Electroanalysis
with
Carbon Paste Electrodes

ANALYTICAL CHEMISTRY SERIES

Series Editor
Charles H. Lochmüller
Duke University

Quality and Reliability in Analytical Chemistry, *George E. Baiulescu, Raluca-Ioana Stefan, Hassan Y. Aboul-Enein*

HPLC: Practical and Industrial Applications, Second Edition, *Joel K. Swadesh*

Ionic Liquids in Chemical Analysis, *edited by Mihkel Koel*

Environmental Chemometrics: Principles and Modern Applications, *Grady Hanrahan*

Quality Assurance and Quality Control in the Analytical Chemical Laboratory: A Practical Approach, *Piotr Konieczka and Jacek Namieśnik*

Analytical Measurements in Aquatic Environments, *edited by Jacek Namieśnik and Piotr Szefer*

Ion-Pair Chromatography and Related Techniques, *Teresa Cecchi*

Artificial Neural Networks in Biological and Environmental Analysis, *Grady Hanrahan*

Electroanalysis with Carbon Paste Electrodes, *Ivan Švancara, Kurt Kalcher, Alain Walcarius and Karel Vytřas*

ANALYTICAL CHEMISTRY SERIES

Electroanalysis
with
Carbon Paste Electrodes

Ivan Švancara • Kurt Kalcher • Alain Walcarius • Karel Vytřas

Featuring a special Foreword by **THEODORE KUWANA**

CRC Press is an imprint of the
Taylor & Francis Group, an **informa** business

Image on the front cover of the book is a historical photo illustrating the detail of the very first prototype of carbon paste holder machined in workshops at the University of Pardubice in the mid-1980s. It should be noted that the extrusion of carbon paste from the body had been arranged just to show the presence of paste and its consistency.

CRC Press
Taylor & Francis Group
6000 Broken Sound Parkway NW, Suite 300
Boca Raton, FL 33487-2742

© 2012 by Taylor & Francis Group, LLC
CRC Press is an imprint of Taylor & Francis Group, an Informa business

No claim to original U.S. Government works

Printed and bound in India by Replika Press Pvt. Ltd.
Version Date: 20111209

International Standard Book Number: 978-1-4398-3019-2 (Hardback)

This book contains information obtained from authentic and highly regarded sources. Reasonable efforts have been made to publish reliable data and information, but the author and publisher cannot assume responsibility for the validity of all materials or the consequences of their use. The authors and publishers have attempted to trace the copyright holders of all material reproduced in this publication and apologize to copyright holders if permission to publish in this form has not been obtained. If any copyright material has not been acknowledged please write and let us know so we may rectify in any future reprint.

Except as permitted under U.S. Copyright Law, no part of this book may be reprinted, reproduced, transmitted, or utilized in any form by any electronic, mechanical, or other means, now known or hereafter invented, including photocopying, microfilming, and recording, or in any information storage or retrieval system, without written permission from the publishers.

For permission to photocopy or use material electronically from this work, please access www.copyright.com (http://www.copyright.com/) or contact the Copyright Clearance Center, Inc. (CCC), 222 Rosewood Drive, Danvers, MA 01923, 978-750-8400. CCC is a not-for-profit organization that provides licenses and registration for a variety of users. For organizations that have been granted a photocopy license by the CCC, a separate system of payment has been arranged.

Trademark Notice: Product or corporate names may be trademarks or registered trademarks, and are used only for identification and explanation without intent to infringe.

Library of Congress Cataloging-in-Publication Data

Electroanalysis with carbon paste electrodes / Ivan Švancara ... [et al.].
    p. cm. -- (Analytical chemistry series)
"A CRC title."
Includes bibliographical references and index.
ISBN 978-1-4398-3019-2 (alk. paper)
1. Electrodes, Carbon. 2. Electrochemical analysis. I. Švancara, Ivan. II. Title. III. Series.

QD572.C37E44 2012
543'.4--dc23
                                                                2011031819

Visit the Taylor & Francis Web site at
http://www.taylorandfrancis.com

and the CRC Press Web site at
http://www.crcpress.com

Contents

Foreword xi
Preface...xv
Acknowledgments..xvii

Chapter 1 Introduction to Electrochemistry and Electroanalysis with Carbon Paste–Based Electrodes... 1

    1.1   Historical Survey and Glossary... 1
        1.1.1   1958: The Very First Report on Carbon Paste Electrodes 2
        1.1.2   1959–1963: Proposals of Carbon Pastes, Their Characterization, and Initial Applications 2
        1.1.3   1964, 1965: First Modifications of Carbon Pastes 2
        1.1.4   1965–1975: Expansion of Carbon Paste Electrodes in Electrochemical Laboratories 3
        1.1.5   1976–1980: Pioneering Chemical Modifications of Carbon Paste.............3
        1.1.6   1981–1988: Era of Chemically Modified Carbon Paste Electrodes..............4
        1.1.7   1988–1995: Worldwide Spread of Carbon Pastes with Enzymes as a Novel Type of Biosensor 5
        1.1.8   1996–2000: Starting Competition of Traditional Carbon Pastes with Screen-Printed Sensors and Other Carbon Composites 5
        1.1.9   2000–2001: Carbon Pastes and New Technologies 6
        1.1.10 2002–2003: Carbon Pastes Follow the Concept of Green Chemistry ... 7
        1.1.11 2003–2010: Brand New Carbon Pastes and Their Immediate Success 7
    1.2   Field in Publication Activities and Literature 9

Chapter 2 Carbon Pastes and Carbon Paste Electrodes... 11

    2.1   Carbon Paste as a Binary Mixture 11
        2.1.1   Definitions of Carbon Paste 11
        2.1.2   Carbon Powder/Graphite 12
            2.1.2.1   Spectroscopic (Spectral) Graphite 12
            2.1.2.2   Other Carbonaceous Materials......................... 13
            2.1.2.3   New Forms of Carbon 14
        2.1.3   Pasting Liquid/Binder 16
            2.1.3.1   Paraffin (Mineral) Oils 17
            2.1.3.2   Aliphatic and Aromatic Hydrocarbons............ 18
            2.1.3.3   Silicone Oils and Greases................................. 18
            2.1.3.4   Halogenated Hydrocarbons and Similar Derivatives 19
            2.1.3.5   Other Pasting Liquids and Mixed Binders 19
            2.1.3.6   New Types of Carbon Paste Binders 20
        2.1.4   Mixing and Preparation of Carbon Pastes 20
            2.1.4.1   Carbon-to-Pasting Liquid Ratio 20
        2.1.5   Handling and Storage of Carbon Pastes............................. 22
    2.2   Classification of Carbon Pastes and Carbon Paste Electrodes 23
        2.2.1   Traditional Types of Carbon Paste–Based Electrodes 23

		2.2.2	Special Types of Carbon Paste–Based Electrodes 26
			2.2.2.1 Carbon Paste Electroactive Electrodes 26
			2.2.2.2 "Solid," "Solid-like," and "Pseudo" Carbon Paste Electrodes 26
		2.2.3	New Types of Carbon Paste Electrodes 27
			2.2.3.1 Diamond as the Electrode Material and Diamond Paste Electrodes 28
			2.2.3.2 Carbon Paste Electrodes and Carbon Nanotubes 29
			2.2.3.3 Carbon Paste Electrodes and Ionic Liquids 32
	2.3	Construction of Carbon Paste Holders 35
		2.3.1	Tubings and Rods (Plugs) with Hollow Ends 36
			2.3.1.1 Rotated Disc Electrodes for Hydrodynamic Measurements 38
		2.3.2	Piston-Driven Electrode Holders 38
		2.3.3	Commercially Available Carbon Paste Electrode Bodies 40
		2.3.4	Carbon Paste–Based Detectors 40
		2.3.5	Planar Constructions of Carbon Paste Electrodes 42
		2.3.6	Miniaturized Variants of Carbon Paste Electrodes 43
		2.3.7	Special Constructions of CPEs 45

Chapter 3 Carbon Paste as an Electrode Material 49

	3.1	Physicochemical Properties of Carbon Pastes 49
		3.1.1 Microstructure of Carbon Pastes 49
		3.1.2 Ohmic Resistance 51
		3.1.3 Instability of Carbon Pastes in Organic Solvents 52
		3.1.4 Aging of Carbon Pastes 52
		3.1.5 Hydrophobicity of Carbon Pastes 52
	3.2	Electrochemical Characteristics of Carbon Pastes 52
		3.2.1 Very Low Background 52
		3.2.2 Individual Polarizability of Carbon Pastes 53
		3.2.3 Specific Reaction Kinetics at Carbon Pastes 56
	3.3	Testing of Unmodified CPEs 59
	3.4	Interactions at Carbon Pastes 62
		3.4.1 Interactions at the Carbon Paste Surface 63
		3.4.1.1 Electrode Reactions with Charge Transfer (Electrolytic Processes) 63
		3.4.1.2 Adsorption and Related Electrode Reactions of Nonelectrolytic Character 63
		3.4.1.3 Ion Exchange and Ion-Pair Formation 64
		3.4.1.4 Electrocatalysis-Assisted Detection 65
		3.4.1.5 Synergistic Processes and Interactions 66
		3.4.2 Interactions in Carbon Paste Bulk 68
		3.4.2.1 Extraction and Reextraction 69
		3.4.2.2 Chemical Equilibria in the Bulk and Transport Models 70

Chapter 4 Chemically Modified Carbon Paste Electrodes 73

	4.1	Classifications of Chemical Modification 73
		4.1.1 Intrinsic Modification 74

Contents vii

		4.1.2	Extrinsic Modification	74
			4.1.2.1 Surface Modification	74
			4.1.2.2 Bulk Modification	76
	4.2	Reasons and Strategies for Modification		76
		4.2.1	Criteria Required for Modifiers	76
		4.2.2	Activation and Restoration of the Modifier	77
		4.2.3	Immobilization and Preconcentration: Concepts and Strategies	77
		4.2.4	Electrocatalysis	78
		4.2.5	Other Purposes for Modification	81
	4.3	Types of Chemical Modifiers		81
		4.3.1	Inorganic Materials	82
			4.3.1.1 Prussian-Blue Derivatives and Polyoxometallates	82
			4.3.1.2 Clays and Zeolite-Based Molecular Sieves	83
			4.3.1.3 Metal Oxides and Sol-Gel-Derived Inorganic Materials	86
			4.3.1.4 Other Inorganic Modifiers	87
		4.3.2	Organic and Organometallic Compounds	89
			4.3.2.1 Organic Ligands	89
			4.3.2.2 Organic Catalysts	91
			4.3.2.3 Organometallic Complexes	95
			4.3.2.4 Surfactants, Amphiphilic and Lipophilic Modifiers	97
			4.3.2.5 Organic Polymers and Macromolecules	98
		4.3.3	Other Possibilities and Recent Approaches to Chemical Modification	100
			4.3.3.1 Organic–Inorganic Hybrid Materials	100
			4.3.3.2 Nanomaterials	101
			4.3.3.3 Surface Treatments and Alterations	103
	4.4	Modeling and Testing of Chemically Modified CPEs		104
		4.4.1	Basic Procedures and Specifics	104
		4.4.2	Chemically Modified CPEs versus Bare CPEs	105
			4.4.2.1 Sorption Processes at the Bare (Unmodified) Carbon Pastes	105
			4.4.2.2 Chemically Modified CPEs	106

Chapter 5 Biologically Modified Carbon Paste Electrodes 109

	5.1	Biosensors and Electrodes for Biological Analysis		109
	5.2	Modifiers and Working Principles		110
		5.2.1	Electrocatalysis and Mediation	110
		5.2.2	Enzymes	111
			5.2.2.1 Oxidases	112
			5.2.2.2 Dehydrogenases	116
			5.2.2.3 Hydrolases	117
			5.2.2.4 Auxiliary Enzymes	117
			5.2.2.5 Enzyme Kinetics	118
		5.2.3	Nucleic Acids	118
		5.2.4	Immunosensors	120
		5.2.5	Tissues and Cells	120
		5.2.6	Other Biomolecules	121
	5.3	Modification Strategies (for Biosensors)		121

Chapter 6 — Instrumental Measurements with Carbon Paste Electrodes, Sensors, and Detectors ... 123

- 6.1 Electrochemical Techniques: Fundamentals and Basic Principles ... 123
    - 6.1.1 Double Layer Concept and Capacitive Current ... 123
    - 6.1.2 Mass Transport ... 124
    - 6.1.3 Faradic Processes ... 125
    - 6.1.4 Diffusion Layer ... 127
        - 6.1.4.1 Diffusion Coefficients ... 127
        - 6.1.4.2 Macrosized Stationary Electrodes ... 127
        - 6.1.4.3 Rotating Disk Electrodes ... 128
        - 6.1.4.4 Microelectrodes ... 129
- 6.2 Voltammetry ... 130
    - 6.2.1 Direct Voltammetry ... 130
        - 6.2.1.1 Voltammetry at Slow Scan Rates ... 130
        - 6.2.1.2 Voltammetry at Faster Scan Rates ... 131
    - 6.2.2 Stripping Voltammetry ... 133
- 6.3 Amperometry ... 134
- 6.4 Potentiometry ... 135
    - 6.4.1 Direct Potentiometry and Potentiometric Titrations ... 135
    - 6.4.2 Chronopotentiometry and Stripping Chronopotentiometry ... 143
- 6.5 Nonelectrochemical Techniques ... 147
    - 6.5.1 Microscopic Observations ... 147
    - 6.5.2 Other Physicochemical Techniques for Special Characterizations ... 149

Chapter 7 — Electrochemical Investigation with Carbon Paste Electrodes and Sensors ... 151

- 7.1 Studies on Electrode Reactions and Mechanisms of Organic Compounds in the Early Era of Carbon Pastes ... 151
    - 7.1.1 Historical Circumstances ... 151
    - 7.1.2 Adams's Investigation of Aromatic Compounds ... 153
    - 7.1.3 Adams's Legacy ... 158
- 7.2 Electrochemistry of Solids ... 158
    - 7.2.1 Historical Introduction ... 158
    - 7.2.2 Main Principles, Configurations, Possibilities, and Limitations ... 159
    - 7.2.3 Survey of Practical Applications ... 161
- 7.3 Electrochemistry *in Vivo* ... 162
    - 7.3.1 Rise of the Field and Its Position ... 162
    - 7.3.2 Basic Characterization of *in Vivo* Measurements ... 163
    - 7.3.3 Survey of Applications and Some Prospects ... 165
- 7.4 Special Studies with Carbon Pastes ... 165
    - 7.4.1 Role of Minor Research ... 165
    - 7.4.2 Selected Themes across the Field ... 166
        - 7.4.2.1 Carbon Pastes as Substrates for Electropolymerized Films ... 166
        - 7.4.2.2 Carbon Pastes as Electronic Tongues ... 166
        - 7.4.2.3 Carbon Pastes as CPEEs for Electrochemical Characterization of New Electrocatalysts ... 167
        - 7.4.2.4 Carbon Pastes in Industrial Use ... 167
    - 7.4.3 Further Examples ... 168

Contents

Chapter 8 Electroanalysis with Carbon Paste–Based Electrodes, Sensors, and Detectors 169

    8.1 Determination of Inorganic Ions, Complex Species, and Molecules 170
        8.1.1 Noble Metals 171
        8.1.2 Heavy Metals... 172
        8.1.3 Metalloids.. 175
        8.1.4 Metals of the Iron, Manganese, Chromium, and Vanadium Groups..175
        8.1.5 Platinum Metals and Uranium 178
        8.1.6 Metals of the Fourth and Third Groups, Metals of Rare Earths...... 179
        8.1.7 Metals of Alkaline Earths and Alkaline Metals........................... 180
        8.1.8 Non-Metallic Ions, Complexes, and Neutral Molecules 181
        8.1.9 Concluding Remarks... 184
    8.2 Determination of Organic Substances and Environmental Pollutants.......... 185
    8.3 Pharmaceutical and Clinical Analysis 188
    8.4 Determination of Biologically Important Compounds 191
        8.4.1 Alcohols... 192
        8.4.2 Aldehydes, Ketones, and Acids.. 193
        8.4.3 Amino Compounds 194
            8.4.3.1 Amides and Amines 194
            8.4.3.2 Amino Acids... 194
        8.4.4 Antioxidants and Phenolic Compounds 196
        8.4.5 Carbohydrates and Related Compounds 197
        8.4.6 Coenzymes, Enzymes, Proteins, and Related Compounds..............200
        8.4.7 Hormones, Phytohormones, and Related Compounds 202
        8.4.8 Neurotransmitters..202
        8.4.9 Nucleic Acid, Nucleic Bases, and Related Compounds 204
        8.4.10 Purines, Pyridines, and Pyrimidines 205
        8.4.11 Vitamins 206
        8.4.12 Whole Cells, Microorganisms, Tissues, and Tissue Extracts as Modifiers..208
        8.4.13 Survey of Applications of Carbon Paste Mini- and Microelectrodes in Brain Electrochemistry/ *in Vivo* Voltammetry 209
        8.4.14 Miscellaneous..209

Chapter 9 In Place of a Conclusion: Carbon Paste Electrodes for Education and Practical Training of Young Scientists 447

Appendix A: RNA: A Profile of the Carbon Paste Inventor and a Great Scientist 449

Appendix B: List of Dissertation Theses Defended in the Authors' Countries and Dealing with Carbon Paste–Based Electrodes, Sensors, and Detectors.............................. 453

Appendix C: Alternate Titles of Chinese, Japanese, and Korean Journals or Periodicals........... 461

References...463

Authors..601

Author Index..603

Subject Index...643

Foreword

A TRIBUTE TO THE LATE PROFESSOR RALPH N. ADAMS, INVENTOR OF THE CARBON PASTE ELECTRODE

During the early days of his research, Professor Ralph N. Adams (RNA) was interested in the electro-oxidation of organic compounds, mainly aromatic amines. The two principal materials used as electrodes for such electro-oxidations were platinum and carbon in various physical configurations (e.g., Pt as wire or planar foil and carbon as rods or sheets). A major problem with such solid electrodes, however, was the contamination and fouling of the electrode surface due to the product(s) of these electro-oxidations. He kept thinking that it would be nice to have an electrode like dropping mercury electrode (DME), which had a renewed surface with each new drop. Of course, Hg is not used in the anodic potential range since it itself readily undergoes oxidation.

In analogy to DME, Adams envisioned a dropping carbon electrode consisting of finely powdered, conductive carbon, like carbon black, with an inert organic liquid, mixed to be sufficiently fluid to flow through a thin capillary tube. Thus, the idea of a carbon paste electrode (CPE) was born, but implementation as a dropping electrode was still a problem. The organic liquid would separate from the carbon particles when the composition was sufficiently fluid to flow in the capillary, thus leading to an electrical disconnect.

At the time, I was a graduate student working on the oxidation of aromatic amines with an electrochemical cell that had a small circular cavity at the bottom containing Hg. The Hg surface was oxidized first in the presence of chloride ion to form a thin passivating film of mercurous chloride prior to introduction of the organic compound of interest. Adams thought that the cavity of this cell could be filled with the carbon paste (CP) to comprise a CPE in a stationary format. This CPE configuration worked well but was impractical since easy replacement of the CPE was inconvenient. So he machined an electrode that had had a small diameter cavity at the end of a Teflon rod which could be easily filled with the CP (carbon particles mixed with bromonaphthalene to a consistency like peanut butter). A Pt wire through the center of the rod extending into the cavity made electrical contact with the CP. This CPE could be readily removed from the cell for refreshing the surface by "scraping" the surface with a spatula or replacing the CP.

The advantages of the CPE (see Refs. [1–3]) were the low residual background current, the reproducibility of successive voltammetric scans, applications to both oxidations and reductions (hence the large applicable potential range), and low cost. For reductions, it was not necessary to remove oxygen because of the large overpotential for oxygen reduction. Adams and his students subsequently developed several configurations (see Refs. [4,5]), including a rotating ring-disc electrode, and studied the physical, chemical, and electrochemical characteristics of CP mixtures with different carbon particles, pasting liquids, and when using various model redox systems for both oxidations and reductions. Acheson graphite mixed with Nujol (a mineral oil) became the standard CPE for the lab.

As suggested by Adams, the reduction of metal ions to their metallic state would produce a reverse oxidative wave similar in shape to the stripping of metals from a Hg electrode. This idea of partitioning into CPE gave rise to experiments where organic compounds, insoluble in water, could be dissolved in the pasting liquid and oxidized or reduced to the water-soluble ionic state, thereby exhibiting voltammetric waves with diffusional characteristics (see Refs. [6,7]). An early application of the CPE was in a flow cell as a detector of electroactive species for liquid chromatography (LC), which subsequently gave rise to many forms of LC-EC.

Adams was an "idea scientist," unafraid to take risk to pioneer innovations. Examples include the placement of a small electrochemical cell with a metal electrode (Pt) in the cavity of the magnet of an ESR spectrometer to obtain spectra of electrogenerated free radicals, and the insertion of microelectrodes in brain tissues for *in situ* electrochemical monitoring of neurotransmitters—all the aforementioned were "firsts," albeit many peer electrochemists at the time ridiculed his ideas when he first discussed them. He would come to the lab before sunrise (e.g., 4:00 a.m.) to do experiments and then return home to have breakfast with his family before coming back to meet with student and teach his courses. His hobby was flying, a skill he learned as a World War II bomber pilot (see Figure F.1), and his students often had memorable (stomach turning) flights of acrobatics with "Buzz"—a nickname for obvious reasons.

FIGURE F.1 R. N. Adams, ace chemist and pilot. A reproduction of the front cover of a contemporaneous local magazine. (From personal archives of Prof. Kuwana, with his kindly permission.)

FIGURE F.2 Adams making carbon paste. A cartoon drawing. (From personal archives of Prof. Kuwana, with his kindly permission and with courtesy of Dr. Donald W. Leedy.)

Foreword   xiii

FIGURE F.3 Professor Adams with his students. From left to right: Theodore "Ted" Kuwana, Ralph "Buzz" Adams, and Donald W. Leedy, the last-mentioned being responsible for the drawing "Adams making carbon paste." This picture was taken in 1999 at a reunion of Professor Adams with his students. (From personal archives of Prof. Kuwana and with his kindly permission.)

FIGURE F.4 From a scientific discussion (about carbon paste electrodes?). From left to right: Professor Kurt Kalcher, Professor Theodore "Ted" Kuwana, and Professor Karel Vytras. The photo was shot in 1999 at the 7th International Seminar on Electroanalytical Chemistry (held in Changchun, China). (From authors' archives.)

His enthusiasm for science and life was infectious (see Figure F.2). Adams was just a wonderful scientist and human being. Many stories abound about him, including his habit from his days when he grew up on the beach in New Jersey—that is, not wearing sox…

Congratulations to the authors of this book for documenting the developments and applications of CPE and allowing this introduction about its inventor, the late Professor Ralph N. Adams (Figures F.3 and F.4). I end with what he told his students: "If you love what you are doing, the rewards will take care of themselves…".

Theodore "Ted" Kuwana
Lawrence, Kansas

REFERENCES

1. Adams, R.N. 1958. Carbon paste electrodes. *Anal. Chem.* 30:1576.
2. Adams, R.N. 1963. Carbon paste electrodes. A review. *Rev. Polarog.* 11:71–78.
3. Olson, C. and R.N. Adams. 1960. Carbon paste electrodes: Application to anodic voltammetry. *Anal. Chim. Acta* 22:582–589.
4. Olson, C.L. 1962. Investigation concerning the development and use of carbon paste electrodes for voltammetric measurements. PhD dissertation, University of Kansas, Lawrence, KS.
5. Galus, Z., C. Olson, H.Y. Lee, and R.N. Adams. 1962. Rotating disk electrodes. *Anal. Chem.* 34:164–166.
6. Kuwana, T. and W.G. French. 1964. Carbon paste electrodes containing some electroactive compounds. *Anal. Chem.* 36:241–242.
7. Schultz, F.A. and T. Kuwana. 1965. Electrochemical studies of organic compounds dissolved in carbon-paste electrodes. *Electroanal. Chem.* 10:95–103.

Preface

This book, *Electroanalysis with Carbon Paste Electrodes*, aims, for the very first time, to cover the long-established position of carbon paste in electrochemistry and electroanalysis.

Indeed, except for a series of contemporary reviews and a recent entry in the *Encyclopedia of Sensors* (*EOS*), there has been no specific textbook since the late 1950s (when carbon paste was invented) devoted exclusively to electroanalysis with carbon paste electrodes (CPEs) that attempts to cover the field in its entirety. The classic monograph, *Electrochemistry at Solid Electrodes*, by Ralph N. Adams—the inventor of carbon paste—published in 1969, cannot be regarded as a comprehensive book on CPEs because it deals with only basic facts and merely covers the early era of the field, whereas most of the milestones and principal innovations took place much later.

The idea of compiling this book came to us sometime in the mid-2000s, shortly after we had finished working on the aforementioned contribution to the *EOS*. We had been provided a wide scope for our overview and, thanks to such benevolence, we were encouraged to explore a still wider concept, resulting eventually in this book. As time went by, our team also published a series of new review articles that, in retrospect, were like actual presentations of the updated information we had been systematically gathering (to complete our archives) for the book.

The actual work on this book started in the winter of 2009/2010 and took us a year and a half to finish. It can be noted here that our original intention to use material from the aforementioned recent reviews was soon abandoned, and the book was written anew. Thus, current "hot topics" were integrated with the latest developments in the field, including novel types of CPEs (associated with the massive use of nanomaterials, ionic liquids, and other newly synthesized products), actual reflections of environment-friendly ("green") chemistry, and all the features illustrating flexible adaptation of CPEs, as well as the latest trends in biological and clinical analysis. (Herein, it should be emphasized that the latter area, especially, has been dealt with particular care as it was reviewed for the first time after many years.)

Moreover, all the actualities included were discussed and analyzed in the overall context considering numerous historical aspects as well as countless relations and links among the individual categories, themes, and phenomena. (This is also the reason why individual sections and paragraphs are closely linked and frequently navigate to the previous or, vice versa, following text.)

This book covers an enormously wide area of electrochemical and electroanalytical measurements with carbon paste–based electrodes, sensors, and detectors and required a balanced approach to compile. This was accomplished by dividing the major topics and distributing the respective themes among the four authors, who worked on a joint schedule, but in their own independent capacity. However, despite our best efforts, we have to admit certain overlaps that can be found in some chapters and sections. However, this has been left as it is so as not to spoil the integrity and the overall presentation of the individual chapters or sections.

On the other hand, we have paid special attention to avoid duplicities in the literature cited. Our decision to gather all the references (3300 in all) at the end of the book rather than citing them at the end of every chapter resulted in tremendous work. Thus, without the invaluable help of a special computer program (developed by one of us), which, at its optimum, could perform more than 3,000,000 mutual comparisons in one run, we would have continued to check and renumber all the references manually till today! Regarding the references as such, the "deadline" for having them ready was set as the summer of 2010, and articles officially published beyond this period are cited only occasionally.

Acknowledgments

Herein, we would like to acknowledge all the people who have been involved with this book. We are grateful to Professor Charles H. Lochmüller, who first suggested CRC Press/Taylor & Francis Group as a possible publisher, and to Barbara Glunn, David Fausel, Vinithan Sethumadhavan, and Joette Lynch with their production teams at Taylor & Francis for kindly helping us whenever we needed and for tremendous editorial work on our textbook. We are particularly delighted that Theodore "Ted" Kuwana, emeritus distinguished professor at the University of Kansas, who witnessed the very beginnings of the field, agreed to write a foreword for this book. He also supplied us with some personal material, which gave impetus to our work, and provided genuine encouragement.

Further, we are indebted to the following collaborators, colleagues, and friends for kindly providing literature material we could use and include in our reference list: Professor Jiří Zima (Charles University, Prague, Czech Republic), Dr. Miroslav Fojta (Institute of Biophysics, Brno, Czech Republic), Professor Ján Labuda (Slovak Technical University, Bratislava, Slovakia), Dr. Zuzana Navrátilová (Ostrava University, Czech Republic), Dr. Jana Skopalová (Palacky University, Olomouc, Czech Republic), and Dr. Libuše Trnková (Masaryk University, Brno, Czech Republic). Thanks also to a number of distinguished scientists who have contributed to our book through fruitful personal discussions or correspondence. They include Professor Joseph "Joe" Wang (University of California San Diego, California), always interested in our opinion on the news in the field; Professor Robert D. O'Neill (University College Dublin, Ireland), who willingly revised the chapter on *in vivo* electrochemistry; Professor Karel Štulík (Charles University, Prague, Czech Republic) with his valuable glosses; and Professor Robert Kalvoda (Heyrovský Institute, Prague, Czech Republic), who sadly passed away during the preparation of this book, leaving us his fond memories of Professor R. N. Adams. We would also like to thank all our coworkers who did research with us and under our supervision, as well as the electrochemists working with CPEs for publishing their results. It is indeed their work that made the realization of this project possible.

We would like to extend our thanks to Professor Jiří Barek (Charles University, Prague, Czech Republic), Professor Andrzej Bobrowski (AGH University, Krakow, Poland), Dr. Anastasios Economou (University of Athens, Greece), and Dr. Samo B. Hočevar (National Institute of Chemistry, Ljubljana, Slovenia), for their kind assessment and critical comments on the preliminary draft. Finally, Ivan Švancara and Karel Vytřas gratefully acknowledge financial support from the Ministry of Education, Youth, and Sports of the Czech Republic (project # MSM0021627502).

In particular, all of us express our gratitude and thankfulness to our relatives and friends and to our beloved family members, wives, and children, all of whom gracefully accepted our prolonged absences due to our constant engagement with the work related to the preparation of the book.

Last but not least, we would appreciate comments from readers, as some contributions or useful details could have escaped our attention and did not appear in the final text. Such a feedback is always welcome and will be reflected in our future work—including a possible updated edition of this book.

Ivan Švancara
Kurt Kalcher
Alain Walcarius
Karel Vytřas

1 Introduction to Electrochemistry and Electroanalysis with Carbon Paste–Based Electrodes

1.1 HISTORICAL SURVEY AND GLOSSARY

"During an investigation of the properties of a dropping carbon electrode, a new paste electrode has been developed which possesses unique advantages in anodic polarography. ..." says an opening sentence of the first report on carbon paste electrode (CPE) with a laconic title *Carbon Paste Electrodes* [1]. The author of this text, Ralph N. Adams (Appendix A), had thus stepped into the history as the inventor of carbon paste for electrochemical and electroanalytical measurements.

The aforementioned formulation reveals that it happened more or less coincidentally, when the background of the discovery was mentioned in the inventor's first review on CPEs [2] and, some years later, revealed fully in his monograph [3]. Therein, on page 280, one can read as follows:

> ... Carbon paste electrodes developed from an attempt to prepare a fluid of suspended carbon particles to be used in the sense of a dropping electrode for anodic oxidations. Such a mixture can be prepared and dropped from a capillary of internal bore slightly greater that that of the DME. However, the conditions for successful operation of the dropping carbon paste electrode are far from ideal, and it soon became apparent that a thicker paste, packed in a pool configuration and used either stationary or rotated, would be far more advantageous ...

Finally, these circumstances were also confirmed by *Theodore Kuwana*, Adams's former student [4,5], the author of a special foreword in this book, and one of the first electrochemists who had ever worked with carbon paste. During a short meeting with the authors a decade ago [6], he said that he was "the slave" who had been working in the laboratory with the dropping carbon electrode (DCE) during the experimental testing of this device, as well as with the very first carbon paste mixtures made afterward. Despite some initial optimism (see the Adams's concluding words in his pioneering report [1]), the whole concept of the DCE failed; nevertheless, the unsuccessful work with fluid graphite suspensions had resulted in an unexpected side product—the carbon paste. In addition, the mixtures of this consistency soon became so popular that carbon pastes as such represented the classic electrode material for laboratory preparations of electrodes, sensors, and detectors of innumerable types, sizes, and configurations.

Before having achieved such a position, however, the carbon paste–based electrodes had undergone a very interesting genesis, comprising a number of significant developments and achievements. It may even be stated that there is scarcely another type of electrode material whose employment would illustrate more faithfully the overall progress—or, on the contrary, the individual movements and trends—in electrochemical and electroanalytical measurements over half a century—across the time period of existence of carbon paste [7].

The era of carbon paste–based electrodes can then be summarized by means of the following periods and key moments.

1.1.1 1958: The Very First Report on Carbon Paste Electrodes

As already noticed, the end of the 1950s saw an unsuccessful attempt at constructing an alternative to Heyrovský's dropping mercury electrode for anodic oxidations [1]. Instead, the electrochemists obtained a brand new electrode material—carbon paste—whose potentialities in electrochemical research and applied measurements could only be assumed at that time. This was pretty well formulated by *Peter T. Kissinger*, another Adams's student, in his dedicated profile [5]: "… I don't think anyone could have predicted a number of carbon paste publications from reading the very first carbon paste article …" (Herein, we can add that this number is currently around 2500 scientific publications, increasing by 20–30 new items each month.)

The first carbon paste was a mixture of 1 g carbon powder with 7 mL bromoform, whose properties were demonstrated on the oxidation of iodide in a solution of 1×10^{-4} M I$^-$ in 1 M H_2SO_4; the second paste was reportedly tested as a mixture of graphite and carbon tetrachloride [1]. In retrospect, these pioneering experiments had been quite atypical for research activities of the carbon paste inventor and his group whose domain was mainly organic electrochemistry.

1.1.2 1959–1963: Proposals of Carbon Pastes, Their Characterization, and Initial Applications

The early years of using CPEs can be classified as the "Era of Adams's Group" or "Era of the Electrochemists from Kansas University (KU)." At that time, several selected types of carbon pastes were subjected to a fundamental characterization in anodic and cathodic voltammetry [8,9], using both inorganic and organic compounds as model species or systems. Of interest were also self-made carbon paste holders applicable in several different electrode configurations [10,11].

As time went by, the initially preferred halogenated liquids were replaced by less harmful and commercially available mineral oils such as Nujol®, which was finally recommended as the pasting liquid of choice. (It is notable that this substance is still one of the most popular binders for the preparation of carbon pastes and its use is being regularly reported in research articles until now [7].) Satisfactory experience from the testing measurements with CPEs soon initiated their systematic employment in investigations of the electrode behavior of various substances, predominantly biologically important aromates from the family of substituted phenols, aromatic amines, and aminophenols. The results of these studies were first summarized in Adams's review [2] and later, in more detail, in the aforementioned textbook [3]. Moreover, after a lengthy period of four decades, these invaluable investigations are again the central point of interest within this book—in Section 7.1.

Extensive studies of both anodic oxidations and cathodic reductions of organic compounds were typically qualitative in nature, and therefore, the first quantitative analysis with carbon paste–based electrodes can be attributed to Jacob's simultaneous determination of gold and silver [12] who was using atypical (bell-shaped) tubings filled with carbon paste. The author of the corresponding procedure, offering the limits of detections at the nanomolar level despite the use of direct voltammetry (i.e., without accumulation step), is also an early user of CPEs who had not originated from the research team at the KU, the other two being Davis and Everhart [13].

1.1.3 1964, 1965: First Modifications of Carbon Pastes

In the mid-1960s, two other breaking events happened in Adams's laboratories. First, it was shown that carbon pastes made of the two main constituents might contain another component and the resultant system offered new attractive applications in electrochemical research [14,15]. The intention of such an approach was to use the binary nature of carbon pastes for obtaining a simple approach to study the electrode behavior of compounds completely insoluble in aqueous solutions (see also the foreword to this book). Because the whole concept "bare carbon paste + substance admixed in" could later be adapted for studying inorganic solids, the pioneering studies by Kuwana et al. are often regarded to be the initial step in the development of the so-called carbon paste electroactive electrodes (CPEEs) that are discussed separately, in Section 7.2.

A similar reason—i.e., combination of CPEs with measurements in organic solvents—was also behind the second key work [16] that introduced a new type of CPE containing up to 50% (w/w) of solid surfactant dissolved in the binder, acting as the stabilizer against disintegration of the native paste in nonaqueous solutions of a majority of organic solvents (e.g., MeOH, EtOH, ACN, DMSO, or DMFA). In this case, one can also refer to the purposely performed change in the typical properties of carbon paste mixtures in an effort to improve their resistivity in aggressive conditions. In other words and generally, this study was the first actual modification of carbon pastes, leading to the desired behavior of a particular CPE for a given application.

1.1.4 1965–1975: Expansion of Carbon Paste Electrodes in Electrochemical Laboratories

Whereas the activities of Adams's school had continued (see, e.g., [17–25]), representing—with a few exceptions [26,27]—the research activities in Northern America, electrochemists and electroanalysts in the Old Continent came rapidly to the fore [28–40]. In the late 1960s, especially, researchers from the Central Europe were active in developing and characterizing various types of CPEs [28–32], including newly proposed mixtures with silicon polymer–based binders such as highly viscous oils and greases [28] or even rubber [32]. (The latter was apparently the first example of the "solid-like" CPE whose originally soft consistency had been turned to a compact composite by solidifying the binder, here, by cold vulcanization.) Within these studies, some innovative constructions of carbon paste holders were designed [28,31], featuring also an unusual device for renewing the carbon paste surface with the aid of a stretched piano string [33] or new filling procedures [34]. Finally, there were some pioneering reports on the use of a CPE in equilibrium potentiometry [35,36], in analysis of biologically important compounds [37], or in voltammetry of solids [38].

The 1970s brought several valuable contributions into the advanced characterization of CPEs [39–41], including a sophisticated procedure for protecting the graphite particles against sorption of air oxygen [39], which had attracted the carbon paste inventor himself [3]. The same time period saw also the decline of Adams's dominance in the field [42–44], more specifically, his reorientation toward pharmaceutical and biological sciences—pharmacotoxicology and *in vivo* monitoring [45–50]. (The last contribution is a review, cementing Adams's position within the clinical research—namely, in contemporary brain electrochemistry—and it has also commented in detail on the first successful assays with carbon paste–based microelectrodes implanted in living laboratory animals.)

1.1.5 1976–1980: Pioneering Chemical Modifications of Carbon Paste

The late 1970s showing some stagnation in publication activities (see Figure 1.1) had spawned two studies whose significance was soon recognized as a breakthrough in the field. At first, Cheek and Nelson ([51]; the second author being further absolvent of Adams's school [4]) reported on an intentionally made modification with amino groups chemically immobilized onto the graphite particles by means of rather complicated laboratory procedure. However, the effort made had been compensated by the result, and the sensor was found to exhibit a high affinity to Ag(I) ions, enabling their determination far below the sub-nanomolar level. The second milestone happened in Japan, where Yao and Musha [52] introduced the first prototype of a carbon paste biosensor, although, according to the authors themselves, the respective configuration was a chemically modified CPE applied to the determination of a biologically important compound. Apart from the proper classification, both reports predetermined the turbulent movement in the subsequent decades signaling two dominant areas: (i) chemically modified CPEs and (ii) biologically modified CPEs (carbon paste–based biosensors).

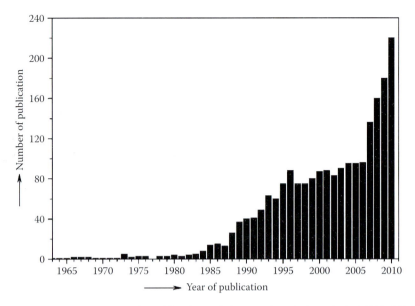

FIGURE 1.1 Carbon paste electrodes in publication activity over the whole period of 1958–2010. Diagram. (From authors' archives.)

1.1.6 1981–1988: Era of Chemically Modified Carbon Paste Electrodes

All the previous results from measurements with carbon pastes containing additional constituents could not remain unnoticed. The specific combination of the two major components in solid and liquid states with evident potential for multicomponent concept and in contrast with simple laboratory preparation gave altogether such an attractive proposition that it was a question of time, when these features would be reflected in wider applications. The real boom was apparently ignited by a report by Ravichandran and Baldwin [53] who had used the advantages of "mechanical modification" (i.e., direct mixing of both main carbon paste components with a modifier) and who were probably the first to introduce the term *chemically modified carbon paste electrode* (CMCPE) [53,54]. Their method of modification by means of incorporation of modifying agent into the carbon paste is now regarded as revolutionary [7], although the concept of direct modification had not been new and resembled the preparation of Kuwana's ferrocene-containing CPEs [14,15] or Adams's "non-aqueous CPE" [16], all appearing in the early 1960s. On the other hand, there were some nuances in the proper making, when the modifier could be delivered into the bulk through (i) its dissolution in the binder [14,15] or loaded as a solid [16,53,54] by (ii) mixing it with both main components as well as into the ready-made carbon paste.

Although the entire 1980s have been influential mainly in association with the development and applications of CMCPEs (for comprehensive surveys, see, e.g., [55,56]), electrodes made of traditional carbon paste mixtures were still of interest. Such unmodified CPEs were successfully employed in analyses of pharmaceuticals [57], biologically important compounds [58], or organic pollutants [59]. In parallel, there also were studies continuing in the characterization of unmodified carbon pastes [60], including their testing in newly introduced modern instrumentation [61,62], some of them trying to improve their surface characteristics by means of a pretreatment by intensive electrolysis [63] or through addition of a surfactant [64]. In addition, there were attempts to replace the pasting liquid by more resistant high-molecular PVC binder [65], chlorofluorinated polymeric fluid, *Kel-F*® [66], or via the total alteration of the CPE surface by its spraying with solid graphite layer [67], all these configurations documenting further applicability of "solid-like" CPE configurations. At the end of the commentary on this period, one should not forget to quote Adams's farewell

Introduction to Electrochemistry and Electroanalysis with Carbon Paste–Based Electrodes   5

to the world of carbon pastes in his classical contribution to the reaction kinetics of CPEs [68], having achieved a remarkable citation index and competing, in this respect, with the introductory report itself (see [7] and data analysis therein).

1.1.7 1988–1995: Worldwide Spread of Carbon Pastes with Enzymes as a Novel Type of Biosensor

Hand in hand with the development of CMCPEs [69], carbon pastes were also examined as a substrate for incorporating biological modifiers—enzymes, co-enzymes, or suitable tissues containing such natural catalysts. Hitherto the most comprehensive review on CP-based biosensors [70] quotes that the earliest examples of such modifications go back to the 1980s or even before, which is the case of the already mentioned electrode [52] with chemically immobilized species of NAD^+ (oxidized form of NADH). Similar attempts followed soon, where enzyme(s) had been anchored onto the protective membranes of carbon paste–based substrates with admixed mediators (see, e.g., [71,72]). However, the direct incorporation of enzyme into a carbon paste—i.e., mechanical modification without any other linking or coating elements—was described for the first time by Matuszewski and Trojanowicz [73]. Their simple sensing device for amperometric detection of glucose contained the paste with glucose oxidase embedded inside and loaded into a small well at the bottom of a common electrode shaft. Reportedly, excellent stability of such a sensor (min. 3 weeks) together with surprisingly rapid response due to the functioning of carbon paste as arrays of miniature membranes [69,70] seems to be a sufficiently stimulating factor that had initiated a real boom in the development and subsequent applications of biologically modified carbon paste electrodes, commonly known as CP-biosensors.

Thus, this type of biosensors came rapidly to the fore, forming in a short time another major area within the electrochemistry with CPEs. Again, one of the reasons for this widespread use was the uniqueness of carbon paste materials in their flexibility to be modified. Similarly as with CMCPEs, carbon paste–based biosensors offer a wider choice in modification strategies than those available for common solid electrode materials. Among others, the substrates from heterogeneous carbon pastes allow one to incorporate an enzyme and a mediator in one matrix, which improves the kinetics of the corresponding enzymatic reaction and fastens the biosensor response. Also, carbon pastes can be simply modified with two different enzymes forming bifunctional (dual) biosensors and may contain suitable stabilizing additives as well as some biological tissues or microbial cells (particular examples and further details are given in Chapter 5). These carbon paste–based dual sensors were one of the models for the future constructions of multifunctional sensors known as electronic tongues (see the following).

It can be concluded that biologically modified carbon pastes had been in a lead position throughout the 1990s, holding the same position up to the mid-2000s, when laboratory-made carbon paste material stepped into an uneven confrontation with commercially marketed carbon inks as substrates for serial machining or manufacturing of screen-printed (carbon) electrodes, SP(C)Es [74]. This hypothetical battle had its beginnings in the mid-1990s [75], thus characterizing the upcoming period of our chronicle.

1.1.8 1996–2000: Starting Competition of Traditional Carbon Pastes with Screen-Printed Sensors and Other Carbon Composites

Probably the first analysis devoted to this competition was included in the review [75] introduced earlier, positioning CPEs in the role of laboratory predecessors of SP(C)Es. Because the respective literature source is not so easily accessible, the entire rather visionary discussion can be briefly presented herein—through interpretation of carbon pastes and carbon inks as the electrode material close in nature. Both are heterogeneous substrates and both can be classified as dispersions of carbon or graphite particles in suitable binders. Regarding CPEs, the latter is represented by pasting liquid that stays in the liquid state even after preparation of the carbon paste mixture; whereas, in SPEs, the original fluid agent is solidified during the machining of the final product. Thus, although the resultant properties of a CPE and the selected SPE can be quite different, the method of eventual modification is nearly the same, including the direct mixing of the two main components with the modifying agent. If such modification is to be used for the first time and hence tested in advance, it is evident that the whole process for SPEs (including model measurements) can be simulated with CPEs, which is advantageous because the corresponding procedure with a CPE is usually much simpler. Yet another factor in favor of this strategy is the fact that SPEs are typically made in special machines as integrated series (sets) of many electrodes [74] and their repeated fabrication just for testing would became unacceptable from the economical point of view.

Besides SP(C)Es, there are a number of other carbon composite materials for which the respective sensing units are called "carbon paste–like" or "pseudo-carbon paste" electrodes, often, by the authors themselves. Some of them have already been introduced in the previous sections [33,62,63]); further types and prototypes have been proposed newly [76–79]. One of such composites was a mixture of carbon powder with melted phenanthrene [76], when the hot liquid homogeneous mixture was cooled down, thus giving rise to a granular mass that could be stored for later use. Then, before use, a portion of such granules was melted up and—as a fluid—packed into an ordinary carbon paste holder. Alternatively, this solid-like carbon paste material could be modified like common carbon paste mixtures, either by incorporating a modifier into the mixture—again, in melted form—or *in situ* if the modifying agent(s) tended to decompose at higher temperatures.

Other types of "solid-like" CPEs were polyethylene-based carbon paste solid-like mixture [77], with the modifier added, and the relatively popular paraffin wax–based carbon composites (see, e.g., [78,79]), continuing in the concept of a pioneering prototype by Crow and his "carbon wax electrode" (CWE [80]). In these cases, the electrode material itself is obtainable in the same way as for the previously described phenanthrene-s-CPE, and its later modification can also be performed analogically.

Similar to the previous decade, the 1990s with CPEs were also surveyed in some contemporary reviews [81,82], highlighting the significance of new technologies in the area and the flexibility of carbon paste as being "ready for things to come."

1.1.9   2000–2001: Carbon Pastes and New Technologies

The arrival of a new millennium coincided with rapidly increasing applicability of materials that had been developed with the aid of new technologies. In electroanalysis with carbon pastes, they are represented mainly by two major groups of modifiers [83]: (a) complexants and (b) redox electrocatalysts, including mediators for biosensing, when—from the chemical point of view—the individual compounds typically belong to the so-called inorganic or organic hybrids; e.g., various (I) cations of the $[Me^N(A)_a(B)_b]^{m+}$ type (where "Me" is a metal of the iron or platinum groups, "A" inorganic ligand, and "B" organic molecule or vice versa) or (II) heteropoly-anions based on Si^{IV}, P^V, and As^V as central atoms, Mo^{VI}, W^{VI}, and V^V as metals in oxo-anionic ligands, both in combination with voluminous organic and organometallic cations as well as neutral organic molecules.

These complex structures offer numerous attractive properties owing to their multifunctional nature, coupling in one the mechanical stability of the inorganic lattice, chemical reactivity of the organic moiety, and its lipophilicity, which ensures good adhesion of the voluminous molecule to the hydrophobic carbon paste surface or bulk, respectively [84].

Incorporations of these substances and materials into various CPE configurations cannot, of course, be confined to a couple of years as evoked by the title of this section, because their applications have been or are of interest for many years—before, as well as after, the commencement of the year 2000. This is also emphasized in two recent reviews [83,84] analyzing for the first time the phenomenon called "carbon paste vs. new technologies" and mapping in detail the respective research activities that accompany the electrochemistry and electroanalysis with carbon paste–based electrodes until now.

In order to describe the earlier stated and to illustrate the situation on the verge of the centuries, some examples can be given. An inorganic or organic hybrid was of interest in studies [85,86], whereas the use of a traditional analytical reagent in a new role of mediator is reported in [87], demonstrating also a special immobilization of the sol-gel type. Such procedures, as cross-linking of an enzyme [88], were usually connected with applications of silica and titanica porous matrices, representing yet another example of "modern" modifiers whose boom had peaked mainly in the 1990s [89].

Among typical trends of the early 2000s, one can quote the advanced investigations of nucleic acids [90], with the roots in the previous decade (see, e.g., [91] or [92], the former being also notable for the lowest detection limit ever reported for a CPE: 4×10^{-16} mol L^{-1} [7]); the increasing use of various nanomaterials (e.g., colloidal gold [93], renamed soon to nanogold [94], apparently, under the influence of contemporary nomenclature—see also a critical note in [84]), continuing efforts in seeking novel types of CPEs (such as "glassy carbon paste electrode" [GCPE] [95], having an almost identical predecessor in the productive 1990s [96]) or further expansion of SP(C)Es ([97], in this case, sorted among CPEs by the authors themselves). Finally, there are also some rarities; for instance, (probably) the first prototype of a reference electrode based on carbon paste [98], cyanobacteria-containing carbon paste applied to accomplish the photoelectrochemical oxidation of water [99] or a novel solar cell [100], documenting once again the potentialities of carbon pastes in new technologies.

1.1.10 2002–2003: CARBON PASTES FOLLOW THE CONCEPT OF GREEN CHEMISTRY

This point reflects the growing intensity of the ecologic aspects of modern instrumental analysis and the associated research, giving rise to a new field of "green chemistry" [101].

In this respect, carbon pastes as such are quite a fine material, when traditional mixtures are practically nontoxic. Nevertheless and despite this, the investigations with CPEs, CMCPEs, and CP-biosensors were further directed toward the green-chemistry concept. One of the most typical activities of the last decade was the search for an alternative to controversial mercury electrodes, which would be applicable in electrochemical stripping analysis (ESA) to the determination of heavy metal ions, with comparable performance. Moreover, the great success with the bismuth film electrode (BiFE), introduced under these circumstances in 2000 [102], opened a door for a new electroanalytical area with nonmercury metallic electrodes, where carbon pastes had been among the firsts [103,104], still the most frequently used electrode substrates thanks to the diversity with which they can be modified with bismuth and bismuth compounds.

Other contributions that were ecologically useful [105–110] appeared in parallel with the early reports on bismuth-film plated CPEs (BiF-CPEs). Thus, there was an interesting method for speciation of Hg$_2^{2+}$ and Hg^{2+} [105]; determination of Hg(II), Pb(II), and Cd(II) [106,107]; or differentiation and subsequent quantification of both As^{III} and As^V [108], all representing highly toxic species in the environment. Finally, similar methods have been proposed for notorious organic pollutants such as phenol [109] and two commercial herbicides [110], which were also phenolic derivatives.

1.1.11 2003–2010: BRAND NEW CARBON PASTES AND THEIR IMMEDIATE SUCCESS

The last period of our chronicle concerns again a number of the previously pursued activities, such as the continuing employment of various CMCPEs and solid-like CPEs in inorganic and organic analyses, pharmaceutical and clinical analyses, or for the investigation and determination of numerous biologically important compounds. At still growing intensity, there have been further applications of new materials, novel types of modifiers, as well as various innovative procedures. All these trends are also documented in the retrospective literature [7,83,84], including the latest two reviews [111,112], with a collection of review articles since the appearance of the very first contribution [2] of a kind compiled nearly a half a century ago by R.N. Adams, the inventor of carbon paste.

However, the most significant movement of the late 2000s is undoubtedly associated with the commencement of newly synthesized forms of carbon (i) besides or (ii) instead of ordinary graphite and with (iii) conventional or (iv) alternate binders, as well as (v) both in one mixture. These five possible combinations have logically resulted in a qualitatively variable family of new carbon paste mixtures.

Within this area, (i) glassy carbon powder with spherical particles and a specially pretreated surface [95,96,113] came again to the fore, shown newly as the electrode material with extraordinary resistivity in media with a higher content of organic solvents (such as solutions containing up to 80%–90% (v/v) MeOH [114]). Better performances due to electrocatalytic effect on reaction kinetics compared with traditional carbon or graphite materials have then been reported for carbon pastes prepared from (ii) acetylene black [115], (iii) template carbon, (iv) porous carbon foam, (v) porous carbon microspheres [116], and (vi) fullerene "C-60" [117]. Interestingly, the carbon moiety in a paste mixture can also be (vii) diamond powder [118] despite its almost zero conductivity. In the last few years, intensively popularized new forms of carbon have been completed with (viii) ordered mesoporous carbon [119]), (ix) carbon nanoparticles [120], (x) electrospun carbon nanofibers (ES-CFs [121]), or with (xi) carbon nanotubes [122] and their two basic variants: single wall carbon nanotubes, SWCNT(s) and multiwall carbon nanotubes, MWCNT(s).

The CNT(s) deserve a special attention as their introduction into electrochemistry with carbon pastes [123,124] has started a little revolution in the field, with the unprecedented boom of qualitatively new carbon paste mixtures and the respective electrodes, known soon as "carbon nanotube paste electrodes" (CNTPEs). A similar situation occurred after the first successful testing of the (room-temperature) ionic liquids, (RT)ILs, in carbon paste configurations [125]. It was shown soon that ILs may effectively replace traditional pasting liquids [126], giving rise to another new type of carbon paste–based electrodes, the so-called "carbon ionic liquid electrodes" (CILEs). In accordance with the aforementioned variability of configurations with alternate carbon paste constituents, the mutual combinations of CNT or ES-CFs with ILs came as well, giving rise to further new mixtures [127,128], where one does not find any of the two traditional components.

Otherwise, the last years of the 2000s have also seen some new trends in electrochemical investigation, for instance, the rapidly increasing popularity of carbon pastes as substrates for electropolymerization or their compatibility to form integrated systems with modifiers loaded inside such polymeric layers (see [129–138]). Another example of such activities is a rapidly growing area of continuing syntheses and testing of various inorganic or organic hybrids, especially, of new catalysts. Within the respective procedures, one can also find quite sophisticated methods of direct modification of carbon during electrospinning [139] or a long string of analogically oriented studies, originating from various Chinese regions and having appeared regularly each year (see, e.g., [140–147]).

For the latter, it is typical that the newly synthesized solids undergo an intimate characterization by means of the whole arsenal of instrumental techniques, namely, CHN-elemental analysis (EA); thermogravimetric analysis (TGA); microscopic techniques (SEM, STEM, and AFM); UV/vis-spectrophotometry; fluorimetry (F); roentgen spectroscopy of solids (XPS, XRF, and XRD); Fourier-transform infrared (FT-IR), Auger-, and mass spectroscopy (MS); and up to four variants of nuclear magnetic resonance (H^1-NMR, C^{13}-NMR, Si^{29}-NMR, and P^{31}-NMR). The electrochemistry with carbon pastes is then represented by cyclic voltammetry (CV), potentiostatic coulometry (COU), equilibrium potentiometry (POT), chronoamperometry (CA), electrochemical impedance spectroscopy (EIS), and mainly solid-state voltammetry (SPV) with the CPEE, where the material studied is admixed into the carbon paste bulk and, in this form, studied to evaluate its electrocatalytic capabilities in the redox behavior of selected model substance(s). Regarding all the nonelectrochemical and electrochemical techniques listed earlier, the individual applications have also been surveyed (see [83,84] and references therein).

To summarize the 2000s with all the events, it should be added that the end of the decade saw a further decline of interest in CP-biosensors—at least, compared with the preceding periods [70,111]—and practically absolute fade-out of solid phase voltammetry with CPEEs, when the respective report(s) had appeared only seldom (e.g., [148,149]). Regarding the overall activities in the field, however, there was an abrupt increase in the last 3 years, for which there are two evident explanations: (i) stormy development of both CNTPEs and CILEs (representing around 150 publications) and (ii) the establishment of new online journals that have quickly become a pivotal platform for several productive authors' teams.

1.2 FIELD IN PUBLICATION ACTIVITIES AND LITERATURE

Research activities in the development and applications of CPEs, CMCPEs, CP-biosensors, or CPEEs within a period of more than half a century have affected practically all the areas of theoretical and applied electrochemistry. In some aspects, the results obtained have also contributed substantially to a successful competition of modern electrochemical measurements with rapidly expanding separation and spectral techniques. Since the discovery of carbon paste until now, it can be estimated that the terms "carbon paste" or "carbon paste electrode," respectively, have appeared in 2500–3000 original papers published in renowned international journals or local periodicals.

Publication intensity had run in several waves reflecting the individual periods of increasing interest, the times of the real boom, as well as some poorer years. To illustrate this, one obtains a diagram shown in Figure 1.1.

It can be deemed from that the exponential growth in the last three decades corresponds to (i) dynamic progress in the area of chemically modified carbon pastes during the 1980s and in the first half of the 1990s [55,56,59,69], (ii) similar boom in the introduction of various carbon paste–based biosensors through the 1990s and in the early 2000s [70,81,82], and (iii) maybe the most dynamically growing attention toward the new carbon paste mixtures, hand in hand with reflections of newly synthesized materials in all the CPE configurations [83]. The latter is particularly typical for the past few years [84,111,112], when the renowned Internet databases (e.g., "Web of Science" [150]) annually offered more than 150 abstracts of freshly published papers—see again Figure 1.1 as well as the closing remark in the previous section.

Regarding the spectrum of international journals publishing a majority of contributions on CPEs, CMCPEs, and CP-biosensors, summarizations of publication activities with the chart of most productive authors and teams, the highlights among the original papers ("the classics of the field") with actual citation indexes and many other trivia, such material has been reviewed a couple of years ago [7] and the reader(s) is advised to consult directly this original title full of never-before published data, surveys, and statistics.

To date, CPEs in all known modifications, configurations, and variants were the central subject of interest in more than 20 review articles [2,7,55,56,59,69,70,75,81,82,84,111,112,151–160]. Among them and together with one book chapter [161], solely a minor part represents general compilations that have covered the topic as a whole, which is obvious for the first review from the early 1960s [2] but remarkable for the others [7,55,56,69,70], when the field had already become a worldwide phenomenon, covering hundreds of papers. Thus, it is quite logical that a majority of reviews have been arranged more specifically and framed either by the time period(s) [75,82,84,111,155] or by the particular (sub)area chosen, the latter being, for example, the case of a survey of physicochemical processes at CPEs and CMCPEs [157], food [112] and water [161] analyses, solid phase voltammetry with CPEEs [151], *in vivo* electrochemistry [152], potentiometry [159], pharmaceutical analysis [158], and electrochemistry with enzymatic CP-biosensors [156]. Most of the reviews [59,81,152–156,158–160], including the already mentioned ones [75,151], target local audiences, often referring to the regional publications and presenting the papers of the authors themselves.

Thus, until now, a chapter in the *Encyclopedia of Sensors* (EOS) book series [83] is the only text that can be regarded as comprehensive and contemporary study covering electrochemistry and electroanalysis with CPEs in its entirety regardless that a certain part has been devoted to related screen-printed electrodes, SP(C)Es. (Such a generous scope could also be chosen thanks to the extraordinary space in the book with an allowance to insert a set of extensive tables and a list with more than 2200 references.)

Finally, information about CPEs is also included in various textbooks and monographs [3,57,58,162–170], most of which is today being considered as the classical electrochemical literature. The selection of these books has not been fortuitous, and they are highlighted here because during the 1970s and 1980s most of them had to substitute the missing review, having provided contemporaneous information on CPEs, when Adams's early article [2] had become old and nonactual.

Furthermore, this collection also mirrors the geography of traditionally strong electrochemical regions: (i) North America [57,58,163,166,168], (ii) Middle Europe [162,164], and (iii) the former U.S.S.R. [169], although practically each of these books has become popular worldwide. Some of them obtained a good reputation after being translated into English (see [164] vs. [165] and [169] vs. [170], the former representing a paradox situation, when the translated version had come to the market 1 year earlier than the original), the other could then be sold in the reeditions ([163,166,168]; the second one reappearing in considerably revised form, where, interestingly, the chapter about CPEs has been updated and rewritten by other authors—compare [166] with [167]).

2 Carbon Pastes and Carbon Paste Electrodes

2.1 CARBON PASTE AS A BINARY MIXTURE

2.1.1 Definitions of Carbon Paste

Similarly as in the previous chapter, the initial definitions of what is understood under the term "carbon paste" can be given by means of the authentic citation of descriptions formulated by its inventor, Adams, in his three principal texts:

(i) Introductory report "Carbon Paste Electrodes" [1] "… The (carbon) paste is prepared by stirring together the carbon and organic liquid until the mass appears uniformly wetted. …" (followed by specification of the first carbon paste of the 7:1 type) "… This is a moderately thick paste, well suited for pool application. …"
(ii) Review [2], p. 71 … (with an introductory paragraph reasoning the introduction of CPEs for anodic voltammetry) "… The carbon paste itself contains a mixture of carbon (graphite) with organic liquid, forming a typical consistency of peanut butter. …"

Note: In the context of this rather idiosyncratic comparison, see also the cartoon picture in the Foreword to this book and look upon its right upper corner.

(iii) Monograph [3], p. 281 (with an introductory paragraph that has been cited in this book, within the opening sentences of Chapter 1) "The carbon paste itself is prepared by simple hand mixing of carbon (usually graphite) with any liquid that is sufficiently nonmiscible with water to keep the electrode matrix from dissolving when immersed in the solution …"

It can be added that, from a traditional point of view, a mixture of carbon (graphite) powder with a suitable binder ("pasting liquid") is a special type of *solid carbon electrode* [2,7,56]. The function of the graphite moiety in carbon pastes may give rise to some specific mixtures, when the resultant electrodes are similar in behavior to related compact carbon materials, such as *pyrolytic graphite electrode* (PyGE) or *glassy carbon electrode* (GCE), the latter representing apparently the most frequently used solid electrode in electrochemical and electroanalytical measurements. Nevertheless, due to the presence of liquid binder and because of its interaction with graphite particles, carbon pastes are usually reported as a specific electrode material with numerous unique properties.

Newer definitions categorize the respective electrodes, carbon paste electrodes (CPEs), among *heterogeneous carbon electrodes* [55,69,82,83]. Such classification then comprises almost all variations of carbon paste–based electrodes and sensors that have appeared in the electrochemistry during the past decades, including those being mentioned in the introductory historic section as particular kinds or configurations.

When considering classical carbon pastes, the two main (major) components represent the moieties with rather contradictory character. This is given by the fact that the conductive carbon (graphite) usually serves as the proper electrode material, whereas liquid binder is of insulating character and acts as an inert medium, binding the individual graphite particles into a compact mixture. However, such an obvious function of the binder is accompanied by numerous effects and side effects, making the role of pasting liquids as important as that of the carbon (graphite) material itself. Thus, one has to pay comparable attention to the choice of both main constituents of carbon pastes.

2.1.2 Carbon Powder/Graphite

2.1.2.1 Spectroscopic (Spectral) Graphite

A great majority of binary (unmodified) carbon pastes are prepared from commercially available graphite powders. Usually, they are produced as conductive substrates for spectroscopy and in this way declared in annual catalogues of renowned companies and globally operating dealers with fine chemicals and laboratory equipment, such as *Merck*, *Sigma*, *Aldrich*, *Fluka*, or *Riedel-de-Haen*. Relatively frequent are also local products from specialized manufacturers marketed under the respective trademarks and abbreviations. Here, for example, one can quote the products denoted by "CR" (Czech Republic), "RW" (Germany, Austria), "BDH" (Scandinavia and United Kingdom), "SMMC" (China), "Acheson," "UCP," "GP," and "SP" (United States), the first two named being popular during the early era of CPEs (for further details, see Adams's reports [2,3,8,9]).

All the aforementioned graphite powders usually offer the features that must be considered when selecting the proper carbon material for making of carbon pastes. Among them, the following criteria should be highlighted:

i. *Particle (grain) size and distribution.* Typically, particle size of common spectroscopic graphites ranges in micrometers or in tens of micrometers, respectively, when most of manufacturers are able to provide an official certificate with the specification of the overall particle size distribution (see, e.g., [171]).

Otherwise, relatively reliable information about the average distribution can also be obtained on one's own with the aid of microscopic imaging. Based on empirical experience, it is better if the distribution is not very diverse and the grain size of pulverized graphites is as uniform as possible. Regarding standard spectral powders, their particle distribution is satisfactorily uniform, typically ranging within 5–20 μm, which is also documented by the bar diagram or in a table inside the review [69], quoting for some individual products of the aforementioned graphites the following average mesh: "CR-5": 5 μm, "RW-B": 5–10 μm, "Acheson G(P)-38": 10–20 μm, "UCP-1-M": 1 μm, and "Graphon GP": 25 μm.

Naturally, carbon powders used for preparation of carbon pastes are of interest with respect to their microstructure (see [96,172,173] and references therein). Spectroscopic graphites do not exhibit any characteristic structure or pattern and their typical morphology can be approximated as a tangle of geometrically irregular objects with sharp edges and rather scratched surface—see Figure 2.1.

Besides ordinary micrometer-sized carbons, there have been reports on some carbon paste mixtures containing graphite with a particle size of about 100 nm [174] or even 30 nm [120]. Such a fine material can nowadays be purchased but is also obtainable in laboratory by intimately grinding some of suitable carbon powders in a minimill [174].

ii. *Low adsorption capabilities.* In some cases, only appropriate experiments reveal that a graphite powder intended to be used for preparation of carbon paste exhibits a high adsorption activity [175,176]. This undesirable feature can be identified via the enhanced content of oxygen entrapped in pores of graphite or absorbed additionally during mechanical homogenization of the paste.

When using CPEs made of such graphite powders in faradic (current-connected) measurements, higher concentrations of oxygen are then registered as a noticeable plateau-like response (resembling the increased background within the potentials from −0.2 to −0.6 V vs. Ag/AgCl) during the cathodic scanning [2,175].

Carbon Pastes and Carbon Paste Electrodes

FIGURE 2.1 Microstructure of two typical spectroscopic graphite powders. (A) "RW-B"[1] and (B) "CR-5".[2] Other specification: (1) spectral coal (series "X-69-257"; Ringsdorff Werke, Bonn, Germany) and (2) gear lubricant (Maziva Týn, Týn nad Vltavou, Czech Republic). Scanning electron microscope (SEM, model "JSM-5500LV"; JEOL, Tokyo, Japan), magnification used: 1:3000. (From authors' archives.)

Note: Some authors had tried to solve the problems associated with the presence of oxygen in graphite powder by its pretreatment with the aid of special procedures. One of such purifying operations utilized the effect of thermal desorption of oxygen at approximately 400°C in the atmosphere of inert gas passed over a small portion of graphite. Reportedly, the whole process taking 3–5 h allowed them to remove irreversibly a significant part of adsorbed oxygen [177]. A similar operation, yet improved by thoroughly impregnating the refined graphite with ceresin wax, was found to be effective as well [39]. Otherwise, however, further attempts to purify graphite powders from oxygen appeared only rarely (see, e.g., [178–181]), when some of these operations had even had different motivation (e.g., improvement of kinetic characteristics [179]). The unwillingness to follow similar procedures can be explained by their unease and sophistication. Finally, from the present day's point of view, it seems that such additional deoxygenation and impregnation is not so inevitable since spectroscopic graphites produced by modern technologies usually exhibit satisfactory low adsorption capabilities [175]. Also, carbon pastes are still being prepared by manual hand mixing and, during this operation, the powder may get back a part of previously removed oxygen by absorbing ambient air.

iii. *High purity.* It is obvious that carbon powder should not contain any impurities interfering with measurements. This is the particular case of experiments connected with detection of very low currents (at the low nA level [152,182]), where even negligible traces of electroactive impurities may release unwanted and overlapping signal(s). Again, the manufacturers usually offer the purity profile of their goods [171,183,184], including specification of the remaining traces from the production process.

Note: The latter is also the case of the aforementioned "RW" spectral coal powder, where the corresponding certificate declares the following content of trace impurities (in $\mu g \cdot g$): B < 0.01, Ca < 0.20, Cu < 0.08, Fe = 0.20, Mg = 0.05, Si = 0.6, Ti < 0.50, and V < 0.20 [171].

2.1.2.2 Other Carbonaceous Materials

Occasionally, carbon pastes are also made from less common carbon powders. The first example is (i) acetylene black (AB) [115,185,186]), obtainable by controlled combustion of acetylene in inert atmosphere or, better, by chemical decomposition. Reportedly, the AB has a crystalline-like consolidated structure, which contributes to a more reproducible behavior of the corresponding carbon paste that also exhibits markedly enhanced adsorption capabilities [185].

Atypical forms of carbon used by Kauffmann et al. [187] were also (ii) carbon black (an amorphous material obtainable by the incomplete combustion of heavy petroleum fractions) and

(iii) colloidal graphite (hexagonal carbon with extremely fine flakes and enhanced conductivity), both having been presented as the constituents of new "homemade" carbon paste mixtures and used in CPEs for pharmaceutical analysis.

In the early 2000s, Stefan-van Staden has published her first contribution on the use of (iv) diamond [118], represented by both natural and synthetic forms of diamond and applied as fine powders with particle diameter from 1 to 50 μm. The initial report was soon followed by a series of further reports, suggesting a versatile applicability of this quite surprisingly chosen carbon moiety (see later, in Section 2.2.3.1).

Also, there have been some single attempts with other carbonaceous materials that are known rather from the industry and common life than as the product of fine chemistry. At first, it is (v) soot formed by incomplete combustion of hydrocarbons or, generally, from flammable materials of organic origin, mentioned already by Adams [2] and, after collecting a small amount of this black material inside a chimney, examined also in carbon pastes [188]. (This study made more or less in an intention "just to see" gave the anticipated result: the tested mixture had been totally inapplicable due to a very high background [189].)

Similar carbonaceous material is also (vi) activated charcoal, widely used either for medicinal purposes (against flatulence, for its capability of absorbing gaseous by-products formed in digestion tract) or in the filters of gas masks. Also in this case, one cannot expect excellent properties of the respective carbon paste due to an extreme adsorptibility, as well as insufficient purity, which is, of course, the case of formerly mentioned soot. Despite this, a commercially available activated charcoal ("USP" powder [190]) was once reported as the material of choice for a CPE that could be successfully employed for the quantitative determination in clinical analysis.

Finally, when considering also (vii) coal ("black coal"), this most common form of natural carbon has so far been used as the modifier of an ordinary carbon paste [191]. And, to complete the list of naturally occurring carbons, there is also lignite ("brown coal") whose use in a CPE configuration has never been reported, however.

2.1.2.3 New Forms of Carbon

Most of physicochemical properties—primarily, distinct catalytic capabilities—of such materials are due to a specific microstructure, which is also very characteristic for each of these carbons. This is illustrated by Figure 2.2, presenting microscopic images of three typical representatives, the first one in two different variants.

One of the first representatives of a material that can be called a new form of carbon is apparently (i) glassy carbon (GC) powder, having for the first time been tested in the mid-1990s [96,192] and, later, successfully used in various glassy carbon paste electrode (GCPE) configurations (see, e.g., [95,113,114,193–197]), including chemically and biologically modified variants [193,194,196,197]. The GC powder is a specially treated graphite with spherical particles produced by pyrolytic degradation of highly molecular resins [184], when the respective forms may differ from each other in the final surface treatment, which is the case of two GC powders "Sigradur K" and "Sigradur G" (see again Figure 2.2, images "A" and "B").

Note: The reports on GC powders have emphasized—besides their unique microstructure—some attractive characteristics like minimal adsorption or excellent polarization capabilities despite somewhat "muddy" consistence of some GCPE mixtures [189], making their filling into the CPE bodies with a narrower cavity quite difficult [192]. A particularly notable feature of the GC powders and of the resultant pastes is the already mentioned resistivity against "aggressive" organic solvents used alone or in mixtures with aqueous solutions in the supporting media or mobile phases for analyses by LC-EC. This phenomenon whose physicochemical background is not still fully explained has been confirmed in many experiments with *Sigradur* powders, when it is evident that the effective protection comes from the graphite alone. Therefore, some newer interpretations explain this phenomenon via a relatively stable interphase formed in between the spherical carbon particles partially solvated by mobile phase [114] or by the benefit of the surface treatment of the individual particles and their repelling effect toward the molecules of an organic solvent [84].

Carbon Pastes and Carbon Paste Electrodes 15

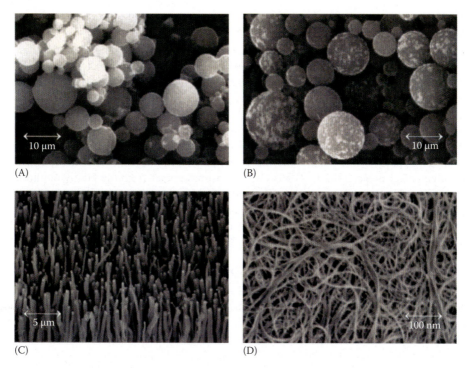

FIGURE 2.2 Microstructure of some new carbon and graphite materials. (A) Glassy carbon powder, variant with untreated surface;[1] (B) with pretreated surface;[2] (C) carbon nanofibers;[3] (D) single-wall carbon nanotubes.[4] Other specification: (1, 2) with spherical particles ("Sigradur G & K"; both HTW Maitingen, Germany), (3) reproduced from http://www.nibib.nih.gov/gallery/images/nibib_092605_082714.jpg; downloaded on January 15, 2011; (4) L-SWNTs (NTP, Shenzhen, China). Scanning electron microscope (SEM, model "JSM-5500LV"; JEOL, Tokyo, Japan), magnification: 1:3000 (A, B) and 1:10,000 (D). (From authors' archives, except (3).)

Chronologically, the next new material is (ii) fullerene "C-60," a representative of molecular carbons that form—together with related "carbon nanotubes" (CNTs) and "carbon nanohorns" (CNHs)—three major classes of carbon nanomaterials. In the carbon paste mixtures, fullerene "C-60" (whose structure of a hollow sphere and building units from hexagonal aromatic rings resembles faithfully a soccer ball) was firstly used in China [198], followed soon by some other reports [117,199,200]. In all cases, fullerene "C-60" was applied as the additional modifier with distinct electrocatalytic properties, whereas a new binary mixture of the "C60-PE" type has not been still reported. (Similarly, there is no mention on the use of an ellipsoidal fullerene, "C-70.") Otherwise, regarding the aforementioned carbon nanomaterials, (iii) carbon nanohorns have been examined in a single study only [201], whereas CNTs dominate among all new forms of carbon (see later and especially Section 2.2.3.2) and some other carbon nanoforms are to come yet.

Next in the order are three novel carbonaceous materials, namely, (iv) template carbon, (v) porous carbon foam, and (vi) porous carbon microspheres (together with a self-made GC powder), all proposed recently by Nossol and Zarbin [116].

The authors see the possibilities and limitations of these materials, synthesized by means of pyrolysis of different poly(furfurylalcohol)–based precursors, in their employment in electrocatalysis, as well as in practical electroanalysis. In their study, all the materials were tested in the corresponding binary mixtures and with Nujol oil as the standard binder.

In the same year, similar investigations were carried out with (vii) ordered mesoporous carbon (OMC [119]) and the OMCPE, offering also improved electron transfer kinetics and catalytic capabilities in connection with the electrode transformations of selected redox systems. Moreover, the OMCPE was also recommended anodic stripping voltammetry of heavy metal ions, offering a superior sensitivity compared to common CPEs.

Finally, the late 2000s witnessed the use of the remaining carbon nanomaterials. Firstly, they arrived in the form of (viii) carbon nanoparticles (of about 30 nm in diameter [120]) in a carbon paste microelectrode (nCµPE) that, like in previous cases, underwent more or less routine characterization with model redox systems of both inorganic and organic nature. The survey continues with (ix) carbon nanofibers (CNF; [121,128,139,202,203]; see Figure 2.2, image "C"), obtainable either by machine-controlled growth during electrospinning or with the aid of chemical vapor deposition (CVD), and with (x) graphene (GR; [204]). The latter is another allotrope of carbon, occurring in one-atom-thick sheets with a honeycomb-like structure and, in fact, represents the limiting case of an indefinite polyaromate. To date, GR is also the last newcomer from the family of alternate carbons being tested in the carbon paste configurations.

And the most popular (xi) CNTs ([122]; see again Figure 2.2 and image "D") that can be titled as "the carbon of the new millennium" also in the electrochemistry with CPEs [84] are already embodied in a special group of CNT-based paste electrodes [123,124]) requiring to be discussed in more detail.

2.1.3 Pasting Liquid/Binder

As already emphasized, mechanical connection of the individual carbon particles into a uniform mass is not the only role of carbon paste binders. Each pasting liquid, including highly chemically inert substances, codetermines principal physicochemical and electrochemical properties of each CPE, including all newly described types (see, e.g., [83,84,111] and references therein). Typical parameters of a suitable pasting liquid can be summarized in a similar way like for carbon materials, when pointing out the following features:

i. *Chemical inertness and electroinactivity.* Liquid binders used for the preparation of carbon pastes are highly stable substances and their interactions at the electrode interface or in the electrode bulk, respectively, are usually of purely physical nature [3,83]. However, some liquid binders may also interact chemically, for instance, via active participation in acido–basic equilibria. This is the case of some organophosphate esters as shown in a pioneering study with otherwise electroinactive and water-insoluble liquid binders [205], forming voluminous protonated cations with high affinity toward ion pairing.

   Electrochemical inertness of the binder is required in experiments based on the current flow, that is, in faradic measurements such as various voltammetric modes (e.g., [8,9,25,34,41,61]), amperometry [62–65], constant current chronopotentiometry [190], or coulometry [206]. This demand is not the case of carbon paste electroactive electrodes (CPEEs) (i.e., CPEs with electrolytic binder [38,151]) or other CPEs with electroactive binders if such constituents do not interfere with the analytical signal of interest [207]. Electrochemical activity is of lesser importance in potentiometric indication [159], when carbon paste mixtures may contain even highly reactive compounds like organic nitro esters such as 2-nitrophenyl-octyl ether (NPOE) (see, e.g., [208] and references therein) or in stripping potentiometry with chemical oxidation [209,210].

ii. *Low volatility.* Apart from attitudes associated with direct physical contact with potentially harmful vapors, the stability and lifetime of CPEs prepared from more volatile pasting liquids is rather limited. Due to slow evaporation of the binder, the respective carbon paste mixtures become desiccated, which is accompanied—besides visible changes in consistency—by irreversible alterations of physicochemical and electrochemical properties [83,177]. These trends can be well documented on CPEs made of liquid tricresyl phosphate (TCP) that, despite their relatively high viscosity, was found to evaporate from carbon paste mixture during 1 or 2 weeks only [205]. Regarding common binders, the requirement of minimal volatility is obeyed by most of the liquids so far proposed.

iii. *Minimal solubility in water.* Except special measurements, both CPEs and CMCPEs are designed as sensors for aqueous solutions and hence, also heterogeneous carbon pastes have to be sufficiently stable in water, without undesirable disintegration and dissolution in consequence of miscibility of the binder with water and, in general, any aqueous solution.

Note: This obvious aspect is of particular significance in measurements with bulk-modified carbon pastes. There are indications that some modifying agents—despite their minimal solubility in water—may tend to leach out of the paste. Although this phenomenon known as "bleeding" [55,75] concerns usually negligible amounts of modifier, it may have a considerable effect, when considering "micro"relations at the electrode interface and the function of modifier itself in lipophilic phase of binders compared to that in aqueous solutions. Then, even essentially insoluble modifiers have to be anchored into the carbon paste bulk either mechanically or, better, via chemical immobilization [51] (for details, see Chapters 4 and 5). A slight solubility of the binder is the case of ordinary carbon pastes and, for instance, even extremely hydrophobic molecules of highly molecular oils are capable of penetrating to the aqueous phase [20] and vice versa, when a little amount of water is entrapped in the carbon paste bulk [157,175]. Such saturation effects can be observed after recommended storage of prefilled electrode holders in a beaker with distilled water, which—after a longtime exposure—may give rise to an opalescent appearance, serving as an experimental proof for such a leakage [189]. The whole process can be revealed during electrochemical experiments when the saturation-affected carbon paste layer suffers from the markedly increased background [175,183].

iv. *Controlled miscibility with organic solvents.* If a pasting liquid is to be sufficiently hydrophobic in order to repel molecules of water in aqueous solutions, such a binder is then miscible with all the solvents of the same character [20,157]. It means that, in principle, carbon pastes are inapplicable in nonpolar solvents and in mixed solutions containing such solvolytic systems [2,3,83]. Unfortunately, as found, common CPEs can also be seriously damaged in some polar solvents, such as MeOH, EtOH, ACN, DMSO, and DMFA [192,210]. In this respect, the use of CPEs in the presence of organic solvents (which can be brought into the system also via the solution of *in situ* modifier spiked into the sample) always require particular care and—if needed—even special precautions and/or maintenance.

Note: The first successful attempt to keep a carbon paste stable in organic solvents was based on addition of highly lipophilic substance into the paste mixture [16]. (Properly chosen surfactant and its substantial amount in the paste, of about 30% w/w, had then been capable of effective repelling some polar solvents like methanol, acetonitrile, or diethylcarbonate, thus hindering their molecules to penetrate onto the carbon paste interior.) The second possibility of how to stabilize carbon pastes in polar solvents is the choice of highly viscous pasting liquid [211] or even resistant solid-like binder (see, e.g., [66]). Finally, there is yet another way of stabilizing the carbon paste mixture during the permanent contact with organic solvent—it is the selection of a special carbon material, such as the GC powder of *Sigradur*® type, reportedly withstanding up to 90% (v/v) MeOH [114].

Although none of pasting liquids obeys all the criteria for an ideal binder, there is a relatively wide offer of organic compounds or mixtures that may—more or less—provide carbon pastes with satisfactory quality. Specifically, one can quote several particular substances described in the following sections.

2.1.3.1 Paraffin (Mineral) Oils

These liquids are undoubtedly the most frequently chosen carbon paste binders. A typical paraffin oil (PO) is formed by a mixture of liquid aliphatic hydrocarbons and usually being marketed under trademark.

The most popular product of this kind is Nujol® (oil or mull, respectively; originally solvent for IR-spectroscopy) obtainable from every renowned dealer; see, for example, [212]. Very similar in nature, as well as in the basic properties, are mineral oils used as solvents in UV-spectroscopy, for example, the product supplied by *Merck* being well known and declared as the mineral oil of Uvasol® purity grade [213]. (Perhaps, under the influence of the name "Nujol," some authors call this product "Uvasol oil" being unaware of the original meaning.)

Despite some classifications as heavy mineral oils, both Nujol and the solvent for UV-spectroscopy are liquids with distinctly liquid consistency (almost like water), completely transparent, odorless, and—if coming into a short contact with skin—essentially harmless. These features are not so obvious (see the following Note) and surely belong among good reasons for their constant popularity and widespread global use [7,83]. Their drawback may be a potential flammability and, especially, their capability to act as solvent of some less resistant materials—potentially vulnerable compartments of measuring cells or of the electrode itself.

2.1.3.2 Aliphatic and Aromatic Hydrocarbons

The dominance of commercial POs with often anonymous composition has also resolved the previous dilemmas on which particular hydrocarbon would be the best as the binder for carbon pastes. As documented in studies with a series of either aliphatic or aromatic hydrocarbons, even a very thorough testing did not lead to any conclusive answer. In this context, for example, one can remember the comparative studies with homologues of aliphatic hydrocarbons C_8–C_{20} [8,9,11,34,60,68] or various derivatives of benzene, naphthalene, and phenanthrene [8,9,40]. Occasionally, some of these compounds have been used again in later electroanalytically oriented measurements, such as hexadecane [60,214,215] or selected items of shorter chain aliphatic hydrocarbons (C_{10}–C_{14}; [216]).

2.1.3.3 Silicone Oils and Greases

The binders of this kind belong among polymerized siloxanes with organic side chains. If the substance of interest is still being fluid, it is usually called silicone oil (SO), whereas a thicker and already gelatin-like derivative is silicone grease (SG). The latter may either be a representative of polysiloxanes with high molecular weight or even silicon oil combined with a thickener (e.g., fumed silica).

In general, both SO and SG share numerous valuable properties of POs, such as chemical inertness, insulating character, and health harmlessness. Moreover, unlike these carbon analogues, silicones are nonflammable and friendlier to some plastic materials. And compared to Nujol or the oil of *Uvasol* grade, typical SOs are often more viscous and usually forming denser and more compact carbon paste mixtures [175].

Both SO and SG represent the second major group of binders used for traditional carbon pastes. The pioneering reports on their use can be traced up as early as in 1960s [20,28,217], including some capacitance testing measurements in Adams's laboratories [2], and various SO-based carbon pastes appear regularly over the upcoming decades (see, e.g., [41,60,61,202,209,218,219,224,225]), when the functioning of numerous SOs or greases was also the subject of extensive comparative studies [2,40,192,218,220–222].

In general, silicone fluids form carbon pastes with properties very close to those of mixtures from mineral oils; nevertheless, there are also some differences in behavior of these two fundamental types. Besides the already mentioned consistency, the most evident dissimilarity can be seen in resistivity of both types of pastes in media with organic solvents, where silicone fluid–based mixtures exhibited better stability compared to the counterparts made of mineral oils [211].

Note: For instance, a CPE of the C/SO type could be reliably operated in a solution of 2 M H_2SO_4 containing 25% (v/v) MeOH, when the C/Nj counterpart at identical experimental conditions had already exhibited the extremely high background. And, after increasing the content of solvent up to

75%, the C/SO was still usable with slightly distorted baseline, whereas its Nujol-based test partner started to leak out from the electrode in consequence of total disintegration. Otherwise, the original study [192] had been performed with a sextet of *Lukoil®* polysiloxanes of the [–O–Si(-CH$_3$)$_2$–O–)]$_n$ and [–O–Si(-CH$_3$)(-CH=CH$_2$)–O–]$_n$ type with the average molecular weight of 500–15,000 g · mol^{-1}, revealing a certain increase in the stability of the individual carbon paste mixtures in relation with the growing molecular weight of the corresponding SO used [188]. Yet another unpublished study concerned the examination of a CPE prepared from SG purchased in a sport shop as a lubricant for fly-fishing sinking lines. Despite quite a fine consistency, the resultant paste ultimately failed during electroanalytical characterization.

The remaining differences in properties of both PO- and SO-based carbon pastes are less evident and can be noticed mainly during advanced electrochemical characterizations (see, e.g., [221–223,226]) or some special studies with the aid of electrochemical impedance spectroscopy (EIS; [227]) or ohmic-resistance measurements (RM; [228,229]). Nevertheless, the resultant nuances ascertained during these experiments are also worthy of mentioning and most of them are of interest within the individual sections in Chapter 3.

2.1.3.4 Halogenated Hydrocarbons and Similar Derivatives

Historically, these substances represent third important group of pasting liquids; and, for example, bromoform, carbon tetrachloride, and α-bromonaphthalene were among the binding agents that had formed the very first types of carbon pastes [1–3,7]. Since then, halogenated derivatives as binders for carbon pastes were selected only occasionally, which is the case of α-bromonaphthalene [40], *p*-dichlorobenzene [230], and *trans*-1,2-dibromocyclohexane [231], the latter two described as electroactive or reactive carbon pastes, respectively.

Note: Except for Adams's laboratory, halogenated hydrocarbons and similar substances are not very popular among electrochemists working with CPEs, which can be understood if one considers that these compounds are often highly toxic; some of them are also proven carcinogens. In context with such carbon paste mixtures, the authors of this text may share their own and rather terrible experience with carbon pastes prepared from halogenated aliphatic hydrocarbons. Apart from rather uncomfortable work in a ventilation box in an atmosphere of intensively aromatic vapors, the resultant CPEs exhibited markedly incompact consistency with unusually rough surface and their electrochemical performance was also very average. Mainly, however, these pastes were found extremely aggressive toward to the electrode bodies used for their housing, in that case a less resistant plastic material (*Silon®*) having been totally destroyed in few hours of use [188].

2.1.3.5 Other Pasting Liquids and Mixed Binders

Occasional use of atypical pasting liquids can be documented on mixtures made of (i) TCP [205,207]), (ii) dioctyl phthalate (DOP; [205,207,208]), or (iii) di-iso-nonyl phthalate (DINP; [232]). Due to a chemical activity of these binders exhibiting noticeable ion-exchanging capabilities, the resultant CPEs can be classified to be a *transient element* between unmodified and modified carbon pastes [75,84]. Similar examples are also the carbon paste configurations based on (iv) NPOE [207,208] in the role of pasting liquid instead of its typical employment as a plasticizer for electrode membranes [159]. Due to the presence of nitro group, this binder is electrochemically highly active, and hence, its applicability seems to be limited to current-free potentiometry and related measurements; nevertheless, it has been shown that typical faradic experiments with current flow are also feasible if the polarization range is properly set and the eventual electrode reduction of the binder prevented [207].

Another liquid binder chosen for preparation of two-component carbon paste was (v) diphenyl ether [233]. Further, there are two related reports on the use of (vi) glycerol [234,235], in this case, as additional stabilizer and in mixture with ordinary mineral oil. (The authors themselves have specified their configuration as carbon paste with "mixed binder.")

Finally, the verge of two millennia saw also two attempts to abandon traditional oils in favor of alternatives, namely, (vii) castor oil (an oily substance of vegetable nature and produced from castor beans of *Ricinus communis* sp.) applied as a special CPE in biological analysis [236] and (viii) vaseline oil (a greasy mass, obtainable by industrial fractionation of petroleum jelly and having the name after a trademark product) mixed together with an avocado tissue in a CP-based biosensor for clinical analysis [237].

2.1.3.6 New Types of Carbon Paste Binders

In the late 1990s, the same authors published two related papers in which they introduced the called "binder paste" (BP; [238,239]), based on a polycationic electrolyte instead of traditional oil. Reportedly, their new composite material offered significant improvements compared to common carbon paste mixtures, especially, for construction of biosensors.

Maybe that some results with the "binder paste electrodes" (BPEs) were also motivating for the upcoming activities on testing (room temperature) ionic liquids, (RT)ILs, and the establishment of the aforementioned, brand new subarea of the so-called "carbon ionic liquid electrodes" (CILEs) [84]. Whereas the BPEs had remained as more or less a single attempt, the CILE configurations went rapidly to the fore since its introduction in 2001 [125,126]. And like the equally popular "carbon nanotubes paste electrodes" (CNTPEs) [123,124], also CILEs represent the new phenomenon in the field, deserving to be discussed separately—in Section 2.2.3.3.

2.1.4 Mixing and Preparation of Carbon Pastes

2.1.4.1 Carbon-to-Pasting Liquid Ratio

In most cases, graphite powder and pasting liquid are mixed together in quantities chosen according to the empiric experience, a typical "carbon-to-pasting-liquid ratio" varying in the interval of 1.0 g:0.4–1.0 mL [2,3,59,69,82,83,111]. Some mixtures contain even a higher percentage of liquid binder since the resultant ratio of the two main components depends also on their mutual adherence [177,205]. Thus, for example, 1 g of graphite may "swallow" up to 7 mL pasting liquid, which is the case of the very first paste mixture [1].

Note: Appropriate consistency can also be a virtual measure of how to adjust the optimal carbon-to-binder ratio; however, one should be aware that the resultant pastes may vary quite much—they can be finely thick, but they can also be rather hard or soft or even sticky [2,3,83,111]. In this context, it has to be pointed out that especially "carbon pastes of the new generation," that is, the new alternate mixtures, may form the masses with atypical consistency. These features are due to specific surface characteristics of new carbons, as well as often very tight adhesiveness of some highly lipophilic binders, both contributing to the resultant compactness of such pastes. Here for example, one can remember again "muddy" and almost fluidic mixtures made of some GC-powders (such as Sigradur with pretreated surface of carbon particles [96,192]) or, oppositely, extremely hard, difficult-to-handle, and notably deep black masses obtained by homogenizing Nujol with higher percentages of multiwalled CNTs [229].

Thus, it can be recommended to seek and find a suitable ratio of both pasting constituents experimentally—for each individual carbon paste mixture intended to be prepared and used. Such evaluation can advantageously be accomplished by means of special testing measurements with a series of CPEs [176,177], where the corresponding carbon paste mixtures have been prepared from identical components and differed solely from the ratio chosen. During this examination, one can utilize a number of special tests with standard inorganic and organic electrode systems; a comparison with measurements carried out by using traditional solid electrodes like GCE or Pt-disc is also advantageous.

Carbon Pastes and Carbon Paste Electrodes

Strategies of such measurements have already been described in numerous special studies (e.g., [8,9,28,31,34,40,41,60,68,221,227–229]) and summarized a decade ago in a pair of practical guides [176,177], instructing how to choose, carry out, and interpret all the assays with freshly tested carbon pastes made of traditional components. Some principles of such testing measurements are included in Section 3.3.

A mixture of graphite powder and pasting liquid chosen is usually handmade using ordinary laboratory equipment—see illustrative photo in Figure 2.3—comprising tools for dosing and later filling of electrode body, scraper (e.g., a small spatula and thin glass bar for more viscous liquids) and, of course, porcelain mortar with a pestle, both of appropriate size and sufficient purity.

Note: Due to a possible risk associated with contamination of the paste during the proper stirring, there are some doubts [240] about the suitability of porcelain mortar (depicted on image), especially, of those that have unglazed surface with rough and potentially crumbly contact layer. Despite a legitimacy of such objection, similar pollution of the paste with the porcelain particles has never been noticed [188] or even reported [7]. Regarding possible use of agate equipment, our own experiences [175,189,208] have shown that these grinding tools cannot be recommended so much. The reason is that the contact zones are too smooth and the mixing itself quite difficult to be performed thoroughly. Moreover, commonly marketed agate mortars are relatively small in order to apply a sufficient pressure during the homogenization.

The proper homogenization by manual mixing of the two components (loaded in the mortar) is advisable to perform in two or three consecutive steps, when—after 2–3 min intensive homogenization—the seemingly compact mixture is scraped down from the rubbing wall with a spatula, collected at the bottom of mortar, and then intimately stirred again [175].

Some distributors marketing with electrochemical instrumentation and related equipment offer their ready-made mixtures, usually of the Nujol oil type together with two types of carbon paste holder (see Table 2.1 [241]).

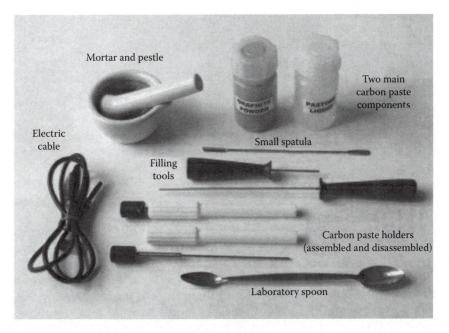

FIGURE 2.3 Equipment and accessories for the preparation of carbon pastes and the corresponding carbon paste electrodes. An illustrative photo. (From authors' archives.)

TABLE 2.1
Commercially Marketed "Ready-to-Use" Carbon Paste with Accessories

6.2801.000 Carbon paste
Used with 6.0802.000 CPE; material for three complete fillings (up to 17 g)

6.2801.020 Carbon paste
Used with 6.0807.000 mini-carbon paste electrode (up to 1.5 g)

6.0807.000 Mini-carbon paste electrode
Diameter of active zone 3 mm (suitable for 656 and 791 ELCD apparatus)

CPE-holder
6.0807.000

Source: Metrohm, Carbon electrodes & accessories, Specification data http://sirius.metrohm.ch/accessories/FMPro (accessed on January 10, 2002; site now discontinued).

As can be ascertained in the literature, some research teams were regularly working with such pastes (see, e.g., [242,243] and CP *Metrohm*, type "EA 267C"); nevertheless, a majority of carbon pastes is preferably made by electrochemists themselves. The popularity of laboratory preparation of carbon paste mixtures can be explained simply—the hand mixing is advantageous because one can choose one's own favorite components, as well as their mutual ratio, both being based on one's previous experience or recommendations from the literature. In this way, specially pretreated carbon paste components can also be used, including procedures for preparation of chemically and biologically modified CPEs. Last but not least such a laboratory way also allows one to prepare the whole series of related carbon pastes, which is arguably the best way for finding the optimal mixture [175].

2.1.5 Handling and Storage of Carbon Pastes

According to practical remarks [2,3,59,152,175], the ready-made carbon paste can either be collected and stored in a suitable container for later use or packed immediately into an electrode body (holder).

In accordance with some former observations, our previous reviews have stated that a long-time storage of carbon pastes does not seem to be very practical due to rather limited lifetime of some types of carbon pastes (see, e.g., [69,82,83,153]).

However, based on a recent finding on a 3-year-stored and still ready-to-use carbon paste mixture—described as a curiosity in the recent retrospective review [7]—one can agree with Adams [2] that the pastes made from traditional components can survive lengthy months if kept in tightly closed containers at a suitable place. But immediate filling of the freshly made carbon paste into an electrode body is also practical and if such a setup is designed properly, a portion of about 1 cm^3 of carbon paste is sufficient for several weeks of intensive measurements; see also [175] and references therein. (This, of course, does not apply to special experiments with accumulations into the carbon paste bulk, requiring a regular and/or frequent paste renewal by cutting the whole layers; see also Sections 2.3.1, 2.3.2, and 3.4.2.)

2.2 CLASSIFICATION OF CARBON PASTES AND CARBON PASTE ELECTRODES

2.2.1 TRADITIONAL TYPES OF CARBON PASTE–BASED ELECTRODES

Regarding classical carbon paste mixtures with two main constituents, the classification of the hitherto existing types can be made according to three different criteria: (i) the physicochemical character of the binder, (ii) consistence, and via (iii) the status whether or not the respective carbon paste is anyhow modified.

In the first category, one has the following:

- *Common (classical) carbon paste electrodes.* Mixtures made of spectral graphite powders and of liquid binders represented by chemically and electroinactive organic substances, such as paraffin or SOs. These CPEs are the most frequent type, representing around 80%–90% carbon paste–based electrodes, sensors, and detectors reported to date and belonging to the two-component configurations [2,3,56,82,83,69,83,166,167].

With respect to the consistence, the carbon pastes can be sorted in this way:

- *"Dry" or "wet" CPEs.* Such carbon paste mixtures differ significantly from the content of the binder, that is, from the "carbon-to-pasting-liquid ratio" used, which is schematically depicted in Figure 2.4 showing also the corresponding cyclic voltammograms for a model redox pair that confirm that too dry mixtures may suffer from less favorable signal-to-noise ratio (see "Dry CPE"). Although this classification is not very common, it can occasionally be found in some special studies [68,175,220,244] or in the early papers (see, e.g., [190]).

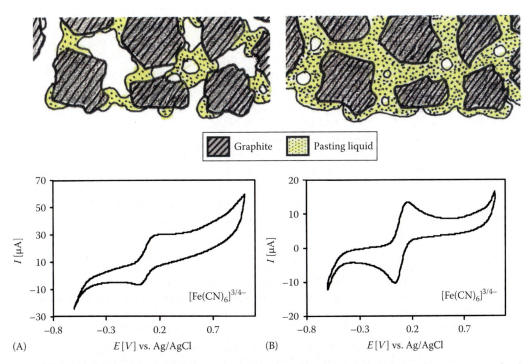

FIGURE 2.4 Microstructure of (A) "dry" and (B) "wet" carbon paste mixtures (a scheme, cross sections) with the corresponding cyclic voltammograms for the $[Fe(CN)_6]^{3/4-}$ redox pair. *Experimental conditions*: Carbon paste from spectral carbon powder (C) and silicone oil (SO), dry mixture: 0.5 g C + 0.2 mL SO, wet mixture: 0.5 g C + 0.4 mL SO; CV, polarization potential: from −1.0 to +1.0 V vs. Ag/AgCl and back, scan rate: 20 mV·s⁻¹; 0.2 M KCl; $c[K_4Fe(CN)_6]$ = 0.005 mol·L⁻¹. (Drawings and voltammograms from authors' archives.)

- *Soft or hard (desiccated) CPEs.* This is yet less common categorization appearing in literature only rarely and is rather subjective. Both terms were sometimes used in context with description of ageing effects [69,83,175,192,205].

For a long time, undoubtedly the most frequent classification is the approach to divide carbon paste–based electrodes, sensors, and detectors into two fundamental categories:

- *Unmodified CPEs.* These electrodes consist of binary carbon pastes, that is, mixtures made solely of the two main constituents—graphite powder and a binder [55,83]. Alternatively, such types of CPEs are called as "bare," "native," or "virgin" CPEs.
- *Modified CPEs.* Original binary mixtures contain additional component, usually, the modifier itself. According to the character of modifying agents, modified CPEs can further be subdivided as *chemically modified carbon paste electrodes*, CMCPEs [55,59,83] or biologically modified CPEs, falling into a large group of *carbon paste biosensors* [70].

Among a myriad of CMCPEs and CP-biosensors, there have been described numerous variants that contained—besides the native paste—even more modifying agents in one mixture. Besides common configurations "paste/modifier" and "paste/enzyme/mediator" [69,70], there were also mixed configurations such as "modifier/stabilizer (additive)" [234,235,245] or "two enzymes/mediator(s)" [246–248] and, eventually, "multienzyme systems" [249,250], enabling to accomplish and combine consecutive analyses of three different biological substances with the aid of bioreactors (containing the corresponding enzymes in banana, horseradish root, and mushroom tissues [249]) or simultaneous detection of two related compounds (saccharides [250]).

After such a juggling with the number of possible constituents in carbon pastes, it is fair to note that there also are binary mixtures that may represent a modified electrode material. For instance, some CPEs contain the pasting liquid formed by chemically active ion exchangers (originally for chromatographic columns), such as *Aliquat 336* [251], *Amberlite LA "IR1 20"* [252], and TCP [210,253,254]. In these cases, the classification of two-component carbon paste as an unmodified mixture is not so clearly defined and, for similar configurations, it is better to specify the active role of pasting liquid (according to previously cited reports [230,231]).

Almost all reviews present the two fundamental categories: (I) chemically and (II) biologically modified carbon pastes, via the bulk modifications by classical analytical reagents, complexants, ion exchangers, electrocatalysts, or well-known enzyme or mediator systems. This traditional approach is also followed within this book, when the corresponding information material is discussed in detail inside the relevant sections—in Chapters 4 and 5.

In this glossary, rather attractive than typical examples are collected in an effort to illustrate the topic alternatively, but with the same goal: to show how diverse modifications can be proposed and often carried into practice in the form of both CMCPEs and CP-biosensor. Moreover, in a debut, (III) physically modified carbon pastes are also presented here as a supplemental classification group, comprising less typical procedures used from time to time to accomplish and/or enhance some physicochemical processes.

I. *Chemical modifications.* The possibilities and limitations of chemical modification can, for example be documented on a group of the so-called metallic film-plated carbon paste electrodes (MeF-CPEs) that have passed through a challenging pathway [255]: from initially poor experience with mercury film-plated CPEs [67,256] and their later overcome [257–259], via somewhat indecisive results with gold-film CPEs [108,260,261]), up to the successful arrival among rapidly growing area of environmentally friendly bismuth film electrodes (BiFEs; [102,262–264]). In this case, CP-configurations were coming consecutively in the alternate variants: (i) BiF-CPE (with Bi-film generated either *in situ* or from special plating solutions [103]), (ii) Bi_2O_3-CPE (film formed *in nascenti* from solid oxide [265]), (iii) Bi(s)-CPE (with dispersions of fine Bi-powder [266]), and (iv) BiF-CMCPE (with *in situ* formed Bi-film and having CP-support additionally modified with clay [267]),

Carbon Pastes and Carbon Paste Electrodes

FIGURE 2.5 Some curious modifiers for carbon pastes. An illustrative sketch. (From authors' archives.)

which is an unprecedented variety within the "green family" of all the BiFEs [268]. And, very recently, this spectrum has been enlarged by introducing (v) BiF_4^--CPE [269] as the newest representative of bismuth-modified CPEs.

II. *Biological modifications.* If one defines this type of modification as the intentional use of material(s) of biological origin, it is possible to reproduce the survey [7] of unusual materials applied as the bulk modifiers (for illustration, see Figure 2.5): plant tissues or substrates (e.g., alga [270], lichen [271], moss [272], tobacco [273], pea seedlings [274], or grass weed [275]); tissues from fruits and vegetables (as natural sources of enzymes); e.g., banana [249,276], orange slice [277], apple [278], pear and peach [279], avocado [237], annona [280], pineapple [281], coconut [282], potatoes [279], or horseradish root [249]; mushrooms [249,279,283,284]; natural macromolecules (e.g., chitin [285], chitosan [286], and keratin [287]) or even living microorganisms (bacteria [288] and viruses [289]).

III. *Physical modifications.* Last but not least and on the occasion of debuting introduction of this category, one can present also some special studies, in which the role of modifier has been undertaken entirely or partially by: (i) daily light (to initiate a photoreaction [99] or for functioning of a solar cell [100]), (ii) UV-irradiation (to activate the surface of solidified CP-SPE [290]), (iii) elevated temperature (heating; in order to enhance interactions with DNA [291–293], diffusion transport of the analyte(s) [294], or mass transfer in the solution as such [295,296]), (iv) a short "shock" freezing (as a way of drying prior to preparing carbon paste in batches and with desired reproducibility [297]), (v) ultrasound (to activate the CPE surface by physical "erosion" [188]), and—finally—(vi) magnetic field (to reorient "by force" some macromolecules such as DNA for their easier oxidation [298,299], to increase again the mass transport [300], or to accomplish an immunoassay [301–303]). Regarding the first cases [298,299], the respective sensing systems have also been introduced under a special term—as a "magnetic carbon paste (electrode)."

2.2.2 SPECIAL TYPES OF CARBON PASTE–BASED ELECTRODES

2.2.2.1 Carbon Paste Electroactive Electrodes

A group of carbon paste–based electrodes appearing firstly in its typical form in the mid-1970s [38,304], but having had the principally similar predecessors a decade ago [14,15,29] or even essentially identical configurations yet before [305], comprises the carbon paste mixtures with qualitatively different compositions. Namely, in the CPEE configurations, the originally insulating and chemically inert pasting liquid is being intentionally replaced by a strong inorganic electrolyte, such as concentrated solutions of mineral acids or alkaline hydroxides [83].

A rather specific area of CPEEs, concentrated in a collection of more than 100 scientific papers [7] and already widely reviewed [151,170,306,307], is not only worthy of mentioning here, but deserves special attention—thanks to a variety of attractive, as well as unique applications—which is accomplished by means of a separate section (see Section 7.2).

2.2.2.2 "Solid," "Solid-like," and "Pseudo" Carbon Paste Electrodes

It is not surprising that the growing diversity of carbon pastes has also motivated numerous scientists to prepare, test, and popularize further related mixtures, in which the traditional oils were replaced by alternate binders. Many of the resultant electrode substrates had exhibited consistency markedly thicker than that of classical CPEs (being like peanut butter as specified by Adams [2]), giving rise to the terms used as headline of this section. Here, the individual examples can be briefly presented, when—except for a few cases—the denotation "solid" or "solid-like" CPE is authentic and come from the authors' reports.

The oldest prototypes of an s-CPE were those combining purposely graphite powder with (i) silicone rubber [32,65,308–310] that had gotten their final consistency after cold vulcanization, having been recommended for measurements in both stationary and hydrodynamic conditions. Similar s-CPEs designed as amperometric sensors or detectors were then made of (ii) highly molecular polyvinylchloride (PVC; [65]), (iii) chloroprene rubber + alkyl-phenyl resin (at the 1:1 ratio [310]), or related (iv) butadiene rubber (caoutchouc; [311]). Other types of s-CPEs were based on (v) paraffin wax [36] or (vi) ceresin wax [37], which can be easily melted (both beneath 100°C [38,80]), mixed with graphite powder, and then left to cool down into fine solid dispersions. At present, wax-impregnated electrodes can be crowned as the most frequently employed s-CPEs (see also [78,79,222,312–317]). Notably high resistance in organic solvents has been shown for a solid-like paste mixture prepared from (vii) poly(chlorotrifluoroethylene) oil (or wax, respectively), known under a trademark *Kel-F*® [66,318,319]. Reportedly, the respective detectors could be used in HPLC being immersed in mobile phase containing up to 70% ACN.

The remaining types of s-CPEs have already made an ephemeral appearance. Among them, one can present electrodes made from the melted and solidified (viii) phenanthrene [76], (ix) polyethylene ([77], yet with a modifying agent as the third component), (x) polypropylene (forming a very dense, sticky, and difficult-to-handle mass [188,320]), and, finally, (xi) 3,3′,5,5′-tetramethylbenzidine [321]—a polymeric binder capable of hosting either bulk-incorporated or surface-attached enzymes.

Note: In association with carbon paste–like mixtures, one should not forget a large family of *screen printed (carbon) electrodes*, *SP(C)Es* [74,83], belonging to most typical *carbon heterogeneous electrodes* or, else, *carbon composite electrodes*. Besides some related characteristics with CPEs, their sensing proper—carbon ink, formed by graphite powder and a polymeric resin (hardener)—has been occasionally called "carbon paste" (see, e.g., [83,97,290,322,323]). (Similar materials seem to be also (i) carbon cement [324] that can be packed as a carbon paste into a special electrode holder, or recently quite popular (ii) carbon films; see, e.g., [325].) As noted in historical introduction, carbon inks and SPEs are connected with carbon pastes and CPEs mainly in the developmental stage—as practical realization of laboratory testing of CPEs and CMCPEs (see also [326,327]). Today, SPEs represent a well-established and large group of modern sensors with a number of specific features (planar configuration, miniature size, possible industrial production, disposable character, and high promise for routine analysis), positioning their presentation already beyond the scope of this book.

In contrast to efforts in seeking new alternate binders, there is a series of related reports [328–331] on "pseudo-carbon paste electrodes" (p-CPEs) whose configuration misses the usual liquid moiety, replacing it with a polymer. More specifically, such porous p-CPEs can be fabricated by mixing graphite powder, pyrrole (as a monomer actually forming the paste-like mixture), and polystyrene microspheres (as a template). After catalyzed polymerization, the latter is removed to yield the resultant porous structure (with pores of 2–5 μm in diameter). Reportedly [328,331], the surface area of such (n)P-p-CPE is greatly increased (from ca. 10 up to $60\,m^2 \cdot g^{-1}$), benefiting—similarly as for typical CPEs—from an easy and simple preparation (when no binding agents needed), easy preservation and renewal, and mainly form being highly hydrophobic, thus exhibiting a substantial affinity for selective accumulations of lipophilic organic compounds [329].

2.2.3 NEW TYPES OF CARBON PASTE ELECTRODES

In fact, each CPE prepared from the materials that have been classified earlier as "new types of carbon" or "new binders" (see Sections 2.1.2 and 2.1.3 or [95,96,113,114,116,117,119–121,139, 193–204] and [238,239]) can also be considered as novel CPEs. In these cases, however, the respective electrodes represent more or less the first prototypes existing so far in a few exemplars and mostly undergoing the basic characterization only, that is, without wider appearance and typical applications.

Into this category, it is possible to add yet another quintet of reports [332–336] on newly proposed carbon paste–based electrodes that can be jointly called *carbon paste film electrodes* (*CPFEs*) and for which the resultant configuration has been reduced to a very thin film/layer of carbon (paste) attached to a compact electrode support, specifically, Ni-Cr alloy [332] and common GCEs [204,333,334]. The paste itself is either "super thin film" (STF; [332]) accomplished by inlaying a mixture of two traditional constituents (and the corresponding electrode denoted as STF-CPE) or composed entirely from alternate CP-components, which is the case of two mixtures MW-CNTs + (RT)IL [333,334] and GR + (RT)IL [204], the latter containing besides GR and ionic liquid (of the $Q^+PF_6^-$ type) also an additional solid binder with a biological sensing element.

A specific variant of these film arrangements is also concerned in two fresh contributions, describing novel two-dimensional carbon materials, "graphite oxide" [335] or "graphene oxide" [336], respectively. The first one was prepared by chemical oxidation of natural graphite powder and then used to modify the surface of the supporting electrode. In approximation, this approach leads to the same effect like electrolytic activation of the surface of some carbon electrodes, including CPEs (see [83,175] and also the following text, in the next chapter), when one can also obtain a very thin and sensitive layer of oxidized carbon particles whose presence totally alter the surface states (characteristics) of an electrode treated in this way [68]. With respect to the fact that graphene-oxide film could be electrolytically deposited directly onto the GCE surface—that is, without any inter-step needed for preparation of oxidized graphite particles and its subsequent attachment [336]—it is questionable, however, whether or not these specific configurations are to be classified as carbon paste– or carbon paste–like electrodes. Rather, it seems that their right ranking is among chemically modified carbonaceous substrates.

Within a wide variety of the new configurations, there are three significant categories of (i) DPEs, (ii) CNTPEs, and (iii) CILEs that have been applied more extensively, which is particularly valid for the latter two. In the late 2000s, they both have achieved the status of dominant groups among novel carbon paste–based electrodes, sensors, and detectors. The following sections bring mainly the basic characterization of all three groups, whereas the details on proper functioning, ways of modification, and practical applicability are only sketched. Again, all this information material is gathered in Chapters 4 and 5 and, then, in Chapter 8, where the individual electroanalytical methods are specified by means of typical data gathered in a series of summarization tables.

2.2.3.1 Diamond as the Electrode Material and Diamond Paste Electrodes

Besides natural and synthetic forms of hexagonal graphite, or various new forms of carbon, there is also a cubic allotrope of the same element—diamond. Apart from other differences between the individual forms of carbon, it is well known that the latter is an excellent electric insulator (with resistivity of about 10^{20} $\Omega \cdot cm^{-1}$ [337]). Thus, the pure forms of diamond were for a long time believed to be inapplicable in electrochemical measurements.

However, this hindrance was overcome by the appearance of synthetically doped and sufficiently conductive modifications that rapidly occupied a position in the corresponding electrode configurations [337,338], especially those based on boron-doped diamond films [339], configured as a very thin crystalline layer material immobilized onto a suitable electrode supports, and giving rise to a large family of *boron-doped diamond electrodes* (*BDDEs*).

If one considers the diamond, as well as its boron-doped variant, to be an alternative to carbon or graphite, respectively, there is yet another example of such carbonaceous material. It is boron carbide (B_4C, a *Norbide* product with satisfactory conductivity), representing—in a certain context—a historic predecessor of carbon pastes in Adams's laboratories [340,341]. This crystalline and extremely compact substance, not dissimilar to diamond, was tested in the form of a *boron carbide electrode* (*B_4CE*) in the bulk configuration, being described as "a solid electrode inert to a much higher degree than do the noble metal counterparts" [3]. From today's point of view, it is quite logical to speculate on why the same material had not been examined in the powdered form and as a paste, which would have given rise to a true alternative to "normal" carbon pastes at the time, when they were just coming to the fore. The answer can likely be withdrawn from Adams's book, where he mentioned "difficulties to mount the electrode," which may imply comparable problems with inevitable grinding of this extraordinarily hard material.

However, the early 2000s had seen such an attempt with pure diamond, when the resultant powders were immediately presented in a form of diamond paste electrodes (DPEs; [118]). Since then, energetic propagators of this new type of electrodes have published a full dozen contributions [342–354] that were to convince the electrochemical audience about realness of incorporating nondoped diamonds into the functioning electrode assembly.

Reportedly [118], the pastes containing either (i) natural or (ii) synthetic diamond powder (both available as especially fine powders with average particle size of 1 µm or even rougher fractions with mesh of 50 µm) and mixed with ordinary POs typically exhibit a very high sensitivity in faradic measurements, giving rise to remarkably high currents during the detection of analytes at "common" concentrations. Maybe that such high absolute values of the overall signal-to-noise ratio could be behind achieving some amazing LODs down to 1×10^{-12} mol·L^{-1} for the single ions [344,347], as well as for some analytes of organic and biological nature [345,352].

To date, DPEs have been applied to the determination of a series of inorganic ions, namely, Fe^{2+} and Fe^{3+} [118,347], Cr^{3+} and Cr^{VI} [344,346], Pb^{2+} [349], Ag^+ [353], and iodide, I^- [342]. Of interest were also some biologically important compounds and pharmaceuticals: L- and D-pipecolic acids [343,350,351], creatine and creatinine [345,352], *Azidothymidine* ([348], one of the effective therapeutic agents for retarding AIDS progress), and *Sildenafil citrate* ([354], known worldwide as *Viagra*® for treating man's erectile dysfunctions). In association with the individual method proposed, it is interesting that all the procedures with DPEs had employed solely *two-component diamond pastes* or, else, that there was no adaptation to a chemically or biologically modified variant.

Last but not least, a rather mysterious task on what is the right reason for the proper functioning and factual conductivity of diamond pastes in electrochemical measurements has been answered by the inventors of DPEs themselves [355], admitting that it would have been and be the presence of ubiquitous trace impurities inside the crystalline structures of both natural and synthetic diamond powders.

2.2.3.2 Carbon Paste Electrodes and Carbon Nanotubes

Since the introduction of CNTs into the modern industry and research (originally under the term "helical graphite" in the already classical report by Iijima [122] having harvested to date more than 10,000 citations [150]), this new form of carbon occupies the prominent position also in the today's electrochemistry and electroanalysis [356–358].

Nearly the same can be stated with respect to the role of the two basic categories of this cylindrical allotrope of carbon, *single-walled* and *multi-walled carbon nanotubes* (*SW-CNTs and MW-CNTs*, respectively) among the carbon paste mixtures. The priority in the use of CNTs in the carbon paste configuration is sometimes being thought to belong to a pair of contributions being published—almost in parallel—in the early 2000s [123,124]. However, after a more intimate literature search, one can trace the paper by Britto et al. [359], having appeared yet a half a decade ago and, in fact, reporting for the first time about the use of CNTs in measurements with CPE.

Specifically, when loosely citing the authors themselves [359], "... New carbon nanotube electrodes were constructed using self-made polyhedral nanotubes mixed with bromoform as binder into a homogeneous blend. The resultant mass was filled into a glass tube, and devised with a Cu- (or Pt-)wire as electric contact by inserting into the paste and fixing on a tube with glue tape ..." Nevertheless, a notable boom is really seen later and connected with the aforementioned pioneering studies in Rivas's [123] and Palleschi's [124] laboratories. Since then and up until now, CNTs in the proper paste mixtures can be found in ca. 100 reports published in the new millennium [119,127,229,333,334,360–449].

In the individual configurations, CNTs can be applied as either SW-CNTs (see, e.g., [124,334,362,396,434]) or MW-CNTs [333,366,383,438,439], when the selection of the two basic forms of CNTs is almost equal despite some differences in behavior, e.g., electric characteristics or resistance to aggressive chemicals [356,357]. On the other hand, both share the most valuable property of CNTs—the electrocatalytic activity—and can be manufactured in similar sizes.

Note: Usually, SW- and MW-CNTs have a diameter in the order of nanometers and length in tens of nanometers [358]; but there are some "extreme" variants, prepared by the CVD method, with abnormal dimensions. For the realm of CPEs and CMCPEs it is interesting and characteristic at once that even such atypical CNTs have already been examined in the paste mixtures—see, e.g., [406] and the use of "ultralong nanotubes" with average length of ca. 1000 nm.

In faradic (current-flow) measurements, the CNT-based pastes are typical for their enhanced signal-to-noise ratio that is given by the specific surface geometry at the carbon tubules, along with their catalytic capabilities. This is demonstrated in Figure 2.6, comparing such an electrode with a conventional CPE.

CNTs can be applied as a substitute of graphite powder (see [360–365,376,387,437]), giving rise to *two-component mixtures* of the "CNTs + liquid binder" type [123,124] known as the so-called *(I) Carbon Nanotube Paste Electrodes (CNTPEs)*. At the same time, this configuration is one of the two major categories of the CNT-containing paste electrodes and sensors [84]. In order to prepare such an electrode, CNTs are commonly mixed with (i) mineral oils [360–364, 423,425,435], occasionally with (ii) SO [229,380], (iii) bromoform [359], or an (iv) electroactive binder (e.g., *N,N*-didodecyl-*N′,N′*-diethylphenylenediamine [391]), and still more frequently with (v) (room temperature) ionic liquids, (RT)ILs [127,333,394,406,414,426,448]. Again, except for the latter, all these combinations can be classified inside the same family as the individual variants of the CNTPE fundamental type [84].

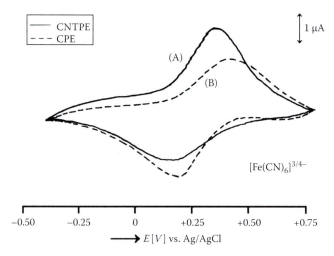

FIGURE 2.6 Cyclic voltammetry of the $[Fe(CN)_6]^{3/4-}$ redox pair at carbon nanotube paste electrode (CNTPE) and common carbon paste electrode (CPE). (A) CNTPE: 0.3 g MW-CNTs (NTP, Shenzhen, China) + 0.20 mL Nujol oil (Merck); (B) CPE: 0.3 g graphite powder ("CR-5," Maziva Tyn, Czech Rep.) + 0.15 mL Nujol oil. *Experimental conditions*: CV, polarization potential from −0.3 to +0.7 V vs. Ag/AgCl and back, scan rate: 25 mV·s^{-1}; 0.25 M KCl; $c[K_4Fe(CN)_6] = 0.005$ mol·L^{-1}. (Voltammograms from authors' archives.)

The second possible configuration is then symbolized by *(II) Carbon Nanotube-modified Carbon Paste Electrodes* (*CNT-CPEs*), where CNTs themselves play the role of an additional constituent, typically, of an electrocatalytic agent. These *three-component systems* consisting of (i) traditional graphite powder, plus (ii) CNTs added in, and (iii) mineral oil can thus be considered as the basic variant of the type "II." And similar to CNTPEs, also CNT-CPEs belong to relatively often recommended carbon paste mixtures (e.g., [372,378,384,397,431,441]).

Yet more complex variants had come to the fore afterward, when originally two- or three-component arrangements were combined with additional constituents. This trend logically led to further *subvariants of CNTPEs and CNT-CPEs*, which was particularly typical for laboratory-made carbon pastes and inconceivably expanded their application potential in the field. Such a deduction can simply be documented by the respective scientific papers, especially those having appeared within the last 2 years [410–449], but known also from the early contributions that represent the first actual chemical [360,368,370] and biological [364,372,375] modifications of CNT–based pastes. Among typical subvariants, one has

(I-a) *Chemically modified carbon nanotube paste electrodes (CMCNTPEs)*. This type can yet be divided into two different subconfigurations. Firstly, the original *two-component mixtures* have the modifying agent ("X") attached on the surface ("X/CNTPE" types); see, for example, those with chemically immobilized fluorine compound [396], electrolytically deposited Ni(II) complexes [411,424] and inorganic/organic hybrid polymer [433], or, eventually, *in situ* adsorbed chelates [397,407]). Secondly, the modifier is manually embedded into the carbon paste bulk ("X-CNTPE"), forming *three-component systems* with admixed Cu^{II}-molecular wires [373] or Cu^{II}-hexacyanoferrate [436].

The respective electrodes benefit from the presence of CNT as the more effective carbon moiety, as well as from the agent "X" bringing the desired chemical selectivity. In some cases, however, the function of both CNT and X can be mutually integrated, when the latter is capable of further amplifying the electrocatalytic effect of CNTs alone (see, e.g., [373,432,444]).

(I-b) *Biologically carbon nanotube paste electrodes (CNTP-biosensors)*. Similar to CP-biosensors in the early 1990s [70], also their CNT-analogues have gained particular attention a decade later, giving rise to a number of new adaptations of already existing configurations.

The CNTP-biosensors typically consist of (i) CNTs, (ii) mineral oil, and an enzymatic system, that is, (iii) enzyme [364,385,405,433,475], microbial cells [409], or biological tissue [359], and a (iv) mediator (or, instead, a binder capable of the same functioning [409]).

(II-a) *Carbon paste electrodes modified with carbon nanotubes and other reagent(s)* that are usually denoted as *X(Y)/CNT-CPEs* can be split again into two related variants with the modifier (i) on the surface [393] or (ii) admixed—together with CNTs—into the original graphite paste [431]. These systems are also *four-component composites* that can be obtained by mixing, for instance, (1) carbon powder + CNTs + hemoglobin + liquid paraffin; (2) graphite powder + CNTs + ferrocene + mineral oil [432] or, more specifically, (3) 60% (w/w) graphite powder + 5% MW-CNTs + 20% PO + 15% ionophore (complexant) [446]; the latter containing CNTs due to their benefit on the overall sensitivity.

And in some applications, the respective mixture can be truly multicomponential, which was the case of two nearly identical mixtures [419,447] containing *five constituents*: (i) 52%–57% graphite powder + (ii) 5% MW-CNTs + (iii) 20% PO + (iv) 15%–20% ionophore + (v) 3% nanosilica. These "hyper-heterogeneous" carbon paste composites, differing mainly in the ionofore used, were the appropriate ones of two ion-selective electrodes developed for direct potentiometric indication in samples with complex matrix.

(II-b) *Carbon paste electrodes modified with carbon nanotubes and an enzyme/mediator system (CNT-CP-Biosensors)*. According to one selected report [381], this setup can be of interest if the CNTs serve as the substrate for immobilizing microbial cells by means of a polymer that effectively shuttles the electrons between redox enzymes in the cell and, at the same time, promotes a stable binding to the surface of a common carbon paste completing the whole sensing system.

(III) *Carbon nanotube paste (Thin) film electrodes (CNTFEs)*. In fact, these atypical configurations have already been discussed in association with *new carbon paste electrodes* (see Section 2.2.3 and [127,204,333,334]). It can be briefly repeated that the CPE is reduced to a carbon paste film hosting at "foreign" electrode substrate. Here, the four remaining examples can be supplemented with other two (previously omitted). First, it is a "MW-CNTs + 6,7-dihydroxy-3-methyl-9-thia-4,4a-diazafluoren-2-one" film at the surface of a CPE [439], the second is *dual carbon-paste-film electrode* configured as an *inlaying ultrathin carbon paste electrode* coated by functionalized (chemically modified) CNTs [446], when the inner carbon paste layer has been spread onto a nichrome–alloy substrate and the whole configuration can be schematized as "CMCNTs/ut-CP/(Ni-Cr)E."

With respect to the *construction size*, a great majority of both CNTPEs and CNT-CPEs were employed in the usual macrodesign (with the electrode diameter in the order of millimeters) and there was only one special series of miniaturized patch- and needle-sensors developed for *in vivo* monitoring and with dimensions balancing between mini- and microdesign [377,382,390,396,409].

And regarding CNTs-based CPEs and their *compatibility with electrochemical techniques*, it can be stated that there are no marked limitations. This can be documented on brief survey comprising (i) cyclic voltammetry (CV; [360–363,390,398,439]) and (ii) linear scan (sweep) voltammetry (LSV; [425]), ramp-modulated modifications like differential pulse (DP; [365,384,416,442]), reverse differential pulse (RDP; [374]), square wave (SW; [369,376,402]), and the second derivative (2ndDe; [393,407]) voltammetry, potentiometric indication [399,447], amperometric detection in flowing streams (FIA; [367,370,436]), and capillary electrophoresis [367,370,401], or electrochemiluminescence detection [386], the latter two also in a mutual combination (ECL-CE; [389]). Furthermore, useful information from the characterization of CNTs-based carbon pastes and the selected modifying systems could be enriched with some specific data obtained with other electrochemical techniques, such as chronoamperometry [405,422,439,442], chronopotentiometry [366], EIS [405,421,440], galvanostatic charge/discharge [406], and ohmic-RM [229,406]. Finally, valuable observations could also be obtained by means of nonelectrochemical measurements, for example, by SEM [446] and STM [405]), x-ray diffraction (XRD; [405]), or Raman spectroscopy [380].

Over the years, both CNTPEs and CNT-CPEs, together with their chemically and biologically modified subvariants, have been the central object of wide *electrochemical investigation*, comprising the initial characterization of basic types of nanotubes and the respective CNTPEs (e.g., [123,124,127,360–364]), their comparison with traditional CPEs [359,436], other new types of CPEs [201,434], or even with further kinds of CNTs-based pastes [417]. Of interest were the specific electrocatalytic properties of both SW-CNTs and MW-CNTs [361,368,403,427], reaction kinetics for various (model) compounds at the corresponding electrodes [368,411,444], specific ion transfer at liquid/liquid phase boundaries [391], or their treatment at elevated temperature [389,406] or in rotated arrangement [127].

Most of CNT-based electrodes have also been successfully applied in practical electroanalysis, where the individual configurations could be employed for a wide spectrum of various inorganic and organic species. This can be documented by the following survey:

- *Inorganic ions and molecules. Single metal ions*: Hg^{2+} [369,448], Cu^{2+} [382,393,425], Cd^{2+} [377], Pb^{2+} [382,447], $Sb^{III,V}$ [428], Zr^{4+} [397,407,408], Er^{3+} [415], and Ho^{3+} [419,446]. *Anions*: sulfite, SO_3^{2-} [436], bromate, BrO_3^- [380], and iodate, IO_3^- [380].
  Molecules: hydrogen peroxide, H_2O_2 [201,431,449] and hydrazine, N_2H_4 [411].
- *Organic substances and biologically important compounds.* Methanol [424], ethanol [364,372], other aliphatic alcohols [424], phenol and its derivatives [362,364,381], quinones [362], nitromethane [204], phenylhydrazine [441], thiolic derivatives [363,370] (including amino acids like acetylcysteine [403], homocysteine [363], glycine [429], glutathione [445], tryptophan [384] and others [401]), ricin [435], flavonoids (quercetin, rutin [366,374], and bergenin [398]), glucose [138,360,371,375,381,385,386,400,405,433], uric acid [361,399,422,427], xanthin and hypoxanthin [399], tocopherols [402] and folic acid [440], ascorbic acid [361,416,422,438], norepinephrine [442], dopamine [123,359,361,376,416,427,443], DOPAC [122,361,367], NAD(H) [364,368,439], and DNA [378,388,394,395,412,421,440].
- *Pharmaceuticals.* Acetaminophen (syn.: Paracetamol [423,438,443]), Caffeine [430], Carbidopa [432], Cisaprid [392], Isoniazid [383,438], Levodopa [420], Methamphetamine [426], Metformin [373], Oxytetracycline [390], Piroxicam [387], and Urapidyl [379].
- *Environmental pollutants. Organophosphate insecticides*: Acephate [389], Dimethoate [389], EPN (*O*-ethyl-*O*-(4-nitrophenyl)-phenyl phosphonothioate [409]), and *herbicide*: Amitrole (3-amino-1,2,4-triazole [365]).

2.2.3.3 Carbon Paste Electrodes and Ionic Liquids

Since publication of the pioneering paper by Tiyapiboonchaiya et al. [450], the so-called (room-temperature) ionic liquids, (RT)ILs, have soon attracted electrochemists and electroanalysts worldwide, initiating a new extensive research focused on possibilities and limitations of incorporating various (RT)ILs into the electrode configurations (see, e.g., [451,452] and references therein). And similar to almost the same move around CNTs, this newly coming trend could not escape attention of experimenters preferring and popularizing carbon paste–based electrodes.

As already stated [452], carbon paste mixtures with their compositional and functional basis offer an almost ideal platform for using (RT)ILs, when a variability in the use of different pasting liquids and their mutual substitutability had opened a particular position for popular ionic liquids. Thus, since the report by Liu et al. [125], in which (RT)ILs were first described to be intentionally added into the carbon paste bulk (when the respective CPE had served as *the probe of choice with a low substrate current and ohmic resistance* for studying the capacitive characteristics of some imidazolium-based ILs) and its direct continuation [126], in which an ionic liquid replaced the conventional binder, the (RT)IL-based CPEs have been the central subject of interest in ca. 80 scientific papers appearing up until now [127,128,204,333,334,394,407,411,413–416,418,419,427,449,453–515].

The individual contributions have shown clearly that electrochemical measurements with CPEs may benefit from a number of specific properties of (RT)ILs [451,452], namely, (i) their excellent solvating properties, (ii) high conductivity, (iii) nonvolatility, (iv) electrochemical stability with (v) wide polarizability, (vi) specific chemical activity (mainly ion-exchanging properties and electrocatalytic effect), as well as (vii) typically low toxicity. And to complete this characterization, there is also (viii) favorable surface microstructure of IL-containing carbon pastes with their markedly smooth pattern (see Figure 3.1), with mutual comparison of various carbon paste mixtures.

Most of these features are also highlighted by Maleki et al. (other pioneers and propagators of ILs in the carbon paste mixtures [453,456,457]), having concluded in their special study [459] that the addition of an ionic liquid into a CPE gives rise to (a) the increased conductivity of the binder with the subsequent (b) decrease of the electrode resistance, (c) increase of ion-exchange capabilities, and (d) inherent catalytic activity of the resultant electrode material containing such an (RT)IL.

In their concluding remarks, however, the authors could have also pointed out the characteristic enhancement of the overall signal-to-noise-ratio at CILEs and related configurations, when the resultant effect in faradic experiments is very similar to that shown earlier for the CNT-based CPEs. Then, also an experiment with the model redox pair yields nearly the same result—see Figure 2.7 and compare it with the previous one, Figure 2.6.

Regarding the individual types of ionic liquids used, their choice was often routine and relatively conservative [452], preferring largely the ion associates of the (i) [n-R–py]$^+$A$^-$, (ii) [R,R'–pyr]$^+$A$^-$, and (iii) [R,R'–Im]$^+$A$^-$ type. Specifically, they are formed by the quaternary cations with alkyl/heterocyclic structure, namely, [n-alkyl-pyridium]$^+$, [N-butyl-N-methyl-pyrrolidinium]$^+$, and [R,R'-imidazolium]$^+$), the latter representing the most frequent case. The anionic moieties are then some voluminous anions, especially hexafluorophosphate, PF$_6^-$ (see, e.g., [394,453,454,464,475,505]), tetrafluoroborate, BF$_4^-$ [410,415,486,494,510], less often bis(trifluoromethylsulfonyl)imide, NTf$_2$ [127,454], ethylsulfate, Et-O-SO$_3^-$ [508], or bromide, Br$^-$ [468]. In such configurations, the (RT)ILs of choice are usually declared under the specific abbreviations, when—in the order of the

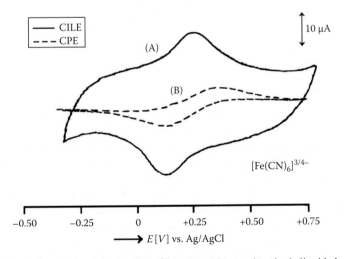

FIGURE 2.7 Cyclic voltammetry of the [Fe(CN)$_6$]$^{3/4-}$ redox pair at carbon ionic liquid electrode (CILE) and common carbon paste electrode (CPE). (A) CILE: 0.3 g graphite powder ("CR-5," Maziva Tyn, Czech Rep.) + 0.2 mL BMIMPF6 (Sigma/Aldrich); (B) CPE: 0.3 g graphite powder + 0.2 mL Nujol oil (Merck). *Experimental conditions*: CV, polarization potential from −0.4 to +0.8 V vs. Ag/AgCl and back, scan rate: 25 mV · s^{-1}; 0.25 M KCl; c[K$_4$Fe(CN)$_6$] = 0.005 mol · L^{-1}. (Voltammograms from authors' archives.)

aforementioned classification—one can quote (i) BPPF6, HPPF6, OPPF6, (ii) C4MPyNTf2, (iii) BMIMPF6, HMIMPF6, OMIMPF6, PMIMPF6, EMIMBF4, BMIMBF4, etc., where "M" denotes "methyl"; "E" ethyl; "B" *n*-butyl; "H" *n*-heptyl; "O" *n*-octyl; "P" phenyl, as well as pyridinium; "Py" pyrrolidinium and "IM" imidazolium. Despite this variability, there are two ionic liquids whose use substantially prevails over the others: BPPF6 and BMIMPF6.

In carbon paste mixtures, an (RT)IL can be used as (i) additional component (usually as a modifier enabling some of the facilities pointed out earlier) or (ii) special binder replacing the original pasting liquid and contributing with its own typical properties. Here, one should note that some (RT)ILs can be nearly solid substances (e.g., OPPF6), thus giving rise to a *solid-like carbon ionic liquid composite* [478]. Otherwise, the two basic approaches (i) and (ii) can be mutually combined, which resembles the variability among CNTs-based electrodes and the classification of (RT)IL-based counterparts can therefore be made in a similar way:

(I) *Ionic liquid-modified carbon paste electrodes (IL-CPEs)*. This *configuration with three components* in the paste mixture opens the list as "historically" first [125] and to date quite frequent category [126,415,453,459,465,468,505,513].

(II) *Carbon ionic liquid electrodes*. A qualitatively new type of CPEs introduced under this name—or as *Carbon Composite Ionic Liquid Electrodes*, respectively—by Maleki et al. [453] represents, at present, the largest group of IL-based CPEs (see, e.g., [126,448,454, 455,457,477,485,492,503,514]). Similar to traditional carbon pastes, they also form *two-component* (*binary*) mixtures, but with the alternate pasting liquid moiety.

(III) *Carbon nanotubes ionic liquid electrodes (CNTILEs)*. The respective composites are "younger" than both previous types, but also popular (e.g., [127,410,412,448,471]), being truly a new kind of CPEs, where both major constituents are replaced by alternate materials.

(IV) *Carbon-nanotubes ionic-liquid film electrodes*. Besides standard configurations of CNTILEs, it is possible to mention again the constructions, where a very thin layer of the CNTs + IL mixture is attached to the surface of a foreign electrode (either the CILE itself [481], a common CPE [479], or—more often—the polished disc of the GCE [204,333,334]), all serving as the support and current collector. In some cases, the composite layer contains also additional constituent(s)/stabilizer(s) in one integrated system [204,479,503,512].

(V) *Chemically and biologically modified IL-CPEs, CILEs, and CNTILEs*. Of course, also these configurations allow one to modify—upon request—the respective native mixtures, when the concrete examples can easily be found in the existing databases. Thus, it is possible to have (i) the "CM-IL-CPE" type (see, e.g., [508]), (ii) "CM-CILE" [126,482,504], and (iii) "CM-CNTILE" [394,509], or CIL- and CNTIL-biosensors (e.g., [475,497] and [414]).

(VI) *Combinations of new types of carbon (graphite) with ionic liquids*. The paste mixtures that do not contain any of the two traditional components form a small group of unique and interesting electrodes but with marginal significance when compared to the types (I-IV) [452]. Hitherto proposed prototypes of mixtures with (RT)ILs have contained the following carbon/graphite materials: (i) glassy carbon microspheres ("GC-ILE" type [486]), (ii) carbon nanofibers ("CNF-ILE"; [128]), and (iii) graphene ("GR-ILE"; [507]). Because these carbon pastes can also be chemically modified (see, e.g., [128] and a reagent attached *in situ* to provide a chemiluminescent signal), the overall variability of IL-based CPEs seems to be yet wider than that described earlier for both CNT-CPEs and CNTPEs.

Finally, there is also a unique application of an (RT)IL in measurements with carbon paste–based electrodes—a study, in which the ionic liquid had been preheated up to 60°C and then used as the supporting medium for testing a temperature-resistant CNTPE [406].

As instrumental techniques of choice to be coupled with IL-based CPEs, very frequent were voltammetric measurements (see, e.g., [126,453–459,469,488,500,513]), including the effective detection with the rotated disc electrode (RDE; [127]), followed by amperometric detection in flowing streams [463,472,479,492], combinations with potentiometric indication [410,415,418,448,457] or electrochemiluminescence detection [462,486,489]. Some characterization and additional studies have then been performed with the aid of EIS [125,466,471]) and scanning electron microscopy (SEM; [466,469,490]) Finally, to evaluate the effectiveness of some surface modifications, also spectral techniques were employed (namely, UV/Vis spectrometry [417,496], FT-IR [413,467,514], and RA-IR [490]). By considering the electrochemical investigations and more theoretical applications, the newly constituted area of IL-based CPEs covers an extensive and a very diverse experimental work—from the initial characterizations [126,453–456], including comparative studies with classical CPEs [394,453,454,457,487] and a report [507], in which an ordinary CPE was presented also as atypical *counter electrode*, via studies on the specific ion/charge transfer in ionic liquids [454,455,512], conductivity changes [127,448,471], or electrocatalytic effect [453–456,459,479,499,514], up to electrochemical characterization of two novel microvariants [472,490] or electrically heated assemblies [473,489].

Practically oriented *electroanalytical applications of IL-CPEs, CILEs, and CNTILEs* are minimally as diverse as the individual variants and subvariants of ionic liquid-based carbon pastes [452]. Again, their applicability to the determination of inorganic species, organic compounds and pollutants, pharmaceuticals, as well as numerous biologically active compounds can be illustrated by means of the following survey:

- *Inorganic ions and molecules. Single metal ions*: Cd^{2+} and Pb^{2+} [495], Cd^{2+}, Pb^{2+}, and Cu^{2+} [513], Hg^{2+}/Hg^{II} [448], Ce^{3+} [418], Yb^{3+} [511], Pr^{3+} [410], Er^{3+} [415,509], and Tb^{3+} [510]. *Anions*: NO_2^- [126,464,467,481], SO_3^{2-} [471], BrO_3^- [515], IO_3^- [482], ClO_4^-, PF_6^- [454]. *Molecules*: H_2O_2 [413,463,464,470,479,487,488,492,499,512,514], O_2 [479], N_2H_4 [472].
- *Organic compounds.* Phenol and 2,4-dichlorphenol [461], hydroquinone and anthraquinone [465,480,505], dihydroxybenzenes [504], metol [493,494]), catechol [461,476], dihydroxy-benzoic acid [501], and trichloroacetic acid [413,427,470,479,481,488,496, 499,503,506].
- *Organic pollutants.* p-Aminophenol [468], tripropylamine [462], nitromethane [204]; *p*-nitrophenol [500], TNT (2,4,6-trinitrotoluene [507]), and methylparathion (herbicide [334]).
- *Pharmaceuticals. Commercial products*: Dobesilate (calcium [460]), Fentanyl (citrate [486]), Paracetamol [477]; *abuse drugs*: Heroin [502] and Methamphetamine [426].
- *Biologically important compounds. Aminoacids*: cysteine [487], *purines*: guanosine [412,484], uric acid [333,456]; *vitamins*: ascorbic acid [453,456,458]; *saccharides*: glucose [475,478,414,497]; *neurotransmitters*: dopamine [453,456,466], adenosine [483]; *enzymes and coenzymes*: NADH [453,128]; *proteins and metaloproteins*: α-fetoprotein [485]; hemoglobin [413,417,464,467,479,496,499,503,512,514], and myoglobin [483,488]; *antioxidants*: rutin [498]; *macromolecules*: DNA (and the respective forms [474,491,508]).

A remarkable collection of nearly 50 papers, appearing all during the last one and a half years ([204,411,413–416,418,419,427,449,477–515], when the authors' literature search for this book had to be stopped in the middle of 2010), is the best indication that the recently established area of CILEs and related electrodes rapidly grows, reflecting faithfully the newest achievements in synthesis of further ionic liquids and similar fluids.

2.3 CONSTRUCTION OF CARBON PASTE HOLDERS

In order to employ soft, mechanically less compact carbon pastes in electrochemical measurements, it is necessary to choose a suitable support, that is, properly designed electrode holder (body). Such an assembly then represents *CPE* from a tradition point of view [3,69,83,166,167].

A task whether a special support for carbon pastes is to be considered as an advantage or as a drawback can be answered that there are pros and cons, depending on the fact which aspects are considered. Undoubtedly, a lack of integrity in the construction of CPEs may be a handicap, for instance, if one is used to work with classical sensors that do not rely on any maintenance prior to measurements and which are almost compatible with commonly used electrochemical instrumentation. Here, especially routine users usually prefer conservative constructions of commercially available electrodes. Nevertheless, the assortment of these electrodes is not so wide in order to cover all specific needs, among others, a freedom in choosing the electrode material and unusual electrode designs for special purposes.

In this respect, it seems that CPEs, CMCPEs, and CP-biosensors may offer practically endless possibilities, which is documented in the following sections surveying the most popular constructions of CPEs, sensors, and detectors, including less common or even rather curious designs.

2.3.1 Tubings and Rods (Plugs) with Hollow Ends

These very simple constructions can be made in every laboratory. Often, they represent a *provisory solution* of how to obtain quickly an electrode for some occasional measurements (see, e.g., [189,206]) or how to use carbon pastes as an alternate electrode material in some comparative studies [34,192,516]. As seen in Figure 2.8 and in Scheme "A," the CPE of this type consists of a glass tubes with appropriate diameter partially filled with a portion of carbon paste, into which a Cu- or Pt-wire is immersed through the upper part of the tube [517–519]. Electrical contact can also be made by adding a small portion of graphite powder with droplet of mercury [520], or even carbon paste alone in some s-CPEs configurations [521].

If needed, the lower part of tubing can further be enlarged into a bell-like shape in order to obtain a capsule with a sufficient capacity for larger portions of carbon paste [10,12]. Instead of glass, tubes (or small columns) from plastic materials that are chemically resistant (e.g., PTFE) can also be used [36,522–525], sometimes in the form of medicinal syringes [525,526], where the piston may serve for intimate fixing of the contact wire [527].

The authors who were employing carbon paste–based electrodes more frequently had usually proposed their *own constructions* (e.g., [3,28]) that—more or less—were genuine copies of the overall design of traditional solid electrodes with respect to shape, dimensions, as well as the manner of connecting the electrode with the apparatus. Moreover, in some cases, these home-made constructions had also been made in several variants. In this respect, the authors of this text are delighted to introduce here a very rare and authentic image from Adams's laboratories, illustrating such variability and so far presented only in a scarcely accessible PhD thesis [11]—see Figure 2.9.

Despite the bad quality, this unique photo depicts clearly the individual differences between all four constructions, two of which have had rather special use and are commented again in the following. The joint image also comprises the pioneering construction of Adams's first special CPE holder—Teflon "plug" (Design "B") whose cross section with detail view into the filling well has already been sketched in the previous drawing (Figure 2.8, Scheme "B").

Regarding the proper procedures of carbon paste filling, refilling, and the renewal as such, the container-like constructions do not offer a very high comfort. The wells/hollows have to be often refilled and, when using constructions with lower capacity, even quite frequently. This is also the case of the aforementioned drilled plug [2,3,11]. Rather time-consuming refilling is apparently the most serious drawback of these otherwise popular constructions and, in some experiments, it could represent unpleasant complications. For instance, anodic oxidations of some organic compounds give rise to reaction intermediate products with strong adsorption capabilities that—if not removed mechanically—may block active sites at the electrode surface, thus hindering the detection of the following electrode transformations [3,82,111].

Carbon Pastes and Carbon Paste Electrodes 37

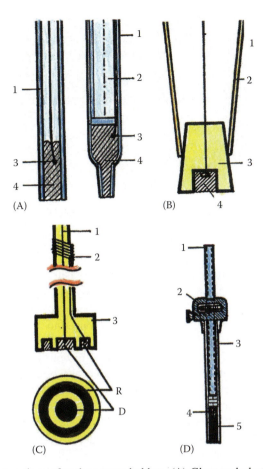

FIGURE 2.8 Basic constructions of carbon paste holders. (A) Glass and plastic tubes: 1, tube (syringe); 2, ... piston, 3, contact (metal wire); and 4, filling of carbon paste (CP). (B) Adams's construction: 1, Cu-wire and Pt-contact; 2, glass tube (conically narrowed); 3, Teflon® plug; and 4, well filled with CP. (Newly redrawn from Adams, R.N., *Electrochemistry at Solid Electrodes*, Marcel Dekker, New York, 1969.) (C) Rotating ring-disc CPE. 1 and 2, inner and outer electrical contact and 3, Teflon body ("P": ring, "D": disc). (D) Monien's construction (Redrawn from Monien, H. et al., *Fresenius Z. Anal. Chem.*, 225, 342, 1967): 1, screw (piston); 2, turning head; 3, electrode body; 4, Pt-contact; and 5, CP. (Redrawn from Kalcher, K. et al., Heterogeneous electrochemical carbon sensors, in *The Encyclopedia of Sensors*, Vol. 4, Eds. C.A. Grimes, E.C. Dickey, and M.V. Pishko, American Scientific Publishers, Stevenson Ranch, CA, pp. 283–429, 2006 and newly colorized.)

However, the main motivation for future innovations of CPE holders was the growing importance of practical electroanalysis, when the respective methods co-opened once more the problems with filling wells of limited capacity, hindering packing larger portions of carbon paste [31,32,34]. For example, still more popular electrochemical stripping analysis with CPEs and utilizing extractive accumulation required the regular carbon paste renewal—typically, after each measurement. The reason was that a certain amount of the substance to be analyzed might have remained entrapped in the surface layer even after rediffusion during the stripping/measuring step [20,175,219,226]. And if the penetration of extracted molecules into the CP-bulk was particularly deep, solely a mechanical removal of the so-affected layer—up to 0.5–1 mm [217,254]—could assure the acceptable reproducibility of measurements. In other words, such a thorough carbon paste renewal had also meant a considerable consumption of the paste and for that kind of experiments, another type of CPE holder had to be sought—ideally, any other sophisticated construction enabling the carbon paste extrusion and its immediate renewal.

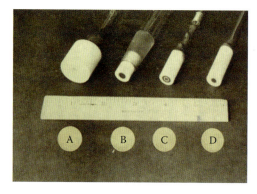

FIGURE 2.9 Four variants of carbon paste electrodes used in Adams's laboratories. (A) Shielded CPE in Teflon; (B) planar CPE with ground glass fitting; (C) rotated disc CPE; (D) planar CPE. (A photo from Prof. Kuwana's personal archives; presented with his kind permission.)

2.3.1.1 Rotated Disc Electrodes for Hydrodynamic Measurements

Yet before going to such more flexible CPE-holders and in accordance with chronological order, it is necessary to present rotated/rotating variants of carbon paste–filled electrode bodies (plugs, cylindrical rods with drilled end holes, etc.) that had belonged to a regular equipment of electrochemists in the early era of carbon paste [2,3,10,18,528–530]. (Two basic variants of RD-CPEs are also included in Figures 2.8 and 2.9 as documented in both legends.) However, the respective constructions had been of interest not only in theoretically oriented studies of electrode reactions, but could also be highly effective in contemporary electroanalysis [530], as well as later on [63,249,519,531–535]. In the meantime, the respective designs underwent slight variations—from common cylindrical [10,528] to bell-shaped variants [12,18], via unshielded and shielded modifications [3,11], up to special rotated assemblies that have been described and used recently [536,537], including studies with new carbon paste mixtures made of CNTs and (RT)ILs [127].

A rather unique variant of rotated electrode was the so-called *rotating ring-disc carbon paste electrode*, *RRD-CPE* (see again Figures 2.8 and 2.9 and the Scheme/Design "C") devised and successfully used for studying some multistep electrode mechanisms [10,11]. Its dual working surface allowed one to apply different potentials at both ring and disc parts separately, thereby even parallel or consecutive electrode processes could be registered one after the other and without mutual interferences.

2.3.2 Piston-Driven Electrode Holders

Apparently the most elegant way of how to renew the used carbon paste was introduced by Monien et al. [31] who had smartly utilized the principle of a micrometric screw for quick and very effective extrusion of the paste out of the body. As seen in Figure 2.8 (Scheme "D"), the cavity inside a glass body with a Teflon sleeve was formed by the respective position of the screw and, reportedly, could contain up to 3.5 g carbon paste. The paste pulled out of the holder could then be simply cut off and polished by a soft paper tissue—for example, wet filter paper is also very convenient [175]—the entire operation lasting a few seconds only. In fact, Monien's design was the predecessor of piston-driven constructions that came to the fore soon after [33], supplemented yet with a special tool for cutting the extruded paste with a stretched piano string [33,538]. Later constructions had also combined the material used, which was the case of a CPE consisting of a glass body and plastic piston [539].

Laboratory-made carbon paste holders with piston had been of continuing interest and the realized constructions brought numerous innovations, offering useful and, in some cases, completely new construction details.

Carbon Pastes and Carbon Paste Electrodes

In this direction, quite a large effort was made by the authors of this book, having proposed a whole series of various CPE-holders, assisting their proper manufacturing since the mid-1980s. The genesis of the individual constructions in our workshops is depicted in Figure 2.10, presenting all important designs of the piston-operated carbon paste holders and assemblies appearing as time went by.

After a provisory use of a small syringe with cut-off end-tip [537], the first piston-driven body (image "A") was machined from Teflon according to Monien. Its weakness was thread carved directly into the soft plastic and, mainly, the piston itself made of ordinary steel and thus susceptible to corrosion. The second design (image "B") had the main body machined from much harder plastic mass (Silon®) with more resistant thread, as well as with piston from a noble stainless steel. This somewhat robust holder ("thick" or "T" type [540]) came as a special electrode kit devised with exchangeable thread-devised endings (caps) machined again from Teflon and having four different end holes allowing one to select three different electrode diameters for voltammetric experiments (1, 2, or 3 mm), or even a "huge" surface (with diameter up to 10 mm) for some potentiometric or coulometric measurements [541]. Later on, in the early 1990s [56,153], the function of thread in the Silon corpus was improved by a special stainless steel inset with contra thread. Meanwhile, however, the Silon material itself had been found inapplicable in some particularly aggressive chemicals (see a note in Section 2.1.3) and the next series of plastic bodies were again machined wholly from Teflon. The resultant CP-holders ("C") were notably thinner and, in overall, less robust compared to

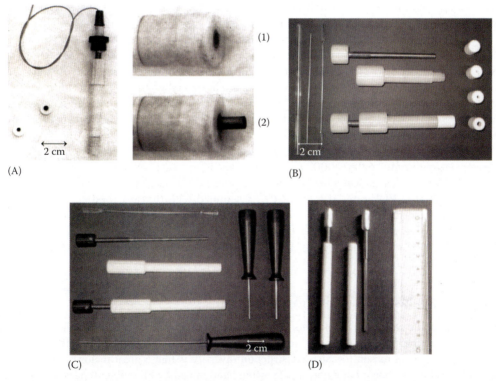

FIGURE 2.10 Piston-driven carbon paste holders and the individual construction variants. (A) First prototype (with two spare end tips and the connecting cable), detail on the right: end tip prior to (1) and after extrusion of carbon paste (2); (B) the "T" type: disassembled (above) and assembled body, with a set of exchangeable end tips (right), and accessories for filling with carbon paste and cleaning (left); (C) "S" type: disassembled (above) and assembled body, with a set of tools for pressing the paste (right and bottom) and filling (a spatula on top); (D) "M" type: disassembled (left) and assembled body, with an office ruler as a measure. (*Note*: All constructions presented are the designs used by the authors (Ivan Švancara and Karel Vytřas), being the property of the University of Pardubice, Pardubice, Czech Republic variants "T" and "S," including principal construction elements of the "M," are the patented material [544]). (Photos from authors' archives.)

the previous type. (Such a reduction in size was inevitable also due to progressive miniaturization of electrochemical instrumentation, including the electrode-cell compartments [175].) And, when abandoning advantageous concept of exchangeable tips—because of very difficult machining of smaller parts from quite elastic Teflon [188]—these holders ("slim" or "S"-type [540]) have been and still are being made in two variants.

Both are nearly identical, but differ from the size of the holes drilled in and through, making these slim holders to have the actual surface diameter either 2 or 3 mm, respectively. Whereas both T- and S-types were manufactured in fairly large series (maybe 100 pieces of each) and then successfully used for years, the newest "ultraslim" or "mini" prototype ("M" [84,111]; see image "D") exists in a few exemplars only, designed and tested as the wall-jet CP-detector for HPLC (see [114] and references therein) and representing, in fact, the construction limit from mechanical and functional points of view, as well as with respect to the material and machinery used.

2.3.3 COMMERCIALLY AVAILABLE CARBON PASTE ELECTRODE BODIES

Both carbon paste–filled drilled rods and piston-equipped holders had also become the inspiration and pattern for some manufacturers of electrochemical instrumentation offering in official catalogues—or, now websites—their own products: CPE bodies (usually together with a ready-to-use carbon paste; see also Section 2.1.4 and Table 2.1). These products have found indisputable popularity among some renowned research groups (see, e.g., [242,243,542]); nevertheless, homemade carbon paste holders still prevail due to a wider versatility.

To the authors' knowledge, commercially available CP-bodies have appeared in both tubing and piston-driven constructions. Regarding the latter, for instance, there has been a "giant" CP-holder (about 25 cm long and weighing some hundreds of grams) with the corpus made of transparent *Plexiglass*® and marketed approximately 15 years ago. The first type—that is, plastic tubes with filling end holes—seems to be more frequent, being designed as small (5–10 cm in length) or even miniaturized (<5 cm) variants that are fully compatible with the electrode-cell compartments offered by the respective manufacturer(s).

To date, products of this kind can be found via the official merchandise, for example, (1) mini-carbon paste electrode "6.0807.000" (see [241] and the updated website [543]); (2) electrode body "MF-2010" with a *carbon paste* "*CF 1010*" supplied extra (see [544] with the link for getting a special manual in PDF); or (3) *a set of CPEs* supplied with the so-called *carbon-paste oil base* [545], altogether forming four different variants—single CPE "010251" (planar body with central well designed as enzymatic reactor), two standard/vertical CPE "002223" and CPE "002210" (with ID = 1.6 and 3 mm, longer bodies), and CPE for RDE "010800" (with ID = 3 mm and a short body mountable to a rotator).

2.3.4 CARBON PASTE–BASED DETECTORS

This rather specific category is typical for a great variety of the individual constructions. It can be even stated that practically there is not a design that would be completely the same like any other [83,84,111]. The main reason for such construction diversity is that a great majority of flow cells with CP-detectors are being laboratory made, often hand in hand with unique separation systems, such as those proposed for analysis of gaseous samples [546,547].

The first types of CP-detectors appeared in sensing units for column separations by HPLC (see, e.g., [46,65,548]). Later on, similar constructions found quite a wide use in flow injection analysis (FIA; [64,73,549]), as well as sequential injection analysis (SIA; [550–552]). Whereas the "golden era" of carbon paste detectors through the 1990s and the early 2000s was undoubtedly associated with the dynamic development of CP-biosensors in flowing streams (see [70,82,111] and references therein), including sophisticated assemblies operated in dual regime [299,553,554]. Since the mid-1990s, the applications of CP-detectors under flow conditions are also connected with capillary electrophoresis (CE) [555–557] and this trend stays on [558–562], adopting yet further related techniques [563]. Among CE setups, one can highlight an unprecedented combination of a chemically modified carbon paste with a dual electrode/dual channel detection system [560] consisting of two working electrodes and two amperometric detectors, the whole assembly enabling the simultaneous determination of four different analytes.

With regard to the electrochemical techniques so far combined with CP-detectors, dominant is of course the amperometric mode (see, e.g., [46,97,280,436,548,564–567]), but there are also occasional reports on the choice of voltammetric ramps [65,310,568], potentiometry [569,570], or electrochemiluminescence measurements [389,462,571,572].

Carbon paste–based detectors are appreciated for (i) easy regeneration and renewal of the electrode material, as well as for (ii) flexibility in modifications manageable with various classical reagents, newly synthesized materials, or even various biological tissues. Both these advantages are fully exploited in a construction shown in Figure 2.11 (Scheme "A," redrawn from illustrations published in [573]).

The figure depicts the design of a dual enzymatic detector for FIA, where a well (3) served as a reactor with carbon paste modified with a mushroom tissue, whereas the second hole (2)—placed in parallel—was the working electrode containing the native paste with immobilized redox mediator. The paste in both filling wells could be fully exchanged and thus effectively renewed [573]. Moreover, such regeneration can be accomplished directly during measurements if the construction employs two parallel reactors; when the first one is in use, the second, at the same moment, is appropriately regenerated [574]. On the other hand, some constructions of flow detectors can also be quite simple with a possibility to incorporate the standard piston-driven CP-holder—see, e.g., [114] and Figure 2.11, Scheme "B."

Relatively serious drawback of CP-detectors is their limited stability in media with organic solvents, such as mobile phases in HPLC and CE. However, usually helpful would be the choice of more resistant mixtures, such as those made of new carbon materials (e.g., GC powders [114]) or some solid-like carbon pastes [78,308,310,318,575].

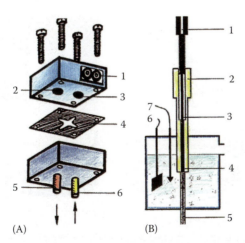

FIGURE 2.11 Typical constructions of carbon paste–based detectors. (A) Dual detector with enzyme-modified carbon paste reactor (Redrawn from Bonakdar, M. et al., *J. Electroanal. Chem.*, 266, 47, 1989): 1, electric plug; 2, working electrode; 3, enzymatic reactor; 4, Teflon seal; 5, outlet; and 6, inlet; (B) electrochemical wall-jet detector employing the piston-driven CPE (After Zima, J. et al., *Rev. Chim. (Bucharest)*, 55, 657, 2005): 1, piston head and electrical contact; 2, Teflon electrode body; 3, steel piston; 4, carbon paste filling; 5, inlet from HPLC; 6, Pt-counter electrode; and 7, reference electrode. (Redrawn and colorized from Kalcher, K. et al., Heterogeneous electrochemical carbon sensors, in *The Encyclopedia of Sensors*, Vol. 4, Eds. C.A. Grimes, E.C. Dickey, and M.V. Pishko, American Scientific Publishers, Stevenson Ranch, CA, pp. 283–429, 2006.)

The individual achievements and other details about CP-detectors can be found in numerous contemporary reviews on *electrochemical detection in flowing streams* (see, e.g., [576–581] and references therein), while more general information and instrumental aspects of various detection systems are yet commented in Chapter 6.

2.3.5 PLANAR CONSTRUCTIONS OF CARBON PASTE ELECTRODES

Besides the aforementioned planar configurations of numerous CP-detectors, biosensors, and bioreactors applicable in FIA, SIA, or EC-LC arrangements, there are also a number of constructions with typically planar configuration that are not necessarily intended for use in flowing streams or hydrodynamic measurements in the batch arrangement.

An apparent priority of such a planar CPE can be attributed to a construction reported by the carbon paste inventor and his team [21] who had used it in combination with triangular-ramp voltammetry for one of their typical studies within organic electrochemistry. A few years later, a similar study—divided into two parts [582,583]—was performed with another planar CPE whose geometry allowed the periodical renewal of the diffusion layer. Of plane constructions was also an s-CPE-based potentiometric ion-selective minielectrode [521] or similarly designed and, again, zero-current-operated multi-electrode array [584]. A planar construction is also a typical feature of a series of electrically heated carbon paste electrodes (EH-CPEs; [90,292–294]) commented in more detail overleaf.

Furthermore, plane configuration is very characteristic for the so-called groove electrodes (GrEs; [295,585–587]) developed as the true carbon-paste alternative to *screen-printed electrodes* with solidified carbon ink, having copied their typical geometry, as well as the overall size. As illustrated in Figure 2.12 on drawing "A," GrEs are designed as *Teflon*®-made prismatic bars with horizontal cavity for carbon paste filling (ca. 1 mm layer), for a metal contact, and additional plastic inserts for mechanical demarcation of the surface area.

Due to rather small dimensions and a plane configuration, the GrE can be inserted into miniaturized flow-through cells, where the carbon paste filling was found sufficiently stable also at relatively high flow rates as found during their advanced testing with modified carbon pastes [586,587]. The kit-like

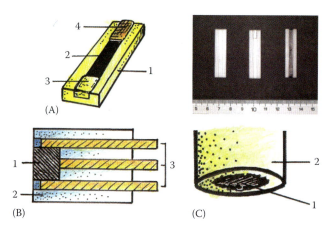

FIGURE 2.12 Schematic constructions of three different planar carbon paste electrodes. (A) Groove electrode (After Švancara, I. et al., Plastic bars with carbon paste: A new type of the working electrode in electroanalysis, in *Monitoring of Environmental Pollutans*, Vol. IV (in Czech), Eds. K. Vytřas, J. Kellner, and J. Fischer, University of Pardubice, Pardubice, Czech Republic, pp. 145–158, 2002); *left*: 1, prismatic bar (made of PTFE); 2, cavity with carbon paste filling; 3, plastic insert; and 4, brass contact; *right*: disassembled kit; (B) electrically heated carbon paste electrode (Redrawn from Flechsig, G.-U. et al., *Electroanalysis*, 14, 192, 2002): 1, carbon paste filling; 2, PTFE corpus; and 3, three Cu-wires; (C) carbon paste film electrode (simplified sketch): 1, carbon paste film/layer and 2, supporting electrode in the disc configuration. (Drawings from authors' archives.)

character of plastic bars was also inspiring for a development of electrically heated groove electrode (EH-GrE; [295,296]) that—dependent on the temperature changes—manifested notable enhancement of the diffusion transport during the electrolytic transformation of the single ions, as well as a certain control of nonelectrolytic processes involving complex species and based on the distribution equilibria (adsorption, extraction, and ion pairing [296]).

Regarding a potential electroanalytical utility, however, the functioning of EH-GrE has remained behind the performance of the original constructions of EH-CPEs ([292–294]; see also Scheme "B"), when comparing the signal stability and the overall benefit of elevated temperature. With EH-GrE, it was quite difficult to control the actual temperature across the large electrode surface (ca. 10 mm^2) as it was in direct contact with ambient solution and its cooling effect [296]. On the other hand, other constructions of electrically heated CPEs reportedly did not suffer from these problems and could also be successfully used in practical analysis, for example, for the determination of selected heavy metal ions [292], organophosphate insecticides [389], flavonoid antioxidants [293], or during studies with [Ru(bpy)$_3$]$^{2+}$ [489] and of DNA damage [292].

Finally, the last example of planar configurations is a group of CPFEs ([127,204,332–334,439]; Scheme "C") discussed already in association with new types of carbon pastes (in Section 2.2.3).

2.3.6 MINIATURIZED VARIANTS OF CARBON PASTE ELECTRODES

Naturally, a myriad of various configurations of CMCPEs, CPEs, CP-biosensors, together with their new alternatives, comprise also numerous designs with overall size being significantly smaller than "normal" electrodes. Thus, if one defines the size of such normal CPE to be in the order of centimeters in overall length (l) and in millimeters when measuring its surface diameter (Ø), the miniaturized variant of a CPE is then each construction whose dimensions fulfill the criterion that l ≪ 5–10 cm and Ø < 1 mm.

Indeed, within the electrochemistry of carbon paste–based electrodes, this rough classification according to the approximate size is much more frequent than more exact definition "macroelectrode vs. microelectrode," depending on the mutual ratio between the surface diameter and the actual size of diffusion layer formed at such an electrode. By accepting the first simplified approach, all the miniaturized versions of CPEs can be divided into three fundamental groups.

(I) *Carbon paste minielectrodes (CPmEs).* The first variants of miniaturized CPEs came to the fore in the mid 1970s [50]; however, their typical inner diameter of maximum 300 μm corresponded better to microelectrodes, despite their contemporaneous description as "mini-," "semi-micro," or even "very small" electrodes [21,548,588].

Thus, some typical CP-minielectrodes can be presented on a quartet of relatively recent designs. At first, it is a miniature enzyme-based sensing device [589] whose premodified carbon paste can be incorporated in the various configurations, including further diminishing to a micro-variant. Or, it is a CP-minibiosensor [590] that, despite a very small size, exhibits surprisingly robust characteristics, such as lifetime for ca. 4 months. A minivariant can also be integrated in a three-electrode microcell, through which μL-sample volumes are passed with the aid of streaming (bubbling) with inert gas [591]. And a fourth example is a CPmE obtainable from disposable plastic tips of common (transfer) pipettes [592]. As shown in Figure 2.13 ("A") with properly inserted electrical contact, this simple construction also allows one to select the surface size with the aid of careful cross-cutting the previously filled tip using a sharp knife or, better, a razor.

Last but not least, some commercial CP-holders are purposely constructed so small that the resultant CPEs balance at the edge of macro- and minielectrode—see, for example, [543] and the respective product advertized as *mini–carbon paste electrode* "6.0807.000."

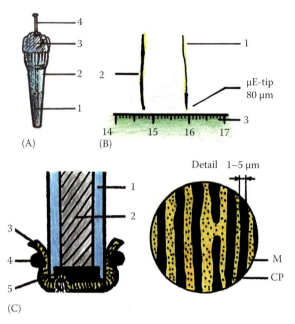

FIGURE 2.13 Schematic constructions of carbon paste mini-, micro-, and ultramicroelectrodes. (A) Carbon paste mini-electrode (a schematic sketch according to Baldrianova, L. et al., *Anal. Chim. Acta*, 599, 249, 2007): 1, carbon paste filling; 2, plastic tip; 3, provisory seal (from Parafilm®); and 4, metal pin (electrical contact); (B) Adams's microelectrodes for *in vivo* brain electrochemistry (Redrawn from a photo in Adams, R.N., *Anal. Chem.*, 48, 1126A, 1976): 1, CPµE (capillary with carbon paste); 2, Ag-wire (as reference electrode), both with connecting wires; and 3, ruler (as a measure); (C) ensembles of carbon paste ultramicroelectrodes with macroelectrode collector (Newly redrawn from Wang, J. and Zadeii, J.M., *J. Electroanal. Chem.*, 249, 339, 1988): 1, PTFE body; 2, steel rod; 3, polycarbonate membrane; 4, rubber ring; and 5, glassy carbon disc, "M", membrane pores; "CP", carbon paste. (Drawings from authors' archives.)

(II) *Carbon paste microelectrodes (CPµEs).* As already mentioned, the early constructions were proposed by agile Adams [48–50], having electrodes with radii of approximately 3–300 µm. Such tiny sensors had been well compatible with miniature detection units for HPLC [46,548] and, mainly, applicable—as the working microelectrodes—for *in vivo* monitoring in brain electrochemistry [48–50,593–599].

In these measurements, the CPµE and a tiny AgCl-coated Ag-wire (see Figure 2.13 and Scheme "B") had been implanted into the brain of a rat, where the cerebrospinal fluid served as the inner electrolyte for the Ag/AgCl reference, as well as the proper supporting medium [50]. For similar purposes, a 45-cm long tubing with an electrode tip filled with carbon paste has also been described being found fully functioning also after intravenous implantation into an anaesthetized monkey (Rhesus Macaque [600]). During the 1980s and 1990s, the research activities within *in vivo* electrochemistry culminated and the attempts to monitor negligible current signals in living organisms have become one of the most challenging areas for practical applications of CPµEs, which can also be documented on principal papers and contemporary reviews [152,594–606].

Otherwise, out of *in vivo* measurements, CP-microelectrodes have been used only seldom. In fact, there was a short recovery with the commencement of the new millennium [607–612], but—since then—the entire 2000s saw only a few contributions on this topic. The first two examples of amperometric CP-biosensors have already been presented in the previous category [590,591] because both balance between mini- and microconstructions. And hitherto last report on a CPµEs has emphasized mainly a new

Carbon Pastes and Carbon Paste Electrodes

carbon nanomaterial used [120], the microelectrodes themselves devised with a special piston mechanism for extruding the fine paste from tiny glass capillaries (with ID from 5 to 50 µm).

(III) *Carbon paste ultramicroelectrodes (CP-UMEs)*. Similarly like CPµEs, also ensembles of these especially miniature electrodes had shortly attracted electrochemists' attention in the late 1980s as documented in three interesting reports [174,613,614]. Within these investigations, it was hoped that CP-UMEs would further improve some physicochemical and electrochemical characteristics of the already existing ensembles of UMEs and such superiority was thanks to unique mechanical properties of heterogeneous carbon pastes.

In this respect, some anticipations have been fulfilled (e.g., those speculating on a lesser ohmic resistance or enhanced capacitance), some predictions mostly failed (e.g., more favorable reaction kinetics compared to the macroelectrode arrangements).

Figure 2.13 and Scheme "C" depict how to accomplish and prepare ensembles of CP-UMEs. According to the authors of this assembly [613], it was necessary to use a special carbon paste (CP) made of very finely grinded graphite with particles of ca. 100 nm. (The resultant powder was a nanocarbon at that time, when similar materials had not been yet called with the modish "nano" prefix [8].) The respective and very elastic paste-like mass was then thoroughly pressed into a polymeric membrane (3) with micrometric pores (M; see also cross section). After careful removal of the exceeding paste, the filled membrane was stretched over the tip of an electrode body (1) and fixed firmly with a rubber ring (4). The supporting electrode area was the GC disc (2) serving as a contact site, collecting the signal from all the individual carbon paste ultramicrodiscs isolated inside the membrane pores. As shown for a similar arrangement of CP-UMEs [174,614], also "normal" CPE could properly serve as the collector.

Note: Although ensembles of CP-UMEs had made more or less an ephemeral appearance, they contributed to the understanding of some processes associated with the behavior of carbon paste–based electrodes. Namely, the studies with CP-UMEs helped to define the character of the standard carbon paste (macro) surface being classified, at that time, to be rather an array of microelectrodes than a uniform area of a pool electrode [276]. When using cyclic voltammetry, it was demonstrated that whereas common CPE configuration had exhibited typical voltammograms with normal peaks, ensembles of CP-UMEs gave rise to sharp sigmoidal curves. Hence, the latter indirectly confirmed that miniature regions of carbon paste had been separated mechanically in the pores, obeying the fundamental criterion of each microelectrode to have comparable dimensions of the diffusion layer and of the electrode surface beneath. In contrast to this and according to latest interpretations [83], the surface of ordinary CPE can now be imaged via a model, where the individual electrode miniregions formed by binder-coated graphite particles are merging into a uniform area and the corresponding diffusion layers are completely overlapped. The result is then a behavior known for traditional solid electrodes, including well-developed peaks in cyclic voltammograms.

A decade ago, ensembles of CP-UMEs had surfaced once more [615] in two different configurations of a (i) CP-disc shaped electrode (with overall diameter, $\varnothing_{CP}$ = 100–150 µm), and a (ii) CP-band electrode, both exhibiting characteristic properties of microelectrodes but suffering from rather poor reproducibility (of about ±20%).

2.3.7 Special Constructions of CPEs

In this closing section, three unique constructions of carbon paste–based electrodes are presented as very atypical configurations standing beyond any established classification. The first one has been reported as carbon paste "U"-shaped electrode [616] filled with mercury and having one end loaded with carbon paste linked to the Hg-reservoir through a solid carbon insert. Due to the presence

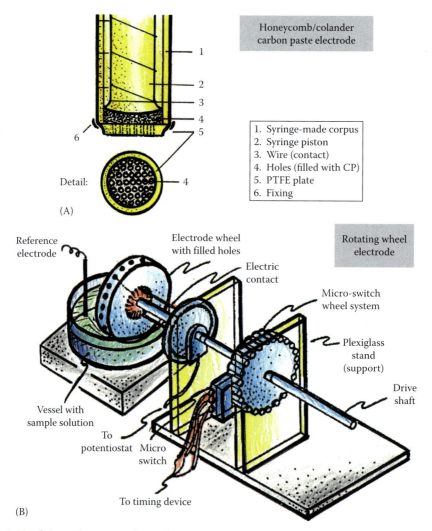

FIGURE 2.14 Schematic constructions of two unusual carbon paste electrodes. (A) Honeycomb carbon paste electrode (Newly sketched after Yao, C.-L. et al., *Anal. Chem.*, 61, 279, 1989): see inscription; (B) rotating wheel electrode (Redrawn and rearranged from Lawrence, R.J. and Chambers, J.A., *Anal. Chem.*, 39, 134, 1967): see inscription. (From authors' archives.)

of liquid mercury, the whole assembly had to be arranged in a reverse configuration, that is, with the carbon paste surface turned upside. Despite this or, rather, thanks to this, the CPUE has been described as quite practical for use.

A loose adaptation of a CP-microelectrode array is the design of a CMCPE shown in Figure 2.14 (Scheme "A") and is reported to offer particularly large and easy-to-renew surface [526]. The tubular corpus of this CPE was made from a disposable syringe by cutting its end tip and connecting the obtained tube to a *Teflon*® plate with 44 holes in a honeycomb arrangement. The carbon paste modified with electroactive compound of interest was then packed into the hollow tube and—upon request—extruded through all the holes. Freshly renewed surface of this CMCPE whose working site resembles also a colander (kitchen sieve) could be obtained by joint polishing of all miniature carbon paste discs.

Probably the most curious design of a CPE ever proposed was a (carbon paste) rotating wheel electrode (CP-RWE; [27]), looking like a miniature water mill (Figure 2.14, Scheme "B"). Its function relies on a large wheel consisting of 24 holes drilled around the periphery and each being filled with carbon paste. The wheel was slowly rotated by electromotor while partially immersed in a bowl with the sample solution. Such construction had allowed one to regenerate the paste during measurements, when the surface of carbon pastes in the holes—being currently outside the solution—could be renewed mechanically.

Note: The purpose of the whole mechanism was to obtain an electrode with periodically renewable surface—a certain analogue to Heyrovsky's dropping mercury electrode. In contrast to unsuccessful experimentation with a dropping carbon electrode [1–3,8], such intention had been attained. This was demonstrated on model polarographic waves with oscillations, corresponding to the periodical renewal of each hole with carbon paste [27]. (In fact, short interruptions of the current flow had been recorded during replacement of one hole by another, as they emerged/submerged while rotating the wheel.)

3 Carbon Paste as an Electrode Material

In the previous chapters, carbon paste was introduced as an electrode material consisting typically of two major building components and possessing numerous specific features. In this chapter, the survey of mechanical and physicochemical properties is carried out, which includes other characteristics of carbon paste that have a direct impact on the resultant behavior of carbon paste electrode (CPEs), chemically modified carbon paste electrodes (CMCPES), and carbon paste (CP) biosensors, and all the remaining sensors, including their electrochemical characteristics.

3.1 PHYSICOCHEMICAL PROPERTIES OF CARBON PASTES

The characteristics in which carbon pastes differ from similar electrode materials, that is, from solid carbon (graphite) and some noble metals, including their variants with additional membranes, can be summarized as follows:

3.1.1 Microstructure of Carbon Pastes

Due to the presence of liquid binder, carbon paste is typically a *multicomponent material*, exhibiting a rather unique microstructure. In order to define the structural character of carbon pastes, basic types of CPEs made of mineral and silicone oils were subjected to some special investigations. Surprisingly, the very first attempts had not been microscopic observations but indirect electrochemical experiments based on measurements of conductance [2] or electrolytic and polarization phenomena [68]. Nevertheless, the most comprehensive insights into the CP microstructure were obtained later, with the aid of modern microscopic techniques, such as scanning tunneling microscopy (STM; [172]), optical microscopy (OM; [617]), scanning electron microscopy (SEM; [96,618,619]), and atomic force microscopy (AFM; [620]). SEM can also be advantageously combined with x-ray microanalysis [173] or profilometer to specify the surface roughness [327]. Regarding microscopic imaging of carbon pastes and their surfaces, further details are available in Section 6.5. Recently, SEM imaging has become very popular, often as an inevitable instrumental tool for the characterization of novel types of electrodes (see, e.g., [96,173] and references therein) and completely new types of carbon pastes, such as those made of carbon nanotubes [124] or ionic liquids [453]. And it should also be mentioned that there were some studies on the effect of chemical modifiers [243,618] or structural alterations after immobilization of biological tissues [619].

Of considerable interest were also preplated metallic films and deposits where attention was focused on diversities in the structure of palladium deposits [621], mercury [189], gold [261], and bismuth films [173,622], and related structures (like, for instance, "solid" dispersions of bismuth and antimony obtained by manual homogenization of the respective fine powders in the native carbon paste [188]).

Based on the observations from OM, STM, and mainly SEM studies, the first *theoretical models on the carbon paste microstructure* have been proposed with an attempt to interpret relations between the quality and mutual ratio of both main carbon paste constituents and the resultant behavior of CPEs, thus supporting observations from electrochemical measurements.

One of the first models of this kind was discussed in detail in the fundamental report of Adams's group [68], and some conclusions can be summarized as follows. The model assumed consequent changes in the surface layer for three different carbon pastes containing a decreasing amount of liquid binder in the mixture. In situation (i), the content of the binder was the highest, resulting in almost total coverage of all graphite particles with the film of liquid. The phase (ii) was then expected to form a mixture, where the amount of liquid binder was somewhat lowered, indicated by the increasing number of "empty spaces" inside the paste, representing air gaps—or even molecular oxygen—entrapped in the carbon paste layer. Finally, scheme (iii) reflected a situation where carbon paste contained only a minimal percentage of pasting liquid, yet sufficient for mechanical binding or sticking of the carbon particles together. Although models (i) and (ii) were presented to approximate the structure of real pastes, the last one (iii) illustrated a hypothetical mixture at the "dry limit" (see also Section 2.2.1), when the resultant CPE exhibited behavior nearly identical to that of solid carbon electrodes. It can be added that Adams's structural models (later redrawn into a set of illustrations and presented in this form [83,153]) were more or less confirmed by later microscopic imaging.

This is shown in Figure 3.1, revealing typical structures of a sextet of different carbon pastes, including carbon paste mixtures prepared from new kinds of carbon, namely, glassy carbon powder with spherical particles or of CNTs and an (RT)IL exhibiting particularly smooth paste composites (see images at bottom).

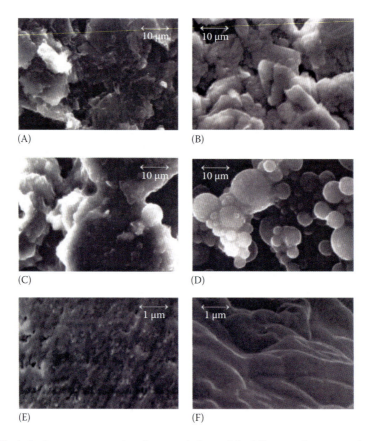

FIGURE 3.1 Typical microstructures and surface morphology of six different carbon paste mixtures. (A) (C/Nj) carbon powder[1] + Nujol oil,[2] (B) (C/SO) + silicone oil,[3] (C) (C/TCP) + tricresylphosphate,[4] (D) (GC/Nj) glassy carbon powder[5] + Nujol oil, (E) (CNT/Nj) carbon nanotubes[6] + Nujol oil, and (F) (C/IL) carbon powder + ionic liquid.[7] *Other specification*: (1) spectroscopic graphite "CR-5" (Maziva Tyn; Czech Republic), (2) mineral oil (Merck), (3) highly viscous product ("Lukoil MV 12,500"; Lucebni zavody Kolin, Czech Republic), (4) mixture of isomers (Fluka), (5) with spherical particles ("Sigradur G"; HTW Maitingen, Germany), (6) L-SWCNTs (NTP, Shenzhen, China), and (7) BMIMPF6 (Sigma/Aldrich). SEM (model "JSM-5500LV"; JEOL, Tokyo, Japan), magnification used: 1:1000 (A–D) and 1:10000 (E,F). (Alternate or unpublished images from authors' archives.)

The first two carbon pastes (imaged in upper row) that have been prepared from the same spectroscopic graphite and two most common pasting liquids "copy" faithfully the original relief of the graphite alone. (Such structure morphology was nicely approached by Wang et al. [172] whose "romantic idyll at the carbon paste surface" is an unprecedented description and is paraphrased in Section 6.5.) In contrast to both C/Nj and C/SO carbon pastes, the third mixture (in the central row, on left) manifests a completely different structure and a very intimate covering of graphite by liquid, giving rise to compact aggregates that might be behind rather specific electrochemical behavior of the respective carbon paste (C/TCP; [205,254]). Another image (central row, right) is apparently the best proof for the legitimacy of Adams's model, showing how the individual graphite spheres are coated by the binder; many of them are "glued" together into the whole clusters. (This phenomenon is yet more evident if one compares this image with the original surface relief of the glassy carbon powder alone; see [96] and SEM photos therein or Figure 2.2 in this text, images "A" and "B.") Finally, specific structures of CNT/Nj and C/IL pastes have already been commented.

3.1.2 Ohmic Resistance

Attractive properties of carbon pastes in electrochemical measurements are also given by their high conductivity, especially, potentiometric indication merits from this feature with respect to the stable and rapid response [157,159]. Ordinary mixtures with paraffin or silicone oils exhibit the ohmic resistance of about 10 Ω or even less [7,26,175,228,229], and somewhat higher values of about 50–200 Ω can then be measured with less compact carbon pastes, such as those made of liquid tricresyl phosphate (TCP) [205,254] or based on glassy carbon powders [95,96]. In experiments with analytical concentrations at normal levels, variations in ohmic resistance can be neglected; nevertheless, some occasional attempts to lower the carbon paste resistance have also been made to minimize undesirable effects of the ohmic drop [26]. In context with this report, it is worth mentioning that a trend of the increasing resistance in carbon pastes has been explained via the quality of graphite powder used without emphasizing the role of liquid binder. The prevailing effect of the carbon moiety to the total ohmic resistance was also confirmed in later studies, again, when testing glassy carbon powders with special surface characteristics [96,192,229].

Such findings were quite surprising as one would expect a more important role of the binder. In fact, for a long time, there was no satisfactory explanation on why (almost totally) insulating binders contribute so little to the overall resistance of carbon paste mixtures, including blends with a rather high content of liquid phase. For example, some hypotheses published so far had speculated on a tunnel effect known for semiconductors or on a special role of the binder acting as a totally permeable membrane [82,157].

It seems that these uncertain interpretations were definitely "buried" recently, when two new studies [228,229] built upon an extensive database of specially performed experiments with two standard carbon pastes (C/Nj and C/SO types) and one novel mixture (SW-CNTs/SO) revealed the close relation between the carbon paste composition(s) and the resultant ohmic resistance(s). Thus, a new hypothesis for the carbon paste microstructure could be postulated and verified, explaining the resultant behavior of both CPEs and CNTPE by means of the *model of the tightest arrangement of solid spheres*, in approximation applicable also to compact graphite particles, and—rather surprisingly—to snake-like carbon nanotubes (for further details, go into Section 3.3).

The task of ohmic resistance is also closely associated with some latest trends in the electrochemistry of modern sensors. It represents an example when some advantages of classical carbon pastes cannot be used during the transformation of the respective procedures with CPEs to a concept employing functionally identical screen-printed electrodes (SPEs) [75,83]. Compared with carbon pastes, solidified carbon inks in SPEs typically possess a much higher resistance (ranging from kΩs up to some MΩs [323,327,623]), which elucidates why some experiments with CPEs are still superior to those performed using SPEs of very similar or even the same construction [295,623].

3.1.3 INSTABILITY OF CARBON PASTES IN ORGANIC SOLVENTS

Tendencies to disintegrate in contact with most of the organic solvents—apart from if being used alone or in mixed media with water—were already commented on in the previous sections—in context with stabilization of carbon pastes by additional agents [16] or through choice of more resistant CP constituents such as highly viscous silicone oils [211], wax-like chlorotrifluoroethylene polymer (*Kel-F*®; [66,318]) or related solid-like materials [308,310], and powdered glassy carbons [82,111,114].

3.1.4 AGING OF CARBON PASTES

According to some special studies (e.g., [192,254]), carbon paste mixtures exhibit a series of *typical periods of changes in time* that can be attributed to the aging effect. Namely, (i) unstable or almost irreproducible response of *freshly made* carbon paste mixtures, (ii) relative stable signal in the course of 2–3 weeks (up to several months [8,9,11]), and (iii) gradual deterioration of the response due to *starting desiccation*. These individual stages indicate a certain *self-homogenization* process in a newly prepared paste (ca. 24–48 h [175,176,192]).

And the lifetime of a CPE ends by reaching the critical limit, when the mixture turns into a desiccated and incompact mixture. The overall duration of such a lifetime is closely connected with the overall quality of both main constituents [175] and, mainly, with the stability of the liquid binder. Then, in extreme cases, the lifespan of a CPE may reach several months [8,11] or even years (see one curiosity in [7]).

3.1.5 HYDROPHOBICITY OF CARBON PASTES

Lipophilic character of the carbon paste material is indisputably the most important physicochemical feature, having a principal significance for the resultant electrochemical and electroanalytical characteristics of CPEs [2,3,68,83,175]. Hydrophobicity is reflected in the so-called "repelling effect" [55,69] at the carbon paste surface toward hydrophilic ions and molecules as well as in *specific interactions* with some substances in the carbon paste bulk (interior).

It can be stated that the *hydrophobic character* of carbon pastes codetermines practically all electrochemical characteristics of CPEs and related sensors and their resultant behavior in practical measurements. Thus, the *lipophilicity of carbon pastes* given by *the presence of liquid binder* is of continuing interest in the following sections, including the discussion on the individual interactions, potentially occurring at various types of carbon pastes, and is summarized in Section 3.4.

3.2 ELECTROCHEMICAL CHARACTERISTICS OF CARBON PASTES

A unique constellation of physicochemical and mechanical properties of the carbon paste electrode material is largely reflected in various specific features that are surveyed and briefly commented on in the following paragraphs:

3.2.1 VERY LOW BACKGROUND

In faradic measurements, satisfactorily low background currents (residual current, parasitic signal, noise) are of principal importance as they contribute directly—in positive or negative sense—to the analytical signal of interest, which, for *current intensity* (I_i), can be expressed by the following relations:

$$I_\Sigma = I_{ANAL} \pm I_{RES} \quad (3.1)$$

where subscript "Σ" means overall or total. The resultant current intensity, I_Σ, then depends on the mutual polarity of both analytical and residual signals being the same physical variable (i.e., electric current) measured in the same units, usually in µA or nA. Reliable electroanalytical measurements with CPEs, say, about ±5%–10% rel. dev. [175], then require suppressing the background below the µA level, which is not so obvious and simple to accomplish when considering some carbonaceous materials (e.g., pyrolytic graphite or more porous carbons) and their potentially higher noise [3,165,167–170].

As early as in Adams's first reports [2,3], "extremely low" *background currents* of carbon pastes had repeatedly been emphasized as one of the most significant benefits when choosing CPEs for current-flow measurements. Especially the polarizability over the whole anodic potential range was classified as extraordinarily good and practically nonconcurrent compared to other solid electrode materials such as platinum, gold, or various compact carbons and graphites.

Note: In typical voltammetric measurements, the background-current level at CPEs usually does not exceed a value of 1 µA. In the anodic range, the residual currents are yet lower and if some special pretreatments of the electrode surface are used, they may drop down to the nanoampere level even for common macroelectrodes that have the surface area of several mm^2 [182]. The absolute magnitude of the background is given by a variety of factors such as the quality of graphite used [176], the carbon-to-pasting-liquid ratio (the aforementioned pastes at the "dry limit" exhibit as higher a background as the amount of graphite increases [68]), and to a certain extent individual polarizability in a supporting medium given. In order to specify the absolute level of background currents, it is advisable to estimate an approximate value as an average noise registered within the potential range of interest, that is, between cathodic and anodic potential limits [3,175].

Regarding CPEs, a specific kind of the undesirable background is the reduction signal of oxygen dissolved in carbon pastes [3,7,26,61,68,69,83]. Oxygen is brought in either as molecules adsorbed in the pores of graphite particles or as air entrapped in carbon paste mixtures during their manual preparation and homogenization (see also Section 2.1.4). Irreversible two-step reduction of O_2 [61,516] is registered during cathodic scanning, and it can be seen as a broad response appearing typically between −0.5 and −1.0 V vs. Ag/AgCl/3 M KCl.

The existence of the parasitic signal of oxygen is generally undesirable because of the deformations of the base line or even of the proper analytical signal. In some situations, the signal of interest can still be evaluated as a peak superimposed at the plateau-like response of oxygen [2,3]; however, if one operates at very low concentration levels, the corresponding peaks of interest may already be partially or even completely overlapped. This is particularly valid in situations, when the paste mixture is made of graphite, manifesting undesirably high adsorption capabilities [176].

In measurements whose output is not electric current, carbon pastes usually exhibit a satisfactorily low background, which can be demonstrated on two typically "zero-current" measuring techniques: (i) *equilibrium potentiometry* (in direct arrangement [35,36] or as titrations [159,208]) and, in particular, (ii) *computer-controlled stripping potentiometry* (in both of its main variants: PSA and CCSA [209,210]), where a very low background can be frequently achieved (thanks to the highly selective character of the electrode potential, E_i, as the measure of choice) as the principal demand for efficient and reliable detections at extremely low concentrations [157,168].

3.2.2 Individual Polarizability of Carbon Pastes

Analogically to other electrodes, polarization capabilities of a CPE define its operability or applicability in practical measurements connected with (faradic) current flow or similar analytical signals. Thus, polarization studies usually represented the initial steps in characterization of various CPEs in the early era of carbon pastes [8,9,28,31] or even later [34,40,41,60,61,63,64,68], including metallic-film-plated variants [625–627], measurements at extremely low concentrations [628–630], or recently proposed carbon paste mixtures and formulations made of new carbon materials [115–120,123,124,128,332–336,361,362,631–633] or alternate liquid binders [125–127,192,205,221,233,238,453,454,517].

Polarization studies conducted by the individual authors and research teams mentioned earlier have been carried out in a wide spectrum of supporting electrolytes, and the results can be summarized in the following points:

(I) Both cathodic and anodic polarization limits of common carbon pastes made of spectroscopic graphites and paraffin or silicone oils can be compared with those offered by other graphite electrodes. The resultant *potential range* ("potential window") of a CPE depends on the supporting electrolyte composition, especially, upon the total acidity of the solution used. Thus, polarization data typical for measurements with carbon paste–based electrodes can be surveyed accordingly, best through the respective potential limits.

- *Highly acidic media* [pH < 1; e.g., solutions containing mineral acids such as HCl, $HClO_4$, H_2SO_4, or HNO_3 in a concentration range of 0.1–1.0 mol L^{-1}; cathodic limit, E_C usually lower than −1.0 V (vs. SCE), max. −1.2 V; anodic limit, E_A: from +1.0 to +1.4 V (in dependence of the acid strength and concentration)]. It is obvious that a low pH of mineral acids significantly limits the use of CPEs for cathodic reductions. On the contrary, polarization of carbon paste in the anodic direction can also be performed in extremely acidic solutions such as 2 M H_2SO_4 [113], 2 M HCl [192], or even 15 M (conc.) H_3PO_4 [634]. The latter documents, among others, why CPEs are attractive for organic electrochemistry, where such solutions are often needed for studying the reaction mechanisms [2,3].

- *Acidic and mild acidic media* [pH 1–5; considerably diluted solutions of mineral acids, acetate buffer, citrate buffer, etc.; E_C: up to −1.5 V; E_A: max. +1.3 V]. Due to the lower acidity of these solutions, the potential range is markedly extended in cathodic direction, which can be exploited in anodic stripping analysis with electrolytical accumulation of some unreadily reducible metals [165,168,257,625]. Since the corresponding measurements are usually characterized by very low background currents over the whole potential range, mild acidic media are apparently the most frequent supporting media in electroanalysis with CPEs (see also Chapter 8 and go through Tables 8.1 through 8.28).

- *Neutral and basic media* [pH 6–10; solutions of nonhydrolyzing salts (KCl, Na_2SO_4), phosphate or ammonia buffers; E_C: up to −1.7 V; E_A: from +0.8 to +1.2 V]. Despite fair cathodic polarizability and a wide potential range, solutions of this type exhibit a somewhat higher background, which is the case of nonbuffered mixtures of neutral salts [625]. In more basic media, the respective anodic limits are already notably lowered, usually, below +1.0 V.

- *Alkaline and highly alkaline media* [pH 10–14; solutions of hydroxides (NaOH or KOH), borate or carbonate buffers; E_C: max. −1.5 V; E_A: from +0.4 to +0.8 V]. Although these solutions are regularly included in polarization studies, practical measurements in alkaline media are seldom performed, for instance, when oxidizing H_2O_2 [635,636] or within special studies with metallic-film-plated CPEs [637].

Note: The reason can be the fact that measurements in highly alkaline media often suffer from substantially deteriorated base line. This also corresponds to some recent observations [637], where the respective polarization curves obtained in more concentrated solutions of NaOH and KOH have exhibited an undesirably high background, including peaks of unknown origin, which might indicate an undesirable interaction of highly aggressive hydroxides with the active sites of the carbon paste surface. Such aggressiveness of more concentrated alkaline hydroxides—and, in fact, severe risk when working, for example, without gloves and glasses—can be documented on the authors' own experience with these solutions being able, in a short time, to etch CPE bodies made of less resistant plastics [188].

(II) There is a general agreement that both cathodic and anodic limits of CPEs are influenced mainly by the *quality of graphite powder* used. For instance, graphite particles impregnated by ceresin wax (i.e., protected against oxygen) may yield a carbon paste with a markedly widened potential window in the anodic range (up to +1.8 V vs. SCE [39]). In addition, relative content of graphite in the paste mixture is an important factor [2,3,68,83,175]. In a limited interval of the carbon-to-pasting liquid ratio (e.g., 1 g C: 0.3–0.6 g PO/SO), it can be even stated that the lesser the percentage of carbon, the wider the polarizability attained. Furthermore, the increasing content of graphite usually results in enhanced background, involving increasing signals of both limiting electrode reactions in aqueous solutions, that is, cathodic reduction of protons and the release of hydrogen or anodic oxidation of water, respectively [3,61,516,622]). Nevertheless, changes in polarization through the carbon-to-pasting liquid ratio are insignificant, and the proper selection of graphite material is more relevant.

(III) With regard to the *type of binder* used, the role of different pasting liquids in the polarization of CPEs is less known and understood. The first attempts to correlate the polarization characteristics of carbon pastes with the properties of a binder were made in a study dealing with seven different liquids [40], however, without decisive results and addressed interpretation. In other studies focused on the function of pasting liquids, certain explanations had already been formulated, namely, possible effect of a more thorough coating of the graphite particles by voluminous molecules of some binders such as highly viscous silicone fluids [211,219–221] or liquid plasticizers and ion exchangers [205,233]. Compared with ordinary carbon pastes, the potential window of these mixtures was extended mainly in the cathodic direction; for example, dioctyl phthalate–based carbon paste could be polarized in 0.1 M HCl yet at a potential of −1.2 V vs. Ag/AgCl, whereas the Nujol paste was polarized at −0.9 V [205].

Note: The mechanism assuming the "coating or packaging effect" of a TCP–based carbon paste (see Figure 3.1) was also applied to explain extreme polarizability of the TCP-CPE that could be operated under scanning from −2.0 to +2.0 V vs. Ag/AgCl, that is, within an interval of 4 V [96]. However, the subsequent experiments performed with CPEs made of alternate TCP [108,182,638,639] have revealed that the resultant polarizability was only slightly better than that of common CPEs. Then, abnormal behavior of the first TCP-CPE should be attributed rather to a catalytic action of some impurities in the original chemical (of technical grade) than to a benefit of the liquid itself [189]. In addition, typically high currents observed during measurements with TCP-CPE (e.g., [182,254]) have indicated a larger surface area, which did not concur with the previous hypothesis. In other words, the interpretation through the inhibition of electrolytic decomposition in aqueous solutions seems to explain better the variability in the polarization of TCP-CPE, particularly, if the same effect can be achieved by special modifiers or catalysts added into the paste (see later and in Sections 4.1 through 4.4).

(IV) Besides modifications mentioned in the footnote, the polarization characteristics of carbon pastes can also be improved by some surface treatments and alterations, which applies, of course, to any electrode or sensor. Here, one can quote an interesting approach based on spraying the native carbon paste surface with an additional graphite layer [67,256] or the already mentioned family of metallic-film-plated CPEs [103,257,260].

(V) In accordance with the information gathered in points I–IV, it can be concluded that *each carbon paste exhibits more or less specific polarization characteristics*, although the differences in both cathodic and anodic potential limits are usually minor or even negligible [8,9,40,68,83,175,227]. Nevertheless, it is always useful—and with newly made pastes strongly advisable—to examine the polarizability of the individual CPEs with the aid of suitable testing experiments ([175,176] and Section 3.3).

3.2.3 SPECIFIC REACTION KINETICS AT CARBON PASTES

One of the many characteristic features of lipophilic liquid binders that influence the properties of carbon pastes is their direct impact on the electrode reactions connected with the charge transfer as well as mass transport [3,68,83]. This effect was of permanent interest since carbon paste–based electrodes had appeared in electrochemical laboratories [2,3]. In a brief definition derived from Adams's publications, atypical kinetics at CPEs can be approached as a phenomenon arising from the ability of binders to moderate numerous faradic processes taking place in the proximity of the active graphite sites [3,528]. Thus, redox processes reversible or nearly reversible at compact electrodes such as GC, Pt, or Au are converted into a "pseudo-reversible" pathway when studied with a CPE. Moreover, analogously, those reactions that are moderated at conventional solid electrodes are slower or even totally irreversible when employing CPEs [3].

The reasons for such a broadening of the overpotential have been the subject of speculations in numerous early reports (e.g., [13,22,26,29,34,60,640–642]); nevertheless, the ultimate explanation was given as late as in the early 1980s by Adams et al. [68]. As already mentioned in the historical survey, this excellent study—followed soon by some other authors [60,61,63,220]—belongs among the classics within the field (see survey of such papers in [7]) and comprises up-to-date valuable discussion material, the most important facts of which are highlighted as follows:

 (i) At the CP surface, hydrophobic molecules of binder form numerous *hydrophobic sites and regions capable of repelling hydrophilic species* from the "electrode/solution" interface, having thus moderated their redox transformation and electron transfer as such.
 (ii) As a rule, the more hydrophobic the carbon paste surface and the more the hydrophilic species involved in the reaction at the interface, the more moderated the electrode process.

Note: Among typical hydrophilic species, one can classify practically all inorganic ions and molecules, but there are also numerous organic compounds whose molecules offer some fragments willing to participate in the protonation, hydroxylation, or in some inductive and mesomeric transmutations, which is the case of functional groups such as $-OH$, $=O$, $-NO_2$, $-N=O$, or $-NH_2$ [220].

 (iii) At the carbon paste surface, the abundance and degree of consolidation of *hydrophobic regions* are in direct relation with the type of liquid binder and its content in the CP mixture. Although the amount of the binder increases, the surface becomes more hydrophobic, and the whole *repelling effect*—that is, inhibition of the electrode process—is yet more pronounced. A *degree of hydrophobicity (lipophilicity)* then determines whether the overall electrode process proceeds reversibly, "pseudo-reversibly," or irreversibly. (In electrochemistry, in contrast to thermodynamics, the irreversible process does not mean the ultimate status, and the respective electrode reaction can still be converted to the reversible process. The key parameter is, however, the actual reaction rate [3,528].)
 (iv) By treating the carbon paste surface by means of special chemical or electrolytic procedures, the typically *hydrophobic surface of a CPE can be converted into a more hydrophilic surface state*, and thereby numerous moderated and irreversible electrode processes may be accelerated. In favorable cases, the respective redox pathways then exhibit similar or even nearly identical reaction rates as those taking place at ordinary solid electrodes.

Note: All the key findings and conclusions summarized earlier have been made thanks to the studies with carbon pastes at the limit of "dry graphite" (see Section 2.2.1). In such experiments, the hydrophobic layer of liquid binder could be removed by intensive chemical oxidation by 30% H_2O_2 or with $Ce(SO_4)_2$, K_2CrO_4, and $K_2S_2O_8$ in 1 M H_2SO_4 at 80°C for about 20 h. Yet more effective oxidation could then be attained with the aid of the so-called *anodization*, that is, controlled electrolytic activation at very high potentials (see the following). The *alteration of the surface states*

Carbon Paste as an Electrode Material

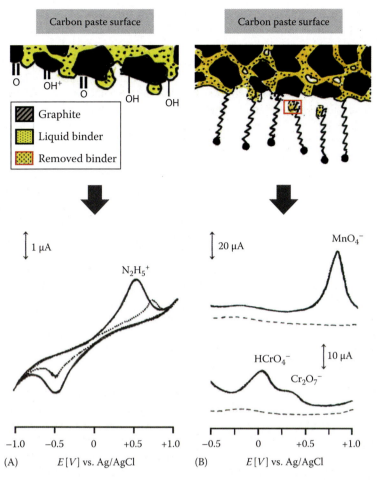

FIGURE 3.2 Carbon paste surface, the effect of electrolytic activation (A) and chemical erosion (B) with the corresponding voltammograms for oxidation (a) and reduction (b) of model compounds. Schematic view and experimental records. (1) before (2) after the surface treatment; (a) hydrazinium ion ($N_2H_5^+$); (b) MnO_4^- and CrO_4^-. *Experimental conditions*: CPE made of silicone oil ("C/SO"); (a) CV, potential scan: from −1.0 to +1.0 V vs. ref. and reversely, scan rate: 20 mV·s^{-1}; 0.1 M phosphate buffer (pH 7), c(Hy) = 5 × 10^{-4} mol·L^{-1}; (b) DPCSV, deposition: E_{DEP} = +1.2 V vs. ref., t_{DEP} = 30 s; stripping limits: +1.2 and −0.3 V vs. ref.; s.e.: 0.05 M HCl + 0.15 M NaCl (pH ∼ 2), c(anion) = 1 × 10^{-5} mol·L^{-1}. (Assembled and redrawn from Švancara, I. and Vytřas, K., *Chemija (Vilnius)*, 11, 18, 2000; Švancara, I. and Schachl, K., *Chem. Listy*, 93, 490, 1999; Švancara, I. et al. 1985–2010, Unpublished results; Digua, K. et al., *Electroanalysis*, 6, 451, 1994; Digua, K. et al., *Electroanalysis*, 6, 459, 1994.)

at CPEs [68] as a consequence of such aggressive chemical or electrochemical oxidation leads to a partial or complete transformation of elemental carbon to higher oxidation states, thus giving rise to various *functional groups containing O and H atoms* hydrophilic in nature, for example, −C-OH, −C=O, −C-OH$_2^+$, and −C=OH$^+$ (see Figure 3.2, top and left). Again, the whole phenomenon is accompanied by an intense *repelling effect* when the molecules of lipophilic binder are removed from the hydrophilic sites. Finally, if one considers "drastic" conditions of activations, the binder itself may be oxidized and hydrophilized as well.

Although the interpretation of specific kinetics of electrode processes at CPEs by means of the altered surface states has been generally accepted [60,63,157,172,220], there were some attempts to postulate alternate theoretical models.

Note: For instance, a research group dealing with modeling the processes at CP biosensors introduced [574,643,644] and later eagerly advocated [645] a hypothesis based on the so-called *pinhole mechanism*. This theory assumes that the binder forms a uniform thin layer or film coating intimately the individual graphite particles and consisting of many miniature holes (with diameter in the micrometer range). Reportedly, such a perforated structure then enables effective charge and mass transfer when the species involved in electrode reactions may pass freely through the holes to reach the active sites at the graphite particles and beneath the liquid layer. The model also tries to explain the effect of surface treatments and the proper alteration to more hydrophilic status by means of enlargement of the individual holes after activation. It is speculated that the entire process would depend primarily on the activation conditions selected, whereas, for example, the carbon-to-pasting liquid is of lesser importance. In addition, it is elucidated that the more intensive the activation, the larger the holes formed, which may expose the hydrophilic graphite surface to action in more extent. Despite the logical deductions behind the whole model, the pinhole mechanism based mainly on mathematic modeling has not been confirmed experimentally, and only a few substances were found to match the theoretical predictions (e.g., highly hydrophilic ferricyanide ion, $[Fe(CN)_6]^{3-}$). Such insufficient verification, as well as the fact that, for example, none of the powerful microscopic techniques was able to reveal and confirm the existence of perforation in the binder layer, seems to be the reason why the pinhole theory has not gained further support and remained abandoned. On the other hand, there have been some interpretations based on spontaneous diffusion or penetration through the binder layer that may remind the pin-hole mechanism used to define the role of the binder in the behavior of CP-based amperometric and potentiometric sensors assuming either an ensemble of miniature membranes [276] or—more simply—a liquid membrane [159], both ensuring efficient and relatively rapid mass transport.

Specific kinetics at carbon pastes was experimentally verified for a number of organic compounds (see [2,3] and Section 7.1), which resulted in some practical advice on how to accomplish the measurements in electrochemical practice. As a consequence of moderated reaction kinetics and the corresponding *overpotential* (*overvoltage*, ΔE), oxidation peaks of various organic substances appear at highly positive potentials, often close to the anodic limit (with the starting electrolyte decomposition). In such cases, the respective signals suffer from a very high background, deteriorating the quality of measurements with respect to unfavorable signal-to-noise ratio and resulting, naturally, in lower selectivity and sensitivity. Moreover, as shown earlier, such unfavorable behavior can be principally changed—and, in certain limits, also controlled—by means of *activation procedures* that enable to substantially lower the overpotential at the carbon paste surfaces.

Then, due to altered reaction kinetics at hydrophilized CPEs, the signals of interest are shifted toward less positive potentials and out of rather noisy potential regions nearby the anodic limit, which is sufficient to bring the desired effect—a marked improvement in the overall signal-to-noise characteristics.

In electroanalysis with CPEs, beneficial effect of activation—after Adams's classical study [68] investigated in detail, for example, by Ravichandran and Baldwin [63])—is well known and examined in numerous studies (e.g., [182,254,600,628–630,642,646]).

The most frequent electrolytic treatment is *anodization*, when highly positive potentials (typically from +1.25 up to +1.75 V vs. SCE) are applied for a period of 30–120 s. After anodic oxidation, the activated carbon paste surface can be "equilibrated" by a short *cathodization* (e.g., at −1.0 V for 15 s). Occasionally, the CP surface is treated by cathodization only, which can be advantageous in measurements requiring a negatively charged active surface [182] or a lower background during the reduction by constant current in the stripping potentiometric mode [254]. In contract to anodic activation, however, only little is known about the chemistry behind cathodic treatments. Perhaps, in analogy and by following some interpretations for solid carbon surfaces [3,167], it can be deemed that the reductive activation would release fragments such as $-C-O^-$ or $=O^-$, accompanied by the effect of the nascent hydrogen, H^0 [173,622]. The entire activation process is usually performed in a separate solution under quiescent conditions; that is, without stirring. Some authors have recommended special "activation media," such as a mixture of $0.1\,M\,KNO_3 + 0.02\,M\,HNO_3 + 0.01\,M\,Na_2HPO_4$ [63] or diluted solutions of mineral acids (e.g., 0.001 or $0.005\,M\,HClO_4$ [108,261]). A very effective activation can also be made with the aid of *potential cycling*, when the CP surface is exposed to periodical changes of highly positive and highly negative potentials (e.g., [646]).

Alternate to electrolytic activation and friendlier with respect to possible overloads in electric circuit due to too intensively anodized electrodes [188] is the chemical treatment of the CP surface by "erosion" *with surfactant(s)*, studied for the first time by Albahadily and Mottola [64] and elaborated in more detail by Kauffmann et al. [243]. The respective procedure relies on one important condition—the solution with the surfactant (e.g., quaternary ammonium salt like CTAB) should be *free of micelles* (not having the characteristics of a colloid). Otherwise, the CP surface is simply immersed into such a solution and—with or without stirring, depending on experience—left for some minutes to be treated. The "alchemy" behind this is depicted in Figure 3.2 (on the right side), showing that lipophilic chains of surfactant gradually withdraw the binder from the surface and that this erosion continues until the treated site gets "naked" and exposes the hydrophilic surface of graphite particle(s) to the solution. In addition, this activation has become popular in electroanalysis, with CPEs applied in a number of practical methods—see again Figure 3.2 and voltammograms given, as well as Chapter 8 and the table surveys with inorganic, organic, or even biological analytes.

There is yet another way of treating the carbon paste to lower the undesirable overpotential. This elegant approach is the *addition of a modifier* into the paste, in this case, a suitable *electrocatalyst*. Among substances with such an effect, one can quote organo-complexes with the central atom of a platinum metal ($Rh^{II/III}$ [526,647], $Ru^{II/III}$ [648], $Os^{III/IV}$ [649,650] or all metals from the iron group ($Fe^{II/III}$ [618], Ni^{II} [651], Co^{II} [652,653]). As confirmed in numerous studies with CPEs, these special modifiers (discussed in detail in Section 4.3) may lower the overpotential down to 400 mV [654], 500 mV [655], or even 600 mV [656], which is scarcely attainable by electrolytic activations. Moreover, the use of such modified carbon pastes eliminates two major drawbacks of electrolytic activations: (i) rather time-consuming procedures and (ii) the already mentioned overloading of the apparatus, especially, of some older instruments that do not have adequate protection against abrupt current changes and, in general, a high-current flow.

Finally, the activation procedure can be avoided and no special modifier or electro-catalyst needed if one uses *commercial activated carbon* [631,632], which is normally available (e.g., *Norit R3EX*) and ready for the preparation of the paste mixture. (The chemical activity of such carbons, namely, the presence of hydrophilic functional groups at the "industrially" preactivated surface, can additionally be characterized by means of pH-metric neutralization titration in combination with FT-IR and the low-temperature nitrogen adsorption method [632].)

3.3 TESTING OF UNMODIFIED CPEs

Electrochemical characteristics as well as some physicochemical properties of carbon pastes surveyed earlier can be, to a certain extent, defined using empirically proposed sets of experiments—by means of the *testing of carbon pastes* [83,175,176]. The corresponding assays usually help to reveal typical properties of carbon paste(s) under test, thus allowing one the basic orientation in a rather variable behavior of various CPEs, CMCPEs, CP biosensors, and other related sensors. In addition, such an approach often represents the only way of characterizing freshly made carbon paste mixtures, including brand new formulations with constituents of hitherto unknown properties.

Figure 3.3 illustrates how to arrange the strategy of such testing measurements in the individual sequences in a step-by-step manner. This *flowchart* has been assembled based on previous experience and recommendations of numerous authors, namely, [2,8,9,20,26,28,29,31,40,41,60,61,63, 68,219,220], and styled for *traditional (binary, unmodified) carbon pastes*. The whole algorithm has

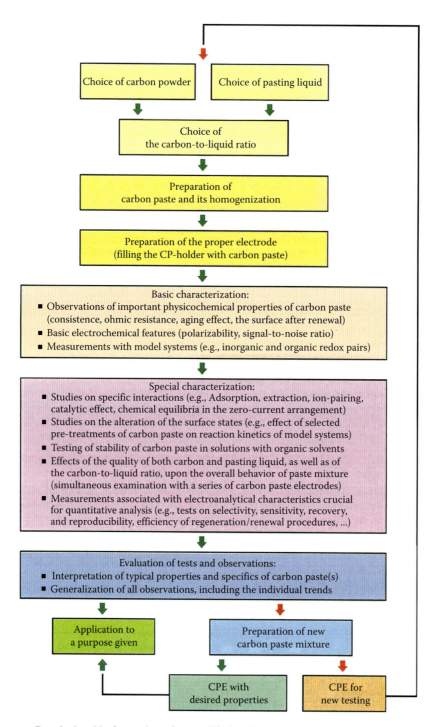

FIGURE 3.3 Practical guide for testing of unmodified carbon pastes. A schematic flowchart. (Retyped, rearranged, additionally colorized after Švancara, I. and Schachl, K., *Chem. Listy*, 93, 490, 1999; Švancara, I. et al. 1985–2010, Unpublished results.)

Carbon Paste as an Electrode Material

been structured as a kit, and the overall scheme can be flexibly adapted according to one's requirement, for instance, by adding new testing subprocedure(s) evaluating the performance of a particular instrumental technique (e.g., potentiometry, amperometry, or measurements in flowing streams), on the contrary—also by removing those assays and procedures that are irrelevant for a given application.

In order to make the individual tests systematic and the respective results comparable, numerous *model systems* have been proposed. Among them, one can recommend the following: $Ag^{+/0}$, $Fe(CN)_6^{3/4-}$, both representatives of reversible inorganic redox pairs; I_2/I^-, essentially reversible system whose oxidation form exhibits marked tendencies to be extracted onto the carbon paste bulk; *quinone* or *hydroquinone*, typical "quasi" reversible organic redox system; *phenol, phenothiazine*, fairly electroactive substances yielding the oxidation products (or intermediates) with strong adsorption capabilities; and $N_2H_5^+$, *ascorbic acid (AA)*, reduced forms of typically irreversible systems whose reaction rate can be controlled *by* electrocatalysis (i.e., suitable for testing of activation procedures and their efficiency).

Besides the specification of various properties and parameters, testing measurements may also help in assessing the quality of both main carbon paste constituents or the optimal carbon paste composition, here, through simultaneous experiments with a series of carbon pastes differing solely in the carbon-to-pasting-liquid ratio. Furthermore, suitably chosen procedures may also help in *testing the electroanalytical performance* of a CPE (e.g., reproducibility and repeatability, operational potential range and the background level, detection capabilities, etc.). A more detailed discussion on the individual tests and assays is beyond the scope of this paragraph, and the reader is advised to consult some special literature (see, e.g., [175,176] and references therein).

Finally, the testing can also be tailored for both *chemically and biologically modified carbon paste electrodes*, with special attention paid to the quality and proper function of the modifying substance(s). Such measurements have already been unified in the *testing and modeling of CMCPEs/CP biosensors* as shown in Sections 4.4 and 5.3.

Note: In the past, including the early era of CPEs, there were some efforts made to unify the individual electrodes via a "standardization of carbon pastes." Besides initial attempts by Adams himself (mentioned in [68]), followed by other authors [40,60], there were also repeated attempts to propose addressed *extraction quotients* [20,219,220] to quantify specific bulk processes at CPEs for the determination of selected organic compounds. Especially the latter seems to be a finely elaborated work, and it is quite surprising that it has not attracted more attention.

Recently, the term "standardization" of CPEs has surfaced again in association with our new activities. After a preliminary sketch of the problem (see [83] and pp. 302–303 and 317–318 therein), our research group finally attempted to elaborate a new study focused on postulating and experimental verification of a simple *qualitative parameter* that could objectively define the behavior of CPEs. One of the respective measures was called "CPE-index," χ_{CPE}, which reflected a simple two-step assay with the ferri or ferrocyanide model redox system, expressed by two alternate formulas:

$$\chi_{CPE} = \frac{\Delta E_P[Fe(CN)_6^{3/4-}]_{CPE}}{\Delta E_P[Fe(CN)_6^{3/4-}]_{TEOR}} \quad (3.2a)$$

$$\chi'_{CPE} = \frac{\Delta E_P[Fe(CN)_6^{3/4-}]_{CPE}}{\Delta E_P[Fe(CN)_6^{3/4-}]_{GCE}} \quad (3.2b)$$

where
$\Delta E_P = E_P(Ox) - E_P(Red)$
$\Delta E_{P\,TEOR} = 59\,mV$ [657]
$\Delta E_{P(GCE)} \geq 59\,mV$
$\Delta E_{P(CPE)} \gg 59\,mV$

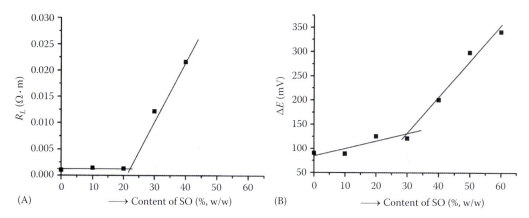

FIGURE 3.4 Dependence of resistivity (R, related to ohmic resistance) on the amount of binder in the carbon paste mixture (A) and dependence of the cathodic-to-anodic peak difference (ΔE, obtained from CVs) upon the amount of binder (B). *Experimental conditions*: carbon paste: graphite powder ("CR-5" type) + highly viscous silicone oil (SO); CV, potential range: from +1.0 V to −0.1 V vs. Ag/AgCl and back, scan rate: 50 mV · s^{-1}; sample solutions: 0.1 M KCl + 0.005 M K$_3$Fe(CN)$_6$. (Based on Mikysek, T. et al., *Anal. Chem.*, 81, 6327, 2009; taken and rearranged from Mikysek, T. et al., Relation between the composition and properties of carbon nanotubes paste electrodes (CNTPEs), in *Sensing in Electroanalysis*, Vol. 5, Eds. K. Vytřas, K. Kalcher, and I. Švancara, University Press Center, Pardubice, Czech Republic, 2010, pp. 69–75.)

the latter two values mirror the essentially reversible behavior at the glassy carbon electrode (GCE), the moderated reaction kinetics at the CPE, and the individual parameters ΔE_P, $\Delta E_{P(GCE)}$, and $\Delta E_{P(CPE)}$ estimable from the corresponding cyclic voltammograms (CVs). Although the first parameter, CPE-*index* "χ_{CPE}," has been related to the theoretical value, its alternative (denoted as χ'_{CPE}) then represents a value correlated to the GCE, which appears to be more convenient as it considers the actual experimental conditions. Both "theoretical" and "practical" CPE indexes can be expressed as dimensionless numbers, defining the *degree of irreversibility* of each CPE and, consequently, the quality of the carbon paste mixture examined.

A couple of years ago, this characterization was combined with the already-mentioned studies [228,229] on the ohmic resistance of CPEs with a direct relation to the carbon paste composition. By using a new *model of the tightest arrangement of carbon particles*, it was shown that the degree of irreversibility (expressed via one of the indexes) is also related to the actual CP composition and the actual ohmic resistance. This is shown in Figure 3.4, illustrating that, up to the ratio of the two main components, $C/L = 70:30\%$ (w/w), the ohmic resistance is still fairly low and nearly constant, starting to grow when the increasing content of liquid has exceeded this limit.

And the respective boundaries—seen as break points on both plots—have exactly fitted the theory of the tightest arrangement of carbon particles that revealed about 65%–70% (w/w) carbon for the two standard CP mixtures tested, "C/Nj" and "C/SO" ([228]; in Figure 3.4, only the latter is shown), as well as two novel paste composites of the CNTs/Nj and CNTs/SO type [229]).

3.4 INTERACTIONS AT CARBON PASTES

Due to its particular composite nature, the carbon paste matrix is likely to induce interactions that can be different from other conventional electrodes. In particular, the binder intrinsically offers new possibilities of interactions not only on the CPE surface but also within the bulk matrix. Hereafter, a classification is proposed to describe the typical behavior of CPEs with respect to where the respective process (or several processes) can take place. Thus, one can distinguish between the surface and bulk processes, as well as their combination should there be considered as the interaction likely to

Carbon Paste as an Electrode Material

occur in a small thickness of "bulk" paste located at the electrode surface (which is often the case for chemically modified CPEs, which are described in detail in Chapter 4).

3.4.1 INTERACTIONS AT THE CARBON PASTE SURFACE

Several electrolytic and nonelectrolytic phenomena can occur at the CPE surface, and most common examples are presented as follows.

3.4.1.1 Electrode Reactions with Charge Transfer (Electrolytic Processes)

As with other electrodes, straightforward processes likely to occur at a biased CPE surface are oxidation and reduction transformations (Equation 3.3, [3,658]). They may involve charge transfers at both inorganic and organic substances as well as organometallic species:

$$[Ox] + ne^- \Leftrightarrow [Red] \tag{3.3}$$

These reactions, and especially *the rate of charge transfer processes*, are affected by the state of the CPE surface and therefore by the type of carbon used to prepare the electrode. As already discussed in Chapter 2, graphite is the most widely used carbon source, but CPEs can also be prepared using other carbonaceous materials such as, for example, activated carbon [632], glassy carbon powder [392,659], carbon nanofibers [121,660], fullerenes [117,200], ordered mesoporous carbon [119], and a wide range of carbon nanotubes [661]. The chemical state of the surface of such carbon-based materials (i.e., presence of surface functional groups, which can be tuned by appropriate treatments [632]) has great influence on the electrochemical properties of the resulting electrodes in terms of variable overpotentials for the detection of redox-active analytes.

After a debut report [360], *carbon nanotube carbon paste electrodes* (*CNTPEs*) have generated a huge amount of work because of the extremely attractive properties of carbon nanotubes that are particularly useful to improve the performance of electroanalytical devices [662]. Compared with the "classical" CPEs, CNTPEs usually exhibit faster charge transfer kinetics and higher sensitivities ([362,363,366,379,398,399]; for other details, see Section 2.2.3.2). Even more recent, and of comparable interest, are *carbon ionic liquid electrodes* (*CILEs*) consisting of homogeneous mixtures of graphite particles with room-temperature ionic liquids acting as the active binder [126,451,663]; combinations of ionic liquids with carbon nanotubes are also feasible [417,418]. In brief, CILEs offer higher sensitivity, faster electron transfer, and better reversibility than ordinary CPEs [126,456,459,461,474,495, 663–665] due to the attractive properties of the room-temperature ionic liquids, such as the increased ionic conductivity, lower resistance, enhanced ion-exchange capacity, or inherent catalytic properties. (For example, the rate constant for oxidation of hydroquinone, H_2Q, was reported to follow the order "CILE > IL > CMCPE > CPE" [465,480]. Again, more information is given in Section 2.2.3.3.)

Electron transfer reactions are also likely to occur when using CPEs coated with polymeric membranes [666–668] or preplated metallic films [103,617,621,637,669–672]. If such metal film-coated CPEs are combined with electrochemical stripping analysis, the electrode process involves the deposition and subsequent dissolution of intermetallic compounds such as $Me(Hg)_x$ amalgams [255] or Hg-Au and Me-Bi adducts (expected to be formed at gold- and bismuth-plated CPEs [260,673]). If metal films are generated *in situ*, these processes can also be accompanied by additional effects as a consequence of mutual competition between the corresponding film and the bare carbon paste surface [260].

3.4.1.2 Adsorption and Related Electrode Reactions of Nonelectrolytic Character

Several redox-active analytes can be accumulated by adsorption (or related nonelectrolytic phenomena) onto CPE surfaces before their electrochemical detection. This has resulted in the development of *adsorptive stripping voltammetry* (*AdSV*) as a particular variant of the electrochemical stripping techniques [674], which has also influenced electrochemistry with CPEs. Several kinds of adsorption processes at CPEs can be distinguished, and, depending on the accumulation mechanism, various experimental parameters are likely to affect the effectiveness of the preconcentration event. The accumulation mechanisms at CPEs have been classified in several categories (Equations 3.4 through 3.8; according to [675]):

Direct adsorption

$$ORG \rightarrow (ORG)_{CPE} \quad (3.4a)$$

$$MeX_m \rightarrow (MeX_m)_{CPE} \quad (3.4b)$$

Competitive adsorption

$$(MeX_m)^{m-} + (X^{m-})_{CPE} \rightarrow (MeX_m)_{CPE} + X^{m-} \quad (3.5a)$$

$$(MeX_m)^{m-} + (Y^{m-})_{CPE} \rightarrow (MeY_m)_{CPE} + X^{m-} \quad (3.5b)$$

Synergistic adsorption

$$MeX + (L)_{CPE} \rightarrow (MeXL)^{m+}_{CPE} \quad (3.6)$$

Electrosorption

$$Me^{m+} + (L)_{CPE} \rightarrow (MeL)^{n+}_{CPE} + (n-m)e^- \quad (3.7)$$

Chemisorption

$$Me^{m+} + (-Z)_{CPE} \rightarrow (Me-Z)^{m+}_{CPE} \quad (3.8)$$

where
the subscript "*CPE*" means the carbon paste surface
other abbreviations are explained in the following

A straightforward way to *direct adsorption* is the *physical sorption* or *chemisorption* of organic molecules ("ORG"; [676–681]). It can also be the case of numerous neutral metal complexes ("MeX_m," e.g., Be^{II}–o-(2-hydroxy-5-methyl-phenylazo) benzoate [682], Ni^{II}- or Co^{II}-dimethylglyoximates [683,684], a series of metal ions—Alizarin Red complexes (with metals = Ce [685], Zr [397,407,686], Cu [393,687], V [688], or Ga [689]), Ag^I-Alizarin violet [690] or Y^{III}(Sc^{III}, Sm^{III}, Eu^{III}, etc.)—*Alizarin*ates [691], and several metal ions-bromopyrogallol (with metals = Bi [692], Sb [693], or Sn [694]). Voluminous inorganic anions or anionic complexes ("$(MeX_m)^{m-}$," e.g., $HCrO_4^-$ [695], $P(Mo_3O_{10})_4^{3-}$ [696], and I^- or I_3^- [639,697]) can also enter into competitive adsorption schemes. One can speak about synergistic adsorption when an uncharged complex MeX formed in the solution interacts with another ligand ("*L*") strongly adsorbed at the CPE surface to form an MeXL adduct tending to be adsorbed as well (e.g., $[Hg^{II}(CN)L_3]^{2-}$ [698]).

One can also observe an *electrosorption* mechanism in which immobilization of the analyte is driven by an electron transfer reaction (e.g., I^- precipitation in the form of CuI at Cu_2O-CPE under cathodic conditions [699]). Chemisorption of cationic species can also occur at CPEs modified with minerals or organic–inorganic hybrids containing functional groups ("–Z") that are likely to be complex metal ions, "Me^{m+}" [700–702], but this relies primarily on chemically modified electrodes (see Chapter 4).

3.4.1.3 Ion Exchange and Ion-Pair Formation

Both processes are based on *electrostatic interactions* between charged analytes and a suitable charged reagent, either dissolved in solution (and with some affinity for the CPE surface) or introduced into the carbon paste as an electrode modifier.

Often, ion exchange and ion pairing act together, when one may be substituted by the other, or they mean the same process, which is also evident from various mechanisms based on and reported in literature. Compared with adsorption, the resultant effectiveness of the accumulation based on ion pairing depends primarily on the reagent involved, whereas the "quality" of the paste used is of marginal importance.

A straightforward approach of ion exchange accumulation is the elegant *in situ* modification of the electrode surface by charged amphiphile moieties, creating thereby an ion exchanger directly during the measurement. Subsequently (or simultaneously), ionic analytes of opposite charge can be immobilized and preconcentrated by ion pairing, if they show sufficient affinity toward the modifier. Typical amphiphile moieties used for that purpose are long-chain alkylsulfonic acids (e.g., sodium dodecyl sulfate, SDS; [703]) or bulky organic ammonium salts (e.g., CTAB; [704–707]), which are lipophilic enough to adsorb strongly at the carbon paste. Since the resultant ion pairs (or ion associates, respectively) show hydrophobic characteristics again, immobilization is often not only restricted to the surface layer but the species may migrate into the bulk of the electrode material; see Section 3.4.2.

Other strategies concern the preparation of chemically modified CPEs displaying ion exchange or ion-pairing properties. This can be achieved by various ways: (i) the direct incorporation of a solid ion exchanger into the paste (first example in early 1984 by Wang et al. [707]); (ii) carbon pastes where liquid anion exchangers are mixed with the paste; and (iii) carbon pastes where the liquid binder takes the role of ion exchanger (discovered in 1993 [205]). The first method was very popular, and many types of organic, inorganic, or organic–inorganic hybrid ion exchangers have been added in carbon paste matrices (a detailed description is given in Section 4.3). Liquid ion exchangers were also widespread, and they were often described as preferable to solid ones because they ensure better homogeneity of the modified paste and, therefore, easier surface renewing and better reproducibility of measurements [82]. The third approach demonstrates some new functions of the carbon paste binder that was mainly made of TCP or dioctyl phthalate (DOP), known as plasticizers in polymeric membrane–based ion-selective electrodes [709]. Of particular interest is the more polar TCP because it has a high tendency to be protonated (Equation 3.9a) in the form of a large (and lipophilic) cation ready to be involved in the ion-pair formation (Equation 3.9b):

$$\{(C_7H_7O)_3 \equiv\}_{CPE}\ P=O + H^+ \rightarrow \{(C_7H_7O)_3 \equiv\}_{CPE}\ P=OH^+ \qquad (3.9a)$$

$$\{(C_7H_7O)_3 \equiv\}_{CPE}\ P=OH^+ + X^- \rightarrow \{(C_7H_7O)_3 \equiv\}_{CPE}\ P=OH^+, X^- \qquad (3.9b)$$

The symbol "$\{(C_7H_7O)_3 \equiv\}_{CPE}$" sketches the highly lipophilic organic moiety of the TCP molecules in the carbon paste surface layer; thereby, the hydrophilic cationic fragments "P=OH$^+$" are oriented into the solution where they act as highly effective ion-pairing sites for interaction with a counter anion. For example, the whole mechanism can be advantageously exploited in electrochemical stripping analysis to accumulate selectively some inorganic anions, such as I$^-$ [254,639,638] or AuX$_4^-$ (where "X" … Cl, Br, SCN, CN [205,710,711]), in measurements using either differential pulse cathodic stripping analysis [205,639,638] or constant current potentiometric stripping analysis [254,710,711]. Other anions can then be determined with the aid of automated potentiometric titration, which is the case for ClO$_4^-$ and BF$_4^-$ [719] or As(Mo$_3$O$_{10}$)$_4^{3-}$[712]. Besides the classical, commercially available organic resins exhibiting ion-exchange properties, which have been used in CPEs (see, e.g., [707,713,714], the most commonly used solid ion exchangers in CPEs were clays [715,716], zeolites [717,718], and porous silica materials bearing charged moieties [718,719]. In all the aforementioned approaches, however, the respective accumulation mechanisms were often synergistic processes, also involving extraction principles or even adsorption.

3.4.1.4 Electrocatalysis-Assisted Detection

Electrocatalysis can be defined as a catalytic process involving oxidation or reduction through the direct transfer of electrons (see Chapter 4 for a detailed presentation of this concept). This usually implies the use of an electrocatalyst, that is, a catalyst that participates in an electrochemical reaction, which modifies and increases the rate of electron transfer reactions without being consumed in the process, resulting in significant improvement in the performance of electrochemical sensors [720]. Facilitating the electron transfer reaction is usually achieved by the appropriate choice of a suitable modifier that can be either added directly in the carbon paste mixture or adsorbed—eventually, chemically immobilized—at the carbon paste surface. The versatility of the carbon paste matrix enables one to use a wide range of mediators, from solid particles to molecular systems (organic molecules or organometallic complexes), including supra and macromolecular compounds. Several hundreds of electrocatalysts have been used to modify either the surface or (mostly) the bulk of CPEs; they are not mentioned here as they are described in Section 4.3. In some cases, relatively high electrocatalytic activity can also be achieved through electrolytic activation of the unmodified (bare) carbon paste [600] or, as largely developed in recent years, by merely choosing a proper carbon material (e.g., carbon nanotubes [123,661] or ordered mesoporous carbons [119]) or, alternatively, by using an active binder such as room-temperature ionic liquids [456,471] or a combination of the two approaches [418].

For instance, the simple replacement (even partial) of graphite by carbon nanotubes was likely to dramatically enhance the sensitivity of the composite electrode, as illustrated for the detection of organic molecules [398,389,423,721], for direct electrochemistry of redox proteins (e.g., hemoglobin [417,431]), or in biosensors [722,421]. Imidazolium- or alkylpyridinium hexa-fluorophosphate–based CILEs were reported to exhibit intrinsic electrocatalytic properties in the voltammetric detection of a long line of organic molecules [480,458,466,477,484,493,494,498,500,505,723]. Sometimes, additional electrocatalysts (e.g., polyoxo-metallates [126,515]) can be introduced into CILE to further improve its catalytic behavior.

Interpretations of electrocatalytic phenomena taking place at the CPE—and, in fact, the other processes as well—are typically based on empirical observations [68], and theoretical formulas are seldom needed. Nevertheless, in association with *surface characteristics* of CPEs, there have been some attempts to define such models and equations for calculations. This is the case of a (i) formula describing the specific adsorption of some organic compounds at the CPE [24] or a (ii) relation for calculating the *effective surface area for electrolytic deposition* [724] with the aid of the geometric parameters of graphite particles and the actual carbon-to-pasting-liquid ratio used. Since the latter could be of some interest today, in connection with the boom of metallic-film-plated CPEs [7,83,84,255,268], the respective equation can be given here. Thus, in somewhat rearranged form, one has

$$A_{eff(CPE)} = 0.97\,\pi \frac{(d_{CPE}/2)^2 \times (V_L/V_C)}{[d_C^{0.14} \times (V_L/V_C)^{1.9}]} \tag{3.10}$$

where
  d_{CPE} is the geometric diameter of the CPE surface
  d_C is the average diameter of carbon particles (that must be in an interval of 10–90 μm)
  (V_L/V_C) is the (pasting) liquid-to-carbon (powder) ratio in volume %, v/v [724] Otherwise, some more theoretical approaches with the mathematical background are occasionally used in modeling of test measurements with CMCPEs and CP biosensors; see Sections 4.4 and 5.3.

3.4.1.5 Synergistic Processes and Interactions

Some of the aforementioned processes are not restricted to the CPE surface only, and the carbon paste bulk (interior) may also be involved in the overall interactions. This is the case of *extraction*—as one of the most typical processes taking place inside the carbon paste and therefore being the subject of special interest in the following Section 3.4.2, where it is shown that extractive processes can occur in addition to the afore-mentioned adsorption, ion pairing, complexation, or precipitate formation.

Carbon Paste as an Electrode Material

Even the electron transfer reaction itself can lead to *incorporation/ejection* of matter in or from the paste. This is notably the case when a hydrophobic redox probe (i.e., ferrocene derivative) is introduced into the CPE bulk (via solubilization into a binder *iso*-nitrophenyl-octylether), the electrochemical oxidation of which into positively charged ferricinium derivative gives rise to anion transfer into the organic phase (or ejection of the electrogenerated ferricinium derivative) to maintain the charge balance [455,725]. A similar example is the ion transfer forced by generation of tert-butylferrocenium cation from the neutral substrate present in a hydrophobic room-temperature ionic liquid (1-decyl-3-methylimidazolium bis(trifluoromethylsulfonylimide)) used as the CPE binder [454].

The combination of *adsorption/extraction* was applied to accumulate organic substances [82,69], environmental pollutants [59,726], BIC [70,156] or some pharmaceuticals [727,728], in view of their subsequent electrochemical detection. For typical examples in the latter category, for example, some analgesics [729], antimycotics [730], antipsychotics [731], including a group of phenothiazines [219,732], extraction may dominate over surface adsorption, which can be exploited for their detection in complex matrices with strongly adsorbing constituents. This is the case of determining chlorpromazine (CLP) in the presence of uric acid (UA) [732] based on simultaneous accumulation of both compounds with the subsequent mechanical removal of the adsorbed UA (for details on this procedure, see Section 8.3).

A combination of *extraction with ion pairing* is typical in procedures employing some commercially available ion exchangers (e.g., *Aliquat 336* [251], *Dowex 50W* [522], *Amberlite LA-2* [733,734]) and surfactants from the family of quaternary ammonium salts (CTAB [735,736], *Septonex*® [695,736,737]) or alkyl sulfonates [182,703]. Of related interest is the interaction of organic molecules with surfactants to form hydrophobic assembles that are likely to be accumulated at CPEs and subsequently detected in the same way with great sensitivity (examples are available for L-dopa [738], β-nitrostyrene [739], *Ciprofloxacin* [740], or theophylline [741]. Some rather sophisticated schemes comprising several physicochemical and electrochemical interactions were occasionally reported. For example, electrolytic deposition can be performed at the electrocatalytically active carbon paste surface, both additionally combined with effective ion pairing [182]. Alternatively, accumulation through ion pairing can be accompanied by *intercalation* [742] or *precipitation* [743]. The methods that use such sophisticated processes usually offer high analytical performance with respect to both selectivity and sensitivity [157], thanks to the multiplicity of the individual processes combinable in a single sensing device.

This has led to the determination of trace amounts of target analytes in the presence of a large excess of potentially interfering species (e.g., I$^-$ vs. Cl$^-$ in a ratio 1:10^6 [639,737]) or to the achievement of extremely low detection limits [182]. The latter is the case of a method that employs the whole sequence of mutually combined and supporting processes; see Figure 3.5 and its upper part. Namely, it is (i) specially chosen carbon paste with TCP as chemically active binder; (ii) cathodic activation of this C/TCP giving rise to negatively charged functional groups at its surface; (iii) electrolytic accumulation of the analyte (the reduction of single ions), when the overall efficiency is further amplified by (iv) additional transfer of the target ions toward the electrode (via ion pairing with modifier *in situ*). Moreover, these processes were likely accompanied by (v) extraction/reextraction of the ion associate formed before reduction (not shown), making the entire mechanism selective prior to final (vi) reoxidation at a favorable potential around 0.0 V vs. ref. with minimal background. As a result [182], the procedure based on a series of these consecutive and simultaneous steps (ii) through (vi) offered an extraordinary sensitivity, being capable of operating at the picomolar level or even below—see Figure 3.5, lower part.

Some synergistic mechanisms can also be found in electroanalysis of biologically important compounds. For example, within the series of momentarily popular modifications by electrogenerated polymeric films (see the following, in Section 7.4), there is a report on a monomer-modified bulk CPE [744] that, after *electropolymerization* by potential cycling, gave rise to a poly(1-naphthylamine)-coated electrode surface with electrocatalytic activity toward glucose oxidation. Another interesting example is the combination of the so-called *enantioselectivity* of some special modifiers with electrochemical detection [745], when the corresponding method may include yet another process (e.g., adsorption). Enantioselective sensors act as chiral differentiators, and the respective CP biosensors

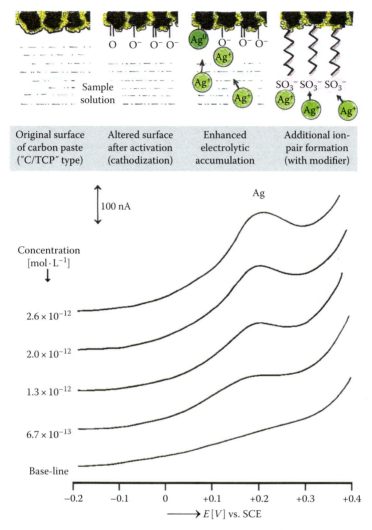

FIGURE 3.5 Voltammetric method for the determination of Ag(I) at tricresyl-phosphate-based CPE: Scheme of its synergistic accumulation mechanism (upper part) and calibration of Ag(I) at the picomolar concentrations (lower part). Base-line: blank supporting electrolyte (s.e.) *Experimental conditions*: C/TCP electrode; DPASV, accumulation conditions: $E_{ACC} = -0.2$ V vs. SCE, $t_{ACC} = 120$ min, $t_{EQ} = 30$ s; stripping limits: $E_{INIT} = -0.2$ V, $E_{FIN} = +0.5$ V vs. ref., scan rate: 20 mV·s^{-1}; s.e.: 0.02 M acetate buffer + 8×10^{-5} M n-C$_7$H$_{15}$-SO$_3$Na + 0.003 M EDTA; each sample purged for 5 min. (Assembled and redrawn according to Švancara, I. et al., *Electroanalysis*, 8, 336, 1996.)

enable the simultaneous analysis of two [746] or even more [747] optically active isomers. Finally, there are also some combinations based on attracting the substance of interest by means of *immunoassay* principles (see, e.g., [286,301–303]) with the subsequent electrochemical detection that enables, among others, to analyze even some microorganisms like bacteria and viruses (see also Chapter 8, the commentary, and table survey).

3.4.2 Interactions in Carbon Paste Bulk

CPEs differ from other solid electrodes by the presence of a (usually) liquid binder. Thereby, most processes occurring in the carbon paste interior are mainly dependent on the presence and properties of the liquid binder. In the following text, one considers solely the process at traditional

Carbon Paste as an Electrode Material

carbon pastes with nonconducting pasting liquids and not the faradic transformations in the bulk of strong electrolyte-based CPEEs, except for special carbon paste mixtures of the CILE type, also containing conductive ionic liquids and which have occupied rapidly the significant position within electroanalysis with CPEs.

3.4.2.1 Extraction and Reextraction

The hydrophobic character of the liquid binder makes possible the extractive transport associated with penetration of species into the bulk of carbon paste matrix, which has become a typical process occurring with this kind of electrode material [20,748]. Extraction capabilities represent perhaps the most distinct difference between CPEs and traditional solid electrodes that are, in principle, inapplicable to similar experiments (except for special configurations with additional membranes and to some extent with mercury electrodes; see discussion in [157]).

Typically, extraction at carbon pastes (see Figure 3.6) is appreciated as quite an attractive feature in practical measurements [732,740], and only exceptionally one can meet some critical comments considering this process as a complicating or even unwanted factor [211]. Basically, the distinction can be made between processes restricted to the electrode surface only and "genuine" extractive behavior, where species migrate appreciably from the surface into the electrode bulk.

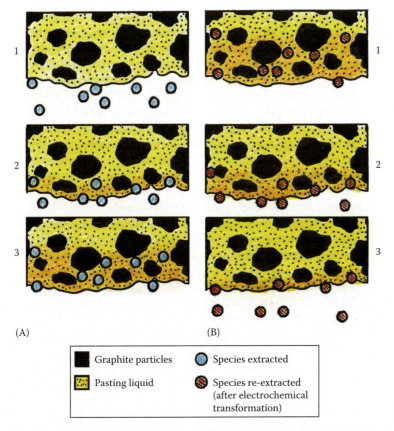

FIGURE 3.6 Typical pathway of extraction (A) and the subsequent re-extraction (B) at carbon paste-based electrodes. A schematic view. 1–3, three consecutive situations. (Redrawn and colorized according to Kalcher, K. et al., Heterogeneous electrochemical carbon sensors, in *The Encyclopedia of Sensors*, Vol. 4, Eds. C.A. Grimes, E.C. Dickey, and M.V. Pishko, American Scientific Publishers, Stevenson Ranch, CA, 2006, pp. 283–429.)

Practically, these two scenarios can occur successively or concomitantly. As shown in Figure 3.6, the overall pathway can be illustrated in two sequences, showing first the transfer of extracted objects into the liquid binder (schemes A—2,3) from a solution in which they have previously been dissolved (scheme A—3). Afterward, a reverse process of rediffusion takes place (schemes B—1,2), resulting in the re-release of objects of interest out of the paste and back to the solution (schemes B—3) where they can be detected.

The figure also tries to illustrate the *dynamic electrode/solution interphase*, depicting that a certain amount of species are likely to remain entrapped in the carbon paste bulk although the rediffusion process has been completed. Moreover, Figure 3.6 depicts the *actual thickness* of the CP layer involved in extraction, which is shown by different tone of color used for the binder moiety. The whole mechanism sketched earlier corresponds to a real situation when extraction is the principle of an electrochemical stripping method.

The extracted objects can be, for example, ion pairs [205,639] or organic molecules [219,221,749]; any of them must however be neutral (with compensated charges or noncharged, respectively) since charged ions cannot be extracted [157,748]. The driving force for the rediffusion process is then a concentration gradient in the electrode bulk with the direction to the electrode surface due to an electrochemical transformation of the species there during the measurement; as this rediffusion occurs during the measurement, some broadening of the resulting signal may be explained by this phenomenon. Rough attempts to determine the thickness of the diffusion layer in the electrode bulk are the controlled removal of surface layers of the paste before starting the measurement; through the removed volume (as calculated by the removed mass, e.g., by wiping off with a wet paper, and the specific mass of the paste) and the electrode area, it is possible to estimate the thickness of the corresponding layer.

Nevertheless, after rediffusion, a certain portion of extracted species may be irreversibly captured inside the paste [175,220]. This is a typical phenomenon for extractive accumulations at CPEs, requiring a frequent or even periodical removal of the used layer to prevent memory effects [175], which is accomplished comfortably with piston-driven CPEs [31,219,243]. Extractive accumulation and the subsequent stripping step can be performed in two different solutions, which is advantageous for optimal adjustment of pH for both processes. Using medium exchange, one can further enhance the overall electroanalytical performance of a procedure given, especially, its selectivity [750,751].

3.4.2.2 Chemical Equilibria in the Bulk and Transport Models

The effectiveness of extraction processes is driven by *distribution equilibria* (i.e., by the ratio between the solubility of a substance and its reaction product in the pasting liquid and in the phase used for the measurement [157,220]). Because the latter is usually an aqueous solution, it becomes obvious that even common carbon pastes made of Nujol®, oil of Uvasol® grade, or silicone fluids offer nearly ideal conditions for extractive accumulation due to a very high hydrophobicity of all these binders.

More exact definitions of extraction processes at CPEs have been of continued interest (e.g., [20,220,221,226,639,748]). It was confirmed that the efficiency and rate of extraction depend on the mutual affinity between the binder and extracted substance, whose chemical structure also plays an important role. For example, the presence of one markedly hydrophilic functional group (e.g., $-NH_2$, $-NO_2$, or $-OH$) in the molecule resulted in lowering the efficiency down to a few tens of percent, and even partial protonation was able to terminate the extraction process completely [20,220,732,748]. The extraction can also be controlled by varying the carbon paste composition. Besides the trivial benefit of an increasing binder ratio in the paste, one could observe somewhat lower extraction capabilities of silicone fluid–based carbon pastes compared to mixtures made from paraffin oil [220], which was explained by a lower adherence of the highly viscous silicone grease to the graphite particles, thus enabling a concurrent adsorption at the uncovered graphite surface [20].

Based on the experiments performed with a series of CPEs containing various alkane homologues, it was found that the extraction process was less satisfactory when using hydrocarbons with shorter chains (<C_{10}), which was explained by means of a certain miscibility of these liquids with aqueous solutions, resulting in "washing-out" effects and losses of the analyte escaping together with the hydrocarbon [220].

The *modeling of transport processes* in the bulk of CPEs is not an easy task because of the heterogeneous nature of the composite. Anyway, some interactions inside the carbon pastes have been interpreted with the aid of the "liquid membrane model" or the "pinhole mechanism" depending on the redox probe used [643] (for the latter, see also special note in Section 3.2.3). The analogy to membrane-like transport allows one to explain the rapid response of CPE-based sensors due to the effective mass transport through miniature membranes—the individual graphite particles coated by binder.

A special case of extraction is that of CILEs in which some redox species can be "dissolved" to further undergo electron transfer reactions, as illustrated for direct electrochemistry of hemoglobin [664] or myoglobin [752]. Another uncommon example is that of zeolite-modified CPEs for which the hydrophilic character of the modifier led to the possible permeation of the solution into micrometer to millimeter thick regions of the paste [753,754]. Using solid paraffin instead of mineral oil in zeolite-modified (solid-like) carbon paste electrodes (ZMCPEs) thus led to more hydrophobic composites, thereby preventing the impregnation by the electrolyte solution in the bulk electrode [754].

4 Chemically Modified Carbon Paste Electrodes

The functionality of all electrode materials is limited with respect to their potential window available. On the cathodic side, a reduction in the cation of the supporting electrolyte results in large currents, whereas at the anodic side, the oxidation processes with the respective responses—either of the electrode material itself or of the anion of the electrolyte—would overlap the signal of interest. Thus, in voltammetric or amperometric measurements, the analytes can basically be monitored if they are electroactive and their responses obtainable within the operational potential range. Apart from sorption effects, the reactivity of the analyte is then limited mainly to electrochemical transformations at the electrode surface, that is, reactions connected with the electron transfer.

The concept of "chemically modified" electrodes (CMEs) appeared in the 1970s, propagated by Murray et al. [756–758] and later also by Wang [759]. In fact, chemical modification of carbon paste had already been performed in the 1960s by Kuwana and French [14], and the term *chemically modified electrode* was coined around 1975 with different types of electrode materials; first publications on explicitly "modified" graphite electrodes dealt with the electrosynthesis of optically active compounds [760,761]. One of the first publications dealing with such *modified carbon paste electrodes* concerned with experimentally rather complicated chemical immobilization of the amino functional groups into the carbon paste bulk; nevertheless, the resultant electrode was found particularly powerful for the determination of silver at subnanomolar level [51].

Initially, the term was quite liberally applied also to the electrodes that could somehow be treated chemically. In today's understanding, CMEs are electrodes also possessing some chemical reactivity besides their ability to transfer the electrons to or from the analyte, which becomes possible by the presence of chemically reactive groups or substances at the electrode surface. Consequently, besides electrochemical transformations, the analytes may undergo some chemical reactions that principally widen the applicability of each electrode material.

Although the modification of solid homogeneous electrode materials is usually a labor-intensive task, heterogeneous materials—in particular, carbon paste mixtures—can be easily modified not only at the electrode surface but also in the whole bulk with comparable effectiveness. This is one of the reasons why heterogeneous electrodes have become so popular over the past three decades.

4.1 CLASSIFICATIONS OF CHEMICAL MODIFICATION

Chemically modified carbon paste electrodes may be classified by different criteria such as (a-i) type and method of modification and (a-ii) location of the modifier. Basically, one can differentiate between *intrinsic* and *extrinsic modification*, depending on whether the components of the paste (conductive particles or pasting liquid) bear modifying functional groups or the modifier has been added to the material as a further component. Another way is to discriminate between (b-i) surface- and (b-ii) bulk modification based on the placement of the modifying agent at or in the electrode. In the first case, the presence of chemically active components is restricted to the electrode surface, whereas the latter employs the whole bulk (interior) of electrode material modified.

A third aspect is (c) time of modification, whether it occurs *during the measurement* or such a process *precedes* it; the latter is the case of the electrolytic formation of a metal film on the support or substrate electrode prior to its use in the MeFE configuration. In this respect, we may distinguish also (c-i) *in situ* and (c-ii) *ex situ* modification. In the former case, the modifier is added to the measurement solution, and modification takes place upon exposure of the unmodified electrode surface to the solution. All other types may be regarded as *ex situ* methods.

4.1.1 INTRINSIC MODIFICATION

Here, the components that constitute the carbon paste bear the modifying functional groups. Thus, intrinsic modification may proceed via the particles or the pasting liquid. In the former case, usually the particles are covalently modified with reactive groups. A convenient starting point is the oxidation of the carbon surface by strongly oxidizing agents (e.g., nitric acid) producing oxygen containing groups on the surface (quinone structures, phenols). As a result, carbonyl groups can be chemically reduced to alcohols and phenols, which in consequence may be exploited directly or may act as anchoring groups for the actual modifier, which can be bound directly or via spacers depending on its chemical structure. A typical example was given by Cheek and Nelson who determined silver by accumulation via a complexing moiety at the carbon surface [51].

A special type of intrinsic modification is preanodization (see Section 3.2), which is usually done *in situ* at the electrode surface. A highly positive potential (beyond +1 V vs. the reference electrode) is applied to the working electrode in order to oxidize the surface of the exposed carbon particles to some extent. In fact, the procedure is most commonly used to improve the electron transfer with the analyte; a positive side effect is the stripping of a layer of pasting liquid from the surface, which otherwise reduces the effective electrode area. An illustrative example is the improvement of the voltammetric responses of uric acid in the presence of ascorbic acid [762].

More seldom, the pasting liquid is the subject of modification. Indeed, we may consider pastes prepared from liquids that may react chemically with the target analyte as this kind of modified material. Examples are carbon pastes where the binder can be protonated or deprotonated and acts as an ion exchanger (e.g., [639]) or where the component is an ionic liquid (see [7] and references therein).

4.1.2 EXTRINSIC MODIFICATION

With extrinsic modification the modifiers are added to the electrode material as additional components. A common way to classify extrinsic modification is the placement of the modifiers in the electrode.

4.1.2.1 Surface Modification

The modifier is located only at the surface of the electrode and may be present in thin layers, in films, or in membranes.

Submonolayers. The modifying agent is randomly distributed over the electrode surface in an amount less than necessary to form a monolayer. Typical examples are *in situ* modifications or direct preconcentration of the analyte with adsorptive stripping analysis, for example [763]. High concentrations of the modifier in the sorbent solution may then result in the formation of mono- or multilayers.

Monolayers. The modifier covers the electrode surface with a single layer usually attained as Langmuir–Blodgett films [764,765] or as self-assembled monolayers (SAMs; see Figure 4.1) [766,767]. Problems may arise from irregularities in the monolayers, that is, holes in the layer structure ("pinholes") [768]. SAMs are formed spontaneously from compounds with functionalized alkyl chains if one end of the molecule sorbs strongly to the surface, whereas the other, usually hydrophilic, may contain another functional group that interacts with the ends of other sorbed molecules to build a stabilized structure. Stabilization may also be effected by van-der-Waals interaction of the alkyl chains. Typical applications comprise cetyltrimethylammonium salts. Often SAMs are

Chemically Modified Carbon Paste Electrodes

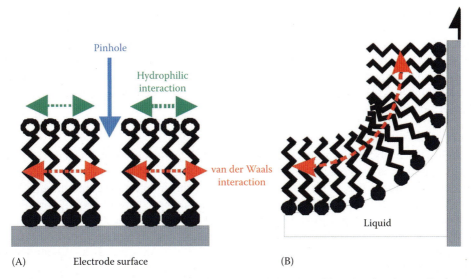

FIGURE 4.1 Self-assembled monolayers (A) and Langmuir-Blodgett films (B). A scheme. (Redrawn and rearranged from Arrieta, A. et al., *Sens. Actuat. B*, B95, 357, 2003.)

formed on silica [551,769–772] or dextrin substrates, which are used as modifiers for carbon pastes [773]. SAMs on gold may be used for potentiometric sensors [774].

Polylayers. These structures consist of many layers of the modifier and can be homogeneous or heterogeneous in nature. In the latter case, modifiers are embedded in a polymeric three-dimensional network. Such layers may be produced by electrodeposition, such as electropolymerized organic polymers (see Figure 4.2), by application of solutions after evaporation of the solvent (typically polymeric solutions), or by direct application of a (usually polymeric) membrane. The deposited layer may interact with the target analyte directly, serving as anchoring backbone or trap for the actual modifier, or may act via its size-exclusion characteristics.

Typical examples for polymeric membranes are dialysis membranes, commonly used for immobilization of biological entities at the electrode surface [156,288,775–788], Nafion®, and other ion exchangers [789–791], redox polymers [783–794], or conductive organic polymers [795–798]. Membranes can be applied directly as sheets to cover the electrode surface or may be generated by casting a solution of the corresponding polymer on the electrode surface and evaporation of the solvent.

Another possibility is in-place polymerization from monomers either by electrolysis or by other common polymerization processes. A special form of the latter is *cross-linking*, frequently applied

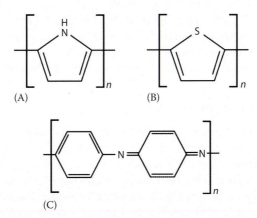

FIGURE 4.2 Some electrolytically generated polymers: polypyrrole (A), polythiophene (B), and polyaniline (C).

to immobilization of proteins, where smaller units are linked with each other by an agent with two reactive groups, which bind covalently to the target molecules, yielding a huge polymeric network with very low solubility in the solvents to which the electrode surface will be exposed. Often this form of immobilization is combined with an additional polymeric backbone structure (membrane) such as Nafion®. Typically, enzyme molecules are cross-linked either to each other or to bovine serum albumin with glutaraldehyde. In fact, cross-linking is possible not only with surface but also with bulk modification.

4.1.2.2 Bulk Modification

Although surface modification can be effected also with homogeneous solid electrodes, heterogeneous materials offer the advantage of simple modification of the whole electrode material under mild conditions due to its composite nature. In the simplest case, the modifier—either liquid or solid—is added directly to the carbon paste (the term "direct mixing" for this procedure being coined by Baldwin et al. [53,54,683]) and is indeed the mildest and simplest way to modify an electrode. Alternatively, the modifier may be mixed to or adsorbed on the carbon powder during the preparation of the carbon paste [799]; another way is its dissolution in the pasting liquid [800].

The *quality of modification* (*homogeneity*) with particulate matter depends on the particle size of the modifier and the care with which it is mixed into the electrode material. In these cases, for example, sonication may improve the results. Due to better homogeneity, liquid modifiers yield more reproducible results than solid ones.

4.2 REASONS AND STRATEGIES FOR MODIFICATION

The general reason for modification of an electrode surface is to improve either the mass transport or the electron transport limitations at an electrode surface.

This can be achieved either via generation or immobilization of electroactive species or reactants on or close to the electrode-solution interface or electron-transfer improving compounds or functional groups. Immobilization of electroactive species directly at the electrode surface decouples mass transport from the electrochemical transformation process, that is, oxidation or reduction, during the measurement and is commonly accompanied by a considerable decrease in the detection capabilities of the system given.

Thus, the main reasons for modification of CPEs involving modifier-substrate interactions are the following [801]:

- Immobilization and preconcentration (accumulation) of the analyte at the electrode surface
- Exploitation of catalytic (mediating) electrochemical reactions, or placement of electron transfer accelerating or improving groups at the electrode surface
- Immobilization of reactants generating electrochemically active products
- Protection and stabilization of the electrode surface
- Change of physical or physicochemical properties of the CP-surface in a favorable way

4.2.1 CRITERIA REQUIRED FOR MODIFIERS

Modifiers should meet some criteria to be useful for the intended purposes: they should be insoluble in solutions to which they will be exposed or at least they should adsorb strongly on the carbon particles; otherwise a "bleeding" of the electrode can occur, yielding poorly reproducible results. Improvements can be made by making modifier molecules more lipophilic [801].

Except if electrocatalytic effects are to be exploited the modifier should not be electroactive at the potential of interest because otherwise high background currents deteriorate or even impair electrochemical responses [802,803]. With bulk modification, the usual concentrations of the modifier are between 5% and 10% (w/w) of the paste. Too small concentrations do not provoke the desired effect, whereas too high concentrations may change the physical and physicochemical properties of the electrode material in an unfavorable way by increasing its ohmic resistivity and the background current. (However, there are also the paste mixtures containing 50% (w/w) modifier or even more (see, e.g., [7] and references therein), where the resultant CPEs are reported to be operating satisfactorily.)

4.2.2 Activation and Restoration of the Modifier

Sometimes, it is necessary to create the actually reactive groups or species of the modifier by conditioning, a procedure that can be *(physico)chemical* or *electrochemical*. Typical examples are protonation of weak amines or deprotonation of weak acids for creating ion exchangers [55,69] or electrolytic generation of the active form [804].

Due to electrochemical reactions during the measurement, the modifier may be consumed or products may adsorb strongly at the electrode surface. In case of repetitive use of the electrode, the original functionality of the modifier must be restored if it does not happen spontaneously. The methods for doing so are straightforward: chemical, electrochemical, and mechanical.

Chemical regeneration means that the electrode surface is exposed to a solution where sorbed species are removed or the modifier is reconstituted by a chemical reaction. Sometimes these effects can be achieved by the application of a potential (electrochemical regeneration). In tedious cases, none of these procedures is successful, and mechanical removal of the outmost layer of the bulk-modified paste, usually done by wiping off with a filter paper, remains the only alternative. With respect to repeatability, chemical or electrochemical regeneration is preferable to mechanical, particularly when particulate modifiers are used. An ensuing conditioning step, with or without a potential, is recommended; both regeneration and conditioning should be performed also before the first measurement to provide the same conditions for all measurements.

4.2.3 Immobilization and Preconcentration: Concepts and Strategies

Preconcentration of an analyte can be accomplished in many ways. In the simplest form, the target molecules show high affinity to the unmodified paste already and accumulation occurs via *adsorption* [226]. If the affinity of an analyte toward the carbon paste material is not high enough due to lack of lipophilicity, other strategies can be applied, such as *ion exchange* [252,522,719,734,800, 805–809], *insoluble salt formation* [810,811], *complexation* [234,549,683,799,812–816], or *ion-pair formation* [182,704–707,817,818].

All these types represent some form of chemical accumulation, where the driving force for the preconcentration of the analyte comes from the chemical reactivity of the target species with the modifier.

Chemical accumulation is often an essential part of stripping analysis with modified electrodes, where a chemical equilibrium reaction replaces the electrochemical deposition step (see Figure 4.3). As the chemical reaction proceeds spontaneously, in many cases the application of a deposition potential is obsolete (open circuit preconcentration); the accumulated amount is controlled by the exposure time of the modified electrode surface to the analyte solution. Nevertheless, an optional potential can be applied if it has a positive effect on the migration of the species to the electrode, if an oxidation state of the analyte that is not present in the sample is involved, or if the product should undergo electrochemical changes (e.g., electrochemical deposition of the analyte in modifier-assisted accumulation [182]).

Very common is a medium exchange (approach) between accumulation and actual measurement (unusual with classical stripping analysis with electrochemical deposition), so that the conditions for preconcentration and determination can be optimized separately and independently. Additionally, interfering species that do not have an affinity to the electrode material are separated off.

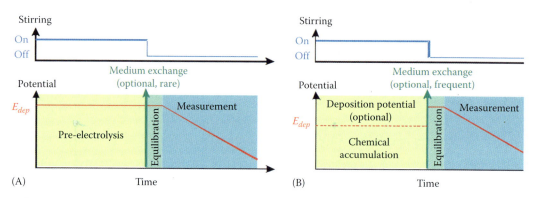

FIGURE 4.3 Comparison of "classical" stripping analysis including electrolytic deposition of the analyte (A) with stripping analysis employing CMCPEs and chemical accumulation (B).

Preconcentrations are referred to as *in situ* accumulations if the surface of an unmodified electrode is altered during the measurement itself by adsorption or electrodeposition of a modifier from the measurement solution [819,820]. In fact, if the reagents are sufficiently water insoluble or adsorbing strongly to the paste, they may also be added directly to the paste.

Ion exchangers can be used not only as solid organic resins [252,522,719,805–808] but also in liquid form [734,800,802,809]. Most popular types are *Amberlite LA2, Aliquat 33, Dowex 50 B, Triton X100*, or tricresyl phosphate, which can function as binder as well. Counterions immobilized with ion exchangers can also be used for potentiometric determinations (see, e.g., [569,821,822]).

Organic ligands for complexation of the analyte are quite frequently employed as modifiers for CPEs, such as dimethyl glyoxime for the determination of Co, Ni, and Pd [234,549,683,812,813]; dithizone for gold and lead [809,814]; or thioridazine for Pd [816].

Formation of insoluble salts can be a successful strategy for accumulating ions forming sparingly soluble agglomerates with corresponding counter-ions immobilized on the electrode surface. In this way, bismuth can be preconcentrated with alkyl mercaptans [810]. Nevertheless, it is possible to use insoluble salts as modifiers and to exploit the different solubility products to accumulate the analyte; thus, oxalate will replace sulfate in $PbSO_4$ and will be immobilized at the electrode surface of a correspondingly modified electrode [811]. The concept of ion pair formation comprises the *in situ* formation of an ion exchanger at the electrode surface [182,704–706,817,818].

Adsorption-capable species, generally sorbents, that show high affinity toward certain molecules are another frequently used group of modifiers for CPEs to accumulate the substrate. Such substances can be sepiolites [823–827], zeolites [828–830] and other types of clay minerals of natural or synthetic origin as well as silica of inorganic–organic hybrid type and other mesoporous materials [717,718,831]. But naturally occurring organic polymers may also be suitable for this purpose. Typical examples for the latter are chitin and humic acids; the former is able to preconcentrate aglutinin, nitrite, copper, or complexed iron [832–836], the latter heavy metals [837]. In fact, the chemical mechanism for accumulation with biopolymers is mostly ion exchange, complexation, or sorption.

4.2.4 Electrocatalysis

The term "electrocatalysis" is often used in combination with modified electrodes if effects of a modifier are involved other than of preconcentrative nature. Most cases of this type represent mediation of the electron transfer to the substrate provoked by a modifying agent in the paste because of some reasons the target molecule or ion is not electrochemically converted directly at the electrode surface at the operation potential. Such a mediation is sketched in Figure 4.4.

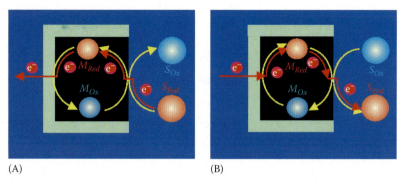

FIGURE 4.4 Mediation of the oxidation (A) and reduction (B) of a substrate S by a modifier M.

The reason for designing an electrocatalytically active electrode surface is mainly the improvement of the electron transfer toward the analyte, which can be accomplished with some envisaged follow-up effects, such as

- Generation of electroactivity of the substrate at the applied potential
- Improvement of the electrochemical signal, either in shape, height, or potential
- Separation of otherwise overlapping signals and elimination of interferences
- Improvement of linear dynamic ranges of the response or of the detection limit
- Improvement of the general analytical situation by decreasing matrix effects

Modifiers that are able to facilitate the aforementioned tasks may be quite different in their chemical nature. Most frequently used mediators are metals, metal complexes and salts, polymers with redox properties, and recently nanosized materials, such as nanoparticles (NPs), nanorods, and nanotubes. In fact, the electrochemical generation of functional groups at the surface of carbon particles by application of a highly positive potential may also be classified as electrocatalytically active modification.

Metals. Compact metallic elements used to enhance the catalytic activity of carbon paste electrodes belong often to the platinum metal group and comprise Pt, Pd, Ru, or Rh [838]; they can be used for the catalytic oxidation of hydrogen peroxide, hydrazines, or related compounds [839]. The metals may be applied in dispersed form but can also be plated on the carbon particles. Nevertheless, metals of less noble electrochemical character can also be used for specific applications [840], such as nickel for the oxidation of methanol [841]. More recently applications with NPs are becoming dominating.

Metal oxides. These binary compounds are useful catalytic mediators if they exist in various oxidation states; thus, MnO_2, FeO, Fe_2O_3, SnO_2, RuO_2, ReO_2 [842], and similar compounds are useful for the determination of corresponding substrates. In many cases, the target is hydrogen peroxide, which shows high overpotentials to both oxidation and reduction. Hydrogen peroxide is of particular interest because it may occur as intermediate in enzymatic reactions of many oxidases. However, other substrates such as amino acids can also be detected [556]. Oxides as modifiers attract increasing attention in the form of NPs with usually improved action over the macro- or microsized powder.

Metal complexes and salts. The biggest group of electrocatalytically active modifiers is made by metal complexes or salts. Hexacyanoferrates and analogues with platinum metals or nickel (for the detection of peroxides) [196,843–852], also coupled synergistically with carbon nanotubes [360,853], ferrocenes (as electron transfer mediator in biosensors mainly) [785,854–881], and phthalocyanines (Co^{II}, Fe^{II}, Ni^{II}, Mo^{VI} salts for the detection of peroxides, thio-compounds, oxalic acid, phenols, ephedrine, vitamin B1, pesticides, and sugars [557,651,652,655,765,882–919]) are most frequently

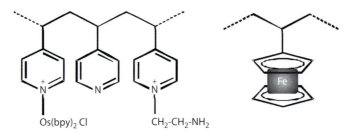

FIGURE 4.5 Some redox polymers.

used, particularly, for the construction of biosensors to decrease the overpotential for intermediates or co-factors. This action may be supported by nanosized materials, such as carbon nanotubes and gold NPs [124,403,420,432,441,920,921], but the redox-active mediator can also be bound to a polymeric backbone [782,856,862,870,922–925]. Ferrocene-modified CPEs were also investigated for their use as reference electrodes [98,634,926–928]. Phthalocyanine-based sensors can be also used as potentiometric sensors for ascorbic acid and thio-compounds [929–931]; the modifier itself can be employed in nanosized form [932–934].

Redox polymers. They can be a very elegant way to wire redox enzymes [935,936]. In this way, the direct electrochemical conversion of the enzyme's active center becomes possible without the need of reagents (reagentless biosensors).

Such polymers may contain ferrocenes, phthalocyanines [782,856,870,922–925], toluidine blue [312,937–939], or osmium [381,385,649,940]. Some osmium- and ferrocene-containing polymers are sketched in Figure 4.5 (after [768]).

Nanosized materials. In the past two decades, these catalytically acting modifiers are gaining importance. Due to an enormously huge surface area, nanosized particles may exert catalytic effects, which are not observable with macrosized form. They may also improve existing catalytic activities of the macro- or micropowder.

In electroanalysis at carbon paste electrodes, modifications with metallic NPs are of long tradition; the most prominent representative is gold (AuNP). AuNPs show very high affinity to thio-compounds, which can be exploited for the generation of self-assembling monolayers, and improve the direct electron transfer to many enzymes, which facilitates the construction of reagentless biosensors [941–946]. They may also act directly in the electrochemical transformation of substrates, such as acetaminophen [567] or homocysteine [947]. In combination with hemoglobin or myoglobin, they may, for example, be used for the detection of nitrite [948,949].

Gold in nanosized form is advantageous for the immobilization of antibodies in the design of immunosensors [485,950–952], but it can also be exploited for hybridization studies of DNA [953–955]; it even shows immobilizing effect on tumor cells [956]. AuNPs have also been used to design potentiometric sensors [774,957–960]. Other metallic NPs with catalytic effects are copper [960–966], platinum metals [967–971], nickel and cobalt (frequently loaded onto electrospun carbon fibers) [139,972–975], iron [976], and silver [416,977].

Among the oxidic NP modifiers, magnetic NPs based on iron oxides are most widely used [850,978–982]; they often consist of a magnetic core to simplify handling operations (with permanent or electric magnets), which is surrounded by a shell of effectively active catalyst or support for further immobilized components, such as antibodies (core-shell NPs) [301–303,983–990]; the shell is often a gold [301,984–986] or silica layer [302,303,987–990]. Other oxidic NPs that can act as modifiers are TiO_2 in combination with others for various neurotransmitters and drugs [499,920,991–997]; ZrO_2 for *Methylparathion* [998]; SiO_2 for hexacyanoferrate, hemoglobin studies, and biosensors [479,503,999,1000]; Fe_2O_3 for hydrogen peroxide [1001]; Al_2O_3 for nitrobenzaldehyde thiosemicarbazone [1002]; CdO for dopamine and ascorbic acid [1003]; and ZnO [496].

Occasionally sulfides or selenides in nanosized shape are used for studying hybridization of DNA [1004–1006]. Or, hexacyanoferrates, sometimes grown under enzymatic control [1007–1010], show increased catalytic activity when present as NPs compared with macrosized manifestations.

Nanoparticulate montmorillonite and zeolite improved the detection of serotonin, dopamine, and tryptophan [1011,1012] similar to phthalocyanines [933,934]. Inorganic or hybrid inorganic–organic heteropolyacids have been applied mainly for the determination of nitrite [143,1013–1016]; silicotungstate in combination with a ruthenium-bipyridyl complex was also used for the electrogeneration of chemiluminescence [1017].

At this point it should be also mentioned that *carbon nanotubes* (see Section 2.2.3.2) can be used as electrically conductive components in the preparation of corresponding pastes; nevertheless, they can also be simply added to common CPs as modifiers [388,721].

Rarely, *nanosized polymeric resins* are used to modify CPEs, such as pegylated (derived from polyethylene glycol, PEG) polyurethane to improve the detection of hydrogen peroxide and dopamine [1018,1019]. Finally, occasionally NPs are used for accumulation, such as silver sulfate for the preconcentration of iodide, where the silver salt of the latter possesses a lower solubility product than Ag_2SO_4 [1020].

4.2.5 OTHER PURPOSES FOR MODIFICATION

The immobilization of reactants at the electrode surface is an important aspect. One of the generated products must exhibit electroactivity, either directly or via mediators. Due to the restriction of the reaction space to the electrochemical monitor, the amount of reactant can be kept small (an important aspect with expensive modifiers and mass production of electrochemical sensors), and also the reaction time can be significantly shorter than with bulk solution processes. Basically, most of the biosensors with an immobilized biological entity at the electrode–solution interface (e.g., an enzyme) exploit this principle. Further details are discussed in the corresponding sections of Chapters 5 and 8.

Another interesting example of immobilizing reactants at the electrode surface was shown by Abruna's group; immobilized amines reacted with aldehydes from the bulk solution, the product of which was electroactive [1021]. Alterations in the physicochemical properties of the electrode surface may occasionally lead to exploitable effects though in fact some other reasons such as electrocatalysis may be responsible for the improved electrochemical behavior. Thus, stearate facilitates the monitoring of dopamine, particularly, *in vivo* measurements in the brain ([1022–1029]; see also Section 7.3). Lipids influence positively the preconcentration of promethazine [1030].

At the end of the chapter, it should be mentioned that any modification of the carbon paste changes indeed its physicochemical and electrochemical properties somehow. Therefore, as a rough rule of thumb, addition of modifiers to the bulk of the paste should not exceed an extent of 5%–10% (w/w), which will provoke usually only small and tolerable alterations in the characteristics of the paste compared with the unmodified material. (However, also such an empirical experience does not have general validity as there are some special types of reliably operating CMCPEs that have contained up to 50% modifier or even more; see, e.g., [7] and examples and references therein.)

4.3 TYPES OF CHEMICAL MODIFIERS

Due to the versatility of the carbon paste matrix to host almost any material (from molecular to macromolecular compounds, including all types of solids), an extremely wide range of modifiers have been used to extend the properties of carbon paste electrodes. Basically, one can define modifiers for CPE as any chemicals used in addition to the carbon source and the liquid binder to form a final *composite paste*. The modifier is most often mixed with the basic carbon paste constituents, resulting in a bulk modified CPE, with the exception of some surface treatments, such as the film formation onto the surface (in an approach that is typical for solid electrodes).

The primary role of a modifier is to bring new or additional features to the CPE (preconcentration of analytes, rejection of interferences, electrocatalytic properties) to improve both the sensitivity and the selectivity of the electrode. Notwithstanding the very high number of modifiers used to date, most of them have been selected for their accumulation capability and electrocatalytic properties. Due to the wide variability of these modifiers, it is quite difficult to provide a universal classification. They are presented hereafter according to three main categories: (i) inorganic materials, (ii) organic compounds, and (iii) organometallic compounds and other "novel" materials (such as nanostructured solids, NPs, hybrids, or other substances used in the recent approaches for modification). For each family, subcategories have also been considered. Note that biomolecules constitute a special category that is described in Chapter 5 and largely also in Section 8.4.

4.3.1 Inorganic Materials

Inorganic materials are usually manufactured in the form of powders or particulate solids, making their immobilization onto a solid electrode surface rather difficult (especially if long-term mechanical stability is required). Therefore, their physical entrapment or embedding into a CPE matrix constitutes an ideal way of immobilization.

4.3.1.1 Prussian-Blue Derivatives and Polyoxometallates

- *Prussian Blue and Related Compounds.* The complex structure of $Fe_4^{III}[Fe^{II}(CN)_6]_3$) and related metal hexacyanoferrates (where Fe^{III} is replaced by another transition metal ion) offer attractive electrocatalytic and ion-exchange properties that can be advantageously exploited in electrochemistry [1031]. After the pioneering work by Boyer et al. [845], the incorporation of metal hexacyanoferrates in CPEs has been largely applied, especially for electrocatalytically oriented determinations. They were mainly based on Prussian Blue itself [196,847,848,850,1032,1033] and also on copper hexacyanoferrate [436,610, 1034–1037], cobalt hexacyanoferrate [1038], or hybrids based on copper and cobalt hexacyanoferrates [1039–1041]. Ru^{III}- [1042], Rh^{III}- [849], Sn^{IV}- [1043], and Th^{IV}- [1044] hexacyanoferrates have also been used, and a comparison between hexacyanoferrates containing Fe^{III}, Cu^{II}, Cu^{I}, Co^{II}, or Ni^{II} with respect to hydrogen peroxide detection is available [843]. CPEs modified with such compounds were applied to electrocatalytic detection of various species, including, for example, amino acids [1036,1040–1042] or other biomolecules [1034,1038,1043,1044], nitrite [847,1032,1037], sulfite [436,1041], persulfate [945], and hydrogen peroxide [196,843,849,] as well as in biosensing applications [610,848,1035]. Ion-exchange properties of Prussian Blue derivatives were also exploited for potentiometric detection of cations [584,1045].

- *Polyoxometallates.* Polyoxometallates (POMs) are heteropolyanions that are likely to form Keggin-type ($XM_{12}O_{40}^{n-}$), Dawson-type ($X_2M_{18}O_{62}^{n-}$), mixed-addenda, or transition metal-substituted structures. They can exhibit well-defined electrochemical responses that have been largely exploited for electrocatalytic purposes [1046]. Most POM structures used in connection with electrochemistry are Keggin-type and transition metal-substituted heteropolyanions. Except the first report on a phosphomolybdate film on CPE [696], pioneering works on POM bulk-modified CPEs have been reported by Xiu Li Wang and coworkers [140,1047,1048]. The most widely used Keggin-type POM-based solids to modify CPEs involved phosphomolybdate (PMo_{12} [126,142,1049,1050]) and silicomolybdate ($SiMo_{12}$ [1049,1051]) but less commonly phosphotungstate (PW_{12} [1052]), silicotungstate (SiW_{12} [1053]), or vanadomolybdate (VMo_{12} [487]). The use of 1:8-type heteropolymolybdate with tetravacant Keggin structure has also been reported [1054]. Sometimes, organic cations (pyridine- or bipyridine-based derivatives) acted as counterions of the heteropolyanions [1049–1051], leading to more sophisticated crystal structures (see two examples in Figure 4.6).

Chemically Modified Carbon Paste Electrodes

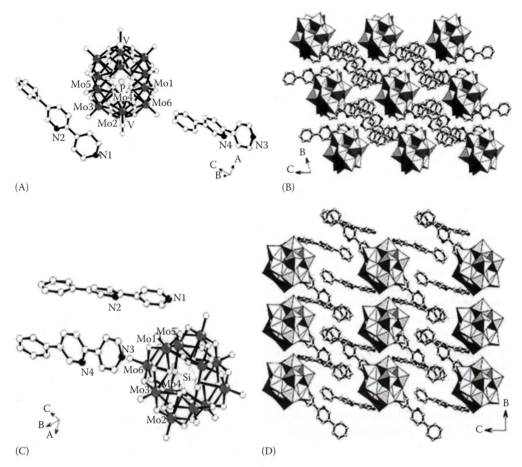

FIGURE 4.6 Schematic representation of two polyoxometallate-based hybrids (pbpy)$_4$H [PMo$_{12}$O$_{40}$(VO)] (1) and (pbpy)$_4$H$_4$[SiMo$_{12}$O$_{40}$] (2), where pbpy … 5-phenyl-2-(4-pyridinyl)pyridine. *Notes*: (A) A unit drawing of 1; polyanion [PMo$_{12}$O$_{40}$(VO)]$^-$ is completely drawn for clarity. Vanadium atom and the corresponding terminal oxygen atoms are only half-occupied with 50% probability; (B) 3D stacking view of 1 along the a axis; the dark-colored polyhedra represent {VO5}; (C) a unit drawing of 2; polyanion [SiMo$_{12}$O$_{40}$]$^{4-}$ is completely drawn for clarity; (D) 3D stacking view of 2 along the a axis. All hydrogen atoms are omitted. (Loosely reproduced from Mazloum Ardakani, M. et al., *Int. J. Electrochem. Sci.*, 4, 308, 2009.)

POM-modified CPE were always used for electrocatalytic detection (e.g., nitrite [1050–1052], bromate [140,1049], or hydrogen peroxide [140,1050,1054]).

4.3.1.2 Clays and Zeolite-Based Molecular Sieves
4.3.1.2.1 Natural Clays

These materials are layered aluminosilicates that exhibit ion-exchange capabilities (most often for cations) and adsorption properties. These attractive features have been largely exploited in electrochemistry to preconcentrate target analytes (usually cations by ion-exchange or organic compounds by adsorption) before their voltammetric detection [715,1055]. They can also be used to immobilize biomolecules, with promising applications in the field of electrochemical biosensors [1056]. Due to their layered structure, clay sheets tend to organize in rather stable layers on solid surfaces, so numerous studies have been performed on the basis of clay-film–modified electrodes [1055–1057].

Nevertheless, many clay-modified electrodes have been prepared by the bulk dispersion of clay particles within the carbon composite matrix.

Several types of clays have been used as modifiers for carbon pastes, including mainly montmorillonites [716,1011,1058–1070] and sepiolite [823–827,1071–1073] but also bentonites [1074–1079], vermiculites [542,1080–1082], kaolinites [1069,1083], kaolin [1084,1085], (fluoro)hectorites [826,1086], muscovite [1087], and some other natural smectites [110,1033,1088–1090]. Another family of layered materials are anion-exchanging clays exhibiting a hydrotalcite-like structure, called "layered double hydroxides" (LDHs), have been exploited in electroanalysis, mostly as thin films [1091,1092] but also in CPEs [1092–1095]. Friedel's salt is a special case of this category [1096]. Crude soil samples were also incorporated directly into CPEs [1097]. Although Wang with Martinez acted as pioneers [1058], the groups of Kula and Navratilova, L. and H. Hernandez, and Kalcher were the most active in the field of clay-modified CPEs and that of Mousty in using LDH anion exchanging clays.

Heavy metal cations such as Hg^{II} [1061,1064,1082], Cu^{II} [1072,1080], Ag^{I} [1080,1082], Fe^{III} [1058], or Bi^{III} [1065], alone or in mixture [1060], have been preconcentrated at clay-modified CPEs to enhance the sensitivity of their subsequent detection. Interestingly, cation exchanging clays were also likely to accumulate anionic species (as exemplified for $AuCl_4^-$ [1062,1063]) as a result of positive charges on the edges of clay sheets inducing both cation- and anion exchange properties [716]. On the other hand, organic molecules (especially pesticides [110,825–827,1071,1072, 1084–1086] but also other pollutants [823,824,1086], biologically-relevant molecules [1011,1066,1073], pharmaceuticals [1098], or dyes [1068]) can be determined after accumulation at clay-modified CPEs. Improvement of their analytical characteristics can be achieved by modifying the clay before use. For example, doping the clay by cationic electrocatalysts such as methylviologen, porphyrins, or Methylene blue was found to enable detection of molecular oxygen [1099], manganese [1079], or ascorbic acid [1087], respectively. Another approach consists in doping the clay with long-chain alkylammonium moieties (that are easily incorporated in the interlamellar space by ion exchange), resulting in highly hydrophobic accumulation centers likely to preconcentrate organic analytes (see, e.g., phenol derivatives accumulated at cetyltrimethylammonium exchanged montmorillonites [1100,1101]). Using organically functionalized alkyl-ammonium-based surfactants further improved the sensor selectivity, as exemplified for clays intercalated with 2-mercapto-5-amino-1,3,4-thiadiazole- or 2-thiazoline-2-thiol-functionalized hexadecylammonium, which were then applied to selective detection of Hg^{II} [1102,1103].

4.3.1.2.2 Zeolite-Based Molecular Sieves

These materials are microporous crystalline aluminosilicates characterized by a regular spatial arrangement of TO_4 tetrahedra (T = Si, Al), leading to well-defined structures made of uniform cages, cavities, or channels of defined dimensions (typically in the 4–15 Å range) with minimum and maximum diameters that can readily discriminate against molecules with dimensional differences less than 1 Å (size selectivity). Substitution of Si^{IV} by Al^{III} in the framework makes it negatively charged and requires the presence of charge-compensating "extra-framework" cations to maintain electroneutrality. These cations are mobile enough to be exchanged by other ones, contributing thereby to the rich ion-exchange chemistry of these materials. Ion-exchange capacity is directly related to the Al/Si ratio in the framework. Zeolites are unique materials exhibiting both size selectivity and ion-exchange capacity in a single solid. This attractive feature has attracted the huge development of zeolite-modified electrodes [1104], and most advances made in electroanalysis are based on zeolite-modified CPEs [717,718,1105]. The reason is probably related to the particulate nature of zeolites, making difficult their immobilization as mechanically stable films onto electrode surfaces (other than in the presence of a polymeric binder or overcoating), contrary to their simple incorporation into CPE matrices.

By far, the most widely used molecular sieve to modify CPEs was the large-pore (8 Å in diameter) faujasite-type zeolite Y [533,753,755,830,850,876,1012,1106–1126]. Its isostructural zeolite X [700,1115,1127–1130] and the small-pore (4 Å in diameter) zeolite A [700,830,1106,1115,1116,1131] were also popular. Sometimes, all of them were used in the same studies to discuss the effect of pore size and ion-exchange capacity of the behavior or performance of zeolite-modified CPEs [736,1115,1121]. Other zeolites include mordenite [829,1132–1134], clinoptilolite [1135–1137], titanium silicalite [1138,1139], ZSM-5 [1122], zeolite beta [1140], some natural zeolites [828,1141–1144], or zeolite mixtures [1145–1147].

The analytical applications of zeolite-modified CPEs can be summarized as follows. Heavy metal cations (Hg^{II} [828], Cu^{II} [1120,1148], Ag^{I} [1131], Co^{II} [1121], or Cd^{II} and Pb^{II} [1123]) have been determined after accumulation by ion exchange, as well as the herbicide *Paraquat* (selectively with respect to *Diquat* as a result of rejection of the latter due to molecular sieving [1117]). Discrimination between dopamine (positively charged at physiological pH) and ascorbic acid has been achieved [1108,1136], and other organic compounds have been determined (e.g., phenol derivatives [1145–1147]). CPEs containing zeolites doped with selected reagents by ion exchange have also been widely used. This was notably the case of the indirect amperometric detection of nonelectroactive species (i.e., species that cannot be detected through direct electron transfer) using zeolites exchanged with redox-active cations [1115], which probably represents the most elegant application of zeolite-modified CPEs as it exploits both ion exchange and size selectivity properties of the modifier. The principle of such a method is illustrated in Figure 4.7, showing that a Cu^{2+}-doped ZMCPE polarized at a potential value likely to reduce Cu^{2+} ions did not give rise to any current response in a supporting electrolyte made of a large cation (i.e., tetrabutylammonium, TBA^+). This was due to the fact that TBA^+ species had not been allowed to enter the zeolite pore aperture, whereas each time a sample containing a small cation (i.e., K^+) was injected onto the electrode surface, it resulted in the amperometric response of Cu^{2+} exchange for K^+ and subsequent Cu^{2+} reduction.

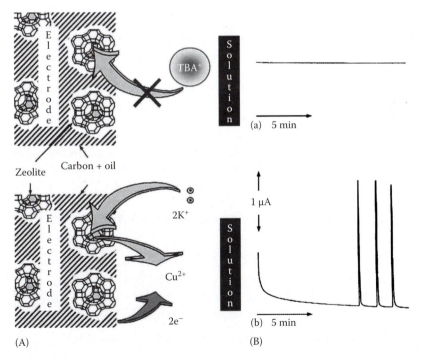

FIGURE 4.7 Schematic representation of the detection principles for the indirect amperometry at the copper-doped zeolite-modified carbon paste electrode (the case of K^+) (A) and flow injection responses obtained at (a) an undoped zeolite Y–modified electrode and (b) a copper-exchanged zeolite Y–modified electrode (B). Experimental conditions: three successive injections of 0.001 M Na^+ (in 0.01 M tetrabutylammonium bromide); carrier, 0.01 M tetrabutyl-ammonium bromide; flow rate: $5\,mL \cdot min^{-1}$; potential: −0.4 V. (Reproduced from Walcarius, A., *Anal. Chim. Acta*, 388, 79, 1999.)

This concept was then extended to the analysis of alkali metal and alkaline earth metal cations after separation by ion chromatography [1111,1116]. Zeolites can also be doped with cationic redox mediators such as Methylene blue [1129,1130], Methylene green [1128,1141], Fe^{III} [1012,1124,1144], Cu^{II} or Cu^{II} porphyrin [1085,1112,1113], methylviologen [1106], or trinuclear ruthenium amine complexes [1126], and the resulting zeolite-modified CPEs have been applied to electrocatalytic detection of various analytes, for example, H_2O_2, O_2, NADH, amino acids, and other biologically active species; such systems are promising for biosensors [876,1118]. Other applications are potentiometric determinations (i.e., $HAsO_4^{2-}$ at Fe^{2+}-doped clinoptilolite [1137] or PO_4^{3-} at surfactant-modified zeolites [1149]) or electrogenerated chemiluminescence (i.e., heroin detection at $Ru(bpy)_3^{2+}$-doped zeolite [1119]).

4.3.1.3 Metal Oxides and Sol-Gel-Derived Inorganic Materials
4.3.1.3.1 Silica-Based Materials

Silica gels (or fumed silicas) are rather old materials exhibiting adsorption and catalytic properties. Such attractive features, associated with usually large specific surface areas (>100 m^2 g^{-1}), are at the origin of their wide use as CPE modifier [89,831]. At first, silica particles were simply introduced into the carbon paste matrix to enhance its electroanalytical performance [419,682,1000,1150–1163]. Examples are available for preconcentration and detection of organic pollutants [682,1152,1154–1156], for the accumulation of heavy metal cations before their voltammetric detection [1157,1158,1160–1162], as well as to improve the response of amperometric biosensors [1000,1153], or for potentiometric determinations [419]. Rapidly after, the versatility of the sol-gel process (i.e., room-temperature synthesis of silica gels by hydrolysis and condensation of alkoxysilane reagents [1164]) was exploited to encapsulate biomolecules (enzymes or redox proteins), and the resulting biocomposite was introduced into CPE as a way of durable immobilization with promising applications in the field of biosensors [1165–1170]. Electrocatalysts (e.g., metal porphyrins [1171,1172]) have also been entrapped in sol-gels and then applied as CPE modifiers. Silica was then exploited to adsorb electrocatalysts acting as electron transfer cofactors, with applications mainly in electrochemical biosensors [1173–1176].

A significant advance in the field was made by the groups of Kubota and Gushikem (among some others). They proposed to extend the properties of silica gels by modifying their surface by very thin reactive layers of metal oxides (i.e., M_xO_y monolayers, which could eventually be further transformed into corresponding metal phosphates) [1177]. The synthetic pathway involves the reaction of surface silanol groups, $\equiv$SiOH, or siloxy bonds, $(\equiv Si)_2O$, with a metal halide, MX_n (Equations 4.1a and b), and their subsequent hydrolysis (Equations 4.2a and b), giving rise to monolayers of metal oxide after drying:

$$m \equiv SiOH + MX_n \rightarrow (\equiv SiO)_m MX_{n-m} + m\, HX \quad (4.1a)$$

$$(\equiv Si)_2 O + MX_n \rightarrow\, \equiv SiX + \equiv SiOMX_{n-1} \quad (4.1b)$$

$$(\equiv SiO)_m MX_{n-m} + (n-m)H_2O \rightarrow (\equiv SiO)_m M(OH)_{n-m} + (n-m)\, HX \quad (4.2a)$$

$$\equiv SiOMX_{n-1} + (n-1)H_2O \rightarrow\, \equiv SiOM(OH)_{n-1} + (n-1)\, HX \quad (4.2b)$$

These metal oxide layers can be further transformed into metal phosphates by treatment with phosphoric acid, as illustrated in Equation 4.3 for titania on silica (($\equiv SiO)_2Ti(OH)_2$):

$$(\equiv SiO)_2 Ti(OH)_2 + H_3PO_4 \rightarrow (\equiv SiO)_2 Ti(OPO_3H)_2 + 2H_2O \quad (4.3)$$

Many examples are available for such metal oxide or metal phosphate thin films grafted on silica gel and incorporated into CPEs, including titanium dioxide, TiO_2 [852,923,1178–1186], zirconium dioxide, ZrO_2 [1187–1190], niobium oxide, Nb_2O_5 [535,1190–1203], antimony oxide, Sb_2O_3 [1204–1206], tin oxide, SnO_2 [1207–1209], alumina, Al_2O_3 [1210,1211], as well as titanium phosphate [87,534,1184,1212,1213], tin phosphate [1207–1209], or zirconium phosphate [1214–1216]. Generation of acid sites is possible in aqueous solution on the surface of SiO_2 or M_xO_y.

Many of the oxides have an amphoteric character, so they can adsorb negatively charged species at solution pH below the point of zero charge ("pzc") and positively charged species above pzc. This has been especially exploited to immobilize various organic or organometallic mediators, such as Meldola's blue [87,1185,1194,1197,1200,1205,1209,1212,1217], Methylene blue [1178,1184,1190,1200,1205], Methylene green [1196], toluidine blue [1182,1195,1204,1205], ferrocene derivatives [535,923], metal phthalocyanines [1179,1180,1183], and metal porphyrins [534,1191,1192,1206,1207,1208] among others [852,1186–1189,1199,1211]. The corresponding modified electrodes were thereby especially applied in electrocatalytically acting biosensors.

4.3.1.3.2 Other Metal Oxides

Not only silica but also other metal oxides have been widely used to modify CPEs because of their attractive properties such as catalysis, host/support for further modification, or source of metallic deposits. Ruthenium oxides (mainly RuO_2) [1218–1224], copper oxides (Cu_2O, CuO) [77,556,699,703,401,1225–1229], and to a less extent cobalt and nickel oxides [1230,1231] have been exploited for their catalytic behavior, mainly with respect to the determination of carbohydrates [77,1218,1221,1222,1225,1226], amino acids [401,556,1223,1230], or hydrogen peroxide [1224,1227,1230]. CPEs modified with manganese oxides (mainly MnO_2) were applied either to the catalytic detection of hydrogen peroxide [326,1232–1235] or in potentiometric indication during the determination of alkali metal ions [742,1236–1240].

Other examples of metal oxides in CPEs are PbO_2 [556,1241], Fe_2O_3 or Fe_3O_4 [1001,1242], Al_2O_3 [1243,1244], ZrO_2 [1245], La^{III}-doped TiO_2 [1246], and some others [556,1247]. In particular, Labuda et al. have provided a comparison between performance observed at CPEs modified with various metal oxides [556]. Perovskite-based oxides [1248–1250] and ceramic powders [1251–1253] have also been reported as CPE modifiers.

Finally, HgO [617] and Bi_2O_3 [103,265,623] are also worth mentioning as they were used as a potential reservoir for the formation of mercury and bismuth films on CPE surfaces.

4.3.1.4 Other Inorganic Modifiers
4.3.1.4.1 Metal Elements

Since the discovery of metal-dispersed carbon paste electrodes by Wang et al. in 1992 [838], a huge amount of work has been performed in this field [94]. The main motivation relies on the ability of small metallic particles to induce catalytic behaviors by lowering overpotentials (see Figure 4.8). This is especially attractive for improving the selectivity of amperometric biosensors by shifting the operational potentials to lower values [1254,1255], that is, where interferent species are no longer likely to affect the transduction of the biocatalytic process [1254].

At first, metalized graphite particles were used as the carbon source for CPE preparation for which various metals (Ru [838,1256–1258], Pt [838,1259], Pd [838], Rh [647,1260–1262], Ir [1263–1265], and even bimetallic Ru-Pt [1266]) have proven to be efficient electrocatalysts for hydrogen peroxide and NADH detection, which have generated a novel family of highly efficient first-generation biosensors [1267]. Later on, metal microparticles or NPs were directly dispersed in conventional carbon paste matrices, and their catalytic properties were exploited as such (direct detections) or associated to redox proteins in biosensing devices. Examples are available for various (usually noble) metallic particles, such as Pd [635,839,940,971,1268,1269],

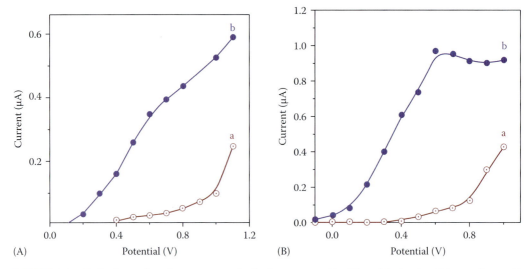

FIGURE 4.8 Hydrodynamic voltammograms for hydrogen peroxide (A) and hydrazine (B) at ordinary (a) and Pt-dispersed CPEs (b). *Experimental conditions*: Batch assays with a 400 rpm solution stirring; electrolyte: 0.05 M phosphate buffer (pH 7.4). (Reproduced from Wang, J. et al., *Anal. Chem.*, 64, 1285, 1992.)

Au [485,567,774,946,948–950,958,959,1270–1280], Cu [960–964,1281], Pt [968,970,1282], Rh [1035], Fe [1972], or Ag [416].

Sometimes, the performance of these metal particles was compared for the same application (e.g., glucose biosensor [1283]). Especially, gold NP-modified CPEs [1284] have attracted considerable attention because they provide a good environment for biomolecules (i.e., possibility for direct electron transfer, as illustrated for hemoglobin [948,1280], myoglobin [949], or cytochrome c [1281]), they can be easily functionalized (through thiol chemistry [774,959]) with promising applications as immunosensors [950,1276,1277], and they can be dispersed into polymeric materials (e.g., chitosan [950] or sol-gels [958]) to facilitate their immobilization onto the CPE. They can also be dispersed into CILEs [485].

4.3.1.4.2 Sparingly Soluble or Insoluble Complexes and Salts

In addition to the poorly soluble heteropolyanions-based complexes of POM types discussed in Section 4.3.1.1, other sparingly soluble or insoluble complexes and salts have been used as CPEs modifiers. Among them, zirconium phosphate particles dispersed into CPEs were largely exploited in biosensor applications because they were found to be attractive substrates for a wide range of redox mediators commonly used in bioelectrochemistry, such as Methylene blue [1285–1288], Nile blue [1286,1289–1291], riboflavin [1287,1292–1294], Methylene green [1287,1294], nitrofluorenone derivatives [1295,1296], and some others [1289,1290,1297–1300]. All of them were applied to NADH oxidation, either for its determination or in biosensors, except one example of ascorbic acid detection [1285].

Another large family is that of calcium phosphates (mainly apatite, $Ca_{10}(PO_4)_6(OH)_2$), which were largely used as CPE modifiers for the analysis of pesticides paraquat and diquat [1301–1304], for *p*-nitrophenol [1305] or for Pb^{II} [1306] or Cd^{II} [1307] subsequent to preconcentration by adsorptive extraction. Other modifiers used for preconcentration purposes include aluminophosphates for Pb^{II} [1308], silver nitrate for selenoamino acids [1309], silver sulfate for iodide [1020], as well as some organometallic complexes (copper(II) diethyldithiocarbamate for cysteine [1310], copper-quinine complex for iodide [1311], or tetrazolium-triiodomercurate for potentiometric analysis of Hg^{II} [1312]).

Chemically Modified Carbon Paste Electrodes 89

Insoluble or sparingly soluble salts were also used for that purpose (e.g., lead sulfate for oxalic acid [811], sodium pyruvate for lactate dehydrogenase [1313], mercury oxalate for heavy metal ions [325], magnesium oxinate for Cu^{II} [1314]). On the other hand, copper(II) hexacyanoferrate [1315] or copper(I) oxide [1316] and Meldola blue intercalated in calcium or barium sulfates [1317,1318] were exploited for electrocatalytic purposes.

4.3.2 ORGANIC AND ORGANOMETALLIC COMPOUNDS

4.3.2.1 Organic Ligands

A large variety of organic ligands have been used for the chemical modification of CPEs, and it is not straightforward to provide a classification. Hereafter are considered molecular ligands, macrocyclic/supramolecular compounds, and polymeric materials (synthetic macromolecular chelating resins and natural polymers (or biopolymers) with anchoring properties).

Organic ligands were essentially exploited for their (expected) recognition properties in order to preconcentrate target analytes (mainly metal ions) at the CPE surface to enhance the sensitivity of their electrochemical detection. The vast majority of applications involved voltammetric or amperometric detection schemes, but potentiometric determinations have also been reported.

4.3.2.1.1 Molecular Ligands

An extraordinary high number of molecular ligands have been used to modify CPEs. Most of them contain N- and S-based complexing centers. They have been classified here on the basis of target analytes.

- Mercury species (i.e., Hg^{II}) have been determined at CPEs modified with 1,5-diphenylcarbazide [1319], 2-mercapto-4(3H)-quinazolinone [1320], imidazole [1321], the Schiff base glyoxal bis(2-hydroxyanil) [1322], diethyldithiocarbamate [1323], tetraethyl thiuram [1324], the Schiff base benzylbisthiosemicarbazone [1325], ethyl-2-(benzoylamino)-3-(2-hydroxy-4-methoxyphenyl)-2-propenoate [1326], and 4-nitrophenyldiazoamino-azobenzene (Cadion A) [1327].
- Silver (Ag^I) was accumulated at CPEs using 2-iminocyclopentanedithiocarboxylic acid [1328], 2,2'-dithiodipyridine [1329], 2-mercaptoimidazole [1330], 2,3-dicyano-1,4-naphthoquinone [1331], the Schiff base glyoxal bis(2-hydroxyanil) [1333], N-benzoyl-N',N'-diisobutylthiourea [1332], N,N'-diphenyl oxamide [1333], (di)thiosalicylic acids [1334], Alizarin violet [690], 3-amino-2-mercaptoquinazolin-4(3H)-one [1335], and 1,11-bis(thiophenoxyl)-3,6,9-trioxaundecane [1336].
- Copper species (i.e., mainly Cu^{II} but also Cu^I) were analyzed after accumulation at CPEs containing, for Cu^{II} detection, 2,9-dimethyl-1,10-phenanthroline [1337], diquinolyl-8,8'-disulfide [1338], salicylideneamino-2-thiophenol [1339], benzoin oxime [1340], 1,2-bismethyl(2-aminocyclopentene-carbodithiolate)ethane [1341], pyruvaldehyde bis(N,N'-dibutylthiosemicarbazone) [1342], 3,4-dihydro-4,4,6-trimethyl-2(1H)-pyrimidinethione [1343], tetraphenylporphyrin [750], salicylaldoxime [1344], 2-mercaptobenzothiazole [1345], tetraethyl thiuram [1324], naphthazarin (5,8-dihydroxy-1,4-naphthoquinone) [1346], and, for Cu^I detection, di(2-iminocyclopentylidinemercaptomethyl)disulfide [1347] and a classical analytical reagent—rubeanic acid [1347].
- Lead species (Pb^{II}) were preconcentrated using benzoin oxime [1349], diphenylthiocarbazone [814], Schiff base N,N-bis[2-(2-iminoantipyrinephenoxy)ethyl]-p-toluene-sulfonamide [1290], pyruvaldehyde bis(N,N'-dibutylthiosemicarbazone) [1342], bis[1-hydroxy-9,10-anthraquinone-2-methyl]sulfide [1351], p-phenylenediamine [1352], and dithizone [1353].
- Cobalt (Co^{II}) has been analyzed at CPEs containing 1,10-phenanthroline [1354], 2,2'-bipyridyl [1355], 1-nitroso-2-naphtholate [1356], 1-(2-pyridylazo)-2-naphthol [1357],

2,4,6-tri(3,5-dimethylpyrazoyl)-1,3,5-triazine [1358], and 5-[(4-chlorophenyl)azo-N-(4′-methyl phenyl)]salicyl aldimine [1359].
- Nickel (Ni^{II}) was essentially complexed at dimethylglyoxime-modified CPEs [683,549,1360,1361].
- Gold species (i.e., Au^{III}) have been determined at CPEs modified with thiobenzanilide [1362] or Rhodamine B [801,818].
- Phenanthroline and 2,2′-bipyridyl derivatives were used to determine total iron [1363], Fe^{III} [1364], or Fe^{II} [1365].
- Chromium species (Cr^{VI} and Cr^{III}) have been analyzed at 1,5-diphenylcarbazide-modified CPE [1366], whereas the separate detection (differentiation) of Cr^{VI} and Cr^{III} was achieved using the respective trioctylamine [1367] and 1-[(2-hydroxy ethyl) amino]-4-methyl-9H-thioxanthen-9-one [1368] ligands.
- Other ligands were 1-(2-pyridylazo)-2-naphthol for Mn^{II} and Mn^{VII} [1369], phenylfluorone for Sb^{III} [1370], tropolone for tin [1371], 8-hydroxyquinoline for Tl^{III} [1372,1373], benzoin oxime for molybdenum [1374], 2,3-diaminonaphthalene for Se^{IV} [1375], 1-furoylthioureas for Cd^{II} [1376], N'-[(2-hydroxyphenyl)methylidene]-2-furohydrazide for Ce^{III} [1377], cinchonine for iodide [1378], propyl gallate for uranium [1379], thenoyltrifluoroacetone for Tc^{IV} [1380], Alizarin derivatives for scandium [1381,1382], or 1,4-diaza-2,3;8,9-dibenzo-7,10-dioxacyclododecane-5,12-dione for potentiometric analysis of Ca^{2+} [1383].

It is also worth mentioning that the same organic ligand can be used for detection of several cations (e.g., benzoin oxime for Cu^{II}, Pb^{II}, and Mo^{III} [1340,1349,1374] or N-phenylcinnamo-hydroxamic acid derivatives for Cu^{II}, Pb^{II}, Co^{II}, Cd^{II}, V^{V}, U^{VI}, and Fe^{III} [1384]). This has opened the way for simultaneous determinations at the same modified electrode, as reported for Hg^{II}, Co^{II}, Ni^{II}, and Pd^{II} at CPE with dimethylglyoxime [234] or Mn^{II}, Cu^{II}, and Fe^{III} using 2-(5′-bromo-2′-pyridylazo)-5-diethylaminophenol in CPE [1389]. Note that a multielement accumulation ligand can also be used to eliminate unwanted interference, as illustrated for the immobilization of toxic metal species with ethylenediaminetetraacetate incorporated into a CPE-based biosensor [1390].

4.3.2.1.2 Macrocyclic Compounds

In addition to the aforementioned derivatives, a number of macrocyclic compounds and cage molecules (see some examples in Figure 4.9) have been used to modify CPEs and to extend their scope of application, notably the cup-like structures that exhibit unique hosting properties (Figure 4.9B), mimicking biological receptors.

The "simplest" structures include crown ethers (mainly 18-crown-6 [525,1391–1400] and dibenzo-18-crown-6 [525,1396,1397,1401]), applied to the detection of Hg^{II}, Pb^{II} or Cu^{II} [579,1391–1393,1399,1400], uranyl [1402], aminoacids [1394 1395,1401], other organics [1396–1398], thiacrown compounds (for Ag^{I} [1403]) and macrocyclic thiohydrazone (for Cu speciation [1404]), and a series of octaaza- and tetra-azamacrocyclic compounds [1405–1408] used for various purposes (i.e., electrocatalysis, voltammetric or potentiometric detection of metal ions, analysis of organics).

Of special interest are ligands with cage structures as γ-cyclodextrins or calixarenes. Cyclodextrins are a family of compounds made up of sugar molecules bound together in a ring (cyclic oligosaccharides, see one example, β-cyclodextrin, in Figure 4.9). Calixarenes are macrocycles or cyclic oligomers based on the hydroxyalkylation products of a phenol and an aldehyde (again, see second example, calix[4]arene, in Figure 4.9). Both of them exhibit a very rich host/guest chemistry, which has already been exploited for electroanalytical purposes (see, e.g., [773,1409,1410]).

Numerous investigations report the incorporation of cyclodextrins (mainly β-cyclodextrin [870,1411–1424] and some others [1423–1427]) into CPEs and their subsequent use as sensors for various organic compounds. Calixarene derivatives were also used as CPE modifiers for sensing metal ions and organics [1428–1433].

Chemically Modified Carbon Paste Electrodes

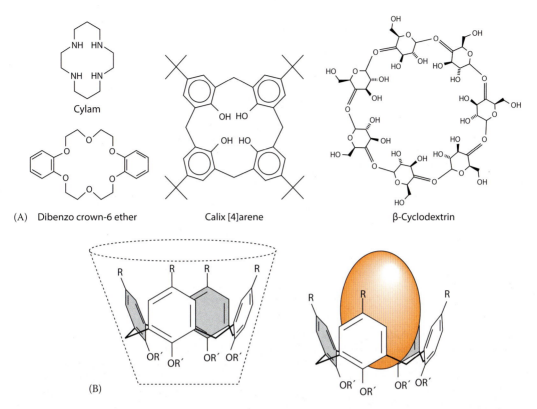

FIGURE 4.9 Examples of macrocyclic compounds used to modify CPEs (A) and an example highlighting the chemical structure of a simple calixarene from which it is easy to see the cup shape of the molecule likely to act as a host for guest species (B). *Notes*: The symbol R is not a single element but a "group" of atoms. The structure of the R group can be varied to give the basic calixarene structure a more selective action rather than simply working on the basis of size.

4.3.2.2 Organic Catalysts

Charge transfer mediators are extremely important in electroanalysis because they enable order to reduce the overpotential usually observed for numerous electroactive analytes. This contributes to increasing the sensitivity of the electrochemical detection as well as to enhance the selectivity by lowering the working potential at values below those characteristic of interferences (with eminent interest in the field of amperometric biosensors, see Chapter 5). The general principle of the so-called "mediated electrocatalysis" is illustrated in Figure 4.10 and described by Equations 4.4 through 4.6, for example, of the specific oxidation of a substrate S_{Red} into a product P_{Ox}.

Assuming a high overpotential for the electrochemical oxidation of S_{Red} ($E_S^{O'} \gg E_S^O$, Equation 4.4; the subscript S indicates the substrate, M the mediator; the prime symbol designates the observed signal with the corresponding overpotential), the addition of a mediator M_{Red} (from a fast M_{Ox}/M_{Red} redox couple) enables lowering the overpotential barrier by electrogeneration of M_{Ox} at $E_M^{0'} = E_M^0$ (Equation 4.6), which then reacts chemically with S_{Red} to produce P_{Ox} with concomitant regeneration of the catalyst M_{Red} (Equation 4.5).

The overall process results therefore in the electrochemical transformation of S_{Red} at a much lower overpotential than the initial substrate in the absence of mediator ($E_M^{O'} \gg E_S^{O'}$):

$$S_{Red} - ne^- \rightarrow P_{Ox} \quad (E_S^{O'} \gg E_S^O) \quad (4.4)$$

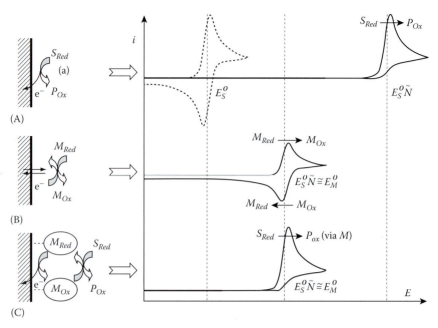

FIGURE 4.10 Scheme illustrating the general principle of "mediated electrocatalysis" at a mediator-modified electrode. (A) overpotential observed for the direct oxidation of a reductant substrate S_{Red} into an oxidized product P_{Ox} (dotted line represents the curve that would have been obtained in the absence of kinetic limitations); (B) electrochemical behavior of a redox mediator couple (M_{Red}/M_{Ox}) characterized by fast electron transfer kinetics; (C) electrocatalytic transformation of S_{Red} into P_{Ox} by means of the mediator immobilized at the electrode surface.

$$S_{Red} + M_{Ox} \rightarrow P_{Ox} + M_{Red} \quad \text{(occurring if } E_S^O < E_M^O\text{)} \tag{4.5}$$

$$M_{Red} - ne^- \rightarrow M_{Ox} \quad (E_M^{O'} \cong E_M^O) \tag{4.6}$$

From a practical point of view, it is important to immobilize the mediator at the electrode surface (reagent-free (bio)sensors), and CPE is a particularly valuable host to achieve this goal. This has been exploited in numerous applications (especially in the biosensors field [70]), and most electron transfer cofactors that have been incorporated in carbon paste matrices (i.e., in small amounts, typically a few w:w %) are described in the following paragraphs.

4.3.2.2.1 Quinone Compounds

This is the first family of mediators that have been incorporated into CPEs (see pioneering work by Ravichandran and Baldwin in 1981 [53]). Quinone derivatives (see Figure 4.11) have become, amongst others, the most widely used electrocatalysts in carbon paste sensors.

The most "popular" is 1,4-benzoquinone (or *p*-benzoquinone), which has been used in numerous CPE-based biosensors (mostly enzyme electrodes [666,788,953,1434–1440] and some others [1441–1443]) and less often in chemical sensors [1444,1445]. Functionalized or substituted benzoquinones [780,855,1445–1447] and *o*-benzoquinone [1298,1440] have also been used for these purposes, as well as the reduced form of the mediator, *p*-hydroquinone [1438,1448].

The quinone-hydroquinone redox couples have been used as mediators for both electroreduction and electrooxidation processes, being especially sensitive for the determination of hydrogen peroxide using horseradish peroxidase. Besides, 1,4-naphthoquinone [1449–1451], 9,10-anthraquinone [1452], and some of their derivatized forms [1451–1455] have been incorporated in CPEs and applied

Chemically Modified Carbon Paste Electrodes

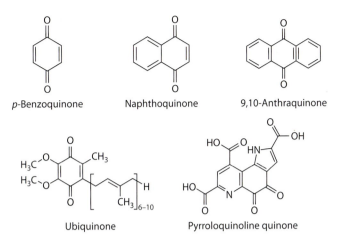

FIGURE 4.11 Some quinone derivatives used as carbon paste modifiers and redox mediators. (Chemical structural formulas.)

to direct catalytic determinations or enzyme-based biosensing. More sophisticated structures are ubiquinone (coenzyme Q_{10}) [1456,1468], pyrrolo-quinoline quinone [1458], menaquinone (vitamin K_2) [1457], quinizarin [1459], and a series of [alcanediylbis(nitriloethylidyne)]-bis-hydroquinone derivatives [422,993,1460–1462].

4.3.2.2.2 Phenothiazine, Phenoxazine, and Phenazine Derivatives

These redox-active dyes constitute another extremely popular class of electrocatalysts; the most widely used in CPEs are depicted in Figure 4.12. In the phenothiazine family, CPE was often modified with Methylene blue [289,1463–1467], Methylene green [244,656,873,1468–1472], Toluidine blue [321,1473–1477], and to a lesser extent *Promazine* (an antipsychotic agent [323,1478,1479]); the behavior of these mediators was sometimes mutually compared to find one leading to the best performance [314,873,1478–1480].

While Methylene blue was often applied as an electrochemical hybridization indicator in DNA sensors, Methylene green and Toluidine blue acted in oxidase- and dehydrogenase-based biosensing devices. These mediators were often immobilized onto inorganic particles (silica-supported metal oxides or metal phosphates), again, with examples of Methylene blue [1087,1178,1184,1190], Methylene green [1196], or Toluidine blue [1204,1205,1249].

In the phenoxazine family, Meldola Blue holds a prominent position to modify CPEs, either dispersed as powder in the matrix [783,863,1481–1483] or incorporated after adsorption onto inorganic particles [87,1106,1173–1175,1185,1205], while some examples also report the use of Nile blue [783,870,1193,1289,1492–1494]. These compounds have been widely used for the determination of the coenzyme NADH generated during analytical methods employing certain dehydrogenase enzymes. A mini-review compares the performance of phenoxazine- and phenothiazine-modified CPEs [784]. Finally, CPEs modified with the title derivatives (i.e., phenazine methosulfate [873,1484,1485]) have also been reported.

4.3.2.2.3 Other Organic Mediators

Within these substances, one can select widely used molecular electrocatalysts, such as tetrathiafulvalene [1486–1491] and tetracyanoquinodimethane [280,1492–1494], alone or together [1495,1496], riboflavin [1198,1294], as well as the cationic methylviologen and some other 4,4′-bipyridyl derivatives (mostly immobilized on inorganic particles before incorporation in CPE [830,1099,1106,1497,1499]; see Figure 4.13). Finally, the catalytic properties of some less common compounds were also exploited, including (nitro) fluorenone derivatives [439,1295,1296], aminobenzoic acid [1021,1500],

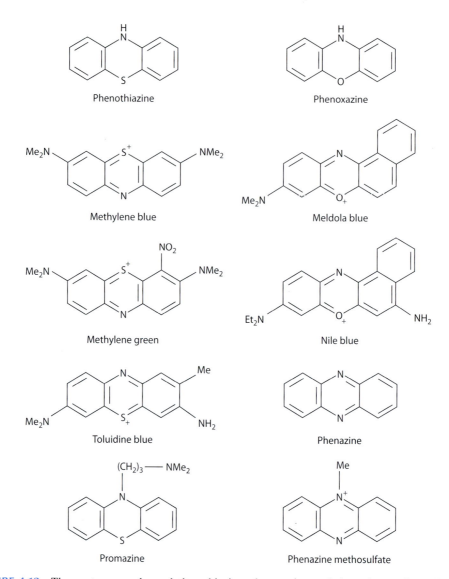

FIGURE 4.12 The most commonly used phenothiazine, phenoxazine, and phenazine mediators in carbon paste electrodes.

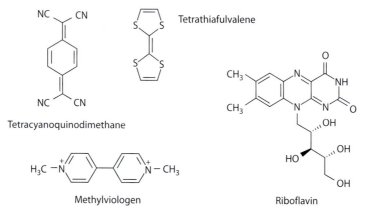

FIGURE 4.13 Some common mediators used as carbon paste modifiers.

Chemically Modified Carbon Paste Electrodes

hematin [636,1501], hemin [1502], hematoxylin [443,1503], hydrazine derivatives [225,1504,1505], flavin [1188], indophenol derivatives [1298], *p*-aminophenol [1506], *p*-bromanil or *p*-chloranil [1447,1507], 1-(*N,N*-dimethyl-amine)-4-(4-morpholine) benzene [1507], thionine [438,1509,1510], 4-nitrophthalonitrile derivatives [1511,1512], Congo Red [1513,1514], Pyrogallol Red [1515], or phenylenediamine [1516].

4.3.2.3 Organometallic Complexes

Besides these organic catalysts, another big group of charge transfer mediators largely used to modify CPEs is that of organometallic complexes. Among them, one can find four major families (ferrocene and ferrocene derivatives, metal phthalocyanines, metal porphyrins, and Schiff bases; see one example of each in Figure 4.14), as well as some others. They are briefly described here.

- *Ferrocene and ferrocene derivatives.* Ferrocene has been widely used as CPE modifier, probably because its hydrophobicity enables easy incorporation into the carbon paste matrix, and this has been particularly important in the development of enzyme-based electrochemical biosensors applicable in a variety of sample matrices [53,88,854, 857–860,864–866,869,871–873,875,921,1517–1527] as well as for some chemical sensors with electrocatalytic response [420,432,877,881,1528,1529]. Dimethyl-ferrocene [420,432,881,1528–1533] or hydroxy-methyl-ferrocene [1534] was also used in CPE-based biosensors, and a wide range of functionalized ferrocene derivatives (i.e., ferrocene

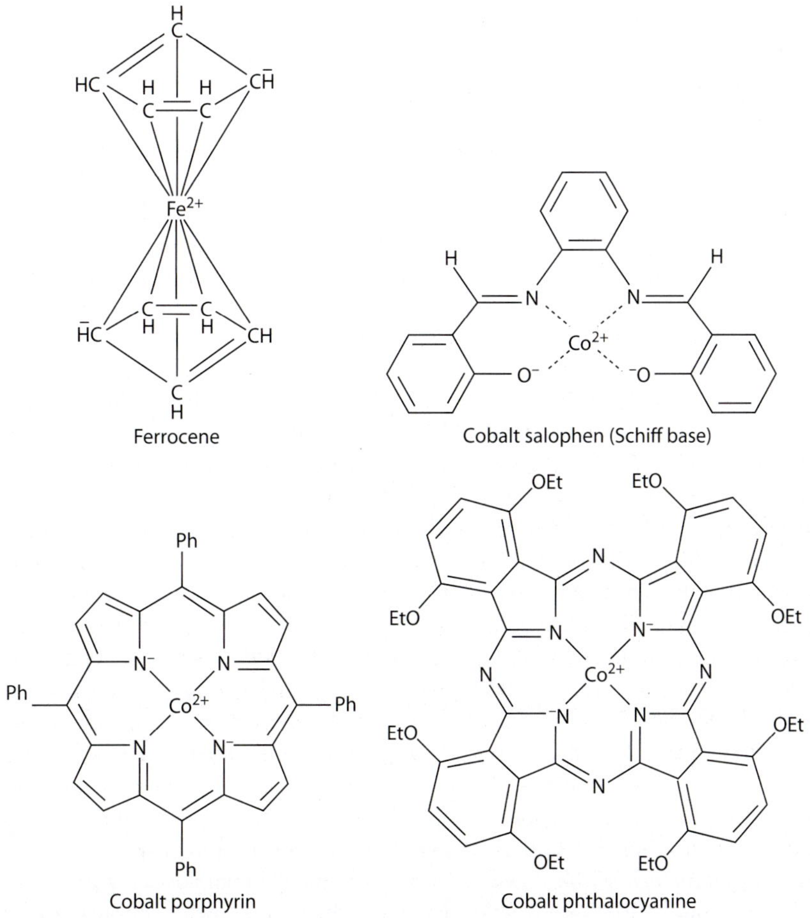

FIGURE 4.14 The most commonly used organometallic complexes as mediators in carbon paste electrodes.

(di)carboxylic acid [441,874,1535–1540], ferrocenyl surfactants [207,1541], 1-[4-(ferrocenylethynyl)phenyl]-1-ethanone [1542–1547], 2,7-bis(ferrocenyl ethyl)fluoren-9-one [879,1548–1550], and some others [1551–1553]) have been exploited for the electrocatalytic detection of various biologically relevant analytes or pollutants. These mediators could be immobilized onto inorganic particles (e.g., zeolites [876,1118] or SiO_2/Me_xO_y [923,1103]) before incorporation into CPEs or associated to organic or inorganic polymers before CPE modification [782,785,856,862,922,924,1554,1555]. Finally, other metallocenes have been exploited in CPE-based immunoassays with electrochemical detection or for electrocatalysis (i.e., cobaltocene derivatives [1556–1559]).

- *Complexes of phthalocyanine and porphyrin derivatives.* Phthalocyanines are intensely colored macrocyclic compounds that form coordination complexes with many elements of the periodic table. Following pioneering works by the Baldwin et al. [882–885], transition metal complexes of phthalocyanine macrocycles are well-known electrocatalysts, and cobalt phthalocyanine (CoPc) in particular has been widely used as CPE modifier for the electrocatalytic determination of biologically important compounds and some other analytes [324,555,560,652,653,886,889–894, 897,898,901,902,904,905,908,914,915,917,929,931,933,1040,1560–1565]. Related CoPc derivatives used in CPEs include cobalt octaethoxyphthalocyanine [1566,1567]. This extraordinary high number of investigations was notably applied to the detection of thiol compounds, saccharides, amino acids, hydrazine, or hydrogen peroxide. CoPc-modified CPEs led not only to direct electrocatalytic determinations but they also prove to be useful detectors for online detection of various analytes, in flow injection analysis [324,902,905], or following their separation by high-performance liquid chromatography [882,884,886,889,890,892] or capillary electrophoresis [555,556,1565]. Besides CoPc, other metal complexes of phthalocyanine macrocycles have been introduced into CPEs for (bio) sensing purposes, including complexes of iron [1562,655,778,887,896,903,909,911,916,1568–1570], nickel [651,895,900,1179,1180], copper [519,900], molybdenum [899], manganese [912], and rare-earth elements [1571]. Of related interest are metalloporphyrins, that is, metal complexes of heterocyclic macrocycles composed of four modified pyrrole subunits interconnected at their α carbon atoms via methane bridges, which have been used as CPE modifiers for electrocatalytic determinations. For species used in CPEs, the porphyrinic ring (functionalized or not) contained metal species such as iron [1171,1572–1575], copper [1112,1113,1574,1576–1578], cobalt [534,1176,1191,1195,1574,1579], manganese [1572,1580], as well as of nickel and zinc [1574]. These catalysts can be incorporated (in the original form) as such in the CPE matrix or previously immobilized in/on inorganic particles such as zeolites [1112,1113] or silica gels coated with thin films of metal oxides or metal phosphates (see, e.g., [206,534,1171,1176,1191]).

- *Schiff base complexes.* Schiff bases are functional groups that contain a carbon-nitrogen double bond with the nitrogen atom connected to an aryl or alkyl group (probably the most well-known chelating ligand of this family is salen, a name originating from the contraction for salicylic aldehyde and ethylenediamine). Some of their metal complexes are attractive mediators that have been used to modify CPEs. The most popular are cobalt salophen (i.e., cobalt(II)-*N*,*N*′-bis(salicylidene)-1,2-phenylenediamine, the salophen name being the contraction for salicylic aldehyde and phenylenediamine) [384,403,1581–1584] and cobalt salophen derivatives [1585–1592], while some other metal Schiff base complexes [536,1593–1596] were also of interest. They were often exploited for electrocatalytic purposes, to detect some principal biologically relevant molecules (mainly, ascorbic acid, dopamine, and uric acid) as well as various pharmaceuticals and drugs.

- *Others/miscellaneous.* Other two families of metal complexes that were widely incorporated into CPEs are metal phenanthroline compounds (iron(II,III) phenanthroline derivatives [68,546,547,1597–1599] or the ruthenium(II) phenanthroline complex [1299,1600,1601]) and bipyridine complexes of osmium(II,III) [1602,1603] and ruthenium(II) [140,1299,1601,1604]. Finally, several other less common organometallic electrocatalysts have been incorporated in CPEs, including Co(II)-, Ni(II)- and Cu(II)-cyclohexylbutyrates [1605–1607], Ru(III)-diphenyl-dithiocarbamate [1608,1609], Cu(II)-diethyldithiocarbamate [1310], Ru(III)-piperidine-dithiocarbamate [1610], Mn(II)-pyrrolidinedithiocarbamate [1611], Ni(II)-baicalein [429] and Ni(II)-quercetin [424], Ru(III)-Rupic [1612], Ru(II)-ethylenediaminetetraacetate [86], ruthenium purple [1613], rhodium acetamide [1614], Rh(II)/Rh(III)-tetrakis(m-2-anilino-pyridinato) [526], iron(II)-nitroprusside [1615], Ni(II)-[N,N'-bis(2-pyridine carboxamido)-1,2-benzene] [1616], Fe(II)-bis(4'-(4-pyridyl)-2,2':6',2'-terpyridine) thio-cyanate [1629], and Zn(II)-, Cu(II)-, or Co(II)-1-alkyl-1H-benzo[d][1,2,3]triazoles [1618–1621].

4.3.2.4 Surfactants, Amphiphilic and Lipophilic Modifiers

Amphiphilic and lipophilic compounds are attractive CPE modifiers as they interact preferably with the hydrophobic matrix, and, once immobilized in or on the composite electrode, they can be further applied to either adsorptive stripping voltammetry of organic compounds (via favorable hydrophobic interactions) or to the preconcentration analysis of ionic species via open-circuit accumulation by ion-pair formation. Various kinds of surfactants [1622] and some lipids [1623] have been exploited for those purposes. It is also noteworthy that treating CPEs with surfactants considerably lowers uncompensated resistance by removing (or decreasing) the oily insulating layer usually produced during the process of smoothing the electrode surface [64].

4.3.2.4.1 Surfactants

Cationic, anionic, and nonionic surfactants have been used to modify CPEs [1622]. The cationic surfactant series is the most popular, and examples are available for cetyltrimetylammonium bromide (CTAB, or its chloride homologue CTAC) [704–706,817,1624–1634], cetylpyridinium [253,709,1635], Septonex® (1-ethoxycarbonyl-pentadecyl-trimethylammonium) [695,736,737,1636], and some others [207,1583,1622]. These species can be incorporated directly into the CPE matrix or also applied *in situ* to the CPE surface (see, e.g., [704–706]).

They have been notably used to detect anionic forms or complexes of metal species [1636]. Neutral alkylamines (mainly octadecylamine [1637–1640] and also triisooctylamine [518] or 1-amino-12-(octadecanoylamino)dodecane [1641]) are intrinsically neutral surfactants, but depending on pH they can be protonated and thereby act as cationic amphiphiles. CPEs modified with other nonionic surfactants have been reported, such as Triton X-100® [317,1634,1642–1649] and some others (of *Triton®*, *Tween®*, *Brij®*, or *Span* families) (see, e.g., [1622,1643,1644,1646]).

Triton X-100-modified CPEs were especially applied to dopamine analysis [318,1645,1648,1649]. 2,9-dichloro-1,10-phenanthroline-surfactant was also reported [1650]. In the anionic surfactant series, sodium dodecyl sulfate (SDS) was by far the most widely used with CPEs [1651–1660] as well as hexadecyl sulfonic acid (sodium salt) [243,1661] and some others [1622]. They were exploited as preconcentration centers or as masking agents (e.g., selective detection of positively charged dopamine in the presence of negatively charged ascorbic acid [1654]). Some studies also compared the performance of CPEs modified with cationic, anionic, or nonionic surfactants in the same conditions and for the same analyte in order to define the most suitable modifier [1622,1634,1643,1656,1658].

4.3.2.4.2 Lipids and Related Compounds

Natural lipids have been incorporated into CPEs, including stearic acid [1027–1029,1662–1664], lauric acid [1665], as well as other fatty acids and phospholipids [1642,1030,1666–1676]. This was carried out either to preconcentrate lipophilic analytes or to mimic the living cell membrane structure with the goal of investigating the drug–membrane interactions. These modified electrodes also found applications in enzyme-based biosensors [1623].

4.3.2.5 Organic Polymers and Macromolecules

Functional macromolecular compounds such as organic and organometallic polymers or organic–inorganic hybrid copolymers have been widely applied to the chemical modification of electrode surfaces, most often in the form of films deposited on solid electrodes [1676].

They have also been used to modify CPEs via two strategies: film deposition onto the CPE surface or direct incorporation in the bulk of the CPE matrix. They were exploited for their intrinsic properties (ion exchange, adsorption, redox/mediator activity, electronic conductivity, etc.) and also as protecting coatings (permselective layers, barriers to retain active materials in or on CPE). They are briefly described here according to their particular features, leading to five categories: chelating resins, ion exchangers, permselective or protecting coatings, redox polymers, and (semi)conducting polymers. Some examples of materials associated with CPEs are illustrated in Figure 4.15), as well as some others. They are briefly described here.

4.3.2.5.1 Chelating Resins and Similarly Functioning Polymers

Synthetic organic polymers bearing chelating ligands or other reactive groups are attractive preconcentration agents. Those used to modify CPEs were either resins containing N- and S-based

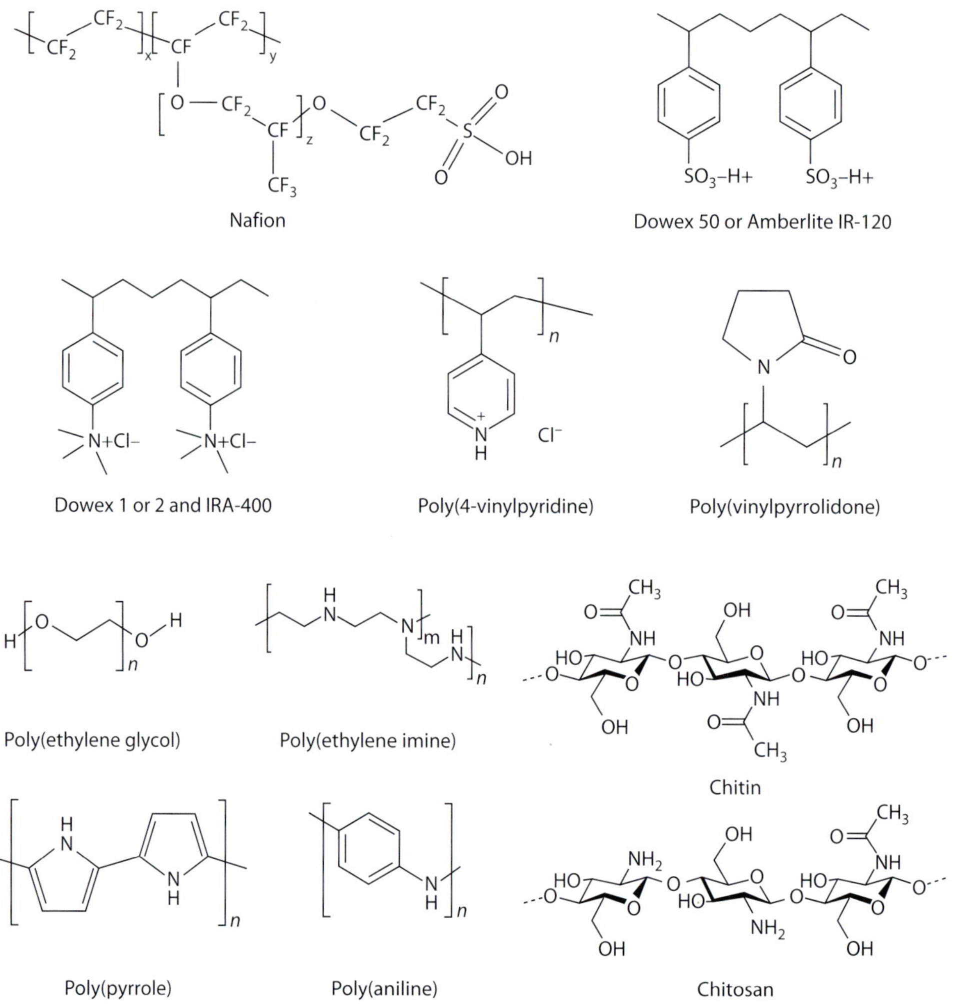

FIGURE 4.15 Some examples of commonly used polymeric materials in/on carbon paste electrodes.

functional groups likely to bind metal ions before their electrochemical detection [808,1676–1684] or functionalized polymers (e.g., poly(*N*-vinylpyrrolidone); [186,1685–1687]) with adsorption properties toward organic target analytes.

A special category also associated with CPEs belongs to molecularly imprinted polymers (i.e., macromolecular compounds formed in the presence of a molecule that is extracted afterward, thus leaving complementary cavities behind, which can then be exploited to the selective accumulation of the original molecule) [1688–1692]. Finally, several natural macromolecular compounds were modifiers of CPEs. The most popular were humic substances [297,837,1336,1693–1704], which were mainly applied to preconcentrate heavy metals or even to get information on humate–metal species interactions [837,1694,1696,1701]. The polysaccharide chitin [285,832,835,1705–1707] and its deacetylated form, chitosan [86,88,499,835,1273,1465,1708–1711], are other attractive CPE modifiers applied to the accumulation of various analytes or as hosts for NP catalysts [86,417, 499,950,1273,1465,1711]. Other less common natural modifiers are lichens [271,1712], ionic polysaccharides (e.g., alginate derivatives [464,479,1713]), or keratin [287].

4.3.2.5.2 Ion Exchangers

Wang and coworkers were the first to report the incorporation of an ion-exchange resin into CPEs for Cu^{2+} preconcentration and detection [708]. Thereafter, numerous cation exchange and anion exchange resins have been used for electroanalytical purposes. Cations exchangers involve most often a polymeric support functionalized with sulfonate or carboxylate groups [708,1714]. The most popular cation exchange polymer is the perfluorinated-sulfonate ionomer, Nafion, which was largely applied in connection with CPEs. Examples are available for adsorptive stripping voltammetry of cationic analytes such as metal ions [708,1354,1355,1364,1365,1714–1716] and several (usually positively charged) organic species [1717–1724].

Nafion was also used in CPE-based biosensors (enzyme electrodes [668,865,1725–1727] and immunosensors [1556,1558,1559,1728]). Nafion was sometimes exploited as a hosting matrix for ligands [1354,1355,1364,1365], charge transfer mediators [1590,1729], or NP catalysts [467,496,1730]. Other cation exchangers include the *Dowex 50W-X8* resin [522,713,807,1731–1734], polyester sulfonate ionomers (e.g., *Eastman Kodak AQ*) [922,1735], and several commercially available resins of the *Amberlite*® trademark (e.g., *IR-120* [252,904,1736], *IRC-718* [1737–1740], or *XAD-2* [1741]). The anion exchange resins were essentially nitrogen polymers, *that is*, quaternized poly(4-vinylpyridine) [1742–1744] and *Amberlite LA2* [800,802,804,1736,1745,1746].

4.3.2.5.3 Permselective or Protective Coatings

CPE-based biosensors were often covered with polymeric layers to protect the biomolecule from the external solution, to prevent from surface fouling effects, or to reject interferences. *Nafion*® was mainly used for this purpose [1725,1727], but *Dextran* was also reported to improve the stability of hemoglobin immobilized on CILE [512]. Ion-exchange polymers are also attractive electrode modifiers when used as thin films because they are likely to induce permselective properties (based on charge exclusion), as exploited for the selective determination of uric acid [1722]. Selective detection of dopamine in the presence of ascorbic acid was achieved on the basis of a poly(vinyl alcohol)-coated CPE [1747].

Other permselective or protective coatings are semipermeable membranes—mostly based on size-exclusion polymers—made of poly(*o*-phenylenediamine) [131,135,648,649,654,841,935, 938–940,1261,1748–1779], or its *m*-derivative [921,1780], poly(ethylene oxides) as poly(ethylene glycol) [319,862,1492,1519,1531,1566,1673,1781–1785], and poly(*o*-aminophenol) [247,1786]. Other macromolecular stabilizing additives in CPE-based biosensors include poly(ethylene imine) [246,321,939,1494,1749,1787], poly(dimethyl-siloxane) [20,1788,1789], silicone-based [1790], or a melanin-type polymer [1282].

4.3.2.5.4 *Redox Polymers and Other Electrocatalytic Systems*

Redox polymers are macromolecules containing redox centers, which support the electron transfer by electron hopping. Similar to organic and organometallic electrocatalysts, the interest of redox polymers relies on their ability to reduce overpotential in the detection of target analytes.

In fact, there are two major reasons why redox polymers can be preferred over "molecular" electrocatalysts: first, the catalytic center is covalently bonded to a macromolecule that is expected to result in more durable immobilization when incorporated into the carbon paste matrix (leaching can occur in case of "molecular" modifiers) and, second, the polymer can act as a stabilization matrix for biomolecules. Both effects contribute to improving the (bio)sensor operational stability and can also contribute to enhancing its analytical performance.

Actually, the electrocatalytic properties of redox polymers have been especially exploited in bioelectrochemistry [935]. The redox polymers involved in the preparation of CPE-based biosensors are the following: ferrocene-containing polymers (polyacrylamide-based [782,785,1555], poly(vinyl-ferrocene) derivatives [924,1748] or ferrocene-modified poly(ethylene imine) [1749]), phenothiazine-based polymers [938,939,1473,1750–1753], osmium or ruthenium 2,2′-bipyridine chloro poly(4-vinylpyridine) complexes [648,649,940] (or related compounds as osmium phenanthrolinedione [1754] or poly[1-vinyl imidazole osmium (4,4′-dimethyl-bipyridinium)] [1755]), poly(xylylviologen) [1498,1756], poly(ether amine quinones) [1757,1758] and related poly(quinone) systems [1759], as well as poly(*N*-vinylimidazole) [1760,1761]. Some redox polymers used for CPE modification have been applied to the direct electrocatalytic detection of neurotransmitters [135,1762,1763], ascorbic acid [135,1763,1764], or nitrite [648]. Siloxane polymers with covalently attached ferrocene derivatives have also been reported [856,922,1554]. Finally, one should mention that polymers supporting catalytic centers (as ions or solid particles) have been used to modify CPEs for electrocatalytic purposes, as reported for nickel in various polymers or redox polymers [131,841,1765–1780] or for metal oxides in neutral resins [1229,1241].

4.3.2.5.5 *Conducting Polymers*

Conducting polymers are attractive electrode modifiers, yet mainly used as films coated onto solid electrode surfaces, because they ensure good electronic conductivity. These materials, adhering very well on hydrophobic carbon pastes and produced directly at the surface by potential cycling, have found important applications with amperometric biosensors, because they enable efficient enzyme immobilization while ensuring good electronic transduction [1791–1794]. Few examples are available for these conducting polymers associated to CPEs, including mainly polypyrrole [584,1776,1795–1803] and polyaniline derivatives [129,130,1497,1804–1806], with applications in the field of biosensors [1497,1798,1800] or for direct electrocatalytic detection of biologically relevant molecules [129,130,1795,1797,1799,1802,1806]. The resort to poly(1,8-diaminonaphthalene) [1807] and poly(phenylenediamine) [1516] was also reported.

4.3.3 OTHER POSSIBILITIES AND RECENT APPROACHES TO CHEMICAL MODIFICATION

This section describes modifiers that are not purely organic or inorganic (such as organic–inorganic hybrids), novel materials (notably those arising from nanotechnology and (nano)materials sciences), as well as some CPE surface treatments intended to modify the electrode properties (electrolytic modifications, metal-film-plated electrodes).

4.3.3.1 Organic–Inorganic Hybrid Materials

These materials, largely arising from the new technologies, combine the mechanical stability of a rigid inorganic structure with the particular reactivity of organic functions.

They are usually classified into two categories, the hybrids of class I characterized by weak interactions between the organic and inorganic counterparts and hybrids of class II exhibiting organofunctional groups covalently bonded to the inorganic component. Here, only hybrids of class II are considered as those of class I usually overlap with inorganic materials bearing organic or organometallic species adsorbed on—or ion exchanged in—the solid phase (see examples in Sections 4.3.1.1 through 4.3.1.3 for organic moieties entrapped in polyoxo-metallates, natural clays, and zeolites or adsorbed on metal oxides and phosphates).

Actually more durable immobilization organofunctional groups onto the inorganic matrix can be achieved by covalent bonding. This has been especially developed with silica-based materials in electrochemistry [89,831,1808]. Chemical modification of the silica surface can be performed by postsynthesis grafting or by the direct (one-pot) sol-gel synthesis involving the use of organosilane reagents. The first examples of CPEs incorporating organically functionalized silicas dealt with long-chain alkyl-grafted silica gels (C_{18} [646,1809–1813] or C_8 [1814,1815]), which were applied to the detection of drugs and pharmaceuticals. Then, silica samples functionalized with more reactive groups have appeared as CPE modifiers, with organofunctional groups such as propylamine (for metal ions accumulation and detection [701,702,1816], as support for electrocatalysts such as PMo_{12} [1817–1819], or to immobilize enzymes [1167]), propylpyridinium (to immobilize negatively charged redox mediators [1820–1824]), a wide range of functions exhibiting high affinity for Hg(II) species (i.e., mercaptopropyl [106,702,1825], benzothiazolethiol [1826], 3-(2-thiobenzimidazoyl)propyl [1827], 2,5-dimercapto-1,3,4-thiadiazole [1828], or 2-aminothiazole [1829]), several groups binding preferably Cu(II) species (i.e., aminopropyl [701,1816], 2-aminothiazole [1830], the carnosine dipeptide [1831,1832], propylsulfonate [1833], salicylidine [1834], or dipyridyl groups [1835]), some functions leading to Pb(II) detection (mercaptopropyl [106] or 2-aminopyridine [1836], as well as some others likely to recognize other metal ions, alone or in mixture [107,771,1837,1838].

Toluidine blue [1476] and a salicylaldehyde-based Schiff base ligand [1211] have also been grafted onto metal oxide particles before incorporation into CPEs. Besides, organoclays started to become attractive CPE modifiers for electroanalytical purposes; they can be obtained by grafting (e.g., amino- or mercaptopropyl groups [1088,1089]) or via incorporation of long-chain amphiphilic ligands into the clay interlayer region [1102,1103]. Finally, self-assembled organic–inorganic hybrid materials belonging to the polyoxometalate family have been prepared and incorporated in CPEs to be exploited for the electrocatalytic detection of hydrogen peroxide or nitrite [1051,1839–1842].

4.3.3.2 Nanomaterials

Nanotechnology is extraordinarily a wide field involving contributions from the physical sciences, life sciences, engineering, and medicine. It is thereby not so surprising to find some traces of the "nanoworld" also in electroanalytical chemistry with CPEs. At present, this has essentially appeared through the incorporation of porous materials ordered at the nanoscale or NPs in CPEs.

4.3.3.2.1 Ordered Mesoporous Materials

Substances of this kind are highly porous solids (pore volume > $0.7\,mL\,g^{-1}$) exhibiting high specific surface areas (500–1500 $m^2\,g^{-1}$) due to a periodic and regular arrangement of well-defined mesopores defined by amorphous inorganic walls. Such regular framework structures are usually prepared by the sol-gel process, involving the hydrolysis and condensation of a tetraalkoxysilane ($Si(OR)_4$), in the presence of a supramolecular template (surfactant or water-soluble polymer). The most widely used are ordered mesoporous silica-based materials prepared by the surfactant template route (see illustration for a hexagonal mesostructure in Figure 4.16A).

These materials have attracted the attention of electroanalysts as their regular channels, high specific surface areas, and consequently high number of easily accessible active sites have enabled improving the performance of silica-modified electrodes [89,816,1843], as reported for the detection of several organic species (mainly drugs) [999,1844–1852] and some metal ions [1853,1854] at mesoporous silica-modified CPEs. Al-doped mesoporous silica in CPEs was also exploited for its catalytic properties in epinephrine and catechol sensing [1855,1856]. Even more interesting is

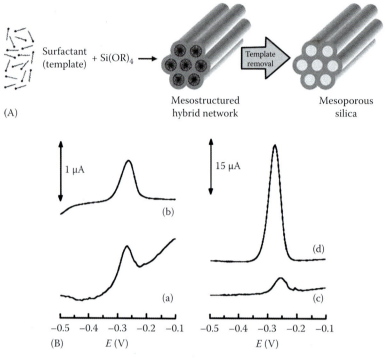

FIGURE 4.16 Schematic pathway for preparing surfactant-templated mesoporous silica from tetralkoxy-silane (A). Comparison between thiol-functionalized silica gels and ordered mesoporous silicas (MCM-41 and SBA-15 types) in CPEs applied to preconcentration electroanalysis of Hg^{II} species (B) *Experimental conditions*: Electrochemical curves obtained after 2 min accumulation in 1 µM Hg^{II}, with various mercaptopropyl-grafted silica samples incorporated in carbon paste electrodes: (a) small pore (~4 nm) silica gel, (b) large pore (~6 nm) silica gel, (c) small pore (~3 nm) MCM-41, and (d) large pore (~6 nm) SBA-15. After transfer to an analyte-free solution (0.1 M HCl + 5% TU), the curves were recorded in the DPASV mode.

the possibility to functionalize mesoporous silica materials. The previous section has underlined the very rich silica chemistry and the interest of silica-based organic–inorganic hybrids as CPE modifiers, and it is therefore not so surprising that an increasing number of studies are dealing with the incorporation of organically functionalized mesoporous silica into CPEs for electroanalytical purposes. The main interest of ordered mesoporous organosilica in comparison to their nonordered homologues is the easier access to binding sites and faster mass transfer processes, inducing dramatic increases in sensitivity of the resulting modified CPEs (see illustrative example for Hg(II) detection at thiol-functionalized silicas in Figure 4.16B [1857]).

Other examples are available for mesoporous silica functionalized with aminopropyl [701,1858,1859], carbamoylphosphonic acid [770,771], 2-benzothiazolethiol [1860–1862], 5-mercapto-1-methyltetrazole [1863], propylsulfonic acid [1833], thiomorpholine [1864], carnosine [1832], acetylacetone [1865], 2-acetylpyridine [1866], thiourea derivatives [1867,1868], or ferrocene [878]. The vast majority of these works used hexagonal mesostructures of MCM-41 or SBA-15 types. Attention should be drawn, however, to the fact that all claimed nanoporous silicas are not obligatory ordered as those formed by the surfactant template route (see, e.g., dipyridyl-functionalized nanoporous silica gels in CPEs [1869–1871]). MCM-41 was also used as a host for electrochemiluminescence reagents as reported for ionic liquid-based CPE [462]. Finally, it is worth mentioning that the use of mesoporous materials other than silica in CPEs, such as nanostructured mixed-valence manganese oxides [1872–1874], mesoporous TiO$_2$ [994,1875] or ZrO$_2$ [1876], as well as ordered mesoporous carbons [116].

Chemically Modified Carbon Paste Electrodes

4.3.3.2.2 Nanoparticles

The second category of "nano" additives to CPEs concerns the nanoparticles (NPs) for which the physicochemical properties can differ from those of the bulk materials. It was mentioned in Section 4.3.1.4 that metal particles can be advantageously exploited in electrocatalysis and biosensors (due to their ability to lower usually observed overpotentials) and some of them were indeed NPs (i.e., Au [485,567,774,946,950,958,959,1277–1280], Cu [960–964,1280,1281], Pd [969,971], Pt [970], or Fe [976]. Note that AuNPs are also attractive for immobilization of thiol-containing groups (e.g., cysteamine [1877,1878]). In addition to them, several other NPs have attracted the attention of electroanalysts, mainly because of their electrocatalytic properties. This is the case of various metal oxides or hydroxides, the most used in connection with CPEs being TiO_2 because of its intrinsic catalytic properties [499,992,995,996,1879] or as support for mediators [920,991,993,997,1880].

Other metal oxides include nano Fe_2O_3 [1001], Fe_3O_4 [980], CoO [1881], SiO_2 [419], ZrO_2 [998], and CdO [1003]. CPEs containing dysprosium or dysprosium oxide nanowires [1882–1887] and $Cu(OH)_2$ nanowires [1888] have been applied to electrocatalytic sensing of drugs. The CdS nanorods [1889], CdSe quantum dots [953], and neodymium(III) hexacyanoferrate(II) NPs [1008] have been exploited in biosensors, while cobalt and iron phthalocyanine NPs in CPEs exhibited electrocatalytic behavior in some organics [932–934]. A last category is that of core-shell iron oxide-based NPs ($MgFe_2O_4/SiO_2$ and $MgFe_2O_4/SiO_2$ [302,990], Fe_3O_4/SiO_2 [988,979], Fe_3O_4/Au [986], and Fe_2O_3/Co hexacyano-ferrate [983]), which were mostly used in CPE biosensors.

4.3.3.3 Surface Treatments and Alterations

Qualitatively new properties can be brought to the CPE surface by its "total" modification by means of electrolytic deposition of some metals in the form of a thin layer. (Momentarily popular *lithographic sputtering* [263,264] or *screen printing* [323] of such layers could also be applicable, but for extraordinarily resistant solid-like CPEs only, whereas common types of incompact, liquid-containing, and temperature-sensitive carbon pastes cannot be used.)

4.3.3.3.1 Metallic-Film-Plated Carbon Paste Electrodes (MeF-CPEs)

Carbon pastes covered with preplated metal films have gained considerable attention for several reasons: (I) this can increase much the cathodic polarization of the electrode, as hydrogen evolution occurs at much lower potentials on, for example, mercury or bismuth surfaces than on carbon ones; (II) avoiding the use of bulk metallic electrodes is often cost-effective (i.e., with respect to noble metals) and might be safer (e.g., in comparison to mercury drop); they can be applied to electrochemical stripping analysis. The so-called *metallic film carbon paste electrodes* can be basically prepared by *in situ* electrolytic deposition from a solution containing the metal ion precursors or *in nascenti* by reducing solid precursors embedded in the paste [103,265,268,269,617]. The second method offers the advantage of operating in conditions of reagent-free sensors for which successive experiments can be easily performed by mechanical polishing of the electrode surface and subsequent metal film deposition between each measurement without the need for an external plating solution. (Remember that morphological transformations can occur in various successive phases of the film formation, when metal films obtained *in situ* can suffer from nonuniformity due to local nucleation (see, e.g., Section 6.5 and Figure 6.11).

Following the worldwide success and use of *mercury film electrodes* [1890], the story of metallic-film-plated CPEs has started with mercury-film-plated carbon paste electrodes (MF-CPEs; [257–259,624]), but the main boom of MeF-CPEs is undoubtedly connected with the rediscovery of bismuth film electrodes (BiFEs; [102,262–264,268]), when the BiF-CPEs [255,626] were the first configuration to employ a support really concurrent to the glassy carbon disc (see also Section 2.2.1). Besides these mercury- [257–259,623,624,1891,1892] and bismuth- [265,626,637,669,671] plated CPEs, the family of related electrodes gathers also gold [108,1893], antimony [1894], palladium [621], and lead

[1895] deposits, when almost all of them have been successfully applied in ESA of metal ions and related species [83,255,262–264,268]. If the preplated metallic film configurations are combined with electrochemical stripping analysis, the electrode process involves the deposition and subsequent dissolution of intermetallic compounds such as $Me(Hg)_X$ amalgams [255,1890], Hg–Au adducts [255,260], and $Me_A Bi_B$ alloys [262,263], expected to be formed in the respective metallic layer.

As shown in experiments with metal films generated *in situ*, these processes can also be accompanied by additional effects as a consequence of mutual competition between the corresponding film and the bare carbon paste surface [189,260,812], which would also occur at CPEs modified with metal powders (e.g., finely pulverized bismuth [266] and antimony [1896]). Finally, the identical metallic layer can also be accomplished in the configurations of CPEs modified with solid oxides, namely, HgO (as yellow allotrope [617]), Bi_2O_3 [268], Sb_2O_3 [1897], and a related hydrolytic product SbOL [1898]), or some inorganic salts—NH_4BiF_4 [269] and BiF_3 or SbF_3 [1899]—being "pseudo-covalent" in nature (forming polymeric structures), insoluble in aqueous media, and applicable as "normal" bulk modifier.

4.3.3.3.2 Electrolytic Modifications and Other Surface Treatments

By referring back to Section 3.2 and highlighting again some fundamentals here, the pretreatments by electrolytic activation (mainly *anodization*, which may sometimes be followed by *cathodization*) and chemical oxidation are applicable to CPEs before use [63,181,254,629,646,675,868,1448,1900–1904]. CPEs treated by such procedures can also be classified as CMEs, as both treatments lead to an alteration of surface states [68]. (As mentioned earlier, this happens through the formation of various functional groups, such as –C-OH, –C=O, –C-OH$_2^+$, and –C=OH$^+$, after intensive oxidation of carbon particles at or near the CP-surface layer, leading to its partial or even total *hydrophilization* [3,68,83,157].) Beneficial effects of anodization (e.g., at −1.5 V for 180 s [176]) include enhancements in both heterogeneous charge transfer rates [868] and preconcentration efficiencies [1902], both improving the overall analytical performance of CP-based electrodes and sensors. After anodic oxidation, the activated carbon paste surface can (or even must) be "equilibrated" by a short cathodization (e.g., at −1.0 V for 15 s [176]), or alternatively, by successive *potential cycling* between the properly set positive and negative potential limits [63,646,868], where the "shock from aggressive anodization" is compensated yet more effectively.

Finally, there are some occasional treatments of the CP-surface, e.g., ultrasonication (to facilitate mediator immobilization onto CNT-modified CPE [1511,1590]) or magnetic field (for increasing the electrochemical flux [1905] or improving the DNA detection [299]); for other examples of these rather unusual pretreatments, refer to Section 2.2.1).

4.4 MODELING AND TESTING OF CHEMICALLY MODIFIED CPEs

Carbon paste electrodes are heterogeneous systems in which *electrically conductive sites or areas* and *insulating zones* are both present. And if there is a modifier on the surface or in the bulk, the constellation becomes yet more complicated. Hence, modeling the behavior of such electrodes is not easy, and some basic procedures that are available have been developed from empirical experience with modeling and testing of binary or unmodified CPEs (see Section 3.3) or from assays for distinguishing between unmodified and modified variants. Mostly, the respective models assumed that such sensors would behave like arrays of microelectrodes, nevertheless, considering that—due to the overlay of the diffusion layers of hypothetical microelectrodes—electrochemical responses are like those of common "macroelectrodes."

4.4.1 BASIC PROCEDURES AND SPECIFICS

Due to the presence of pasting liquid, the voltammetric signals at CPEs are not ideal even for electrochemically reversible systems, when showing deviations as (i) flattening of sigmoidal curves in

DC-voltammetry, (ii) broadening of the peaks in DP-and SW-voltammetry, and (iii) widening the distance between the signal maxima or minima in cyclic voltammetry (CV).

Two methods can be suggested to characterize and quantify the deviation from the ideal behavior as well as the influence of the insulating matrix. For ideal electrochemical behavior, the Nernst equation and all derivations from it should be somehow rigorously valid. As a consequence, the potential difference of the peak maxima of the oxidation and reduction in cyclic voltammetry should correspond to some 59/n mV, where n represents the number of transferred electrons. Deviations from the ideal behavior by whatsoever reasons could be simply described by Equation 4.7:

$$\chi_{CPE} = \frac{(E_{p,a} - E_{p,c})_{CPE} * n}{0.059} \quad (4.7)$$

where
  χ_{CPE} is a dimensionless factor describing the extent of deviation of the carbon paste electrode
  $E_{p,a}$ and $E_{p,c}$ are the anodic and cathodic peak potentials of the system (i.e., CPE in this case) in the cyclic voltammograms in Volts

In order to quantify the influence of the matrix, a similar approach can be chosen, where instead of the ideal behavior, the voltammetric responses of a corresponding homogeneous solid electrode (graphite or glassy carbon, GC) are related with the observed values of the CPE (Equation 4.8):

$$\chi_{CPE/GC} = \frac{(E_{p,a} - E_{p,c})_{CPE}}{(E_{p,a} - E_{p,c})_{GC}} \quad (4.8)$$

4.4.2 Chemically Modified CPEs versus Bare CPEs

As very often carbon pastes represent somehow individuals changing their properties slightly from each batch preparation to the next, a simple tool to quantify the electrochemical characteristics is provided here, by distinguishing the bare CPE (where sorption processes are expected to occur) and chemically modified CPEs (by choosing an illustrative example, that is, accumulation at clay-modified CPE, which could be extended to other CPEs modified with solid particles exhibiting some affinity for target species.

4.4.2.1 Sorption Processes at the Bare (Unmodified) Carbon Pastes

Due to the presence of carbon particles and a more or less lipophilic matrix, CPE exhibits enhanced affinity to organic molecules with the tendency to be spontaneously adsorbed onto the electrode surface, which can be exploited for preconcentrating the target compounds, either in the actual form or as various metal complexes, via which numerous single metal ions can also be accumulated.

When sorption takes place at carbon paste electrodes, similar conditions can be found as with preconcentration of metals on liquid mercury surfaces. Depending on the lipophilicity of the adsorbed species and the composition of the electrode material, sorption can be mainly restricted to the surface of the electrode, or migration of the sorbed compound into the electrode bulk can occur ("extractive accumulation"; see Section 3.4 and Figure 3.6). This provokes a loss of accumulated species at the electrode surface for long time accumulations and a concentration gradient of the analyte of interest into the electrode bulk. On the other hand, during the measurement, the gradient is reversed due to consumption of the electroactive analyte at the electrode surface, which will prompt a migration of accumulated species back from the interior of the electrode to the solution interface with the consequences of additional signal broadening and memory effects.

The thickness of the mean reaction (diffusion) layers inside the electrode bulk can be experimentally roughly estimated by wiping off layers of the carbon paste on paper until no parasitic signal from the memory effect can be monitored. By determining the mass of the stripped paste and via the electrode surface and the specific mass of the paste, the diffusion layer thickness can also be evaluated.

4.4.2.2 Chemically Modified CPEs

Let us consider the example of the accumulation of mono- and divalent metal ions via ion exchange at a vermiculite-modified CPE [1080]. The concept developed in this work is rather general for CPEs, which have been bulk modified with solid modifiers so that it could be applied also to other systems accordingly, as notably demonstrated for ion exchange at zeolite-modified CPEs [1114], for metal ions binding to humic acid–modified CPEs [1694,1701], or for electrodes containing functionalized silica-based organic–inorganic hybrids [719].

For the respective modeling, two distinct cases have to be considered: (i) the one-to-one reaction (binding of monovalent ions) and (ii) the two-to-one reaction for divalent species. If a solid modifier M (i.e., vermiculite) is to be added to carbon paste, it will also appear on its surface at a certain surface concentration, $[M]_{surf}$. And if the analyte "A" reacts in a 1:1 ratio as (i) with the modifier, the reaction can be described as shown in Equation 4.9:

$$A_{sol} + M_{surf} \leftrightarrow AM_{surf} \qquad (4.9)$$

where the subscripts sol and surf designate the solution and the electrode surface, respectively. The corresponding concentration equilibrium constant, K', is defined as

$$K' = [AM]_{surf,eq}/[A]_{sol}/[M]_{surf,eq} \qquad (4.10)$$

The subscript *surf* indicates surface concentrations and *eq* chemical equilibrium.

The dependence of the voltammetric current response, $i_{A,eq}$, on the concentration $[A]_{sol}$ after attaining chemical equilibrium (after long interaction time between analyte and modifier) is given by Equation 4.11:

$$\frac{1}{[A]_{sol}} = \left(\frac{1}{i_{A,eq}}\right) * K' * i_{A,max} - K' \qquad (4.11)$$

where $i_{A,max}$ corresponds to the current with complete saturation of all modifier on the surface with the analyte. A plot of the inverse concentration of the analyte in the solution, $1/[A]_{sol}$, versus the inverse voltammetric current at chemical equilibrium, $1/i_{A,eq}$, should give a straight line with intercepts of $i_{A,max}$ on the inverse current axis and $-K'$ on the inverse concentration axis. Thus, K' values can be estimated.

The dependence of the current $i_{A,t}$ on the accumulation or reaction time t is given by

$$i_{A,t} = i_{A,eq} * [1 - \exp(-k_v * t * [A]_{sol})] \qquad (4.12)$$

where k_v represents the velocity constant for the bimolecular kinetics of the accumulation or reaction, which can also be expressed according to Equation 4.13:

$$\frac{di_{A,t}}{dt} = k_v * [A]_{sol} * (i_{A,eq} - i_{A,t}) \qquad (4.13)$$

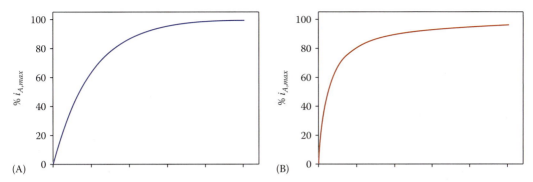

FIGURE 4.17 Dependence of voltammetric signal on the accumulation or reaction time. (A) "1:1" reaction; (B) "1:2" reaction between modifier and analyte (x-axis: time in arbitrary units).

This mathematical function for $i_{A,t}$ represents an exponential curve with a quasi-linear range at short times, because in this case the exponential function $\exp(-x)$ can be approximated with $1 - x$ (see illustrative curve shown in Figure 4.17A).

This model also allows an estimation of the thickness ϑ of the sorption layer involved with the reaction between modifier and analyte, if a few other parameters are known or have been determined (Equation 4.14):

$$\vartheta \geq \frac{[AM]_{max}}{([AM]_{surf,max} * \rho * m_M)} \qquad (4.14)$$

where

ρ is the specific mass of the modified paste
m_M is the mass ratio of the modifier in the paste (mass of modifier/mass of paste)
$[AM]_{max}$ represents something like maximum reaction capacity, that is, the maximum amount of analyte A (in moles) that can react with a certain mass of modifier M (in kg)

This can be simply theoretically evaluated in the case of stoichiometric chemical reactions or can be practically determined in other cases (e.g., ion-exchange capacity for ion-exchange reactions). $[AM]_{surf,max}$ is the maximum possible concentration of the analyte on the surface; its evaluation requires usually coulometric measurements with the same sensors to determine the relation $f_{i \to c}$ between surface concentration $[AM]_{surf}$ and voltammetric current i_A (Equations 4.15a and b):

$$[AM]_{surf} = f_{i \to c} * i_A \qquad (4.15a)$$

$$[AM]_{surf,max} = f_{i \to c} * i_{A,max} \qquad (4.15b)$$

The situation is however more complicated for a 1:2 reaction (*two-to-one reaction*) between analyte and modifier (Equation 4.16):

$$A_{sol} + 2M_{surf} \leftrightarrow (AM_2)_{surf} \qquad (4.16)$$

The corresponding equilibrium constant K'' is defined in Equation 4.17:

$$K'' = [AM_2]_{surf,eq} / [A]_{sol} / [M]^2_{surf,eq} \qquad (4.17)$$

Unfortunately, in this case the surface concentrations are in different powers in the numerator and devisor of the fraction with the consequence that, when expressing them as current responses or charges, the corresponding conversion factors (e.g., $f_{i \to c}$ from voltammetric current to surface concentration) are not eliminated, when replacing the concentrations by the signals (Equation 4.18):

$$K'' = i_{A,eq}/(i_{A,max} - i_{A,eq})^2/(4 * f_{i \to c} * [A]_{sol}) \tag{4.18}$$

In this respect, K'' has a direct meaning only when comparing the same or similar systems with voltammetric measurements and is not directly comparable to formation constants as given in the literature, because surface concentrations are involved rather than volume concentrations. For the determination of $f_{i \to c}$, coulometric measurements with the same sensors are indispensable because they provide the necessary relation between the surface concentration and the voltammetric signal. Additionally, the value for K'' cannot be obtained as simply as in the 1:1 case, but it has to be estimated by corresponding fits [1080].

The dependence of the voltammetric current on the accumulation or reaction time (see Figure 4.17B) is mathematically shown in Equation 4.20, which is in fact rather different to Equation 4.12 but is similar in shape to the previous graph (once again, compare both parts of the figure):

$$i_{A,t} = i_{A,eq} * \left(1 - \frac{1}{(4 * f_{i \to c} * i_{A,eq} * k_v * t * [A]_{sol} + 1)}\right) \tag{4.19}$$

The proper application of this procedure then allows one to estimate the thickness of the sorption layer using the formula of Equation 4.14, which can be formally rewritten as follows (as Equation 4.20):

$$\vartheta = [AM]_{max} * ([AM]_{surf,max} * \rho * m_M)^{-1} \tag{4.20}$$

when the individual variables and symbols are otherwise identical.

5 Biologically Modified Carbon Paste Electrodes

5.1 BIOSENSORS AND ELECTRODES FOR BIOLOGICAL ANALYSIS

A chemical sensor is a simple, small device that provides information about the chemical composition of a system [1906]. Biosensors are chemical sensors that contain a biological component in the recognition layer of the sensing device [1907]. Though the biological entity is not clearly defined as such, it is usual to name devices as biosensors, if the biocomponents are biological polymers (proteins or enzymes, polysaccharides, or nucleic acids), cell organelles or even whole cells (microorganisms), and cell aggregates (tissues). Sensors where antigens or antibodies are immobilized in the recognition layer are categorized as biosensors as well.

The term "biosensor" is sometimes used more liberally in the literature to just indicate that a *biological reaction is involved*, which is not in accordance with the IUPAC suggestions briefly outlined earlier. There is also no strict classification of biosensors. Usually, categories are distinguished according to the modifier, that, the biological component present in the receptor of the sensor.

Regarding the most conventional classifications, there are four major categories, all having respective carbon paste variants:

- *Enzyme electrodes* contain enzymes; most applications deal with oxidases or dehydrogenases yielding electrochemically active products or cofactors, which form the bases of detection. Nevertheless, hydrolases may also be used if they yield an electroactive product by hydrolysis; they are frequent subjects of inhibition studies.
- *Tissue electrodes* contain more or less homogenized tissue, usually from plants. The tissue itself serves normally as a source of an enzyme without the necessity to isolate it in a pure form. In fact, care must be taken that the target enzyme controls the biochemical recognition of the substrate, and other components of the tissue exert no or only a minor effect.
- *DNA electrodes* are electrochemical sensors where deoxyribonucleic acid is involved in the recognition layer of the sensor. This kind of biological modifiers can be used for different purposes, such as accumulation of interacting and intercalating compounds, for interaction studies for the sake of characterizing reactions between a substrate and a nucleic acid including DNA-destructing substances, and for specific hybridization when using single-stranded DNA as a modifier.
- *Electrochemical immunosensors* represent electrochemical devices where one partner of an immunochemical reaction is immobilized at the electrode surface.

As this classification, when looking to these terms which are frequently used in the literature, is neither unique nor covers all the ways of biological modification, it is more informative about the individual types of electrodes—here, the biologically modified variants.

In the following paragraphs, an overview of biological modifiers as well as the corresponding working principles is briefly sketched. As many of the examples are summarized in Section 8.4 (in the form of tables with commentary and the corresponding literature), global citations of all literature referring to special modifiers are omitted here, and only special features are highlighted with the concrete reference. Anyway, the fundamentals of biosensors and bioanalysis are available in an excellent overview by Gorton ([70]; moreover, in close association with carbon pastes), or one can go through the monograph by Hart ([58] devoted exclusively to biologically important [active] compounds [BIC]; again, with numerous examples of CP-based detection systems).

5.2 MODIFIERS AND WORKING PRINCIPLES

5.2.1 ELECTROCATALYSIS AND MEDIATION

Enzymes make up the biggest number of modifiers used to design electrochemical biosensors based on carbon paste. When considering electrochemical conversion potentials of biological substrates, biopolymers, or compounds involved in biological reactions, it is clear that only a few meet the requirement of undergoing oxidation or reduction at reasonably low operating potentials (see Figure 5.1).

The higher the applied potential (positive or negative), the higher the risk for cooxidation or coreduction of components present in the complex matrix of biological samples, such as whole blood, plasma, or urine.

The reason for large positive or negative potential to prompt electron transfer to the analyte is often not a very high normal potential but kinetic implications at the electrode surface that cause so-called overpotentials, which means that an increased electrochemical potential must be applied to transfer electrons between the electrode and the substrate. Additionally, it must be considered that an amperometric current or a voltammetric signal caused by a direct electron transfer between the analyte and the electrode is usually not very specific as usually quite a few compounds could undergo oxidation or reduction.

Many studies with electrochemical sensors and biosensors deal with the problem to reduce the overpotential—in fact, a shift of the signal(s) of interest into the less favorable position within the operational potential range—and this applies either to the analyte itself or to the intermediate(s) of the biological reaction given. Such attempts by means of adding a modifier, either directly to the paste or to the solution, are generally called electrocatalysis. A closer insight into the problem reveals that, basically, two possibilities should then be discriminated as shown in Figure 5.2.

In sketch A, a mediator is incorporated in or on the surface of the paste; in considering the oxidation of the substrate S_{Red}, which shows an overpotential at the electrode material, the mediator M_{Ox} must be chosen in such a way that (i) it is capable of oxidizing the substrate, as a rule of thumb: $E_0^{Ox}(M) > E_0^{Ox}(S)$, and that (ii) it can be reoxidized at a potential significantly lower than the direct oxidation potential of the substrate ($\eta^{Ox}(M) \ll \eta^{Ox}(S)$ where η denominates the overpotential). A detection event of this type can be considered as mediation and is typically characterized such that the substrate increases the electrochemical signal of the mediator, but usually does not change its potential.

In a closer meaning of electrocatalysis (see Figure 5.2B), the substrate is directly converted (e.g., to the oxidized form if initially present in the reduced) at the electrode surface and is then reconstituted to its original state by a reagent (R_{Red}). Thus, the potential of the direct oxidation remains

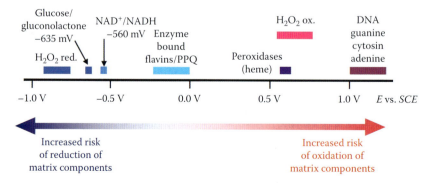

FIGURE 5.1 Some potential ranges of the reduction and the oxidation of biologically significant compounds.

Biologically Modified Carbon Paste Electrodes

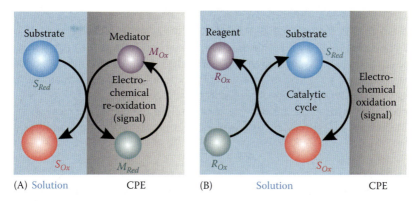

FIGURE 5.2 (A) Electrochemical mediation and (B) electrochemical catalysis ("electrocatalysis") in a closer sense.

unchanged, but the signal is magnified, and apparent effective cycle numbers may be evaluated. Both cases may improve the analytical characteristics of the method significantly with respect to interferences or detection limit.

Nevertheless, direct mediation of the analyte will still show quite a high degree of nonspecificity because any substances reacting with the meadiator will influence the response of the target compound. Electrocatalysis may increase the current, but will still be the subject of interferences if the corresponding potentials for the direct measurement are to be set rather positive or negative.

5.2.2 Enzymes

The unsatisfying analytical situation where the direct or even electrocatalytic electrochemical conversion of the analyte lacks specificity can be very much improved by the use of an enzyme that specifically attacks one single or a group of similar target compounds. The resulting sensors are termed as enzyme electrodes and may show quite high specificity for certain analytes depending on the nature of the enzyme used.

As most of the biosensors described in the literature are amperometric or voltammetric devices, it is evident that the preferred types of enzymes used are the ones that change the oxidation state of the substrate, that is, oxidases and dehydrogenases. Less frequently, peroxidases are used and only occasionally catalases. In fact, all these enzymes belong to the group of oxidoreductases. Their way of action is depicted in Figure 5.3.

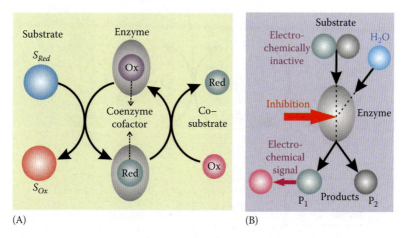

FIGURE 5.3 Way of action of oxidoreductases (A) and hydrolases (B).

The sketch (Figure 5.3A) shows the oxidation of a reduced substrate, S_{Red}, with the aid of the enzyme. The enzyme, in fact its oxidized cofactor or coenzyme, is reduced; and in biological systems it must be reconstituted by a cosubstrate in the oxidized form, which is reduced as a consequence. If the cosubstrate is oxygen, the enzymes are called oxidases with the possibilities to reduce O_2 to hydrogen peroxide or water; in case the cofactor is a biological mediator, such as nicotine adenine dinucleotide, NAD^+, they are classified as dehydrogenases.

On the other hand, hydrolases are enzymes that catalyze the hydrolytic cleavage of a substrate (Figure 5.3B). To some extent they are also exploited for preparing enzyme-modified carbon paste sensors.

5.2.2.1 Oxidases

Oxidases use elemental oxygen, O_2, as an electron acceptor for the oxidation of the substrate. The simplest way to regenerate the initial state of the enzyme (Figure 5.3A) would be a direct electrochemical conversion of the cofactor or coenzyme that is located at the active center. Practically, it is very difficult in most cases because the active center is usually buried underneath the protein shell of the apoenzyme, which prevents or deteriorates the electron transport inside the enzyme.

When exploiting oxidases for designing biosensors, there are quite a few possibilities to obtain the amperometric signals being proportional to the concentration of the substrate (see Figure 5.4).

Going back in history, it is found that the first biosensor was designed by L. Clark in the early 1960s [1908] and, indeed, it was based on the detection of depletion of oxygen, measured by the oxygen sensor (also developed by him) and which can be estimated as the very first amperometric sensor [1909–1911]. This design is one possibility to correlate the substrate concentration to the current signal obtained by the reduction of O_2 (see Figure 5.4, pathway "A").

Many oxidases produce hydrogen peroxide as an intermediate: it is electroactive and can be detected either by reduction to water (pathway "B") or by oxidation to oxygen (pathway "C"). The possibilities to detect the intermediate are sketched in Figure 5.5.

Direct electrochemical conversion of H_2O_2 imposes some problems because both electrochemical reactions, oxidation (Figure 5.5, pathway "A") and reduction (Figure 5.5, pathway "B"), show very high overpotentials with all the risks of detecting a lot of other compounds.

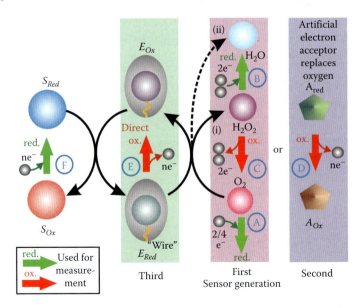

FIGURE 5.4 Way of action of oxidases, electrochemical detection, and sensor generations.

Biologically Modified Carbon Paste Electrodes

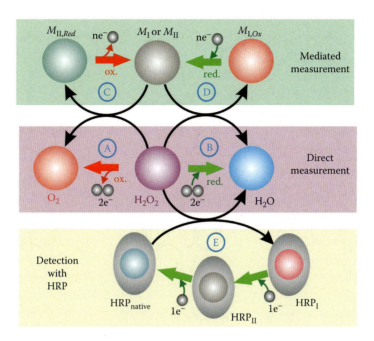

FIGURE 5.5 Possibilities to detect hydrogen peroxide.

Therefore, some alternatives can be applied—mediated oxidation (Figure 5.5, pathway "C") or reduction (Figure 5.5, pathway "D") yielding the corresponding amperometric currents due to the electrochemical reconstitution of the mediator. Convenient methods are mediated electrochemical transformations with Prussian blue and derivatives, such as osmium purple and other types of hexacyanoferrates (for oxidation and reduction), metal oxides (for oxidation and reduction), or carbon nanotubes (for reduction). Another possibility includes horseradish peroxidase (HRP; Figure 5.5, pathway "E"), which is oxidized by H_2O_2 to compound I (HRP_I) and which can be reduced via compound II (HRP_{II}) to the native state either directly or again by mediation. HRP offers even the possibility of accumulating hydrogen peroxide by incubation over a certain period, which facilitates the detection of low concentrations [57,58,70].

Devices where either the depletion of oxygen or the formation of hydrogen peroxide is detected are sometimes called biosensors of the first generation, mostly applied for glucose sensors but sometimes also used for other biosensors relying on oxidases (Figure 5.4). Oxygen, and in consequence hydrogen peroxide, may impose problems on the analytical detection. The cosubstrate O_2 must be always present in sufficient excess not to deteriorate the amperometric signal. Mediation requires an additional modifier. Thus, one intention was to replace oxygen by other artificial electron acceptors (Figure 5.4, pathway "D"). In this case it is possible to measure in oxygen-free media; such types of oxidase-based biosensors (mainly glucose biosensors) are occasionally termed as second-generation biosensors. The signal is obtained by reoxidizing the reduced form of the acceptor.

Similarly as in the previous chapter concerning chemically modified carbon paste electrodes (CMCPEs), it is useful to give here some examples of how certain substances may act in such roles. And in this context, it is quite logical that some items will be presented again, because the corresponding substances are applicable in the configuration of a CMCPE, as well as in a CP-biosensor. Schemes 5.1 through 5.4 thus summarize the most commonly used electron transfer cofactors that have been incorporated into the carbon paste matrices of the respective CP-biosensors and related sensing devices.

As can be seen across Schemes 5.1 through 5.4, a great number of them belong to the classes of phenothiazine I, phenoxazine II, or phenazine III, including, for example, methylene blue IV,

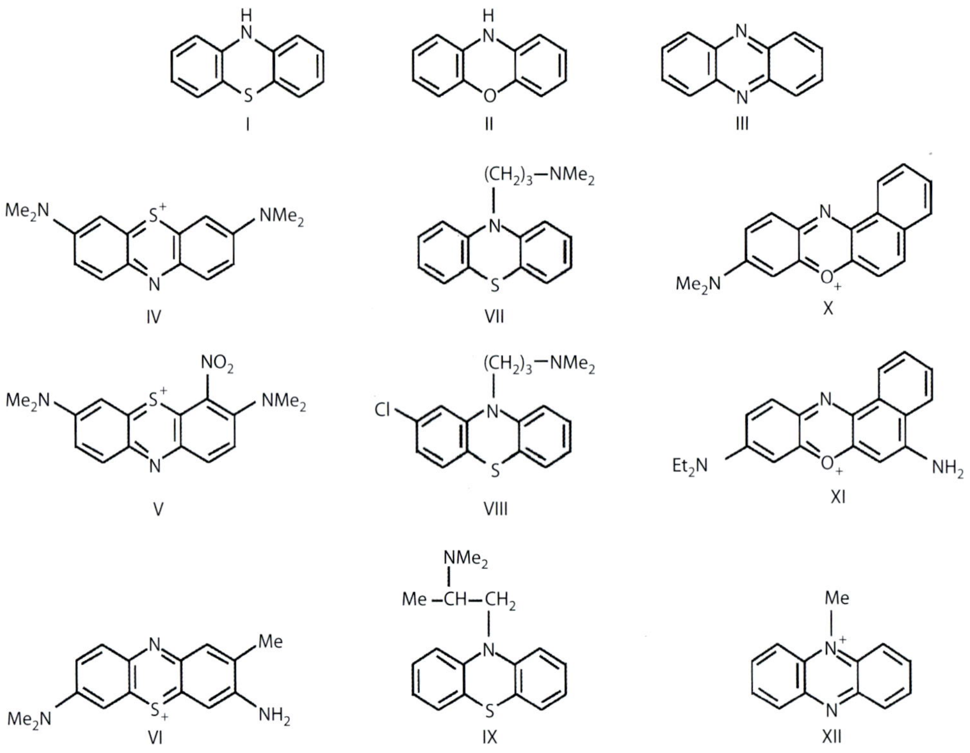

SCHEME 5.1 Phenothiazines, phenoxazines, and phenazines as artificial electron acceptors for use with oxidase-based biosensors.

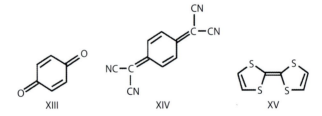

SCHEME 5.2 Parabenzoquinone, tetracyanoquinodimethane, and tetrathiafulvalene.

methylene green V, toluidine blue VI, promazine VII, chloropromazine VIII, promethazine IX, Meldola blue X, Nile blue XI, and phenazine-methosulfate XII (see Scheme 5.1). These mediators were often incorporated directly into the paste, alone or together with enzymes (in case of second-generation biosensors), but many efforts have been directed to improve their immobilization at the electrode surface and to prevent their leaching into the solution. This was especially achieved by means of their adsorption onto inorganic particles (e.g., silica or other metal oxides), which were subsequently dispersed in the carbon paste matrix.

Other organic electrocatalysts are *p*-benzoquinone XIII and related compounds, tetracyanoquinodimethane XIV, and tetrathiafulvalene XV (Scheme 5.2). Finally, an important class of mediators widely used as carbon paste modifiers is illustrated in Scheme 5.3 and represents organometallic compounds, such as ferrocene XVI and ferrocene-based derivatives, osmium–bipyridine complexes (e.g., osphendione XVII, Scheme 5.3). Metal porphyrins and especially those based on cobalt or copper (e.g., cobalt-tetraphenyl-porphyrin XVIII), and cobalt phthalocyanines, are also very frequently used (Scheme 5.4).

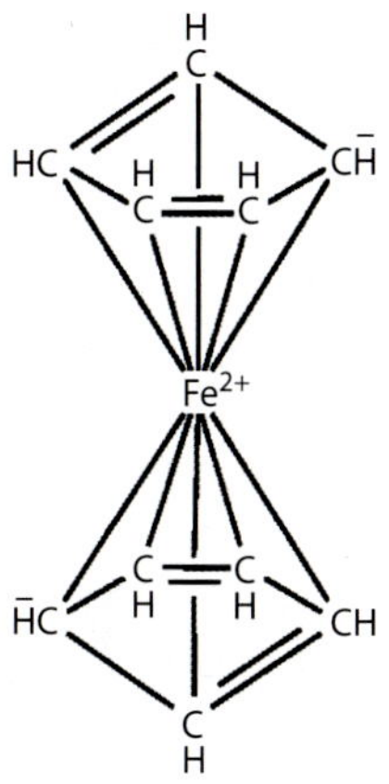

Biologically Modified Carbon Paste Electrodes

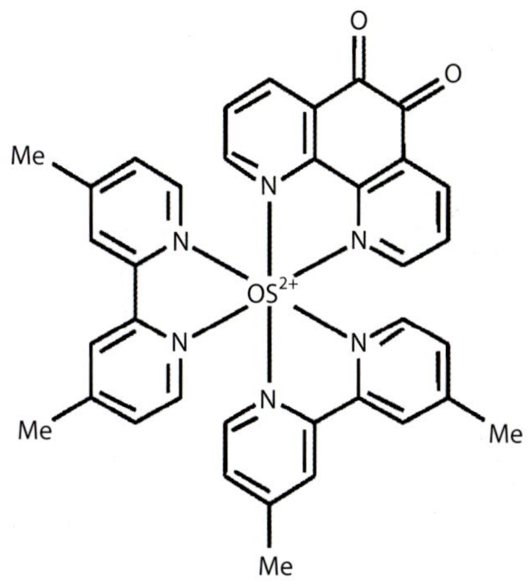

SCHEME 5.3 Ferrocene and osmium bipyridyl.

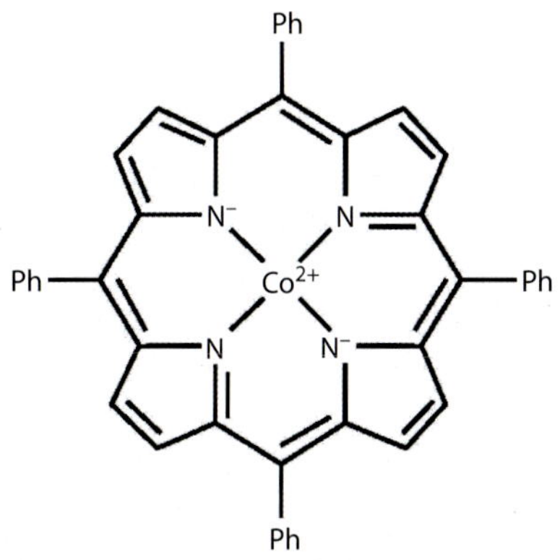

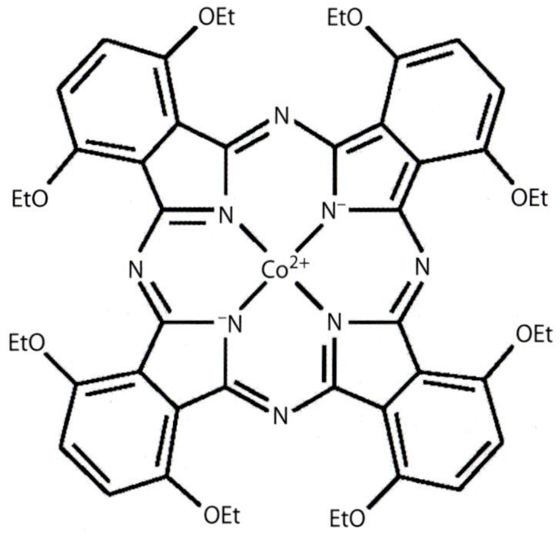

SCHEME 5.4 Co-tetraphenylporphyrin and cobalt phthalocyanine.

The simplest way from an analytical point of view would be a direct electrochemical reconstitution of the enzyme (Figure 5.4, pathway "E"). The main problem for this was already briefly sketched, and lies in the fact that the active center is usually buried in the protein shell. To circumvent the problem, it is attempted to "wire" the active center with the electrode surface (see Figure 5.6). Sensors with an underlying concept of direct electron transfer are sometimes called third-generation sensors.

The oldest approach of wiring is the use of redox polymers (Figure 5.6, part "A"). The electrons are shuttled via redox groups bound to a polymeric backbone into the active center of the enzyme. Useful redox groups are ferrocene and Os-bipyridine complexes (Scheme 5.3). Other assays have been developed recently and rely mainly on a conductive "wire" (carbon nanotubes) or gold nanoparticles attached to the coenzyme (FAD in case of oxidases) [1912,1913]. The attachment is attained normally via covalent bonding (CNT) or sulfide sorption (AuNP). The apoenzyme is reconstituted with the modified FAD where the nanosized material acts as a wire. Apart from the three "generations" of oxidase-based sensors another detection pathway is open if the product of the enzymatic reaction is electroactive. This is often the case with phenols and polyphenols in combination with polyphenol oxidase, which produces quinoid structures that can be reduced again.

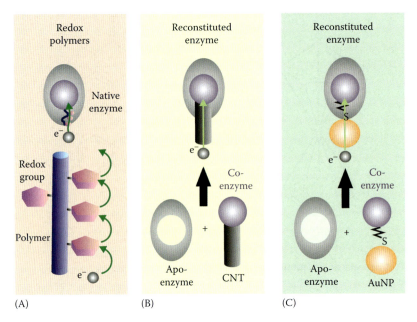

FIGURE 5.6 Wiring of the active center of enzymes with redox-polymers (A), or by reconstitution of the enzyme from the apo-enzyme with the co-enzyme bound either to carbon nanotubes, CNT (B), or to gold nanoparticles, AuNP (C).

The first and the most widely used oxidase [73], and, in fact, the most popular enzyme of all incorporated in CPEs, has been glucose oxidase (GOD, most frequently from *Aspergillus niger*), either alone or together with a redox mediator (Chapter 8, Table 8.19). GOD is dependent on a flavin cofactor (i.e., FAD, flavin adenine dinucleotide) strongly bound inside the enzyme structure, as for all the oxidases. The most often utilized mediators were ferrocene and its derivatives, as well as quinones and benzoquinones or HRP.

Beside GOD, many other oxidases were used to design CP-biosensors (see Section 8.4 and the respective tables). They include tyrosinase, polyphenol oxidase, and laccase, which were widely applied to the detection of phenolic and catechol compounds, xanthine oxidase, L- or D-amino acid oxidase, alcohol oxidase, L-glutamate oxidase, pyruvate oxidase, cholesterol oxidase, pyranose oxidase, amine oxidase, choline oxidase, and others. These enzymes react similarly to GOD, with an intimately bound cofactor, with or without mediator (using molecular oxygen as electron acceptor in the latter case), and several of them can be used in bienzyme configurations.

5.2.2.2 Dehydrogenases

Contrarily to oxidases, dehydrogenases do not require molecular oxygen as electron acceptor. One may distinguish two groups of dehydrogenases: those with a soluble cofactor acting as a cosubstrate in the enzymatic reaction (nicotinamide adenine dinucleotide $NAD^+/NADH$ or nicotinamide adenine dinucleotide phosphate $NADP^+/NADPH$) or bound cofactors (flavin adenine dinucleotide, $FAD^+/FADH$, or pyrroloquinoline quinone) (see Scheme 5.5).

A review surveys CP-biosensors based on immobilized dehydrogenases [156]. Again, the literature is surveyed in Chapter 8. The most widely used enzymes are alcohol dehydrogenase and those related to sugar detection: glucose dehydrogenase, fructose dehydrogenase, oligosaccharide dehydrogenase, aldose dehydrogenase, and pyrroquinoline quinone glucose dehydrogenase.

Other related enzymes include lactate and glutamate dehydrogenase, D-gluconate dehydrogenase, glycerol dehydrogenase, and others. Usually the cofactor $NAD(P)^+$ is added as a modifier at the electrode surface; the reoxidation of the reduced intermediate $NAD(P)H$ is commonly exploited as a signal for the substrate; it can be monitored directly, mediated, or with the aid of diaphorase.

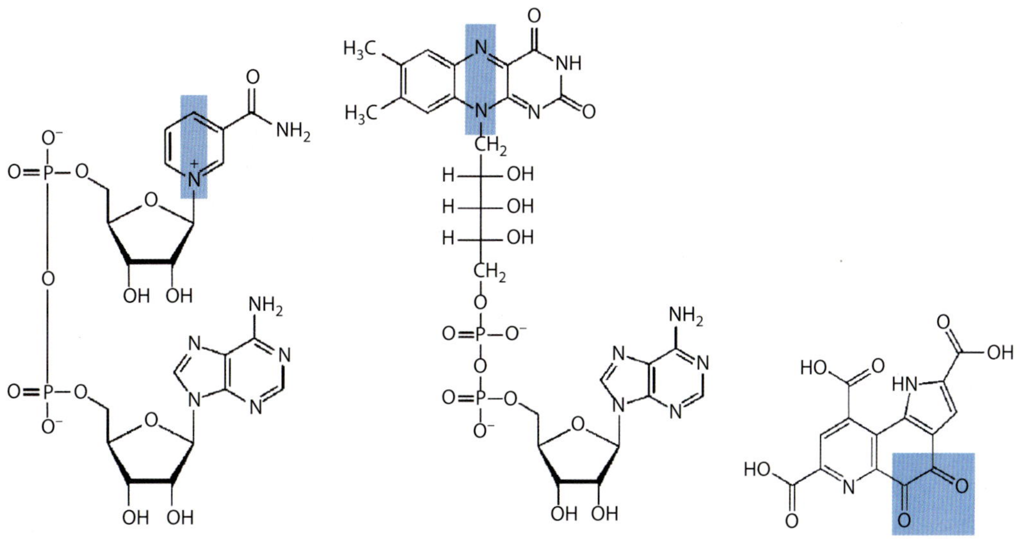

SCHEME 5.5 Enzyme cofactors: nicotinamide adenine dinucleotide, NAD⁺ (XX), flavine adenine dinucleotide, FAD (XXI), and pyrroloquinoline quinone (PQQ, XXII). *Note*: Blue rectangles mark the electrochemically active sites of the individual molecules.

5.2.2.3 Hydrolases

Occasionally hydrolases are used as modifiers for carbon paste electrodes; their way of action is depicted in Figure 5.2. In most cases hydrolases are used for inhibition studies. To do so, the enzyme immobilized at the electrode surface is supplied with a corresponding substrate, and one of the cleavage products is electrochemically active. This system was widely applied to detect organophosphorus compounds that inhibit (alkaline) phosphatase (see Section 8.2 and the tables therein). In fact the method is not very specific because also heavy metals inhibit the enzyme. Another hydrolase used for modification of electrochemical sensors is acetylcholine esterase.

5.2.2.4 Auxiliary Enzymes

Sometimes enzymes are immobilized in the recognition layer not with the intention to generate an electroactive compound or intermediate from the substrate, but to assist and improve the analytical detection process. In this respect peroxidases (in particular horse radish peroxidase, HRP) are widely used; they can be oxidized by hydrogen peroxide, which appears often as an intermediate with oxidases. Therefore, HRP is frequently used to mediate the reduction of H_2O_2.

Therefore, peroxidases were often combined with oxidases, to enhance the response to hydrogen peroxide, since particularly horseradish peroxidase may act as a mediator for the reduction of H_2O_2. Nevertheless, peroxidases were also applied as indicator enzymes to the direct detection of hydrogen peroxide and small organic peroxides, notably in flow injection analysis. HRP can be combined with charge transfer mediators, with the advantage that almost any reducing agent (e.g., ferrocyanide, phenol, *o*- and *p*-phenylenediamines, iodide, ascorbate, etc.) may act as electron donor. These mediators can be used either as soluble species in solution or immobilized in the carbon paste matrix (or even on its surface). Microperoxidase-11 (MP-11) was also described [1115].

Other helper enzymes are catalase in combination with an oxidase (e.g., ascorbate oxidase) to remove interferences (ascorbate), mutarotase to facilitate the equilibrium formation between different glucose species when formed freshly in solution, and diaphorase to detect NADH molecule.

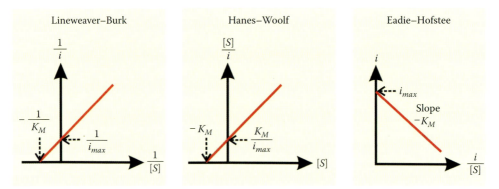

FIGURE 5.7 Usual evaluation procedures for evaluating the apparent Michaelis–Menten constant with enzyme electrodes.

5.2.2.5 Enzyme Kinetics

Enzyme kinetics are usually described by the Michaelis–Menten model. In a simplified form the enzyme, *E*, forms a complex with the substrate, *S*, yielding *ES*, which is followed by the conversion to the product and the dissociation of the product–enzyme complex, *EP*. In case the enzyme concentration and the enzyme–substrate complex (steady state) do not change with time, the reaction rate, *v*, tends to approach a maximum value, v_{max}, with increasing substrate concentrations. The *Michaelis–Menten constant*, K_M, corresponds then to the substrate concentration, where half of the maximum reaction rate is attained (see Equation 5.1).

$$v = \frac{v_{max} \cdot [S]}{K_M + [S]} \quad (5.1)$$

In order to estimate the *apparent Michaelis–Menten constant* with respect to the situation present at the electrode surface, the usual evaluation procedures are applied, that is, *Lineweaver–Burk*, *Hanes–Woolf*, and *Eadie–Hofstee plots* (Figure 5.7). Instead of reaction rates, the signals obtained for the substrate, that is, the corresponding currents, are used for the approach.

5.2.3 Nucleic Acids

Within the past 20 years, numerous publications have appeared that deal with carbon paste electrodes modified with nucleic acids, oligonucleotides, and base constituents of nucleic acids. Deoxyribonucleic acid, DNA, the main center of research, shows high affinity toward carbon paste and strongly adsorbs to its surface. DNA yields a signal below +1.0 V vs. SCE due to the oxidation of guanine; adenine moieties being also oxidized at slightly more positive potentials. In fact, the other bases typically produce oxidation currents at even higher potentials (see Figure 5.8).

Studies of nucleic acids with carbon paste electrodes may be mainly categorized into three topics (see Section 8.4, Table 8.23):

(i) *Preconcentration of nucleic acids* with anodic stripping voltammetry; as the accumulation is not specific, different types of DNA cannot be discriminated.
(ii) *Accumulation and interaction studies* with DNA as modifiers; many substrates show high affinity toward DNA and can be accumulated; the process is often intercalation of the target species into the DNA, frequently into the minor groove of the double helix. Interaction studies include also investigations on damaging effects of the substrate on the modifier.

Biologically Modified Carbon Paste Electrodes

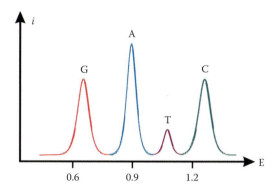

FIGURE 5.8 Schematic baseline corrected responses of DNA, where "G" is guanine, "A" adenine, "T" thymine, and "C"- cytosine. (Newly drawn after Diculescu, V.C. et al., *Sensors*, 5, 377, 2005.)

(iii) *Hybridization studies* are probably one of the most promising direction of applying DNA and oligonucleotide-modified electrodes for base sequencing of analytes. The sensors are modified with single-stranded DNA (ssDNA) and exposed to the analyte solution; if complementary ssDNA is present, hybridization occurs (see Figure 5.9).

Hybridization can be monitored with reporter or indicator molecules. Two cases may be differentiated. If the affinity of the indicator is significantly higher to the single strands than to the double helix, the blank signal will correspond to a complete mismatch, but will decrease correspondingly to the extent of hybridization; for example, methylene blue.

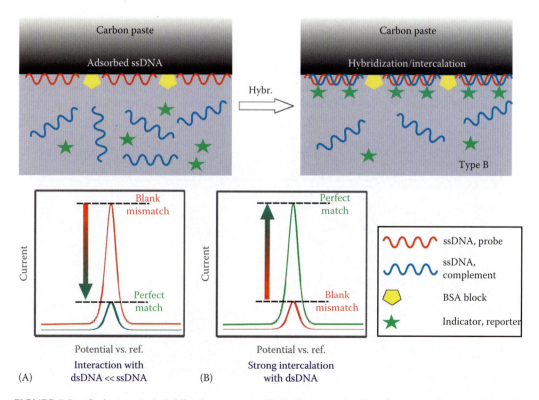

FIGURE 5.9 Carbon paste hybridization sensors with indicator molecules where the affinity to the single stranded DNA is higher (A) or lower (B) than to the dsDNA.

On the other hand, a signal increase with hybridization will be observed, if the reporter molecule intercalates strongly into the double helical structure, as is the case with Co(II)-bipyridyl complex. Bovine serum albumin (BSA) may be used to prevent nonspecific adsorption. Peptide nucleic acids (PNA), in which the deoxyribose-phosphate-backbone is substituted by a proteinic structure, may replace ssDNA with hybridization studies. Substitution of guanine by electroinactive inosine in single strands allows monitoring hybridization via the guanine moiety of the complementary strand.

Occasionally, aptamers—DNA or RNA oligomers with less than 100 mers (i.e., replicates) that can bind specifically to certain biomolecules—are used to specifically bind proteins or antibodies to the electrode surface [58,90–92].

5.2.4 Immunosensors

Immunoreactions involve the combination of an antibody with an antigen, which can be realized with carbon paste electrodes. Antibodies are usually immobilized on the electrode surface; two assays are very common to determine antigens from the sample: sandwich and competitive assay (Figure 5.10).

For the sandwich assay, the antibodies are reacted with the antigens from the sample; in an ensuing step the antibody–antigen complex is incubated with a labeled antibody; the label can be an enzyme producing an electroactive product from an inactive substrate (e.g., alkaline phosphatase with a phenolic phosphate as a substrate) or metal nanoparticles that can be determined by dissolution and stripping voltammetry. With sandwich complexes the signal increases with an increasing amount of antigen.

In competitive assays a labeled antigen is added to the sample, which competes with the unlabeled sample constituent for the antibodies. As a consequence, the higher the signal of the label, the lower the concentration of the antigen in the test solution. Carbon paste immunosensors will also be discussed in Section 8.4.

5.2.5 Tissues and Cells

Tissues and whole cells were sometimes used as such to modify CPEs for two goals: (i) to circumvent the need to isolate the corresponding enzyme in its pure form while even preserving its stabilizing environment, and (ii) to bioaccumulate analytes (e.g., metal ions) prior to their voltammetric detection. Mostly plant *and* mushroom tissues or their crude extracts were used as sources of

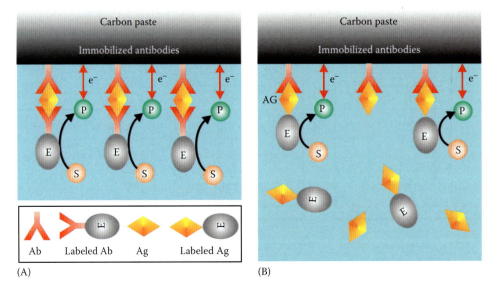

FIGURE 5.10 Immunosensors: sandwich (A) and competitive assay (B).

Biologically Modified Carbon Paste Electrodes

oxidases, the latter for phenol oxidase (see Section 8.4, Table 8.26). Yeasts may serve as a source of alcohol dehydrogenase. Algae, mosses, and lichens were dispersed in carbon pastes and exploited for the preconcentration of metal species. *Mitochondria* have been immobilized on carbon paste as well [1640].

5.2.6 OTHER BIOMOLECULES

Other biological materials used to modify CPEs include cytochrome C, hemoglobin, myoglobin (see Section 8.4, Table 8.20), melanins, and heparin [1777]. Polysaccharides (cyclodextrins) and chitosan are used for accumulations and immobilization.

5.3 MODIFICATION STRATEGIES (FOR BIOSENSORS)

Biomolecules have often a high affinity toward carbon paste so that they can be either admixed directly to the bulk of the paste or adsorbed to the surface. Enzymes often maintain their activity in the carbon paste matrix, and can even show increased temperature and acid resistance [58,90–92]. In order to prevent leaching, different strategies can be applied to immobilize biomolecules.

(i) *Cross-linking* with glutaraldehyde is a common procedure to connect biomolecules via aminogroups to larger molecular networks with lower solubility. Addition of bovine serum albumin (BSA) provides even a large pool of $-NH_2$ groups to which the molecules can be attached. Other possibilities are the entrapment in porous membranes (e.g., *Nafion*®), electropolymerized layers (e.g., polypyrrole), or behind dialysis membranes. Usually membranes deteriorate the kinetics of the accompanying chemical reactions and diffusion. Membranes can be applied from the corresponding solutions via drop or spin coating, by electrolysis from the monomer, or by preformed layers (e.g., dialysis membrane).

(ii) *Electrostatic immobilization* via surfactants, functionalized nanoparticles, and ion exchangers is convenient if the modifier has an electric net charge. For instance, popular *ionic liquids* as binders or modifiers may exert a very positive effect (see also Section 2.2.3.3).

(iii) *Magnetic nanoparticles* enable special interactions that can often be accomplished by using Fe_3O_4 for instance. Outside the core, they may be covered with a functionalized shell, to which biomolecules can be bound (for other details, go to Sections 2.2.1 and 4.2).

(iv) *Avidin–biotin complexation* represents another important way of immobilization. Streptavidin modified surfaces (e.g., nanoparticles) and many biotinylated biomolecules and biocomponents are commercially available. As shown in Section 8.4, the avidin–biotin pair plays a particular role with immunoassays.

Other possibilities of modification. (v) Surfactants change the character of the carbon paste surface from lipophilic to more hydrophilic and can be used as *in situ* ion exchangers in case they are charged; (vi) nucleic acids can be immobilized at positive potentials (around +0.5 V vs. reference electrode) after activation of the electrode surface at very positive potentials (preanodization). Such (vii) preanodization of carbon pastes or also (viii) thermal activation of the graphite moiety prior to preparation of the paste may significantly improve the analytical signals of interest.

Finally, (ix) carbon nanotubes and (x) other nanosized particles (both as plain or functionalized, see also Section 4.2) possess catalytic properties for many chemical and electrochemical reactions (e.g., reduction of H_2O_2) and can be therefore useful modifiers or even main constituents of the past. Here, it should be repeated that modifiers in the form of nanoparticles usually manifest considerably higher activity than macro- or microsized particles due to their substantially larger surface in relation with the volume.

6 Instrumental Measurements with Carbon Paste Electrodes, Sensors, and Detectors

6.1 ELECTROCHEMICAL TECHNIQUES: FUNDAMENTALS AND BASIC PRINCIPLES

In general, heterogeneous carbon electrodes and sensors are very versatile and can be employed in connection with all conventional electrochemical techniques—of course, except polarography—such as voltammetry, amperometry, potentiometry, and all their variants, as well as many others. Compared with the use of homogeneous electrodes, the combinations with carbon paste electrode (CPEs), chemically modified carbon paste electrodes (CMCPES), and carbon paste (CP)-biosensors show various advantages and disadvantages, which are sketched briefly in the following sections. Otherwise, the significant characteristics of each technique are summarized in a condensed form only, as they can be referred to in the corresponding handbooks [3,162–165,1915–1925].

6.1.1 Double Layer Concept and Capacitive Current

CPEs, as most of the electrodes for electrochemical analysis, are mainly exposed to solutions containing a supporting electrolyte. All methods relying on the measurement of currents (voltammetry, amperometry, but indirectly also many other methods such as coulometry or measurements of resistivity) apply a working potential (static or changing with time).

The application of a potential to an electrode surface exposed to a solution causes the formation of a double layer. Basic concepts have already been established by Helmholtz; in the course of improvements, the model was refined to describe properly the behavior of charged particles from the surface into the solution. The double layer is caused by the distribution of charge on the electrode surface by enhancing or depleting electrons from the current source (potentiostat). This charge is compensated by mobile charged particles in the solution, which results in the formation of a double layer corresponding to a charged capacitor. The structure of the solution–electrode boundary in the liquid phase close to the electrode surface may be divided into a rigid compact layer directly in contact with the surface (caused by specifically adsorbed ions [inner Helmholtz-plane] and by nonspecifically adsorbed species, for example, hydrated ions [outer Helmholtz plane]), and an adjacent diffuse layer caused by thermal agitation of the solution and extending from the compact layer into the solution bulk (Stern layer) according to the Gouy–Chapman–Stern model ([1915,1918,1919]; see Figure 6.1).

The current necessary to charge the interface is called capacitive current. When applying a constant potential (potential step), it decreases rapidly to small values according to an exponential decay with practically vanishing magnitude after some time, assuming a simplified circuit of the system with a serial arrangement of a resistor (solution) and a capacitor (solution–electrode interface; Equation 6.1):

$$i_C = \frac{E}{R_s} e^{-t/(R_s \cdot C_d)} \quad (6.1)$$

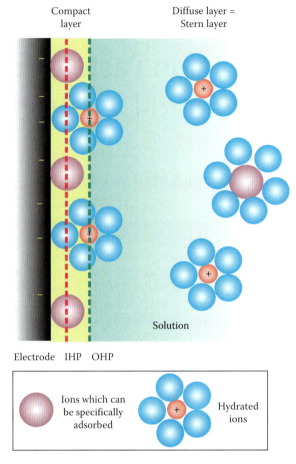

FIGURE 6.1 Guy–Chapman–Stern model of the double layer. IHP, inner Helmholtz plane and OHP, outer Helmholtz plane.

where
- i_C is the capacitive current
- E is the applied potential
- R_s is the resistance of the solution
- t is the time
- C_d is the capacity of the double layer

Typical values for the capacitance of electrode surfaces are 10 to 50 μF cm^{-2}, but they may differ significantly for CPEs particularly when modifying the interface structure. For a capacitance C_d of 20 μF and a resistance R_s of 1 Ω, the time constant τ ($=R_s \cdot C_d$) is around 20 μs. Within 3τ the capacitive current drops by 95% of its value. The thickness of the diffuse double layer depends on the type of electrolyte and its concentration: according to the Gouy–Chapman–Stern model for a 1:1 electrolyte it is around 10 nm for a total ion concentration of 10^{-3} mol dm^{-3}, 100 nm for 10^{-5} mol dm^{-3}, and 1 μm for 10^{-7} mol dm^{-3}. When applying a linear voltage ramp rather than a constant potential, i_C drops to a constant value of $v \cdot C_d$, where v is the scan rate (V·s^{-1}).

6.1.2 Mass Transport

With an electric field applied to a solution, there are basically three circumstances that may provoke the movement of charged particles in this field. The displacement of particles is in general called

Instrumental Measurements with Carbon Paste Electrodes, Sensors, and Detectors

mass transport or *mass transfer*. A basic description of the mass transfer in an electric field is given by the Nernst–Planck equation whose one-dimensional form is shown in Equation 6.2:

$$J_i(x) = -D_i \frac{\partial C_i(x)}{\partial x} - \frac{z_i F}{RT} D_i C_i \frac{\partial \phi(x)}{\partial x} + C_i v(x) \tag{6.2}$$

where
  $J_i(x)$ is the flow of species "*i*" (mol s^{-1} m^{-2})
  x is the distance from surface (m)
  D_i is the diffusion coefficient (m^2 s^{-1})
  C_i is the concentration of species i (mol m^{-3})
  z_i is the charge
  ϕ is the potential
  v is the velocity of moving volume element (m s^{-1})

Reasons for the mass transfer of the ionic species may be as follows:

(i) *Diffusion* along a concentration gradient (chemical potential) is described by the first term in Equation 6.2 and corresponds to the first Fick law. Concentration gradients occur in electrochemical measurements if a transfer of electrons occurs at the electrode surface due to reduction or oxidation (faradic process), which consumes the species of interest. The depletion zone around the electrode surface is called diffusion layer. In most of the voltammetric and amperometric measurements, the aim is to determine the diffusion current.
(ii) *Migration* is represented by the second term in the Nernst–Planck equation and is caused by the gradient of the electric field; it can be suppressed by the addition of a supporting electrolyte, which also decreases the resistivity of the solution.
(iii) *Convection* (expressed in the third term of Equation 6.2) is usually the forced movement of the solution relative to the electrode, caused by streaming liquids or stirring of the solution but also by displacement of the electrode (e.g., rotation). The situation where convection has a significant influence on the measurement is usually referred to as hydrodynamic condition.

6.1.3 Faradic Processes

Faradic processes involve an electron transfer of the target species at the electrode surface; the corresponding current is called faradic. The reaction can, in general, be represented by Equation 6.3:

$$O + ne^- \leftrightarrow R \tag{6.3}$$

where
  O means the oxidized species
  R means the reduced form
  n means the number of electrons transferred

A reduction proceeds from the left to the right, an oxidation in the opposite direction. Electron transfers of this type provoke a depletion of the target species, which as a consequence launches a mass transfer of the bulk species toward the electrode. Thus, the mass transfer and the electron transfer determine the kinetics of a faradic process (see Figure 6.2).

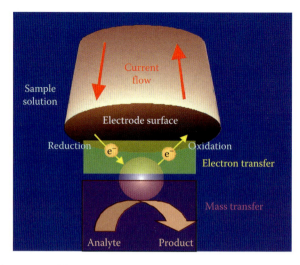

FIGURE 6.2 Faradaic process with mass and electron transfer.

With many ideal and quasi-ideal systems, the electro-transfer is much quicker than the mass transfer. In such a case, the system is called diffusion or mass-transfer controlled. A good indication if diffusion is controlling the kinetics is performing linear sweep or cyclic voltammetry at different scan rates and to correlate the peak current with the square root of the scan rate according to the Randles-Sevcik equation (Equation 6.4 for reversible reactions, where the diffusion coefficients for oxidized and reduced species, D, are practically identical):

$$i_p = (2.69 \cdot 10^5) \cdot n^{2/3} \cdot A \cdot D^{1/2} \cdot C^* \cdot v^{1/2} \tag{6.4}$$

where
  i_p is the peak current
  A is the electrode area
  D is the diffusion coefficient
  C^* is the bulk concentration of the electroactive species
  v is the scan rate ($mV \cdot s^{-1}$)

Apparent sluggish electron transfer, which will consequently control the kinetics rather than mass transfer, can be described with the Butler–Volmer equation, which considers forward and backward transfers (Equation 6.5):

$$i = i_0 \cdot \left(e^{-\alpha \cdot n \cdot F \cdot (E - E_{eq})/RT} - e^{-(1-\alpha) \cdot n \cdot F \cdot (E - E_{eq})/RT} \right) \tag{6.5}$$

where
  α is the transfer coefficient
  E_{eq} is the equilibrium potential
  i_0 is the exchange current

The left term is the forward reaction, the right the backward. Curve analysis can be made from voltammetric sigmoidal curves. When considering that far away from the equilibrium potential, E_{eq} (potential where the net current flow is 0, that is, forward and backward current are of the same

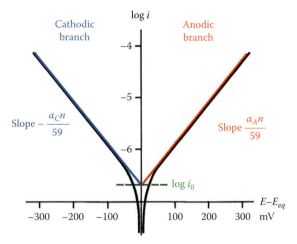

FIGURE 6.3 Tafel plot for symmetric cathodic and anodic branches (with $C_O = C_R$ and $\alpha = 0.5$).

size, i_0, but of different sign), the counter reaction can be neglected. Thus, the logarithmation of the corresponding term of the Butler–Volmer equation yields the Tafel equation (Equation 6.6):

$$\ln i = \ln i_0 - \frac{\alpha \cdot n \cdot F \cdot \eta}{R \cdot T} \quad (6.6)$$

where "α" means the transfer coefficient, either for the reduction (α_C) or for the oxidation (α_A) with $\alpha_C + \alpha_A = 1$. An example of a Tafel plot is shown in Figure 6.3. The product $\alpha \cdot n$ can be evaluated from the slope, i_0, from the extrapolated intercept.

6.1.4 Diffusion Layer

6.1.4.1 Diffusion Coefficients

Diffusion can be described by the Fick laws (see Equation 6.2), where D represents the diffusion coefficient. In aqueous solutions at room temperature, it is very similar for common ions and ranges between 6×10^{-10} and 2×10^{-9} m² s⁻¹. Biological molecules, which are usually larger and clumsier, show D-values of 10^{-11} to 10^{-10} m² s⁻¹.

6.1.4.2 Macrosized Stationary Electrodes

In case the electrode surface is large with respect to the solution volume, a diffusion layer is formed increasing with time. With disk-shaped electrodes, planar diffusion with a direction perpendicular to the electrode surface is the dominating type.

When applying a potential to macroelectrodes, where a diffusion-controlled faradic process occurs to its full extent (potential step experiment), the measured current (chronoamperometry; i.e., measurement of current in dependence of time) decreases according to the Cottrell equation (Equation 6.7):

$$i(t) = n \cdot F \cdot A \cdot C_O^* \sqrt{\frac{D_O}{\pi \cdot t}} = \mathrm{const} \cdot t^{-1/2} \quad (6.7)$$

Here, the subscript "O" stands for the oxidized species in case of a reduction, and the asterisk means the bulk concentration. With sufficiently long periods, the current will drop to neglectible values, which means practically to 0. Cottrell behavior can be exploited for estimating "D" with known "n" or vice versa.

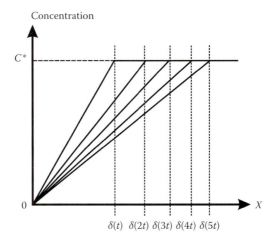

FIGURE 6.4 Diffusion layer thickness and concentration gradient in the diffusion current regime at constant potential with increasing time. x is the distance to the electrode surface.

The thickness of the diffusion layer δ can then be estimated according to Equation 6.8, when assuming Nernst's model of concentration gradient in the diffusion layer (Equation 6.9, Figure 6.4):

$$\delta(t) = \sqrt{\pi D_O t} \tag{6.8}$$

$$\frac{\partial C}{\partial x} = \frac{C_O^* - C_O}{\delta} \tag{6.9}$$

where
C_O^* denotes the concentration of (oxidized) species in the bulk
C_O at the electrode surface

Usually under conditions of measurement the diffusion layer ranges between 10 and a few hundred micrometers depending on the experimental circumstances.

6.1.4.3 Rotating Disk Electrodes

When rotating the electrode, there is an additional mass transfer due to convection, which decreases the diffusion layer (Figure 6.5).

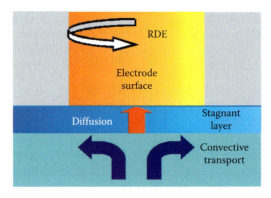

FIGURE 6.5 Convective mass transport due to the rotation of the electrode.

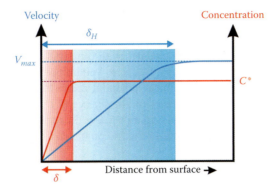

FIGURE 6.6 Diffusion layer (δ) and hydrodynamic (Prandtl) boundary layer (δ_H), at a rotating electrode.

In fact, the hydrodynamic boundary layer, δ_H, also called Prandtl layer, that is, the layer with a velocity gradient, is connected with the diffusion layer δ (Equation 6.10):

$$\delta \approx 2\left(\frac{D}{v}\right)^{1/3} \delta_H \qquad (6.10)$$

where "v" means the kinematic viscosity in this formula. Assuming some standard values ($D = 10^{-5}$ cm² s⁻¹, $v = 10^{-2}$ cm² s⁻¹), the formula gives an estimation of $\delta_H \approx 5\delta$ (see Figure 6.6).

When performing DC voltammetry at a rotating electrode, the limiting current of a mass transport–controlled system depends on the angular velocity, $\omega = 2\pi N$ (where N is the number of revolutions per second) as expressed in the Levich equation, which is given here for a reduction process (Equation 6.11):

$$i_{l,c} = 0.620 \cdot n \cdot F \cdot A \cdot C_O^* \cdot D_O^{2/3} \cdot v^{-1/6} \cdot \omega^{1/2} \qquad (6.11)$$

In this equation, "v" is the kinematic viscosity.

6.1.4.4 Microelectrodes

Microelectrodes are electrodes where the diffusion layer thickness $(Dt)^{1/2}$ is much larger than the radius r_o of the electrode. As a consequence, the electrode area is too small to produce a continuously increasing depletion of its surrounding and therefore leads to steady-state conditions where the current does not drop to zero but to a constant value. Typical microelectrode diameters are in the rang of 2–10 µm. When applying a constant potential (potential step experiment) to a microelectrode, chronoamperometric measurements show current decays at the beginning with a constant value after some time (Equation 6.12). The term $1/r_o$ that is responsible for this is called spherical correction:

$$i_t = n \cdot F \cdot A \cdot D \cdot C^* \left(\frac{1}{\sqrt{\pi \cdot D \cdot t}} + \frac{1}{r_o}\right) \qquad (6.12)$$

For hemispherical electrodes, Equation 6.13 is valid, and for planar disks, the factor 2π has to be replaced by number 4:

$$i_{t \to \infty} = 2\pi \cdot n \cdot F \cdot D \cdot C^* \cdot r_o \qquad (6.13)$$

Microelectrodes with small diffusion layers allow voltammetry, with rapid scan rates still providing sigmoidal curves rather than peak shapes. Very fast scan rates (usually larger than a few hundred V s⁻¹) will eventually produce peak-type curves. As the ohmic drop is of small influence only (usually less than 1 mV for 1 mM of electroactive species), the use of a supporting electrolyte is often obsolete. For similar reasons, even the use of a reference electrode can be avoided.

6.2 VOLTAMMETRY

Voltammetry, as a shortening of voltamperometry, is in principle the monitoring of a current in dependence of an applied potential, which changes with time (potential sweep). The methods are referred to as controlled potential techniques. Different voltammetric techniques differ in the way of establishing and modulating the basic linear potential ramp. Electrochemical transformations (oxidations, reductions) of electroactive species on the surface of the electrode provoke an electrolytic current, named as faradic current. This is in most cases the relevant component for the voltammetric measurements though some techniques rely also on the capacitive current.

6.2.1 Direct Voltammetry

In linear potential sweep techniques, a potential is applied to the electrode, starting at an initial potential, and changed linearly with time to a final value with a constant scan rate.

6.2.1.1 Voltammetry at Slow Scan Rates

With low scan rates, the shape of the current response is sigmoid, characterized by the limiting current i_d, which is caused and usually controlled by diffusion ("diffusion current"), the half-wave potential $E_{1/2}$, where half of the diffusion current is attained, and which is somehow connected to the normal reduction potential, and the steepness of the curve, which is related to the number of electrons transferred with one molecule or ion of the electroactive component. The curves can be described by the Heyrovsky-Ilkovic equation (Equation 6.14):

$$E = E_{1/2} + \frac{RT}{nF} \ln \frac{i_d - i}{i} \quad (6.14)$$

where
  E is the applied potential (V)
  R is the general gas constant (8.314 J K^{-1} mol^{-1})
  F is the *Faraday constant* (96,485 C · mol^{-1})
  i is the current at the potential E
  i_d is the limiting current
  n is the number of electrons transferred

Substituting the constant and referring to the temperature for the standard conditions as well as conversion of the natural to the decadic logarithm yields Equation 6.15:

$$E = E_{1/2} + \frac{0.059}{n} \ln \frac{i_d - i}{i} \quad (6.15)$$

Voltammetric methods with slow or moderate scan rates (usually ≤10 mV) are the following:

- *Direct current voltammetry (DCV)* continuously monitors the current during the potential scan. As the scan rate is usually slow, sigmoidal curves are obtained as explained earlier.
- *Sampled DCV* samples data points discontinuously; otherwise, the curve looks comparable to DCV. It was a tremendous advantage in polarography because oscillations of the response were eliminated.
- *Staircase voltammetry (SCV)* changes the potential incrementally, not continuously; the data are sampled after a waiting period to allow the capacitive current to vanish.

- *Normal pulse voltammetry (NPV)* switches the potential off between the measuring intervals, leading to a pulse-like form of the excitation potential; the corresponding limiting current is given according to the Cottrell equation (6.16):

$$i_{d,NP} = n \cdot F \cdot A \cdot D^{1/2} \cdot C^* \cdot \sqrt{\frac{1}{\pi \cdot t_p}} \quad (6.16)$$

where
  $i_{d,NP}$ designates the diffusion limiting current of this technique
  t_p designates the pulse time after which the measurement of current is launched

- *Differential pulse voltammetry (DPV)* maintains still the basic linear voltage ramp, but a potential pulse is superimposed to the ramp and the current is measured before and after the application of the pulse (usual values are some 20–100 mV) and the decay of the capacitive current as a difference; thus, the current response somehow corresponds to the first derivative of the fundamental sigmoid response curve and consequently resembles a peak. The peak height is again proportional to the bulk concentration of the electroactive species. If the pulse height is less than 59/n mV, then the peak current can be estimated according to Equation 6.17:

$$i_p = \frac{n^2 F^2}{4RT} A \cdot D^{1/2} \cdot C^* \cdot \Delta E_p \cdot \sqrt{\frac{1}{\pi \cdot t_p}} \quad (6.17)$$

- *Semiderivative and semi-integral voltammetry.* Occasionally 2.5th derivative voltammetry is used by some Chinese electrochemical groups [1926–1928]. Semidifferential electroanalysis was introduced by Goto and Ishii [1929] based on the idea of semi-integral electroanalysis as developed by Oldham [1930,1931]. 2.5th derivative voltammetry was first described in 1978 again by Goto et al. [1932].

  Semiderivative or semi-integral response curves can be done by measuring with semidifferential electronic circuits [1929,1932] or by mathematical treatment of linear sweep (scan) voltammetries (LSVs) [1933–1935]. Compared with linear sweep voltammetry, semiderivative methods show lower detection limits and better resolution capabilities for overlapping peaks [1932].

- *Elimination voltammetry (ELV).* This type of voltammetry tries to eliminate kinetic and charging currents by recording a voltammogram at the reference scan rate v as well as at the half and the double scan rate $v_{1/2}$ and v_2, respectively. The voltammogram in linear scan mode is reconstituted by the relation (Equation 6.18 after [1936,1937]):

$$f(i) = -11.657 i_{1/2} + 17.485 i - 5.8284 i_2 \quad (6.18)$$

The subscript of i refers to the scan rate with which they were recorded. ELV was applied in connection with the determination of deoxyribonucleic acid and its components [1938–1945]. ELV should be applied with care in order to avoid artefacts in the results.

6.2.1.2 Voltammetry at Faster Scan Rates

If the scan rate is increased (>10 to 1000 mV s^{-1}), peaks are formed rather than sigmoidal curves due to the fact that the potential is changed already before the current characteristic for slow scan rates is reached. For reversible electrochemical reactions, the **Randles-Sevcik equation** can be used to describe the dependence of the peak current "i_p" on the scan rate "v" (Equation 6.4). Thus, a plot of i_p vs. the square root of the scan rate should give a linear relation.

- *Linear sweep (scan) voltammetry* is, in fact, DCV at higher scan rates, which can go up to 1000 mV s^{-1} and even more. Sometimes the expressions "linear sweep voltammetry" and "direct current voltammetry" are used or referred to identically in the literature.
- *Cyclic voltammetry (CV)* is probably the most important technique for studying electrochemical reactions proceeding at the electrode surface. It represents an extended form of LSV. With CV a scan in one potential direction (half-cycle) is combined with a scan in the counter-direction, yielding one full cycle. The voltammograms are not necessarily started at one of the vertex potentials but usually at a potential with negligible faradic current. Therefore, it is a powerful tool to study whether reduction–oxidation couples appear in the voltammetric scan and whether they are reversible or not; in case of reversibility (quick electron transfer for both components of the redox couple), the potential difference between corresponding cathodic and anodic peaks should be 59/n mV (see Figure 6.7). In fact, CPEs rarely meet the criteria of ideal reversibility due to the presence of pasting liquid in the surface; the deviation can serve as a characterization criterion for the paste (see Sections 3.3 and 4.4.1). Cycles can be repeated ad libidum and may yield information on adsorption–desorption characteristics or chemical changes of reactants or products.
- *Square-wave voltammetry (SWV)*: A square wave is superimposed over the linear potential ramp yielding a peak-shaped current response. The current is measured as a difference from the forward and the backward pulses; SWV often yields -higher responses than DPV in comparable shorter times; the rapidness even allows its use with electrochemical detectors. The method was developed by Barker based on older works from Kalousek [1925]; the method was popularized by J. Osteryoung and is therefore also sometimes referred to as "Osteryoung SWV" (similar to DC-polarography, known as "Heyrovsky polarography").
- *Alternating current voltammetry (ACV)* superimposes an alternating current on the potential ramp and measures the corresponding harmonic tone or overtones. Out-of-phase measurements allow detailed studies of adsorption processes via capacitive currents apart from the faradic reactions.

For investigating electrochemical reactions with heterogeneous carbon electrodes, the considerations shown earlier are basically applicable, but it can be almost always noticed that deviations from the theoretical behavior occur. The main reason is the presence of the pasting liquid

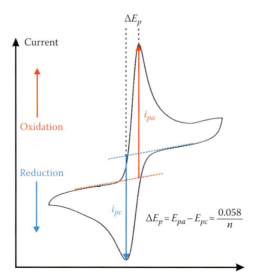

FIGURE 6.7 A typical cyclic voltammogram. *Note*: For a reversible redox couple, the ΔE_p value should be about 59/n [mV] and $i_{pa}/i_{pc} \sim -1$.

Instrumental Measurements with Carbon Paste Electrodes, Sensors, and Detectors

or the polymeric matrix, which worsens the electron transfer between electrode surface and analyte, resulting in signal broadening.

6.2.2 Stripping Voltammetry

Stripping voltammetry, in the former U.S.S.R. and some Central European countries also known as inverse voltammetry, employs in its classical arrangement a preconcentration (preelectrolysis or deposition, respectively) to accumulate the target species on the electrode surface before measurement, which can be DCV, DPV, or any other voltammetric method.

Most typical applications are the preconcentration of zinc, cadmium, lead, or copper, alone or in combination, as amalgams at the hanging mercury drop electrode. The most important is the immobilization of the target species at the electrode surface, resulting in an accumulation and consequently in a decrease in the detection limit. In order to avoid diffusion effects, the solution is usually stirred during this step. For settling the solution, stirring is usually stopped for some 10–20 s (with the potential still applied) before the measurement starts to provide quiet current responses. The so-called medium exchange approach is possible but is hardly performed in the classical technique. The ensuing measurement can be performed either in cathodic or anodic direction depending on the preconcentrated system, named consequently as cathodic or anodic stripping voltammetry (CSV, ASV).

Characteristic for stripping analysis is the decoupling of the mass transfer process in the direction of the electrode from the measurement; nevertheless, reaction products are still subject to mass transfer away from the surface in case they are not sorbed. Modified electrodes have significantly broadened the applicability of stripping analysis to analytes, which cannot be preconcentrated by means of electrolysis. In such cases, chemical preconcentration can improve the analytical situation. Here, a chemical reaction is the basis of the immobilization of the analyte at the electrode surface.

Often, it can be done under open circuit conditions without a potential applied; sometimes an electric field may also support the accumulation process. Chemical preconcentration is often combined with an exchange of the analyte medium by the measurement medium with obvious advantages: the interaction time can be well controlled, and both media can be optimized with respect to their composition independently from each other. Electroactive components from the analyte solution not showing an affinity for the carbon paste or the modifier will be separated.

- *Adsorptive stripping voltammetry (AdSV)*, introduced by Kalvoda (see, e.g., [168] and references therein), is—in its simplest form—the accumulation of the analyte by adsorption to the electrode surface. In a more common variant, an adsorbant or reactant present on the surface immobilizes the target species by interacting with it. Reactants can also be added directly to the analyte solution modifying the electrode surface *in situ*; that is, from the sample solution, which is added directly to the analyte solution and modifies *in situ* the surface of the electrode in order to accumulate the target species. Typical examples are the preconcentration of Ni or Co as oxime complexes at the electrode surface with the subsequent reduction of the Me^{II} atom during measurement. Also here, the potential applied controls the adsorption process.
- *Adsorptive catalytic stripping voltammetry (AdCtSV)*, propagated by Bobrowski et al. (again, see [168]), is an interesting variant with the measurement combined with a catalytic process, where an additional reagent, oxidant or reductant, converts the product of the electrochemical reaction at the electrode surface again to its initial, preconcentrated state yielding an increased current response via a catalytic cycle.
- *Adsorptive transfer stripping voltammetry (AdTSV)* is in fact AdSV with a medium exchange between the accumulation and the measurement step. At least some authors use this term in such a context. It was originally suggested by Palecek's group [1921] in DNA detection for the technique to add a small volume of a solution to the surface of the electrode; after some time, during which the target analytes adsorb at the interface, the solution can be washed off, and the electrode inserted to the electrochemical cell.

6.3 AMPEROMETRY

Amperometry is the *monitoring of the current* in dependence of a parameter, such as volume of a reagent (titration) or time (chronoamperometry). In a more general sense, amperometry is a *special case of voltammetry*, when the potential (ramp) is set constant, at a fixed value that is to be chosen in a way to enable the full electrochemical transformation of the redox species analyzed. In principle, amperometry can be applied to static (batch analysis, stop-flow systems) and dynamic solutions (stirred solutions, flow systems).

Static conditions usually prevail if the current is measured and used directly for the evaluation of the concentration of the analyte, either by external calibration curves or by the standard addition method. Dynamic conditions are found when the solution is moved relatively to the electrode, which is the case in stirred solutions or in flow systems. The instrumental setup for amperometry is usually equivalent to that of voltammetry, usually comprising three electrodes (working electrode, reference electrode, and auxiliary electrode). As mentioned earlier, the operating potential is usually set to a constant value, where the limiting current of the target species can be monitored, which is directly proportional to the concentration of the analyte in the solution.

In chronoamperometry, considerations of the potential step experiment (see Section 6.1) can be applied. A chronoamperometric potential step measurement is either the application of a single constant potential starting with open circuit conditions or it includes two constant potentials (double step experiments), jumping from one potential to another.

- *Hydrodynamic amperometry (HA)* is the monitoring of currents at a constant potential under hydrodynamic conditions, usually in stirred solutions. In fact, rotated electrodes or moving streams of solutions are also possible. Hydrodynamic amperometric measurements can be useful for investigating sensor response times and stabilities or to perform quantitative determinations.
- *Hydrodynamic voltamperometry (HVA)* is the recording of (usually steady state) currents under hydrodynamic conditions in dependence of a potential; such voltamperograms are very useful to find optimal operating potentials.
- *Amperometric detection (AD)* is a general term, where the amperometric measuring system is combined with a system under hydrodynamic conditions, for example, flow systems such as flow injection analysis (FIA), sequential injection analysis (SIA), batch injection analysis (BIA), or analytical separation (column chromatography, capillary zone electrophoresis, etc.).
- *Flow injection analysis*, designed by Ruzicka and Hansen [1947], exists in its simplest form as a device consisting of a pump, an injector, and a detector. The task of the pump can be even taken over by gravitation. A carrier is transported constantly through the system; injection of an electroactive component produces a transient signal at the detector. The system is versatile because reagents can be admixed and reactors can be inserted. It is an ideal method for testing and optimizing analytical conditions and is very often used for investigations with biosensors.
- *Sequential injection analysis*, considered as a second-generation flow injection technique, consists of a high precision pump that can be operated in both directions, a holding coil, and a multidelivery valve that is connected to various reagents and to the sample solution. At the beginning, the system is filled with the carrier; then sequentially sample and reagents are aspired as a stack and eventually brought to the holding coil, where due to the parabolic zone profiles and supported by a fore and back movement caused by the pump, mixing occurs. The sample-reagent cocktail is finally pumped through the valve to the detector. The indisputable advantage of SIA over FIA is a considerably lesser consumption of solvents and reagents.

- *Batch injection analysis (BIA)* is operated under dynamic conditions [1948–1954], where the analyte solution is injected directly onto the surface of the sensor in a wall-jet configuration; the sensor is submerged in a large volume of solution. A transient signal occurs, similar to FIA, due to the stream of injected analyte solution toward the electrode and ensuing dilution in the bulk, yielding a peak-shaped signal.

  Manual BIA usually yields worse reproducibility than FIA due to variations when injecting the solutions. Therefore, automated systems [1948] or capillary injection in combination with small-sized electrodes [1954] are preferable. BIA can also be combined with voltammetry and even with stripping voltammetry [1951,1953].
- *Liquid chromatography (HPLC)* offers similarly favorable conditions for AD employing often the *glassy carbon electrode* (*GCE*) in the detection system. Carbon paste can provide a simple alternative particularly if the CMCPE variants are to be used.
- *Capillary (zone) electrophoresis, C(Z)E,* is a relatively new method of employing some microsized CPEs applicable as amperometric detectors, including dual configurations (see Sections 2.3.4 through 2.3.6 and some examples in Section 8.4).

  With AD, the peak current, i_p, is dependent on the geometry of the detector electrode and its position toward the solution stream.
- *Wall-jet configuration* is a special arrangement of the working electrode inside a detection cell, where the electrode surface perpendicular to the incoming stream of the carrier i_p can be approximated by the limiting current i_d at disk electrodes in wall-jet configuration and continuous flow (Equation 6.19 according to [1955]):

$$i_p \approx i_d = 1.43 \cdot n \cdot F \cdot (R \cdot V_f)^{3/4} \cdot D^{2/3} \cdot v^{-5/12} \cdot a^{-1/2} \cdot C^* \qquad (6.19)$$

where
  R is the radius of the disk electrode
  V_f is the volume flow rate of the solution
  v is the viscosity of the solution
  a is the diameter of the jet stream

Other symbols have the same meaning as before. For some other configurations, see Table 6.1 [1915].

6.4 POTENTIOMETRY

6.4.1 Direct Potentiometry and Potentiometric Titrations

(Equilibrium) potentiometry is another common electrochemical technique for measurements with carbon paste–based electrodes. It can be estimated that (i) direct potentiometry and (ii) potentiometric titrations are both on the third rank with respect to the abundance of their use, immediately after voltammetry and amperometry. This has been documented in a wide variety of publications mentioned earlier (e.g., [36,159,208,232,251,253,448,457,517,538,569,570,655,709,712]), within this section (together with associated citations [178,1956–1976]), as well as later in Chapter 8 and throughout the table surveys.

Regarding direct potentiometry, the respective measurements are employed to complete chemical analyses of species for which an indicator electrode is available. The technique is simple, requiring only a comparison of the voltage developed by the measuring cell (consisting of two half-cells—indicator and reference electrodes) in the test solution with its voltage when immersed in a

TABLE 6.1
Limiting Current Responses of Electrodes of Various Geometry in Flow-Through Cells

Electrode Geometry	Limiting Current
Tubular	$1.61 \cdot n \cdot F \cdot \left(D \cdot \dfrac{A}{R}\right)^{2/3} \cdot V_f^{1/3} \cdot C^*$
Planar (parallel flow)	$0.68 \cdot n \cdot F \cdot \left(\dfrac{D \cdot A}{d}\right)^{2/3} \cdot V_f^{1/3} \cdot C^*$
Thin-layer cell	$1.47 \cdot n \cdot F \cdot \left(\dfrac{D \cdot A}{d}\right)^{2/3} \cdot V_f^{1/3} \cdot C^*$
Planar (perpendicular)	$0.93 \cdot n \cdot F \cdot D^{2/3} \cdot v^{-1/6} \cdot A^{3/4} \cdot u^{1/2} \cdot C^*$

Source: Assembled after Wang, J., *Analytical Electrochemistry*, 1st edn., New York: VCH publishers, 1994.

d, thickness of the channel; u, velocity (cm s^{-1}); C^*, concentration (mM); other symbols with their usual meaning.

standard solution of the analyte. If the indicator electrode response is specific for the analyte and independent of the matrix, no preliminary steps are required. In addition, although discontinuous measurements are mainly carried out, direct potentiometry is readily adapted to continuous and automated monitoring.

In pH measurements, an electrochemical pH-cell consists essentially of a measuring (indicator) electrode together with a reference electrode, both in contact with the solution to be studied. For precise measurements, it has been suggested to use two standard buffers, pH(S_1) and pH(S_2), which straddle the pH(X) value. The pH(X) value is then given as

$$\frac{[\text{pH}(X) - \text{pH}(S_1)]}{[\text{pH}(S_2) - \text{pH}(S_1)]} = \frac{[E(X) - E(S_1)]}{[E(S_2) - E(S_1)]} \tag{6.20}$$

This equation represents the so-called practical definition of the pH scale [1956–1958]. Evidently, in practice the voltage differences expressed on the right side of the aforementioned equation are not measured inasmuch as the present meters provide direct pH readings. Hence, instead of the $E(S)$ values, the pH(S) values are adjusted directly on the instrument scale. The Nernst equation contains a temperature term that can be corrected by automatic temperature compensation; detailed instructions are attached to each pH meter.

For direct concentration measurements with ion-selective electrodes, measurements of pX, different methods are available, although none is as well organized as pH measurements. The simplest procedure is to measure the voltage of the cell containing an ion-selective electrode in solutions of graduated concentrations, usually between 10^{-1} and 10^{-6} mol L^{-1} (or similar to the pH scale, in an interval of pX 1–6). A typical calibration graph is usually linear between pX 1 and 5, defined for p$X = -\log a_i$ (see Figure 6.8).

In practice, however, the determination of concentration is more frequently requested; it is possible and more often used. In this case, the cell voltage values are plotted against the logarithms of the concentration of the ion determined, p$X = -\log c_i$. Such a calibration graph, however, differs from that obtained by measuring the activity at higher concentrations, when y_i (the activity

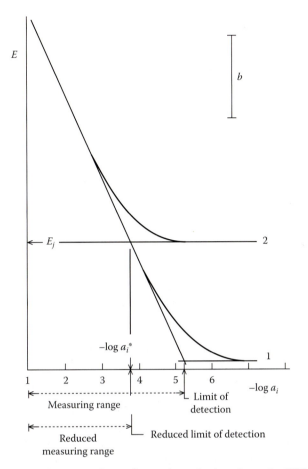

FIGURE 6.8 Typical calibration dependence for an ion-selective electrode (ISE) and its evaluation. (1) Calibration E vs. $-\log a_i$ in solutions of a free determinand "i." The practical limit of detection may be taken as the activity (or the concentration) at the point of the extrapolated lines as shown. (2) Calibration E against $-\log a_i$ in the presence of the interferant "j," the activity of which is of a known and constant value. The response E_j is obtained for $-\log a_i^* = -\log k_{i,j} a_j z_i/z_j$. The resulting interference can restrict the measuring range. (b) The abscissa approximately equal to $60/z_i$ in millivolts. (Reproduced from authors' archives.)

coefficient for recalculations between activities and concentrations, $a_i = c_i y_i$) is less than 1. A series of standard solutions with a composition as close as possible to that of the sample is employed, and the conditions are maintained identical with those used for the measurement of the sample (pH, ionic strength adjustment, screening of interfering ions, etc.). The best results are obtained with simulated standards, where the effects of other components of the sample solution are included in the calibration curve. Curvature occurs at higher and especially at lower concentrations, and eventually the electrode no longer responds to concentration changes.

The most frequently recommended mode is calibration with two standard solutions. It is appropriate for any analysis by direct potentiometry in the Nernstian range of an ion-selective electrode, particularly for analyses carried out at varying temperature. The calibration is performed with two standard solutions in each sample batch. The first gives the value of the cell constant (imitating standard potential E^0 of the sensing electrode], and the second the calibration slope. The two standards should span the concentration range expected in the samples, as any error is magnified by extrapolation. In the non-Nernstian region, the concentrations are obtained from a calibration graph rather than by calculation, and because the graph is curved, more (at least four) standard solutions are needed to define it.

Addition and subtraction modes can also be used; both require a knowledge of the calibration slope but not of the cell constant. The simplest method includes two voltage readings, E_1 before and E_2 after the addition of a volume V_s of a standard solution to V_x volume of the sample:

$$E_1 = \text{Constant} + \text{Slope} \log c_x \quad (6.21a)$$

$$E_2 = \text{Constant} + \text{Slope} \log \frac{(c_x V_x + c_s V_s)}{(V_x + V_s)} \quad (6.21b)$$

where
c_x denotes the concentration of the sample
c_s denotes that of the standard solution

From the cell voltage change $\Delta E = E_2 - E_1$, the unknown c_x concentration can be calculated as

$$c_x = \frac{c_s V_s}{[(V_x + V_s) \, 10^{\Delta E/\text{Slope}} - V_x]} \quad (6.22)$$

In the well-known addition method, the slope factor is determined simultaneously with the concentration by interactive calculation. If a multiaddition method is used, the unknown c_x concentration can be evaluated graphically. The most important application of the addition method is the measurement of the total concentration of the substance to be determined in the presence of a complexing agent (assuming that the matrix effect is mostly eliminated).

In direct potentiometry arranged as continuous or flow measurements, often, in fully automated mode, many conditions must be met in order to obtain a reliable sensor. First, the sensing electrode must respond selectively to the species required and must not be attacked by any other component of the system. Secondly, the electrode must measure reliably for prolonged time intervals, have a reasonably long life, and be easily recalibrated when necessary. The response time must be as short as possible, the accessible concentration range must be sufficiently wide, and the determination limit sufficiently low. A useful and rapid method of such automated analysis is the technique of FIA, in which the sample or a reagent is injected into the stream of a solution of constant composition. Calibration of FIA systems requires the injection of standard solutions, equal in volume to that of sample, into the carrier stream. The background chemical composition of the standards should be equal, as nearly as possible, to that of the samples. Frequent standardization is not necessary, because the measurement of peak heights, albeit on a sloping base line, is relatively not affected by cell voltage drift. Also, the time response is not so important because all the cell voltage values are recorded after fixed time periods.

A weakness of all direct techniques is that the electrodes are not specific but only selective. It means that these may respond to certain other ions in addition to the selected ion "i." Interferences by such ions "j" are expressed in Nikolskii–Eisenman equation:

$$E = \text{Constant} + \text{Slope} \log\{a_i + k_{i,j}^{pot} a_j^{z_i/z_j}\} \quad (6.23)$$

where
E is the voltage of the potentiometric cell in which an ion-selective electrode participates
"Constant" includes all the constant potential contributions mentioned
z_i and z_j denote the charge of the ions i and j
$k_{i,j}^{pot}$ is the potentiometric selectivity coefficient (in case of the presence of more interfering species, the last product is summarized through all j)

The $k_{i,j}^{pot}$ can be experimentally determined; one of the methods is demonstrated in Figure 6.8. When calibration dependence is measured in the presence of the known and constant activity a_j of the interfering j ion (curve 2), both measuring range and limit of detection are limited correspondingly. Moreover, response E_j is obtained for $-\log a_i^* = -\log k_{i,j}^{pot} a_j^{zi/zj}$ from curve 1; thus, this can also represent one method of the evaluation of this coefficient, $k_{i,j}^{pot} = a_i^*/a_j^{zi/zj}$. Various methods were recommended in practice to establish selectivity coefficients, but for many reasons, the values obtained by different methods can also differ but never by orders of magnitude. For many electrodes, especially liquid-membrane systems, selectivity coefficients are also not highly reproducible or precise quantities because they are time dependent. Nevertheless, they can serve to estimate measurement errors in the presence of interfering species.

In contrast to direct potentiometry, potentiometric titrations offer the advantage of high accuracy and precision, although at the cost of increased time and increased consumption of titrants (it should be mentioned that using modern automatic titrators, this is solved by using precise measuring microburettes). A further advantage is that the potential break at the titration end point must be well defined, but the slope of the sensing electrode response need be neither reproducible nor Nernstian, and the actual potential value at the end point is of secondary interest. Only in case of titrations to a fixed potential, the correct voltage value must be chosen by some means of calibration. The errors involved are usually much smaller than in direct potentiometry. This allows the use of simplified sensors [1967]. In the case of CPEs, it means that depending on the chemical reaction on which the determination is based. In precipitation titrations, for example, it is well known that the magnitude of the potential break is governed by the solubility product of the precipitate formed. However, when the precipitate is extractable into the membrane mediator (concretely, the pasting liquid), the role of the extraction parameters must also be taken into account. A quantitative expression for the precipitation equilibrium is given by the solubility product, $K_s(QX)$, which is defined by

$$K_s(QX) = [Q^+]_w [X^-]_w \tag{6.24a}$$

In the presence of another, immiscible organic solvent (a pasting liquid), another equilibrium attained because the two ions will form an associated ion pair in a phase of lower polarity, characterized by the extraction constant, $K_{ex}(QX)$:

$$K_{ex}(QX) = [QX]_o [Q^+]_w^{-1} [X^-]_w^{-1} \tag{6.24b}$$

The distribution ratio of Q^+ ion, $D(Q^+)$, can then be expressed by

$$D(Q^+) = [QX]_o [Q^+]_w^{-1} = K_{ex}(QX)[X^-]_w^{-1} \tag{6.25}$$

It is expected that during conditioning in an aqueous suspension of the appropriate ion pair, an organic solvent (a pasting liquid, analogous to plasticizers of polymeric membrane electrodes) becomes gradually saturated with the ion-pair QX. In the absence of side reactions, the ion-pair concentration in the organic phase, which is in contact with such an aqueous suspension, is determined by the products of the two aforementioned constants, that is,

$$[QX]_o = K_{ex}(QX) \cdot K_s(QX) \tag{6.26}$$

This concentration can be considered as that of ion-exchanging sites in such a Q^+- or X^--selective electrode. It follows that this concentration increases with increasing values of both the extraction constant and the solubility product of the ion pair formed [1967,1973]. When the titration system (titrant, determinand, or both) is changed, usually two titrations are needed to recombine the ion pair in the organic phase for the new one. In analytical practice, these principles may be used in titrations of compounds of pharmaceutical importance [1972] and, especially, in determinations of surfactants of different types, thus eliminating the health-unfriendly or even hazardous two-phase titrations [1966].

Similar explanations can be offered for titrations based on acid-base, complex-forming, or oxidation–reduction equilibria [1967]. Generally, more detailed information on various potentiometric techniques can be found in specialized literature [1956–1961].

Carbon paste–based electrodes, when compared with traditional and marketed ion-selective electrodes, offer several advantages. One of them is significantly lower Ohmic resistance (less than 10 Ω instead of up to MΩ values for electrodes equipped with polymeric membranes). Thus, the experimental work with CPEs is more convenient, and simpler potentiometers not allowing measurements using potentiometric cells with such a high inner resistance may be applied. Historically, it is necessary to mention first the universal carbon-based electrodes called "Selectrodes," introduced by Ruzicka et al. [35,1962,1963]. These were made of Teflon tubing machined so that inside the lower end a thick carbon rod could be fitted very tightly. This spectral-grade porous carbon rod was first hydrophobized and impregnated with pure organic solvent (CCl_4, $CHCl_3$, xylene, benzene, etc.) and then mounted into the electrode body (for more details, see [1962]). Finally, it was immersed in a solution of a proper ion exchanger in the organic solvent with which the carbon rod was originally impregnated. With this method, a nearly invisible layer was formed at the carbon surface, which was water repellent and could be renewed either by dipping the electrode in the organic phase, by scratching off the electrode, or by a combination of both.

Mesaric and Dahmen were probably the first who used electrodes filled with carbon pastes [36]. Later, an increasing number of papers dealing with similar electrode constructions appeared [36,178,251,517,538,1964,1965,1974]. Evidently, the more polar solvents (the same as those used as plasticizers of PVC membranes) can be used in preparing carbon pastes [709]. Because of their good extraction ability against the ion-associated carbon paste composed of lipophilic ionic species, they may also be applied to monitor titrations based on ion-pair formation [253,709,712, 1966–1968]. Mixing of carbon powder with epoxy resin, the ion exchanger of choice, plasticizer, and hardener then led to the construction of *carbon paste*–coated *ion-selective electrodes* (*CP-ISEs*; [1969–1971]).

From the viewpoint of equilibrium potentiometry, the composition of carbon pastes makes it possible to classify the CPEs as ion-selective liquid membrane type electrodes. Pasting liquids usually exhibit good extraction ability against neutral electroactive species of the type of nondissociated weak acids, neutral metal chelates, or ion associates, and the potential of the electrode containing such an organic solvent extract is predominantly governed by an ionic exchange at the interface between the organic phase of the electrode and the sample solution, resulting in the so-called Donnan potential:

$$\Delta \varphi_D = \frac{RT}{z_i F} \ln \frac{a_i(w)}{a_i(o)} \tag{6.27}$$

where
- $a_i(w)$ is the activity of the ionic species ($i = H^+$, M^{z+}, L^{z-}, etc.) in the aqueous sample phase
- $a_i(o)$ is the activity of these ionic species present in an extractable form (HL, ML, etc.) in the organic phase
- z_i is the corresponding ion charge (including the positive or negative sign)
- R, T, F have the usual meaning

If the discussion is confined to dilute solution of ionic species in the aqueous phase as well as to dilute solutions of extractable species in the organic phase, activities can be replaced by concentrations, and the theory can be applied for evaluation of electrode properties by means of extraction data.

Instrumental Measurements with Carbon Paste Electrodes, Sensors, and Detectors

Thus, one can obtain a *hydrogen ion-selective electrode* (if the organic phase forming the electrode surface contains a weak organic acid, HL):

$$\Delta\varphi_D = \frac{RT}{z_i F} \ln \frac{[H^+]_w}{[HL]_o} \tag{6.28}$$

where
 $[H^+]_w$ is the hydrogen ion concentration in the measured aqueous solution
 $[HL]_o$ is an equilibrium concentration of weak organic acid in the paste, which can be equilibrated with the starting amount of the compound, c_{HL}, i.e., $[HL]_o = c_{HL}$

Then, assuming that all other potentials contributing to complete the potentiometric cell voltage E are constant, one can write

$$E = \text{Constant} - 0.059\,\text{pH} \tag{6.29}$$

As can be seen, such an electrode can serve as a pH sensor (the slope of which, of course, may be far from the theoretical value of 0.059 mV per decade), and its potential is more negative, the higher the concentration of organic acid in the paste. Another pH electrode can be prepared by mixing carbon paste with either bismuth or antimony powder [587]. As known, robust electrodes of these metals can function as pH sensors thanks to the surface redox reaction

$$M_2O_3 + 6H^+ + 6e^- \Leftrightarrow 2M + 3H_2O \tag{6.30}$$

in which hydrogen ions participate. The corresponding Nernst equation, assuming unit activities of M, M_2O_3, and H_2O, gets a form

$$E = E^0(M_2O_3/M) - 0.059\,\text{pH} \tag{6.31}$$

which is analogous to that mentioned earlier.

If an extractable metal chelate is incorporated to the paste (or, generally, to the liquid membrane), we have to assume the following reaction:

$$M^{z+}(w) + z\text{HL}(o) = ML_z(o) + H^+(w) \tag{6.32}$$

and the corresponding expression of the interface potential:

$$\Delta\varphi_D = \frac{RT}{z_i F} \ln \frac{[M^{z+}]_w}{[ML_z]_o} \tag{6.33}$$

resulting in the cell potential dependence expressed as

$$E = \text{Constant} + \frac{0.059}{z} \log [M^{z+}]_w \tag{6.34}$$

in case the pH of the aqueous sample solution is kept constant, and all the reagent in the organic phase is converted to the form of a metal chelate, i.e., $z[ML_z]_o = c_{HL}$.

A similar situation occurs in the case of carbon pastes (or, liquid membranes) containing extractable ion pairs, formed according to the overall reaction

$$Q^+(w) + X^-(w) = QX(o) \tag{6.35}$$

The principle for design is as follows [1972]: to build an electrode responsive to anion X^-, for example, the salt Q^+X^- should be incorporated into a nonvolatile solvent using as a membrane plasticizer or, in the case of carbon pastes, as pasting liquid, and the Q^+ ion must be highly lipophilic. Similarly for an electrode responsive to cation Q^+, an oil-soluble salt Q^+X^- is used, where the X^- ion is lipophilic. Thus, the quaternary log-chain alkyl and aryl ammonium salts and high molecular weight cationic dyes, etc., are known to behave as liquid anion exchangers suitable for preparation of anion-selective electrodes. Tetraphenylborate and high molecular anionic surfactants such as dodecylsulphate, etc., show good selectivity for heavier univalent inorganic cations and are also used in electrodes for other "onium" ions.

If such an ion-pair compound (for simplicity, both moieties are univalent) is dissolved in organic phase, one has an electrode (and a potentiometric cell) responding to both cationic and anionic species:

$$E = \text{Constant} + 0.059 \log[Q^+]_w = \text{Constant} - 0.059 \log[X^-]_w \tag{6.36}$$

In contrast to paste electrodes used for electrolytic accumulations, the potentiometric sensors may be prepared even with the use of polar liquids containing reducible groups. Thus, for example, nitrobenzene extracts of ion pairs such as $Ni(phen)_3(ClO_4)_2$ or $Fe(phen)_3(BF_4)_2$ (where phen = 1,10-phenanthroline) were reported as pasting liquids and ion exchangers for perchlorate or fluoroborate ion-selective CPEs [709].

In Figure 6.9, a set of potentiometric curves is shown (in the upper part), together with the following corresponding plot, both concerning the automated ion-pair formation-based titrations of the homologous series of quaternary ammonium salts that varied in the length of their alkyl chains. The titrant containing lipophilic counter ion was tetraphenylborate (used as $NaBPh_4$, where Ph = C_6H_5 [1975]).

According to this, the extractability of the respective ion pairs formed has increased, which is well reflected in the growing magnitude of the end-point breaks of individual potentiometric titration curves (the previous set). As illustrated by the following "ΔE vs. n_C" plot, the phenomenon observed has exhibited a fairly linear trend when being related to the number of carbon atoms in the varied alkyl chain, R.

A brand new experiment is then depicted in Figure 6.10, revealing that some carbon pastes may manifest a notable activity in the solutions with different pH. Although these measurements have primarily been focused on pH-sensing with CPEs modified with bismuth [587] and, newly, with mixtures of Bi + Bi_2O_3 ([1976]; both in powdered form), the initial studies with bare CPEs have shown clearly that even carbon paste alone may act as a pH-electrode with limited operational range.

Among the examples shown in the graph and tested in an interval of pH 5–13, this was the case of a CPE made of ordinary graphite powder (originally produced as gear lubricant), the respective "CR-5/SO" carbon paste exhibiting fairly reproducible and relatively stable response with sub-Nernstian slope of about 40 mV/pH (full line). Compared with this, the second "RW-B/SO" carbon paste prepared from a special spectroscopic graphite of high grade—and, in fact, representing "standard" carbon paste component [82–84]—has shown considerably lower activity in dependence of pH (dashed line), moreover, with a very long time needed to attain the steady-state response.

Such a behavior, together with other comparisons, has implied that the observed pH-activity is associated with the quality of the graphite used and specifically, with the feasibility of carbon powder particles to have partially altered surface states [68], containing various oxygen-containing functional groups that can readily be protonized or deprotonized in dependence of pH (see also Sections 3.2 and 3.4).

Instrumental Measurements with Carbon Paste Electrodes, Sensors, and Detectors

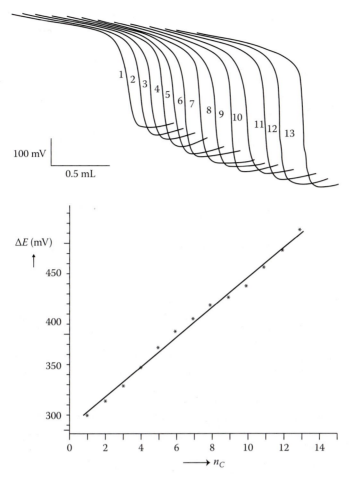

FIGURE 6.9 Potentiometric ion-pair formation-based titrations of quaternary ammonium salts of the $[CH_3(CH_2)_{10}COOCH_2CH_2N(CH_3)_2–R]^+Br^-$ type using $NaBPh_4$ as the titrant showing the dependence of the titration end-point break on the lipophilicity of the cation. A set of titration curves (above) and the "ΔE [mV] vs. n_C" plot (below). 1–13, actual number of titration curves corresponding to the number of carbon atoms (n_C) in the alkyl chain R of the salt. *Experimental conditions*: titrations in the automated mode; tricresylphosphate-containing carbon paste; c(salts) = 1×10^{-5} mol·L^{-1}, c(titrant) = 0.01 mol·L^{-1}. (Redrawn from Vytřas, K. et al., A low ohmic resistance sensor for potentiometric titrations of tensides, in *XXVII Seminar on Tensides and Detergents. Proceedings (in Czech)*, Novaky, Slovakia, 1994, pp. 35–48.)

6.4.2 Chronopotentiometry and Stripping Chronopotentiometry

These techniques whose stripping modifications have been recently in the centre of interest in passionate debates on their proper classification and terminology [1977] offer an interesting alternative to other electrochemical measurements, which is also reflected in electroanalysis with CPEs—see the already cited studies (e.g., [13,26,91,92,108,190,209,210,254,341,366,588,626,629, 630,684,710,711]), contributions within this section (plus related items [1978–1999]), and some others in Chapter 8, surveyed in the respective tables.

Chronopotentiometry uses a three-electrode configuration in which a small constant current is impressed between the auxiliary and working electrodes, and the potential of the working electrode vs. reference electrode is measured as a function of time, that is, $E = f(t)$ and known as E–t curve(s). The operation is done in unstirred solution, and thus, an electroactive substance is transported by diffusion only [1978–1980]. From the corresponding chronopotentiogram, the so-called *transition time*, "τ," can be evaluated. During this interval, the concentration of the reacting substance on the

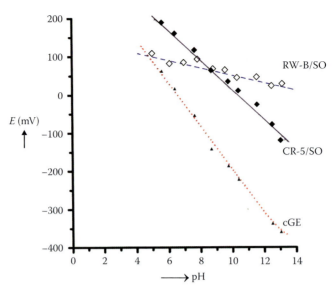

FIGURE 6.10 Calibration curves for pH-measurement obtained with two different bare carbon pastes and with the commercial pH-probe. CR-5/SO, full line: carbon paste from carbon gear lubricant and silicone oil, RW-B/SO, dashed line: ... from high-grade spectral graphite, cGE, dotted line: combined glass sensor (*Metrohm*). *Note*: Due to unstable response of the RW-B/SO, light points were plotted as "rough" average values. (According to Rejhonová, H., Development and testing of the dual sensors based on carbon pastes modified with Bi + Bi$_2$O$_3$ and Sb + Sb$_2$O$_3$ mixtures. MSc dissertation (in Czech), University of Pardubice, Pardubice, Czech Republic, 2011.

electrode surface decreases from the starting c^0 value to zero. Under conditions of linear diffusion with no kinetic complications, the relationship between the transition time and concentration is given by Sand's equation:

$$\tau^{1/2} = \frac{\pi^{1/2} \cdot n\, F \cdot A \cdot D^{1/2} \cdot c^0}{2 \cdot I} \tag{6.37}$$

where
  I is the applied current
  n is the number of electrons transferred at electrode reaction

$$\text{Ox} + ne^- \rightleftarrows \text{Red} \tag{6.38}$$

where
  F is Faraday's constant
  A is the electrode surface area
  D is the diffusion coefficient of the reacting substance
  c^0 is the starting concentration of the reacting substance

As evident from the equation, measurement of transition time can be used for determination of electroactive substances in the analyzed sample, but the method found a limited application only in practical electroanalysis [1981,1998]. Nevertheless, five papers employing CPEs in chronopotentiometric studies appeared in the literature [13,20,26,190,1999].

Much more important and frequent is stripping chronopotentiometry, introduced by Jagner and Graneli in 1976 [1982]. At that time, this novel analytical technique for the determinations of heavy metal traces was called potentiometric stripping analysis (PSA), because analyses based on the oxidation of species previously deposited on an electrode by oxidant currents convectively to the

Instrumental Measurements with Carbon Paste Electrodes, Sensors, and Detectors

electrode surface had not yet been included among electroanalytical techniques by the *International Union of Pure and Applied Chemists* (*IUPAC*). However, as admitted by its proponents themselves, the technique should be referred to more accurately as chronopotentiometric stripping analysis (again, see in [1977]). This alternative arose from ASV and, in both techniques, metals in a sample are electrolytically deposited on an electrode. The two, however, differ in the way in which deposited metals are stripped off and the analytical signal obtained. In PSA, no control is made of the potential of the working electrode during metal stripping, which is accomplished by using a chemical oxidant in solution (originally Hg(II) or dissolved oxygen). The working conditions are set in such a way that the rate of oxidation of deposited metals remains constant throughout the stripping process; such a rate is determined by the oxidant diffusion from the solution to the electrode surface [1982–1987]. Under these conditions, the analytical signal is recorded by monitoring the potential of the working electrode as a function of time; the distance between the two consecutive inflex points in a curve (stripping time) is proportional to the metal concerned in solution, whereas the potential of the central zone (formal redox potential, E^{0f}) is a characteristic measure for that.

Modern PSA instruments use microcomputers to register fast stripping events and convert the wave-shaped response to a more convenient peak over [1983–1985,1987]. Then, the number of counts can be registered in differential units, dt/dE (s·V^{-1}), when the peak area is evaluated in seconds, corresponding to the stripping time (see Figure 6.11 [1991]).

According to Jagner [1982] and some of his continuators (e.g., [1983–1985]), stripping potentiometry can be classified into two major modifications (i) PSA with chemical oxidation and (ii) constant-current stripping analysis (CCSA), both existing in a number of further variants proposed later on. Namely, they are as follows: *reductive PSA, adsorptive PSA, derivative PSA, differential PSA, kinetic PSA, multichannel-monitoring PSA, oxidative CCSA, reductive CCSA*, and, finally, *constant-current-enhanced PSA*. In this survey, however, some terms mean only the different naming for the already existing technique, which is the case of the latter being one of the CCSA variants.

In measurements, where the species accumulated at the working electrode are to be reduced, either chemical reductants or a constant current are applied in the stripping step [1982,1984,1987]. Whereas the first approach is seldom used (see, e.g., [1986]), the constant current with negative

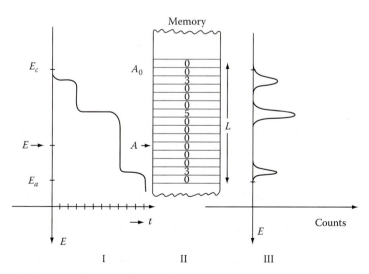

FIGURE 6.11 Example of a stripping potentiogram. (I) Potential vs. time dependence of working electrode during redissolution of three amalgamated metals in the potential window of $E_a - E_c$. (II) Computer memory section; the data storage area starting at address A_0 holds a record of accumulated clock pulse counts. Right (III): the resultant multichannel chronopotentiogram. (Reproduced from Mortensen, J. et al., *Anal. Chim. Acta*, 112, 297, 1979. With permission.)

polarity and very low intensity (at the µA-level) can be set and varied as required via the software of modern electrochemical analyzers [1983].

In principle, CPEs are applicable in both PSA and CCSA without significant limitations. However, compared with extensive voltammetric applications, CPEs have been combined with stripping potentiometry only occasionally [13,190,684,1988], and some valuable properties of these electrodes remained practically unnoticed. Among them, one can quote very favorable signal-to-noise characteristics of CPEs within a wide anodic potential range utilizable mainly for oxidation of organic compounds or unique interaction between the carbon paste and an analyte. As confirmed recently [254,710,711,1989,1990], for example, extraction and ion-pairing processes at CPEs are highly effective even in the PSA.

Perspectives of CPEs in stripping potentiometry have been discussed and some examples have been given for the accumulation of analytes using both electrolytic and nonelectrolytic deposition [210]. CPEs covered with mercury films are applicable for the determination of gold and some heavy metals [108,209,1992,1993], whereas with gold films for the determination of mercury, copper, and arsenic [108,209], the latter in both As^{III} and As^V forms (see also Section 8.1 and Figure 8.6).

A decade ago, a novel approach to the PSA measurements was presented based on total substitution of mercury by bismuth [626], and this intentional replacement had been innovative in two respects. First, bismuth(III) salts may be applied for forming bismuth films, and these films can be generated on such supports like CPEs. Secondly, Bi(III) salts may substitute Hg(II) in its role of an oxidant. This has offered new possibilities for determinations of trace heavy metals like Cd, Pb, etc., with a higher sensitivity, because Bi(III) is a "softer" oxidant than both oxygen and mercury(II). Taking into account all side reactions, the lowest value of the formal redox potential $E^{0f}(Bi^{3+}/Bi^0)$ was obtained in an acidic medium containing bromides as complexing ions. Thus, it was in accordance with the known relation that the weaker the oxidant, the higher the sensitivity of measurements in PSA. Then, the utilization of the bismuth(III) species as an "especially gentle" oxidant would lead to further improvements in the detection abilities of PSA and its overall analytical performance [626].

In 2007, antimony film electrodes were introduced to electrochemical stripping analysis [1995,1996], and analogous new procedures were presented utilizing antimony(III) salts, which can serve as an oxidant after electrodeposition of metals at antimony films generated onto CPEs. Compared with similar total substitution of traditionally used mercury(II) by bismuth(III), the use of antimony(III) offers even higher sensitivity in the detection of heavy metals [1994] as the weakest chemical oxidant from the three Hg^{II}, Bi^{III}, and Sb^{III} species. Of course, the corresponding constant current–based stripping chronopotentiometric technique was examined as well [1997].

Regarding PSA and its potentialities in practical analysis, yet another feature should not be omitted herein and is illustrated in Figure 6.12, making a comparison between the electroanalytical performance of both ASV and stripping potentiometry. This confrontation of the two related techniques was made with the same sample solution and identical working electrode, a bismuth film-plated CPE. As seen, both voltammograms (in the DPASV mode) had exhibited distinct deformations of the baseline with a large maximum (of so far unclear origin [84]), and the signals of interest were badly developed. In contrast, the respective PSA curves were much favorable and easily evaluable.

This test is a textbook case of how the result of an analysis may rely on the technique chosen for the stripping step. Since voltammetric measurements are based on the detection of electric current whose origin is quite diverse (e.g., possible trace electroactive impurities and their residual signal, non-faradic phenomena releasing also the electric current, electrical noise, and the ohmic current, etc.), the resultant signal may suffer—and usually does so—from numerous interferences. In contrast to this, the potentiometric stripping regime registers the E–t dependence (or its mathematical transformed analogue, respectively), where the equilibrium potential, E_{EQ}, represents a highly selective signal independent of any current-releasing disturbance [711,1984].

As a result, the interferences under comparable conditions were severe for voltammetric measurements but almost none in PSA. And this "insensitivity" of stripping potentiometry is also the

Instrumental Measurements with Carbon Paste Electrodes, Sensors, and Detectors

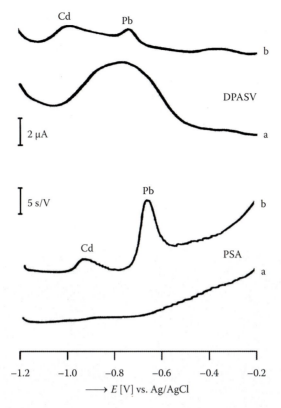

FIGURE 6.12 Anodic stripping voltammograms and stripping potentiograms of Pb(II) and Cd(II) obtained at bismuth film–plated carbon paste electrodes by analyzing a model sample of mineral acid–digested petroleum oil. DPASV, differential pulse anodic stripping voltammetry; PSA, stripping potentiometric analysis (with chemical oxidation); sample base-lines (a), sample + 50 μL 1×10^{-4} M Pb^{2+} + Cd^{2+} [a spike corresponding to $c(Me) = 2 \times 10^{-7}$ mol·L^{-1}] (b). *Experimental conditions*: silicone oil–based carbon paste; sample: digested crude oil in 0.9 M ammonia buffer + 1×10^{-5} M BiIII (pH 4.3); t_{ACC} = 120 s, t_{EQ} = 15 s, E_R from −1.2 to −0.5 V vs. Ag/AgCl; PSA: sampling frequency: f_{PSA} = 90 MHz, DPASV: v = 20 mV s^{-1}, ΔE = −50 mV. *Note*: Y-axis expressed as dt/dE [s/V] for PSA; and ΔI [μA] for DPASV. (Redrawn and rearranged from Švancara, I. et al., *Cent. Eur. J. Chem.*, 7, 598, 2009; Švancara, I. et al., *Sci. Pap. Univ. Pardubice Ser. A*, 10, 5, 2004.)

main reason why methods employing this technique or its CCSA variant, respectively, are very convenient for analysis of samples with extremely complex matrices. This was also the case of the experiment depicted in Figure 6.12, where the crude petroleum decomposed by microwaves had yielded a sample digest with extreme acidity about 3 M HNO_3 that had to be neutralized by intense buffering when containing also numerous residua of inorganic and organic origin.

6.5 NONELECTROCHEMICAL TECHNIQUES

6.5.1 Microscopic Observations

CPEs differ from other electrodes as carbon pastes are typical multicomponent materials exhibiting a rather unique microstructure due to the presence of a liquid binder. The straightforward way to observe the CPE surface is to resort to optical microscopy (OM; [617]), but the most comprehensive insights into the carbon paste microstructure were obtained with the aid of modern microscopic techniques, such as scanning electron microscopy (SEM; [96]) occasionally combined with x-ray microanalysis (XRM; [618,622]) scanning tunneling microscopy (STM; [172]), or atomic force microscopy (AFM; [620]). OM was mainly used to evidence macroscopic surface changes upon modification, as the effect of surfactant incorporation into carbon paste upon the coverage degree of graphite particles [243,1639,1661], or the generation of mercury films onto HgO-modified CPEs [617].

SEM observations were first devoted to the analysis of the surface state of unmodified electrodes to characterize the relations between the quality and mutual ratio of both main carbon paste constituents and the resulting electrochemical behavior of CPEs. In spite of the heterogeneous nature of the matrix, some "models" have been proposed (see [175] and references therein), which were mainly based on the influence of the carbon-to-binder ratio, considering two limiting situations and one intermediary case: (L1) at higher contents of binder, essentially all graphite particles are coated with a thin film of liquid; (Im) as the content of binder decreases, some "empty spaces" appear that are attributed to expansive air/oxygen gaps inside the paste; (L2) an extreme situation when carbon paste contains only a minimal amount of pasting liquid yet sufficient to stick the carbon particles together (the so-called "dry" limit; see also discussions in Chapters 2 and 3, as well as Figure 2.4).

The usefulness of SEM imaging is documented in Figure 6.13 and in its legend, which specifies the individual changes in the overall pattern of all the bismuth film microstructures. Here, it could be

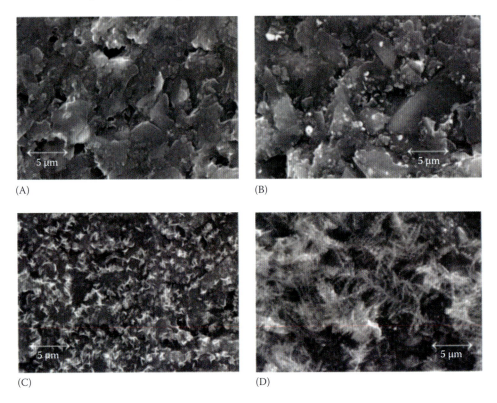

FIGURE 6.13 Three phases of the bismuth film microstructure and the respective changes in the overall morphology in dependence of the deposition intensity. SEM images. (A) CP substrate: the surface-microstructure pattern of carbon paste (CP) made of silicone oil and used for electrolytic deposition of bismuth films; (B) BiF, phase I: unconsolidated assembly of tiny crystallites forming a bismuth film (BiF, noticeable as small whitish objects); (C) BiF, phase II: initial crystalline structure growing from small agreggates; (D) BiF, phase III: the fully developed crystals with distinct cross-chaining growth and the spruce-branch structure. *Experimental parameters*: Deposition conditions: 0.05 M HCl + 0.15 M KCl (pH 2.5), E_{DEP} = −1.0 V vs. Ag/AgCl (I–III), t_{DEP} = 30 s (I), 60 s (II), and 180 s (III); scanning electron microscope (model "JSM-5500LV"; JEOL, Japan), magnification used: 1:3000. (Unpublished images from authors' archives; for further details, see Švancara, I. et al., Microscopic studies with bismuth-modified carbon paste electrodes: Morphological transformations of bismuth microstructures and related observations, in *Sensing in Electroanalysis*, Vol. 2, Eds. K. Vytras and K. Kalcher, University of Pardubice, Pardubice, Czech Republic, 2007, pp. 35–58; Švancara, I. et al., *Electroanalysis*, 17, 120, 2005.)

added that the respective study [173] has for the first time supported experimentally the formation of three consecutive structural models as well as a hypothesis on direct relation between the structure of Bi-films and their electroanalytical performance. It can briefly be stated as "the more developed the bismuth crystallites the worse the properties of the resultant BiFE with respect to its applicability in ESA."

The microdistribution of conductive and insulation regions on CPE surfaces can also be valuably characterized by means of STM. The scanning probe technique was indeed likely to characterize the very complex relief of carbon paste as "… valleys with hills and mountains (carbon particles) hiding numerous brooks, rivers, and waterfalls (pasting liquid)" [172].

Note that observing the carbon paste surface by STM resulted in somewhat "smoothed" images compared with the higher resolution of SEM imaging. This is most probably due to the fact that the STM detector collects only the impulses from conductive substrates, and the overall image reflects the structure of the individual microregions.

On the other hand, SEM can be an effective way to characterize the effect of chemical modifiers [117,243,399,618] or structural alterations after immobilization of biological tissues [619]. This was also considerably useful to follow the formation preplated metallic films to evidence diversities in the structure of palladium deposits [621], mercury-[189,617], gold-[1893], and bismuth films [172,622]. CPEs alone and with electrolytically deposited bismuth-[620] or palladium-[971] films have also been characterized by AFM, which was proven to be effective for scanning such delicate surfaces. Both intermittent contact and phase-shift scan modes provide useful information about nanodistribution and topography as well as "hard" and "soft" areas on the surface. For instance, phase shift scans of the same sections of a Bi-plated CPE depict regions of high and uniform phase shift that cannot be found on the bare CPE and have to be attributed to the newly formed Bi-film (see Figure 6.14).

6.5.2 Other Physicochemical Techniques for Special Characterizations

Various spectroscopic techniques such as *UV-visible spectrometry and fluorimetry, Fourier-transform infrared and Raman infrared spectroscopy (FT-IR and RA-IR), nuclear magnetic resonance (NMR), mass spectrometry (MS), Auger spectroscopy (AGS), x-ray photoelectron spectroscopy (XPS)*, as well as some other physicochemical methods like *microbalance weighing*, identification by *controlled crystallization, CNS-elemental analysis, nitrogen adsorption–desorption method, thermogravimetric analysis (TGA), x-ray diffraction (XRD), and x-ray fluorescence (XRF)* have been extensively used to characterize newly synthesized CPE modifiers (see concrete examples in [140–147,986,1049,1186,1801,1824,1833,1858]), but this is not the purpose of this section as this relies more on the intrinsic features of these modifiers (rather than those of the CPE itself), which might be very different depending on the modifier type (see Section 4.3).

Note that these techniques can also be useful to provide information on the (bio)chemical stability of modifiers once incorporated into or onto CPEs with respect to those of the pristine materials and, thereby, to help at optimizing the modified CPE preparation in order to maintain as good as possible the modifier properties that are intended to be exploited for electroanalytical purposes. For instance, UV/vis-S and FT-IR were often applied to show that redox proteins as haemoglobin or myoglobin kept their native structure when immobilized in CPEs or CILEs [99,664,1168,1730]. In addition to the aforementioned techniques, electrochemical impedance spectroscopy (EIS) was useful to characterize changes in charge transfer resistance; for example, enhancement of charge transfer rates has been reported for CPE modified with TiO_2 nanoparticles [920] or treated with surfactants [2000].

Optical spectroscopic techniques were also used in addition to electrochemical ones to study interactions between one component in the paste and another one in solution; see, for example, reaction of Cu(II) with humic acid [297] or interactions between colchicine and bovine serum albumin [978]. These works combine the results of optical and electrochemical data usually obtained separately. By contrast, examples of real combination of these techniques are available through spectroelectrochemistry (SPEC; [149,1093] or voltabsorptometry (VAb; [2001]) applied to CPEs, most often to elucidate the mechanisms involved in the electrochemical response of the electrode. Combination of spectrophotometric

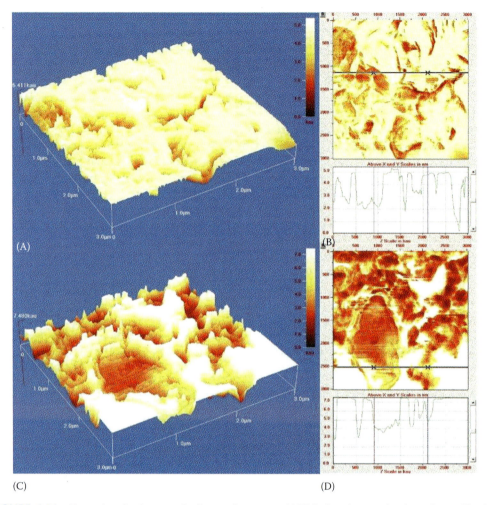

FIGURE 6.14 Example of using atomic force microscopy (AFM) for characterization of metallic-film modified CPE. PhaseMode scans of a bare (A, B) and a Bi-film-modified (C, D) CPE, same 3 µm × 3 µm section, z-scale 5.4 kau (A) and 7.5 kau (C); 3D images (A, C) and 2D images with profiles (B, D). *Experimental conditions*: scan frequency 1 Hz, Bi-film deposition in a 0.05 M H_2SO_4 + 1 ppm Bi^{3+} at −1.0 V vs. Ag/AgCl (3 M KCl) for 10 min after 10 s conditioning at 1.5 V vs. ref. (Reproduced from Flechsig, G.U. et al., *Electrochem. Commun.*, 7, 1091, 2005. With permission.)

and electroanalytical methods has been otherwise reported for the detection of biologically relevant molecules or drugs [1723,1724,2002]. Another application concerns electrochemiluminescence (ECL) sensors and CP-biosensors (e.g., [386,571,572,1119,2003]) for which the carbon paste matrix (or surface), including eventual modifier, was exploited to immobilize the ECL active reagents, for example, *o*-aminophthaloyl hydrazide (known as "luminol") and *tris*(2,2'-bipyridine)ruthenium.

The last category relies on carbon paste electroactive electrodes (i.e., CPEs with conductive binders such as a strong inorganic electrolyte or concentrated solutions of mineral acids or alkaline hydroxides), which were usually applied to study the behavior of redox-active solids (e.g., metal sulfides or metal oxides and hydroxides). Often such investigations were performed using additional characterization techniques before, in the course of, or after, the electrochemical transformation of the sample, including XPS [2004,2005], XRD and direct reflectance IR [2006], absorbance spectrometry [2007], and some others [2008].

7 Electrochemical Investigation with Carbon Paste Electrodes and Sensors

When looking back through the five decades of existence of the field [1,7], one can classify all hitherto reported research activities with carbon paste electrodes (CPEs), chemically modified carbon paste electrodes (CMCPES), and carbon paste (CP)-biosensors—including some work with carbon paste alone—into *two major categories* [83,188]:

(I) *Electrochemical investigations* that comprise either some theoretically orientated experimentation or highly specialized areas for which the instrumentation and applicability are so specific that the respective area is sorted separately
(II) *Electroanalytical applications* that concern more practically focused measurements associated with qualitative identification and quantitative determination of the individual groups of inorganic, organic, and biologically active compounds

Although the latter is largely reviewed in Chapter 8, the first group is represented by several particular topics whose significance within the electrochemistry with carbon paste–based electrodes is indisputable. The first is the extensive research on the electrode reactions of organic compounds reviewed for the first time here (in Section 7.1) as the authors' homage to the carbon paste inventor. Of particular attention should also be the solid-state electrochemistry with carbon paste electroactive electrodes (CPEEs) as well as voltammetry *in vivo*, covered in Sections 7.2 and 7.3 in honor of the fact that both disciplines have spawned together 250 remarkable scientific papers, that is, almost one tenth of all publications concerning carbon pastes. Finally, there are also special studies that, as such, are difficult to be unambiguously categorized and of which typical examples are gathered in Section 7.4.

7.1 STUDIES ON ELECTRODE REACTIONS AND MECHANISMS OF ORGANIC COMPOUNDS IN THE EARLY ERA OF CARBON PASTES

7.1.1 HISTORICAL CIRCUMSTANCES

In the early years of carbon paste in electrochemical laboratories, Adams and his coworkers carried out tremendous work on investigations of electrode behavior of various organic compounds, and a substantial part of the respective measurements were carried out with carbon paste–based electrodes.

This can be well documented by Table 7.1, offering an almost complete survey of the organic compounds that had been of interest in the early era of carbon paste, together with all the major topics that have been in the focus of the inventor's team over one and a half decade [1,8–10, 14–25,42–50,530,2009–2016], including the authors, trained within his group or collaborating with him, when some contributions appeared during their professional career [14,15,20,23,24,27,530].

Most importantly, the table shows clearly the concrete contribution of RNA's group to organic electrochemistry, which had its logical genesis (see historical introduction in Section 1.1). As can be seen, the research interest of pioneering propagators of CPEs had comprised mainly the electrode reactions

TABLE 7.1
Organic Compounds Studied with Carbon Paste–Based Electrodes and Sensors in the 1960s and Early 1970s

Compounds/Groups of Compounds (Abbreviation, Trivial Name) (a)	Subject(s) of Interest (Specification, Selected Details)	References
Native and substituted hydrocarbons: anthracene; 1-nitropropane; 2-nitropropane; *p*-nitrostyrene; nitrocyclohexane	Oxidations and reductions; Aq. soln (HCl, HCl + 25% AcT); Eval. of $E_{p(1/2)}$; reaction kinetics ("k_0"); adsorpt. ef.; Eld. conf.: std. CPE; CMCPE; Elchem. tech.: CV, LSV.	[9,14,2017]
Substituted aromatic amines: *p*-hydroxydiphenylamine; *N,N*-dimethyl aniline; *N,N*-dimethyl-*p*-phenyldiamine; *p*-methylene-bis-(*N,N*-dimethylaniline; *N,N*-dimethyl-*p*-phenylenediamine; 2,4-diaminodiphenylamine; *p*-ethoxy-diphenylamine; *o*-dianisidine; *N*-methyl-dianisylamine; *p*-anisidine; *o*-toluidine; *p*-toluidine; *N,N*-dimethyl-*p*-toluidine; *p*-nitroaniline; *p*-nitro-dimethylamine; *p*-chloro-nitrobenzene; *N,N*-dimethyl-*p*-nitrosoaniline; 4-amino-benzophenone; aminobenzene-sulfonic (orthanilic, metanilic sulfanilic) acids	Anodic oxidations; follow-up reactions (e.g., hydrolysis, hydroxylation, reoxidation); St. reversibility vs. quasi-reversibility/irreversibility; "$E_{P(1/2)}$ vs. pH" plots; Eval.of "z," "D_x," "k_0"; St. on extraction and adsorption; Exp. in aq. soln. (H_2SO_4, HCl, BRB, AcB, AmB) MeOH, AcN, DMSO, PLC; comp. with PtE, GCE, DME; Eld. conf.: std. CPE, µE, RDE, RRDE, RWE; Elchem. tech.: CV, LSV; PS, RCCP; Other tech.: EPR, ESR, UV-S, TLC	[1,8–10,14,16,20,25,27, 530,533,583,640,2009, 2013,2015, 2016,2018]
Substituted phenols, quinones, catechols (1): phenol, *p*-aminophenol; 3-methyl-*p*-aminophenol; *p*-dimethylaminophenol; *p*-methoxyphenol; hydroquinone (H_2Q); 2-methylnaphtho-hydroquinone diphosphate; 2-Cl- and 2-Br-4-aminophenols; 3-methyl-*p*-aminonaphthalene; catecholamines; 6-hydroxydopamine (6-OHDA)	Anodic oxidations; follow-up reactions (e.g., hydrolysis, hydroxylation, intratransfer r.); reversibility/irreversibility; St. of intermediates; St of. adsorption.; $E_{P(1/2)}$ - pH" plots; Eval. of "z," "D_x," and "k_0"; Comp. with PtE; Exp. in aq. soln. (HCl, H_2SO_4, $HClO_4$, HCl, BRB, AcB), in MeOH; Eld. conf.: std. CPE, RDE, RRDE, µE; Elchem. tech.: CV, LSV, PS, CP, CA, COU, *in vivo* EL; Other tech.: UV/VIS, EPR, ESR, MS, NMR	[16,17,21–25,42,43,45, 47,50, 2012]
Substituted naphtols and indols: 1-naphtol; 1-naphthylamine; 1-amino-7-naphthol; 1,5-dihydroxy-naphthalene; 1,4,5-trihydroxynaphthalene; 5-amino-2-naphthalene-sulfonic acid; 7-anilino-1-naphthol -3-sulfonic acid, 5-amidoindole	Anodic oxidations; follow-up reactions (e.g., hydrolysis); degree of irreversibility, eval. of "k_0"; exp. with periodic surface renewal; Eld. conf.: std. CPE, RWE; Elchem. tech.: CV, LSV	[8,27,2014]
Other substituted aromates: 1-nitro-biphenyl; 3,4-diamino toluene; 2,4-diaminotoluene; amino-azobenzene	Oxidations, reductions; follow-up reactions (hydrolysis); degree of irreversibility, $E_{P(1/2)}$ Eld.conf.: std. CPE; Elchem. tech.: CV, LSV	[8,9]

TABLE 7.1 (continued)
Organic Compounds Studied with Carbon Paste–Based Electrodes and Sensors in the 1960s and Early 1970s

Compounds/Groups of Compounds (Abbreviation, Trivial Name) (a)	Subject(s) of Interest (Specification, Selected Details)	References
Substituted carboxylic acids (1): p-nitro-benzoic acid, 2,5-dihydroxybenzoic and 2,4,5-trihydroxy-benzoic acid; 3,4-dihydroxyphenylacetic acid (DOPAC); dihydrooxymandelic acid (DHMA); homovanillic acid (HVA); vanillomandelic acid (VMA); EDTA (*Chelaton III*)	Anodic oxidations; follow-up reactions like hydrolysis; "$E_{P(1/2)}$ vs. pH" plots; Aq. soln. (BRB, AmB); Comp. with PtE, GCE; Eld. conf.: std. CPE(s), μE; Elchem. tech.: CV, LSV, PS, CP	[9,42,47,641]
Miscellaneous (2): triphenylmethane dyes (e.g., *Crystal Violet*, *Malachite Green*); dihydroxyphenylethanol (DHPE); dihydroxyphenylglycol (DHPG); 3-methoxy-4-hydroxyphenylethanol (MH-PE); 3-methoxy-4-hydroxy-phenylglycol (MHPG); ferrocene (Fc/Fc$^+$ redox pair)	Anodic oxidations; follow-up reactions (e.g., hydrolysis, intramolecular r.); St. of various intermediates; Eval. of $E_{P(1/2)}$; Comp. with PtE; Eld. conf.: std. CPE, RDE, CMCPE; Elchem. tech.: CV, LSV, PS	[15,47,2010,2011]

Remarks: (a) The individual compounds, the respective abbreviations, and classification groups are given in accordance with the original papers; (1) including BICs and (2) in chronological order.

Abbreviations and symbols *(in alphabetical order)*: AcB, acetate buffer; AcN, acetonitrile; AcT, acetone; AmB, ammonia buffer; BRB, *Britton-Robinson* buffer(s); conf., configuration(s); aq., aqueous; CA, chronoamperometry; Comp., comparison(s); COU, coulometry; CP, chronopotentiometry; CPE, carbon paste electrode; CMCPE, chemically modified carbon paste electrode; CV, cyclic voltammetry; DME, dropping mercury electrode; DMSO, dimethylsulfoxide; D_x, diffusion coefficient(s); e.g., for example; EL, electrochemistry; ef., effect(s); Elchem., electrochemical; Eld, electrode; E_P, peak potential; $E_{P(1/2)}$, half-peak potential; EPR, electron paramagnetic resonance; ESR, electron spin resonance; Eval., evaluation; Exp., experiment(s); GCE, glassy carbon electrode; k_0, rate constant; M, mol L^{-1}; MeOH, methanol; MS, mass spectrometry; NMR, nuclear magnetic resonance; LSV, potential scan (sweep) voltammetry; pH, measure of acidity, $-\log a(H^+)$; PLC, propylene carbonate; PS, potential stepping (voltammetry); PtE, platinum electrode; r., reaction; RCCP, reverse current chronopotentiometry; RDE, rotated disc electrode; RRDE, rotated ring-disc electrode; RWE, rotating wheel electrode; SCE, (reference) saturated electrode; soln., solution(s); St., study/studies; std., standard (normal); tech., technique(s); TLC, thin-layer chromatography; UV/VIS, spectrophotometry (in ultraviolet and visible spectrum region); V, volt.; z, the number of electrons; μE, microelectrode.

of aromatic systems, namely, four major groups: (i) aromatic amines, (ii) phenolic compounds (substituted hydroquinones and catechols), (iii) aminophenols, and (iv) catecholamines (aromatic compounds combining the structure of a catechol with a fragment of amino acid). Of comparable interest were also (v) imines, (vi) quinone imines, (vii) quinones, and (viii) related structures, as the individual oxidation forms, intermediates, or the subsequent hydrolyzation or hydroxylation products.

7.1.2 Adams's Investigation of Aromatic Compounds

Even from the aforementioned brief listing—and when considering that the respective activities had covered more than 50 different substances—it is practically impossible to describe comprehensively all the individual observations and results as well as all the nuances from measurements with CPEs. Instead, the principal conclusions from extensive investigations by Ralph N. Adams (R.N.A.) within the previously highlighted groups of organic compounds can be summarized in the following way:

(i) Aromatic amines. Presumably, anodic oxidation of aniline and some of its ring-substituted derivatives at CPEs in acidic media takes place according to the scheme [3]:

$$Ph-NH_2 \underset{+e^-}{\overset{-e^-}{\rightleftharpoons}} Ph-NH_2^{+\bullet} \quad (7.1a)$$

Formation of a radical ion

$$Ph-NH_2^{+\bullet} \rightleftharpoons \text{(quinoid radical cation)} \quad (7.1b)$$

Intramolecular rearrangement: dearomatization of the C-atom in position "4"

$$\text{(quinoid radical cation form A)} \rightleftharpoons \text{(quinoid radical cation form B)} \quad (7.1c)$$

Intramolecular rearrangement: transfer of the H-atom from pose "4" to "2"

continuing by condensation reactions:

$$2 \text{ (quinoid radical cation)} \longrightarrow Ph-NH-C_6H_4-NH_2 \quad (7.2a)$$

Dimerization I

$$2 \text{ (quinoid radical cation)} \longrightarrow H_2N-C_6H_4-C_6H_4-NH_2 \quad (7.2b)$$

Dimerization II

$$\text{(quinoid radical cation)} + Ph-NH_2^{+\bullet} \longrightarrow Ph-NH-C_6H_4-NH_2 \quad (7.2c)$$

Alternate coupling

In addition, the product of the alternate tail coupling (3.2c) can be further oxidized, forming the following:

$$Ph-NH-C_6H_4-NH_2 \underset{+2e^-}{\overset{-2e^-}{\rightleftharpoons}} Ph-N=C_6H_4=NH + 2H^+ \quad (7.3a)$$

Subsequent oxidation I

Finally, benzidine formed by the dimerization "II" (3.2b) also yields an oxidation product:

$$H_2N-C_6H_4-C_6H_4-NH_2 \underset{+2e^-}{\overset{-2e^-}{\rightleftarrows}} H_2N^+{=}C_6H_4{=}C_6H_4{=}NH_2^+ \quad (7.3b)$$

Subsequent oxidation II

Despite possible occurrence of these subsequent oxidations, both *p*-aminodiphenylamine and benzidine are typically final products of the respective pathway for oxidation of aniline, representing an ECE mechanism.

Here, the intermediate processes (7.1b) and (7.1c) are classified as rapid follow-up reactions whose pathway at CPEs can somewhat be moderated; nevertheless—reportedly—the overall reaction kinetics is still comparable to that known for standard Pt-electrode [8,9]. Regarding the derived structures, for example, *p*-substituted anilines, $X–C_6H_4–NH_2$, after oxidation they give $X–C_6H_4–N=C_6H_4=NH$—that is, the partially oxidized form of *p*-substituted aminodiphenylamine [3]. At carbon paste–based electrodes, the corresponding cyclic voltammograms (CVs) of such derivatives at CPEs—matching almost perfectly the individual nuances—then exhibit large single peaks whose $E_{P(1/2)}$ varies in an interval from +0.60 to +0.90 V vs. SCE [2015].

(ii) Phenolic compounds (substituted hydroquinones and catechols). In general, substituted hydroquinones (H_2Q) undergo notable anodic transformation to the corresponding quinone (Q) only if the respective substituent is a strongly electrophilic group, facilitating the effective attack by the molecule of water. Overall, the respective ECE mechanism can be drawn in the following way [42]:

$$HO-C_6H_3(R)-OH \underset{+2e^-}{\overset{-2e^-}{\rightleftarrows}} O{=}C_6H_2(R){=}O + 2H^+ \quad (7.4)$$

Oxidative formation of *p*-quinone

$$O{=}C_6H_2(R){=}O + H_2O \longrightarrow HO-C_6H_2(OH)(R)-OH \quad (7.5a)$$

Reaction rate controlled hydrolysis

$$HO-C_6H_2(OH)(R)-OH \underset{+2e^-}{\overset{-2e^-}{\rightleftarrows}} O{=}C_6H(OH)(R){=}O + 2H^+ \quad (7.5b)$$

Likely reverse reoxidation to *p*-quinone

Regarding the carbon paste electrode material, the studies with hydroquinone and its derivatives had also confirmed specific reaction kinetics at CPEs [2,68], when the redox systems studied exhibited moderate—or even markedly slowed—electron/mass transfer.

The whole phenomenon can be illustrated on two CVs depicted in Figure 7.1 and making a comparison of the electrode behavior of the quinone/hydroquinone (Q/H$_2$Q) redox system at two different CPEs (dashed and full lines) that have been made from identical carbon paste components but differed from the carbon-to-pasting-liquid ratio used (see also Section 2.2.1 and Figure 2.4).

As known, under normal conditions, the electrode transformation $Q + 2e^- + 2H^+ \rightarrow H_2Q$ exhibits markedly irreversible behavior at most solid electrodes [3]. In addition, both CPEs tested

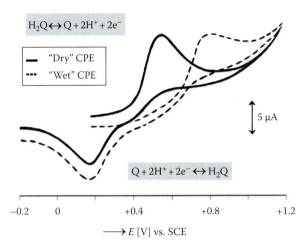

FIGURE 7.1 Cyclic voltammetry of the quinone/hydroquinone (Q/H$_2$Q) redox pair in acidic media and its behavior at two different CPEs. dry CPE: 0.5 g graphite powder (C) + 0.2 mL Nujol oil, wet CPE: 0.5 g C + 0.4 mL Nujol. *Experimental conditions*: CV: potential scan, +0.2 V → +1.2 V and reversely up to −0.2 V vs. SCE, scan rate: 50 mV · s^{-1}; model solution: 0.1 M H$_2$SO$_4$ (pH ≈ 0.5), c(H$_2$Q) = 5 × 10^{-3} mol · L^{-1}. (Unpublished data, an outtake from Švancara, I. and Schachl, K., *Chem. Listy*, 93, 490, 1999.)

behaved in a similar way, but the degree of irreversibility was incomparable. If this parameter is expressed as the difference of both anodic and cathodic peak potentials, $\Delta E = E_{P(A)} - E_{P(C)}$ (theoretically 59 mV for ideally reversible reaction with one-electron transfer), the experiment with "dry" CPE has given $\Delta E = 340$ mV, whereas CPE$_{(W)}$ resulted in a $\Delta E = 590$ mV. Although the Q/H$_2$Q redox pair is not an ideal model (due for some anomalies in the dependence of pH [3]), the whole assay has demonstrated that the increasing content of binder—represented by "wet" CPE—is mirrored in higher irreversibility and vice versa.

Compared with hydroquinone and its substituted homologues, the anodic oxidation of catechols leads to the *o*-quinones, which, in aqueous solutions, are known as unstable intermediates transforming themselves into the corresponding hydrolytic products such as 3-alkyl-4-hydroxy-*o*-hydroquinone:

(7.6)

Oxidative formation of unstable *o*-quinone

(7.7)

Reaction rate–controlled hydrolysis

The anodic hydroxylation reaction (7.7) is completed solely after a prolonged electrolytic oxidation, again, giving rise to a peak whose shape is markedly broader than that of the previous oxidation signal (corresponding to the *o*-quinone formation).

Apart from a potentially slowed reaction rate for these electrode systems at CPEs, the measurements with carbon pastes can also result in some irregularities, including the so-called quasi-reversibility (i.e., reversibility at higher scan rates [2,3]), associated with the already mentioned anomalies in the electrode behavior of the native hydroquinone.

(iii) Aminophenols. The chemical steps assumed to be involved in the oxidation of these compounds can be formulated by two different EC processes according to the scheme [17]:

$$HO\text{-}C_6H_4\text{-}NH_2 \underset{+2e^-}{\overset{-2e^-}{\rightleftharpoons}} O=C_6H_4=NH + 2H^+ \qquad (7.8)$$

Formation of quinone imine

$$O=C_6H_4=NH + H_2O \xrightarrow{+H^+} O=C_6H_4=O + NH_4^+ \qquad (7.9a)$$

Acidic hydrolysis (likely pathway)

$$O=C_6H_4=NH + H_2O \longrightarrow HO\text{-}C_6H_4\text{-}NH\text{-}OH \qquad (7.9b)$$

Neutral hydrolysis (alternate pathway)

Naturally, there were efforts to distinguish between the two possible mechanisms, which has also been examined with the aid of cyclic voltammetry at carbon paste–based electrodes, but the details are beyond the scope of this schematic description and can be found elsewhere [17]. In brief, the corresponding CVs at CPEs exhibit two well-developed peaks, the latter being markedly broader, thus implicating the reaction rate–controlled hydrolysis.

(iv) Catecholamines. As biologically active compounds whose physiological metabolisms can be simulated by electrochemical studies, catecholamines had been of particular interest in Adams' group over the lengthy years of their involvement in bioelectrochemistry. Basically, there are two major ECE pathways: (I) Oxidation of aliphatic amine side chain:

$$(HO)_2C_6H_3\text{-}CHR\text{-}CH_2\text{-}NH_2 + 2H^+ \xrightarrow{-e^-} (HO)_2C_6H_3\text{-}CHR\text{-}CHO + NH_4^+ \qquad (7.10)$$

when the unstable aldehyde formed is either further oxidized to acids ("acidic metabolites") or reduced to the corresponding alcohols ("neutral metabolites"):

$$(HO)_2C_6H_3\text{-}CHR\text{-}CHO + H_2O \xrightarrow{-e^-} (HO)_2C_6H_3\text{-}CHR\text{-}COOH \qquad (7.11a)$$

$$(HO)_2C_6H_3\text{-}CHR\text{-}CHO + H^+ \xrightarrow{+2e^-} (HO)_2C_6H_3\text{-}CHR\text{-}CH_2OH \qquad (7.11b)$$

Alternatively, (II) the respective catecholamine may undergo catalyst-assisted O-methylation in the position "3," when the resultant product, methoxycatechol

$$(HO)_2C_6H_3\text{-}CHR\text{-}CH_2\text{-}NH_2 \xrightarrow{\text{Cat.}} (CH_3O)(HO)C_6H_3\text{-}CHR\text{-}CH_2\text{-}NH_2 \qquad (7.12)$$

can again be converted by oxidation to an unstable intermediate in a reaction analogical to (7.10) and the aldehyde formed further oxidized or reduced according to the scheme (7.11a,b).

In living organisms, both mechanisms (I) and (II) are accomplished by means of specific enzymatic catalysis. In electrochemical measurements, the first pathway can be performed under more or less ordinary conditions, whereas the methoxylation has to be imitated by synthesis of the respective semiproducts. In addition, their subsequent electrode or reduction is more difficult, requiring higher voltages to be applied.

The electrochemical simulation of the proper transformation includes the use of aqueous media at physiological pH (phosphate buffers) and the presence of NaCl, ensuring the desired salinity and constant ionic strength [47].

7.1.3 Adams's Legacy

After Adams's ultimate anchoring in bioelectrochemistry and pharmacobiochemistry (see [45–50, 548,578,593,594], theoretically oriented electrochemical investigations of organic compounds appeared to a less extent. Nevertheless, a "fingerprint" of the carbon-paste inventor's research can be traced in the work of numerous continuators who had preferred CPEs in the organic electrochemistry contemporaneously or even later, during the rather stagnating 1970s. Among these activities, one should not forget to quote the studies conducted within scientific groups headed by Kitagawa et al. [640,641], Grandi et al. [2017], Farsang et al. [2018], Desideri, Lepri, and Heimler [582,583], Lindquist [40], Bobbitt et al. [251,517], and Sternson and Hes [2019–2021], the latter falling into the employment of CPEs in biological and medicinal research.

It can be summed up that the earlier presented information proves the unquestionable contribution of Adams' group to organic electrochemistry. In addition, within the field of electrochemistry with CPEs, the research of such an extent and focus is unprecedented and unique up until now. It can even be stated that every organic electrochemist interested in working with CPEs and newly confronted with the field should start with his experimentation by following the fundamentals from Adams's classical research.

7.2 ELECTROCHEMISTRY OF SOLIDS

As mentioned in the previous sections, CPEEs are defined as CPEs containing redox-active compound(s) inside the carbon paste matrix (or in the bulk of a carbon paste), embedded in the form of sparingly soluble or insoluble solids. CPEEs are particularly suited to (i) the electrochemical characterization of redox properties of solids, (ii) the analysis (qualitative and quantitative determination) of an element in the solid, and (iii) a combination of (i) and (ii) to investigate the oxidation states and relative amounts of selected elements in complex matrices as well as to study their interaction with other species (i.e., those likely to react or adsorb on their surface). Most works have been performed using an electrolytic binder, but some extensions to nonelectrolytic binders are also described.

7.2.1 Historical Introduction

The concept as such was introduced by Kuwana in the early 1960s [14,15] who studied organic compounds dissolved in the binder of traditional CPEs made of bromonaphthalene [14] or paraffin [15]. Some years later, the same idea was extended to insoluble minerals (sphalerite, galena, and covellite) in a bromonaphthalene-based CPE [305,2022]. The real breakthrough in the research came in 1974, when a group of French electrochemists, Gaillochet et al. [38,304], described the first CPEE prepared from an electrolytic binder and containing an insoluble electroactive compound as the sample. (Some interesting and unpublished notes behind the discovery of CPEEs can be found in the thesis of the named author [2023], defended a year later and summarizing his activities in the new area.)

Electrochemical Investigation with Carbon Paste Electrodes and Sensors

Note: As indicated in Section 2.2.2, the priority in the very first publication on electroactive carbon pastes is also being questioned [2022], as some review articles consider [151,169] or, at least, admit [189] that the aforementioned paper by Barikov et al. [305], originating from the extensive solid-phase research in the former U.S.S.R., was the true pioneering report in the area.

In such a way, the electrochemical response is no more dependent on diffusion but only on the volume of paste and amount of electroactive material (at sufficiently low potential scan rates; see the following), and the advantage of total transformation of the sample (easily achievable with CPEEs) makes the method quantitative on the basis of the Faraday law without the need of further calibration. Later on, another progressive method was developed by Sanchez Batanero et al. [2024] enabling to distinguish between the electrochemical reactions involving solids (i) admixed or (ii) dissolved in the electrolyte of a CPEE. Nowadays, many kinds of solids (mostly metal oxides, metal sulfides, and related minerals, as well as metal-based catalysts) are studied using CPEEs (see the following), and one can state that the method significantly extended the electrochemical characterization of poorly conducting and insoluble solids [306], thus contributing to broaden the scope of solid-state electrochemistry (along with the recently developed "voltammetry of microparticles" method [307]).

7.2.2 Main Principles, Configurations, Possibilities, and Limitations

Studying the electrochemical behavior of insulating and insoluble compounds requires a triple contact between these solid compounds, the electrode surface, and the electrolyte solution to enable electron transfer reactions.

This can be achieved by CPEEs made of a mixture of graphite powder (or carbon black), the particulate compound of interest, and a conducting binder (typically an electrolyte solution [mainly made from mineral acids or even salt solutions] or a conducting polymer). Sometimes, a nonconducting binder can be also used, but in that case the electrochemical reactions are most often restricted to a restricted depth of the outermost electrode surface. The original and most common electrode configuration is illustrated in Figure 7.2. The paste containing all the components (graphite, sample solid, and binder) in a mixture is introduced into a glass or PTFE tube equipped with a current collector and compacted by pressing onto a glass frit, which then serves to establish the contact between the electroactive electrode and the external solution while preventing leaching of the paste constituents.

Two other configurations, yet less widespread, have also been described (see Figure 7.3). The first one involves the dispersion of carbon powder and the sample electroactive solid along with an

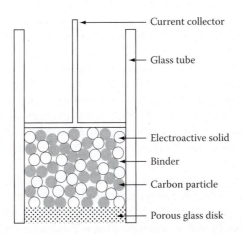

FIGURE 7.2 Classical configuration of carbon paste electroactive electrode (CPEE). A schematic view.

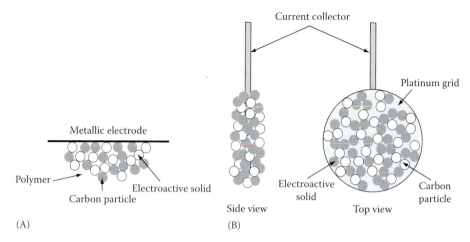

FIGURE 7.3 (A) Two less common configurations for CPEE. (B) A schematic view. (For details, see inscriptions.)

insoluble polymer in an appropriate organic solvent (usually tetrahydrofuran or dichloromethane) and the deposition of a drop of this suspension onto a solid electrode surface, and the resulting evaporation of the solvent leads to the formation of a CPEE as illustrated in Figure 7.3A. Of course, the use of such an electrode is restricted to solvents in which the polymeric binder is not soluble.

The second configuration is a binder-free CPEE (see Figure 7.3B) obtained by pressing a carbon + solid sample mixture onto a metallic grid (Pt or Au) under 2 or 3 tons of pressure for some minutes to form a composite pellet likely to be used in any solvent. This configuration suffers however from some lack of mechanical stability when used for a prolonged time in solution.

Traditional cyclic voltammetry (CV) with linear-scan potential ramp is often used to characterize the redox properties of incorporated solid samples. By adopting careful operational conditions (low amount of electroactive material, i.e., a few percent, potential scan rates typically lower than 0.1 mV s^{-1}), the typical shape of cyclic voltammograms obtained with CPEEs in the simplest cases (reversible electron transfer, strictly solid-state transformations, no soluble species) is relevant to thin-layer voltammetry (symmetric anodic and cathodic peaks located at the same potential; see peaks A_1 and C_1 in Figure 7.4 as an illustrative example for chalcocite (Cu$_2$S) oxidation [2025]). Integration of these peaks enables via the Faraday law to determine the amount of electroactive centers. In most cases, however, voltammograms of complex shape are obtained as a result of contributions not only from the solid but also from soluble electroactive species [2024]; dissolution kinetics can also affect the cyclic voltammograms, making the interpretation of the data somewhat complex.

Basically, one knows the amount of electroactive material present in the paste and, in the presence of an electrolytic binder, this quantity is likely to be transformed. This is not always true, as the charge efficiency of CPEE depends on various parameters such as the paste composition (relative ratios between carbon, solid, and binder), the nature, the structure and morphology of the redox-active compound, and the electrode configuration. It has been reported, for instance, that some iron oxides can be totally transformed whereas some others may suffer from lack of charge transfer efficiency [2026,2027]. Too large loadings of CPEEs with electroactive solids (e.g., 10%) also cause decreases in reaction yields, as reported for the analysis of ferrocene [38,2028].

The quantitative aspects have been fully studied by Lamache, and conditions have been defined for which characteristic voltammetric and chronoamperometric curves can be used in qualitative and quantitative analyses of solid compounds using CPEEs [2029]. Fundamentals and limitations for the application of CPEEs in the electroanalysis of solids are also discussed using Cu and TiC as model solids, showing the critical influence of the carbon powder type, the quality and content of the binder, the size and shape of carbon and solid sample particles, as well as the paste preparation with respect to homogeneity and reproducibility [2030].

Electrochemical Investigation with Carbon Paste Electrodes and Sensors

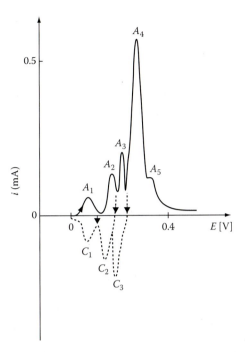

FIGURE 7.4 Current–voltage curve of chalcocite (Cu_2S) in the bulk of a CPEE. *Experimental conditions*: Cu_2S (0.6 mg), H_2SO_4 1 M (40 mg); carbon (50 mg); scan rate: 10^{-4} V · s^{-1}. (For further parameters and details, see Lecuire, J.-M., *J. Electroanal. Chem.*, 66, 195, 1975.)

7.2.3 Survey of Practical Applications

With respect to the electrochemical behavior, there has been a whole series of various electroactive solids characterized by using CPEEs. They include the following:

- *Single metal oxides*, such as iron oxides (Fe_2O_3 [2027,2031–2033], Fe_3O_4 [2033,2034], or several Fe_xO_y in the same study [148,2035,2036]); copper oxides (CuO [2033,2037–2039], Cu_2O [402,2033,2038,2039]); lead oxides (PbO [2040], PbO_2 [2031,2040,2041], mixture of PbO_x for ceramic analysis [432]); manganese oxides (MnO [2042], MnO_2 [616,2031,2034,2042], Mn_3O_4 [2042], or various Mn oxides [2044,2045] or a manganese ore [2046]); vanadium pentaoxide (V_2O_5 [2047], or several others (V_2O_3, V_2O_4, $NaVO_3$, $VOSO_4$) in one joint study [2048–2051]); In_2O_3 [2052]; Bi_2O_3 [2053,2054]; Ag_2O [2037]; SnO_2 [2055]; Co_3O_4 [2056]; PtO_2 [2057]; or metal oxides in mixture (V_2O_5 + MoO_3 [2058])
- *More "Sophisticated" metal oxides*, such as metal ferrites (containing copper [2033,2034,2059,2060], manganese [2034], lead [2061], cadmium [2062], or nickel [2077]); supraconductors (e.g., in order to define the state of Cu-atom in $Y_ABa_BCu_CO_D$ [539,2038], Bi-state in $Bi_ASr_BCa_CCu_DO_E$ [539,2053], Pb- and Cu-state in $Bi_APb_BSr_CCa_DCu_EO_F$ [2064,2065]); $BaPbO_3$ [2040]; vanadium-doped zirconia $VZrO_2$ [2066]
- *Sulfide-based minerals*, such as sphalerite ZnS [305,2067–2070] or marmatite ZnS [2071]; galena PbS [305,2007,2072–2075]; covellite CuS [305] and chalcocite Cu_2S [616,2076]; chalcopyrite $CuFeS_2$ [2077–2080]; arsenopyrite FeAsS [2081]; arsenic sulfides (As_2S_3 and As_4S_4 [2075]) or nickel sulfides [2082,2083]; cinnabar HgS [2084]; metal sulfide concentrates [2069,2085]; or the even more complex sulfide $AgCrTiS_4$ [2086]
- *Other chalcogenides*, such as PbSe [2087]; CuSe [2088]; Na_2SeO_3 [2089]; Ag_2Se, Ag_2Te, and Ag_2S [2090]; $AuTe_2$ [616]; or several Mo-chalcogenides [2091]

- *Other less common solids*: $NiAl_3$, Ni_2Al_3, and Al [2092]; a series of rare earth borides [2093]; iron carbide Fe_3C [2094]; $BaCrO_4$ [2095]; As^0, As^{III} [2096], and Rh^{III} [2097]; *Ilmenite* ($FeTiO_3$ [2098,2099]); Zn powder [2100,2101]; or iron nitroprusside [2101]

Most studies were devoted to the basic characterization of the redox properties of selected elements in solid samples as well as to the observations and interpretation of their dissolution behavior in various media.

Examples of the first category are, for example, the study of metal sulfides in mineral processing [2007,2067,2072] or in hydrometallurgy [2075], characterization of metal species, and the elucidation of oxidation states of metallic centers in complex solids such as metal ferrites [2033,2034,2060,2061,2063] or supraconductors [539,2053,2064,2065]. Examples of dissolution studies involve iron and chromium oxides [2027,2059], other metal oxides (e.g., Mn_3O_4 [2043] or Co_3O_4 [2056]), various metal sulfides [2069,2078,2084] for which the effect of chloride on their dissolution was often investigated [2073,2077,2080], or less common solids such as $AuTe_2$ [616].

Another category of redox-active substances is that with organic compounds, such as 4-aminobenzophenone [14], hydroquinone [2102], naphthoquinone derivatives [1449], anthraquinone derivatives [1455,1809], insoluble quinones [2104]), or organometallic (ferrocene [14,2028,2105]) species, which were introduced into the paste as solids often likely to be solubilized in the CPE binder. In addition, CPEEs can also be exploited for the analysis of interactions between electroactive solids and selected species in solutions (e.g., Ag^+ and Hg^{2+} on pyrite [2106] or corrosion inhibitors in Bronze patina [2107]). Note that if most investigations were based on conducting binders, the extension of the CPEE approach to nonconducting binders (e.g., polyethylene [77], Nujol, or silicone oil [2105]). Finally, the method was also compared with the so-called "abrasive voltammetry" (in which microparticles of the sample are mechanically immobilized onto the surface of a solid electrode [307]), showing similar behavior (e.g., for lead oxides [2040]), the main difference being the leaching of soluble species in solution for the latter.

CPEEs are also attractive from the analytical point of view as they enable quantitative determinations, usually of one element or one compound in a solid sample. Examples are available for U in ores [304], TiO_2 in solid cassiterite samples [2108], free CuO in $Y_ABa_BCu_CO_D$ [2109], Pd in oxidation automotive catalysts [2110], Ge in semiconductor materials [2111], $V^{(III,IV,V)}$ in chloride melts [2050], Cd in bentonite [1078], $Pb^{(II,IV)}$ in medieval glazes [1252], Ni in fly ash [2082], Mn in ores [2046], Fe^{III}/Fe^{II} ratio in archaeological ceramic materials [1253], iron oxides in geological samples [148,2036], or As in contaminated soils [2112].

Even if most applications rely on solid-state analysis, an example of preconcentration electroanalysis is also available: Pb^{II} ions have been accumulated from a solution (in the 10^{-5}–10^{-9} M range) by ion exchange at a CPEE containing $HClO_4$ as the binder before their voltammetric analysis [2113].

7.3 ELECTROCHEMISTRY *IN VIVO*

7.3.1 RISE OF THE FIELD AND ITS POSITION

According to a recent review by one of the leading research teams in the area of electrochemical measurements in living organisms [606], the priority in such experimentation can be attributed to two studies from the late 1950s by Clark et al. [2114,2115], who had been using linear sweep voltammetry for *in vivo* detection of oxygen and ascorbic acid in brain tissue. Nevertheless, the overall concept and strategy for voltammetric *in vivo* monitoring were first laid by Adams with collaborators in the early and mid-1970s [48–50], which is—besides the earlier reviewed investigations of electrode mechanisms of organic compounds—another heritage of the carbon paste inventor for modern electrochemistry summarized in his last two publications and an edited textbook [2116–2118].

The running time soon saw his continuators but not necessarily from the KU bioelectrochemical base. Surprisingly, abundant research within the area of *in vivo* electrochemistry with CPEs has also been established thanks to extensive investigations by Blaha et al. (who had begun in the late 1970s [595] regularly publishing up to the early 2000s [603]), O'Neill et al. (starting in the early 1980s [598] and active up until now [2119]), or newly by Ly et al. [376,390,409].

Thus, with a collection of contributions from other authors (see, e.g., [611,1108,1717,2120–2125]), the specific area of *in vivo* applied CPEs has already spawned around 100 papers, a major part of which is also cited herein—see the reference list and the entries [48–50,152,376,377,390,396,409,430, 593–606,611,1022–1028,1108,1629,1642,1662,1717,2116–2154]. Others can then be anticipated in the near future in connection with the dynamic progress of new carbon pastes and brain neuroscience.

7.3.2 Basic Characterization of *In Vivo* Measurements

For *in vivo* measurements, various carbon paste–based mini- or microelectrodes (CPm/µEs) are used, with the surface diameter in the range of 50–300 µm [50,152,606], which is also the case of a microtip of otherwise nearly half a meter long electrode described by Wang et al. [600]. In fact, the variability of implantable CPm/µEs is rather in construction than in the carbon paste composition.

Thus, traditional mixtures of graphite powder + Nujol have largely prevailed (see, e.g., [152,600,606] and references therein), and only a limited series of reports quote other mixtures, such as those made of silicone oil [598,2143], fluorocarbon oil [611], or new kinds of paste-like composites from carbon nanotubes [376,377,390,396,409]. With respect to the construction, a simple electrode can be made from a miniature glass capillary, where the carbon paste is additionally fixed with epoxy resin ([50]; see also Section 2.3.6 and Figure 2.13).

More sophisticated design is an implantable CPm/µE (shown on a schematic picture in [605]) constructed from PTFE-coated thin wire. One end of this electrode is devised with soldered gold clip (for connection with potentiostat), whereas the other hides a small cavity for carbon paste obtained by peripheral cutting of the plastic sleeve near the gold contact, followed by careful pulling of the insulation out for 0.5 mm over the wire (from silver or Ir-Pt alloy). The carbon paste can then be packed into the obtained cavity using a piece of the same wire, again, with removed insulation at one end.

CPm/µEs for *in vivo* measurements could be operated via the native carbon pastes as well as various chemically and biologically modified variants. Native (unmodified) pastes have been used as such [50,2126] or treated electrolytically by anodization (see [63,68] and Section 3.2) as well as chemically via "surface erosion" [64,243] by surfactants and lipids (e.g., stearate [1028,2121,2133]), or even by the measured tissue alone [1027]. Yet another surface modification was the case of Nafion protective layer [1717], whereas the most typical approach for carbon pastes—the bulk modification—was successfully accomplished with natural clay (zeolite [1108]), some enzymes [1450,2124,2125], albumin [2134], or a piece of spinach leaf [2123].

Experimentation with implanted electrodes is traditionally connected with *in vivo* voltammetry (IVV) [605,606]. Besides voltammetric measurements with a potential ramp in the linear sweep/scan (LSV; e.g., [2135]), staircase (SCV; [2143]), and differential pulse (DPV; [597]) modes, amperometry [2145] and chronoamperometry (CA; [2124]) complete the techniques of choice for *in vivo* applications. Due to extremely low currents to be measured (in the nA range or even below [2135]) with possible signal oscillations and baseline drift, less favorable signal-to-noise ratio due to higher background in complex matrices of biological fluids and tissues, and mainly because of very close peak-potential characteristics of principal compounds of interest (e.g., catechols vs. ascorbate, DOPAC, or uric acid), "current vs. potential applied" ("I_i vs. E_{APP}") measurements are often combined with computer processing of the resultant signal(s) by means of the so-called deconvolution [2155,2156], using mathematical transformation of the experimental data by using semidifferentiation (or semiderivation, respectively; [2157–2159]) and, eventually, the baseline subtraction (i.e., correction or even total elimination of undesirable background [1937]). For other information on these techniques, go to Section 6.2.1.1). In IVV assays (e.g., [596,598,2121,2131]), these operations enable one to (i) evaluate abnormal responses, such as overlaps, background-obscured and badly developed peaks, and otherwise deformed signals) or (ii) suppress the unwanted background or residual noise. On the other hand, there are studies in which nearly the same is achieved by finding the optimal constellation of the measuring technique chosen, some key instrumental parameters, and inevitable data treatment (see [599] and the authors' selection of LSV, slow scan rate, and PC-controlled background subtraction).

In general, the recording characteristics of CPm/μEs applied *in vivo* are quoted to be quite different from those known for common measurements *in vitro*. In brain electrochemistry, these differences are attributed to (i) chemical modification of the carbon paste and (ii) the processes that deliver the substrate to the electrode surface [2138]. Chemical changes due to absorption of lipophilic organic molecules are seen in shifts of oxidation potentials of monitored compounds, whereas physical effects are mirrored in the changes in diffusion conditions (in the proximity of the electrode) and in reaction kinetics.

All these factors explain why *in vitro* characterization or even calibration of a CPm/μE is not possible and the quantification of the analyte has to be made *in vivo*, best, by the standard-addition method combined with time-controlled injections during real-time monitoring. A typical recording from the IVV measurements is depicted in Figure 7.5, representing a simplified redrawing of the already published experiment [2137].

The upper part of the figure shows the selected cut of the time-course recording obtained by monitoring the current response for HVA, a metabolite of dopamine (DA), in the striatum (subcortical part of the forebrain) of a rat whose motor activity has been registered in parallel. The whole assay took over a week, from which three daily cycles are shown here, simulated so as to keep the same time span for days and nights (see legend of Figure 7.5). Among others, the current flow

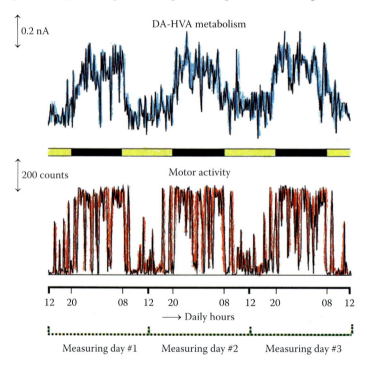

FIGURE 7.5 Time-course of changes in the linear-sweep voltammetric signal of homovanillic acid (HVA)* with a CPμE implanted in the brain of a freely moving rat recorded at regular intervals (12 min, blue plot) and with simultaneous measuring of the roden's motor activity (red plot). *Notes and experimental conditions*: (*) metabolite of dopamine (DA). Each day's recording started at 12.00 h, with lights off in an interval of 20.00–08.00 h, which is indicated by the yellow and black bands between both recordings; the period plotted on the X-axis is a time-span of day № 3–№ 6; the current intensity on the Y-axis was measured as the LSV peak for HVA; the measure for the lower record is the number of counts of the animal's movement measured *via* a low-intensity Doppler-shift microwave device. For details, go to the original literature [2137]. (In simplified form redrawn from O'Neill, R.D., *Sensors*, 5, 317, 2005. With permission of the author.)

monitored during the assay has confirmed that rats are nocturnal animals, showing a significantly higher level of motor activity in the darkness.

Regarding the detection as such, this experiment has also demonstrated the stability of the µE used and its proper function inside the brain [605]. Especially the long-term stability of CPm/µEs, which may attain—under favorable circumstances—up to several months of permanent use [601], is quoted to be maybe the most valuable feature of CPEs in connection with IVV and related applications.

As implied by the relatively frequent term "brain electrochemistry" [50], a great majority of measurements required—either acute (short-time) or chronic (long-term) [2129]—implantation of a CPm/µE into the respective parts of the brain of anesthetized animals, although experimentation without anesthesia is reportedly also feasible [2126,2140]. Mostly, the laboratory animals examined are rats (e.g., [598,1025,2152]), occasionally monkeys [600,2128], cats [2128], or even fish [37]. Other parts of the body for implantation were also reported (e.g., spinal cord [600] or rat tail [396]). In contrast to this variety, "nonanimal" applications of implanted sensors are very rare and, in fact, represented to date by a lone study with a CNTPmE penetrating the skin of an orange or an apple [409].

7.3.3 SURVEY OF APPLICATIONS AND SOME PROSPECTS

In the individual reports, a logical domain of *in vivo* measurements is the monitoring of biologically important compounds (BIC), and all typical applications are presented in detail in Chapter 8—with the typical experimental data gathered in Table 8.27.

Here, only an informative survey is given, specifying mainly which substances are of interest. One can start with catechols and catecholamines, that is, dopamine (DA; e.g., [48,376,602,605,1022,1108, 2130,2137,2153]) plus its metabolite homovanillic acid (HVA; [2119,2151]) and related L-DOPA [2119] or 6-OHDA (6-hydroxydopamine or oxidopamine [49]), and norepinephrine (syn. noradrenaline [2117,2120]) with their interfering "companions" ascorbic acid or ascorbate (AA; [597,1642,2135,2142,2151]), uric acid (UA; [2136,2141,2145,2151]), or 5-HIAA (hydroxy-indoleacetic acid [2141]).

The list continues with serotonin (5-hydroxytryptamine, 5-HT; [593,2146,2147]), glutamate (amino acid [2152]), choline [2152], glucose [611,1774,2152], and some pharmaceuticals such as *Acetaminophen* (syn. *Paracetamol* [600]), *Flunitrazepam* (a benzodiazepine anxiolytic [2139]) with hypnotics and abuse drugs (D-*Amphetamine* [603,2149]), caffeine [430], cocaine [2132], or morphine [1023]), the latter two studied via their pathway in the brain and their role in the metabolism of principal neurotransmitters.

The list of substances and species detectable *in vivo* can be completed by quoting the real-time monitoring of oxygen [2142,2145,2154], or recently introduced environmental monitoring with electrodes of the new generation, CNTPm/µEs (surveyed in Section 2.2.3), so far applied to some bioaccumulative pollutants, namely, insecticide "EPN" (i.e., ethyl-*O*-*p*-nitrophenyl-phenyl phosphonothionate [409]) and two heavy metal ions: Cd^{2+} [377] and Cu^{2+} [396]. Within the area of *in vivo* measurements, these less typical applications are important with respect to future prospects, showing that there are other promising routes for investigations with implanted electrodes that may lead to a wider diversity of the field, which is still anchored mainly in neuroscience.

7.4 SPECIAL STUDIES WITH CARBON PASTES

7.4.1 ROLE OF MINOR RESEARCH

Among the "remaining" research activities with carbon pastes, there are classified *special studies* whose impact on the field is ambivalent as having crossed both large categories, thus allowing one to classify such measurements additionally. By taking into account the overall abundance, they surely fall into a *minor research*; nevertheless, their outputs and achievements cannot be considered to be of marginal importance. On the contrary, numerous results manifest pretty well some less-obvious features of CPEs and the carbon pastes, which is a sufficient reason for the statement that even minor and specially focused investigations deserve adequate attention.

7.4.2 Selected Themes across the Field

7.4.2.1 Carbon Pastes as Substrates for Electropolymerized Films

As already emphasized in the historical survey and in association with the newest trends, the currently abundant applications of carbon pastes in the form of supports for electrolytically deposed polymeric layers cannot be overlooked. Since a pioneering report of this [2160], especially the last 2 years have seen a whole series of such studies [129–138,433], confirming—among others—the good adhesive property of hydrophobic carbon pastes for firmly attaching polymeric structures that alter the carbon paste surface with respect to its overall nature and quality.

The macromolecular structures are obtained by controlled polymerization of substituted anilines (e.g., *N*-methylaniline [132], *N,N*-dimethylaniline [129,133], and *o*-anisidine [2160]), including the native aniline [138], or other related compounds (e.g., *o*-aminophenol [131], 1,5-diaminonaphthalene [130], 5-amino-1,10-phenanthroline [433], indole [134], or an amino ester [136]). Compact films can be formed by means of *potentiostatic deposition* [129] or, more often, in the potentiodynamic mode—by *repetitive potential cycling* [132,133,2160] in solutions containing the respective monomer and, usually, more concentrated mineral acid (e.g., 5% H_2SO_4 [135] or 1 M HCl [138]).

In most cases, the polymerized films formed at the CP-surface serve to house the modifier(s) of choice (e.g., $Fe^{II/III}$ [129], Ni^{II} [130], Pt^0 [134], or *Patton & Reeder*'s modifier with $-SO_3^-$ and $-COO^-$ functional groups [135]). However, the same effect can also be accomplished by selecting the proper monomer—a compound that already contains the reactive functional group(s) in built inside the molecule (see [137] and 2-amino-5-mercapto-1,3,4-thiadiazole used therein).

In conclusion, the presence of polymerized layers at the CP-surface enables to increase both selectivity and sensitivity due to intentionally introduced electrocatalytical sites [129–133,136]), which can be realized by the polymer itself (via its functional groups [135,137] or via immobilization of special modifiers [130–132,134], including some more resistant enzymes (e.g., GOD [138,433]).

7.4.2.2 Carbon Pastes as Electronic Tongues

Some advantageous properties of carbon pastes could not remain unnoticed during the development of multichannel sensing assemblies, the so-called ETs or noses, respectively. The recent years have spawned a quintet of reports concerning carbon paste–based configurations [584,765,2161–2163].

As illustrated in Figure 7.6, sophisticated sensing systems of ETs [2164] are based on measurement and subsequent evaluation of multiple signals, when some specific features can be monitored, mathematically or statistically analyzed and correlated, for example, by means of the method of *partial least squares* (*PLS*; [584]) or using *principal component analysis* (*PCA*; [2162]), and thus ultimately distinguished.

Regarding the constructions with carbon paste, the respective assemblies have served for (i) qualitative analysis of potable waters, some soft drinks, and beers [584]; (ii) recognition of five fundamental tastes (namely, sweet, bitter, salty, acidic, and umami [765]); (iii) classification of a series of red wines through the release of specific compounds that characterize the process of aging of oak barrels and whose different properties has allowed one to identify also the country of origin [2163]); (iv) analysis of white wines depending on the local climate and different quality of grapes [2162], enabled by a sextet of carbon pastes modified with three rare-earth phthalocyanines (of the $Me^{III}(Pc)_2$ type) and three different perylenes. The last construction could then be employed for (v) evaluation of the overall bitterness of extra-virgin olive oils [2163], when the assembly of CPEs bulk modified with the individual olive oils differing in their bitterness reflect the redox behavior of the compounds of interest—in this case, electroactive polyphenols—present in the individual olive oils. These oils were mechanically admixed into a carbon paste of otherwise identical carbon-to-pasting-liquid composition and used for preparation of nine different CPEs forming the proper sensing array.

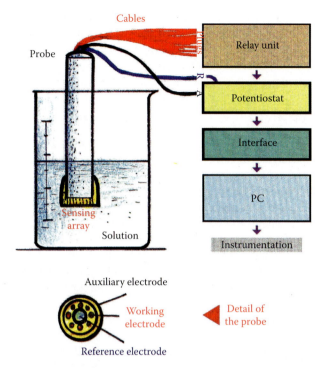

FIGURE 7.6 A possible construction of an electronic tongue/nose with the respective instrumentation and accessories. A schematic view. See inscriptions. (Sketched based on Kalvoda, R., Electronic noses and tongues, in *Modern Electroanalytical Methods—XIX. Book of Abstracts (in Czech), 8*, SES Logis, Usti nad Labem, 1999.)

7.4.2.3 Carbon Pastes as CPEEs for Electrochemical Characterization of New Electrocatalysts

Similarly noticeable activities within special studies have resulted in a collection of ca. 50 papers (see, e.g., [150] and the abstracts available therein) concerning the same topic—and, in fact, presented in almost identical style—that are devoted to the physicochemical and instrumental characterization of various newly synthesized substances. In many cases, these substances are inorganic or organic hybrids with more or less distinct electrocatalytic activity for the electrode transformations of various redox species, such as inorganic ions and molecules (namely, NO_2^-, ClO_3^-, BrO_3^-, IO_3^-, and H_2O_2) or some organic substances (trichloroacetic acid, CCl_3COOH) and compounds of biological importance (L-cysteine, dopamine, and ascorbic acid).

As can be seen in a series of typical publications [140–147,2165–2167], the sequence of instrumental characterizations by means of numerous optical and spectral techniques is usually ended by a cyclic voltammetric study employing carbon paste(s) with the new substance of interest incorporated in the bulk, that is, by means of solid-phase voltammetry (SPV) in combination with the CPEE. When considering the decline of SPV with CPEEs in the recent years (again, see the introductory part, as well as Section 7.2), these studies can be understood as the new chance for a wider applicability of a once popular kind of carbon paste–based electrodes.

7.4.2.4 Carbon Pastes in Industrial Use

Rather curious applications can be traced in the literature concerning the preparative electrolysis or catalyst-assisted industrial production. For instance, one contribution has dealt with *thin-layer graphite paste anode* [2168] in a rotated electrode chamber, where the surface of the paste could be permanently reactivated, thus enabling the electrolysis of highly anode-deactivating systems. Another article reports on *carbon paste briquettes* [2169] continuously consumed during the process—electrolytic production of fine Al, Si, and CaC_2. The last example is a special *carbon paste substrate* [2170] for a Cu-alloy catalyzing electrochemical reduction of CO_2. In all cases, there is only minimal association with traditional carbon pastes, the relation is mainly through the name itself.

7.4.3 Further Examples

A short survey of special studies can be completed with 10 studies [99,100,404,2171–2177] also documenting a very diverse use of carbon pastes. Thus, one can mention a (i) solar cell, based on naturally occurring *Chlorophyll* analogue, molecular oxygen (dissolved in aqueous solution), and a CPE yielding the measurable photocurrent [100]. We can also introduce a (ii) photosynthetic system based on a cyanobacteria and a benzoquinone-derived mediator, giving rise to the steady-state current signal for photoelectrochemical oxidation of water [99]. The next two examples are theoretical studies focused on (iii) surface inclusion [2171] and (iv) biosorption phenomena [2173]. Other two contributions (v) and (vi) concern the construction of electrostatic quadrupole lens systems with carbon nanotube paste composites in the role of emitters [404,2172]. A completely different topic is the (vii) multiinstrumental analysis of the so-called "environmental stress" [2174] reflected by a sunflower and its growth depression, changes in color, or deformities of root system, all correlated with the cascade of natural processes connected with photosynthesis. There are also two particular studies (viii) and (ix) that can be considered as the actual contributions to the use of new materials [2175] or dealing with the development of a novel CP-based biosensor [2176].

Finally, to come to a full circle, this section can be ended as it has begun: with a brief presentation of (x) another solar cell [2177]. This last example also mirrors the momentary preference of new materials because the proper operational unit—a solid-like carbon paste composite—has been made exclusively of such constituents, namely, carbon black, a conducting polymer, and an ionic liquid.

8 Electroanalysis with Carbon Paste–Based Electrodes, Sensors, and Detectors

Within this chapter, all hitherto-reported electroanalytical applications of unmodified carbon paste electrodes (CPEs), chemically and biologically modified carbon paste electrodes (CMCPEs and CP-biosensors, respectively), as well as other related sensing units, including those made of new carbon paste components, are surveyed as a whole. It can be revealed here that the originally planned concept—a systematic presentation of the individual methods for all major categories of analytes and via the selection of typical procedures, their commentary and discussion on possible applications—had to be abandoned in confrontation with the enormous information material needed to be processed in a comprehensive way.

Instead, but with comparable effectiveness, the same has been prepared by means of a series of summarizing tables, gathering all the typical data characterizing each method reported within half a century of CPEs in electroanalysis [1,7]. Thus, inside the tables, one can find the following: (i) the *target analyte* of interest, (ii) specification on *carbon paste composition* and *configuration* of each CPE employed, (iii) the *principle(s)* of every method, (iv) *instrumental technique* used, including possible *variants and measuring modes*, (v) basic *electroanalytical characterization,* (vi) specification of *samples in practical applications*, (vii) selected *experimental* and (viii) *instrumental parameters*, (ix) numerous useful *remarks*, and (x) the corresponding *reference(s)*. And all this information material is yet accompanied with the retrospective glossaries, highlighting the key facts, achievements, and contemporary trends, including actual examples and interesting trivia.

With respect to inorganic analytes, the individual methods have been gathered in the set of tables comprising all so-far concerned elements being divided into 11 groups, mostly in accordance with the periodical system of the elements. Although these surveys—that is, Tables 8.1 through 8.11*—have been built up exclusively for this issue, they inevitably include numerous information material and data that were already published in the previous, similarly structured reviews [55,69,82–84,112,161].

Regarding organic analytes and applications of CPEs, CMCPEs, and CP-biosensors for their determination, selected examples were of interest in some recent reviews devoted to organic electrochemistry in a wider perspective [2178], electroanalysis of organic pollutants [2179–2181], and exclusively to organic electroanalysis with CPEs [111,114,2182]. In this text, organic compounds as the second major group within the electrochemistry with CPEs are surveyed again in a series of summarizing tables:

1. Table 8.12 comprising *organic compounds* of environmental significance with various *native derivatives* and the respective *homologues*, plus *industrially important substances*.
2. Table 8.13 with *organic environmental pollutants* known under commercial names and/or trademarks (mainly, various *pesticides* and related substances used in agriculture).
3. Tables 8.14 summarizing pharmaceutical preparations, formulations, and abuse drugs; all of them forming a very diverse category of commercial products for which it is appropriate to them in alphabetical order with a brief specification of main therapeutic use.

---

* Tables 8.1 through 8.28 are at the end of Chapter 8.

4. Tables 8.15 through 8.28 with all the remaining organic compounds that can be classified as *biologically important compounds* (*BIC* [58]) and that have been predominantly—but not necessarily—determined with the aid of various CP-biosensors. For technical reasons, the respective table had to be divided into a series of sub-tables.

Yet before going to the individual areas of practical electroanalysis, it is useful to comment briefly on those features that make some distinct differences in practical analysis within the two main categories. Thus, under the optics of comparing inorganic analysis versus organic analysis, distinct differences clearly prevail over insignificant nuances. Whereas the determination of inorganic ions, complexes, and molecules with the bare (unmodified) CPEs is rather sporadic, both unmodified and modified CPEs are equally applicable in organic analysis. Unmodified CPEs are popular mainly in situations where a hydrophobic binder acts as a certain type of modifier, attracting lipophilic molecules of many organic and biological compounds, including various pharmaceuticals or drugs. Furthermore, due to a very close behavior of numerous organic substances, special pre-separation steps—by using HPLC, capillary electrophoresis (CE), and liquid–liquid extraction and solid phase extraction (LLE and SPE, resp.)—are often inevitable prior to the electrochemical detection. Or, in contrast to inorganic analysis, the substances of organic and biological origin are nearly equally analyzed in the batch arrangement and in flowing streams—with the aid of electrochemical detectors in flow injection analysis, liquid chromatography, capillary electrophoresis, and electrochromatography; i.e. in arrangements of the FIA-EC, LC-EC, CE-EC, and ECC-EC types. Also, in organic electrochemistry, practical measurements suffer more from the undesirable "poisoning" of the CPE surface by the interfering species whose voluminous molecules may totally block the electrode surface [3,82]. Whereas the protocols for inorganic analysis solve these problems with the use of protective membranes or via electrode regenerations [55], organic electroanalysis requires more radical solution, total isolation of the target analyte from the original matrix being the best. Finally, organic electroanalysis often does not require extreme performance of detection systems, except some special applications in environmental monitoring [2179,2180].

A particular case is then pharmaceutical and clinical analysis with CPEs (recently reviewed in [158,2183]) where the aforementioned problems are yet more pronounced due to the specific matrix of biological fluids, representing commonly analyzed samples. Although the respective procedures can be "relatively simple" with respect to the presence of interfering species, the electroanalytical performance as such is usually quite limited due to the disastrous effect of major matrix constituents. They can be electroinactive, but their excessive content is the source of serious troubles, including the aforementioned physical blocking of the electrode surface. Due to this, pharmaceutical and clinical analysis has always been a domain of a very narrow spectrum of research teams whose scientific reports are relatively well recognizable via their own characteristic, often empirical or even routine approaches to the problematics. Last but not least, pharmaceutical analysis is the only area where unmodified CPEs are regularly applied to date.

Nearly the same can be stated when one tries to specify analysis of biologically important (active) compounds. Of course, in electroanalysis with CPEs, there are different proportions and a wider impact because the category of BIC dominates among the analytes of interest for lengthy years. Typical is also the vitality of various specially oriented investigations (e.g., electrochemistry with/of DNA [2184,2185], *in vivo* monitoring [152,606], or the research on electrochemical immunosensors [2186,2187], with applicability rather in biology, toxicology, and medicine than in electroanalysis.

8.1 DETERMINATION OF INORGANIC IONS, COMPLEX SPECIES, AND MOLECULES

Inorganic analysis with various types of CPEs, CMPEs and CP-biosensors belongs undoubtedly among the largest areas of applications of carbon pastes in the electrochemistry. This can be documented in a myriad of publications mentioned earlier and within this section, throughout the Tables 8.1 through 8.11; the new ones being numbered [2188–2408].

Electroanalysis with Carbon Paste–Based Electrodes, Sensors, and Detectors

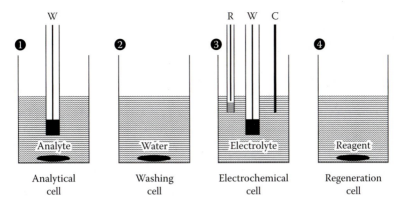

FIGURE 8.1 Schematic illustration of the four successive steps usually involved in a typical voltammetric analysis performed at a chemically modified carbon paste electrodes after accumulation. W, R, and C refer to the working, the reference, and the counter electrodes, respectively. ❶ open-circuit accumulation; ❷ rinsing; ❸ electrochemical detection; ❹ chemical regeneration.

Up until now, more than 70 chemical elements across the whole periodic table were of interest [7,83] and determined via the corresponding single ions, complexes, and molecules when using various CPEs, CMCPEs, and CP-biosensors in combination with voltammetric, amperometric, or potentiometric techniques.

Typically, the data presented in tables concern the sensing (practical) applications, while omitting investigations in which the inorganic analytes had served as model systems for testing some new devices or elaborating novel methodologies. Special attention was given to quantitative determinations and much less to qualitative aspects (e.g., demonstrating a catalytic effect, characterizing a new material as electrode modifier, etc.); some of the latter is presented in Section 4.3.

In inorganic analysis, the methods for quantification have predominantly employed various CMCPEs, allowing one to increase both sensitivity and selectivity of the respective method(s) by means of a special preconcentration step. Usually, it is an integral part of the respective electrochemical stripping method, involving the chemical accumulation of the target analyte by a suitable electrode modifier—see general operational scheme in Figure 8.1.

After such accumulation, the electrochemical detection can be accomplished by directly reducing the preconcentrated species, but the "electrolytic deposition/anodic reoxidation" scheme is often preferred due to the generally poorer pathways for cathodic reductions suffering from the undesirable interferences from molecular oxygen contained in the paste. In some cases, the preconcentration and the detection step can also be performed in different solutions, when the sample solution acts in the first phase for depositing the analyte either as such or in a form of suitable compound, whereas the proper stripping/detection takes place in a new medium—the so-called "blank"—being free and protected from the original matrix constituents and other unwanted residua.

And if needed, a special rinsing solution can also be added between the two main steps. Both these variants are known as "medium exchange" (MEx [55]; see also Section 6.2.2) which is quite a popular approach in inorganic electrochemistry, and much more frequent and almost obligate in pharmaceutical and clinical analysis, operating in particularly complex matrices of biological fluids, which is reviewed in Section 8.3.

8.1.1 Noble Metals

Au^{III}, Ag^{I}, Hg^{II} (or even Hg_2^{II}, resp.), and Cu^{II} species in aqueous solutions of definite composition (inducing selected speciation to the element) can be electroactive at CPEs (Au^{III} mainly as $AuCl_4^-$;

Ag^I as Ag$^+$; Hg^{II} as Hg^{2+}, Hg(OH)$_2$, or HgCl$_x^{[x-2]-}$; and Cu^{II} as Cu^{2+}). They can thus be directly detected via electrolytic reduction of the metal ion/complex M^N into a metal deposit M^0. As already explained, this reductive deposition is normally followed by re-oxidation, M$^0 \rightarrow$ M^{N+}; therefore, the entire "electrolytic deposition—anodic reoxidation" scheme has been the main principle of many methods and numerous examples are also included in Table 8.1 (with methods for determination of gold), Table 8.2 (silver), Table 8.3 (mercury), and Table 8.4 (copper).

Compared to the widespread applicability of CMCPEs, the use of anodic stripping voltammetry with unmodified CPEs is marginal due to the problems with closely lying stripping responses (i.e., with similar peak potentials) for all the four precious metals. Regarding the accumulation schemes, they usually involve the ion-exchange and ion-pairing, adsorption, and complex formation, the latter by using selective sorbents/ligands. Less common are then further processes such as bioaccumulation and extraction. Matrix samples were mostly model solutions for Au and Ag (+ photographic baths as particular case for Ag) while Hg and Cu have been often analyzed in real samples (such as natural or polluted waters, wastes or environmental samples, or biological matrices). As an example, one can cite the speciation of mercury as a function of chloride concentration [1061], which is likely to distinguish between the main mercury species that occur in the natural environments—from drinking, via lake, to sea waters, that is, from Hg^{2+} or Hg(OH)$_2$, via HgCl$_2$, to HgCl$_3^-$ or HgCl$_4^{2-}$. Interestingly, such speciation procedures led also to a possible joint detection of Ag$^+$ + Hg^{2+}, as demonstrated in a method [1322] utilizing glyoxal bis(2-hydroxyanil) as preconcentration agent. Thiol-containing modifiers also enabled simultaneous determination of Hg^{2+} and Cu^{2+} (along with Pb^{2+}; [1683,2223]). As diffusion is often the rate-determining step of the whole accumulation/detection process, attempt to enhance mass transport contributes to improving the sensitivity of the sensor.

In the particular case of CPEs modified with porous organic–inorganic hybrid materials (i.e., organically functionalized silica), it was shown that significant improvement can be obtained when using regularly structured mesoporous solids in comparison to their non-ordered homologues, especially at trace levels, as illustrated in Figure 8.2 for copper analysis at cyclam-functionalized silica-modified CPEs [2248].

A completely different study concerns the voltammograms depicted in Figure 8.3 that was obtained during the initial testing of a tricresyl phosphate–based CPE [205]. Some years later, the same set of DPV curves served as the experimental material for evaluating the stability constants, β_{MLn}, of the respective complex halido- and pseudohalido-aurates, [AuX$_4$]$^-$ (where "X" means F, Cl, Br, SCN, and CN).

Since equilibrium constants could be found in specialized handbooks (e.g., [2405,2406]), the aim of the entire study was to compare the experimental values with literature and to prove whether or not carbon paste–based electrodes with rather poorly defined surface can be used in such theoretical investigations. By adapting a formula from classical polarography that defines the shift of polarographic waves depending on the complexing ligand present [2407] and when considering the actual experimental conditions (0.1 M H$_2$SO$_4$ with pH 0.5, c(X$^-$) = 0.1 mol L^{-1}, v = 20 mV s^{-1}, and ΔE = +50 mV) and the known value of E^0(Au^{III}/Au^0), the following stability constants could be calculated (for details, see [189]) and successfully compared to the literature data (withdrawn from [2405] and given in parentheses):

$$\log \beta[\text{AuCl}_4^-] = 24.6 (\text{lit.}: 26); \quad \log \beta[\text{AuBr}_4^-] = 26.7 (27)$$
$$\log \beta[\text{Au(SCN)}_4^-] = 43.5 (43); \quad \log \beta[\text{Au(CN)}_4^-] = 57.7 (56)$$
(8.1)

8.1.2 Heavy Metals

In addition to Hg and Cu, other heavy metals (mainly Pb, Cd, and Zn, but also Tl, Sn, In, Bi, and Sb) have been determined using CPEs and related sensors—see Tables 8.5 and 8.6.

Electroanalysis with Carbon Paste–Based Electrodes, Sensors, and Detectors 173

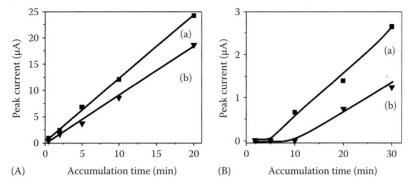

FIGURE 8.2 Influence of the silica structure incorporated into a carbon paste electrode on the voltammetric response of cyclam-functionalized silica samples to Cu(II), recorded as a function of accumulation time. (a) Mesostructured SBA15-cyclam and (b) non-ordered silica gel K60-Cyclam. Preconcentration was carried out in (A) 1.0×10^{-6} M and (B) 1.0×10^{-8} M Cu(II) solutions (0.1 M citrate buffer, pH 6.4). Square-wave voltammograms recorded in 3 M HNO_3 after 60s-electrolysis at −0.5 V. (Redrawn and rearranged from Stephanie, G.R. et al., *Electroanalysis*, 21, 280, 2009.)

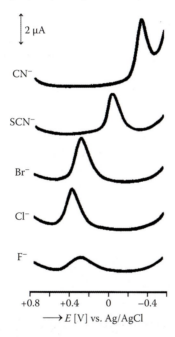

FIGURE 8.3 A set of voltammograms obtained by analyzing the $[AuX_4]^-$ complexes for subsequent evaluation of the corresponding stability constants, $\beta_{(AuX)}$. See inscriptions. *Experimental conditions*: CPE containing tricresyl phosphate as the binder ("C/TCP" type); DPCSV, deposition conditions: E_{DEP} = +1.0 V vs. ref., t_{DEP} = 30 s, scan rate: 20 mV·s^{-1}, ΔE = +50 mV; stripping limits: E_{INIT} = +1.0 V and E_{FIN} = −0.5 V vs. ref.; s.e.: (a) 0.1 M H_2SO_4 + 0.1 M KX (where "X," F, Cl, Br, and SCN; equimolar mixture (1:1), pH ~ 0.5–1.0); (b) 0.01 M KOH + 0.1 M KCN (pH ~12); c[Au(III)] = 1×10^{-5} mol·L^{-1}. (Assembled after Švancara, I., Carbon paste electrodes in electroanalysis, Hab dissertation (in Czech), University of Pardubice, Pardubice, Czech Republic, 2002 when using some data from Švancara, I. and Vytřas, K., *Anal. Chim Acta*, 273, 195, 1993.)

When possible, the detection by direct voltammetry or anodic stripping voltammetry, whether associated or not to a preceding chemical accumulation step, was applied but this was not always possible due to "poor electrochemistry" of some of these species at carbon surfaces and/or the respective large overpotential. For instance, reduction of Hg^{2+} can be quite easily performed on CPE, that of Cd^{2+} is still possible but the corresponding signal is situated close to the cathodic barrier, and that of Zn^{2+} cannot be detected at all on unmodified CPE, whereas all of these metal ions are easily determined using mercury electrodes (because of the extended cathodic potential range).

An elegant way to overcome this limitation is to resort to metal-plated CPEs (with mercury, bismuth, or even antimony films), enabling the reduction of metal ions in an amalgam form. As shown in Tables 8.5 and 8.6, the approach was sometimes used for Pb determinations, more often for the amperometric detection of Cd, and systematically for Zn electroanalysis (except in the case of potentiometric detection [1974]). Regarding the individual configurations, the heterogeneous character and a kit-like concept of carbon pastes naturally led to a variety of applicable configurations. The most common metallic-film electrodes as well as some related arrangements are shown in Figure 8.4.

Besides the most frequent variants of BiF-CPE, Bi-CPE, and SbF-CPE described in several reports, this figure also includes a rather curious configuration of *bismuth paste electrode* (*BiPE* [266]) that is—to the authors' knowledge—the only example of functioning paste-like mixture that does not contain any carbon/graphite moiety. (Despite its rather poor performance, the BiPE seems to be the first—but still very little—step in seeking the way how to devise a bismuth dispersion–based sensor [268] that could be an environmentally friendly alternative to Heyrovsky's dropping mercury electrode [2408].)

A huge amount of work has been made on lead (and cadmium; see Table 8.5) because of the presence of these toxic species at nonnegligible levels in various natural samples. Methods are available for analysis of several aqueous media, including natural (tap, river, lake, sea, ground, and well) waters and wastewaters (polluted waters, industrial effluents, and electroplating baths), solid samples (sediments and food products), or biological matrices or pharmaceuticals. More specific samples are atmosphere (especially for Cd), human hair (Pb and Cd), and gunshot residues (Pb) [807,1836,2259,2275]. Again, as for Hg and Cu, better sensitivity and selectivity was achieved when applying a preconcentration step prior to the electrochemical detection. This was made possible by means of electrode modifiers exhibiting complexation, sorption, or ion exchange properties. Molecular ligands or chelates can be used in principle but better long-term stability was obtained when incorporating solid macromolecular compounds (e.g., organic resins and organic–inorganic hybrid solids). Many methods were applicable to the simultaneous detection of lead and cadmium, sometimes also with zinc [103,617,623,2283].

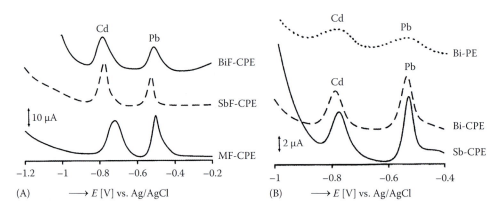

FIGURE 8.4 Square-wave anodic stripping voltammograms of Cd^{2+} + Pb^{2+} model mixtures at different types of metal film–plated and metal-modified carbon paste and paste electrodes. MeF-plated (A) and Me-modified (B) electrodes; the individual types as inscribed. *Experimental conditions*: silicone oil–based carbon paste [the same for all, except BiPE made of bismuth powder (50%, m/m) + silicone oil (SO); content of both metals in Bi-CPE and Sb-CPEs, 17% (m/m)]; SWASV: f_{SW} = 25 Hz, ΔE = 50 mV, s.i. = 4 mV; s.e.: 0.01 M HCl (pH 2); c(Hg,Bi,Sb) = 1 × 10^{-5} mol·L^{-1} (for A); c(Cd,Pb) = 50 ppb (for A), 25 ppb (for B); accumulation conditions: E_{ACC} = −1.2 V (A), −1.0 V vs. ref., t_{ACC} = 120 s, t_{EQ} = 15 s (B); potential limits, from −1.2 V to −0.2 V vs. ref. (Redrawn and rearranged from Švancara, I. et al., *Cent. Eur. J. Chem.*, 7, 598, 2009.)

In addition, other metallic species (Tl, Sn, In, Bi, and Sb; see Table 8.6) have been successfully determined at chemically modified CPEs by exploiting the selective recognition and/or electrocatalytic properties of the modifier. Bi^{III} and Sb^{III} species can be accumulated and then detected by anodic stripping voltammetry because their electrochemistry on CPE is compatible with the "electrolytic deposition—anodic reoxidation" scheme, whereas the application of this scheme for In^{III} was only possible on metal-plated CPEs through alloy formation in the electrolytic deposition step. Studied samples were water solutions, soils, human hairs, and pharmaceuticals.

Two main strategies were usually used for the analysis of Tl^{III} and Tl^{I} as well as Sn^{II} and Sn^{IV} species, the first one similar as the one mentioned earlier (but requiring metal-plated CPE and alloy formation in the electrolytic deposition step) and the second one based on direct electron transfer between soluble forms of these elements. A specificity of thallium determination at CPEs is that a conversion of Tl^{I} into the Tl^{III} state offers an attractive eventuality to pair thallate anions with a suitable counter-ion, offering improved performance with respect to the "simple" Tl^{I} determination via electrolytic reduction.

Otherwise, in mixtures with other heavy metal ions, there is also a possibility to utilize a well-known approach from the electrochemistry of thallium—its determination in the presence of EDTA that firmly masks all the interfering ions whereas the Tl(I) themselves remain uncomplexed and thus available for detection. A study on such a masking is depicted in Figure 8.5, in this case, performed atypically in basic media of ammonia buffer.

The individual combinations (A–D) document that the chelating of both Cd^{II} and Pb^{II} has been effective enough, at least, for concentrations and their mutual ratios chosen. Less evident from these sets of voltammograms was an observation that, at higher concentrations, stripping analysis of thallium at the Bi-CPE usually led to a couple of peaks, Tl(1) and Tl(2) [672], whose occurrence might be due to mutual alloying between the metallic thallium and bismuth.

Finally, one has to note that most of the elements presented in Table 8.6 are sensitive to hydrolysis, requiring to be analyzed in the complexed form (either by adding a suitable ligand in the solution or by exploiting the ligand properties of the electrode modifier).

8.1.3 Metalloids

Only few applications of CPEs for electroanalysis of arsenic and selenium species have been reported (see the bottom of Table 8.6), which contrast with the tremendous works that appeared in the analysis of highly toxic arsenic compounds in many kinds of polluted media.

Already in the first reported method for the determination of arsenic at a CPE, the differentiation of both As^{III} and As^{V} forms was successful [261] and, despite serious interferences from the Cu^{II} ions, the corresponding method could operate at satisfactorily low concentrations, especially, when coupled with CCSA [108]. This is also documented in Figure 8.6, presenting the experimental records used for estimating the respective limits of detection (LODs) having been found at the low ppb (µg L^{-1}) level for both As^{III} and As^{V}. Notable was also rather different character of the corresponding potentiograms with a considerable shift in the peak potentials, reflecting maybe some effect of solid α-cysteine that had to be added directly in sample solutions in order to accomplish the inevitable chemical reduction of electrochemically almost inert As^{V} to more reactive atoms of As^{III}.

Among commonly occurring selenium species, only tetravalent Se^{IV} (selenite) can be detected since other soluble forms of selenium are electroinactive on carbon electrodes. Otherwise, this interesting element (bioessential in traces and highly toxic at slightly higher concentrations) is also determinable in the form of $SeCN^-$ as shown in a lone report [652].

8.1.4 Metals of the Iron, Manganese, Chromium, and Vanadium Groups

In contrast to metal species discussed in Sections 8.1.1 and 8.1.2, the determination of metals Fe, Co, and Ni (Table 8.7), and Mn, Tc, Cr, Mo, W, and V (see the top of Table 8.8), is not straightforward

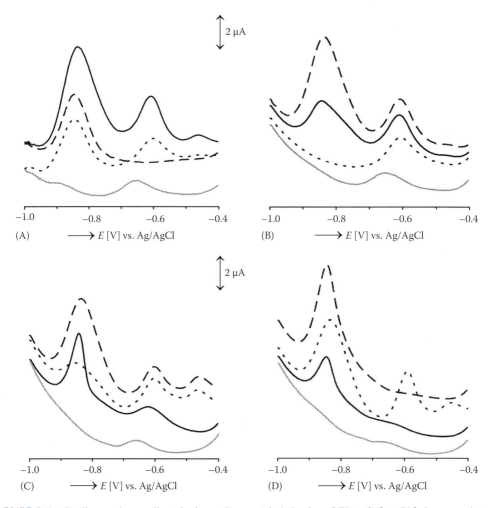

FIGURE 8.5 Studies on the anodic stripping voltammetric behavior of Tl$^+$ + Cd^{2+} + Pb^{2+} ions as mixtures in 0.1 M ammonia buffer and the effect of EDTA. (A) order of addition: Cd, Pb, Tl, EDTA; (B) Pb, Tl, Cd, EDTA; (C) Tl, Pb, Cd, EDTA; (D) Tl, Cd, Pb, EDTA; black full-line: mixture with Tl, dashed line: with Cd, dotted line: with Pb, grey full-line: with EDTA. *Experimental conditions*: Bi$_{17\%}$-CPE: carbon paste with 17 % bismuth powder (w/w); SWASV: f_{SW} = 25 Hz, ΔE = 50 mV, s.i. = 4 mV; accumulation conditions: E_{ACC} = −1.8 V vs. ref., t_{ACC} = 120 s, t_R = 10 s; stripping limits: E_{INIT} = −1.4 V, E_{FIN} = −0.4 V vs. ref.; s.e.: 0.1 M NH$_3$ + NH$_4$Cl (pH 8.8); c(Tl) = 100 μg L^{-1}, c(Cd) = 25 μg L^{-1}, c(Pb) = 50 μg L^{-1}; c(EDTA) = 10 mg L^{-1}. (Redrawn and rearranged from Švancara, I. et al., *Sci. Pap. Univ. Pardubice Ser. A*, 12, 5, 2007.)

at CPEs and usually requires rather sophisticated procedures. This is mainly due to the existence of several oxidation states and the tendency to transform from one valence to another depending on the medium conditions (e.g., sampling and storage conditions and/or moving from anoxic to oxic environment can contribute to speciation changes, which could thus not reflect the actual forms in the target medium). These ions also form stable complexes with many ligands, contributing also to rendering difficult the distinction between the individual forms, including more complicated speciation, also attainable by means of electroanalysis at CPEs.

Iron, cobalt, and nickel are the most widely studied elements of group VIII using CPEs. The main strategies to analyze iron species were directed to exploit the Fe^{II}/Fe^{III} and electrochemistry in complexing media (using, e.g., Cl$^-$, CN$^-$, or phenanthroline ligands) where this redox couple can undergo more or less reversible transformations. An elegant way to avoid the addition of selected

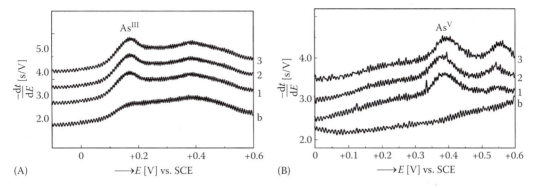

FIGURE 8.6 Stripping potentiograms used to estimate detection limits for arsenic analyzed in the form of As(III) (A) and chemically reduced As(V) (B). (b) base-line (blank), 1–3 5 ppb As^{III} * or 2 ppb As^{V} ** (three replicates). *Experimental conditions*: Gold film–plated carbon paste electrode, AuF-CPE; CCSA, accumulation conditions: $E_{ACC} = -0.4$ V vs. SCE, $t_{ACC} = 15$ s (A), 300 s (B), $t_{EQ} = 10$ s; stripping limits: $E_{INIT} = -0.3$ V and $E_{FIN} = +0.6$ V vs. ref., constant current, +5 µA; s.e.: 1 M HClO$_4$ + 0.2 M HCl. *Note*: the analyte added as aliquiot of (*) As^{3+}, (**) H$_2$AsO$_4^-$ pre-reduced with solid α-cysteine. (Taken and rearranged from Švancara, I. et al., *Talanta*, 58, 45, 2002.)

ligands in the sample is the use of CPEs modified with such ligands (reagentless detection), offering also the advantage of further preconcentration.

Contrary to iron, Co^{II} and Ni^{II} species can be directly reduced to their elemental state, but this transformation was often performed in complexing media to improve the analytical performance of the method (by enhancing charge transfer kinetics and/or accumulation of the target analytes, or even to avoid precipitation of the analyte). Especially dimethylglyoxime (DMG), and also cyclam derivatives and some other ligands, have been used for these purposes, DMG being moreover likely to determine both Co^{II} and Ni^{II} in mixture. Due to a certain solubility of DMG in aqueous solutions, CPEs prepared by simple admixing of the modifier into the paste can suffer from poor long-term stability, which can be somewhat got round by *in situ* approaches (i.e., DMG in solution to form the metal-ligand complex which is then accumulated by adsorption on CPE). These elements have been analyzed in various natural water samples, ethanol fuel, electroplating baths, or even biological fluids.

Strategies exploiting CPEs to detect other metal species, that is, of groups V, VI, and VII, have also been developed, for Mn, Tc, Re, Cr, Mo, W, and V (Table 8.8). Vanadium was mainly detected in the V^V form after the accumulation at modified CPEs, which is also the case of a method depicted in Figure 8.7 by means of a set of calibration voltammograms and operated via a complex with oxalate.

A huge amount of work has been performed to determine chromium at both Cr^{III} and Cr^{VI} oxidation states, using various detection techniques (anodic and cathodic stripping voltammetry with linear scanning, in differential pulse mode, or with potentiometric indication), usually associated to accumulation procedures involving complexation or ion-pairing with suitable modifiers. One has to confess, however, that the proposed methods to date do not reach the analytical performance of those employing mercury electrodes. Other elements of the same group VI can also be detected, including molybdenum (either Mo^{IV}O^{2+} or Mo^{VI}O$_4^{2-}$ forms) and tungsten (the only example to date [2324]). Finally, dealing with elements of group VII, some methods have been developed for manganese species and, to less extent, for technetium, which were likely to distinguish between various redox states. The usual way to determine Mn^{II} species relies on their oxidation at a carbon paste anode to form insoluble MnO$_2$ which can then be reduced quantitatively (reduction of Mn^{II} into metallic manganese is also an alternative, yet much less used). To date, direct detection of chromate species has not been reported (although this should be

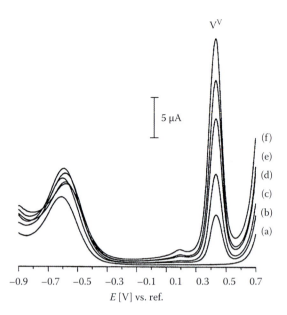

FIGURE 8.7 Differential pulse voltammetry of vanadium (V) at a carbon paste electrode modified "*in situ*" with oxalic acid. Model calibration. (a) 0, (b) 20, (c) 40, (d) 60, (e) 80, and (f) 100 μg·L^{-1} V (V). *Experimental conditions*: Nujol oil-based carbon paste; DPV, accumulation conditions: t_{ACC} = 1 min, E_{ACC} = −0.9 V vs. Ag/AgCl, scan rate: 50 mV·s^{-1}; s.e.: 0.01 M H$_2$C$_2$O$_4$ + 0.25 mM CTAB. (For other parameters and details, see Stadlober, M. et al., *Electroanalysis*, 9, 225, 1997; Stadlober, M. et al., *Sci. Pap. Univ. Pardubice, Ser. A*, 3, 103, 1997.)

possible by potentiometry subsequent to ion-pairing), Mn^{VI} being usually reduced into Mn^{II} prior to determination at CPEs. Detection of technetium involved accumulation as Tc^{IV} and subsequent oxidation into Tc^{VII}.

8.1.5 PLATINUM METALS AND URANIUM

In addition to iron, cobalt, and nickel, other elements of group VIII (Pt, Ir, Os, Ru, Rh, Pd) have been analyzed using CPEs (see bottom of Table 8.8). The electrochemistry of these species has been largely investigated but essentially for exploiting their electrocatalytic properties (notably in the form of metal deposits or nanoparticles), and only few analytical determinations have been published (mainly for Cd^{II} and seldom for the others). All are based on adsorptive accumulation (via complexation or ion-pairing) and subsequent voltammetric detection of anionic complexes.

In contrast to numerous investigations involving osmium and ruthenium derivatives in electrocatalysis, very limited examples are available for their study as the target analytes [736,1636,2330,2334]. Figure 8.8 shows such a procedure that could be generally used for all platinum metals, except palladium.

The respective procedure based on the effective ion-pairing with a quaternary ammonium ion (added as *in situ* modifier) was particularly good for some typical anionic forms of three heavy platinum metals (i.e., [Os^{IV}Cl$_6$]$^{2-}$, [Ir^{III}Cl$_6$]$^{3-}$, and [Pt^{IV}Cl$_6$]$^{2-}$), enabling also their simultaneous determination (see again Figure 8.8 and a voltammogram in the lower part).

Also not so widespread are the methods devoted to U$^{IV/VI}$ determination at CPEs (see middle of Table 8.8), all of them being based on the analysis of UO$_2^{2+}$ after accumulation at open circuit. This can be notably achieved by resorting to nano-engineered adsorbents designed to selectively bind actinides or lanthanides (such as carbamoyl-phosphonic acid functionalized mesoporous

Electroanalysis with Carbon Paste–Based Electrodes, Sensors, and Detectors

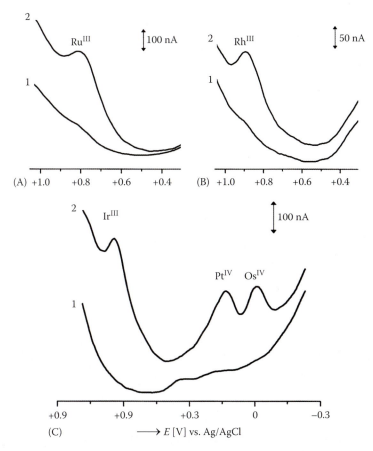

FIGURE 8.8 Voltammetric analysis of model solutions containing Ru(III) (A), Rh(III) (B), and a mixture of Pt(IV), Ir(III), and Os(IV) (C) at a carbon paste electrode modified with surfactant of the $R_3R'N^+X^-$ type. (1) Base-line (supporting electrolyte, s.e.; see the following) (2) 5×10^{-6} M Ru^{III} (A), 5×10^{-6} M Rh^{III} (B), and 5×10^{-7} M Pt^{IV} + 4×10^{-6} Ir^{III} + 5×10^{-8} M Os^{IV} (C). *Experimental conditions*: DPCSV; accumulation conditions: E_{ACC} = +0.9 V vs. Ag/AgCl; t_{ACC} = 30 s, t_{EQ} = 15 s; stripping limits: E_{INIT} = +0.9 V, E_{FIN} = −0.3 V vs. ref., scan rate, 10 mV s^{-1}; pulse height, ΔE = −25 mV; s.e.: 0.10 M acetate buffer + 0.15 M KCl (pH 4.5) + 1 × 10^{-5} M Septonex® (1-(ethoxycarbonyl)pentadecyl-trimethylammonium bromide). (Assembled and redrawn from Švancara, I. et al., *Talanta*, 72, 512, 2007; Galik et al., *Sensing in Electroanalysis*, Eds. K. Vytřas, K. Kaleher, University of Pardubice Press, Pardubice, Czech Republic, pp. 89–107, 2005.)

silica [770]). An original approach is the adaptation of carbon paste as the sensing element of a "lab-on-a-cable" device which was then applied to decentralized uranium analysis in undergroundwaters [1379].

8.1.6 Metals of the Fourth and Third Groups, Metals of Rare Earths

Several elements belonging to groups III and IV (Ti, Zr, Al, Ga, and Sc) and some of their related lanthanides or actinides (light and heavy rare earths) have been analyzed using CPEs (see second half of Table 8.9), although most of them do not exhibit attractive electrochemical features. This probably contributes to explaining why a relatively wide panel of methods has been proposed for their determination.

Dealing with Ti, Zr, Al, Ga, or Sc, thanks to their ability to form stable complexes with electroactive ligands (e.g., Alizarin derivatives), a strategy based on indirect voltammetric detection (i.e., monitoring changes in the electrochemical behavior of the ligand probe upon complexation with the

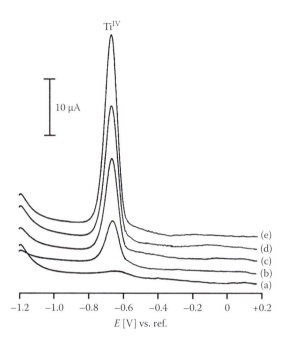

FIGURE 8.9 Differential pulse voltammetry of titanium(IV) at a carbon paste electrode modified 'in situ' with oxalic acid. Model calibration. (a) 0, (b) 25, (c) 50, (d) 75, and (e) 100 µg · L⁻¹ Ti(IV). *Experimental conditions*: Nujol oil-based carbon paste; DPV, accumulation conditions: t_{ACC} = 2 min, E_{ACC} = −1.2 V vs. Ag/AgCl, scan rate: 50 mV · s⁻¹; s.e.: 0.05 M acetate buffer + 0.01 M $H_2C_2O_4$. (For other parameters and details, see Stadlober, M. et al., *Talanta*, 43, 1915, 1996; Stadlober, M. et al., *Sci. Pap. Univ. Pardubice Ser. A*, 3, 103.)

target metal species) was often applied. On the other hand, direct electrochemical methods involved potentiometry, exploiting electrostatic interactions between charged forms of the analyte and suitable modifiers, and sometimes also the direct voltammetric detection mode (restricted to date to $Ti^{IV}O_2^{2+}$ and Ga^{3+} species). The studied matrices were mainly water solutions and ore specimens, which were also the cases of a method for determination of Ti(IV) illustrated in Figure 8.9 (on model calibration) and based on *in situ* modification with oxalate that, under slightly altered conditions, had also been applicable to V(V) and Mo(IV).

Recently, efforts have been made to detect rare earth metals. In addition to the aforementioned strategy involving electroactive ligands such as Alizarin derivatives, novel approaches have exploited the attractive features of room-temperature ionic liquid–based CPEs (as such or "doped" with carbon nanotubes) for the potentiometric detection of lanthanide or actinide cations, even in complex mixtures (see, e.g., [2361]). Ionophores were sometimes added into the composite electrode to improve sensitivity and selectivity of the method via strong interaction effects [415,419,510,511].

8.1.7 METALS OF ALKALINE EARTHS AND ALKALINE METALS

This family (namely, Li, Na, K, Cs, Be, Mg, and Ca) constitutes a special class of analytes as most alkali metal and alkaline earths metal ions are not electroactive in the sense that they cannot undergo electron transfer reactions at electrode surfaces, at least under usual conditions. It is therefore not so surprising to find potentiometry at the top place among electrochemical techniques applied to determine these species using CPEs (see first half of Table 8.9). To this end, CPEs were modified with selected ionophores (e.g., macrocyclic compounds) or ion-exchangers to induce preferable recognition via host–guest interactions. The only exception is the Be^{2+} ion, which can be detected by cathodic reduction [2342].

Nevertheless, various strategies based on indirect amperometric detection of these electroinactive cations have appeared in the literature. They include (1) the resort to electrodes modified with intercalation materials (i.e., metal oxides) for which cation insertion–ejection phenomena result in change in their electrochemical behavior, which can be related quantitatively to the cation concentration in solution, (2) the exploitation of the electrochemical activity of an electroactive ligand likely to interact with the metal ion (making the resulting complex likely to undergo electron transfer via its ligand part), or (3) the use of CPEs containing organic or inorganic ion exchangers doped with an electroactive cation which is likely to be released from the solid (and thereby detected at the electrode surface) by ion exchange with the cationic analyte. The latter approach was particularly developed using zeolites exchanged with redox-active cations [1115], as these solids feature both ion exchange and size selectivity properties, enabling the indirect amperometric detection of non-size-excluded cations while using a supporting electrolyte containing cations larger than the zeolite pore aperture (size-excluded cations), which did not contribute to the amperometric response of the electrode (see method principles in Figure 4.7, in Section 4.3).

An attractive extension of this approach concerns its possible use in flowing streams (batch or flow injection conditions) *in the absence* of supporting electrolyte, the cationic analyte playing the role of analyte by liberating the redox probe from the zeolite and the role of (transient) electrolyte, with a resulting amperometric response being the sum of two contributions: a faradic one arising from the probe reduction and a capacitive one resulting from conductivity changes at the electrode/solution interface [1116]. In such conditions, any kind of ion exchanger (not only molecular sieves) doped with a redox active ion (either cation or anion) can be used, and this has opened the door to applications as amperometric detectors in ion chromatography operating with the suppressor technology with analytical performance of the same order as the classical conductivity detector—see illustration in Figure 8.10, depicting the analysis of some alkali metal and alkaline earths cations, together with ammonium (according to [1736].

8.1.8 Non-Metallic Ions, Complexes, and Neutral Molecules

This last category dealing with non-metallic species is characterized by a wide diversity of physicochemical and electrochemical properties of the target analytes. They have been classified, more or less arbitrarily, on the basis of their charge. Data relative to non-metal cations and inorganic molecules are gathered in Table 8.10 while inorganic anions, such as single and oxo-anions, plus some anionic complexes, dwell in Table 8.11. The latter ones include halide species (Cl^-, Br^-, I^-, and I_n^-), halo-anions (ClO_3^-, ClO_4^-, BrO_3^-, and IO_3^-), or pseudo-halides (CN^-, SCN^-, $SeCN^-$, and N_3^-), as well as other anionic species formed from nitrogen (i.e., NO_2^- and NO_3^-), sulfur-containing anions (S^{2-}, SO_3^{2-}, SO_4^{2-}, and $S_2O_8^{2-}$), phosphate anions, or even some other, that is: $[Si(Mo_3O_{10})_4]^{4-}$ and $B(O_2)(OH)_2^-$.

Figure 8.11 shows the initial study on an ion-exchanger-modified carbon paste developed for sensitive detection of nitrite, NO_2^-. As documented by the set of CV cycles given, such CMCPE had exhibited strong sorption capabilities toward nitrite—the CVs themselves having illustrated the opposite situation: its rapid desorption—enabling the determination of this toxic anion down to the low ng mL^{-1} level [734].

The second category comprises the proton, H^+, and protonated amino-derivatives (NH_4^+, NH_3^+OH, $N_2H_5^+$) as cations, and several molecules as oxygen, hydrogen peroxide, nitrogen oxide and dioxide, hydroxylamine, and hydrazine. As a result of such diversity, a wide variety of methodological approaches, modifiers, and measuring techniques has been considered. Among detection modes, one can mention direct potentiometry, automated potentiometric titrations, a series of modifications of electrochemical preconcentration techniques with various stripping modes, voltammetry associated or not to electrocatalysis, amperometric detection in flowing streams, or electrochemiluminescence measurements. The selection of the modifier was of course the function of the target analyte, as briefly discussed hereafter.

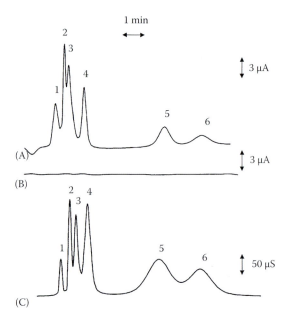

FIGURE 8.10 Chromatograms obtained using Cu^{2+}–Amberlite IR-120* modified electrode. (A), unmodified (bare) carbon paste electrode (B), and a classical conductimetric detector (C). *Note*: Amberle IR-120 is commercially available cation-exchange resin. *Experimental conditions*: Injection of 25 μL of a solution containing in a mixture of (1) 1.0 mM Li$^+$, (2) 1.0 mM Na$^+$, (3) 2.0 mM NH$_4^+$, (4) 1.0 mM K$^+$, (5) 2.0 mM Mg^{2+}, and (6) 2.0 mM Ca^{2+}. Mobile phase: 3 × 10^{-3} M methyl sulfonic acid. Flow rate: 0.75 mL · min^{-1}. Applied potential: −0.4 V vs. ref. [*Note*: only for indirect amperometric detection (A) and (B)]. (Adapted and redrawn from Mariaulle, P. et al., *Electrochim. Acta*, 46, 3543, 2001.)

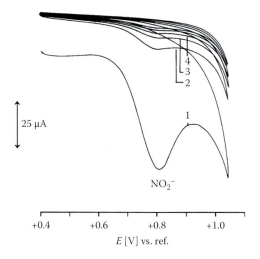

FIGURE 8.11 Cyclic voltammetry of nitrite anion at a carbon paste electrode modified with an anion exchanger. Model calibration. 1, 2, 3, 4: number of cycles. *Experimental conditions*: Nujol oil-based carbon paste containing *Amberlite*; CV, potential scan from E_{INIT} = +0.4 V vs. SCE to +1.05 and back, scan rate: 50 mV · s^{-1}; s.e.: 0.1 M HCl. (For other details, see Kalcher, K., *Talanta*, 33, 489, 1986.)

If excluding the few examples of CPE-based potentiometric sensor applicable to pH-measurements (because they are not expected to compete with the conventional glass electrodes), most cations of non-metallic character which have been determined using CPEs are nitrogen-containing species such as ammonium, NH_4^+, hydrazinium, $N_2H_5^+$, and hydroxylammonium, NH_3OH^+. Ion-pairing and amperometric detection at electrodes modified with ion-exchangers was especially used for ammonium, while the detection schemes for hydrazinium and hydroxylammonium were always based on the use of oxidation electrocatalysts such as phthalocyanines, porphyrins, and quinone derivatives, among others, as applied to basically native unprotonated compounds (i.e., NH_3, N_2H_4, and NH_2OH).

Other molecular analytes were essentially oxygen, O_2, hydrogen peroxide, H_2O_2, and nitrogen oxides, NO_X. Data presented in Table 8.10 for these species are related only to their direct detection while omitting methods in which they are reagents or products of enzymatic reactions at CPE-based biosensors (i.e., the analyte of interest being another substance). The most commonly used modifiers were electrocatalysts, most preferably phthalocyanines, porphyrins, and quinone derivatives, but also methylviologens, phenothiazines, or hexacyanoferrates (see again Table 8.10).

The effect of H_2O_2 concentration in enhancing or suppressing enzymatic activities was also exploited for sensing this molecule and one example of electrochemiluminescence detection was also reported [2366]. At present, most applications remained restricted to model solutions (and biological/pharmaceutical samples). Gaseous samples were sometimes analyzed on their content of NO_X (see [547,568,2372]).

Dealing with non-metallic anions and oxy-anions, almost each of them requires rather specific conditions for sensing at CPEs, so it is rather difficult to draw a general trend or to get some "rare" procedures. A basic distinction between the analytes reported in Table 8.11 can however be made on the basis of their intrinsic or mediated electroactivity or their non-electroactive character. When looking across the table rows, it cannot be overlooked that iodide, I^-, was one of the most frequently determined anions, which would not be so surprising, however, if one considers its easy transformation to elemental iodine, as well as a fine electroactivity in higher valencies, or its tendency to form stable ion-pairs and precipitates.

Figure 8.12 shows the fundamentals of one exemplary method that could be used in several adaptations for ESA [210,254,638,639,697,737] without losing its outstanding selectivity over other halides (expressed, e.g., as I^-: Cl^- = 1: 2,000,000) being superior to commercially marketed ion-selective electrodes for iodides.

Voltammograms included in the scheme of Figure 8.12 mirror some results from the recent studies focused on the method performance for particularly high concentrations of iodide [296]. It has been revealed—confirming also some results from the "prehistory" of the field [1,8,13,40,217]—that the $2I^-/I_2$ redox system may participate in both extraction and adsorption processes, when iodide–iodine or chloride–iodine adducts exhibit particularly strong sorption capabilities. Otherwise, the iodide species could also be detected directly via anodic oxidation without the preceding accumulation step or by potentiometric indication of the changes in chemical equilibria involving the I^- anion or even triiodide, I_3^-. Most of other anionic species have then been determined via the amperometric/voltammetric detection, which is the special case of oxy-anions, requiring the assistance of electrocatalysts such as polyoxometallates, Schiff bases, $Me^{I/II}$-bipyridines, Me^{II}-phenan-throline complexes, and macromolecular myoglobin. In many procedures, special efforts were directed to synthesize first the new modifiers whose improved performance has enabled the desired determination. Examples of potentiometric or indirect amperometric detections are also available (see throughout the tables). Among the analytes that are rarely determined by electrochemical methods, one can cite secondary *ortho*-phosphate or *ortho*-silicate, persulfate and perborate, or azide moieties, $S_2O_8^{2-}$, $BH_2O_4^-$, and N_3^-, the latter directly in solid lead azide, after its stabilization with dextrin [2385].

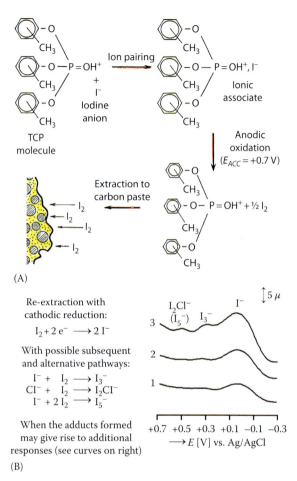

FIGURE 8.12 Principles of a method for the determination of iodine at the CPE with chemically active binder and employing the synergictic accumulation mechanism. An overall scheme (above); Typical stripping voltammograms for determination of iodine at a higher concentration level. (1) 5×10^{-6}, (2) 1×10^{-5}, and (3) 5×10^{-5} M KI. *Experimental conditions*: tricresyl-phosphate based carbon paste ("C/TCP" type); DPCSV, accumulation conditions: $E_{ACC} = +0.7$ V vs. Ag/AgCl, $t_{ACC} = 30$ s, $t_{EQ} = 15$ s; stripping limits: $E_{INIT} = +0.7$ V, $E_{FIN} = -0.3$ V; scan rate: $20\,mV \cdot s^{-1}$, pulse height: -50 mV; s.e.: 0.2 M NaCl + 0.1 M HCl (pH ~ 1.5). (Redrawn and adapted according to Švancara, I. et al., Temperature-controlled processes at carbon paste-based electrodes: Possibilities and limitations in electroanalytical measurements, in *Sensing in Electroanalysis*, Vol. 4, Eds. K. Vytřas, K. Kalcher, and I. Švancara, University Press Center, Pardubice, Czech Republic, 2009, pp. 7–26; Švancara, I. et al., *Electroanalysis*, 10, 435, 1998; Švancara, I. and Vytřas, K., *Sci. Pap. Univ. Pardubice Ser. A*, 7, 5, 2001.)

8.1.9 Concluding Remarks

In order to come full circle and finish the whole chapter where it started, naturally occurring chemical elements and their compounds that have not been yet analyzed using carbon paste electrodes, sensors, and detectors must be surveyed; that is, the proper methods for their quantification is still missing. Specifically, such absence is the case of Rb(I), Sr(II), Ba(II), Hf(IV), Te(IV/VI), Nb(V), Ta(V), and Re(IV/VII).

In a footnote of our recent review [7], mapping such "white sites" within the applicability of CPEs, fluoride has also been quoted as hitherto undetermined species. As ascertained during the completion of literature archives for the book, this uneasy-to-determine anion had already been successfully determined at a CPE, which is now documented by additionally found report [2380].

Electroanalysis with Carbon Paste–Based Electrodes, Sensors, and Detectors

Among others, this is another proof that any archive is never completed and, from an electroanalytical point of view, it documents the versatility in which one can use carbon paste–based sensing systems for quite difficult practical tasks.

8.2 DETERMINATION OF ORGANIC SUBSTANCES AND ENVIRONMENTAL POLLUTANTS

Also organic pollutants attract considerable attention for a long time, which can be illustrated in a plethora of publications mentioned earlier and within this section, throughout the Tables 8.12 and 8.13; including [2409–2641] as further entries.

Tables 8.12 and 8.13 summarize all the methods hitherto used in electrochemical studies and determinations of organic substances that can appropriately be sorted as (i) traditional organic compounds with derivatives with environmental significance (Table 8.12) and (ii) various synthetic preparates, including industrially important substances (Table 8.13).

As seen, among organic substances that have been already determined or at least studied in combination with CPEs, the only missing important category is the family of aliphatic and aromatic hydrocarbons (plus some basic heterocyclic derivatives), all representing almost non-electroactive compounds with non-ionic character. (For electroinactive but ionic-structure-forming organic compounds, the corresponding ion-selective electrodes have been developed and already successfully applied when coupled with potentiometric determinations [2491].)

In determinations of alcohols, phenols, and related compounds (e.g., [216,381,480,504,915, 1101,1628,2409–2416]), the catalytic properties of tyrosinases, laccases, and peroxidases are often exploited for the construction of sensors with narrow or broad selectivity [222,554,2411]. Moreover, the microbial biosensor was also prepared by using phenol-adapted bacteria and then calibrated to phenol [381]. Evidently, such modifications are noneffective if readily oxidizable compounds from the hydroquinone family are to be determined [405,504,979,2415,2416]. The same is valid for readily reducible emodin (6-methyl-1,3,8-trihydroxy-anthraquinone) [505,2417,2418], Alizarin (1,2-dihydroxy-anthraquinone) [1647], or naphthoquinones [2419]. CPEs modified with Ni [130] and Pd [1021]) or organic modifiers were used in determination of aliphatic [130,1268] and aromatic aldehydes [233,1021] or ketones (namely, coumarins and psoralens [2421]).

A rather special category comprises aliphatic amines. It should be mentioned that these compounds exist in aqueous solutions to a considerable extent as geminal diol derivatives. Their dehydration and formation of the electroactive unhydrated carbonyl is base catalyzed; but, in more alkaline solutions, the reactive OH$^-$ ions add in a nucleophilic attack to the aldehyde, forming a geminal diol, which is not reducible, undergoing oxidation only.

The chemical reactions involved [2178] can be described as follows:

$$R - CH(OH_2) \rightleftarrows R - CH = O + H_2O \tag{8.2}$$

$$R - CH = O + OH^- \rightleftarrows R - CH(OH)O^- \tag{8.3a}$$

$$R - CH = O + 2H^+ + 2e^- \rightleftarrows R - CH_2OH \tag{8.3b}$$

Readily reducible organic peroxides were determined at CMCPEs as well [894,896,2422,2423]; interaction of peroxynitrite with DNA is studied at a biosensor [282].

A lot of contributions deal with determinations of aliphatic and aromatic amines and their derivatives (e.g., [565,1413,2021,2426,2429]), aminofluorenes [2431,2432], aminopyrene [1424], aminoquinones [2435], and other amine compounds. Similarly, numerous organic nitro-compounds were studied and determined using CPEs [198,500,680,823,1075,1611,1690,1849,2440,2445]. In fact, many of these amino- and nitroaromates belong among well-known polyaromatic derivatives—see again Table 8.12 and the rows full of data on the methods for determination of PAHs, APAHs, and NPAHs—that represent highly toxic substances, including potentially genotoxic, as well as the already proved carcinogens and mutagens [2179–2181].

The last named nitro derivatives and the nitro group itself are one of the most important reducible species among organic compounds whose electrode transformations via nitroso, hydroxylamine, and finally amine group [2178],

$$-NO_2 + 2H^+ + 2e^- \rightleftarrows -N=O + H_2O \qquad (8.4a)$$

$$-N=O + 2H^+ + 2e^- \rightleftarrows -NH-OH \qquad (8.4b)$$

$$-NH-OH + 2H^+ + 2e^- \rightleftarrows -NH_2 + H_2O \qquad (8.4c)$$

are strongly dependent on the experimental conditions, in particular, upon the actual pH. Furthermore, it can be recalled here that the ultimate reaction mechanism and the resultant product(s) depend also on the chemical structure of the compound of interest because other substituents may participate in the corresponding redox reactions and/or structural rearrangements of the molecule(s) involved.

A similar variability is also the case of compounds containing nitrogen in other readily oxidizable and/or reducible groups, such as hydroxylamines [2019], hydrazines [2449], organic azo-compounds (commercial azo-dyes [2450]), and *N*-heterocyclic systems [2451], including some alkaloids [215,2452] that have also been largely subjected to electroanalytical treatment at various CPEs, CMCPEs, and CP-biosensors.

An illustrative analysis of a derivative containing both amino- and nitro-functional groups is presented in Figure 8.13, documenting also the usefulness of bismuth-modified CPEs (see Sections 2.2.1, 3.4, and 8.1).

In this case, an extremely high content of powdered bismuth in one of the three electrode materials tested and offering the most favorable electroanalytical performance for 4A3NP is notable. According to the authors' archives and a recent review [7], such 85% content of a modifier in the CP-mixture is maybe the highest percentage ever reported as being used for the bulk modification of carbon paste.

Whereas industrially important macromolecules and polymers have been analyzed with CPEs seldom [435,2464], the determination of surfactants of various types has been shown to offer a surprisingly wide applicability, thanks to potentiometric titrations. They represent an ecologic alternative to the official two-phase titrimetric method which has given a basis to the international standard method described in the respective norms ISO 2271, ISO 2871-1, and ISO 2871-2, when all utilize almost identical procedures [2492]. In this method, chloroform as the organic phase and a mixed indicator containing Disulphine Blue VN and 3,8-diamino-5-methyl-6-phenylphenan-thridinium (dimidium) bromide are used, both representing quite harmful substances. Possibilities for potentiometric end-point indication have been extended simultaneously with the development of membrane ISEs [1959,1960]. The main advantages of potentiometry as opposed to two-phase titration are as follows: (i) it permits the use of much higher concentrations of titrant, resulting in sharper end-points and thus better reproducibility; (ii) it is less subjective; (iii) it eliminates the use of chloroform, whose vapor is toxic and may be carcinogenic; (iv) it readily lends itself to automation; (v) it is much less fatiguing for the operator.

Till this time, procedures involving potentiometry and potentiometric titrations have been described which allow reliable determination of surfactants of various types—anionic, cationic, ampholytic, as well as non-ionic based on poly(oxyethylene) or poly(oxypropylene) chains. As shown [208,2461,2462], CPEs can advantageously serve as potentiometric sensors to monitor the

Electroanalysis with Carbon Paste–Based Electrodes, Sensors, and Detectors

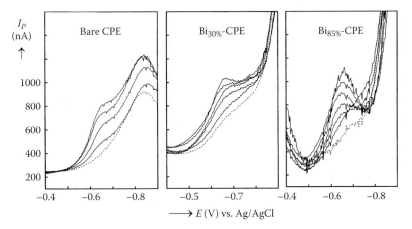

FIGURE 8.13 From the final optimization of a method for determination of 4-amino-3-nitro phenol (4A3NP) at three different carbon paste electrodes. Sets of voltammograms obtained by analyzing the compound of interest at the lowest detectable concentrations. Bare CPE made from Nujol oil–based carbon paste; Bi-CPE, carbon pastes with bismuth powder admixed at 30% and 85% (w/w). *Experimental conditions*: DPV, scan rate: $20\,mV \times s^{-1}$, pulse height: 50 mV, pulse width: 80 ms; supporting electrolyte, B-R buffer (pH 4, with 10% MeOH; dashed line); $c(4A3NP) = 4–10 \times 10^{-6}$ (for bare CPE), $2–10 \times 10^{-6}$ (Bi$_{30\%}$-CPE), and $1–10 \times 10^{-5}$ mol × L^{-1} (Bi$_{85\%}$-CPE); full line. (Redrawn from Dejmkova, H. et al., Application of carbon paste electrodes with admixed bismuth powder for the determination of 4-amino-3-nitrophenol, in *Sensing in Electroanalysis*, Vol. 3, Eds. K. Vytřas, K. Kalcher, and I. Švancara, University Press Center, Pardubice, Czech Republic, 2008, pp. 83–89. With kind permission of the authors.)

titrations of surfactants as well; corresponding procedures were reviewed [159,1966]. In voltammetric procedures, similar ion-pairing principles as in potentiometry can be utilized; the respective CPEs modified with compounds containing lipophilic counter-ions and the analytes—that is, corresponding surfactants—are accumulated at their surfaces [207,2463].

Organic pollutants known under *commercial names* and/or *trademarks* are then listed in Table 8.13, documenting that these substances are represented mainly by various pesticides from the families of carbamate and dithiocarbamate derivatives, organo-phosphates, triazines, or polychloro- and polynitro benzenes. As can be seen, the electrochemistry with CPEs has significantly contributed to developing a lot of new procedures for these traditionally determined organic pollutants [59,83].

In association with momentarily popularized "green-chemistry" approaches, the actual impact of some widely used pesticides on the environment is intended to be studied via *online monitoring* and evaluated by means of the *time-controlled biodegradability* [2180–2182]. For such purposes, it is necessary to have a reliable method capable of determining these compounds at sufficiently low concentrations, including their residua in natural matrices. Some new studies have shown that the combination of choice could be *electrochemical stripping analysis (ESA)* with a CPE based on tricresyl phosphate–containing carbon paste (C/TCP) whose electroactive binder can effectively interact with lipophilic molecules of some pesticides [2493]. One of the initial stages of such environmental investigation is captured in Figure 8.14, demonstrating why the C/TCP electrode is of such interest. Regarding the already existing collection of organic pollutants determinable at CPEs gathered in Tables 8.12 and 8.13, there are some joint characteristics that can be briefly commented on at the end of this section. First, the individual methodological procedures employ still popular ESA techniques, often, in combination with effective preconcentration via adsorption, extraction, or ion-pairing.

Of course, should they be successful also in determination of residual concentrations and analysis of samples with complex matrices, the corresponding methods must offer both *high selectivity and sensitivity*, which demands not only the proper working electrode, but also strictly defined experimental conditions, mainly, the continuous pH-control.

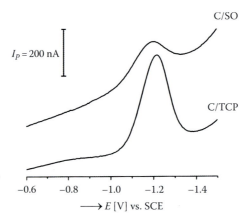

FIGURE 8.14 Comparison of differential pulse voltammetric signals for imidacloprid (insecticide) obtained at two different carbon paste electrodes. C/SO, carbon paste mixture made of highly viscous silicone oil; C/TCP, made of tricresyl phosphate (mixture of isomers). *Experimental conditions*: DPV, scan rate: $25\,mV \cdot s^{-1}$, pulse height: 50 mV, width: 50 ms; phosphate buffer (pH 7); $c(Imc) = 33.3\,\mu g \cdot mL^{-1}$. (Redrawn and rearranged from Papp, Zs. et al., New applications of tricresyl phosphate-based carbon paste electrodes in voltammetric analysis, in *Sensing in Electroanalysis*, Vol. 4, Eds. K. Vytřas, K. Kalcher, and I. Švancara, University Press Center, Pardubice, Czech Republic, 2009, pp. 47–58.)

Despite the high standard of many CPEs, CMCPEs, and CP-detectors, sometimes, the carbon paste surface has to be additionally treated by *mechanical regeneration* (see, e.g., [1268,2430,2466]) or *electrolytic activation* [2451,2472] prior to accumulation of the analyte. These operations may complicate the respective analytical procedures, but their benefit is so profound that both are included quite often (see, again, Tables 8.12 and 8.13). On the other hand, there are also some specific features of carbon pastes that can be exploited for highly effective accumulation without any pretreatment. It is the hydrophobic CP-surface [3,56,68,83] and its affinity toward more lipophilic compounds [59,157]. Such compounds are usually poorly soluble in aqueous media, which can be solved either by *solubilizing the analyte by its complexation* [55,2490] or, more often, by adding an organic solvent into the solution. If the latter cannot be avoided, one has to always consider that common carbon paste mixtures can be seriously damaged by a direct contact with such a solvent [3,56,83,211]. Thus, it is advisable to avoid this contact by the previously introduced *medium exchange* [55,1071,1395] based on short-time accumulation of the analyte in the original solution, followed by its transferring into another solution which is solvent-free and sufficiently safe for performing the entire stripping step and electrochemical detection in the selected mode.

Last but not least, if the analyte(s) of interest is completely insoluble in aqueous media (e.g., the already-mentioned PAHs, APAHs, and NPAHs) and requires the *presence of organic solvent*, there are several possible ways as to how to accomplish this without resigning on the use of a CPE or CMCPE. It can be (i) selection of a special CP-mixture with enhanced resistivity to organic solvents [197,2433–2435], (ii) choice of more stable solid-like CPEs [317,329,2480], or protection of the carbon paste material with the aid of (iii) polymeric membrane [560,565,2480] or even chemically, by adding (iv) a modifier [561,2488] with such protective function.

8.3 PHARMACEUTICAL AND CLINICAL ANALYSIS

Since the first reports in the early and mid-1980s (see, e.g., [219,242,2127,2499]), this specific area of applied analytical chemistry have firmly anchored in the electrochemistry with CPEs, CMCPEs, and CP-biosensors, proving soon a particular suitability and high effectiveness of using carbon paste–based electrodes for the determination of various pharmaceuticals and related substances.

In the area that offers extraordinarily *wide employment for traditional (unmodified) CPEs*, this can be shown in a long line of publications mentioned earlier and within this section, throughout the Table 8.14.

Among others, this specific area offers many attractive applications for potentiometry. In the 1970s—the "golden age" for ion-selective electrodes—their development was quickly followed by applications not only in inorganic analysis but also in biomedical analysis and in the pharmaceutical field. In first books and reviews dealing with the matter (see, e.g., [1972,2636,2637]), however, CPEs did not appear very frequently. Although a great majority of pharmaceuticals and drugs are organic substances in nature, their lipophilic moieties can often be converted into typically ionic species and, in this form, involved in potentiometric measurements with electroactive ion-exchangers (as appropriate modifiers of pasting liquids), thus being determinable by applying the ion-pairing principles described earlier (in Section 6.4.1).

Similarly, voltammetric procedures for pharmaceutical analysis with carbon paste–based sensors possess some similarities or even identical features described in previous chapters. At first, it is somewhat surprising that unmodified CPEs were so popular, relegating the CMCPEs to a spectrum of modifiers narrower than those exploited for analyses of organic compounds and pollutants or biological substances. This might be due to the fact that pharmaceuticals and related substances have often been determined in formulations or model solutions where the target analyte was the sole substance in the absence of interfering species.

Secondly, many methods imply open-circuit accumulation and MEX. The reason is that numerous drug substances can be preconcentrated by adsorption and/or extraction, that is, via open-circuit processes that do not require the application of any potential. These non-electrolytic depositions then minimize the eventual interference likely to arise from various electroactive species present, for example, in very complex matrices of biological tissues and fluids, representing maybe the largest group of samples in pharmaceutical analysis. The MEX approach may bring even double benefit: (i) additionally enhanced selectivity after replacing the sample solution (a real matrix) by a suitable blank electrolyte; (ii) effective control of the deposition step in the new supporting medium, thanks to carefully adjusted experimental conditions (possible masking reagents, selection of buffer and the respective pH, optimal ionic strength, etc.). Finally, owing to the character of typical real samples mentioned earlier, pharmaceutical analyses with CPEs usually require more sophisticated procedures involving (i) additional steps such as extractive isolation from interfering species and (ii) frequent electrolytic pretreatments of the electrode surface, with an advantage of quick mechanical renewal if neither anodization nor potential cycling led to the desired effect (e.g., in samples with very complex matrix or containing trace dispersions).

Pharmaceuticals of very diverse therapeutic use and action, together with some well-known abuse drugs, are listed in Table 8.14, surveying in alphabetical order the individual representatives of various inhibitors and stimulants of the central nervous system, antiseptics and disinfectants, antipyretics, diuretics, hypoglycemic agents, chemotherapeutics, hormonal supplements, vitamins, etc., always, with the key data characterizing the corresponding methods and referring to the original literature source(s).

Regarding some "more popular" pharmaceuticals, besides frequently studied and/or determined antipyretics *Acetaminophen* (syn. *Paracetamol*) or *Piroxicam*, also anti-psychotics from the family of substituted phenothiazines, the so-called *major tranquilizers* have attracted considerable attention. The reason could also be their very fine electroactivity as illustrated in Figure 8.15. At carbonaceous electrodes, these compounds are readily oxidized [187,211,2638], giving rise to two distinct signals (or even more [211,314]) of which practically either can be selected for identification and subsequent calibration (see the set of DPV curves in part "A").

Moreover, some more lipophilic phenothiazines, such as *Promethazine*, *Phenazine*, *Fluphenazine*, and mainly *Chlorpromazine* (*Clp*)—the latter being shown in Figure 8.15—exhibit distinctive extraction capabilities, which can be purposely exploited for their determination in the presence of some interfering species.

An exemplary case of such differentiation was described by Wang et al. [220,226], having studied—among others—the electrochemical behavior of Clp in model mixtures with potentially interfering uric acid (UA). Thanks to a spontaneous extractability of the former at appropriately chosen conditions (i.e., by keeping Clp in the form of neutral and uncharged molecules), the analysis could be performed in a sophisticated way, when both substances were preconcentrated at a CPE; Clp being extracted into the bulk, whereas UA adsorbed at the surface only. Then, the surface layer with accumulated species was cut off and UA removed as such. Because the molecules of Clp remained entrapped in the carbon paste (see also Section 3.4 and Figure 3.6), this drug could be re-extracted out of CPE and subsequently detected, both performable during the voltammetric scanning. The individual sequences can then be formulated as follows:

1. *Accumulation*

$$(Clp)_{SOLN} + (UA)_{SOLN} \rightarrow (Clp)_{CP\text{-}BULK} + (UA)_{CP\text{-}SURF} \quad (8.5)$$

2. *Surface removal and voltammetric detection*

$$(Clp)_{CP\text{-}BULK} + c(UA)_{CP\text{-}SURF} \rightarrow (Clp)_{SOLN} \quad (8.6a)$$

$$(Clp)_{SOLN} \rightarrow (OxP)_{SOLN} \quad (8.6b)$$

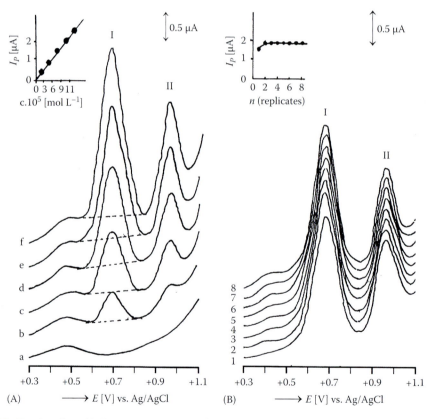

FIGURE 8.15 Anodic oxidation of chlorpromazine (Clp*) at a carbon paste electrode in highly acidic media. (A) Typical calibration voltammograms: (a) base-line (s.e.), (b–f) 2–10 × 10^{-5} M *Clp*; (B) reproducibility study: (1–8) eight replicates. *Experimental conditions*: CPE made of highly viscous silicone oil; direct DPV, scan rate: 50 mV · s^{-1}, pulse height: −50 mV; s.e.: 2 M H_2SO_4 + MeOH [30 % (w/w); pH ∼ 0] (*) *Chlorpromazine*, antipsychotic agent; 3-(2-chloro-10*H*-phenothiazinyl)-*N*,*N*-dimethyl-propan-1-amine hydrochloride, $C_{17}H_{19}N_2SCl$. (From authors' archives, an outtake from Švancara, I. et al., *Electrochim. Acta*, 37, 1355, 1992.)

The whole assay, where OxP means an oxidation product of the original pharmaceutical resulted in highly selective procedure for practical determination of Clp ([226] with a certain tolerance to the precision due to slight losses of Clp in the removed CP-layer. Nearly the same result was achieved in an alternate procedure, when both UA and Clp had been left to be accumulated together and, without any surface renewal, the corresponding (and almost merged) response was registered.

In the second step, the CP-surface was carefully removed and the measurement repeated. Thus, solely Clp as the compound entrapped in the carbon paste bulk might give the respective signal and the difference of both responses from the experiments No 1 and No 2 led to the desired separation:

1. *Accumulation*

$$(Clp)_{SOLN} + (UA)_{SOLN} \rightarrow (Clp)_{CP\text{-}BULK} + (UA)_{CP\text{-}SURF} \quad (8.7)$$

2. *Voltammetric scan No 1*

$$(Clp)_{CP\text{-}BULK} + (UA)_{CP\text{-}SURF} \rightarrow (Clp + UA) \quad (8.8)$$

3. *Surface removal and voltammetric scan No 2*:

$$(Clp)_{CP\text{-}BULK} + (UA)_{CP\text{-}SURF} \rightarrow (Clp) \quad (8.9a)$$

$$(Clp + UA) - (Clp) \rightarrow (UA)^* \quad (8.9b)$$

where $(UA)^*$ is the differentiated signal (or concentration, resp.) for interfering species UA.

Although both approaches described may lead to somewhat different results, all these experiments have clearly demonstrated a specific feature of CPEs—their operability (or functioning) by means of electrochemically active bulk.

Last but not least, some contemporary reviews on electroanalysis of pharmaceuticals should also be mentioned [187,2183,2639–2641], including a report devoted to the CPEs themselves [158], all being recommended as the primary source of expertise information.

8.4 DETERMINATION OF BIOLOGICALLY IMPORTANT COMPOUNDS

The last large group of the analytes in electroanalysis with CPEs, CMCPEs, and CP-biosensors is gathered in the following sections, in Tables 8.15 through 8.28, including those cited in the previous text and via a long string of new references [2642–3269].

BICs have long been the center of interest in analytical chemistry and biochemistry, either to monitor biochemical pathways or to detect the fate of exogenic or endogenic compounds and toxins. After the construction of the first (electro)chemical sensor by Cremer—who was, in fact, biologist—and its description by Haber and Klemensiewicz [3256,3257], it took roughly half a century until the first biosensor for glucose was invented by L. Clarke, earning thus the merit of being the "father of biosensors" [1908]. First commercialization of amperometric glucose biosensors started in the early 1970s by *Yellow Springs Instruments*. In fact, the real breakthrough was achieved in 1987 by *MediSense* when launching screen-printed electrodes as glucose biosensors for medical application. Yet before, Guilbault was the first to use a purified enzyme (urease) to construct a potentiometric biosensor [3261], Palecek started with electrochemistry of DNA in 1958 [3028,3258,3259], and Janata must be accredited with describing the first immunosensor [3260]. These are just some milestones in the history of the electrochemistry of BIC and of biosensors. In general, electrochemical sensors have a much longer history than optical sensors, whose development started eventually in the second half of the twentieth century. A brief overview has been given in [1911].

The following sections with Tables 8.15 through 8.28 summarize electrochemical methods for determination of BIC at carbon paste–based electrodes, except for a few classes of compounds that have already been reviewed—in the Sections 8.2 and 8.3.

8.4.1 Alcohols

Analytical methods of alcoholic compounds involving carbon paste electrodes are summarized in Table 8.15.

Many alcohols can be oxidized with the aid of Cu(I)-oxide; the actual oxidation mechanism proceeds via higher oxidation states [1225]. Also Ru, as RuO_2 or its complexes, can be exploited for this purpose [1220,2642–2644]. Here, again, the reaction proceeds via higher oxidation states of the metal with assumed tetra- to hexavalencies.

Choline can be detected with biosensors containing choline oxidase and ferrocene or phenothiazine as electron acceptors and membranes to immobilize the bioconstituent [1531,2645]. Phosphatidyl choline in rat saliva can be determined with this type of sensors.

Ethanol is the most frequently described alcoholic analyte in the literature. Its non-enzymatic oxidation may be catalyzed by Ni nanoparticles [139] or Ni(II) or Ru(II) complexes, where again an involvement of higher oxidation states of the metal ion can be assumed [1299,2646].

The majority of analytical ethanol assays employ oxidizing enzymes, most frequently alcohol dehydrogenase (ADH) [52,214,319,364,372,783,784,1000,1197,1471,1474,1599,1601,1750, 1752,1772,1786,2647–2651] and to a lesser extent alcohol oxidase [1755,2652–2659]. The enzymes are not specific to ethanol but oxidize also methanol and other aliphatic alcohols.

The enzyme ADH can be present in the solution [52,319] but is usually immobilized as a modifier at the electrode surface. ADH needs NAD^+ as a cofactor which is usually added to the solution or to the carbon paste itself; immobilization of NAD^+ with an aldehyde is also possible [52]. The metabolite NADH can be oxidized directly or with a mediator; the latter can be polymerized *o*-aminophenol or *o*-phenylenediamine [1772,1786], Ru complexes [1601], phendione complexes of Fe or Re [1599], or polymerized vinyl-ferrocene [2649]. Organic mediators can be toluidine blue [784,1474], Meldola blue [372,783,1197], phenothiazine derivatives [1750,1752,2650,2651] or Methyl green [1471].

Alcohol oxidase is usually combined with horseradish peroxidase (HRP) to detect hydrogen peroxide [1755,2652–2659]; HRP can be reconstituted directly via wiring with an Os-polymer or mediated with ferrocene [2652]. Yeast can be used as a source of ADH in corresponding biosensors [2660–2662], and plant tissue (tomato seeds) [2663] or whole cells of *Acetobacter* spp. may also serve as a pool of the enzyme [780,2664]. As a matter of fact, alcohol sensors are mostly applied for the determination of ethanol in beverages, such as wine, whiskey, etc.

For glycerol, glycerol dehydrogenase was used to construct a biosensor with NAD^+ as a co-factor [654]. In this assay NAD^+ was partly overoxidized, and the oxidation products strongly adsorbed to the carbon paste and showed catalytic effect on the re-oxidation of NADH formed by the enzymatic reaction.

Methanol can be catalytically oxidized by various metals mainly from the iron-platinum group [134,841,2666–2668]. Ni(II)-complexes can be used as well; probably they act via NiO(OH) which is formed in the strongly alkaline medium [131,424,2669–2671]; multi-wall carbon nanotubes (MWCNT) may support the oxidation [424]. Sulfur analogues of alcohols, thiols (organic mercaptans), can be oxidized easily with catalysts such as cobalt phthalocyanine (see Section 8.4.5); thiols can also be determined by the use of sulfhydryl oxidase as amperometric detector for chromatographic separations [2672].

8.4.2 Aldehydes, Ketones, and Acids

Table 8.16 gives an overview of analytical methods based on carbon paste sensors applicable to the determination of aldehydes, ketones, and carboxylic acid which are important in biological processes. Other methods can be found in Section 8.2.

Aliphatic aldehydes can be reduced to alcohols with palladium-modified carbon paste electrodes in alkaline medium [1268]. An interesting approach for the determination of aromatic aldehydes was presented by Abruna's group: the carbon paste was modified with a complexed aminopyridine, which accumulates aromatic aldehydes as Schiff bases; the latter can be electrochemically reduced [1021]. Ceresin wax as a binder offers increased stability against organic solvents.

A biologically important ketone (and carbonic acid) is pyruvate; it can be detected with corresponding biosensors based either on pyruvate oxidase [781,1468] with HRP or methyl green for the detection of H_2O_2 or on pyruvate dehydrogenase with vitamin K3 as a co-factor [2673].

Glycolate is enzymatically oxidized with glycolate oxidase using dimethylferrocene as an electron acceptor [854,856]. Instead of the purified enzyme, biosensors can be also constructed from tissues, such as spinach or sunflower leaves [2674–2676]. The sensor with spinach leaves can be also designed for electrogenerated chemiluminescence (ECL) with luminol (phthaloyl-hydrazide [2675]).

Lactate is an important intermediate in energy metabolism. Carbon paste sensors for lactate are usually designed as enzymatic biosensors. Most frequently lactate dehydrogenase (LDH) is used for this purpose which oxidizes lactate to pyruvate [52,180,855,875,937,1000,1217,1475,1672,1771, 1786,2677–2680]. Sensors have been developed for the D- or L-enantiomers using the corresponding enzymes. Catalysts for the reoxidation of NADH are Meldola blue [1217], ferrocene or naphthoquinone in combination with diaphorase, an enzyme which supports the reoxidation of NAD(P)H [855,875], and toluidine blue O, preferentially as redox polymer [937,1475,1786,2677–2679].

Addition of alanine aminotransferase in the presence of glutamate only insignificantly suppresses the interference from pyruvate, but is effective if used in an enzyme reactor [2679]. Glutamic pyruvic transaminase is also effective for the removal of the intermediate pyruvate with glutamate to yield alanine and α-oxoglutarate [1771]. Cytochrome c shows electrochemical activity in the presence of negatively charged lipids (asolectin) and acts as an electron acceptor of LDH [1672]; it can be replaced by phenazine methosulfate Reineckate [2680]. Diaphorase in combination with a naphthoquinone can be used to reoxidase NADH as an enzymatic intermediate [855]. Cells or membranes of *Paracoccus denitrificans* can be used as a source for membrane-bound dehydrogenases and respiratory chain enzymes [1520].

Lactate oxidase can be used as an alternative to LDH in biosensors for detecting lactic acid [246,364,647,661,863,1527,1567,2681–2684]. Mediators for the detection of hydrogen peroxide are Rh [647], Pt [1567], MWCNT [364,661], Prussian blue [196], and peroxidases [246,1527,2681,2683,2684]. As electron acceptors, Co-phthalocyanines [1567], ferrocene [863,1527,2681,2682], phenothiazines [863], or methylene green [1470] was used. Yeasts (baker's yeast or *Hansenula anomala*) can be the source of the enzyme and of cytochrome b2 [1483,2685,2686]. Malate dehydrogenase can recognize malate in corresponding biosensors with toluidine blue O-polymers [938]. Pd nanoparticles were used as modifiers of CPEs in an amperometric detector of capillary zone electrophoresis [2687], showing a promising way also for other platinum metals.

Oxalic acid can be oxidized with Pd nanoparticles as modifiers; the catalytic reaction proceeds via Pd(II) [971,2688]; Os (spongy) acts similarly [2689]. Lead sulfate as a carbon paste modifier can preconcentrate oxalate, because the solubility product of lead oxalate is significantly lower than that of $PbSO_4$; the accumulated oxalate can be oxidized to carbon dioxide [811]. Co-phthalocyanine is a useful catalyst for the oxidation of oxalate via Co(III) [557,885,1172,1561,1822,2690–2692]. Ruthenium bipyridyl silicotungstate can be used for detection of oxalic acid with ECL [1017]. Biosensors for oxalic acid involve oxalate oxidase via detection of H_2O_2 [1182,2693].

Salicylate (or salicylic acid, a close derivative to acetylsalicylic acid forming well-known *Aspirin*, being of interest also in the previous Section 8.3) can be determined with the aid of some carbon paste biosensors containing salicylate hydrolase, which releases catechol that can subsequently be oxidized to the corresponding quinone, either directly or via Meldola blue as mediator [1174,2694].

8.4.3 Amino Compounds

8.4.3.1 Amides and Amines

Table 8.17 comprises analytical methods for the determination of amines with carbon paste electrodes. Amines with neurotransmitter function will be discussed in Table 8.22.

Acrylamide can be detected indirectly with a biosensor containing hemoglobin; it forms an adduct with the biological recognition element and decreases its reduction current [2420]. Cu(I) oxide can be used as a catalyst for the oxidation of amines in alkaline medium, similar to alcohols [1225]. The principle of formation of a Schiff base for the detection of aldehydes [1021] can be also applied to amines when using aldehyde pentacyano-Fe(II) complexes; the Schiff base is irreversibly oxidized at around 0.9 V vs. SCE [1743]. Co(II)-phthalocyanine can oxidize amines and ephedrines in alkaline medium via Co(III) [902].

Synthetic and biogenic amines can be oxidized by amine oxidase either in purified form [2695] or in pea seedling tissue [274]; the intermediate H_2O_2 is detected either via direct oxidation or with HRP as a mediator. Aniline can be preconcentrated at sepiolite-modified carbon paste electrodes and determined in beverages without need to prior isolation [2424]. Benzocaine and lidocaine react chemically with *p*-chloranil and decrease its electrochemical response [1507].

Monoamine oxidase (MAO) can oxidize benzyl amine and can be exploited to determine the analyte in mouthwashes [1167]. Biogenic amines are usually freshness indicators of fish or meat. They can be detected via simultaneous monitoring of their oxidation by different phthalocyanine-modified CPEs and chemometric evaluation [2696]. Other possibilities involve HRP [2697], carbon nanotubes [2698], amine oxidase with HRP on sepharose and ferrocene [589], or specific oxidases for the individual amines (sarcosine, putrescine, and cadaverine). Such types of sensors are often designed as thick film sensors (screen-printed electrodes). Histamine can be detected with amine dehydrogenase and ferrocene [2699].

Phenylethylamine and tyramine can be catalytically oxidized with CNT [2700], penicillamine and tryptophan with quinizarine on TiO_2 nanoparticles [1880]. Oat seedling tissue contains polyamine oxidase and peroxidase; it can be used in biosensors for polyamines which are detected via released hydrogen peroxide [864].

Aromatic amines and cancerogens can be determined by adsorptive anodic stripping with CPEs modified with cyclodextrins and direct electrochemical oxidation [726,1413].

A biosensor for urea was designed with urease and glutamate dehydrogenase [2701], when the respective ammonium ions are indirectly detected via the decrease of the NADH response which is consumed by the dehydrogenase to form glutamate from ketoglutarate.

8.4.3.2 Amino Acids

Amino acids were frequently the subject of studies with carbon paste electrodes. The majority deal with voltammetric and amperometric investigations, although potentiometric detection is also possible.

In amperometry, detection may proceed either via the oxidation of the amino-moiety or via the chemical and/or electrochemical reactivity of a structural element of the compound. *N*-acetyl cysteine is a mucolytic agent and is also used as a food supplement. It can be determined with catalysts, such as caffeic acid [2702], iron oxide-Co(III)-hexacyanoferrate core-shell nanoparticles [983], Co(III)-salophen in combination with carbon nanotubes [403], or methylphenothiazine [2703].

Many amino acids can be electrocatalytically oxidized by copper micro- and nanoparticles and powder [963,966,2704,2705], Cu(I)-oxide [401,556,1225,2706] and various other oxides [556,1223], usually in strongly alkaline media; Cu(II)-cyclohexal-butyrate can be used as well [1607].

Co-phthalocyanine is also a catalyst [558,902] and can be employed in combination with Cu(II) for electrochemical detection in CE at a dual electrode, where the first oxidizes Cu(II) to Cu(III) [558].

D- or L-Amino acids can be also quantified with biosensors containing amino acid oxidase, which produce the corresponding keto-acid by oxidative deamination [179,854,1264,2708–2710]. Electrospun carbon nanofibers catalyze the oxidation of cysteine, tryptophan, and tyrosine [121]. Crown ethers form host–guest complexes with cysteine, homocysteine, and glutathione and support their electrochemical oxidation [1394].

Cysteine has been a very frequent subject of electrochemical studies with CPEs [324,487,555, 877,883–886,899,995,1040,1042,1230,1310,1316,1459,1535,1536,1543,1581,1586,1594,1608,1821, 2711–2721]. Ionic liquids exert a favorable effect on its oxidation and determination in soy milk [2711]. Nanocomposites from iron intercalated into graphite oxide and reduced catalyze the oxidation of cysteine [2712]. Bismuth can preconcentrate the analyte as Bi-cysteinate which can be reduced electrochemically by square wave voltammetry [2713]. Cu(I) forms strong complexes with R–SH, which can be reduced at around −0.65 V vs. SCE [1316].

The oxidation of cysteine can be catalyzed by some metal oxides [1230,2714], vanadium in different oxidation states [487,1594], and various hexacyanoferrates [1040,1042,1821,2715] and dithiocarbamates [1310,1608]. Cu(II)-cyclohexylbutyrate accumulates cysteine as Cu(I)-complex at negative potentials, which can be reoxidized in anodic stripping voltammetry [2716]. Two catalysts have been widely employed for the oxidative determination of cysteine: ferrocene [877,1535,1536,1543,2717] and cobalt-phthalocyanine derivatives [324,555,883,884,886,2718,2719]; the latter may be also used for the determination of glutathione and other thiols. Oxomolybdenum(V)-phthalocyanine is also able to oxidize cysteine [1581]. Co(II) salophen complexes can be exploited for potentiometric (via Co(II)/Co(I)-couple) and voltammetric determinations [1581,1586]. Quinones act as catalysts for the oxidation of cysteine [995,1459]; the hydroquinone derivative immobilized on TiO_2 nanoparticles allows simultaneous determination with dopamine UA and (DA) [995]. The system tyrosinase (immobilized at the carbon paste surface) and catechol can be used for the indirect determination of cysteine and other thiols, for example, glutathione [2721]. The enzymatically generated quinine is electrochemically recycled to the hydroquinone yielding a reduction current; thiols add to the quinine to the aromatic ring forming thiol-substituted hydroquinones not undergoing reduction.

Glutamate has been mainly determined with biosensors using either glutamate dehydrogenase [939,1751,2722,2723] or glutamate oxidase [1488,1773,2710,2724]. In the case of dehydrogenase, mediators are used and usually the co-factor NADP is added; with oxidases, either hydrogen peroxide is detected with peroxidase or an electron acceptor is added. The product with both enzyme is α-ketoglutarate by oxidative deamination.

Histidine can be accumulated via adsorption at the carbon paste surface and reduced by cathodic stripping voltammetry [2725]. Sorption can be improved by modifying the paste with alumina [2726] whereas semi-differential pulse voltammetry increases the signal even more. Cyclodextrins and fullerenes can be used as chiral selectors for L-histidine to check the enantiopurity of the amino acid in pharmaceuticals [2727,2728]. Tetra-3,4-pyridinoporphyr-azinatocopper(II) is an ionophore for histidinate, which can be exploited for potentiometric sensors [1577].

Homocysteine and other thiols are catalytically oxidized on multiwall carbon nanotubes [2728]; similarly gold nanoparticles immobilized with cysteamine improve its oxidation; the sensor can serve as a detector in high-performance liquid chromatography (HPLC) [947]; the system works also with methionine, as well as with some other thio-compounds [1877,1878].

A method for the determination of leucine and other amino acids was proposed by EC with Ru(III)-bipyridyl on Pd nanoparticles after derivatization with acetaldehyde [2704]. Lysine oxidase oxidizes lysine; liberated hydrogen peroxide is detectable with Prussian blue [196]. Lanthanum hydroxide nanoparticles catalyze the oxidation of mefenamic acid [2585].

β-N-Oxalyl-1,3-diaminopropionic acid (β-ODAP) is the toxic principle of *Lathyrus sativus* (grass peas) causing chronic poisoning (lathyrism). It is metabolized by glutamate oxidase with slower kinetics than glutamate; interferences from the latter are eliminated by decarboxylation with glutamate decarboxylase [842,2729–2731]. The detection proceeds via hydrogen peroxide with manganese dioxide as mediator.

Biosensors for phenylalanine contain phenylalanine dehydrogenase which oxidizes it to phenyl pyruvate and releases NADH; the latter can be reoxidized with dihydroxy benzaldehyde [2732]. Oxygenation removes interferences from ascorbic acid (AA) and uricase from UA. A trienzyme biosensor was suggested by Scheller's and Kuwana's groups [2733]. Besides phenylalanine dehydrogenase, salicylate hydroxylase and tyrosinase were co-immobilized; the hydroxylase converts salicylate to catechol in the presence of oxygen and NADH from the dehydrogenase. The other enzyme, tyrosinase, then oxidizes the catechol to o-quinone, which is electrochemically detected by the reduction to catechol at an operating potential of around −50 mV vs. Ag/AgCl. The analyte can be detected in blood, serum, and urine, which is important to uncover phenylketonuria, the condition of not being able to metabolize phenylalanine that leads to mental retardation.

L-Proline interacts with cyclodextrins which is exploited for enantioselective determination with potentiometry [2734]. Complexation studies of selenoamino acids with Ag^+ were performed with cyclic voltammetry and chronocoulometry [1309].

Tryptophan shows a high affinity for carbon paste and is electroactive; it can be oxidized at relatively high positive potentials. Therefore, it can be detected with adsorptive stripping voltammetry [223,2735–2738]. Preanodization or over-oxidized polypyrrole films improve the sorption [181,1797,2739]. Mediated oxidation can be achieved with quinizarine on titanium dioxide nanoparticles (with the possibility of simultaneous determination of penicillamine) [1880], ferrocenes [920,1545], or Co-salophen with carbon nanotubes [384]. Potato juice as a carbon paste modifier increases the selectivity [2741]; polyamide [2740] and clay minerals [1012,1122,1140,2742] assist accumulation. Doping with Fe^{III} increases the anodic peak current.

Tyrosine can be oxidized at unmodified CPEs [2743,2744]; an effect similar to tryptophan was observed with clay minerals and with Fe^{3+} [1124,2745]. Polypyrrole molecularly imprinted with L-phenylacetic acid increases the selectivity [2746]; polyvinylpyrrolidone with cetyltrimethyl ammonium bromide (CTAB) supports accumulation [2747].

8.4.4 Antioxidants and Phenolic Compounds

Table 8.18 gives an overview of the determination of antioxidants with carbon paste electrodes. AA will be reviewed with vitamins in Table 8.25; phenolic amino acids can be found in Table 8.17. An overview of the antioxidant activity of natural products is given in [2748].

Ni-porphyrin on phosphate silica-niobia was used as a catalyst for the oxidation of phenolic compounds, in particular, p-aminophenol and hydroquinone [1201]. The electrochemical behavior of antioxidants was critically compared with different electrodes in [2749]. Mn(II)-hexacyanoferrate catalytically oxidizes butylated hydroxyanisole via hexacyanoferrate(III) [226,2750]; Ni-phthalocyanine is probably reduced to Ni(I) by the analyte and can be electrochemically re-oxidized [651,895]. Butylated phenols are used as antioxidants in food industry. Caffeic acid (3-(3,4-dihydroxyphenyl)-acrylic acid) in the presence of hydrogen peroxide is catalytically oxidized by peroxidase (in green bean tissue) to the quinone which can be electrochemically reduced [2751]. β-Cyclodextrin supports the extraction of (+)-catechin from tea; the analyte can be oxidized to an o-quinone [1419].

Catechol is oxidized by tyrosinase and polyphenol oxidase and also by laccase (monophenol oxidase and blue copper enzyme) [222,279,364,554,987,2752–2758]. The enzymes can be applied in purified form, and also with plant tissues [279,2756–2758]. DNA improves the stability of tyrosinase-modified carbon pastes and prolongs the shelf lifetime [2755]. The detection proceeds via the reduction of the enzymatically produced quinone.

DA pyrocatechol, and flavonols can be determined with plant tissue electrodes (banana and green apple); the tissue acts as a source for polyphenol oxidase [278,2759,2760]. Dihydroxybenzenes [41] and flavonoids can be oxidized with unmodified CPEs [2761–2763]; flavonoids have been also investigated with respect to their interaction with double-stranded and single-stranded deoxyribonucleic acid (dsDNA and ssDNA) [293,2764]. They show some protective effect against dsDNA cleavage by Cu-phenanthroline and hydrogen peroxide.

Hydroquinone is oxidized by Co-phthalocyanine [633] or by laccase (on magnetic core-shell nanoparticles) [979] or polyphenol oxidase from apple tissue [2765]. *Moraxella* spp. can degrade *p*-nitrophenol to hydroquinone, which can be oxidized electrochemically; this system was used to construct a microbial carbon paste biosensor for *p*-nitrophenol [2440].

Phenol can be oxidized directly at carbon paste electrodes at rather positive potentials [548,1903,2766,2767] and its redox transformation is often accompanied by distinct adsorption effects [3,27]. Ionic liquids [461], polyamide [2768], CTAB [1101,2769], silica gel [1150], montmorillonite [1101,2770], sepiolite [824], mesoporous TiO_2 [994], bentonite [2696], and β-cyclodextrin [1411] improve the conditions for accumulation and oxidation. Procaine hydrochloride [2413] and over-oxidized polymers [2412] improve the electron transfer. RuO_2 and colloidal Pt [2771] and co.phthalocyanine [2772] act as electrocatalysts for oxidation. HRP catalytically oxidizes phenol with the aid of hydrogen peroxide [1181,1192,1525,2785]. This system also works with resveratrol, a "magic" stilbene compound with life prolonging and anti-aging effects in the popular literature [775]. Extracts of soybean seed hulls [2777] or eggplant ("gilo" for rutin and H2Q) [2791] can be a source of the peroxidase. Phenols and its compounds can be enzymatically oxidized by tyrosinase, often in combination with supporters and electron acceptors [194,384,573,898,979,1258,1265,1275,1798,1800,2125,2410,2773–2776,2778–2784]. Mushroom or plant tissue may serve as pools for polyphenol oxidase [194,283,898,1258,1265,2564,2783].

Cu(II)- and Mn(II)-phthalocyanines were used as biomimetic agents which should mimic in combination with histidine and hydrogen peroxide dopamine β-hydroxylase to oxidize phenolic compounds [519,912]. As a pathway oxidation of the phthalocyanine is assumed which in consequence oxidizes the phenol; electrochemical monitoring is achieved by reducing the generated quinoid. The sensors are called enzymeless biosensors, but in fact they are not biosensors.

p-Aminophenol, hydroquinone, and catecholamines, such as DA, epinephrine, and norepinephrine, can be detected with quinoprotein glucose dehydrogenase; electrochemical detection is achieved by oxidation of the phenols at 0.5 V vs. Ag/AgCl; the signal is amplified by electron transfer from the enzyme to the quinoid as an electron acceptor [2786]. Quercetin adsorbs on carbon paste and can be determined with adsorptive differential pulse anodic stripping voltammetry [366].

Redox compounds in olive oil and wines can be used for bitterness or adulteration criteria [918,1571,2163,2787,2788]. For wines, the array with different phthalocyanines in combination with statistical evaluation may act as an electronic tongue (see Section 7.4.2).

Rosmarinic acid [2789] and rutin [2790] are catalytically oxidized with transition metal complexes in the presence of hydrogen peroxide; the sensors should somehow mimic the action of peroxidase and red kidney purple acid phosphatase. Thymol, zearalenone, and zearalenole can all be oxidized directly at unmodified carbon paste electrodes used as amperometric detectors in flow systems [2627,2792].

8.4.5 Carbohydrates and Related Compounds

Table 8.19 surveys analytical methods with carbon paste electrodes for the determination of sugars and related compounds.

Alditols and saccharides can be catalytically oxidized with Co-phthalocyanine [882,889,890]. Aldoses can be determined enzymatically with pyrroloquinoline quinone (PQQ)-dependent aldose dehydrogenase, where the reduced form $PQQH_2$ is electrochemically reoxidized with ferrocene as a mediator [1735,2793,2794].

Metal oxides can catalyze the oxidation of carbohydrates; typical representatives are Co-oxides [1230,2706], copper oxides and hydroxides [1225,1226,1229,2706], NiO (nanoparticles) [1231,2706], $Ni(OH)_2$ [497], and RuO_2 [1218,1221,2706]. The metals can be present also in complexed form, for example, Cu(II)-porphyrin [1576], Ni(II)-*o*-amino-phenol [1766], Ni(II)-1-naphthylamine [2687], or Ni(II)-quercetine [414], nickel atom acting via the redox couple Ni(II)/Ni(III). Biosensors with activity toward a few carbohydrates contain oligosaccharide dehydrogenase, which can be also used for the determination of starch in combination with amylase [1437], or glucose oxidase (GOD) in combination with invertase and mutarotase [1523,2795]. GOD needs β-D-glucose as a substrate, which is in consequence oxidized to D-glucono-1,5-lactone with the aid of oxygen. Invertase hydrolyzes saccharose to glucose and fructose; mutarotase is an epimerase accelerating mutarotation from α-D-glucopyranose to β-anomer. Fructose dehydrogenase oxidizes specifically D-fructose to 5-ketofructose; ferrocene (oxidized) can be used as an electron acceptor [872].

Valine is the N-terminal amino acid of hemoglobin; it is glycosylated in hemoglobin HbA_{1c} by elevated glucose levels in the blood; HbA1c is a factor reflecting exposition to elevated glucose levels in diabetes patients for over 8 weeks. Fructosyl valine is a model compound for HbA1c; it contains fructose bound via the C1-atom to the aminogroup of valine, yielding the structural element $-CH_2-NH-$, which can be oxidized to the imine either directly electrochemically or by fructosyl amine oxidase from a marine yeast (*Pichia* spp.) [2796–2798]. The imine is then hydrolyzed to valine and glucosone.

Glucose, as apparently the most frequently determined analyte at CP-based electrodes and sensors, can be oxidized by Cu_2O (via Cu^{III}) and by RuO_2 in strongly alkaline medium [633,1222]. Ni nanoparticles on carbon nanofibers and Ni(II) on a poly(*o*-aminophenol) film act similarly [975,2800]. Determination of glucose is possible via its Cu(II)-complex which can be reduced at lanthanum hydroxide nanowires [2799].

Nevertheless, most of the studies and analytical methods for the determination of glucose deal with biosensors, which are mostly based on GOD [2801]. The enzyme has flavin adenine dinucleotide, Flavin adenine dinucleotide (FAD), as a coenzyme, which oxidizes glucose to δ-gluconolactone (which is further hydrolyzed to D-gluconate and H^+) with simultaneous formation of the reduced form of FAD, that is, $FADH_2$. The latter is re-oxidized with the aid of oxygen (yielding hydrogen peroxide) or with other oxidized species as electron acceptors. The enzyme itself is inexpensive and user-friendly with respect to storage and handling; therefore, GOD is being the template enzyme of choice to test new biosensing configurations (see Figure 8.16). Hydrogen peroxide can be monitored directly electrochemically [73,2802]; carbon nanotubes [123,371,661,2803] or ordered mesoporous carbon [119] supports detection. It is also possible to transfer electrons directly to the enzyme with covalently bound FAD and reconstitution of the native enzyme with the apo-enzyme [2804]. Platinum metal oxides mediate the electrochemical detection of hydrogen peroxide [1949,2805]. Quinones have been frequently used as electron acceptor; they are reduced and electrochemically oxidized again [72,666,1434,2806–2808].

Osmium complexes work as electron acceptors via Os(III)/Os(II); in combination with electrolytic binders without any oil component they yield high signals [238,2809]; osmium redox polymers can wire the enzyme for direct electron transfer [649]. Ferrocenes, either in solution [1250,2810] or as modifiers of the carbon paste, are the most popular artificial electron acceptors for GOD [88,119,202,589,785,857,859,860,862,865,869,871,922–924,942,988,1118,1153,1260, 1492,1519,1522,1524,1526,1530,1532,1533,1551,1554,1555,1725,1727,1748,1749,1776,1783,2811–2828,2846,2847]; moreover, they can be present as pasting liquid [2827,2828].

In fact, ferrocene, bis(cyclopentadienyl)-Fe(II), must be oxidized to the ferricenium cation, [bis(cyclopentadienyl)-Fe(III)]$^+$, which acts as the effective acceptor for one electron and is reoxidized electrochemically afterward, providing the current signal. Nickelocene in combination with platinum particles may be used as well [2847]. Co-phthalocyanine is frequently employed as an electron acceptor; similar to ferrocene, oxidation of $FADH_2$ of GOD proceeds via Co(III) [1566,2830–2832]. Electropolymerized Fe(III)-aminophenanthroline may be used for the mediated reduction of hydrogen peroxide via the couple Fe(III)/Fe(II) [433]. Hexacyanoferrate (III) is a soluble electron acceptor to re-oxidize $FADH_2$ from the enzyme [643,1788]. Prussian blue, Fe(III)-hexacyanoferrate, its Ru-analogue, ruthenium purple, Ni(II)-hexacyanoferrate, and Cu(II)-hexacyanoferrate are mediators for the electrochemical detection of H_2O_2 [196,360,848,1035,1613,2813].

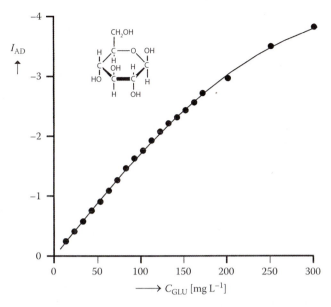

FIGURE 8.16 Testing of a new CP-biosensor with GOD immobilized onto the electropolymerized film. Calibration of glucose (GLU) at the mg·L^{-1} level. *Experimental conditions*: Nujol oil–based carbon paste coated with polyaniline film; hydrodynamic amperometry (HA), detection potential, −300 mV; phosphate buffer (pH 7.5), c(GLU) = 10–300 ppm (introduced to the solution as the adequate additions of stock standard solution). (An outtake from Stoces, M. et al., *Int. J. Electrochem. Sci.*, submitted paper, 2010.)

Enzyme-induced precipitation of Cu(II)- and Nd(III)-hexacyanoferrate(II) starts with ferricyanide, which is an electron acceptor from FADH$_2$ of GOD (in solution) forming [Fe(CN)$_6$]$^{4-}$ which then precipitates with the counterion [1008,1009]. CuO catalyzes the oxidation and the reduction of hydrogen peroxide [2833]. Mn-perovskite [1249] and manganese oxides [1234,1874,2834] mediate the oxidation of hydrogen peroxide to oxygen; manganese oxides in lower oxidation states can be electrochemically reoxidized to MnO$_2$ again. Metal oxides as mediators for detecting H$_2$O$_2$ were studied in [2835]. Platinum metals, their alloys, and copper were widely used [285,291,375,405,609,611,612,754,838,873,1035,1256,1260,1262,1263,1266,1267, 1281–1283,1711,1789,1854,1949,2733,2837]; they can be applied in dispersed form of metalized graphite particle, as powder and as micro- and nanoparticles. At positive potentials, platinum metals mediate the oxidation of H$_2$O$_2$ to O$_2$, at negative the reduction to H$_2$O. The studies also revealed high thermal [291,2707] and acid [1854] stability of the enzyme GOD. Efforts to avoid oxygen depletion by the enzymatic action resulted in addition of myoglobin as an oxygen pool, or in the use of *Kel-F* or siloxane as pasting liquids that can dissolve O$_2$ [611,873,1789,2837]. Iron nanoparticles on carbon nanotubes (carbon-iron nanocomposite) may transfer electrons directly to GOD in a third-generation biosensor [2712], similar to gold nanoparticles [941,1270,1274]; here catalytic effects of the Au-nanoparticles on the reduction of oxygen are also assumed [1274]. A nanostructured perovskite immobilizes GOD well, but the catalytic action is done by ferrocene [1250]. Tetrathiafulvalene (TTF, 2,2′-bis-1,3-dithiolylidene) can be oxidized, acting as an electron acceptor [1486,2841–2843,2846,2847]. Tetracyano-quinodimethane (TCNQ, 2,5-cyclohexadiene-1,4-diylidene) is readily reduced to a blue anionic radical and is therefore also a useful oxidant for FADH$_2$ in GOD [1493,2844–2846]. Both TTF and TCNQ together form a blue salt with high conductivity, "organic metal"; it can be used to build a third-generation glucose biosensor which is less stable than the one with TTF alone [2846]. Quinones are another group of organic electron acceptors in biosensors with GOD [1438,1440,1448,1450,1456,1757, 1759,2848,2852]. One assay uses a silicagel particle with immobilized modifiers as a microreactor implanted on the CPE surface [1440,2850]. Other organic electron acceptors used with glucose biosensors were methylene green [244], viologen derivatives [1498,1756], 1-(*N*,*N*-dimethylamine)-4-(4-morpholine) benzene [574,2857,2858], Meldola blue [1482], and poly(*o*- or *m*-phenyl-enediamine films [1261,1780]. GOD bound to poly(ethylene glycol) gives access to direct electron transfer [1781,1782,860,2853]. HRP detects hydrogen peroxide which oxidizes the enzyme that can be electrochemically reduced directly or via a mediator, tetramethylbenzidine [321]. Fibrinogen increases the sensitivity of a glucose biosensor with direct oxidation of H_2O_2 [2855,2856].

Glucose can be also oxidized with glucose dehydrogenase [385,935,938,1481,1601,1754,2850,2860,2862]. It needs a co-factor, NAD^+, which is reduced to NADH and can be electrochemically reoxidized, either directly or via mediators. Typical representatives are Meldola blue [1481,2801,2859], phenazine methosulfate [2861], toluidine blue O [935,938,2860], ruthenium complexes [1601], and Os-polymers [385,1754]. Glucose biosensors were also constructed from microorganisms which serve as a source of glucose dehydrogenase: *Gluconobacter* sp. [1441,2863,2864], *Escherichia coli* [288], *Lactobacillus fructivorans* [1443], and *Aspergillus niger* [1521].

Occasionally special designs with carbon paste were used for determining glucose + other carbohydrates. Carbon paste served only as a support for GOD, where the released hydrogen peroxide was detected with a platinum electrode placed downstream [2865]. An optical fiber sensor based on luminescence from H_2O_2, enzymatically released by GOD, and electrooxidized luminol was described in [2866]. Dual electrodes were employed for the simultaneous determination of glucose and other sugars, such as lactose [2867] and sucrose [250]. For the lactose, sensor galactosidase was used additionally to GOD, for sucrose GOD, invertase and mutarotase; catalase was placed between the electrodes to avoid cross-talks via hydrogen peroxide. In both cases, another electrode was a glucose sensor based on GOD; the disaccharide concentration was evaluated as difference.

Pyranose oxidase can metabolize a few carbohydrate, such as glucose, xylose, galactose, and is the main ingredient of a pyranose biosensor [2868]. It detects hydrogen peroxide with HRP and contains other additives as supporters and stabilizers, mutarotase for quick formation of the beta-anomer of glucose, polyethyleneimine for its stabilizing effect on the oxidase, and streptomycin/dihydrostreptomycin for its positive action on the peroxidase. Hydrogen peroxide from pyranose oxidase can be detected also with carbon nanotubes [400]. Sixteen mono- and oligosaccharides were observed to give response with a biosensor containing oligosaccharide dehydrogenase and benzoquinones [1439].

Pectins, mainly poly(galacturonic acid) esterified with methanol, can be quantified via the alcoholic component with a tissue-enzyme hybrid biosensor: orange peel as a source of pectin esterase on top of an alcohol sensor (alcohol oxidase) will release methanol which is monitored by the enzyme electrode [277].

8.4.6 COENZYMES, ENZYMES, PROTEINS, AND RELATED COMPOUNDS

Table 8.20 reflects extensive investigations with carbon paste electrodes on proteins, enzymes, and coenzymes.

FAD, flavin mononucleotide (FMN; riboflavin-5′-phosphate), and riboflavin were adsorbed on silica gel modified with zirconium oxide; their voltammetric behavior was studied with cyclic voltammetry [1188]. Lipoic acid can be catalytically oxidized with Ni(II)-cyclohexyl butyrate [1606] or, like other thiols and disulfides, with cobalt phthalocyanine [886].

Most studies were performed with nicotine adenine dinucleotide in reduced and oxidized forms (NADH, NAD^+) as well as on its phosphate (NADPH, $NADP^+$) with a phosphate group on the 2′-C atom of the ribose, because NAD^+ is a very frequent co-factor with dehydrogenases [64,119, 124,203,368,395,439,453,472,535,654,656,782,875,876,938,1130,1189,1190,1193,1194,1195,1196, 1199,1205,1212,1286,1287,1290–1293,1295,1296–1298,1299,1469,1473,1476,1503,1512,1779,1786, 1904,2869–2890]. The reduced form can be oxidized at plain carbon paste, though surfactants [64], carbon nanotubes [124,395,2871], carbon nanofibers alone [203] or with Pd nanoparticles [2875], and ordered mesoporous carbon [119] support the electron transfer. Poly(methyl methacrylate) [2872] and ionic liquids [453,472] as binders decrease the overpotential. Electrochemical pretreatment and oxidation products of NAD⁺ exert a catalytic effect on the oxidation of NADH [654,1904,2873,2874]; electrooxidized products from ATP show similar characteristics [1779]. Ferrocenes were used as mediators for the oxidation, either on a niobium oxide-silicagel [535], in combination with diaphorase (an oxidoreductase which catalyzes the oxidation of NADH and NADPH) [875], as a redox gel with direct electron transfer [782], or with synergistically active zeolite [876]. An Ru(II)-phenanthroline complex on zirconium phosphate also catalyzes the oxidation of NADH [1299]. Other mediators are guanine, oxoguanine [568], hematoxylin [1503], quinones and indophenol [1189,1298], coumestan [2877], dihydroxy benzaldehyde [368,2878,2879], methylaminophenol [368], and o-phenylenediamine [1199,1786]. Some redox dyes, Meldola blue, methylene green, methylene blue, fluorenones, Nile blue, and toluidine blue often on acidic or phosphate supports catalyze the oxidation of NADH [439,656,1130,1109,1193,1194,1196,1205,1212,1286,1287,1290,1291,1295,1296,1469,1476,2880–2886]. A few other organic compounds, including riboflavin, were tested to act successfully as mediators [1292,1293,1297,1512,2886–2888]. NAD(P)H is catalytically oxidized with a phenazine and oxygen as a co-substrate, yielding hydrogen peroxide; in a cascadic process HRP and ferrocene are oxidized yielding ferricinium, which is electrochemically reduced at the electrode surface [2889]. Diaphorase and glucose-6-phosphate dehydrogenase metabolize NADH and NAD⁺ in the presence of glucose-6-phosphate with vitamin K3, a quinone, as an electron acceptor [2890].

Glutathione is an oligopeptide with –SH function, which can undergo oxidation; it was a frequent subject for electrochemical studies with carbon paste electrodes [324,445,555,777,883,884, 920,947, 1394,1496,1511,1529,1550,1594,1608,2497,2540,2665,2672,2721,2891–2893]. As most methods rely on the oxidation of the thiol group, cysteine and other thiols are usually co-determined. Gold nanoparticles immobilized on cysteamine [947], Cu(OH)$_2$ (probably via Cu^{III}) [2891], V^{IV}O-salen (via V^V) [1594], and Ru(III)-diphenyldithiocarbamate [1608] support catalytic oxidation. Other mediators are ferrocenes [920,1540,1550], Co-phthalocyanine [324,555,883,884,2892], tetrathiafulvalene-tetracyanoquinodimethane (TCQM [1496]), Clp (in combination with carbon nanotubes [445], and 4-nitro-phthalonitrile [1511]. Crown ethers form host–guest complexes with glutathione and lower the detection limit down to around 50 nmol L^{-1} [1394]. Biosensors are based on the direct action of an enzyme, sulfhydryl oxidase [2672], or on inhibition, such as of pyranose oxidase [2665] or of HRP [2497,2540,2893]; another strategy is substrate blocking, where the substrate of tyrosinase is deactivated by reaction with R–SH [2721]. Oligopeptides (and amino acids) can be detected chromatographically by post-column derivatization with Cu(II) to form complexes, which can be oxidized to Cu(III) and detected with a Co-phthalocyanine-modified carbon paste electrode [558].

Carbon paste electrodes were also used to estimate the activity of the enzymes [291,779,1224, 1313,1779,2125,2707,2775,2776,2836,2894–2898]. Thermal stability of enzymes, particularly GOD [291,2707], and stability in acidic media [2836] are supported by the carbon paste matrix environment. Biotinylated alkaline phosphatase can be coupled to the electrode with streptavidin, and the activity was monitored via indigo from 3-indoxyl phosphate [2894]. Catalase activity can be determined with ruthenium dioxide–modified CPEs with hydrogen peroxide as a substrate [1224]. A size-exclusion membrane facilitates the estimation of γ-glutamyl transferase with glutathione, which is metabolized to glycyl-cysteine and can permeate through the membrane to the electrode surface [2895]. Isocitrate dehydrogenase activity can be followed with isocitrate via NADPH [1779]. LDH is determined indirectly with pyruvate and NADH by the signal decrease of the latter [1313]. Lysozyme attaches to anti-lysozyme aptamer and decreases its guanine oxidation signal [2896]. Neuropathy target esterase releases phenol from phenyl valerate detectable with tyrosinase and phenazine mediators [2125,2775,2776]. Pepsin is monitored with polyphenol oxidase through released tyrosine [779], plasmin [2897], and thrombin [2898] by liberated electroactive components.

Many proteins exhibit electroactivity due to oxidizable amino acids, mainly tyrosine and tryptophan, such as agglutinin [2899], albumin [832, 2704], angiotensin and vasopressin [92], avidin [2902], cucurbitin [2705], and α-synuclein [2930]; oxidation may proceed either directly or with the aid of modifiers, such as Cu-microparticles. Proteins modified with Au-nanoparticles can be determined by the electroactivity of the latter, for example, by deposition and re-oxidation of silver [2900]. Biotin is detectable with alkaline phosphatase labels and phenyl phosphate by the release of phenol and its oxidation with tyrosinase [2901]. Carbon paste is a very convenient electrode material for characterizing proteins and their reactions and interactions with target molecules, such as azurine [1499], chromatophores [2904], cytochromes [1271,1499,2905], fibrinogen [2856,2907], proteins and polycations [644,868], riboflavin binding protein [2928], and rusticyanine [2929]. Many studies deal with hemoglobin and myoglobin, their characterization and use as an electrode modifier; both proteins were frequently used for the catalytic reduction of oxygen, hydrogen peroxide, nitrite, and trichloroacetic acid [307,417,431,463,464,470,481,496,499,503,512,514,664,949,974,996,1168–1170, 1730,1889,2369,2797,2908–2918,2924–2953]. Proteins were frequently subject to immunosensors and immunoassays, so for carcinoembryonic antigen [302,985,2903], α-1-fetoprotein [301], immunoglobulin G [989,1276,1277,2824,2919–2950,3294], pneumolysine [2955,2927], protein A [925] and streptavidin [93].

8.4.7 HORMONES, PHYTOHORMONES, AND RELATED COMPOUNDS

In Table 8.21, an overview is given on investigations with carbon paste electrodes on hormones and phytohormones, or related substances.

Estrogens can be electrochemically oxidized, directly or with mediators [66,318,1674,2931,2932]; some xenoestrogens are detected amperometrically after capillary electrophoretic chromatographic separation [563]. Insulin is detectable via Zn after mineralization [2933], or via electroactive amino acids in the protein structure [612,2934]. The latter principle can be also applied to LH releasing hormone, bombesin, and neurotensin [630]. Melatonin can be accumulated at carbon pastes and is oxidized primarily irreversibly to the hydroquinone, which gives a reversible response hydroquinone-quinone; preanodization supports the electron transfer [236,2935–2939]. Thyroxine can be accumulated and oxidized electrochemically with supportive effect from surfactants [2000,2940–2943]; it is detectable also with immunosensors containing anti-T4 [2944]. Bile acids can be catalytically oxidized with the aid of cobalt phthalocyanine [633]. Cholesterol is usually determined by hydrogen peroxide (mediated redox reaction) released by the enzymatic reaction with cholesterol oxidase [871,1534,2945–2948].

Indole-3-acetic acid, a plant growth factor, is oxidized at carbon paste and is detectable after chromatographic separation [2949]; it can be accumulated in acidic medium electrostatically [2950] or with OV-17 stationary phase [1790]; mung bean leaves may serve as a source for indole acetic acid oxidase [2951]. Kinetin can be also preconcentrated and determined by oxidation [2952]. Roxarsone, an artificial growth factor for chicken, may be accumulated by the liquid anion exchanger Amberlite LA2; the electrochemical measurement is either oxidation or reduction of the analyte [1746].

8.4.8 NEUROTRANSMITTERS

Electrochemical methods for neurotransmitters and with detection by carbon paste electrodes are contained in Table 8.22.

Acetylcholine responds to a biosensor containing acetylcholine esterase and choline oxidase; the cofactor $FADH_2$ can be directly oxidized by TTF [1487]. Crude extracts of tissues or tissues themselves serve as a source of laccase which oxidizes catecholamines such as DA and epinephrine (adrenaline) [2953,2954]; even dual-tissue modification is possible (oyster mushroom for laccase and zucchini for peroxidase) [2954].

L-Dopa can be oxidized directly at carbon pastes [2743,2744], supported by surfactants [738], but usually it is done with mediators like dysprosium nanowires [1882], chloranil [2955], lead dioxide [1241], ferrocene with carbon nanotubes [420], and ruthenium red [1126]. Enzymatically oxidized dopa melanins were also characterized [2956]. Dopac gives improved signals with nanosized carbon black [2752].

DA has been one of the central topics with electrochemical investigations of neurotransmitters since the Adams' era [3,50]. It can be detected by differential pulse voltammetry at carbon pastes with other phenols and polyphenols like epinephrine, homovanillic acid, and also ascorbic and uric acids which usually overlap with the signal [2957]; carbon composites are applicable as amperometric detectors in CE [559]. Ionic liquids and nanostructured carbon materials catalyze the oxidation or improve the signal of the analyte [119,124,200,203,362,367,453,456,466,2958–2961]. Some catalysts were used to improve the electrochemical oxidation [2962–2964]. Some of the main interference in biological sample are AA and UA which overlap with the signal of DA with unmodified electrodes. Many efforts were made to improve the analytical situation in this respect: negatively charged barriers reject the anions of the acids if the pH is sufficiently high, such as Nafion membranes [2961], silica gel [1163], silica gel with phosphates [1213,127], zeolites [700,1108], zeolite with Nafion [1136], and anionic surfactants [243,1654,1659,1661,2966,2967]. Polyalcohols [195,1747,2965] improve the resolution of the signals by shifting the potentials of electrochemical responses. Surfactants in general have a strong influence on the oxidation signal of DA. Stearic acid has long been used as a modifier for carbon paste electrodes for *in vivo* brain electrochemistry due to its signal-resolving effect [1029]; nevertheless the application seems problematic (see the paragraph on brain electrochemistry). Neutral surfactants (*Triton X100*), often in combination with some acid, increase the signal [317,1500,1634,1648,1649]; cationic surfactants improve the resolution of overlapping signals [767,1632,2969–2972]. β-Cyclodextrin forms inclusion complexes with DA and supports its accumulation [2974–2976], and so do electropolymerized films of L-dopa [2977] and nanoparticles of SiO_2 [1008]. Various types of nanoparticles catalyze the oxidation and decrease the overpotential, such as Au [2978,2979], Pd [969], Ag [416], Fe [976], and Ni [972]. Ni(II) [1616,2980], SnO [2981], Fe(III) [1012], cobalt oxides [1230], ferrocenes [881,986], hexacyanoferrates [1043,1044,1799,1802], and even CdO nanoparticles [1003] can act as oxidation catalysts.

Some complexes of Fe(II) [1617], Cu(II) [2982], Co(II) [1583], and Ru(III) [1612,2000] are also active in this respect. Phthalocyanines of various metals, such as Cu [519,2983], Fe(II) [933,1570,2984], Co(II) [933,2984], and Mn(II) [912], can be also exploited for this task. The latter in combination with histidine should bio-mimic dopamine β-monooxygenase which hydroxylates DA; it is assumed that hydrogen peroxide oxidizes the phthalocyanine, which in turn extracts electrons from the analyte whereupon the quinone is electrochemically reduced. Analogous reaction is assumed for copper phthalocyanine [519]. Porphyrins of Fe(III) [1202] are also catalysts for the oxidation of DA. Apart from inorganic compounds and metallo-organic complexes, a variety of organic substances have been used as electrode modifiers. Acetone [2986] is questionable if it is really a "modifier," or if it just dissolves pasting liquids and improves the physical characteristics of the surface. Catalytically active useful substances are 2,4-dinitrophenyl hydrazine [225], *N*-hydroxysuccinimide [2987], phenylenediamines [992,1516], quinones and hydroquinones [992,995,1446,1462], tetracyanoquinodimethane [2988], Eriochrome black T in monomeric [2989] or polymeric form [1763], pyrogallol red [1515], hematoxylin [443], thionine [1729], Lamotrigine [1645], and Eperisone [1633]. Electropolymerized films [135,1762,1763,1806,2991,2992] seem to have a positive effect on the analytical behavior. Biosensors for DA contain DNA (interaction) [376], HRP (catalytically oxidizing the analyte with the aid of H_2O_2) [2697], quinoprotein dehydrogenase (using oxidized phenols as electron acceptors) [2786], polyphenol oxidase [194,280,661,1495], and tyrosinase [78,1258,2779,2993,2994]. Quite a few tissues can be used as a source of polyphenol oxidase [276,279,1018,2123,2564,2756,2759,2995–3004,3266].

DA metabolites have been detected chromatographically with CPEs as amperometric detectors [3005,3006]. Epinephrine, also known as adrenaline, can be adsorbed and electrochemically oxidized [553,607,3007–3009]; surfactants improve the signal [3010–3012]. Mesoporous aluminum oxide-SiO_2 supports the accumulation [1855]; iodine oxidizes the analyte which can be detected by reduction [3013]. Quite a few compounds were found to exert a catalytic effect on the oxidation of adrenaline, such as Fe(III) on zeolite [1124], phthalocyanines [913,3014], Methylene green [3015], hydroquinones [993,3016], and a carbothiamide [3017]. Biosensors for epinephrine are based on laccase or tissues (crude extract) as a pool of polyphenol oxidase [967,3018,3019].

Neurotransmitters on catechol base can be enzymatically oxidized with HRP and hydrogen peroxide [2697]. Noradrenaline (*syn.* norepinephrine) is detectable with unmodified CPEs by differential pulse voltammetry [3020] or by amperometric detection in HPLC [3021]. Hydroquinones (after oxidation) and chloranil catalytically oxidize the analyte [442,991,3022].

Serotonin was determined either with unmodified carbon paste electrodes [659,2737,2738,3023] or with modified dodecyl sulfate [1657], mesoporous silicagel [1852], or a clay mineral [1011] to facilitate accumulation and oxidation.

8.4.9 Nucleic Acid, Nucleic Bases, and Related Compounds

Table 8.23 surveys electrochemical studies with nucleic acids or their constituents. These bioessential substances have attracted some attention also by electrochemists [3024–3028]. Apart from mercury, they adsorb strongly to carbon; thus, carbon paste electrodes have been widely applied in studies with nucleic acids, because they allow simple addition of additional modifiers to characterize the target molecules and interactions with them.

Apart from studies on the constituents of nucleic acids (bases, nucleosides, and nucleotides), investigations may be categorized mainly into three major topics:

1. Adsorption studies including stripping analysis
2. Interaction studies with adsorbed nucleic acids (as modifiers) to
    a. Accumulate other analytes
    b. Study interactions of substances to evaluate protective or damaging effects
    c. Investigate possibilities for substances as reporter (indicator) molecules by intercalation
3. Hybridization studies with single-stranded DNA or oligomers

Adenine, guanine, cytosine, and thymine can be oxidized directly, but the corresponding potentials are rather in the positive range starting at around +0.7 V vs. SCE with guanine and increasing in the order guanine < adenine < thymine < cytosine ([328,723,1901,3029–3031,3268]; see also Figure 5.8). Binding to ribose or deoxyribose and polymerization to nucleic acids does not change the electrochemical behavior significantly. Therefore, nucleotides and oligonucleotides can be accumulated and directly oxidized [3033]. Direct oxidation of nucleic acid relies on the oxidation of the bases, but mainly the guanine signal is monitored because it appears lowest in potential.

Often background current correction is performed to improve the signal evaluation; carbon pastes with a wide positive potential window are preferable; this is particularly necessary if also other bases are monitored by direct oxidation.

Preanodization in an oxidizing medium creates strongly oxidizing groups at the carbon paste surface which facilitate the accumulation of oxidized adenine under open circuit conditions [3030]. Ionic liquids usually improve the resolution of signals [723] and even facilitate the distinction between guanine and guanosine [412,484]. Additives, such as cyclodextrin [1421] or montmorillonite [1066], improve the signal. It is also possible to use electrocatalysts, such as cobalt phthalocyanine [907,1043], Mo(VI) complexes [997], and hexacyanoferrates [1033,1038]. Inosine, a nucleoside with hypoxanthine bound to ribose and able to pair with adenine, cytosine, and uracil, is not electroactive but can be detected as a copper complex with La-hydroxide nanowires; the same is possible with uracil [3032]. A biosensor for nucleosides with nucleoside oxidase has been developed [788].

DNA, single or double stranded, and RNA easily adsorbs on the surface of carbon paste electrodes [361,474,491,629,661,3034–3036]. Usually the electrodes are preanodized, and adsorption is performed at a positive potential around 0.5 V vs. SCE. Ionic liquids or carbon nanotubes improve the analytical conditions by decreasing the oxidation potentials [444,661,918,394]. Dissolved DNA has been used to detect the voltage-dependent release of ethidium from ethidium-tetracyanoquinodimethane by complex formation and laser-induced fluorescence [3037,3038]. Sorption of DNA on clay [1069] or magnetic particles [3039] can be monitored with methylene blue. Ru(II)-bipyridyl improves the oxidation of guanine moieties in DNA [1604].

Oligonucleotides are detectable via Cu(I)-complexes [591], via accumulation and direct oxidation [3040], via preferential inclusion of oligonucleotides in electropolymerized polypyrrole films and discrimination from ssDNA and dsDNA [3041], or via handling with magnetic particles [299]. Heated electrodes improve the accumulation process [90].

DNA dendrimers are highly branched DNA molecules where subregions of more than two strands form double helical structures, thus interconnecting the different strands; they adsorb strongly to CPE and show very low detection limits [3042].

Peptide nucleic acids (PNA), introduced in 1992 by Egholm, Burchardt, Nielsen, and Berg [3262], are completely synthetic ssDNA mimicries with a peptide backbone (rather than ribose-phosphate): H–(NH–CH$_2$–CH$_2$–NR–CH$_2$–CO)$_n$–lys–NH$_2$ where "R" means –CO–CH$_2$– Base with Base as a common nucleic base [3263]. PNAs can hybridize with ssDNA. PNA can be accumulated at carbon paste electrodes and determined with stripping analysis, for example, potentiometric stripping analysis, similar to DNA [1988]. In fact, not only DNA but also RNA can be adsorptively preconcentrated and determined by means of ESA with either voltammetric or (chrono)potentiometric detection [91].

Many investigations deal with DNA-modified carbon paste electrodes in order to exploit them for accumulation of analytes or to uncover interactions between target compounds and the nucleic acid [85,289,292,293,376,390,430,608,1424,1463,1464,1467,1480,1602,1625,1921,2422,2451,2535,2538, 2541,2546,2549,2553,2565,2580,2582,2595,2596,2611,2622,2869,3043–3074]. Many substances intercalate with DNA, that is, they penetrate into the DNA helical structure. Interactions can be monitored by changes in the guanidine signal (increase or decrease) or by a stripping behavior of an electroactive analyte. By immobilizing DNA it is easy to perform damage and protection studies [292,293,2764,2422,3051,3066,3067] or to discover new electroactive intercalating compounds which can advantageously be employed as reporter molecules for hybridization [289,1463,1464,1467,1625,2565,3069].

One very promising field for many future applications are hybridization and sequencing sensors [289,298,378,388,421,766,953–955,1004,1005,1465,1466,1603,1638,1801,3075–3120,3269]. Usually a single-stranded DNA or oligonucleotide is adsorbed on the electrode surface; on exposition to the sample solution, hybridization occurs if complementary ssDNA is present. Hybridization can be monitored with hybridization indicators, also called reporter molecules. They are electroactive substances which have either a higher affinity to the free bases or which show a dominating intercalating effect with the double helix. In the former case a signal decrease compared to the ssDNA-modified electrode will be observed, in the latter an increase.

An interesting approach is the replacement of electroactive guanine by electroactive inosine (hypoxanthine as a base) in the single strands, which still can pair with adenine, cytosine, and uracil; the signal comes from guanine from the complementary strand in the hybrid [3084,3087,3088]. ssDNA-modified biosensors have been used to determine HIV virus [3093,3106], hepatitis B virus [289,3103–3105], cauliflower mosaic virus [954,1005,1025], aflatoxigenic *Aspergillus* spp. [3075], *Cryptosporidium* [3083,3084], *Mycobacterium tuberculosum* [3094], *Microcystis* spp. [3095], and *Salmonella* [3115] and also to reveal genetic diseases and mutations, such as cystic fibrosis (mucoviscidosis) [1638], factor V Leiden-mutation (hypercoagulability of the blood) [3087], achondroplasia (a form of dwarfism with normal torso and short limbs) [3088], point-mutations in genes [3089], and genetically modified food [298,954,1004,3098,3099,3107,3108,3110,3112–3114,3116,3117]; even single base mismatches can be detected [3120].

8.4.10 PURINES, PYRIDINES, AND PYRIMIDINES

Purine and pyrimidine bases which are important for nucleic acids have already been considered in previous section. Table 8.24 gives an overview of other "-ines," that is, pyridines and pyrimidines.

Caffeine is preconcentratable on carbon pastes, and can be directly oxidized, but at rather positive potentials [39,1691,3121–3122], whereas 1,4-benzoquinone forms an electroinactive complex with caffeine, which allows its indirect determination via the quinine signal [1444]. DNA enhances the accumulation [430]. Mesoporous TiO_2 improves the oxidation current of hypoxanthine [1875]; this analyte is frequently detected with xanthine oxidase [572,858,945,1497], where the electrochemical signals can be improved by Au nanoparticles or mediated by ferrocene or methyl viologen.

Nicotine can be quantified indirectly by inhibition of choline oxidase with choline as a substrate and benzoquinone as an electron acceptor; the enzyme converts choline into betaine [2452]. Some pyrimidine derivatives can be directly oxidized on carbon paste electrodes, which can be exploited for amperometric detection in chromatography [3123].

Theophylline also undergoes direct oxidation at around +1.0 V vs. Ag/AgCl [3125]; cetyl trimethylammonium bromide supports accumulation [741]; and cobalt phthalocyanine and K-hexachloroplatinate serve as catalysts [908,932,2626]. Thiouracil like many other thio-compounds is catalytically oxidized by cobalt phthalocyanine [908].

UA is another most frequently studied compound when using various carbon paste–based electrodes and sensors [119,135,203,226,247,422,427,442,453,456,590,659,660,676,916,969,995,1122, 1420,1446,1509,1513,1590,1632,1722,1847,1900,2738,2826,2876,3009,3016,3022,3126–3133]; mainly, because of the fact that UA in biological matrices AA is quite a common interferent. UA is a final product of the purine metabolism in many animals. It shows a poorly developed direct oxidation signal which can be improved with glassy carbon powder [659,2738,3126], carbon nanofibers [203,660], carbon nanotubes [3128], or ionic liquids [453,456]. Preanodization improves the electron transfer and allows simultaneous determination with xanthine, hypoxanthine, or AA [1900].

Nafion eliminates AA interferences by electrostatic repulsion [1722]. Cationic surfactants [1632,2971,2972,3129], cyclodextrin [1420], and Pd-nanoparticles [969] improve the resolution with other compounds yielding overlapping peaks (particularly AA). Mesoporous SiO_2 increases the oxidation signal by adsorption [1847].

Quite a few mediators may catalyze the oxidation of UA: Fe(III) on zeolites [1122], chloromercuri ferrocene [3130], Fe(II)-phthalocyanine [916], Co(II)-5-nitrosalophen [1590], 4-nitrophthalonitrile [3131], quinones and hydroquinones [422,397,442,995,1446,3016], thionine [1509], Congo red [1513], poly(calconcarboxylic acid) [135], and chloranil [3022]. Uricase, an enzyme occurring in many animals which metabolize (oxidize) UA to allantoin, can be exploited for corresponding biosensors [247,590,3132]; the enzyme can be also placed on a precolumn reactor in flow injection analysis [2826].

Xanthine is also a metabolite of purine degradation; it occurs in caffeine-containing plants and many animal tissues, in plasma, and in canned fish. It can be oxidized directly on carbon paste with poorly established signal [1900]; the electron transfer is significantly improved with carbon nanotubes and ionic liquids [3128]. The most common way of determination is with xanthine oxidase [722,943,3134–3139].

8.4.11 Vitamins

A myriad of analytical methods for vitamins [2748] that have been employing various types of carbon paste electrodes are summarized in Table 8.25.

One of the most frequently studied substances is AA [53,63,86,123,135,144,416,422,438,453 ,456,458,653,655,660,783,874,916,930,933,969,1003,1029,1043,1044,1087,1094,1110,1122,1133 ,1186,1202,1209,1229,1285,1288,1316,1318,1430,1446,1460,1461,1472,1491,1509,1516,1528 ,1542,1548,1549,1570,1583,1586,1590,1609,1610,1612,1617,1661,1704,1729,1742,1763,1764, 1799,1802,1806,1820,2002,2681,2748,2957,2964,2969–2972,2985,2989,3009,3131,3140–3177]). It is the main water-soluble antioxidant and is widely distributed in the animal and plant kingdom; it is also widely employed in food industry as a preservative.

AA is readily oxidized at low positive potentials already and can be determined directly [453,456, 458,1029,2957,3140,3145]; hexadecyl sulfonate shifts the potential to more anodic direction by electrostatic repulsion [1661], cationic surfactants [783,2969–2972], stearic acid [1029,2002], ionic liquids [453,456,458,3147], carbon ceramic [3009], carbon nanotubes [123], and carbon nanofibers [660] improve the signals and in many cases also the resolution from other analytes, in particular from DA and UA. Enhancement can also be achieved by intimate anodization of the CP-surface [63,210], regardless of which detection technique is used; see Figure 8.17 with the choice of stripping potentiometry.

Cyclodextrine and mesoporous silicagel facilitate adsorptive accumulation [3148–3150]. Cu(II) rather than Cu(I) forms complexes with AA, which can be reduced at slightly negative potentials [1316]. At positive potentials Cu(I)-oxide can be used to oxidize the analyte [1229]; catalytic oxidation also occurs with Cu(II)-phosphate [3151], Cu(II)-phthalocyanine [3152], hexacyanoferrates [1043,1044,1094,1742,1799,1802,1820], immobilized Fe(III) [1110,1122,1704,3153], pentacyano nitrosyl ferrate(II) [3154,3155], ferrocenes [870,874,1528,1542,1548,1549,3130,3155–3160] and other Fe-complexes [1202,1617], phthalocyanine [653,655,916,930,933,1570,3161,3162], Co(II)-complexes [1583,1586,1590], Ru(III)-complexes [86,1609,1610,1612,2985], Zn-azamacrocycles [3163], vanadates [144,3164], phosphomolybdate [3165], CdO nanoparticles [1003], and Pd and Ag nanoparticles [416,969]. Also organic compounds catalyze oxidation, such as quinones and hydroquinones [53,422,1446,1460,1461], tetramethyl phenylenediamine [1516], chloranil [3166], 4-nitrophthalonitrile [3131], thionine [438,1509,1729], TCQM [3167], TTF [1491], calixarene with Pb(II) [1430], various organic polymers [135,1764,1806,2964,3168,3169], monomeric and polymeric Eriochrome black T [1763,2989], Methylene blue [1087,1285,1288,1318,3170–3172], Methylene green [1472], Meldola blue [1209,3173], Brilliant yellow [1186], and Congo red [3174]. HRP [2681] acts catalytically via Fe(II) in heme; other biosensors can be made from plant tissue [3175,3176]. AA can also be determined indirectly by suppressing the electrochemiluminescent signal of luminol [3177].

Vitamins of the B group, such as folic acid [1663], nicotinamide, pyridoxine, riboflavin [1397], and nicotinic acid [1246,3178], give electrochemical responses at carbon pastes. The latter can be also determined with *Pseudomonas fluorescens* whole cells; they hydroxylate the

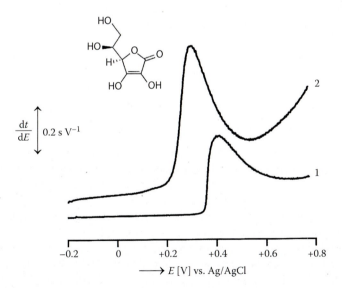

FIGURE 8.17 Oxidation of L-ascorbic acid (AA) at unmodified carbon paste electrode (1) and at the same electrode with electrolytically activated surface (2). *Experimental conditions*: silicone oil–based CPE; constant current stripping analysis (CCSA), deposition potential, -0.5 V vs. ref.; deposition time, 15 s; equilibration time, 10 s; stripping step: $-0.2 \rightarrow +0.8$ V vs. ref.; stripping current, $+1\,\mu$A; 1 M acetate buffer (pH 4.5), $c(AA) = 5 \times 10^{-4}$ mol·L^{-1}; activation procedure (prior to CCSA), $E_{AND} = +1.75$ V (for 120 s) + $E_{CATH} = -1.0$ V (15 s), 0.1 M Na$_2$HPO$_4$ + 0.05 M H$_3$PO$_4$ (pH 9). (Redrawn and rearranged from Švancara, I. et al., *Anal. Sci.*, 18, 301, 2002.)

analyte and use benzoquinone or phenazine methosulfate as an electron acceptor which provides an anodic re-oxidation current [786,1442].

Panthenol, used in skin ointments, is catalytically reduced, where probably the carbonyl group of the amide moiety is reduced to the alcohol, which can be further reduced to the amine by eliminating H_2O [1622]. Phylloquinone strongly adsorbs to carbon paste and is determined with adsorptive stripping voltammetry with open circuit accumulation [3179].

Pyridoxal can be bound with its aldehyde function to a lipophilic amine yielding a Schiff base –CH=N– which can be reduced to the amine [1637,1641].

Pyridoxine is oxidized either directly [1398,3180] or catalytically with hexacyanoferrate(III) or V(IV)-Salen (via V^V) [1034,1585]; crown ether may act as an accumulator [1398]. Riboflavin is the main component of the co-factor of flavoproteins, FAD and FMN; it can be immobilized on unmodified carbon paste or on silicagel or phosphate sorbents [39,1188,1198,1292–1294,3182] and is also used for the catalytic determination of NADH [1292,1293]. Crown ethers form supramolecular complexes with the analyte [1398,1408]. Thiamine is catalytically oxidized by Mn(II)-phthalocyanine (via Mn^{III}) [906].

Also, tocopherols can be studied [3184,3185] and determined [3183] with carbon paste electrodes. By the way, α-, β-, and γ-tocopherols are "the pioneers" among vitamins and the BIC in general, due to the fact that they were successfully recognized in a mixture and subsequently quantified already in the early 1970s [37].

Quite a few other vitamins are detectable with CPEs, such as vitamin A [3186–3188], vitamins of the B-group [232,1386,1486,3189], vitamin E [3190–3192], and vitamin K3 [1651]. One method for vitamin B12 is very straightforward by mineralization of the organic compound and determination of Co via adsorptive stripping with a hydroxamic acid [1386].

8.4.12 WHOLE CELLS, MICROORGANISMS, TISSUES, AND TISSUE EXTRACTS AS MODIFIERS

It is interesting to notice that various microorganisms and tissues have relatively often been used as modifiers for carbon pastes, as documented in an overview given in Table 8.26.

The various purposes of using tissues rather than purified single compounds are as follows:

1. *Sorption character* of the surface of microorganisms or tissues, which can be exploited for accumulations or as support for mediators
2. Tissue or its extract as a *source of an enzyme* or another important biological component without the need of time-consuming clean-up processes
3. *Monitoring* of biological processes

A disadvantage of using tissues and microorganisms as sources of biological modifiers is the variation of the target compound in different individuals, and frequently also within the tissue. Individuals of this type of biosensors usually have individual characteristics.

Algae have been used to accumulate metal ions, such as Au [270,3193], Cd [2260], Cu [2226, 2236,2245,2260,3194], and Pb [3196]. Some algae have been employed as a support for hexacyanoferrate for the catalytic determination of hydrogen peroxide [3195].

Bacteria rarely serve accumulation purposes [2198,2199]. Otherwise, they are exploited as a source of corresponding dehydrogenases or oxidases, like for ethanol [780,2664,3197] and glucose [288,381, 1441,1443,1521,2863,2864]. Nicotinic acid is hydroxylated by *P. fluorescens* with quinones as electron acceptor [786,1442]. *Moraxella* metabolizes *p*-nitrophenol to *p*-quinone and reduces it to *p*-hydroquinone which is more electroactive for oxidation than *p*-nitrophenol [2440]. Surface-expressed organophosphorous hydrolase on *Pseudomonas* hydrolyzes organophosphorous herbicides with ensuing degradation of the released nitrophenol [2476]. The respiratory activity of *P. fluorescens* on eggshell membrane attached to carbon paste can be monitored by the oxygen consumption [3198].

Lichens and mosses are occasionally used to design biomass biosensors for sorptive accumulation of inorganic species, such as copper, lead, and mercury [271,1712,3199]. Tissues were widely used as electrode modifiers, mainly as sources of certain enzymes [70,2564,3001,3200,3201]. (Notably, the first type of this sensor was called a "bananatrode" (with banana tissue on traditional Clark electrode [3264].)

Only a few applications of tissue electrodes are devoted to inorganic analytes; fluoride (inhibition of peroxidase) [2380], Pb(II) (sorption on banana tissue) [2270], and in fact hydrogen peroxide (usually as substrate for peroxidase) [311,2368,3202–3207].

Tissues most frequently serve as a source for peroxidase, such as horseradish (root) [3202,3203,3208], spinach root [3204], cabbage [311,3205,3206], green beans [2751], spinach leaves, and sunflower leaves [2123,2674–2676]. Most applications deal with tissues containing various phenol oxidases, for example, mushrooms [194,279,283,284,898,2564,2783,2840,2953,2954,3210,3211], palm trees and their fruits [2756,3019], coconuts [2754], potatoes [278,279,1754,2757,2760,2999,3000], zucchini [3001], eggplants [2791,3002,3266], fruits [278–280,2760,2765,2998], and, in particular, bananas [276,278,1265,2758–2760,2995–2997,3002]. Spinach leaves and sunflower leaves contain also glycolate oxidase [2674–2676], pea seedlings diamine oxidase [274], oat seedlings polyamine oxidase [864], and mung beans indole acetic acid oxidase [2951]. Orange peel can be used for the determination of pectin with an alcohol sensor, because the peel contains pectin esterase releasing methanol from pectin [277]. (For some further modifiers of such a kind, see also Section 2.2.1 and Figure 2.5.)

Yeast is a source of ADH [2660–2662] for the determination of ethanol and cytochrome b2 [1483,2685,2686] for the oxidation of lactate. Some special applications mimic bacteria counting, where bacteria [3212] or bacterial lysates [3213] are exposed to carbon paste surfaces, where they adsorb and decrease the electrochemical signal of a probe by blocking the active surface area.

Salmonella can be detected by an immune assay with phosphatase-linked anti-salmonella and phenyl-phosphate as a substrate by the release of electroactive phenol [3214,3215]. Wang with coworkers have suggested a carbon paste as the entrapping substrate for biological structures (enzymes, tissues, and cells) where biological processes can be monitored with microelectrodes [619].

8.4.13 SURVEY OF APPLICATIONS OF CARBON PASTE MINI- AND MICROELECTRODES IN BRAIN ELECTROCHEMISTRY/*IN VIVO* VOLTAMMETRY

An interesting area of *in vivo* measurements with CPm/μEs has already been reviewed in the previous chapter (see Section 7.3) with a note that the individual applications would be surveyed in more detail herein. Thus, Table 8.27 summarizes such measurements offering yet deeper insight into the individual methods and a database with typical experimental and instrumental conditions. This is also the case of experimentation with several macrosized constructions [152,605,2124,3216,3217]; nevertheless, numerous mini- and micro-variants have always prevailed [152,430,498,599,1022,1027,1028,1642,1674,1664,1717,1774,2119,2120,2122,2125–2141,2143–2153,3218,3122,3218–3251]. Exceptionally, the table includes also *in vitro* measurements (e.g., [2142,2154]) belonging, however, to the field as a useful comparative material. Finally, all the principal reviews are also cited.

8.4.14 MISCELLANEOUS

A few compounds which were determined with carbon paste electrodes and did not fit in the schemes displayed earlier are listed in Table 8.28 [952,2460,3252]. Again, the table also contains a series of selected review articles that could be of some interest for detailed studies [70,156,842,935,936,1254,2710,3200,3253–3255].

TABLE 8.1
Determination of Gold at Carbon Paste Electrodes, Sensors, and Detectors: Survey of Methods and Selected Studies

Form, Species	Type of CPE (Modifier)	Technique (Mode)	Measuring Principles (Method Sequences)	Linearity Range (LOD; t_{ACC}/t_R)	Sample(s)	Other Specification (Remarks)	Ref(s)
Au^{III}, Ag^{I}	C/Nj (unm.)	LSV	Electrolytic deposition Anodic reoxidn.	5×10^{-9}–3×10^{-7} M (5 nM; 15 min)	Model solns.	Simultaneous detn. of Au^{III} and Ag^{I} CPE: unusual bell-shaped constr.	[12]
Au^{III}, AuX_4^-	C/Nj (unm.)	LSV	Electrolytic deposition Anodic reoxidn.	25–500 ng mL^{-1} (1 ng mL^{-1}; 20 min)	Model solns.	pH-study included; quantification via peak shifts; X … Cl, Br	[2188]
Au^{III}, Me^N	C/PO (unm.)	LSV	Electrolytic deposition Anodic reoxidn.	(µM level)	Model solns.	detn. in presence of Pt-metals; Me^N: Pt^{IV}, Ru^{III}, Rh^{III}, Pd^{II}, Ir^{IV}	[2189]
Au^{III}, Au^{I}	C/Nj (Ce^{4+}, in situ)	LSCP (CCSA)	accum. assisted by modif. oxidn. by constant current	2×10^{-6}–1×10^{-5} M (80 nM; 15 min)	Drugs, human serum	Samples digested in min. acids Quantif. via E-t curves	[190]
Au^{III}, $AuCl_4^-$	C/Nj (unm.)	LSV	Electrolytic deposition Anodic reoxidn.	(0.1 ppm)	Metallic copper	s.e.: 0.1 M HCl	[2190]
Au^{III}, $AuCl_4^-$	C/PO (unm.)	DCV	Electrolytic deposition Anodic reoxidn.	(0.1 ppb)	Elemental silicon	s.e.: 0.1 M HCl + 0.001 M Fe^{3+}	[2191]
Au^{III}, $AuCl_4^-$	C/Nj (unm.)	ASV, CV	Electrolytic accum. Anodic reoxidn.	0.1–10 µM (0.1 µM)	Feed, raws, cleaner, post-cyanide baths	interfs. from Fe^{3+} suppressed by solvent extraction; no eff. of Ag^+	[2192]
Au^{III}, AuX_4^-	C/Nj (four diff. ion-exs)	CSV	accum. via ion-pairing Cathodic redn.	100–300 µg L^{-1}	Model solns.	interfs. from I^-, As^{III}, CrO_4^{2-}, and Pb^{2+}; Me … Cl, Br	[802]
Au^{III}, $AuCl_4^-$	C/Uv (dithizone)	AdSV	o.c. accum. via sorption Cathodic redn.	0.1–1.0 mg mL^{-1} (2 min)	Model solns.	Dithizone: reagent with $-N=$, $-OH$, and $-SH$ funct. groups	[799]

Au^{III}, AuX_4^-	C/Uv (TOPO, TBP)	AdSV	accum. via ion-pairing Cathodic redn.	$(0.02-1.0\,\mu g\,L^{-1})$	Model solns.	TOPO: trioctylphosphine oxide, TBP: tributyl phosphate; Me ... Cl, Br. accum. in open circuit Regeneration in NaCN solns.	[809]
Au^{III}, $AuCl_4^-$	C/Nj (algae *Chlorella*)	CSV	accum. via bio-affinity; Cathodic redn.	(μM level)	Model solns.		[270]
Au^{III}	C/PO (terc. alkyl amine)	DPV	accum. via ion-pairing	$0.1-20\,mg\,L^{-1}$ $(0.07\,mg\,L^{-1})$	Ores (ctng. gold)	Modifier: comm. compd. "N_{235}" incl. interf. study	[2193]
Au^{III}	C/PO (chelating resin)	DPV	accum. via $-NH_2$ and $-SH$ funct. Groups	$3\times10^{-8}-1\times10^{-6}\,M$ (10 nM; 15 min)	Minerals, copper, anode mud	incl. characterization by CV Acc. under open circuit conds.	[1678]
Au^{III}, $AuCl_4^-$	C/PO (flower *Datura*)	CSV	o.c. accum. by bio-affinity effect; cathodic redn.	$2.5-25\,mg\,L^{-1}$ $(0.5\,mg\,L^{-1})$	Model solns.	With MEx interfs. from Cd^{2+}, Tl^+, and V^V	[2194]
Au^{III}, $AuCl_4^-$	C/Uv (Rhodamine "B")	DPV	Electrostatic effect Cathodic redn.	$0.2-50\,mg\,mL^{-1}$	Model solns.	Modifier attached with surfactant acc. under open circuit conds.	[801,818]
Au^{III}, $AuCl_4^-$	C/PO (unm.)	CSV	accum. via adsorption at highly positive E_{acc}; redn.	$2-100\,ng\,mL^{-1}$ $(0.2\,ng\,L^{-1};$ 5 min)	Ores ctng. gold	s.e.: 0.1 M HCl; E_{acc} = +1.4 V. vs. Ag/AgCl; exp. parameters opt.	[2195]
Au^{III}, AuX_4^-	C/TCP (unm.)	DPCSV	accum. by extraction onto special liquid binder	$20-200\,\mu g\,L^{-1}\,(-)$	Model solns.	s.e.: 0.1 M HCl + 0.1 M KX (mixt.) (where X: F,Cl,Br,SCN,CN); eval. of stability constants, β_{AuXn}	[189,205]
Au^{III}, $AuCl_4^-$	C/Uv (thiobenzanilide)	DPCSV	accum. via $-N-$ and $-SH$ funct. groups; redn.	$0.05-0.6\,mg\,mL^{-1}$ (2 min)	Model solns.	Acc. under open-circuit conds. interfs from Hg^{2+} and Pt-metals	[1362]
Au^{III}	C/PO (tri-*iso*-octylamine)	DPV	Extractive accum. Cathodic redn.	$0.1-10\,\mu g\,L^{-1}$ $(30\,ng\,L^{-1};$ 15 min)	Ores ctng. gold	No sample pretreatment needed incl. interf. Studies	[518]

(continued)

TABLE 8.1 (continued)
Determination of Gold at Carbon Paste Electrodes, Sensors, and Detectors: Survey of Methods and Selected Studies

Form, Species	Type of CPE (Modifier)	Technique (Mode)	Measuring Principles (Method Sequences)	Linearity Range (LOD; t_{ACC}/t_R)	Sample(s)	Other Specification (Remarks)	Ref(s)
Au^{III}, $AuCl_4^-$	C/Nj (crown-ether)	DPV	o.c.-accum. by compl. and intercalation effect; redn.	$0.1–1\ \mu g\ mL^{-1}$ ($20\ ng\ mL^{-1}$)	Ores ctng. gold	modif.: 8-crown-6 Proc. involving MEx	[2196]
Au^{III}, $AuCl_4^-$	C/TCP (unm.)	DPCSV	accum. via ion-pairing (enabled by binder itself) Cathodic redn.	$1 \times 10^{-7} – 1 \times 10^{-5}\ M$ ($5\ min$)	Electrotechnical components (ec)	TCP: tricresyl phosphate (mixt. of isomers); interfs. of Fe^{3+} elim. by EDTA; ec. digested in *aqua regia*	[2197]
Au^{III}, $AuCl_4^-$	C/Nj (humic acid + en)	CV, DPSV	bioaccum. Process Cathodic redn.	$5 \times 10^{-8} – 4 \times 10^{-6}\ M$	Model solns.	en: ethylene diamine; s.e.: $0.4\ M\ HCl$ (pH 1.5); method with MEx	[1699]
Au^{III}	C/Nj (bacteria)	CSV	bioaccum. process Cathodic redn.	$10–100\ \mu g\ mL^{-1}$ ($1\ ng\ mL^{-1}$)	Model solns.	Modifier dried, grinded prior use Electrochem. pretreatm. of CPE (anodic activation)	[2198,2199]
Au^{III}, $AuCl_4^-$	C/Nj (SWy-2)	DPCSV	accum. by sorption/ion-exchange; cathodic redn.	$(8 \times 10^{-7}\ mol\ L^{-1})$	Tap water (spiked), mineral water	SWy 2: Na-montmorillonite; s.e.: $NaCl + HCl$; interf. study with Hg^{II}, SCN^- ($c_{Hg}:c_{Au} = 100:1$, $c_X:c_{Au} = 10^{-2}:1$)	[1062]
Au^{III}, $AuCl_4^-$	C/Nj (SWy-2)	DPCSV	accum. by sorption/ion-exchange; cathodic redn.	$(8 \times 10^{-7}\ mol\ L^{-1})$	pharms. forms (drug *Tauredon*)	SWy-2: Na-montmorillonite (natur. clay); pH-studies; tests on detn. in simulated sample of whole blood	[1063]
Au^{III}, ($AuCl_4^-$)	C/TCP (binder as modif.)	CCSA	accum. via ion-pairing redn. by const. current	($1 \times 10^{-7}\ M$; $5\ min$)	Gold-rich soils	interfs. from Fe^{III} supprs. by KF Samples mineralized by MWD	[710]

TABLE 8.2
Determination of Silver at Carbon Paste Electrodes, Sensors, and Detectors: Survey of Methods and Selected Studies

Form, Species	Type of CPE (Modifier)	Technique (Mode)	Measuring Principles (Method Sequences)	Linearity Range (LOD; t_{ACC}/t_R)	Sample(s)	Other Specification (Remarks)	Ref(s)
Ag^I, Au^{III}	C/Nj (unm.)	LSV	Electrolytic depos. Anodic reoxidn.	5×10^{-9}–3×10^{-7} M (5 nM, 15 min)	Model solns.	Simultaneous detn. of $Ag^+ + Au^{III}$ CPE: unusual bell-shaped constr.	[12]
Ag^+	C/BRF (unm.)	CV, ASV	Electrolytic depos. Anodic reoxidn.	(1×10^{-8} M; 5 min)	Model solns.	s.e.: 0.1 M ammonia buffer incl. interference studies	[217]
Ag^+	C/Nj (unm.)	CV, ASV	Electrolytic depos. Anodic reoxidn.	1–100 ng mL^{-1} (2×10^{-9} M; 5 min)	Model solns.	s.e.: 0.2 M KNO_3, RSD = ±9% Use of new CPE-piston holder	[31]
Ag^+	C/PO (unm.)	LSV	Electrolytic depos. Anodic reoxidn.	1×10^{-10}–1×10^{-8} M; (0.1 nM; 20 min)	Ores, wastewater, semiconductors	s.e.: 0.01 M ammonia buffer No interfs. from Fe^{III}, Mn^{II}, Hg^{II}, Cu^{II}, Bi^{III}, Pb^{II}, and Zn^{II}	[2200]
Ag^+	C/Nj; C/PW (AgX; Ag_2S)	POT	Equilibrium potential at liquid membrane ISE	(Ag^+: 5×10^{-5} M)	Model solns.	Applicable also to detn. of X^- ions (X … Cl, Br, I); strong oxidants and reductants interf.	[36]
Ag^+	C/Nj (en, dien ders.)	FIA-EC, LC-EC	compl. via $-NH_2$ groups prcp. AgCl and cath. redn.	1×10^{-11}–1×10^{-7} M (10 pM; 180 min)	Model solns.	Prior to chem. immobilization, C-particles were ox. by min. acid.	[51]
Ag^+	C/Nj (unm.)	DCV	Electrolytic depos. Anodic reoxidn.	(5 ppm; 5 min)	Metallic copper, alloy ctng. Cu	interfs. suppressed by adding compl. DCTA into s.e.	[2201]
Ag^+	s-CPE (C/Wx) (AgI)	POT	Equilibrium potential at liquid membrane ISE	1×10^{-4}–0.1 M	Model solns.	CP of semisolid character (with paraffin wax diss. in chloroform incl. evaluation of select. coeffs.	[1964]
Ag^+	C/Nj (zeolites)	DPV	accum. via ion-exchange Cathodic redn.	(<1 mg L^{-1}; 3 min)	Model solns.	Combined with MEx simult. detn. of Hg^{II} possible	[1131]
Ag^+	C/PO (thiacrown cmpds)	DCV	compl. via chelating redn. and anodic reoxidn.	0.5–2.5 mM (0.5 mM; 5 min)	Model solns.	Open circuit + MEx No interfs. from Cu^{II}, Cd^{II}, Ni^{II}, Co^{II}	[14403]
Ag^+	C/Nj (chelating resin)	DPV	compl. by $-NH-$ groups Cathodic redn.	5×10^{-10}–1×10^{-7} M (0.3 nM; 5 min)	Wastewater	s.e.: opt. pH: 6.5–7.5 No interfs. from common metals	[1677]

(continued)

TABLE 8.2 (continued)
Determination of Silver at Carbon Paste Electrodes, Sensors, and Detectors: Survey of Methods and Selected Studies

Form, Species	Type of CPE (Modifier)	Technique (Mode)	Measuring Principles (Method Sequences)	Linearity Range (LOD; t_{ACC}/t_R)	Sample(s)	Other Specification (Remarks)	Ref(s)
Ag^+	C/Nj (chelating cmpd)	CV, LSV, DPV	compl. by –S– groups Cathodic redn.	1×10^{-9}–1×10^{-6} M (1 nM; 20 min)	CRMs (–)	Regeneration in min. acid	[1328]
Ag^+	C/PO (AgI)	POT	Equilibrium potential Liquid membrane ISE	5×10^{-7}–0.01 M (0.1 μM)	Fixing and waste electroplate solns.	No interfs. from Hg^{II} (at pH 5–6) Response slope: 63 ± 2 mV	[2202]
Ag^+	C/Nj (2,2'-dithiopy.)	CV, LSV, DPV	compl. by –S– groups Cathodic redn.	—	Model solns.	comp. with other subst. pys incl. interf. studies	[1329]
Ag^+	C/PO (unm.)	LSV	Electrolytic dep. Anodic reoxidn.	1×10^{-8}–1×10^{-7} M (<10 nM; —)	Model solns.	s.e.: 0.5 M KNO_3 detn. in presence of Hg^{2+}	[2203]
Ag^+	C/Nj (phen drv. + surf.)	DPV	compl. via ion-pairing Cathodic redn.	8×10^{-8}–5×10^{-7} M (0.5 nM; —)	Wastewater	s.e.: HNO_3-based medium incl. interf. Studies	[1660,2204]
Ag^+	C/Nj (cinchonidine)	CV, DPV	compl. via ion-pairing Cathodic redn.	8×10^{-9}–1×10^{-6} M (0.8 nM; 20 min)	Model solns.	s.e.: neutral medium (pH 7) interfs. from Hg^{II} and Mn^{II}	[2205]
Ag^+	C/Nj (CYNQ)	LSV	compl. in open circuit redn. after MEx	1×10^{-7}–8×10^{-5} M (50 nM; 10 min)	Model solns.	CYNQ: 2,3-dicyano-1,4-naphthoquinone	[1331]
Ag^+	C/PO (unm.)	CV, ASV	Cathodic electrodep. Anodic reoxidn.	5×10^{-9}–1×10^{-5} M (5 nM; —)	Model solns.	no interfs. from Cr^{III}, Ni^{II}, Mn^{II}, Zn^{II}, and Pb^{II}	[2206]
Ag^+	C/Uv (MI)	DPASV	o.c.-accum.; MEx redn.; anodic reoxidn.	0.5–1000 mg L^{-1} (0.1 mg L^{-1}; 5 min)	pharms.	MI: 2-mercaptoimidazole: interfs. from other ions elim. by EDTA	[1330]
Ag^+, Hg^{2+}	C/Nj (GHA)	CV, DPV	accum. by compl. Cathodic redn. + regnt.	(1×10^{-10} M; 40 min)	CRM (urine)	GHA: glyoxal bis(2-hydroxyanil) simult. detn. of Ag^+ + Hg^{2+}	[1322]
Ag^+	C/TCP ($C_7H_{15}SO_3Na$)	DPASV	Ion-pairing via modif. in situ; redn. + anodic detc.	3×10^{-12}–2×10^{-5} M (<1 pM; 120 min)	Tap water	elchem. cathodic pretreat. of CPE interfs. from Hg^{II} and Cu^{II} elim. by EDTA, from Au^{III} by column sep.	[182]
Ag^+, Cu^{2+}	C/PO (–)	ASV	Electrolytic accum. Anodic reoxidn.	0.05–150 μg L^{-1} (10 min)	Model solns.	Regeneration of CPE in dil. HNO_3 incl. studies on pH of s.e.	[2207]

Electroanalysis with Carbon Paste–Based Electrodes, Sensors, and Detectors

Analyte	Electrode	Method	Principle	Range (LOD)	Matrix	Notes	Ref.
Ag^+	C/PO (BTT)	CSV	accum. by modif. (donor podant); cathodic redn.	$1 \times 10^{-7}–1 \times 10^{-6}$ M (–)	Model solns.	BTT: 1,1′-bis(thiophenoxy)-3,6,9-trioxaundecane; studies on effect of pH and various interf. ions	[1336]
Ag^+, Ag_mL_m	C/PO (unm.)	SWSV	accum. of free and compl. forms and their detc.	(<0.2 µg L^{-1})	River, lake water	incl. speciation study (Ag^+ vs. Ag^I-compl. with natur. ligands)	[2208]
Ag^+, Hg^{2+}	C/Nj (vermiculite)	SWV	accum. by ion exchange MEx; redn.; ASV detc.	$1 \times 10^{-7}–8 \times 10^{-6}$ M (6×10^{-8} M; 10 min)	Model solns.	CMCPE regnt. and actv. simult. detn. of $Hg^{2+} + Ag^+$	[1082]
Ag^+	C/PO (keratin)	DPASV	compl. by –S, –SH groups redn. and anodic reoxidn.	(2×10^{-8} M; 10 min)	Photographic developer	s.e.: 0.1 M acetate buffer No interfs. from common ions	[287]
Ag^+	C/PO (S_2O_2-compd.)	DPCSV	accum. via donor podant Cathodic redn.	$5 \times 10^{-7}–6 \times 10^{-6}$ M (2×10^{-7} M; 10 min)	Model solns.	S_2O_2: donor podant (added *in situ*) No interfs. from Me^{n+}, except Pb^{2+}	[2209]
Ag^+	C/PO (EDTA)	ASV	accum. (electrolytic redn.) Anodic reoxidn.	$2 \times 10^{-9}–1 \times 10^{-6}$ M (1×10^{-9} M; 10 min)	Model solns.	EDTA used as modif. agent elim. interfs. from other ions	[2210]
Ag^+	C/PO (DPO)	DPASV	o.c. accum. via compl./redn.; anodic reoxidn.	$5 \times 10^{-9}–1 \times 10^{-7}$ M (0.7 nM; 10 min)	Model solns.	DPO: N,N'-diphenyl oxamide No interfs. from common ions	[1333]
Ag^+, Pb^{2+}	C/PO (SWy-2)	DPASV	accum. via ion-pairing redn. + anodic reoxidn.	$8 \times 10^{-9}–1 \times 10^{-7}$ M (0.1 nM; 4 min)	Model solns.	SWy-2: Na-montmorillonite (clay); in presence of CTAB; possible join detn. of Pb^{2+} investigated	[2211]
Ag^+	C/PO (Alizarin violet)	DPASV	accum. via complexation el. redn./MEx/reoxidn.	$3 \times 10^{-10}–1 \times 10^{-7}$ M (0.1 nM; 3 min)	Wastewater, zinc alloy	s.e.: (a) 0.1 M AcB (pH 5.2), (b) 0.1 M H_2SO_4 + 1×10^{-4} M KBr No interfs. from Me^I and Me^{II}	[690]
Ag^+	C/DINP (Ag-*Thimerosal*)	POT, FIA (dir., titr.)	Chemical equilibrium and steady-state potential for $Ag^+ \leftrightarrow Ag^I$-compl.	$1 \times 10^{-8}–5 \times 10^{-7}$ M (3×10^{-7} M)	Cosmetics, pharms.; radiology films	DINP: di-*iso*-nonyl phthalate der.; pharms: *Thiopental*, *Thimerosal* (both by potentiometric titration)	[222]
Ag^+	C/Nj (+BiF; *in situ*)	SWASV	accum. via alloy form. Anodic reoxidn.	5–50 µg L^{-1} (–)	Model solns.	Study of initial character (detn. of metals nobler than Bi); also Hg^{2+}	[2212]

(*continued*)

TABLE 8.2 (continued)
Determination of Silver at Carbon Paste Electrodes, Sensors, and Detectors: Survey of Methods and Selected Studies

Form, Species	Type of CPE (Modifier)	Technique (Mode)	Measuring Principles (Method Sequences)	Linearity Range (LOD; t_{ACC}/t_R)	Sample(s)	Other Specification (Remarks)	Ref(s)
Ag^I	C/PO + memb. (BNSAO)	POT (dir., titr.)	Chemical equilibrium and steady-state potential for $Ag^+ \leftrightarrow Ag^I$-compl.	9×10^{-7}–0.03 M (4×10^{-7} M)	Cosmetics, radiology films	BNSAO: bis 5-(4-nitrophenyl-azo) salicylaldimine (*Schiff base*); CPE compared with similar CWE	[2213]
Ag^+	C/PO, (DPSG)	POT (dir.)	Chemical equilibrium and steady-state potential for $Ag^+ \leftrightarrow Ag^I$-compl.	5×10^{-7}–0.1 M (1×10^{-7} M)	Wastewater	DPSG: dipyridyl-functionalized silica gel; s.s.: AcB (pH 5.5) CPE long stable (>6 months)	[1869]
Ag^+	C/PO (AMQ)	DPASV	accum. via ion-pairing redn. + anodic reoxidn.	1–300 µg L^{-1} (0.4 µg L^{-1}; 12 min)	Natural waters, photographic films	AMQ: 3-amino-2-mercapto quinazolin-4-one; comp. with GF-AAS	[1335]
Ag^+	C/PO (Tu-SBA-15)	DPASV	accum. via complexation redn. + anodic reoxidn.	8–80 pmol L^{-1} (5×10^{-12} M)	Tap and sea water	mdf.: phenylthiourea-nanoporous silica gel (10% w/w in the paste) Results compared with GF-AAS	[1868]
Ag^+	C/PO (p-IPCX)	DPASV	accum. via complexation redn. + anodic reoxidn.	5×10^{-8}–2×10^{-6} M (5×10^{-8} M)	X-ray photographic film	pIPCX *p*-isopropylcalix[6]arene; s.e.: 0.1 M H$_2$SO$_4$ + 0.02 M KBr	[1433]
Ag^+	DPEs [D/PO] (unm., three types)	DPASV	accum. (electrolytic redn.) Anodic reoxidn.	0.5–10.0 µg L^{-1} (<0.5 µg L^{-1})	Residual water	DPEs made of nat. and two synth. diamond powders; $E_P = +0.08$ V vs. ref.; interfs. study (Mn, Cr, Fe, Cu)	[353]

Electroanalysis with Carbon Paste–Based Electrodes, Sensors, and Detectors 217

TABLE 8.3
Determination of Mercury at Carbon Paste Electrodes, Sensors, and Detectors: Survey of Methods and Selected Studies

Form (Species)	Type of CPE (Modifier)	Technique (Mode)	Measuring Principles (Method Sequences)	Linearity Range (LOD; t_{acc}/t_R)	Sample(s)	Other Specification (Remarks)	Ref(s)
Hg^{2+}	C/Nj	LSV	Electrolytic accum.	$0.8–2.5\,\mu g\,L^{-1}$	Model solns.	s.e.: $3.6\,M\,Li_2SO_4$ incl. interf. study	[2214]
Hg^{II} $(Hg(SCN)_4)^{2-}$	C/Nj	LSV	Electrolytic accum.	$1 \times 10^{-9}–1 \times 10^{-7}\,M$	Human urine	s.e.: $0.1\,M\,KSCN + 0.03\,M\,HCl$ In the presence of Cu^{2+}	[2215]
Hg^{2+}	C/Nj (zeolite)	SWV, CV	accum. by ion-pairing Cathodic redn. of Hg^{II}	$0.1–2.2\,mg\,L^{-1}$ ($0.1\,mg\,L^{-1}$, 6 min)	Model solns.	s.e.: KNO_3-based solns. incl. pH studies	[828]
Hg^{2+}	C/Nj (crown ethers)	DPASV	accum. by compl. and adsorption; cathodic redn.	10–60 mM	Model solns.	modifs.: two diff. crown ethers s.e.: acetate buffer (pH 4)	[1391]
Hg^{2+}	C/Nj (1-spartein)	DPV	accum. by compl. Cathodic redn. of Hg^{II}	$5 \times 10^{-7}–2 \times 10^{-6}\,M$	Waters	CPE regnt. in acetate buffer Ag^+ interf., suppr. by adding KCl	[2216]
Hg^{II} $(Hg(SCN)_4)^{2-}$	C/PO (1,5-DPC)	DPV	accum. by compl. Cathodic redn.	$4 \times 10^{-7}–3 \times 10^{-5}\,M$ ($5 \times 10^{-9}\,M$)	Model solns.	modif.: 1,5-diphenyl carbazide s.e.: $0.1\,M\,KSCN$, Cu^{2+} interf.	[1319]
Hg^{2+}	C/PO (humic acid)	DPV	accum. by compl. Cathodic redn.	$5 \times 10^{-8}–5 \times 10^{-7}\,M$	Model solns.	Cu^{2+} interf. (c_{Cu} : c_{Hg} = 10 : 1)	[1693]
Hg^{2+}	C/Nj (NaTPB, in situ)	DPASV	det. via decrease of peak of TPB (undirect method)	$1 \times 10^{-8}–1 \times 10^{-6}\,M$ ($5 \times 10^{-9}\,M$; 10 s)	Natural water (spiked)	NaTPB: sodium tetraphenyl borate No interf. from Ag^+, Cu^{2+}, Bi^{3+}	[2217]
Hg^{2+}	C/Nj (SiO_2 + TBIM)	DPASV	accum. by compl. redn.; anodic reoxidn.	$0.1–2.0\,ng\,L^{-1}$	Model solns.	TBIM: 3-(2-thiobenzimidazolyl)-funct. group attached to silica gel	[1827]
$(HgCl_4^{2-}, HgC_6H_5^+)$	C/PO (Amberlit LA-2)	DPASV	o.c.-accum. by ion pairing MEx; redn.; reoxidn.	$1–1000\,\mu g\,L^{-1}$ ($1\,\mu g\,L^{-1}$, 15 min)	pharms.	Applicable to detn. phenylmercury compds.; Ag and Pt-metals interfs.	[800]
Hg^{2+}	C/PO (TQZ)	DPASV	o.c.-accum. via compl. MEx; redn.; reoxidn.	$5–6000\,\mu g\,L^{-1}$ ($0.1\,\mu g\,L^{-1}$, 15 min)	Sewage sludge, plant	TQZ: 2-thio-4(3H)-quinazoline	[1320]
Hg^{2+} (+Me^{2+})	C/PO (Amberlite IRC-718)	DPASV	accum. via chelating with modif.; redn. + reoxidn.	(mg L^{-1} level)	Model solns.	Me: Zn, Cd, Pb, and Cu; study on selective retention with the aid of modif. and possible speciation	[1738]

(continued)

TABLE 8.3 (continued)
Determination of Mercury at Carbon Paste Electrodes, Sensors, and Detectors: Survey of Methods and Selected Studies

Form (Species)	Type of CPE (Modifier)	Technique (Mode)	Measuring Principles (Method Sequences)	Linearity Range (LOD; t_{ACC}/t_R)	Sample(s)	Other Specification (Remarks)	Ref(s)
Hg^{2+}, Pb^{2+}, Cu^{2+}	C/Nj (humic acid, HA)	CV, DPASV	accum. by complexation with HA; redn. + reoxidn.	$(8 \times 10^{-9}$ M; 20 min)	CRMs (urine samples)	HA-CPE regnd.; interf. by Ag^+ ion minimized by precipitating with KCl; simult. detn. with Pb^{2+}, Cu^{2+} tested.	[1695]
$(Hg^{2+} HgCH_3^+)$	C/Nj (thiolic resin)	CV, CSV	accum. by compl. Cathodic redn.	$(4 \mu g\ L^{-1}$ for Hg^{2+}; $2 \mu g\ L^{-1}$ for $HgCH_3^+)$	Aquatic samples	Applicable to speciation of inorg. mercury and organomercury	[1680]
Hg^{2+}, Ag^+	C/Nj (Schiff bs.: GHA)	CV, DPV	accum. by compl. Cathodic redn. + regnt.	$(1 \times 10^{-9}$ M; 40 min)	CRM (urine)	GHA: glyoxal bis(2-hydroxyanil) simult. detn. of Hg^{2+} + Ag^+	[1322]
Hg^{2+}, Hg_2^{2+}	C/Nj (AHA)	CV, LSV	o.c.-accum. via compl. MEx; cathodic redn.	5×10^{-8}–5×10^{-6} M	Model solns.	AHA: amidised humic acid Speciation of Hg^I + Hg^{II} possible	[1698]
Hg^{2+}	C/SO (+AuF, in situ)	DPASV	Electrolytic accum. Anodic reoxidn.	1×10^{-9}–5×10^{-7} M (3×10^{-10} M; 30 min)	Natural water (polluted)	s.e.: 0.1 M HNO_3 + 0.02 M KCl No interfs. By Cu^{II} (Cu:Au <100:1) little interfs. from Hg^{II}, Se^{IV}, Ag^I	[260]
Hg^{2+}	C/PP (Au-colloid)	DPASV	Electrolytic accum. Anodic reoxidn.	(ca. 20 ng L^{-1})	Model solns.	Preliminary study Mechanical regnt. of "solid" CP	[320]
Hg^{2+}, Ag^+	C/Nj (vermiculite)	SWV	accum. by ion exchange MEx; redn.; ASV detc.	1×10^{-7}–8×10^{-6} M (6×10^{-8} M; 10 min)	Model solns.	CMCPE regnt. and actv. simult. detn. of Hg^{2+} + Ag^+	[1082]
Hg^{2+}	C/PO (Zn^{II}-DDC)	DPASV	accum. By compl. + MEx redn. + anodic reoxidn.	5×10^{-8}–1×10^{-7} M (8×10^{-10} M; 15 min)	Natural water, human urine	DDC: diethyldithiocarbamate Combined with MWD	[1323]
Hg^{II} ($HgCl_4^{2-}$)	C/Nj (SWy-2)	DPV	accum. via ion-exchange Cathodic redn.	5×10^{-8}–2×10^{-7} M (15 min)	Drinking, sea waters (spiked)	s.e.: chloride-based media Speciation of $HgCl_4^{2-}$ vs. Hg^{2+}	[1061]
Hg^{2+}	C/PO (SWy-2 clay)	DPASV	Ion-exchange + adsorption redn. + anodic reoxidn.	1×10^{-9}–5×10^{-7} M (1×10^{-10} M; 6 min)	Water samples	SWy-2: sodium montmorillonite	[2218]

Hg_2^{2+}	C/PO (PA-NO)	ASV	Specific interact. with Hg^I. Anodic reoxidn.	3×10^{-8}–1×10^{-6} M	Human urine	PA-NO: picolinic acid N-oxide. Analyzed as $Hg^I + Hg^{II}$ mixtures	[2219]
Hg^{II}, HgI_3^-	C/PO (TZ^+; $[HgI_3]^-$)	POT (dir.)	Chemical equilibrium: $Hg^{2+} \leftrightarrow Hg^{II}$-compl.	6×10^{-6}–0.001 M (4×10^{-6} M)	Wastewater, alloy, dental amalgam	TZ: tetrazolium (quaternary) ion incl. studies on select. coeff.	[1312]
Hg^{2+}, Pb^{2+}	C/PO (SH/ms-SiO$_2$)	SWV	accum. via compl. redn.; anodic reoxidn.	20–1600 ppb Hg (10 ppb; 5 min)	Model solns.	SH/ms.SiO$_2$: thiol-ctng. mesoporous silica; simult. detn. of Hg+Pb	[772]
Hg^{2+}	C/SO, C/PO (+AuF, ex situ)	SP (CCSA)	accum. by electrolytic redn. reoxidn. with I_{CONST}	(2 ppb; 10 min)	Model solns.	AuF plated in situ or preplated simult. detn. of $Hg^{2+} + Cu^{2+}$	[1993]
Hg^{II} (var. sp.)	C/PO (SH-, NH$_2$-PSXL)	LSV, DPCSV	Specific sorption processes o.c. accum. + cathodic redn.	— (not specified)	Soils and water samples	Species studied: $HgCl_4^{2-}$, $HgCl_3^-$, $HgCl_2$, $Hg(OH)_2$ (speciation st.) PSXL: polysiloxane ligand	[702]
Hg^{2+}, Hg^{II} (HgX_n^{m-})	C/PO (BbTSC)	CV, SWV	Specific complexation Anodic reoxidn.	10–50 ppb (8 ppb; 15 min)	River water samples	BbTSC: benzylbisthiosemicarbazone Speciation of Hg^{2+}, $HgCl_3^-$, and $HgCl_4^{2-}$ via the ligand competition	[1325]
Hg^{2+}	C/PO (DTTD-SG)	CV, DPCSV	Sorption + ion-exchange accum./cathodic redn.	3×10^{-8}–1×10^{-8} M (1×10^{-8} M; 3 min)	Polluted water	DTTD-SG: 2,5-dithio-1,3,4-thiadiazole functionalized silicagel CMCPE regenerated chemically	[1628]
Hg^{2+}	C/PO (APSC, MPSC)	DPASV	o.c. accum. (ion-exchange) Anodic reoxidn.	1×10^{-8}–7×10^{-8} M (8×10^{-8} M; 5 min)	Model solns.	A(M)PSC: amino(mercapto) phyl-lo silicate clay (synth. by grafting)	[1089]
Hg^{2+}	C/PO (TZT-HDTA-C)	CV, DPASV	Selective sorption + redn. Anodic reoxidn.	—	Model solns.	TZT-HDTA-C: 2-thiazoline-2-thiol-hexadecyltrimethylammonium-clay No interfs. from Pb, Cd, Cu, or Zn	[1102]
Hg^{2+}	CNTPE (unm.)	CV, SWASV	Electrocatalysis-assisted electrolysis; reoxidn.	1–25 + 40–200 μg L^{-1} (0.4 μg L^{-1}; 12 min)	Water samples	CNT-PE: carbon nanotube paste electrode; s.e.: AcB (pH 4.0)	[369]

(continued)

TABLE 8.3 (continued)
Determination of Mercury at Carbon Paste Electrodes, Sensors, and Detectors: Survey of Methods and Selected Studies

Form (Species)	Type of CPE (Modifier)	Technique (Mode)	Measuring Principles (Method Sequences)	Linearity Range (LOD; t_{ACC}/t_R)	Sample(s)	Other Specification (Remarks)	Ref(s)
Hg^{2+}	C/PO (EBHMP)	POT (dir.)	Chemical equilibrium and steady-state potential for $Hg^{2+} \leftrightarrow Hg^{II}$-compl.	3×10^{-7}–0.01 M (1×10^{-7} M)	Wastewater (spiked); amalgam	EBHMP: ethyl-2-(benzoylamino)-3-(2-hydroxy-4-methoxyphenyl)-2-pro-penoate; s.e.: pH 1–4, t_R: <5 s	[1326]
Hg^{2+}	C/PO (SIAMT)	CV, DPASV	accum. via sorption and complex.; redn. + reoxidn.	1–20 μg L^{-1} (0.1 μg L^{-1}; 15 min)	Wastewater (spiked); amalgam	SIAMT: silica gel functionalized with 2-aminothiazole; var. s.e.	[1829, 1837]
Hg^{2+}	C/PO (SWy-12/MATD)	LSV, LC-EC	accum. via sorption and complexation; redn.	(10 μg L^{-1}; —)	Model solns.	SWy-12/MATD: Na-montmorillonite + 2-mercapto-5-amino-1,3,4-thiadiazole; clay of natural origin (deposit)	[2220]
Hg^{2+}	C/PO (DTSA)	POT (dir.)	Chemical equilibrium and steady-state potential for Hg^{II}-HA ↔ Hg^{II}-compl.	1×10^{-7}–3×10^{-4} M (2×10^{-8} M)	Model solns.	DTSA: dithiosalicylic acid; HA: humic acid (competitive ligand) Evaluation of stability constants	[2221]
Hg^{2+}	C/Nj (chitosan, CHS)	CV, DPASV	accum. via sorption and complex.; redn.+ reoxidn.	1×10^{-6}–4×10^{-5} M (6×10^{-7} M; 4min)	Water samples (spiked)	CP-composition: 60% (m/m) C + 20% PO (Nujol) + 20% (m/m) CHS s.e.: 0.2 M KNO_3 (pH 6.2)	[1709]
Hg^{2+}	C/PO (Cryptofix)	ASV	accum. via complexation el. redn./anodic reoxidn.	— (0.1 μg L^{-1})	Model solns.	Cryptofix: commercial reagent (with ion-exchange capabilities)	[2222]

Hg^{2+}	C/PO (Cadion A)	CV, LSV	accum. via complexation el. redn./anodic reoxidn.	0.1–20 µg L^{-1} (0.1 µg L^{-1}; 1 min)	Polluted waters	Cadion A: 4-nitrophenyldiazo-aminoazobenzene; s.e.: 0.2 M NaAc (pH 7) accum. performed in open circuit	[1327]
Hg^{2+}, Cu^{2+}, Pb^{2+}	C/PO (AMT/plex-µS)	SWASV	accum. via complexation el. redn. + reoxidn.	12–70 µg L^{-1} (10 min) (5 µg L^{-1}; 15 min)	Model solns.	AMT/plex-µS: 2-aminothiazole in plexi-polymer made microspheres Simultaneous detn. with Cu^{2+}, Pb^{2+}	[1683]
Hg^{2+}, Cu^{2+}, Pb^{2+}	C/PO (SBA-15)	CV, DPASV	accum. via sorption and compl.; redn.+ reoxidn.	2 × 10^{-6}–1 × 10^{-5} M (5 × 10^{-7} M; 4 min)	Natural water; alcoholic drink (from sugar cane)	SBA-15: nanostructured silica functionalized with 2-benzothiazolethiol Simultaneous detn. with Cu^{2+}, Pb^{2+}	[2223]
Hg^{2+}	C/PO (DPSG)	CV, LSV	accum. via complexation el. redn./anodic reoxidn.	20–100 nM (8 nM)	Wastewater	DPSG: dipyridyl functionalized nanoporous silica gel (15=); 0.02 M KNO$_3$ results compared with CV-AAS	[1871]
Hg^{2+}	C/PO (FTU-LUS-1)	POT (dir.)	Chemical equilibrium and steady-state potential for Hg^{2+} ↔ Hg^{II}-compl.	1 × 10^{-7}–0.1 M (7 × 10^{-8} M; 35 s)	Wastewater, fish tissue	mdf.: thiourea-functionalized nanoporous silica, its special characteris. by CNS, TG, 4x NMR, FTIR, XRD	[1867]
Hg^{2+}	C/PO (Salophen)	POT (dir., titr.)	Chemical equilibrium and steady-state potential for Hg^{2+} ↔ Hg^{II}-compl.	3 × 10^{-7}–3 × 10^{-4} M (2 × 10^{-7} M; 35 s)	Wastewater, fish tissue	mdf.: bis-(salicylaldehyde)-phenylene Diamine; pH 3.8–7.8; study on the role of Hg(OH)$^+$ ion; interf.	[2224]

TABLE 8.4
Determination of Copper at Carbon Paste Electrodes, Sensors, and Detectors: Survey of Methods and Selected Studies

Form, Species	Type of CPE (Modifier)	Technique (Mode)	Measuring Principles (Method Sequences)	Linearity Range (LOD; t_{Acc}/t_R)	Sample(s)	Other Specification (Remarks)	Ref(s)
Cu^{2+}, Cu^+	C/PO (four diff. modifs.)	DCV, ASV	accum. by compl. *in situ* Anodic reoxidn.	(ca. 1 ppb, 3 min)	Drinking water	modifs.: dithiooxamide, biquinoline, Cuproin, Bathocuproin High selectivity (c_{Cu} : c_{Me} = 1:10⁵)	[2225]
Cu^+	C/Nj (1,10-phen der.)	CV, ASV	accum. by compl. Cathodic redn.	(3×10^{-7} M)	Natural waters	modif.: 2,9-dimethyl-1,10-phen. Requires pre-redn. step: $Cu^{II} \rightarrow Cu^I$	[1337]
Cu^{2+}	C/Nj (alga *Eisenia*)	CV, FIA	bioaccum. by compl. MEx; cathodic redn.	4×10^{-9}–1×10^{-8} M	Model solns.	Dry alga dissolved in binder incl. interference studies	[2226]
Cu^{2+}	C/PO (Cu^{II}-zeolite)	CV, CSV	accum. by ion-exchange Cathodic redn.	(3×10^{-8} M)	Model solns.	incl. study with various s.es. (KCl, KBr, LiCl, K_2PtF_6)	[829]
Cu^{2+}	C/Nj (*Dowex 50W*)	DPV	o.c.-accum. (ion exchange) MEx + cathodic redn.	(18 μM)	pharms.	incl. studies on mechanism of preconcentration	[522]
Cu^{2+}	C/PO (unm.)	FIA	Column pre-separation Re-elution; anodic oxidn.	—	CRM (human serum)	CP-det.: wall-jet configuration s.e.: 1 M acetate buffer + 0.05 M NH_3OH^+ (no interfs. from Fe^{III})	[2227]
Cu^+	C/PO (benzoin oxim)	DPV	accum. by chelating effect and adsp.; cathodic redn.	(ca. 1 ng mL⁻¹)	Tap, river, lake waters	s.e.: 0.01 M HNO_3 No interfs. from Ag^I, Bi^{III}, Sb^{III}	[2228]
Cu^{2+}	C/Nj (DQDS)	DPCSV	accum. by chelating effect Cathodic redn.	3×10^{-9}–2×10^{-6} M	Soils, minerals	DQDS: diquinolyl-8,8'-disulfide s.e.: 0.1 M acetate buffer	[1338]
Cu^{2+}	C/Nj (SATP)	DPCSV	o.c.-accum. via compl. Cathodic redn.	2×10^{-9}–1×10^{-7} M	CRMs (pepper bush, hair)	SATP: salicylideneamino-2-thiophenol; s.e.: 0.01 M acetate buffer	[1339]
Cu^+	C/Nj (2,2'-BQ + Nf.)	CV, DPV	accum. via compl. Cathodic redn.	(ca. 1 nM, 2 min)	Model solns.	BQ + Nf: biquinoline + Nafion® CMCPE regnt. anodically	[1715]

Electroanalysis with Carbon Paste–Based Electrodes, Sensors, and Detectors 223

Analyte	Electrode	Technique	Procedure	Range/Conditions	Matrix	Notes	Ref.
Cu^{2+}	C/PO (*Duolite ES 467*)	DPCSV	accum. via ion-exchange by chelating resin; redn.	$(6 \times 10^{-9}$ M)	Human hair	Resin pre-protonated in min. acids Coupled with MWD	[2229]
Cu^{2+}	C/PO (benzoin oxime)	FS-LV	accum. via chelating by *Cupron*; cathodic redn.	$1 \times 10^{-8} – 1 \times 10^{-6}$ M $(5 \times 10^{-9}$ M, 10 min)	Anodic muds, polluted water	s.e.: ammonia buffer (pH 8.5) incl. studies on interfs.	[1340, 2230]
Cu^{2+}	C/Nj (PAN)	DPCSV	o.c.-accum. via compl. Cathodic redn.	$2 \times 10^{-7} – 1 \times 10^{-6}$ M $(5 \times 10^{-9}$ M, 10 min)	Model solns.	PAN: 1-(2-pyridylazo-2-naphthol) s.e.: 0.1 M KNO_3; regnt.: HCl	[2231]
Cu^{2+}	C/Nj (hectorit)	CV, DPV	o.c.-accum. via compl. Cathodic redn.	0.1–0.7 mg L^{-1} (7 µg L^{-1})	Mineral ores	s.e.: 0.05 M KCl; incl. studies on reaction mechanism and interfs.	[2232]
Cu^{2+}	C/MO (oxin)	DPV	accum. by adsopt.; redn. (with catalytic effect)	$1 \times 10^{-9} – 2 \times 10^{-6}$ M	Human hair, blood	Sensitivity of detc. enhanced by catalytic wave of H_2	[2233]
Cu^+	C/Nj (DIDD)	CV, DPV	accum. via compl. after chem. redn. $Cu^{II} \rightarrow Cu^{I}$	$(5 \times 10^{-11}$ M, 30 min)	CRM (urine)	DIDD: diiminocyclopentylidine-thiomethyl disulfide CMCPE regnt. by anodization	[1347]
Cu^{2+}	C/Nj (PAN)	DPCSV	o.c.-accum. via compl. Cathodic redn.	$8 \times 10^{-7} – 8 \times 10^{-6}$ M	Anodic muds	PAN: 1-(2-pyridylazo-2-naphthol) s.e.: acetate buffer (pH 4.5)	[2234]
Cu^{2+}	C/PO (vermiculite)	SWV	o.c.-accum. (ion exchange) redn. + reoxidn.	$1 \times 10^{-8} – 8 \times 10^{-5}$ M $(5 \times 10^{-9}$ M, 10 min)	CRM (soil)	s.e.: BR-buffer + 0.1 M $NaNO_3$ CMCPE regnt. by anodization	[542]
Cu^{2+}	C/PO (unm.)	POT	Selective extr./adsorp. Equilibrium potential with super-Nerstian response	$1 \times 10^{-6} – 0.001$ M	Model solns. (in acetate buffer)	Unclear mechanism of interaction of Cu^{2+} and paraffin binder in CP	[2235]
Cu^{2+}	C/PO (Amberlite IRC)	ASV	o.c.-accum. (ion exchange)—cathodic redn.	$1 \times 10^{-5} – 3 \times 10^{-4}$ M	Model solns.	incl. studies on CP-composition, pH-effect and other variables	[252]
Cu^{2+}	C/Nj + Nf (alga *Anabaena*)	DPASV	o.c.-bioaccum. redn. + anodic reoxidn.	$2 \times 10^{-4} – 0.001$ M (8×10^{-5} M; 10 min)	Model solns.	Serious interfs. from compl. agents (e.g.,: CN^-, Ox^{2-}, EDTA)	[2236]
Cu^{2+}	C/PO (*Chelite P*)	CV, DPV	o.c.-accum. (ion exchange)—cathodic redn.	(2 µg L^{-1}; 10 min)	River water	*Chelite P*: chelating resin with NH_2–$P(=O)$– groups	[1681]

(continued)

TABLE 8.4 (continued)
Determination of Copper at Carbon Paste Electrodes, Sensors, and Detectors: Survey of Methods and Selected Studies

Form, Species	Type of CPE (Modifier)	Technique (Mode)	Measuring Principles (Method Sequences)	Linearity Range (LOD; t_{ACC}/t_R)	Sample(s)	Other Specification (Remarks)	Ref(s)
Cu^{2+}	C/PO (SWy-2)	DPASV	accum. by ion-exchange and sorption; elec. redn. anodic reoxidn.	sub-μM level (4×10^{-8} M)	Model solns.	SWy-2: Na-montmorillonite (clay) interfs. of Pb^{2+} ($c_{Pb}:c_{Cu} = 10:1$)	[1059]
Cu^+	C/Nj (rubeanic acid)	CV, DPV	accum. via compl. after chem. redn. $Cu^{II} \to Cu^{I}$	$1 \times 10^{-7} - 5 \times 10^{-6}$ M (5×10^{-8} M; 15 min)	CRM (human urine)	CMCPE regnt. by anodic cycling in $0.1 M HNO_3$	[1348]
Cu^{2+}	C/PO (ADE)	FIA-EC	accum. via compl. Under FIA conds.; elchem. detc.	$2 \times 10^{-6} - 2 \times 10^{-4}$ M (1×10^{-6} M, 4 min)	Model solns.	ADE: 1,2-bismethyl(2-aminocyclo pentenecarbodithiolate) ethane	[1341]
Cu^{2+}	C/Nj (Sal-TSC)	FIA-EC	accum. via compl. under FIA conds.; volt. detc.	(1×10^{-10} M)	CRM (sea water), tap water	Sal-TSC: salicylaldehyde thiosemi-larbazone	[2237]
Cu^{2+}	C/Nj (SiO_2)	CV, SWV	o.c.-accum. via compl. MEx; redn. + reoxidn.	$5 \times 10^{-9} - 5 \times 10^{-6}$ M (2×10^{-9} M, 10 min)	Model solns.	s.e.: ammonia buffer, $0.1 M HNO_3$ incl. study on accum. mechanism	[1157]
Cu^{2+}	C/PO (chitin)	ASV	accum. via compl.; MEx redn. + anodic reoxidn.	$3 \times 10^{-9} - 9 \times 10^{-7}$ M (1×10^{-9} M)	Model solns.	accum. in open circuit incl. study on accum. mechanism	[834]
Cu^{2+}	C/PO (L_1, L_2)	POT	Chelating effect Equilibrium potential	$1 \times 10^{-5} - 0.01$ M (pCu = 4.8–5.6)	solns. with humic acids	L_1, L_2: macrocyclic thiohydrazone and thiosemicarbazone ligands	[1406]
Cu^{2+}, Pb^{2+}	C/PO (TPSB)	ASV	accum. via chelatation redn.; anodic reoxidn.	$1 \times 10^{-9} - 1 \times 10^{-7}$ M (2×10^{-10} M; 20 min)	Natural waters	modif.: pyruvaldehyde bis-dibutyl thiosemicarbazone; some interfs. by Se^{IV}, Sr^{2+}, Ga^{3+}; Pb^{2+} also detn.	[1342]
Cu^{2+}	C/PO (crown ethers)	DPASV	accum. via chelating and intercalation effect; redn. Anodic reoxidn.	1–100 ppb	Alcoholic beverages	Three diff. sp. of crown ethers incl. studies on interfs. from Ni^{II}, Co^{II}, Mn^{II}, Fe^{III}	[1399]

Analyte	Electrode	Method	Principle	Range (LOD)	Sample	Notes	Ref.
Cu^{2+}, Cu^+	CP-SPE (Cu_2S, CuS)	POT	Ion-exch. equilibrium in Cu-doped thick layer	—	Model solns.	Disposable sensor made by screen-printed techn. ("thick film-CP")	[2238]
Cu^{2+}	C/PO (AP-GS)	CV, SWV	o.c.-accum. via compl. MEx; redn. + reoxidn.	$5 \times 10^{-8} - 2 \times 10^{-7}$ M (3×10^{-9} M, 10 min)	Tap water	AP-GS: aminopropyl-functionalized grafted silica; studies on accum. Mechanism; interfs. Studies	[1816]
Cu^{2+}	C/PO (TH)	CV, LSV	o.c.-accum. via compl. + redn./cathodic reoxidn.	(0.5 µg L^{-1}, 10 min)	River water	TH: macrocyclic thiohydrazone incl. studies on interfs.	[1404]
Cu^{2+}	C/PO (DTPT)	POT (dir.)	Chemical equilibrium and steady-state potential for $Cu^{2+} \leftrightarrow Cu^{II}$-DTPT	$1 \times 10^{-6} - 0.08$ M (7×10^{-7} M)	Electronics (waste solns.)	modif.: (3,4-dihydro-4,4,6-trimethyl-2(1H) pyrimidine thione	[1343]
Cu^{2+}	C/PO (TPP)	DPASV	o.c.-accum. by compl. MEx/redn. + reoxidn.	$9 \times 10^{-8} - 5 \times 10^{-5}$ M (2×10^{-9} M, 12 min)	Minerals	s.e.: BR-buffer (pH 6) Sample extracted prior to detn.	[750]
Cu^{2+}	C/PO (CAR-SG)	DPASV	o.c. accum. by compl. redn.; anodic reoxidn.	$5 \times 10^{-8} - 1 \times 10^{-6}$ M (4×10^{-9} M)	Model solns.	CAR-SG: Carnosine immobilized onto silica (solid phase extractant)	[1831]
Cu^{2+}	s-CPE (C/wx) (zeolite)	CV, FIA-EC	o.c. accum./ion-exchange Cathodic redn.	—	Model solns.	s-CPE: with solid paraffin matrix whose viscosity is temperature controlled; initial accum. study	[755]
Cu^{2+}	C/PO (humic acid)	CV, DPV	o.c. compl.+ redn. Anodic reoxidn.	$3 \times 10^{-8} - 1 \times 10^{-5}$ M	Model and real samples	Study on accum. mechanism and possible speciation	[1702]
Cu^{2+}	C/PO (MDPT)	ASV	o.c. accum. by compl. + redn.; anodic reoxidn.	$1 \times 10^{-7} - 1 \times 10^{-4}$ M (1×10^{-7} M, 6 min)	Coal ash	MDPT: 4-methoxy-2,6-bis(3,5-di-methylpyrazoyl)-1,3,5 triazine s.e.: tartrate buffer (pH 4)	[2239]
Cu^{2+}	C/PO (MBTZ-SG)	LSV, ASV	o.c. accum. via compl. + sorption/MEx/el. redn. Anodic reoxidn.	$1 - 10 \times 10^{-7}$ and $1 \times 10^{-5} - 0.01$ M (0.1 µM)	Model solns.	MBTZ-SG: 2-mercaptobenzothiazole functionalized silica gel; two linear ranges for calibration	[1345]

(continued)

TABLE 8.4 (continued)
Determination of Copper at Carbon Paste Electrodes, Sensors, and Detectors: Survey of Methods and Selected Studies

Form, Species	Type of CPE (Modifier)	Technique (Mode)	Measuring Principles (Method Sequences)	Linearity Range (LOD; t_{ACC}/t_R)	Sample(s)	Other Specification (Remarks)	Ref(s)
Cu^{2+}	C/SO, C/PO (+AuF)	SP (CCSA)	accum. by electrolytic redn. reoxidn. with I_{CONST}	(5 ppb; 10 min)	Model solns.	Au-film plated in situ or preplated. Simultaneous detn. of Cu^{2+} + Hg^{2+}	[1993]
Cu^{2+}	C/PO (salicylaldoxime)	CV, ASV	o.c. accum. via compl. el. redn./anodic reoxidn.	0.1–10 ppm (0.1 ppm; 3 min)	Wastewater, wine	s.e.: diluted HNO_3; intermediate reduction for 100 s	[1344]
Cu^{2+}	C/PO (Z)	DPCSV	o.c. accum. via sorption MEx/cathodic redn.	5×10^{-8}–5×10^{-6} M (2×10^{-8} M, 3 min)	Model and real samples	Z: natural zeolite (modif.); accum./regeneration scheme during anal.	[1142]
Cu^{2+}	C/SO (+BiF/BiP)	SWASV	accum. onto Bi via Cu_aBi_b alloy formation; reoxidn.	5–100 µg L^{-1} (2 ppb; 2 min)	Model solns.	CPE either with BiF or dispersed Bi-powder; s.e.: 0.2 M AcB (4.5), 0.2 M AmB (pH 10)	[2240]
Cu^{2+}	C/PO (ARS/$S_2O_8^{2-}$)	CtAdSV, 2nd DLSV	o.c. accum. via compl. + adsorption/el. redn./el. catalyst-assisted reoxidn.	8×10^{-10}–3×10^{-8} M (2×10^{-10} M, 3 min)	Natural water, soil samples	s.e.: BRB (pH 4.6); modif. in situ; no regeneration required	[687]
Cu^{2+}	C/PO [calix[4]arene]	DPASV	o.c. accum. via intercalation el. redn./anodic reoxidn.	(0.1 µg L^{-1}; 10 min)	Tap water	s.e.: buffer (pH 6.5–7.5); analysis compared to the reference (AAS)	[1431]
Cu^{2+}, Cu^+	C/PO (nano-Pt)	CV, ASV	el. accum. $Cu^{II} \to Cu^I \to Cu^0$ with intermediate catalysis Anodic reoxidn.	4×10^{-8}–2×10^{-6} M (4×10^{-9} M, 10 min)	Model solns.	Pt-particles, 20 nm characterized by SEM and XRF; LOD lowered by the effect of some surfactants	[968]
Cu^{2+}	s-CPE/C/Wx (AMTZ-S)	CV, DPASV	o.c. accum. via complex. el. redn./anodic reoxidn.	8×10^{-8}–3×10^{-6} M (3×10^{-8} M, 20 min)	Commercial ethanol fuel	Solid-like CPE cctng. paraffin wax (stable in EtOH); AMTZ-S: amino-thiazole-functionalized silica	[1830]

Analyte	Electrode	Method	Principle	Range (LOD)	Sample	Note	Ref.
Cu^{2+}	C/PO (DPN-SG)	POT (dir.)	Chemical equilibrium and steady-state potential for $Cu^{2+} \leftrightarrow Cu^{II}-DPN$	1×10^{-7}–0.01 M (8×10^{-8} M, 50 s)	Wastewater	DPN-SG: dipyridyl group-functionalized nanoporous silica gel Slope: Nenstian, ca 28.5 mV/dec.	[1835]
Cu^{2+}	C/PO (Cu-SALHMN)	POT (dir., titr.)	Chemical equilibrium and steady-state potential for $Cu^{2+} \leftrightarrow Cu^{II}-SALHMN$	4×10^{-7}–0.01 M (6×10^{-8} M, 12 s)	Tap water, multi-vitamin tablets	SALHMN: N,N'-disalicylidene-hexameythylenediaminate; s.s.: buffer (pH 4–6.5); interfs. study	[2241]
Cu^{2+}	C/PO (chitosan)	ASV	o.c. accum. via adsorption + el. redn.; anodic reoxidn.	2×10^{-7}–7×10^{-6} M (8×10^{-8} M, 4.5 min)	Wastewater samples	modif. content in CP: 25% (m/m); s.e.: 0.1 M KNO_3 (pH 6.5)	[2242]
Cu^{2+}	C/PO (Zincon)	DPASV	o.c. accum. via compl. + sorption/anodic reoxidn. of modif. (indirect detn.)	2–220 μg L^{-1} (1 μg L^{-1}; 5 min)	Natural waters, human hair	Zincon: 2-carboxy-2-hydroxy-5-sulfo formazyl benzene; s.e.: PhB (pH 6.4)	[2243]
Cu^{2+}, Hg^{2+} + Pb^{2+}	C/PO (AMT/plex-μS)	SWASV	accum. via complexation el. redn. + reoxidn.	3–50 μg L^{-1} (10 min) (1 μg L^{-1}; 15 min)	Tap water (spiked) wastewater	AMT/plex-μS: 2-aminothiazole in plexi-polymer made microspheres Simultaneous detn. with Hg^{2+}, Pb^{2+}	[1683]
Cu^{2+}, Hg^{2+} + Pb^{2+}	C/PO (SBA-15)	CV, DPASV	accum. via sorption and complex.; redn. + reoxidn.	8×10^{-7}–1×10^{-5} M (2×10^{-7} M; 3 min)	Natural water; alcohol drink (from cane sugar)	SBA-15: nanostructured silica functionalized with 2-benzothiazolethiol Simultaneous detn. with Hg^{2+}, Pb^{2+}	[2223]
Cu^{2+}	C/PO (s.a. n-Au + three diff. modifs)	POT (dir.)	Chemical equilibrium and steady-state potential for $Cu^{2+} \leftrightarrow Cu^{II}-SA-nAu/M$	8×10^{-9}–0.001 M ($3-4 \times 10^{-8}$ M; 5 s)	Water samples, human hair	s.a.: self-assembled n-Au-particles modifs.: MMN-IT, MNFT, MNTT (SH-ctng. heterocyclic derivatives)	[959]
Cu^{2+}, Pb^{2+}	CNTPE (+immob. DNA)	CV, SWV	accum. via the $Cu^{II}/-SH$ affinity (provided by DNA modif.); cathodic redn.	0.5–3.2 ng L^{-1} (0.4 ng L^{-1}; 200 s)	Fish tissue	s.e.: 0.1 M AmB (pH 10); RSD <0.1% ($n=15$); simultaneous detn. of Cu^{2+} and Pb^{2+}	[382]
Cu^{2+}	CNT(F)-PE, CNT(F) PE	SWV, CA (in vivo)	o.c. accum. via compl. + sorption/anodic reoxidn.	0.01–0.10 μg L^{-1} (0.005 μg L^{-1})	Tap water, blood, rat tail	CNT(F): fluorinated carbon nano-tubes; electrode lifetime in appl. in vivo, t_{iv} > 1 month	[396]

(continued)

TABLE 8.4 (continued)
Determination of Copper at Carbon Paste Electrodes, Sensors, and Detectors: Survey of Methods and Selected Studies

Form, Species	Type of CPE (Modifier)	Technique (Mode)	Measuring Principles (Method Sequences)	Linearity Range (LOD; t_{ACC}/t_R)	Sample(s)	Other Specification (Remarks)	Ref(s)
Cu^{2+}	C/PO (TMTDS)	POT (dir.), FIA-EC	Chemical equilibrium and steady-state potential for $Cu^{2+} \leftrightarrow Cu^{II}\text{-TMTDS}$	$(5 \times 10^{-8}$ M for stat.; 2×10^{-7} M for FIA)	Model and real samples	TMTDS: tetramethyl-thiuram di-sulfide; optimization studies for both stac. and FIA modes	[2244]
Cu^{2+}	C/PO (SAL/PSX)	POT (dir., titr.)	Chemical equilibrium and steady-state potential for $Cu^{2+} \leftrightarrow Cu^{II}\text{-SAL}$	2×10^{-7}–0.001 M (3×10^{-8} M; 8 s)	Water samples, urine	SAL/PSX: salicylidine-functionalized polysiloxane; opt.: pH 2–5.5; titr. performed with EDTA	[1834]
Cu^{2+}	C/PO (naphthazarin)	AdSV, FIA POT (dir.)	Chemical equilibrium and steady-state potential for $Cu^{2+} \leftrightarrow Cu^{II}\text{-TMTDS}$	$(2 \times 10^{-6}$ M for stac.; 3×10^{-5} M for FIA) ($t_R < 50$ s)	Metal alloys with Cu-traces	modif.: 5,8-dihydroxy-1,4-naphtho-quinone; s.s.: 0.1 M AmB/AcB; lifetime, $t > 60$ days	[1346]
Cu^{2+}	C/PO (microalgae)	DPCSV	o.c. accum. via bio-sorption Cathodic redn.	5×10^{-8}–1×10^{-6} M (5×10^{-10} M, —)	Model and real samples	Microalgae type: *Tetraselmis Chuii*; added in content of 3%–20% (m/m)	[2245]
Cu^{2+}	C/PO (SPADNS)	ASV	Adsorptive accum. via the dye (used as modif). redn. Anodic reoxidn.	2–800 µg L^{-1} (0.4 µg L^{-1})	Natural water, human hair	modif.: 2-(p-sulfophenylazo)-1,8-di-hydroxy-3,6-naphthalene disulfate in surfactant-modified zeolite	[2246]
Cu^{2+}	C/PO (H-BDBTU)	POT (dir.)	Chemical equilibrium and steady-state potential for $Cu^{2+} \leftrightarrow Cu^{II}\text{-BTU}$	1×10^{-5}–0.001 M (1×10^{-5} M; 10 s)	Model solns.	H-BDBTU: *N*-benzoyl-*N'*,*N'*-di-*n*-butyl-thiourea; studies on reaction mechanism and eq. conditions	[2247]
Cu^{2+}	CNTPE (ARS)	2nd D-LSV	Adsorptive accum. via the complex; cathodic redn.	2×10^{-11}–4×10^{-7} M (4×10^{-12} M; 4 min)	Biological tissues	ARS: Alizarin Red "S"; 0.04 M AcB (pH 4.2); $R_R = 96\%$–102%	[393]

Analyte	Electrode	Method	Principle	Range (LOD; t_acc)	Sample	Note	Ref.
Cu^{2+}	C/PO (Cyclam/mS)	ASV	accum. via macrocycle redn.; anodic reoxidn.	2–100 nM (2 nM)	Tap water	modif.: Cyclam ders. grafted via silylation to mesoporous silica; study on effect of modif. prepared	[2248]
Cu^{2+}	C/PO (Cyclam/mS)	POT, AM, SECM	Dual function of the CPE (as ISE or a tip for SECM)	1×10^{-6}–0.001 M (1 μM; 15 s)	Model solns.	CP-ISE microelectrode (with tip of 25 μm); SECM: scanning electro-chemical microscopy	[2249]
Cu^{2+}	C/PO (SB/ SiO_2–Al_2O_3)	CSV	accum. via complexation Anodic reoxidn.	3×10^{-10}–3×10^{-7} M (0.2 nM; 5 min)	Tap water, CRMs	modif.: Schiff base covalently bound with mixed oxide, its identification by CHN, TG, FTIR; CPE regnt. needed	[1221]
Cu^{2+}	C/Nj (DAN)	ASV (with RDE)	accum. via complexation Anodic reoxidn.	0.1–250 ppb (—)	Tap water, orange juice	DAN: 1,8-diaminonaphthalene (as conductive polymer); acidic media accum. in open circuit; reproducibility of CPE < 4.3%	[537]
Cu^{2+}	C/PO (zeolite)	SWCSV	accum. via ion-exchange Anodic reoxidn.	0.1–65 mg L^{-1} (0.05 mg L^{-1})	Model solns.		[1148]
Cu^{2+}	C/PO (HBP)	LSV	accum. via ion-exchange Anodic reoxidn.	1×10^{-9}–3×10^{-5} M (0.4 nM)	Human hair	HBP: hyperbranched polymer (as amine-ester); s.e.: 0.25 M KCl	[136]
Cu^{2+}	CNTPE (cCTS)	LSASV	accum. by compl. with modif.; anodic reoxidn.	8×10^{-8}–2×10^{-5} M (8×10^{-8} M; 300 s)	Natural and wastewater, human urine	cCTS: cross-linked chitosan; 65% CNTs + 20% MO + 15% modif.; s.e.: 0.05 M KNO_3 + HNO_3 (pH 2.3)	[425]
Cu^{2+}	C/PO (PYTT-nAu)	POT (dir.)	Chemical equilibrium and steady-state potential for $Cu^{2+} \leftrightarrow Cu^{II}$-PYTT(nAu)	4×10^{-9}–0.07 M (1 nM; < 5 s)	Natural water (spiked)	modif.: 5-(pyridin-2-yl-methyleneamino)-1,3,4-thiadiazole-2-thiol at gold nanoparticles; pH 3.0–6.5	[957]
Cu^{2+}	C/PO (Rhodotomula sp.)	CV, DPCSV	accum. via bio-affinity to Rhodotomula sp. (fungi); cathodic redn.	1×10^{-7}–1×10^{-5} (0.1 μM; 15 min)	Natural water (spiked)	CPE: microbial biosensor; modif. obtainable from pigmented yeast; s.e.: 0.05 M $NaNO_3$ (pH 5)	[2250]
Cu^{2+}; Cd^{2+}, Pb^{2+}	IL-CPE (quercetin, Qc)	AdSV	accum. by adsorption of the complex with modif. Cathodic redn.	—	Model solns.	Qc-IL-CPE used for comparative studies only; s.e.: 0.1 M HCOONa + HCl (pH 4.7)	[513]

TABLE 8.5
Determination of Cadmium and Lead at Carbon Paste Electrodes, Sensors, and Detectors: Survey of Methods and Selected Studies

Ion, Species	Type of CPE (Modifier)	Technique (Mode)	Measuring Principles (Method Sequences)	Linearity Range (LOD; t_{ACC}/t_R)	Sample(s)	Other Specification (Remarks)	Ref(s)
Pb^{2+}, Fe^{3+}	C/PO (unm.)	LSV	Electrolytic accum. Anodic reoxidn.	$1-8 \times 10^{-8}$ M (0.001%Pb/Fe)	Cast iron, stainless steel	s.e.: 1 M HCl (pH 0); selectivity Pb: Fe = 1: 5×10^4 (by masking Fe)	[2251]
Pb^{2+}, Hg^{2+}	C/PO (CEs, CR)	DPCSV	o.c. accum. via compl.; MEx; cathodic redn.	Micromolar level (1 µM; 20s)	CRMs (water, urine)	CEs: crown ethers; CR: cryptand CE(R)-CPE also sensitive to Hg^{II}	[525]
Cd^{2+}	C/PO (Dowex 50W)	LSV, DPV	o.c.-accum. (ion-exchange) MEx + cathodic redn.	(1 ng mL^{-1}, 5 min)	Sampled air (atmosphere)	s.e.: 0.1 M KNO_3 Air sampled in smoking area	[807]
Pb^{2+}	C/Nj (chelating resin)	ASV	o.c. accum. via compl.; redn.; anodic reoxidn.	4–200 ng mL^{-1} (1.2 ng mL^{-1}; 10 min)	Rain water, human urine	Chelating resin: Amberlite IRC-75 (a) AmAc (pH 5), (b) HCl (pH 1)	[808]
Pb^{2+}	C/SC [MeF–s-CPE]	DPV	elec. accum. + amalgam formation; anodic reoxidn.	0.01–250 µM	Motor oil (used)	SC: silica gel composite (as solid-like CPE); without removal of O_2	[2252]
Cd^{2+}	C/PO (Amberlit IRC)	DPV	o.c.-accum. (ion-exchange) MEx + cathodic redn.	(5 ng mL^{-1})	River water (polluted)	s.e.: (1) ammonia buffer (pH 9); (2) 0.1 M HCl	[1737]
Pb^{2+}	C/PO (crown ether)	CSV	accum. via compl. Cathodic redn.	Nanomolar level (0.2 ng L^{-1})	Natural waters, sediments	No interfs. from Mn^{II}, Zn^{II}, Cd^{II}, Tl^{I}, Hg^{II}, and Ag^{I}	[2253]
Pb^{2+}	C/PO (moss Sphagnum)	DPASV	o.c.-bioaccum. + redn. Anodic reoxidn.	(2 ng mL^{-1}; 15 min)	Natural waters, tap water	s.e.: acetate buffers (pH 5 and 6) CMCPE regnt. in 0.05 M $HClO_4$	[272]
Pb^{2+}	C/PO (benzoin oxime)	DPASV	o.c.-accum. + MEx redn.; anodic reoxidn.	1×10^{-8}–1×10^{-7} M (8×10^{-9} M; 10 min)	Wastewater	Studies on accum. with Cuproin No interfs. from common ions	[1349,2254]
Pb^{2+}, Me^{n+}	C/PO (CE)	DPASV	Intercalation effect of CE for Pb^{2+} ion; redn., reoxidn.	Nanomolar level (3×10^{-9} M; 5 min)	Model solns.	CE: dibenzo-18-crown-6; study on structural mechanism with diff. Me	[2255,2256]
Pb^{2+}	C/PO (dithizone)	DPCSV	o.c.-accum. + MEx Cathodic redn.	20–520 µg L^{-1} (17 µg L^{-1}, 6 min)	Natural waters	interfs. from some ions elim. by masking except sulfide, S^{2-}	[814]

Electroanalysis with Carbon Paste–Based Electrodes, Sensors, and Detectors 231

Analyte	Electrode	Method	Principle	Concn. range (LOD; t_acc)	Sample	Remarks	Ref.
Cd^{2+}	C/PO (oxin + glycerine)	CSV	accum. via *compl.* Cathodic redn.	$9 \times 10^{-8} - 1 \times 10^{-5}$ M (5×10^{-8} M) ($5 \mu g L^{-1}$; 5 min)	Natural water	oxin.: 8-hydroxyquinoline Glycerin served as adhesive	[2257]
Cd^{2+}, Pb^{2+}	C/Nj (+MF, *in situ*)	ASV	Electrolytic accum. via amalgamation; reoxidn.		Natural water, biol. mat., food stuff	MF generated *in situ* from Hg^{2+} ions (added to sample soln.)	[2258]
Pb^{2+}, Cd^{2+}	C/Nj + TBP (PMBP *in situ*)	CSV	o.c.-accum. + MEx Cathodic redn.	$1 \times 10^{-8} - 8 \times 10^{-6}$ M Pb $5 \times 10^{-8} - 2 \times 10^{-6}$ M Cd	Wastewater, human hair	s.e.: acetate buffer (pH 5) No or little interfs. from common	[2259]
Cd^{2+}, Pb^{2+}	C/PO (alga *Anabaena*)	DPASV	bioaccum. by compl. redn.; anodic reoxidn.	$1 \times 10^{-6} - 8 \times 10^{-6}$ M (8×10^{-6} M)	Model solns.	incl. bioaccum. mechanism study	[2260]
Pb^{2+}	C/Nj + TBuP (PMBP *in situ*)	CSV	o.c.-accum. Cathodic redn.	$1 \times 10^{-9} - 3 \times 10^{-7}$ M	Natural water, powdered alloy	CMCPE regnt. in 0.1 M HCl modif: 1-phenyl-3-methyl-4-benz-oyl-pyrazolone; no serious interfs.	[2261]
Pb^{2+}, Cu^{2+}	C/PO (TPSB)	ASV	accum. via chelating redn.; anodic reoxidn.	$8 \times 10^{-8} - 2 \times 10^{-6}$ M (2×10^{-9} M; 20 min)	Natural waters	modif: pyruvaldehyde bis-dibutyl thiosemicarbazone; interfs. from Se^{IV}, Sr^{2+}, Ga^{3+}; also detn. of Cu^{2+}	[1342]
Cd^{2+}, Pb^{2+}, Cu^{2+}	C/PO, C/SO (+MF, *in situ*)	PSA	Electrolytic accum. chem. reoxidn. by Hg^{II}	5–150 ppb Me^{2+} (0.5–1 ppb; 3 min)	Model solns.	compr. with MF(GCE); studies on simult. detn. of Me^{2+} (Cd,Pb,Cu)	[1992]
Cd^{2+}, Pb^{2+}	C/PO (CCHA)	DPASV	accum. by compl.; MEx redn.; anodic reoxidn.	$4 \times 10^{-8} - 3 \times 10^{-6}$ M (1×10^{-9} M; 2 min)	Municipal and mineral waters	CCHA: *N-p*-chlorophenylcinnamo-hydroxamic acid; CP: anodic regnt.	[1388]
Pb^{2+}	C/Nj (bPE-AQ)	DPASV	o.c. accum. by complex., followed by cathodic accum. Anodic reoxidn.	$2 \times 10^{-9} - 1 \times 10^{-5}$ M (1×10^{-9} M; 30 s)	Wastewaters	modif.: 1,4-bis(propen-nyloxy)-9,10-anthraquinone; mech. regnt.	[1454]
Pb^{2+}	C/PO (crown ethers)	DPASV	accum. via compl. redn.; anodic reoxidn.	20–100 ppb (1 ppb; 30 s)	Alcoholic beverages	s.e.: aqueous solns. + 40% MeOH	[1400]
Pb^{2+}	C/SO (+MF, *in situ*)	PSA	Electrolytic accum. chem. reoxidn. by Hg^{II}	5–150 μg L^{-1} (1 μg L^{-1}; 3 min)	Tap, natural, wastewater; snow	simult. detn. of Cu^{II} also tested Hg^{II} added acted as (i) source of MF and (ii) the chemical oxidant	[209, 2262]

(*continued*)

TABLE 8.5 (continued)
Determination of Cadmium and Lead at Carbon Paste Electrodes, Sensors, and Detectors: Survey of Methods and Selected Studies

Ion, Species	Type of CPE (Modifier)	Technique (Mode)	Measuring Principles (Method Sequences)	Linearity Range (LOD; t_{acc}/t_R)	Sample(s)	Other Specification (Remarks)	Ref(s)
Pb^{2+}, Zn^{2+}, Cd^{2+}	EH-CPE (+BiF, in situ)	SWASV	accum. onto Bi via Me_aBi_b alloy formation; reoxidn.	20–150 µg L^{-1} Pb^{2+} (3 µg L^{-1}; 120 s)	Model solns.	EH: electrically (pre)heated CPE Optimization of detn. only for Pb^{2+} (but tested in presence of both Me)	[294]
Cd^{2+}, Pb^{2+}	C/SO (+BiF, in situ)	PSA	accum. onto Bi via Me_aBi_b alloy formation; reoxidn.	5–100 ppb Me^{2+} (0.5 ppb; 10 min)	Model solns.	Bi^{3+} ions added into soln. serve as (i) source of BiF, (ii) as chemical oxidant; s.e.: 0.2 M HCl + KCl/KBr	[626]
Cd^{2+}, Pb^{2+}, Zn^{2+}	C/SO (Bi$_2$O$_3$)	DPASV	accum. via electrolytic redn.+ Me_aBi_b alloy form. Anodic reoxidn.	10–100 µg L^{-1} Me^{2+} (3 µg L^{-1}; 5 min)	Model solns.	Way of plating: (i) in situ, (ii) ext., (iii) in-nascenti (via Bi$_2$O$_3$ redn.); detn. of Zn with high background	[103]
Cd^{2+}, Pb^{2+}	C/PO (Bi$_2$O$_3$)	DPASV	accum. via electrolytic redn.+ Me_aBi_b alloy form. Anodic reoxidn.	5–100 µg L^{-1} Cd, Pb (0.5 µg L^{-1}; 6 min)	Tap and mineral waters, h. urine	BiF generated from oxide in nascenti; s.e.: 0.1 M acetate buffer (pH 4.5) simult. detn. of Cd+Pb^{II} possible	[265]
Cd^{2+}, Pb^{2+}, Zn^{2+}	C/SO [HgO or Bi$_2$O$_3$ (s)]	DPASV	accum. via electrolytic redn. at in-nascenti formed MF or BiF; anodic. reoxidn.	1–50 ppb Cd, Pb (1–2 ppb Me; 10 min)	Tap and natural waters	HgO, Bi$_2$O$_3$ added in CP (5% m/m); tests on reproducibility ($R_\pm < 5\%$); compared with MF- and BiF-SPE	[623]
Cd^{2+}	C/PO (BTT + am. SiO$_2$)	DPASV	adsor. accum. via compl. redn.; anodic reoxidn.	6×10^{-7}–4×10^{-5} M (1×10^{-7} M; 2 min)	Natural water (spiked)	BTT: benzothiazole-thiol s.e.: phosphate buffer (pH 7.5)	[107]
Cd^{2+}, Pb^{2+}	C/PO (DMG)	DPCSV	accum. by compl. + adsort. Cathodic redn.	1×10^{-7}–2×10^{-5} M Pb; 3×10^{-7}–3×10^{-5} M Pb	Waters	DMG: dimethyl glyoxime interfs. from Ni^{II} and Hg^{II}	[2263]

Analyte	Electrode	Method	Principle	Concentration range	Sample	Notes	Ref.
Pb^{2+}	C/PO (plant tissue)	DPASV	bioaccum. (ion-exchange) redn.; anodic reoxidn.	(0.01 ppb)	Natural waters	modif.: grass weed (*Pennisetum*) s.e.: acetate buffer (pH 5)	[2756]
Pb^{2+}, Hg^{2+}	C/PO (SH/ms-SiO$_2$)	SWV	accum. via compl. redn.; anodic reoxidn.	10–1500 ppb Pb 20–1600 ppb Hg	Model solns.	SH/ms.SiO$_2$: thiol-cntg. mesoporous silica; simult. detn. of Hg^{II}	[772]
Pb^{2+}	C/PO (1,8-DAN)	DPASV	accum. via compl.; redn. Anodic reoxidn.	50–2000 ppb (30 ppb; 10 min)	Model solns.	DAN: diaminonaphthalene + CP acts as conducting polymer	[1807]
Cd^{2+}, Pb^{2+}	C/PO (α-CD, β-CD)	ASV	accum. by ion-inclusion effect; cathodic redn.	(6×10^{-7} M Pb) (2×10^{-5} M Cd)	Model solns.	CDs: cyclodextrins; incl. studies on diff. prfm. of α-CD and β-CD	[1416]
Cd^{2+}, Pb^{2+} Tl^+	C/SO (+BiF, *in situ*)	DPASV	accum. onto Bi via Me$_a$Bi$_b$ alloy formation; reoxidn.	50–500 μg L^{-1} Me^{n+} (–)	Model solns.	s.e.: 0.1 M NaOH/KOH (pH 12) Study on complexation by OH$^-$	[637]
Cd^{2+}, Pb^{2+} Me^{n+}	C/SO (+BiF)	CV, DPASV	accum. onto Bi via Me$_a$Bi$_b$ alloy formation; reoxidn.	50–1000 μg L^{-1} Me^{n+} (–)	Model solns.	Initial study with BiF-CPEs made *in situ*/*ex situ*; Me^{n+}: Mn, Zn, Tl, Sn, In, and Sb.	[625, 2264]
Pb^{2+}	D/PO (DPEs, three types)	DPASV	accum. via electrolytic redn.; anodic reoxidn.	1×10^{-10}–1×10^{-6} M, depending on DPE; (10–100 pM; —)	Natural water, tea samples	D: natural or synthetic diamond (particle size: 1 or 50 μm); DPE: diamond paste electrode	[349]
Cd^{2+}, Pb^{2+} Cu^{2+}	C/PO (CPA + ms. SiO$_2$)	SWV	accum. via compl./adsorp. redn./anodic reoxidn.	10–200 ppb (0.5 ppb; 20 min)	Model solns.	CPA: carbamoyl-phosphonic acid as self-assembled monolayer on mesoporous silica; also detn. of Cu	[771]
Pb^{2+}	C/PO [(OH-AQ-Me)$_2$S]	DPASV	accum. via complexation e. redn./anodic reoxidn.	6×10^{-10}–6×10^{-6} M (4×10^{-10} M; 11 min)	Wastewater	modif.: bis[1-hydroxy-9,10-anthra-quinone-2-methyl] sulfide	[1351]
Pb^{2+} Me^{n+}	C/SO (+BiF, *in situ*)	DPASV, SWASV	accum. onto Bi via Me$_a$Bi$_b$ alloy formation; reoxidn.	2–20 ppb Pb (0.8 ppb; 20 min)	Tap and sea water	Initial study with BiF-CPE *in situ*; s.e.: 0.1 M AcB (pH 4.5) or 0.1 M NaOH (12); Me^{n+}: Zn, Cd, and Tl	[2265]
Cd^{2+}, Pb^{2+}	C/SO (+BiF, *ex situ*)	DPASV	accum. onto Bi via Me$_a$Bi$_b$ alloy formation; reoxidn.	50–200 ppb Me^{2+} (–)	Model solns.	Investigation of Bi-film deposition by SEM; Cd^{2+} + Pb^{2+} as model ions	[622]

(*continued*)

TABLE 8.5 (continued)
Determination of Cadmium and Lead at Carbon Paste Electrodes, Sensors, and Detectors: Survey of Methods and Selected Studies

Ion, Species	Type of CPE (Modifier)	Technique (Mode)	Measuring Principles (Method Sequences)	Linearity Range (LOD; t_{ACC}/t_R)	Sample(s)	Other Specification (Remarks)	Ref(s)
Pb^{2+}	C/PO (o-, m-, p-PD/E)	DPASV	accum. via complexation e. redn./anodic reoxidn.	$5 \times 10^{-8} - 1 \times 10^{-5}$ M (1×10^{-9} M; 10 min)	Model solns.	PD/E: phenylenediamine (isomers) mixed in CP/ electropolymerized interfs. of various Me^{2+} evaluated	[1352]
Pb^{2+}	C/PO [wj-D] (APA-MMS)	AD, FIA-EC	Pb^{II}-vs.-APA affinity el. redn. by E_{CONST}	1–25 ppb Pb (1 ppb)	Model solns.	APA-MMS: acetamide-phosphonic acid functionalized mesoporous SiO_2; wj-D: wall-jet detector	[547]
Cd^{2+}, Pb^{2+}	C/SO (Bi-powder)	SWASV, CCSA	accum. onto Bi via Me_aBi_b alloy formation; reoxidn.	10–100 ppb Me^{2+} (1 ppb; 3 min)	Tap water (spiked)	Carbon paste modified with fine Bi-powder (15% w/w); compared with BiF-GCE and BiF-CPE	[266]
Cd^{2+}, Pb^{2+}	CP-GrE (+BiF, in situ)	DPASV	accum. onto Bi via Me_aBi_b alloy formation; reoxidn.	50 ppb Me^{2+} (–)	Model solns	GrE: "groove" CPE based on mini planar holder with a cavity for CP; initial characterization with Me^{2+}	[585]
Cd^{2+}, Pb^{2+}	C/SO (+MF, in situ)	DPASV	accum. on Hg via Me_aHg_x amalgamation; reoxidn.	30–300 µg L^{-1} Me^{2+} (–)	Tap water (spiked)	Statistical analysis of about 600 detns. performed in framework of students' lab exercises per decade	[671]
Cd^{2+}, Pb^{2+}, Me^{n+}	C/SO (+BiF/BiP)	DPASV, SWASV	accum. onto Bi via Me_aBi_b alloy formation; reoxidn.	10–100 ppb (<1 ppb; 5 min)	Tap and sea waters (spiked samples)	Advanced studies with BiF-CPEs and Bi-CPEs; s.e.: 0.1 M AcB, 0.1 M NaOH; Me: Mn, Zn, Tl	[670]

Pb^{2+}	C/PO (ZrP$_2$O$_7$-SG)	DPASV	o.c. accum. (ion-exchange) el. redn./anodic reoxidn.	3×10^{-9}–5×10^{-6} M (4×10^{-10} M; 2 min)	Wastewater	SG: functionalized silica gel interfs. study with Zn, Cd, Sn, Tl	[1214]
Pb^{2+}	AB/PO (I$^-$, in situ)	SWASV	accum. via spec. sorption el. redn./anodic reoxidn.	2×10^{-8}–4×10^{-6} M (6×10^{-9} M; 10 min)	Water samples	AB: acetylene black (replacing here graphite powder); addition of I$^-$ ion further enhances the sensitivity	[115]
Cd^{2+}, Pb^{2+}	C/SO (+BiF, in situ)	SWASV	accum. onto Bi via Me$_a$Bi$_b$ alloy formation; reoxidn.	(0.15 μg L^{-1} Cd 0.10 μg L^{-1} Pb)	Model solns.	Studies on conc. relations among Bi^{3+} and Cd^{2+} + Pb^{2+}; compared with BiF-AuE; imaging by SEM	[2266, 2267]
Pb^{2+}	C/PO (OF-S)	DPASV	accum. by adsorption to S el. redn./anodic reoxidn.	5–1000 ng L^{-1} (5 ng L^{-1}; 1 min)	Hair sample (spiked)	OF-S: amino-functionalized silica; CP-lifetime >4 months	[1836]
Pb^{2+}	C/PO (Dithizone)	DPASV, FIA	o.c. accum. via complex. redn./anodic reoxidn.	8×10^{-8}–1×10^{-5} M (5–8×10^{-8} M; 8 min)	Soil samples	interfs. study with Zn, Cd, Cu, Hg Samples collected near metallurgic plant; analyses compared to AAS	[815]
Pb^{2+}	C/PO (fruit tissue)	DPCSV	o.c. accum. via biosorption; MEx/cathodic redn.	1–10 ng mL^{-1} (5 ng mL^{-1}; 1 min)	Water (spiked), laboratory waste	Fruit: dried/pulverized pineapple interfs. study with 15 Me^{n+} ions; Bio-CPE regeneration by EDTA	[281]
Cd^{2+}	C/PO (ZrP$_2$O$_7$-SG)	DPASV	o.c. accum. (ion-exchange) el. redn./anodic reoxidn.	3–1400 ng mL^{-1} (3 ng L^{-1}; 2 min)	Wastewater	SG: functionalized silica gel incl. optimization studies	[1215]
Pb^{2+}	C/Pos (+DTBA/ MBA)	POT (dir., titr.)	Chemical equilibrium and steady-state potential for Cu^{2+} ↔ Cu^{II}–TMTDS	(5×10^{-8} M for CPE1, 4×10^{-8} M for CPE2)	Humate extracts	Two modifs: dithiodibenzoic acid, mercaptobenzoic acid (added in: 25%, m/m); evaluation of β_{Pb-TBA}	[2268]

(continued)

TABLE 8.5 (continued)
Determination of Cadmium and Lead at Carbon Paste Electrodes, Sensors, and Detectors: Survey of Methods and Selected Studies

Ion, Species	Type of CPE (Modifier)	Technique (Mode)	Measuring Principles (Method Sequences)	Linearity Range (LOD; t_{ACC}/t_R)	Sample(s)	Other Specification (Remarks)	Ref(s)
Cd^{2+}, Pb^{2+} Cu^{2+}, Zn^{2+}	s-CPE (C/wx) ($Hg_2C_2O_4$)	DPASV	accum. via ion-exchange redn.; anodic reoxidn.	—	Medicinal plants, pharms.	CP-binder: solidified paraffin wax pharm.: ayurvedic tablets; results compared with ref. AAS	[316]
Cd^{2+}, Pb^{2+} Me^{n+}	C/SO (Bi-powder)	SWASV	accum. onto Bi via Me_aBi_b alloy formation; reoxidn.	50–1000 ppb (—)	Model solns.	Studies on use of Bi-CPE in AmB (pH 9–10); Me: Mn, Zn, Tl, and In	[672]
Cd^{2+}	CNTPE (+MF; in situ)	SWASV (in vivo)	accum. at highly negative pot. by Hg-amalgamation Anodic reoxidn.	10–80 μg L^{-1} (5×10^{-8} M, 400 s)	Plant and fish tissues	E_{acc} = −1.6 V vs. ref.; RSD < 6% Real-time monitoring (in vivo)	[377]
Pb^{2+}	C/PO (SiO_2/Al_2O_3)	CV, DPASV	o.c. accum. via sorption el. redn./anodic reoxidn.	2×10^{-9}–5×10^{-5} M (1×10^{-9} M; 5 min)	Real samples (—)	modif.: mixed oxide, characterized by TG, XRF, FTIR; added in 5%	[1210]
Cd^{2+}, Pb^{2+}	mCPE (+BiF, in situ)	SWASV	accum. onto Bi via Me_aBi_b alloy formation; reoxidn.	1×10^{-9}–5×10^{-6} M (1.3 nM Cd, 0.8 nM Pb)	Model solns.	Initial characterization of a mini-CPE plated with BiF; s.e.: AcB	[592]
Pb^{2+}	C/PO (Ca^{II}-MMT)	CV, DPASV	o.c. accum. via ion-exch. el. redn./anodic reoxidn.	—(6×10^{-9} M, —)	Water samples	MMT: montmorillonite; s.e.: 0.01 M HCl; comparison with bare CPE	[1067]
Cd^{2+}, Pb^{2+}	C/PO (+BiF/ Fbg)	SWASV (BIA)	accum. onto Bi via Me_aBi_b alloy formation; reoxidn.	(0.2 μg L^{-1} for Cd, 0.1 μg L^{-1} for Pb)	Tap water tea samples	Fbg: Fibrinogen (a protective layer against surfactants); BIA mode: batch injection analysis in microvolume	[2269]
Cd^{2+}	C/PO (BTZT/ SBA-15)	DPASV	accum. via complex and sorption; redn./reoxidn.	$1–10 \times 10^{-6}$ M (5×10^{-7} M; 2 min)	Natural water	modif: 2-benzothiazolethiol + com nanostructured silica; PhB (pH 3) interfs. from other Me^{n+} tested	[1860]

Electroanalysis with Carbon Paste–Based Electrodes, Sensors, and Detectors 237

Analyte	Electrode	Technique	Principle	Range (LOD; t_{acc})	Sample	Notes	Ref.
Cd^{2+}, Pb^{2+}	C/SO (+SbF/SbP)	SWASV	accum. onto Sb via Me_aSb_b alloy formation; reoxidn.	10–250 ppb (—)	Model solns.	Basic characterization of SbF- and Sb-CPE with Cd^{2+} + Pb^{2+} as model ions; s.e.: 0.01 M HCl (pH 2)	[1996]
Pb^{2+}, Cu^{2+}	CNTPE (+immob. DNA)	CV, SWV	accum. via the Pb^{II}/-SH affinity (to DNA-compl.)	0.5–2.2 ng L^{-1} (0.4 ng L^{-1}; 200 s)	Fish tissue	s.e.: 0.1 M AmB (pH 10); RSD < 0.1% ($n = 15$); detn. of Pb^{2+}+ Cu^{2+}	[382]
Pb^{2+}	C/PO (banana tissue)	DPV, AdCSV	bioaccum. via modif. + redn.; anodic reoxidation (el. called "bananatrode")	1–20 mg L^{-1} (0.1 mg L^{-1}; 6 min)	Polluted nat. water (in industrial area)	CP in PET-tube with Cu-contact; modif.: 20% (w/w) in CP, 0.01 M HCl; t_{ACC} = 6 min, t_{DEP} 3 min	[2270]
Pb^{2+}	C/PO (NFR, in situ)	DPV, AdCSV	accum. via adsorption of complex; cathodic redn.	0.5–200 ng mL^{-1} (0.2 ng mL^{-1}; 1 min)	Mineral water, powdered drinks	NFR: Nuclear fast red; s.e.: buffer (pH 3) + 5×10^{-5} M modif.; intefs. s.	[2271]
Cd^{2+}	C/PO (MNT/n-Au)	POT (dir.)	Chemical equilibrium and steady-state potential for $Cd^{2+} \leftrightarrow Cd^{II}$-MNT/ nAu	3×10^{-8}–3×10^{-4} M (2×10^{-8} M; 6 s)	Water samples, human hair	MNT: 2-mercapto-5-(3-nitrophenyl) 1,3,4-thiadiazole; slope: Nernstian, ca 29.5 mV/dec; s.s.: pH 2–4	[774]
Pb^{2+}, Hg^{2+}, Cu^{2+}	C/PO (AMT/plex-μS)	SWASV	accum. via complexation el. redn.+ reoxidn.	10–70 μg L^{-1} Pb (4.5 μg L^{-1}; 10 min)	Model solns.	AMT/plex-μS: 2-aminothiazole in plexi-polymer-made microspheres Simultaneous detn. with Hg^{2+}, Cu^{2+}	[1683]
Pb^{2+}, Hg^{2+}, Cu^{2+}	C/PO (SBA-15)	CV, DPASV	accum. via sorption and complex.; redn. + reoxidn.	3×10^{-7}–7×10^{-6} M (4×10^{-8} M; 3 min)	Natural water; alcoholic drink (from sugar cane)	SBA-15: nanostructured silica functionalized with 2-benzothiazolethiol Simultaneous detn. with Hg^{2+}, Cu^{2+}	[2231]
Cd^{2+}, Pb^{2+}	C/SO + Ze (+BiF)	DPASV	accum. via electrolytic redn. + alloy formation Anodic reoxidn.	(0.10 μg L^{-1} for Pb, 0.08 μg L^{-1} for Cd; 2 min for both)	Natural water	Ze: natural zeolite added into CP (serving for more effective plating with BiF); s.e.: AcB (pH 4.5)	[267]

(continued)

TABLE 8.5 (continued)
Determination of Cadmium and Lead at Carbon Paste Electrodes, Sensors, and Detectors: Survey of Methods and Selected Studies

Ion, Species	Type of CPE (Modifier)	Technique (Mode)	Measuring Principles (Method Sequences)	Linearity Range (LOD; t_{ACC}/t_R)	Sample(s)	Other Specification (Remarks)	Ref(s)
Pb^{2+}	C/SO (chitosan)	DPCSV	o.c. accum. via biosorption Cathodic redn.	$10–110$ ng mL^{-1} (2 ng mL^{-1})	Tap water, pharm., human blood, pre treated urine	s.e.: 0.5M HCl; no interfs. from Cd, Tl, Sn, Cu (at 10-fold excess)	[1710]
Pb^{2+}	C/PO (unm.)	DPASV	o.c. accum. by biosorption Cathodic redn.	2×10^{-8}–1×10^{-7} M (5×10^{-9} M Pb)	environm. samples, human urine	Results of analyses compared with those obtained by spectral method	[2272]
Cd^{2+}	C/PO (phen-on; in situ)	2nd-DeLV	accum. via adsorption of complex; redn. and anodic reoxidn.	6×10^{-9}–2×10^{-7} M (3×10^{-10} M Cd)	Water samples	Phen-on: 1,10-phenanthroline-5,6-dione; s.e.: 0.05 M AcB (pH 4.7)	[2273]
Pb^{2+} (Cd^{2+})	C/SO (QPu-TU)	CV, R_{tr} SWASV	o.c. accum. via chelating el. redn./anodic reoxidn.	0.005–$5(0)$ mg L^{-1} (0.2 mg L^{-1}; 10 min)	Tap, lake, and wastewaters	QPu-TU: thiourea-functionalized macroporous resin "QuadraPure"; 30% (w/w) in CP; s.e.: 0.1 M AcB	[1682,1684, 2274]
Pb^{2+}	CP-μD (+BiF, in situ)	DPASV, FIA	accum. onto Bi via Me_aBi_b alloy formation; reoxidn.	0.3–10.0 mg L^{-1} Pb (0.2 mg L^{-1}; 60s)	Gunshot residues (coll. from hands)	μD: micro-detector (as flow cell); FIA mode with MEx and automated standard addition method	[2275]
Cd^{2+}, Pb^{2+} Zn^{2+}, Tl^+	C/SO (+SbF, in situ)	PSA	accum. onto Bi via Me_aBi_b alloy formation; reoxidn.	10–100 ppb Cd, Pb (–)	Model solns.	Sb^{III} species added to soln. serve as (i) source of SbF (ii) for chemical oxidn.; s.e.: 0.01 M HCl + KCl/KBr	[1994]
Pb^{2+}	C/PO (TETAM)	ASV	o.c. accum. by recognition chelating; MEx + redn. Anodic reoxidn.	1×10^{-8}–1×10^{-5} M (3×10^{-9} M Pb; 120 s)	Tap water	TETAM: 2-(4,8,11-tris-carbamoyl-methyl-1,4,8, 11-tetraaza-cyclo-tetra-decyl) acetamide; s.e.: pH 10	[2276]

Analyte	Electrode	Technique	Procedure	Concentration range	Matrix	Notes	Ref.
Pb^{2+}	mCPE (+MF, in situ)	SIA-DPASV (mLoV)	Electrolytic accum. via amalgamation; reoxidn.	(Low ppb level, ultratrace conc.)	Natural water	mLoV: mini-lab-on-valve regime; analysis efficiency: up to 50 cycles	[552]
Cd^{2+}, Pb^{2+}	C/SO (+SbF, in situ)	DPASV, FIA	accum. onto Bi via Me$_a$Bi$_b$ alloy formation; reoxidn.	5–50 ppb Cd, Pb (0.8 ppb Cd, 0.2 ppb Pb)	Lake water (spiked)	Advanced study with SbF-CPE in comp. with MF-, BiF-, and SbF-GCE; s.e.: 0.01 M HCl (pH 2)	[1894]
Cd^{2+}, Pb^{2+}	C/SO [RDE] (+BiF, in situ)	SWASV	accum. onto Bi via Me$_a$Bi$_b$ alloy formation; reoxidn.	5–150 ppb Cd, Pb (0.3 ppb Cd, 0.4 ppb Pb)	Canned food (meat paste)	Reproducibility of BiF/CP-RDE: $R = 3\%$–6%; analyses validated via comp with ICP-OES	[2277]
Cd^{2+}, Pb^{2+}	C/PO (ZeY)	DPCSV	accum. by ion-exchange Cathodic redn.	5–100 ppb Cd, Pb (1 ppb Cd, 3.6 ppb Pb)	Groundwater, industrial effluent	ZeY: NH$_4$+Y-modif. zeolite; opt.; results compared with ref. AAS	[1123]
Pb^{2+}	C/PO (HAP)	CV, SWCSV	accum. by ion-exchange Cathodic redn.	Nanomolar level (8×10^{-10} M; 5 min)	Environmental samples	HAP: Ca$_{10}$(PO$_4$)$_6$(OH)$_2$, hydroxy-apatite; s.e.: 1 M HClO$_4$	[1306]
Cd^{2+}	C/PO (HAP)	CV, SWCSV	accum. by ion-exchange Cathodic redn.	2×10^{-8}–3×10^{-5} M (4×10^{-9} M Pb; 5 min)	Natural water	HAP: hydroxyapatite (fine powder) opt. of mdf. content; $R_R = 104\%$	[115]
Cd^{2+}	C/PO (DNTPMBA)	POT (dir.)	Chemical equilibrium for Cd^{2+} ↔ Cd^{II}–BA	3×10^{-7}–0.1 M (2×10^{-7} M Pb; 50 s)	Well water	modif: 3,5-dinitro-N-(tri-2-pyridyl methyl) benzamide; $t_E > 3$ months	[2278]
Pb^{2+}	C$_{ACT}$/PO (virgin)	ASV	Electrolytical deposition redn.; anodic reoxidn.	8×10^{-9}–2×10^{-6} M (2×10^{-9} M Pb; 5 min)	Water samples	C$_{ACT}$: carbon powder activated by oxidn. with 30% H$_2$O$_2$ ("AC6")	[2279]
Cd^{2+}, Pb^{2+}	CILE (HAP)	SWASV	accum. by ion-exchange redn. + anodic reoxidn.	1×10^{-9}–1×10^{-7} M (0.5 nM Cd, 0.2 nM Pb)	Model solns.	HAP: hydroxy-apatite; accum. enhanced by IL chosen (MMIMPF$_6$)	[495]
Cd^{2+}, Pb^{2+}	EH/HS-CPE (+BiF, in situ)	SWASV, PSA	accum. onto Bi via Me$_a$Bi$_b$ alloy formation; reoxidn.	50–200 μg L^{-1} Cd, Pb (–)	Model solns.	EH: electrically pre-heated; HS: heated by solution ($t = 25°C$–$55°C$); initial study of temperature effect	[296,2280]

(continued)

TABLE 8.5 (continued)
Determination of Cadmium and Lead at Carbon Paste Electrodes, Sensors, and Detectors: Survey of Methods and Selected Studies

Ion, Species	Type of CPE (Modifier)	Technique (Mode)	Measuring Principles (Method Sequences)	Linearity Range (LOD; t_{acc}/t_R)	Sample(s)	Other Specification (Remarks)	Ref(s)
Cd^{2+}, Pb^{2+} Me^{2+}	C-TCP (+BiF; *in situ*)	SWASV	accum. via alloy form. Anodic reoxidn.	25 μg L^{-1} Cd, Pb (–)	Model solns.	The surface of TCP-CPE activated study of initial character	[2281]
Cd^{2+}, Pb^{2+}	C/SO (NH$_4$BiF$_4$)	SWASV	accum. via Me$_a$Bi$_b$ alloy Anodic reoxidn.	20–10 ppb Cd, Pb (10 ppb Cd, 1 ppb Pb)	Model solns.	Initial study on BiF$_4$-CPE; comp. to BiF–, Bi-CPE; s.e.: 0.1 M HCl	[269]
Pb^{2+}	C/PO (HMS/MMT)	SWAdSV	o.c. accum. via complex MEx, cathodic redn.	1–100 μg L^{-1} Pb (<μg L^{-1}; 120 s)	Tap water and groundwater	modif.: mesoporous silica funct. with SH-; also tests with Cd + Cu	[1863]
Cd^{2+}	CP-ISE (TMMCM-41)	POT (dir.)	Chemical equilibrium for Cd^{2+} ↔ Cd^{II}–TM/nM	1 × 10^{-6}–0.01 M (6 × 10^{-7} M Pb; 10 s)	Polluted tap water, electroplating bath	modif.: thiomorpholine-ctng nano/mesoporous substr.; t_L > 4 months	[1864]
Cd^{2+}	C/PO (AANH$_2$-HMS)	SWASV	o.c. accum. via complex redn., anodic reoxidn.	80–2000 ppb (50 ppb; 120 s)	Natural and residual water	modif.: mesoporous silica functionalized with HO–; interf. studies	[1865]
Pb^{2+}	CNT-CPE (n-SiO$_2$)	POT (dir.)	Chemical equilibrium and steady-state potential for Pb^{2+} ↔ Pb^{II}–n-SiO$_2$	1 × 10^{-7}– 0.01 M (–)	Model solns.	n-SiO$_2$: 3% nanosilica in mixt. of 57% C. + 5% CNTs + 20% PO + 5% ionofor (Thiram); t_L > 60 days	[447]
Cd^{2+}, Pb^{2+} Cu^{2+}	IL-CPE (quercetin, Qc)	AdSV	accum. by adsorption via modif.; cathodic redn.	—	Model solns.	IL-CPE for comparative studies; s.e.: 0.1 M HCOON a (pH 4.7)	[513]
Cd^{2+}, Pb^{2+} Me^{n+}	C/SO (Sb$_2$O$_3$, SbO-L)	SWASV	accum. via Me$_a$Bi$_b$ alloys Anodic reoxidn.	10–100 ppb Cd, Pb (5–10 ppb; 300 s)	Model solns.	L: OH– or Cl– (ligand); 0.1 M HCl; tested also for Zn^{2+}, Sn^{2+}, In^{3+}, Tl$^+$	[1898]

TABLE 8.6
Determination of Zinc, Thallium, Tin, Indium, Bismuth, Antimony, Arsenic, and Selenium at Carbon Paste Electrodes, Sensors, and Detectors: Survey of Methods and Selected Studies

Ion, Species	Type of CPE (Modifier)	Technique (Mode)	Measuring Principles (Method Sequences)	Linearity Range (LOD; t_{ACC}/t_R)	Sample(s)	Other Specification (Remarks)	Ref(s)
Zn^{2+}							
Zn^{2+}	C/PO (BG + $Zn(SCN)_4^{2-}$)	POT (dir.)	Ion-pairing with modif. Equilibrium potential	1×10^{-4}–0.1 M (50 μM; 60 s)	Model solns.	BG: *Brilliant Green* (dye) Slow nearly Nerstian response	[1974]
Zn^{2+}	C/TCP (+MF, *in situ*)	DPASV	Electrolytic accum. Anodic reoxidn.	2×10^{-7}–1×10^{-5} M (2×10^{-7} M, 3 min)	Drinking w., deionized w.	s.e.: 0.1 M ammonia buffer (pH 9) In presence of Cu^{II} (Cu:Zn < 5:1)	[257]
Zn^{2+}, Pb^{2+}, Cu^{2+}	C/PO [ZeMS(s)]	DPCSV	accum. via intercalation effect; subs. redn.	2×10^{-7}–5×10^{-6} M (145 nM; 4 min)	Model solns.	ZeMS(s): zeolite-based molecular sieves(s); study on Zn^{II} species in AmB; extensive interf. studies	[2382]
Zn^{2+}, Cd^{2+}, Pb^{2+}	EH-CPE (+BiF, *in situ*)	SWASV	accum. onto Bi via Me_aBi_b alloy formation; reoxidn.	ca. 5 μg L^{-1} Zn^{2+} (–)	Model solns.	EH: electrically (pre)heated CPE; detn. of Zn tested only in presence of Me^{2+} (in 10 × conc. excess)	[294]
Zn^{2+}, Cd^{2+}, Pb^{2+}	C/SO (Bi_2O_3)	DPASV	accum. via electrolytic redn. + Me_aBi_b alloy form. Anodic reoxidn.	50–200 μg L^{-1} Zn (20 μg L^{-1}; 5 min)	Model solns.	Way of plating: (i) *in situ*, (ii) ext., (iii) *in-nascenti* (via Bi_2O_3 redn.); detn. of Zn with high background	[103]
Zn^{2+}, Cd^{2+}, Pb^{2+}	C/SO [HgO, Bi_2O_3 (s)]	DPASV	accum. via electrolytic redn. at *in-nascenti* formed MF or BiF; anodic. reoxidn.	50–1000 ppb Zn (25 ppb; 10 min)	Tap and natural waters	HgO, Bi_2O_3 added as 5% (m/m); tests on reproducibility ($R_{\pm} < 5\%$); compared with MF- and BiF-SPE	[623,2283]

(continued)

TABLE 8.6 (continued)
Determination of Zinc, Thallium, Tin, Indium, Bismuth, Antimony, Arsenic, and Selenium at Carbon Paste Electrodes, Sensors, and Detectors: Survey of Methods and Selected Studies

Ion, Species	Type of CPE (Modifier)	Technique (Mode)	Measuring Principles (Method Sequences)	Linearity Range (LOD; t_{ACC}/t_R)	Sample(s)	Other Specification (Remarks)	Ref(s)
Zn^{2+}, Me^{2+}	s-CPE (C/wx) ($Hg_2C_2O_4$)	DPASV	accum. via ion-exchange redn.; anodic reoxidn.	ca. 100 μg L^{-1} Zn^{2+} (–)	Medicinal plants, ayurvedic tablets	CP-binder: solidified paraffin wax. Compared to AAS; Me: Cd, Pb, Cu	[316]
Zn^{2+}, Me^{2+}	C/SO (+SbF; *in situ*)	PSA	accum. onto Bi via Me_aBi_b alloy formation; reoxidn.	ca. 50 μg L^{-1} Zn^{2+} (–)	Model solns.	Sb^{III} added serves as a (i) source of SbF, (ii) for chemical oxidn.; s.e.: acidified KCl, KBr; Me: Cd, Pb, Tl	[1994]
Zn^{2+}, Tl^+, In^{3+}	C/SO (+SbF; *in situ*)	SWASV, CCSA	accum. via electrolytic redn., Me_aBi_b alloy form. Anodic reoxidn. by Sb^{3+}	0.1–0.5 mg L^{-1} Zn (–)	Model solns.	comp. to MF-, BiF-, and SbF-CPE, as well as CCSA with SWASV s.e.: 0.01 M HCl (pH 2).	[1894, 1996]
Zn^{2+}, Me^{2+}	C/SO (Sb_2O_3, SbOL)	SWASV	accum. via electrolytic redn., Me_aBi_b alloy form. Anodic reoxidn.	20–200 ppb Zn (19 ppb Zn; 300 s)	Model solns.	L: OH- or Cl- (ligand); comp. with SbF- and Sb_2O_3-CPE; 0.1 M HCl; Me: Cd, Pb, Sn, In, and Tl	[1898]
Tl^+, Tl^{III}							
Tl^{III}, $TlCl_4^-$	C/PO (*Amberlite LA-2*)	CV, DPASV	accum. via ion-pairing; elelyt. redn. ($Tl^{III} \rightarrow Tl^0$) Anodic reoxidn.	0.1–1000 μg L^{-1} Tl (ca. 0.1 μg L^{-1})	Fly ash	o.c. and MEx; s.e.: 0.001 M HCl; CMCPE regnt. in 1 M AmB	[1745]
Tl^{III}, TlX_4^-	s-CPE (C/phen) (R_4N^+, Ar_4P^+; *in situ*)	DPASV	o.c.-accum. by ion-pairing MEx; redn.; reoxidn.	10–1000 μg kg^{-1} (10 μg kg^{-1})	Zinc alloys	Phen: solidified phenanthroline; X: Cl, Br; s.e.: ammonia buffer, 0.01 M HCl + 0.15 M KCl/KBr	[76]

Analyte	Electrode	Method	Principle	Concentration range / LOD	Sample	Notes	Ref.
Tl^I, Tl^{III}	C/PO (oxin)	DPV	Electrolytic accum. of Tl^I (or chem. red. Tl^{III}); oxidn.	1×10^{-7}–1×10^{-6} M (5×10^{-9} M, 120 s)	Model solns. (with Tl^I + Tl^{III})	interfs. from Bi^{III}, Pb^{II}, and Cu^{II} supprd. by masking with EDTA	[1372]
Tl^{III}, Tl^{3+}	C/PO (oxin)	DPV	o.c.-accum. + MEx redn. (Tl^{III} → Tl^0); reoxidn.	5×10^{-10}–2×10^{-5} M (2×10^{-10} M, 120 s)	See water, human urine	s.e.: (1) BR-buffer (pH 4.6), (2) ammonia buffer (pH 10)	[1373]
Tl^{III}, Tl$_x$Cl$_n^{m-}$	C/TCP (CPy+Tl^{III}Cl$_4^-$)	POT	Ion-associate formation; equilibrium potential	6×10^{-6}–0.003 M (6×10^{-6} M; 20 s)	Model solns.	CPy: cetylpyridinium ion (titrant) Studies on compl. stoichiometry	[253]
Tl$^{I, III}$, TlCl$_4^-$	C/TCP (unm.)	CCSA	accum. via ion-pairing as {H-TCP$^+$; TlCl$_4^-$}; redn. with $I_{CONST} = -3$ µA	0.1–1.0 µM Tl^{III} (1×10^{-7} M; 3 min)	Model solns.	TCP: tricresyl phosphate (binder); s.e.: 0.1 M HCl + 2 M KCl: study on differentiation of Tl^I and Tl^{III}	[2384, 2285]
Tl$^+$, Cd^{2+}, Pb^{2+}	C/SO (+BiF, in situ)	DPASV	accum. onto Bi via Me$_a$Bi$_b$ alloy formation; reoxidn.	ca. 200 µg L^{-1} Tl (–)	Model solns.	s.e.:0.1M NaOH/KOH(pH 12) Study on complexation by OH	[637]
Tl$^+$, Cd^{2+}, Pb^{2+}	C/SO (+BiF, in situ)	DPASV	accum. via complexation, el. redn. + alloy formation Anodic reoxidn.	ca. 200 µg L^{-1} Tl (–)	Model solns.	s.e.: 0.1–1.0 M NaOH (pH > 12); studies on Tl-deposition at BiFE plated from Bi(OH)$_4^-$; effect of Me^{2+}	[670]
Tl$^+$, Cd^{2+}, Pb^{2+}	C/SO (+Bi-powder)	DPASV, SWASV	accum. via complexation, el. redn. + alloy formation Anodic reoxidn.	100–250 µg L^{-1} Tl^I (–)	Model solns.	s.e.: 0.1 M AmB (pH 9.5); study on behavior of Tl + Pb + Cd mixt. in atypical supporting media	[672]
Tl$^+$ Me^{2+}	C/SO (+SbF, in situ)	PSA	accum. onto Bi via Me$_a$Bi$_b$ alloy formation; reoxidn.	ca. 50 µg L^{-1} Zn^{2+} (–)	Model solns.	Sb^{III} added serves as (i) source of SbF, (ii) for chemical oxidn.; s.e.: acidified KCl, KBr; Me: Zn, Cd, Pb	[1994]
Tl$^+$, Zn^{2+}, In^{3+}	C/SO (+SbF, in situ)	CCSA, (SWASV)	accum. via electrolytic redn., Me$_a$Bi$_b$ alloy form. Anodic reoxidn. by Sb^{3+}	5–50 µg L^{-1} Tl (1.4 µg L^{-1} Tl; 120 s)	Model solns.	comp. to MF-, BiF-, and SbF-CPE, as well as CCSA with SWASV s.e.: 0.01 M HCl (pH 2)	[1894, 1996]

(continued)

TABLE 8.6 (continued)
Determination of Zinc, Thallium, Tin, Indium, Bismuth, Antimony, Arsenic, and Selenium at Carbon Paste Electrodes, Sensors, and Detectors: Survey of Methods and Selected Studies

Ion, Species	Type of CPE (Modifier)	Technique (Mode)	Measuring Principles (Method Sequences)	Linearity Range (LOD; t_{Acc}/t_R)	Sample(s)	Other Specification (Remarks)	Ref(s)
Tl^+, Me^{2+}	C/SO (SbO-L)	SWASV	accum. via electrolytic redn., Me_aBi_b alloy form. Anodic reoxidn.	50–250 ppb Tl (35 ppb Tl; 300 s)	Model solns.	L: OH^- or Cl^- (ligand); comp. with SbF^- and Sb_2O_3–CPE; 0.1 M HCl; Me: Zn, Cd, Pb, Sn, and In	[1898]
Sn^{2+}, Sn^{IV}							
$Sn^{II,VI}$, Pb^{2+}	C/PO (Hematein, in situ)	LSASV	o.c.-accum. via compl. as Sn^{II}; anodic reoxidn.	10–100 ng mL^{-1} (–)	Lead-tin alloy (with 0.005% Sn)	No interfs. from $Pb^{I,II,V}$, Cu^{II}, Sb^{II}, As^{III} (up to conc. ratio of 1:10^4) simult. detn. of Pb^{2+} also tested	[2286]
Sn^{IV}, SnO^{2+}	C/SO (Tropolone)	CV, ASV	accum. via compl. with modif.; cathodic redn.	4–50 mg L^{-1} (0.2 mg L^{-1})	Model solns.	Mechanical regnt. of CMCPE incl. interf. studies	[1371]
Sn^{II}, Me^N	C/PO (Metrohm®) (Oxin)	CV, DPV	Catalyzed oxidn. of $Sn^{II} \rightarrow Sn^{IV}$ (by modif.); oxidn. of the Sn^{II}–Ox complex	Micromolar level	Model solns.	Oxin: 8-hydroxyquinoline; com. marketed carbon paste mixture Other Me^N also tested	[2287]
Sn^{IV}, SnO^{2+}	C/PO (Alizarin violet)	ASV	o.c. accum. via complex; MEx/redn./an. reoxidn.	8×10^{-9}–1×10^{-6} M (4×10^{-9} M; 2 min)	Canned food	s.e.: AcB (pH 4.5); studies on adsorpt. mechanism of accum.	[2288]
Sn^{2+}	C/SO (BiF; in situ) C/SO (Bi-powder)	SWASV	accum. via electrolytic redn. + alloy formation Anodic reoxidn.	0.2–1 mg L^{-1} (0.1 mg L^{-1}; 5 min)	Model solns.	s.e.: 0.2–1.0 M HCl (pH < 2) + 0.005 M $N_2H_5^+$ (stabilizing agent); studies on detn. of tin as Sn^{II}	[2289]
Sn^{2+}, Sn^{IV}	C/PO (BPR)	DPASV	o.c. accum. via complex. el. redn. ($Sn^{II} \rightarrow Sn^0$) MEx/anodic reoxidn.	0.1–50 µg L^{-1} (0.1 µg L^{-1}; 2 min)	Wastewater, canned food	BPR: bromo-pyrogallol red; s.e.: (a) 0.1 M AcB, (b) 4 M HCl; pre-redn. $Sn^{IV} \rightarrow Sn^{II}$ by chem. agent	[694]

Analyte	Electrode	Method	Procedure	Range (accum.)	Sample	Notes	Ref.
In^{3+}, In^{+}							
In^{3+} (In^{+}) $Cd^{2+} + Pb^{2+}$	C/SO (+BiF, *in situ*)	SWASV	accum. via electrolytic redn. + alloy formation. Anodic reoxidn.	25–250 µg L^{-1} in (10 µg L^{-1}; 5 min)	Water samples (spiked); CRM (soil)	s.e.: 0.1 M AcB + 0.2 M KBr; studies on simultaneous detn. of In + Cd + Pb and detn. of In^{III} alone	[2290]
In^{3+}, Me^{2+}	C/SO (+SbF; SbOL)	CCSA, SWASV	accum. via electrolytic redn., Me_aBi_b alloy form. Anodic reoxidn. by Sb^{3+}	100–1000 ppb in (75 ppb in; 300 s)	Model solns.	L: OH$^-$ or Cl$^-$ (ligand); comp. with Sb_2O_3-CPE; s.e.: 0.01–0.1 M HCl; Me: Zn, Cd, Pb, Sn, and Tl	[1894, 1898]
Bi^{3+}, Bi^{III}							
Bi^{3+}	C/PO (unm.)	LSV	Electrolytic accum. Anodic reoxidn.	(ca. 0.1 µM)	Model solns.	s.e.: 0.15 M KSCN + 0.05 M HCl. Study of preliminary character	[2291]
Bi^{3+}	C/PO + SpG (+MF, *ex situ*)	DPASV	Electrolytic accum. Anodic reoxidn.	(2×10^{-9} M, 10 min)	Model solns.	SpG: graphite layer sprayed onto the CPE surface as aerosol	[67]
Bi^{3+}	C/Nj (decan-1-thiol)	DPASV	o.c.-accum. + redn. Anodic reoxidn.	0.02–4 ppm	Model solns.	eld. regnt. in 0.01 M HCl; s.e.: 0.2 M HCl; oxidn. agents interf.	[810]
Bi^{3+}	C/Nj (*Bismuthol I*)	DPCSV	o.c.-accum. Cathodic redn.	0.01–4 mg L^{-1} (5 µg L^{-1}, 2 min)	Model solns.	incl. study on accum. mechanism and interfs. from various ions	[2292]
Bi^{III} BiI_4^{-}	C/TCP (unm.)	DPCSV, DPASV	o.c.-accum. (ion-pairing + extraction; cathodic redn. Electrolytic deposition + anodic reoxidn.	3×10^{-7}–2×10^{-5} M (1×10^{-7} M, 3 min)	Model solns.	TCP: tricresyl phosphate acting as ion-pairing agent; ASV mode: no separation of Bi and Sb; CSV: Bi detn. in presence of Sb (c/c = 1:10)	[188, 205]
Bi^{3+}	C/PO (*Bismuthol II*)	LSV	o.c.-accum. by complex form.; cathodic redn.	3×10^{-8}–6×10^{-6} M (2×10^{-8} M, 5 min)	Actual samples (–)	modif.: 5-thio-3-(*p*-carboxyphenyl-1,3,4-thiadiazoline-2-thione	[2293]

(*continued*)

TABLE 8.6 (continued)
Determination of Zinc, Thallium, Tin, Indium, Bismuth, Antimony, Arsenic, and Selenium at Carbon Paste Electrodes, Sensors, and Detectors: Survey of Methods and Selected Studies

Ion, Species	Type of CPE (Modifier)	Technique (Mode)	Measuring Principles (Method Sequences)	Linearity Range (LOD; t_{ACC}/t_R)	Sample(s)	Other Specification (Remarks)	Ref(s)
Bi^{3+}	C/PO (sodium humate)	CV, DPV	o.c.-accum. + MEx redn. + anodic reoxidn.	$5 \times 10^{-8} - 2 \times 10^{-5}$ M	Waters, alloy	s.e.: (1) 0.05 M KNO_3 + 0.01 M HNO_3, (2) 0.5 M HNO_3	[1697]
Bi^{III} $BiCl_4^-$	C/Nj (Pyrogallol red)	2nd DeLV	accum. by compl. redn. + anodic reoxidn.	$1 \times 10^{-9} - 6 \times 10^{-7}$ M (5×10^{-10} M, 5 min)	Model solns.	s.e.: 0.3 M HCl incl. study on accum. mechanism	[2294]
Bi^{3+}	C/PO (SWy-2)	DPASV	o.c. accum. via ion-exch. el. redn./anodic reoxidn.	$4 \times 10^{-9} - 1 \times 10^{-6}$ M (1×10^{-10} M; 5 min)	Water sample, nickel alloy	SWy-2: Na-montmorillonite; ion-exchange + strong sorption of Bi^{III}	[2295]
Bi^{3+}	C/PO (BPR; in situ)	DPASV	o.c. accum. via chelating el. redn./anodic reoxidn.	$1 \times 10^{-9} - 5 \times 10^{-7}$ M (5×10^{-10} M; 3 min)	Water sample, human hair	BPR: bromopyrogallol red; s.e.: 0.3 M HCl + 2×10^{-5} M BPR	[692]
Bi^{3+}	C/PO type (1),(2) (BiI_4^- + POE/OPE + $Fe(phen)_3$)	POT (dir.)	Chemical equilibrium and steady-state potential for $Bi^{3+} \leftrightarrow (POE/OPE)^+BiI_4^-$	(4×10^{-6} M (1) 2×10^{-6} M (2) $t_R = 20-40$ s)	Suppositories, ointment dosage	POE: polyoxy ethylene, OPE: octyl- phenyl ether; "phen": 1,10-phenan-throline; s.e.: buffers (pH 3–9)	[2296]
Sb^{3+}, Sb^{III}, Sb^V							
Sb^{3+}	C/PO	DPASV (with RDE)	accum. via electrolytic redn.; anodic reoxidn.	0.005–0.500 ng mL^{-1} (500 pg mL^{-1}; 10 min)	Model solns.	RDE: rotated disc electrode (2500 rpm); s.e.: diluted HCl (pH 2)	[529]
Sb^{3+}	C/Nj (phenylfluorone)	DPASV	o.c.-accum. via compl. redn. + anodic reoxidn.	$3 \times 10^{-8} - 1 \times 10^{-7}$ M (9×10^{-9} M, 10 min)	Soil, human hair	incl. interf. studies on various ions and org. substances	[1370]
Sb^{3+}	C/PO (BPR)	DPASV	o.c. accum. via chelating el. redn./anodic reoxidn.	$2 \times 10^{-9} - 5 \times 10^{-7}$ M (1×10^{-9} M; 2.5 min)	Water sample, human hair	BPR: bromo-pyrogallol red; s.e.: 0.1 M HCl + 3×10^{-5} M BPR (chel. ag. preventing hydrolysis of Sb^{III})	[693]

Electroanalysis with Carbon Paste–Based Electrodes, Sensors, and Detectors 247

Analyte	Electrode	Method	Principle	Concentration range (detection limit)	Sample	Notes	Ref.
Sb^{3+}, Cu^{2+}	C/PO	DPASV (with RDE)	accum. via electrolytic redn.; anodic reoxidn.	10–50 ng mL^{-1} Sb (5 ng mL^{-1}; 5 min)	CRMs (iron and steel)	s.e.: HCl + KI + ascorbic acid (for separation of Sb from Cu); interfs. other ions tested (Bi^{3+}, Hg^{2+}, Pb^{2+})	[2297]
Sb^{3+}	C/PO type (1), (2) (SbI_4^- + CPy/TPhT)	POT (dir.)	Chemical equilibrium and steady-state potential for $Sb^{3+} \leftrightarrow (CPy/TPhT) + SbI_4^-$	4×10^{-6} M (1) 5×10^{-6} M (2) $t_R = 20–30$ s	Wastewater, anti-bilharzial comp.	CPy: cetyl pyridinium, TPhT: tri-phenyl tetrazolium, s.m.: pH 4–10; little interfs. from Cd^{2+}, Hg^{2+}, Bi^{3+}	[428]
Sb^{III}, Sb^V	CNTPE (unm.)	PSA (ox.: Hg^{II})	accum. via electrolytic redn.; catalytic reoxidn.	10–50 + 100–500 ppb (6 ppb Sb; 3 min)	pharms. (*Meglumine-Sb.*)	Multiwalled CNTs mixed with Nj s.e.: diluted acid (pH 3.6); cysteine for chem. redn. of Sb^V; interf. stud.	[2298]
As^{III}, As^V							
$As^{III} + As^V$	C/SO (+AuF, *ex situ*)	DPASV	Electrolytic deposition; anodic reoxidn.	5–50 µg L^{-1} (1 µg L^{-1}, 10 min)	Model solns.	Serious interfs. from Cu^{II} (0.5:1) With possible detn. of As^{III} + As^V (chem. redn. with α-cysteine)	[261]
$As^{III} + As^V$	C/SO (+AuF, *ex situ*)	CCSA ($I_C = 1$ µA)	accum. (electrolytic redn.) oxidn. with const. current	As^{III}: 3 ppb (15 s) As^V: <1 ppb (5 min)	River water (polluted)	Interf. from Cu^{II} (Cu:As >5:1) With diffn. of As^{III} vs. As^V; CPE regnt. mech. and in 0.001 M $HClO_4$	[108]
As^V ($HAsO_4^{2-}$)	C/TCP (unm.)	POT (titr.)	Chemical equilibrium with ion-pairing: $HAsO_4^{2-} \leftrightarrow CPyB \cdot H_2; [As(Mo_3O_{10})_4]^-$	(0.2 mg L^{-1} As^V)	Mineral water, org. substances	TCP: tricresyl phosphate (as binder) CPyB: cetylpyridinium bromide	[712]
As^V ($HAsO_4^{2-}$)	C/PO (Fe^{II}-clinoptite)	POT (dir.)	Ion-pairing equilibrium: $HAsO_4^{2-} \leftrightarrow (ZE)_2$-$HAsO_4^{2-}$	2×10^{-8}–0.001 M (3×10^{-8} M; 5–10 s)	Natural water, wastewater	Dry ashing of solid samples modif.: natural zeolite; t_{CP-ISE} >2 months; evaluation of k^{pot}	[1137]

(continued)

TABLE 8.6 (continued)
Determination of Zinc, Thallium, Tin, Indium, Bismuth, Antimony, Arsenic, and Selenium at Carbon Paste Electrodes, Sensors, and Detectors: Survey of Methods and Selected Studies

Ion, Species	Type of CPE (Modifier)	Technique (Mode)	Measuring Principles (Method Sequences)	Linearity Range (LOD; t_{ACC}/t_R)	Sample(s)	Other Specification (Remarks)	Ref(s)
As^{III}, As^V	C/PO (hematite)	ASV	As^{III} reduced at pH 4, As^V at pH 0 (both at electrode); anodic reoxdn.	5–50(70) ppb $As^{III/V}$ (As^{III}: 5, As^V: 2 ppb)	Drinking water, compost lixiviate	Speciation of As^{III} + As^V via the way of electrode reduction; Cu^{II} and Bi^{III} interfs. with detn. of As^V	[2299]
Se^{IV}, Se^{-II}							
Se^{IV} (SeO_3^{2-})	C/PO (diamino-2, 3-naphthalene)	LSV	Adsorp. accum.; MEx redn. + anodic reoxidn.	(1×10^{-7} M; 20 min)	Model solns.	s.e.: (1) 0.01 M HCl + 0.1 M KCl (2) 0.01 M HNO_3 + 0.1 M KNO_3	[1375]
Se^{IV} (SeO_3^{2-})	GTE (with CP-layer)	LSV	Electrolytic accum. and redn.; anodic reoxidn.	1×10^{-7}–1×10^{-6} M (8×10^{-8} M; 5 min)	Model solns.	GTE: graphite thick-layer el. with carbon paste-like material	[2300]
Se^{IV} (SeO_3^{2-})	CP-ISE (CoTMeOPP)	POT (dir.)	Chelate formation with modif.; equilibrium pot.	5.2×10^{-5}–0.012 M ($3–5 \times 10^{-5}$ M; <15 s)	Na_2SeO_3 (s), mixt. of Se^{IV}/Se^{VI}, pharm.	modif.: Co^{II}-porphyrin der.; pH-dependent response (for $HSeO_3^-$ and SeO_3^{2-}); other two ISEs also tested.	[2301]
Se^{-II} ($SeCN^-$)	C/PO (Co^{II}-Pc)	AD, FIA	Anodic oxidation (with cathodic reactivation)	0.2–20 mM (0.1 mM; >5 s)	Model solns. (saliva for SCN^-)	Pc: phthalocyanine; PhB (pH 7) Detn. together with SeO_3^{2-}, SCN^-, SO_3^{2-}, $S_2O_3^{2-}$, NO_2^-, and $HAsO_3^{2-}$	[652]

Electroanalysis with Carbon Paste–Based Electrodes, Sensors, and Detectors

TABLE 8.7
Determination of Iron, Cobalt, and Nickel at Carbon Paste Electrodes and Sensors: Survey of Methods and Selected Studies

Ion, Species	Type of CPE (Modifier)	Technique (Mode)	Measuring Principles (Method Sequences)	Linearity Range (LOD; t_{ACC}/t_R)	Sample(s)	Other Specification (Remarks)	Ref(s)
$Fe^{2+}, Fe^{3+}, Fe^{III}$							
$Fe^{3+}, FeCl_4^-$	C/PO (unm.)	LSV	adsort. accum. Cathodic redn.	$1 \times 10^{-6} – 1 \times 10^{-3}$%	conc. HCl (model solns.)	In presence of 5–6 M Cl$^-$ interfs. from Cu^{II}, Au^{III}, Sb^{III}	[2302]
$Fe^{3+}, Fe(CN)_6^{3-/4-}$	C/Uv (*Amberlite LA-2*)	DPV	o.c. accum. by ion-pairing oxidn./redn.	6–6000 µg L^{-1} (0.3 µg L^{-1}, 3 min)	Wines	incl. interfs. study from typical constituents in wine matrices	[803]
Fe^{3+}	C/PO (*SWy*)	DPV	o.c. accum. (ion-exchange) MEx + cathodic redn.	0.5–6.0 ppm (0.2 ppm, 3 min)	Model solns.	SWy: sodium montmorillonite CMCPE regnt. in Na$_2$CO$_3$	[1058]
Fe^{3+}	C/Nj (chitin)	CV, DPV	accum. via compl. Cathodic redn.	$2 \times 10^{-6} – 7 \times 10^{-6}$ M	River, tap water	s.e.: 0.01 M KCl + 10^{-4} M EDTA No interfs. from common ions	[1705]
Fe^{2+}	C/PO (1, 10-phen + Nf)	CV, DPV	accum. by chelating Cathodic redn.	$1 \times 10^{-8} – 5 \times 10^{-7}$ M (5 min)	CRMs (water), natural water	Fe^{III} chem. red. with NH$_3$OH$^+$ incl. study on accum. mechanism	[1364]
Fe^{3+}	C/PO (1, 10-bipy + Nf)	CV, DPV	accum. by form. of chelate Cathodic redn.	$1 \times 10^{-8} – 2 \times 10^{-6}$ M (3 min)	CRMs (water), natural water	Fe^{III} chem. red. to Fe^{II}; CMCPE regnt. in acidic soln.	[1365]
Fe^{3+}	C/PO (diff. chelators)	CV, DPV	accum. by chelating redn. + anodic reoxidn.	6–60 ppb (2 ppm)	Biological fluids	modifs.: pyridine, hydrazone, and semicarbazone derivatives.; tests on simult. detn. of Hg^{2+} + Cu^{2+}	[1679]
$Fe^{3+}, Hg^{2+}, Cu^{2+}$	C/Nj (*Desferal*)	CV, DPV	accum. via compl. oxidn. of ligand + redn.	Down to 1 µM (0.5 µM)	Model solns.	modif: trihydroxamic acid (THA) s.e.: AcB or AmB (pH 4 or 10)	[2303]
$Fe^{3+}, Fe(CN)_6^{3-}$	C/PO (ODA)	CV, LSV AD (stat.)	Effect of higher coverage of C-particles; el. redn.	(µM level)	Model solns.	ODA: octadecyl amine; special microscopic study; selectivity is adjustable via the modif. content	[1639]
Fe^{3+}	C/PO (+EDTA, *in situ*)	POT, (OGI-titr.)	Equilibrium potential with Super-*Nernstian* response	$5 \times 10^{-8} – 2 \times 10^{-5}$ M	Model solns.	OGI: oscillographic indication Soln.: 0.01 M HCl (pH 2)	[2304]
Fe^{3+}	C/PO (unm.)	DPCSV	compl. with analyte (AIP) sample; electrolytic redn.	$3 \times 10^{-7} – 5 \times 10^{-5}$ M (3×10^{-7} M; 3 min)	pharm. (AIP-A)	AIP-A: 5-aminoisophthalic (acid) No sample pretreatment needed	[2305]

(continued)

TABLE 8.7 (continued)
Determination of Iron, Cobalt, and Nickel at Carbon Paste Electrodes and Sensors: Survey of Methods and Selected Studies

Ion, Species	Type of CPE (Modifier)	Technique (Mode)	Measuring Principles (Method Sequences)	Linearity Range (LOD; t_{ACC}/t_R)	Sample(s)	Other Specification (Remarks)	Ref(s)
Fe^{2+}, Fe^{3+}	DPEs [D/PO] (unm., three types)	DPASV	accum. (electrolytic redn.) Anodic reoxidn.	$1 \times 10^{-12} - 1 \times 10^{-6}$ M, [$1-10 \times 10^{-13}$ M]	(a) pharm. forms. (multivitamin) (b) Natural waters	D: natural or synthetic diamond (p. size: 1 or 50μm); interf. study	(a) [118] (b) [347]
Fe^{2+}, Fe^{3+}	C/PO + Nf (1,10-phen)	SWASV, AD (stat.)	accum. by electrolytic redn. Anodic reoxidn.	$6 \times 10^{-6} - 2 \times 10^{-5}$ M (2×10^{-6} M; 5 min)	Fuel ethanol samples	Nf: Nafion® (stabilizer of carbon paste in EtOH; comp. to ref. AAS)	[1363]
Fe^{3+}, $Fe(CN)_6^{3-}$	C/PO (APS)	CV, DPV	o.c. accum. by electrostatic attraction; cathodic redn.	$2 \times 10^{-7} - 1 \times 10^{-5}$ M (8×10^{-8} M; 5 min)	Electroplating waste	APS: aminopropyl-grafter mesoporous silica (characterized by FT-IR) APS-CPE anodized (charged to +)	[1859]
Fe^{3+}, $Fe(CN)_6^{3-}$	C/PO (SDS)	CV, LSV	Electrostatic attraction by the surfactant monolayer	(1×10^{-4} M)	Model solns.	SDS: Na-dodecylsulfate; comp. to bare CPE; accum. studies	[1659]
Fe^{3+}, $Fe(CN)_6^{3-}$	(a) C/PO, C/SO, (b) GC/PO, CNTPE	CV, R_{tr}	Basic redox behavior Ohmic resistance effect	(mM level)	Model solns.	comp. study with various CPEs, analyte as model ion, evaln. of CP indexes; new theory verification	(a) [228] (b) [229, 2306]
Co^{2+} (+Ni^{2+})							
Co^{2+}	C/PO (1,10-phen.)	CV, DPV	accum. by chelating redn. + anodic reoxidn.	$1 \times 10^{-6} - 3 \times 10^{-4}$ M	Model solns.	incl. studies on compl. effect and interfs. from Fe^{II} and Cu^{I}	[2307]
Co^{2+}, Ni^{2+}	C/SO [+MF] (DMG, in situ)	DPCSV	accum. by chelating with DMG and adsorption; cathodic redn.	$1 \times 10^{-8} - 1 \times 10^{-6}$ M Co (5×10^{-9} M; 10min)	Sea water (synthetic)	DMG: dimethylglyoxime; detns. made with bare CPE as well as at MF-CPE; s.e.: AmB + TEA (pH 9)	[2308]
Co^{2+}	C/PO [+Nf] (1,10-bipy, -phen)	DPASV	accum. by chelating oxidn. of Co^{II} in compl.	$7 \times 10^{-7} - 1 \times 10^{-5}$ M (3×10^{-7} M, 5 min)	CRMs (alloys, anodic mud)	CMCPE regnt. in acidic solns. No interfs. from common ions	[1354, 1355]
Co^{2+}, Ni^{2+}	C/SO [+MF] NN, in situ or DMG, in situ)	DPCSV	accum. by chelating with NN or DMG, adsorption; cathodic redn.	$1 \times 10^{-8} - 1 \times 10^{-6}$ M Co (5×10^{-9} M; 10min)	Sea water (sampled in 100–1500 m depth)	NN: 1-nitroso-2-naphthol; s.e.: AmB + TEA (pH 9); analyses under expedition conditions at ship-board lab	[812]

Electroanalysis with Carbon Paste–Based Electrodes, Sensors, and Detectors 251

Analyte	Electrode	Method	Principle	Range (LOD)	Sample	Note	Ref.
Co^{2+}	C/PO (1-spartein)	CV, DPV	accum. via compl. oxidn. of ligand + redn.	2×10^{-7}–5×10^{-5} M (1×10^{-7} M, 20 min)	Model solns.	incl. studies on compl. effect of other ions	[2309]
Co^{2+}	C/Nj (PAN)	DPV	accum. by chelating Cathodic redn.	2×10^{-7}–1×10^{-4} M (1×10^{-7} M, 15 min)	Real samples	PAN: 1-(2-pyridylazo)-2-naphtol CMCPE regn. in acidic soln.	[2310]
Co^{2+}	C/PO (+phen.) (HTTA, in situ)	2.5-DPV	accum. via form. of ternary compl.; redn.	8×10^{-9}–7×10^{-7} M (1×10^{-9} M, 7 min)	Natural water	HTTA: thiophenecarboxylic trifluoroacetate; s.e.: acetate buffer	[2311]
Co^{2+}	C/Nj (CPCHA)	DPASV	accum. by compl.; redn. Anodic reoxidn.	1×10^{-6}–4×10^{-5} M (3×10^{-7} M, 5 min)	pharm. (vitamin)	CPCHA: N-p-chlorophenylcinna-mo hydroxamic acid	[1386]
Co^{2+}	C/PO (PAN)	DPCSV	o.c.-accum. + MEx oxidn. + cathodic redn.	1×10^{-8}–1×10^{-8} M (6×10^{-9} M, 3 min)	CRM, human hair, pig liver, spinach	PAN: 1-(2-pyridylazo)-2-naphtol interfs. from V^{II}, Ce^{III}, and EDTA	[1357]
Co^{2+}	C/PO (Cyclam)	POT (dir., titr.)	Chemical equilibrium and steady-state potential for $Co^{2+} \leftrightarrow Co^{II}\text{-}Cyclam$	6×10^{-6}–0.1 M (3×10^{-6} M; —)	Wastewater (from electroplating baths)	Cyclam: agent entrapping Co^{2+} via its ion-diameter; solns. cntg. 25% EtOH; Nerstian slope: 28.4 mV/dec.	[1407]
Co^{2+}	C/SO [+ PbF] (Nx, in situ)	CtAdSV	accum. via compl.; electrocatalysis-assisted el. redn. (nioxime/ NO_2^- system)	1×10^{-9}–5×10^{-7} M (4×10^{-10} M; 2 min)	Model solns.	s.e.: 0.1 M AmB + 5×10^{-4} M Nx (nioxime) + 0.25 M KNO_2; detn. of Co^{II} in the presence of Ni^{II} and Zn^{II} at high excess	[1895]
Co^{2+}	C/PO (ZEs: A,X,Y types)	ASV	accum. via ion-exchange + sorption/el. redn. Anodic reoxidn.	(3 ppm; 15 min)	Model solns.	ZEs: zeolites; characterization by XRF + particle size analysis	[1121]
Ni^{2+} ($+Co^{2+}$)							
Ni^{2+}	C/Nj (DMG, in bulk)	DPCSV, EC-FIA	accum. by chelating and adsorpt.; cathodic redn.	0.05–5 ppb (10 min)	Natural water, fly ash	DMG: dimethylglyoxime; CPE regnt. in 1 M HNO_3; s.e.: AmB (pH 8); no interfs. from O_2	[549,683]
Ni^{2+}, Co^{2+}	C/SO [+MF] (DMG, in situ)	DPCSV	accum. by chelating with DMG and adsorption; cathodic redn.	1×10^{-8}–1×10^{-6} M Co (5×10^{-9} M; 10 min)	Sea water (synthetic)	DMG: dimethylglyoxime; detn. performed with the bare CPE as well as MF-CPE; s.e.: AmB + TEA (pH 9)	[2308]

(continued)

TABLE 8.7 (continued)
Determination of Iron, Cobalt, and Nickel at Carbon Paste Electrodes and Sensors: Survey of Methods and Selected Studies

Ion, Species	Type of CPE (Modifier)	Technique (Mode)	Measuring Principles (Method Sequences)	Linearity Range (LOD; t_{ACC}/t_R)	Sample(s)	Other Specification (Remarks)	Ref(s)
Ni^{2+}	C/PO (DMG)	PSA (chem. ox.)	accum. by chelating with modif.; redn. + reoxidn. by O_2 (diss. in soln.)	0.01–3 mg mL^{-1} (8 ng mL^{-1})	Model solns.	DMG: dimethylglyoxime incl. study on chem. oxidn.	[684]
Ni^{2+}, Co^{2+}	C/SO [+MF] DMG, in situ	DPCSV	accum. by chelating with NN or DMG, adsorption; cathodic redn.	1×10^{-8}–1×10^{-6} M Co (5×10^{-9} M; 10 min)	Sea water (sampled in 100–1500 m depth)	s.e.: AmB + TEA (pH 9); analyses under expedition conditions at ship	[812]
Ni^{2+}, Co^{2+}	C/PO (Cyclam & ders.)	CSV	accum. by ion-inclusion MEx + cathodic redn.	(4×10^{-8} M; 5 min)	Model solns.	modif(s).: azacrown ether compds. s.e.: (a) 0.1 M KCl, b) 0.1 M KOH; test on simult. detn. of Co^{2+}	[2312]
Ni^{2+}	C/PO (DMG)	POT (dir.)	compl. with modif. Equilibrium potential	5×10^{-7}–0.001 M	Steel sample	High stability of CMCPE; spec. dissolved in HNO_3 and buffered	[2313]
Ni^{2+}	C/Nj (DCPAB)	POT (dir.)	compl. with modif. Equilibrium potential	3–6000 ppm	CRMs (alloys)	modif.: 4-(3,5-dichloro-2-pyridyl)-1,3-diamino benzene; titr. with EDTA in 6M HCl or 2M NaOH	[2314]
Ni^{2+}	C/Nj (Dowex 50W)	AdSV	o.c. accum. by ion-pairing redn. + reoxidn.	5–6000 µg L^{-1} (1 µg L^{-1}, 12 min)	Tap and mineral waters	s.e.: 0.005 M HCl (pH 3); interfs. from Hg^{II} and Ag^+	[1734]
Ni^{2+}	C/SO + MF (DMG, in situ)	DPCSV, CCSA	Ads. accum. by chelating redn. (cathodic or I_{CONST})	(5×10^{-7} M; 60 s)	Crude oil digested (as spiked sample)	Coupled with MWD; high acidity buffered with 30% NH_3 (to pH 9)	[627]
Ni^{2+}	C/PO (DMG)	DPCSV, AD (HA)	accum. via chelating + adsorption; cathodic redn.	5×10^{-9}–5×10^{-7} M; (3×10^{-9} M; 25 min)	Commercial ethanol fuel	DMG: dimethyl glyoxime; modif. added in paste; compared to AAS	[1361]
Ni^{2+}	s-CPE/C/Wx (AMTS + DMG, in situ)	CV, DPCSV	accum. via adsorption (as Ni^{II}-AMT)/MEx + DMG; deposition/cathodic redn.	8×10^{-9}–1×10^{-6} M; (2×10^{-9} M; 20 min)	Commercial ethanol fuel	Solid-like CPE ctng. paraffin wax (stable in EtOH); AMTZ-S: amino-thiazole-functionalized silica	[1838]

TABLE 8.8
Determination of Metals of the V, VI, VII, and VIII Group at Carbon Paste Electrodes, Sensors, and Detectors: Survey of Methods and Selected Studies

Ion, Species	Type of CPE (Modifier)	Technique (Mode)	Measuring Principles (Method Sequences)	Linearity Range (LOD; t_{ACC}/t_R)	Sample(s)	Other Specification (Remarks)	Ref(s)
$V^{IV,V}$							
V^V (VO^{3+})	C/PO (SRV RS, in situ)	LSV	adsort. accum. as compl. Anodic oxidn. of ligand	0.1–1 µg L^{-1}	CRM (basic slag)	SRV RS: Solochrome violet RS simult. detn. with Sb^{III} possible	[529]
V^V (VO_2^-/Ox_2^{3-}) Ti^{IV}, Mo^{VI}	C/U v (CTAB, in situ)	DPASV	accum. via complexes of oxalate, ion-pair formation Stat. redn. of V^V → V^{IV} and anodic reoxidn.	5–200 µg L^{-1} (0.07 µg L^{-1}; 10 min)	Model solns.	s.e.: 0.01 M H$_2$C$_2$O$_4$ for in situ complexation with MeN; simult. detn. of V^V, Mo^{VI}, and Ti^{IV}, incl. test measurements in mixtures	[706,817]
V^V, Me^N	C/PO (ARS)	2nd DeLV	accum. via adsorption of VO$_2$-ARS; cathodic redn.	0.1–15 µg L^{-1} (0.04 µg L^{-1}; 2 min)	Water samples	ARS: Alizarin red S; s.e.: AmAc; detn. at high scan rate (200 mV/s) Other Me^N: Fe^{III}, Cr^{VI}, Mn^{VII}	[688]
V^V	AB/PO (AV, in situ)	LSASV	o.c. accum. by compl. and adsorption; anodic oxidn.	8 × 10^{-10}–1 × 10^{-7} M, (6 × 10^{-10} M; 90 s)	Natural water	AB: acetylene black; AV: Alizarin violet; s.e.: urotropin/HCl buffer (pH 4.4); E_R = 0–1.0 V vs. ref.	[2315]
$Cr^{III,VI}$							
Cr^{VI} (CrO$_4^{2-}$)	C/PO (chitin)	DPV	accum. by ion-pairing with modif.; cathodic redn.	3 × 10^{-8}–1 × 10^{-5} M	Model solns.	simult. detn. of NO$_2^-$ + MnO$_4^-$ incl. studies on other interfs.	[1706]
Cr^{III} + CrO$_4^{2-}$	C/Nj (DPC)	CSV	synerg. accum. via redox process + compl.; redn.	(1 × 10^{-7} M; 5 min)	Soil extract	DPC: diphenyl carbazide; incl. studies on accum. mechanism	[1366]

(continued)

TABLE 8.8 (continued)
Determination of Metals of the V, VI, VII, and VIII Group at Carbon Paste Electrodes, Sensors, and Detectors: Survey of Methods and Selected Studies

Ion, Species	Type of CPE (Modifier)	Technique (Mode)	Measuring Principles (Method Sequences)	Linearity Range (LOD; t_{ACC}/t_R)	Sample(s)	Other Specification (Remarks)	Ref(s)
Cr^{VI} ($Cr_2O_7^{2-}$)	C/PO (unm.)	POT	Ion-pairing with titrant Equilibrium potential	(3×10^{-8} M; 20 s)	Model solns.	pot. titration in 2 M H_2SO_4 Super-Nernstian slope (180 mV/dec.)	[2316]
$Cr^{III,VI}$	D/PO (DPEs, three types)	DPASV	accum. via electrolytic redn.; anodic reoxidn.	$1 \times 10^{-10} – 1 \times 10^{-7}$ M, (1×10^{-11} M)	Vitamins (dosage tablets)	DP: diamond paste (prepared from nat./synth. diamond ($p = 1/50$ μm)	[344,346]
Cr^{VI} ($HCrO_4^-$)	C/PO (CTAB or SPX, both in situ)	DPCSV	accum. via ion-pairing as {$RR'N_4^+$; $HCrO_4^-$} + extr. Cathodic redn.	$5 \times 10^{-7} – 5 \times 10^{-5}$ M (5×10^{-8} M; 5 min)	CRMs (plants), tea extracts	SPX: Septonex ($RR'N_4^+Br^-$); s.e. 0.1 M HCl + KCl (pH 1); Cr^{III} in samples oxidized by H_2O_2 + NaOH	[695]
Cr^{VI}	C/PO (modif.)	CSV	accum. via the reaction with modif.; cathodic redn.	$1 \times 10^{-8} – 1 \times 10^{-6}$ M (9×10^{-9} M)	Water samples	modif.: org. complexants of the Dithizon type (with –N, =C–OH$^+$, and –SH functional groups)	[2317]
Cr^{VI}	C/PO (DFC)	CV, DPASV	accum. via Cr^{III}–DFC' (red. form of DFC); anodic oxidn.	$1 \times 10^{-7} – 5 \times 10^{-6}$ M (7×10^{-8} M; 120 s)	Sea water	DFC: diphenylcarbazide, DFC': diphenylcarbazone; no effect of O_2	[2318]
Cr^{III}	C/PO (DHPIE)	POT (BIA, FIA)	Complexation with modif. Equilibrium potential	(1.4×10^{-7} M; BIA) (5.4×10^{-7} M; FIA)	Soil extracts, electroplating baths	modif.: di-hydroxy-phenyl-imino ethane; Cr^{VI} reduced chemically	[2319]
Cr^{III}	C/PO (APy + n-SiO_2)	POT (dir.), ACV	Complexation with modif. Equilibrium potential	$1 \times 10^{-8} – 0.001$ M (8×10^{-9} M; 55 s)	Food stuff (coffee, tea leaves)	modif.: 2-acetylpyridine + nanoporous silica gel; modif. also characterized by UV-vis and ACV	[1866]

Analyte	Electrode	Method	Principle	Range (LOD; time)	Matrix	Notes	Ref.
Cr^{III}	C/PO (Cr^{III}-Salophen)	POT (dir.+ titr.)	Complexation with modif. Equilibrium potential	7.5×10^{-6}–0.01 M (2×10^{-6} M; 8s)	River water, drug (Salophen)	modif.: N,N'-bis(salicylidene)-o-phenediaminate-Cr(III); pH 4.5–7.7; titr.: EDTA; comp. with PVC-ISEs	[2320]
Cr^{VI} ($HCrO_4^-$)	C/Nj (TOA)	CV, ASV	Complexation with modif. el. redn. + anodic reoxidn.	nM level (3×10^{-9} M; 10min)	Electronic material and components	TOA: trioctylamine; the selectivity Cr^{III} vs. Cr^{VI} (600:1); $E_{ac} = +0.45$ V	[1367]
Cr^{III}	C/PO ($NaBPh_4$ + AMTX)	POT (dir. + titr.)	Complexation with modif. Equilibrium potential	3×10^{-7}–0.1 M (2×10^{-7} M; <10s)	Model solns.	modif.: 1-[(2-hydroxy-ethyl)amino]-4-methyl-9H-thioxanthen-9-one; evaln. of k_i^{POT} by various methods	[2321]
Mo^{IV-VI}							
Mo^{VI} (MoO_4^{2-})	C/Nj (DEDC, in situ)	ASV	accum. via redox compl. Anodic reoxidn.	5–50 ng mL^{-1}	$NaWO_4$ (s)	DEDC: diethyldithio carbamide s.e.: acetate buffer (pH 4.8) No interfs. from W^{VI} (up to 1:10^5)	[2322]
Mo^{VI} (MoO_2^{2+})	C/Nj (αBO + ClO_3^-)	DPV	accum. by compl. Catalytic redn. of ClO_3^-	2×10^{-6}–8×10^{-5} M (1×10^{-7} M; 60s)	Model solns.	BO: benzoinoxime; s.e.: acetate buffer; interfs. from Sn^{IV}	[1374]
Mo^{VI} (MoO_4^{2-})	C/PO (chitin)	ASV	accum. by redox. compl. reoxidn. ($Mo^{IV} \to Mo^{VI}$)	2×10^{-7}–5×10^{-6} M (1×10^{-7} M; 20min)	Sea water	s.e.: 0.1 M acetate buffer (pH 4.3) No interfs. from common ions	[1707]
Mo^{VI} (MoO_3/Ox^{2-}) Ti^{IV}, V^V	C/Uv (CTAB, in situ)	DPASV	Ion-pairing accum. + redn. ($Mo^{VI} \to Mo^V + Mo^{IV}$); oxidn.	0.5–500 μg L^{-1} (0.4 μg L^{-1}; 10min)	Model solns.	CTAB: cetyltrimethylammonium bromide; s.e.: 0.01 M $H_2C_2O_4$; simult. detn. of Mo^{VI}, V^V, and Ti^{IV}	[705,817]

(continued)

TABLE 8.8 (continued)
Determination of Metals of the V, VI, VII, and VIII Group at Carbon Paste Electrodes, Sensors, and Detectors: Survey of Methods and Selected Studies

Ion, Species	Type of CPE (Modifier)	Technique (Mode)	Measuring Principles (Method Sequences)	Linearity Range (LOD; t_{ACC}/t_R)	Sample(s)	Other Specification (Remarks)	Ref(s)
Mo^{VI} (MoO_4^{2-})	C/PO (*Luminol*; in situ)	LSV-ECL	Amplifying effect of Mo^{VI} on ECL (via modif. redn.)	5×10^{-4}–0.01 mg L^{-1} (–)	Model solns.	*Luminol*: reagent (org. hydrazide); providing blue luminescent signal in presence of H_2O_2 (intermediate) interfs. studies with sel. MeO_m^{n-}	[2323]
W^{VI}						
W^{VI} (WO_4^{2-})	C/Uv (impregnated with SH-*oxin* in DMSO)	CSV	accum. via compl. with modif.; cathodic redn.	5×10^{-4}–0.001 M (–)	Model solns. (complex mixts.)	SH-*oxin*: 8-thioquinoline; s.e.: HCl-based solns; study on reaction kinetics of irreversible elec. redn.	[2324]
$U^{IV/VI}$						
U^{VI} (UO_2^{2+})	C/SO (Pt^{III}-gallate)	DPV (online)	accum. by chelating Anodic reoxidn.	(<1×10^{-7} M)	Groundwater (sampled in caves)	CPE designed for remote sensing (connected to shielded long cable)	[1379]
U^{VI} (UO_2^{2+})	s-CPE (C/Wx) (CPHA-MS)	AdSV (SWV)	accum. via adsorption of the complex UO_2-CPH Cathodic redn.: $U^{VI} \to U^{IV}$	5–50 ppb (1 ppb; 20 min)	Model solns.	s-CPE: solid-like el. made of paraffin wax; CPHA-MS: carbamoyl-phosphonic acid in mesoporous silica	[770]
U^{VI} (UO_2^{2+})	C/PO (B15C5)	DPASV, CV, EIS	accum. Via intercalation effect of modif.; redn. Anodic reoxidn.	0.002–0.2 μg mL^{-1} (1.1 ng mL^{-1}; 5 min)	Real samples (nat. waters)	modif.: benzo-15-crown-5 (as 5% in CP-bulk); comp. with ICP-AES	[1402]
Mn^{II-VII}						
Mn^{2+}, Mn^{IV}	C/SO (umm.)	ASV, CSV	Electrolytic accum. as Mn^0 or MnO_2; reoxidn. (redn.)	(5×10^{-8} M; 10 min)	Model solns.	s.e.: 0.1 M HCl + 0.1 M KCl compr. of detn. as Mn^{II} or Mn^{IV}	[2325]

Electroanalysis with Carbon Paste–Based Electrodes, Sensors, and Detectors

Analyte	Electrode	Method	Principle	Concn. range (LOD; t_{acc})	Sample	Note	Ref.
Mn^{2+}, Co^{2+}, Mg^{2+}	C/BN, C/Nj, C/SO (azodyes, in situ)	LSV	Adsorpt. accum. by compl. oxidn. of org. ligand	$0.1–1\ \mu M$ (in AmB)	Model solns.	modifs.: *Eriochrom Black* and *Blue*; s.e.: AmB (pH 9–10); interf. study, simult. detn. of $Mn^{2+} + Co^{2+}$ tested	[2326]
Mn^{2+}	C/Nj (HRP + 1,2-NQ)	AM-S	Catalytic effect of Mn^{II} in enzym. reaction; redn.	$(5 \times 10^{-7}\ M)$	Model solns.	HRP: horseradish peroxidase, NQ: 1,2-naphtoquinone; interf. by Co^{II}	[2327]
$Mn^{2+} + MnO_4^-$	C/PO (PAN)	DPCSV	accum. as Mn^{II}-PAN; oxidn. redn.: $Mn^{IV} \to Mn^{II}$	$1 \times 10^{-8}–1 \times 10^{-7}\ M$ $(7 \times 10^{-9}\ M;\ 200\ s)$	CRM, sea water	Mn^{VII} chem. red. with H_2O_2 interfs. supprd. by masking by Co^{II}	[1307]
Mn^{2+}	C/PO (BE/PO)	ASV	accum. via compl. form.; electrocatalysis-assisted el. redn.	$6 \times 10^{-7}–5 \times 10^{-4}\ M$ $(1 \times 10^{-7}\ M;\ 4\ min)$	Wheat flour, wheat rice, vegetables	BE/PO: Benthonite/ porphyrin system (characterized by TG, SPF); no interfs. of Me^{2+} at 1000x-excess	[1079]
Mn^{2+}	C/PO (bare)	DPCSV	anod. deposition as MnO_2. Cathodic redn. $Mn^{IV} \to Mn^0$	$1 \times 10^{-6}–1 \times 10^{-5}\ M$ $(1 \times 10^{-7}\ M;\ 2\ min)$	Pharmaceuticals	s.e.: 0.1 M PhB (pH 7.4); samples: diet supplement dosage tablets	[2328]
$Tc^{IV/VII}$							
Tc^{IV}, Tc^{VII}	C/Uv (TTA)	DPASV	accum. via redox. process and complex formation of Tc^{IV}-$(TTA)_n$; reoxidn.	$(<1 \times 10^{-7}\ M)$	Model solns.	TTA: thenoyltrifluoroacetone incl. study on diffn. $Tc^{IV} + Tc^{VII}$	[1380]
Tc^{IV}, Tc^{VII} $(TeCl_n^{m-})$	C/Uv (TOA)	DPASV	o.c. accum. (as $Te^{IV}Cl_n^m$) Potentiostatic reoxidn. of $Tc^{IV} \to Tc^{VII}$	$(4 \times 10^{-8}\ M;\ 3\ min)$	Model solns.	TOC: tri-n-octylamine s.e.: 2 M HCl Radioactive labeling tested	[2329]
Platinum metals							
Ru^{III}, Rh^{III} (Pd^{II})	C/SO (CPyB; in situ)	DPCSV	Ion-pairs: CPy^+ $MeCl_6^{3(4)-}$ and its adsorp.; el. redn.	$5–10 \times 10^{-7}$ for Ru,Rh; $1 \times 10^{-6}\ M$ for Pd	Model solns.	CPyB: cetylpyridinium bromide; initial study on detn. of $Me^{III(II)}$	[2330]

(continued)

TABLE 8.8 (continued)
Determination of Metals of the V, VI, VII, and VIII Group at Carbon Paste Electrodes, Sensors, and Detectors: Survey of Methods and Selected Studies

Ion, Species	Type of CPE (Modifier)	Technique (Mode)	Measuring Principles (Method Sequences)	Linearity Range (LOD; t_{ACC}/t_R)	Sample(s)	Other Specification (Remarks)	Ref(s)
Pd^{II}	C/PO (azomethine)	CSV	Interaction with modif. Cathodic redn.	(2×10^{-7} M Pd)	ores (Cu-Ni-S)	modif.: comp. with properties of liquid crystals; $E_R = 0 \rightarrow -1.5$ V No interfs. of Pt- and Fe-metals	[2331]
Pd^{II}	C/PO (TR, in situ)	DPASV	adsorpt. accum. by compl. Anodic reoxidn.	1–450 µg L^{-1} (1 µg L^{-1}, 120 s)	Fresh water, catalysts	TR: thioridazine incl. interfs. studies	[2332]
Pd^{II}, Me^{II}	C/LP+Gl (DMG, in situ)	DPCSV	accum. by chelating Cathodic redn.	(1×10^{-8} M, 120 s)	Tea extract, rice, human hair	LP + GL: liquid paraffin + glycerol; Me^{II}: simult. detn. of Hg,Co & Ni	[234]
Pd^{II}	C/PO (DMG)	EC-FIA	accum. by chelating Amperometric detc.	5×10^{-7}–1×10^{-4} M	Model solns.	modif.: embedded in the CP-bulk	[2333]
Pd^{II} ($PdCl_4^{2-}$)	C/PO (Na-humate)	LSV	o.c. accum. by ion-pairing MEx; anodic reoxidn.	9×10^{-8}–5×10^{-6} M (5×10^{-8} M)	Noble metal solns., waste catalysts	s.e.: (a) BR-buffer (pH 2.8), (b) 1 M HCl; quant. via a couple of peaks	[1700]
Os^{IV} Ir^{III}, Pt^{IV}	C/SO (CTAB or SPX; applied in situ)	DPCSV, POT (titr.)	accum. via ion-pairing as CTA(Sept)$^+$; $MeCl_6^{3(4)-}$/extraction; cathodic redn.	1–50 × 10^{-8} for Os (5 × 10^{-9} M Os; 1 min)	Wastewater (spiked)	SPX: Septonex: 1-(ethoxycarbonyl)-pentadecyltrimethyl-ammonium bromide; stoichiometry studied by SPX-titration and UV–vis	[736,1636, 2334]

Ir^{IV} ($IrCl_6^{2-}$)	C/PO (*Prussian blue*)	CV, HV	Electrolytic redn. Anodic reoxidn.	(Millimolar conc.)	Model solns.	Study of the modif. function; other species: $Fe^{II/III}$ $(CN)_6$ and $Ru^{III/IV}$ Hydrodynamic studies with RDE	[531]
Ir^{IV} ($IrCl_6^{2-}$)	C/Uv (ion-exchangers)	DPV	accum. by ion-pairing Cathodic redn.	0.2–5 ppm	Model solns.	modifs.: *Amberlite LA-2, Aliquat* Noticeable interfs. from Pt-metals, except Ru	[2335]
Ir^{III}	C/DCB (Ir^{III}-PTQ)	ASV	accum. via compl. Anodic reoxidn.	(3×10^{-10} M)	Model solns.	PTQ: 4-phenyl-8-thioquinolinate No interfs. from other Pt-metals, Ni, and Au; CPE: chem. regnt.	[230]
Ir^{III} Os^{IV}, Pt^{IV}	C/SO (CTAB or SPX; applied *in situ*)	DPCSV, POT (titr.)	accum. by ion-pairing and extraction; cathodic redn.	$1-10 \times 10^{-6}$ M Ir (1×10^{-6} M; 1 min)	Wastewater (spiked)	SPX: *Septonex*: 1-(ethoxycarbonyl)-pentadecyltrimethyl-ammoniumBr^- Stoichiometric study ion-pair titration	[1636,2330]
Pt^{IV} ($PtCl_6^{2-}$)	C/PO (umm.)	ASV	accum. via compl. form.; anodic reoxidn.	(0.002 ppm Pt)	Waste solns. (ctng. Cu+Ni)	s.e.: 2 M HCl + 0.02 M EDTA + 0.03 M Sn^{IV}.	[2336]
Pt^{IV} Os^{IV}, Ir^{III}	C/SO (CTAB or SPX; applied *in situ*)	DPCSV, POT (titr.)	accum. by ion-pairing and extraction; cathodic redn.	$1-10 \times 10^{-6}$ M Pt (1×10^{-6} M; 1 min)	Wastewater (spiked)	s.e.: 0.1 M HCl; stoichiometric study by ion-pair formation titration	[1636,2330]

TABLE 8.9
Determination of Metals of the I, II, III, and IV Group and of Rare Earths at Carbon Paste Electrodes, Sensors, and Detectors: Survey of Methods and Selected Studies

Ion, Species	Type of CPE (Modifier)	Technique (Mode)	Measuring Principles (Method Sequences)	Linearity Range (LOD; t_{ACC}/t_R)	Sample(s)	Other Specification (Remarks)	Ref(s)
Alkaline metals							
Li^+	C/PO [λ-MnO_2 (spinel)]	CV, ASV, POT (dir., FIA-ind.), SP (CCSA)	accum.: redn. $Mn^{IV} \to Mn^{III}$ followed by intercalation (insertion) of Li^I into the spinel structure; re-oxidn. changes in equilibria pot.	3×10^{-6}–0.002 M (ASV) 8×10^{-5}–0.01 M (POT) 5×10^{-7}–1×10^{-5} (CCSA) (2×10^{-7} M; 30/5 s)	Model solns., natural waters, pharm. prepn. (comm. tabs.)	modif.: 25% λ-MnO_2 (m/m) in CP s.e.: borate buffer/*TRIS* (pH 7–10) No interferes by Me^+/Me^{2+} ions POT: Nerstian slope, 79 mV/dec. CCSA: E_{ac} = +0.3 V, I_C = +10 µA	[742,1236, 1237,1238, 1240]
Na^+	C/PO (12-crown-4)	POT (dir.)	Inclusion compl. effect Equilibrium potential	1×10^{-4}–0.01 M	Model solns.	modif. added in CP in 5% (w/w) Lower selectivity vs. H^+ and K^+ ions	[2337]
Na^+, Cu^{2+}	s-CPE/C/Wx (natural zeolite)	AD, (FIA-EC)	accum. via ion-exchange equivalent to Na^I-conc.	1–50 ppm Na^+	Model solns.	s-CPE: mixture ctng paraffin wax Indirect detn. (via another analyte)	[755]
Na^+	C/PO (MnO_2)	CV, LSV	oxidn. of $Mn^{III} \to Mn^{IV}$ inclusion of Na^+ in MnO_2	7.9×10^{-5}–3.5×10^{-5} M (3.5×10^{-5} M)	Human urine	15% (m/m) modif. (nat. *Birnessite*) in CP-mixture; TRIS buff. (pH 8)	[1241]
K^+ Cs^+	s-CPE/C/SR (unm.)	LSASV, AD (stat.)	Indirect detn. (via electro-oxidn. of $NaBF_4$)	(mM level)	Model solns.	SR: silicone rubber; the end-point detection at E_{const}; RSD < ±0.9%	[309]
K^+, Li^+, Na^+ and NH_4^+	s-CP (C/*Kel-F*) (Ce^{3+}; *in situ*)	AD (LC-EC)	Indirect elchem. detection s-CPD stable in mobile ph. with org. solvents (up 80%)	— (down to 2 ppm)	Model solns.	Solid-like CP-detector; Kel-F: commer. available fluorinated polymer Detn. of Na^+, Li^+, NH_4^+ also tested.	[2338]
K^+, Li^+, Na^+ Cs^+, Me^{2+}	C/Nj (a) s-CPE/C/Wx (b) (ex-Z)	AD: (a) FIA-EC (b) BIA-EC (c) LC-EC	accum. by the electrostatic interaction/ion-exchange; equivalent to Me^I-conc.	1–50 ppm Li^+ (ca. 1 ppm)	(a,b) Model solns. (c) Pure water, free of electrolyte	s-CPE: mixture ctng. paraffin wax with modif. dispersed in the bulk (a,b) Indirect detn.; (b) separation by HPLC; Me^{2+}: Mg^{2+}, Ca^{2+}, Cu^{2+}	(a) [755,2339] (b) [1116] (c) [1115]

Analyte	Electrode	Method	Principle	Range (LOD)	Matrix	Notes	Ref.
K^+	C/PO [MnO_2 (hollandite)]	CV, AD	oxidn. of $Mn^{III} \to Mn^{IV}$ inclusion of K^+ in MnO_2	5×10^{-5}–9×10^{-4} M (2.5×10^{-5} M)	Human urine	Hollandite: nanostructured type of MnO_2; $E_{AMP} = +0.8$ V vs. SCE	[2340]
K^+	C/PO [$KSr_2Nb_2O_{15}$]	POT (dir.)	Intercalation effect of new modif. and equilibrium pot.	1.3×10^{-5}–0.002 M (7.3×10^{-5} M; 10 s)	Model solns.	modif.: triple oxide with special tetragonal structure (TTB type)	[2341]
Cs^+, Li^+, Na^+, K^+, Me^{2+}	C/Nj (ex-Z)	AD: FIA-EC, BIA-EC	accum. by the electrostatic interaction/ion-exchange; equivalent to Me^i-conc.	1–50 ppm Cs^+ (ca. 1 ppm)	Model solns.	ex-Z: methyl-viologen (MV^{2+}) exchanged-zeolite "Y"; BIA: batch injection analysis; indirect detn.	[1116,2339]

Alkaline earths metals

Analyte	Electrode	Method	Principle	Range (LOD)	Matrix	Notes	Ref.
Be^{2+}	C/PO (HMPB, in situ)	LSV	accum. by compl. Cathodic redn.	(1×10^{-7} M; 5 min)	Model solns. (+Al^{III}, 3000:1)	HMPB: o-(2-hydroxy-5-methyl-phenylazo) benzoic acid; s.e.: 1 M ammonia buffer (pH 8–9);	[2342]
Mg^{2+}	C/BN, C/Nj, C/SO (azodyes, in situ)	LSV	adsorpt. accum. by compl. oxidn. of org. ligand	0.1–1 μM (in AmB)	Model solns.	modifs.: *Eriochrom Black and Blue*; s.e.: AmB (pH 9–10); interf. study, negative effect of Mn^{2+}, Co^{2+}, Fe^{3+}	[2326]
Mg^{2+}	C/Nj [+MF] (Na^L-TP, in situ)	SWAdSV	accum. via adsorption of the $Mg(OH)_2$–TP adduct; cathodic redn. of ligand	6×10^{-9}–9×10^{-8} M (5×10^{-9} M; 60 s)	Tap water, human urine	TP: thiopentone; s.e.: PhB (pH 11) No interfs. of Al, Ca, Fe, Zn, and Pb or biol. matrix; indirect detn.	[1891]
Ca^{2+}	C/DOPhP (Ca^{II}-DOP)	POT (dir.)	Ion-pairing with modif. Equilibrium potential	0.001–0.1 M	Model solns. (+Mg^{II}, Ba^{II}, 10:1)	DOPhP: di-n-octylphenylphosphonate, DDP: didecylphosphate; Samples ctng. Mg^{2+}, Ba^{2+} (as 10:1)	[251]
Ca^{2+}	C/EtB (Ca^{II}-TTFA)	POT (titr.)	compl. reactions; changes in equilibrium potential	1×10^{-5}–0.1 M	Model solns.	EtB: ethyl benzoate, mdf.: thenoyl trifluoroacetone; titrant: EDTA	[178]
Ca^{2+}	C/PO (MDA + D^4)	POT (dir.)	compl. reactions; changes in equilibrium potential	1×10^{-6}–0.001 M (8×10^{-7}; <10 s)	pharm. prepn.	MDA: macrocyclic diamide, "D^4": 1,4-diaza-2,3;8,9-dibenzo-7,10-di-oxa-cyclododecane-dione; evaln. of selectivity coeff., k^{POT}	[1383]

(continued)

TABLE 8.9 (continued)

Determination of Metals of the I, II, III, and IV Group and of Rare Earths at Carbon Paste Electrodes, Sensors, and Detectors: Survey of Methods and Selected Studies

Ion, Species	Type of CPE (Modifier)	Technique (Mode)	Measuring Principles (Method Sequences)	Linearity Range (LOD; t_{ACC}/t_R)	Sample(s)	Other Specification (Remarks)	Ref(s)
Ca^{2+}	C/PO (ARS; *in situ*)	2nd DeLVs	adsorp. accum. by compl. Cathodic redn.	3×10^{-8}–2×10^{-6} M (9×10^{-9}; 90 s)	Tap water; milk, human serum	s.e.: 0.02 M KOH + 20 mM ARS (i.e., *Alizarin Red 'S'*), $E_{ac} = -0.1$ V regnt. of bare CPE in 0.2 M HCl	[2343]
Al^{3+}, Ga^{3+}							
Al^{3+}, V^V	C/PO (SRV *RS*, *in situ*)	LSV	adsorp. accum. as compl. Anodic oxidn. of ligand	0.1–1 µg L^{-1} Me (<0.1 µg L^{-1}; 5 min)	CRM (silica brick)	SRV *RS*: Solochrome violet *RS* s.e.: 0.2 M CH$_3$COONa; method for simult. detn. of Al^{III} + V^V and indirect detc. of org. moiety	[2344]
Al^{3+}, Be^{2+}	C/PO (SRV *RS*, *in situ*)	LSV	adsorp. accum. as compl. Anodic oxidn. of ligand	(ca. 1 µg L^{-1})	Model solns. (ctng. Be^{2+})	Indirect method; high selectivity toward Be^{2+}, but serious interfs. from Pb^{2+} ions	[2345]
Al^{3+}	C/PO (oxin)	DPASV	accum. by compl. Anodic reoxidn.	2×10^{-8}–6×10^{-6} M (1×10^{-8} M, 30 s)	Model solns.	incl. studies on interfs. from metal ions and redn. agents	[2346]
Al^{3+}	C/PW (s) (β-CD clp + ARS)	POT (dir.)	accum. by inclusion effect and compl.; equilibrium pot.	1×10^{-4}–0.1 M (8×10^{-5} M; —)	pharms.	modif.: β-cyclodextrin polymer mixed with *Alizarin Red S*	[2347]
Al^{3+}	C/PO (SG-nAu-dNNMA)	POT (dir.)	Complex formation (1:1); changes in equilibrium pot.	5×10^{-10}–0.05 M (2×10^{-10} M; 5 s)	Real samples (water, soils)	modif.: Sol/Gel Au-nanoparticles + 2,2′-dihydroxy-1-naphthylidine-1′-naphthyl methyl amine; pH 3–7	[958]
Ga^{3+}, Zn^{2+}	C/PO (unm.)	LSV	Direct measurement with cathodic redn.	(µM level)	Industrial solns. (nat. minerals of Al(OH)$_n$ type)	s.e.: 0.1–0.6 M monoethanolamine + (Et)$_4$N$^+$Br$^-$ (acc. to conc. of Ga); simult. detn. with Zn^{II}	[2328,2349]
Ga^{3+}	C/PO (ARS)	CV, 2nd DeLV	adsorp. accum. via compl. Cathodic redn.	0.02–6.00 µg L^{-1} (0.01 µg L^{-1}; 3 min)	Food	ARS: *Alizarine red S*; 0.1 M AcB studies on complex. mechanism(s)	[689,2350]

Electroanalysis with Carbon Paste–Based Electrodes, Sensors, and Detectors 263

Ti^{IV}, Zr^{IV}						
Ti^{IV} (TiO/ Ox_2^{2-})	C/Uv (CTAB, in situ)	DPASV	accum. by ion-pairs form.; $Ti^{IV} \rightarrow Ti^{III}$ redn. + reoxidn.	5–160 µg L^{-1} (0.1 µg L^{-1}, 10 min)	Model solns.	CTAB: cetyltrimethylammonium bromide; s.e.: 0.01 M H$_2$C$_2$O$_4$	[704,817]
Zr^{IV}	C/PO (NN + Zr^{IV} salt)	POT (dir.)	compl. and ion-pairing Change in equilibrium pot.	1×10^{-6}–5×10^{-4} M	Ceramic glass	NN: 1-nitroso-2-naphthol; solns. buffered to pH 2; sample dissolved in a mixture with HF	[2351]
Zr^{IV} (ZrO^{2+})	C/PO (ARS, in situ)	2nd DeLV	adsorp. accum. as compl. Cathodic redn. of ligand	2×10^{-10}–4×10^{-7} M (1×10^{-10} M, 4 min)	CRMs (ores)	ARS: Alizarin Red S; s.e.: mixed acetate/phthalate buffer (pH 4.8)	[2352]
Zr^{IV} (ZrO^{2+})	C/PO (ARS + Ca^{2+})	AdSV, DPCSV	adsorp. accum. as Zr^{IV} Ca^{II}-ARS compl.; ligand redn.	1×10^{-10}–2×10^{-7} M (4×10^{-11} M; 4 min)	CRMs (ores)	modif. in situ; 0.04 M phthalate buffer + 0.1 M glycine (pH 4.0)	[2353]
Zr^{IV} (ZrO^{2+})	C/Nj (Morin)	2nd DeLV	adsorp. accum. as compl.; el. redn. + anodic reoxidn.	0.6–6 $\times 10^{-8}$–3 $\times 10^{-6}$ M (3 $\times 10^{-9}$ M; 3 min)	CRMs (ore specimens)	Morin: an anthraquinone der.; s.e.: 2 M HCl + 10 mM Mo; no extraction and other pretreatment of samples	[2354,2355]
Zr^{IV} (ZrO^{2+})	C/PO (a), CNT-CPE (b–d) (ARS, in situ)	2nd DeLV	adsorp. accum. as compl. (with transduction effect by CNTs); anod. oxidn.	1×10^{-9}–2×10^{-7} M (a) 2×10^{-11}–10 nM (b–d) (1×10^{-11} M; 3 min)	Ore specimens	ARS: Alizarin Red S + Ca^{II} (b,c) s.e.: AmAc (pH 4–5); a,b: fast scan rate, 200 mVs^{-1}; E_R = 0–1.0 V vs. ref.	[3977,407,408, 686]
$Ce^{3+/4+}$							
$Ce^{3+/4+}$	C/Nj (ALC + CTAB)	CV, 2nd DeLV	adsorp. accum. as complex with Ce^{III} (with chem. redn. of Ce^{IV}); anodic reoxidn.	8×10^{-10}–3×10^{-9} M (6×10^{-10} M; 2 min)	CRM (nodular cast iron)	ALC: Alizarin complexone; CTAB: cetyltetraammonium bromide; AcB + PhB (pH 5); complexation study	[1629,2356]
Ce^{3+}	C/PO (BPy/n-SG)	ASV	accum. via complexing of Ce^{III}-BPy; anodic oxidn.	1–30 ng mL^{-1} (1 ng mL^{-1})	Real samples (ore specimens)	BPy/n-SG: bipyridyl-functionalized nanoporous silica gel; s.e.: AcB (4.5); RSD < 3%	[1870]
Ce^{3+}	C/PO (PMFH)	ASV	accum. via complexing of Ce^{III}-NHMF; anodic oxidn.	5–90 nmol L^{-1} (0.8 nmol L^{-1})	Wastewater phosphate rock	PMFH: N'-[(2-hydroxy-phenyl) methylidene]-2-fluorohydrazide s.e.: AcB + PhB (pH 4–5)	[1377]
Ce^{3+}	CNT-ILE (NHMF)	POT (dir.)	compl. and ion-pairing Change in equilibrium pot. (at pH 3.5–9.0)	8×10^{-7}–0.1 M (3 $\times 10^{-7}$ M; 20 s)	Model solns.	modif.: semicarbazide derivative el. opt. comp.: 44% C + 15% CNTs + 25% [BuMIm]BF$_4$ + 16% NHMF	[418]

(continued)

TABLE 8.9 (continued)
Determination of Metals of the I, II, III, and IV Group and of Rare Earths at Carbon Paste Electrodes, Sensors, and Detectors: Survey of Methods and Selected Studies

Ion, Species	Type of CPE (Modifier)	Technique (Mode)	Measuring Principles (Method Sequences)	Linearity Range (LOD; t_{ACC}/t_R)	Sample(s)	Other Specification (Remarks)	Ref(s)
Rare earth metals							
Th^{4+}	C/PO (*ALC*, *in situ*)	AdSV	accum. by compl./ads. Cathodic redn. of ligand	$3 \times 10^{-9} – 8 \times 10^{-7}$ M $(1–9 \times 10^{-10}$ M; 3 min)	Clays, CRMs (ore specimens)	*ALC*: Alizarin complexon reagent; interf. studies (with 10 various metals, incl. Me$_{RE}$)	[2357,2358]
Sc^{3+}	C/PO (*ARS*, *in situ*)	2nd DeLV	adsort. accum. as compl. Cathodic redn. of ligand	$1 \times 10^{-9} – 4 \times 10^{-7}$ M $(6 \times 10^{-10}$ M; 3 min)	Minerals, CRMs (ore specimens)	*ARS*: *Alizarin Red 'S'*; s.e.: mixed acetate/phthalate buffer (pH 4–6) interf. studies with Me$_{RE}$	[1381,1382]
Er^{3+}	CILE, CNTPE, CNT-CILE (a–c) (dMNSP$_3$CA)	POT (dir.)	Strong interaction effect Changes in equilibrium potential (at pH 3.5–9)	$1 \times 10^{-7} – 0.1$ M $(5 \times 10^{-8}$ M) [for CNT-CILE]	Model solns.	modif: 5-(dimethylamino) naphtha- lene-1-sulfonyl-4-ph-semicarbazide opt. comp.: 45% C + 20% CNTs + 20% [BuMIm]BF$_4$ + 20% modif.	[415]
Er^{3+}	CNT-C-ILE (HDEBH)	POT (dir.)	Strong interaction effect Changes in equilibrium potential (at pH 3–8)	$8 \times 10^{-8} – 0.01$ M $(6 \times 10^{-8}$ M; 15 s)	Tap and river water	modif.: N'-(2-hydroxy-1,2-di-phenethylidene)-benzohydrazide; (rt)IL: [BMIM]BF$_4$ + 3% n-SiO$_2$ added	[2359]
Eu^{3+}	C/PO [Ze(nat.)]	SWCSV	accum. via adsorption to modif.; cathodic redn.	$1 \times 10^{-7} – 2 \times 10^{-5}$ M $(4 \times 10^{-8}$ M; –)	Stream sediments	modif.: natural zeolite (montmorillonite); high selectivity over Me$_{RE}$	[2360]

Ho^{3+}	CNT-CPE [+n-SiO$_2$ (a)] [+TBA (b)]	POT (dir.)	Transduction effect and changes in equilibrium potential (at pH 3.8–8)	1×10^{-7}–0.01 M (a) 1×10^{-8}–0.01 M (b) (7×10^{-9} M; 10 s)	Model solns., alloys (ctng. Ho), human serum	opt. comp.: (a) 52% C + 5% CNTs + 15% ionophore (PO) + 3% n-SiO$_2$. (b) 60% C + 5% CNTs + 20% PO + 15% TBA (as ionophore); $t_L > 60$ d	[419,446]
Pr^{3+}	CNT-C-ILE	POT (dir.)	Transduction effect and changes in equilibrium potential (at pH 3–9)	1×10^{-8}–0.001 M (–)	Model solns.	Three-components paste mixture (with C(G) + MW-CNTs + (RT)IL; the latter of the [BuMIM]BF$_4$ type.	[410]
Tb^{3+}	CNT-C-ILE (NdMNSH = L)	POT (dir.)	Strong interaction effect Changes in equilibrium potential (at pH 3–8)	8×10^{-8}–0.01 M (6×10^{-8} M; 15 s)	Tap and river water	L: N'-(2-naphthoyl)-8-(di-methyl-amino)napht.-2-sulfono-hydrazide; 60% C + 10% CNTs + 15% IL + 15% L	[510]
Yb^{3+}	CNT-C-ILE (NOPFCH = L)	POT (dir.)	Strong interaction effect Changes in equilibrium potential (at pH 4–9)	1×10^{-8}–0.01 M (8×10^{-9} M; 15 s)	Model solns.	L: N'-(1-oxoacena-phthylen-2(1H)-ylidene) furan-2-carbo-hydrazide; 45% C + 5% CNTs + 25% IL + 25% L; new modif. studied by fluorimetry	[511]
Me$^{3+}_{(L)}$, Me$^{3+}_{(H)}$*	C/PO (ALC, in situ)	AdSV, 2nd DeLV	adsorpt. accum. by compl. Anodic reoxidn.	ca. 1×10^{-7} M Me$_{(L)}$ ca. 5×10^{-8} M Me$_{(H)}$	CRM (nodular cast iron)	ALC: *Alizarin* complexon (com.) s.e.: AcB + PhB (pH 5.0–5.5) ind. rare earths not identified, but quantified as a sum (total conc.) Me$_{(L)}$ and Me$_{(H)}$ can be separated	[2361]

* Where: Me$^{3+}_{(L)}$ ... Y, Sc, Sm, Eu, Gd, and Tb (Light rare rarths); Me$_{(H)}$ ... Dy, Ho, Er, Tm, Yb, and Lu (Heavy rare earths).

TABLE 8.10
Determination of Non-Metal Cations and Inorganic Molecules at Carbon Paste Electrodes, Sensors, and Detectors: Survey of Methods and Selected Studies

Ion, Species	Type of CPE (Modifier)	Technique (Mode)	Measuring Principles (Method Sequences)	Linearity Range (LOD; t_{ACC}/t_R)	Sample(s)	Other Specification (Remarks)	Ref(s)
H^+							
H^+	C/PO [(a) TDA, (b) D/P]	POT, pH-sensing	Ion-exchange effect Equilibrium potential	pH (a) 5–8, (b) 3–11	Model solns.	(a) mod.: tridodecylamine, no interfs. from Na^+, K^+; (b) funct. dendrimer on plexi attached to (CP)biosensor	(a) [2337], (b) [2362]
H^+	C/SO (Bi-powder) [two diff. designs standard + GrE]	POT, pH-sensing (FIA)	Redox reaction involving Bi, Bi_2O_3, and H^+ ions Equilibrium potential	pH 2–10	Model solns.	CPE: standard and planar configuration; Bi_2O_3 released from Bi(0); sub-Nernstian slope (−41.2 mV/pH)	[587]
O_2, O_2^{2-}							
O_2	C/Nj (unm. + GR)	CV, LSV	Cathodic. redn. $O_2 \rightarrow H_2O_2$ via superoxide radical HO_2^-	(mM level)	Model solns., el. material	Study on two-step redn. and effect of el. catalysis from graphite added	[516, 2363]
O_2	C/PO (Co-PC)	LSV	Electrocatalytic effect Cathodic redn.	0.2–28 g L^{-1}	Model solns.	modif.: Co^{II}-phthalocyanine s.e.: 2 M HCl	[2364]
O_2	C/Nj (SW_y + MV)	DCV	Electrocatalytic effect Cathodic redn.	(2 µg L^{-1})	Aqueous solns. (pH 7)	MV: methylviologen; SW_y: Na-montmorillonite; 0.1 M $LiClO_4$	[1106]
O_2	C/Nj (SW_y + MV)	LSV	Electrocatalytic effect Cathodic redn.	—	Aqueous solns. with diss. O_2	Commercially made CP-holder s.e.: phosphate buffer (pH 7)	[1099]
O_2	C/PO (MPP/ $SiO_2 + Sb_2O_3$)	CV, CA, LSV	Electrocatalytic reduction of O_2 via modif. effect	1–13 mg L^{-1} (<1 mg L^{-1})	Water solns. (with dissolved O_2)	MPP: methyl-4-pyridyl)-porphine immobilized by sol–gel method	[1206]

O_2	C/PO (CTAB; *in situ*)	CV, LSV	Electrocatalytic reduction of O_2 via effect of modif.	1–10 mM	Water solns.	modif. attached via hydrophobic affinity and adsorption; pHB	[1626]
O_2	C/PO (SiO$_2$/Nb$_2$O$_5$)	CV, CA, LSV	Electrocatalytic reduction of O_2 via effect of modif.	1–14 mg L^{-1} (<1 mg L^{-1}; 5 s)	Model solns.	Immobilized by sol–gel method; 2-el. ex. process; 1M KCl (pH 6)	[1203]
O_2	C/PO (AQ ders.)	CV, CA	Electrocatalytic reduction of O_2 via effect of modif.	(mM level)	Model solns.	AQ: anthraquinone; pH 7–8; over-potential lowered down to 570 mV	[1453]
O_2	C/PO (Co^{III}-SB)	CV, HV, CA	Electrocatalytic reduction of O_2 via effect of modif.	(mM level)	Model solns.	SB: *Schiff base*; AcB (pH 4); over-potential lowered down to 800 mV Kinetic and hydrodynamic studies with RDE	[536]
O_2	C/PO (DABQ)	CV, LSV	Chemical reaction of O_2 with modif.; indirect detn.	250–1250 μM (<250 μM)	Human urine	modif.: diamino-*o*-benzoquinone; s.e.: citrate buff (pH 2); quantif. of diss. O_2 via the peak decrease(s)	[2365]
O_2	C/PO (FeTCPP-Cl)	CV, HV, CA	Electrocatalytic reduction of O_2; el. oxidn.	(mM level)	Model solns.	modif.: Fe^{III}-tetracyanophenylporphyrin chloride; kinetic and hydro-dynamic studies with RDE (unm.)	[1575]
H_2O_2							
H_2O_2	C/PO (Pd-powder)	LSV	Electrocatalytic effect by modif.; anodic oxidn.	0.1–600 mg L^{-1} (50 μg L^{-1})	Model solns.	Pd dispersed in the bulk; CMCPE activated by potential cycling; s.e.: 0.001 M NaOH (pH 11)	[635]
H_2O_2	C/Uv (MnO$_2$)	CV, LSV, FIA-EC	detn. via catalytic oxidn. currents	0.5–350 μg L^{-1} (45 μg L^{-1})	Model solns.	modif. added into the CP-bulk	[1233]
H_2O_2	C/Uv (MnO$_2$)	CV, LSV, FIA-EC	detn. via catalytic oxidn. currents	5–450 μg L^{-1} (4.5 μg L^{-1})	Model solns., rain water	modif. dep. electrolytically as film s.e.: 0.2 M ammonia buffer (pH 9)	[326, 1232]
H_2O_2	s-CPE [C/Wx] (pTMB + HRP)	CV, AM	Enzymatic reactivity in redn. of H_2O_2; el. oxidn.	1×10^{-8}–5×10^{-5} M (1×10^{-8} M)	Model solns.	Solid-like CPE (ctng. paraffin wax) modif.: polymerized 3,3′,5,5′-tetramethyl benzidine with enzyme	[321]

(continued)

TABLE 8.10 (continued)
Determination of Non-Metal Cations and Inorganic Molecules at Carbon Paste Electrodes, Sensors, and Detectors: Survey of Methods and Selected Studies

Ion, Species	Type of CPE (Modifier)	Technique (Mode)	Measuring Principles (Method Sequences)	Sample(s)	Linearity Range (LOD; t_{ACC}/t_R)	Other Specification (Remarks)	Ref(s)
H_2O_2	C/Nj (Co-PC)	CV, DPV	Electrocatalytic effect by modif.; el. oxidn.	Model solns.	$(6 \times 10^{-7}$ M$)$	Co^{II}-phthalocyanine; Ni^{II}- and Cu^{II} ders. also studied; NaOH (pH 11)	[900]
H_2O_2	C/PO (hematin)	ECL (FIA)	ECL effect via oxidn. of the intermediate	Model solns. (with luminol)	4×10^{-6}–0.001 M	ELC: electrochemiluminescence s.e.: carbonate buffer (pH 10)	[2366]
H_2O_2	C/PO (Rh-*Blue*)	CV, FIA-AD	Electrocatalytic effect by modif.; el. oxidn.	Model solns.	5×10^{-5}–9×10^{-4} M (3×10^{-5} M)	Rh-*Blue*: Rh$^{III}_4$[Ru^{II}(CN)$_6$]$_3$ modif. deposited electrolytically	[849]
H_2O_2	C/PO (Ni^{II}-L-calix[4]arene)	CV, DPV	Electrocatalytic effect by modif.; el. oxidn.	Rain water	2×10^{-6}–1×10^{-4} M (1×10^{-6} M)	L: 5,11,17, 23-tetra-*tert*-butyl-25, 2,7-bis (diethylcarbamoylmethoxy): 0.05 M NaClO$_4$ + 0.001 M NaOH	[2367]
H_2O_2	C/PO (HRP + M on Nb$_2$O$_5$/SiO$_2$)	CV, AD (stat.)	Electrocatalytic effect by "M"; el. oxidn.	Model solns.	1–700 µM (0.5 µM)	M: Methylene Blue, Meldola Blue, or phenazine Methosulfate; pH 6.8 Up to 300 detns. (with R_R = 92%)	[1200]
H_2O_2	C/PO (nCM/Mn-oxides)	CV, AD (stat.)	Electrocatalysis-assisted oxidn. of H_2O_2 (E_{CONST})	Model solns.	1×10^{-5}–7×10^{-4} M (2×10^{-6} M; 10 s)	nCM/Mn-oxides: nanostructured Mn(II,IV) oxides of Cryptomelane type (added as 5% (m/m) in CPE)	[1872]
H_2O_2	C/PO [Me$_n$O$_m$ or Me^N(OH)CO$_3$]	CV, AD (stat.)	Electrocatalytic effect by "M-comp."; el. oxidn.	Model solns.	3×10^{-5}–0.009 M (2×10^{-5} M; < 3 s)	modifs: a-Fe$_2$O$_3$, β-Fe$_2$O$_3$, Fe$_3$O$_4$, FeCO$_3$, and FeO(OH); pH 7	[1001]
H_2O_2	C/PO (coconut tissue)	CV, AD (stat.)	Enzymatic reactivity of peroxidase in fruit tissue Electrolytic redn.	pharms. (hygienic cleaning solns. and cosmetic ointment)	2×10^{-4}–0.003 M (4×10^{-5} M; 7 s)	modif.: as dried fibers; pH 5.2 Samples: antiseptic, opthalmo and dental solns., hair-blonding cream	[2368]
H_2O_2	C/PO (Mb-TATP)	CV, AD (stat.)	Enzymatic reactivity in redn. of H_2O_2; el. oxidn.	Model solns.	80–1140 µM (55 µM; 7 s)	modif.: Myoglobin-triacetone tri-peroxide composite/film; pH 6.9 (PhB); kinetic and stability studies	[2369]

Electroanalysis with Carbon Paste–Based Electrodes, Sensors, and Detectors

Analyte	Electrode	Method	Principle	Range (LOD; t_R)	Medium	Notes	Ref.
H_2O_2	C/PO (Ga^{III}HCF)	CV, CA, AD (stat.)	Electrocatalytic effect by modif. (redn.); el. oxidn.	5×10^{-6}–4×10^{-4} (1×10^{-6} M; 7 s)	Model solns.	HCF: hexacyanoferrate (synth. and characterized by FTIR); opt. s.e.: 0.05 M PhB (pH 6.8); evaln. of D_x	[2370]
H_2O_2	IL-CPE (heme proteins)	CV, AD (stat.)	Catalytic effect by modif. modif. (in redn. of HP)	1×10^{-6}–0.01 M (1×10^{-6} M)	Environmental specimens	IL-CPE: C + PO + [BMIM][PF6]; enhanced stability and operational pH-range due to presence of RTIL	[463]
H_2O_2	C/PO (NZe or SZe)	CV, LSV	Electrocatalytic effect by modif. (redn.); el. oxidn.	0.005–0.05 M (0.5 mM)	Model solns.	N/SZe: natural/synthetic zeolite of NaX-type; studies on pH-effect	[1128]
H_2O_2	C/PO (pTF/Hrp + nAu)	CV, CA	Electrocatalytic effect by modif. (redn.); el. oxidn.	1×10^{-5}–0.001 M (8×10^{-7} M; 2 s)	Model solns.	modif.: electropolymerized multi-porous polythionine film with immobilized enzyme and nano-gold	[944]
H_2O_2	s-CPE [C/cpst] (HRP + @Fe$_3$O$_4$)	CV, AD (stat.)	Enhanced enzyme activity (stabilizing effect of Fe$_3$O$_4$) followed by electrode redn.	1.2×10^{-7}–8.3×10^{-5} (5×10^{-8} M; <5 s)	Model solns.	Cpst: solid composite; @-Fe$_3$O$_4$; lab-synthesized magnetic granules s-CPE easily renewable and stable	[981]
H_2O_2	C/PO (MB/Ze)	CV, AD (stat.)	Electrocatalytic effect by modif. (redn.); el. oxidn.	8×10^{-5}–0.001 M (1×10^{-4} M)	Model solns.	MB: Methylene blue, Ze: zeolite pH- and kinetic studies (k_R, α, D_x)	[1129]
H_2O_2	C/PO (Cnp + Cat (im.))	CV	Enzymatic reactivity in redn. of H_2O_2; el. oxidn.	5×10^{-6}–0.001 M (8×10^{-7} M)	Milk specimens	Cnp: Clinoptilotite, Cat: catalasa; RSD < 2%; t_{LT} = >60 days	[2371]
H_2O_2	C/PO (PESs)	CV, AD (stat.)	Enzymatic reactivity in oxidn. redn. of H_2O_2	(sub-mM level)	Real samples (waters)	PES: perovskites (La$_{(1-x)}$A$_{(x)}$MnO$_3$ type); calcination of modif. studied	[1249]
H_2O_2	C/SO (Co-PMA-SDS)	CV, CA, DPV, SWV	Electrocatalytic effect of modif. (oxidn.); el. redn.	(sub-mM level)	Cosmetic products	modif.: Co^{II} in electropolymerized film of N-methylaniline and Na-dodecyl sulfate; cat. process study	[1660]

(continued)

TABLE 8.10 (continued)
Determination of Non-Metal Cations and Inorganic Molecules at Carbon Paste Electrodes, Sensors, and Detectors: Survey of Methods and Selected Studies

Ion, Species	Type of CPE (Modifier)	Technique (Mode)	Measuring Principles (Method Sequences)	Linearity Range (LOD; t_{ACC}/t_R)	Sample(s)	Other Specification (Remarks)	Ref(s)
H_2O_2	s-CPE [C/BR] (plant tissue)	CV, AD (stat.)	Enzymatic activity in irrev. (oxidn.); potentiostatic redn.	5×10^{-5}–0.001 M (3×10^{-5} M)	Model solns.	Solid-like CPE (made of butadiene rubber); enzyme from gr. cabbage	[311]
H_2O_2	CNT-CPE (Cu^{II}-Ze)	CV, HV AD (stat.)	Electrocatalytic activity of modif. (in redn. of HP)	0.5–10.0 mM (0.3 mM; 1 s)	Model solns.	CP with four components: C + CNT + PO + modif. (Cu^{II}-ctng nat. zeolite) Evaln of k_R(HP) with RDE (unm.)	[449]
NO_X							
NO_2	C/HD + Gr [Fe^{II} (bipy)$_3$]	CV, FIA-EC	Electrocatalytic effect of modif.; el. oxidn.	(0.5 ppm; v/v)	Model air samples	HD: hexadecane (I). Gr: vacuum grease; serious interfs. from Cl_2^-; CPE surface regnt. by Triton X100	[2372]
NO_2	C/PO [Fe^{II} (phen′)$_3$]	EC-FIA, LC-EC	Electrocatalytic effect of modif.; el. oxidn.	(ca. 1 ppm; v/v)	Model air samples	phen′: 4,7-diphenyl-1,10-phen simult. detn. SO_2, H_2S, Cl_2 also examined and evaluated	[547,568]
NO_2, NO	C/PO (Ni-DFTAA)	CV, FIA-EC	Electrocatalytic oxidn./redn. via modif. effect	(Down to 1 μM)	Aqueous solns.	modif.: 6,17-diferrocenyldibenzo-5,9,14,18-tetraaza[14]annulen-Ni^{II} BR-B (pH 5–8); regnt. of CPE	[1529]
NH_3, NH_4^+							
NH_4^+	C/PO + PCM (PQQ/ MeOH-DE)	AD (stat.)	enzym. reaction (mediated by phenazine-SO_4 + NH_4Cl)	(3–60 mM; v/v)	Model solns.	PCM: polycarbonate membrane; DE: dehydrogenase enzyme (isolated from Pseudomonas sp.)	[2373]

Electroanalysis with Carbon Paste–Based Electrodes, Sensors, and Detectors

Analyte	Electrode	Method	Principle	Range (LOD)	Medium	Notes	Ref.
NH_4^+	C/Nj (Cu-ex-Ze)	AD, FIA-EC	accum. by the electrostatic interaction/ion exchange; equivalent to NH_4^+-conc.	2×10^{-5}–1×10^{-3} M (5×10^{-6} M)	Groundwater (synth. s.)	modif.: Cu(II)-exchanged *Clino-ptilolite* (of the nat. zeolite type); indirect amperometric detection	[1135]
NH_4^+	C/PO (+NH_4HPO_4/ SiO_2 + SnO_2)	POT (dir.titr.)	Chemical reaction involv.: NH_4^+(soln.) $\leftrightarrow$ NH_4^+(bulk) Equilibrium potential	8×10^{-7}–0.04 M (2×10^{-7} M, 60 s)	Natural waters	modif.: prepared by the sol-gel procedure; evaln. of coeff. k_i^{POT}; sub-Nerstian response for CP-ISE	[2374]

NH_2OH, NH_3^+OH

Analyte	Electrode	Method	Principle	Range (LOD)	Medium	Notes	Ref.
NH_2OH	C/PO (Pd-powder)	FIA-EC	Electrocatalytic effect of modif.; el. oxidn.	0.1–10 ng (abs.) (20 pg)	River water	Pd-powder dispersed in the bulk and activated by potential cycling	[1269]
NH_2OH, N_2H_4	C/PO (coumestan)	CV, BA DPV	Electrocatalytic oxidation of $N^{-II/-I}$ via modif. effect	0.2–15 mM NH_2OH (–)	Model solns.	Mechanism, reaction kinetics, catalytic, and activity ($y_\pm$) aspects studied and evaluated	[2375]

N_2H_4, $N_2H_5^+$

Analyte	Electrode	Method	Principle	Range (LOD)	Medium	Notes	Ref.
N_2H_4	C/Nj (Ti-SiO_2 + Ni-SPC)	LSV	Electrocatalytic effect of modif.; el. oxidn.	1×10^{-4}–6×10^{-4} M (1×10^{-5} M)	Aqueous solns. (pH 7)	modif.: titanised silicagel + Ni^{II}-tetrasulfonated phthalocyanine	[1180]
N_2H_4	C/PO ($Co^{II,III}$-compl.)	CV, LSCP	Electrocatalytic effect; electrocatalytic effect E vs. t detc.	(1.5–2 mM)	Model solns.	modifs.: $Co^{II,III}$-8-hydroxyquinolinate, Co^{III}-diethylthiocarbamate	[2376]
N_2H_4	C/PO (SWy + Cu-TPP)	LSV, FIA-EC	Electrocatalytic effect of modif.; el. oxidn.	5×10^{-6}–6×10^{-5} M	Model solns.	modif.: Cu^I-tetraphenylporphyrin, SWy: sodium zeolite	[1113, 1123]
N_2H_4	C/PO (Nb-SiO + H_2TCPP)	CV, FIA	Electrocatalytic effect of modif.; el. oxidn.	1×10^{-5}–5×10^{-4} M	Aqueous solns. (pH 7)	modif.: meso-tetracarboxyphenylporphyrin + Nb_2O_5-grafted silica	[2377]
N_2H_4	C/PO (+SX-pol.) (Fe^{III}-tSPhP/SiAl)	CV, CA	Electrocatalytic effect of modif.; el. oxidn.	5×10^{-5}–6×10^{-4} M (3×10^{-5} M, –)	Model solns.	tSPhP: 2,6-difluoro-3-sulfonatophenyl porphyrinate immobilized on SiO_2/Al_2O_3 in siloxan polymer	[1823]

(continued)

TABLE 8.10 (continued)
Determination of Non-Metal Cations and Inorganic Molecules at Carbon Paste Electrodes, Sensors, and Detectors: Survey of Methods and Selected Studies

Ion, Species	Type of CPE (Modifier)	Technique (Mode)	Measuring Principles (Method Sequences)	Linearity Range (LOD; t_{ACC}/t_R)	Sample(s)	Other Specification (Remarks)	Ref(s)
N_2H_4	C/PO (Co^{II}-Pc)	CV, CE-AD	Electrocatalytic effect of modif.; el. oxidn.	20–200 µM N_2H_4 (10 µM in PhB)	Model solns., water samples	Pc: phthalocyanine; with "on-chip" separation of R- or Ar-hydrazines	[1565]
N_2H_4	C/PO (Cu^{II}/Co^{II}-HCF)	CV, CA, CCP	Electrocatalytic effect of modif.; el. oxidn.	0.1–12 mM N_2H_4 (–)	Model solns.	HCF: hexacyanoferrate; s.e.: PhB (pH 7); mechanism, react. kinetics and diffusion processes studied	[1937]
N_2H_4	C/SO (Q/HQ)	CV, CA, DPV	Electrocatalytic effect of modif.; el. oxidn.	7×10^{-6}–8×10^{-4} M (5×10^{-6} M, for DPV)	Wastewaters	s.e.: buffer (pH 10); overpotential lowered to 500 mV; samples colld. nearby a wood&paper factory	[2378]
N_2H_4	C/PO (Co^{II}-SB)	CV, HV (+RDE)	Electrocatalytic effect of modif.; el. oxidn.	5×10^{-5}–0.01 M (3×10^{-5} M)	Model solns.	SB: *Schiff base* (hydroxy-3-OMe-benzyliden)-hydrazide; kinetic st.	[2379]
N_2H_4	C/SO (AcFc)	CV, DPV	Electrocatalytic effect of modif.; el. oxidn.	3×10^{-5}–0.001 M (1×10^{-5} M, in DPV)	Natural water	AcPc: acetylferrocene (dissolved in pasting liquid); s.e.: pH 7.5; kinetic studies (evaln. of k_R, D_X)	[1553]

Analyte	Electrode	Method	Process / Reaction	Range (LOD / t_R)	Sample	Notes	Ref.
N_2H_4	µ-CILEs [various ILs]	CV, LSV	Electrocatalytic oxidn. of N_2H_4 via function of IL	0.01–1.00 mM (<0.01 mM)	Model solns.	(RT)ILs: Bu-, He-, Oc-MIM[PF_6] µ-CILEs also tested for H_2O_2, AA	[472]
N_2H_4	C/PO (Co^{II}-Pc)	CV, LSV	Electrocatalytic effect of modif.; el. oxidn.	$1 \times 10^{-4} – 1 \times 10^{-5}$ M (7×10^{-5} M)	Industrial water (boiler feed w.)	Pc: phthalocyanine; redn. requires highly alkaline solns. (pH 13); no interfs. from common Me^{n+} ions	[917]
N_2H_4	CNTPE (Ni^{II}-BA)	CV, CA	Electrocatalytic effect of modif.; el. oxidn.	2.5 µM–0.2 mM (0.8 mM; 5 s)	Model solns.	BA: *Baicalein* (soil) deposited on the el. surface; kinetic study (evaln. of k_R(hy), α, and D_X.	[411]
SO_2, SO_3							
SO_2	C/PO (unm.) [+Fephe, *in situ*]	AD, CL-FIA	chem. redn. and elchem. oxidn. of redox system $Fe^{2+}(phe)_3 \leftrightarrow Fe^{3+}(phe)_3$	0.3–14.0 ppm (0.3 ppm, <3 s)	Air samples	Phe: phenanthroline; CL: close loop system; up to 35 samples per hour, working CPE regnt. by surfactant	[546]
CO, CO_2							
CO	C/PO (+DMFc) [+m/CO-OR]	CV, AD (stat.)	Enzymatic and mediated oxidation of $CO \rightarrow CO_2$ (at pH 7.5 and $T = 80°C$)	20 nmol–65 µmol (20 nmol; $t_R < 15$ s)	Model spec., ambient air	mdf.: dimethylferrocene, CO-oxido reductase (obt. from microbial cells *Pseudomonas thermocarboxydovorans*) membrane; kinetic st. (evaln.of k_R)	[71]

TABLE 8.11
Determination of Non-Metal Anions and Complex Structures at Carbon Paste Electrodes, Sensors, and Detectors: Survey of Methods and Selected Studies

Ion, Species	Type of CPE (Modifier)	Technique (Mode)	Measuring Principles (Method Sequences)	Linearity Range (LOD; t_{ACC}/t_R)	Sample(s)	Other Specification (Remarks)	Ref(s)
Halides							
F^-	C/PO (*Asparagus*, Fc)	CV, AD	detn. via inhibition of enzyme (peroxidase)	(0.5 mg L^{-1})	Fluoride tablets	Fc: phthalocyanine; s.e.: 0.05 M PhB, pH 7 (for CV) or pH 5 (AD) + 0.1 mM H$_2$O$_2$ (added to s.e.)	[2380]
Cl^-, Br^-, I^-	(a) C/Nj (b) s-CPE/CWx (AgX + Ag$_2$S)	POT (dir.)	Membrane ion-exchange Equilibrium potential	(5 × 10^{-5} M Cl, Br; 5 × 10^{-7} M I; <10 s)	Model solns.	s-CPE: solid-like paraffin-CPE interfs. of strong oxidn. and redn. agents; simult. detn. of Cl$^-$, Br$^-$, I$^-$	[36]
Cl^-, NO_3^-	C/*Aliquat* 336 (R$_3$R'N$^+$Cl$^-$)	POT	Membrane ion-exchange Equilibrium potential	0.07–4.0 pCl	Model solns.	Binder served as modif.; interfs. from X$^-$ when Br, I:Cl ≥10:1	[251]
Cl^-	s-CPE [C/Wx] (R$_4$N$^+$Cl$^-$ + P + H)	POT	Membrane ion-exchange Equilibrium potential	2.0–4.7 pCl$_{TEOR}$ 2.0–4.0 pCl$_{PRACT}$	Tap, groundwater	Solid-like CP (Wx: paraffin wax, P: plasticizer, H: hardener); s.e.: buffered medium (pH 3–9)	[1971]
Cl^-, Br^-	C/PO (Fc-CXP)	SWV	modif.: redox receptor, binding X$^-$ anion; oxidn.	10–100 µM Cl$^-$	Model solns.	Fc-CXP: ferrocene functionalized calix[4]pyrrole; interfs. of F$^-$ ions	[1803]
Br^-	C/Nj, C/BN	CP, RCP, R$_{ir}$	3-step oxidn., where chem. oxidant is intermediate Br$_2^-$	1 mM NaBr (+1 M H$_2$SO$_4$)	Model solns.	RCP: reverse chronopotentiometry Study on 2 Br$^-$ → Br$_2$; evaln. of τ_t	[13]
Br^-	C/Nj (AgBr)	CV, LSCP, CCSA	Inhibition effect of Br$^-$ on AgBr redn.	(µM level)	Photographic developers	Strong adsorpt. agents interfere Quantification via potential shift	[1999]
Br^-	C/PO (Hg^{II}-Py/PTC)	POT (titr.)	Chemical equilibrium for H$^+$ ↔ PTC-HBr	1 × 10^{-5}–0.03 M (4 × 10^{-6} M; —)	Tap water	Py: pyridine, PTC: proton-transfer and carrier for Br$^-$ ion; pH 4.0–8.3	[2381]
I^-	C/Nj (*Amberlite* LA-2)	DPASV	o.c. accum. by ion-pairing Anodic oxidn.	(ca. 1 × 10^{-7} M)	Model solns.	s.e.: 0.02 M KCl No interfs. from Br$^-$ and Cl$^-$	[806]
I^-	C/Nj (CTAB, *in situ*)	DPCSV	Ion-pairing + extr. after oxidn.; cathodic redn.	0.05–0.25 mg L^{-1} (0.03 mg L^{-1})	Table salts	CTAB: cetyltrimethylammonium bromide; s.e.: BR-buffer (pH 3)	[1624]
I^-	C/Nj (cinchonine)	CV, DPV	accum. via ion-pairing Anodic oxidn.	(8 × 10^{-8} M, 20 min)	Iodine-based disinfectant	No interfs. from X$^-$ or pseudo-X$^-$; some interf. from S$_2$O$_3^{2-}$	[1378]

Electroanalysis with Carbon Paste–Based Electrodes, Sensors, and Detectors 275

Analyte	Electrode	Method	Principle	Conc. range	Sample	Notes	Ref.
I⁻, I₃⁻, (IO₃⁻)	C/TCP (binder as modif.)	CV, DPCSV	accum. via ion-pairing/extr. (after electrode oxidn.) Cathodic redn.	3×10^{-7}–5×10^{-5} M (1×10^{-7} M; 5 min)	Mineral waters, table salts (ctng. I⁻ or IO₃⁻)	TCP: tricresyl phosphate (ion-ex.) IO₃⁻ chem. pre-redn. with $N_2H_5^+$ -High selectivity (Cl: I = 10⁵:1)	[638, 639]
I⁻	C/SO (Septonex, in situ)	DPCSV	accum. via ion-pairing + oxidn.; cathodic redn.	4×10^{-7}–8×10^{-5} M (2×10^{-7} M, 15 min)	Mineral waters	SPX: Septonex: 1-ethoxycarbonyl pentadecyltrimethylammonium Br⁻ interfs.: Au^{III}, none by Cr^{VI}, Mn^{VII}	[737]
I⁻, I$_n$ (n = 3, 5)	C/TCP (binder as modif.)	DPCSV	accum. via ion-pairing/extr. (after el. oxidn. I⁻ → I₂) Cathodic redn. (I₂ → I⁻)	5×10^{-7}–5×10^{-5} M (3×10^{-7} M; 5 min)	Potassium iodide-ctng. tablets	KI-tbs: distributed to inhabitants living nearby nuclear powerplants incl. study on interfs. from $S_2O_3^{2-}$	[697]
I⁻, I₃⁻	C/TCP (binder as modif.)	SP (CCSA)	accum. via ion-pairing/extr. (after electrode oxidn.) redn. with I_{CONST} = 1–5 μA	8×10^{-8}–5×10^{-5} M (1×10^{-8} M; 15 min)	Table salts (with I⁻) human urine	0.5 M NaCl + 0.1 M HCl (pH 1.5) Sample only acidified and diluted (without any further pretreatment)	[210, 254]
I⁻	C/Nj (Cu^{II}quinine-DN)	DPASV	accum. by ion-exchange MEx + anodic oxidn.	1×10^{-8}–1×10^{-6} M	Iodine-based disinfectant	s.e.: CHCl₃ + MeOH, 0.1 M KNO₃ No interfs. from X⁻ or pseudo-X⁻	[1311]
I⁻	C/PO (Cu₂O)	DPV	accum. by precip. as Cu₂I₂ Anodic/cathodic detc.	1×10^{-6}–2×10^{-5} M (5×10^{-7} M)	Model solns.	Electrolytic regnt. of CMCPE interfs. from Br⁻ and Cl⁻ studied	[743]
I⁻	C/PO (dIH-Friedel salt)	DPASV	accum. via ion-pairing/MEx; anodic oxidn.	50 μM–1 mM (0.05 μM; –)	Ground and sea waters (synth.)	dIH: double-layer hydroxide; s.e. Chloride-based medium	[1096]
I⁻	C/PO (CTAB, in situ)	LSSV	accum. via ion-pairing oxidn. and extraction; cathodic redn.	8×10^{-9}–5×10^{-6} M (2×10^{-9} M; 3 min)	Table salts	CTAB: cetyltrimethylammonium bromide; s.e.: 0.1 M NaCl + modif. incl. interf. studies from X⁻ and Y⁻	[735]
I⁻	DPE [D/PO] (unm.; three types)	DPASV	accum. via electrolytic redn.; anodic reoxidn	(1×10^{-7} M; –)	Vitamins, table salts	D: natural or synthetic diamond, DPE: diamond paste electrode; no intefs. from Cl⁻ and Br⁻ ions	[342]
I⁻, Q⁺I₃⁻	C/PO (CTAI)	POT (dir.)	Ion-pairing of membr. type Chemical equilibrium for I⁻/I₃⁻ ↔ {CTA⁺; I⁻/I₃⁻}	5×10^{-5}–0.1 M (4×10^{-5} M, 30 s)	Pharmaceuticals (relaxant drug)	Q⁺: org. cation (ion-pair moiety) CTAI: cetyltrimethyl ammonium iodide; Nerstian slope: −55 mV/dec.	[707]
I⁻, R⁻, Ar–I	s-CPE [C/cmps] (μ-Ag/n-Ag)	CV, HV IC-EC	Interaction Ag and I⁻; redox behavior of I⁻/I₂ el. oxidn. at E_{CONST}	0.64–64 μg L⁻¹ (0.47 μg L⁻¹ = 4 nM)	CRM (milk powder)	cmps: composite ctng. micro- and nano-silver powder; selected org. compounds ctng. I also studied	[2382]

(continued)

TABLE 8.11 (continued)
Determination of Non-Metal Anions and Complex Structures at Carbon Paste Electrodes, Sensors, and Detectors: Survey of Methods and Selected Studies

Ion, Species	Type of CPE (Modifier)	Technique (Mode)	Measuring Principles (Method Sequences)	Linearity Range (LOD; t_{acc}/t_R)	Sample(s)	Other Specification (Remarks)	Ref(s)
I^- (IO_3^-)	C/PO (n-Ag_2SO_4)	SFnE-AD	Ion exchange: $2I^- \leftrightarrow SO_4^{2-}$; el. redn.: $Ag^I \to Ag(0)$ at constant potential	5×10^{-12}–4×10^{-9} (5 pM)	Table salt (ctng. IO_3^-)	modif.: nanosized salt; IO_3^- chem. redn. prior to analysis; solid phase nano-extraction; $R_R = 10\%$–110% no interfs. of Cl^-, BrO_3^-, and IO_3^-	[10020]
Halo-anions							
ClO_3^-, BrO_3^-	C/PO (MMS + PMo_{12})	CV, DPV	elchem. activity of modif. + catalytic effect	(1 μM XO_3^-)	Model solns.	PMo_{12}: $H_3[P(Mo_3O_{10})_4] \cdot n\text{-}H_2O$ MMS: mesopore molecular sieve	[1817]
ClO_4^-	C/Nj + *Amberlite* ($TlCl_4^-$, *in situ*)	DPV	o.c. accum.; catalytic effect—chem. reoxidn. of modif.	(50 μg L^{-1}, 12 min)	Drinking water (spiked)	quantif. via signal of modif. incl. interf. studies	[733]
ClO_4^-	C/TCP+NB [Ni^{II}-$CPyClO_4$]	POT	Pairing with counter ion Equilibrium potential	$pClO_4$ 1–5	Explosives, rocket propellant extrs.	Titrant: cetylpyridinium chloride Applicable also to detn. of BF_4^-; NB: nitrobenzene	[709]
IO_3^-	C/PO (SGAm + PW_{12})	CV, FIA-AD	Electrocatalytic effect of modif. (with its renewal)	5×10^{-6}–0.001 M (3×10^{-6} M)	Model solns. (in stream)	SGAm-PW_{12}: phosphato-tungstate-functionalized silica gel derivatized by amino-funct. group (3D-modif.)	[1818]
IO_3^-	CILE (PW_{12})	CV, AD	Electrocatalytic effect of modif.; el. redn.	1.0–50.0 mg kg^{-1} (0.5 mg kg^{-1})	comm. table salt (ctng. IO_3^-)	PW_{12}: 1–12 phosphomolybdic acid (RT)IL used: n-$OcPy^+PF_6^-$; pH 5–6 Kinetic studies (evaln. of k_H, α)	[482]
Pseudo-halides							
CN^-	C/PO (Cu^{II}-azomethine)	ASV	accum. by compl. Anodic oxidn. of analyte	(6×10^{-9} M)	Model solns.	s.e.: ammonia buffer (pH 10) No interfs. from X^-, S^{2-}, SCN^-	[523]
CN^-	C/PO (Hg^{II} + L, *in situ*)	ASV	Competitive compl. form. between Hg^{II}, L, and CN^-	3×10^{-7}–2×10^{-6} M	Model solns.	Indirect method based on decrease in signal during $Hg^{II} \to Hg^0$ redn.	[698]

CN⁻	C/Nj (AL+Cc-C, Cc-Ox)	FIA-EC	Inhibition of enzymatic reaction; el. oxidn.	(6×10^{-9} M)	Model solns.	AL: *Asolectin*; Cc-C, Cc-Ox: cytochrome "C", cytochrome oxidase	[2383]
SCN⁻	C/PO (Co^{II}-PC)	FIA-EC	Catalytic effect of modif. Alternate oxidn./redn.	(ca. 1 μM)	Diluted solns. of saliva	PC: phthalocyanine; det. of SeCN⁻, SO_3^{2-}, $S_2O_3^{2-}$, SeO_3^{2-} also tested	[652]
CN⁻, I⁻	C/PO [+membr.] (AgI/Ag$_2$S mixt.)	POT (dir.)	Chemical equilibrium for X⁻ ↔ AgX⁻; eq. potential	pX = 2.6–4.7 for CN⁻, pX = 2.0–6.5 for I⁻ (t_R = 30s)	Model solns.	Indication of CN⁻ requires regular regeneration of CP in acidic media	[2384]
CN⁻	C/PO (Co^{II}-TPPA)	POT (dir.)	Chemical equilibrium for CN⁻ ↔ CoTTPA-CN adduct	2×10^{-5}–0.01 M (9×10^{-6} M; 5 s)	Mineral water (commercial p.)	TPPA: 3,4-tetra pyridinoporphirazinate; no interfs. of Cl⁻, Br⁻, I⁻, SCN⁻	[1148]
N-anions							
N$_3^-$	C/Nj (unm.)	LSV	Uncatalyzed anodic oxidn. at high positive potentials	(0.05 mg m^{-3} NaN$_3$)	Air specimen, Pb(N$_3$)$_2$ (s)	s.e.: 0.1 M AcB (pH 4.6); air coll. in column with KOH; solid lead azide stabilized with *Dextrin*	[2385]
N$_3^-$	CP-ISE [C/PO] (Fe^{III}-SB)	POT (dir.)	Ion-exchange Fe^{III}-SB ↔ Fe^{III}-N$_3$; equilibrium pot.	1×10^{-6}–0.05 M (9×10^{-7} M; <15 s)	Model solns.	SB: *Schiff base*; s.e.: pH 4.3–10.2 Nernstian slope: ca. 59 mV/dec^{-1}; no interfs. from X⁻ and pseudo-X⁻	[2386]
NO$_m$-anions							
NO$_2^-$	C/Nj (unm.)	LSV	Anodic oxidn.	1×10^{-5}–5×10^{-5} M (ca. 1×10^{-6} M)	Model solns.	s.e.: 1 M KNO$_3$; compared with GCE; serious interfs. from S^{2-}, CN⁻, and I⁻	[2387]
NO$_2^-$	C/Nj (unm.)	CV, LSV	Anodic oxidn.	(ca. 1×10^{-5} M)	Industrial products (TNT, nitrocellul.)	s.e.: 0.5 M AcB (pH 4); detn. in the presence of HNO$_3$ + H$_2$SO$_4$	[2388]
NO$_2^-$	C/Nj (*Amberlite LA-2*)	CV, DPV	o.c. accum. by ion-pairing Cathodic redn.	(1 ng mL^{-1})	Model solns.	s.e.: 0.01 M KCl incl. studies with related modifs.	[734]
NO$_2^-$	C/PO (Ru^{II}-BPP)	FIA-EC	Electrocatalytic effect of modif.; el. redn.	5×10^{-8}–5×10^{-4} M (3×10^{-8} M)	Model solns.	modif.: [Ru^{II}(bipy)$_2$(PVP)$_{10}$-Cl] Cl polymer	[648]
NO$_2^-$	C/Nj (*Prussian Blue*)	CV, DPV	Electrocatalytic effect of modif.; el. redn.	(200 ppb)	Model solns., wastewater	s.e.: acidic buffers incl. studies on accum. mechanism	[847]
NO$_2^-$	C/PO [chitin, CT]	CV, DPV	Deposition and oxidn. of CT-NO$_2$ complex	2×10^{-8}–2×10^{-7} M (1×10^{-9} M)	Model solns., natural water	s.e.: BR buffer (pH 3.9) R_R = 96%–110%	[833]

(continued)

TABLE 8.11 (continued)

Determination of Non-Metal Anions and Complex Structures at Carbon Paste Electrodes, Sensors, and Detectors: Survey of Methods and Selected Studies

Ion, Species	Type of CPE (Modifier)	Technique (Mode)	Measuring Principles (Method Sequences)	Linearity Range (LOD; t_{ACC}/t_R)	Sample(s)	Other Specification (Remarks)	Ref(s)
NO_2^-	C/PO (Ru^{II}-BPD)	FIA-EC	Electrocatalytic effect of modif.; el. redn.	—	Model solns.	modif.: [Ru^{II}(bipy)$_2$dppz]$^{2+}$ polymer s.e.: HCl-based medium	[2389]
NO_2^-	C/Nj {Mb + colloid-Au}	CV	Electrocatalytic effect of modif.; el. redn.	10–110 μM	Model solns.	Mb: myoglobin s.e.: 0.1 m BR-buffer (pH 7)	[948]
NO_2^-	C/PO [+membr.]	LSV	Anodic oxidn.	(14 ppb; i.e., 3 ppm in sample analyzed)	Cured meat (preserved by NO_2^-)	Membrane from cellulose-acetate; stabilization soln.: AA + m-PhA + EDTA; compared with UV-method	[2390]
NO_2^-	C/PO (MPP/SiO$_2$ + SnO$_2$)	CA, HV (+RDE)	Electrocatalytic effect of modif.; el. redn.	(50 μM; —)	Model solns.	MPP: *tetrakis*(1-methyl-4-pyridyl) 21H, 23H-porphine ion; mechanism and kinetic parameters investigated	[1208]
NO_2^-	C/PO (DAN)	CV, DPV	Electrocatalytic effect of modif.; el. redn.	(0.2 mM; —)	Real samples	DAN: 2,3-diammino-naphthalene; up to 120 detns./h	[2391]
NO_2^-	C/PO Fe(CN)$_6^{3-}$; *in situ*	CV, CA	Electrocatalytic effect of modif.; el. redn.	5×10^{-5}–0.001 M (3×10^{-5} M, —)	Model solns.	Overpotential lowered for 700 mV; reaction at higher pH; mechanism, kinetic, and activity coeffs. ($\gamma_\pm$)	[2392]
NO_2^-	C/PO (poly-o-TO)	CV, CA	Electrocatalytic effect of modif.; el. redn.	5×10^{-4}–0.02 M (3×10^{-4} M, —)	Real samples	Poly-o-TO: poly(ortho-toluidine); redn. requires highly acidic solns.	[2393]
NO_2^-	C/PO (PW-12)	CV, CA	Electrocatalytic effect of modif.; el. redn.	3×10^{-5}–0.001 M (3×10^{-5} M, 2 s)	Water samples	PW-12: H$_3$[P(Mo$_3$O$_{10}$)$_4$]; s.e.: 1 M H$_2$SO$_4$; kinetic studies, evaln. by Nicholson-Shain method	[1052]
NO_2^-	C/PO (V^{IV}O-SBC)	CV, LSV	Electrocatalytic effect of modif.; el. redn.	4×10^{-6}–0.004 M (6×10^{-7} M, —)	Model solns.	SBC: *Schiff base* complex; react. mechanism, diffusivity, kinetics, and electrocatalysis studied and calct.	[1592]
NO_2^-	C/SO (DQBQ)	CV, CA DPV	Electrocatalytic effect of modif.; el. redn.	6×10^{-6}–8×10^{-4} M (4×10^{-6} M; DPV)	Weak liquor (from w-&-p factory)	modif.: *p*-duroquinone(tetramethyl *p*-benzoquinone; optimal pH: 1.0 w-&-p: wood and paper	[1445]

Electroanalysis with Carbon Paste–Based Electrodes, Sensors, and Detectors

Analyte	Electrode	Method	Principle	Range	Sample	Notes	Ref.
NO_2^-	C/PO (Cu^{II}-MPS)	CV, CEC–µED	Extraction, CE separation; amperometric oxidn.	1–160 ppm NO_2^- (0.3 ± 0.05 ppm)	Ham, sausage	modif.: Cu^{II}-(3-mercaptopropyl)-trimethoxysilane (characterized by UV–vis, XPS, FTIR); µC: microchip	[2394]
NO_2^-	C/SO (PNMA+DDS)	CV, LSV	Electrocatalytic effect of modif.; el. redn.	1×10^{-4}–0.02 M ($<1 \times 10^{-4}$ M)	Model solns.	modif.: electropolymerized N-Me-aniline in presence of Na-dodecyl sulfate; poly-film char. by SEM	[132]
NO_2^-	C/PO (Zn^{II}-PBT)	CV, LSV	Electrocatalytic effect and complex form. with modif. Subsequent el. redn.	10–50 ppm	Model solns.	modif.: $Zn(C_{11}H_{15}N_3)_2]Cl_2$ compl. CMCPE applicable also to H_2O_2; CP-surface regnt. mechanically	[1618]
NO_3^-	C/Aliquat 336 ($R_3R'N^+NO_3^-$)	POT	Membrane ion-exchange Equilibrium potential	0.07–2.5 pNO_3	Model solns.	Binder served as mdf.; interfs. Of X^- and ClO_4^- when $c(A) \geq 10:1$	[251]
NO_3^-	C(CW)/ionex (ionex: Orion 92-07-02)	POT	Membrane ion-exchange Equilibrium potential	1×10^{-5}–0.1 M	Model solns.	CW: ceresin wax (used for treating graphite particles); pH 3–8; surface of CPE regnt. by special cutting pr.	[538]
NO_3^-	CP-CWE ($R_4N^+Cl^-$+P+H)	POT	Membrane ion-exchange Equilibrium potential	1–5 pNO_3	Model solns.	"Solid-like" CP (P: plasticizer; H: hardener); optimal pH 2.5–10.3	[1970, 2395]
NO_3^-	C/Nj (+LA-2) (Tl^{III}, in situ)	DPASV	o.c. accum. + oxidn. Electrocatalytic effect	0.5–60 µg mL^{-1}	Drinking water	LA-2: comm. ion-exchange resin, s.e.: HCl-based medium (pH 2–3) Mechanical regnt. of CMCPE	[804]

S-anions

Analyte	Electrode	Method	Principle	Range	Sample	Notes	Ref.
S^{2-}	C/PO (DMA+LS, in situ)	CV, ACV	accum. via MB formt. by ion-pairing + adsort.; redn.	(5×10^{-11} M)	Model solns.	Indirect detn. via Methylene blue formation; modif.: p-dimethyl-N-aniline + lauryl sulfate	[2396]
S^{2-}	C/PO (FePC)	CV, POT (dir.)	Catalytic effect + oxidn. Equilibrium potential	1×10^{-6}–0.005 M (—)	Model solns.	FePC: Fe^{II}-phthalocyanine; detn. of S^{II}-org. compds. also studied	[2397]

SO_m-anions

Analyte	Electrode	Method	Principle	Range	Sample	Notes	Ref.
SO_3^{2-}	C/Nj (PbO_2, in situ)	LSV	Electrocatalytic effect of modif.; el. oxidn.	(ca. 1×10^{-6} M)	Model solns.	modif. dep. from solns. ctng. Pb^{2+} s.e.: 0.5 M H_2SO_4	[520]
SO_3^{2-}	C/PO (AG + HCF + S-Ox)	FIA-EC	Mediated enzym. reaction Amperometric detn.	(nM level)	Model solns.	AG: agarose gel, HCF: $Fe(CN)_6^{4-}$, S-Ox: sulfite oxidase; $SO_2(g)$ detn.	[2398]

(continued)

TABLE 8.11 (continued)

Determination of Non-Metal Anions and Complex Structures at Carbon Paste Electrodes, Sensors, and Detectors: Survey of Methods and Selected Studies

Ion, Species	Type of CPE (Modifier)	Technique (Mode)	Measuring Principles (Method Sequences)	Linearity Range (LOD; t_{ACC}/t_R)	Sample(s)	Other Specification (Remarks)	Ref(s)
SO_3^{2-}	s-CPE [C/SBM] (three diff. enzymes)	AD	Mediated enzym. reactions Amperometric detn.	(nM level)	White and red wines	SBM: solid-like binding matrix s-CPE-multibiosensor applicable also to selected electroactive BICs	[575]
SO_3^{2-}	C/PO (BFEFM)	CV, CA DPV	Electrocatalytic effect of modif.; el. oxidn.	4×10^{-6}–0.01 M (2×10^{-7} M for DPV)	Real sample (unspecified)	BFEFM: bis (ferrocenyl-ethyl)-fluoren-9-one; pH 8.0; evaln. of kinetic parameters (k_S, α, D_X)	[879]
SO_3^{2-}	C/PO (Fc-EPhE)	CV, DPV	Electrocatalytic effect of modif.; el. oxidn.	4×10^{-6}–1×10^{-4} M (2×10^{-7} M for DPV)	Model solns.	Fc-EPhE: 1[4-(ferrocenyl-ethynyl) phenyl] ethanone; pH 8; evaln. of kinetic parameters (k_S, α, D_X)	[1546]
SO_3^{2-}	s-CPE [C/Wx] [Ni^{II}-Fe(aq)(CN)$_5$]	CV, CA LSV, FIA	Electrocatalytic effect of modif.; el. oxidn.	3×10^{-6}–0.003M (9×10^{-7} M; LSV)	Real samples	modif. characterized by XRD,UV; s.e.: PhB (pH 8–10); long lifetime	[2399]
SO_3^{2-}	CILE (unm.)	CV, DPV	Electrocatalytic oxidn. (with min. deactivation)	5–1000 μM (4 μM)	Mineral water, grape juice, non-alcoholic beer	(RT)IL of the Q$^+$PF$_6^-$ type; study on r. kinetics (evaln. of k_S, α, D_X); comparison with CPE and GCE	[471]
SO_3^{2-}	C/PO (ETF)	CV, LSV, AD	Electrocatalytic effect of modif.; el. oxidn.	5 μM–5 mM (1 μM; <5 s)	Real samples (wines)	ETF: electrodeposited thin film of Cu^{II}/Co^{II}-Fe(CN)$_6$; s.e.: pH 8–10	[1041]
SO_3^{2-}	CNT-CPE [Cu$_2$Fe(CN)$_6$]	CV FIA-PV	Electrocatalytic effect of modif.; el. oxidn.	0.5–50.0 mg L^{-1} (0.4 mg L^{-1}; <5 s)	Food products	PV: pervaporation technique; s.e.: 0.1 M KNO$_3$; E_A = +0.55 V vs. ref. Comparison with DPP-method	[436]
SO_3^{2-}	CNTPE (FcDCA)	CV, EIS DPV	Electrocatalytic effect of modif.; el. oxidn.	0.5–100.0 μM (0.3 μM)	Real samples	mdf.: ferrocene-dicarboxylic acid neutral solns. (pH 7); kinetic study (evaln. of k_H and α)	[444]
SO_4^{2-}	s-CPE/C/Wx (R$_3$R'N$^+$HSO$_4^-$ + P + H)	POT	Membrane ion-exchange Equilibrium potential	1–4 pSO$_4$	Model solns.	Solid-like CP: P: plasticizer (DOP) H: hardener (epoxy resin); pH 2–3 Indirect method (detc. of modif.)	[1969]
SO_4^{2-}	C/PO (Cr^{III}-SBC)	POT (dir.,titr.)	Chemical equilibrium for $SO_4^{2-} \leftrightarrow$ Cr-SB-SO$_4$ adduct	2×10^{-6}–0.05M (9×10^{-7} M; 10 s)	Mineral water	SBC: *Schiff base* (N,N'-ethylene-bis(5-hydroxy-salicylideneiminate; Ba^{2+} as titrant; optimal pH 4–9	[2400]

Analyte	Electrode	Method	Notes	Application	Ref.
p-Aminophenol PAP [C_6H_7ON]	IL-CPE (with HMIMPF6)	CV, DPV	Study on el. oxidn, comp. to ordinary CPE; evaln. of ΔE; s.e.: 0.1 M PhB (pH 7); lin.r.: 2×10^{-6}–3×10^{-4} mol L^{-1}	Wastewater (synt. soln.)	[468]
Pyrocatechol 1,2-Dihydroxy-benzene [$C_6H_6O_2$]	C/PO (unm.)	CA, DPV, FIA	PhB (pH 6.5), sub-µM level; sampled in collector with 0.1 M HCl or NaOH; tested via drying paper satrd. with Pcl.	Ambient air (indoor pollution)	[2414]
Hydroquinone 1,4-dihydroxy-benzene [$C_6H_4(OH)_2$]	C/PO (+Pd and extr.)	CV, DPV	Extr.: plant extract from *Zucchini* (ctng. enzyme); detn. via oxidn. HQ → Q and re-redn. Q → HQ; lin.: 6×10^{-5}–0.001 M	Photographic developers	[2415]
Hydroquinone H$_2$Q [$C_6H_4(OH)_2$]	[CILE, IL-CPE, CPE] (both ctng. HMIMPF6)	CV, DPV	comp. of three CP-configurations; st. on redox behav. of H$_2$Q, effect of IL and its content; LOD: 1×10^{-6} M	Skin cosmetics (medicated cream)	[405]
Hydroquinone H$_2$Q [$C_6H_4(OH)_2$]	CILE (with BPPF$_6$)	CV, DPV	Study on redox behavior and el-catalytic effect of BPPF6; s.e.: PhB (pH 2.5); lin.r.: 5×10^{-6}–0.005M + RSD <4%	Wastewater (synth. soln.)	[480]
Hydroquinone H$_2$Q [$C_6H_4(OH)_2$]	C/PO (+*Lacc*/n-Fe$_3$O$_4$/S) [CP-bio. with enzyme]	CV, AM	modif.: *Laccase* enzyme attached onto magnetic core-shell with nano-Fe$_3$O$_4$/silica; lin.: 1×10^{-7}–2×10^{-4} M, t_R < 60 s	Compost extracts	[979]
Hydroquinone and HQ-ethers H$_2$Q, H$_2$Q-BeE, H$_2$Q-DMeE	C/Nj (unm.)	DPV	Be: benzyl; s.e.: 0.1 M PhB (pH 2.1); lin.rs: 0.1–8.8, 0.2–31.4, 0.2–33 mg L^{-1}; comp. with Pt-, GCE, and HPLC.	Skin lighteners, bleaching creams	[2416]
Pyrocatechol, Resorcinol, Hydroquinone 1,2-, 1,3-, 1,4-Dihyd.-benzenes [$C_6H_6O_2$]	CNT-CILE (+β-CD) [+CPyB, *in situ*]	CV, DPV	modifs: β-cyclodextrin (in bulk) and cetylpyridinium Br- (at surface); E_P = +0.03, +0.14, +0.52V vs. ref.; LODs: 4–9 $\times 10^{-8}$ M; simult. detn. of three ders.	Wastewater (synth. solns.)	[504]

(continued)

TABLE 8.12 (continued)
Determination of Organic Compounds at Carbon Paste Electrodes, Sensors, and Detectors: Survey of Methods and Selected Studies: Part 1: Environmental Pollutants, Synthetic Substances, and Industrially Important Products

Analyte/Substance (Org. Compound)	Specification [Chem. Formula]	Type of CPE (Modifier)/ [Configuration]	Technique (Mode)	Experimental and Electroanalytical Characterization, Selected Data; Notes	Sample(s)	Ref(s)
Metol	2-Me-1,4-dihydroxybenzene [$C_7H_8O_2$]	(a) CILE (with BPPF6) (b) CILE (with EMIMBF4)	CV, EIS (SEM)	Study on quasi-rev. redox behavior of Mt, evaln. of kinetics data (k_S, α, D_X) and effect of IL; lin.: 5×10^{-6}–0.001 M	Model samples, photographic baths	(a) [493] (b) [494]
Bisphenol A	BPA [$C_{15}H_{16}O_2$]	(a) C/PO (+CTAB, in situ) (b) C/PO (+Co^{II}-PC) (c) C/PO (unm., as [ec-D]) (d) C/PO (+thionine-Tyr.)	(a) CV, DPV (b) CV, cCOU (c) AD-pCEC (d) CV, AD	(a) enhanced hydrophobicity of CPE (b) catalyzed el. oxidn., (c) pre-sepn. and detn. with DCP and PCP; (d) enzymatic effect of Tyrosinase; LODs = (a) 8, (b) 10nM, (c) 2 μg L^{-1}, (d) 0.15 μM; evaln. of biodegradation process of BPA	(a,b) Waste plastic materials (c) Chicken egg, milk powder (d) Polluted water	(a) [1628] (b) [915] (c) [553] (d) [1510]
Tannic acid	Polyphenol, [$C_{76}H_{52}O_{46}$]	C/PO (unm. + SBA-15) C/PPyrr (unm.) [trad. CPEs, ps-CPE]	CV, DPV	PPyrr: polypyrrole, SBA-15: ordered mesoporous silica; ps-: pseudo; A_{CPE}: 60m^2 g^{-1}; LOD = 0.01 mM	Wood-staining baths	[329]

Quinones and carboxylic acids

Anthraquinone	In aloe-latex (aloe-emodin) [$C_{14}H_8O_2$]	(a,b) C/PO (unm.) (c) IL-CPE (with BMIMPF6)	(a,b) CV, LSV, COU (c) CV, DPV	(a,b) Catalytic rev. redox reaction in the presence of O_2, (c) catalysis by RTIL; (a,b) AmB (pH 8.9), (c) BR-B (pH 1.5); LODs: a,b) 1×10^{-10} M, (b) 3×10^{-9} M	(a) Medicinal forms (Radix Rhei®) (b) Chinese verbs (c) Model solns.	(a) [2417] (b) [2418] (c) [505]
1,2-Dihydroxy-anthraquinone	Alizarin, ALZ [$C_{14}H_8O_4$]	C/PO (unm.), C/PO (+TX-100, in situ)	CV, LSV	Study on redox behavior, pH- and catalytic/enhancing effect of surfactant; s.e.: 0.2 M AcB (pH 4); sub-mM level	Model solns.	[1647]

Hydroxy-1,2-Naphthoquinones	3 NQ-isomers [$C_{10}H_6O_3$]	C/PO (unm.)	DPV	NQs: lawsone, juglone, and plumbagin; CPE used only in comparative studies	Model solns.	[2419]
Acrylamide (acrylic-acid amide)	AcA [C_3H_5ON]	C/PO [+Hb (attached)]	CV, DPV	Interaction of AcA with modif., detn. via redn. Hb-Fe^{III}/Hb-Fe^{II} (decrease); LOD: 1.2×10^{-10} M; AcA: carcinogen	Foods (potato chips extracts)	[2420]
3,4-Dihydroxy-benzoic acid	DBA [$C_7H_6O_4$]	CILE (+poly-MBF)	CV, EIS	modif.: Methylene blue-film obtained by electropolymerization; el.-catalytic effect; linear range: 5×10^{-4}–0.003 M	Model solns.	[501]
Other O-compounds (e.g.,: –CH=O; –O–COOR; –O–O–)						
Formaldehyde	[HCH=O]	C/PO (+Ni^{II}-p-MeAF)	CV, CA	modif.: *in situ* electropolymerized film of N-methylaniline funct. with Ni^{III}, cat. oxidn. and r. kinetics studied (k_R, α, D_X)	Model solns.	[130]
Aliphatic and aromatic aldehydes	R–CH=O, Ar–CH=O	C/PO (+AN, AmBA, or PCF-AmPy)	CV, DPV	modifs.: aniline, 4-aminobenzoic acid, [Fe(CN)5(4-aminopyridine)], elchem. detn. via imine formed; sub-mM level	Model solns.	[1021]
Aliphatic and aromatic aldehydes	R–CH=O, Ar–CH=O	C/Nj (+Pd-powder)	CV, LSV	Electrodeposited Pd acts as el. catalyst; detn. via cathodic redn. and desorption lin.r.: µM-mM; mech. regnt. of CPE	Model solns. (of HCHO)	[1268]
Vanillin	Subst. benzaldehyde [$C_8H_8O_3$]	C/DPhE (unm.), C/Nj (unm.)	CV, Ad/ExSV	Binder: diphenyl ether; s.e.: BR-B (pH 5) (in 50% PrOH); lin.r.: 0.6–4mM	Biscuits, canned custard powder	[233]
Coumarins, Furocoumarins	Hetero-compds. (1-benzopyran-2-ones ders.)	C/PO (unm.)	CV, Ad/ExSV	Study on oxidn., also using CFi-uµE*, BR-B (pH 2–12), five diff. ders.; LODs: 0.05–0.20 mg L^{-1}	Essential oil (from citrus)	[2421]
Aliphatic peroxides	BuP, BuHP, CmP [R–O–O–R]	C/PO (+CoPC) [ft-D]	(a) AD-FIA (b) AD-HPLC	Cm: cumene, HP: hyperoxide, Co^{II}-phthalocyanine catalyzes the oxidn. of $R_2(H)O_2$; (b) lin. r.: 1–100 pmol L^{-1}	(a) Drinking water (b) Model solns.	(a) [894] (b) [896]
Benzoyl peroxide	BPO ($C_6H_5CO)_2O_2$	C/PO (+coconut fibers)	HA, AD	modif.: source of peroxidase enzymes; s.e.: 0.1 M PhB (pH 5), redn. with $E_P = 0$ V vs. ref.; lin: 5–55 µmol L^{-1}; $t_R < 5$ s	Facial creams, dermatol. shampoos	[282]

(continued)

TABLE 8.12 (continued)
Determination of Organic Compounds at Carbon Paste Electrodes, Sensors, and Detectors: Survey of Methods and Selected Studies: Part 1: Environmental Pollutants, Synthetic Substances, and Industrially Important Products

Analyte/Substance (Org. Compound)	Specification [Chem. Formula]	Type of CPE (Modifier)/ [Configuration]	Technique (Mode)	Experimental and Electroanalytical Characterization, Selected Data; Notes	Sample(s)	Ref(s)
Peroxynitrites (superoxides)	$R-NO(O_2)-$	C/PO (+MSIN + o/dG15) [CP-bio. with ss-ds-DNA]	DPV	modif.: morpholinosydnonimine with immobilized 15 guanine oligo-n-bases; study on reactivity of ONOO vs. DNA	Model substrates	[2422]
Lipid peroxide(s)	(Form. by oxidn. degradation of lipid compds.)	C/PO (+disolv. analyte) [CPEE]	CV, LSV	Oxygenation performed at 120°C and then followed in the SPV mode (12 h)	Linseed oil	[2423]
Aliphatic N-compounds						
Triethylamine, tri-n-propylamine	$[C_6H_{15}N]$ $[C_9H_{21}N]$	(a) C/PO (+RuC, *in situ*) [EH-CPE] (b) CILE (+RuC-*MCM-41*)	(a) ECL-CE (b) ECL-FIA	modif.: $[Ru^{II}(bpy)_3]^{2+}$ in soln. (a), immobilized on sulfonic-funct. silica sb.; LODs: <1 μM at 40°C; (a)	Model solns. and mixts.	(a) [571] (b) [462]
Nitromethane	$[CH_3NO_2]$	GNS/IL (+chitosan + Hb) [CP-film on the GCE]	CV, CA AD	7.2 nM. (b) Hb: hemoglobin; special paste made of graphene nano-sheets and ionic liquid attached to the GCE; LOD: 6×10^{-10} M	Fresh water (spiked)	[204]
1-Nitropropane	$[C_3H_7NO_2]$	C/Nj (unm.)	CV, LSV	Study on el. redn. (affected by reaction products and their desorption); s.e: aq. soln. of KCl, HCl, H_2SO_4; LOD: 5 mM	Model solns.	[2017]
Aromatic amines						
Aniline	$[C_6H_5NH_2]$	C/Nj (+HRP) [CP-bio + enzyme/Nf]	CV, AD-FIA	Nafion-membrane to immob. enzyme; low ppb level; $tD < 30$ s, fs = 20h^{-1}	Vegetable oils (spiked s.)	[2021]

Aniline	[C$_6$H$_5$NH$_2$]	C/Nj (unm.) [+enzyme, in situ]	CV, DPV	Indirect detn. via oxidn. of metabolite (NH$_2$-PhOH); LOD: 1 μM, RSD = 3%–5%	Liver homogenates	[2424]
Aniline	[C$_6$H$_5$NH$_2$]	C/Nj (+HRP) [CP-bio + enzyme/Nf]	CV, AD-FIA	Nafion-membrane to immob. enzyme; low ppb level; $tD < 30$ s, fs = 20 h^{-1}	Vegetable oils (spiked s.)	[565]
N-subst. anilines	Ar-NRR'	C/Nj (unm.)	CV, DPV	accum. via adsorp./extract. process; study on sel.: (Me)$_2$AN vs. AN (10:1)	Model solns.	[2425]
Aromatic amines, diamines	R–Ar–(NH$_2$)$_n$ ($n = 1,2$)	C/SO (unm.)	CV, DPV	Study on oxidn. of diff. subst. amines; pH-effect, tests on detn. in mixtures	Natural waters (spiked)	[2426]
m-Phenylendiamine, p-phenylendiamine	PDA [C$_6$H$_8$N$_2$]	C/PO (unm.)	DPV, DPAdSV	el. oxidn. and study of intermediates; AcB (pH 4.8), sub-μM level; interfs./simult. detn. of other NH$_2$–, NO$_2$– ders.	oxidative hair dyes (com. products)	[2427]
APAHs						
1-Aminonaphthalene, 2-aminonaphthalene, 2-aminobiphenyl	[C$_{10}$H$_9$N] [C$_{12}$H$_{11}$N]	(a) C/PO (+α, β, γ-CDs) (b) C/PO, GC/PO (unm.) [wall-jet-D]	(a,b) CV,DPV, DPAdSV (b) AD-HPLC	CDs: cyclodextrins, acc. by inclusion and adsorp. effect; pH-study; LOQs: 1–3 × 10^{-8} M; (a) only 1-AN, (b) all ders.	(a,b) Model mixts. (a,b) Drinking and river waters	(a) [1413] (b) [2428, 2429]
1,5-Diaminonaphth., 1,8-diaminonaphth.	[C$_{10}$H$_{10}$N$_2$]	GC/PO (unm.) [wall-jet-D]	AD-HPLC	s.e.: 0.01 M PhB (pH 7) + MeOH (2:3) lin. r.: 5 × 10^{-9}–1 × 10^{-4} M; studies on simult. detn. of two DANs, mech. regnt.	Model solns.	[2429,2430]
2-Aminofluorene, 2,7-diaminofluorene	[C$_{13}$H$_{11}$N] [C$_{13}$H$_{12}$N$_2$]	C/PO (unm.)	DPV, DPAdSV	el. oxidn.; BR-B: pH (i) 13, (ii) 1012: lin. r.: 2 × 10^{-7}–1 × 10^{-7} M (t_{ACC} = 200 s)	Model solns.	[2431]
2-Acetamidofluorene, fluoren-9-ol	[C$_{15}$H$_{13}$ON] [C13H10O]	C/PO (unm.)	DPV, SWV	el. oxidn.; (i) BR-B (pH 7), (ii) 0.1 M H$_2$SO$_4$ (pH 0.5); LODs: (i) 0.05 μM, (ii) 1.00 μM; simult. detn. also tested	Model solns.	[2432]
1-Aminopyrene, 1-hydroxypyrene	[C$_{16}$H$_{11}$N] [C$_{16}$H$_{10}$O]	C/PO (+βγ-CDs) C/PO (+ds-DNA)	DPV, DPAdSV	CDs: cyclodextrins; s.e.: BR-B; lin.r.: 2 × 10^{-8}–4 × 10^{-7} and 2 × 10^{-7}–4 × 10^{-6} M	Model solns.	[1424]

(continued)

TABLE 8.12 (continued)
Determination of Organic Compounds at Carbon Paste Electrodes, Sensors, and Detectors: Survey of Methods and Selected Studies: Part 1: Environmental Pollutants, Synthetic Substances, and Industrially Important Products

Analyte/Substance (Org. Compound)	Specification [Chem. Formula]	Type of CPE (Modifier)/ [Configuration]	Technique (Mode)	Experimental and Electroanalytical Characterization, Selected Data; Notes	Sample(s)	Ref(s)
Aminoquinolines (3-, 5-, 6-, and 8-)	AQLs [$C_9H_8N_2$]	(a–c) C/PO (unm.) GC/PO (unm.) (a) [wall-jet-D]	(a) AD-HPLC (a–c) DPV (as AdSV)	s.e.: 0.1 M H_3PO_4 (dir.), 0.1 M NaOH (with accum.); LOQs: $1–5 \times 10^{-7}$ M (for AdSV); simult. detn. tested on (a,c) 3-, 5- and 6-AQLs and (a) 5- and 8-AQLs	Model solns. and mixtures	(a) [2433] (b) [2434] (c) [2435]
3-Aminofluoranthene	3-AFT [$C_{16}H_{11}N$]	C/PO (unm.)	DPV, DPAdSV	Study on irrev. oxidn.: BR-B (pH 2–11) LOD: 2×10^{-8} M; comp. with GCE	Model solns.	[2436]
Aromatic nitro-compounds						
Nitrobenzene	[$C_6H_5NO_2$]	(a) C/PO (+*Sepiolite*) (b) C/Nj (+BiF) [BiF-CPE + CTAB, *in situ*]	(a) CV, DPV (b) SWV	modif.: (a) clay (Mg^{II}-hydroxysilicate) (b) bismuth film; (a) o.c. accum., (b) el. redn.; LOD: (a) 0.4 mg L^{-1}, (b) 8×10^{-7} M	(a) Beer, wines, cider (apple wine) (b) Model solns.	(a) [823] (b) [2437]
2-Nitrophenol	2-(o-)NP [$C_6H_5O_3N$]	(a) GC/Nj (b) C/PO (+*Bentonite*) (c) [ft-D] (d) C/PO (+"C-60") (e) C/PO (+hmp-S) (f) GC/PO (+SWy-2) GC/PO (+*Sepiolite*)	(a,d) CV, DPV EC-LC (b,c) AD-FIA (e) CV, EIS (f) DPV (dir.)	modifs.: (b,c) natur. clay (Al^{III}-silicate) (d) fullerene (new form of C), (e) hexagonal mesoporous silica; (f) nat. clays (b,c) NaCl + HCO_2Na, (d) Bu_4NOH + ACN, (e,f) BR-Bs; LODs: (a–c) 0.05–10 mg L^{-1}; (c,e) 10 μM; (f) 7×10^{-7} mol L^{-1} (a–c) studies and tests on simult. detn. (d–f) studies on total redn. $NO_2 \rightarrow NH_2$	(a) River water (spiked) (b,c) Sea water (synth., real, spiked with 0.5–2.5 mg L^{-1}) (d,f) Model solns. (e) Wastewater	(a) [2438] (b) [1077] (c) [1076] (d) [198] (e) [1849] (f) [197]

Analyte	Electrode	Method	Notes	Sample	Ref.
4-Nitrophenol 4-(p-)NP [C$_6$H$_5$O$_3$N]	(a) C/PO (+Zeolite Y) (b) C/Nj (+Anthrobacter sp) (c) C/PO (+Moxarella cells) (d) C/PO (+MIP) (e) C/PO (+HAP) (f) CILE (with BPPF6) (g) GC/PO [+nat. clay(s)]	(a) CV, DPV (b) AD (c) CV (d) CV, LSV (e) CV, SWV (f) CV (g) DPV (dir.)	modifs.: (a) natur. clay; (b,c) microbial mater., (d) molecular imprint polymer (a) BR-B; (b,c) PhB + tartrate (pH 7) (d) extract. accum., LODs.: (a) 10 μM (b) 0.7 ppb, (c) 2×10^{-9} M, (d) 3×10^{-9} M modifs.: (d) molecular imprint polymer (e) hydroxy-o-apatite, (f) (RT)IL; (e) o.c. accum., (f) electrocatalytic effect by IL LODs.: (d) 3×10^{-9}, (e) 8×10^{-9}, 1×10^{-6} M simult. detn. of 2-NP and 4-NP tested	(a) Model solns. (b–d) Lake waters (e) River water (f) Model solns. (g) Model mixts.	(a) [1145] (b) [2439] (c) [2440] (d) [1689] (e) [1305] (f) [500] (g) [197]
Nitrophenols (2-, 3-, 4-) [C$_6$H$_5$O$_3$N]	(a) C/PO (+"C-18," 50%) (b) C/Nj (2:1, +α-CD) (c) C/Nj (+PVPy) [DEDC for detn. of 4 sp.]	(a) DPV (b) LSV (c) AD-CE	modifs.: (a) com. ionex resin, (b) cyclodextrin; c) polyvinyl-pyrrolidone film (c) conf.: dual-electrode/dual-channel; LODs: (a) 2–8 ng mL^{-1}, (b) μM; (c) 25 μM simult. detn.: (a,c) 2-,3-,4- and (b) 2-,4- NPs together with other APh-ders.	(a,c) Natural water (from lake, rain, river) (b) Model mixtures	(a) [1611] (b) [2441] (c) [560]
Nitrophenols, nitroanilines (both 2-, 3-, 4-); o-nitrobenzoic acid [C$_6$H$_5$O$_3$N] [C$_6$H$_6$O$_2$N$_2$] [C$_7$H$_5$O$_4$N]	C/HC, ArHC, SO, OE** Bi-CPE (30% and 85% Bi)	CV, DPV DPAdSV	Pasting liquids (binders): aliphatic and aromatic hydrocarbons, silicon oil, org. esters; Bi-CPE: with bismuth powder; study on el. redn. and adsorption; interf. of isomers; aq.buffers, sub-mM level	Model solns. and mixtures	[223, 2442–2445]
2-Methyl-3-nitro-aniline [C$_7$H$_8$O$_2$N$_2$]	C/PO (+"C-18," ionex)	CV, DPAdSV	o.c. accum.; s.e.: 0.01 M KOH (pH 10) LOD: 0.3 mg L^{-1}; simult. detn. with herbicide Ioxynil (diiodo-benzonitrile)	Drinking water	[1813]
1,3,5-Trinitrotoluene (industrial explosive) T.N.T. [C$_7$H$_5$O$_6$N$_3$]	(a) Gr-ILE (as WE) CPE (as Aux-E) (b) C/PO (+MIP)	(a) CV, DPV (b) DPV, SWV	Gr: graphene (new type of 3D-carbon) MIP: molecularly imprinted polymer functd. to selective detn. of TNT; pH-effect; LODs: (a) 0.5 ppb, (b) 2×10^{-9} M	(a) Model solns. (b) Natural water, soil extracts	(a) [517] (b) [1690]

(continued)

TABLE 8.12 (continued)
Determination of Organic Compounds at Carbon Paste Electrodes, Sensors, and Detectors: Survey of Methods and Selected Studies: Part 1: Environmental Pollutants, Synthetic Substances, and Industrially Important Products

Analyte/Substance (Org. Compound)	Specification [Chem. Formula]	Type of CPE (Modifier)/ [Configuration]	Technique (Mode)	Experimental and Electroanalytical Characterization, Selected Data; Notes	Sample(s)	Ref(s)
NPAHs						
2-Nitrofluorene, 2,7-dinitrofluorene	[$C_{13}H_9O_2N$] [$C_{13}H_8O_4N_2$]	GC/PO (umm.)	LSV, DPV	s.e.: BR-B (pH 7–10) + MeOH (1:1); lin. r.: 1×10^{-6}–1×10^{-4} M; mech. regnt.	Model solns.	[2446]
5-Amino-6-nitro quinoline	ANQ [$C_9H_7N_3O_2$]	GC/PO (umm.) [trad., wall-jet-D]	DPV, AdSV AD-HPLC	s.e.: BR-B (pH 7.0) + MeOH (1:9); lin. r.: 1×10^{-7}–1×10^{-4} M; mech. regnt.	Model solns.	[680,2447]
Other N-ctng. compounds (i.e..: –NH–OH; –NH–NH$_2$; –N=N–; heterocycles; alkaloids; etc.)						
Arylhydroxyl-amines	[Ar–NHOH]	C/Nj (umm.)	DPAdSV	*In vivo* enzymatic redn. of nitro-ders. and el. oxidn. of metabolite (NH$_2$-phenols); E_A = 0.0 vs. SCE; lin.r.: sub-μM level	(a) Liver ctng. suspension, (b) liver homogenate	(a) [2019] (b) [2020]
Methylhydrazine, dimethylhydrazine	[CH_3NH–NH_2] [$(CH_3)_2NNH_2$]	(a) C/SO (+Tyr) (b) C/SO (+DNA)	CV, DPV	Study on interaction with: (a) *Tyrosinase*, (rev. competitive inhibition), (b) DNA (damage); (a,b) lin. r.: low ppb level	(a,b) River water (a) Drinking water	(a) [2448] (b) [2449]
Phenylhydrazine	[$C_6H_5NHNH_2$]	CNT-CPE (+Fc-mCA)	DPV	modif.: ferrocene monocarboxylic acid; el.-catalytic oxidn.; lin.r.: 20–800 μM; simult. detn. with other HZs also tested	Model solns. (spiked s.)	[441]
Azobenzene, nitroazobenzenes (2-,3-,4-)	[$C_{12}H_{10}N_2$] [$C_{12}H_9O_2N_3$]	C/Nj (umm.) C/SO (umm.)	CV, DPV	Studies on redox behavior of sel. ders.; pH-effect (BR-B, 1–12); lin: μM level; detn. of total content tested	Wastewater (spiked s.)	[2450]

Acridine (and its ders.)	[C$_{13}$H$_9$N]	C/PO (+DNA)	DPASV	accum. & interaction with DNA; effect of CPE activation; LOD: nmol L^{-1} level simult. detn with catechin, poly-PhOH	Model solns.	[2451]
Nicotine	Alkaloid [C$_{10}$H$_{14}$N$_2$]	C/PO (+CHOD/BSA)	DPASV	mdf.: Choline-oxidase + bovine serum albumin; microbial oxidn. to BQ + redn. 0.07 M PhB (pH 7.4); LOD: 1×10^{-5} M	Tobacco leaves	[2452]
Brucine	Alkaloid of the *Strychnine* gr. [C$_{23}$H$_{26}$N$_2$O$_4$]	C/HD (unm.) C/HD (+CaC$_2$O$_4$)	(a,b) CV, CCOU (b) LSASV	Studies on el. oxidn. and form. of intermediates (three peaks: E_{P1} = 0.9, E_{P2} = 0.47, E_{P3} = 0.27 vs. SCE), catalysis by OxA; study on indirect detn. of nitrate, NO$_3^-$	Model solns.	[215, 2453]
Amino-poly-carboxylic acids	NTA, EDTA, DTPA [various sp.]	(a) C/Nj (unm.) (b) C/PO	(a) CV, CA (b) AD-LC	Study on anodic oxidn. & (a) pH-effect (pH < 2.5: one step, pH 4–8: two steps) (b) Direct oxidn. (one step), LODs: 0.1–0.2 ppm; LC with reversed phase mode	(a) Model solns. (b) Natural and wastewaters (from wat. treatment station)	(a) [640, 641] (b) [564]
Halogen-compounds						
2-Chlorophenol	2(*o*)-CP [C$_6$H$_5$OCl]	C/PO (+Tyr/sg-SiO$_2$) [CP-bio with enzyme]	HA, AD	mdf.: *Tyrosinase* enzyme immobilized onto sol–gel silica layer, lin.: μM level; simult. detn. with *o*-, *m*-, *p*-cresols	Model solns.	[1166]
4-Chlorophenol	4(*p*)-CP C$_6$H$_5$OCl	(a) C/PO (+CTAB/*Mmt*) (b) C/PO (+ms-TiO$_2$)	DPASV	mdf.: (a) *Ca-Montmorillonite* (clay) + TAB, (b) ms: mesoporous, studies on at. oxidn.; lin.r.: 5×10^{-8}–1 to 5×10^{-5} M	Natural water	(a) [1100] (b) [2454]
2,4-Dichlorophenol	2,4-DCP [C$_6$H$_4$OCl$_2$]	C/PO (unm.) [end-col. D]	AD-pCEC SPE)	Sepn. and detn. of five ders. (DCP, PCP, bis-Ph-A, Oc- and No-Ph); pre-sepn. by SPE; LODs: 2–50 ng mL^{-1}; R = 80%–100%	Chicken egg, milk powder	[563]

(*continued*)

TABLE 8.12 (continued)
Determination of Organic Compounds at Carbon Paste Electrodes, Sensors, and Detectors: Survey of Methods and Selected Studies: Part 1: Environmental Pollutants, Synthetic Substances, and Industrially Important Products

Analyte/Substance (Org. Compound)	Specification [Chem. Formula]	Type of CPE (Modifier) [Configuration]	Technique (Mode)	Experimental and Electroanalytical Characterization, Selected Data; Notes	Sample(s)	Ref(s)
2,4,6-Trichlorophenol	TCP [$C_6H_3OCl_3$]	(a) C/PO (+Zeolite-Y) [ft-D] (b) C/PO (+Caldariomyces s) [CP-bio with enzyme]	(a) DPV-FIA (b) HA, AD	Studies on (a) irrev. oxidn., (b) microbial catalysis by chloroperoxidase; (a) AcB; lin.: 0.05–1.0 mg L^{-1}; 1–10×10^{-7} M Tcp	(a) Synth. sea water (b) Model solns.	(a) [1147] (b) [2455]
2,3,4,5,6-Pentachlorophenol	PCP [C_6HOCl_5]	(a) C/PO (+silica gel) (b) C/PO (+HAs) (c) C/PO (unm.; as [ec-D])	(a) DPV, SWV (b) AD-pCEC	Studies on irrev. oxidn.; (a) PhB (pH 3) (b) modif.: humic acids; c) sepn./detn. of PCP + four other ders; LODs: 10 nM	(a) Fungicides, soils (b) Model solns. (c) Chicken egg, milk	(a) [682] (b) [1703] (c) [563]

S-, Se-, P-, As-ctng. org. compounds
– (none)

Organic dyes and coloring agents

Analyte/Substance (Org. Compound)	Specification [Chem. Formula]	Type of CPE (Modifier) [Configuration]	Technique (Mode)	Experimental and Electroanalytical Characterization, Selected Data; Notes	Sample(s)	Ref(s)
Azo- and diazo-dyes (synthetic products)	(Commercial products)	C/SO (unm.)	CV, DPV	Studies on behav. and detn. of com. dyes: (a) F3-B, Saturn Red, Saturn Blue LBR (b) Chromolan Red G, Egacid Orange, Orange GG, and Tartrazine O	Industrial wastewater (sampled near chem. factory Ostacolor®)	[2450]
Triphenylmethane dyes, subst. bis-indole sulfonate	Food coloring additives (comm. preps.)	C/PO (unm.) [trad.: wall-jet D]	CV, LSV AD-FIA	Studies on redox behav. and the detn. of sel. products: Green S, Patent Blue V, Brilliant Blue FC & Indigo Carmine	Model solns.	[2456]
Indigo White [aka Leuko-indigo] (bis-hydroxy-indole)	Dye: red. form. [$C_{16}H_{10}N_2O_2$]	C/PO (pretreated) [+$Na_2S_2O_4$, in situ]	CV, ACV, AdSV-FIA	redn. Indigo Blue → Indigo White by $S_2O_4^{2-}$ in NaOH; adsopt. accum. + rev. oxidn. in Tris/HCl; LOD: nM (2 min)	Model solns.	[2457]

Compound	Composition	Method	Notes	Matrix	Ref.
Bromothymol Blue (subst. bromo-tri-phenylmethane der.) [$C_{27}H_{28}O_5SBr_2$]	C/PO (unm.), C/PO (+surf., *in situ*)	CV	Study on redox behavior and pH-effect, adsorpt. and accum. in presence of diff. surfactants (CTAB, SDS, and TX-100)	Model solns.	[1658]
Acid Blue 120 (poly-naphthalene-di-Ph-diazo sulfonate) [$C_{33}H_{23}N_5O_6S_2$]	C/PO (unm.), C/PO (+surf., *in situ*)	CV, SWV	Study on irrev. oxidn., adsorption, and uptake efficiency; lin.: 2 – 20 + 60 – 520 mM; interf. study (inorg ions, fatty a.)	Silk-dyeing control	[2458]
Methylene Blue (phenothiazine der.) [$C_{16}H_{18}N_3SCl$]	(a) C/PO (+ZrP$_2$O$_7$) [ft-D] (b) C/PO (+MPTMS-smc)	(a) CV, SPE (b) CV	Studies on redox and adsorpt. behavior; modif: (b) SH-Pr-Me$_3$-SX immob. onto smectite-clay; LOD: (a) 1.0, (b) 0.4 µM	Model solns.	(a) [1288] (b) [1090]
Methyl Red (subst. azo-comp.) [$C_{15}H_{15}N_3O_2$]	C/PO (unm.)	LSV	Studies on redox behavior and form. of intermediates (in presence of O$_2$); AcB (pH 4.4); tests on detn. ($c_{MR} \cong 1$ µM)	Model solns.	[2459]
Malachite Green (trialkyl-methane anilinium chloride) [$C_{23}H_{25}N_2Cl$]	C/PO (+SDBS, *in situ*)	DPASV	Enhanced accum. via anionic surfactant and oxidn. of MG; s.e.: PhB (pH 6.5); lin.: 8 × 10^{-9}–5 × 10^{-7} M (t_{ACC} = 5 min)	Fish tissues	[2460]
Sudan I (1-phenylazonaphthol) [$C_{16}H_{12}N_2O$]	(a) C/PO (unm., +Ca-*Mmt*) (b) AB/PO (+PVP + SDS)	(a) LSV (b) CV, DPV	AC: acetylene black; modifs: (a) natur. clay (ion-pairing agent), (b) polyvinyl-pyrrolidone + surfactant (*in situ*); lin.r.: (a) 0.05–1 mg L^{-1}, (b) 2 × 10^{-7}–8 × 10^{-6} M	Food stuff (chilli products)	(a) [1068] (b) [186]
Surfactants and related substances					
Surfactants (anionic type) Alkyl-sulfates, alkyl-sulfonates [RR'–(O)SO$_3^-$]	C/NPOE (unm.) C/DOP (unm.) [+Fc(Me)$_3$N$^+$, *in situ*]	CV, AdSV	CP-binders: NO$_2$-Ph-Oc-ether, di-Oc-phthalate; mdf: Ferrocenyl-trimethyl ammonium ion.; detn. uses ion-pairing between mdf. + analyte; LOD: 0.1 µM	Model solns.	[207]

(*continued*)

TABLE 8.12 (continued)
Determination of Organic Compounds at Carbon Paste Electrodes, Sensors, and Detectors: Survey of Methods and Selected Studies: Part 1: Environmental Pollutants, Synthetic Substances, and Industrially Important Products

Analyte/Substance (Org. Compound)	Specification [Chem. Formula]	Type of CPE (Modifier)/ [Configuration]	Technique (Mode)	Experimental and Electroanalytical Characterization, Selected Data; Notes	Sample(s)	Ref(s)
Surfactants (cationic, anionic, and non-ionic types)	(R3R′N+X−) (R−(OCH$_2$)$_n$−OH [R′R−(O)SO$_3$−]	C/TCP, C/DOP, C/NPOE [CP-ISEs with Q+Y−]	POT (dir., titr.)	TCP: tricresyl phosphate; Q+Y−: ion-associate (as modif. or for saturation *in situ*); LODs: down to 1×10^{-6} M	Model solns., natural/industrial wastewater, (polluted or spiked)	[2461,2462]
Triton X-100 (non-ionic surfactant)	comm. product [C$_{14}$H$_{22}$O−(C$_2$H$_4$O)$_n$]	C/PO (unm.)	CV, DPV	Study on adsorption affinity of T-X at diff. conc. ranges (c_{LIM} = 3 μM, when above/below diff. monolayer structure	Model solns.	[317]
Dodecylbenzene-sulfonate, sodium (anionic surfactant)	com. p.: SDBS [C$_{18}$H$_{29}$NaO$_3$S]	C/PO (+Bu-*Rhodamine-B*)	CV, ECL	Extraction of ion-associate with modif. (1) and ECL-detn. (2) at +1.3 V vs. ref.; s.e.: (1) HCl, (2) NaOH; LOD: 5×10^{-7} M	Model solns. (wastewater)	[2463]

Macromolecular and polymeric structures

Ricin (oil) (aka "*Castor oil*")	Toxic protein (glycosylated hetero-dimer)	CNTPE (with MW-Ts) C/PO (unm.)	AImD	Immunodetection utilizing crevice-like structure of CNTs (confirmed by SEM); 0.6–25.0 ng mL^{-1}; trad. CPE also tested	Water samples	[435]
Polystyrene sulfonate (PSS)	Anionic poly-electrolyte(s) [RR′−(O)SO$_3$−]	C/PO (unm.) [+Fc(Me)$_3$N+, *in situ*]	CV, AdSV	Indirect detn. based on ion-pairing of Fc(Me)$_3$N+ and PSS with oxidn. of el. active Fc-group; LOD: 1 μM; usable to some bio-polymers (e.g.: heparin)	Model solns.	[2464]

* uμ ...ultramicro-. ** All mixts. unm. and of diff. compositions.

TABLE 8.13
Determination of Organic Compounds at Carbon Paste Electrodes, Sensors, and Detectors: Survey of Methods and Selected Studies: Part 2: Preparations and Products Marketed under Commercial Names

Analyte (Form/ Alter. n.)	Classification (Add.) [Chem. Specification]	Type of CPE (Modifier) [Configuration]	Technique (Mode)	Selected Experimental Conditions and Analytical Parameters; Notes	Sample(s)	Ref(s)
Acephate	(Foliar) insecticide [methoxy-sulfonyl-phospho-acetamide]	CNTPE (+Ru(bpy)$_3^{2+}$) [EH-μD]	CE-ECL	EH: electrically heated electrochemiluminescence detection; lin.r.: μM level simult. detn. with *Dimethoate*	comm. forms	[389]
Amitrole	(Grass) herbicide [3-amino-1,2,4-triazole]	(a) C/PO (CoPC) (b) CNTPE (with MW-Ts) (c) C/PO (+*n*-FePC)	(a) FIA-EC (b) AdSV (c) CV, CA	Study on oxidn., el.-catalytic processes (a) PC: phthalocyanine; 0.1 M NaOH; (b) 0.05 M PhB; (c) PhB + Na$_2$SO$_4$ (12) detn. In presence of interf. a. NH$_4$SCN LODs: (a,b) 40–50 μg L^{-1}, (c) 0.3 μg L^{-1}	(a,c) comm. forms (b,c) Tap water (c) River water	(a) [904] (b) [365] (c) [934]
Amoben (*Chloramben*)	Herbicide [2,5-dichloro-amino-benzoic acid]	C/PO (Co-DMDP)	DPV	Study on H$_2$-wave el. catalytic effect; DMDP: dimethyldithiophosphate (s); MEx. AcB (pH 3); LOD: 0.7 mg L^{-1}	Model solns.	[2465]
Aziprotryne	Herbicide [methyl-thio-triazine]	C/PO (CoPC)	CV, FIA-EC	Studies on catalytic redn., pre-extract. step, pH 7; LOD: 9 μg L^{-1} (0.9 ng)	Environ. samples, tap water	[905]
Bendiocarb (*Ficam*)	(Antimalaric) insectic. [dimethyl-benzo-di-oxo-ethylcarbamate]	C/Nj (+"C-18")	CV, DPV	modif.: ionex resin (50% in CP); o.c., Ex, BR-B (pH 5); LOD: 0.7 μg mL^{-1}	Model solns., soils	[1812]
Benomyl (*Benlate*)	(Selective) fungicide [benzimidazole der.]	C/SO [+analyte(s)]	LSV	MEx, BR-B (pH 5–7); 4–40 μg mL^{-1}; CPEE-configuration (with CP-regnt.)	Model solns., samples (s)	[2466]
Bentazon	(In-crop) herbicide [subst. benzothiadi-azin-3*H*-one-dioxide]	C/PO [+pANF, pPPF]	CV, UV–vis	modifs.: polyaniline-, polypyrrole conducting polymers (films for sorption); detn. with *Glyphosate* and other pests.	Model solns.	[2467]

(continued)

TABLE 8.13 (continued)
Determination of Organic Compounds at Carbon Paste Electrodes, Sensors, and Detectors: Survey of Methods and Selected Studies: Part 2: Preparations and Products Marketed under Commercial Names

Analyte (Form/Alter. n.)	Classification (Add.) [Chem. Specification]	Type of CPE (Modifier) [Configuration]	Technique (Mode)	Selected Experimental Conditions and Analytical Parameters; Notes	Sample(s)	Ref(s)
BM-Ts thiazole	(Cold) vulcanization accelerator/catalyst [SH-benzothiazole]	C/PO (+AgSCN)	DPV	modif. dispersed in the pasting liquid; s.e.: 0.1 M LiClO$_4$ (pH 6.5–7); LOD: 0.08 mg L^{-1}	Wastewater (near rubber factory)	[2468]
Bromodialone	(Anti-rat) rodenticide [subst. bromo-phenyl-hydroxy-coumarin]	C/PO (+FePC) [CP-bio with enzymes]	CV, DPV	Study on BDL transport inside the food chain (via various metabolite products) 0.2 M AcB (pH 4.2); LOD: 0.5 ng mL^{-1}	Animal tissues (pheasant and fox liver)	[2469]
Captax (syn. *Kaptax*)	(Cold) vulcanization accelerator/catalyst [SH-benzothiazole]	C/PO (+Co-DDTP)	LSV, DPV	modif.: Co^{II}-dimethyl-dithiophosphate dil. AcB (pH 3–4); LOD: 0.005 mg L^{-1}; simult. detn. with *BM-Ts* also tested	Wastewater (near rubber factory)	[2468]
Carbaryl	(Home garden) insecticide [1-naphthyl-methyl-carbamate]	C/PO (+polyamide) [ft-D]	CZE-AD	Hydrolysis to phenolic ders. and detn.; simult. detn. with *Fenobucarb*, *Isopro-carb* and *Metolcarb*; LODs: 3–6 × 10^{-8} M	Environmental specimens	[561]
Carbofuran	(Field) insecticide [subst. benzofuran methylcarbamate]	C/PO (+FePC) [CP-bio with enzymes]	AD (stat.)	Inhibition effect of analyte; two diff. enzymes; LOD: 1 × 10^{-10} M	Model solns.	[778]
Carbathion	Herbicide [methylaminomethane thioate; Na$^+$-salt]	(a) C/PO (Co-DMDP) (b) C/PO (crown-ethers)	CV, DPV	DMDP: dimethyldithiophosphate, MEx; (a) Study on H$_2$-wave el. catalytic effect; LODs: (a) 0.8 mg L^{-1}, (b) 5 × 10^{-9} M	Model solns.	(a) [2465] (b) [1395]
Carbophos (*Malathion*)	Insecticide [di-MeO-phosphino-thio-butanoic acid e.]	C/PO (Co^{II}-2,2′-dipy)	DPAdSV	Studies on accum. mechanism, adsopt. pH-effect, interfs.; LOD: 1.2 × 10^{-9} M	Model solns.	[2470]
Chloridazolne	Herbicide *N*-phenyl-2-chloro-3-amino-pyridazine)	C/PO (+SiO$_2$)	DPV, SWV	Study on irrev. oxidn. (el.-catalyzed by modif.) and adsopt.; BR-B (pH 11.3); lin.r.: 4 × 10^{-8}–9 × 10^{-7} M; RSD < 1.5%	Soil extracts	[1154]
Chlorpyrifos	Insecticide [diethyl-trichloro-pyridyl-phosphoro-thiolate]	C/PO (+Zeolite Y)	DPAdSV	Study on ion-pairing accum. and effect of pH, modif.: nat. clay; lin.: 0.0001–2 ppm (t_{ACC} = 80 s); interf. studies	Environmental specimens	[2471]

Clothianidin	Insecticide [chlor-thiazol-methyl Me-2-nitroguanidine]	C/SO (unm.), C/TCP (unm.)	DPV (dir.)	s.e.: BR-B (pH 7.0), lin.r.: 3–33 mg L^{-1}; detn. with *Imidacloprid* and *Nitenpyram*	Model solns.	[2472]
Cythioate	Insecticide/anthelmintic (anti-worm a.) [organo-S-phosphate]	C/PO (crown-ethers)	CV, DPV	Studies on accum., re-oxidn. processes; lin.r.: 4×10^{-9}–9×10^{-8} M; RSD < 4%	Model solns.	[1395]
2,4-D	(Non-spec.) herbicide [2,4-dichlorophenoxy acetic acid]	(a) C/PO [+F-HE + an.] (b) C/PO (+elp-ANFs)	(a,b) CV, LSV (b) UV–vis	modifs: (a) fluoro-hectorite (nat. clay) analyte (s)/CPEE-arr., (b) electropolymerized aniline films; LODs: ca 1 μM	(a) Herbicide solid sp.(b) biodegradability tests (model s.)	(a) [1086] (b) [2473]
2,4-DCP	(Non-spec.) herbicide [2,4-dichlorophenol]	C/PO [+F-HE + an.]	CV, LSV	Affinity/sorption effect of mdf (fluoro hectorite); analyte added into CP-mixt.; s.e.: PhB (pH 6.5); LOD: ca. 1 μM	Herbicide preps.	[1086]
Dimethoate	(Contact) insecticide [dimethyl-amino-2-oxo-dithiophosphate]	CNTPE (+Ru(bpy)$_3^{2+}$) [EH-μD]	CE-ECL	EH: electrically heated electrochemiluminescence detection; lin.r.: μM level; simult. detn. with *Acephate*	comm. forms.	[389]
Dinoseb	Herbicide [2,4-dinitro-6-butyl (*sec*)-phenol]	C/PO (unm.; +clay)	CV, DPV	modif: nat. clay of the *Bentonite* type; studies on pH-effect, accum. process, modif. effect, interfs.; LOD: 5×10^{-10} M	Environmental samples	[110]
Dinoterb (acetate)	Herbicide [2-*tert*-butyl-4,6-dinitrophenol]	C/PO (unm.; +clay)	CV, DPV	modif: nat. clay of the *Bentonite* type; studies on pH-effect, accum. process, modif. effect, interfs.; LOD: 5×10^{-10} M	Environmental samples	[110]
Diquat (dibromide)	Herbicide [quaternary ammonium salt, R'RN$^+$X$^-$]	(a) C/Nj (+*Zeolite Y*) (b) C/PO (+FAP) (c) C/PO (+KA)	(a) SWV (b,c) CV, SWV	(a) Detn. after wet digestion + extraction (b,c) modifs.: fluoro-apatite and kaolin, 2 redn. peaks; LOD: (a) 2×10^{-8} M, (b,c) p1: 3×10^{-8}, p2: 8×10^{-8} M, lin.: 1–80 μM	(a) Potatoes (b,c) River water (c) (spiked s.)	(a) [2474] (b) [1303] (c) [1084]
Disulfiram (*Antabus*)	Herbicide/fumigant [dimethydithiocarba-moyl-disulfide]	C/Nj (Co-PC)	DPV, FIA-EC	Studies on adsorpt. accum. mechanism, pH-effect, interfs.; LOD: 2×10^{-8} M	Preps. (under initial testing)	[897]

(*continued*)

TABLE 8.13 (continued)
Determination of Organic Compounds at Carbon Paste Electrodes, Sensors, and Detectors: Survey of Methods and Selected Studies: Part 2: Preparations and Products Marketed under Commercial Names

Analyte (Form/ Alter. n.)	Classification (Add.) [Chem. Specification]	Type of CPE (Modifier) [Configuration]	Technique (Mode)	Selected Experimental Conditions and Analytical Parameters; Notes	Sample(s)	Ref(s)
Endosulfan (EDS)	Insecticide/acaricide [subst. hexachloro-bis benzo-oxa-thiepine]	C/PO (+ionex "C-18") [trad. CPE + Cu^{2+}, in situ]	DPV	Indirect detn. based on interaction of electroinactive EDS with Cu(II)/via the peak decrease; LOD: 40ng L^{-1}	Model solns.	[2475]
EPN	Insecticide [ethyl-O-nitro-Ph-Ph-phosphonothioate]	(a) CNTPEs (+MeF, DNA) [μEs: needle type] (b) C/PO (+P Pseudomonas putida sp. + OPH-enzyme)	(a) ASV (trad. and in vivo) (b) AD (stat.)	(a) Study with a series of w. electrodes, incl. GCE,AuE). MeF: mercury film, 0.1 M $NH_4H_2PO_4$; lin.r.: 10–210 ng L^{-1} (b) CP-bio with modif. as biodegrader, LOD: 1.6 ppb; detn. with Fenitrothion	(a) Orange, apple, and skin tissues (b) Lake water	(a) [409] (b) [2476]
Fenitrothion	Insecticide [diMeO-4-nitro-PhO-thioxo-phosphorane]	C/PO (+Pseudomonas sp. + OPH-enzyme) [CP-bio + Co^{2+}, in situ]	AD (stat.)	modif: microbial cells (as biodegrader) 0.05 M PhB (pH 7.5), E_w = 0.6 V/ref.; LOD: 1.4 ppb; simult. detn. with EPN	Lake water	[2476]
Fenobucarb	(Agricult.) insecticide [2-butan-yl-2-phenyl methyl-carbamate]	C/PO (+polyamide) [ft-D]	CZE-AD	Hydrolysis to phenolic ders. and detn.; simult. detn. with Carbaryl, Isoprocarb, and Metolcarb; LODs: $3–6 \times 10^{-8}$ M	Environmental specimens	[561]
Glyphosate	(Perennials) herbicide [N-(phosphonomethyl) glycine]	C/PO [+pAN, pPP]	CV, UV–vis	modifs.: polyaniline-, polypyrrole conducting polymers (sorption substrates); detn. with Bentazon and other pests	Model solns.	[2467]
Hostaquick	Pesticide [chloro-bicyclo-di-methyl phosphate]	C/Nj + m. (enzyme + Co-PC)	AD, HA	Detn. based on analyte inhibition/bio-enzymatic effect; PC: phthalocyanine LOD: $3 \times 10^{-4} – 0.1$ g L^{-1}	Model solns.	[893]

Imazaquin	(Weeds) herbicide [imidazolinone der., Na⁺-salt]	C/PO (+pANF)	CV	modif.: polyaniline film with sorption effect; pH 2.5, lin.r.: mid-μM range	Model solns.	[1805]
Imidacloprid	Insecticide [subst. chloro-nitro-neonicotinoid der.]	(a) C/SO, C/nTD, C/TCP (b) C/SO, C/TCP [all CPEs as unm.]	(a,b) DPV (dir.)	(a) comp. of three diff. CPEs (made of SO, tetradecane, and tricresyl phosphate) s.e.: BR⁻B (pH 7.0), lin.r.: 3–33 mg L⁻¹; detn. with *Clothianidin* and *Nitenpyran*	(a) River water, comm. preps. (b) Model solns.	(a) [2477] (b) [2472]
Ioxynil (*Toxynil*)	Herbicide [4-hydroxy-3,5-di-iodo-benzonitrile]	(a) C/Nj (+"C-18") (b) C/PO (crown-ethers)	(a,b) DPV	(a) o.c., 0.01 M HCl, LOD: 0.1 μg mL⁻¹ (b) MEx, pH 3; lin.: 4 × 10⁻⁹–9 × 10⁻⁸ M	(a) Drinking water (b) Model solns.	(a) [1813] (b) [1395]
Isoprocarb	(Agricult.) insecticide [iso-propan-2-phenyl methyl-carbamate]	C/PO (+polyamide) [ft-D]	CZE-AD	Hydrolysis to phenolic ders. and detn.; simult. detn. with *Carbaryl*, *Fenobu-carb*, *Metolcarb*; LODs: 3–6 × 10⁻⁸ M	Environmental specimens	[561]
Linuron	(grass) herbicide [3,4-dichloro-phenyl-methoxy-diurea-on]	C/PO (+*Sepiolite*)	DPV	o.c., MEx, pH 1.7–2.0 (0.01 M KNO₃); LOD: 0.75 mg L⁻¹; interfs. from Cl⁻	River water	[1071]
MCPA	(Weed) herbicide [mono-chloro-phen-oxy acetic acid]	C/PO (+MnO₂)	AD (stat.)	Study on el. catalytic redn. by modif.; (also with MnF-GCE); LOD: 0.2 ppm	Soils and soil extracts	[2478]
Maleic hydrazide	Herbicide [3,6-dihydroxo-pyri-dazine; oxidn. form]	C/PO (unm.) C/PO (+Pd) [ft-D]	AD-FIA	Study on catalytic effect of Pd-powder; s.e.: PhB (pH 7); E_w = +0.7 V vs. ref., v_F = 2 mL min⁻¹; lin.: 2 × 10⁻⁷–1 × 10⁻⁴ M	Drinking and natural water	[839]
Metamitron	(In-crop) herbicide [amino-3-methyl-Ph-1,2,4-triazin-one]	(a,b) C/PO (+SiO₂) (c) C/PO (+BiF-, *ex situ*)	(a) DPV (a–c) SWV (c) DPA, FIA	(a,b) Studies on el. oxidn. (c)+ Bismuth film, 0.1 M AcB (pH 4.5), (a,b) LOD: 4–40 × 10⁻¹⁰ M, (c) lin.r.: 10–200 μM	(a) Natural water (b) Soils (c) Model solns.	(a) [1155] (b) [1156] (c) [2479]
Met(h)am sodium	Fungicide/fumigant [methylaminomethane dithioate, Na⁺-salt]	C/Wx (+PVC-layer) [s-CP-bio with enzyme]	AD (stat.)	Enzyme immobilized at CPE by photo-cross-linkage with PVC; LOD: 100 ppb	Model solns., soil samples	[2480]

(*continued*)

TABLE 8.13 (continued)
Determination of Organic Compounds at Carbon Paste Electrodes, Sensors, and Detectors: Survey of Methods and Selected Studies: Part 2: Preparations and Products Marketed under Commercial Names

Analyte (Form/ Alter. n.)	Classification (Add.) [Chem. Specification]	Type of CPE (Modifier) [Configuration]	Technique (Mode)	Selected Experimental Conditions and Analytical Parameters; Notes	Sample(s)	Ref(s)
Methylparathion (syn. *Metathion*)	(Contact) insecticide, acaricide [subst. *p*-nitro-phen-oxy-organophosphate]	(a) C/PO (+FePC) [CP-bio with enzymes] (b) C/Nj (+"C-18") (c) C/MO (+enzyme) (d) CNT-IL (BMIMPF$_6$) [CP-film at the GCE] (e) C/PO (MBD + OPH) (f) C/PO (unm.) (g) C/PO (+mep-ZrO$_2$) (h) C/PO (+nano-ZrO$_2$)	(a) AD (stat.) (b,d) DPV (c) FIA-EC (e) LSV, AD (stat.) (f) SWAdSV (g) DPASV (h) SWASV	(a) Nal. inhibition/bioenzymatic effect (b) interfs. studies; LOD: 7.9 ng mL^{-1} (c) catalyzed hydrolysis; 0.2–1.0 μM (d) catalytic effect of both CNTs and IL, o.c. accum., PhB (pH 7), LOD: 1 nM (e) modif. microbial cells (degrader) + enzyme; LODs: 0.3–5.3 ppb, simult. detn. with *Paraoxon* and *Parathion* (f) accum. by adsorpt., LOD: 0.05 μM (g,h) o.c. accum. with strong affinity of P=O to mdf., DLs: (a) 5 nM, (b) 5 μg L^{-1}	(a) Model solns. (b) lake water (c) well water (spiked) (d) nat. water, soil (e) soil extracts (f) model mixts. (g) apple tissue (h) diff. water specimens	(a) [893] (b) [1611] (c) [2481, 2482] (d) [3334] (e) [2483, 2484] (f) [2485] (g) [1876] (h) [998]
Metolcarb	(Agricult.) insecticide [methyl-hydroxy-2-Ph-methylcarbamate]	C/PO (+polyamide) [ft-D]	CZE-AD	Hydrolysis to phenolic ders. and detn.; simult. detn. with *Carbaryl*, *Fenobucarb*, *Isoprocarb* LODs: 3–6 × 10^{-8} M	Environmental specimens	[561]
Nitenpyram	Insecticide [subst. chloro-nitro-neunicotinoid der.]	C/SO, C/TCP [both CPEs as unm.]	DPV (dir.)	s.e.: BR-B (pH 7.0), lin.r.: 3–33 mg L^{-1}; detn. with *Clothianidin* & *Imidacloprid* electrolytic activation of CPEs tested	Model solns.	[2472]
Nogos (50EC)	Insecticide [dimethyl-2,2-dichlorovinyl phosphate]	C/PO (+Co-PC) [CP-bio with enzymes]	AD (stat.)	Detn. based on analyte inhibition/ bio-enzymatic effect; PC: phthalocyanine, LOD: 2 × 10^{-4}–0.1 g L^{-1}; RSD < ±3%	Model solns.	[893]

Paraoxon	Insecticide [subst. p-nitro-phen-oxy-organophosphate]	(a) C/PO (+OPH-enzyme) (b) C/PO (FePC + m-ezs) (c) C/PO [+Arthrobacter sp. + OPH-enzyme (m)] (d) C/PO [+Pseudomonas putida JS444 sp + OPH]	(a) FIA-EC (b) AD (stat.) (c) AD (stat.) (d) LSV	OPH: phosphorus-hydroxylase; m-ezs add. membrane + 2 enzymes; (a,b) PhB (pH 7–8), (c) PhB + Na-citrate (pH 7.5) LODs: (a) 0.1 µM, (b) 1×10^{-10} M, (c) 2.8 ppb; (c) Detn. with *Methylparathion* (d) mdf: microbial degrader + enzyme, detn. with *Parathion*, *Methylparathion*	(a) Well water (spiked) (b) Model solns. (c) Soil extracts (d) Model solns.	(a) [2481, 2482] (b) [778] (c) [2483] (d) [2484]
Paraquat	(Non-sel.) herbicide [N,N-dimethyl-bis-pyridine dichloride]	(a) C/Nj (Amberlite XAD-2) (b) C/PO (+zeolite) (c) C/PO (+HAP) (d) C/PO (+FAP)	(a) DPV (b,c) SWV (d) CV, SWV	(a) LOD: 0.1 mg mL^{-1}; (a,b) studies on accum. mechanism (b) LOD: 2×10^{-8} M detn. after wet digestion and extraction (c) Detn. via oc.-acc. by hydroxy-apatite & redox behav. of PA; l.l.: 0.02–20 µM (d) modif.: fluoro-apatite, 2 redn. peaks; s0.1 M K_2SO_4, DL: 4×10^{-8} and 7×10^{-8} M	(a) River water (b) Potatoes (c) Model solns, natural water (d) Food stuff (model s.)	(a) [2486] (b) [2474] (c) [1301, 1302] (d) [2487]
Parathion	Insecticide/acaricide [subst. p-nitro-phen-oxy-organophoshate]	(a) C/PO [+*Pseudomonas putida JS444* sp + OPH] (b) C/PO (+MIP)	(a) LSV (b) CV, DPV	(a) mdf: microbial degrader + enzyme; detn. with *Paraoxon*, *Methylparathion* (b) mdf: molecularly imprinted polymer (sel. recogn./extr. effect), l: 2–900 nM.	(a) Model solns. (b) Groundwater, vegetables	(a) [2484] (b) [2488]
Phosalone	Insecticide/acaricide [chloro-diethoxy-thio phosphine-benzoxazole]	C/PO (+Co^{II}-2,2′-dipy)	DPAdSV	Studies on pH-effect, adsorpt. and accum. mechanism, interfs.; LOD: 5.5×10^{-9} M	Model solns.	[2470]
Prefix	(Soybean) herbicide [2,6-dichloro-thio- benzamide]	C/PO (+crown-ethers)	CV, DPV	Studies on accum. and reoxidn. process; lin.r.: 4×10^{-9}–9×10^{-8} M	Model solns.	[1395]
Sayphos (Menazon)	(Anti-aphids) insectic. [subst. diamino-phos-phino- thioyl-triazine]	C/PO (+crown-ethers)	CV, DPV	Studies on accum. and reoxidn. process; lin.r.: 5×10^{-9}–1×10^{-7} M	Model solns.	[1395]

(continued)

TABLE 8.13 (continued)
Determination of Organic Compounds at Carbon Paste Electrodes, Sensors, and Detectors: Survey of Methods and Selected Studies: Part 2: Preparations and Products Marketed under Commercial Names

Analyte (Form/ Alter. n.)	Classification (Add.) [Chem. Specification]	Type of CPE (Modifier) [Configuration]	Technique (Mode)	Selected Experimental Conditions and Analytical Parameters; Notes	Sample(s)	Ref(s)
Seedox	Fungicide [1,3,4-trichloro-phe-noxy acetic acid]	C/Nj (+CoPC) [CP-bio with enzyme]	AD (stat.) AD (FIA)	Inhibition/bio-enzymatic mechanism; PC: phthalocyanine; LOD: 3 × 10^{-4}–0.1 g L^{-1}; tests in flowing streams	Model solns.	[893]
Tetramethrin	Insecticide [subst. cyano-phen-oxy-benzylalcohol]	C/PO (+*Sepiolite*)	DPV	Studies on pH, accum. process, interfs. o.c.. MEx, pH 5.3 + 12, LOD: 45 µg L^{-1}	Water samples, soils	[825]
Thiamethoxam	Insecticide [neonicotinoid der.]	C/TCP (unm.)	DPV (dir.)	Detn. without accum., s.e.: BRB (pH 7) lin.r.: (a) 4–40µg mL^{-1}, RSD = ± 1.3%	River water, comm. prep. (*Actara 25-WG*)	[511]
Thiram	Ectoparasiticide [bis-*N*,*N*'-dimethyl-carbamoyl disulfide]	C/Nj (+Co-PC)	DPV, FIA-EC	Studies on pH-effect, adsorpt. accum. process, interfs.; o.c.. LOD: 7 × 10^{-8} M	Strawberries (spiked samples)	[897]
Zineb	(Foliate) fungicide [Zn^{II}-ethane-1,2-diyl bis-dithiocarbamate]	sCI (+Pt-sputtered) [SPCP-bio + NADH/OD]	AD (stat.)	sCI: solidified carbon ink (+modif.) and covered by PVC-layer (ctng. enzyme); s.e.: Na_2EDTA soln., LOD: ca. 8 ppb	Model solns., soil samples	[2490]

TABLE 8.14
Determination of Pharmaceuticals and Drugs at Carbon Paste Electrodes, Sensors, and Detectors: Survey of Methods and Selected Studies

Analyte (Chemical Form)	Classification (Chem. Specification)	Type of CPE (Modifier)/ [Configuration]	Technique (Mode)	Experimental and Electroanalytical Characterization, Selected Data; Notes	Sample(s)	Ref(s)
Part 1: Substances "A–H"						
Aceclofenac	Antiinflammatory ag. (glycolic acid)	(a) C/Nj (+Triton X, SDS) (b) C/PO (unm.)	(a) AdSV (b) CV, DPV	modifs.: non-ionic/anionic surfactant (a) ppb level; studies on pH, CP comp. (b) 2.8 ppb; MeOH (20%) based s.e.	(a) Tablets (b) Commercial forms	(a) [1644] (b) [2494]
Acetaminophen syn. paracetamol [N-(4-hydroxyphen-yl)-acetamide]	Wide-range analgesic (i.e., pain reliever) or antipyretic (i.e., fever reducer)	(a) C/PO (unm.) (b) s-CPE [C/PWx] (c) C/VO (avocado tissue) (d) C/PO (pumice, 6% w) (e) CP-bio (+mSiO$_2$ + HRP) (f) CILE (ctng. BMIMPF$_6$) (g) CNTPE (unm.) (h) C/NJ, C/Nj (+n-Au) (i) C/PO (+Cu$_3$[Fe(CN)$_6$]$_2$) (j) C/PO (+Dy^{III}-NWs) (k) CNT-CPE (+thionine)	(a) LC-AD (b) CV, DPV (c) CV, CA (d) DPV (e) CV, DPV (f) CV, DPV (g) 2nd DeLV (h) FIA-AD (i) CV (j) SWAdSV (k) CV, DPV	(a) Special study with Acp-metabolites (b) CP from paraffin wax, lin: nM level (c) VO: vaseline oil, BRB, LOD: 90 μM (d) modif.: "sea foam," se: 0.1 M H$_2$SO$_4$ (e) Magnetized nanoporous μ-particles (f) s.e.: AcB (pH 4.6), lin.: 1 μM–2 mM (g) lin.: 0.1–125 μM, compd. with CPE (h) Catalytic effect; lin.: 0.1–80 mg L^{-1} (i) AcB + 0.05 M NaCl; LOD: 1 × 10^{-5} M (j) NW: nanowire, detn. with *Naproxen* (k) LOD: 0.1 mM; detn. with *Isoniazid*	(a) Model solns. (b) Dosage forms (vials, tablets, suppositories) (c,d) com. tablets, urine (e) Model solns. (μM) (f) com. tablets, urine (g) Effervescent dosage (h) com. drugs (six types) (i, j) pharm. forms. (k) comm. drugs, blood plasma	(a) [2495] (b) [2002] (c) [237] (d) [2496] (e) [2497] (f) [477] (g) [423] (h) [567] (i) [2498] (j) [1887] (k) [438]
n-Acetylcysteine	Supplemental ag. for chronic bronchitis (der. of aminoacid)	CNTPE (+Co^{III}-SLP; diff. types)	CV, DPV	SLP: Schiff bases; study on catalyzed oxidn.; other –SH ders. also examined lin.: 1 × 10^{-7}–1 × 10^{-4} M; RSD < 4% (n = 10)	Model solns.	[403]

(continued)

TABLE 8.14 (continued)
Determination of Pharmaceuticals and Drugs at Carbon Paste Electrodes, Sensors, and Detectors: Survey of Methods and Selected Studies

Analyte (Chemical Form)	Classification (Chem. Specification)	Type of CPE (Modifier)/ [Configuration]	Technique (Mode)	Experimental and Electroanalytical Characterization, Selected Data; Notes	Sample(s)	Ref(s)
Acyclovir (syn. *Zovirax*)	Synth. antiviral drug (amino-OH-EtO)-Me-1H-purin-6,9H-one)	C/PO (+μ-Cu-Ps) C/PO (+n-Cu-Ps)	CV, AD	Elec.-catalytic oxidn. (via $Cu^{II} \rightarrow Cu^{III}$), μ-Ps, n-Ps: micro- and nanoparticles, lin.: low micromolar level; RSD < 5%	comm. products (tablets, topical cream)	[961]
Adriamycin	Antitumor drug (anthracycline gr.)	(a) C/Nj (unm.) (b) C/PO (+CTAB, *in situ*)	(a) CV, DPV (b) AdSV	(a) pH 4.5, LOD: 0.1 μM; test EC-LC (b) 3×10^{-8}–5×10^{-6} M, CTAB: surf. for erosive activation of CP-surface	biol. fluids (human urine)	(a) [2499, 522] (b) [2501]
Amantadine	Anti-Parkinsonian ag. (amino-adamantane)	C/PO (+β-CD)	POT (dir.)	Inclusion effect; CD: cyclodextrin; s.e: 0.02 M AcB (pH 4.7), LOD: 5×10^{-10}	pharm. forms (tablets)	[2502]
Ambroxol (hydrochloride)	Bronchosecretolytic agent (cyclohexanol)	C/Nj (unm.)	CV, AV	s.e.: 0.5 M H_2SO_4 (pH 0); 4×10^{-7}–6×10^{-5} M (LOD: 2×10^{-7} M)	Model solns.	[2503]
Amikacin	Antibioticum (aminoglycoside)	C/PO (+n-CuO)	CV, AD (stat.)	modif: nanosized oxide; lin. range: 1–200 mM; comp. to the bare CPE	pharm. forms.	[1228]
Amikhellin	Antiasthmatic agent (chromone der.)	C/Nj (*Metrohm*)	CV	Commercially available CP-mixture; oxidn. process studied: LOD: 1 μM	Model solns.	[2504]
Amiodarone	Antiarrhythmic agent (benzofuran diiodo-phenoxydiamine der.)	(a) C/Nj (*Metrohm*) (b) C/PO (+*Trit.*, *in situ*)	(a) CV, COU (b) 2.5nd DeLV	(a) Studies on electrode mechanism and reaction kinetics, lin: 5×10^{-6}–5×10^{-5} M (b) modif: *Triton X-100*; LOD: 0.1 nM	Model solns.	(a) [2505] (b) [2506]
Aminophylline	Bronchodilator (bis-dimethyl-purine-one thylendiamine)	C/PO (+Pt-F)	CV, DPV	PtE: electrodeposited Pt-film; study on el. oxidn.; s.e.: BR-B, AcB, PhB (pH 2–8); lin.: μM level; GCE also tested.	Medical prod. (anti-asthma thigh creams)	[2507]
Amitriptyline (hydrochloride)	Antidepressant (tricyclic der.)	C/PO (+PV-N-imidazole)	CV, DPV	Studies on electrode oxidn. and pH; interfs.; compared to the bare CPE	pharms. (dosage forms)	[2508]
Amlodipine (besylate)	Antihypertensive ag. (dihydropryridine der)	C/PO (unm.)	CV, DPAdSV	Studies on el. irrev. oxidn. and adsorp.; s.e.: pH 11; lin.r.:1×10^{-8}–2×10^{-7} M	com. drugs (e.g., *Amlopres-A®*)	[2509]

Amoxicillin	Antibioticum (from *Penicillium* antibiotics group)	(a) C/PO (PV-pyridine) (b) C/PO (PV-N-imidazole) (c) C/Nj (+V^{IV}O-*Salen*)	(a,b) CV, ASV (c) CV, LSV, DPV, SWV	(a,b) PV: poly-4-vinyl; study on oxidn, LOD: 1×10^{-6} M; (c) detn. via signal form V^{III}/V^{IV}; LODs: 8.5–16.5 µM	(a,b) Tablets, capsules, oral suspensions (c) comm. Tablets	(a) [1744] (b) [1760] (c) [2510]
Ampicillin	Bacterial antibiotic (from β-lactam gr.)	C/PO (+FcDCA)	LSV, CA, DPV	modif: ferrocene dicarboxylic acid, el.-catalyzed oxidn. (pH 10); $E_A = +0.48$ vs. ref; lin.: 2.3–30.0 and 40–700 µM	com. drugs, human urine	[1538]
Apomorphine	Anti-Parkinsonian ag., anti-heroin addiction (di-benzo-quinoline)	C/Nj (umm.)	CV, LSV	Study on irrev. oxidn. (incl. 2e$^-$, 2H$^+$); PhB, simult. detn. with AA; µM level	Model solns.	[2511]
Aspirin	Principal antipyretic, antiinflammatory dr. (acetylsalicylic acid)	C/PO (umm.; 70:30%) [ft-D]	CV, SWV, FIA-ASD	Indirect detn. based on hydrolysis of ASA → SA; simult. detc. with group of salicylates; LOD: 1 fmol (in 5 µL)	pharm. forms (tabs. *Acylpyrin*®)	[2512]
Artemisinin	New anti-malarian d. (subst. sesquiterpene endoperoxide)	C/PO (+Co^{II}-PC)	CV, DPV	PC: phthalocyanine (el. cat.), s.e.: PhB (pH 7); linear range: 2×10^{-5}–5×10^{-4} M	Plant materials (*Artemisia* sp.)	[914]
Atenolol	Antihypertensive dr. (subst. amino-propo-xy-phenoacetamide)	(a) C/PO (umm.) (b) C/PO (+*Mordenite*) (c) C/PO (+*n*-Au-Ps)	(a–c) CV, DPV (c) CCOU	(a) Study on irrev. oxidn. and mechanism; s.e.: PhB (pH 10.4); lin: sub-µM level (b) mdf.: zeolite, AcB (5), LOD: 0.1 µM (c) Ps: particles; pH 10, LOD: 7×10^{-8} M	(a–c) pharm. forms. (com. tablets) (a,c) Human urine	(a) [2513] (b) [1134] (c) [1279]
Azithromycin	Antibioticum (macrolide gr.)	C/Nj (+PV-N-imidazole)	SWV	s.e.: AcB + AcN (30% v/v, pH 4.6) lin.r.: 0.5–1.5 ppb; RSD < 5 %.	pharm. solns.	[2514]
Aztreonam	Antibioticum (monobactam gr.)	C/PO (umm.) C/PO (+gelatine)	DPV, SWV	LOD: 2×10^{-8} M; comp. with other electrodes (DME, HMDE, GCE)	Model solns., urine	[2515]
Azidothymidine	Anti-HIV/AIDS drug (subst. tetrahydroxy-pyrimidin-one azide)	(a) C/PO (+anti-AZT) (b) DPE (umm.)	SIA-AD	DP: diamond paste; immunosensing principles (modifier of CPE: antibody) E_A ca. +0.2 V; lin.: micromolar level	Raw pharm. products (quality control)	(a) [550] (b) [348]
Benorilate	Antiinflammatory and antipyretic agent (der. of *Asp* + *Par*)	(a) C/PO (umm.) (b) C/PO (+*n*-Ag)	CV, DPV	Studies on irrev. oxidn, incl. metabolite PhB (pH 6.9), $E_A = +1.02$ V vs. ref; LOD: 50 nM; (b) m.: Ag-nanoparticles	(a,b) pharm. forms. (com. tablets) (b) Human urine	(a) [2516] (b) [977]
Benperidol	Antidepressivum (F-alkylketone)	C/PO (umm.)	CV, DPV	0.1 M H$_2$SO$_4$, LOD: 1×10^{-6} M; studies on irreversible oxidn.	pharm. forms.	[242]
Benserazide	Anti-Parkinsonian ag. (trihydroxy-benzyl-propenyl hydrazide)	C/PO (umm.)	CV, DPV	Studies on electrode process and pH-effect; BRB (2–12); lin.: 5–500 µM	Model solns.	[2517]

(continued)

TABLE 8.14 (continued)
Determination of Pharmaceuticals and Drugs at Carbon Paste Electrodes, Sensors, and Detectors: Survey of Methods and Selected Studies

Analyte (Chemical Form)	Classification (Chem. Specification)	Type of CPE (Modifier)/ [Configuration]	Technique (Mode)	Experimental and Electroanalytical Characterization, Selected Data; Notes	Sample(s)	Ref(s)
Bentazepam	Anxiolytic agent (benzodiazepine)	(a) C/PO (+*Sepiolite*) (b) C/PO (+C-18, 50%)	CV, DPAdSV	Studies on adsoprt. accum. mechanism; modifs.: (a) clay, (b) ion-exchange resin; interfs. from diff. antipsychotic drugs	Human urine	(a) [1072] (b) [1810]
Benziodarone	Vasodilator (benzofuran-di-iodo-phenolic der.)	C/Nj (*Metrohm*)	CV, COU	Studies on electrode mechanism and reaction kinetics, iodinated ders. also of interest; c(a) = 5×10^{-6}–5×10^{-5} M	Model solns.	[2505]
Benzocaine	Local anesthetic and topical pain reliever (Et-amino-benzoate)	C/PO (+*Chloranil*)	CV, LSV	Undirect method (via the decrease in signal of modif.), lin: sub-μM level; simult. detn. with *Lidocaine*	Model solns.	[1507]
Bergenin	Immuno-modulatory ag. (glycoside of 4-*o*-methyl-gallic acid)	CNT-CPE (with MW-CNTs)	CV, DPV	Studies on electrode mechanism and reaction kinetics; s.e.: PhB (pH 7.0); LOD: 7×10^{-8} M, R_R = 99.8%–100.2%	pharm. forms. (com. tablets)	[398]
Bumetanide	Diureticum (sulfonamide der.)	C/Nj (*Metrohm EA207C*)	DPASV	pH 2.5 and 7 (MEx), LOD: 5×10^{-6} M; MeOH (10%) ctng. s.e. (prior to MEx)	Commercial forms. (tabs *Fordiuran*®)	[2518]
Buprenorfin	Diureticum (sulfonamide der.)	C/PO (unm.)	CV, DPV	pH 9.0 (BR buff.), LOD: 1×10^{-8} M: CPE tested as thin-layer cell for FIA	pharm. forms. (tablets, vials)	[2519, 2520]
L-*Butaclamol*	Antipsychotic drug (subst. benzocyclo-hepta-iso-quinoline)	C/PO (+M$_{I-III}$-CDs)	CV, DPV	Enantioanalysis of two R- & L-forms, modif.: 3 types of malto-cyclodextrin, LODs: 1×10^{-10}–1×10^{-8} M; R_R > 90%	Model solns., human urine	[2521]
Buzepide (methiodide)	Antidepressant agent (Me-azepine-di-Ph-dimethan-amide)	C/PO (unm.) C/PO (+*n*-TiO$_2$-Ps)	CV, DPV	el-catalyzed oxidn. peak (by modif.); PhB (pH 6.5), lin.r.: 5×10^{-8}–1×10^{-5} M	Human blood, human serum	[1879]
Caffeine	Psychoactivity stimulating agent/drug (natural alkaloid)	C/PO (unm.)	CV, SWV	Indirect method (based on suppr. of 1,4-benzo-quinone peak); s.e. PhB (pH 7.4) lin.: 0.1–0.5 and 0.5–8.0 mM (diff. sens.)	Coffee extracts	[1444]

Electroanalysis with Carbon Paste–Based Electrodes, Sensors, and Detectors 307

Captopril	Antihypertensive ag. (subst. sulfonyl-pyrrolidin-carboxylic a.)	(a) C/PO (+Co-NSLF) (b) C/PO (+Fc-DCA)	(a) CV, DPV (b) CV, CA, EIS, SWV	(a) mdf.: Co^{II}-5-nitro-*Salophen* (*Schiff b*) study on catalytic effect; LOD: 10nM (b) mdf.: subst. ferrocen, LOD: 9 × 10^{-8} M	(a,b) Human urine (b) comm. Drug	(a) [1587] (b) [1539]
Carbidopa	Anti-Parkinsonian ag. [(OH)$_2$-Ph-Me-propanoic hydrazide]	(a) CPE (+PbO$_2$/PER) (b) CNTPE (+Fc)	(a,b) CV, DPV (b) CA	(a) PER: polyester resin, LOD: 3 × 10^{-7} M simult. detn. of *Carbidopa* and L-dopa (b) Study on el.-catalytic oxidn. (evaln. of kinetics: α, k_{ft}, D_X), lin.: 5–600 μM	(a) pharm. forms. (com. tablets) (b) Human urine	(a) [1241] (b) [432]
Carboplatin	Anticancer agent (organometallic der.)	C/Nj (*Metrohm*)	CV, DPV	pH 6.0, 10–300μM; studies on redn. process and Pt-plating processes, CPE pre-activated by cycling	Model solns.	[2522]
Carminomycin	Antitumor antibiotic (methyl-transferase)	C/PO (unm.)	CV, DPV	Study of electroxidn. process at pH 1; comp. to GCE and CCE; lin.: 1 × 10^{-6}–1 × 10^{-4} M, LOD: 6 × 10^{-8} M	Model solns.	[2523]
Efalosporines	Antibiotics (six diff. drugs from *Cephalosporine* gr.)	(a) C/Nj (unm., fatty acids) (b) C/PO (+Co^{II}-*Salophen*)	(a) CV, AdSV (b) CV, DPV	(a) *Cefatrexyl*, *Cefobid*, *Claforan*, *Roceph-in* and *Velosef*; s.e.: dil. H$_3$PO$_4$, LOD: 10nM, effect of AA,UA (b) *Cefalexin*, *Cefazolin*; pH 3, lin.: 1 × 10^{-5}–0.001 M	(a) pharm. forms., biol. fluids (b) Human serum	(a) [2524, 2525] (b) [1584]
Cefotaxime	Antibioticum (*Cephalosporine* dr. of the third generation)	(a) C/Nj (unm.) (b) C/PO (+Zn^{II}-*SB*)	(a) CV (b) CV, DPV (EQCM)	s.e.: 0.3 M H$_2$SO$_4$ (pH 2.3); (a) CPE used in reference studies; (b) SB: *Schiff base*, lin: 1–500nM, study with electrochemical quartz crystal microbalance	(a) Model solns. (b) Human blood (acidified s.)	(a) [2526] (b) [1596]
Celiptium	Antineoplastic (ellipticinium acetate)	C/Nj (*Metrohm*)	CV, AdSV	MEx, LOD: 1 × 10^{-8} M; incl. studies on oxidn. mechanism and interfs.	biol. fluids (human urine)	[1669]
Cetirizine (hydrochloride)	Anti-allergic and anti-hay-fever agent (subst. Cl-Ph-acetate)	C/PO (+Cez-αCD) C/PO [+Cez$^+$TFPB$^-$] [CP-ESE, CP-ISE]	POT (dir.)	mdf.: inclusion and ion-exchange effects TFPB: [B(CF$_3$-C$_6$H$_4$)$_4$], S = 57–60mV/d, lin.r.: 5 × 10^{-6}–0.1 M; R_R = 100.1%	Human urine (spiked s.)	[2527]
Cisatracurium (besilate)	Neuromusc. blocking drug (tetrakis-MeO-benzo-polyether)	C/PO (unm.)	CV, DPV	Study on irrev. oxidn. and el. processes LOD: 0.4 μg mL^{-1}, LOQ: 1.3 μg mL^{-1}	biol. fluids (human urine and serum)	[2528]
Cis-platin	Anticancer agent [Pt(II) derivative]	(a) C/Nj (*Metrohm*) (b) C/Nj (+DNA)	(a) CV, DPV (b) CV, HA	(a) pH 6.0; 0.01–1.00mM; studies on redox process and hydrolytic effects (b) Interactions with ssDNA	Model solns.	(a) [2529] (b) [2530]

(*continued*)

TABLE 8.14 (continued)
Determination of Pharmaceuticals and Drugs at Carbon Paste Electrodes, Sensors, and Detectors: Survey of Methods and Selected Studies

Analyte (Chemical Form)	Classification (Chem. Specification)	Type of CPE (Modifier)/ [Configuration]	Technique (Mode)	Experimental and Electroanalytical Characterization, Selected Data; Notes	Sample(s)	Ref(s)
s-Cilazapril	Antihypertensive a., (Py-carboxamide)	C/PO (AAOd)	SIA-AD	modif: L-aminoacid-oxidase; enantioselectivity principles; s.e.: pH 6.8–7.4; 0.001–100 μM (LOD: 1 nM)	Model solns. (quality control)	[747]
Ciprofloxacin	Antibacterial agent (synth. chemotherp.) (subst. fluoroquinol-one-carboxylic acid)	(a) C/Nj (unm.) (b) C/PO (+Na-DBS) (c) C/PO (+CTAB)	(a) CV, AdSV (b,c) DPV	(a) BR-B (pH 8.5), CPE s. activated by electrolysis; lin.r.: 6×10^{-7}–4×10^{-6} M (b) modif.: surfactant; LOD: 2×10^{-8} M (c) PhB (pH 7); lin.: 2×10^{-7}–2×10^{-5} M	(a) Urine (spiked s.) (b,c) pharm. forms. (com. tablets)	(a) [2531] (b) [2532] (c) [740]
Cisapride	Gastroprokinetic ag. (subst Cl,F-piperidin methoxy-benzamide)	CNTPE (ctng. MW-CNTs)	CV, DPV	Study on irr. oxidn. (2e−, H+, adsorp.); BR-B (pH 6.1), E_A = +0.89 V vs. ref.; lin.: 4×10^{-8}–2×10^{-5} M; RSD < 4%	pharm. forms. (tablets)	[392]
Clenbuterol (hydrochloride)	Bronchospasmolytic (subst. aminobenzen-methanol)	(a,b) C/Nj (+Nafion) (c) C/PO (+subst. CD)	(a) CV (b) DPAdSV (c) POT (dir)	(a) Study on oxidn. mechanism, modif., interfs. (b) 0.1 M NaOH, LOD: 1 nM (b) CD: cyclodextrin, lin.: 1 μM–1 mM	(a) Model solns. (b) Bovine urine (c) Human serum	(a) [1720] (b) [1721] (c) [2533]
Chloram-phenicol	Synth. antibioticum (Cl-Ph-acetamide d.)	C/PO, GC/PO (unm.)	LSV, DPV	GC: glassy carbon powder (globules); BR-B (pH 7), lin.: 6×10^{-7}–1×10^{-4} M	Model solns	[2534]
Chlordiaz-epoxide	Antipsychotic agent (benzodiazepin-oxide)	C/PO + Gl [+SLS, in situ]	CV, CSV	Gl: glycerol, SLS: Na-lauryl sulfate 0.01 M HCl (pH 2), LOD: 5×10^{-10} M	biol. fluids (human serum)	[235]
Chloroquine	Antimalarial agent (4-amino-quinoline)	(a) C/Nj (+dsDNA) (b) C/PO (+Cu^{II}-NWs)	CV, DPV	(a) Study on oxidn (b) NWs: nanowires LODs: 10 μM, par. detn. of *Primaquine*	(a) Human serum (b) comm. tablets	(a) [2535] (b) [1888]
Chlorphenir-amine (maleate)	Antiallergic and anti-histaminic dr. (used also in veterinary p.)	C/DOP [+Cm+ (SiW$_{12}$)−] C/DOP [+Cm+ B(Ph$_4$)−] [CP-ISEs]	POT (dir.)	Ion-exchange & ion-pairing principles; DOP: dioctylphthalate (as binder); s.e.: pH 4.5–7.7; lin.r.: 1×10^{-6}–0.01 M, 10 s	pharm. forms, serum, urine	[822]
Chlorpromazin (hydrochloride)	Antipsychotic agent, neurolepticum (phenothiazine der.)	(a) C/SG, S/SO (unm.) (b,c) C/PO (unm.) (d) C/SO, C/Nj (unm.)	(a–d) CV, DPAdSV (c) CP, CM	(a) pH 7.4 (BR-buff.), LOD: 1×10^{-7} M, no interfs. from AA, UA (b) Th. Study (c) o.c., 0.02–1 mg L^{-1}, study of adsorp. (d) s.e.: 2 M H$_2$SO$_4$ + MeOH (pH 0)	(a–c) Model solns. (a,b) biol. fluids (human urine, serum)	(a) [218,219, 220,732] (b) [2536] (c) [731,2537] (d) [211]

Chlor-tetracycline	Tetracycline antibiotic (from *Streptomyces* sp. antibiotic group)	(a) C/PO [+dsDNA+ Cu^{2+}] (b) C/PO (+Cel/dsDNA)	(a) DPASV (b) CE-AD	s.e.: PhB; LODs: 1–5 nM; simult. detn. with *Chlortetracycline*, *Oxytetracycline* and *Doxycycline*; (b) on-chip μ-detector	(a) Test solns. (in AcB) (b) Beef meat	(a) [2538] (b) [562]
Clofibric Acid	Anti-high cholesterol a. (Chlorophenol der.)	C/PO (unm.), CNTPE	CV, DPV (+SPEx)	lin.: 4.7 μM; 100× enrichment by solid-phase extraction made for real samples; simult. detn. with *Dcf*, *Ofx*, and *Prp*.	River water (spiked s.)	[2539]
Clozapine syn. *Clozaril*, *Azaleptin*	Antipsychotic agent, treat. schizophrenia (Cl-Me-piperazin-yl di-benzodiazepine)	(a) C/Nj (*Metrohm*), GC/ PO, CB/PO (b,c) C/PO (unm.) (d) s-CPE [C/Wx + HRP] (e) C/PO (+m-nS + HRP) (f) C/PO (+ds-DNA)	(a) CV (b) AdSV (c) Ad/ExSV (d–f) DPV	(a) Study with three types of CPE (made of carbon, colloidal graphite, carbon black) (b) s.e.: pH 3.2, MEx, 0.03–026 μg mL^{-1} (c) MEx, LOD: 4 × 10^{-9} M (d) BRB (4.5) (e) mdf.: magnetized n-silica + enzyme (f) ds: double stranded; LOD: 2 × 10^{-9} M	(a,e) Model solns. (b,f) Human serum (c) Human urine (c,d) pharm. tablets	(a) [187] (b) [827] (c) [731, 2537] (d) [315] (e) [2540] (f) [2541]
Cocaine	Narcoticum, abuse drug (alkaloid from *Eryhroxylon* sp.)	C/PO (unm.)	(a) CV, CM (b) FIA	(a) pH 9, LOD: 1 μM; CPE activated (b) LOD: 0.2 μM; CPE not regnt	Studies on irrev. oxidn. mechanism	(a) [2542] (b) [2543]
Codeine	Narcoticum (alkaloid, *Opium* sp.)	C/Nj (*Metrohm EA207C*) CG/PO (colloidal graphite) CB/PO (carbon black)	CV, LSV	Studies with various types of CPEs, incl. comm. marketed CP-mixture; inv. on reaction pathway, pH effect	Model solns.	[187]
Colchicine	Antirheumatic agent (alkaloid *Colchicum*)	(a) C/PO (unm.) (b) C/PO (+mg-*n*-Fe$_3$O$_4$)	(a) CV, ASV (b) CV, DPV	(a) Study on oxidn. (b) Albumin accum. 0.05 M H$_2$SO$_4$, lin.: 4 × 10^{-7}–0.001 M	(a) Tabs as raw material (b) Model solns.	(a) [2544] (b) [978]
Curcumin	Occasional antifever ag. (MeO-Ph-dye)	C/PO (unm.)	CV, DPV	Studies on oxidn. and redn. processes (also with HMDE), LOD: ca 1 × 10^{-6} M	Model solns.	[2545]
Cyclophosph-amide	Anticancer prodrug (of *Oxazophorine* gr.)	C/PO (+ss-, dsDNA)	CV, DPV	Studies on interaction with DNA; via oxidn. of guanine and adenine bases	Model solns.	[2546]
Dacarbazine	Antineoplastic drug (triazen-carboxamid)	C/PO (unm.)	CV, DPV	0.1 M HClO$_4$, pH 4.4; studies on reaction mechanism and kinetics	Model solns.	[2547]
Daunorubicin	Antileukemic agent (*Streptomyces* sp.)	(a) C/PO (unm.) (b) C/PO (+DNA)	(a) CV, AdSV (b) CV, CCSA	(a) pH 4.4 (acetate buff.); studies on (a) oxidn., (b) interaction with DNA	(a) Urine (b) Model solns.	(a) [2548] (b) [2549]

(continued)

TABLE 8.14 (continued)
Determination of Pharmaceuticals and Drugs at Carbon Paste Electrodes, Sensors, and Detectors: Survey of Methods and Selected Studies

Analyte (Chemical Form)	Classification (Chem. Specification)	Type of CPE (Modifier)/ [Configuration]	Technique (Mode)	Experimental and Electroanalytical Characterization, Selected Data; Notes	Sample(s)	Ref(s)
Dexamethasone	Antiinflammatory ag. (synt. cortico-steroid)	C/PO (unm.) C/PO (+β-cyclodextrin)	CV, DPV	Study on the inclusion effect of modif.; lin.r.: $4 \times 10^{-7} - 2 \times 10^{-5}$ M; simult. detn. with *Hydrocortisone* and *Prednisone*	pharm. forms, biol. fluids	[2550]
Dextro-methorphan	Anti-tussive drug and cough suppressant (MeO-Me-morphinan)	C/PO [+Q⁺ B(Ph)₄⁻]	POT, COU (dir.); FIA	Q: *Dex*-cation; evaln. of K_s(Dex) by COU; lin.r.: $1 \times 10^{-5} - 0.01$ M; $t_R < 2$ s; comparison with PVC-based ISE	pharm. forms.	[2551]
Diazepam	Antidepressivum (benzodiazepine)	(a) C/PO (unm.) (b) C/Nj (+DNA)	(a) CV, AdSV (b) DPV	(a) 0.05 M HCl, LOD: 5×10^{-8} M; redn. (b) 1 nM; biosensing comb. with accum.	(a) Human serum (b) Model solns.	(a) [2552] (b) [2553]
Diclofenac	Antiinflammatory a. (di-chlor-Ph-amino-phenyl-acetic acid)	C/PO (unm.) CNTPE	CV, DPV (+ SPEx)	Lin: 0.8 μM; 100 × enrichment by solid-phase extraction made for real samples; simult. detn. with *Cla*, *Ofx*, and *Prp*.	River water (spiked s.)	[2539]
Dicyclomine (hydrochloride)	Anticholinergic drug (bicyclo-carboxylate)	C/DOP (+IEx₍₁₎ + IEx₍₂₎)	POT, FIA (dir./titr.)	DOP: dioctyl phthalate; l.: (a) $1 \times 10^{-5} - 0.02$ M (b) 1 nM; accum. biosensing	Pure drug, biological fluids	[569]
Diethyl-stilbestrol	Antitumor drug (synthetic estrogen)	(a) C/Nj (*Metrohm*) [ft-D] (b) C/Nj (CPyB)	(a) LC-EC (b) DPASV	(a) Pre-extraction step, CH₃CN (50%)-based eluent, 2–5 ng mL; (b) pH-studies	Model solns. (injection samples)	(a) [2554] (b) [2555]
Diflunisal	Antiinflammatory d., pain reliever (subst. di-F-di-Ph-carb. acid)	C/PO (+Ca^II-MM)	SWASV	MM: *Montmorillonite* clay, s.e.: AcB (pH 5); LOD: 3×10^{-9} M; no extraction prior to analysis (three comm. drugs)	pharm. forms (e.g., *Dolozal®*), human serum	[1070]
Diphen-hydramine	Anti-histaminic drug (subst. di-Me-EtO-di-Me-ethanamine)	C/PO (+Dy^III-NWs) [ft-D]	FFT-SWV + FIA	NWs: nanowires, FFT: fast Fourier-tr. (increasing S/N ratio and suppr. noise); E_A = +1.1 V vs. ref.; LOD: 4×10^{-11} M	comm. drug, urine (spiked s.)	[1883]
Dipyrone	Antipyreticum (subst. aminomethane sulfonic acid)	(a) C/PO (unm.) (b) C/PO [+VO(*Salen*)] (c) C/PO (+Cu₃[Fe(CN)₆]₂)	(a) FIA-EC (b) CV, DPV (c) CV	(a) pH-effect; lin.: $5 \times 10^{-6} - 3 \times 10^{-4}$ M (b) modif.: bis-(salicylidene-iminato) oxovanadium(IV); LOD: 7×10^{-6} M (c) AcB+0.05M NaCl; LOD: 2×10^{-4} M	(a) pharm. forms. (b) Raw product (quality control) (c) pharm. forms.	(a) [566] (b) [2556] (c) [2498]

Dipyridamole	Anti-thrombus form: agent (di-piperidin-pyrimido-piperidine)	C/PO (+CTAB, in situ)	CV, COU, EIS, DPV	Study on irr. oxidn. and intermediates enhanced accum. by the modif. effect, $E_A = +0.53$ V vs. ref.; LOD: 0.01 μg L^{-1}	pharm. forms. (dosage tablets)	[1630]
Dobesilate (calcium)	Capillary/vein stab. (dihydroxy-benzene-sulfonate calcium)	(a) CILE [with (PMIM) PF$_6$] (b) C/PO (+La^{III}-nWs)	(a) CV, DPV (b) 2nd DeLV	(a) Catalytic effect of (RT)IL; s.e.: 0.05 M H$_2$SO$_4$; lin.: 8×10^{-7}–1×10^{-4} mol L^{-1} (b) nWs: nanowires, LOD: 5×10^{-11} M	(a) Capsules, dil. urine (b) Model solns.	(a) [460] (b) [2557]
Dobutamine (hydrochloride)	Cardiotonicum (subst. pyrocatechol)	C/PO (unm.)	AdSV	0.1 M H$_2$SO$_4$; 3×10^{-9}–1×10^{-6} M; studies on reaction mechanism	Model solns.	[2558]
Doxazosin	Antihypertensive a. subst. benzodioxin-carbonyl piperazine	(a,b) C/PO (Tenax, "C-8") (c) C/PO (+Nafion)	DPAdSV, SWAdSV	Studies on irrev. oxidn. and adsorpt. behavior; s.e.: (a,b) BR-B (pH 6.6) (c) PhB + citrate (pH 3.0); (a–c) LODs: 2–5×10^{-11} M; (a,b) CPE surface regnt.	(a–c) pharm. Forms. (tablets), human urine	(a) [2559] (b) [1723] (c) [1815]
Doxorubicin	Antineoplastic, anticancer ag. (antibiotic of Streptomyces gr.)	(a) C/Nj (unm.) (b) C/PO, GC/PO (unm.)	(a) AD-FIA (b) DPAdSV, LC-AD	(a) 0.2M AcB (pH 4), lin.r.: 1–1000nM (b) BR-B (pH 7), LODs: 3×10^{-9} M (DP) 4×10^{-7} M (AD), e..: PhB+ MeOH (75%)	(a) Human urine (b) Model. Solns	(a) [2560] (b) [681]
Doxycycline	Tetracycline antibiotic (from Streptomyces sp. antibiotic group)	C/PO (+dsDNA/cellulose [lab-on-chip μD]	CZE-AD	Interaction with dsDNA; LOD: 5 nM; simul. detn. with Chlortetracycline, Oxy-tetracycline, Tetracycline; met.st.add.	Beef meat	[562]
Droperidol	Antidepressivum (F-alkylketone)	C/PO (unm.)	CV, DPV	s.e.: 0.1 M H$_2$SO$_4$, LOD: 1×10^{-6} M; studies on irrev. oxidn. and adsorpt.	pharm. forms.	[242]
Drotaverine (hydrochloride)	Antispasmodic drug (analkaloid from the Papaverine group)	C/PO [+Q-Si(W$_3$O$_{12}$)$_4$] C/PO [+Q-B(Ph)$_4$] [CP-ISEs]	POT, FIA (dir., titr.)	Q.: Dro-cation; $S = 59.2 \pm 2$ mV dec^{-1}; interfs. of Men+, sugars, and AAs also studied; lin. range: 5×10^{-7}–0.01 M	pharm. forms. (turbid solns.)	[570]
s-Enalpril	Antihypertensive a. (subst. L-proline)	(a) C/PO (L-AmA oxidase) (b) C/PO (β-cyclodextrin) (c) CP-biosensor	(a) AD (stat.) (b) CP (c) AD-SIA	(a) Study on enantioselectivity effect (b) 4×10^{-5}–0.06 M; interfs. studies (c) 5×10^{-6}–0.01 M; simult. analysis	Forms., raw material (synthesis control)	(a) [2561] (b) [2562] (c) [2563]
Ephedrine (derivatives)	Sympathomimetic drugs (alkaloids from Ephedra sp.)	(a) C/Nj (+banana tissue) (b) C/VO (+C$_{18}$/SiO$_2$)	(a) EC-LC (b) CV, CA, SWAdSV	VO: vaseline oil (a) studies of enzyme processes; (b) pH 10, MEx, CPE activ. (b) LOD: 270 mg L^{-1} (c) LOD: 3 mg L^{-1}	(a) Model. solns. (b) Clinical samples (human urine)	(a) [2564] (b) [646, 1073]

(continued)

TABLE 8.14 (continued)
Determination of Pharmaceuticals and Drugs at Carbon Paste Electrodes, Sensors, and Detectors: Survey of Methods and Selected Studies

Analyte (Chemical Form)	Classification (Chem. Specification)	Type of CPE (Modifier)/ [Configuration]	Technique (Mode)	Experimental and Electroanalytical Characterization, Selected Data; Notes	Sample(s)	Ref(s)
Epirubicin	Antineoplastic (antibioticum from *Streptomyces* sp.)	C/PO (unm.) C/PO (+ss-, dsDNA)	CV, DPV	Interaction with double-stranded calf thymus DNA (hybridization indicator)	Model solns.	[2565]
Ethinylestradiol	Orally active medic. (semisynth. steroid)	C/PO (unm.) C/PO (+CPyB)	CV, LSV, DPV	Surfactant enhances hydrophobicity of the CPE surface; LOD: 3×10^{-8} mol L^{-1}	comm. tablets (*Levonorgestrel*®)	[1635]
Etoposide	Antineoplastic (antibioticum from *Streptomyces* sp.)	C/PO (unm.)	CV, LSV, DPV	Study of rev. elec. oxidn., BR-B (pH 3; pH > 4 gives to more compl. process) $E_A = +0.51$ V vs. ref.; LOD: 1×10^{-7} M	Human serum (spiked s.)	[2566]
Etidronate (sodium)	Skin cosmetic agent (biphosphonate der.)	C/PO (+n-Cu)	CV, LSV, DPV	el.-catalyzed oxidn. (via Cu^{III} → Cu^{II}) evaln of kinetic data (k_S, α, and D_X); lin.r.: 200–2500 mM; $R_R = 95\%$–105%	Model solns.	[960]
Fenoterol (hydrobromide)	Sympathomimetic a. (subst. phenyl amino-ethyl benzendiol)	(a) C/Nj (unm.; *Metrohm*) (b) C/Nj (+Nafion)	(a) COU (b) CV, DPV	(a) Pool-configured CPE ($d = 2$ cm) used for coulometric evaln. of number of e$^-$ (b) LOD: 1×10^{-8} M; simult.detn. of 3 d.	Model solns.	(a) [1719] (b) [1718]
Fentanyl (citrate)	Narcotic analgesic ag. (Ph-Et-piperidinyl-N-Ph-propanamide)	CILE ctng. GCP + OPy+BF$_4^-$)	CV, ECL	GC: glassy carbon powder, OPy: octyl pyridinium; ECL reaction with Ru^{III} sp. lin: 1×10^{-8}–1×10^{-4} M; RSD < 2% ($n = 10$)	Model solns., pharm. forms.	[486]
Flaxedil	Muscle relaxant (G-triethiodide)	C/PO (CTA$^+$I$^-$) [CP-ISE]	POT (dir., titr.)	pH 5–9, 2×10^{-6}–0.1 M; G: gallamine, titrants: BF$_4^-$, W(Mo$_3$O$_{10}$)$_4^{3-}$; no interfs.	Injection samples (ampoules)	[707,2567]
Flunitrazepam	Hypnoticum (benzodiazepin-one)	C/Nj (+*Bentonite*, clay)	CV, DPV	o.c., MEx: (1) AcB (pH 3.8) + (2) 0.5 M KNO$_3$ (pH 7); lin.: 0.2–4 µg mL^{-1}	Tablets (*Rohypno*®), human serum, urine	[1074]
Fluphenazine	Antipsychotic drug (phenothiazine der.)	C/PO (unm.)	DPV, CP, CM	o.c., 0.02–1 µg mL^{-1}; LOD: 5 ng mL^{-1} studies on adsorp. accum. mechanism	pharm. forms. (dosage tablets)	[731,2537]

s-Flurbiprofen	Antiinflammatory d. (aryl-propionic acid)	(a,b) C/Nj (α, β, γ subst-CDs) (c) C/Nj (+two diff. mcAbs) [CP-ESE]	POT, (dir.)	(a,b) accum. via inclusion and chiral effect by cyclodextrins; subst.: HO-N(Me)$_3$, LODs: $1{-}10 \times 10^{-9}$ M (c) mdf: macrocyclic antibiotics (ch. selectors)	(a,b) pharm. product (tablets Froben®) (c) Raw product (drug)	(a) [2568] (b) [2569] (c) [2570]
Gatifloxacin	Antibacterial agent (fluoroquinoline der.)	(a) C/PO (unm.) (b) C/PO (+β-cyclodextrin)	CV, DPV	(a) simult.detn. with *Lof* + *Spf*; (b) with *Lof*, *Nof*, and *Ofl*.: (a) LOD: 1×10^{-7} M (b) LOD: 2×10^{-8} M, s.e.: BRB (pH 4.0)	Dosage forms., human urine (dil. and spiked)	(a) [678] (b) [1418]
Glibenclamide	Antidiabetic agent (SO$_2$-urea hydrazide)	C/PO (unm.), C/PO (+*Sephadex*)	CV, DPV	Studies on adsor. accum. mechanism o.c., s.e.: 0.04M BRB (pH 2); 1×10^{-9}–5 $\times 10^{-7}$ M, LOD: 0.4nM ($t_{AC} = 180$ s)	pharm. forms. (tablets) human urine	[2571]
Gliclazide	Antidiabetic agent (SO$_2$-urea hydrazide)	C/PO (unm.)	DPV, CP, CM	Study on electrode oxidn. (two peaks), pH 2–12; lin.range: 1×10^{-8}–5 $\times 10^{-6}$ M	pharm. forms. (dosage tablets)	[2572]
Heroin	Narcoticum, abuse drug (alkaloid sp. from *Morphin* gr.)	(a) C/PO (unm.) (b) C/PO (+Ru(bpy)$_3$–*ZeY*) (c) CILE [ctng. (MIM) PF$_6$ + Ru^{III}-complex as reagt.]	(a) CV, LSV (b) FIA-ECL (c) ECL	(a) Study on oxidn. mechanism and pH-effect; lin. range: 1×10^{-7}–1 $\times 10^{-5}$ M (b) mdf.: altered zeolite Y, LOD: 1 µM (c) lin.: 2×10^{-9}–2 $\times 10^{-5}$ M, RSD < 5%	(a) Seized drug s. (b) Model solns. (c) Human serum (R_R = 94%–101%)	(a) [2573] (c) [502] (c) [502]
Honokiol	Antibacterial agent (nat. biphenol ctnd. in *Magnolia* flower)	C/PO (+mpSiO$_2$ and surf.)	CV, 2ndDeLV	Enhanced oxidn. peak (E = +0.3 vs. ref.) by cationic surfactant; pH 6.5, LOD: 0.5 µg L^{-1}; simult. detn. with *Magnolol*	pharm. preps, (Chinese med.)	[1671]
Hydrocortisone	Antiinflammatory ag. (nat. cortico-steroid)	C/PO (unm.) C/PO (+β-cyclodextrin)	CV, DPV	Study on the inclusion effect of modif.; lin.r.: 4×10^{-7}–3 $\times 10^{-5}$ M; simult. detn. with *Dexamethasone* and *Prednisone*	pharm. forms, biol. fluids	[2550]
Hydrochloro-thiazide	Antihypertension and first-line diuretic agent (subst. sulfonamide)	C/PO (+Fc-DCA)	CV, CA, SWV	Study on el.-catalytic effect of modif. on oxidn.; DCA: dicarboxylic acid.; pH 9; lin.r.: 8.0×10^{-8}–5.8 $\times 10^{-6}$ M	comm. drug, human urine	[1537]
Part 2: Substances "I–Z"						
Imipramine	Antidepressant (tricyclic der.)	(a) C/PO (PV-N-imidazole) (b) C/PO (fatty acid)	(a) CV, DPV (b) DPASV	(a) pH-, opt. and interfs. studies (b) pH 9, 0.1–8 µM; opt. + interfs.	(a,b) Dosage forms (b) Biological fluids	(a) [2508] (b) [1675]

(continued)

TABLE 8.14 (continued)
Determination of Pharmaceuticals and Drugs at Carbon Paste Electrodes, Sensors, and Detectors: Survey of Methods and Selected Studies

Analyte (Chemical Form)	Classification (Chem. Specification)	Type of CPE (Modifier)/ [Configuration]	Technique (Mode)	Experimental and Electroanalytical Characterization, Selected Data; Notes	Sample(s)	Ref(s)
Indomethacin	Antiinflammatory a. (subst. indole acetate)	(a) C/Nj (unm.) (b) C/Nj, C/SO (+castor oil)	(a) LSV, DPV (b) Ad/ExSV	(a) pH 8, s.e.: AmB + EtOH(40% w); LOD = 2ppm; (b) pH 7; 3×10^{-8} M	(a) Model solns. (b) Diluted urine	(a) [2574] (b) [729]
Indapamide	Diureticum (subst. sulfamylindole benzamide)	C/Nj (+castor oil)	CV, DPASV	pH 4.0, MEx, $5 \times 10^{-8}-1 \times 10^{-7}$ M; incl. studies on oxidn. process, exp. conds.	Model solns, spiked urine	[2575]
Inosine	Muscle functionality reactivator (O$_2$ transf.) (purin-one-furan der.)	C/PO (+La^{III}-NWs) [+Cu^{2+} + S$_2$O$_8^{2-}$, *in situ*]	CV, 2nd DeLSV	NWs: nanowires; ele.-catalytic oxidn., $E_A = +0.2$ V vs. ref., LOD: 8×10^{-10} M	pharm. forms., human serum	[2576]
Iproplatin	Antitumor agent [Pt(II) derivative]	C/Nj (*Metrohm*)	CV, DPV	Studies on reaction mechanism and pH, CPE compared with GCE and PtE	Model solns.	[2577]
Isoniazid	Anti-tuberculosis ag., DNA-intercalator (semisynthetic der., *Amycolatopsis* sp.)	(a) C/PO (unm.) (b) C/PO (unm.) (c) CNTPE (unm.) (d) CNT-CPE (+thionine)	(a) CV, SWV (b) CV, LSV (c,d) CV, DPV	(a,b) Study on irrev. oxidn., involv. 2e$^-$ + 2H$^+$; (a–d) lin: $1 \times 10^{-6}-0.001$ M; simult. detn. with: (a,b) *Rifampicin*, (c) AA,DA (d) AA and *Acetaminophen*	(a,b) com. tablets (*Rimactazid*®) (c) Model mixts. (d) Drug, blood plasma	(a) [2578] (b) [2579] (c) [383] (d) [438]
Isoprenaline	Sympathomimetic ag. (subst-R-benzen-diol)	C/PO (+Cu$_2$Fe(CN)$_6$)	CV, DPV	Studies on reaction mechanism and pH, $5 \times 10^{-8}-2 \times 10^{-4}$ M: comp. with UV-vis	pharm. forms.	[1315]
Khellin	Antiasthmatic agent (chromone der.)	(a) C/Nj (*Metrohm*) (b) C/Nj (+DNA)	CV	(a) Study of oxid. mech.; LOD: 0.8 μM (b) o.c., MEx, pH 7; l: $1 \times 10^{-8}-1 \times 10^{-7}$ M	(a) model soln. (b) Dosage form, human serum	(a) [2504] (b) [2580]
Ketoconazole	Antimycoticum (subs. piperazine)	C/PO (unm.)	CV, DPV	Studies on adsorp./extr. accum.; pH effect; lin.: $3 \times 10^{-8}-1 \times 10^{-5}$ M	pharm. forms., human urine	[730]
Lanzoprazole	Antibioticum (*Pivampicilin* sp.)	C/Nj (unm.)	CV, DPV	pH 6.0 (BR-buff.), LOD: 1×10^{-8} M; studies on oxidn. process, exp. conds.	pharm. forms. (capsules)	[2581]
Levodopa (syn. L-*Dopa*)	Anti-Parkinsonian ag. and dietary supplement (3,4-dihydroxy-L-phenylalanin)	(a) CPE (+PbO$_2$/PER) (b) CNTPE (+Ferrocene) (c) C/PO (+Dy^{III}-NWs)	(a) DPV (b) AD-FIA (c) FFT-SWV	(a) PER: polyester resin, LOD: 3×10^{-7} M simult. detn. of L-*dopa* and *Carbidopa* (b) Study on anodic and catalyzed oxidn. lin.: $2 \times 10^{-6}-5 \times 10^{-4}$ M, no s. treatment (c) FFT: fast Fourier-tr. (supprs. noise)	(a,c) pharm. forms. (com. tablets) (b,c) Human urine (c) Human serum	(a) [1241] (b) [420] (c) [1882]

Name	Description	Electrode	Technique	Notes	Sample	Ref.
Levofloxacin	Antimicrobial drug (fluoroquinolone)	C/Nj (+dsDNA)	CV, FIA-AD	5×10^{-7}–5×10^{-6} M; anodic-activated CPE; studies on interaction with DNA	Human urine	[2582]
Lidocaine	Local anesthetic and topical pain reliever (Et-amino-benzoate)	C/PO (+*Chloranil*)	CV, LSV	Undirect method (via the decrease in signal of modif.), lin: sub-μM level; simult. detn. with *Benzocaine*	Model solns.	[1507]
Lomefloxacin	Antibacterial agent (fluoroquinoline der.)	(a) C/PO (unm.) (b) C/PO (+β-cyclodextrin)	CV, DPV	simult. detn. with *Gaf* and *Spf*; (b) with *Gaf*, *Nof*, and *Ofl*.: (a) LOD: 1×10^{-7} M (b) LOD: 2×10^{-8} M, s.e.: BRB (pH 4.0)	Dosage forms., human urine (dil. and spiked)	(a) [678] (b) [1418]
Loprazolam	Hypnoticum (benzodiazepin-one)	C/Nj (unm.)	CV, DPV	pH 3–10, LOD: 2×10^{-7}, LOQ: 2×10^{-6} M; studies on irrev. oxidn. + adsorp.	pharm. forms. (*Somnovit*®)	[2583]
Magnolol	Antibacterial agent (nat. biphenol ctnd. in *Magnolia* flower)	C/PO (+mpSiO$_2$ and surf.)	CV, 2nd DeLV	Enhanced oxidn. peak ($E = +0.4$ vs. ref.) by cationic surfactant; pH 6.5, LOD: 10 g L^{-1}; simult. detn. with *Honokiol*	pharm. preps. (Chinese med.)	[1671]
Marcellomycin	Antineoplastic anti-bioticum (*Actino- sprangium* sp.)	C/Nj (lipids)	CV, DPV	Studies on accum. process, pH-effect, CPE compared with GCE and PtE; MEx, lin. range: 1×10^{-8}–2×10^{-6} M	Model solns. (spiked urine)	[1667,2584]
Mefenamic acid	Antiinflammatory a. (2,3-dimethyl-phenyl aminobenzoic acid)	C/PO (+La^{III}-NWs)	CV, 2nd DeLSV	NWs: nanowires; study of oxidn. (via 1e$^-$ and H$^+$) at effectively porous surf. given by NWs; lin.: 2×10^{-11}–4×10^{-9} M	Model solns.	[2585]
Meloxicam	Antiinflammatory a. (subst. benzothiazine carboxamide dioxide)	C/Nj (unm.)	CV, LSV	Studies on oxidn. mechanism and pH; BR-B (pH 3.0), lin: 5×10^{-7}–5×10^{-5} M	Dosage forms.	[2586]
Metaproterenol (sulfate)	Sympathomimetic a. (subst. benzendiol)	(a) C/Nj (unm.; *Metrohm*) (b) C/Nj (+Nafion)	(a) COU (b) CV, DPV	(a) Pool-configured CPE ($d = 2$ cm) used for coulometric evaln. of a number of e$^-$ (b) LOD: 2×10^{-8} M; simult.detn. of 3 d	Model solns.	[1719]
Metformin	Anti-diabetic drug for oral intake (biguanide der.)	(a) CNTPE (+Cu^{II}-MWs) (b) CNTPE (+Cu^{2+}, *in situ*) (c) C/PO [+Mt$^+$B(FPh)$_4$] [CP-ISE]	(a) 2nd DeLV (b) 2nd DeLV (c) POT (dir)	(a) MWs: molecular wires: Cu$_x$[bis(Py) propane]$_4$(2-methylacrylic acid)$_8$ H$_2$O, studies on irrev.oxidn. and mechanism LOD: 5×10^{-7} M; (b) AmB (pH 8.9); lin: 0.2–10 μM; (c) pH 4–8, LOD: 1×10^{-5} M	(a–c) Model solns.	(a) [373] (b) [2587] (c) [2588]
Met(h)adone	Anti-opiate agent (dimethylamino-di-phenyl-heptanone)	C/PO (unm.)	CV, DPV	Studies on el. reaction mechanism and r. kinetics, evaln. of k_S, α, D_X; LOD: 1×10^{-8} M; anodically pretreated CPE	Model solns.	[2589]

(*continued*)

TABLE 8.14 (continued)
Determination of Pharmaceuticals and Drugs at Carbon Paste Electrodes, Sensors, and Detectors: Survey of Methods and Selected Studies

Analyte (Chemical Form)	Classification (Chem. Specification)	Type of CPE (Modifier)/ [Configuration]	Technique (Mode)	Experimental and Electroanalytical Characterization, Selected Data; Notes	Sample(s)	Ref(s)
Methamphet-amine (hydrochloride)	Psycho-stimulant ag. (N-methyl-1-phenyl-propan-2-amine)	CNT-ILE (composite el.) [ctng. CNT + (MIM)PF$_6$]	CV, ECL	Studies on irrev. oxidn. and adsorption carbon moiety: multi-wall CNTs; both constituents in reaction with Met give ECL-signal; LOD: 8×10^{-9} M; RSD = 3%	Model solns.	[426]
Methimazole	Antihyperthyroidism ag. (subst. thioamide)	C/PO (+Co^{II}-SB)	CV, DPV	SB: Schiff base (new type), studies on catalytic effect; pH 6; LOD: 1×10^{-6} M	Clinical prep., pharm., human serum (synth.)	[1591]
Methotrexate	Antileukemic and anticancer/tumor agent (subst. aminopteroyl-L-glutamic acid)	(a) C/PO (unm.) (b) C/PO (+HPρCD)	(a) CV, DPV (b) SWASV	(a) pH 3 (BR-buff.), 2nM; (a,b) studies on redn. mechanism, pH-effect (b) Use of L- and D-enantioactivity; LOD: μM	(a) Model solns. (b) Com. tablets, injections (CRM)	(a) [2590, 2591] (b) [2592]
Metoclopramide (hydrochloride)	Gastroprokinetic and anti-migrene agent (sb. OMe-benzamide)	C/PO (+gly-PP)	CV, SWASV	Studies on oxidn. mechanism and accum. modif.: glycosylated porphyrin; s.e.: 0.4 M HCl + AcB; lin.: 0.07–270 μg L^{-1}	pharm. forms. (dosage tablets)	[2593]
Metronidazole (benzoate)	Antiprotozoal agent (imidazole-ethanol)	(a) C/PO (+gly-PP) (b) C/PO (+Zn^{II}-MHP-PP)	(a) SIA-AD (b) DPCV	(a) Studies on oxidn. mechanism, lin.: 6×10^{-8}–0.003 M; (b) MHP-PP: tetra- MeO-OH-Ph porphyrin, LOD: 4.4 μM	(a,b) pharm. forms. (c) Human urine	(a) [1580] (b) [2594]
Mifepristone	Glucocorticoid-based receptor antagonist (synth. steroid)	C/PO (unm.) C/PO (+ss-, dsDNA) [ft-D]	SIA-AD	2×10^{-7}–2×10^{-6} M; analytically useful interaction with ds-DNA only	pharm. forms. (tablets)	[2595]
Minodixyl	Hair-restoring agent (piperidin-pyrimidine)	C/Nj (+lipids)	CV, taV, pseudo-DPV	1×10^{-8}–1×10^{-6} M study on pH-effect, pKi evaluation, oxidn. process	pharm. forms	[1670]
Mitomycin-C	Chemotherapeutic ag. (subst. quino-aziridin methyl carbamate)	C/PO (+ss-, dsDNA)	DPV, SWV	detn. via the guanine signal; LOD: μM level; DNA/ C-SPE also tested.	Model solns.	[2596]
Mitoxanthron	Antibioticum (anthraquinone der.)	C/PO (unm.)	CV, DPV	pH 10 (BR-buff.), LOD: 2×10^{-7} M; studies on redn. mechanism, pH-effect	Model solns.	[2597]

Electroanalysis with Carbon Paste–Based Electrodes, Sensors, and Detectors

Moexipril (hydrochloride)	Antihypertensive dr. (subst. MeO-isoquinoline carboxylate)	C/PO (unm.) [+SDS, *in situ*]	CV, DPV	Enhanced accum. by surfactant SDS, ir. oxidn., BR-B (Ph 2–11); lin.: 4×10^{-7}–5×10^{-5} M; $R_R = 99.7$–100.8%, RSD $< 1\%$	pharm. forms (com. tablets) [703]
MPATD, MRATD	2-mercapto-5-(*R*)-phenylamino-1,3,4-thiadiazoles	C/PO (unm.) s-CPE (C/Wx)	CV, LSV	pH 1–10, 3×10^{-9}–1×10^{-4} M; studies on pH-effect, irrev. oxidn. process	Model solns., dosage forms, biol. media [79,2598]
Naltrexone	Antiopiate agent (morphin hydrazone)	C/PO (unm.)	(a) CV, DPV (b) FIA-EC	(a) 7×10^{-6}–1×10^{-4} M; pretreated CPE; (b) s.e.: 0.1 M HClO$_4$, lin. range: 2×10^{-8}–1×10^{-5} M (LOD: $< 1 \times 10^{-8}$ M)	pharm. forms (a) [2599] (b) [2600]
Naproxen (sodium)	Antiinflammatory ag. (methoxy-naphthalen 2-yl propanoic acid)	C/PO (+Dy^{III}-NWs)	CV, SWAdSV	NWs: nanowires (catalytic effect), s.e.: PhB (pH 7), lin.: 1×10^{-9}–5×10^{-4} mol L^{-1}; simult. detn. with *Acetaminophen*	pharm. forms (com. tablets) [1887]
Natamycin	natur. antifungal drug (of *Streptomyces* gr.)	C/Nj (unm.)	CV, DPV	(a) 7×10^{-6}–1×10^{-4} M; pretreated CPE (b) s.e.: 0.5 M H$_2$SO$_4$ (pH 1.8); lin. r.: 2×10^{-6}–8×10^{-5} M; comp.to UV/vis-S	pharm. forms (capsules) [2601]
Neomycin	Antibioticum (from *Streptomycin* sp.)	C/Nj (+RuO$_2$)	CV, LC-EC	Studies on electrode mechanism and catalytic effect; CP-detector compr. to GCE; lin.range: 1×10^{-6}–0.001 M	Model solns. [1219]
Nicarpidine	Cerebral vasodilator (subst. pyridindicarboxylate ethylester)	C/SO (unm.)	CV, DPV	pH 7.1 (BR-buff.), LOD: 2×10^{-8} M, MEx, studies on extr. accumulation	Model solns. [2602]
Nicergoline	Cerebral vasodilator (subst. Br-nicotin carboxylate lester)	C/PO (unm.)	CV, AdSV	Study on oxidn. process; o.c., MEx; pH 7; lin. range: 1×10^{-6}–1×10^{-5} M; LOD: 0.8 µM;	(a) Model soln. (b) Dosage forms. (doped serum) [2603]
Nifuroxazide	Antibacterial agent (furanyl-hydrazide)	(a) C/PO (unm.; + polymer) (b) C/Nj (unm.; + *Sephadex*)	(a) AdSV (b) CV, DPV	(a,b) Studies on pH-effect, irrev. oxidn., modif. effects; (a) LOD: 10 ng mL^{-1}	(a) Human urine (b) Model solns., urine (a) [2604] (b) [2605]
Nitrendipine	Antihypertensive a. (pyridindicarboxylate dialkylester der.)	C/PO (+β-cyclodextrin)	CV, DPV	Studies on, oxidn. process, pH-effect, opt. of modif. content; LOD:	Model solns. [1414]

(continued)

TABLE 8.14 (continued)
Determination of Pharmaceuticals and Drugs at Carbon Paste Electrodes, Sensors, and Detectors: Survey of Methods and Selected Studies

Analyte (Chemical Form)	Classification (Chem. Specification)	Type of CPE (Modifier)/ [Configuration]	Technique (Mode)	Experimental and Electroanalytical Characterization, Selected Data; Notes	Sample(s)	Ref(s)
Norfloxacin	Antibacterial agent, chemotherapeutic (subst. F-quinoline)	(a) C/PO (unm.) (b) C/PO (+β-cyclodextrin)	CV, DPV	accum. effect of β-CD, BRB (pH 4.0); simult. detn. with *Gaf, Lof, Oft,* & *Spf*. (a) LOD: 12×10^{-7} M; (b) LOD: 2×10^{-8}	Dosage forms., hum. urine (spiked)	[1418]
Novobiocin	Antibioticum (saccharide-rel. gr.)	C/Nj (+RuO$_2$)	CV, LC-EC	Studies on catalytic effect; CP-detector compr. with GCE; r: 1×10^{-6} – 0.001 M	Model solns.	[1219]
Ofloxacin	Antibacterial agent, chemotherapeutic (subst. F-quinoline)	(a) CNTPE (b) C/PO (unm.) and C/PO (+β-cyclodextrin)	CV, DPV (+SPEx)	(a) $100 \times$ acc. by solid-phase extraction, simult. detn. with *Cla, Dcf,* and *Prp*. (b) simult. detn. with *Gaf, Lof, Nof, Spf*.	(a) River water (spiked) (b) hum. urine (spiked)	(a) [2539] (b) [678, 1418]
Omeprazole	Secretion inhibitor (sulfinyl-H-benzimi-dazole der.)	C/Nj (unm.)	CV, DPV	pH 6.0 (BR-buff.), LOD: 3×10^{-8} M; studies on oxidn. process, exp. conds.	pharm. forms. (capsules)	[2581]
Oxytetracycline	Tetracycline antibiotic (from *Streptomyces* sp. antibiotic group)	(a) CNTPE [+dsDNA] (b) C/PO [+dsDNA + Cu^{II}] (c) C/PO (+Cel/dsDNA)	(a) SWASV (b) DPASV (c) AD-CE	(a–c) interaction with dsDNA/cellulose + Cu^{2+}, LOD: 1–5 nM; (a–c) simult.detn. with *Chlortetracycline, Doxycycline* & *Tetracycline*; (c) on-chip μ-detector	(a) Hatchery fish tissue (b) Model solns. (AcB) (c) Beef meat	(a) [390] (b) [2538] (c) [562]
Paracetamol	syn. *Acetaminophen*					
Pantoprazole	Secretion inhibitor (SH-benzimidazole)	C/Nj (unm.)	DPAdSV	Studies on accum. mechanism, interfs.; o.c.: BR-B (pH 4); LOD: 2×10^{-8} M	pharm. forms. (dosage tablets)	[2606]
s-Perindopril	Antihypertensive a. (subst. ethoxy-indol carboxylic acid)	(a) C/Nj (β-cyclodextrin) (b) C/Nj (αβ-CD) [CP-ESEs]	POT, CP	accum. via inclusion and chiral effect by CD: lin.: 1×10^{-5} – 0.01 M; mech. Renewal of CPE, interfs. studies	(a) Model solns. (b) pharm. forms.	(a) [2607] (b) [2608]
s-Pentopril	Antihypertensive a. (subst. N-indol pentanoic acid ethylester)	C/PO (AAOd)	AD-SIA	modif.: L-aminoacid oxidase; enantio-selectivity principles; pH 6.8–7.4; lin. range: 0.001–100 μM	Model samples (quality control)	[747,2563]

Electroanalysis with Carbon Paste–Based Electrodes, Sensors, and Detectors 319

Analyte	Electrode	Technique	Conditions / Notes	Sample	Refs.
D-Penicillamine	(a) C/PO (+CoSLf) (b) C/PO (+Fc-CA) (c) C/PO (+Fc-subst FL) (d) C/PO (+HCF, in situ) (e) C/PO (+n-TiO₂ + Qz)	(a–c) DPV (b–d) CV, CA (d) DPV (e) SWV	(a) modif.: Co^II-Salophen (Schiff base); s.e.: 0.1 M PhB (pH 3); LOD: 1 × 10⁻⁷ M; (b–d) modif.: subst. fluoren-ferrocenes, study of catalytic oxidn. (pH 7); LOD: 6 × 10⁻⁶ M (d) lin.r.: 8 × 10⁻⁶–2 × 10⁻⁴ M (e) Qz: quinizarine; simult.detn. of Try	Antirheumatic drug with chelator funct. (amino-3-Me-3-sul-fanyl-butanoic acid) (a) Drug (raw product) (b,c) pharm. capsules (c) Human serum (d) pharm. preps. e) Model mixts. (at sub-μM c. level)	(a) [1582] (b) [880] (c) [1544,2609] (d) [2610] (e) [1880]
Pefloxacin	C/Nj (+dsDNA)	CV, DPASV	Interaction of double-stranded DNA + Pex; study on irrev. oxidn. lin: 10 nA–10 μM, RSD < 5 % (n = 5)	Chemotherapeutic ag. (subst. fluoro-piperazine-quinolinic acid) Human urine (dil. s.)	[2611]
Perphenazine	C/MO (unm.)	DPV, CP, CM	o.c., 0.02–1.0 μg mL⁻¹; LOD: 5 ng mL⁻¹ studies on adsorp. accum. mechanism; rare use of CPE for CM-measurement	Antipsychotic drug (phenothiazine der.) Dosage forms. (tablets)	[731,2537]
Phenothiazine, Phenoxazine and related derivatives*	(a) C/SG, C/PO (unm.) (b) C/SO, C/Nj (c) C/PO (+DNA) (d) C/PO (Zr-phosphate) (e) C/Nj, s-CPE (C/Wx) (both as CPEE conf.)	(a,b) CV, DPV, AdSV (c) CV, SP (CCSA) (d,e) CV, LSV, DPV	SG: silicone grease, Wx: paraffin wax; (a,d) Study on adsorp./extr. mechanism. interfs. by UA and DA; (b) 2 M H₂SO₄ + 30% MeOH; (c) interactions with DNA; (e) exps. by solid state electrochemistry*	Antipsychotic drugs (neuro transquilizers) (native phenothiazine, luoromethyl phenot- hiazine, phenoxazine) (a) Human urine (d) Model substs. (in solid state) (d) Model substs. (in solid state) (e) Model solns.	(a) [187,732, 2537,731] (b) [211] (c) [1480, 2553] (d) [1286, 2612] (e) [314]
Phenacetine	(a) C/PO (unm.) (b) C/PO, C/PE	CV	PE: polyethylene; studies on pH, irrev. oxidn. process; form. of metabolites.	Analgeticum (acet-p-phenetidine) Model solns.	(a) [2495] (b) [187]
Phenazopyridine (hydrochloride)	C/Nj (unm.)	CV, SWV	Study on el. redn.; s.e.: 0.5 M H₂SO₄ (pH 0.5), lin.r.: 3 × 10⁻⁸–3 × 10⁻⁶ M	Local analgesic effect (with selective use) Dosage tablets, urine (spiked)	[2613]
Phenylephrine	C/PO (+ion-exchanger)	CV, LSV	Study on electrode oxidn. and accum. mechanism; s.e.: 0.1 M HCl (pH 1)	Sympathomimetic a. (subst. aminomethyl-benzenemethanol) Model solns.	[1714]
Phenobarbital	C/PO (1-octanol + Nafion®)	CV, SWV	E_p = −0.92 V vs. ref; LOD: 25 nM; studies on analyte immunoassay via Coc⁺-labeled analyte (as tracer)	Antiepileptic/hypnotic (5-ethyl-phenyl-pyrimidine-trione) Clinical samples (human serum)	[1558,1559]

(continued)

TABLE 8.14 (continued)
Determination of Pharmaceuticals and Drugs at Carbon Paste Electrodes, Sensors, and Detectors: Survey of Methods and Selected Studies

Analyte (Chemical Form)	Classification (Chem. Specification)	Type of CPE (Modifier)/ [Configuration]	Technique (Mode)	Experimental and Electroanalytical Characterization, Selected Data; Notes	Sample(s)	Ref(s)
Phenytoine	Antiepileptic agent (diphenyl-imidazolidine-dione)	C/PO (1-octanol + Nafion®)	CV, SWV	$E_p = 0.39$ V vs. ref. LOD. 50 nM; studies on analyte immunoassay, incl. dual detection of PhB and PhT	Clinical samples (human serum)	[1556,1558, 1559]
Physcion	Antifungal agent (subst. methoxy-anthraquinone der.)	C/PO (unm.)	CV, SWASV	Anodic oxidn. in presence of ambient O_2; s.e.: AmB (pH 10.5); lin.: 2×10^{-10} – 4×10^{-9} M ($t_{ACC} = 120$ s); RSD < 3%	Medicinal plant (*Polygonum multi* sp.)	[677]
Piribedil	Antifungal agent (subst. piperazine- pyrimidine der.)	C/TCP (+Pir-PMo$_3$O$_{12}$) [CP-ISE]	POT (dir.) (stat./FIA)	Ion-exchange and ion-pairing in liquid phase; $S = 58.4 \pm 0.6$ mV dec^{-1}; lin.: 8×10^{-7} – 0.001 M; interf. st. (Ss, AAs)	pharm. forms., human urine	[2614]
Piroxicam	Antiinflammatory, antirheumatoid ag. (subst. benzothiazine carboxamide dioxide)	(a) C/Nj (*Metrohm*) (b) C/PO (+fatty acids) (c) C/Nj (surfactants) (d) C/PO [+analyte(s)] (e) CNTPE (+ MW-CNTs)	(a) CV, DPV (b,c) DAdSV (d) CV (e) CV, DPV	(a) PhB + MeOH(10%), LOD: 2 μM (b) LOD: 1 nM; interf. study (AA,UA) (c) pH 2, r.: 40–100 μM, modifs. effect (d) LOD: 0.5 g kg; CPEE-arrangement (e) st. on catal. oxidn.: r: 0.15–5 μg mL^{-1}	(a–c) Model solns. (d,e) Dosage tablets (*Feledene* capsules, *Piroxicam* gel)	(a) [187, 2615] (b) [2616] (c) [1643] (d) [2617] (e) [387]
Prazosine (hydrochloride)	Antihypertensive a. (quinazolin-pip. de.)	(a) C/PO (+Nafion) (b) C/PO (+"C-8")	(a) DPAdSV (b) SWAdSV	(a) pH 6.0 (BR-buff.), LOD: 3×10^{-11} M; (b) pH 2.6, 8×10^{-10} M: accum. studies	Tablets (*Minipres*®); human urine	(a) [1724] (b) [1814]
Prednisolone	Antiinflammatory ag. (nat. cortico-steroid)	C/PO (unm.) C/PO (+β-cyclodextrin)	CV, DPV	Study on the inclusion effect of modif.; lin.r.: 6×10^{-7}–1×10^{-5} M; simult. detn. with *Dexamethasone* and *Hydrocortisone*	pharm. forms, biol. fluids	[2550]
Primaquine	Antimalarial agent (4-amino-quinoline)	C/PO (+Cu^{II}-NWs)	CV, DPV	NWs: nanowires, LOD: 0.25 μg mL^{-1}, pH 5.5, simult.detn. with *Chloroquine* (with R_R = 95% and RSD < ±5%)	comm. tablets	[1888]
Procarbazine (hydrochloride)	Synth. anti-brain cancer drug (iso-Pr-Me-hydrazino benzamide)	C/PO (+unm.) [as flow-through det.]	LC-AD	detn. via electroactive degradation pr.; m.p.: PhB + MeOH (pH 7); LOD: 2 ng (in 1 μL-cell), comp. with UV-detector	Human urine, plasma (untreated)	[2618]
Procaine (hydrochloride)	Anesthesicum (local) (subst. aminoethanol)	C/PO (+pumice)	DPV	Studies on catalyzed electrode oxidn.; s.e.: PhB (pH 6.9); LOD: 5×10^{-8} M	Injection samples, human urine	[2618]

Proflavine	Bacteria disinfectant (diaminoacridine der.)	C/Nj (unm.)	CV, SWAdSV	Study on electrode oxidn. and redn.; also HMDE; $E_A = +0.2$, $E_C = -0.9$ V vs. ref. lin.r.: $1.3 \times 10^{-6} - 1.2 \times 10^{-5}$ M; $r^2 = 0.998$	Bovine serum	[679]
Promethazine	Antihystaminicum (phenothiazine der.)	(a) C/PO (lipids, fatty acids) (b) C/Nj, s-CPE [+anal(s)]	(a) CV, DPV (b) LSV	(a) pH 9, LOD: 1×10^{-10} M; (a,b) studies on pH, adsorp./extr. accum., modifs. (b) Studies in the CPEE configuration	(a) Human serum, urine (b) Model substances (in solid state)	(a) [1030, 1666] (b) [314]
Propranolol	Anti high-cholesterol agent (chlor-phenol-phenothiazine der.)	(a,b) C/PO (unm.) (b) CNTPE (unm.) (ctng. MW-CNTs)	(a) CV, DPV (b) (+SPEx)	(a) Study of pH effect (BR-B, pH 2–11) lin.: $6 \times 10^{-7} - 5 \times 10^{-5}$ M; (b) s. enrichment by solid-phase extraction, lin: 0.5 μM (b) simult. detn. with *Cla*, *Dcf*, and *Ofx*.	(a) Human serum, urine (b) River water (spiked)	(a) [2620] (b) [2539]
Propylthio-uracil	Antihyperthyroidism (a thioamide der.)	C/PO (+Co-Cl-*SLF*)	CV, DPV	modif.: Co^{II}-4-chlor-*Salophen* (*Schiff* base); study on el.-catalyzed oxidn.; lin.r.: $8 \times 10^{-6} - 8 \times 10^{-4}$ M; RSD < 4%	Human serum	[1588]
s-Ramipril	Antihypertensive ag. (subst. indol carboxylic acid alkylester)	C/PO (+AAOd)	SIA-AD	modif.: L-aminoacid oxidase; enantioselectivity principles; pH 6.8–7.4; lin. range: 0.001–100 μM	Model solns.	[2561, 2563]
Repaglinide	Antidiabetic agent (subst. peperidin-Ph-Bu-EtO-benzoic acid)	C/PO (unm.)	SIA-AD	Study on irrev. oxidn. (BR-B, pH 2–11) lin.r.: $8 \times 10^{-7} - 3 \times 10^{-6}$ M; RSD = 0.6%–1.4%	com. tablets, human serum	[2621]
Rifampicin	Anti-tuberculosis and anti-leprosy ag. (der. of *Amycolatopsis* sp.)	(a) C/PO (+ss, dsDNA) (b) C/PO (unm.) (c) C/PO (+Dy^{III}-NWs)	(a) CV, DPV (b) CV, SWV (c) SWAdSV	(a) Studies on interaction of Rif + DNA, MEx: AcB (pH 5) → PhB (pH 7.4) (b) simult.stud. and detn. with *Isoniazid* (c) NWs: nanowires; pH5, LOD: 0.5 nM	(a) Model solns. (b) com. tablets (*Rimactazid*®) (c) Capsules, urine	(a) [2622] (b) [2578, 2579] (c) [1884]
Salbutamol	Muscle relaxant (phenylethanol der.)	(a) C/Nj (unm.; *Metrohm*) (b) C/Nj (+Nafion)	(a) COU, (b) CV, DPV	(a) Pool-configured CPE ($d = 2$ cm) used for coulometric evaln. of number of e^- (b) LOD: 3×10^{-8} M; simult.detn. of 3 *d*	Model solns.	[1719]
Sildenafil (citrate)	For treating erectile dysfunction (subst. pyr-pym-piperazine)	DPEs (three types)	CV, SWV	DPs: diamond pastes made of nat. and two synth. powders ($p = 1$ and 50 μm) $E_A = +0.18$ V; lin.: $1 \times 10^{-12} - 1 \times 10^{-8}$ M	pharm. forms. (tablets *Viagra*®)	[354]

(continued)

TABLE 8.14 (continued)
Determination of Pharmaceuticals and Drugs at Carbon Paste Electrodes, Sensors, and Detectors: Survey of Methods and Selected Studies

Analyte (Chemical Form)	Classification (Chem. Specification)	Type of CPE (Modifier)/ [Configuration]	Technique (Mode)	Experimental and Electroanalytical Characterization, Selected Data; Notes	Sample(s)	Ref(s)
Sparfloxacin	Antibacterial agent (fluoroquinolone de.)	(a) C/PO (unm., +β-CD) (b) C/PO (unm.)	CV, DPV	(a) CD: cyclodextrin; acc. Mechanism lin.: 0.04–60 μM; (b) r.: 2×10^{-7}–6×10^{-5} M; simult. detn. with Gaf and Lof	pharm. forms, diluted urine	(a) [1417] (b) [678]
Spiperon	Antidepressivum ag. (fluoro-alkylketone)	C/PO (unm)	CV, DPV	Study on irrev. oxidn. and adsorption; s.e.: 0.1 M H_2SO_4, LOD: 1×10^{-6} M	pharm. forms.	[242]
Spiroplatin	Antitumor drug [platinum(II) der.]	C/Nj (*Metrohm*)	CV, DPV	Study on electrode oxidn. and redn., incl SEM; sub-millimolar level of analyte	Model solns.	[2623]
Streptomycin	Antibioticum (from *Streptomyces* sp.)	C/Nj (+RuO$_2$) [ft-D]	CV, LC-EC	1×10^{-6}– 0.001 M: studies on catalytic effect, CP-support compr. with GEE	Model solns.	[1219]
Sumatriptan (succinate)	Anti-migraine agent (Me$_2$-amino-Et-indol methan-sulfonamide)	C/Nj (unm.)	COU	Studies on electrode reaction and evaln. of n(e$^-$); "giant" CPE (Ø = 20 mm), detn. with GC-D (in LC-EC mode)	Pure. preps. (dosage tablets)	[206]
Tamoxifen (citrate)	Anti-breastcancer ag. (di-Ph-PhO-N,N′-di-methyl-ethanamine)	C/PO (unm.)	LSV, 2nd DeLV	Study on irrev. oxidn., AcB + MeOH (15%v), $|E_A|$ = +1.1 V vs. SCE; lin.r.: 7×10^{-10}–3×10^{-8} M: RSD < 3%	pharm. forms. (com. tablets)	[2624]
Tanshinone II-A	Anti-atherosclerotic a. (trad. nat. Chinese d.)	C/PO (unm.)	CV, 2nd DeLSV	Study of rev. eld. reaction (via: e$^-$ + H$^+$) s.e.: BR-B + EtOH (40% v/v; pH 2.4); lin.: 1×10^{-8}–2×10^{-7} M; CP: mech.rgnt	pharm. forms. (pills *Danshen*®)	[2625]
Tenoxicam	Antiinflammatory a. (subst. thiazinecarbo- xamide dioxide)	C/PO (unm; +fatty acids)	CV, DPAdSV	0.1 M BR-buff., LOD: 1 nM; adsor., modif., and interfs. studies (AA, UA)	Model solns.	[2616]
Tetracycline	Tetracycline antibiotic (from *Streptomyces* sp. antibiotic group)	(a) C/PO [+dsDNA + Cu^{2+}] (b) C/PO (+Cel/dsDNA)	(a) DPASV (b) AD-CE	(a,b) Interaction with dsDNA/cellulose + Cu^{2+}, LODs: 1–5 nM; simult. detn. with Chlortetracycline,Oxytetracycline and Doxycycline; (b) on-chip μ-detector	(a) Test solns. (in AcB) (b) Beef meat	(a) [2538] (b) [562]

Electroanalysis with Carbon Paste–Based Electrodes, Sensors, and Detectors   323

Theophylline	Anti-asthmatic agent (methyl-xanthine rel. to *Caffeine*)	(a) C/PO (+*n*-CoPC) (b) C/PO (+K$_2$PtCl$_6$)	(a) CV, DPV (b) AD-LC	Study on catalytic effect and formation of three metabolites; E_A = +0.17 V vs. ref. LODs: 1.1 µg L^{-1} (Thp), 11 µg L^{-1} (mbs)	(a) [932] (b) [2626]
Thioridazine	Antidepressant drug (tricyclic heteroder.)	C/PO (+b-CD)	CV, DPV	Inclusion ("host–guest") accum. Effect by cyclodextrin; lin.: nM; also simult. detn. with *Imipramine, Trimipramine*	[1412]
Thymol	Disinfectant agent (isopropyl-*m*-cresol)	GC/Nj (unm.)	DPV, AD-HPLC	Pastilles (*Septolete*®), thyme syrup	[2627]
Tifluadome	Analgeticum (opiod benzodiazepine der.)	C/PO (+"C-18")	CV, DPV	modif.: com. ion-pair/exchange resin o.c., BR-B (pH 6), 2 × 10^{-7}–5 × 10^{-6} M; studies on adsor.; no sample treatment	[1809]
Tizanidin	Muscle relaxant benzothiazole gr.)	C/Nj (unm.; *Metrohm*)	CV, COU, DPV	Study on irrev. oxidn; e.: PhB + MeOH (pH 7); lin.r. 2 × 10^{-5}–1 × 10^{-4} M (DPV)	[2628]
Todralazine (hydrochloride)	Antihypertensive *a*. (subst. phthalazine-hydrazine der.)	C/Nj (+silica gel)	CV, DPV	o.c., pH 1.8 (0.1 M KNO$_3$), 3 × 10^{-8} M; studies on adsor, mechanism, modif.	[1152]
Tramadol hydrochloride)	Analgesic, severe pain reliever (synth. der. from *Opioid* group.)	C/NPOE (+Tr- PW$_3$O$_{12}$) C/-": [+Tr-H$_3$(SiMo$_{12}$)] [CP-ISEs]	POT (dir.)	Ion-association and chem. equilibrium NPOE: nitrophenyl-octyl ether; t_R < 5 s, lin.r.: 6 × 10^{-6}–0.1 M; R_R = 90%–95%	[2629]
s-Trandorapril	Antihypertensive *a*. (subst. *N*-indol carboxylic acid alkylester)	C/PO (+AAOd)	AD-SIA	modif.: L-aminoacid oxidase; enantio-selectivity principles; pH 6.8–7.4; lin. range: 0.001–100 µM	[747]
Trazodone (hydrochloride)	Antidepressant (piperazinyl]-propyl-1,2,4-triazol-pyridine)	C/PO (unm.)	CV, COU, DPV	Study on pH, pK$_s$, k$_s$ (for irrev. oxidn), form. of metabolites; lin.range: 1 × 10^{-6}–1 × 10^{-4} M (for DPV mode)	[2630]
Trifluperazine	Antipsychotic drug (phenothiazine der.)	C/PO (unm.)	DPV, CP, CM	o.c., 0.02–1 µg mL^{-1}; LOD: 5 ng mL^{-1} studies on adsor. accum. mechanism	[731,2537]
Trimipramine	Antidepressant (subst. dibenzoaze- pine propanamine)	(a) C/PO (unm) (b) C/Nj (+β-CD)	DPV	Dosage forms. (tablets) Model solns.	(a) [2631] (b) [1412]
					(continued)

TABLE 8.14 (continued)
Determination of Pharmaceuticals and Drugs at Carbon Paste Electrodes, Sensors, and Detectors: Survey of Methods and Selected Studies

Analyte (Chemical Form)	Classification (Chem. Specification)	Type of CPE (Modifier)/ [Configuration]	Technique (Mode)	Experimental and Electroanalytical Characterization, Selected Data; Notes	Sample(s)	Ref(s)
Triprolidine (hydrochloride)	Antihistaminic agent (methyl-phenyl-pyr-rolidin-enyl-pyridine)	C/PO (+Trp⁺BF₄⁻) [CP-ISE]	POT (dir.)	CP-ISE together with other two ISEs; S_{CP} = 54 mV/dec; pH 4.7–8.5, 2×10^{-5}–0.01 M; detn. by st.add. method	pharm. forms, human urine	[2632]
Troponin I	Cardiac stimulator (a part of macromolecular complex)	C/PO (+three diff. MSs)	CV, ASV	MSs ... mol. sieves: *Zeolite-Y*, *SBA-15*, and *MCM-41*; antibody "sandwich" immuno-reaction; lin.: 0.5–5.0 ng mL⁻¹	pharm. forms, human urine	[1844,1846, 2633,2634]
Troxerutin	Vein dilatation drug (nat. flavonoid der.)	C/PO (+cl-PV-Pyrr)	CV, DPV	modif.: cross-linked electrodeposited polyvinyl-pyrrolidone; s.e.: 0.1 M KCl, LOD: 5×10^{-9} mol L⁻¹ (o.c., t_{ACC} = 300 s)	pharm. forms, (tabs *Cilkanol®*)	[1686]
Urapidil	Sympatholytic, antihypertensive agent (piperazo-pyrimidine)	CNTPE (cntg. MW-CNTs)	CV, DPAdSV	Study on oxidn., involving 2e⁻ + 2H⁺ and adsorp., BR-B (pH 6.8), E_A = +0.6 V vs. ref.; LOD: 4×10^{-8} M, t_{ACC} = 60 s.	comm. tablets	[379]
Xipamide	Diureticum agent (sulfoamide der.)	C/Nj (*Metrohm*)	CV, DPV	MEx, pH 2.5, MeOH(10%)-ctng. s.e., LOD: 5×10^{-7} M; with pre-separation	pharm. forms,. urine-spiked sample	[2635]
Zovirax	syn. *Acyclovir*					

* Chlorpromazine, fluphenazine, perphenazine, and promethazine given extra.

TABLE 8.15
Analysis of Biologically Important Compounds at Carbon Paste–Based Electrodes: Alcohols

Analyte	Modifier	Carbon Paste	Techniques	Comments	Ref(s)
Alcohols, Diols	Cu_2O	G/mo	CV, FIA (+0.55V)	0.1 M NaOH; LOD 0.2–28 ng	[1225]
Alcohols Methanol Ethanol Propanol-1 Propanediol Glycerol Ethylene glycol	RuO_2	G/mo	CV, HDV, A, FIA 80.4 V vs. Ag/AgCl	Catalytic oxidation, maybe via Ru(VI); NaOH (1 M)	[1220]
Alcohols	$[Ru(4,4'-Me_2bpy)_2(PPh_3)(H_2O)](ClO_4)_2$	C/mo	CV	PBS $\mu = 0.2$ M, pH 6.8	[2642]
Benzyl alcohol					
Benzyl alcohol	Diaqua-M-oxotetrakis (2,2'-bipyridine) diruthenium(IV)/molecular sieve	C/Nj	CV	Electrocatalytic oxidation via Ru(V); NaOH (1 M)	[2643]
Benzyl alcohol	Ru-terpyridine phenanthroline complex/Dowex W50x8		CV	Electrocatalytic oxidation via Ru/IV); NaOH (1 M)	[2644]
Choline	Poly(ethylene glycol)-choline oxidase + 1,1'-dimethyl Fc + membrane		A (0.3 V vs. SSE)	Biosensor, ox., regeneration of enzyme via Fc^+, Tris buffer 0.1 M, pH 8	[1531]
Choline	HRP + BSA + GA + phenothiazine + dialysis membrane + choline oxidase	G/p wax (solid)	A (0 V vs. SSE)	det. via H_2O_2; PBS 0.1 M, pH 7.4; LOD 0.1 μM; s: phosphatidyl choline in rat saliva (with phospholipase D)	[2645]
Ethanol	Ni-NP on carbon fibers (electrospun from PAN and carbonization)/mo	Carbon fiber paste	CV; A (0.55 V vs. SSEI)	Enzyme-free sensor; electrocatalytic oxidation; LOD 0.25 mM; s: liquors	[139]
Ethanol Methanol	Ni-DMG	G	CV	Catalytic oxidation 1 M NaOH	[2646]
Ethanol	Bis(1,10-phenanthroline-5,6-dione) (2,2'-bipyridine) ruthenium(II) in zirconium phosphate + ADH	G/mo	CV	PBS ($\mu = 0.1$); ADH and NAD^+ in solution; via catalytic oxidation of NADH	[1299]

(continued)

TABLE 8.15 (continued)
Analysis of Biologically Important Compounds at Carbon Paste–Based Electrodes: Alcohols

Analyte	Modifier	Carbon Paste	Techniques	Comments	Ref(s)
Ethanol	—	C/Kel-F wax	FIA (0.75 V vs. SSE)	Carrier: 0.05 M tris-HCl buffer (pH 8) + ADH + PEG + NAD$^+$ LOD 0.44 mM; s: wines	[319]
Ethanol	NAD$^+$ immobilized with octanal	—	LSV	Enzymatic reduction of the co-substrate and oxidation of ethanol	[52]
Ethanol	ADH + NAD$^+$	—	A (0.7 V vs. SSE)		[2647]
Ethanol	ADH + NAD$^+$	MWCNT/mo	A (0.4 V vs. SSE)	Amperometric biosensor PBS 0.05 M, pH 7.4; s: wines	[364]
Ethanol	ADH + NAD$^+$ + BSA	G/po, dodecane, hexadecane		Study of pasting liquids; s: beer, wine	[214] [2648]
Ethanol	ADH + NAD$^+$ + poly (o-phenylene diamine)	G/po	V, A (0.15 V vs. SSE)	Amperometric biosensor PBS 0.1 M, pH 8.5; s: cider, wine, whiskey	[1772]
Ethanol	ADH + NAD$^+$ + poly (o-aminophenol), poly (o-phenylenediamine)		CV, FIA (0.15 V vs. SSE)	Amperometric biosensor PBS 0.1 M, pH 8.5; LOD 0.51 µM	[1786]
Ethanol	ADH + NAD$^+$ + fumed SiO$_2$		A		[1000]
Ethanol	QH-ADH + ruthenium complexes	G/so	CV, CA	Different biosensor designs PBS 0.05 M, pH 7.0	[1601]
Ethanol	ADH + NAD$^+$ + phen-dione complexes of Fe(II) and Re	G/po	CV, A (0 V vs. SSCE)	Amperometric biosensor PBS 0.1 M, pH 8.0 LOD 45 µM	[1599]
Ethanol	ADH + NAD$^+$ + poly-(vinylferrocene)	G/po	A (0.7 V)	Amperometric biosensor LOD 0.39 mM; s: wines	[2649]
Ethanol	ADH + toluidine blue + NAD$^+$ + dialysis membrane		A (0.05 V vs. SSE); FIA	Amperometric biosensor; LOD 10 µM; s: wines	[784] [1474]
Ethanol	ADH (GA/BSA) + NAD$^+$ + meldola blue adsorbed on silica gel/Nb-oxide	G/mo	A (0 V vs. SCE)	Biosensor; PBS 0.1 M, pH 7.8; LOD 8 µM; s: wine, beer, spirits	[1197]
Ethanol	ADH + NAD$^+$ + meldola blue + dialysis membrane	CNT paste	A (0.05 V vs. SSE)	Amperometric biosensor LOD 0.1 µM; s: beverages	[372] [783]

Ethanol	ADH/PEG + NAD⁺ + 7-dimethylamine-2-methyl-3-β-naphtamido-phenothiazinium chloride		A (0.3 V vs. SSE)	Biosensor PBS 0.1 M, pH 7.5 LOD 50 µM s: wine, beer, sake, soy sauce	[1752] [2650] [2651]
Ethanol	γ-ADH + NAD⁺ + PEI+phenothiazine polymer	G/po	FIA (0.1 V vs. SSE)	Reagentless amperometric biosensor LOD 2 µM	[1750]
Ethanol	ADH + NAD⁺ + methylene green + poly(ester sulfonic acid) cation exchange membrane		A	Reagentless amperometric biosensor	[1471]
Ethanol	Alcohol oxidase + HRP + ferrocene		FIA (−0.05 V vs. SSE)	Biosensor	[2652]
Ethanol	Alcohol oxidase + HRP + PEI	G(heat treated)/ phenylmethyl so	FIA; detector in HPLC	Amperometric biosensor	[2653–2657]
Methanol Ethanol Methanol n/i-Propanol Butanol	ADH + HRP + GA + PEI + poly(o-phenylene diamine) + cation exchange membrane		FIA; detector in HPLC	PBS 0.05 M, pH 8; s: plasma, urine Reagentless biosensor	[2658]
Ethanol	Ethanol oxidase			Review	[2659]
Ethanol Methanol	Alcohol oxidase + Os-wired HRP + PEI	G/so	FIA (−0.05 V vs. SSE)	Amperometric biosensor 0.1 M KCl; LOD 10 µM	[1755]
Ethanol	Yeast			Biosensor via ADH LOD 0.14 mM s: beer	[2660]
Ethanol Alcohols	Yeast		FIA	Amperometric biosensor LOD 2 µM; s: alcoholic beverages	[2661]
Ethanol Methanol t-Butanol i-Amylol	Yeast			Biosensor via ADH; s: fermentation broth	[2662]
Ethanol Propanol Butanol	Tomato seeds (soaked, crushed)	G/mo	CA (0.6 V vs. SSE)	Amperometric biosensor; PBS 0.05 M, pH 7.4, NAD⁺, K₃[Fe(CN)₆] 1 mM lod 7 µM (allyl alcohol); 54 µM (ethanol)	[2663]

(continued)

TABLE 8.15 (continued)
Analysis of Biologically Important Compounds at Carbon Paste–Based Electrodes: Alcohols

Analyte	Modifier	Carbon Paste	Techniques	Comments	Ref(s)
Allyl alcohol					
Ethanol	*Acetobacter pasteurianus* whole cells + nylon net	G/p	CV, A (0.5 V vs. SSE)	Biosensor McIlvaine buffer pH 6.0 + $K_3[Fe(CN)_6]$ 20 mM	[2664]
Ethanol	*Acetobacter aceti* + dialysis membrane		A (0.5 V vs. SSE)	Biosensor; McIlvain buffer (pH 6) + 0.5 mM benzoquinone	[780]
Ethanol Glutathione	Pyranose oxidase		A (0.9 V vs. SSE)	Biosensor via detection of H_2O_2 and inhibition of pyranose oxidase by ethanol or glutathione	[2665]
Glycerol	Glycerol dehydrogenase + NAD^+ + poly(o-phenylene diamine) (electropolym.)	G/silicone grease	A (0.15 V vs. SSE)	Catalytic effect by ox. products from NAD^+; PBS, pH 10	[654]
Methanol	NiNb/Pt, Pt-Sn, Pt-Ru amorphous alloys		CV, CA	Electrocatalytic oxidation	[2666, 2667]
Methanol	Ni in poly(m-toluidine)	G/p	CV	0.1 M NaOH; electrocatalytic oxidation probably to formate	[841]
Methanol	Pt or Pt/Ni microparticles in poly(indole) film	G/p	CV, IS	Electrocatalytic oxidation; $HClO_4$ (1 M)	[134]
Methanol	Pt-doped Co hexacyanoferrate	G/p	CV	Electrocatalytic oxidation; KH_2PO_4 (0.1 M)	[2668]
Methanol ethylene glycol	Poly(o-aminophenol) or poly(1,5-diamino–naphthalene) + Ni(II)	G/so	CV, CA	electrocatalytic oxidation via NiOOH NaOH (0.1 M)	[131, 2669–2671]
Methanol Alkyl alcohols	Ni(II)-quercetin	MWCNT/po	CV, LSV	0.1 M NaOH: catalytic oxidation LOD 5 mM	[424]
Thiols	Sulfhydryl oxidase on gelatine	G/mo	A (0.7 V vs. SSE), AD with HPLC	Biosensor; PBS 0.05 M, pH 7.5	[2672]

TABLE 8.16
Analysis of Biologically Important Compounds at Carbon Paste Electrodes: Aldehydes, Ketones, and Carboxylic Acids

Analyte	Modifier	Carbon Paste	Techniques	Comments	Ref(s)
Aldehydes					
Aldehydes, aliphatic	Pd	G/po	SE, CV, DCV, A (−0.32 V vs. SCE)	Catalytic reduction at −0.3 V vs. SCE; NaOH (0.006 M) + NaClO$_4$ (0.1 M)	[1268]
Aldehydes, aromatic	[Fe(CN)$_5$L]$^{3-}$ (L = 4-aminopyridine)	G/ceresin wax	oc, acc, DPV	Formation of Schiff base; acc in etOH/water, measurement in CH$_3$CN/bu$_4$NClO$_4$ (0.1 M); red. (benzaldehyde imine) at −0.63 V vs. SSCE	[1021]
Ketones					
Pyruvate	Pyruvate oxidase + HRP + trehalose/lactitol + poly(amino acids)	G/po	FIA (−0.05 V vs. SSE)	Bienzyme biosensor Phosphate/citrate puffer 0.1 M, pH 7 LOD 5 µM	[781]
Pyruvate	Pyruvate oxidase + methylene green		A	Biosensor; PBS 0.1 M, pH 7 s: human sweat	[1468]
Pyruvate	Pyruvate dehydrogenase + vit. K3 + dialysis membrane		A (0.1 V vs. SSE)	Biosensor; enzymatic oxidation PBS, pH 7.0	[2673]
Acids					
Glycolate	Glycolate oxidase + dimethylferrocene + polycarbonate membrane	Synthetic g/p	A (0.22 V vs. SSE)	Biosensor; Tris buffer, pH 8.3	[854]
Glycolate	Glycolate oxidase + dimethylferrocene on siloxane polymer	G/po	CV, A (0.3 V vs. SCE)	Biosensor	[856]
Glycolate	Spinach leave tissue + ferrocene	G/mo	CV, A (0 V vs. SSE)	Plant tissue biosensor via glycolate oxidase and peroxidase; PBS 0.05 M, pH 7.5 LOD 1 µM; s: human urine	[2674]
Glycolate	Spinach leave tissue	G/mo	Potentiostatic 0.8 V vs. SSE	ECL biosensor via luminol; luminol (0.25 mM) + KNO$_3$ (0.05 M), pH 9; LOD 15 µM	[2675]
Glycolate	Sunflower leave tissue + ferrocene	G/mo	FIA (0 V vs. SSE)	Plant tissue biosensor via glycolate oxidase and peroxidase; PBS 0.05 M, pH 7.5 LOD 1 µM; s: human urine	[2676]

(continued)

TABLE 8.16 (continued)
Analysis of Biologically Important Compounds at Carbon Paste Electrodes: Aldehydes, Ketones, and Carboxylic Acids

Analyte	Modifier	Carbon Paste	Techniques	Comments	Ref(s)
Lactate	NAD+ immobilized with octanal		LSV	Enzymatic reduction of the co-substrate and oxidation of lactate	[52]
Lactate	Lactate dehydrogenase + NAD+ + fumed SiO$_2$		A		[1000]
Lactate	Lactate dehydrogenase + GA + BSA + NAD+ + meldola blue on silica gel/Nb-oxide	G/mo	A (0 V vs. SCE)	Reagentless amperometric biosensor; PBS 0.1 M, pH 7; LOQ 0.1 mM; s: blood plasma	[1217]
L-Lactate	L-Lactate dehydrogenase + ferrocene + diaphorase + NAD+ + BSA + GA	G/vaseline	CV, FIA (0 V vs. SSE)	Biosensor; PBS 0.1 M + KCl 0.1 M, pH 7	[875]
L-Lactate	L-Lactate dehydrogenase + GA + BSA + toluidine blue O/+ NAD+	G/po	FIA (0 V vs. SSE)	Biosensor; PBS 0.1 M + KCl 0.1 M (+ NAD+ 1%), pH 7	[1475]
D-Lactate	PEI + D-lactate dehydrogenase + NAD+ + toluidine blue O redox polymer	G/po	FIA (−0.05 V vs. SSE)	Biosensor; PBS 0.1 M, pH 7 LOD 20 µM	[2677]
Lactate	Lactate dehydrogenase + NAD+ + poly(o-aminophenol), poly(o-phenylenediamine)		CV, FIA (0.15 V vs. SSE)	Amperometric biosensor PBS 0.1 M, pH 8.5 LOD 0.75 µM	[1786]
D-Lactate	D-Lactate dehydrogenase + polym.-toluidine blue O	G/po	A	Reagentless biosensor for monitoring fermentation process	[937,2678]
D/L-Lactate	D-Lactate dehydrogenase + mediator + NAD+/L-lactate oxidase + HRP	G(act)/so	FIA	Biosensors; simultaneous determination with flow-split	[180]
D-Lactate	PEI + D-lactate dehydrogenase + alanine aminotransferase + NAD+ + toluidine blue O redox polymer	G/po	FIA (−0.05 V vs. SSE)	Biosensor; PBS 0.1 M, pH 7	[2679]
Lactate	Lactate dehydrogenase + glutamic pyruvic transaminase + NAD+	G/po	CV, A (0 V vs. SSE)	Biosensor PBS 0.1 M + glutamate 0.01 M, pH 9.5 s: cider	[1771]
L-Lactate	Lactate dehydrogenase + cytochrome c + asolectin		A (0.15 V vs. SCE)	Reagentless biosensor; LOD 1 µM	[1672]

D-Lactate	Lactate cytochrome c oxidoreductase + Reinecke salt phenazine methosulfate	G/po	A (0 V vs. SCE)	Biosensor; PBS 0.05 M + NaCl 0.06 M, pH 7 LOD 56 μM; s: light beer	[2680]
Lactate	Lactate dehydrogenase + diaphorase + 2-methyl-1,4-naphthoquinone + dialysis membrane		A	Biosensor	[855]
Lactate Succinate	Cells or membrane vesicles of Paracoccus denitrificans + mediator + dialysis membrane	G/po	A	Biosensor via mediated ox. of enzymes of respiratory chain; soluble or insoluble mediators; membrane-bound dehydrogenase in vesicles; PBS 0.05 M, pH 7.3	[1520]
Lactate	Lactate oxidase + PEI + Rh	Rh-C/mo	CV, A, FIA (0 V vs. SSE)	Biosensor; PBS 0.05 M, pH 7.4 LOD 15 μM	[647]
Lactate	Lactate oxidase + CoPC or CoPC(OEt)$_8$ or Pt		A (0.65 or 0.45 or 0.6 V vs. SSE)	Biosensor; PBS 0.1 M, pH 7	[1567]
L-Lactate	L-lactate oxidase + HRP + PEI	G(act)/so	FIA (0 V vs. SSE)	Biosensor; PBS 0.1 M, pH 7; LOD 3 μM	[246]
L-Lactate	Lactate oxidase + (HRP)ox+ ferrocene	G/po	FIA (−0.1 V vs. SSE)	Biosensor s: fermentation reactor of milk	[1527]
Lactate	Lactate oxidase + (HRP)ox + ferrocene	G/po	FIA	Biosensor; PBS 0.1 M + KCl 0.1 M, pH 7.2: ascorbate interference study	[2681]
Lactate	Lactate oxidase + GA + BSA + ferrocene derivatives	G/po	A (0.7 V vs. Pt pseudoreference)	Biosensor; PBS 0.1 M, pH 8.9	[2682]
L-Lactate	L-Lactate oxidase + ferrocenes or phenothiazines (Meldola blue) (+ Nafion membrane)		A(0.2–0.4 V vs. SCE [ferrocenes], 0.05–0.25 V [Mb])	biosensor pH optimum, pH 7.9 s: serum	[863]
Lactate	Lactate oxidase + NAD⁺	MWCNT/po	CV, A (−0.1 V vs. SSE)	Biosensor; PBS 0.05 M, pH 7.4; via reduction of H$_2$O$_2$; LOD 0.3 mM	[364,661]
L-Lactate	L-lactate oxidase in poly (1,3-diaminophenylene) rescorcinol + TTF		A	Amperometric biosensor LOD 56 μM s: milk, yogurt	[1489]
Lactate	Lactate oxidase + PB	GC(act)/mo	A (0 V vs. SSE)	Biosensor PBS 0.1 M + KCl 0.1 M, pH 7	[196]
L-Lactate	Lactate oxidase + methylene green	G/po	A (0.15 V vs. SSE)	Biosensor; PBS + KCl (0.1 M), pH 7.0 s: goat whole blood	[1470]
Lactate	Lactate oxidase + peroxidase + trehalose/lactitol + PEI/polylysine	G(act)/po	FIA	Reagentless biosensor	[2683, 2684]

(continued)

TABLE 8.16 (continued)
Analysis of Biologically Important Compounds at Carbon Paste Electrodes: Aldehydes, Ketones, and Carboxylic Acids

Analyte	Modifier	Carbon Paste	Techniques	Comments	Ref(s)
Lactate	Baker's yeast + nile blue, methylene blue, toluidine blue, prussian blue, meldola blue	G/po	CV, A (0.1/0.3 V vs. SSE)	Biosensor; yeast as a source for cytochrome b2 PBS, pH 7.2	[1483]
L-Lactate	Baker's yeast (pretreated with meOH or etOH) + ferricyanide	G/po	CV, A (0.3 V vs. SSE)	Biosensor with yeast as a source of flavocytochrome b2 PBS + LiCl 0.1 M, pH 7.3; LOD 3–6 μM	[2685]
Lactate	Yeast *Hansenula anomala* + mediators	G/Nj or po	A (0.05–0.3 V vs. SCE)	Biosensor via cytochrome b2 PBS, pH 7.6–7.7	[2686]
Malate	Toluidine blue O PEI polymer + malate de-hydrogenase + NAD$^+$	G/po	CV, FIA (0 V vs. SCE)	Biosensor, redox polymer, direct e−-transfer,; PBS 0.25 M, pH 7	[938]
Organic acids Oxalic Tartaric Malic Fumaric Citric Succinic	Pd NP		CZE	As detector; LOD < 2 μM s: fruit juices	[2687]
Oxalic acid	Pd NP electrodeposited		CV	Catalytic ox. via Pd(II); H$_2$SO$_4$ 0.1 M; LOD 20 μM	[971]
Oxalic acid	Carbon nanofibers + Pd NP	G/mo	CV, DPV	Decrease of overpotential via ox. of Pd; HClO$_4$ 0.1 M; LOD 0.2 mM; s: spinach	[2688]

Analyte	Modifier	Paste*	Technique	Notes	Ref.
Oxalic acid	spongy Os			Catalytic ox.	[2689]
Oxalic acid	PbSO$_4$	G/po	acc. (o.c.), me, SWV	acc. due to lower solubility product; BRP 0.04 M, pH 8; LOD 0.3 mg L^{-1} (acc. 5 min); s: fruit juices	[811]
α-Keto acids	CoPC		A (0.75 V vs. SSE); as detector in HPLC	LOD 0.3 pmol (oxalic acid) s: serum, urine	[885,2690]
Oxalic acid	CoPC	G/wax (solid)	CV	Catalytic ox.; PBS 0.05 M, pH 7.4	[2691]
Oxalic acid	CoPC		AD in CE		[557]
Oxalic acid	CoPC in situ on SiO$_2$		CV, DPV, CA (0.71 V vs. SCE)	Catalytic ox.; KCl 1 M; LOD 1 mM	[1172]
Oxalic acid	CoPC on SiO$_2$/3-n-propylimidazole	G/mo	CV, DPV	Catalytic ox. via Co(III); ABS, pH 3.5 (+NaCl 0.1 M)	[15561]
Oxalic acid	Sulfonated CoPC on silsesquioxane on Al-oxide or phosphate	G/mo	CV, DPV, CA	Catalytic ox. via Co(III); KCl 1 M; LOD 18 μM (Al$_2$O$_3$) or 10 μM (AlPO$_4$)	[1822, 2692]
Oxalic acid	[Ru(bpy)$_3$]$_2$SiW$_{12}$O$_{40}$ · 2 H$_2$O NP	G/silicone grease	CV, ECL	H$_2$SO$_4$ (CV), ABS 0.1 M, pH 4.5 (ECL)	[1017]
Oxalic acid	Oxalate oxidase		A (−0.1 V vs. SCE)	Biosensor via det. of H$_2$O$_2$; s: urine, beer	[2693]
Oxalic acid	Oxalate oxidase + HRP on SiO$_2$/TiO$_2$/toluidine blue + GA	G/po		Biosensor, det. via H$_2$O$_2$; succinate buffer 0.1 M + KCl 0.4 M, pH 3.6; s: spinach	[1182]
Salicylate	Salicylate hydrolase	G/po	A (0.35 V vs. SCE)	Biosensor via release of salicylic acid, ox. to quinine; PBS, pH 7.6	[2694]
Salicylate	Salicylate hydrolase + Meldola blue on SiO$_2$		A (0.2 V vs. SCE)	Biosensor, det. by ox. of released catechol by ox. via mediator	[1174]

TABLE 8.17
Analysis of Biologically Important Compounds at Carbon Paste Electrodes: Amines and Other Amino-Compounds

Analyte	Modifier	Carbon Paste	Techniques	Comments	Ref(s)
Amides and Amines					
Acrylamide	Hemoglobin (adsorbed from DDAB vesicles)	G/p	CV, SWV	Biosensor; via adduct with Hb; decrease of red.curr.of Hb by acrylamide; LOD: 13 pM s: food	[2420]
Amines Ethylenediamine Diethylenetriamine	Cu_2O	G/mo	CV, FIA (+0.55 V vs. Ag/AgCl)	0.1 M NaOH LOD 0.2–0.3 ng	[1225]
Amines Amino naphthalene	Polyvinylpyridine hexafluorophosphate+$[Fe(CN)_5L]^{3-}$ L = pyridine-4-carboxaldehyde	C/ceresin wax + po	SWV	via Schiff base; irrev. ox. at around +0.9 V vs. SCE	[1743]
Amines Ephedrines	CoPC		FIA Pulsed potential (0.3 V 220 ms, −0.3 V 100 ms)	0.1 M NaOH	[902]
Amines Biogenic Synthetic	Modified triazine-Silasorb + immob. amine oxidase (pea seedlings) + HRP		A (0 V vs. Ag/AgCl)	Biosensor H_2O_2 detection via HRP	[2695]
Amines Biogenic amines spermidine	Pea seedling tissue (diamine oxidase)	G/mo		Biosensor H_2O_2 detection at 0.9 V vs. AgCl LOD 7.1 nmol spermidine	[274]
Aniline	Sepiolite		Preconcentration (pH 6.9), DPV (pH 1.5)	LOD 15 ng mL^{-1} s: beverages	[2424]
Benzocaine, lidocaine	p-Chloranil	G/mo	SWV	BRB, pH 1 Indirect det. via decrease of p-chloranil signal	[1507]
Benzylamine, benzydamine	Monoamine oxidase in sol–gel layer	C/epoxy	CV, FIA	Biosensor PBS, HEPES buffer (pH 7.4) s: mouthwashes	[1167]

Biogenic amines	Phthalocyanine CoPC, LnPC, LuPC, GdPC	G/Nj	CV, SWV	Comparison with SPEs, GC chemometric evaluation s: fish	[2696]
Biogenic amines	HRP cross-linked with GA and CDI			Biosensor s: rat blood	[2697]
Biogenic amines	CNT/polyaniline				[2698]
Cadaverine spermine	Amine oxidase and HRP on Sepharose + Fc	G/silicone grease	A (0.2 V vs. SCE)	Biosensor; HEPES 0.05 M + KCl 0.05 M, pH 7	[589]
Histamine	Amine dehydrogenase + ferrocene			Biosensor	[2699]
Phenylethylamine, tyramine	CNT		CV	Catalytic oxidation of amines	[2700]
Penicillamine	TiO_2 NP + quinizarine	G/p	CV, CA, SWV	PBS 0.1 M, pH 7.0; peak at 0.5 V vs. Ag/AgCl; simultaneous det. of try possible; LOD 0.76 µM s: drugs	[1880]
Polyamines	Oat seedlings + Fc	G/mo	A, FIA (0 V vs. SSE)	Tissue as a source of polyamine oxidase and peroxidase, Fc for red. of enzyme; PBS 0.1 M, pH 7.4	[864]
Polycyclic aromatic amines	α-, β-, γ-cyclodextrin	G/vaseline	CV, DPV, adASV	BRB, pH 7	[726,1413]
Urea	Urease + glutamate dehydrogenase		A (1.1 V vs. SSE)	Indirect det. via decrease of ox. of NADH; PBS 0.1 M + α-ketoglutarate 0.4 mM + NADH 0.4 mM, pH 7.2	[2701]
Amino acids					
N-acetyl cysteine	Caffeic acid		CV, DPV, A	LOD 0.17 µM	[2702]
N-acetyl cysteine	Fe(III)-oxide (core) Co(III)-hexacyanoferrate (shell) nanoparticles	C/Nj	CV, LSV, A (890 mV vs. Ag/AgCl)	LOD 205 nM (LSV), 21 nM (A); in tablets	[983]
N-acetyl cysteine	CNT + Co(III)-salophen	G/Nj	CV, DPV	LOD 50 nM; in pharmaceutical preparations	[403]
N-acetyl cysteine	10-methylphenothiazine		CV, LSV	LOD 0.8 µM (LSV)	[2703]

(continued)

TABLE 8.17 (continued)
Analysis of Biologically Important Compounds at Carbon Paste Electrodes: Amines and Other Amino-Compounds

Analyte	Modifier	Carbon Paste	Techniques	Comments	Ref(s)
Amino acids Glycine Aspartic acid Glutamic acid Cysteine Tyrosine	Copper micro- and nanoparticles		CV, FIA, BIA	Electrocatalytic oxidation via Cu(III); 0.1 M NaOH; operating pot. 720 mV vs. Ag/AgCl	[963]
Amino acids	Cu nanoparticles			PBS, pH 8	[966]
Amino acids	Cu micropartices	CNT/mo	SWV A	0.05 M PBSr, pH 7.4; 0 V vs. Ag/AgCl; LOD 10^{-4}–10^{-5} M	[2704]
Amino acids Glutamine Proline Tyrosine Valine	Cu powder	C/po	FIA 0.12 V vs. Ag/AgCl	PBS 0.002 M; pH 7.0	[2705]
Amino acids	Cu_2O	G/MO	CV, FIA (+0.55V)	0.1 M NaOH LOD 0.4–4 ng	[1225]
Amino acids	MWCNT + Cu_2O	C/po	CV; amperometric detection in CZE	PBS, pH 7.6; 0.75 V vs. SCE	[401]
Amino acids	Cu_2O, RuO_2, NiO, CoO	G/mo	FIA (0.45 V vs. SSE)	Four sensor array; statistical evaluation of responses; NaOH 1 M	[2706]
Amino acids	CuO, Cu_2O, NiO, CoO, PbO_2, Ag_2O_2, MnO_2, Bi_2O_3, PdO	G/vaseline	CV; detector in CE	0.05 M NaOH	[556]
Amino acids	RuO_2	G/oil	FIA; 0.45 V vs. Ag/AgCl	KOH 1 M	[1223]
Amino acids	Co_3O_4	G/oil	FIA; 0.55 V vs. Ag/AgCl	KOH 1 M	[1223]
Amino acids	Cu(II)-cyclohexylbutyrate			LOD 10^{-6} M	[1607]
Amino acids Phenylalanine	L-Amino acid oxidase	C-Ir/mo	FIA	Biosensor Improved thermal stability of enzyme by CP entrapping	[291,2707]
D-, L-Amino acids	D-, L-Amino acid oxidase, Ir-dispersed			Biosensor via H_2O_2; LOD 10^{-5} M	[1264]

D-, L-Amino acids	D-, L-Amino acid oxidase + HRP + PEI		FIA	Biosensor; via H_2O_2 and HRP	[2708]
Amino acids	CoPC		FIA; pulsed potential (0.45 V 220 ms, −0.3 V 100 ms)	0.1 M NaOH	[902]
Amino acids	CoPC	G/po	A detector in microchip CE; dual electrode	Post-column derivatization to Cu(II) complex; ox. to Cu(III) at first el., red. at second; oligopeptides can be detected as well	[558]
L-Amino acids	L-Amino acid oxidase + Fc derivatives + dialysis membrane		A	Biosensor	[854]
D-Amino acids	D-Amino acid oxidase + ferrocene + PVC membrane	G/fatty acid FAD on top of G electrode	A	Biosensor 700 mV vs. Ag/AgCl 0.05 M PBS, pH 8.4	[2709]
D-Amino acids	D-Amino acid oxidase + peroxidase Electrospun carbon nanofibers			Biosensor	[179, 2710]
Amino acids		G/mo	CV, DPV	Electrocatalytic oxidation at 0.65 zo 0.75 V vs. Ag/AgCl; PBS 0.1 M, pH 7	[121]
Tryptophan Tryptophan Tyrosine					
Amino acids Cysteine Homocysteine Glutathione	Crown ethers		SV	Host–guest complex LOD 2–5 × 10⁻⁸ M	[1394]
Cysteine		CILE	CV	PBS, pH 7.0 ox. at +49 V vs. Ag/AgCl LOD 2 μM s: soy milk	[2711]
Cysteine		FeC–nanocmposite + po	CV		[2712]
Cysteine	Bi-powder	G/so	SWV; SWCSV	Acetate buffer 0.1 M, pH 4.7 LOD 0.3 μM (t_{ACC} 600 s)	[2713]
					(continued)

TABLE 8.17 (continued)
Analysis of Biologically Important Compounds at Carbon Paste Electrodes: Amines and Other Amino-Compounds

Analyte	Modifier	Carbon Paste	Techniques	Comments	Ref(s)
Cysteine	Cu_2O	G(arc)/mo	CV, LSV	Peak at −650 mV (probably reduction of the Cu(I)-SR-complex) Borate buffer, pH 9.2 LOD 2 nM	[1316]
Cysteine	Co(II,III)-oxide	G/mo	FIA (0.4 V vs. SSE)	Catalytic ox.; PBS 0.1 M, pH 7 or NaOH 0.1 M	[1230]
Cysteine Cystine	RuO_2				[2714]
Cysteine	$V^{IV}O$-salen	G/mo	CV, LSV, A (0.8 V vs. SCE)	LOD $1.7 \cdot 10^{-4}$ M (A); also for N-acetylcysteine, glutathione and 6-thiopurine	[1594]
Cysteine	12 molybdovanadate(V)	CILE	CV, A (0.4 V vs. Ag/AgCl)	Pharmaceutical samples	[487]
Cysteine	Poly(N,N-dimethylaniline) + ferrocyanide	G/p	CV, A (0.2 V vs. Ag/AgCl)	LOD 6.17×10^{-5} M (CV), 6.38×10^{-6} M (A)	[2715]
Cysteine	Hexacyanoferrate/SiAl-ion exchanger	G/hydrocarbon oil	CV	ox. of cysteine	[1821]
Cysteine	Cu(II)/Co(II)-hexacyanoferrate	G/po	CV, LSV	PBS, pH 2, 0.1 M LOD 5 μM	[1040]
Cysteine Cystine Methionine	Ru(III)-hexacyanoferrate(II) film		FIA		[1042]
Cysteine	Cu(II)-diethyldithiocarbamate		DPASV	LOD 2 μM	[1310]
Cysteine-glutathione	Ru(III)-diphenyldithiocarbamate	G/p	CV, A	ox at +0.38 V vs. SCE; KNO_3/HNO_3, pH 3; LOD 9.16 mg L^{-1}	[1608]
Cysteine	Cu(II)-cyclohexylbutyrate		ASV (E_{acc}, −0.9 V vs. Ag/AgCl)	acc. as Cu(I)-complex, ox at 0.03 V to Cu(II); LOD 2 nM	[2716]
Cysteine	Ferrocene	C/po	CV, DPV	PBS, pH 7.0 LOD 4.7 μM (DPV); in soya protein powder, human serum	[877]

Electroanalysis with Carbon Paste–Based Electrodes, Sensors, and Detectors 339

Analyte	Modifier	Technique	Notes	Ref.
Cysteine	Ferrocenecarboxylic acid	CV, DPV	PBS, pH 7.0; LOD 25 nM (DPV); in soya protein powder, human serum	[1535]
Cysteine	Ferrocenedicaroxylic acid	C/po	CV, DPV, CA	PBS 0.1 M, pH 8.0; LOD 1.4 μM (DPV); in plasma and pharmaceuticals	[1536]
Cysteine	Acetylferrocene	G/po	CV, CA	LOD 2.5 μM; kinetic studies	[2717]
Cysteine	1-[4-(Ferrocenyl ethynyl) phenyl]-1-ethanone	G/po	CV, DPV	PBS, pH 7.0; LOD5 μM (DPV); in plasma and pharmaceuticals	[1543]
Cysteine Homocysteine N-Acetylcysteine Glutathione	Co(II)-PC	G/Nj	CV, A detector in LC	ox at 0.75–0.85 V vs. Ag/AgCl	[883]
Cysteine	Co(II)-PC		CV, POT	POT via Co(II)/Co(I) LOD 0.5 μM (POT)	[2718]
Cysteine	Co(II)-PC	G/mo	Microelectrodes for CE; A 400 mV vs. Ag/AgCl	ox via Co(III)	[555]
Glutathione Cysteine N-Acetylcysteine Glutathione	Co-PC	Carbon cement	FIA, HPLC (0.7 V vs. Ag/AgCl)	LOD 31 nM; in urine	[324]
Cysteine Glutathione	Co-PC	G/Nj	A, det in LC (0.75 V vs. Ag/AgCl)	In serum	[884]
Cysteine Thiols Disulfides	Co-PC		HPLC	LOD 0.2 μM	[886]
Cysteine	MWCNT/α-(2,4-di-tert-butylphenoxy) phthalocyaninatocobalt		CV, A	LOD 5 μM	[2719]
Cysteine	Oxomolybdenum(V)-PC, oxomolybdenum(V)-tetrasulfoPC	G/mo	CV	ox via Mo(VI) at 0.26/0.28 V vs. Ag/AgCl; LOD 10^{-3}–10^{-2} M	[899]

(continued)

TABLE 8.17 (continued)
Analysis of Biologically Important Compounds at Carbon Paste Electrodes: Amines and Other Amino-Compounds

Analyte	Modifier	Carbon Paste	Techniques	Comments	Ref(s)
Cysteine	Co(II)-salophen	G/Nj	CV, A, DPV, POT	Catalytic oxidation, red. to Co(I) salophen E(POT) determined by Co(II)/Co(I) ratio LOD(POT) 1 μM	[1581]
Cysteine Ascorbic acid	Co(II)-4-methylsalophen	G/Nj	CV, DPV	Acetate buffer 0.1 M, PH 5.0; LOD 0.5 μM; s: synthetic serum	[1586]
Cysteine Dopamine Uric acid	2,2′-[1,7-Heptanediylbis (nitriloethylidine)]-bis-hydroquinone on TiO2 nanoparticles	G/po	CV, DPV, SWV	Simultaneous det.; PBS, pH 7; LOD 0.84 μM; s: ampoules	[995]
Cysteine	Quinizarine	G/po	CV, DPV	PBS, pH 7.0; LOD 2.2×10^{-7} M; in the presence of tryptophan; serum, blood, pharmaceuticals	[1459]
Cysteine	4-Nitrophthalonitrile	G/mo	CV, A (0.33 V vs. Ag/AgCl)	Reduction of mediator to nitroso- and hydroxyl amine comp.; PBS, pH 7.0; LOD 250 nM; in food supplements	[2720]
Cysteine Glutauthion	Tyrosinase	G (activated) po	CV, A, FIA	Biosensor via blocking of substrate (catechol) recycling by thiols	[2721]
Glutamate	Glutamate dehydrogenase, NADP, polymeric toluidine blue (+[Ru(NH3)6]Cl3)	C/po (+p wax)	FIA	Biosensor; regeneration of NADPH with polymeric toluidine blue; LOD 0.3 mM	[1751]
Glutamate	Glutamate dehydrogenase + NADP + polymeric toluidine blue + lactitol + diethylaminoethyl dextran	C/mo	A (0.1 V vs. Ag/AgCl)	Biosensor; improved stability	[939]
Glutamate	Glutamate dehydrogenase + octadecylamine			Biosensor; mediation via hexacyanoferrate(III) and phenazine methosulfate	[2722]
Glutamate	Glutamate dehydrogenase + dialysis membrane		BIA (0.05 V vs. SSE)	Biosensor; Dulbecco buffer + phenazine methosulfate 0.5 mM + NAD⁺ 1 mM	[2723]

Analyte	Modifier	Binder/type	Method	Notes	Ref.
Glutamate	Glutamate oxidase/poly(o-phenylenediamine) + NAD	G/po	A (0.15 V vs. Ag/AgCl)	Biosensor; buffer, pH 8; LOD 3.8 μM; in chicken bouillon soup	[1773]
Glutamate	Glutamate oxidase + peroxidase + polyethyleneimine	G/po	A (−0.15 V vs. Ag/AgCl); HPLC	Biosensor; in plasma, urine	[2724]
Glutamate	Glutamate oxidase + peroxidase			Biosensor	[2710]
Glutamate	Glutamate oxidase (cross-linked with GA) + electropolymer + tetrathiafulvalene		A (+0.15 V vs. Ag/AgCl)	Biosensor; TTH reoxidizes NADH; LOD 2.6 μM; food samples	[1488]
Histidine			CV, DPV; acc at −0.1 V vs. Ag/AgCl	Acetate buffer, pH 4.1; peak at −0.45 V vs. Ag/AgCl; LOD 1 mg L^{-1}; injection solutions	[2725]
Histidine	Alumina		2,5th DPV	Succinic acid–borax buffer 0.05 M, pH 3.5; LOD 0.4 μM	[2726]
Histidine	Tetra-3,4-pyridinoporphirazinatocopper(II)	G/liquid p	Pot	Borax buffer 0.05 M, pH 9.5; LOD: 20 μM; in serum	[1577]
L-Histidine	Cyclodextrins	G/po	Pot	Cyclodextrins as chiral selectors; CPE as membrane; LODs: 61.7 nM to 47.7 pM; enantiopurity in pharmaceuticals	[2727]
L-Histidine	Fullerenes C60, C60 carboxylic acids	G/po	Pot	Fullerenes as chiral selectors; CPE as membrane; LODs: 6.3 nM to 2.2 pM; enantiopurity in pharmaceuticals	[2728]
Homocysteine Cysteine N-Acetylcysteine Glutathione		MWCNT/mo	CV, A	ox. at approx. 0.5 V vs. Ag/AgCl; PBS 0.05 M, pH 7.4; LOD 4.6 μM	[883]
Homocysteine	AuNP-cysteamine	G/mo	CV, A (600 mV vs. Ag/AgCl); HPLC	PBS, pH 7.0; responses also from cys, N-acetyl-cys, glutathion, penicillamine; in serum samples	[947]
Leucine Amino acids	Tris(2,2′-bipyridyl) dichlororuthenium(II) hexahydrate on dendritic Pd-NP	G(heat act.)/po	CV, ECL-FIA	Deriv. of amino acids with acet aldehyde; electrogenerated chemiluminescence	[2704]
Lysine	Lysine oxidase + PB	GC(act)/mo	A (0 V vs. SSE)	Biosensor: PBS 0.1 M + KCl 0.1 M, pH 8.0	[196]

(continued)

TABLE 8.17 (continued)
Analysis of Biologically Important Compounds at Carbon Paste Electrodes: Amines and Other Amino-Compounds

Analyte	Modifier	Carbon Paste	Techniques	Comments	Ref(s)
Mefenamic acid	Lanthanum hydroxide nanowires	G/po	CV, LSV	PBS 0.1 M, pH 5.8; ox. ~0.85 V vs. SCE; LOD 6 pM; s: pharmaceuticals	[2585]
Methionine	Colloidal Au + cysteamine	G/po	CV, FIA	Suitable for other S-organic compounds	[1877,1878]
β-N-Oxalyl diaminopropionic acid	Glutamate oxidase/Nafion + MnO_2	G/po	CV, FIA	Toxic principle of *Lathyrus sativus*; glutamate eliminated by decarboxylation; pH 7.4; s:grass peas (extract)	[842, 2729–2731]
Phenylalanine	Phenylalanine dehydrogenase, uricase, NAD^+ + 3,4-dihydroxybenzaldehyde	C(heat act.)/po	CV, LSV	Biosensor; PBS 0.1 M, pH 7.0; LOD: 0.4 mM; s: urine	[2732]
Phenylalanine	Phenylalanine dehydrogenase + salicylate hydroxylase + tyrosinase + $NAD+$	G(heat act.)/po		Trienzyme biosensor; oxidative deamination, oxidative decarboxylation and oxidation to quinone; LOD 5 μM; s: serum, whole blood	[2733]
L-Proline	Cyclodextrins	G/po	Pot	Enantioselective det.; LOD 0.1–0.9 nM	[2734]
Selenoamino acids	$AgNO_3$	G/liquid p	CV, chrono-coulometry	via complexation of surface-bound Ag^+; complexation studies	[1309]
Tryptophan	Unmodified	Unmodified	V	LOD 0.25 μM	[2735]
Tryptophan	Unmodified	G/mo	CV, DPV	ox. ~0.7 V vs. Ag/AgCl BRP, pH 7.4; LOD 1.7 μM s: pharmaceuticals	[2736]
Tryptophan Serotonin	Unmodified	G/Nj	adSV with me	det. as indole after incubation with tryptophanase; s: serum	[2737]
Tryptophan Serotonin Uric acid	Unmodified	GC paste	DPV	via redox peak of ox product; PBS; improvements compared to CPE	[659,2738]
Tryptophan	Preanodization	C/solid p	Adsorptive stripping	Acetate buffer, pH 3.5 ox. at 1.01 V vs. Ag/ AgCl; LOD 2 ng; s: synthetic serum	[181]
Tryptophan	Preanodization	C/solid p	Adsorptive stripping	Acetate buffer, pH 3.5 ox. at 1.8 V vs. Ag/AgCl for 10min; LOD 2 μg/L; s: spiked serum, amino acid injection	[2739]

Electroanalysis with Carbon Paste–Based Electrodes, Sensors, and Detectors

Analyte	Modifier	Electrode	Method	Conditions	Ref.
Tryptophan	Overoxidized ppy film	G/p	ASV(oc)	pH 2 (HCl/KCl); ox. at ~1 V vs. Ag/AgCl	[1797]
Tryptophan	TiO$_2$ NP + quinizarine	G/p	CV	PBS 0.1 M, pH 7.0; peak at 0.75 V vs. Ag/AgCl; simultaneous det. of penicillamine possible	[1880]
Tryptophan Glutathion	TiO$_2$ NP + ferrocene carboxylic acid	G/p	CV, CA	PBS 01, M, pH 7; LOD 98 nM; s: blood	[920]
Tryptophan	1-[4-(Ferrocenyl ethynyl) phenyl]-1-ethanone	G/po	CV, DPV	PBS, pH 7.0; LOD 0.56 μM (DPV); s: serum	[1545]
Tryptophan	MWCNT + Co-salophen	G/Nj	CV, DPV	Acetate buffer, pH 4.0; LOD 0.1 μM (DPV); s: pharmaceuticals	[384]
Tryptophan Tyrosine	Polyamide		2,5th DPV		[2740]
Tryptophan	Potato juice		CV, adSV	PBS, pH 7.4; LOD 5·10^{-6} M	[2741]
Tryptophan	Montmorillonite		DPV, acc. (oc)	ox. +0.75 V vs. SSE	[2742]
Tryptophan	Zeolites		CV	LOD 70 ng mL^{-1}	[1140]
Tryptophan	Zeolite/Fe^{3+}	G/mo	V	LOD 0.1 μM; s: pharmaceuticals	[1122]
Tryptophan Uric acid Ascorbic acid			CV, DPV; E_{acc} −0.3 V vs. SSE, t_{ACC} 30 s	PBS 0.15 M, pH 3.5; LOD 0.06 μM; s: serum	
Tryptophan Dopamine	Zeolite NP/Fe^{3+}		DPV	PBS, pH 5; submicromolar levels; s: serum	[1012]
Tyrosine	Unmodified	CPE, rotating CPE		ca 1 N H$_2$SO$_4$; ox	[2743,2744]
Tyrosine	zeolite/Fe^{3+}	G/mo	CV, DPV	Simultaneous det.; PBS, pH 3.0; LOD 0.32 μM; s: serum	[1124]
Epinephrine			E_{acc} −0.2 V vs. SSE; t_{ACC} 40 s DPV, acc. (oc)		
Tyrosine	Montmorillonite		CV	KCl/HCl (pH 2); ox. +0.9 V LOD 63 ng mL^{-1}	[2745]
D-Tyrosine	Molecular imprinted polypyrrole membrane			Imprinting with L-phenylacetic acid	[2746]
Tyrosine	Polyvinylpyrrolidone + CTAB *in situ*		acc (oc); ASV	irrev.ox. at ~0.43 V vs. SCE; LOD 0.1 μM (3 min acc); s: alcoholic beverage, urine	[2747]

TABLE 8.18
Analysis of Biologically Important Compounds at Carbon Paste Electrodes: Phenolic Compounds and Antioxidants

Analyte	Modifier	Carbon Paste	Techniques	Comments	Ref(s)
Various				Review—antioxidant activity of natural products	[2748]
4-Aminophenol hydroquinone	Nb_2O_5-SiO_2/$H_2PO_4^-$/ Ni-porphyrin	G/mo	CV	KCl 0.1 M	[1201]
Antioxidants t-Butyl phenols Irganox Propyl gallate		G/p or Teflon	CV, DPV, FIA	Critical comparison of electrodes	[2749]
Butylated hydroxyanisole	Mn(II)-hexacyanoferrate(II)	G/p wax	CV, DPV, A (0.45 V vs. SCE)	Catalytic oxidation to quinone via Fe(III); 0.05 PBS + 0.1 M NaCl, pH 6; s: potato chips	[2750]
Butylated hydroxyanisol	Mn(II)-hexacyanoferrate(II)	G/p	Extractive preconcentration	Competitive extractive preconcentration with other aromatic compounds	[226]
t-Butylhydroxy-anisole	NiPC		CV; DPV	Probably via reoxidation of Ni(I) produced from the antioxidant; $HClO_4$ 0.1 M; LOD 3.6 µg L^{-1}; potato flakes	[895]
t-Butylhydroxy-toluene	NiPC		CV; DPV	Probably via reoxidation of Ni(I) produced from the antioxidant; methanol-water (30%), pH 2; LOD 0.17 mg L^{-1}	[651]
Caffeic acid	Green bean tissue + chitin (from squid) + GA + epichlorhydrin	G/Nj	CV, A (0.1 V vs. SSE)	Tissue as a source of peroxidase, catalyzes ox. to quinone, det. by red.; PBS 0.1 M + H_2O_2 2 µM, pH 7; LOD 2 µM; s: white wine	[2751]
(+)-Catechin	β-Cyclodextrin	G/po	SWV	Oxidation of o-diphenol group to o-quinone; phosphate/borate buffer, pH 7.4; LOD 1.35 mg L^{-1}; s: fresh and commercial tea	[1419]
Catechol	Tyrosinase	Nanosized carbon black	CV	Biosensor; comparison with CPE	[2752]
Catechol Hydroquinone	Polyphenol oxidase	MWCNT/mo	A (−0.05 V vs. SSE)	Biosensor; catalytic effect of CNT on reduction of quinones; PBS 0.05M, pH 7.4; s: rd wine	[364]

Electroanalysis with Carbon Paste–Based Electrodes, Sensors, and Detectors 345

Analyte	Electrode material	Technique	Notes	Ref.
Catechol	Laccase (extract from *Pycnoporus sanguineus*)	G/mo	CV, DPV	Biosensor; responds also to hydroquinone and cresol; LOD 4.5 µM	[2753]
Catechol	Laccase on core-shell Fe$_3$O$_4$-SiO$_2$/NH$_2$ NP + GA		A	Biosensor with magnetic NP; LOD: 75 µM; s: compost extracts	[987]
Catechol Phenol	Tyrosinase	g/so	AD detection (−0.2 V vs. SSE) in RP-HPLC	Study on influence of viscosity of binder; improvement by pre-ox. in serial dual cell	[222,554]
Catechol Hydroquinone Phenol	Tyrosinase		FIA, ED in HPLC	Biosensor; comparison with laccase and coconut tissue; comparison with reactors	[2754]
Catechol	DNA + tyrosinase + membrane Cuprophan	G/mo	CV, FIA (−0.15 V vs. SSE)	Biosensor, improved stability and shelf life by DNA; LOD 1 µM	[2755]
Catechol Dopamine	*Latania* spp., fresh or dry tissue + Nafion film	g/mo	FIA	Tissue as a source of polyphenol oxidase; PBS 0.05 M, pH 5.7	[2756]
Catechols Dopamine	Fresh tissue of potato, peach, mushroom, pear		A (0 V vs. SCE)	Tissue as a source of polyphenol oxidase; McIlvaine buffer, pH 6.75	[279]
Catechol	Potato tissue	G/N	A	s: beer	[2757]
Catechol	Banana tissue + Au NP or MWCNT	GC	A (−0.2 V vs. ref.)	s: wine	[2758]
Dihydroxy benzenes	CP impregnated with so or ceresin wax		LSV, DPV	LOD 0.2–0.4 µM (hydroquinone)	[41]
Dopamine Pyrocatechol	Banana tissue		CV, DPV, FIA	Biosensor with tissue as a source for polyphenoloxidase; comparison with unmodified CPE	[2759]
Flavanols	Apple (green or red) or banana or potato tissue	g/Nj	A (−0.25 V vs. SSE)	Biosensor; tissue as a source of polyphenol oxidase; oxidation of *o*-diphenols to quinone and electrochemical reduction; PBS 0.05 M, pH 7.4; s: beer	[278,2760]
Flavonoids Diosmin Diosmin Quercetin Quercitrin Rutin		G/mo	SWV	PBS 0.2 M, pH 7; s: urine	[2761]
Flavonoids		G/Nj	CV, adDPSV	BRP ph 5 (rutin); s: rutin in capsules	[2762]

(continued)

TABLE 8.18 (continued)
Analysis of Biologically Important Compounds at Carbon Paste Electrodes: Phenolic Compounds and Antioxidants

Analyte	Modifier	Carbon Paste	Techniques	Comments	Ref(s)
Flavonoids		G/Nj, G/diphenylether + TCP	adSV in FIA	BRP ph 5; s: rutin in capsules	[2763]
Flavonoids	DNA adsorbed	G/mo, electrically heated	DPV	Biosensor; protective antioxidative properties toward cleavage of DNA by Cu(II)-phen complex + H_2O_2 using Co(phen)$_3^{3+}$ as reporter	[293]
Flavonoids Quercetin Rutin		G/mo	SWV	Direct interaction study with ssDNA; PBS, pH 7	[2764]
Hydroquinone	Core-shell Fe_3O_4-SiO_2/NH_2 NP + GA + laccase	g/p (solid)	CV, A (−0.232 V vs. SCE)	Biosensor; PBS 0.067 M, pH 5.5; LOD 0.67 μM; s: compost extract	[979]
Hydroquinone	CoPC	Conductive carbon cement	CV	PBS 0.1 M, pH 7	[633]
Hydroquinone	Apple tissue		DPV	Biosensor; PBS 0.05 M, pH 6	[2765]
p-Nitrophenol	*Moraxella* spp.	g/mo	A (0.3 V vs. SSE)	Microbial biosensor via p-hydroquinone which is oxidized; PBS 0.02 M, pH 7.5; LOD 20 nM; s: lake water	[2440]
Phenol			A (0.895 V vs. SSE)	Direct ox., phenol released from phenyl phosphate in a chromatographic immunoassay	[2766]
Phenol	Unmodified		ads. acc. (−0.5 V), stripping (0.5 V)	Borate buffer; LOD 80 nM; s: tomato sauce	[2767]
Phenol	Unmodified		CV, adASV (second-order diff.), acc. (0.1 V vs. SSE)	NH_3/NH_4Cl buffer, pH 9.25; LOD 1.3 μM; s: water	[1903]
Phenol		Semimicroelectrode		Sample volumes 1–300 μL; extracts from thin-layer chromatography	[548]
Phenol 2,4-Dichloro-phenol		CILE g/octylpyridinium hexafluoro-phosphate	CV, A (1.2 V vs. SSE)	BRP, pH 2	[461]

Electroanalysis with Carbon Paste–Based Electrodes, Sensors, and Detectors 347

Analyte	Modifier	Method	Notes	Ref.
Phenol	C/polyamide composite	o.c. acc., 2,5th order DPV	BRP, pH 2; s: cola drinks	[2768]
Phenol	CTAB	LSV	PBS 0.1 M, pH 8; LOD: 50 nM; s: industrial wastewater	[2769]
Phenol	Silica gel	CV, LSV	Na-citrate 0.05 M + acetonitrile (15%); ox. at 0.85 V vs. SSE	[1150]
Phenol		po		[2770]
Phenol	Montmorillonite	DPV	oc accumulation; ox. at +0.9 V vs. SSE; acetate buffer 0.1 M; LOD: 40 ng mL^{-1}; s: soft drinks	[2770]
Phenol	Montmorillonite-CTAB	ASV	Improvement of oxidation of phenol; NaOH 0.1 M; LOD: 60 nM; s: water	[1101]
Phenol	Sepiolite	DPV, oc acc.	acc at pH 1.5; meas. at pH 2; s: soft drinks	[824]
Phenol	Procaine HCl; cycling	CV	Signal improvement by procaine; BRP, pH 12; s: decaying leaves	[2413]
Phenol	G/po Polypyrrole, polyvinyl pyrrolidone, overoxidized	LSV; o.c. acc. (10 min in 0.1 M KCl)	ox., PBS, pH 10; LOD 0.1 μM; s: wastewater	[2412]
Phenol	RuO$_2$/Pt colloid		catalytic ox. of phenol; LOD 1 nM	[2771]
Phenol	CoPC		Catalytic ox. of phenol; PBS, pH 8.7; LOD 10 μM; s: wastewater	[2772]
Phenol	HRP immobilized on SiO$_2$/Nb$_2$O$_5$ with GA	A (0 V vs. SCE), FIA	Biosensor with mixed oxide via sol–gel; PBS 0.1 M + H$_2$O$_2$ 0.05 mM, pH 7; LOD 300 nM	[1192]
Phenol	HRP immobilized on SiO$_2$/TiO$_2$ with GA	LSV, A (0 V vs. SCE)	Biosensor; block of direct electron transfer to H$_2$O$_2$ by HRP due to immobilization; PBS 0.1 M, pH 6.8; LOD 1 μM	[1181]
Phenol	Tyrosinase + poly-pyrrole + PVA/cyclodextrine membrane	A (−0.2 V vs. SCE)	Biosensor via detection of quinones produced from tyrosinase; PBS 0.1 M + NaClO$_4$ 0.1 M, pH 7; s: salmon, impregnated textiles (surface contact with sensor)	[2773]
Phenol	Tyrosinase + ferrocene ormosil + coating (polyvinyl alcohol cross-linked with GA)	CV, A (0.2 V vs. NHE)	Biosensor; reduction of o-quinone formed; PBS 0.1 M, pH 7; LOD 0.5 μM	[2774]
Phenol	Tyrosinase + 1-methoxyphenazine methosulfate	CV	Biosensor; also used to determine neuropathy target esterase activity via phenol from phenyl valerate; PBS 0.05 M + 0.1 M NaCl, pH 7.0; LOD 25 nM; s: NTE assay in hen brain and blood	[2125, 2775, 2776]

(continued)

TABLE 8.18 (continued)
Analysis of Biologically Important Compounds at Carbon Paste Electrodes: Phenolic Compounds and Antioxidants

Analyte	Modifier	Carbon Paste	Techniques	Comments	Ref(s)
Phenol	Tyrosinase + AuNP	g + AuNP/po	CV, A (−0.15 V vs. SCE)	Biosensor; AuNP for tyrosinase immobilization; reduction of o-quinone PBS 0.1 M, pH 7	[1275]
Phenol	Core-shell MgFe$_2$O$_4$–SiO$_2$/NH$_2$ NP + GA + tyrosinase	g/p (solid)	CV, A (−0.15 V vs. SCE)	Biosensor with magnetic NP; red. of generated quinone; PBS 0.05 M, pH 7; LOD 0.6 µM	[979]
Phenol	Extract soy bean seed hulls		A	Biosensor via peroxidase; pH 7.4	[2777]
Phenols			A (0.8 V vs. SSE); detector in pCEC	LOD 2–50 ng mL^{-1}	[563]
Xenoestrogens				s: chicken egg, milk powder	
Phenols	Mesoporous TiO$_2$ NP	g/po	CV	Enhanced sensitivity by TiO$_2$ NP; 0.1 M NaCl	[994]
Phenols	Bentonite	g/mo	DPV in FIA with stopped flow	s: sea water	[2696]
Phenols	Tyrosinase (+ carbodiimide/+ GA/+ BSA)	G/po	AD in HPLC (−0.05 V vs. SSE)	Biosensor by oxidation of phenols to quinines via catechols and detection of quinone; PBS 0.1 M + Tween 80 (0.025%), pH 6.0; comparison with graphite electrode; s: wastewater (pulp industry)	[2778]
Phenols Phenol Epicatechin Ferulic acid	Polyphenol oxidase + polypyrrole in situ	G/Nj/pyrrole	CV, A (−0.2 V vs. SCE)	Biosensor; polypyrrole created in situ in the paste; PBS 0.1 M	[1798, 1800]
Phenols Catechols	Tyrosinase + sodium hexadecanesulfonate			Biosensor for solvents with high organic liquid content (methanol, acetonitrile); s: pharmaceuticals	[2779]
Phenols	β-Cyclodextrin	G/po	CV, DPV; oc acc	acc. via complex with dextrin	[1411]
Phenols	Tyrosinase + maltodextrin + PEI + Nafion membrane		A (−0.1 V vs. SCE)	Biosensor; LOD 50 nM; pH 5.4; s: wastewater	[2780, 2781]
Phenols	Tyrosinase + Os-mediator	G/Nj	CV, FIA	Biosensor; comparison of commercial tyrosinases	[2782]
Phenols	CuPC + histidine	G/mo	A (static and as RDE)	Based on chemistry of dopamine β-hydroxylase; PBS 0.25 M + H$_2$O$_2$ 0.12 mM, pH 6.9 LOD 9 µM	[519]

Phenols	MnPC + histidine	G/mo	A (0 V vs. SSE)	Biomimetic sensor; PBS 0.1 M, pH 7 + H_2O_2 0.25 mM; assumed ox. of MnPC and ensuing catalytic ox. of phenols; electrochem.red. of the quinoid; LOD 0.01 mM (catechol); s: pharmaceuticals	[912]
Phenols Ascorbic acid Dopamine	Polyphenol oxidase or mushroom tissue	GC/mo	CV, A (−0.1 V vs. SSE)	Biosensor; PBS 0.05 M, pH 7.4; LOD 0.4 µM (catechol); s: red wine, acetaminophen in pharmaceuticals	[194]
Phenols	Mushroom		AD in HPLC (−0.2 V vs. SSE)	Biosensor; PBS 0.05 M/acetonitrile (70:30), pH 5	[2564]
Phenols	Mushroom tissue (partially purified tyrosinae) + dialysis membrane		A (−0.1 V vs. SSE)	Biosensor; detection via red. of o-quinone; pH 6 or 6.5 (two isozymes of tyrosinae); LOD 10 nM	[283]
Phenols	Tyrosinase + hexacyanoferrate/PVP	G/mo	CV, DPV, FIA (−0.2 V vs. SSE)	Biosensor via red. of o-quinone mediated by $[Fe(CN)_6]^{4-}$; KCl 0.1 M LOD 14 ng/mL	[573]
Phenols	Mushroom tissue + CoPC or $K_4[Fe(CN)_6]$	g/mo	A (−0.22 V vs. SSE)	Biosensor; tissue as a source of tyrosinase; catalytic red. of o-quinone; PBS 0.05 M, pH 7.4	[898,2783]
Phenols	Tyrosinase + Ru (dispersed on C)	G/mo	A (−0.1 V or 0 V vs. SSE)	Biosensor; PBS 0.05 M, pH 7.4; LOD 25 µM (0 V); s: groundwater	[1258]
Phenols Catechol	Polyphenol oxidase or banana tissue + Ir	G/mo	A (−0.1 V vs. SSE)	Biosensor; catalytic reduction of the quinone; PBS 0.05 M, pH 7.4	[1265]
Phenols	Tyrosinase		ED with HPLC (−0.2 V)	s: soil leachate, sludge extrac.	[2410,2784]
Phenols	Tyrosinase	g/mo or hydrocarbons	FIA (−0.05 V vs. SSE), AD in HPLC	Biosensor, study of influence of hydrocarbons as liquid binder on extraction; PBS 0.05 M, pH 7.4	[384]
Phenols	Tyrosinase in sol–gel SiO_2 layer		CV, A (0 V vs. SSE)	Biosensor; electrochem. reduction of the quinones; PBS 0.02 M, pH 6.5	[1166]
Phenols (+)-catechin	Ferrocene + HRP		A	Biosensor via decreased amount of H_2O_2 consumed by phenols; LOD 0.3 µM (catechin); s: wine, tea	[1525]
Phenols	Lactitol + HRP	g/po	FIA (−0.05 V vs. SCE)	Biosensor; reduction of phenoxy-radicals; PBS 0.1 M, pH 7 + 0.9 mM H_2O_2	[2785]
Phenols	Quinoprotein glucose dehydrogenase + PEI	g(act)/po	A (0.5 V vs. SSE)	Biosensor; oxidized phenols as electron acceptor; PBS + 15 mM glucose + 1 mM $CaCl_2$	[2786]
Quercetin		G/Nj, G/CNT	CV, adDPASV	ox. of o-diol to o-quinone; acetate buffer + NaCl (1%); s: spiked blood	[366]

(continued)

TABLE 8.18 (continued)
Analysis of Biologically Important Compounds at Carbon Paste Electrodes: Phenolic Compounds and Antioxidants

Analyte	Modifier	Carbon Paste	Techniques	Comments	Ref(s)
Redox compounds	—	G/vegetable oil	CV, SWV	Determination of antioxidants in vegetable oils; discrimination of oils via statistics; discrimination of bitterness in olive oils	[2163, 2787]
Redox compounds	Bis-PC of Lu(III) or Gd(III) or Pr(III)	G/Nj	CV, SWV	Sensor system for the discrimination of wines; statistical evaluation	[1571, 2788]
Redox compounds	Sensor array with phthalocyanines	G/Nj	CV, SWV	Electronic tongue for adulterations in wine; also polypyrrole electrodes doped with anions	[918]
Resveratrol	Peroxidase basic isoenzymes + dialysis membrane		A (0V vs. SCE)	Biosensor, indirect method via H_2O_2; PBS, pH 7 + 0.1 mM H_2O_2	[775]
Rosmarinic acid	Fe(III)Zn(II)-complex	G/Nj	SWV, CV; AD in CZE	Biomimetic red kidney purple acid phosphatase sensor, catalytic ox. with H_2O_2, det. via red. of o-quinone; PBS 0.1 M + H_2O_2 0.119 mM, pH 7.5; s: plant extracts	[2789]
Rutin	Mn(II)/Mn(III) dinuclear complex	G/Nj	SWV	Biomimetic peroxidase sensor, catalytic oxidation of rutin with H_2O_2; PBS 0.1 M + H_2O_2, $4 \cdot 10^{-5}$ M, pH 6; LOD 0.18 µM; s: pharmaceuticals	[2790]
Rutin Hydroquinone	Eggplant (gilo) crude extract + chitosan + GA + epichlorhydrin or carbodiimide	G/Nj	A (0.124 V vs. SSE), SWV	Tissue as a source of peroxidase, catalytic ox., det.by red. of quinone; PBS 0.1 M + H_2O_2 2 µM, pH 7; LOD 20nM (ru) or 2 µM (hy)	[2791]
Thymol	—	GC	A (1.1 V vs. SSE)	Amperometric detection after HPLC separation; s: thyme syrup, pastilles	[2627]
Zearalenone Zearalenole	—	G/mo	AD with electrophoresis	Phenolic *Fusarium* toxin; s: maize flour (supercritical fluid extraction)	[2792]

TABLE 8.19
Analysis of Biologically Important Compounds at Carbon Paste Electrodes: Carbohydrates and Related Compounds

Analyte	Modifier	Carbon Paste	Techniques	Comments	Ref(s)
Alditols Acidic sugars Mono-, oligo-saccharides Deoxysugars Polyalcohols	CoPC	G/po	CV, AD in AIEC (0.4–0.5 V vs. SSE), pulsed detection (−0.3 V/+0.39 V vs. SSE)	Catalytic oxidation via Co(III); NaOH 0.15 M	[882,889,890]
Aldoses Xylose Glucose	Aldose dehydrogenase + polymer-Fc + PQQ	G/po	A, FIA (0.2 V vs. SSE)	Biosensor with PQQ-dependent dehydrogenase via polymer-bound Fc-mediated reoxidation of $PQQH_2$; s: yeast fermentation	[1735,2793, 2794]
Carbohydrates	Co(II,III)-oxide	G/mo	FIA (0.5 V vs. SSE)	0.1 M NaOH	[1230]
Carbohydrates	Cu_2O	G/mo	CV, FIA (+0.55 V vs. AgAgCl)	0.1 M NaOH LOD 0.2–3.2 ng	[1225]
Carbohydrates Glucose Sucrose Fructose	Cu_2O + polyethylene glycol		CV, AD in CZE (0.06 V vs. SCE)	Simultaneous det. of ascorbic acid; NaOH 0.03 M; LOD 0.12–0.49 µM; s: black tea beverage	[1229]
Carbohydrates	Cu_2O, CuO, $Cu(OH)_2$		CV, FIA, AD in HPLC (0.45 V vs. SSE)	Oxidative degradation to formic acid; NaOH 0.3 M	[1226]
Carbohydrates	Cu(II)-porphyrin	G/p	CV	via oxidation of reduced Cu(0?) at 0.25 V vs. SSE; HNO_3 0.025 M	[1576]
Carbohydrates Sucrose Glucose Fructose	NiO NP	G/po	CV, AD in CZE (0.55 V vs. SCE)	NaOH 0.05 M; LOD 0.3–0.6 µM; s: honey	[1231]
Carbohydrates Glucose Maltose Lactose Sucrose	Ni(II)/o-amino phenol	G/so	CV, CA	Electropolymerization of the o-aminophenol; mediated oxidation via NiO(OH); NaOH 0.1 M; LOD 8–90 µM	[1766]

(continued)

TABLE 8.19 (continued)
Analysis of Biologically Important Compounds at Carbon Paste Electrodes: Carbohydrates and Related Compounds

Analyte	Modifier	Carbon Paste	Techniques	Comments	Ref(s)
Carbohydrates Glucose Galactose Lactose Maltose Sucrose Sorbitol	Ni(II)/1-naphthyl amine	G/so	CV	Electropolymerization of 1-naphthyl amine; LOD 6–8.8 µM	[2687]
Carbohydrates	RuO$_2$	G/mo	CV, FIA (0.4 V vs. SSE)	NaOH 1 M	[1218,1221]
Carbohydrates	Cu$_2$O, RuO$_2$, NiO, CoO	G/mo	FIA (0.45 V vs. SSE)	Four sensor array; statistical evaluation of responses; KOH 0.5 M	[2706]
Carbohydrates	p-Benzoquinone + oligosaccharide dehydrogenase + dialysis membrane		CV	Biosensor; PBS 0.1 M, pH 7; amylase activity determination possible with starch substrate	[1437]
Carbohydrates Glucose Sucrose	Glucose oxidase + HRP + Fc			Biosensor in combination with invertase and mutarotase; s: fruit juice	[1523]
Carbohydrates Sucrose Sucrose Lactose Fructose Fructose	Osphendione + GDH + NAD$^+$ Mutarotase + invertase + GDH + NAD$^+$ GDH + NAD$^+$ GDH + NADP$^+$ FDH + Os(bpy)$_2$ + PEI	G(act.)/po	AD (0.15 V vs. SSE, 0.1 V for fructose)	Biosensors in multianalyte flow cell, detection via osmium complexes; LOD 0.1–0.8 mM; s: fruit juices, milk derivatives	[2795]
Carbohydrate antigen 19–9	AuNP + CA 19–9 antibody labeled with HRP	G(act.)/po	CV, DPV, AC impedance spectroscopy	Reagentless immunosensor via decrease of HRP activity due to antigen–antibody complex formation; s: human serum	[951]
D-Galactose	Galactose oxidase + 1,1'-dimethyl Fc	G/epoxy re	CV	pH 9	[854]

Electroanalysis with Carbon Paste–Based Electrodes, Sensors, and Detectors 353

Analyte	Modifier	Electrode	Technique	Notes	Ref.
Fructose	D-Fructose dehydrogenase + Fc or hydroxymethyl-Fc	G/po	CV, CA	Mediated ox. by oxidized Fc; PBS 0.1 M, pH 7.2	[872]
Fructosyl valine	—	GC/mo on ITO, 200°C for 24 h	CV, A (1.0 V vs. SSE)	Oxidation of the amide to the imide; PBS, pH 7.4	[2796]
Fructosyl valine	Poly(vinyl imidazole)	—	A (0.1 V vs. SSE)	ox. as model compound for glycosylated Hb; artificial fructosyl valine oxidase; PBS 0.01 M + 1-methoxyphenazine methosulfate 1 mM, pH 7	[2797]
Fructosyl valine	Fructosyl valine oxidase	—	—	Oxidase from marine yeast, det. via peroxide	[2798]
Glucose	Cu_2O	Conductive carbon cement	CV	Oxidation of glucose via Cu(III); NaOH 0.1 M	[633]
Glucose	RuO_2	—	—	RuO_2 in alkaline medium as inorganic enzyme analogue	[1222]
Glucose	$La(NO_3)_3$ in H_2O	G/po	CV	Reduction of Cu(II)–glucose complex at $La(OH)_3$ nanowires; NH_3/NH_4Cl-buffer, pH 9.8 + 0.5 μM Cu(II); LOD 0.35 μM; s: injection solutions	[2799]
Glucose	NiNP (on C nanofibers)	C nanofibers (electrospun)/mo	CV, A (0.6 V vs. SSE)	Nonenzymatic sensor; NaCl 0.1 M + NaCl 0.1 M; LOD 1 μM	[975]
Glucose	Poly(o-aminophenol)–Ni(II) film (electropolymerized)	—	CV	Catalytic oxidation at 0.9 V; NaOH, 0.09 M; LOD 10 μM	[2800]
Glucose	—	—	—	Review—glucose oxidase biosensors	[2801]
Glucose	Glucose oxidase in paste	G/so	FIA (+0.9 V vs. SSE)	Biosensor without mediator; pH 6.5; LOD 0.5 mM	[73]
Glucose	Glucose oxidase	G/epoxy resin	LSV, CV, A (1.15 V vs. SSE)	Biosensor with direct oxidation of H_2O_2; PBS 0.1 M + KCl 0.1 M, pH 7	[2802]
Glucose	Glucose oxidase	MWCNT/mo	A (−0.1 V vs. SSE)	Biosensor based on detection of H_2O_2; PBS 0.05 M, pH 7.4; LOD 0.6 mM	[123,661]
Glucose	Glucose oxidase	Bamboo-structured CNT/po	CV, A (−0.1 V vs. SSE)	Biosensor; PBS 0.05 M, pH 7.4	[371,2803]
Glucose	Glucose oxidase (apo or native) + FAD (covalently immob.)	GC/so	CV, FIA	Biosensor; reconstitution of apo-enzyme and covalently bound FAD with direct e-transfer; KCl 0.1 M	[2804]

(continued)

TABLE 8.19 (continued)
Analysis of Biologically Important Compounds at Carbon Paste Electrodes: Carbohydrates and Related Compounds

Analyte	Modifier	Carbon Paste	Techniques	Comments	Ref(s)
Glucose	Glucose oxidase (+RuO$_2$)	G/mo	BIA (1.0V vs. SSE, 0.4V with RuO$_2$)	Biosensor; BIA compared with FIA	[1949]
Glucose	Glucose oxidase + IrO$_2$ or PtO$_2$ or PdO	G/po	FIA	Biosensors; PBS 0.1M, pH 7.4; LOD: 0.83–2 mg L^{-1}	[2805]
Glucose	Glucose oxidase (+DMFc)	G/po	CV, A (1.1V vs. SCE, 0.16V with DMFc)	Biosensor; PBS, pH 7 s: plasma	[1517]
Glucose	Glucose oxidase + p-benzoquinone + nitrocellulose or dialysis membrane	G/p	CV, A (0.5V vs. SCE)	Biosensor with p-BQ as e-acceptor from the enzyme; acetate buffer 0.1 M, pH 5–7; LOD 0.5–10 mM (dependent on membrane)	[72,666,2806, 2807]
Glucose	Glucose oxidase + p-benzoquinone + dialysis membrane + Au minigrid + Nylon net	G/po, Nj	A (0.2V vs. SCE), 0.5V vs. SCE (minigrid)	Biosensor; gold minigrid eliminates interference from ascorbic acid by ox.	[1434]
Glucose	Glucose oxidase + siloxane polymers with benzo- or naphthoquinone	G/po		Biosensor; pH 7	[2808]
Glucose Fructose	Glucose oxidase + lipophilic or hydrophilic Os-complex	G/po or poly(vinylpyridine) derivatized with bromoethylamine	CV, A (0.1 [hydrophobic] or 0.4 [hydrophilic] V vs. SSE)	Biosensor; perchlorate buffer 0.25 M, pH 6	[238,2809]
Glucose	Glucose oxidase + BSA + GA + Os-redox polymer	G/so	CV, A (0.4V vs. SSE)	Biosensor via Os(II,III) e-transfer from enzyme; comparison of different electrode designs; PBS 0.1 M + KCl 0.1 M, pH 7.2	[649]
Glucose	Glucose oxidase	G/PVC	A (+0.3V vs. SSE)	Biosensor; Fc monocarboxylic acid in solution; pH 7.4; LOD 0.32 mM	[2810]
Glucose	Glucose oxidase	Ordered mesoporous carbon	CV, A (0.4V vs. SSE)	Biosensor; PBS 0.05M, pH 7.4; LOD 72 µM	[119]
Glucose	Glucose oxidase + Fc	G/po	CV, A (+0.7V vs. SCE)	Biosensor; pH 7; LOD 3 mM; s: must	[2811]
Glucose	Glucose oxidase (bound to Nylon net) + Fc + cellulose triacetate membrane	G/cellulose triacetate	FIA (0.16V vs. SSE)	Biosensor; LOD 0.01 mM	[859]

Analyte	Composition	Type	Method	Notes	Ref.
Glucose	Fc + glucose oxidase in poly-phenol films				[869]
Glucose	Glucose oxidase + 1,1′-dimethyl Fc	G/po	A (0.15 V vs. SCE)	Biosensor; LOD 0.5 mM	[2812]
Glucose	Recombinant *Microdochium navale* carbohydrate oxidase + 1,1′-dimethyl Fc + AMB + dialysis membrane	G/po	A (0.04 V vs. SCE)	Biosensor; detection via oxidized mediator; PBS, pH 6.6	[1532,1533]
Glucose	Glucose oxidase + GA + dimethyl Fc (+ poly(pyrrole) layer + poly(o-phenylene diamine), electropolymerized)	G/so	A, FIA (0.35 V vs. SSE)	Biosensor via re-oxidation of Fc; comparison with corresponding Pt- and organic conductive salt (TCQN-TTF) electrodes; PBS 0.05M + KCl 0.1 M, pH 7; LOD 0.3 (without bilayer) or 0.6 (with) mM; s: synthetic serum	[1776]
Glucose	Glucose oxidase + stearic acid + Fc derivatives	G/mo	CV, A (0.3 V vs. SSE)	Biosensor, CPE microelectrode possible; pH 6.5	[857]
Glucose	Glucose oxidase + heptyldimethyl aminomethyl ferrocene	G/po	A (0.6 V)	Biosensor; pH 7; LOD 0.5 mM	[2813]
Glucose	Glucose oxidase + zeolite/Fc			Biosensor; improved performance and storage stability	[1118]
Glucose	Glucose oxidase and Fc carboxylic acid on silica/TiO₂	G/mo	CV, A (0.34 V vs. SCE)	Biosensor; PIPES 0.1M, pH 6.8; s: blood	[923]
Glucose	Glucose oxidase + siloxane polymer with Fc and derivatives	G/po	CV, A (0.3–0.4 V vs. SCE)	Biosensors; pH 7; LOD 0.1 mM	[1554, 2814–2817]
Glucose	Glucose oxidase + polymers with Fc	G	CV, A (+0.35 V vs. SCE)	Biosensor; PBS, pH 7.4	[2818]
Glucose	Glucose oxidase + siloxane polymer with Fc	G/po	A (0.3 V vs. SSE)	Biosensor; PBS 0.1 M + KCl 0.1 M, pH 7	[922]
Glucose	Glucose oxidase + Fc on siloxane polymer	G/po	CV, A (0.35 V vs. SSE)	Biosensor; PBS + KCl, 0.1 M, pH 7	[2819]
Glucose	Glucose oxidase + ethyleneoxide polymer with Fc	G/po	CV, A (0.1–0.3 V vs. SCE)	Biosensor; PBS 0.1M, pH 7	[862,2817]
Glucose	Glucose oxidase + siloxane ethylene oxide polymer with Fc	G/po	A (0.3 V vs. SCE)	Biosensor; pH 7	[2820]
Glucose	Glucose oxidase + poly(vinylferrocene-methacrylates)	G/po	CV, A (0.35 V vs. SCE)	Biosensor with redox polymer for e-transfer to enzyme; PBS, pH 7.4	[1748,2821, 2822]

(continued)

TABLE 8.19 (continued)
Analysis of Biologically Important Compounds at Carbon Paste Electrodes: Carbohydrates and Related Compounds

Analyte	Modifier	Carbon Paste	Techniques	Comments	Ref(s)
Glucose	Glucose oxidase + acryl amide polymer + Fc			Reagentless biosensor with hydrogel redox polymer for e-transfer from the enzyme	[1555]
Glucose	Glucose oxidase + 2-hydroxypropyl-β-cyc-Iodextrin + vinyl Fc in poly (acrylamide) gel + dialysis membrane	G/mo	CV, A (0.4 V vs. SSE)	Biosensor with redox gel; PBS 0.1 M, pH 7	[785]
Glucose	Glucose oxidase + Fc on PEI	G/po	CV, A (0.1–0.4 V vs. SSE)	biosensor; PBS 0.1 M + KCl 0.1 M, pH 7	[1749]
Glucose	Glucose oxidase + Fc + Nafion gel layer	G/PBS 0.1 M, pH 7.4; also as RDE	CV, A (0.6 V vs. SCE)	Biosensor based on aqueous CP and hydrophobic Nafion gel layer; kinetic studies; PBS 0.1 M, pH 7.4; LOD 10 μM	[865, 17725]
Glucose	Glucose oxidase + 1,1′-DMFc (+ membrane of Nafion or lipid)	G/po	A (0.45 V vs. SCE)	Biosensor; membrane to diminish interference from ascorbic acid; pH 7; LOD 1 mM	[1530]
Glucose	Glucose oxidase + vinyl Fc (+ dialysis membrane)	G colloidal emulsion on GC	CV, A (0.3 V vs. SCE)	Biosensor; PBS 0.1 M, pH 6	[2823]
Glucose	Glucose oxidase/chitosan/GA + Fc/chitosan		CV, A (0.4 V vs. SCE)	Biosensor; PBS 0.02 M; s: soft drinks	[88]
Glucose	Glucose oxidase on chitosan + Fc monocarboxylic acid on Fe_3O_4/SiO_2 core-shell NP		CV; A (0.35 V vs. SSE)	Biosensor; PBS 0.1 M, pH 7.3; LOD 3.2 μM	[988]
Glucose	Glucose oxidase + DMFc or Fc + tetramethyl ammonium tetraphenylborate			Biosensor; ion-associate	[2824]
Glucose	Glucose oxidase + poly(vinyl Fc) on silicagel	G/po	CV, A (0.2 or 0.35 V vs. SSE)	Biosensor; improved catalytic activity due to silicagel; PBS 0.1 M + KCl 0.1 M, pH 7.4	[924]
Glucose	Glucose oxidase + Fc (+HRP)	G/po	FIA (0 V vs. SSE)	Biosensor; PBS 0.1 M + KCl 0.1 M, pH 7.2; s: wine	[2825]
Glucose	Glucose oxidase bound to poly(ethylene glycol) or glucose oxidase + DMFc		CV, DPV, A (+0.4 V vs. SSE)	Biosensor; K-acetate buffer 0.1 M, pH 5; LOD 0.1 mM	[1519, 860]

Glucose	Glucose oxidase on poly (ethylene glycol) + DMFc or anthraquinone or methylene blue or TCQM or chlorophenol indophenol		A (0.05–0.3 V vs. SSE)	Biosensors; comparisons of various mediators for PEG-bound GOx; citrate buffer 0.1 M, pH 5.5	[1492]
Glucose	Glucose oxidase on poly (ethylene glycol) + DMFc + HRP	G/silicon grease	A (−0.2 V vs. SSE)	Biosensor; citrate buffer 0.1 M, pH 5.5; LOD 20 μM; s: juices, soft drinks	[1783]
Glucose	Glucose oxidase on Sepharose + Fc		A 0.4 V vs. SCE)	Biosensor; PBS 0.1 M + KCl 0.05 M, pH 7	[589]
Glucose	Glucose oxidase + Fc + AuNP			Biosensor; improvement of sensitivity by AuNP	[942]
Glucose	Glucose oxidase + Fc-carboxylic acid	Carbon nanofibers/mo	CV, A (0.4 V vs. SSE)	Biosensor; PBS 0.1 M + KCl 0.5 M, pH 7	[202]
Glucose	Glucose oxidase/BSA/GA + Fc or hydroxymethyl Fc	G/po or castor oil	CV, A (0.9 V vs. Pt quasi-ref.)	Biosensor; thermal simulation of aging and estimation of shelf lifetime	[1524]
Glucose	Glucose oxidase/GA/BSA + peroxidase + Fc	G/po	CV, A (0 V vs. SSE)	Biosensor; Fc+ and oxygen as concurring e-acceptors for GOx; PBS 0.1 M, pH 7; digital simulations	[1526]
Glucose	Glucose oxidase + HRP + Fc	G or GC/po	FIA (−0.05 V vs. SSE)	Biosensor with oxidized (NaIO4) HRP forming –CHO which cross-link with –NH2 from GOx; PBS 0.1 M, pH 7.2	[1522]
Glucose	Glucose oxidase + Fc + HRP in poly (3-aminophenol), electropolym.	C/liquid p	A (0.05 V vs. SSE)	Biosensor; PBS 0.1 M, pH 7	[871]
Glucose	HRP + Fc + membrane (Nafion-coated cellulose acetate)	G/po	FIA (0.1 V vs. SSE)	Biosensor with glucose oxidase on pre-column; PBS 0.1 M + KCl, pH 7; LOD: 0.1 μM; s: serum	[1727]
Glucose	Glucose oxidase/BSA/GA + mutarotase + Fc		A (0.9 V vs. Pt quasi-ref. el.)	Biosensor with improved characteristics due to mutarotase, but less accurate; PBS 0.1 M, pH 7.4	[2826]
Glucose Sucrose	Glucose oxidase horseradish peroxidase + Fc	G/po	FIA (0 V vs. SSE)	Biosensor; determination of sucrose via invertase and mutarotase in solution; PBS 0.1 M + KCl 0.1 M, pH 7.2; s: fruit juice	[1523]
Glucose	Glucose oxidase + Fc-Si dendrimers	G/po	CV, A (0.35 V vs. SCE)	Biosensor; PBS 0.1 M + KCl 0.1 M, pH 7	[1551]
Glucose	Glucose oxidase + fumed silica (+DMFc)	G/mo	A (+0.3 V vs. SSE [with DMFc] or 1 V [without])	Biosensor with improvement of sensitivity by fumed silica; PBS 0.05 M, pH 7.4	[1153]
Glucose	Glucose oxidase	G/t-pentyl Fc or n-butyl Fc	CV, A (0.4 V vs. SSE)	Biosensor with mediator as pasting liquid; PBS 0.05 M, pH 7.4; s: physiological fluids	[2827]

(continued)

TABLE 8.19 (continued)
Analysis of Biologically Important Compounds at Carbon Paste Electrodes: Carbohydrates and Related Compounds

Analyte	Modifier	Carbon Paste	Techniques	Comments	Ref(s)
Glucose	Glucose oxidase	G/n-butyl Fc	CV, LSV	Biosensor; self-catalytic paste; KPF6 0.1 M; pH 7.1–7.4; s: blood	[2828]
Glucose	Glucose oxidase + nickelocene + Pt particles		A (−0.2 V vs. SSE)	Biosensor; s: serum	[2829]
Glucose	Glucose oxidase + CoPC	G colloidal emulsion on GC	CV, A (0.7 V vs. SCE)	Biosensor; PBS, pH 6; s: serum	[2830]
Glucose	Glucose oxidase + CoPC		A (0.7 V vs. SCE)	Biosensor via catalytic oxidation of H_2O_2; pH 7; LOD 0.01 mM	[2831]
Glucose	Glucose oxidase + CoPC or Pt-black		A (0.4–0.7 V vs. SSE)	Biosensor	[2832]
Glucose	Glucose oxidase on poly (ethylene glycol) + Co-octoethoxy PC or CoPC		A (0.45 V [Co-oePC] or 0.65 V vs. SSE [CoPC])	Biosensor; PBS 0.1 M, pH 7	[1566]
Glucose	Glucose oxidase + poly(Fe(III)-5-aminophenanthroline) (electropolymerized)	MWCNT/Nj	CV, A(−0.1 vs. SSE)	Biosensor via catalytic reduction of H_2O_2; PBS 0.05 M, pH 7.4	[433]
Glucose	Glucose oxidase	G/polysiloxane	A (0.2 V vs. SCE)	Biosensor with ferricyanide as e-akzeptor; PBS 0.1 M + KCl 0.1 M + $K_3[Fe(CN)_6]$ 0.5 mM	[1788]
Glucose	Glucose oxidase	G/Nj	CV	Biosensor; PBS 0.01 M + KCl 1 M + $K_3[Fe(CN)_6]$ 1 mM	[643]
Glucose	Prussian blue/Fe_3O_4-NP Aminosilane Glucose oxidase	C/solid p	CV, A (−0.15 V vs. SSE)	Biosensor with magnetic NP containing catalyst for red. of H_2O_2 and covalently bound enzyme; PBS 0.025 M + KCl 0.1 M, pH 6.5; LOD 0.1 µM; s: blood	[850]
Glucose	Glucose oxidase + Prussian blue	GC/po	A (0 V vs. SSE)	Biosensor; PBS, pH 6.0; LOD 50 µM	[196]
Glucose	Glucose oxidase + Prussian blue	SWCNT		Biosensor; LOD 0.1 mM	[360]
Glucose	Glucose oxidase + Prussian blue	G(act. with aqua regia)/mo or paraffin	CV, A	Biosensor; PBS 0.1 M, pH 6; LOD 0.1 mM	[848]

Electroanalysis with Carbon Paste–Based Electrodes, Sensors, and Detectors 359

Analyte	Modifier	Electrode	Method	Notes	Ref.
Glucose	[Ru(CN)$_6$]$^{4-}$ on Fe-enriched cinder + glucose oxidase-BSA-GA in Tosflex layer	G/mo; also as rotating disk	CV, A, FIA (V vs. SSE)	Biosensor, formation of ruthenium purple in cinder; ammonium buffer, 0.1 M, pH 7; LOD 0.35 µM; s: physiological solutions and tablets	[1613]
Glucose	Glucose oxidase + Ni-hexacyanoferrate		A (0.02 V vs. SSE)	Biosensor; LOD 0.1 mM	[851]
Glucose	Glucose oxidase + Cu$_2$[Fe(CN)$_6$]	G/mo as microsensor	CV, A (−0.1 V vs. SSE)	Biosensor; PBS 0.05 M, pH 7.4; LOD 0.1 mM	[2813]
Glucose	Unmodified	G/po	SWV; optical detection	Enzyme-induced growth of Cu$_2$[Fe^{II}(CN)$_6$] NP; citrate buffer 0.1 M, pH 5 + Cu(II) 0.001 M + ferricyanide 0.05 M + glucose oxidase 50 mg L^{-1}	[1009]
Glucose		G/po	CV, SWV	via formation of Nd(III)-hexacyanoferrate(II) NP by enzymatic action; KCl 0.1 M + 1 mM Nd(III) + 5 mM [Fe(CN)$_6$]$^{3-}$ + GOx 20 µg mL^{-1}; LOD 3 µM	[1008]
Glucose	Ni(OH)$_2$ NP	CILE MWCNT/ octylpyridinium iodide	CV, A (0.5 V vs. SSE)	Catalytic oxidation of glucose via Ni(III); NaOH 0.5 M; LOD 6 µM; s: serum	[497]
Glucose	Ni(II)-quercetin complex (electrodeposited)	CILE MWCNT/po + 1-butyl-3-methylimidazolium hexafluorophosphate	CV, A (0.6 V vs. SSE)	Catalytic oxidation of glucose via Ni(III); NaOH 0.1 M; LOD 1 µM	[414]
Glucose	Glucose oxidase + Mn-perovskite	G/mo	A (−0.1 or 0.7 V vs. SSE)	Biosensor via H$_2$O$_2$; PBS 0.05 M, pH 7.4	[1249]
Glucose	Glucose oxidase + CuO	G/mo	CV, A (−0.1 V vs. SSE)	Biosensor; PBS 0.05 M, pH 7.4; LOD 20 µM	[2833]
Glucose	Glucose oxidase + MnO$_2$	G/po	CV, A (0.48 V vs. SSE)	Biosensor via oxidation of H$_2$O$_2$; PBS 0.2 M, pH 7.5; LOD 11 mg L^{-1}; s: wine	[1234]
Glucose	Glucose oxidase (Nafion film) + MnO$_2$	G/po	CV, FIA (0.48 V vs. SSE)	Biosensor via oxidation of H$_2$O$_2$; PBS 0.5 M, pH 7.4; LOD 5 mg L^{-1}; s: wine	[842]
Glucose	Glucose oxidase + manganese oxides		A (0.3 V vs. SSE)	Biosensor; PBS, pH 7.4; LOD 0.05–0.1 mM	[1874]
Glucose	Glucose oxidase + MnO$_2$		Pot	Biosensor; LOD 3.3 µM	[2834]
Glucose	Glucose oxidase + metal oxide	G/po	CV, FIA	Biosensors; comparison of different metal oxides as mediators for detection of H$_2$O$_2$	[2835]
Glucose	Glucose oxidase	Pt-C/mo	A (0.8 V vs. SSE)	Biosensor; PBS 0.05 M, pH 7.4	[838]
Glucose	Glucose oxidase	Ru dispersed C/mo	FIA (−0.15 V vs. SSE)	Biosensor; pH 7.4; LOD 1 mM	[1256]

(*continued*)

TABLE 8.19 (continued)
Analysis of Biologically Important Compounds at Carbon Paste Electrodes: Carbohydrates and Related Compounds

Analyte	Modifier	Carbon Paste	Techniques	Comments	Ref(s)
Glucose	Glucose oxidase	Ru or Pt or Ru-Pt(alloy)-dispersed C/mo	CV, A, FIA (−0.05 V vs. SSE)	Biosensor; PBS 0.05 M, pH 7.4	[1266]
Glucose	Glucose oxidase + PEI	Rh-C/mo	A, FIA (−0.1 or 0 V vs. SSE)	Biosensor	[1262]
Glucose	Glucose oxidase	Rh-C/mo (or silicone grease)		Biosensor; remarkable thermal stability over a long period	[291,2707]
Glucose	Glucose oxidase in paste	Rh-C/mo	A (0.6 V vs. SSE)	Biosensor; resistance toward acidic deactivation; PBS 0.05 M, pH 7.4	[1854]
Glucose	Glucose oxidase + zeolite	Rh-C/mo	A (−0.05 V vs. SSE)	Biosensor; increased sensitivity due to hydrophilic involvement of interior of electrode; PBS 0.05 M, pH 7.4	[754]
Glucose	Glucose oxidase	Rh-C/mo	A (0.6 V vs. steel)	Biosensor; needle-type electrode, dual sensors for glucose and insulin	[612]
Glucose	Glucose oxidase	Rh-C/poly(dimethyl siloxane) or KelF-oil	A (+0.7 V [siloxane] or 0.2/0.6/0.8 V [KelF] vs. SSE) or flow A	Biosensor; dissolved O_2 in siloxane or fluorocarbons as a source for oxygen; PBS 0.05 M, pH 7.4	[611,873,1789, 2837]
Glucose	Glucose oxidase + myoglobin	Rh-C/mo or KelF	A (0.6 V vs. SSE)	Biosensor with myoglobin as oxygen reservoir; PBS 0.05 M, pH 7.4	[609]
Glucose	Glucose oxidase + Rh or dimethyl Fc	G/po	A, FIA (−0.05 V [Rh] or 0.1 [DMFc] V vs. SSE)	Biosensors; comparison between metallized (Rh) and mediated (Fc) CPE glucose biosensors; PBS 0.05 M, pH 7.4	[1260]
Glucose	Glucose oxidase + PEI [+ Cu(II)-hexacyanoferrate(II)]	Rh-C or C (for CuHCF)/mo	A (−0.1 V vs. SSE)	Comparison between Rh- and CuHCF CPEs; PBS, pH 0.05 M, 7.4	[1035]
Glucose	Glucose oxidase + chitin + Pt-powder	GC/di-n-octyl phthalate	CV, A (0.55 or 0.6 V vs. SSE)	Biosensor; electrostatic immobilization of GOD on protonated chitin; acetate buffer, 0.1 M, pH 6.2; LOD 0.5 μM: s: sports drinks	[285,1711]
Glucose	Glucose oxidase on keratin + Pt-powder		A (0.85 V vs. SSE)	Biosensor via detection oh H_2O_2; acetate buffer 0.1 M, pH 6	[2838]

Analyte	Modifier	Electrode	Technique	Notes	Ref.
Glucose	Glucose oxidase	G/oil, electrochem. platinized	CV, A (0.5 V vs. SSE)	Biosensor; PBS 0.05 M, pH 7.4; LOD 20 µM	[2839]
Glucose	Glucose oxidase + Pt + melanin-type polymer layer			Biosensor; decrease of interferences; LOD 60 µM	[1282]
Glucose		PtNP decorated CNT/po	A, EIS		[405]
Glucose	Glucose oxidase	Pt, Pt-Ru, Ru-dispersed G	A (−0.05 V vs. SSE)	Biosensor; PBS 0.05 M, pH 7.4	[1267]
Glucose	Glucose oxidase + mushroom tissue	C-Pt/mo	A (0.6 V vs. SSE)	Biosensor; mushroom tissue as a source of tyrosinase to eliminate interference from acetaminophen; PBS 0.05 M, pH 7.4	[2840]
Glucose	Glucose oxidase + PEI	Ir-C/mo	CV, A (0 V vs. SSE), FIA	Biosensor; PBS 0.05 M + KCl 0.03 M, pH 7.4	[1263]
Glucose	Glucose oxidase + Ir or Pd or Pd/Ir or Cu or Ir/Cu or Ru or Ru/Ir	G/mo	CV, A (−0.1 to −0.1 V vs. SSE)	Biosensors via H_2O_2; PBS 0.05 M, pH 7.4	[1283]
Glucose	Glucose oxidase		A (−0.1 V vs. SSE)	Biosensor; PBS 0.05 M, pH 7.4	[375]
Glucose	Glucose oxidase + Cu microparticles	CNT/mo + Ir or Cu microparticles	CV, A (−0.1 V vs. SSE)	Biosensor; PBS 0.05 M, pH 7.4; LOD 0.15 mM; s: blood	[1281]
Glucose	Glucose oxidase (Nafion membrane) + carbon-iron nanocomposite	G/oil	CV, A (0.05 V vs. SCE)	Reagentless biosensor; PBS 0.067 M + KCl 0.1 M; pH 7; LOD 3.2 µM; s: serum	[2712]
Glucose	Glucose oxidase + Au microparticles	G/p	A, FIA (0.5 V vs. SSE)	Biosensor; PBS 0.05 M + NaCl 0.03 M, pH 7.4	[1270]
Glucose	Glucose oxidase + colloidal Au (NP)	G/mo	CV (peak at −0.43 V vs. SSE)	Biosensor with direct e-transfer to AuNP; i inversely proportional to the particle size; pH 5–7	[941]
Glucose	Glucose oxidase (ads) + AuNP	G(heat act)/po	CV	Biosensor with direct e-transfer; PBS 0.1 M, pH 5; LOD 10 µM; s: serum	[1274]
Glucose	Glucose oxidase + $LaNi_{0.5}Ti_{0.5}O_3$–$NiFe_2O_4$	G/po	CV, A (0.35 V or 0.5 V vs. SCE)	Biosensor with improved immob. of GOD; PBS 0.1 M + Fc-carboxylic acid 0.5 mM, pH 7; LOD 0.05 mM	[1250]
Glucose	Glucose oxidase + BSA + GA + TTF + polycarbonate membrane	G/Nj	CV, FIA (0.2 V vs. SSE)	Biosensor; pH 7.5–8; LOD 30 µM; s: whole blood (without dilution)	[1486, 2841–2843]
Glucose	Glucose oxidase + TCNQ	G/Nj	A (0.22 V vs. SCE)	Biosensor; pH 7; LOD 0.5 mM	[2844]

(continued)

TABLE 8.19 (continued)
Analysis of Biologically Important Compounds at Carbon Paste Electrodes: Carbohydrates and Related Compounds

Analyte	Modifier	Carbon Paste	Techniques	Comments	Ref(s)
Glucose	Glucose oxidase + BSA + GA + tetrabutyl ammonium TCNQ	G/po	CV, A (0.15 V vs. SCE)	Biosensor with e-transfer to enzyme via mediator; PBS 0.1 M + KCl 0.1 M, pH 7; s: serum	[1493,2845]
Glucose	Glucose oxidase + TCNQ/TTF (/Fc) or + Fc	G	CV, A	Comparison of different mediators	[2846]
Glucose	Glucose oxidase bound to Sepharose CL-6B on composite + TTF or Fc	G + epoxy resin powder (pressed)	A (0.27 V for TTF, 0.15 V for Fc)	Biosensor; PBS 0.01 M, pH 6.8; LOD 0.1 mM; s: molasses	[2847]
Glucose	Glucose oxidase (apo- and native) (+FAD) (+p-benzoquinone)	GC/so	Impedance, capacitance measurements	Biosensor; surface and capacitance characterization	[2848]
Glucose	Glucose oxidase + hydroquinone	G/mo, activated	CV, A (0.15 V vs. SSE)	Biosensor; activation by cycling between 0.6 and 2 V vs. SSE; PBS 0.1 M, pH 7 (CV) or NaHCO$_3$ 0.5 M (A); s: Monitrol	[1448]
Glucose	Glucose oxidase + naphthoquinone		CV, A (−0.16 V vs. SCE)	Biosensor with naphthoquinone as e-acceptor, for *in vivo* meas.; PBS 0.15 M, pH 7.4	[1450]
Glucose	Glucose oxidase + benzoquinone or riboflavin	Rotating disc electrode	CV	Biosensor; study of electron transport from enzyme to electrode	[2849]
Glucose	Glucose oxidase + polymers with benzoquinone		CV, A	Biosensor with redox polymer	[1438]
Glucose	Glucose oxidase + phenothiazine or benzoquinone + hexacyanoferrate(III) or NaFeEDTA on silica gel	G/p	A (0.15–0.27 V vs. SSE)	Biosensor with silicagel particle as microreactor encrusted in CPE; PBS 0.05 M + KCl 0.1 M, pH 7	[1440,2850]
Glucose	Glucose oxidase + (poly) aminobenzoquinones	G/po	CV, A (0.1–0.4 V vs. SSE)	Biosensor with quinones as e-acceptors; PBS 0.1 M + KCl 0.1 M, pH 7	[1759]
Glucose	Glucose oxidase + poly (ether amine quinone)	G/po	CV, A (0.3 V vs. SSE)	Biosensor; PBS 0.1 M + KCl 0.1 M, pH 7	[1757]
Glucose	Glucose oxidase + ubiquinone (in po)	G/po	A (0.3–0.5 V vs. SSE)	Biosensor with ubiquinone as e-acceptor; PBS, pH 7	[1456,2851, 2852]
Glucose	Glucose oxidase + methylene green	G/po	A (0.068 V vs. SCE)	Biosensor with MG as e-acceptor; PBS 0.1 M + KCl 0.1 M, pH; s: serum	[244]

Electroanalysis with Carbon Paste–Based Electrodes, Sensors, and Detectors 363

Analyte	Modifier / Enzyme system	Electrode	Technique	Remarks	Ref.
Glucose	Glucose oxidase + viologen derivatives	G/po	CV, A (−0.1 to −0.2 V vs. SCE)	Biosensor; PBS 0.1 M + KCl 0.1 M, pH 7	[1498,1756]
Glucose	Glucose oxidase + poly(m-phenylene diamine) film	G/mo	CV, A (0.65 V vs. SSE)	Biosensor with catalytic activity of polymer to H_2O_2; PBS 0.05 M, pH 7.4	[1780]
Glucose	Glucose oxidase + poly (o-phenylenediamine) film	Rh-C/mo	CV, A (−0.1 V vs. SSE)	Biosensor with permselective characteristics; PBS 0.05 M, pH 7.4	[1261]
Glucose	Glucose oxidase bound to poly(ethylene glycol) or glucose oxidase	G/mo	CV, DPV, A (0 V vs. SSE)	Biosensor with direct electron transfer; citrate buffer 0.1 M, pH 5	[1781,1782,860, 2853]
Glucose	Glucose oxidase + HRP, bound to C and cross-linked with GA	C act. with carbodiimide/so	A (−0.05 V vs. SSE)	Biosensor with direct e-transfer to HRP; pH 5	[2657]
Glucose	Glucose oxidase + HRP + tetramethylbenzidine	G/solid paste		Biosensor via H_2O_2	[321]
Glucose	Glucose oxidase	CNT/mo	CV, ECL	Biosensor via H_2O_2; PBS 7 + luminol 10^{-5} M; s: serum	[386]
Glucose	Glucose oxidase + fibrinogen (ads.)	G/Nj	CV, A (1 V vs. SCE)	Biosensor; improved sensitivity due to fibrinogen; PBS 0.1 M + KCl 0.1 M, pH 7; LOD 0.1 mM	[2855,2856]
Glucose	Wired glucose oxidase cross-linking with poly(ethylene glycol) diglycidyl ether		Microdialysis with FIA (0.15 V vs. SSE)	Biosensor for *in vitro* determination of glucose	[1784]
Glucose	Glucose oxidase + GA + 1-(N,N-dimethylamine)-4-(4-morpholine)benzene	G/different pasting oils, dual CPE or array	A (0.08 V vs. SSE)	Biosensor; long-term stability investigations; PBS 0.1 M + NaCl 0.1 M + PEG 2%, pH 7.4	[2857,2858]
Glucose	Glucose oxidase + GA + 1-(N,N-dimethylamine)-4-(4-morpholine)benzene	G/liquid p	A (0.1 V vs. SSE)	Biosensor; Fourier analysis of periodic variations after slicing off surfaces (0.3 mm)	[574]
Glucose	Glucose oxidase/GA + Meldola blue	G/po	A (0.05–0.1 V vs. SCE)	Biosensor; PBS 0.1 M + KCl 0.1 M + PEG 2%, pH 7.4	[1482]
Glucose	Glucose dehydrogenase + NAD⁺ + Meldola blue + Eastman AQ290 membrane	G/po	FIA (0.1 V vs. SSE)	Biosensor; PBS 0.25 M, pH 7	[1481,2801, 2859]

(continued)

TABLE 8.19 (continued)
Analysis of Biologically Important Compounds at Carbon Paste Electrodes: Carbohydrates and Related Compounds

Analyte	Modifier	Carbon Paste	Techniques	Comments	Ref(s)
Glucose	Glucose dehydrogenase + NAD$^+$ + Toluidine blue O polymers (+Eastman AQ290 membrane)	G/po	CV, A (0 V vs. SSE)	Biosensor; PBS 0.25 M, pH 7	[935,2860]
Glucose	Toluidine blue O PEI polymer + glucose de-hydrogenase + NAD$^+$	G/po	CV, FIA (0 V vs. SCE)	Biosensor, redox polymer, direct e$^-$-transfer.; PBS 0.25 M, pH 7	[938]
Glucose	Ru-complexes + glucose dehydrogenase (mixed, ads. or in membrane) + Nafion membrane (ads.)	C/so	CV, A (0.8 V vs. SSE)	Biosensor; PBS 0.05 M, pH 7	[1601]
Glucose	Glucose dehydrogenase + Os polymers + poly(ethyleneglycol) diglycidyl ether + diaphorase + NAD$^+$	SWCNT/mo	CV; A (0.2 V vs. SSE)	Biosensor; catalytic reoxidation of NADH with diaphorase and Os(III); catalytic oxidation of Os(II) at the CNT; s: wine	[385]
Glucose	Glucose dehydrogenase + Os polymers + NAD$^+$		A (0.1 V vs. SSE)	Reagentless biosensor	[1754]
Glucose	Glucose dehydrogenase + phenazine methosulfate		A (0.1–0.2 V vs. SSE)	Biosensor	[2861]
Glucose	Glucose dehydrogenase			Computer-simulated models	[2850,2862]
Glucose	*Gluconobacter oxydans* + dialysis membrane + mediator (ferrocenes or quinones)	G/mo	CV, A (0.065–0.51 V vs. SSE)	Biosensor; phosphate-citrate buffer 0.15 M, pH 6.0	[2863]
Glucose	*Gluconobacter industrius* + benzoquinone + dialysis membrane	G/liquid p	CV, A (0.5 V vs. SSE)	Biosensor with *Gluconobacter* as a source for dehydrogenase; PBS 0.1 M, pH 7; LOD 10 µM	[1441,2864]
Glucose	*Escherichia coli* + 2,3-dimethoxy-5-methyl-1,4-benzoquinone + dialysis membrane		CV, A (0.2 V vs. SSE)	Biosensor with *E. coli* as a source of glucose dehydrogenase; PBS + KCl 0.3 M + PQQ 1 µM + MgSO$_4$ 5 mM, pH 6.5; s: blood, regeneration with EDTA	[228]
Glucose	*Lactobacillus fructivorans* (Hiochi) + benzoquinone		A (0.5 V vs. SSE)	Biosensor with *Bacilli* as dehydrogenase source; PBS 0.2 M, pH 6; s: sake brewing	[1443]

Analyte	Enzyme / biocomponent	Paste composition	Technique	Note	Ref.
Glucose	*Aspergillus niger* (in paste) + Fc	G/po	CV, A (0.35 V vs. SSE)	Microbial biosensor with A. niger as a source of GOx; PBS 0.2 M, pH 7; s: xanthan fermentation on milk, lactose	[1521]
Glucose	Glucose oxidase	G/po or KelF oil		Fill and flow cell; carbon paste only as an enzyme matrix, H_2O_2-detection with downstream Pt-electrode	[2865]
Glucose	Glucose oxidase + hematin	G/silicone grease	A (−0.4 V vs. SSE)	Chemiluminescence sensor with luminol; potential to prevent monotonous decrease of signal	[1501]
Glucose	Glucose oxidase			Optical fiber biosensor, based on luminescence by enzymic release of H_2O_2 and electrooxidation of luminol	[2866]
Glucose Lactose	Glucose oxidase (+β-galactosidase) + dialysis membrane	Dual electrode G/mo	A (0.9 V vs. SSE)	Biosensor; lactose conc. evaluated as difference; PBS 0.1 M, pH 7.5; s: milk	[2867]
Glucose Sucrose	Glucose oxidase/BSA/GA (el.1) and glucose oxidase + invertase + mutarotase/BSA/GA (el.2) + catalase (between el.1 and 2)	Dual electrode (el.1 and 2) in flow cell; G/light mo	FIA	Biosensors for simultaneous detection of glucose and sucrose; PBS, pH 6.8 s: cola drinks	[250]
Glucose labeled with daunomycin	Chitin	G/di-n-octyl phthalate or Nj	DPV	Indirectly via precipitation and inactivation with wheat germ agglutinin on chitin; electrochem. active label; LOD 1 nM	[832]
Mono-saccharides Glucose Glucose galactose	Pyranose oxidase + HRP + (GA + PEI + lactitol + (dihydro)-streptomycin + mutarotase)	C+carbodiimide/po	FIA (−0.05 V vs. SSE), det. in HPLC	Biosensor; PBS 0.1 M, pH 7	[2868]
Pectin	Alcohol oxidase + Eastman AQ29D + orange peel (or pectin-esterase in membrane) + dialysis/PTFE membrane	G/mo	A (0.9 V vs. SSE)	Hybrid tissue-enzyme biosensor; orange peel as a source of pectin esterase on alcohol sensor; detection via released methanol; PBS 0.1 M, pH 7.5	[277]
Pyranoses	Pyranose oxidase + MWCNT	G/mo	CA	Biosensor; PBS 0.05 M, pH 7.5; s: glucose in wine	[400]
Saccharides	Oligosaccharide dehydrogenase + benzoquinone or poly (ether amine benzoquinone)	G/po	FIA	Biosensor with quinone as e-acceptor; responds to 16 sugars with highest response to glucose; PBS 0.1 M, pH 7; LOD 1.7 μM (glucose)	[1439]

TABLE 8.20
Analysis of Biologically Important Compounds at Carbon Paste–Based Electrodes: Coenzymes, Enzymes, Proteins, and Related Compounds

Analyte	Modifier	Carbon Paste	Techniques	Comments	Ref(s)
Coenzymes					
FAD, FMN, riboflavin	Adsorbed on SiO_2/ZrO_2	G/mo	CV	PBS 0.1 M, pH 7	[1188]
Lipoic acid	Ni(II) cyclohexyl butyrate	G/p (liquid)	acc. (−1.35 V vs. ref., 2 min), LSV	ox.: ABS 0.01 M, pH 4.6; LOD 4 nM	[1606]
Lipoic acid Thiols, disulfides	Co(II)PC			ox.	[886]
NADH			AD in CZE (0.85 V vs. SCE)	Separation from uric acid; PBS, pH 7.5. LOD 0.6 µM	[2869]
NADH	Surfactant	G/po or silicon grease or ceresin wax	V	Improved current, decreased cell resistance	[64]
NADH Ascorbic acid Catechol Cysteine		G/Se-30 (stationary phase in chromatography)	CV, EIS: A (0.6 V vs. SSE), FIA (0.47 V vs. SSE)	Increase of e-transfer rate due to increased hydrophilicity; PBS 0.1 M, pH 7	[2870]
NADH		MWCNT/po	CV, A (0.45 V vs. SSE)	Comparison of MWCNT electrodes, highest catalysis with 20–30 nm diameter; PBS, pH 7.4	[395]
NADH Dopamine		SWCNT/mo	CV	Pretreatment of CNT, comparative study with GC, PT and CPE; PBS 0.2 M, pH 7	[124]
NADH DOPAC Dopamine Catechol Guanine		SWCNT/mo	CV, DPV, CA	Improved response (current, potential) compared to unmodified CPE, PBS 0.1 M, pH 7	[2871]

Analyte	Electrode	Method	Notes	Ref
NADH, Dopamine, Ascorbic acid, Uric acid, Catechol, Catechol, Acetamidophenol	Carbon nanofibers/mo	CV, EIS	Improved current, decrease of overpotential; PBS 0.06M, pH 6.85	[203]
NADH, Uric acid, Dopamine, Acetaminophen, Ascorbic acid, Epinephrine	Ordered mesoporous carbon/mo	CV	Improving effects of ordered mesoporous C, comparison with CNT and CPE; PBS 0.05 M, pH 7.4	[119]
NADH	G/poly(methyl methacrylate)	CV, EIS, A (0.45V)	Composite electrode, decrease of overpotential; PBS 0.1M, pH 7; LOD 3.5 µM	[2872]
NADH, Ascorbic acid, Ascorbic acid, Hydrazine	G/butyl-methyl-imidazolium hexa-fluorophosphate, microelectrodes	CV, A (0.4 or 0.8V vs. SSE)	Comparison of various CILEs; PBS 0.05 M, pH 7.4	[472]
NADH	CILE 1-octyl pyridinium hexa-fluoro phosphate	CV, DPV	Significant decrease of overpotential; PBS 0.1M, pH 6.8	[453]
NADH	G/po	CV	Effect of electrochem. pretreatment (preanodization and precathodization) on signal; catalytic influence of ox. products of NAD⁺; PBS 0.1 M, pH 10	[654,1904, 2873,2874]
NADH, H_2O_2	NAD⁺ modified and unmodified; + electrochemical pretreatment	EIS, CV	Catalytic ox.	[2875]
NADH	Pd NP on C nanofibers (electrospun)			
NADH	Ferrocene carboxylic acid on SiO_2/Nb_2O_5 Also as RDE	CV, A (0.35V vs. SCE)	Catalytic ox.; KCl 0.5M, pH 6.5	[535]
NADH	Ferrocene + diaphorase + Tween 20 + GA + BSA G/vaseline	CV, FIA (0V vs. SSE)	Biosensor, catalytic ox.; PBS 0.1 M + KCl 0.1M, pH 7	[875]

(continued)

TABLE 8.20 (continued)
Analysis of Biologically Important Compounds at Carbon Paste–Based Electrodes: Coenzymes, Enzymes, Proteins, and Related Compounds

Analyte	Modifier	Carbon Paste	Techniques	Comments	Ref(s)
NADH NADPH	Vinyl Fc + acryl amide and N,N'-methylene bisacrylamide + lipo-amide dehydrogenase/glutathione reductase + dialysis membrane	G/mo	CV	Redox gel; enzymes accelerate co-polymerization, catalyze ox. of NAD(P)H, direct e−-transfer via Fc.; PBS 0.1 M, pH 7	[782]
NADH	Zeolite/Fc or zeolite + Fc	G/po	CV, CA (0.3 V), FIA (0.3 V vs. SSE)	Synergetic effect of zeolite + Fc, catalytic ox.; PBS 0.1 M, pH 7	[876]
NADH	Bis(1,10-phenanth-oline-5,6-dione) (2,2′-bipyridine) Ru(II) in Zr-phosphate	G/mo	CV, SWV	Catalytic ox., PBS ($\mu = 0.1$)	[1299]
NADH NADPH 8-Oxoguanine	Guanine or 8-oxoguanine (electroox., +1.1 V vs. SSE)	G/Nj	DPV	Signal improvement; universal buffer, pH 8: LOD 3.3 µM (NADH), 3.7 µM (NADPH)	[2876]
NADH	Hematoxylin	G/po	CV, DPV, CA	Catalytic ox.; PBS 0.1 M, pH 7; LOD 80 nM	[1503]
NADH	Pyrroloquinoline quinone on SiO_2/ZrO_2	G/hydrocarbon oil	CV	Catalytic ox.; Ca^{2+} improves electrode stability; PBS 0.2 M, pH 4.5	[1189]
NADH	Indophenol, o-quinone-derivatives on Zr-phosphate	G/po	CV	Catalytic ox. via intermediate complex, positive effect of Ca^{2+}	[1298]
NADH	Coumestan	G/po	CV, CA (0.3 V vs. SCE), DPV	Catalytic ox.; PBS 0.1 M, pH 7	[2877]
NADH	3,4-Di-hydroxy benzaldehyde	CN/mo		Comparison of mediators with GCE; PBS 0.1 M, pH 7	[2878]
NADH	3,4-Di-hydroxy benzaldehyde (electropolymerized)	G/po	CV, A (0.23 V vs. SSE)	Catalytic ox.; PIPES buffer, pH 6.8	[2879]
NADH	p-Methylaminophenol-sulfate or 3,4-di-hydr-oxybenzaldehyde	SWCNT		Catalytic ox.; small effects of p-maps; PBS 0.1 M, pH 7	[368]
NADH	o-Phenylene diamine on SiO_2/Nb-oxide	G/mo, also as RDE	CV, CA (−0.5 V vs. SCE)	Catalytic ox.; PBS 0.1 M, pH 7	[1299]

NADH NADPH	Poly(o-phenylene diamine) or poly(o-aminophenol) (electropolymerized)	G/po	CV, A (0.15 V vs. SSE)	Catalytic ox.; LOD 0.81 nM; PBS 0.1 M, pH 7	[1786]
NADH	Methylene green	G/po	FIA	Catalytic ox.	[656]
NADH	Diaphorase + methylene green or Meldola blue	G/po	CV	Biosensor, catalytic ox.; PBS, pH 7	[1469]
NADH	Methylene green on SiO_2/Nb-oxide	G/mo	CV, FIA (−0.5 V vs. SCE)	Catalytic ox.; KCl 0.5 M; LOD 8.2 μM	[1196]
NADH	Methylene blue on SiO_2/ZrO_2/Sb_2O_5	G/mo	CV, A (0.05 V vs. SCE)	Catalytic ox.; PBS 0.06 M +KCl 0.2 M, pH 6.8	[1190]
NADH	Methylene blue/zeolite-Ca	G/po	CV, A (0 V vs. SSE)	Catalytic ox.; PBS 0.1 M, pH 7; LOD 0.8 (Ca) and 4.3 μM	[1130]
NADH	Methylene blue on Ba-phosphate	G/mo	CV, CA (−0.08 V vs. SCE)	Catalytic ox.; PBS 0.5 M, pH 7; LOD 3.9 μM	[2880]
NADH	6,7-Dihydroxy-3-methyl-9-thia-4,4a-diazafluoren-2-one/MWCNT film	G/p	CV, DPV, CA (0.35 V vs. SSE)	Catalytic ox.; PBS 0.1 M, pH 7 or 7.5	[439]
NADH	Nitro-fluorenone derivatives on Zr-phosphate	G/po, also as RDE	CV, FIA (0.35 V vs. SSE)	Catalytic ox., influence of Ca^{2+}; Tris buffer, 0.1 M, pH 7; LOD 0.07–0.6 μM	[1295,1296]
NADH	Nile blue or N-methyl phenazonium on Zr-phosphate	G/po	CV	Catalytic ox., complex formation; application as ethanol sensor with ADH	[1291]
NADH	Nile blue or methyl viologen or benzyl viologen on Zr-phosphate	G/po	CV	Catalytic ox.	[1290]
NADH	Nile blue on SiO_2/Nb-oxide	G/mo, also as RDE	CV, A (−0.2 V vs. SCE)	PBS 0.5 M, pH 7	[1193]
NADH	Methylene blue or Nile blue on Zr-phosphate	G/mo	CV	Catalytic ox.; KCl 0.1 M	[1286]
NADH	Toluidine blue O on polymer	G/p	CV, FIA	Redox polymer, catalytic ox.; PBS 0.25 M, pH 7 (CV) or PBS 0.1 M, pH 7.5 (FIA)	[1473]
NADH	Toluidine blue O polymers	G/po	CV, A (0 V vs. SCE)	Redox polymers, catalytic ox.; PBS 0.25 M, pH 7	[938]

(continued)

TABLE 8.20 (continued)
Analysis of Biologically Important Compounds at Carbon Paste–Based Electrodes: Coenzymes, Enzymes, Proteins, and Related Compounds

Analyte	Modifier	Carbon Paste	Techniques	Comments	Ref(s)
NADH	Toluidine blue on SiO_2/Nb-oxide	G/mo	CV	Catalytic ox.; PBS 0.5 M, pH 7; LOD 34 µM	[1195]
NADH	Toluidine blue covalently on SiO_2	G/po, also as RDE	CV, CA (0.1 V vs. SSE), FIA (0.1 V)	Catalytic ox.; PBS 0.1 M, pH 7	[1476]
NADH	Meldola blue	G/po or sol–gel from Zr-propoxide or tetraethyl o-titanate	CV, CA (0.6 V vs. SSE), A, FIA (0.2 V vs. SSE)	Carbon composites also with o-silicates, catalytic ox.; PBS 0.05 M, pH 7.4	[2881]
NADH	Meldola blue on SiO_2/Ti-phosphate	G/po	CV, A (0 V vs. SCE)	Catalytic ox.; PBS 0.1 M, pH 7.4	[1212]
NADH	Meldola blue on SiO_2/Nb-oxide	G/mo	CV	Catalytic ox.; KCl 0.5 M	[1194]
NADH	Meldola blue on Ti-phosphate + PEI	GC/po, also as RDE	CV, A (0.05 V vs. SSE)	Catalytic ox.; Tris buffer, 0.1 M, pH 7	[2882]
NADH	Meldola blue or methylene green or riboflavin on Zr-phosphate	G/po as RDE	CV, A	Catalytic ox.; pH dependence; Tris buffer 0.1 M, pH 6.5 or 7	[1287]
NADH	Meldola blue on Zr-phosphate + PEI	GC/po, also as RDE	CV	Catalytic ox., positive effect of Ca^{2+}; Tris buffer, 0.1 M, pH 7	[2883,2884]
NADH	Meldola blue or toluidine blue on cellulose acetate/TiO_2	G/p (liquid)	CV, A (0.02 V vs. SCE)	Catalytic ox.; KCl, 1 M, pH 7	[2885]
NADH	Meldola blue, methylene blue, toluidine blue on SiO_2/Sb_2O_3	G/hydrocarbon oil	CV, A (0 V vs. SCE)	Catalytic ox.; comparative study; PBS 0.2 M + KCl 0.06 M, pH 7; LOD 7–42 µM	[1205]
NADH	p-Nitrophenylcarboxyl toluidine blue O			Catalytic ox.; LOD 5.1 µM	[2886]
NADH	16H,18H-dibenzo[c,l]-7,9-dithia-16,18-diaza-pentacene (on Zr-phosphate)		CV	Catalytic ox., improvement by Ca^{2+}; PBS 0.1 M, pH 7	[1297]

Analyte	Modifier	Binder	Method	Note	Ref.
NADH	3,7-Di(m-aminophenyl)-10-ethyl phenothiazine on zeolite or bentonite	G/po, also as RDE	CV	Catalytic ox.; PBS 0.1 M, pH 9	[2887]
NADH	4-Nitrophthalonitrile	G/mo, also as RDE	CV, A (0.1 V vs. SSE)	Catalytic ox. via electrogenerated NO-compound; Tris buffer 0.1 M, pH 7	[1512]
NADH	Methylated azopyridine	G/N,N'-dimethyl-4,4'-azopyridinium hexafluorophosphate	CV, A (0.25 V vs. SSE)	Catalytic ox.; PBS 0.1 M, pH 6.8; LOD 2 µM	[2888]
NADH	Riboflavin + (or on) Zr-phosphate	G/po	A (0.05 or 0.15 V vs. SCE)	Catalytic ox.; improvement by Ca^{2+} or Mg^{2+}	[1292,1293]
NADH NADPH	+HRP + Fc + dialysis membrane		CV	ox. with 1-methoxy-phenazine metho-sulfate, reox. with O_2 under formation of H_2O_2, biocatalytic det. of H_2O_2; PBS 0.1 M, pH 7.4	[2889]
NADH NAD^+	Diaphorase + glucose-6-phosphate dehydrogenase + vitamin K3 + membrane			Biosensor; increased response of NADH and NAD^+ in the presence of glucose-6-phosphate	[2890]
NADH NADPH	Electrooxidation products from ATP + poly(o-phenylene diamine) membrane (electropolymerized)	G/silicone grease	CV, A	Catalytic ox., Tris buffer, 0.1 M, pH 7.8; LOD 5 nM (NADH)	[1779]
Oligopeptides					
Glutathione Cysteines	AuNP-cysteamine	G/mo	CV, A (600 mV vs. Ag/AgCl); HPLC	Catalytic ox., PBS, pH 7.0; s: serum	[947]
Glutathione Disulfide	Nanoscale $Cu(OH)_2$	G/octylpyridinium hexafluorophosphate CILE	CV, EIS	Catalytic irreversible ox. of –SH at 0.25 V, of –SS–at 1.1 V, simultaneous det., PBS 0.1 M, pH 7	[2891]
Glutathione Cysteine	$V^{IV}O$-salen	G/mo	CV, LSV, A (0.8 V vs. SCE)	Catalytic ox. of thiol group via V(V)/V(IV)	[1594]
Glutathione Cysteine	Ru(III)-diphenyldithiocarbamate	G/p	CV, A	ox at +0.38 V vs. SCE; KNO_3/HNO_3, pH 3; LOD 9.16 mg L^{-1}	[1608]

(continued)

TABLE 8.20 (continued)
Analysis of Biologically Important Compounds at Carbon Paste–Based Electrodes: Coenzymes, Enzymes, Proteins, and Related Compounds

Analyte	Modifier	Carbon Paste	Techniques	Comments	Ref(s)
Glutathione	Fc	G/p (liquid)	CV, DPV, CA (0.45/0 V vs. SSE)	Catalytic ox.: PBS 0.1 M, pH 7; LOD 18 μM (CV), 2.1 μM (DPV)	[1529]
Glutathione	2,7-Bis (ferrocenyl ethyl) fluoren-9-one	G/p	CV, DPV, CA (0.4/0.2 V vs. SSE)	Catalytic ox.: PBS 0.1 M, pH 7; LOD 14 μM (CV), 0.51 μM (DPV); s: hemolyzed erythrocytes	[1550]
Glutathione Tryptophan	TiO$_2$ NP + ferrocene carboxylic acid	G/p	CV, CA (0.45/0.3 V vs. SSE), EIS, DPV	Catalytic ox.: PBS 0.1 M, pH 7; LOD 98 nM; s: blood	[920]
Glutathione	CoPC + dialysis membrane		A (0.15 V vs. ref.)	Indirect det. by GSH current decrease by addition of N-ethylmaleimide, no deproteinization; s: whole blood	[777]
Glutathione Cysteine	Co(II)-PC	G/Nj	CV, A det. in LC	ox at 0.75–0.85 V vs. Ag/AgCl	[883]
Glutathione Cysteine	Co(II)-PC	G/mo, microelectrodes for CE	A (0.4 V vs. SSE)	ox via Co(III); LOD 31 nM; s: urine	[555]
Glutathione Cysteine	Co-PC	Carbon cement	FIA, HPLC (0.7 V vs. Ag/AgCl)	In urine	[324]
Glutathione Cysteine	Co-PC	G/Nj	A, det in LC (0.75 V vs. Ag/AgCl)	In serum	[884]
Glutathione	CoPC + membrane (better permeability for thionitrobenzoic acid)		A (0.2 V vs. ref.)	det. via increase of current by 5,5'-dithiobis-(2-nitrobenzoic acid)due to formation of TNB; s: yeast extract	[2892]
Glutathione	TTF-TCNQ	G/Nj	CV, A (0.2 V vs. SCE)	PBS 0.1 M + KCl 0.1 M + EDTA 0.5 mM, pH 8	[1496]
Glutathione	MWCNT + chlorpromazine	G/p	CV, CA (0.4/0.8 V vs. SSE)	Catalytic ox.: universal buffer 0.4 M, pH 4; LOD 0.16 μM; s: hemolyzed erythrocytes	[445]
Glutathione	4-Nitrophthalonitrile	G/mo	A (0.4 V vs. SSE)	Catalytic ox.: PBS 0.1 M, pH 7; LOD 2.7 μM	[1511]
Glutathione Amino acids	Crown ethers		SV	Host–guest complex LOD 2–5 × 10^{-8} M	[1394]
Glutathione Thiols	Sulfhydryl oxidase on gelatine	G/mo	A (0.7 V vs. SSE), AD with HPLC	Biosensor; PBS 0.05 M, pH 7.5	[2672]

Electroanalysis with Carbon Paste–Based Electrodes, Sensors, and Detectors 373

Analyte	Electrode	Method	Notes	Ref.
Glutathione Ethanol	Pyranose oxidase	A (0.9 V vs. SSE)	Biosensor via detection of H_2O_2 and inhibition of pyranose oxidase	[2665]
Glutathion Thiols Cysteine	Magnetic silica microparticles/ Au + cysteamine + GA + HRP	G/p (solid)	CV, A (−0.25 V vs. SSE)	Indirect det. by ox. of hydroquinone with H_2O_2/HRP and inhibition of electrochem. red.; PBS + H_2O_2 0.1 mM, pH 6.8	[2893]
Glutathion Thiols	HRP on magnetic silica microparticles	G/p (liquid)	CV, A (0 V vs. SSE)	Indirect det. by ox. of clozapine with H_2O_2/HRP and inhibition of electrochem. red.; PBS + H_2O_2 0.1 mM, pH 7.4	[2540]
Glutathion Thiols	HRP on magnetic silica microparticles	G/p (solid)	CV, LSV, A (−0.1 V vs. SSE)	Indirect det. by ox. of acetaminophen with H_2O_2/HRP and inhibition of electrochem. red.; PBS + H_2O_2 0.1 mM, pH 7.4	[2497]
Glutathion Cysteine	Tyrosinase	G (activated) po	CV, A, FIA	Biosensor via blocking of substrate (catechol) recycling by thiols	[2721]
Oligopeptides Amino acids	CoPC	G/po	A detector in microchip CE; dual electrode	Post-column derivatization to Cu(II) complex; ox. to Cu(III) at first el., red. at second	[558]

Enzymes and enzyme activity

Analyte	Electrode	Method	Notes	Ref.
Acid stability	Glucose oxidase or polyphenol oxidase	Rh-C/mo	A (0.6 V vs. SSE)	Biosensor; resistance toward acidic deactivation; PBS 0.05 M, pH 7.4	[2836]
Alkaline phosphatase biotinylated biotin hydrazide	Streptavidin + BSA (blocking)	G/po	CV	Monitoring of avidin-biotin complex formation, det. via indigo from 3-indoxyl phosphate; Tris buffer 0.1 M, pH 7.2; competitive immunoassay for hydrazide	[2894]
Catalase activity H_2O_2	RuO_2	G/po	CV, A (0.6 V vs. SCE)	Catalase activity by decomposition of H_2O_2; PBS + KCl 0.1 M; LOD H_2O_2 20 μM (pH 7.4), 7 μM (pH 9); s: plants	[1224]
γ-Glutamyl transferase	CoPC + dialysis membrane (cutoff 5000)		A	det. via current increase by degradation of glutathione to cys-gly (more electroactive due to membrane); Tris buffer 0.1 M + GSH 1.2 mM + gly-gly 50 mM + EDTA 5 mM; pH 8.3; s: serum	[2895]
Isocitrate dehydrogenase	Electrooxidation products from ATP + poly(o-phenylene diamine) membrane (electropolymerized)	G/silicone grease	A (0.15 V vs. SSE)	det. via $NADPH$; $NADP^+$, 0.3 mM + isocitrate, 10 mM + Mn(II) 30 mM as substrate, s: serum	[1779]

(continued)

TABLE 8.20 (continued)
Analysis of Biologically Important Compounds at Carbon Paste–Based Electrodes: Coenzymes, Enzymes, Proteins, and Related Compounds

Analyte	Modifier	Carbon Paste	Techniques	Comments	Ref(s)
Lactate dehydrogenase	NADH + pyruvate	G/p (solid)	LSV	Indirect det. via decreased ox. of NADH; PBS 0.15 M, pH 7.4; s: biological fluids	[1313]
Lysozyme	IgG (blocking) + anti-lysozyme-aptamer + lysozyme		SWV	Increased immobilized lysozyme causes decrease of guanine or adenine ox. of aptamer; ABS 0.02 M, pH 5	[2896]
Neuropathy target esterase	Tyrosinase + 1-methoxyphenazine methosulfate	G/po	CV	Biosensor; determine neuropathy target esterase activity via phenol from phenyl valerate; PBS 0.05 M + 0.1 M NaCl, pH 7.0; LOD 25 nM; s: NTE assay in hen brain and blood	[2125,2776]
Neuropathy target esterase	Tyrosinase	G/po	FIA (−0.15 V vs. SSE)	Biosensor; determine neuropathy target esterase activity via phenol from phenyl valerate; PBS 0.05 M + NaCl 0.1 M, pH 7; inhibition studies by organic phosphorous compounds; s: whole blood	[2775]
Pepsin	Polyphenol oxidase (ads) + dialysis membrane	Acetylene carbon black/mo	CV, A (−0.05 V vs. SSE)	det. via released tyrosine; PBS 0.1 M + NaCl 0.15 M, pH 6.65	[779]
Plasmin	Adsorbed on CPE	G/Nj	UV–vis photometry	Activity study with H-D-val-leu-lys-p-nitroanilide, temperature, pH dependence	[2897]
Thermal stability of enzymes in CPE	Rh	G/mo		Remarkable stability of glucose oxidase in CPE environment	[291,2707]
Thrombin			UV–vis or electrochemical	det. via release of amine from H-D-Phe-Pip-Arg-4 MeO β-naphthylamide, 2HCl	[2898]
Proteins and related compounds					
Agglutinin (wheat germ)	Chitin		acc.. V	Selective binding to N-acetyl glucosamine residues of chitin	[2899]
Albumin	Cu microparticles	CNT/mo	acc. (−0.1 V vs. SSE, 10 min), SWV	det. via ox. of amino acids in protein; 0.05 M PBS, pH 7.4	[2704]
Amino acids					

Electroanalysis with Carbon Paste–Based Electrodes, Sensors, and Detectors 375

Analyte	Modifier	Paste	Technique	Notes	Ref.
Albumin–Au NP		G/po	ads; CV	det. via dep. of silver and reox.	[2900]
Alkaline phosphatase	Tyrosinase	G/po	CV, A (0V vs. SSE)	det. via phenyl phosphate and acc. of phenol; det. of biotin via competitive streptavidin alkaline phosphatase-labeled biotin assay; LOD 35 nM (phenol)	[2901]
Biotin					
Angiotensin II		G/mo	acc., PSA	det via ox. of tyrosine residue; comparison with HMDE	[92]
Vasopressin					
Avidin		G/mo	adtSV, SWV	via tyr and trp residues; comparison with CE and electrophoresis; ABS 0.2 M, pH 4; s: transgenic tobacco plants	[2902]
Azurin	4,4′-Dipyridyl	G/oil	CV	Reaction study	[1499]
Cytochromes					
Bovine serum albumin			CV	Direct oxidation of tyrosine and tryptophan residues	[832]
Concanavalin A					
RNAse A					
carcinoembryonic antigen CEA	Fe_3O_4 NP (core) + Ag-shell + anti-CEA	G/po	CV, EIS, Pot	Immunoassay, det. via potential shift upon antigen binding; s: serum	[303]
Carcinoembryonic antigen CEA	Fe_3O_4 NP + Si-thiol or polythionine + AuNP + CEA	G/po	DPV	Competitive immunoassay with HRP-labeled CEA antibodies; det. via o-phenylenediamine–H_2O_2	[2903,985]
Chromatophores from *Rhodospirillum rubrum*	Triton X100		CV	Characterization	[2904]
Cucurbitin	Cu		FIA +0.12V vs. Ag/AgCl	PBS, pH 7.0	[2705]
Cytochrome *c*	Au NP + cytochrome *c* (ads.)	G/po	CV	Characterization; response to H_2O_2; PBS 0.1 M, pH 7	[1271]
Cytochrome *c*	Cellulose + ssDNA	G/mo	acc. (oc, 10min); CV, DPV	det. via red. of cyt c; PBS 0.2 M, pH 7.4; LOD 0.5 μM; s: mitochondria rat liver	[2905]
Cytochrome *c*	Cytochrome reductase + NADH		CV	red.: PBS 0.1 M, pH 7.4	[2906]

(continued)

TABLE 8.20 (continued)
Analysis of Biologically Important Compounds at Carbon Paste–Based Electrodes: Coenzymes, Enzymes, Proteins, and Related Compounds

Analyte	Modifier	Carbon Paste	Techniques	Comments	Ref(s)
α-1-Fetoprotein	Fe_3O_4 (core) + Au (shell) NP + anti-AFP		Pot	det. by change of potential or current after antigen binding	[301]
Fibrinogen	Fibrinogen	G/Nj	CV	Characterization, pH-dependent permselectivity (iep 5.5); probes: hexacyanoferrate, Ru-ammine complexes	[2856,2907]
Hemoglobin	CTAB + hemoglobin	G/po	CV, A (−0.4 V vs. SCE)	Direct e⁻-transfer, catalytic ability for red. of O_2, H_2O_2, NO_2^-; PBS 0.1 M, pH 7.4	[2908]
Hemoglobin	Surfactant + hemoglobin (ads.)	G/po	CV	Direct e⁻-transfer; PBS 0.1 M, pH 7.3; catalytic red. of H_2O_2 and of NO	[2909]
Hemoglobin	Triacetonetriperoxide + hemoglobin	G/Nj	CV	Direct e⁻-transfer, catalytic red. of H_2O_2 and O_2; PBS 0.1 M, pH 7 or 6	[2910]
Hemoglobin	Hb + Nafion	G/p		Nafion to protect from leaching, PBS, pH 7; catalytic red. of H_2O_2 and trichloroacetic acid	[2911]
Hemoglobin	Hb	G/1-butyl-3-methyl-imidazolium hexafluorophosphate CILE	CV, EIS	Direct e⁻-transfer, catalytic ability for red. of O_2, H_2O_2, NO_2^-; PBS 0.1 M, pH 7	[664,464]
Hemoglobin	Hemoglobin in SiO_2-sol–gel film	G/Nj	CV	Direct e⁻-transfer, catalytic ability for red. of O_2, H_2O_2, NO_2^-; PBS 0.1 M, pH 7	[1169,1170, 2797]
Hemoglobin	Hb in sol–gel derived film	G/1-butyl-3-methyl-imidazolium hexa-fluorophosphate CILE		Characterization in methanol, acetonitrile and propanol, peroxidase mimetics	[2912]
Hemoglobin	Hb in Na-alginate film	G/1-butyl-3-methyl-imidazolium hexa-fluorophosphate CILE	CV	red., electrocatalytic activity to H_2O_2 and NO_2^-; BRP, pH 7	[2913]
Hemoglobin	SiO_2 NP-alginate film + hemoglobin	G/1-butyl-3-methyl-imidazolium hexa-fluorophosphate CILE	CV	Characterization, good electrocatalytic activity to trichloroacetic acid; PBS, pH 7	[503]

Electroanalysis with Carbon Paste–Based Electrodes, Sensors, and Detectors 377

Hemoglobin	Hemoglobin + Nafion + CaCO$_3$ NP	G/1-butyl-3-methyl-imidazolium hexa-fluorophosphate CILE	CV, EIS	Characterization, good electrocatalytic activity toward red. of trichloroacetic acid, H$_2$O$_2$, NO$_2^-$; BRP, pH 7	[2910,1730]
Hemoglobin	CNT + Hb		CV	Characterization, reductive dechlorination of trichloroacetic acid	[2914]
Hemoglobin	CNT + hemoglobin			Direct e$^-$-transfer, catalytic red. of H$_2$O$_2$, trichloroacetic acid, nitrobenzene; PBS, pH 7	[2915]
Hemoglobin	CNT + hemoglobin	G/p (liquid)		Direct e-transfer, catalytic red. of H$_2$O$_2$, trichloroacetic acid	[431]
Hemoglobin	SWCNT + hemoglobin + Nafion	G/1-butyl-3-methyl-imidazolium hexa-fluorophosphate CILE	CV, EIS	Characterization, good electrocatalytic activity toward red. of trichloroacetic acid; BRP, pH 7	[2916]
Hemoglobin	MWCNT + chitosan film + hemoglobin	G/CILE	CV	Characterization, good electrocatalytic activity toward trichloroacetic acid; PBS, pH 7	[417]
Hemoglobin	Hb + TiO$_2$ NP	G/CILE N-butyl-pyridinium hexafluorophosphate		Characterization, red. by direct e$^-$-transfer; PBS, pH 7	[996]
Hemoglobin	Hb in chitosan + bentonite film	G/1-butyl-3-methylimidazolium hexafluorophosphate CILE		red., catalytic red. of H$_2$O$_2$ and trichloroacetic acid	[470]
Hemoglobin	Hb in chitosan + TiO$_2$ NP film	G/CILE 1-butyl-3-methylimidazolium hexafluorophosphate		Catalytic red., catalytic effects to H$_2$O$_2$ and trichloroacetic acid; PBS, pH 7	[499]
Hemoglobin	Hb in film 1-butyl-3-methyl-imidazolium tetra-fluoroborate + chitosan + ZrO$_2$ NP	G/N-butylpyridinium hexafluoro-phosphate CILE	CV, A (−0.25V vs. SCE)	Characterization, catalytic red.; PBS 0.1M, pH 7	[2917]
Hemoglobin	Hb in dextran or agarose film	G/1-butyl-3-methylimidazolium hexafluorophosphate	CV	Improved direct e$^-$-transfer; catalytic activity to red. of H$_2$O$_2$	[512,514]

(continued)

TABLE 8.20 (continued)
Analysis of Biologically Important Compounds at Carbon Paste–Based Electrodes: Coenzymes, Enzymes, Proteins, and Related Compounds

Analyte	Modifier	Carbon Paste	Techniques	Comments	Ref(s)
Hemoglobin	Hemoglobin + Nafion + ZnO NP + 1-butyl-3-methylimid-azolium hexafluorophosphate	G/p (liquid)	EIS, CV	Characterization; good electrocatalytic activity toward trichloroacetic acid (reductive de-chlorination, re-ox. of Fe(II) from he); PBS 0.1 M, pH 7	[496]
Hemoglobin	Hb in CdS nanorods + Nafion film	G/1-butyl-3-methyl-imidazolium hexa-fluorophosphate CILE	CV	red.; catalytic effect to trichloroacetic acid; PBS, pH 7	[1889]
Hemoglobin	Cresyl fast violet		CV	Catalytic red.; ABS 0.2 M, pH 5.5	[2918]
IgG human	anti-IgG + BSA (block)	G/p (solid)	LSV	Sandwich immunoassay with Au-labeled antibodies; det. via Au (ox.–red.)	[1276]
IgG human	Fe_3O_4 NP + cysteine + GA + anti-IgG + BSA (blocking)	G/p (solid)	Pot	Immunosensor, PBS, pH 7.4; LOD 23 ng^{-1}L	[980]
IgG human	IgG (ads.) + BSA (blocking) + anti-IgG-alkaline phosphatase	G/po	adSV with ACV	Immunoassay, det. via indigo from -indoxyl phosphate; Tris buffer 0.1 M, pH 7.2	[2919,2920]
IgG human	IgG (ads.) + anti-IgG-alkaline phosphatase	G/po	adSV with ACV	det. via naphthol from naphthyl phosphate; Tris buffer 0.1 M + $MgCl_2$ 1 mM, pH 9	[2921]
IgG alkaline phosphatase	IgG-alkaline phosphatase (ads.) + anti-alkaline phosphatase	G/po	adSV with ACV	det. via inhibition of alkaline phosphatase by antibody, det. via indigo from 3-indoxyl phosphate; Tris buffer 0.1 M, pH 7.2	[2922,2920]
IgG Human Rabbit		G/po	ASV with SWV	Immunochromatographic test strip, sandwich conjugates with IgG and antibodies-AuNP; after separation isolation of zones, dissolution of gold, SV of Au(III)	[1277]

Analyte	Electrode/modifier	Technique	Notes	Ref.
IgG	Anti-IgG on magnetic beads		Sandwich immunoassay with gold labels; det. via stripping analysis of Au	[2824]
IgG		LSV	Immunoassay with two electrodes, sandwich on immunoelectrode with antibody conjugated with alkaline phosphatase; release of ascorbic acid from ascorbic acid 2-phosphate which reduces Ag+ in cathode compartment (30 min); Ag(0) is then stripped	[3265]
Lactoferrin	G/mo	LSV	det. after chromatographic isolation; ABS 0.1 M, pH 5	[2924]
Met-hemoglobin Met-myoglobin		DPV	Retaining of catalytic activity for red. of O_2 and H_2O_2; PBS 0.1 M, pH 7	[2925]
Myoglobin	G/Nj	CV	Direct e−-transfer, catalytic ability for red. of O_2, trichloroacetic acid, NO_2^-; PBS 0.1 M, pH 7	[1168]
Myoglobin	G/po	CV, EIS, A (−0.95 V vs. SCE)	Direct red., catalytic red. of NO_2^-; PBS 0.1 M, pH 7.4	[949]
Myoglobin	G/po	A (−0.45 V vs. SSE)	Sensor for red. of H_2O_2; PBS 0.1 M, pH 7	[463]
Myoglobin	G/p (liquid)	EIS, CV	Characterization; good electrocatalytic activity toward trichloroacetic acid (reductive de-chlorination, re-ox. of Fe(II) from my); PBS 0.1 M, pH 7	[307, 481]
Myoglobin	G/p (liquid)	CV, EIS	PBS 0.1 M, pH 7, $[Fe(CN)_6]^{3/4-}$ (EIS); catalytic activity to H_2O_2 (red.) and trichloroacetic acid (red.)	[974]
Myoglobin	G/Nj	CV	Direct e−-transfer, catalytic red. of H_2O_2; PBS, pH 6.92	[2369]
Pneumolysine	C/po	LSV, ACV, FIA	Adsorptive stripping of liberated indigo from ELISA	[2926, 2927]

(continued)

TABLE 8.20 (continued)
Analysis of Biologically Important Compounds at Carbon Paste–Based Electrodes: Coenzymes, Enzymes, Proteins, and Related Compounds

Analyte	Modifier	Carbon Paste	Techniques	Comments	Ref(s)
Protein A (*Staphylococcus aureus*)	Magnetic NP/Au-shell + 11-mercaptoundec-anoic acid + IgG		CV, ACV, EIS	Immunological reaction between human IgG and protein A	[984]
Proteins Polycations	BSA	G/po	CV, EIS	Characterization as modifiers, effects of pretreatmet, probes: hexacyanoferrate, Fc monocarboxylic acid as probes	[644,868]
Riboflavin binding protein		G/mo	PSA	Characterization, interaction study with riboflavin; PBS 0.05 M, pH 7	[2928]
Rusticyanin	4,4′-Bipyridyl	G/Nj	CV	Characterization, reaction with Fe(II), Cr(II)	[2929]
Streptavidin	Biotinylated albumin	G/po	CV, reox. of deposited Ag on Au NP(0.08 V vs. SSE)	det. of biotin–avidin complex formation via dep. of Ag on Au NP and reox., competitive immunoassay with streptavidin labeled with Au NP	[93]
α-Synuclein		G/mo	adSV with SWV	Native and aggregate, det. via ox. of tyrosine; ABS 0.2M, pH 5	[2930]

TABLE 8.21
Analysis of Biologically Important Compounds at Carbon Paste–Based Electrodes: Hormones, Phytohormones, Steroids, and Related Compounds

Analyte	Modifier	Carbon Paste	Techniques	Comments	Ref(s)
Hormones					
Estradiol valerate	Iron tetrapyridino-porphyrazine	G/mo	CV, A (0.6 V vs. SSE)	Catalytic ox.; acetonitrile: PBS (0.1 M, pH 7) 47:53 (v:v); s: pharmaceuticals	[2931]
Estrogens			acc. (o.c.)	Enhancement effect of CTAB; LOD 0.8 nM (acc. 6 min)	[2932]
Estrogens Estriol		G/Kel-F	AD in HPLC (0.95 V vs. SSE)	Fluorinated Kel-F increases stability in organic solvents; acetonitrile/PBS (0.025 M, pH 3.2)	[66,318]
Estrogens Estriol	Oleic acid		acc.	LOD 0.4 µM; s: pharmaceuticals	[1674]
Estrogens Estriol		G/po	acc., CV, LSV	ox.; PBS 0.067 M, pH 8.04; LOD 30 nM (acc. 250 s); s: tablets	[1635]
Ethinyl estradiol	4-(3,5-Dichloro-2-pyridylazo)-1,3-diaminobenzene	G/Nj	pot. titration	det. via Zn after mineralization	[2933]
Insulin	Preanodization	G/mo	acc., PSA	det. via ox. of tyrosine; Na$_2$CO$_3$ 0.5 M, LOD 2 nM (5 min acc.)	[2934]
Insulin Myoglobin	RuO$_x$ electrodeposited	G/mo needle-type dual microelectrode	FIA (0.6 V vs. SSE)	det. via ox.; dual microelectrode for simultaneous det. of glucose; PBS 0.05 M + NaCl 0.1 M; pH 7.4	[612]
Insulin	Preanodization	G/mo	acc., PSA	det. via ox. of tyrosine or tryptophan; LOD 0.2 nM (10 min acc.); PBS 0.2 M, pH 7	[630]
LH releasing hormone					
Bombesin					
Neurotensin					
Melatonin		G/castor oil	acc.	LOD 1 nM; s: serum	[236]
Melatonin			o.c. acc., me DPASV	s: urine	[2935]
Melatonin			LSV	Extractive acc. (rather than adsorptive); LOD 0.15 µg L^{-1} HClO$_4$ 0.02 M; LOD 2.3 µM; s: capsules	[2936]
Melatonin					[2937]
Melatonin	Preanodization	G/po	CV, DPV, FIA	LOD 1 nM (DPV); HClO$_4$ 0.1 M; s: pharmaceuticals	[2938]
Melatonin	Preanodization	G/po	CV, ACV; acc. (0.15 V vs. SSE, 2 or 5 or 10 min)	ox. prim. to hydroquinone (irrev.), then to quinone (rev.); HClO$_4$ 0.1 M; LOD 90 pM (10 min acc.)	[2939]

(continued)

TABLE 8.21 (continued)
Analysis of Biologically Important Compounds at Carbon Paste–Based Electrodes: Hormones, Phytohormones, Steroids, and Related Compounds

Analyte	Modifier	Carbon Paste	Techniques	Comments	Ref(s)
Thyroxine	Polyvinylpyrrolidone	G/po	CV, LSV, CC	ox. of OH of T4; NaOH 0.1 M + CTAB 60 µM; LOD 80 nM	[2940]
Thyroxine		G/po	CV, EIS, LSV; acc. (0.9 V vs. SCE)	Signal improvement by CTAB; KCl 0.1 M	[2000, 2941, 2942]
Thyroxine			acc.	Activation by anionic surfactant; LOD 60 nM (acc. 5 min); s: urine	[2943]
L-Thyroxine	Anti-L-T4	G/po	SIA	Immunosensor; NaCl 0.1 M; s: pharmaceuticals	[2944]
Xenoestrogens (phenolic)			AD in pressurized CEC	s: chicken egg, milk powder	[563]
Steroids and derivatives					
Bile acids	CoPC	Conductive carbon cement	CV	PBS 0.1 M, pH 7; quick electrode deactivation in flow streams	[633]
Cholesterol	HRP + BSA + GA + hydroxymethyl Fc	G/po	CV, CV (0 V vs. SCE)	Cholesterol oxidase and esterase in solution; PBS 0.1 M + Triton X100, 0.35%, pH 7 s: serum	[1534]
Cholesterol	HRP + GA + Pt	G/po	CV, A (−0.05 V vs. SSE)	PVC-reaction cell with immobilized cholesterol esterase and oxidase, incubation, det. via H_2O_2; PBS 0.1 M, pH 7; s: serum	[2945]
Cholesterol	Cholesterol oxidase + Fc + HRP in poly (3-aminophenol), electropolym.	C/liquid p	A (0.05 V vs. SSE)	Biosensor, det. via H_2O_2; PBS 0.1 M, pH 7	[871]

Cholesterol	CdS-carbon nanofibers or CdS hollow spheres + cholesterol oxidase	G/po	EIS, ECL	det. of H_2O_2 by ECL, improvement by nanofibers; PBS 0.1 M, pH 8; LOD 0.8 µM	[2946]
High density lipoprotein cholesterol	Fc + peroxidase + cholesterol oxidase-polyethylene glycol			Biosensor via H_2O_2; selective activity of PEG-oxidase to HDL cholesterol with dextrane sulfate and Mg(II); s: serum	[2947,2948]
Phytohormones					
Indole-3-acetic acid		G/Nj	AD with HPLC (0.83–0.9 V vs. SCE)	ox., separation with strong anion exchanger, reversed phase or adsorbent; s: beans, cereals, tomatoes	[2949]
Indole-3-acetic acid	Sodium dodecyl sulfate	G/po	acc., CV, SWV	Accumulation via electrostatic interaction, ox.: McIlvain buffer 0.1 M, pH 3; LOD 20 nM (acc. 3 min); s: gladiola and phoenix tree leaves	[2950]
Indole-3-acetic acid		G/OV-17 (stationary phase GC)	a.c. acc., DPV, CV	ox.; s: stems *Echallium elaterium* after extraction	[1790]
Indole-3-acetic acid	Mung bean leave + Fc		CV	Tissue as source of IAA oxidase, catalytic ox.; PBS, pH 5.8; LOD 4.2 mg L^{-1}	[2951]
Kinetin		G/phenylmethyl silicone	acc. (o.c., 600 s), CV, SWV	ox., BRP 0.04 M, pH 7; s: apple extract	[2952]
Synthetic animal growth factors					
Roxarsone	Amberlite LA2	G/po	CV, DPV; o.c. acc. possible	Amberlite LA2 supports acc., red., ox. can also be exploited; BRP, pH 4; LOD 0.1 µM; s: tablets for poultry	[1746]

TABLE 8.22
Analysis of Biologically Important Compounds at Carbon Paste Electrodes: Neurotransmitters and Related Compounds

Analyte	Modifier	Carbon Paste	Techniques	Comments	Ref(s)
Acetylcholine	Acetylcholine esterase + choline oxidase + TTF	G/po	CV, A (0.2 V vs. SCE)	Direct e⁻-transfer of $FADH_2$ to TTF	[1487]
Catecholamines Dopamine Epinephrine	Crude extract of oyster mushroom	G/Nj	DPV	Source of laccase, catalytic ox., det. by red. of quinones as sum; LOD 7.9 (ad) or 9.8 μM (da); PBS 0.1 M, pH 7; s: pharmaceuticals	[2953]
Catecholamines Adrenaline L-Dopa Dopamine Isoprenaline	Laccase (oyster mushroom) + peroxidase (zucchini)	G/po	DPV	Synergistic ox., det. by red. of quinones; PBS 0.1 M, pH 6, LOD 24–27 nM	[2954]
L-Dopa Tyrosine L-Dopa Me-dopa	Unmodified (Ionic surfactants)	CPE, rotating CPE	CV, DPV	ca 1 N H_2SO_4; ox ox. to quinone; s: pharmaceuticals	[2743,2744] [738]
L-Dopa	Dy-nanowires	G/po	acc. (−0.2 V vs. SSE, 30 s), FFT SWV	Catalytic ox.; ABS, pH 7; LOD 4 nM; s: urine, plasma	[1882]
L-Dopa Benserazide	Chloranil	G/p liquid	CV, DPV	Catalytic ox.; KCl 0.1 M, LOD 0.95 μM (do) and 0.65 μM (be); pH 10; s: urine	[2955]
L-Dopa Carbidopa	PbO_2 in polyester	C/p solid	CV, DPV	Perchloric acid, 0.1 M; LOD 25 μM (do) and 3 μM (ca); s: pharmaceuticals	[1241]
L-Dopa	Fc + MWCNT	G/po	CV, CA (0.45 V vs. SSE), DPV	Catalytic ox.; PBS, 0.1 M, pH 7; LOD 1.2 μM; s: urine	[420]
L-Dopa	Ru-red on Na zeolite Y [$(NH_3)_5Ru^{III}$–O–$Ru^{IV}(NH_3)_4$–O–$Ru^{III}(NH_3)_5$]$^{6+}$	G/mo	CV	Catalytic ox. to dopaquinone, det. via reox. of Ru-complex; ABS 0.1 M, pH 4.8, LOD 85 μM; s: pharmaceuticals	[1126]
Dopa-melanins prepared by enzymatic ox. of dopa		G/KCl 0.1 M, pH 5.6		Eumelanins in paste, electrochem. characterization; KCl, pH 5.6	[2956]

Analytes	Modifier	Technique	Notes	Ref.
Dopac, Epinephrine, Catechol	Nanosized carbon black	CV	Comparison with CPE	[2752]
Dopamine, Epinephrine, Homovanilic acid, Ascorbic acid	G/Nj	DPV	Evaluation of DPV at CPEs	[2957]
Dopamine	CILE N-butylpyridinium hexafluorophosphate	CC, EIS, CV	Catalytic ox.; PBS 0.1 M, pH 6; LOD 0.7 μM; s: injection solutions	[466]
Dopamine, Ascorbic acid, Uric acid	CILE 1-octyl pyridinium hexafluoro phosphate	CV, DPV	Simultaneous det. without mediators; PBS 0.1 M, pH 6.8; LOD 1 μM; s: serum, urine	[453,456]
Dopamine, Ascorbic acid, Uric acid	C nanofibers (electrospun) G/mo	CV, DPV	Improved peak resolution; PBS 0.1 M, pH 4.5; LOD 2 μM	[660]
Dopamine	C nanofibers (oxidized with H_2O_2) G/mo	CV, DPV	Oxygen rich groups on NF surface; adsorption; PBS 0.1 M, pH 6; LOD 50 nM; s: injection	[2958]
Dopamine	SWCNT/mo	CV	Pretreatment of CNT, comparative study with GC, PT, and CPE; PBS 0.2 M, pH 7	[124,362]
Dopamine, Dopac, Epinephrine, Nor-epinephrine, Ascorbic acid	CNT/po	CV, FIA 0.4–0.7 V vs. SSE); also det. in CZE	PBS 0.05 M, pH 7.4; LOD 1.8–0.7 μM; CZE: boric acid, PH adj. 9.7	[367]
Dopamine, Ascorbic acid, Uric acid, Epinephrine, Catechol, Acetaminophenol, NADH	Carbon nanofibers/mo	CV, EIS	Improved current, decrease of overpotential; PBS 0.06 M, pH 6.85	[203]

(continued)

TABLE 8.22 (continued)
Analysis of Biologically Important Compounds at Carbon Paste Electrodes: Neurotransmitters and Related Compounds

Analyte	Modifier	Carbon Paste	Techniques	Comments	Ref(s)
Dopamine Acetaminophenol Ascorbic acid Epinephrine NADH Uric acid	Ordered mesoporous carbon/mo		CV	Improving effects of ordered mesoporous C, comparison with CNT and CPE; PBS 0.05 M, pH 7.4	[119]
Dopamine	Rice hull–activated carbon			PBS, pH 7.63; LOD 75 nM	[2959]
Dopamine Adenine Guanine	C60 deposited onto electrode by physical vapor deposition		CV	Catalytic effect; PBS 0.2 M, pH 6.5	[200]
Dopamine Epinephrine		Carbon composite via sol–gel from methyl trimethoxysilane	AD in CE (0.85 V vs. SSE)	PBS 0.025 M, pH 6.5; LOD 30–60 nM (ep)	[559]
Dopamine Ascorbic acid	Stearic acid	G/Nj	CV, ACV	PBS 0.1 M, pH 7	[1029]
Dopamine $K_3[Fe(CN)_6]$	1-Butyl-4-methyl pyridinium tetrafluoro borate	G/so	CV	PBS 0.1 M, LOD 10 μM	[2960]
Dopamine	1-Octyl-3-methylimidazolium bromide (ionic liquid) + Nafion		CV, SWV	Improved response, selective in presence of AA and UA; PBS 0.1 M, pH 7.4, LOD 10 nM	[2961]
Dopamine $K_3[Fe(CN)_6]$	Acrylamide	G/so	CV, DPV	Electrocatalytic ox.; PBS 0.2 M, pH 7.4; LOD 0.5 μM	[2962]
Dopamine $K_3[Fe(CN)_6]$	Losartan	G/so		PBS 0.2 M, LOD 0.1 μM	[2963]
Dopamine Serotonin Ascorbic acid	Poly(2-amino-5-mercapto-thiadiazole) film	G/paraffin wax	CV, DPV; A (0.55 V vs. SSE)	Catalytic ox., no significant acc.; BRP 0.2 M, pH 5; LOD 0.7 nM (da), 0.4 nM (st); s: injections	[2964]
Dopamine $K_3[Fe(CN)_6]$	Poly(ethylene glycol)	G/so	CV	Interaction with positively charged DA; PBS 0.1 M, pH 6.6; LOD 10 μM	[2965]
Dopamine Ascorbic acid	Polyvinylalcohol	G/so	CV, DPV	Signal improvement, separation of signal from ascorbic acid, simultaneous det.; PBS 0.2 M, pH 7	[1747]

Analyte	Modifier	Electrode type	Method	Notes	Ref.
Dopamine	Melanin-type polymer (electropolym. from L-dopa)			Comparison of different C electrodes; PBS 0.05 M, pH 7.4	[195]
Dopamine	Silica gel	G/po	CV, acc. (oc, 90 s); adDPASV	Repression of AA and UA by electrostatic repulsion; PBS 0.1 M, pH 8; LOD 48 nM; s: injection, serum	[1163]
Dopamine Epinephrine	Zeolite Y	G/mo	CV, FIA with DPV	Rejection of AA, acc. possible; PBS 0.01 M, pH 7.4	[1108]
Dopamine	Zeolites	G/Nj	CV	acc. by ion exchange; 0.05 M NaNO$_3$	[700]
Dopamine	Ti-phosphate/silica gel	G/so	CV, DPV	Selective in presence of AA; PBS 0.1 M, pH 7.5; LOD 20 µM; s: injections	[1213]
Dopamine	Zr-phosphate/silica gel	G/Nj	CV, DPV	Improvement of ox. of dopamine, det. possible beside UA and AA; BRP 0.1 M, pH 5.5; LOD 20 nM	[1216]
Dopamine	Nafion on clinoptilolite (zeolite)	G/mo	CV, DPASV, acc. (oc, 120 s)	acc.; Nafion prevents interference from AA; PBS 0.05 M, pH 7.4; LOD 10 nM	[1136]
Dopamine Epinephrine Phenothiazines	Na-hexdecyl sulfonate	G/Nj	LSV, acc (5 min)	Cathodic shift for ox. potential due to surfactant; PBS 5 mM, pH 2.2	[243,1661]
Dopamine	Na-dodecyl sulfonate (monolayer)	G/po	EIS, CV, DPV	Suppression of AA; PBS, 0.1 M, pH 7.4; s: injection solutions	[2966]
Dopamine	Na-dodecyl sulfonate micelles		DPV	Repression of AA ("masking") by SDS; 0.1 M NaCl + SDS 3 mM; LOD 5 µM; s: pharmaceuticals	[1654,2967]
Dopamine K$_3$[Fe(CN)$_6$]	Na-dodecyl sulfonate	G/so	CV	Signal improvement, possibility for acc.; LOD 0.1 µM	[1659]
Dopamine	Na-dodecyl sulfonate (in paste)		DPV	Repulsion of negatively charged interferences; s: deproteinized serum	[2968]
Dopamine Ascorbic acid Uric acid	Cetyl pyridinium bromide		CV, CA, CC	Improvement of resolution and signal; s: tablet, food	[767,2969 – 2972]
Dopamine Ascorbic acid Uric acid	CTAB	G/so	CV	Resolution of overlapping responses; PBS 0.2 M, pH 7.4	[1632]

(continued)

TABLE 8.22 (continued)
Analysis of Biologically Important Compounds at Carbon Paste Electrodes: Neurotransmitters and Related Compounds

Analyte	Modifier	Carbon Paste	Techniques	Comments	Ref(s)
Dopamine	Triton X100	G/so or ceresin wax	CV	T X100 improves slightly current	[317]
Dopamine $K_3[Fe(CN)_6]$	Phthalic acid + Triton X100	G/so	CV	Increased currents by phthalic acid and Triton; PBS 0.2 M, pH 7	[1648]
Dopamine	p-Aminobenzoic acid (+surfactant SDS or CTAB)	G/so	CV	Signal improvement, further enhancement by surfactant; ABS, pH 7	[1500]
Dopamine	Salicylic acid + surfactant	G/so	CV	Catalytic oxidation, improvement by surfactants; PBS 0.2 M, pH 7	[1634]
Dopamine	Mannitol + Triton X100	G/so	CV	Increased currents by mannitol and Triton; ABS 0.2 M, pH 6	[1649]
Dopamine	L-Aspartic acid in Nafion film	CILE G/1-butyl-3-methyl-imidazolium hexafluorophosphate		Electrocatalytic ox.; PBS 0.1 M, pH 7.4; LOD 0.03 μM	[2973]
Dopamine	β-Cyclodextrine/membrane			Inclusion complexes; adsorption control with $HClO_4 > 2$ M	[2974]
Dopamine Ascorbic acid	β-Cyclodextrine, CNT or poly (cyclodextrin) (electropolym.)			Formation constants of inclusion complexes, pH 3; supramolecular complexes	[2975,2976]
Dopamine Hydroxy-tryptophan	Melamine polymer film from electropolym. of L-dopa	MWCNT/po	CV, A (0.2 V vs. SSE), adSDPV (oc, medium exchange)	Accumulation and catalytic effect; PBS 0.05 M (+ ascorbic acid 10^{-3} M: current increase due to red. of dopamine quinone), pH 7.4; LOD 20 nM (5′ acc.)	[2977]
Dopamine Ascorbic acid	Polyglycine (electropolym.)	G/so	CV	Electrocatalytic effect on ox.; simultaneous det.; ABS 0.2 M, pH 5	[1764]
Dopamine	Mesoporous SiO_2 NP	G/po	CV; acc. 2 min	Current enhancement by adsorption; PBS 0.1 M, pH 7; LOD 0.1 μM: s: serum, injections	[999]
Dopamine	AuNP		DPV	No responses from uric and ascorbic acids; universal buffer, pH 7.4; LOD 5.9 nM	[2978]
Dopamine Ascorbic acid Uric acid	AuNP				[2979]

Electroanalysis with Carbon Paste–Based Electrodes, Sensors, and Detectors 389

Analyte	Modifier	Electrode	Technique	Notes	Ref.
Dopamine, Ascorbic acid, Uric acid	Pd NP on C nanofibers (electrospun)	G/mo	CV	Improved peak resolution, simultaneous det.; PBS 0.1 M, pH 7; s: injection solutions	[969]
Dopamine, Ascorbic acid	Ag NP + MWCNT	G/Nj	CV, DPV	Decrease of ox. overpotentials; PBS 0.1 M, pH 2; LOD 0.3 μM; s: Ringer's serum	[416]
Dopamine	FeNP in Nafion film			Electrocatalytic ox. and red., increased currents, no interference from AA; BRP, pH 7; LOD 3 μM	[976]
Dopamine	NiNP + ionic liquid		CV	PBS 0.1 M, pH 6; LOD 6.5 nM	[972]
Dopamine	[N,N'-Bis (2-pyridine carboxamido)-1,2-benzene] nickel(II)		CV	Catalytic ox.; LOD 62 nM; s: pharmaceuticals, biol. s.	[1616]
Dopamine, Ascorbic acid	Ni(II) on poly(o-amino phenol) (electropolym.) + dsDNA		DPV	Catalytic ox.; suppression of ascorbic acid; PBS, pH 5.1	[2980]
Dopamine, Catechols	SnO			Thin-layer cell with plant tissue bioreactor and SnO/CPE	[2981]
Dopamine	CdO NP	G/so	CV	Increase of current for ox. of DA and AA due to CdO NP	[1003]
Dopamine, Ascorbic acid	Fe(III) on zeolite NP		DPV	Catalytic ox.; simultaneous det.; PBS, pH 5	[1012]
Dopamine, Tryptophan	Co(II,III)-oxide	G/mo	FIA (0.6 V vs. SSE)	Catalytic ox.; PBS 0.1 M, pH 7 or NaOH 0.1 M	[1230]
Dopamine	Fc	G/po, also RDE	CV	Catalytic ox.; interference from AA eliminated with tetraphenylborate; PBS 0.1 M + KCl 0.1 M, pH 6; s: serum	[881]
Dopamine	6-Ferrocenylhexanethiol on magnetic core (Fe_3O_4)—shell (Au) NP	G/po	CV, A (0.38 V vs. SSE)	Catalytic ox.; PBS 0.1 M + $KClO_4$ 0.1 M, pH 6.86	[986]
Dopamine, Ascorbic acid	Sn(IV)-hexacyano ferrate (II) + CTAB	G/po	CV	Electrocatalytic ox. enhanced by CTAB; PBS 0.5 M, pH 7; s: injection solution	[1043]
Dopamine, Ascorbic acid	Th(IV)-hexacyano ferrate	G/p	CV, CA (0.8/0 V vs. SSE)	PBS 0.1 M, pH 3	[1044]
Dopamine, Ascorbic acid	Polypyrrole (electropolym.)/ hexacyanoferrate		CV, DPV, LSV	Simultaneous det.; PBS 0.1 M + KCl 0.05 M, pH 6; LOD 15 μM	[1799,1802]

(continued)

TABLE 8.22 (continued)
Analysis of Biologically Important Compounds at Carbon Paste Electrodes: Neurotransmitters and Related Compounds

Analyte	Modifier	Carbon Paste	Techniques	Comments	Ref(s)
Dopamine Ascorbic acid	Bis(4′-(4-pyridyl)-2,2′,6′,2″-terpyridine) iron(II) thiocyanate	G/po	CV	Decrease of ox. overpotential; acetate buffer, LOD 1 µM; simultaneous det. of AA and DA; 0.1 M, pH 5; s: injection solutions	[1617]
Dopamine	3-Mercaptopropyltrimethoxysilane Cu encaps. in mesoporous crystalline molecular sieve			Electrocatalytic ox.; also on microchip CE; LOD 0.2 µM	[2982]
Dopamine Ascorbic acid	Co(II)-5-nitrosalophen + tetraoctyl ammonium bromide	G/Nj	CV, DPV	Catalytic oxidation via Co(III); simultaneous det.; ABS 0.1 M, pH 5; LOD 0.5 µM; s: injections	[1583]
Dopamine Serotonin Phenols	CuPC + histidine	G/mo	A (static and as RDE)	ox. via Cu(II)-OOH(?); det. via red. of quinine; PBS 0.25 M + H_2O_2 0.12 mM, pH 6.9 LOD 11 µM (do) and 24 µM (ser)	[519]
Dopamine	Alcian blue (Cu^{II} PC dye) (+surfactant)	G/so	CV	Catalytic oxidation; improvement by surfactant (Triton X100) 0.2 M PBS, pH 6.6; LOD 0.1 µM	[2983]
Dopamine Ascorbic acid	Fe(II)-tetrasulfo-PC	G/mo	LSV	Simultaneous det. of ratio DA/AA via potential shift; pH 7.4; LOD 0.45 µM	[1570]
Dopamine Ascorbic acid	CoPC NP (+CTAB)	G/po	CV, DPV	Catalytic ox.; simultaneous det. possible with CTAB; PBS 0.025 M, pH 7.4; LOD 1 µM; s: injections	[933]
Dopamine Uric acid	Co(II)-octanitro PC			Catalytic ox.; simultaneous det.	[2984]
Dopamine Serotonin	Fe(II)-tetrasulfo-PC	G/mo	CV, SWV	Catalytic ox. via Fe(III); comparison with Co and Ni analogues; Tris buffer, pH 7.4; LOD ca.1 µM	[903]
Dopamine Ascorbic acid Uric acid	Fe(II)-octanitro PC + CTAB	G/so	CV	Simultaneous det.; PBS + KCl 0.1 M, pH 7	[916]

Analyte	Modifier	Paste type	Method	Notes	Ref.
Dopamine	MnPC + histidine	G/mo	A (0 V vs. SSE)	Biomimetic sensor; PBS 0.1 M, pH 7 + H_2O_2 0.25 mM; assumed ox. of MnPC and ensuing catalytic ox. of phenols; electrochem. red. of the quinoid; s: pharmaceuticals	[912]
Dopamine Ascorbic acid	Fe(III)-tetraphenylporphyrin on SiO_2/Nb_2O_5	G/Nj	CV	Catalytic ox. via Fe(III)/Fe(II); KCl 0.5 M; LOD 10 µM	[1202]
Dopamine Ascorbic acid	mer-ruthenium(III) bis[1,4-bis (diphenyl phosphino) butane] piccoline trichloride	G/Nj	CV	Catalytic ox.; multivariate calibr. for simultaneous det.; KCl 0.5 M, pH 2; LOD 20.8 mM (?)	[1612, 2000]
Dopamine	Acetone (?)	G/so	CV	Improved signal of DA over AA and UA; PBS 0.2 M, pH 7.4	[2986]
Dopamine	2,4-Dinitrophenyl hydrazine			Catalytic ox.; pyrophosphate buffer 0.2 M, pH 7 det. in presence of AA; PBS 0.1 M + NaCl 0.14 M, pH 4; LOD 50 nM	[225]
Dopamine	N-hydroxysuccinimide		CV, SWV		[2987]
Dopamine Ascorbic acid NADH	N,N',N'-Tetramethyl phenylenediamine	G/Nj	CV	Catalytic ox. via quinone diimine; PBS or citric buffer, 0.15 M, pH 7	[1516]
Dopamine	N,N'(2,3-Dihydroxybenzylidene)-1,4-phenylenediamine on TiO_2 NP	G/po	CV, DPV, CC	Mediated ox.; PBS 0.1 M, pH 8; s: injection	[992]
Dopamine	Tetrabromo-p-benzoquinone	G/po	CV, DPV, CA (0.22 V vs. SCE)	Electrocatalytic ox.; simultaneous det.; PBS 0.1 M, pH 7	[1446]
Dopamine Ascorbic acid Uric acid	2,2'-[3,6-Dioxa-1,8-octanediyl bis(nitriloethylidyne)] -bis-hydroquinone	G/liquid p	CV, SWV	Catalytic ox.; PBS 0.1 M, pH 7; LOD 3.2 µM; s: injection solutions	[1462]
Dopamine	TCNQ			Catalytic ox.; s: pharmaceuticals	[2988]
Dopamine Ascorbic acid	Eriochrome Black T		CV, DPV	Catalytic oxidation; KCl 1 M; simultaneous det. possible; LOD 0.18 µM	[2989]
Dopamine Uric acid	Pyrogallol red			Catalytic oxidation; s: drugs, urine	[1515]

(continued)

TABLE 8.22 (continued)
Analysis of Biologically Important Compounds at Carbon Paste Electrodes: Neurotransmitters and Related Compounds

Analyte	Modifier	Carbon Paste	Techniques	Comments	Ref(s)
Dopamine Acetaminophen	Hematoxylin/MWCNT	G/po	CV, CA (0.28 V vs. SCE), DPV	Catalytic improvement by CNT and hematoxylin, combined effect; simultaneous det.; PBS 0.5 M, pH 7; LOD 0.06 μM; s: tablets, injections, oral solutions	[443]
Dopamine	2,2′-(1,3-Propanediyl bisnitrilo ethylidene) bis-hydroquinone on TiO$_2$ NP	G/po	CV, LSV, SWV	Selective det, in presence of try and uric acid; PBS 0.1 M, pH 7; LOD 0.47 μM	[2990]
Dopamine Uric acid	2,2′-[1, 2-Ethanediylbis(nitrilo ethylidyne)]-bis-hydroquinone + CNT	G/po	CV, LSV	Catalytic ox.; PBS 0.1 M, pH 7, LOD 87 nM	[427]
Dopamine Cysteine Uric acid	2,2′-[1,7-Heptanediylbis (nitriloethylidine)]-bis-hydroquinone on TiO$_2$ NP	G/po	CV, DPV, SWV, CA (0.5 V vs. SCE)	Catalytic ox.; simultaneous det. of DA, UA and cys; PBS 0.1 M, pH 7; LOD 0.84 μM; s: ampoules	[995]
Dopamine Ascorbic acid	Thionine/Nafion on MWCNT	G/Nj	CV, DPV	Simultaneous det.; ABS 0.1 M, pH 4; LOD 80 nM	[1729]
Dopamine	Lamotrigine + Triton X100	G/so	CV	Signal improvement; PBS 0.2 M, pH 7	[1645]
Dopamine	Eperisone + CTAB	G/so	CV	Catalytic ox.; PBS 0.2 M, pH 7	[1633]
Dopamine Ascorbic acid	Polyalanine (electropolym.)	G/so	CV	Catalyt. improvement of ox.; ABS 0.2 M; pH 7	[1806]
Dopamine Epinephrine	Poly(tannic acid) film	G/so	CV	Enhanced oxidation; PBS 0.2 M, pH 7	[2991]
Dopamine Epinephrine	Poly(isonicotinic acid)	G/so		Catalytic ox.; simultaneous det.; PBS, pH 5.3; LOD 20 μM	[1762]
Dopamine Ascorbic acid Uric acid	Poly(calmagite) film	G/so	CV, DPV	Simultaneous det. possible; electrocatalytic ox.; PBS 0.2 M, pH 7; LOD 10 nM (DPV); s: injection	[2992]
Dopamine Ascorbic acid	Poly(Eriochrome black T) electropolym.	G/so	CV	Electrocatalyt. ox.; KCl, 1 M	[1763]

Ascorbic acid Uric acid Dopamine	Poly (calconcarboxylic acid), electropolym.	G/so	CV	Electrocatalytic activity, simultaneous det.; acetate buffer, 0.2 M, pH 5.4	[135]
Dopamine	DNA	CNT	CV, SWV; acc 5 s	Biosensor; NH$_4$H$_2$PO$_4$ 0.1 M, pH 3.5; LOD 20 pM	[376]
Dopamine Serotonin Epinephrine Norepinephrine	HRP cross-linked with BSA with GA and CDI	G/mo	A (−0.5 V vs. SSE)	Biosensor, catalytic ox. of catechol, det, by redu. of quinone; PBS 0.1 M, pH 7 + H$_2$O$_2$ 10 µM; LOD 14–33 µg L^{-1}; s: rat blood	[2697]
Dopamine Epinephrine Norepinephrine Phenols	Quinoprotein glucose dehydrogenase + PEI	g(act)/po	A (0.5 V vs. SSE)	Biosensor; oxidized phenols as electron acceptor; PBS + 15 mM glucose + 1 mM CaCl$_2$	[2786]
Dopamine Phenols	AuNP + polyphenol oxidase	G/mo	A (−0.1 V vs. SSE)	Biosensor; PBS 0.05 M, pH 7.4; LOD 0.2 µM	[946]
Dopamine Phenols Ascorbic acid	Polyphenol oxidase or mushroom tissue	GC/mo	CV, A (−0.1 V vs. SSE)	Biosensor; PBS 0.05 M, pH 7.4	[194]
Dopamine Catechin NADH	Polyphenol oxidase	CNT/mo	A (−0.05 V vs. SSE)	Biosensor, improved characteristics by CNT; PBS 0.05 M, pH 7.4; LOD 1 µM	[661]
Dopamine	Polyphenol oxidase from soursop, soursop tissue + TCQN	G/silicone wax	FIA (0.1 V vs. SSE)	Biosensor; PBS 0.3 M, pH 7.8; LOD 0.15 mM; s: pharmaceuticals	[280]
Dopamine	Dopamine polyphenoloxidase + TCQN/TTF + dialysis membrane		A (−0.075 V vs. SSE)	Biosensor; PBS, pH 6.75; LOD 0.25 µM	[1495]
Dopamine Phenols	Tyrosinase + Ru (dispersed on C)	G/mo	A (−0.1 V or 0 V vs. SSE)	Biosensor; PBS 0.05 M, pH 7.4	[1258]
Dopamine Phenols Catechols	Tyrosinase + sodium hexadecanesulfonate			Biosensor for solvents with high organic liquid content (methanol, acetonitrile); s: pharmaceuticals	[2779]
Dopamine Phenols Catechols	Tyrosinase	G/solid p and other pastes	FIA (−0.2 V vs. SSE)	Biosensor; PBS 0.1 M, pH 7; LOD 50 nM	[2993,2994]

(*continued*)

TABLE 8.22 (continued)
Analysis of Biologically Important Compounds at Carbon Paste Electrodes: Neurotransmitters and Related Compounds

Analyte	Modifier	Carbon Paste	Techniques	Comments	Ref(s)
Dopamine Dopac Norepinephrine	Tyrosinase	G/solid p	AD in HPLC (−0.2 V vs. SSE)	Biosensor; det. via red. of quinone; pH 6.5; LOD 0.25–0.8 µM	[78]
Dopamine	Banana tissue	G/mo	DPV, FIA (−0.2 V vs. SSE)	Biosensor, tissue source of polyphenoloxidase; PBS 0.05 M, pH 7.4; LOD 13 nM	[276]
Dopamine Epinephrine Norepinephrine	Banana tissue		AD in HPLC (−0.2 V vs. SSE), C18 column	Biosensor, tissue source of polyphenoloxidase; PBS 0.05 M, pH 6.5; s: urine	[2564]
Dopamine	Banana tissue			For in vitro and in vivo det.	[2995–2997]
Dopamine Pyrocatechol	Banana tissue		CV, DPV, FIA	Biosensor; tissue as a source of polyphenoloxidase	[2759]
Dopamine	Apple		DPV	Biosensor, apple as a source for polyphenol oxidase; det. by red. of quinone; LOD 0.2 µM; s: injections	[2998]
Dopamine	Potato tissue			Biosensor, tissue as a source of polyphenoloxidase; LOD 1.6 µM	[2999]
Dopamine L-Dopa Dopac Pyrocatechol	Potato tissue	G/mo, also as RDE	CV, CA (0.4 V vs. SCE)	Biosensor, tissue as a source of polyphenoloxidase; McIlvaine buffer, pH 6.5; LOD 2.5 µM (5500 rpm)	[3000]
Dopamine Norepinephrine	Zucchini tissue	G/mo	DPV, FIA (0.7 V vs. SSE), CA (−0.1/+0.7 V vs. SSE)	Source of polyphenoloxidase and ascorbate oxidase—eliminates AA interference; PBS 0.05 M, pH 7.4	[3001]
Dopamine	Spinach tissue			Biosensor for in vivo and in vitro measurements; LOD 0.71 µM	[2123]
Dopamine L-Dopa Catechol	Eggplant tissue	G/mo	A (−0.2 V vs. SSE)	Biosensor; tissue as a source for polyphenoloxidase; det. by red. of quinone; PBS 0.1 M, pH 7	[3266]

Dopamine L-Dopa Norepinephrine	Eggplant-banana tissue	Microelectrode		Eggplant to eliminate AA interference; LOD 0.32 μM (da)	[3002]
Dopamine	Pine kernel peroxidase on pegylated polyurethane NP	G/mo	CV, SWV	Biosensor, enzyme catalyzes ox. of dopamine, det. via red. of quinone; PBS 0.1 M, pH 6.5 + H_2O_2 2 mM; LOD 9 μM; s: pharmaceuticals	[1018]
Dopamine Catechol	*Latania* spp., fresh or dry tissue + Nafion film	g/mo	FIA	Tissue as a source of polyphenol oxidase; PBS 0.05 M, pH 5.7	[2756]
Dopamine Catechols	Fresh tissue of potato, peach, mushroom, pear		A (0 V vs. SCE)	Tissue as a source of polyphenol oxidase; McIlvaine buffer, pH 6.75	[279]
Dopamine Epinephrine	Crude extract of cara root		LSV	As a source of polyphenol oxidase; ox. to quinine; LOD 0.75 mM (da) and 0.082 mM (ep)	[3003]
Dopamine	Plant tissue			Separation after extraction	[3004]
Dopamine metabolites			AD in HPLC adSV		[3005,3006]
Epinephrine				Adsorptive acc.; det via ox.; LOD 2.5 nM; s: injections	[3007]
Epinephrine		G/mo as CP needle microelectrode	CV, LSV	PBS 0.01 M, pH 7.4 or no supporting electrolyte	[607]
Epinephrine Epinephrine-Be(II) complex				Electrochem. characterization and quantum mechanical calculations	[3008]
Epinephrine Catecholamines		CPE array (16 electrodes)	AD in FIA and HPLC; AD (0.6 V vs. SCE)	Possible simultaneous monitoring of unmodified and modified electrodes	[553]
Epinephrine Ascorbic acid Ascorbic acid Uric acid		C ceramic electrode, also as RDE	CV, SWV, CA	Prepared via sol–gel from methyl trimethoxysilane as precursor; PBS 0.1 M, pH 5; s: urine (DA), serum (ad)	[3009]
Epinephrine		MWCNT/IL 1-butyl-3-methyl-imidazolium hexa-fluorophosphate	CV	Paste on GC support; improved ox.; PBS, pH 7; LOD 0.1 μM	[3267]
Epinephrine Ascorbic acid	CTAB		DPV	Better signal resolution; NaCl 0.1 M, pH 3	[3010,3011]

(*continued*)

TABLE 8.22 (continued)
Analysis of Biologically Important Compounds at Carbon Paste Electrodes: Neurotransmitters and Related Compounds

Analyte	Modifier	Carbon Paste	Techniques	Comments	Ref(s)
Epinephrine	Sodium dodecyl sulfate	Acetylene black/po	EIS, CV, DPV, acc. 70 s	acc. of protonated adrenaline; H_2SO_4 0.5 M; LOD 10 nM; s: injections	[3012]
Epinephrine	Mesoporous Al-oxide/SiO_2	G/po	CV, SWV	acc. possible, improved ox.; PBS 0.1 M, pH 7; s: urine	[1855]
Epinephrine	Lewatit M500 + I_3^-	G/mo	CV, DPV	ox. by iodine, det. of quinone; electroregenerable I_3^- at 0.65 V vs. SSE; PBS, pH 6; LOD 3.9 µM: s: pharmaceuticals	[3013]
Epinephrine Tyrosine	Zeolite/Fe^{3+}	G/mo	CV, DPV acc. −0.2 V vs. SSE; 40 s	Catalytic ox. and acc.; simultaneous det.; PBS, pH 3.0; LOD 0.3 µM; s: serum	[1124]
Epinephrine	Fe(II)-PC	G/Nj	CV, DPV	Catalytic ox., det. in presence of AA and UA; ABS 0.1 M, pH 4; LOD 0.5 µM	[913]
Epinephrine Dobutamine Dopamine	(Tetrakis(t-butyl)CoPC)	CILEs with ionic liquids	CV, acc (10 min)	IL as ion exchanger, PC still increases signals; H_2SO_4 0.1 M + KCl 0.01 M; LOD 0.13 µ (ad), 1.2 µM (DA), 83 nM (do); s: pharmaceutical	[3014]
Epinephrine	Methylene green		CV, FIA	Catalytic ox.; LOD 10 nM	[3015]
Epinephrine Acetaminophen	2,2′-[1,2-Butanediylbis(nitriloethylidyne)] − bishydroquinone + TiO_2 NP	G/po	CV, CA, DPV	Catalytic ox.; PBS 0.1 M, pH 8; LOD 0.2 µM; s: injections	[993]
Epinephrine Uric acid Folic acid	2,2′-[1,2-Ethanediylbis(nitriloethylidyne)] − bishydroquinone + DWCNT	G/po	CV, CA, DPV	Catalytic ox.; simultaneous det.; PBS 0.1 M, pH 7; LOD 0.22 µM; s: injections, blood	[3016]
Epinephrine Norepinephrine	2-(4-Oxo-3-phenyl-3,4-dihydroquinazolinyl)-N′-phenylhydrazinecarbothioamide + MWCNT	G/po also RDE	SWV	Catalytic ox., simultaneous det.; PBS 0.1 M, pH 7	[3017]
Epinephrine	Pt NP in ionic liquid + laccase	CILE1-butyl-3-methylimidazolium hexafluoro phosphate	SWV	Biosensor, ox. of adrenaline with laccase and oxygen, electrochem. red. of adrenalinequinone; PBS 0.1 M, pH 6.5; s: pharmaceuticals	[967]

Analyte	Modifier	Method	Notes	Ref
Epinephrine	Banana crude extract	G/vaseline	Pot	Polyphenoloxidase; *in vitro* and *in vivo* measurements; PBS, pH 7; LOD 8 nM; s: pharmaceuticals	[3018]
Epinephrine	Palm tree fruits	G/po	CV, AD (−0.1 V vs. SSE) in FIA and bioreactor	Tissue as a source of polyphenol oxidase; ABS 0.1 M, pH 4.4; LOD 15 μM; s: pharmaceuticals for inhalation	[3019]
Neurotransmitter Catechols	HRP + CDI + BSA	G/mo	A (−0.05 V vs. SSE)	Catalytic ox. with HRP and H_2O_2; PBS 0.1 M + H_2O_2, 10 μM, pH 7; s: blood	[2697]
Norepinephrine			Differential double pulse V	LOD 1 μM	[3020]
Norepinephrine			HPLC with AD	s: plasma	[3021]
Norepinephrine Penicillamine Uric acid	2,2′-[1,2-Ethanediylbis(nitriloethylidyne)] – bis-hydroquinone + MWCNT	G/po	CV, CA, DPV	Catalytic ox., simultaneous det.; PBS 0.1 M, pH 7; LOD 82 nM; s: injections	[442]
Norepinephrine Folic acid Acetaminophen	2,2′-[1,2-Buthanediylbis(nitriloethylidyne)] – bis-hydroquinone + TiO_2 NP	G/po	CV, CA (0.3 V vs. SCE), DPV	Catalytic ox., simultaneous det.; PBS 0.1 M, pH 8; LOD 0.5 μM	[991]
Norepinephrine Folic acid Uric acid	Chloranil + MWCNT	G/po	CV, CA (0.15 V vs. SSE), DPV	Catalytic ox., simultaneous det.; PBS 0.1 M, pH 7; LOD 11 nM; s: ampoules	[3022]
Serotonin Tryptophan Uric acid	Unmodified	GC paste	DPV	via redox peak of ox product; PBS; improvements compared to CPE	[659, 2738]
Serotonin 5-OH tryptophan 5-Hydroxy indole-3-acetic acid			CV	Voltammetric characterization, ox.	[3023]
Serotonin Indoles Harmaline Tryptophan	Unmodified	G/Nj	adSV (acc. 60 s) with me	Adsorptive acc.	[2737]
Serotonin	Na-dodecyl sulfate		CV	Improved ox.; pyrophosphate buffer 0.1 M, pH 7	[1657]
Serotonin	Mesoporous SiO_2		acc. (2 min)	LOD 80 nM; s: serum	[1852]
Serotonin	Na-montmorillonite	G/p	CV	Catalytic ox.; LOD 57 nM; s: serum	[1011]

TABLE 8.23
Analysis of Biologically Important Compounds at Carbon Paste Electrodes. Determination of DNA, RNA, Components of Nucleic Acids, and Related Compounds. Investigations with DNA-Modified CPEs

Analyte	Modifier	Carbon Paste	Techniques	Comments	Ref(s)
Various				Reviews	[3024–3057]
Bases, nucleotides, nucleosides					
Adenine Adenosine 5′-AMP			LSV, DPV, CA	Comparative study, direct ox.	[3029]
Adenine	Preanodization (NaOH 0.001 M + NaNO$_3$ 0.1 M)	G/po	acc. (oc, 10 min); CV, SWV	acc. of oxidized adenine; LOD 0.15 µM	[1901]
Adenine dsDNA	Electrochemical pretreatment		acc., DPV	PBS 0.1 M, pH 7	[3030]
Adenine Guanine			SWV	Detection after hydrolysis of DNA, comparison of electrodes	[3268]
Adenine Guanine Oligonucleotides		G/poly(pyrrole) (electropolymerized) Porous pseudo-CP	Adsorptive stripping	Porosity attained with polymethacrylate microspheres and their re-dissolution; LOD 8 nM (ad), 20 nM (gu)	[328]
Adenine Guanine ssDNA (thermally denatured)		G/N-butyl pyridinium hexafluorophosphate CILE	CV	Simultaneous det. by catalytic ox.; also two signals from DNA; BRP 0.2 M, pH 5; LOD 0.25 µM (ad), 0.079 µM (gu)	[723]
Adenine Guanine ssDNA		Carbon nanofibers/mo	DPV	ox., dsDNA yields no signal, only ssDNA; PBS 0.1 M + KCl 0.5 M, pH 7	[202]
Adenine Guanine	Poly(acetyl aniline) (electroox.) + cyclodextrin	G/oil	CV, DPV	Interaction with modifier, ox.; PBS 0.1 M, pH 7; LOD 0.05 µM	[1421]
Guanine			CV, DPV	Comparison with GCE; PBS 0.5 M; s: DNA after hydrolysis	[3031]
Guanine	Na-montmorillonite		acc. (4 min)	Improvement of signal and potential; LOD 20 nM	[1066]

Analyte	Modifier	Electrode configuration	Method	Notes	Ref.
Guanine ssDNA	Co(II)-PC	G/po, also as RDE	CV, DPV	Catalytic ox.; LOD 85 ng mL^{-1} (gu) and 280 ng mL^{-1} (ssDNA)	[907]
Guanine ssDNA	Co(II)-hexacyanoferrate film electrodeposited	G/po, also as RDE	CV	Catalytic ox.; NaCl 0.2 M	[1038]
Guanine	Hexacyanoferrate intercalated in cinder		CV	Improved ox.; KCl 0.1 M + HCl, pH 2	[1033]
Guanine	Bis[bis(salicylidene-1,4-phenylenediamine) molybdenum(VI)] on TiO$_2$ NP	G/p	CV, DPV	Catalytic ox.; PBS 0.1 M, pH 5; LOD 3.4 nM	[997]
Guanosine	N-Butylpyridinium tetrafluoroborate	G/po CILE	acc. (0.6 V vs. SCE, 5 min), CV	Direct ox.; simultaneous det. of guanine; BRP 0.2 M, pH 4.5; LOD 0.26 µM; s: spiked urine	[484]
Guanosine	MWCNT + 1-ethyl-3-methyl imidazolium tetrafluoroborate	G/po	acc. (0.6 V vs. SCE, 5 min), CV, DPV	Simultaneous det. of guanine; BRP 0.2 M, pH 5; LOD 78 nM	[412]
Inosine	La-hydroxide nanowires	G/po	CV	det. via electroactive inosine-Cu-complex, ox.; PBS 0.1 M + Cu^{2+} 6 µM + K$_2$S$_2$O$_8$ 6 µM, pH 6.7; LOD: 0.8 nM; s: serum, pharmaceuticals	[3032]
N6-isopentenyl-adenine (immob.)	[Ru(bpy)$_3$]$^{2+}$	G/1-butyl-3-methylimidazolium hexafluorophosphate CILE, electrically heated (58°C)	ECL	Enhanced ECL by adenine compound, dependent on temp.; PBS 0.05 M, pH 9; LOD 36 nM; s: coconut milk	[489]
Mononucleotides Oligonucleotides		G/mo	acc. (0.5 V vs. SSE, 2 min); SWV	Accumulation and direct ox.	[3033]
Nucleosides	Nucleoside oxidase + p-Benzo-quinone or p-Benzohydroquinone + dialysis membrane + nylon net		CV, CA (−0.2 V vs. SSE)	Biosensor via red. of quinone produced by oxidized nucleoside oxidase; PBS, pH 6.0	[788]
Ribonucleotides	CoPC		As AD in HPLC	Catalytic ox. of ribose containing compounds, not of deoxy ribose; NaOH 0.15 M	[1043]
Uracil	La-hydroxide nanowires			det. via electroactive uracil-Cu-complex; LOD: 0.2 nM; s: uracil-based anticancer drugs	[3032]

(continued)

TABLE 8.23 (continued)
Analysis of Biologically Important Compounds at Carbon Paste Electrodes. Determination of DNA, RNA, Components of Nucleic Acids, and Related Compounds. Investigations with DNA-Modified CPEs

Analyte	Modifier	Carbon Paste	Techniques	Comments	Ref(s)
Nucleic acids					
dsDNA, ssDNA, tRNA		G/mo	FIA (1.0 V vs. SSE)	Flow injection with AD. ox. of guanine; LOD 460–750 pg	[3034]
ssDNA		G/mo	acc., PSA	Adsorptive potentiometric stripping; ABS 0.2 M, pH 4.8	[629]
dsDNA plasmid		G/mo	acc., PSA, DPV, LSV	Adsorptive stripping det. via guanine moiety, denatured yields higher signal; ABS 0.2 M, pH 5	[3035]
dsDNA, oligo(dG)21		GC	PSA	Study on ads. and electroox.; ABS 0.2 M, pH 4 or PBS, 0.02 M + NaCl 0.5 M, pH 7.4; LOD 200 µg L^{-1} (DNA) and 21 µg L^{-1} (odG)	[3036]
dsDNA (herring sperm)		G/N-butylpyridinium hexafluorophosphate CILE	acc. (0.3 V vs. SCE, 5 min)	Direct ox.: BRP 0.2 M, pH 4; LOD 17 µg mL^{-1}	[491]
ssDNA (thermally denatured)		CILE G/N-butyl pyridinium hexafluorophosphate	acc. (0.35 V vs. SCE, 160 s), CV	Direct det.; PBS 0.1 M, pH 7	[474]
dsDNA ssDNA Oligonucleotides		MWCNT/po	acc. (0.2 V vs. SSE); CV, PSA	Enhanced oxidation of guanine moiety by CNT; ABS 0.2 M, pH 5 or PBS 0.2 M + NaCl 0.5 M, pH 7.4; LOD 170 µg L^{-1} dsDNA	[361]
dsDNA Oligonucleotides		MWCNT/mo	CP, acc. (1 or 5 min, 0.2 V vs. SSE)	Stripping potentiometric det.; ABS 0.2 M, pH 5	[661]
ssDNA (oligonucleotides) Dopamine Ascorbic acid Methylene blue		SWCNT/1-butyl-3-methylimidazolium hexafluorophosphate	CV, DPV, EIS	Adenine and guanine ox. at significantly reduced potentials; BRP 0.2 M, pH 7; LOD 9.9 pM (via ox. of gu)	[394]

Electroanalysis with Carbon Paste–Based Electrodes, Sensors, and Detectors

Analyte	Modifier/Electrode	Configuration	Method	Notes	Ref.
dsDNA	Ethidium-TCNQ	G/mo		Fluorimetric det. of ethidium-DNA complex after pulsed release of ethidium at 0.45 V vs. SSE	[3037,3038]
ssDNA dsDNA	Clay minerals (drop-coated suspension)	G/p (solid)		DNA binding capacity study with methylene blue or ethidium bromide as marker	[1069]
DNA	Methylene blue on magnetic G/Fe$_3$O$_4$ NP			det. by decrease of MB signal: NH$_4$Cl, 0.1 M	[3039]
ssDNA	Tris(2,2′-bipyridyl) dichloro-ruthenium(II)/ Nafion	G/mo	CV, FIA (0.9 V vs. SSE)	Improved guanine ox.: ABS 0.2 M, pH 5.1	[1604]
Oligodeoxynucleotides		G/mo; inverted drop microcell	DPASV	det. by ASV of accumulated acid hydrolyzed oligodeoxynucleotide-Cu(I) complexes	[591]
Oligonucleotides Synthetic oligo(dG)n		G/mo	ads. acc. (0.5 V vs. SSE, 1–5 min), PSA	Adsorptive stripping analysis, direct ox. of guanine; ABS 0.2 M, pH 5	[3040]
Oligonucleotides	Poly(pyrrole) electropolymerized	G/mo	Electropolymerization, me, SWV, PSA	Oligonucleotides as dopants in poly(pyrrole), discrimination of dsDNA and ssDNA	[3041]
Oligonucleotides labeled with magnetic microspheres		G/mo	PSA	Direct ox. of guanine moiety; attraction to electrode via magnet	[299]
Nucleic acids dsDNA, ssDNA, tRNA		Electrically heated CPE	Adsorptive acc.: PSA, CV	Improved acc. at heated electrode in quiescent solution; ABS 0.2 M, pH 5	[90]
Nucleic acid dendrimers		G/mo	acc. (0.5 V vs. SSE), PSA, DPV	Improved guanine signals, ABS 0.2 M, pH 5; LOD, 3 pM for 4-layer dendrimer (acc. 15 min)	[3042]
PNAs		G/mo	acc. (−0.1 V vs. SSE); DPV, PSA	acc. and comparison with oligonucleotides	[1988]
RNA		G/mo	Adsorptive acc.: PSA, CV, SWV	ABS 0.2 M, pH 5	[91]

Nucleic acid modified electrodes

Analyte	Modifier/Electrode	Configuration	Method	Notes	Ref.
Acridine derivatives	DNA		acc., DPV	Intercalation	[2451]
Catechin					
1-Aminopyrene	dsDNA	G/vaseline, heated electrode	SWV	Decrease of signal with increasing temperature	[1424]

(continued)

TABLE 8.23 (continued)

Analysis of Biologically Important Compounds at Carbon Paste Electrodes. Determination of DNA, RNA, Components of Nucleic Acids, and Related Compounds. Investigations with DNA-Modified CPEs

Analyte	Modifier	Carbon Paste	Techniques	Comments	Ref(s)
Amodiaquine	ssDNA	G/mo	CV	Biosensor; study of adduct formation, probably with guanine; BRP 0.01 M, pH 4	[3043]
Azepine Phenothiazine drugs	DNA		acc., DPV	Biosensor	[2553]
Caffeine	DNA	CNT		Enhanced acc. at DNA; for brain electrochem.: LOD 0.35 μM (with DNA), 51 μM (without)	[430]
Characterization	Dirhodium(II) tetrakis [methyl-2-oxopyrrolidine-5(S)-carboxylate] + DNA	G/mo	CV	Electrochemical characterization, Doyle catalyst; KCl 0.5 M, pH 7	[85]
Chloroquine	dsDNA	G/Nj	acc. (oc), CV, DPV	acc. with DNA, ox.; LOD 30nM; s: serum	[2535]
Chlorpromazine Phenothiazine Promethazine	dsDNA	G/mo	acc., PSA, CV	Intercalative acc., ox.; ABS 0.2 M, pH 5; LOD 7 nM (ch), 12 nM (ph). 5 nM (pr)	[1480]
Clozapine	dsDNA		acc. (180s), adSV	Interaction study, acc. by intercalation, ABS 0.05 M, pH 4.3; LOD 1.5 nM; s: serum	[2541]
Cu(II), Pb(II)	dsDNA		acc (190s); SWV	Biosensor; PBS 0.1 M, pH 9; LOD 6.3 pM (Cu), 1.9 pM (Pb); s: fish tissue	[382]
Cu(I)	dsDNA	G/Nj	CV, DPSV, ACV	Interaction study on el. surface and in solution, complex formation; ABS 0.2 M, pH 5	[3044]
Cu(II)	dsDNA	G/Nj	CV, DPSV, ACV	Interaction study on el. surface and in solution, complex formation	[3045]
Cu-complexes with thiophens	dsDNA		adDPV	Interaction study: ABS, pH 5	[3046]
Cyclophosphamide	dsDNA or ssDNA	G/mo	acc. (DPV, ACV)	Interaction study; ABS 0.2 M + NaCl 0.02 M, pH 4.8; LOD 30 NM (dsDNA), 8 nM (ssDNA)	[2546]
Cytochrome c	Cellulose + ssDNA	G/mo	acc. (oc, 10min); CV, DPV	det. via red. of cyt c; PBS 0.2 M, pH 7.4; LOD 0.5 μM; s: mitochondria fraction from rat liver	[1921]
o-Dianisidine	dsDNA	G/mo	acc. (oc, 1 min), SWV	det. via decrease of guanine ox. or direct ox. of od; ABS 0.2 M + KCl 0.01 M, pH 4.7	[3047]

Dopamine	DNA	CNT	CV, SWV; acc 5 s	Biosensor; $NH_4H_2PO_4$ 0.1 M, pH 3.5; LOD 20 pM	[376]
Dopamine	dsDNA on poly(o-aminophenol)/Ni^{2+}		DPV	Catalytic ox.; PBS, pH 5.1; ascorbic acid does not interfere	[3048]
Epirubicin	dsDNA, ssDNA		acc.; CV, DPV	Possible hybridization indicator, ox.; ABS 0.5 M, pH 4.8	[2565]
O-Ethyl-O-4-(nitrophenyl) phenyl phosphonothioate	DNA	CNT	CV, adsorptive stripping (30 s) with SWV	Biosensor for pesticide, increased sensitivity due to DNA; PBS 0.2 M, LOD 7.9 pM	[409]
Fluoroquinolones	DNA		V	Review on interaction of fluoroquinolones with DNA	[3049]
Hydrogen peroxide Nitrite Trichloroacetic a.	DNA/hemoglobin/Nafion film	G/1-ethyl-3-methyl imidazolium tetrafluoroborate + p (liquid) CILE	CV, EIS	Characterization of Hb (retains native form); catalytic red.; PBS, pH 7; LOD 70 nM (H_2O_2)	[3050]
Interaction Cd^{2+}	Alternating layers of poly (lysine) and dsDNA on SWCNT (functionalized with COOH) + methylene violet	G/p solid	CV, EIS, DPV	Biosensor; damage study with Cd^{2+}; MV acts as intercalation redox probe; BRP, pH 7	[3051]
Interaction actinomycin D	dsDNA or ssDNA (denatured)		DPV	Intercalation with DNA, study on el. and in solution, ABS 0.2 M, pH 5.2	[3052]
Interaction acridine orange	dsDNA or ssDNA (denatured)		adSV	Intercalation with DNA, study on el. and in solution, ABS 0.2 M, pH 5	[3053]
Interaction As_2O_3	dsDNA or ssDNA	G/mo	DPV	Interaction study at surface and in solution, decrease of guanine oc. after interaction	[3054]
Interaction benzo[a]pyrene	dsDNA		DPV, PSA	Interaction study; decrease of guanine ox., increase of ox. of adduct; ABS 0.5 M + NaCl 0.02 M, pH 4.8	[3055]
Interaction $[Co(phen)_3]^{2+}$	CTAB + dsDNA	G/po	CV	Immobilization of $[Co(phen)_3]^{2+}$ and dsDNA on CTAB-modified CP	[1625]
Interaction $[Cu\ phen_2]^{3+}$	dsDNA	CPE electrically heated	SWV, DPV	Monitoring of damage via $[Co(phen)_3]^{3+}$ or guanine ox.	[292]
Interaction curcumin			CV, SWV	dsDNA and curcumin interaction study in solution	[3056]
Interaction curcumin + Cu^{2+}		G/po	DPadSV; acc. (+0.5 V vs. SSE)	Interaction study of curcumin in presence of Cu(II)	[3057]

(continued)

TABLE 8.23 (continued)
Analysis of Biologically Important Compounds at Carbon Paste Electrodes. Determination of DNA, RNA, Components of Nucleic Acids, and Related Compounds. Investigations with DNA-Modified CPEs

Analyte	Modifier	Carbon Paste	Techniques	Comments	Ref(s)
Interaction daunomycin	dsDNA or daunomycin	G/mo	CV, PSA	Study with ads. DNA and in solution, increase of ox. of guanine; ABS 0.2 M, pH 5	[2549]
Interaction 4,4'-dihydroxy chalcone	dsDNA, ssDNA	G/mo	CV, DPV	Interaction study at el. surface and in solution, decrease of ox. signal of guanine or adenine; ABS 0.5 M + NaCl 0.02 M, pH 4.8	[3058,3059]
Interaction ethidium Br Acridine orange Actinomycin D	Hexadimethrine/dextrane sulfate/hexadimethrine on SiO_2 + DNA	G/mo	adDPSV, ACV	Interaction/intercalation studies; effect of multiple layers, ox. of intercalants; comparison with HMDE; ABS 0.1 M, pH 5	[3060,3064]
Interaction flavonoids			SWV	Some flavonoids interact directly, potentially harmful carcinogens	[293,2764]
Interaction 2-hydroxyquinoline-3-carbaldehyde oxime		G/mo	CV	Interaction study with dsDNA, decreased signal in presence of DNA; Tris buffer 5 mM + NaCl 0.05 M, pH 7.2	[3065]
Interaction I-131 Tc-99 m		G/mo	DPV	Monitoring of radioactive damage, acc. (+0.5 V vs. SSE) of dsDNA after irradiation; ABS 0.5 M + NaCl 0.02 M, pH 4.8	[3066]
Interaction isopomiferin, pomiferin, osajin				Interaction study for antioxidative capacity to prevent damage to radical oxygen species from Fenton reaction; PBS 0.1 M, pH 6.98	[3067]
Interaction isorhamnetin	DNA (natural and denatured)		CV	Interaction study, intercalation and supramolecular complex; UV–vis study	[2869]
Interaction lycorin	dsDNA	G/mo	DPV	Interaction study, det. via ox. of guanine moiety; comparison with pencil graphite el.	[3068]
Interaction Meldola blue	dsDNA or ssDNA	G/mo	CV, DPV; adSV	Signal decrease with dsDNA, hybridization monitoring, comparison with HMDE; PBS 0.05 M + $HCOONH_4$ 0.3 M, pH 6.98	[3069]
Interaction methylene blue $[Ru(bpy)_3]^{2+}$	dsDNA, ssDNA		CV, DPV	Interaction study, hybridization indicator, ox. of Ru and red. of mb; ABS 0.5 M + NaCl 0.02 M, pH 4.8	[289,1463]

Interaction methylene blue	ssDNA, dsDNA		SWV; CC, DPV	Interaction study, interaction via guanine moiety; hybridization indicator	[1464,1467]
Interaction mitomycin C	dsDNA (fish sperm, natural and denatured)	G/mo	DPV; acc. (oc, 2 min)	Interaction det. by decrease of ox. of guanine; ABS 0.5 M + NaCl 0.02 M, pH 4.8	[2596]
Interaction mitoxantrone	dsDNA, ssDNA		CV, DPV	Interaction study: ABS 0.5 M, pH 4.8	[3070]
Interaction peroxynitrite	dsDNA or ssDNA or oligo(dG)$_{15}$	G/mo	DPV	Interaction study for damaging effect, direct ox. of guanine moiety; hybridization study of (dG)$_{15}$ with (dC)$_{15}$	[2422]
Interaction rifampicin	dsDNA or ssDNA (denatured)		adDPSV	Intercalation with DNA, study on el. and in solution, decrease of DNA ox. by ri; ABS 0.2 M, pH 5 and PBS 0.2 M, pH 7.4	[2622]
Tris-bipyridyl osmium(II/III)	ssDNA or dsDNA + stearylamine	Acetylene black/PTFE emulsion	CV, LSV, CC	Interaction study; PBS, pH 7	[1602]
Khellin	dsDNA		acc. (oc), CV, LSV, DPV	PBS 0.02 M, pH 7; LOD 10 nM (DPV, acc. 300 s); s: serum, tablets	[2580]
Levofloxacin	dsDNA	G/mo	acc. (oc), CV	Biosensor; ABS 0.2 M, pH 5; LOD 25 mg L^{-1}	[2582]
6-Mercapto purine	DNA		acc.; DPV	Biosensor; LOD 2 μM	[3071]
Mercury	DNA + CNT		adSV with SWV (acc. 400 s)	NH$_4$H$_2$PO$_4$, 0.1 M, pH 4.0; LOD 0.2 ng L^{-1}; s: fish liver	[3072]
Mifepristone	dsDNA	G/mo	CV, DPV, second harmony ACV	Biosensor, increase of ox. current in ACV; PBS 0.05 M, pH 7; LOD 0.1 μM	[2595]
Ofloxacin		CNT	CV	Biosensor, interaction study with dsDNA, acc. by intercalation; ABS 0.2 M, pH 5; LOD 85 nM	[3073]
Pefloxacin	dsDNA		CV, DPV	acc. on dsDNA, ox.; LOD 50 nM; s: urine	[2611]
Oxytetracycline	dsDNA	CNT/mo	adSV with SWV (acc. −1.7 V vs. SSE, 500 s)	Biosensor, accumulation by intercalation; PBS 0.1 M, pH 5.5; LOD 0.4 ng L^{-1}; s: fish tissue	[390]
Oxytetracycline Tetracycline Chlortetracycline	dsDNA		adSV	Biosensor; interaction with DNA; detection in the presence of Cu(II); ABS 0.2 M + NaCl 0.02 M, pH 4.7; LOD 1–6 nM	[2538]
Rutin	DNA	G/1-butyl-3-methyl imidazoliumhexafluorophosphate + po	CV, ASDPV (acc 240 s)	Biosensor with acc.; BRP, 0.1 M, pH 3; LOD 1.3 nM; s: tablets	[3074]
Release of DNA	dsDNA, ssDNA (ads.)	G/mo	SWV, PSA	Potential controlled release of adsorbed DNA (−1.2 V vs. SSE)	[608]

(continued)

TABLE 8.23 (continued)
Analysis of Biologically Important Compounds at Carbon Paste Electrodes. Determination of DNA, RNA, Components of Nucleic Acids, and Related Compounds. Investigations with DNA-Modified CPEs

Analyte	Modifier	Carbon Paste	Techniques	Comments	Ref(s)
Hybridization and sequencing					
Aflatoxigenic *Aspergillus* sp.	Oligonucleotide (ads.) + BSA (surface blocker)	G/mo	LSV, DPV, SWV	Hybridization with biotinylated DNA and coupled to streptavidin-phosphatase, det. via 1-naphthol from 1-naphthyl phosphate	[3075]
Homopolynucleotides	Homopolynucleotides, poly(G) or poly(C)	G/p wax	PSA	Study of duplex and triplex formation between poly(G) and poly(C) with $[Co(phen)_3]^{3+}$, comparison with HMDE	[3076]
Hybridization	ZrO_2 (in paste) + ssDNA	G/p wax	CV, DPV	ads. due to strong affinity of phosphate to zirconia, methylene blue as indicator	[1245]
Hybridization	Oligonucleotides	G + poly(methyl methacrylate) + polypyrrole (polym. by Fe^{3+})	ASV	Porous pseudo carbon paste; gold NP catalyzed silver enhancement; LOD 50pM	[1278]
Hybridization	Fe_3O_4/methylene blue on NT	SWCNT	V	MB as indicator	[3077]
Interaction Brilliant cresyl blue	ssDNA	G/po	CV, DPV	Interaction with BCB as hybridization label for oligonucleotides; PBS 0.1 M, pH 7	[3077]
Interaction $[Co(phen)_3]^{3+}$	dsDNA, ssDNA		CV, DPV	Co(III)-1,10-phenanthroline complex as hybridization indicator	[3269]
Interaction Cu(II)-dimethyl glyoxime complex	ssDNA/Al(III)/stearic acid	G	CV	Interaction study, $Cu_2(dmg)_4$ as indicator for hybridization; BRP + KNO_3 0.05M, pH 5; s: BAR gene from transgenic corn	[3079]
Interaction ethidium bromide, SYBR green	dsDNA or amplicons	G/mo	DPV, acc. (0.5 V vs. SCE, 5 min)	Intercalation with conformation change and decrease of signal; ABS 0.2 M + NaCl 0.02 M, pH 5	[3080]
Interaction Meldola blue, proflavine	dsDNA, ds oligonucleotides		CV, SWV	Meldola blue or proflavine as intercalation labels	[2981,2982]

ssDNA Oligonucleotides Cystic fibrosis	Octadecyl amine or stearic acid + ssDNA (covalently), oligonucleotides	G/mo	CV	det. via [Co(phen)$_3$]$^{3+}$ or via [Co(bpy)$_3$]$^{3+}$ as indicator; Tris buffer 5 mM + NaCl 20 mM, pH 7	[1638]
ssDNA Cryptosporidium	38-mer	G/mo	PSA	det. via [Co(phen)$_3$]$^{3+}$ as indicator; Tris buffer 0.05 M, pH 7	[3083]
ssDNA Cryptosporidium	29-mer (guanine substituted by inosine)	G/mo	PSA	Indicator-free biosensor due to guanine from complement after hybridization; ABS 0.2 M, pH 5	[3084]
Oligonucleotides	Oligonucleotides	G/mo	cv, SWV, PSA	Evaluation of PSA as det. method for hybridization with indicator [Co(phen)$_3$]$^{3+}$; Tris buffer 0.02 M, pH 7.4	[3085]
Oligonucleotides	Oligonucleotides + silica-titania beads	G/mo	PSA	Polishable and renewable, det. via [Co(phen)$_3$]$^{3+}$	[3086]
Oligonucleotides; factor V Leiden mutation	21-Oligomer (guanine replaced by inosine)	G/mo	DPV	Indicator-free det. via ox. of guanine in hybrid (inosine electroinactive); ABS 0.5 M + NaCl 0.02 M, pH 4.8	[3087]
ssDNA Achondroplasia G380R mutation	12-Oligomer (guanine replaced by inosine)	G/mo	DPV	Label-free det. via ox. of guanine in hybride (inosine electroinactive); ABS 0.5 M + NaCl 0.02 M, pH 4.8	[3088]
Oligonucleotides; point mutation p53 gene	17-Mer PNA or 17/30-mer oligonucleotides	G/mo	PSA	Hybridization indicator [Co(phen)$_3$]$^{3+}$	[3089]
Oligonucleotides	Peptide nucleic cid, oligonucleotides			Hybridization study with indicator [Co(phen)$_3$]$^{3+}$	[3090]
ssDNA Oligonucleotides	PNA	G/mo	DPV	Label-free det. via decreased ox. of guanine after hybridization	[3091]
ssDNA Oligonucleotides	dsDNA, ssDNA, PNA		ACV, DPV	Hybridization monitoring via methylene blue; Tris buffer 0.02 M, pH 7	[1466]
ssDNA Oligonucleotides	Biotin; immobilization of streptavidin-magnetic beads with biotinylated PNAs—after hybridization with ssDNA and acc. of Meldola blue		V	Hybridization det. via Meldola blue; LOD 2 pM (20 min acc. of indicator)	[3092]

(continued)

TABLE 8.23 (continued)
Analysis of Biologically Important Compounds at Carbon Paste Electrodes. Determination of DNA, RNA, Components of Nucleic Acids, and Related Compounds. Investigations with DNA-Modified CPEs

Analyte	Modifier	Carbon Paste	Techniques	Comments	Ref(s)
ssDNA Oligonucleotides		G/mo	adSV using OsO_4 + bipy as DNA marker before or after hybridization	det. via adSV of Os-marked released DNA; hybridization at magnetic beads and release of target DNA independent of measurement	[1603]
Oligonucleotides HIV virus	Oligonucleotides	G/mo	PSA	Tris buffer 0.02 M + indicator $[Co(phen)_3]^{3+}$ 0.05 mM, pH 7.4	[3093]
Oligonucleotides, *Mycobacterium tuberculosum*	Oligonucleotides	G/mo	PSA	Tris buffer 0.02 M + indicator $[Co(phen)_3]^{3+}$ 0.1 mM, pH 7	[3094]
Oligonucleotides *Microcystis* spp.	Oligonucleotides	G/mo	CV, DPV	det. of hybridization via labels: methylene blue (decrease of red. current) or $[Ru(bpy)_3]^{2+}$ (decrease of guanine ox. current)	[3095]
Oligonucleotides		G/mo electrically heated	PSA after hybridization on magnetic beads, release by NaOH; hot acc. at el.	Enhanced acc. of released DNA at elevated T, applicable to small volumes; ABS 0.2 M, pH 5	[3097]
Oligonucleotides (20 mer)	CNT + oligonucleotide	G/p (liquid)	DPV, EIS	Indicator-free hybridization bio-sensor; Tris buffer 0.02 M, pH 7	[421]
Oligonucleotides biotinylated	Streptavidin			det. via decrease of ox. of streptavidin (due to Tyr and Trp) signal by adduct with biotin	[3097]
ssDNA	ssDNA		SWV	Hybridization det. via methylene blue and daunomycin; det. of *Listeria monocytogenes* and genetically modified food	[3098]
Lyseriolysin gene Phosphotransferase neomycin gene					
ssDNA	Membrane (polyaniline NT + polylysine) + ssDNA		EIS	Label-free enhanced hybridization recognition; s: transgenic soy bean	[3099]
ssDNA	ssDNA on polypyrrole film	MWCNT	DPV	Hybridization biosensor with ethidium bromide as indicator; PBS 0.1 M, pH 7; LOD 85 pM	[3100]

Electroanalysis with Carbon Paste–Based Electrodes, Sensors, and Detectors

ssDNA	MWCNT/Ag-TiO$_2$ + ssDNA, DNA		EIS	Increased immobilization capacity by NT; label-free hybridization sensor by EIS	[3101]
ssDNA labeled with Bi$_2$S$_3$ NP/polyvinyl pyrrolidone	AuNP + ssDNA			Hybridization monitored via Bi signal; LOD 38 fM	[3102]
ssDNA labeled with CdSe, cauliflower mosaic virus	ssDNA			Hybridization monitored via Cd after dissolution in HNO$_3$	[953]
ssDNA labeled with PbSe NP	CTAB + ssDNA (electrostatic binding)			Hybridization biosensor with PbSe label; LOD 8.7	[1004]
Phosphinothricin acetyltransferase transgene					
ssDNA	CTAB + ssDNA		CV, DPV	Hybridization biosensor with K$_3$[Fe(CN)$_6$] or methylene blue as indicator; LOD 80 nM	[766]
ssDNA	ssDNA-HRP on Au NP in composite film		CV, DPV	Biosensor via suppression of HRP activity toward H$_2$O$_2$ red. by hybridization; LOD 50 pM	[955]
ssDNA	Chitosan + ss oligonucleotides or ssDNA or PNA, dsDNA	G/mo	DPV	det. of hybridization label-free (ox. of guanine) or with methylene blue	[1465]
ssDNA Hepatitis B virus	Oligonucleotides	G/mo	DPV	via [Co(phen)$_3$]$^{3+}$ as indicator; Tris buffer 0.02 M, pH 7	[3103]
ssDNA Hepatitis B virus TT virus	ssDNA, oligonucleotides		SWV	via methylene blue as indicator; Tris buffer 0.02 M, pH 7; s: serum with PCR amplification	[3104]
ssDNA from hepatitis B virus	Chitosan + ssDNA (electrostatically bound)	G/Nj	DPV	Hybridization biosensor with methylene blue as indicator; s: blood (with PCR)	[289, 3105]
ssDNA—35S promoter from cauliflower mosaic virus	PbSe/chitosan		DPV	Hybridization monitored with methylene violet as indicator; LOD 16 pM	[1005]
ssDNA HIV virus	ssDNA			Hybridization monitored with Co(bpy)$_3^{3+}$	[3106]
ssDNA phosphinothricin acetyltransferase gene	ssDNA (19 mer, covalently attached)	G/mo	SWV	Hybridization monitored with Co(bpy)$_3^{3+}$ (0.5 V vs. SSE, 300 s) or via ox. of guanine; PBS 0.05 M, pH 7; LOD 0.1 μM	[3107]

(continued)

TABLE 8.23 (continued)
Analysis of Biologically Important Compounds at Carbon Paste Electrodes. Determination of DNA, RNA, Components of Nucleic Acids, and Related Compounds. Investigations with DNA-Modified CPEs

Analyte	Modifier	Carbon Paste	Techniques	Comments	Ref(s)
ssDNA Phosphinothricin acetyltransferase transgene	SWCNT + poly(lysine) (electropolymerized) + ssDNA	G/p (solid)	CV, EIS	NaCl 0.1 M + $[Fe(CN)_6]^{3-/4-}$ (1 mM) as probe; LOD 0.3 fM; s: transgenic beans (after PCR)	[3108]
ssDNA	ssDNA (covalently bound)	G/stearic acid	CV	det. via methylene blue	[3109]
ssDNA Bar gene	Stearic acid + ssDNA (covalently via ethylenediamine)	G/mo	SWV	Hybridization biosensor with $Co(bpy)_3^{3+}$ as indicator	[3110]
ssDNA (cucumber)	ssDNA (cucumber) + stearic acid			Hybridization biosensor with $Co(bpy)_3^{3+}$ as indicator	[3111]
ssDNA Phosphinothricin acetyltransferase transgene	Stearic acid + Al(III) + ssDNA	G/p (solid)	CV	Biosensor with MB as indicator via decrease of signal after hybridization; s: transgenic corn	[3112]
ssDNA CP4 epsps gene Bar gene	Stearic acid + SDS + Al(III) + ssDNA	G/p (solid)	CV, DPV	Biosensor with MB as indicator via decrease of signal after hybridization; LOD 23 nM; s: transgenic soy bean (with PCR)	[3113]
ssDNA Phosphinothricin acetyltransferase transgene	Al(III)-poly(L-glutamic acid) (electropolym.) + ssDNA (electrostatic binding)	G/p (solid)	EIS	Label-free hybridization biosensor; KCl 0.1 M + $[Fe(CN)_6]^{3-/4-}$ (0.5 mM) as probe; LOD 3 pM	[3114]
ssDNA—35S promoter from cauliflower mosaic virus Nopaline synthase gene	Au NP/TiO$_2$ hollow microsphere (membrane) + ssDNA		CV, EIS	EIS with hexacyanoferrate probe; LOD 0.23 pM; s: transgenic soy beans (via NOS gene after PCR)	[954]

Analyte	Electrode composition	Method	Notes	Ref.
ssDNA	CNT + streptavidin + ssDNA-biotin	DPV	Hybridization sensor with ssDNA labeled with Au NP, enhancement by multiple layer hybridization; det. via Au	[378,388]
ssDNA invA from *Salmonella*	Avidin gel beads + ssDNA-biotin + PPQ glucose dehydrogenase-avidin	A (0.1 V vs. SSE)	Hybridization sensor and labeling with glucose dehydrogenase and glucose substrate; MOPS buffer 0.01 M, pH 7	[3115]
ssDNA Phosphinothricin acetyltransferase transgene Nopaline synthase gene	Poly(aniline) nanofibers + MWCNT/Au NP gelatin film + ssDNA (immobilized at 0.6 V vs. SSE for 500 s)	CV, DPV, EIS	Hybridization biosensor (0.4 V vs. SSE for 600 s); label-free EIS det. with methylene blue or hexacyanoferrate(II/III); LOD 0.56 pM	[3116]
ssDNA Phosphinothricin acetyltransferase transgene Nopaline synthase gene	MWCNT + poly(aniline) nanofibers + chitosan + ssDNA	CV, EIS	Synergistic effect of MWCNT and nanofibers, label-free EIS det. with methylene blue or hexacyanoferrate(II/III); LOD 27 fM	[3117]
ssDNA	Polyaniline/graphite oxide (+ssDNA)	SWV	Signal at −0.27 V vs. SCE as hybridization signal; Tris buffer, pH 7.2	[3118]
ssDNA	Polypyrrole/graphite oxide (+ssDNA)	DPV	Catalytic red. of ssDNA, useful for hybridization	[1801]
ssDNA	Oligo(dT)$_6$ + MWCNT-oligo(dA)$_6$ + (dT)$_6$-marked tag (T substituted by I) + ssDNA-tag	SWV	Direct ox., hybridization amplification by attachment of 6 mer dA-labeled MWCNT to CPE by hybridization, hybridization with tag and further hybridization with ssDNA with tag	[3119]
ssDNA BRCA1 gene	ssDNA-biotin + magnetic beads-streptavidin	PSA	Magnetic handling of hybrids, ox. of guanine	[298]
Single base mismatch recognition	CPE with magnet	adSV with SWV	via repair enzyme MutS, incubated on immobilized DNA duplex, released and detected via adsorption directly or after partial hydrolysis	[3120]

TABLE 8.24
Analysis of Biologically Important Compounds at Carbon Paste Electrodes: Purines, Pyridines, and Pyrimidines

Analyte	Modifier	Carbon Paste	Techniques	Comments	Ref(s)
Caffeine		G(atc. 100°C)/ceresin wax	LSV	Paste with wide anodic range; H_2SO_4, 0.1 M	[39]
Caffeine		G/mo or graphite pencil	CV, acc. (0 V vs.), SWASV	ox. at around 1.4 V vs. SSE; PBS 0.1 M, pH 9	[3121]
Caffeine Catechol	Bi	CNT/mo	LSV, adSV with SWV	ox.: simultaneous det.: for brain electrochem.: 0.1 M H_3PO_4 or 0.1 M NH_4PO_3; LOD 0.18 μM	[3122]
Caffeine Theophylline	Molecular imprinted poly(methacrylate)	G/eicosane solid	acc. (oc., pH 7; medium exchange)	ox.: LOD 15 nM; phosphoric acid, 0.15 M; s: cola, tea	[1691]
Caffeine Aspirin Acetaminophen	Triton X100 + MWCNT	G/po	CV, adSDPV, acc. (−0.7 V vs. SSE, 300 s); EIS	ox., simultaneous determination; PBS 0.1 M, pH 7; LOD 88 nM; s: pharmaceuticals, coffee-based drinks, cola	[3122]
Caffeine	(+DNA)	CNT		Enhanced acc. at DNA; for brain electrochem.: LOD 0.35 μM (DNA), 51 μM (without)	[430]
Caffeine	1,4-Benzoquinone	G/po	CV, SWV	Suppression of benzoquinone red. peak via π-complex formation; LOD 0.3 μM	[1444]
Hypoxanthine	Mesoporous TiO_2	G/po	CV, SWV	Enhanced ox. current; PBS 0.1 M, pH 7; LOD 50 nM; s: serum	[1875]
Hypoxanthine	Xanthine oxidase + GA + BSA (drop coated)	Heated (35°C)	CV, ECL	Enhancement of CL of luminol by H_2O_2; PBS 0.2 M, +luminol 20 μM, pH 7; LOD 3 μM	[572]
Hypoxanthine	Xanthine oxidase + GA + BSA + Au NP (electrodep.)		A (0 V vs. SSE)	Biosensor; PBS 0.05 M, pH 7.4; LOD 0.22 μM; s: sardines, chicken meat	[945]
Hypoxanthine	Xanthine oxidase on cellulose acetate + hydroxymethyl Fc	G/p liquid	CV, A (0.29 V vs. SSE)	Biosensor; PBS 0.1 M, pH 7.8; LOD 0.6 μM; s: fish	[858]
Hypoxanthine	Methylviologen on montmorillonite + xanthine oxidase in polyaniline film		CV	Biosensor; det. by red. of H_2O_2; PBS, pH 7; LOD 0.8 μM; s: fish meat	[1497]

Analyte	Modifier	Configuration	Technique	Notes	Ref.
Nicotine	Choline oxidase + BSA + GA	G/p (solid)	A (0.45 V vs. SCE)	Biosensor by enzyme inhibition of ox. of choline to betaine; PBS 0.067 M + 1,4-benzoquinone 0.5 mM + choline 0.5 mM, pH 7.4; s: tobacco	[2452]
Pyrimidine derivatives	Polymer		AD after HPLC separation	Direct ox. of oxidizable groups (e.g., $-NH_2$, $-SH$)	[3123]
Theophylline 3-Methyl xanthine			DPV	Extraction with $CHCl_3$/pr-OH, re-extraction with NaOH; ox. at 1 V vs. SSE; PBS, pH 7.4; s: plasma	[3125]
Theophylline	CTAB		DPV	Enhanced acc.; LOD 0.18 µM; s: tablets, urine	[741]
Theophylline	CoPC NP		DPV	Catalytic ox.; LOD 0.14 µM	[932]
Theophylline Methylated purines	K_2-Hexachloroplatinate	G/po	AD in HPLC (0.17 V vs. SSE)	Catalytic ox.; LOD 1–11 ng mL^{-1}; s: serum	[2626]
Thiouracil Thiobarbituric acid	CoPC			Electrocatalytic ox. of thiols	[908]
Uric acid	Unmodified	G/silicone grease, as RDE	DPV, also in FIA; acc. (to 3 min)	Adsorptive acc., ox., competition study; PBS 0.05M, pH 3; s: urine	[226,676]
Uric acid	Unmodified	GC paste	DPV	In presence of ascorbic acid and acetaminophen	[3126]
Uric acid Serotonin Tryptophan	Unmodified	GC paste	DPV	ox.; PBS; improvements compared to CPE	[659,2738]
Uric acid			A	Multiparameter sensor system with linear readout circuit for pNa, pH, and uric acid	[3127]
Uric acid Dopamine Ascorbic acid Epinephrine Catechol Acetaminophenol NADH		Carbon nanofibers/mo	CV, EIS	Improved current, decrease of overpotential; PBS 0.06M, pH 6.85	[203]

(*continued*)

TABLE 8.24 (continued)
Analysis of Biologically Important Compounds at Carbon Paste Electrodes: Purines, Pyridines, and Pyrimidines

Analyte	Modifier	Carbon Paste	Techniques	Comments	Ref(s)
Uric acid Ascorbic acid Dopamine	C nanofibers (electrospun)	G/mo	CV, DPV	Improved peak resolution; PBS 0.1 M, pH 4.5; LOD 0.2 µM	[660]
Uric acid Dopamine Acetaminophen Ascorbic acid Epinephrine NADH		Ordered mesoporous carbon/mo	CV	Improving effects of ordered mesoporous C, comparison with CNT and CPE; PBS 0.05 M, pH 7.4	[119]
Uric acid Ascorbic acid Neurotransmitter		C ceramic electrode, also as RDE	CV, SWV, CA	Prepared via sol–gel from methyl trimethoxysilane as precursor; PBS 0.1 M, pH 5	[3009]
Uric acid Xanthine Hypoxanthine	Preanodization (1.3 V vs. SCE, 10 s)	G/po	CV, DPV	Improvement of signal and resolution; NaClO$_4$ 0.1 M + NaOH 0.002 M, LOD 5–10 µg L^{-1}; s: urine, plasma	[1900]
Uric acid Ascorbic acid Dopamine		CILE 1-octyl pyridinium hexafluoro phosphate	CV, DPV	Simultaneous det. without mediators; PBS 0.1 M, pH 6.8; LOD 1 µM; s: serum, urine	[453,456]
Uric acid		SWCNT/1-butyl-3-methylimidazolium hexfluorophosphate	acc (oc, 180 s), CV, CC	Paste on GC support; PBS, pH 4; LOD 5 nM; s: urine	[3128]
Uric acid	Nafion film (spin coated)	G/mo	SWV, acc. (0.2 V vs. SSE, 30 s)	In the presence of AA: citrate buffer, 0.05 M, pH 4; LOD 0.25 µM; s: urine	[1722]
Uric acid	CTAB		LSV	PBS 0.1 M, pH 7.4; LOD 1 µM; s: urine	[3129]
Uric acid Ascorbic acid Dopamine	CTAB	G/so	CV	Resolution of overlapping responses; PBS 0.2 M, pH 7.4	[1632]
Uric acid Ascorbic acid	Cetyl pyridinium bromide		CV, CA, CC	Improvement of resolution and signal	[2971, 2972]

Electroanalysis with Carbon Paste–Based Electrodes, Sensors, and Detectors

Analyte	Modifier	Electrode	Method	Notes	Ref.
Uric acid	β-Cyclodextrin film (electrogenerated)	G/Nj	CV, LSV, A (0.6 V vs. SSE)	det. in presence of AA; ABS, pH 5; LOD 4.6 μM; s: urine, saliva	[1420]
Uric acid	Mesoporous SiO$_2$		CV, SWV (acc. 3 min)	Increased ox. current; ABS 0.1 M, pH 5; LOD 80 nM; s: serum	[1847]
Uric acid Ascorbic acid Dopamine	Pd NP on C nanofibers (electrospun)	G/mo	CV	Improved peak resolution; PBS 0.1 M, pH 7	[969]
Uric acid Tryptophan Ascorbic acid	zeolite/Fe^{3+}	G/mo	CV, DPV; acc. −0.3 V vs. SSE, (30 s)	PBS 0.15 M, pH 3.5; LOD 0.08 μM; s: serum	[1122]
Uric acid Ascorbic acid	Chloromercuri ferrocene	MWCNT/p	CV	Simultaneous determination of ascorbic and uric acids; PBS 0.1 M, pH 4; LOD 2.6, 0.8 μM; s: serum, urine	[3130]
Uric acid Ascorbic acid Dopamine	Fe(II)-octanitro PC + CTAB	G/so	CV	Simultaneous det.; PBS + KCl 0.1 M, pH 7	[916]
Uric acid Ascorbic acid	MWCNT/Nafion + Co(III)-5-nitrosalophen	G/Nj	CV, DPV	Catalytic ox. via Co(III); simultaneous det.; ABS 0.1 M, pH 4; LOD 60 nM	[1590]
Uric acid Ascorbic acid	4-Nitrophthalonitrile	G/mo; also RDE	CV, A	Electrocatalytic ox.; PBS 0.1 M, pH 7; LOD 1.3 μM; s: urine	[3131]
Uric acid Dopamine Ascorbic acid	Tetrabromo-p-benzoquinone	G/po	CV, DPV, CA (0.22 V vs. SCE)	Electrocatalytic ox.; simultaneous det.; PBS 0.1 M, pH 7	[1446]
Uric acid Norepinephrine Penicillamine	2,2′-[1,2-Ethanediylbis (nitriloethylidyne)]-bis-hydroquinone + MWCNT	G/po	CV, CA, DPV	Catalytic ox., simultaneous det.; PBS 0.1 M, pH 7; LOD 82 nM	[442]
Uric acid Cysteine Dopamine	2,2′-[1,7-Heptanediylbis (nitriloethylidine)]-bis- hydroquinone on TiO$_2$ nanoparticles	G/po	CV, DPV, SWV	Simultaneous det.; PBS 0.1 M, pH 7	[995]
Ascorbic acid Uric acid	2,2′-[1,2-Ethanediylbis (nitriloethylidyne)]-bis-hydroquinone	CNT/p, also RDE	CV, A, DPV	Electrocatalytic ox.; PBS 0.1 M, pH 7; LOD 75 15 μM	[422]
Uric acid Dopamine	2,2′-[1,2-Ethanediylbis(nitrilo ethylidyne)]-bis-hydroquinone + CNT	G/po	CV, LSV	Catalytic ox.; PBS 0.1 M, pH 7, LOD 15 μM	[427]

(*continued*)

TABLE 8.24 (continued)
Analysis of Biologically Important Compounds at Carbon Paste Electrodes: Purines, Pyridines, and Pyrimidines

Analyte	Modifier	Carbon Paste	Techniques	Comments	Ref(s)
Uric acid Epinephrine Folic acid	2,2′-[1,2-Ethanediylbis (nitriloethylidyne)]-bis-hydroquinone + DWCNT	G/po	CV, CA, DPV	Catalytic ox.; simultaneous det.; PBS 0.1 M, pH 7; LOD 8.8 µM; s: blood	[3016]
Uric acid Ascorbic acid	Thionine on Nafion	G/Nj	CV, DPV	Electrocatalytic ox. by thionine; ABS 0.1 M, pH 5; LOD 50 nM; s: urine, serum	[1509]
Uric acid	Congo red on Nafion/MWCNT	G/mo	CV, DPV	ox.; effects of AA masked; ABS 0.1 M, pH 5 and PBS 0.1 M, pH 7; s: urine, serum	[1513]
Uric acid Ascorbic acid Dopamine	Poly (calconcarboxylic acid), electropolym.	G/so	CV	Electrocatalytic activity; acetate buffer, 0.2 M, pH 5.4	[135]
Uric acid NAD(P)H 8-Oxoguanine	Guanine or 8-oxoguanine (electroox., +1.1 V vs. SSE)	G/Nj	DPV	Signal improvement; universal buffer, pH 8; LOD 6.6 µM (ua), 2 µM (og)	[2876]
Uric acid Norepinephrine Folic acid	Chloranil + MWCNT	G/po	CV, CA (0.15 V vs. SSE), DPV	Catalytic ox., simultaneous det.; PBS 0.1 M, pH 7	[3022]
Uric acid	Uricase + TCNQ	G/mo	CV, FIA (0.34 V vs. SSE)	Tris buffer 0.1 M, pH 9; s: serum	[590]
Uric acid	Uricase + β-cyclodextrin + MnO$_2$ NP + Fc			Biosensor, synergistic effect of MnO$_2$ and Fc; LOD 3 µM	[3132]
Uric acid	Uricase + poly(o-aminophenol) (electropolym.)		CV, FIA (0.05 V vs. SSE)	Biosensor; PBS 0.1 M, pH 6.5 or 7.5 (FIA); LOD 3 µM; s: serum	[247]

Analyte	Modifier	Electrode	Technique	Notes	Ref.
Uric acid	Uricase + HRP + Fc + membrane (Nafion-coated cellulose acetate)	G/po	FIA (0.1 V vs. SSE)	Biosensor with uricase on pre-column; PBS 0.1 M + KCl, pH 7; LOD: 0.1 µM; s: serum	[2826]
Uric acid		G/methyl methacrylate	Electro chemiluminescence	Indirectly via inhibition of ECL of luminal; LOD 12 pM; s: body fluids	[3133]
Xanthine		SWCNT/1-butyl-3-methylimidazolium hexfluorophosphate	acc (oc, 210s), CV, LSV	Paste on GC support; small effects by UA and hypoxanthine; PBS, pH 6; LOD 2 nM; s: serum, urine	[3128]
Xanthine	Xanthine oxidase (adsorbed) + membrane	G/Nj	A (0.4 V vs. SSE)	Biosensor via ox. of released UA; PBS 0.1 M, pH 7.2	[3134]
Xanthine Hypoxanthine	Xanthine oxidase	G/Nj	CA	Biosensor via ox. of released UA	[3135]
Xanthine Hypoxanthine	Xanthine oxidase (drop coated)	G/Nj	CV, A (0.4 V vs. SSE); acc. (3 min)	Biosensor via ox. of released UA; PBS 0.1 M, pH 7.2	[3136]
Xanthine	Xanthine oxidase	Dual el.; CPE (mod.) and GCE (unmod.)	A (0 V vs. SCE)	Biosensor; det. via superoxide	[3137]
Xanthine	Xanthine oxidase + SWCNT or DWCNT or MWCNT	G/mo	CV, CA (0.9 V vs. SSE)	Biosensor; PBS 0.05M, pH 7; s: canned tuna fish	[722,3138]
Xanthine Hypoxanthine	Xanthine oxidase	GC/mo	CA (0.9 V vs. SSE)	Biosensor, det. of H_2O_2; PBS 0.05 M, pH 7; LOD 0.1 µM (x), 5.3 µM (hx); s: plasma	[3168]
Xanthine Hypoxanthine	Xanthine oxidase + Au NP	GC/mo	CV, A (0.8 V vs. SSE)	Biosensor; PBS 0.05M, pH 7.5; s: canned tuna fish	[943]

TABLE 8.25
Analysis of Biologically Important Compounds at Carbon Paste–Based Electrodes: Vitamins

Analyte	Modifier	Carbon Paste	Techniques	Comments	Ref(s)
Various				Review	[2748]
Ascorbic acid		G/silicone grease	CV	s: fruit, beverages, vegetables	[3140]
Ascorbic acid			CV	PBS 0.1 M, pH 7; rate constants based on semi-integration	[3141]
Catechols					
Ascorbic acid		G/Nj	CV, AD in chromatography (0.3 V vs. SSE)	PBS, pH 7; enhanced response by preanodization + precathodization	[63]
Hydroquinone					
NADH					
Hydrazines					
Ascorbic acid		G/Nj	DPV	Evaluation of DPV at CPEs	[2957]
Dopamine					
Epinephrine					
Homovanilic acid					
Ascorbic acid			c	det. of AA as student lab experiment	[3142]
Ascorbic acid	Polypyrrole			KNO_3 0.1 M, LOD 5 pM	[3143]
Ascorbic acid		G/di-iso-octyl phthalate	POT	Potentiometric sensor without ionophore	[3144]
Ascorbic acid			LSV	Citrate-phosphate buffer, pH 4.7; s: pharmaceuticals	[3145]
Ascorbic acid	Hexadecyl sulfonate	G/Nj	LSV	Repulsion by negative charge and anodic shift of potential by surfactant	[1661]
Ascorbic acid	CTAB		DPV	Better signal resolution; NaCl 0.1 M, pH 3	[3010]
Adrenaline					
Ascorbic acid	CTAB	G/so	CV	Resolution of overlapping responses; PBS 0.2 M, pH 7.4	[1632]
Uric acid					
Dopamine					
Ascorbic acid	Cetyl pyridinium bromide		CV, CA, CC	Improvement of resolution and signal; s: tablet, food; LOD 2 μM	[783, 2969–2972]
Dopamine or uric acid					
Ascorbic acid	Stearic acid	G/Nj	CV, ACV	PBS 0.1 M, pH 7	[1029]
Dopamine					

Electroanalysis with Carbon Paste–Based Electrodes, Sensors, and Detectors

Ascorbic acid Acetaminophen	Unmodified or stearic acid	G/p (solid)	LSV	ox.; ABS, pH 4.7; s: tablets	[2002]
Ascorbic acid				Determination and calculation of diffusion coefficients	[3146]
Ascorbic acid		Ionic liquid 1-butyl-3-methyl imidazolium hexafluorophosphate		Direct ox.; decrease of ox. overpotential	[3147]
Ascorbic acid Dopamine Uric acid		CILE 1-octyl pyridinium hexafluoro phosphate	CV, DPV	Simultaneous det. without mediators; PBS 0.1 M, pH 6.8; LOD 20 µM; s: serum, urine	[453,456]
Ascorbic acid Dopamine		CILE N-butylpyridinium hexafluorophosphate	CV	Decrease of overpotential for ox.; PBS, pH 6.5; LOD 8 µM; s: tablet	[458]
Ascorbic acid Uric acid Neurotransmitter		C ceramic electrode, also as RDE	CV, SWV, CA	Prepared via sol–gel from methyl trimethoxysilane as precursor; PBS 0.1 M, pH 5	[3009]
Ascorbic acid Dopamine Dopac Uric acid		CNT/mo	LSV, CV	PBS 0.05 M, pH 7.4	[123]
Ascorbic acid Uric acid Dopamine	C nanofibers (electrospun)	G/mo	CV, DPV	Improved peak resolution; PBS 0.1 M, pH 4.5; LOD 2 µM	[660]
Ascorbic acid	β-Cyclodextrine		CV	Adsorption of AA; ABS 0.1 M, pH 4.69	[3148]
Ascorbic acid	Mesoporous SiO_2		adSV; oc acc.	Analyte adsorptively preconcentrated on SiO_2; PBS 0.1 M, pH 5.5; LOD 50 nM	[3149, 3150]
Ascorbic acid Cysteine	Cu_2O	G (act. by electric arc)/mo	CV, LSV	Catalytic red. via Cu(II)-AA complex; PBS, pH 6.9; LOD 1 nM; s: pharmaceuticals	[1316]
Ascorbic acid Sugars	Cu_2O + polyethylene glycol		CV, AD in CZE (0.06 V vs. SCE)	Simultaneous det. of ascorbic acid; NaOH 0.03 M; LOD 0.12–0.49 µM; s: black tea beverage	[1229]
Ascorbic acid	Cu(II)-phosphate in polyester resin	g/mo	CV; A (0.06 V vs. SCE)	Catalytic ox.; ABS 0.1 M, pH 5; LOD 10 µM; s: pharmaceuticals	[3151]

(continued)

TABLE 8.25 (continued)
Analysis of Biologically Important Compounds at Carbon Paste–Based Electrodes: Vitamins

Analyte	Modifier	Carbon Paste	Techniques	Comments	Ref(s)
Ascorbic acid	Cu(II)-tetrasulfo PC encapsulated in n-propylpyridinium silsesquioxane polymer on SiO_2/Al_2O_3		CV	Catalytic ox.; pH 5–8; s: tablets	[3152]
Ascorbic acid	Poly(4-vinylpyridine)/ hexacyanoferrate(II)	G/Nj	CV	Catalytic ox.; glycine buffer 0.5 M, pH 3.2	[1742]
Ascorbic acid Dopamine	Polypyrrole (electropolym.)/ hexacyanoferrate		CV, DPV, LSV	Simultaneous det.; PBS 0.1 M + KCl 0.05 M, pH 6; LOD 33.8 µM	[1799, 1802]
Ascorbic acid	Hexacyanoferrate(II) on silica gel/propylpyridine	G/mo	CV	Catalytic ox. via hexacyanoferrate(III); NaCl 0.7 M	[1820]
Ascorbic acid	Hexacyanoferrate(II) on hydrotalcite or hexacyanoferrate (III) on trioctylamine	G/po or Nj or vaseline or TOA	CV (i), CA (ii) (0.2 V vs. SCE)	Catalytic oxidation; (ii) oc reaction of AA with Fe(III) with ensuing CA; PBS 0.1 M, pH 7.4; LOD 0.2 mM (i) or 0.1 mM (ii); s: tea bag, juice	[1094]
Ascorbic acid Dopamine	Sn(IV)-hexacyano ferrate (II) + CTAB	G/po	CV	Electrocatalytic ox. enhanced by CTAB; PBS 0.5 M, pH 7; s: injection solution	[1043]
Ascorbic acid Dopamine	Th(IV)-hexacyano ferrate	G/p	CV, CA (0.8/0 V vs. SSE)	PBS 0.1 M, pH 3	[1044]
Ascorbic acid Tryptophan Uric acid	Fe(III) on zeolite	G/mo	CV, DPV; E_{acc} −0.3 V vs. SSE, t_{ACC} 30 s	PBS 0.15 M, pH 3.5; LOD 0.21 µM; s: serum	[1122]
Ascorbic acid	Fe(III) on zeolite Y	G/Nj	SWV	Catalytic ox.; PBS 0.1 M, pH 5.4; s: orange and lemon juice	[1110, 3153]
Ascorbic acid	Fe(III) on humic acid		CV, A (0.87 V vs. SCE)	Catalytic ox.; PBS 0.5 M or ABS 0.5 M + KCl 0.5 M, pH 5.4; s: orange juice	[1704]
Ascorbic acid	Fe(II)-pentacyano nitrosyl ferrate film			catalyt. ox.; KNO_3 0.5 M; LOD 4 µM	[3154]

Analyte	Modifier	Electrode	Method	Notes	Ref.
Ascorbic acid	Fe(II)-pentacyano nitrosyl ferrate	CV		Catalytic ox.; LOD 1.4 mM; s: pharmaceuticals	[3155]
Ascorbic acid	Ferrocene carboxylic acid	G/po	CV, CA (0.5/0.2 V vs. SSE)	Catalytic oxidation; PBS 0.1 M + LiClO$_4$ 0.1 M, pH 5	[874]
Ascorbic acid	Fc	G/po	CV, CA, CC	catalyt. ox.; pH 5; LOD 32 µM	[3155]
Ascorbic acid	Fc	G/po	CA, SWV	Catalytic ox.; LOD 1.5 µM	[3157]
Ascorbic acid	Fc	G/p (solid)	A	Catalytic ox.; NH$_3$/NH$_4$Cl-buffer, pH 9.3; LOD 0.5 µM; s: drinks	[3158]
Ascorbic acid Uric acid	Chloromercuri ferrocene	MWCNT/p	CV	Simultaneous determination of ascorbic and uric acids; PBS 0.1 M, pH 4; LOD 2.6 µM (AA), 0.8 µM (UA); s: serum, urine	[3130]
Ascorbic acid	1-[4-(2-Ferrocenyl ethyl)-phenyl]-1-ethanol	G/so	CV, CA	Catalytic ox.; pH 6; LOD 38 µM; s: tablets, ampoules, syrup	[1528]
Ascorbic acid	2,7-Bis (ferrocenyl ethynyl) fluoren-9-one	G/po	CV, DPV, CA (0.85/0 V vs. SSE)	Catalytic ox. via Fc; PBS 0.1 M + LiClO$_4$ 0.1 M, pH 7.4; LOD 4.2 µM; s: tablets, injection solutions, syrup	[1548, 1549]
Ascorbic acid	1-[4-(Ferrocenyl ethynyl) phenyl]-1-ethanone	G/so or po or 1-bromonaphthalene	CV, CA (0.85/0 V vs. SSE)	Catalytic ox. via Fc; PBS 0.1 M + LiClO$_4$ 0.1 M, pH 7.0; LOD 63 µM; s: tablets, injection solutions, syrup	[1542]
Ascorbic acid Catechols	Fc + PVC		CV	PBS, pH 7.4	[3159]
Ascorbic acid	β-Cyclodextrine/Fc inclusion complex	G/p (solid)	CV, A (0.2 V vs. SCE)	Enhanced electrode stability and reproducibility by complex; NH3/NH4Cl buffer 0.1 M, pH 10; LOD 0.12 µM	[877, 3160]
Ascorbic acid Dopamine	Fe(III)-tetraphenylporphyrin on SiO$_2$/Nb$_2$O$_5$	G/Nj	CV	Catalytic ox. via Fe(III)/Fe(II); KCl 0.5M; LOD 51 µM	[1202]
Ascorbic acid Dopamine	Bis(4'-(4-pyridyl)-2,2':6',2"-terpyridine) iron(II) thiocyanate	G/po	CV	Decrease of ox. overpotential; acetate buffer, LOD 2 µM; simultaneous det. of AA and DA; 0.1 M, pH 5; s: injection solutions	[1617]
Ascorbic acid Dopamine Uric acid	Fe(II)-octanitro PC + CTAB	G/so	CV	Simultaneous det.; PBS + KCl 0.1 M, pH 7	[916]
Ascorbic acid	Fe(II)PC dioctyl phthalate			Catalytic ox.; LOD 0.32 µM; s: tablets, orange juice	[3161, 3162]

(continued)

TABLE 8.25 (continued)
Analysis of Biologically Important Compounds at Carbon Paste–Based Electrodes: Vitamins

Analyte	Modifier	Carbon Paste	Techniques	Comments	Ref(s)
Ascorbic acid Dopamine	Fe(II)-tetrasulfo-PC	G/mo	LSV	Simultaneous det. of ratio DA/AA via potential shift; pH 7.4; LOD 0.75 µM	[1570]
Ascorbic acid	Fe(II)-PC	G/Nj	CV, POT	Potentiometric response by Fe(II)/Fe(III)-PC redox couple; PBS 0.05 M, pH 7; LOD 0.5 µM; s: vitamin preparations	[655]
Ascorbic acid	Co(II)-PC	G/Nj	POT	Potentiometric response by Co(II)/Co(III)-PC redox couple; PBS 0.05 M, pH 7; LOD 0.2 mg L^{-1}; s: vitamin preparations	[930]
Ascorbic acid	Co(II)-PC	G/p (liquid)	DPV	Catalytic ox.; ABS 0.1 M + KCl 0.1 M, pH 4.7; LOD 0.8 mg L^{-1}; s: fruit juice	[653]
Ascorbic acid Dopamine	CoPC NP (+CTAB)	G/po	CV, DPV	Catalytic ox.; simultaneous det. possible with CTAB; PBS 0.025 M, pH 7.4; LOD 1.7 µM; s: injections	[933]
Ascorbic acid Dopamine	Co(II)-5-nitrosalophen + tetraoctyl ammonium bromide	G/Nj	CV, DPV	Catalytic oxidation via Co(III); simultaneous det.; ABS 0.1 M, pH 5; LOD 0.7 µM; s: vitamin preparations	[1583]
Ascorbic acid Uric acid	MWCNT/Nafion + Co(II)-5-nitrosalophen	G/Nj	CV, DPV	Catalytic ox. via Co(III); simultaneous det.; ABS 0.1 M, pH 4; LOD 0.1 µM	[1590]
Ascorbic acid Cysteine	Co(II)-4-methylsalophen	G/Nj	CV, DPV	Acetate buffer 0.1 M, PH 5.0; LOD 0.8 µM	[1586]
Ascorbic acid	Ru(III)-EDTA on 4-pyridylmethylchitosan			Catalytic ox. via Ru(III)/Ru(II)	[86]
Ascorbic acid Dopamine	mer-Ruthenium(III) bis [1,4-bis(diphenyl phosphino) butane] piccoline trichloride	G/Nj	CV	Catalytic ox.; multivariate calibration for simultaneous det.; KCl 0.5 M, pH 2; LOD 44.8 mM (?)	[1612, 2985]
Ascorbic acid	Ru(III)-diphenyl dithiocarbamate	G/p (liquid)	CV, HDV, A (0.38 V vs. SCE)	Catalytic ox.; KNO$_3$ 0.1 M, pH 3; LOD 5 mg L^{-1}; s: tablets	[1609]
Ascorbic acid	Penta(N,N-piperidine dithiocarbamate)-diruthenium(III) chloride		LSV	Potassium hydrogen phthalate, pH 4; LOD 7 µM; s: pharmaceuticals	[1610]

Analyte	Modifier	Electrode type	Method	Notes	Ref.
Ascorbic acid	Azamacrocycles + Zn(NO$_3$)$_2$	G/Nj	CV, DPV	Electrocatalytic ox.; BRP, pH 1.5; LOD 0.1 mg L^{-1}; s: tablets, juices, wine	[3163]
Ascorbic acid	unmod. or ammonium vanadate	G/mo + epoxy resin	FIA (0.8 V vs. SSE)	PBS, pH 6.5; s: soft drinks	[3164]
Ascorbic acid	Cd-(phenanthroline or dipyridyl) hexadeca vanadate	G/mo	CV	Catalytic ox.; H$_2$SO$_4$ 1 M	[144]
Ascorbic acid	Piperazinium phosphoromolybdate			Electrocatalytic ox.	[3165]
Ascorbic acid Dopamine	CdO NP	G/so	CV	Increase of current for ox. of DA and AA due to CdO NP	[1003]
Ascorbic acid Uric acid Dopamine	Pd NP on C nanofibers (electrospun)	G/mo	CV	Improved peak resolution; PBS 0.1 M, pH 7; s: injection solutions	[969]
Ascorbic acid Dopamine	Ag NP + MWCNT	G/Nj	CV, DPV	Decrease of ox. overpotentials; PBS 0.1 M, pH 2; LOD 12 μM; s: Ringer's serum	[416]
Ascorbic acid	2,2′-(1,8-Octanediyl bisnitriloethylidyne)-bis-hydroquinone	G/po	CV, A	Electrocatalyt. ox.; PBS 0.1 M, pH 7; LOD 0.6 μM; s: pharmaceuticals	[1461]
Ascorbic acid	2,2′-[3,6-Dioxa-1,8-octanediylbis(nitrilo-ethylidyne)]-bis-hydroquinone	G/po	CV, LSV,CA (−0.1/0.3 V vs. SSE)	PBS 0.1 M, pH 7; LOD 0.38 μM; s: pharmaceuticals	[1460]
Ascorbic acid Uric acid	2,2′-[1,2-Ethanediylbis (nitriloethylidyne)]-bis-hydroquinone	CNT/p, also RDE	CV, A, DPV	Electrocatalytic ox.; PBS 0.1 M, pH 7; LOD 75 nM; s: pharmaceuticals	[422]
Ascorbic acid Dopamine Uric acid	Tetrabromo-p-benzoquinone	G/po	CV, DPV, CA (0.22 V vs. SCE)	Simultaneous det.; PBS 0.1 M, pH 7; LOD 0.62 μM	[1446]
Ascorbic acid	Phenanthrene quinone	G/Nj	CV	Lowering of overpotential	[53]
Ascorbic acid Dopamine NADH	N,N,N′,N′-Tetramethyl phenylendiamine	G/Nj	CV	Catalytic ox. via quinone diimine; PBS or citric buffer, 0.15 M, pH 7	[1516]
Ascorbic acid	Chloranil	G/silicon or bromo naphthalene and po	CV, CA (0.2/−0.2 V vs. SSE)	Catalytic oxidation; PBS 0.1 M + KCl 0.1 M, pH 7; s: tablets, syrup, ampoules	[3166]

(continued)

TABLE 8.25 (continued)
Analysis of Biologically Important Compounds at Carbon Paste–Based Electrodes: Vitamins

Analyte	Modifier	Carbon Paste	Techniques	Comments	Ref(s)
Ascorbic acid Uric acid	4-Nitrophthalonitrile	G/mo; also RDE	CV, A	Electrocatalytic ox.; PBS 0.1 M, pH 7; LOD 1.6 μM; s: urine	[3131]
Ascorbic acid Acetaminophen Isoniazide	Thionine on MWCNT	G/Nj	CV, DPV	Simultaneous determination (ox.); LOD 0.8 μM; acetate buffer, 0.1 M, pH 4; s: serum	[438]
Ascorbic acid Uric acid	Thionine on Nafion	G/Nj	CV, DPV	Electrocatalytic ox. by thionine; ABS 0.1 M, pH 5; LOD 0.5 μM; s: tablets, multivitamin drops	[1509]
Ascorbic acid Dopamine	Thionine/Nafion on MWCNT	G/Nj	CV, DPV	Simultaneous det.; ABS 0.1 M, pH 4; LOD 80 nM	[1729]
Ascorbic acid	Tetracyanoquino dimethane			Catalytic ox.	[3167]
Ascorbic acid	Tetrathiafulvalene		CV	Catalytic ox.; PBS, pH 7	[1491]
Ascorbic acid	Calixarene	G/mo	CV, DPV	Supramolecular metallocatalytic oxidation via complexed Pb(II); CH_3COONH_4 0.01 M + $Pb(NO_3)_2$ 0.005 M, pH 5.1; s: vitamin preparates, fruit juice	[1430]
Ascorbic acid	Polypyrrole within clay		CV	Electrocatalytic ox.	[3168]
Ascorbic acid	Microcrystalline cellulose	Thin electrode	CV	LOD 0.75 μM	[3169]
Ascorbic acid Dopamine	Polyglycine (electropolym.)	G/so	CV	Electrocatalytic effect on ox.; simultaneous det.; ABS 0.2 M, pH 5	[1764]
Ascorbic acid Dopamine	Polyalanine (electropolym.)	G/so	CV	Catalyt. improvement of ox.; ABS 0.2 M; pH 7	[1806]
Ascorbic acid Uric acid Dopamine	Poly (calconcarboxylic acid), electropolym.	G/so	CV	Electrocatalytic activity; acetate buffer, 0.2 M, pH 5.4	[135]
Ascorbic acid Dopamine Serotonin	Poly(2-amino-5-mercapto-thiadiazole) film	G/paraffin wax	CV, DPV; A (0.55 V vs. SSE)	Catalytic ox., no significant acc.; BRP 0.2 M, pH 5; LOD 1.5 nM	[2964]
Ascorbic acid Dopamine	Eriochrome Black T		CV, DPV	Catalytic oxidation; KCl 1 m; simult. det. possible; LOD 0.27 μM	[2989]

Electroanalysis with Carbon Paste–Based Electrodes, Sensors, and Detectors

Analyte	Modifier	Paste type	Technique	Conditions / Notes	Ref.
Ascorbic acid, Dopamine	Poly(Eriochrome black T) electropolym.	G/so	CV	Electrocatalyt. ox.; KCl, 1 M	[1763]
Ascorbic acid	Methylene blue intercalated in Ca-phosphate	G/mo	CV, A (0 V vs. SCE)	Catalytic oxidation; KCl 0.5 M, pH 6; s: tablets	[1318]
Ascorbic acid	Methylene blue on Zr- or Ti-phosphate cond. with n-butyl amine		Photoamperometric FIA (0.1 V vs. SCE)	0.1 M phosphate buffer, pH 7	[1285, 1288, 3170]
Ascorbic acid	Methylene blue on cellulose acetate/TiO$_2$	G/liquid p	CV, CA (0.3 V vs. SSE)	Catalytic ox. via MB; KCl 1 M, pH 7; LOD 15 µM; s: tablets	[3171]
Ascorbic acid	Methylene blue on synth. zeolites		A	Catalytic ox.; PBS, pH 7.0; LOD 17–47 µM	[3172]
Ascorbic acid	Methylene blue on morenite (zeolite)	G/po	CA (0.2 V vs. SCE)	PBS + KCl 0.5 M, pH 6.2; LOD 12.1 µM; s: tablets, ampoules, syrup	[1133]
Ascorbic acid	Methylene blue on muscovite	G/mo	DPV, FIA (0.05 V vs. SCE) with photoelectrochem.	Catalytic ox.; PBS 0.1 M, pH 7; LOD 10 nM; s: tablets	[1087]
Ascorbic acid	Methylene green		FIA (0.5 V)	Catalytic oxidation; LOD 10 nM	[1472]
Ascorbic acid	Meldola blue on SiO$_2$/SnO$_2$/phosphate	G/po	CV, CA (0.04 V vs. SSE)	Catalytic ox.; KCl 0.5 M, pH 7 s: pharm. tablets; fruit juice	[1209, 3172]
Ascorbic acid	Brilliant yellow on silica gel/TiO$_2$ xerogel	G/Nj	CV, CA (0.18 V vs. SCE)	Catalytic ox.; KCl 1 M, pH 6; LOD 15.3 µM; s: tablets, orange juice	[1186]
Ascorbic acid	Congo Red on silica/aniline xerogel		CV, CA (0.22 V vs. SCE)	KCl 0.5 M, pH 7; s: tablets	[3174]
Ascorbic acid	HRP	G/po	CV	Biosensor; catalytic oxidation via Fe(II)/Fe(III) of heme in HRP; PBS 0.1 M + KCl 0.1 M, pH 7.2	[2681]
Ascorbic acid	Green pepper seed		DPV	Biosensor; LOD 9.9 µM	[3175]
Ascorbic acid	Gooseberry		DPV	Biosensor; catalytic ox.; PBS, pH 5.6; LOD 5 µM; s: fruit juice	[3176]
Ascorbic acid		Poly(methyl methacrylate)	CV, EIS, ECL	Inhibition of ECL of luminol; LOD 8.3 nM; PBS, pH 7.5 + luminol 20 µM	[3177]
Folic acid (vitamin B9)	Palmitic acid or stearic acid		acc (60 s), LSV	ox.; PBS 0.04 M, pH 3.4; LOD 5 nM; s: urine, serum	[1663]

(continued)

TABLE 8.25 (continued)
Analysis of Biologically Important Compounds at Carbon Paste–Based Electrodes: Vitamins

Analyte	Modifier	Carbon Paste	Techniques	Comments	Ref(s)
Nicotinamide Pyridoxine Riboflavin	Macrocycles		CV, DPV, EIS	Accumulation; LOD 0.03 mg L^{-1}; s: pharmaceuticals	[1397]
Nicotinic acid			V	red.: 0.15 M HCl; LOD 3.3 μM; for drug analysis	[3178]
Nicotinic acid		MWCNT + TiO$_2$ with La(III)/po	CV, A (0.32 V vs. SCE)	Catalytic ox., KCl 0.1 M, pH 10; LOD 0.27 μM; s: urine	[1246]
Nicotinic acid	*Pseudomonas fluorescens* TN5, whole cells + *p*-benzoquinones or phenazine methosulfate + dialysis membrane		A (0.5 V vs. SSE)	Catalytic hydroxylation of na; quinine or ohenazine as electron acceptor; PBS 0.1 M, pH 7	[786,1442]
Panthenol Panthothenic acid	CoO + surfactants	G/p (liquid)	CV, LSV, CP, DPV	Studies with surfactants and metal oxides, catalytic red.: PBS, pH 6.08; s: cosmetics, pharmaceuticals	[1622]
Phylloquinone (vitamin K1)		G/Nj	LSV, oc acc. (15 min), adSV	s: plasma after extraction (hexane/alc.)	[3179]
Pyridoxal	Octadecyl amine or 1-amino-12-(octadecanoylamino) dodecane	G/mo	oc. acc. (trifluoroacetate buffer), DPV	Detection via Schiff base; Na-perchlorate 0.1 M; LOD 0.5 μM; s: non-fat dry milk	[1637, 1641]
Pyridoxine (vitamin B6)			CV	ox.; interference of ascorbic acid removed by ion exchange	[3180]
Pyridoxine	Cu(II) hexacyanoferrate(III)	G/mo	CV, LSV	Catalytic ox. via Fe^{III}; ABS with 0.05 M Na$^+$, pH 5.5; LOD 0.41 μM; s: pharmaceuticals	[1034]
Pyridoxine	V(IV)-salen	G/mo	CV, LSV	Catalytic ox. via V(V); KCl 0.1 M, pH 7; LOD 37 μM; s: pharmaceuticals	[1585]
Pyridoxine Riboflavin Thiamine	Crown ethers	G/po	CV, DPV, EIS	Complexation of vit. B6 by crown ether; TRIS buffer 0.05 M, pH 10.3; s: pharmaceuticals; LOD 0.2 mg L^{-1}	[1398]
Riboflavin (vitamin B2)		G/liquid paraffin	DPV	NaCl 0.1 M, pH 2.5; LOD 20 nM; s: tablet	[3181]
Riboflavin		G(atc. 100°C)/ceresin wax	LSV	Paste with wide anodic range; H$_2$SO$_4$, 0.1 M	[39]
Riboflavin	Immobilized on SiO$_2$/Nb-oxide grafted with phosphate	G/mo	CV	NaNO$_3$, 0.1 M, pH 7	[1198, 3182]

Electroanalysis with Carbon Paste–Based Electrodes, Sensors, and Detectors 427

Analyte	Modifier	Binder	Method	Notes	Ref.
Riboflavin FMN, FAD	Immobilized on SiO$_2$/ZrO$_2$	G/mo	CV	PBS 0.1 M, pH 7	[1188]
Riboflavin	Zr-phosphate	G/po	Fast scan CV, SECM	Adsorption at neutral pH	[1294]
Riboflavin	Immobilized at unmodified and on Zr-phosphate	G/po	CV	Used for catalytic det. of NADH	[1292, 1293]
Riboflavin	Crown ethers	G/po	CV, DPV, SWV	Complexation; BRP 0.04M, pH 1.5; LOD 0.2 μg L^{-1}; s: pharmaceuticals, milk powder, mango drink	[1408]
Thiamine (vitamin B1)	Mn(II)-PC	G/Nj	CV	Catalytic ox. via Mn(III); Tris buffer, pH 10; LOD 14.6 μM; s: tablets	[906]
Tocopherols		G(wax sealed)/so		s: pharmaceuticals, vegetable oil, food	[37]
α-Tocopherol Ascorbic acid		G/po	CV, DPV	Simultaneous det.; ethanol (50%) + water or acetonitrile (40%) + water + 0.1 M surfactant (Triton X100 or others) + 0.1M NH$_3$; s: multivitamin preparations	[3183]
α-Tocopherol Quinones Fc	Dissolved in Nj	G/Nj	CV	Influence of surface active compounds	[3184, 3185]
Vitamin A		C/so	CV, DPV	ox. in (partly) organic media; s: pharmaceuticals, margarine	[3186, 3187]
Vitamin A Vitamin B1,B2				s: carrots (vit. A), meat (vit. B)	[3188]
Vitamin B p-Amino benzoic acid	Crown ethers	G/po	CV, DPV	BRB, pH 2.0; LOD 0.1 mg L^{-1}; s: ointment, spirulina, yeast	[1486]
Vitamin B6-Fe(III) complex			CV	HEPES buffer, pH 5.1–13.1	[3189]
Vitamin B12	N-p-chlorophenyl cinnamohydroxamic acid		CV, DPASV	ABS 0.2M, pH 6; det. of vit. B12 via Co after mineralization with HNO$_3$/H$_2$O$_2$	[1386]
Vitamin B12		g/trans-1,2-Dibromo cyclohexane	CV, SWV	Electrocatalytic red. by binder; PBS 0.2M, pH 2.5; s: tablets	[231]
Vitamin E		G/so	CV, stripping voltammetry	Extractive preconcentration after ox. with HNO$_3$ to electroactive toco-red; HNO3 1.M in etOH; LOD 6 μM; s: capsules	[3190–3192]
Vitamin K3			CV	Aqueous solutions with supercritical conc. of surfactants (CTAB, SDS…)	[1651]

TABLE 8.26
Analysis at Carbon Paste–Based Electrodes Modified with Whole Cells, Microorganisms, Cell Biomass, Tissues, and Tissue Extracts Plus Related Topics

Analyte	Modifier	Carbon Paste	Techniques	Comments	Ref(s)
Algae					
Au(III)	*Chlorella pyrenoidosa*	G/mo	acc. (o.c.), me	Sorptive biosensor; high affinity for Au	[270]
Au(III)	*Bacillus* sp. DO1		CSV (Au), ASV (Cu)	Simultaneous det; LOD 20 μg L^{-1} (Au) an 0.5 μg L^{-1} (Cu)	[3193]
Cu(II)					
Cd(II)	*Anabaena*		acc., DPSV	Sorptivebiosensor; PBS, pH 6; LOD 0.5 μM	[2260]
Cu(II)	*Bacillus* sp. DO1 + Au-film (electrodep.)			Biosensor; s: water	[3194]
Cu(II)	*Chlorella pyrenoidosa* or others	G/mo	acc. (o.c.), me, DPV	Sorptive biosensor; ABS, pH 5	[2226]
Cu(II)	*Anabaena* + Nafion		acc. (o.c., 15 min), me, DPASV	Sorptive biosensor; buffer, pH 4 (acc.), KCl 0.1 M (meas.); LOD 75 μM	[2236]
Cu(II)	*Tetraselmis chuii*	G/mo	acc. (o.c.), CV, DPCSV	Sorptive biosensor; LOD 0.46 nM; HClO$_4$ 0.05 M (meas.); s: mineral capsules for animals	[2245]
H$_2$O$_2$	*Spirulina platensis* or other algae + [Fe(CN)$_6$]$^{4-}$			Catalytic biosensor, algae as supports for electrocatalytically active metal complexes	[3195]
Pb(II)	*Formidium* spp. (heat dried + formaldehyde) + formaldehyde	G/p (liquid)	acc. (o.c., 10 min), CV, DPASV	Sorptive biosensor; improvement of binding sites by formaldehyde; Tris buffer 0.05 M, pH 8; LOD 25 nM	[3196]
Bacteria					
Au(III)	*Bacillus megatherium* DO1	G/p wax	acc. (0.75 V vs. SCE, 10 min), CSV	Biosensor; KCl 0.1 M	[2198,2199]
Ethanol	*Acetobacter aceti*		CV	Decrease of oxygen wave when offering etOH to cells (alternate O-source)	[3197]
Ethanol	*Acetobacter aceti* + dialysis membrane		A (0.5 V vs. SSE)	Biosensor; McIlvain buffer, pH 6 + 0.5 mM benzoquinone	[780]
Ethanol	*Acetobacter pasteurianus* whole cells + nylon net	G/p	CV, A (0.5 V vs. SSE)	Biosensor; McIlvaine buffer, pH 6.0 + K$_3$[Fe(CN)$_6$] 20 mM	[2664]

Electroanalysis with Carbon Paste–Based Electrodes, Sensors, and Detectors   429

Analyte	Composition	Configuration	Method	Notes	Ref.
Glucose	*Gluconobacter industrius* + benzoquinone + dialysis membrane	G/liquid p	CV, A (0.5 V vs. SSE)	Biosensor with gluconobacter as a source for dehydrogenase; PBS 0.1M, pH 7; LOD 10μM	[1441,2864]
Glucose	*Gluconobacter oxydans* + dialysis membrane + ferrocenes or quinones	G/mo	CV, A (0.065–0.51 V vs. SSE)	Biosensor; phosphate-citrate buffer 0.15 M, pH 6.0	[2863]
Glucose	*Escherichia coli* + 2,3-dimethoxy-5-methyl-1,4-benzoquinone + dialysis membrane	G/mo	CV, A (0.2 V vs. SSE)	Biosensor with *E. coli* as a source of glucose dehydrogenase; PBS + KCl 0.3M + PQQ 1μM + MgSO$_4$ 5 mM, pH 6.5; s: blood, regeneration with EDTA	[288]
Glucose	*Lactobacillus fructivorans* (Hiochi) + benzoquinone	G/mo	A (0.5 V vs. SSE)	Biosensor with bazilli as dehydrogenase source; PBS 0.2 M, pH 6; s: sake brewing	[1443]
Glucose	*A. niger* (in paste) + Fc	G/po	CV, A (0.35 V vs. SSE)	Microbial biosensor with *A. niger* as a source of GOx; PBS 0.2 M, pH 7; s: xanthan fermentation on milk, lactose	[1521]
Glucose Phenol	CNT + poly(1-vinylimidazole)$_{12}$-[Os-(4,40-dimethyl-2,20-dipyridyl)$_2$Cl$_2$]$^{2+/+}$ + *Pseudomonas putida*	G/mo	CV, CA	Direct e-transfer to cell-wall bound enzymes, phenol det. with phenol-adapted bacteria; Tris-buffer, pH 9	[381]
Nicotinic acid	*Pseudomonas fluorescens* TN5 + p-benzoquinones + dialysis membrane	G/mo	A (0.5 V vs. SSE)	Catalytic hydroxylation of na; PBS 0.1M, pH 7	[786,1442]
p-Nitrophenol	*Moraxella* spp.	g/mo	A (0.3 V vs. SSE)	Biosensor via p-hydroquinone which is oxidized; PBS 0.02M, pH 7.5; LOD 20nM; s: lake water	[2440]
Organophosphate Fenitrothion EPN	*Pseudomonas putida* JS444 with surface-expressed organo-phosphorous hydrolase	G/mo	A (0.6 V vs. SSE)	Direct det. after hydrolysis and bacterial degradation of 3-methyl-4-nitrophenol; citrate-phosphate buffer 0.05 M + CoCl$_2$ 005 mM, pH 7.5; LOD 1.4–1.6μg L^{-1}	[2476]
O$_2$ (respiratory activity)	*Pseudomonas fluorescens* + eggshell membrane (+Fc)	G/mo	CA (−0.7 V vs. SSE, +0.4 V vs. SSE with Fc)	Biosensor, PBS 0.05 M, pH 7 + glucose	[3198]
Lichens and mosses					
Cu(II) Pb(II)	*Ramalina stenospora* or *Sphagnum* moss	G/mo	acc., ASV	Sorptive biosensor; s: water	[3199]
Cu(II) Pb(II)	Cell walls from *Cladonia portentosa Lobaria pulmonaria* or *Roccella* spp.	G/mo	DPASV	Biomass biosensor; LOD 20μM Pb(II)	[271]

(*continued*)

TABLE 8.26 (continued)
Analysis of with Carbon Paste–Based Electrodes Modified with Whole Cells, Microorganisms, Cell Biomass, Tissues, and Tissue Extracts Plus Related Topics

Analyte	Modifier	Carbon Paste	Techniques	Comments	Ref(s)
Cu(II) Pb(II) Hg(II)	Cell walls from *Cladonia portentosa*, *Lobaria pulmonaria* or *Roccella* spp.	G/mo	V	Biomass biosensor; det. in mixtures	[1712]
Tissue					
Various	Plant tissues			Reviews and general concepts	[70,2564, 3001,3200, 3201]
F	Asparagus + Fc	G/mo	CV, A (−0.05 V vs. SSE)	det. via inhibition of peroxidase; PBS 0.05 M, pH 7 (CV) or 5 (A) + 0.1 mM H_2O_2; LOD 0.5 mg L^{-1} s: fluoride tablets	[2380]
Pb(II)	Banana tissue		acc. (−1.5 V vs. ref., 300 s)	Sorption/complexation on tissue; pH 5 (acc.), 0.01 M HCl (meas.); LOD 10 mg L^{-1}	[2270]
Hydrogen peroxide	Horseradish root	FIA	FIA (−0.2 V vs. SSE)	PBS 0.05 M + *o*-phenylenediamine 1 mM, pH 7.4	[3202]
Hydrogen peroxide	Horseradish root or HRP	G/paraseal wax	CA (−0.2 V vs. SSE)	Stripping of H_2O_2; PBS 0.05 M, pH 7.4	[3203]
Hydrogen peroxide	Spinach root + Fc		A (−0.3 to 0 V vs. SSE)	LOD 2.3 µM	[3204]
Hydrogen peroxide	Cabbage tissue (+Fc or *o*-phenylene diamine)		A (−0.25 V vs. SSE)	Tissue as a source of peroxidase; PBS 0.1 M, pH 6.5	[3205]
Hydrogen peroxide	Cabbage tissue	C/butadiene rubber	CV, A (0.4 V vs. SSE)	Rubber shows improved antifouling properties and stability, tissue as a source of peroxidase	[311,3206]
Hydrogen peroxide	Peroxidases on coconut fibers	G/mo	CV, A (−0.15 V vs. SSE)	As support, naturally immobilized; PBS 0.1 M, pH 5.2; LOD 40 µM; s: pharmaceuticals	[2368]
Hydrogen peroxide	Chicken small intestine tissue + Fc	G/mo	CV, A (−0.15 V vs. SSE)	ABS 0.1 M, pH 4.75; LOD 50 µM	[3207]
Alternaria mycotoxins	Mushroom tissue or tyrosinase	G/mo		McIlvaine buffer, pH 6.7; LOD 19 µM (alternariol) or 24 µM (methyl ester)	[284]

Electroanalysis with Carbon Paste–Based Electrodes, Sensors, and Detectors 431

Analyte	Matrix/electrode	Composition	Technique	Notes	Ref.
Amines	Pea seedling tissue (diamine oxidase)	G/mo	A (0.9 V vs. SSE)	Biosensor; via H$_2$O$_2$; LOD 7.1 nmol spermidine	[274]
Ascorbic acid	Green pepper seed		DPV	Biosensor; LOD 9.9 μM	[3175]
Ascorbic acid	Gooseberry		DPV	Biosensor; catalytic ox.; PBS, pH 5.6; LOD 5 μM; s: fruit juice	[3176]
Caffeic acid	Green bean tissue + chitin (from squid) + GA + epichlorhydrin	G/Nj	CV, A (0.1 V vs. SSE)	Tissue as a source of peroxidase, catalyzes ox. to quinine, det. by red.; PBS 0.1 M + H$_2$O$_2$ 2 μM; LOD 2 μM; s: white wine	[2751]
Catecholamines Adrenaline L-Dopa Dopamine Isoprenaline	Laccase (oyster mushroom) + peroxidase (zucchini)	G/po	DPV	Synergistic ox., det. by red. of quinones; PBS 0.1M, pH 6, LOD 24–27 nM	[2954]
Catecholamines Dopamine Epinephrine	Crude extract of oyster mushroom	G/Nj	DPV	Source of laccase, catalytic ox., det., by red. of quinones as sum; LOD 7.9 μM (ad) or 9.8 μM (da); PBS 0.1 M, pH 7; s: pharmaceuticals	[2953]
Catechol Dopamine	*Latania* spp., fresh or dry + Nafion film	g/mo	FIA	Tissue as a source of polyphenol oxidase; PBS 0.05M, pH 5.7	[2756]
Catechols Dopamine	Fresh tissue of potato, peach, mushroom, pear		A (0 V vs. SCE)	Tissue as a source of polyphenol oxidase; McIlvaine buffer, pH 6.75	[279]
Catechol	Potato tissue	G/N	A	s: beer	[2757]
Catechol	Banana tissue + Au NP or MWCNT	GC	A (−0.2 V vs. ref.)	s: wine	[2758]
Dopamine	Soursop tissue + TCQN	G/silicone wax	FIA (0.1 V vs. SSE)	Biosensor, tissue as a source of polyphenol oxidase; PBS 0.3M, pH 7.8; LOD 0.15 mM; s: pharmaceuticals	[280]
Dopamine Pyrocatechol	Banana tissue		CV, DPV, FIA	Biosensor with tissue as a source for polyphenoloxidase	
Dopamine	Banana tissue	G/mo	DPV, FIA (−0.2 V vs. SSE)	Biosensor, tissue source of polyphenol-oxidase; PBS 0.05M, pH 7.4; LOD 13 nM	[276]

(continued)

TABLE 8.26 (continued)
Analysis of with Carbon Paste–Based Electrodes Modified with Whole Cells, Microorganisms, Cell Biomass, Tissues, and Tissue Extracts Plus Related Topics

Analyte	Modifier	Carbon Paste	Techniques	Comments	Ref(s)
Dopamine Epinephrine Norepinephrine	Banana tissue		AD in HPLC (−0.2 V vs. SSE)	Biosensor, tissue source of polyphenol-oxidase; PBS 0.05 M, pH 6.5; s: urine	[2564]
Dopamine	Banana tissue			For *in vitro* and *in vivo* det.	[2995–2997]
Dopamine	Apple		DPV	Biosensor, apple as a source for polyphenol oxidase; det. by red. of quinone; LOD 0.2 µM; s: injections	[2998]
Dopamine	Potato tissue			Biosensor, tissue as a source of polyphenoloxidase; LOD 1.6 µM	[2999]
Dopamine L-Dopa Dopac Pyrocatechol	Potato tissue	G/mo, also as RDE	CV, CA (0.4 V vs. SCE)	Biosensor, tissue as a source of polyphenoloxidase; McIlvaine buffer, pH 6.5; LOD 2.5 µM (5500 rpm)	[3000]
Dopamine Norepinephrine	Zucchini tissue	G/mo	DPV, FIA (0.7 V vs. SSE), CA (−0.1/+0.7 V vs. SSE)	Source of polyphenoloxidase and ascorbate oxidase—eliminates AA interference; PBS 0.05 M, pH 7.4	[3001]
Dopamine	Spinach tissue			Biosensor for *in vivo* and *in vitro* measurements; LOD 0.71 µM	[2123]
Dopamine L-Dopa Catechol	Eggplant tissue	G/mo	A (−0.2 V vs. SSE)	Biosensor; tissue as a source for polyphenoloxidase; det. by red. of quinone; PBS 0.1 M, pH 7	[3266]
Dopamine L-Dopa Norepinephrine	Eggplant-banana tissue	Microelectrode		Eggplant to eliminate AA interference; LOD 0.32 µM (da)	[3002]
Dopamine	Pine kernel peroxidase on pegylated polyurethane NP	G/mo	CV, SWV	Biosensor, enzyme catalyzes ox. of dopamine, det. via red. of quinone; PBS 0.1 M, pH 6.5 + H_2O_2 2 mM; LOD 9 µM; s: pharmaceuticals	[1018]

Analyte	Source/Modifier	Electrode	Method	Notes	Ref.
Dopamine Epinephrine	Crude extract of cara root		LSV	As a source of polyphenol oxidase; ox. to quinine; LOD 0.75 mM (da) and 0.082 mM (ep)	[3003]
Dopamine	Plant tissue				[3004]
Dopamine Acetaminophen	Papaya tissue	G/mo	DPV	Papaya tissue for preventing fouling by proteins; PBS 0.05 M, pH 7.4	[3201]
Epinephrine	Banana crude extract	G/vaseline	POT	Polyphenoloxidase; *in vitro* and *in vivo* measurements; PBS, pH 7; LOD 8 nM; s: pharmaceuticals	[3018]
Epinephrine	Palm tree fruits	G/po	CV, AD (−0.1 V vs. SSE) in FIA and bioreactor	Tissue as a source of polyphenol oxidase; ABS 0.1 M, pH 4.4; LOD 15 µM; s: pharmaceuticals for inhalation	[3019]
Flavanols	Apple (green or red) or banana or potato tissue	g/Nj	A (−0.25 V vs. SSE)	Biosensor; tissue as a source of polyphenol oxidase; oxidation of *o*-diphenols to quinone and electrochemical reduction; PBS 0.05 M, pH 7.4; s: beer	[278,2760]
Glucose	Glucose oxidase + mushroom tissue	C-Pt/mo	A (0.6 V vs. SSE)	Biosensor; mushroom tissue as a source of tyrosinase to eliminate interference from acetaminophen; PBS 0.05 M, pH 7.4	[2840]
Glycolate	Spinach leave tissue + ferrocene	G/mo	CV, A (0 V vs. SSE)	Plant tissue biosensor via glycolate oxidase and peroxidase; PBS 0.05 M, pH 7.5; LOD 1 µM; s: human urine	[2674]
Glycolate	Spinach leave tissue	G/mo	Potentiostatic 0.8 V vs. SSE	ECL biosensor via luminol; luminol (0.25 mM) + KNO$_3$ (0.05 M), pH 9; LOD 15 µM	[2675]
Glycolate	Sunflower leave tissue + ferrocene	G/mo	FIA (0 V vs. SSE)	Plant tissue biosensor via glycolate oxidase and peroxidase; PBS 0.05 M, pH 7.5; LOD 1 µM; s: human urine	[2676]
Hydroquinone	Apple tissue		DPV	PBS 0.05 M, pH 6	[2765]
Indole-3-acetic acid	Mung bean leave + Fc		CV	Tissue as a source of IAA oxidase, catalytic ox.; PBS, pH 5.8; LOD 4.2 mg L^{-1}	[2951]
Neurotransmitters Phenols	Tissue or microbes or yeast			Tissue- and yeast-modified CPEs as detectors in HPLC	[2564]

(continued)

TABLE 8.26 (continued)
Analysis of with Carbon Paste–Based Electrodes Modified with Whole Cells, Microorganisms, Cell Biomass, Tissues, and Tissue Extracts Plus Related Topics

Analyte	Modifier	Carbon Paste	Techniques	Comments	Ref(s)
Organic peroxide	Horseradish root or HRP	G/silicon grease	AQ, FIA (−0.2 V vs. SSE)	PBS 0.05 M, pH 7.4; s: water	[3208]
Pectin	Alcohol oxidase + East-man AQ29D + orange peel + dialysis/PTFE membrane	G/mo	A (0.9 V vs. SSE)	Hybrid tissue-enzyme biosensor; orange peel as a source of pectin esterase on alcohol sensor; detection via released methanol; PBS 0.1 M, pH 7.5	[277]
Phenol	Mushroom tissue + EDTA + Nafion film	G/mo	A (−0.24 V vs. SSE)	EDTA prevents effects from metals; ammonium buffer 0.1 M + [Fe(CN)$_6$]$^{4+}$ 30 µM, pH 8.8	[3209]
Phenols Ascorbic acid Dopamine	Polyphenol oxidase or mushroom tissue	GC/mo	CV, A (−0.1 V vs. SSE)	Biosensor; PBS 0.05 M, pH 7.4; LOD 0.4 µM (catechol); s: red wine, acetaminophen in pharmaceuticals	[194]
Phenols	Mushroom		AD in HPLC (−0.2 V vs. SSE)	Biosensor; PBS 0.05 M/acetonitrile (70:30), pH 5	[2564]
Phenols	Mushroom tissue (partially purified tyrosinae) + dialysis membrane		A (−0.1 V vs. SSE)	Biosensor; detection via red. of o-quinone; pH 6 or 6.5 (two isozymes of tyrosinae); LOD 10 nM	[283]
Phenols	Mushroom tissue + CoPC or K$_4$[Fe(CN)$_6$]	g/mo	A (−0.22 V vs. SSE)	Biosensor; tissue as a source of tyrosinase; catalytic red. of o-quinone; PBS 0.05 M, pH 7.4	[898,2783]
Phenols Catechol	Banana tissue + Ir	G/mo	A (−0.1 V vs. SSE)	Biosensor; catalytic reduction of the quinone; PBS 0.05 M, pH 7.4	[1265]
Phenols	Coconut tissue		FIA, ED in HPLC	Biosensor; comparison with laccase and tyrosinase; comparison with reactors	[2842]
Polyamines	Oat seedlings + Fc	G/mo	A, FIA (0 V vs. SSE)	Tissue as a source of polyamine oxidase and peroxidase, Fc for red. of enzyme; PBS 0.1 M, pH 7.4	[864]
Tryptophan	Potato juice		CV, adSV	PBS, pH 7.4; LOD 5 × 10^{-6} M	[2741]

Electroanalysis with Carbon Paste–Based Electrodes, Sensors, and Detectors

Tyrosinase inhibitors	Mushroom tissue		A, FIA	Via inhibition with catechol as substrate; inhibitors: diethyldithiocarbamate, thiourea, benzoic acid	[3210,3211]
Yeast					
Ethanol	Yeast			Biosensor via ADH; LOD 0.14 mM; s: beer	[2660]
Ethanol Alcohols	Yeast		FIA	Amperometric biosensor LOD 2 μM; s: alcoholic beverages	[2661]
Ethanol Alcohols	Yeast			Biosensor via ADH; s: fermentation broth	[2635]
Lactate	Baker's yeast + nile blue, methylene blue, toluidine blue, prussian blue, meldola blue	G/po	CV, A (0.1/0.3 V vs. SSE)	Biosensor; yeast as a source for cytochrome b2; PBS, pH 7.2	[1483]
L-Lactate	Baker's yeast (pretreated with meOH or etOH) + ferricyanide		CV, A (0.3 V vs. SSE)	Biosensor with yeast as a source of flavocytochrome b2; PBS + LiCl 0.1 M, pH 7.3; LOD 3–6 μM	[2685]
Lactate	Yeast *Hansenula anomala* + mediators	G/Nj or po	A (0.05–0.3 V vs. SCE)	Biosensor via cytochrome b2; PBS, pH 7.6–7.7	[2686]
Other					
Bacteria (quantitation)		G/mo	acc. (0.3 V vs. SSE), LSV	Probing: acc. of (2′-(4-hydroxy-phenyl)-5-(4-methyl-1-piperazinyl)-2,5′-bi-1*H*-benzimidazole; PBS 0.01 M, pH 7	[3212]
Bacterial lysates (quantitation of bacteria)		G/mo	acc. (1 V vs. SSE for 900s), LSV	Lysis by heat or lysozyme, probing with Ag⁺, O₂ or [Fe(CN)₆]³⁻ via signal decrease; s: cooling tower water	[3213]
Biological processes	Enzymes, tissues, algae		Scanning electrochemical microscopy	Biological entities incorporated in CP, processes monitored with a microelectrode	[619]
Salmonella typhimurium		G/p (liquid)	DPV	det. via ox. of phenol with immunoassay of alkaline phosphatase-linked anti-*Salmonella* and phenyl phosphate; s: chicken carcass washing water	[3214,3244]

TABLE 8.27
Brain Electrochemistry/*in Vivo* Voltammetry with Carbon Paste–Based Electrodes

Analyte	Modifier	Carbon Paste	Techniques	Comments	Ref(s)
Various				Reviews and related articles	[50,152,601,602,604–606, 2124,3216,3217]
Adrenaline	Banana crude extract	G/vaseline	POT	Source of polyphenol oxidase; PBS, pH 7; s: pharmaceuticals; *in vivo* measurement (jugular vein of rats)	[3018]
Ascorbic acid Homovanilic acid Dopamine				*In vivo* detection in brain (brain electrochemistry), *in vitro* real-time monitoring	[598,599,2126,2135,2138,2139, 2148–2150,2154,3218–3229]
Neurotransmitters and products					
Ascorbic acid Dopamine	Nafion membrane			*In vivo* comparison with uncoated electrodes	[1717]
Ascorbic acid	Unmodified, Triton X100, phosphatidyl ethanolamine, or brain tissue	G/so	CV	Influence of surfactants, lipids, and brain tissue on the voltammetric response	[1642]
Ascorbic acid Dopamine	Stearic acid	G/Nj or p(solid)	LSV	Modification of electrode by brain material eliminates discrimination between AA and DA; not sensitive for det. of DA	[1027,1028]
Ascorbic acid Dopamine	Stearic acid	G/p (solid)	CV, CA	Studies on influence of homogenized brain tissue	[1664]
Ascorbic acid		G/so	DP staircase V	Influence of proteins and lipids on the signal for *in vivo* studies; comparison with carbon fibers	[2143]
Ascorbic acid	Lipid		DP A (0.05–0.25 V vs. SCE)	*In vitro* study for simultaneous det. of AA and O_2	[2142]
Ascorbic acid Uric acid		G/so	LSV	*In vivo* measurements in brain with implanted electrodes	[2151,3230]

Analyte	Modifier	Technique	Notes	Refs
Ascorbic acid	G/so	A (0.25 V vs. Ag/AgCl) with microdialysis	*In vivo* measurements of extracellular AA	[3231–3233]
Ascorbic acid Oxygen			Simultaneous monitoring of glucose	[1774]
Caffeine	CNT (+DNA)		Enhanced acc. at DNA; for brain electrochem.; LOD 0.35 μM (DNA)	[430]
Caffeine Catechol	CNT/mo Bi	LSV, adSV with SWV	Simultaneous det., for brain electrochem.: 0.1 M H_3PO_4 or 0.1 M NH_4PO_3; LOD 0.18 μM	[3122]
Dopamine	G/so	Cyclic staircase voltammetry	Interference study of DOPAC-generated electrochem. from homovanillic acid for brain electrochem.; PBS 0.04 M + 0.15 M NaCl (pH 7.4)	[2119]
Dopamine Neurotransmitter Metabolites	Microsized electrodes		Brain electrochemistry or analysis of perfusions	[2120,2137,2146,2147,3221, 3224,3226–3229,3234–3243]
Dopamine	Microsized electrodes		Influence study of brain tissue on response	[152]
Dopamine Homovanillic acid, uric acid	Microsized electrodes	Computer-based LSV	Control of up to eight electrodes by microcomputer	[3244]
Dopamine	Microsized electrodes	CA	Brain electrochemistry	[1022,1662,1650,3245–3247]
Dopamine	Stearate Banana tissue		Tissue as a source of polyphenol oxidase; *in vitro* and *in vivo* det.	[2995–2997]
Dopamine	Spinach tissue		Tissue as a source for polyphenol oxidase; LOD 0.71 μM; useful for *in vivo* pharmacokinetic studies	[2123]
O-Ethyl-O-4-(nitrophenyl) phenyl phospho-nothioate	CNT DNA	CV, adsorptive stripping (30 s) with SWV	Biosensor for pesticide, increased sensitivity due to DNA; PBS 0.2 M, LOD 8 pM; *in vivo* pesticide assay	[409]
Oxygen		A	Brain electrochemistry with rats, comparison with blood flow	[3248]
Oxygen			Brain electrochemistry, rats, comparison with blood flow, lactate (microdialysis), and glucose (biosensor)	[2144,3249]
Uric acid			Brain electrochemistry	[2122,2136,2140,2141,2165, 3221,3227,3250,3251]

TABLE 8.28
Analysis of Biologically Important Compounds at Carbon Paste–Based Electrodes: Other Compounds + Bonus

Analyte	Modifier	Carbon Paste	Techniques	Comments	Ref(s)
Miscellaneous					
Carcinoma antigen 125	Thionine—AuNP—anti-CA 125 + BSA	G/po	CV, DPV	Direct immunoassay via thionine ox.: ABS, pH7; s: serum	[952]
Malachite green	Dodecyl benzenesulfonate *in situ*		acc., V	Enhancement of signal by surfactant; PBS, pH 6.5; LOD 4 nM (5 min acc.); s: fish	[2460]
Various				Biosensors with functional matrices	[3252]
Survey of selected reviews (on biologically important compounds and with particular focus)					
Modified electrodes as chemical sensors					3253
Carbon paste biosensors					70
Carbon paste biosensors modified with redox enzymes					2710
Metallized carbon biosensors					1254
Biosensor applications with coatings of conductive polymers					3254
Carbon paste biosensors with dehydrogenases					156
Biosensors on catalytic regeneration of NAD + with redox polymers					935
Application of CNT in oxidase biosensors					3255
Biosensors based on manganese dioxide					842
Electrochemical biosensors based on tissues and crude extracts					3200
Electron transfer mechanism in amperometric biosensors					936

ABBREVIATIONS AND SYMBOLS USED [IN ALPHABETICAL ORDER] (TABLES 8.1 THROUGH 8.14)

a., ag.	agent
A⁻	anion (unspecified)
AA	ascorbic acid
AB	acetylene black
absorb.	absorption, absorbed
accum.	accumulation, accumulated
AD	amperometric detection
add.	additional, additionally
Ac(A)	acetate (acetic acid)
AcB	acetate buffer
ACV	alternating current voltammetry
AC-IM	AC-impedance measurements
adsorbd.	adsorbed
adsorpt.	adsorption, adsorbed
AdSV	adsorptive stripping voltammetry
Ag/AgCl	silver/silver chloride electrode
am.	amorphous
AmA	amino acid
AmB	ammonia buffer
an.	analyte
anal.	analysis, analyzed
APAH(s)	amino-(derivative of) PAH(s)
appl.	application, applicable
ASV	anodic stripping voltammetry
Au-E	platinum electrode
AuF	gold film
B(e)	benzo-
BF	bromoform
BIA	batch injection analysis
BiF	bismuth film
biol.	biological
BN	bromonaphthalene
BR-B	Britton Robinson buffer
B(u)	butyl-
calc.	calculation, calculated
c	concentration
C	carbon, carbon powder
CA	chronoamperometry
CB	carbon black
CCOU	chronocoulometry
CCSA	constant current stripping analysis
CD	cyclodextrin
CE, CZE	capillary (zonal) electrophoresis
CEC	capillary electrochromatography

CF	chloroform
char.	characterization, characterized
CILE	carbon ionic liquid electrode
CM	conductometry
CMCPE	chemically modified carbon paste electrode
comm.	commercial, commercially available
comp.	composition, component
compd.	compound
compl.	complexation, complexed with
compr.	comparison, compared to/with
conf.	configuration, configured
constr.	construction, constructed
CNT	carbon nanotube(s)
CNT-CPE	carbon nanotube-modified CPE
CNT-ILE	carbon nanotube ionic liquid electrode
CNTPE	carbon nanotube paste electrode
COU	coulometry
CP	carbon paste
CP-bio	carbon paste–based biosensor
CP(O)	chronopotentiometry
CPE	carbon paste electrode
CPEE	carbon paste electroactive electrode
CP-ESE	carbon paste enantio-selective electrode
CP-ISE	carbon paste–based ion-selective electrode
CP(y)B	cetylpyridinium bromide
CRM	certified reference material(s)
CSV	cathodic stripping voltammetry
CTAB	cetyltriethylammonium bromide
ctng.	containing
curr.	current
CV	cyclic voltammetry
CWE	coated-wire electrode
D	detector
DCV	direct current voltammetry
def.	definition, defined
DeL(S)V	derivative linear (scan) voltammetry
der.	derivative, derived from
detc.	detection, detected
detn.	determination(s)
Dien	diethylenetriamine
diff.	different
dir.	direct
diss.	dissolution, dissolved
DNA	deoxyribonucleic acid
DPASV	differential pulse anodic stripping voltammetry
DPAdSV	differential pulse adsorptive stripping voltammetry
DPE	diamond paste electrode

DPP	differential pulse polarography
DPV	differential pulse voltammetry
ds-	double stranded (DNA)
D_X	diffusion coefficient
E	electrode
E	electrode potential
E_A	anodic potential
E_C	cathodic potential
ECL	electrochemiluminescence
EDTA	ethylenediamine-tetraacetic acid
E_P	peak potential
E_R	potential range ("window")
E^0, $E^{0'}$	standard redox/formal potential
FIA	flow injection analysis
FIA-AD	flow injection analysis with amperometric detection
FIA-EC	flow injection analysis with electrochemical detection
EIS	electrochemical impedance spectrometry
el.	electrode
elec.	electrolytic, electrolysis
elim.	elimination, eliminated
ELISA	enzyme-linked immunosorbent serologic assay
en	ethylenediamine
environm.	environmental
ESP	electro-spinning
evaln.	evaluation
Et	ethyl
EtOH	ethanol
exp.	experiment, experimental
ExSV	extractive stripping voltammetry
extr.	extraction, extract, extracted
Fc	ferrocene
form.	formula, formulation
FT-IR	Fourier-transform infrared spectroscopy
funct.	function, functionalized
G	gram
GCE	glassy carbon electrode
GC	glassy carbon powder
GCPE	glassy carbon paste electrode
gr.	group, family
HA	hydrodynamic amperometry
HC	hydrocarbon(s)
He	hexyl-
HMDE	hanging mercury drop electrode
h(r)	hours
HRP	horseradish peroxidase
HV	hydrodynamic voltammetry
H_2Q	hydroquinone

I, i	current (intensity)
I_P	peak current
IC	ion(ic) chromatography
IL	ionic liquid
IL-CPE	ionic liquid–modified carbon paste electrode
IM	imidazolyl
incl.	including
interf(s)	interference(s) from (of, by)
irrev.	irreversible
ISE	ion-selective electrode
k	kilo- (prefix, 10^3)
k_R	rate constant
k_S	solubility product
Kel-F®	poly-chlorotrifluoroethylene (PCTFE)
L	liter
$L_{(n)}$	ligand (unspecified)
LC-EC	liquid chromatography with electrochemical detection
lin., lin.r.	linearity, linear range
LOD	limit of detection
LOQ	limit of quantification
LSCP	linear sweep chronopotentiometry
LSV	linear sweep/scan voltammetry
m	milli- (prefix, 10^{-3})
m.	membrane
m-	*meta-*
M	molar concentration [mol L^{-1}]
max.	maximum, maximally
M(e)	methyl-
MeO	methoxy-
MEx	medium exchange (experiment with two diff. solns.)
mech.	mechanical, mechanically
MeOH	methanol
mdf.	modifier
MF	mercury film
min.	mineral
mixt.	mixture
modif.	modification, modified with, modifier
ms.	mesoporous
MWD	microwave-assisted digestion
n, n-	nano- (prefix, 10^{-9})
n-	(symbol for linear alkyl-)
nat(ur).	natural
Nf	Nafion®; sulfonated poly-tetrafluoroethylene
Nj	Nujol (mineral oil)
NPAH(s)	nitro- (derivative of) PAH(s)
o-	*ortho-*
Oc	octyl-

o.c.	open-circuit (i.e., without potential applied)
OD	oxidase
opt.	optimum, optimal, optimization
oxid.	oxidant(s), oxidizing
oxidn.	oxidation
p	pico- (prefix, 10^{-12})
p-	*para-*
p	particle
PAH(s)	polycyclic aromatic hydrocarbon
PbF	lead film
pH	acidity unit (def.: $-\log a_{H^+}$)
Ph	phenyl-
PhO	phenoxy-
pharm.	pharmaceutical
PhB	phosphate buffer
Phen	1,10-phenanthroline
PhOH	phenol
pK_i	dissociation constant
PO	paraffin/mineral oil
poll.	(environmental) pollutant
POT	potentiometry (trad., equilibrium pot.)
ppb	parts-per-billion (concentration unit)
ppm	parts-per-million (concentration unit)
prcp.	precipitate, precipitation
prep.	preparative (product)
pretreat.	pretreated, pretreatment
Pr	propyl
PSA	potentiometric stripping analysis
Pt-E	platinum electrode
PV(C)	polyvinyl (chloride)
py	pyridine
Py	pyridinium
Q	quinone
Q^+	lipophilic (large) cation
R_{ir}	ohmic resistance (measurement)
RDE	rotated/rotating disc electrode
red(s)	reductant(s), reducing, reduced
redn.	reduction
regnt.	regeneration, regenerated
ref.	reference
rel.	related (item), relative
reoxidn.	re-oxidation
rev.	reversible
R_R	recovery rate
RRDE	rotating ring-disc electrode
RSD	relative standard deviation
(RT)IL	(room-temperature) ionic liquid

S	second
s.	sample/specimen
(s)	plural (suffix)
(s.)	solid (state)
SbF	antimony film
SCE	saturated calomel electrode
s-CPE	solid-like ("pseudo") carbon paste electrode
SDS	sodium dodecylsulfate
s.e.	supporting electrolyte
SEM	scanning electron microscopy
SG	silicone grease
SIA	sequential injection analysis
simult.	simultaneous, simultaneously
SO	silicone oil
soln.	solution
sp., s.	species/family
SP	(computer controlled) stripping potentiometry
SP(C)E	screen-printed (carbon) electrode
SPCPE	screen-printed carbon paste electrode
SPEC	spectroelectrochemistry
SPEx	solid phase extraction
SPV	solid phase voltammetry
ss-	single stranded (DNA)
stat.	stationary (arrangement)
subst.	substitution, substituted
suppr.	suppression, suppressed
surf.	surfactant (surface active compound)
SWV	square-wave voltammetry
SX	siloxane
synth.	synthetic (artificial, model)
t_{ACC}	accumulation (preconcentration) time
taV	tast voltammetry
TCP	tricresyl phosphate
terc.	tercial
Tf	Teflon®; poly-tetrafluoroethylene (PTFE)
titr.	titration, titrated
t_R	response time
trad.	traditional
UA	uric acid
unm.	unmodified
Uv	mineral oil (for UV-spectroscopy)
UV–vis	ultraviolet–visible spectro(photo)metry
v, v/v	volume (%)
V	Volt
VO	vaseline oil
vs.	versus
w, w/w	weight (%)

Electroanalysis with Carbon Paste–Based Electrodes, Sensors, and Detectors

w.	water
Wx	paraffin wax
X	halogen, halide (if not stated otherwise)
Y⁻	anion (unspecified)
z	charge
2nd-	second order
α	charge transfer coefficient
β, β'	stability constant, conditional st. const.
μ	micro- (prefix, 10^{-6})
/	or
–	not given, not found
*,**	footnote(s)
?	unclear, unsure data

ABBREVIATIONS (TABLES 8.15 THROUGH 8.28)

A	amperometry
ABS	acetate buffer solution
acc.	accumulation
act	activated
AD	amperometric detection
AMB	1-(N,N-dimethylamine)-4-(4-morpholine)benzene
CA	chronoamperometry
CC	chronocoulometry
CDI	carbodiimide
CNT	carbon nanotubes
CV	cyclic voltammetry
DMFc	dimethyl ferrocene
Fc	ferrocene
FIA	flow injection analysis
G	graphite
GA	glutar aldehyde
gc	glassy carbon
GC	gas chromatography
HDV	hydrodynamic voltammetry
HRP	horseradish peroxidase
LOD	limit of detection
me	medium exchange
mo	mineral oil
MWCNT	Multiwall carbon nanotubes
Nj	Nujol
oc.	open circuit
p	paraffin
POT	potentiometry
PBS	phosphate buffer solution
PC	phthalocyanine

pCEC	pressurized capillary electrochromatography
PEG	polyethyleneglycol
PEI	polyethyleneimine
PPQ	pyrroloquinoline quinone
ppy	polypyrrole
po	paraffin oil
QH	quinohemoprotein
s	sample(s)
SCE	saturated calomel electrode
SE	spectroelectrochemistry
SSCE	sodium saturated calomel electrode
SSE	silver/silver chloride reference electrode
SV	stripping voltammetry
SWV	square wave voltammetry
TCQM	7,7,8,8-tetracyanoquinodimethane
TTF	tetrathiafulvalene
?	unclear, unsure data

9 In Place of a Conclusion: Carbon Paste Electrodes for Education and Practical Training of Young Scientists

Here, the authors' team ends the journey across the world of carbon paste–based electrodes, sensors, and detectors. We are not afraid about their future. Instead of speculative prospects that have been outlined recently, for example, in our previous review articles [7,83,84,111,112,161]), it is sufficient to go into the new web databases as all the major topics of the field have spawned a long line of new studies. Among many others and besides two brand new reviews [2182,2185], there are fresh contributions within the methods for determination of inorganic ions (see, e.g., [1314,2493]), for analysis of organic pollutants [2437,2493] and pharmaceuticals [335,1692], as well as biologically active compounds [2992,3270]. New publication activities can also be found within *in vivo* electrochemistry [2154], biosensing [138,3270], or associated with the continuing boom of new technologies and the use of still popular nanomaterials and ionic liquids [331,3271–3275].

Thus, we would like to conclude our book in a different way, showing a possibility to train the new generations of electrochemists and electroanalysts in measurements with carbon paste electrodes (CPEs). At first, imagine that there are the following attributes: (i) easy-to-obtain chemicals, (ii) inexpensive and common laboratory equipment, (iii) simple and quick preparation, (iv) minimal toxicity—with full aspiration to join the concept of "green analytical chemistry" in the form of traditional, as well as new carbon pastes, including many modified variants, (v) massive use of new attractive materials, (vi) compatibility with instrumentation of different kind and quality, and (vii) practically unlimited applicability in both electrochemical and electroanalytical measurements. Second, consider that all these aspects are typical for carbon pastes and the respective CPEs and that these features are always welcome in laboratories of various educational institutions. Then, it appears evident that experimentation with carbon paste–based electrodes is indeed feasible as real laboratory practice of students and young scientists almost everywhere regardless whether the respective institutions offer luxury instrumentation or are equipped with basic apparatus only.

For that, there is one requirement—whether the supervising staff is willing to gain the elementary knowledge about CPEs gathered in each good review, book chapter, as well as in some original papers. If so, the skillfulness in working with carbon pastes and experience to solve problems on the way then come by learning and practicing. This knowledge obtained mainly empirically can be confirmed by the authors of this book who have worked with CPEs for decades, passing initial embarrassment and later routine, of seeking and finding, great expectations and lesser achievements, as well as many failures compensated by heartwarming successes. In this sense, the key chapters of this monograph can be used as the starting point.

The rest is up to the experimenter who intends to work with CPEs apart from whether he or she is an experienced electrochemist or a young scientist at the beginning of his or her career. In addition, this knowledge has been gained by the authors, together with the individual members of their research teams and other collaborating groups worldwide. Especially delightful is then a finding that the experimental work with carbon paste–based electrodes has become, as time goes by, still more challenging for both undergraduates and graduate students. It can be documented by presenting such activities gathered in Appendix B that surveys the BSc, MSc, and PhD dissertation theses elaborated in the course of more than two decades in laboratories of authors' home institutions as well as at other universities involved in various collaboration programs.

When going through the list, it can be seen that there have been no limitations in the topic(s) practiced, starting from fundamental laboratory training of bachelors, via aimed and successful measurements performed by MSc students in the framework of projects focused on theoretical or applied electrochemistry, up to highly specialized and experimentally valuable research by numerous PhD students, contributing thus to the overall scientific profiles of the individual educational and research institutions. Since similar lists of dissertation theses or related reports could be assembled by other propagators of CPEs over the whole world, it can be stated that the area of experimentation with these electrodes would have never arrived into the prominent position where it is now without the enthusiastic work of the beginning scientists. Indeed, despite the existence of a huge database of papers from renowned scientists, the value of such dissertations may be comparable as they often hide numerous useful (or valuable) details without which numerous achievements would have been far less attractive or even impossible.

Since similar details have undoubtedly enriched the content of this book, the presented survey is meant also as the authors' homage and their thanks to all young people who had, in some way, contributed to the field of electrochemistry and electroanalysis with CPEs. To the field that mirrors the state of the art of electrochemical experiments over a half a century in a comparable way like Heyrovsky's polarography did in the preceding years, in the era of measurements with the dropping mercury electrode.

Appendix A: RNA: A Profile of the Carbon Paste Inventor and a Great Scientist

Ralph Norman Adams was born on August 26, 1924, in Atlantic City (NJ, USA). Young Ralph had begun his college education before the World War II, but, in 1942, he was recruited into the *Chemical Warfare Service*. He had always had a desire to be a pilot, and his wish came true after joining the *Army Air Corps* 1 year later. During this eventful period of his life, he piloted military aircrafts "B-17s" and "B-29s" and, as a first lieutenant, operating in the Pacific Territory. At that time, he also earned his well-known nickname, "Buzz."

RNA (1924–2002)

After returning to the academic sphere, Adams received a BS degree in chemistry from the Rutgers University (in 1950). Then, he attended Princeton University, where he worked under the direction of Professor N. H. Furman, receiving a PhD in chemistry (1953). At the Princeton faculty, the scientist at the beginning of his career remained for the next 2 years after which he joined the staff of the department of chemistry at the Kansas University, KU (1955). Since then, his research interests have been centered on electrooxidations in solid electrodes and the mechanisms of organic electrode reactions. In the late 1950s, Adams and his students experimented with a dropping carbon electrode (DCE), but, despite enormous effort invested in the exciting idea, the concept of a carbon-based alternative to Heyrovský's mercury dropping electrode finally failed. Luckily, meanwhile, the tireless Adams discovered a hitherto unknown carbonaceous mixture—the carbon paste—that had been soon shown to replace fully the originally wanted DCE, thus entering electrochemistry as the new and a very progressive type of the electrode material. No need to say that the carbon paste and the respective electrodes (CPEs) made Adams's discovery to be one of the milestones within the electrochemistry of the twentieth century. In fact, Dr. Adams was a leading authority in the development of CPEs during the entire 1960s.

In 1964, he was a J.S. Guggenheim fellow in Zuerich (Switzerland), involved in the EPR electrochemistry in the laboratories of the Varian AG company and the Eldgenoessische Technische Hochschule (ETH). Among others, Adams' research had proven the existence of the so-called free radicals in the human body. In the late 1960s, the inventor of CPEs crowned the era of his activities in organic electrochemistry (with predominantly nonmercury electrodes) by finishing his debut monograph Electrochemistry at Solid Electrodes, which would be appraised worldwide as the classic within electrochemical literature.

Those years also saw Adams's scientific reorientation toward neuroscience, with particular interest in catechols and other neurotransmitters as the primary markers of schizophrenia. In 1970, Adams spent a sabbatical year at the department of psychobiology, *University of California*, in Irvine (CA, USA), where he received fundamental training in neuroanatomy, neurophysiology, etc. Adams's research in this area led to the discovery of the chemical background of the function of the two brain halves, showing that both have different chemical make-ups. In the period of 1978–1982, he served as a professor of neurobehavioral sciences at the *Menninger School of Psychiatry* in Topeka (KS, USA). In the mid-1970s, Professor Adams was also a pioneer in electrochemistry *in vivo*, thus paving the way for later miniaturization of electrodes and sensors with their widespread use in pharmacological and clinical analysis, as well as biological research as such.

Adams's retirement in 1992 was rather formal because his former students and collaborators, together with his wife, hired a small laboratory, where Adams could continue in his research activities. Revived, at the age of 70, he started some new experiments; among others, the testing of novel batteries for cellular phones. Otherwise, he was still interested in everything new that was going on at the KU, being in permanent contact with many of his former students, with regular reunion events (see photo in the Foreword). Professor Adams died on November 28, 2002, at St. Luke's Hospital in Kansas City (Kansas), after a short illness and at the blessed age of 78.

Over the course of his career, Professor Adams was awarded the J. S. Guggenheim Fellowship (1964), Fisher Award in Analytical Chemistry (from the American Chemical Society, 1982), Higuchi Award for Excellence in Basic Sciences (as one of the very first scientists worldwide, 1982), C. N. Reilley Award for Electroanalytical Chemistry (1984), I. M. Kolthoff Gold-Medal Award (from the American Pharmaceutical Society, 1985), Jacob Javits Neuroscience Investigator Award (from the National Institute of Neurological and Communicative Disorders and Stroke, 1986), American Chemical Society National Award in Electrochemistry (1989), and Oesper Award (1996). In 1997, Professor Adams was nominated for a Nobel Prize.

"He was a magnificent scientist and an exceptional human being," recalls T. Kuwana, a distinguished Professor of Chemistry at KU. *"He lived what he believed in. And uppermost was his integrity and belief in values."* *"Ralph was an inspiration to everybody,"* agrees Professor C. Lunte, a colleague of Kuwana. *"I've never known a scientist who was that dedicated to knowledge. Dr. Adams' love for science and enthusiasm for academia influenced countless students during his life. There are Adams' students all over the country. I myself am a second-generation Adams' student,"* he adds.

During his scientific and pedagogical career, Professor Adams supervised ca. 60 students, and his laboratories hosted a number of renowned scientists from the United States and other countries. Ralph N. Adams is the author or co-author of 214 original papers, from which about 40 contributions are associated with carbon pastes and carbon paste–based electrodes.

Ralph N. Adams: Alumni   *(in alphabetical order)*
Jeff Bacon (PhD); C. LeRoy Blank (PhD); Charles W. Bradberry (PhD); Jon Cammack (PhD); Peter Capella (PhD); James Q. Chambers, III (PhD); Willie Chey (MS); James C. Conti (PhD); R. Keith Darlington (UG); Roger J. Dreiling (UG); Benjamin A. Feinberg (PhD); Steve Feldberg; Zbigniew "George" Galus (Postdoc); Greg A. Gerhardt (PhD); Benham Ghasemzadeh (PhD); Cynthia Gouvion

Appendix A: RNA: A Profile of the Carbon Paste Inventor and a Great Scientist

(RA); M. Dale Hawley (PhD); Rita M. Huff; Beb Jeftic (Postdoc); Richard W. Keller, Jr. (PhD); Peter T. Kissinger (Postdoc); Theodore "Ted" Kuwana (PhD); Thomas P. Layloff, Jr. (PhD); Donald W. Leedy (PhD); Yola Liang (PhD); Paul A. Malachesky (PhD); Charles R. "Gus" Manning (PhD); Lynn S. Marcoux (PhD); Charles Marsden; Leslie J. May (UG); Richard L. McCreery (PhD); Ivan N. Mefford (PhD); Kristin H. Milby (PhD); Ellen G. Miller (UG); Terry A. Miller (UG); Kim M. Mitchell (PhD); Bita Moghaddam (PhD); Theodore R. Mueller (PhD); Robert F. Nelson; Arvin Oke (PhD); Carter L. Olson (PhD); Lucien M. Papouchado (PhD); George B. Park (PhD); Glen A. Petrie (PhD); Stanley Piekarski (PhD); Paul Plotsky (UG); Barbara L. Prater (UG); Keith B. Prater (UG); Charles J. Refshauge (PhD); Margaret E. Rice (PhD); Robert W. Sanford, Jr. (MS); James O. Schenk (PhD); Eddie Seo (PhD); Ramin Shiekhattar (PhD); Steve Soper (PhD); Jerzy Strojek; Elaine Strope (PhD); Greg Swain; Shankar V. Tatawadi (PhD); K.V. Thrivikraman (PhD); Dan Tse (UG); Beverly Voran; Mark Wentz (UG); Wesley White (PhD); Mark Wightman (PhD); John Zimmerman (PhD).

Apparently the greatest honor to Professor Adams came posthumously. In the area where R.N.A. spent the majority of his career, within the KU campus, a new research institution has recently been established and named Ralph N. Adams Institute for Bioanalytical Chemistry. The Buzz Adams Institute is an interdisciplinary consortium of researchers dedicated to achieving international leadership in bioanalytical science. More than 200 scientists are involved in a wide variety of activities covering the state-of-the-art sampling, separation, detection, and selected characterization techniques, such as microdialysis, mass spectrometry, molecular imaging and spectroscopy, lab-on-chip analytical devices, or electroanalysis.

Compiled and written by the first author (I.Š.), *the son of a pilot of former Czechoslovak People's Army, when using biographical material and some related information from the following sources, comprising also freely accessible websites*:

1. KU: Ralph N. 'Buzz' Adams—R. N. Adams Institute of Bioanalytical Chemistry, http://www.adamsinstitute.ku.edu/about/r_n_adams.shtml (accessed January 20, 2011).
2. Recollections—Ralph N. Adams, http://electroanalytical.org/Adams.html (accessed January 20, 2011).
3. Harmony, M.D. 2006. History of the KU chemistry department: 1950–2000. Lawrence: Univ. Kansas; http://www.chem.ku.edu/docs/historyofthedepartment.pdf (accessed January 20, 2011).
4. Editorial. 1999. Dedicated to Professor Ralph N. Adams on the occasion on his 75th birthday. *Electroanalysis* 11:284–291.
5. Švancara, I. 2002. Carbon paste electrodes in electroanalysis (in Czech). Hab dissertation, University of Pardubice, Pardubice.

Appendix B: List of Dissertation Theses* Defended in the Authors' Countries and Dealing with Carbon Paste–Based Electrodes, Sensors, and Detectors

UNIVERSITY OF PARDUBICE

t-1 Riha, V., Jr. 1987. Determination of trace concentrations of heavy metals by inversion voltammetry. MSc dissertation (in Czech), University of Chemical Technology, Pardubice.

t-2 Švancara, I. 1988. Application of carbon paste electrodes in voltammetric determination of low concentrations of nickel (II) and cobalt (II) ions. MSc dissertation (in Czech), University of Chemical Technology, Pardubice.

t-3 Suska, M. 1989. A Study on suitability of solid electrodes for the determination of low concentrations of Ni and Co using stripping voltammetry method. MSc dissertation (in Czech), University of Chemical Technology, Pardubice.

t-4 Chybova, O. 1991. Electrochemical reductions of some azo-compounds at carbon paste electrodes; Part I. MSc dissertation (in Czech), University of Chemical Technology, Pardubice.

t-5 Beranova, G. 1991. Electrochemical reductions of some azo-compounds at carbon paste electrodes; Part II. MSc dissertation (in Czech), University of Chemical Technology, Pardubice.

t-6 Srey, M. 1992. Voltammetric and potentiometric determination of gold method. MSc dissertation (in Czech), University of Chemical Technology, Pardubice.

t-7 Hvizdalova, M. 1994. A study on applicability of carbon paste electrodes in voltammetry. MSc dissertation (in Czech), University of Pardubice, Pardubice.

t-8 Jezkova, J. 1994. The use of carbon paste electrodes in potentiometry. MSc dissertation (in Czech), University of Pardubice, Pardubice.

t-9 Musilova, J. 1995. Carbon paste electrodes for potentiometric determination of the anions. MSc dissertation (in Czech), University of Pardubice, Pardubice.

t-10 Strakosova, A. 1995. Determination of trace amounts of iron in raw and fine products from pharmaceutical synthesis. MSc dissertation (in Czech), University of Pardubice, Pardubice.

t-11 Matousek, M. 1996. Voltammetry at electrode plated with a gold film. MSc dissertation (in Czech), University of Pardubice, Pardubice.

t-12 Ticha, I. 1996. Automated potentiometric titrations of surfactants. MSc dissertation (in Czech), University of Pardubice, Pardubice.

t-13 Konvalina, J. 1997. Extraction at carbon paste electrodes and its application to the determination of iodine. MSc dissertation (in Czech), University of Pardubice, Pardubice.

t-14 Novakova, M. 1997. Determination of ascorbic acid in the foodstuff. MSc dissertation (in Czech), University of Pardubice, Pardubice.

t-15 Skorepa, J. 1997. Determination of inorganic ions using potentiometric titration with carbon paste electrodes. MSc dissertation (in Czech), University of Pardubice, Pardubice.

t-16 Cermakova, I. 1998. Applicability of surfactants in voltammetry with carbon paste electrodes. MSc dissertation (in Czech), University of Pardubice, Pardubice.

---

* See notes behind the list.

t-17 Fidlerova, P. 1998. Determination of the elements capable of forming the hetero-polyanions with the aid of automated potentiometric titration. MSc dissertation (in Czech), University of Pardubice, Pardubice.

t-18 Metelka, R. 1998. Possibilities of coupling PA-3 polarographic analyser to a personal computer for modern voltammetric analysis. MSc dissertation (in Czech), University of Pardubice, Pardubice.

t-19 Chadim, P. 1999. Possibilities and limitations of carbon paste electrodes plated with a gold film for voltammetric determination of arsenic. MSc dissertation (in Czech), University of Pardubice, Pardubice.

t-20 Salakova, Z. 1999. Automated potentiometric titrations of the hetero-polyanions; Part I. MSc dissertation (in Czech), University of Pardubice, Pardubice.

t-21 Slavikova, S. 2000. Automated potentiometric titrations of the hetero-polyanions; Part II. MSc dissertation (in Czech), University of Pardubice, Pardubice.

t-22 Stiburkova, M. 2000. Development and applications of carbon paste electrodes modified with mercuric oxide. MSc dissertation (in Czech), University of Pardubice, Pardubice.

t-23 Pazdera, R. 2001. Stripping potentiometry with carbon paste electrodes in practical analysis. MSc dissertation (in Czech), University of Pardubice, Pardubice.

t-24 Ticha, J. 2001. The formation of the heteropolyanions and potentiometric analysis. MSc dissertation (in Czech), University of Pardubice, Pardubice.

t-25 Vitova, V. 2001. The electrochemistry of amines at carbon paste electrodes. MSc dissertation (in Czech), University of Pardubice, Pardubice.

t-26 Czaganova, J. 2002. Screen-printed electrode plated with mercury- and bismuth films in electrochemical stripping analysis. MSc dissertation (in Czech), University of Pardubice, Pardubice.

t-27 Jansova, G. 2002. Carbon paste electrodes plated with bismuth films in electroanalysis. MSc dissertation (in Czech), University of Pardubice, Pardubice.

t-28 Foret, P. 2003. Possibilities of determination and speciation of chromium at carbon paste electrodes. MSc dissertation (in Czech), University of Pardubice, Pardubice.

t-29 Fairouz, M. 2003. Carbon paste electrodes plated with a mercury film. Some contribution to their voltammetric characterisation and applicability to the determination of selected metals in crude oil. MSc dissertation (in English), University of Pardubice, Pardubice.

t-30 Ismail, Kh. 2003. Carbon paste electrodes plated with a bismuth film. Some contribution to their voltammetric characterisation and applicability to the determination of heavy metals in crude oil. MSc dissertation (in English), University of Pardubice, Pardubice.

t-31 Brazdilova, P. 2004. Modifications of carbon electrodes and their applications in the construction of biosensors. MSc dissertation (in Czech), University of Pardubice, Pardubice.

t-32 Mikysek, T. 2004. The electrochemistry of mono- a bis(imidazolyl)pyridines and their possible use as carbon paste modifiers. MSc dissertation (in Czech), University of Pardubice, Pardubice.

t-33 Galik, M. 2005. Electrochemical stripping analysis of platinum metals at carbon paste electrodes. MSc dissertation (in Czech), University of Pardubice, Pardubice.

t-34 Stoces, M. 2006. Stripping potentiometry at bismuth-modified electrodes. MSc dissertation (in Czech), University of Pardubice, Pardubice.

t-35 Zeravik, M. 2008. Application of groove carbon paste electrode in flow injection analysis. MSc dissertation (in Czech), University of Pardubice, Pardubice.

t-36 Hradilova, S. 2009. Non-electrolytic processes at carbon paste electrodes and their utilisation in electro-analysis. MSc dissertation (in Czech), University of Pardubice, Pardubice.

t-37 Nepejchalova, L. 2009. Development of a new type of electrically heated carbon paste electrode and its testing in electroanalysis. MSc dissertation (in Czech), University of Pardubice, Pardubice.

t-38 Oubrechtova, K. 2009. Use of groove carbon paste electrode in designing of amperometric biosensors. MSc dissertation (in Czech), University of Pardubice, Pardubice.

t-39 Bartos, K. 2010. Deposition of bismuth- and antimony films from atypical solutions with possible use in electroanalysis. MSc dissertation (in Czech), University of Pardubice, Pardubice.

t-40 Švancara, I. 1995. Voltammetric applications of carbon paste electrodes. PhD dissertation (in Czech), University of Pardubice, Pardubice.

t-41 Jezkova, J. 1999. Potentiometric determinations with a carbon paste electrode. PhD dissertation (in Czech), University of Pardubice, Pardubice.

t-42 Konvalina, J. 2001. Carbon paste electrodes in stripping potentiometry. PhD dissertation (in Czech), University of Pardubice, Pardubice.

t-43 Metelka, R. 2005. Heterogeneous carbon electrodes with electrochemically deposited metal films. PhD dissertation (in Czech), University of Pardubice, Pardubice.

t-44 Kotzian, P. 2006. Mediators of electron transfer in amperometric enzyme biosensors. PhD dissertation (in Czech), University of Pardubice, Pardubice.

Appendix B: List of Dissertation Theses   455

t-45 Galik, M. 2008. Electrochemistry and electroanalysis of platinum metals and biologically important thiols. PhD dissertation (in Czech), University of Pardubice, Pardubice.
t-46 Mikysek, T. 2008. Carbon paste electrode vs. mercury drop electrode in the present days electrochemistry and electroanalysis. PhD dissertation (in Czech), University of Pardubice, Pardubice.
t-47 Baldrianova, L. 2009. Development and applications of novel types of bismuth-based electrodes. PhD dissertation (in English), University of Pardubice, Pardubice.
t-48 Tesarova-Svobodova, E. 2009. Electroanalytical characterisation of novel types non-mercury electrodes. PhD dissertation (in Czech), University of Pardubice, Pardubice.
t-49 Švancara, I. 2002. Carbon paste electrodes in electroanalysis. Hab dissertation (in Czech), University of Pardubice, Pardubice.
t-50 Navratilova, Z. 2006. Electroanalysis with carbon paste electrodes modified with natural substances. Hab dissertation (in Czech), University of Pardubice, Pardubice.

CHARLES UNIVERSITY, PRAGUE

t-51 Bouzkova, T. 2005. Determination of 2-nitrophenol on carbon paste electrodes. BSc dissertation (in Czech), Charles University, Prague.
t-52 Kocourkova, M. 2006. Study of lomustine reducibility at carbon paste electrodes. BSc dissertation (in Czech), Charles University, Prague.
t-53 Nemcova, L. 2006. Determination of 5-amino-6-nitroquinoline using HPLC with electrochemical detection. BSc dissertation (in Czech), Charles University, Prague.
t-54 Svecova, M. 2007. Determination of ambroxol using carbon paste electrodes. BSc dissertation (in Czech), Charles University, Prague.
t-55 Vokalova, V. 2007. Determination of benzocaine using HPLC-ED. BSc dissertation (in Czech), Charles University, Prague.
t-56 Vlachova, K. 2008. Determination of aminoglutethimide at carbon paste electrodes. BSc dissertation (in Czech), Charles University, Prague.
t-57 Mikes, M. 2009. Determination of metoclopramide using carbon paste electrodes. BSc dissertation (in Czech), Charles University, Prague.
t-58 Muzikova, J. 2010. Determination of carvacrole using HPLC with electrochemical detection. BSc dissertation (in Czech), Charles University, Prague.
t-59 Smutna, K. 2001. Voltammetric determination of 2-hydroxyphenanthrene at carbon paste electrodes. MSc dissertation (in Czech), Charles University, Prague.
t-60 Kacur, M. 2003. Voltammetric determination of fluorene nitroderivatives using carbon paste electrodes. MSc dissertation (in Czech), Charles University, Prague.
t-61 Dvorakova, J. 2004. Determination of 1-aminonaphthalene at carbon paste electrodes using voltammetry and HPLC-EC. MSc dissertation (in Czech), Charles University, Prague.
t-62 Pekarova, Z. 2004. Voltammetric determination of methotrexate using a carbon paste electrode. MSc dissertation (in Czech), Charles University, Prague.
t-63 Zitova, A. 2004. Determination of aminoquinolines at carbon paste electrodes using voltammetry and HPLC-ED. MSc dissertation (in Czech), Charles University, Prague.
t-64 Cerna, P. 2005. Determination of mitoxantron using carbon paste electrodes. MSc dissertation (in Czech), Charles University, Prague.
t-65 Dejmkova, H. 2005. Determination of naphthalene aminoderivatives using HPLC with electrochemical detection. MSc dissertation (in Czech), Charles University, Prague.
t-66 Jemelkova, Z. 2005. Voltammetric determination of doxorubicin on carbon paste electrodes. MSc dissertation (in Czech), Charles University, Prague.
t-67 Pudilova, H. 2006. Determination of chloramphenicol at carbon paste electrodes. MSc dissertation (in Czech), Charles University, Prague.
t-68 Cienciala, M. 2007. Determination of thymol at carbon paste electrodes. MSc dissertation (in Czech), Charles University, Prague.
t-69 Tinkova, V. 2007. Determination of aminoanthraquinone at carbon paste electrode. MSc dissertation (in Czech), Charles University, Prague.
t-70 Kocourkova, M. 2008. Voltammetric determination of benserazide at carbon paste electrodes. MSc dissertation (in Czech), Charles University, Prague.
t-71 Nemcova, L. 2008. Determination of 5-amino-6-nitroquinoline at carbon paste electrode. MSc dissertation (in Czech), Charles University, Prague.

t-72 Svecova, M. 2009. Determination of ambroxol using carbon paste electrodes. MSc dissertation (in Czech), Charles University, Prague.

t-73 Vokalova, V. 2009. Determination of benzocaine using FIA and HPLC at carbon paste electrodes. MSc dissertation (in Czech), Charles University, Prague.

t-74 Hranicka, Z. 2010. Determination of selected nitrophenols using modified carbon paste electrodes. MSc dissertation (in Czech), Charles University, Prague.

t-75 Kalusova, M. 2010. Voltammetric determination of 4-aminobiphenyl at montmorillonite modified carbon paste electrodes. MSc dissertation (in Czech), Charles University, Prague.

t-76 Vlachova, K. 2010. Determination of aminoglutethimide using HPLC-ED with carbon pastes. MSc dissertation (in Czech), Charles University, Prague.

t-77 Vysoka, M. 2010. Determination of propyl gallate at carbon paste electrode. MSc dissertation (in Czech), Charles University, Prague.

MASARYK UNIVERSITY, BRNO

t-78 Fulneckova, J. 1997. Utilization of electrochemical methods for study of specific DNA interaction with small molecules. MSc dissertation (in Czech), Masaryk University, Brno.

t-79 Kovarova, L. 2000. Interaction of tumor suppressor protein p53 with DNA damaged by mutagens, carcinogens, antitumor agents. MSc dissertation (in Czech), Masaryk University, Brno.

t-80 Tomschik, M. 1998. Analysis of nucleic acids and peptides by voltammetry and chrono-potentiometry using electrodes modified with biopolymers. PhD dissertation (in Czech), Masaryk University, Brno.

t-81 Masarik, M. 2005. Study of proteins by electrochemical methods and their application in proteomics, genomics, and biomedicine. PhD dissertation (in Czech), Masaryk University, Brno.

t-82 Vacek, J. 2009. Recent approaches in electrochemical analysis of damage, hybridization, and interactions of DNA. PhD dissertation (in Czech), Masaryk University, Brno.

MENDEL UNIVERSITY, BRNO

t-83 Benova, V. 2006. Using of avidin-biotin technology for studying programmed cell's death. BSc dissertation (in Czech), Mendel University, Brno.

t-84 Hradecky, J. (2007). Study of plants protective signals by electrochemical methods. MSc dissertation (in Czech), Mendel University, Brno.

t-85 Petrlova, J. (2006). Determination of DNA and proteins in transgenic plants using electrochemical and chromatographic techniques. PhD dissertation (in Czech), Mendel University, Brno.

PALACKY UNIVERSITY, OLOMOUC

t-86 Macikova, P. 2007. The study of redox properties of rutin and quercetin. BSc dissertation (in Czech), Palacky University, Olomouc.

t-87 Kulova, P. 1998. Determination of mercury(II) at a carbon paste electrode modified with natural clay. MSc dissertation (in Czech), Palacky University, Olomouc.

t-88 Radova, O. 2003. Voltammetric analysis of 1,2,4-triazines on carbon paste electrodes. MSc dissertation (in Czech), Palacky University, Olomouc.

t-89 Halouzka, V. 2007. Carbon paste and composite electrodes modified by nanoscopic iron(III) oxides and Prussian blue. MSc dissertation (in Czech), Palacky University, Olomouc.

t-90 Macikova, P. 2009. Application of modified carbon paste electrodes for voltammetric determination of rutin. MSc dissertation (in Czech), Palacky University, Olomouc.

UNIVERSITY OF OSTRAVA

t-91 Kudrysova, K. 1997. Problems of oxygen in voltammetry on carbon paste modified electrodes. BSc dissertation (in Czech), University of Ostrava, Ostrava.

t-92 Strnadova, H. 2000. Mercury behavior on the carbon paste electrode modified with clay minerals. BSc dissertation (in Czech), University of Ostrava, Ostrava.

t-93 Stiborova, L. 2001. Study of ion exchange by means of carbon paste electrode modified with clay minerals. BSc dissertation (in Czech), University of Ostrava, Ostrava.

Appendix B: List of Dissertation Theses

t-94 Orosova, L. 2006. Preparation and testing of modified carbon paste electrodes. BSc dissertation (in Czech), University of Ostrava, Ostrava.
t-95 Pavlicek, D. 2007. Preparation and characterization of film clay electrode. BSc dissertation (in Czech), University of Ostrava, Ostrava.
t-96 Patrmanova, M. 2009. Carbon paste electrode modified with coal. BSc dissertation (in Czech), University of Ostrava, Ostrava.
t-97 Marsalek, R. 1997. Voltammetric determination of selected metals on the carbon paste electrode modified with montmorillonite. MSc dissertation (in Czech), University of Ostrava, Ostrava.
t-98 Hranicka, Z. 2007. Study of sorption properties of montmorillonite by modified carbon paste electrode. MSc dissertation (in Czech), University of Ostrava, Ostrava.
t-99 Orosova, L. 2008. Voltammetry of solid phase aimed at carbonaceous materials. MSc. dissertation (in Czech), University of Ostrava, Ostrava.

INSTITUTE OF INORGANIC CHEMISTRY AS CR, REZ (NEARBY PRAGUE)

t-100 Grygar, T. 2000. The electrochemical dissolution of iron(III) and chromium(III) oxides and ferrites under conditions of abrasive stripping voltammetry. PhD dissertation (in Czech), Academy of Sciences of the Czech Republic, Prague.

SLOVAK TECHNICAL UNIVERSITY, BRATISLAVA

t-101 Ozabalova, K. 1997. Selective determination of trace concentration using a modified electrode. BSc dissertation (in Slovak), Slovak University of Technology, Bratislava.
t-102 Fuknova, M. 1998. Determination of DNA using electrochemical methods. BSc dissertation (in Slovak), Slovak University of Technology, Bratislava.
t-103 Baranova, V. 1999. Analysis using DNA biosensor. BSc dissertation (in Slovak), Slovak University of Technology, Bratislava.
t-104 Sivakova, Z. 2000. Use of DNA biosensor. BSc dissertation (in Slovak), Slovak University of Technology, Bratislava.
t-105 Bubnicova, K. 2001. Study of DNA damage and its inhibition. BSc dissertation (in Slovak), Slovak University of Technology, Bratislava.
t-106 Tothova, A. 2001. Analytical use of DNA biosensor. BSc dissertation (in Slovak), Slovak University of Technology, Bratislava.
t-107 Madarasova, I. 2002. Possibilities of use of DNA biosensor for evaluation of food antioxidants. BSc dissertation (in Slovak), Slovak University of Technology, Bratislava.
t-108 Kruzlicova, D. 2003. Evaluation of food antioxidants using DNA biosensor. BSc dissertation (in Slovak), Slovak University of Technology, Bratislava.
t-109 Benikova, K. 2008. The role of nanotechnology in the development of DNA biosensors. BSc thesis (in Slovak), Slovak University of Technology, Bratislava.
t-110 Riecka, J. 2008. *In vitro* redox metabolism and effects of some food additives, drugs and pollutants investigated by DNA electrochemical biosensor. BSc dissertation (in Slovak), Slovak University of Technology, Bratislava.
t-111 Kubricanova, V. 2009. DNA biosensor in flow-through arrangement. BSc dissertation (in Slovak), Slovak University of Technology, Bratislava.
t-112 Zrubak, R. 2010. Electrochemical DNA biosensor based on nanomaterials. BSc dissertation (in Slovak), Slovak University of Technology, Bratislava.
t-113 Zanatova, K. 1992. Determination of trace metals using chemically modified electrodes. MSc dissertation (in Slovak), Slovak University of Technology, Bratislava.
t-114 Suriakova-Krekacova, E. 1993. Analytical use of carbon electrode covered by polymeric film. MSc dissertation (in Slovak), Slovak University of Technology, Bratislava.
t-115 Repaska, M. 1994. Use of chemically modified electrode in clinical analysis. MSc dissertation (in Slovak), Slovak University of Technology, Bratislava.
t-116 Hudakova, M. 1996. Analysis with an electrode modified by redox mediator. MSc dissertation (in Slovak), Slovak University of Technology, Bratislava.
t-117 Hruskova, S. 1997. Chemiluminometric and electrochemical determination of biologically important species. MSc dissertation (in Slovak), Slovak University of Technology, Bratislava.
t-118 Ozabalova, K. 1999. Electrochemical sensors with selective modifier. MSc dissertation (in Slovak), Slovak University of Technology, Bratislava.

t-119 Fuknova, M. 2000. Analysis using DNA biosensor. MSc dissertation (in Slovak), Slovak University of Technology, Bratislava.

t-120 Baranova, V. 2001. Analysis using DNA biosensor. MSc dissertation (in Slovak), Slovak University of Technology, Bratislava.

t-121 Nemcova, R. 2002. Study of DNA damage and its inhibition. MSc dissertation (in Slovak), Slovak University of Technology, Bratislava.

t-122 Sivakova, Z. 2002. Determination of antioxidants using DNA biosensor. MSc dissertation (in Slovak), Slovak University of Technology, Bratislava.

t-123 Bachrata, S. 2003. Use of DNA biosensor for determination of antioxidants. MSc dissertation (in Slovak), Slovak University of Technology, Bratislava.

t-124 Bubnicova, K. 2004. Detection of environmentally risk species using DNA biosensor. MSc dissertation (in Slovak), Slovak University of Technology, Bratislava.

t-125 Kovalova, L. 2004. Use of DNA sensor for the determination of risk species. MSc dissertation (in Slovak), Slovak University of Technology, Bratislava.

t-126 Madarasova, I. 2005. DNA biosensor as sensor of chemical toxicity. MSc dissertation (in Slovak), Slovak University of Technology, Bratislava.

t-127 Novakova, K. 2005. Preparation and use of DNA biosensor for the detection of species bound to DNA. MSc dissertation (in Slovak), Slovak University of Technology, Bratislava.

t-128 Benikova, K. 2010. Nanostructured DNA biosensors for the evaluation of DNA damage. MSc thesis (in Slovak), Slovak University of Technology, Bratislava.

t-129 Riecka, J. 2010. Analysis with DNA biosensor in flow-through electrochemical and impedimetric arrangement. MSc dissertation (in Slovak), Slovak University of Technology, Bratislava.

t-130 Buckova, M. 1999. Development of electrochemical sensors for clinical and environ-mental analysis. PhD dissertation (in Slovak), Slovak University of Technology, Bratislava.

t-131 Ferancova, A. 2001. Development and utilization of sensor based on β-cyclodextrin modified electrodes. PhD dissertation (in Slovak), Slovak University of Technology, Bratislava.

t-132 Ovadekova, R. 2007. Development of biosensors for evaluation of biomedical materials, food additives and environmental risk species. PhD dissertation (in Slovak), Slovak University of Technology, Bratislava.

t-133 Galandova, J. 2008. Development of DNA biosensors as analytical screening devices. PhD dissertation (in Slovak), Slovak University of Technology, Bratislava.

t-134 Sirotova, L. 2010. New sensors and methods for evaluation of selected phytochemicals. PhD dissertation (in Slovak), Slovak University of Technology, Bratislava.

PAVOL JOZEF SAFARIK UNIVERSITY, KOSICE[A]

t-135 http://ais-old.upjs.sk/servlet/javanet/Report?_template=pw/institucie1/charustavu.html.free&cislo=350124&instit=350001&jazyk=250112#350701; download on December 10, 2010.

UNIVERSITIES IN FRANCE[B]

t-136 Gaillochet, P. 1975. Redox properties of oleums and concentrated or pure sulfuric acid. Electrochemical behavior of solids dispersed in carbon paste. Application to the determination of uranium in ores. PhD dissertation (in French), University of Paris 6, Paris, France.

t-137 Mouhandess, M.-T. 1983. Electrochemical study of some powdered metal oxides with the aid of a carbon paste electrode. PhD dissertation (in French), University of Lyon 1, Lyon, France.

t-138 Gobon Robveille, S. 1985. Nucleophilic substitution in aromatic series: Study of the reactivity of cyanide ion on aryl radicals by SN1. Study of nickel-based catalyst by carbon paste electrode. PhD dissertation (in French), University of Paris 7, Paris, France.

t-139 Bennouna, A. 1988. Relationship between the electrochemical reactivity and granulometry characteristics of chromium mixed oxides incorporated into carbon paste electrode. PhD dissertation (in French), University of Lyon 1, Lyon, France.

t-137 Berger, P. 1988. Mechanistic study of the chemical and electrochemical oxidoreductive dissolution of actinides dioxides (UO_2, NpO_2, PuO_2, AmO_2) in acidic aqueous medium. PhD dissertation (in French), University of Paris 6, Paris, France.

t-138 Diaw, M. 1990. Detection and determination of precious metals in automotive post-combustion catalysts by electrochemistry. PhD dissertation (in French), University of Paris 6, Paris, France.

t-139 Cocuaud, N. 1990. Synthesis and study of high critical temperature supraconducting ceramics. PhD dissertation (in French), University of Toulouse, Toulouse, France.

Appendix B: List of Dissertation Theses

t-140 Randiamahazaka, H. 1991. Study and modelling of the behaviour of modified carbon electrodes intended to be used for the voltammetric detection of immobilized protease activities. PhD dissertation (in French), University of Toulouse, Toulouse, France.

t-141 Boujtita, M. 1992. Enzymatic electrodes for determination of ethanol and lactate in food products. PhD dissertation (in French), University of Nantes, Nantes, France.

t-142 Adekola, F.A. 1993. Study of the electrochemical behaviour of platinum, palladium, and rhodium compounds using carbon paste electrode. Application to the characterization of automotive post-combustion catalysts. PhD dissertation (in French), University of Paris 6, Paris, France.

t-143 Bacha, S. 1993. Modelling of electroenzymatic biosensors. PhD dissertation (in French), University of Toulouse, Toulouse, France.

t-144 Avoundogba, N. 1995. Chemical and electrochemical behaviour of silver selenide and telluride in concentrated sulfuric and hexafluorosilicic acids and in water-acetonitrile mixtures. PhD dissertation (in French), University of Nancy 1, Nancy, France.

t-145 Rapicault, S. 1996. Design of a carbon paste electrode impregnated with Nafion. Application to the immunodetection of drugs and indirect determination of an enzyme. PhD dissertation (in French), University of Clermont-Ferrand, Clermont-Ferrand, France.

t-146 Navarro, P. 1996. Study of the electrochemical behaviour of carbon paste electrodes. Application to the study of species adsorbed onto non electroactive solids. PhD dissertation (in French), University of Lyon 1, Lyon, France.

t-147 Godet, C. 1997. Study of electron transfer reactions at the interface between aqueous solutions and carbon paste electrodes modified or not with proteins. PhD dissertation (in French), University of Nantes, Nantes, France.

t-148 Bordes, A.-L. 1998. Optimization of simultaneous determination of several antigens in immunoanalysis with electrochemical detection. PhD dissertation (in French), University of Clermont-Ferrand, Clermont-Ferrand, France.

t-149 Despas, C. 1998. Analysis of sorption properties of silica towards bases and cationic species using dielectric and electrochemical methods. PhD dissertation (in French), University of Nancy 1, Nancy, France.

t-150 Nguyen, N.H. 1999. Study of the electrochemical behaviour of activated carbons with the aid of carbon paste electrodes: Application to the determination of metal ions and organic molecules. PhD dissertation (in French), University of Lyon 1, Lyon, France.

t-151 Rondeau, A. 1999. Carbon paste electrodes modified with one or two enzymes (GOD/GOD-HRP) and a mediator (Fc): Application to glucose determination in flow injection analysis (FIA). PhD dissertation (in French), University of Nantes, Nantes, France.

t-152 Serpentini, C.-L. 2000. Study of electrochemical properties of synthetic eumelanins for a novel approach in the conception of depigmenting or hyperpigmenting agents. PhD dissertation (in French), University of Toulouse 3, Toulouse, France.

t-153 Gueguen, S. 2000. Development of modified carbon paste electrodes for the determination of diholosides in food industry. PhD dissertation (in French), University of Nantes, Nantes, France.

t-154 Ramirez Molina, C. 2000. Amperometric biosensors based on NAD+ cofactor regeneration for the specific determination of L-lactate in flow injection analysis. PhD dissertation (in French), University of Nantes, Nantes, France.

t-155 Guemas, Y. 2001. Elaboration of novel analytical methodologies via biosensing of glucose and saccharose mixtures. Application to the analysis of fruit juice. PhD dissertation (in French), University of Nantes, Nantes, France.

t-156 Zaydan, R. 2003. Elaboration of novel analytical method based on the use of an electrode selective for L-lactic acid coupled to a bioreactor for the determination of non-specific inhibition of lactic fermentation in milk. PhD dissertation (in French), University of Nantes, Nantes, France.

t-157 Benoit, S. 2003. Electrochemical approach for the study of interactions between humic acid and metal cations. PhD dissertation (in French), University of Nantes, Nantes, France.

t-158 Serban, S. 2004. Modified electrodes for improving properties of enzymatic biosensors. PhD dissertation (in French), University of Nantes, Nantes, France.

t-159 Lubbad, I. 2004. Contribution to the electrochemical trace analysis of aromatic compounds of nitrogen, lead and cadmium in natural/industrial waters. PhD dissertation (in French), University of Lyon 1, Lyon, France.

t-160 Goubert Renaudin, S. 2007. Synthesis, characterization and reactivity in aqueous medium of silica-based materials functionalized with dithiocarbamate and cyclam ligands. PhD dissertation (in French), University of Nancy 1, Nancy, France.

KARL-FRANZENS UNIVERSITY GRAZ[C]

p-1 Refera, T., B.S. Chandravanshi, and **H. Alemu**. 1998. Differential pulse anodic stripping voltammetric determination of cobalt(II) with N-p-chlorophenyl-cinnamohydroxamic acid modified carbon paste electrode. *Electroanalysis* 10:1038–1042.

p-2 **Beyene Negussie, W.**, P. Kotzian, K. Schachl, H. Alemu, E. Turkusic, A. Copra, H. Moder-egger, I. Švancara, K. Vytřas, and K. Kalcher. 2004. (Bio)sensors based on MnO_2-modified carbon substrates: Retrospections, further improvements and applications. *Talanta* 64:1151–1159.

p-3 **Cai, X.-H.**, B. Ogorevc, and K. Kalcher. 1995. Synergistic electrochemical and chemical modification of carbon paste electrodes for open-circuit preconcentration and voltammetric determination of trace adenine. *Electroanalysis* 7:1126–1131.

p-4 **Diewald, W.**, K. Kalcher, C. Neuhold, I. Švancara, and X.-H. Cai. 1994. Voltammetric behavior of thallium(III) at a solid heterogeneous carbon electrode using ion-pair formation. *Analyst (U.K.)* 119:299–304.

p-5 **Koelbl, G.**, K. Kalcher, and A. Voulgaropoulos. 1992. Voltammetric determination of gold with a Rhodamine B-modified carbon paste electrode. *Fresenius J. Anal. Chem.* 342:83–86.

p-6 Schachl, K., E. Turkusic, A. Komersova, M. Bartos, **H. Moderegger**, I. Švancara, H. Alemu, K. Vytřas, M. Jimenez Castro, and K. Kalcher. 2002. Amperometric determination of glucose with a carbon paste biosensor. *Collect. Czech. Chem. Commun.* 67:302–313.

p-7 **Neuhold, C.G.**, K. Kalcher, W. Diewald, X.-H. Cai, and G. Raber. 1994. Voltammetric determination of nitrate with a modified carbon paste electrode. *Electroanalysis* 6:227–236.

p-8 **Raber, G.**, K. Kalcher, C. Neuhold, C. Talaber, and G. Koelbl. 1995. Adsorptive stripping voltammetry of palladium(II) with thioridazine in situ modified carbon paste electrodes. *Electroanalysis* 7:137–142.

p-9 **Schachl, K.**, H. Alemu, K. Kalcher, J. Jezkova, I. Švancara, and K. Vytřas. 1997. Determination of hydrogen peroxide with sensors based on heterogeneous carbon materials modified with MnO_2 (a review). *Sci. Pap. Univ. Pardubice, Ser. A* 3:41–55.

p-10 **Stadlober, M.**, K. Kalcher, and G. Raber. 1997. Voltammetric determination of titanium, vanadium, and molybdenum using a carbon paste electrode modified with cetyl-trimethylammonium bromide (a review). *Sci. Pap. Univ. Pardubice, Ser. A* 3:103–137.

p-11 **Talaber, C.**, K. Kalcher, G. Koelbl, G. Raber, and X.-H. Cai. 1992. Voltammetric determination of gold with a carbon paste electrode modified with Rhodamine B hexadecyl ester. *Sci. Int. (Lahore)* 4:347–350.

p-12 **Turkusic, E.**, K. Kalcher, K. Schachl, A. Komersova, M. Bartos, H. Moderegger, I. Švancara, and K. Vytřas. 2001. Amperometric determination of glucose with an MnO_2 and glucose oxidase bulk-modified screen-printed carbon ink biosensor. *Anal. Lett.* 34:2633–2647.

p-13 Begic, S., E. Turkusic, **T. Tadesse Waryo**, E. Kahrovic, K. Vytřas, and K. Kalcher. 2005. Some metal oxides as mediator for the amperometric determination of hydrogen peroxide; In *12th Young Investigators' Seminar on Analytical Chemistry*. Book of Abstracts, p. 30. Sarajevo (BIH): Univ. Press.

ABBREVIATIONS USED AND NOTES

t … Thesis, p … publication (scientific paper); cze … Czech Republic, svk … Slovakia, fra … France, aut … Austria. *The individual theses are listed chronologically and in order: *Bachelor* (BSc), *Master* (MSc), *Doctor* (PhD), and, eventually, *Associate Professor* (Hab) for each institution included in the list. (**a**) The respective survey has not been obtained and is replaced with a web link that specifies the research activities at the faculty within which techniques related to the solid-state electrochemistry with CPEEs are also represented; (**b**) in selected cases, particularly long and composite titles have been shortened, which is indicated with ellipsis; (**c**) for technical reasons (temporarily inaccessible archives), the corresponding survey could not be obtained and is replaced by an alphabetically sorted list of representative publications by the same authors (highlighted in bold) and concerning identical topic(s)—and usually also the year—as those of the original theses.

Appendix C: Alternate Titles of Chinese, Japanese, and Korean Journals or Periodicals*

English Transcription	English Translation (Abbreviated)
Aichi-ken Sangyo Gijutsu, Kenkyusho Kenkyu Hokoku	*Aichi Industr. Technol. Inst. Res. J.*
Bunseki Kagaku	*Jpn. Anal.*
Chuangan Jishu Xuebao	*Chin. J. Sens. Actuators*
Denki Kagaku	*J. Electrochem. Assoc. Jpn.*
Denki Kagaku oyobi Kogyo Butsuri Kagaku	*Electrochemistry*
Dian Huaxue	*J. Chem.*
Dianchi (Shuangyuekan)	*Battery (Monthly)*
Dongbei Shida Xuebao, Ziran Kexueban	*J. Dongbei Norm. Univ., Nat. Sci.*
Fenxi Ceshi Xuebao	*J. Instrument. Anal.*
Fenxi Huaxue	*Chin. J. Anal. Chem.*
Fenxi Kexue Xuebao	*J. Anal. Sci.*
Fenxi Shiyanshi	*J. Anal. Lab.*
Fukuoka Kogyo Daigaku Erekutoronikusu,	*Fukuoka Inst. Technol. Electron. Res.*
Gaodeng Xuexiao Huaxue Xuebao	*Chem. Res. Chin. Univ.*
Guangdong Gongye Daxue Xuebao	*J. Inst. Technol.*
Guangzhou Huagong	*Guanzhou Chem. Ind.*
Gunma Daigaku Kyoikugakubu Kiyo, Shizenkagaku-hen	*Sci. Rep.—Gunma Univ. Fac. Educ.*
Hokkaidoritsu Kogyo Shikenjo Hokoku	*Rep.—Electron.*
Huagong Shikan	*Chem. Ind. Times*
Huanjing Kexue	*Chin. J. Environ. Chem.*
Huaxue Chuanganqi	*Chem. Sensor*
Huaxue Yanji, Yu Yingyong	*Chem. Res., Appl.*
Hubei Minzu Xueyuan Xuebao, Ziran Kexueban	*J. Hubei. Inst. Nat. Nat. Sci.*
Jinan Daxue Xuebao, Ziran Kexueban	*J. Jinan Univ., Nat. Sci.*
Jisuanji Yu Yingyong Huaxue	*Comput. Appl. Chem.*
(Kikai Gijutsu) Kenkyusho Shoho	*J. Mech. Eng. Lab.*
Kexue Xuebao	*J. Chin. Chem. Soc.*
Kobunshi Ronbunshu	*Jpn. J. Polym. Sci. Technol.*
Lihua Jianyan Huaxue Fence	*Phys. Chem. Testing*
Nippon Kagaku Kaishi	*J. Chem. Soc. Jpn.*
Punsok Kwahak	*Korean J. Anal. Chem.*
Qingdao Keji Daxue Xuebao	*J. Quindao Univ. Sci. Technol.*
Shandong Jiancai Xueyuan Xuebao	*J. Shandong Univ. Build. Mater.*

(continued)

---

* Cited in this book and given in alphabetical order.

(continued)	
English Transcription	**English Translation (Abbreviated)**
Shipin Yu Fajiao Gongye	*Food Ferment. Ind.*
Weishengwuxue Tongbao	*Microbiology*
Wuhan Daxue Xuebao, Lixueban	*J. Wuhan Univ., Phys.*
Wuhan Huagong Xueyuan Xuebao	*J. Wuhan Univ., Nat. Sci.*
Wuji Huaxue Xuebao	*Chin. J. Inorg. Chem.*
Wuli Huaxue Xuebao	*Acta Physico-Chimica Sinica*
Xiandai Yiqi	*Open Library*
Xibei Shifan Daxue Xuebao, Ziran Kexueban	*J. Xibei Norm. Univ., Nat. Sci.*
Xinan Shifan Daxue Xuebao, Ziran Kexueban	*J. Xinan Norm. Univ., Nat. Sci*
Yakugaku Zashi	*Pharm. Soc. Jpn.*
Yanbian Daxue Xuebao, Ziran Kexueban	*J. Yanbian Univ., Nat. Sci.*
Yaoxue Xuebao	*Acta Pharm. Sinica*
Yingyong Huaxue	*Chin. J. Appl. Chem.*
Yingyong Kexue Xuebao	*J. Appl. Sci.*
Zhongguo Zuzhi Gongcheng, Yanjiu Yu; Linchuang Kangfu	*J. Clin. Rehab. Tissue Eng. Res.*
Ziran Kexueban	*Nat. Sci.*

SURVEY OF THE REMAINING TITLES (OFFICIALLY NAMED IN ENGLISH)

English Title (Full Text)	**Abbreviated Version**
Analytical Sciences (Japan)	*Anal. Sci. (Jpn.)*
Academic Journal of Xi'an Jiatong University	*Acad. J. Xi'an Jiatong Univ.*
Bulletin of the Chemical Society of Japan (BCSJ)	*Bull. Chem. Soc. Jpn.*
Bulletin of the Korean Chemical Society	*Bull. Korean Chem. Soc.*
Chinese Chemical Letters	*Chin. Chem. Lett.*
Chinese Journal of Chemistry	*Chin. J. Chem.*
Chinese Journal of Structural Chemistry	*Chin. J. Structur. Chem.*
Chinese Science Bulletin	*Chinese Sci. Bull.*
Journal of the Chinese Chemical Society (JCCS)	*J. Chin. Chem. Soc. (Taipei)*
Journal of the Korean Chemical Society	*J. Korean Chem. Soc.*
Journal of Korean Air Pollution Research Association	*J. Korean Air Pollut. Res. Assoc.*
Journal of Tongji Medical University	*J. Tongji Med. Univ.*
Reviews on Polarography (Kyoto/Japan)	*Rev. Polarog. (Kyoto)*
Science in China, Series B: Chemistry	*Sci. China, Ser. B*

References

1. Adams, R.N. 1958. Carbon paste electrodes. *Anal. Chem.* 30:1576.
2. Adams, R.N. 1963. Carbon paste electrodes. A review. *Rev. Polarog. (Jpn.)* 11:71–78.
3. Adams, R.N. 1969. *Electrochemistry at Solid Electrodes*, pp. 26–27, 30, 55, 100, 125–130, 280–283, 330–367. New York: Marcel Dekker.
4. Editorial. 1999. Dedicated to Professor Ralph N. Adams on the occasion on his 75th birthday. *Electroanalysis* 11:284–291.
5. Kissinger, P.T. 1999. Lessons from the work of Professor Ralph N. Adams. *Electroanalysis* 11:292–294.
6. Kuwana, T. 2001. Personal communication.
7. Švancara, I., K. Vytřas, K. Kalcher, A. Walcarius, and J. Wang. 2009. Carbon paste electrodes in facts, numbers, and notes: A review on the occasion of the 50-years jubilee of carbon paste in electrochemistry and electroanalysis. *Electroanalysis* 21:7–28.
8. Olson, C. and R.N. Adams. 1960. Carbon paste electrodes: Application to anodic voltammetry. *Anal. Chim. Acta* 22:582–589.
9. Olson, C. and R.N. Adams. 1963. Carbon paste electrodes: Application to cathodic reductions and anodic stripping voltammetry. *Anal. Chim. Acta* 29:358–363.
10. Galus, Z., C. Olson, H.Y. Lee, and R.N. Adams. 1962. Rotating disk electrodes. *Anal. Chem.* 34:164–166.
11. Olson, C.L. 1962. Investigation concerning the development and use of carbon paste electrode for voltammetric measurements. PhD dissertation, University of Kansas, Lawrence, KS.
12. Jacobs, E.S. 1963. Anodic stripping voltammetry of gold and silver with carbon paste electrodes. *Anal. Chem.* 25:2112–2115.
13. Davis, D.G. and M.E. Everhart. 1964. Chronopotentiometry of the bromide-bromine couple at platinum and carbon paste electrodes. *Anal. Chem.* 36:38–40.
14. Kuwana, T. and W.G. French. 1964. Carbon paste electrodes containing some electroactive compounds. *Anal. Chem.* 36:241–242.
15. Schultz, F.A. and T. Kuwana. 1965. Electrochemical studies of organic compounds dissolved in carbon-paste electrodes. *J. Electroanal. Chem.* 10:95–103.
16. Marcoux, L.S., K.G. Prater, B.G. Prater, and R.N. Adams. 1965. Nonaqueous carbon paste electrode. *Anal. Chem.* 37:1446–1447.
17. Hawley, D. and R.N. Adams. 1965. Homogenous chemical reactions in electrode processes: Studies of the follow-up chemical reactions. *J. Electroanal. Chem.* 10:376–386.
18. Prater, B. and R.N. Adams. 1966. A critical evaluation of practical rotated disc electrodes (RDEs). *Anal. Chem.* 38:153–155.
19. Chambers, C.A.H. and J.Q. Chambers. 1966. Electrochemical oxidation of a quinol phosphate. *J. Am. Chem. Soc.* 88:2922–2923.
20. Chambers, C.A.H. and J.K. Lee. 1967. Studies of the extraction of organic molecules into the carbon-paste electrode. *J. Electroanal. Chem.* 14:309–314.
21. Hawley, D., S.V. Tatawawadi, S. Piekarski, and R.N. Adams. 1967. Electrochemical studies of the oxidation pathways processes of catecholamines. *J. Am. Chem. Soc.* 89:447–448.
22. Malachesky, P.A., K.B. Prater, G. Petrie, and R.N. Adams. 1968. Measurement of chemical reaction rates following electron transfer: An empirical approach by using ring disk electrodes (RDEs). *J. Electroanal. Chem.* 16:41–46.
23. Marcus, M.F. and M.D. Hawley. 1968. The electrochemical oxidation of p-dimethyl-aminophenol in aqueous solution. *J. Electroanal. Chem.* 18:175–183.
24. Meier, E.P. and J.Q. Chambers. 1969. Adsorption of organic molecules at a carbon paste electrode. *Anal. Chem.* 41:914–918.
25. Papouchado, L., J. Bacon, and R.N. Adams. 1970. Potential-step cyclic voltammetry for the study of electrode reaction mechanisms. *J. Electroanal. Chem.* 24:1–5.
26. Beilby, A.L. and B.R. Mather. 1965. Resistance effects of two types of carbon paste electrodes. *Anal. Chem.* 37:766–768.

27. Lawrence, R.J. and J.A. Chambers. 1967. Voltammetry at solid electrodes with renewable surface: A revolving wheel electrode. *Anal. Chem.* 39:134–136.
28. Farsang, Gy. 1965. Voltammetric properties and analytical uses of carbon paste electrodes prepared with silicone oil. *Acta Chim. Acad. Sci. Hung.* 45:163–176.
29. Landsberg, R. and R. Thiele. 1966. On the effect of inactive surface areas upon the diffusion current at a rotating disc electrode and the transition time in galvanostatic measurements (in German). *Electrochim. Acta* 11:1243–1259.
30. Brezina, M. 1966. Estimation of electrochemical activity of carbon using a paste electrode. *Nature* 212:283–283.
31. Monien, H., H. Specker, and K. Zinke 1967. Application of various carbon electrodes for inverse voltammetric determination of silver. *Fresenius Z. Anal. Chem.* 225:342–351.
32. Pungor, E. and E. Szepesvary. 1968. Voltammetric studies with silicone rubber-based graphite electrodes. *Anal. Chim. Acta* 43:289–296.
33. Lindquist, J. 1968. A new carbon paste electrode holder and a simple method for preparing reproducible electrode surfaces. *J. Electroanal. Chem.* 18:204–205.
34. Neeb, R., I. Kiehnast, and A. Narayanan. 1972. On the use of carbon paste electrodes in voltammetry. *Fresenius Z. Anal. Chem.* 262:339–343.
35. Ruzicka, J., E.H. Hansen, and J.C. Tjell. 1973. Selectrode: The universal ion-selective electrode. *Anal. Chim. Acta* 67:155–178.
36. Mesaric, S. and E.M.F. Dahmen. 1973. Ion-selective carbon-paste electrodes for halides and silver(I) ions. *Anal. Chim. Acta* 64:431–438.
37. Atuma, S.S. and J. Lindquist. 1973. Voltammetric determination of tocopherols by use of a newly developed carbon paste electrode. *Analyst (U.K.)* 98:886–894.
38. Bauer, D. and P. Gaillochet. 1974. Behavior of carbon pastes containing an electro-active compound (in French). *Electrochim. Acta*, 19:597–606.
39. Lindquist, J. 1973. Carbon paste electrode with a wide anodic potential range. *Anal. Chem.* 45:1006–1008.
40. Lindquist, J. 1974. A study of seven different carbon paste electrodes. *J. Electroanal. Chem.* 52:37–46.
41. Soderhjelm, P. 1976. A comparison of the analytical utility of three different potential ramp techniques in voltammetry, using a carbon-paste electrode. *J. Electroanal. Chem.* 71:109–115.
42. Papouchado, L., G. Petrie, and R.N. Adams. 1972. Anodic oxidation pathways of phenolic compounds. Part 1: Anodic hydroxylation reactions. *J. Electroanal. Chem.* 38:389–396.
43. Petek, M., S. Bruckenstein, B. Feinberg, and R.N. Adams. 1973. Anodic oxidation of substituted methoxyphenols: MS-identification of methanol formed. *J. Electroanal. Chem.* 42:397–401.
44. Papouchado, L., R.W. Sandford, G. Petrie, and R.N. Adams. 1975. Anodic oxidation pathways of phenolic compounds: Part 2. Stepwise electron transfers and coupled hydroxylations. *J. Electroanal. Chem.* 65:275–284.
45. Adams, R.N., E. Murrill, R. McCreery, L. Blank, and M. Karolczak. 1972. 6-Hydroxy-dopamine: A new oxidation pathway. *Eur. J. Pharmacol.* 17:287–289.
46. Kissinger, P.T., C. Refshauge, R. Dreiling, and R.N. Adams. 1973. Electrochemical detector for liquid chromatography with picogram sensitivity. *Anal. Lett.* 6:465–477.
47. Sternson, A.W., R. McCreery, B. Feinberg, and R.N. Adams. 1973. Electrochemical studies of adrenergic neurotransmitters and related compounds. *J. Electroanal. Chem.* 46:313–321.
48. Kissinger, P.T., J.B. Hart, and R.N. Adams. 1973. Voltammetry in brain tissue: A new neurophysical measurement. *Brain Res.* 55:209–213.
49. McCreery, R.L., R. Dreiling, and R.N. Adams. 1974. Voltammetry in brain tissue: The fate of injected 6-hydroxydopamine. *Brain Res.* 73:15–21.
50. Adams, R.N. 1976. Probing brain chemistry with electroanalytical techniques. *Anal. Chem.* 48:1126A–1138A.
51. Cheek, G.T. and R.F. Nelson. 1978. Applications of chemically modified electrodes to analysis of metal ions. *Anal. Lett.* 11:393–402.
52. Yao, T. and S. Musha. 1979. Electrochemical enzymic determinations of ethanol and L-lactic acid with a carbon paste electrode modified chemically with nicotinamide adenine dinucleotide. *Anal. Chim. Acta* 110:203–209.
53. Ravichandran, K. and R.P. Baldwin. 1981. Chemically modified carbon paste electrodes. *J. Electroanal. Chem.* 126:293–300.
54. Ravichandran, K. and R.P. Baldwin. 1983. Phenylenediamine-containing chemically modified carbon paste electrodes as catalytic voltammetric sensors. *Anal. Chem.* 55:1586–1591.

References

55. Kalcher, K. 1990. Chemically modified carbon paste electrodes in voltammetric analysis. *Electroanalysis* 2:419–433.
56. Švancara, I.K. Vytřas, F. Renger, and M.R. Smyth. 1992/93. Application of carbon paste electrodes in electroanalysis. A review. *Sb. Ved. Pr., Vys. Sk. Chemickotechnol., Pardubice* 56:21–57.
57. Wang, J. 1988. *Electroanalytical Techniques in Clinical Chemistry and Laboratory Medicine*, pp. 3, 31, 35, 39, 40, 97, 113, 124, 133, 150, 157, 158. Weinheim, Germany: VCH Publishers.
58. Hart, J.P. 1990. *Electroanalysis of Biologically Important Compounds*, pp. 29, 34, 68, 69, 139, 169, 179, 183, 184. Chichester, U.K.: Ellis Horwood.
59. Ulakhovich, N.A., E.P. Medyantseva, and G.K. Budnikov. 1993. Carbon-paste electrodes as chemical sensors in voltammetry. *J. Anal. Chem. (Russ.)*/transl. of *Zh. Anal. Khim.* 48:980–998.
60. Urbaniczky, C. and K. Lundstroem. 1984. Voltammetric studies on carbon paste electrodes. The influence of paste composition on electrode capacity and kinetics. *J. Electroanal. Chem.* 176:169–182.
61. Kingsley, E.D. and D.J. Curran. 1988. Some observations on carbon paste electrodes in AC-voltammetry. *Anal. Chim. Acta* 206:385–390.
62. Curran, D.J., M.B. Gelbert, and E.D. Kingsley. 1990. An instrument for alternating current voltammetry featuring a digital phase-sensitive detector: Application to flow-injection analysis using ac amperometry. *Electroanalysis* 2:435–442.
63. Ravichandran, K. and R.P. Baldwin. 1984. Enhanced voltammetric response by electrochemical pretreatment of carbon paste electrodes. *Anal. Chem.* 56:1744–1747.
64. Albahadily, F.N. and H.A. Mottola. 1987. Improved response of carbon-paste electrodes for electrochemical detection in flow systems by pretreatment with surfactants. *Anal. Chem.* 59:958–962.
65. Stulik, K. and V. Pacakova. 1981. Comparison of several voltammetric detectors for high-performance liquid chromatography. *J. Chromatogr. A* 208:269–278.
66. Danielson, N.D., J. Wangsa, and M.A. Targove. 1989. Comparison of paraffin oil and poly(chlorotrifluoroethylene) oil carbon paste electrodes in high organic content solvents. *Anal. Chem.* 61:2585–2588.
67. Kauffmann, J.-M., A. Laudet, G.J. Patriarche, and G.D. Christian. 1982. The graphite spray electrode and its application in the anodic stripping voltammetry of bismuth. *Anal. Chim. Acta* 135:153–158.
68. Rice, M.E., Z. Galus, and R.N. Adams. 1983. Graphite paste electrodes: Effects of paste composition and surface states on electron-transfer rates. *J. Electroanal. Chem.* 143:89–102.
69. Kalcher, K., J.-M. Kauffmann, J. Wang, I. Švancara, K. Vytřas, C. Neuhold, and Z. Yang. 1995. Sensors based on carbon paste in electrochemical analysis: A review with particular emphasis on the period 1990–1993. *Electroanalysis* 7:5–22.
70. Gorton, L. 1995. Carbon paste electrodes modified with enzymes, tissues, and cells. A review. *Electroanalysis* 7:23–45.
71. Turner, A.P.F., W.J. Aston, I.J. Higgins, J.M. Bell, J. Colby, G. Davies, and H.A.O. Hill. 1984. Carbon monoxide-acceptor oxidoreductase from pseudomonas thermo-carboxydovorans strain C2 and its use in a carbon monoxide sensor. *Anal. Chim. Acta* 162:161–174.
72. Ikeda, T., H. Hamada, K. Miki, and M. Senda. 1985. Glucose-oxidase immobilised benzoquinone carbon paste electrode as glucose sensors. *Agricult. Biol. Chem.* 49:541–543.
73. Matuszewski, W. and M. Trojanowicz. 1988. Graphite paste-based enzymatic glucose electrode for flow-injection analysis. *Analyst (U.K.)* 113:735–738.
74. Wang, J. 1994. Decentralized electrochemical monitoring of trace metals: From disposable strips to remote electrodes. *Analyst (U.K.)* 119:763–766.
75. Kalcher, K.K. Schachl, I. Švancara, K. Vytřas, and H. Alemu. 1997. Recent progress in the development of electrochemical carbon paste sensors. *Sci. Pap. Univ. Pardubice Ser. A* 3:57–85.
76. Diewald, W., K. Kalcher, C. Neuhold, I. Švancara, and X.-H. Cai. 1994. Voltammetric behavior of thallium(III) at a solid heterogeneous carbon electrode using ion-pair formation. *Analyst (U.K.)* 119:299–304.
77. Cataldi T.R.I. and D. Centonze. 1996. Development and treatment of a carbon paste (composite) electrode made from polyethylene and graphite powder modified with Cu_2O. *Anal. Chim. Acta* 326:107–115.
78. Pravda, M., C. Petit, Y. Michotte, J.-M. Kauffmann, and K. Vytřas. 1996. Study of a new solid carbon paste tyrosinase-modified amperometric biosensor for the determination of catecholamines by HPLC. *J. Chromatogr. A* 727:47–54.
79. Mirel, S., R. Sandulescu, J.-M. Kauffmann, and L. Roman. 1999. Electrochemical study of some 2-mercapto-5-R-amino-1,3,4-thiadiazole derivatives using solid carbon paste electrode. *J. Pharm. Biomed. Anal.* 18:535–544.

80. Crow, D.R. 1975. Voltammetry with the carbon-wax-based electrode. *Proc. Anal. Div. Chem. Soc.* 12:181–184.
81. Zhang, Z.-Q., H. Liu, and Z.-F. Li. 1998. New developments of carbon paste electrode (a review). *Fenxi Kexue Xuebao* 14: 80–86; 1998. *Chem. Abstr.* 125:175403x.
82. Švancara, I., K. Vytřas, J. Barek, and J. Zima. 2001. Carbon paste electrodes in modern electroanalysis (a review). *Crit. Rev. Anal. Chem.* 31:311–345.
83. Kalcher, K., I. Švancara, R. Metelka, K. Vytřas, and A. Walcarius. 2006. Heterogeneous electrochemical carbon sensors. In *The Encyclopedia of Sensors*, Vol. 4, Eds. C.A. Grimes, E.C. Dickey, and M.V. Pishko, pp. 283–429. Stevenson Ranch, CA: American Scientific Publishers.
84. Švancara, I., A. Walcarius, K. Kalcher, and K. Vytřas. 2009. Carbon paste electrodes in the new millennium (a review). *Cent. Eur. J. Chem.* 7:598–656.
85. Gil, E.S. and L.T. Kubota. 2000. Electrochemical properties of Doyle catalyst immobilized on carbon paste in the presence of DNA. *Bioelectrochemistry* 51:145–149.
86. Rodrigues, C.A. and E. Stadler. 2000. Electrocatalytic oxidation of ascorbic acid on carbon paste electrode modified with [Ru(EDTA)(4PMC)]. *Portugal. Electrochim. Acta* 18:55–61.
87. Ferreira, C.U., Y. Gushikem, and L.T. Kubota. 2000. Electrochemical properties of Meldola's blue immobilized on silica-titania phosphate prepared by the sol-gel method. *J. Solid State Electrochem.* 4:298–303.
88. Miao, Y., L.-S. Chia, N.-K. Goh, and S.-N. Tan. 2001. Amperometric glucose biosensor based on immobilization of glucose oxidase in chitosan matrix cross-linked with glutaraldehyde. *Electroanalysis* 13:347–349.
89. Walcarius, A. 2001. Electroanalysis with pure, chemically modified, and sol-gel-derived silica-based materials (an overview). *Electroanalysis* 13:701–718.
90. Wang, J., P. Gruendler, G.U. Flechsig, M. Jasinski, G.A. Rivas, E. Sahlin, and J.L. Lopez Paz. 2000. Stripping analysis of nucleic acids at a heated carbon paste electrode. *Anal. Chem.* 72:3752–3756.
91. Wang, J., X.-H. Cai, J. Wang, C. Jonsson, and E. Palecek. 1995. Trace measurements of RNA by potentiometric stripping analysis at carbon paste electrodes. *Anal. Chem.* 67:4065–4070.
92. Tomschik, M., L. Havran, M. Fojta, and E. Palecek. 1998. Constant current chronopotentiometric stripping analysis of bioactive peptides at mercury and carbon electrodes. *Electroanalysis* 10:403–409.
93. Gonzalez Garcia, M.B. and A. Costa Garcia. 2001. Silver electrodeposition catalyzed by colloidal gold on carbon paste electrode: Application to biotin/streptavidin interaction monitoring. *Biosens. Bioelectron.* 15:663–670.
94. Hernandez Santos, D., M.B. Gonzalez Garcia, and A. Costa Garcia. 2002. Metal nanoparticles based electroanalysis (a review). *Electroanalysis* 14:1225–1235.
95. Wang, J., U.A. Kirgoz, J.-W. Mo, J.-M. Lu, A.N. Kawde, and A. Muck. 2001. Glassy carbon paste electrodes. *Electrochem. Commun.* 3:203–208.
96. Švancara, I., M. Hvizdalova, K. Vytřas, K. Kalcher, and R. Novotny. 1996. A microscopic study on carbon paste electrodes. *Electroanalysis* 8:61–65.
97. Grennan, K., A.J. Killard, and M.R. Smyth. 2001. Physical characterizations of a screen-printed electrode for use in an amperometric biosensor system. *Electroanalysis* 13:745–750.
98. Hammouti, B., H. Oudda, A. El Maslout, and A. Benayada. 2000. Realisation of ferrocene reference electrode. *Bull. Electrochem.* 16:283–284.
99. Torimura, M., A. Miki, A. Wadano, K. Kano, and T. Ikeda. 2001. Electrochemical investigation of photoreduction catalyzed by cyanobacteria *Synechococcus* sp. PCC7942 in exogenous quinones and photoelectrochemical oxidation of water. *J. Electroanal. Chem.* 496:21–28.
100. Kureishi, Y., H. Shiraishi, and H. Tamiaki. 2001. Self-aggregates of synthetic zinc chlorins as the photosensitizer on carbon paste electrodes for a novel solar cell. *J. Electroanal. Chem.* 496:13–20.
101. Wang, J. 2002. Real-time electrochemical monitoring: Toward green analytical chemistry. *Ass. Chem. Res.* 35:811–816.
102. Wang, J., J.-M. Lu, S.B. Hočevar, P.A.M. Farias, and B. Ogorevc. 2000. Bismuth-coated carbon electrodes for anodic stripping voltammetry. *Anal. Chem.* 72:3218–3222.
103. Krolicka, A., R. Pauliukaite, I. Švancara, R. Metelka, A. Bobrowski, E. Norkus, K. Kalcher, and K. Vytřas. 2002. Bismuth film-plated carbon paste electrodes. *Electrochem. Commun.* 4:193–196.
104. Hocevar, S.B., B. Ogorevc, J. Wang, and B. Pihlar. 2002. A study on operational parameters for advanced use of bismuth film electrodes in anodic stripping voltammetry. *Electroanalysis* 14:1707–1712.
105. Kong, Y.-T., G.-H. Choi, M.-S. Won, and Y.-B. Shim. 2002. Determination of $Hg_2^{(2+)}$ ions using the specific reaction with a picolinic acid N-oxide modified electrode. *Chem. Lett.* 1:54–55.
106. Yantasee, W., Y.-H. Lin, T.S. Zemanian, and G.E. Fryxell. 2003. Voltammetric detection of Pb(II) and Hg(II) using a carbon paste electrode modified with thiol self-assembled monolayer on mesoporous silica. *Analyst (U.K.)* 128:467–472.

References

107. Marino, G., M.F. Bergamini, M.F.S. Teixeira, and E.T.G. Cavalheiro. 2003. Evaluation of a carbon paste electrode modified with organo-functionalized amorphous silica in for the determination of cadmium(II) by using differential pulse anodic stripping voltammetry. *Talanta* 59:1021–1028.
108. Švancara, I., K. Vytřas, A. Bobrowski, and K. Kalcher. 2002. Determination of arsenic at a gold-plated carbon paste electrode using constant current stripping analysis. *Talanta* 58:45–55.
109. Freire, R.S., N. Duran, and L.T. Kubota. 2002. Electrochemical biosensors for continuous phenols monitoring in environmental matrices. *J. Brazil. Chem. Soc.* 13:456–462.
110. Sreedhar, M., L.M. Reddy, K.R. Sirisha, and S.R.J. Reddy. 2003. Differential pulse adsorptive stripping voltammetric determination of Dinoseb and Dinoterb at a modified electrode. *Anal. Sci. (Jpn.)* 19:511–516.
111. Zima, J., I. Švancara, J. Barek, and K. Vytřas. 2009. Recent advances in electroanalysis of organic and biological compounds at carbon paste electrodes. *Crit. Rev. Anal. Chem.* 39:204–227.
112. Kalcher, K., I. Švancara, M. Buzuk, K. Vytřas, and A. Walcarius. 2009. Electrochemical sensors and biosensors based on heterogeneous carbon materials. *Monatsh. Chem.* 140:861–889.
113. Varma, S. and C.K. Mitra. 2002. Low frequency impedance studies on covalently modified glassy carbon paste. *Electroanalysis* 14:1587–1596.
114. Zima, J., J. Barek, and A. Muck. 2005. Monitoring of environmentally and biologically important substances at carbon paste electrodes. *Rev. Chim. (Bucharest)* 55:657–662.
115. Li, G., Z.-M. Ji, and K.-B. Wu. 2006. Square wave anodic stripping voltammetric determination of Pb^{2+} using acetylene black paste electrode based on the inducing adsorption ability of I(–). *Anal. Chim. Acta* 577:178–182.
116. Nossol, E. and A.J.G. Zarbin. 2008. Carbon paste electrodes made from novel carbonaceous materials: Preparation and electrochemical characterization. *Electrochim. Acta* 54:582–589.
117. Miranda Hernandez, A., M.E. Rincon, and I. Gonzalez. 2005. Characterization of carbon-fullerene-silicone oil composite paste electrodes. *Carbon* 43:1961–1967.
118. Stefan, R.I. and S.G. Bairu. 2003. Monocrystalline diamond paste-based electrodes and their possible applications to the determination of Fe(II) in vitamins. *Anal. Chem.* 75:5394–5398.
119. Zhu, L., C. Tian, D. Zhu, and R. Yang. 2008. Ordered mesoporous carbon paste electrodes for electrochemical sensing and biosensing. *Electroanalysis* 20:1128–1134.
120. Hocevar, S.B. and B. Ogorevc. 2007. Preparation and characterization of carbon paste micro-electrode based on carbon nanoparticles. *Talanta* 74:405–411.
121. Tang, X.-F., Y. Liu, H.-Q. Hou, and T.-Y. You. 2010. Electrochemical determination of L-tryptophan, L-tyrosine and L-cysteine using electrospun carbon nanofibers modified electrode. *Talanta* 80:2182–2186.
122. Iijima, S. 1991. Helical microtubules of graphitic carbon. *Nature* 354:56–58.
123. Rubianes, M.D. and G.A. Rivas. 2003. Carbon nanotubes paste electrode. *Electrochem. Commun.* 5:689–694.
124. Valentini, F., A. Amine, S. Orlanducci, M.L. Terranova, and G. Palleschi. 2003. Carbon nanotube purification: Preparation and characterization of carbon nanotube paste electrodes. *Anal. Chem.* 75:5413–5421.
125. Liu, H.-T., P. He, Z.-Y. Li, Y. Liu, J. Li, L.-Z. Zheng, and J-H. Li. 2005. The inherent capacitive behavior of imidazolium-based room-temperature ionic liquids at carbon paste electrodes. *Electrochem. Solid State Lett.* 8:17–19.
126. Liu, H.-T., P. He, Z.-Y. Li, C.-N. Sun, L.-H. Shi, Y. Liu, G.-Y. Zhu, and J.-H. Li. 2005. An ionic liquid-type carbon paste electrode and its polyoxometalate-modified properties. *Electrochem. Commun.* 7:1357–1363.
127. Kachoosangi, R.T., G.G. Wildgoose, and R.G. Compton. 2007. Room temperature ionic liquid carbon nanotube paste electrodes: Overcoming large capacitive currents using rotating disk electrodes. *Electroanalysis* 19:1483–1489.
128. Liu, Y., D.-W. Wang, J.-S. Huang, H.-Q. Hou, and T.-Y. You. 2010. Highly sensitive composite electrode based on electrospun carbon nanofibers and ionic liquid. *Electrochem. Commun.* 12:1108–1111.
129. Ojani, R., J.B. Raoof, and E. Zarei. 2009. Electrocatalytic oxidation and determination of cysteamine by poly-N,N-dimethylaniline/ferrocyanide film modified carbon paste electrode. *Electroanalysis* 21:1189–1193.
130. Raoof, J.B., A. Omrani, R. Ojani, and F. Monfared. 2009. Poly(N-methylaniline)/nickel modified carbon paste electrode as an efficient and cheep electrode for electrocatalytic oxidation of formaldehyde in alkaline medium. *J. Electroanal. Chem.* 633:153–158.
131. Ojani, R., J.B. Raoof, and S. Fathi. 2009. Poly(o-aminophenol) film in the presence of sodium dodecyl sulfate: Application to the Ni(II) dispersion and the electrocatalytic oxidation of methanol and ethylene glycol. *Electrochim. Acta* 54:2190–2196.

132. Ojani, R., J.B. Raoof, and B. Norouzi. 2009. Electropolymerization of N-methylaniline in the presence of sodium dodecylsulfate and its application for electrocatalytic reduction of nitrite. *J. Mater. Sci.* 44:4095–4103.
133. Ojani, R., J.B. Raoof, A. Ahmady, and S.R. Hosseini. 2010. Carbon paste as a candidated electrode substrate for electrochemical polymerization of 2,5-dimethylaniline. *Asian J. Chem.* 22:943–952.
134. Nagashree, K.L., N.H. Raviraj, and M.F. Ahmed. 2010. Carbon paste electrodes modified with Pt and Pt-Ni microparticles dispersed in poly-indole film for electro-catalytic oxidation of methanol. *Electrochim. Acta* 55:2629–2635.
135. Pandurangachar, M., B.E.K. Swamy, U. Chandra, O. Gilbert, and B.S. Sherigara. 2009. Simultaneous determination of dopamine, ascorbic acid and uric acid at poly(Patton and Reeder's) modified carbon paste electrode. *Int. J. Electrochem. Sci.* 4:672–683.
136. Zhou, C.-L., Y.-X. Zheng, H.-W. Chen, C.-Q. Shou, Y.-M. Dong, Z. Liu, and L.-R. Chen. 2009. Preparation and application of the hyperbranched polymer modified carbon paste electrode. *Fenxi Huaxue* 37:111–114.
137. He, J.-B., F. Qi, Y. Wang, and N. Deng. 2010. Solid carbon paste-based amperometric sensor with electropolymerized film of 2-amino-5-mercapto-1,3,4-thiadiazole. *Sens. Actuators B* 145:480–487.
138. Stoces, M., K. Kalcher, I. Švancara, and K. Vytřas. 2011. A biosensor for glucose based on electrochemically generated poly-aniline film and carbon paste. *Int. J. Electrochem. Sci.* 6:1917–1926.
139. Liu, Y., L. Zhang, Q.-H. Guo, H.-Q. Hou, and T.-Y. You. 2010. Enzyme-free ethanol sensor based on electrospun nickel nanoparticle-loaded carbon fiber paste electrode. *Anal. Chim. Acta* 663:153–157.
140. Wang, X.-L., Z.-B. Han, E.-B. Wang, H. Zhang, and C.-W. Hu. 2003. A bifunctional electrocatalyst containing tris(2,2′-bipyridine)Ru(II) and 12-molybdophosphate bulk-modified carbon paste electrode. *Electroanalysis* 15:1460–1464.
141. Wang, L., J. Li, E.-B. Wang, L. Xu, J. Peng, and Z. Li. 2004. Preparation and characterization of ultrathin multilayer films based on polyoxometalate of the $Mo_8V_2O_{28}x_7$-H_2O type. *Mater. Lett.* 58:2027–2031.
142. Han, Z.G., Y.-L. Zhao, J. Peng, A.-X. Tian, Y.-H. Feng, and Q. Liu. 2005. Inorganic/organic hybrid polyoxometalate: Preparation, characterization and electrochemical properties. *J. Solid State Chem.* 178:1386–1394.
143. Wang, X.-L., H.-Y. Zhao, and Y.-F. Wang. 2006. Preparation, electrochemical property and application in bulk-modified electrode of Dawson type phosphomolybdate-doped polypyrrole composite nanoparticles. *Gaodeng Xuexiao Huaxue Xuebao* 22:556–559.
144. Dong, B.-X., J. Peng, A.-X. Tian, J.-Q. Sha, L. Li, and H.S. Liu. 2007. Two new inorganic–organic hybrid single pendant hexadecavanadate derivatives with bifunctional electrocatalytic activities. *Electrochim. Acta* 52:3804–3812.
145. Zhao, Z.-F., B.-B. Zhou, Z.-H. Su, H.-Y. Ma, and C.-X. Li. 2008. A new $[As_3Mo_3O_{15}](3-)$ fragment binded with Cu(I)-imidazole complexes: Synthesis, structure, and electrochemical properties. *Inorg. Chem. Commun.* 11:648–651.
146. Wang, X.-L., B.-K. Chen, G.-C. Liu, H.-Y. Zhao, H.-Y. Lin, and H.-L. Hu. 2009. Hydrothermal syntheses and structural characterization of two new supramolecular compounds: $(H_{(2)}bbi)_{(2)}[Mo_8O_{26}]$ and $(H_{(2)}bbi)_{(2)}$ $[SiW_{12}O_{40}]$ x $2H_{(2)}O$. *Solid State Sci.* 11:61–67.
147. Jin, H.-J., B.-B. Zhou, Z.-F. Zhao, and Z.-H. Su. 2010. Synthesis, structure and electrochemical property of a supramolecular compound $(H_{(2)}en)_{(2)}$ $[Cu(en)_{(2)}(H_2O)_{(2)}]$ $[Mo_8O_{28}]$. *Chin. J. Struct. Chem.* 29:213–218.
148. Memon, M., K.Z. Memon, M.S. Akhtar, and D. Stuben. 2009. Characterization and quantification of iron oxides occurring at low concentration level in soil materials. *Commun. Soil Sci. Plant Anal.* 40:62–178.
149. He, J.-B., G.-H. Ma, J.-C. Chen, Y. Yao, and Y. Wang. 2010. Voltammetry and spectro-electrochemistry of solid Indigo dispersed in carbon paste. *Electrochim. Acta* 55:4845–4850.
150. Web of Science–Science–Thompson Reuters, http://thomsonreuters.com/products_services/science/science_products/a-z/web_of science/ (accessed December 10, 2010).
151. Brainina, Kh.Z. and V. Ashpur. 1979. Phase electrochemical analysis with a carbon paste electrically-active electrode (in Russian). *Zavod. Lab. (U.S.S.R.)* 45:10–20.
152. O'Neill, R.D. 1993. Sensor–tissue interactions in neurochemical analysis with carbon paste electrodes in vivo (a review). *Analyst (U.K.)* 118:433–438.
153. Švancara, I. and K. Vytřas. 1994. Preparation and properties of carbon paste electrodes (a review, Part I; in Czech). *Chem. Listy* 88:138–146; Vytřas, K. and I. Švancara. 1994. Applications of carbon paste electrodes in electroanalysis (a review, Part II; in Czech). *Chem. Listy* 88:412–422.
154. Vytřas, K. and I. Švancara. 1994. Some ways of modern voltammetry: Carbon paste electrodes. *Egypt. J. Anal. Chem.* 3:78–86.

References

155. Kalcher, K., X.-H. Cai, G. Kolbl, I. Švancara, and K. Vytřas. 1994. New trends in voltammetric analysis: Modified carbon paste electrodes. *Sb. Ved. Pr., Vys. Sk. Chemickotechnol. Pardubice* 57:5–27.
156. Lobo Castanon, M.J., S.L. Alvarez Crespo, M.I. Alvarez Gonzalez, S.B. Saidman, A.J. Miranda Ordieres, and P. Tunon-Blanco. 1998. Biosensors based on carbon paste electrodes using immobilized dehydrogenase enzymes. An overview and trends. *Sci. Pap. Univ. Pardubice Ser. A* 3:17–29.
157. Švancara, I. and K. Vytřas. 2000. Physicochemical processes in analytical electrochemistry with carbon paste electrodes. An overview. *Chemija (Vilnius)* 11:18–27.
158. Radi, A.E. 2006. Applications of stripping voltammetry at carbon paste and chemically modified carbon paste electrodes to pharmaceutical analysis (a review). *Curr. Pharm. Anal.* 2:1–8.
159. Vytřas, K. and I. Švancara. 2007. Carbon paste-based ion-selective electrodes. In *Sensing in Electroanalysis*, Vol. 2, Eds. K. Vytřas and K. Kalcher, pp. 7–22. Pardubice, Czech Republic: University Press.
160. Vytřas, K., I. Švancara, and R. Metelka. 2009. Carbon paste electrodes in electroanalytical chemistry. *J. Serbian Chem. Soc.* 74:1021–1033.
161. Zima, J., I. Švancara, K. Peckova, and J. Barek. 2009. Carbon paste electrodes for the determination of detrimental substances in drinking water. In *Progress on Drinking Water Research*, Chapter 1, Eds. M.H. Lefebvre and M.M. Roux, pp. 1–54. Hauppauge, NY: Nova Science Publishers.
162. Neeb, R. 1969. *Inverse polarographie und voltammetrie (Inverse polarography and voltammetry)*, Orig. Edn. (in German), pp. 26, 55, 58, 68, 104–107. Weinheim, Germany: Verlag Chemie.
163. Meittes, L., P. Zuman, and A. Rott. 1976. *Handbook Series in Inorganic Electrochemistry* (Vol. I + II) & *Handbook Series in Organic Electrochemistry* (Vol. III + IV), Orig. Edn., pp.: throughout the tables. Cleveland, OH: CRC Press.
164. Vydra, F., K. Stulik, and E. Julakova. 1977. *Rozpousteci polarografie a voltametrie (Stripping polarography and voltammetry)*, Orig. Edn. (in Czech), pp. 20, 137, 138, 143, 145, 180, 193, 196, 215, 220–227, 237, 242. Prague, Czech Republic: SNTL.
165. Vydra, F., K. Stulik, and E. Julakova. 1976. *Electrochemical Stripping Analysis*, Eng. Edn., Chichester, U.K.: Ellis Horwood.
166. Dryhust, G. and D.L. McAllister. 1985. Carbon electrodes. In *Laboratory Techniques in Electroanalytical Chemistry*, Chapter 10-III, Eds. P.T. Kissinger and W.R. Heineman, Orig. Edn., pp. 294–301. New York: Marcel Dekker.
167. McCreery, R.L. and K.K. Cline 1996. Carbon electrodes. In *Laboratory Techniques in Electroanalytical Chemistry*, Chapter 10-III, Eds. P.T. Kissinger and W.R. Heineman, Revised and Expanded Edn., pp. 311–312, 816, 817. New York: Marcel Dekker.
168. Wang, J. 1985. *Stripping Analysis: Principles, Instrumentation, and Applications*, Orig. Edn., pp. 62, 64, 73–75, 137, 142, 143. Deerfield Beach, FL: VCH Publishers.
169. Brainina, Kh.Z., E.Ya. Neiman, and V.V. Slepuschkin. 1988. Инверсионные електроаналитические методы (*Inverse Electroanalytical Methods*), Orig. Edn. (in Russian), pp. 48, 49, 55, 56, 59, 78, 79, 127, 132, 135, 138–143, 150, 159, 169–171, 174. Moscow, Russia: ХИМИЯ (Chemistry).
170. Brainina, Kh. and E. Neiman. 1993. *Electroanalytical Stripping Methods*, Eng. Edn., New York: John Willey & Sons.
171. Anonymous. 1995. *Spectral coal—powder, "RW-B" product, the highest purity*. Characterization & guarantee certificate (in German). Bonn, Germany: Ringsdorff–Werke GmbH.
172. Wang, J., T. Martinez, D.R. Yaniv, and L. McCormick. 1990. Characterization of the micro-distribution of conductive and insulating regions of carbon paste electrodes with scanning tunneling microscopy. *J. Electroanal. Chem.* 286:265–272.
173. Švancara, I., L. Baldrianova, E. Tesarova, M. Vlcek, K. Vytřas, and S. Sotiropoulos. 2007. Microscopic studies with bismuth-modified carbon paste electrodes: Morphological transformations of bismuth microstructures and related observations. In *Sensing in Electroanalysis*, Vol. 2, Eds. K. Vytřas and K. Kalcher, pp. 35–58. Pardubice, Czech Republic: University of Pardubice.
174. Cheng, I.F. and C.R. Martin. 1988. Ultramicrodisk electrode ensembles prepared by incorporating carbon paste into a microporous host membrane. *Anal. Chem.* 60:2163–2165.
175. Švancara, I. and K. Schachl. 1999. Testing of unmodified carbon paste electrodes. *Chem. Listy* 93:490–499.
176. Švancara, I., J. Zima, K. Schachl. 1998. The testing of carbon paste electrodes: An example on the characterisation of a carbon paste electrode prepared from newly used graphite powder. *Sci. Pap. Univ. Pardubice Ser. A* 4:49–63.
177. Zakharchuk, N.F. 1991. Personal communication.
178. Ali Qureshi, G. 1978. Carbon paste for calcium selective electrode. *Libyan J. Sci.* 8A:37–41.

179. Johansson, E., G. Marko Varga, and L. Gorton, 1993. Study of a reagent- and mediatorless biosensor for D-amino acids based on co-immobilized D-amino acid oxidase and peroxidase in carbon paste electrodes. *J. Biomater. Appl.* 8:146–173.
180. Narasaiah, D., U. Spohn, and L. Gorton. 1996. Simultaneous determination of L- and D-lactate by enzyme modified carbon paste electrodes. *Anal. Lett.* 29:181–201.
181. Wang, H., H. Cui, A. Zhang, and R. Liu. 1996. Adsorptive stripping voltammetric determination of tryptophan at an electrochemically pre-treated carbon-paste electrode with solid paraffin as a binder. *Anal. Commun.* 33:275–277.
182. Švancara, I., K. Kalcher, K. Vytřas., and W. Diewald. 1996. Voltammetric determination of silver at ultra-trace levels using a carbon paste electrode with improved surface characteristics. *Electroanalysis* 8:336–342.
183. Analytical chemistry & chromatography products, Fluka & Supelco/Sigma-Aldrich-com, http://www.sigma-aldrich.com/saws.nsf/FluProducts?OpenFrameset (accessed May 15, 2004).
184. Anonymous. 1995. *Sigradur®* Pulver: Characterization & Guarantee Certificate (in German). Meitingen, Germany: HTW-Hochtemperatur-Werkstoffe GmbH.
185. Li, G., C.-D. Wan, Z.-M. Ji, and K.-B. Wu. 2007. An electrochemical sensor for Cd2+ based on the inducing adsorption ability of I(−). *Sens. Actuators, B* 124:1–5.
186. Yang, C.-H., J. Zhao, J.-H. Xu, C.-G. Hu, and S.-S. Hu. 2009. A highly sensitive electrochemical method for determination of Sudan-I at polyvinyl-pyrrolidone modified acetylene black-based paste electrode with enhancement effect of sodium dodecyl sulphate. *Int. J. Environ. Anal. Chem.* 89:233–244.
187. Kauffmann, J.-M., M.P. Prete, J.-C. Vire, and G.J. Patriarche. 1985. Voltammetry of pharmaceuticals using different types of modified electrodes. *Fresenius Z. Anal. Chem.* 321:172–176.
188. Švancara, I., J. Jezkova, P. Kotzian, R. Metelka, J. Zima, K. Kalcher, K. Vytras et al. 1985–2010. Unpublished results.
189. Švancara, I. 2002. Carbon paste electrodes in electroanalysis. Hab dissertation (in Czech), University of Pardubice, Pardubice, Czech Republic.
190. Schmid, G.M. and G.W. Bolger. 1973. Determination of gold in drugs and serum by use of anodic stripping voltammetry. *Clin. Chem.* 19:1002–1005.
191. Navratilova, Z. 2009. Coal as a new carbon paste electrode modifier with sorption properties. *Electroanalysis* 21:1758–1762.
192. Hvizdalova, M. 1994. A study on applicability of carbon paste electrodes in voltammetry. MSc dissertation (in Czech), University of Pardubice, Pardubice, Czech Republic.
193. Varma, S. and C.K. Mitra. 2002. Bioelectrochemical studies on catalase modified glassy carbon paste electrodes. *Electrochem. Commun.* 4:151–157.
194. Rodriguez, M.C. and G.A. Rivas. 2002. Glassy carbon paste electrodes modified with polyphenol oxidase. Analytical applications. *Anal. Chim. Acta* 459:43–51.
195. Rubianes, M.D. and G.A. Rivas. 2003. Amperometric quantification of dopamine using different carbon electrodes modified with a melanin-type polymer. *Anal. Lett.* 36:329–345.
196. Ricci, F, C. Goncalves, A. Amine, L. Gorton, G. Palleschi, and D. Moscone. 2003. Electroanalytical study of Prussian blue modified glassy carbon paste electrodes. *Electroanalysis* 15:1204–1211.
197. Dejmkova, H., Z. Hranicka, Z. Navratilova, and J. Barek. 2010. Voltammetric determination of nitrophenols at clay-modified glassy carbon paste electrodes. *Chem. Listy* 104:563–566.
198. Qian, H., J.-N. Ye, and L.-T. Jin. 1997. Study of the electrochemical properties of C-60 modified carbon paste electrode and its application for nitrobenzene quantitation based on electrocatalytic reduction. *Anal. Lett.* 30:367–381.
199. Kureishi, Y., H. Tamiaki, H. Shiraishi, and K. Maruyama. 1999. Photoinduced electron transfer from synthetic chlorophyll analog to fullerene C-60 on carbon paste electrode. Preparation of a novel solar cell. *Bioelectrochem. Bioenerg.* 48:95–100.
200. Lokesh, S.V., B.S. Sherigara, Jayadev, H.M. Mahesh, and R.J. Mascarenhas. 2008. Electrochemical reactivity of 'C-60' modified carbon paste electrode by physical vapor deposition method. *Int. J. Electrochem. Sci.* 3:578–587.
201. Zhu, S.-Y., L.-S. Fan, X.-Q. Liu, L.-H. Shi, H.-J. Li, S. Han, and G.-B. Xu. 2008. Determination of concentrated hydrogen peroxide at the single-walled carbon nanohorn paste electrode. *Electrochem. Commun.* 10:695–698.
202. Pruneanu, S., Z. Ali, G. Watson, S.-Q. Hu, D. Lupu, A.R. Biris, L. Olenic, G. Mihailescu. 2006. Investigation of electrochemical properties of carbon nanofibers prepared by the CCVD method. *Part. Sci. Technol.* 24:311–320.
203. Wu, Y.-H., X.-Y. Mao, X.-J. Cui, and L.-D. Zhu. 2010. Electroanalytical application of graphite nanofibers paste electrode. *Sens. Actuators B* 145:749–755.

204. Wang, L., X.-H., Zhang, H.-Y. Xiong, and S.-F. Wang. 2010. Novel nitromethane biosensor based on biocompatible conductive redox matrix of chitosan/hemoglobin/graphene/(room temperature) ionic liquid. *Biosens. Bioelectron.* 26:991–995.
205. Švancara, I. and K. Vytřas. 1993. Voltammetry with carbon paste electrodes containing membrane plasticizers used for PVC-based ion-selective electrodes. *Anal. Chim. Acta* 273:195–204.
206. Sagar, K., J.H. Fernandez Alvarez, C. Hua, M.R. Smyth, and R. Munden. 1992. Differential pulse voltammetric determination of Sumatriptan succinate (1–1) in a tablet dosage. *J. Phar. Biomed. Anal.* 10:17–21.
207. Hattori, T., M. Kato, S. Tanaka, and M. Hara. 1997. Adsorptive stripping voltammetry of anionic surfactants on a carbon paste electrode using ferrocenyl cationic surfactant as an analytical electrochemical probe. *Electroanalysis* 9:722–725.
208. Jezkova, J. 1999. Potentiometric determinations with a carbon paste electrode. PhD dissertation (in Czech), University of Pardubice, Pardubice, Czech Republic.
209. Švancara, I., R. Pazdera, R. Metelka, E. Norkus, and K. Vytřas. 2001. On the applica-bility of stripping potentiometry at carbon paste electrodes plated with mercury and gold films. In *Monitoring of Environmental Pollutants*, Eds. K. Vytřas, J. Kellner, and J. Fischer, pp. 123–134. Pardubice, Czech Republic: University of Pardubice.
210. Švancara, I., B. Ogorevc, S.B. Hocevar, and K. Vytřas. 2002. Perspectives of carbon paste electrodes in stripping potentiometry. *Anal. Sci. (Jpn.)* 18:301–305.
211. Švancara, I., K. Vytřas, F. Renger, and M.R. Smyth. 1992. Application of carbon paste electrodes in highly methanolic solutions. *Electrochim. Acta* 37:1355–1361.
212. Sigma-Aldrich: Products, services, support, http://www.sigmaaldrich.com/cgibin/hsrun/DistributedEU/HahtShop/HahtShop.htx; start = HS_Frameset Main (accessed May 30, 2004; site now discontinued).
213. Merck, http://www.merck.de/english/services/chemdat/english/index.htm (accessed May 30, 2004; site now discontinued).
214. Boujtita, M., M. Chapleau, and N. El Murr. 1996. Biosensors for analysis of ethanol in food: Effect of the pasting liquid. *Anal. Chim. Acta* 319:91–96.
215. Nyasulu, F.W. 1990. Electrochemistry of brucine. Part I: Brucine. *Electroanalysis* 2:327–331.
216. Wang, J., F. Lu, S.A. Kane, Y.-K. Choi, M.R. Smyth, and K. Rogers. 1997. Hydro-carbon pasting liquids for improved tyrosinase-based carbon paste phenol biosensors. *Electroanalysis* 9:1102–1106.
217. Farsang, Gy. 1965. Voltammetric behavior of Ag at C paste electrode, and some problems of the applicability of this electrode to anodic stripping. *Acta Chim. Acad. Sci. Hung.* 45:257–266.
218. Wang, J. and B.A. Freiha. 1983. Selective voltammetric detection based upon adsorptive preconcentration for flow injection systems. *Anal. Chem.* 55:1285–1288.
219. Wang, J. and B.A. Freiha. 1984. Extractive preconcentration of organic compounds at carbon paste electrodes. *Anal. Chem.* 56:849–852.
220. Wang, J., B.K. Deshmukh, and M. Bonakdar. 1985. Solvent extraction studies with carbon paste electrodes. *J. Electroanal. Chem.* 194:339–353.
221. Deshmukh, B.K. 1987. Role of solute in the behavior of carbon paste electrodes. *Indian J. Chem. A* 26:315–319.
222. Rogers, K.R., J.Y. Becker, J. Cembrano, and S.H. Chough. 2001. Viscosity and binder composition effects on tyrosinase-based carbon paste electrode for detection of phenol and catechol. *Talanta* 54:1059–1065.
223. Kuzmina, N.V., F.K. Kudasheva, V.N. Maistrenko, and S.V. Sapelnikova. 2003. Influence of the nature of the pasting liquid on the accumulation of nitroanilines at carbon paste electrode during determination by absorptive stripping voltammetry. *Anal. Bioanal. Chem.* 375:1182–1185.
224. Pauliukaite, R., G. Zhylyak, D. Citterio, and U.E. Spichiger Keller. 2006. L-Glutamate biosensor for estimation of the taste of tomato specimens. *Anal Bioanal. Chem.* 386:220–227.
225. Chandra, U., B.E.K. Swamy, O. Gilbert, S.S. Shankar, K.R. Mahanthesha, and B.S. Sherigara. 2010. Electrocatalytic oxidation of dopamine at chemically modified carbon paste electrode with 2,4-dinitrophenyl hydrazine. *Int. J. Electrochem. Sci.* 5:1–9.
226. Wang, J. and D.-B. Luo. 1984. Competition studies in voltammetric measurements based on extractive accumulation into carbon paste electrodes. *J. Electroanal. Chem.* 179:251–261.
227. Mikysek, T., A. Ion, I. Švancara, K. Vytřas, and F.G. Banica. 2005. Carbonaceous materials for single-use metal ion sensors. Quality assessment by electrochemical impedance spectrometry. In *Sensing in Electroanalysis*, Eds. K. Vytřas and K. Kalcher, pp. 19–27. Pardubice, Czech Republic: University of Pardubice.
228. Mikysek, T., I. Švancara, M. Bartos, K. Kalcher, K. Vytřas, and J. Ludvik. 2009. New Approaches to the characterization of carbon paste electrodes based on ohmic resistance and qualitative carbon paste indexes. *Anal. Chem.* 81:6327–6333.

229. Mikysek, T., M. Stoces, I. Švancara, and J. Ludvik. 2010. Relation between the composition and properties of carbon nanotubes paste electrodes (CNTPEs). In *Sensing in Electroanalysis*, Vol. 5, Eds. K. Vytřas, K. Kalcher, and I. Švancara, pp. 69–75. Pardubice, Czech Republic: University Press Center.
230. Ulakhovich, N.A., E.P. Medyantseva, and S.V. Mashkina. 1997. Electroactive carbon-paste electrode for determining iridium(III). *J. Anal. Chem. (Russ.)*/transl. of *Zh. Anal. Khim* 52:331–333.
231. Tomcik, P., C.E. Banks, T.J. Davies, and R.G. Compton. 2004. A self-catalytic carbon paste electrode for the detection of vitamin B12. *Anal. Chem.* 76:161–165.
232. H. Ibrahim. 2005. Carbon paste electrode modified with silver thimerosal for the potentiometric flow injection analysis of silver(I) ion. *Anal. Chim. Acta* 545:158–165.
233. Cookeas, E.G. and C.E. Efstathiou. 1992. Preconcentration of organic compounds at a diphenyl ether graphite paste electrode and determination of vanillin by adsorptive-extractive stripping voltammetry. *Analyst (U.K.)* 117:1329–1334.
234. Zhang, Z.-Q., H. Liu, H. Zhang, and Y.-F. Li. 1996. Simultaneous cathodic stripping voltammetric determination of mercury, cobalt, nickel, and palladium by mixed binder carbon paste electrode containing dimethylglyoxime. *Anal. Chim. Acta* 333:119–124.
235. Li, Y.-F., Z.-L. Xiao, X.-Y. Liu, and Z.-Q. Zhang. 1998. Voltammetric characteristics of chlordiazepoxide at the mixed binder carbon paste electrode. *Fenxi Kexue Xuebao* 14:106–109.
236. Radi, A. 1999. Electroanalysis of melatonin using castor oil-graphite paste electrode. *Anal. Commun.* 36:43–44.
237. Fatibello Filho, O., K.O. Lupetti, and L.D. Vieira. 2001. Chronoamperometric determination of paracetamol using an avocado tissue (*Persea americana*) biosensor. *Talanta* 55:685–692.
238. Parellada, J., A. Narvaez, E. Dominguez, and I. Katakis. 1997. A new type of hydrophilic carbon paste electrodes for biosensor manufacturing: Binder paste electrodes. *Biosens. Bioelectron.* 12:267–275.
239. Parellada, J., E. Dominguez, and I. Katakis. 1997. Binder paste: New composite material for biosensors and electrochemical bioreactors. *Polym. Mater. Sci. Engineer.* 76:511–512.
240. Wang, J. 2002. Personal communication.
241. Metrohm—Carbon electrodes & accessories, http://sirius.metrohm.ch/accessories/FMPro (accessed January 10, 2002; site now discontinued).
242. Pap, S., J.-M. Kauffmann, J.-C.Vire, G.J. Patriarche, and M.P. Prete. 1984. The electrochemistry of butyrophenones studied with the aid of modified carbon electrodes (in French). *J. Pharm. Belg.* 39:335–340.
243. Digua, K., J.-M. Kauffmann, and J.L. Delplancke. 1994. Surfactant-modified carbon paste electrode. Part 1: Electrochemical and microscopic characterization. *Electroanalysis* 6:451–458; Digua, K., J.-M. Kauffmann, and M. Khodari. 1994. Surfactant-modified carbon paste electrode. Part 2: Analytical performances. *Electroanalysis* 6:459–462.
244. Kulys, J., L. Wang, H.E. Hansen, T. Buch Rasmussen, J. Wang, and M. Ozsoz. 1995. Methylene green-mediated carbon paste glucose biosensor. *Electroanalysis* 7:92–94.
245. Lutz, M., E. Burestedt, J. Emneus, H. Liden, S. Gobhadi, L. Gorton, and G. Marko Varga. 1995. Effects of different additives on a tyrosinase based carbon paste electrode. *Anal. Chim. Acta* 305:8–17.
246. Spohn, U., D. Narasaiah, L. Gorton, and D. Pfeiffer. 1996. A bienzyme modified carbon paste electrode for the amperometric detection of L-lactate at low potentials. *Anal. Chim. Acta* 319:79–90.
247. Miland, E., A.J. Miranda Ordieres, P. Tunon Blanco, M.R. Smyth, and C.O. Fagain. 1996. Poly(o-aminophenol)-modified bienzyme carbon paste electrode for the detection of uric acid. *Talanta* 43:785–796.
248. Huang, T., A. Warsinke, O.V. Koroljova Skorobogatko, A. Makower, T. Kuwana, and F.W. Scheller. 1999. A bienzyme carbon paste electrode for the sensitive detection of NADPH and the measurement of glucose-6-phosphate dehydrogenase. *Electroanalysis* 11:295–300.
249. Wang, J. and M.-S. Lin. 1989. Dual electrode configurations with a plant tissue as generator reactor. *Anal. Chim. Acta* 218:281–290.
250. Zhang, X. and G.A. Rechnitz. 1994. Simultaneous determination of glucose and sucrose by a dual-working electrode multienzyme sensor flow-injection system. *Electroanalysis* 6:361–367.
251. Sapio, J.P., J.F. Colaruotolo, and J.M. Bobbitt. 1973. Universal ion-selective electrode based on graphite paste. *Anal. Chim. Acta* 67:240–242.
252. Wu, W.-S., M.S. Uddin, and C. Hua. 1995. Cation exchange based chemically modified electrodes. *Bull. Electrochem.* 11:402–406.
253. Vytřas, K., E. Khaled, J. Jezkova, H.N.A. Hassan, and B.N. Barsoum. 2000. Studies on the potentiometric thallium(III)-selective carbon paste electrode and its possible applications. *Fresenius J. Anal. Chem.* 367:203–207.
254. Švancara, I., B. Ogorevc, M. Novic, and K. Vytřas. 2002. Simple and rapid determination of iodide in table salt by stripping potentiometry at a carbon-paste electrode. *Anal. Bioanal. Chem.* 372:795–800.

References

255. Švancara, I., K. Kalcher, and K. Vytřas. 1997. Solid electrodes plated with metallic films. *Sci. Pap. Univ. Pardubice Ser. A* 3:207–225.
256. Kauffmann, J.-M., A. Laudet, and G.J. Patriarche. 1982. The modified carbon paste electrode. On its use by differential pulse anodic stripping voltammetry in the presence of a mercury film (in French). *Anal. Lett.* 15:763–774.
257. Švancara, I., M. Pravda, M. Hvizdalova, K. Vytřas, and K. Kalcher. 1994. Voltammetric investigations on carbon paste electrodes as supports for mercury films. *Electroanalysis* 6:663–671.
258. Guo, S.X. and S.B. Khoo. 1997. Formation of a mercury plated carbon paste electrode by electroreduction of a mercury(II) diethyldithiocarbamate modified carbon paste. *Environ. Monitor. Assess.* 44:471–480.
259. Navratilova, Z. and Vaculikova. 2006. Electrodeposition of mercury film on electrodes modified with clay minerals. *Chem. Pap.* 60:348–352.
260. Švancara, I., M. Matousek, E. Sikora, K. Schachl, K. Kalcher, and K. Vytřas. 1997. Carbon paste electrodes plated with a gold film for the voltammetric determination of mercury(II). *Electroanalysis* 9:827–833.
261. Chadim, P., I. Švancara, B. Pihlar, and K. Vytřas. 2000. Gold-plated carbon paste electrodes for anodic stripping determination of arsenic. *Collect. Chem. Czech. Commun.* 65:1035.
262. Wang, J. 2005. Stripping analysis at bismuth electrodes: A review. *Electroanalysis* 17:1341–1346.
263. Economou, A. 2005. Bismuth-film electrodes: Recent developments and potentialities for electroanalysis. *Trends Anal. Chem.* 24:334–340.
264. Kokkinos, C. and A. Economou. 2008. Stripping analysis at bismuth-based electrodes (a review). *Curr. Anal. Chem.* 4:183–190.
265. Pauliukaite, R., R. Metelka, I. Švancara, A. Krolicka, A. Bobrowski, K. Vytřas, E. Norkus, and K. Kalcher. 2002. Carbon paste electrodes modified with Bi_2O_3 as sensors for the determination of cadmium and lead. *Anal. Bioanal. Chem.* 374:1155–1158.
266. Hocevar, S.B., I. Švancara, B. Ogorevc, and K. Vytřas. 2005. Novel electrode for electrochemical stripping analysis based on carbon paste modified with bismuth powder. *Electrochim. Acta* 51:706–710.
267. Cao, L.-Y., J.-B. Jia, and Z.-H. Wang. 2008. Sensitive determination of Cd and Pb by differential pulse stripping voltammetry with in-situ bismuth-coated zeolite doped carbon paste electrodes. *Electrochim. Acta* 53:2177–2182.
268. Švancara, I., C. Prior, S.B. Hočevar, and J. Wang. 2010. A decade of bismuth-modified electrodes in electroanalysis (a review). *Electroanalysis* 22:1405–1420.
269. Sopha, H., L. Baldrianova, E. Tesarova, G. Grinciene, T. Weidlich, I. Švancara, and S.B. Hocevar. 2010. A new type of bismuth electrode for electrochemical stripping analysis based on the ammonium tetrafluorobismuthate bulk-modified carbon paste. *Electroanalysis* 22:1489–1493.
270. Gardea Torresdey, J., D. Darnall, and J. Wang. 1988. Bioaccumulation and voltammetric behavior of gold at alga-containing carbon paste electrodes. *J. Electroanal. Chem.* 252:197–208.
271. Connor, M., E. Dempsey, M.R. Smyth, D.H. Richardson. 1991. Determination of some metal ions using lichen-modified carbon paste electrodes. *Electroanalysis* 3:331–336.
272. Ramos, J.A., E. Bermejo, A. Zapardiel, J.A. Perez, and L. Hernandez. 1993. Direct determination of lead by bioaccumulation at a moss-modified carbon paste electrode. *Anal. Chim. Acta* 273:219–227.
273. Navaratne, A. and G.A. Rechnitz. 1992. Improved plant tissue-based biosensor using in-vitro cultured tobacco callus tissue. *Anal. Chim. Acta* 257:59–66.
274. Wijesuriya, D. and GA. Rechnitz. 1991. Mixed carbon paste-pea seedling electrochemical sensor for measuring plant growth-regulating activity of amines. *Anal. Chim. Acta* 243:1–8.
275. Ouangpipat, W., T. Lelasattarathkul, C. Dongduen, and S. Liawruangrath. 2003. Bioaccumulation and determination of lead using treated-Pennisetum-modified carbon paste electrode. *Talanta* 61:455–464.
276. Wang, J. and M.-S. Lin. 1988. Mixed plant tissue carbon paste bioelectrode. *Anal. Chem.* 60:1545–1548.
277. Horie, H. and G.A. Rechnitz. 1995. Hybrid tissue/enzyme biosensor for pectin. *Anal. Chim. Acta* 306:123–127.
278. Cummings, E.A., P. Mailley, S. Linquette Mailley, B.R. Eggins, E.T. McAdams, and S. McFadden. 1998. Amperometric carbon paste biosensor based on plant tissue for the determination of total flavanol content in beers. *Analyst (U.K.)* 123:1975–1980.
279. Forzani, E.S., G.A. Rivas, and V.M. Solis. 1997. Amperometric determination of dopamine on vegetal-tissue enzymic electrodes. Analysis of interferents and enzymic selectivity. *J. Electroanal. Chem.* 435:77–84.
280. Bezerra, V.S., J.L. de Lima Filho, M.C. Montenegro, A.N. Araujo, and V. Lins da Silva. 2003. Flow-injection amperometric determination of dopamine in pharmaceuticals using a polyphenol oxidase biosensor obtained from soursop pulp. *J. Pharm. Biomed. Anal.* 33:1025–1031.

281. Mojica Elmer, R.E., S.P. Gomez, J.R.L. Micor, and C.C. Deocaris. 2006. Lead detection using a pineapple bioelectrode. *Philipp. Agricult. Sci.* 89:134–140.
282. Kozan, J.V.B., R.P. Silva, S.H.P. Serrano, A.W.O. Lima, and L. Angnes. 2010. Amperometric detection of benzoyl peroxide in pharmaceutical preparations using carbon paste electrodes with peroxidases naturally immobilized on coconut fibers. *Biosens. Bioelectron.* 25:1143–1148.
283. Skladal, P. 1991. Mushroom tyrosinase-modified carbon paste electrode as an amperometric biosensor for phenols. *Collect. Czech. Chem. Commun.* 56:1427–1433.
284. Moressi, M.B., A. Zon, H. Fernández, G.A. Rivas, and V. Solis. 1999. Amperometric quantification of Alternaria mycotoxins with a mushroom tyrosinase modified carbon paste electrode. *Electrochem. Commun.* 1:472–476.
285. Sugawara, K., T. Takano, H. Fukushi, S. Hoshi, K. Akatsuka, H. Kuramitz, and S. Tanaka. 2000. Glucose sensing by a carbon-paste electrode containing chitin modified with glucose oxidase. *J. Electroanal. Chem.* 482:81–86.
286. Lei, C.-X., F.-C. Gong, G.-L. Shen, and R.-Q. Yu. 2003. Amperometric immunosensor for *Schistosoma japonicum* sp. antigen using antibodies loaded on a nano-gold monolayer modified chitosan-entrapped carbon paste electrode. *Sensors Actuators B* 96:582–588.
287. Sugawara, K., H. Matsui, S. Hoshi, and K. Akatsuka. 1998. Voltammetric detection of Ag(I) at carbon paste electrode modified with keratin. *Analyst (U.K.)* 123:2013–2016.
288. Ito, Y., S. Yamazaki, K. Kano, and T. Ikeda. 2002. *Escherichia coli* and its application in a mediated amperometric glucose sensor. *Biosens. Bioelectron.* 17:993–998.
289. Erdem, A., K. Kerman, B. Meric, U.S. Akarca, and M. Ozsoz. 2000. Novel hybridization indicator methylene blue for the electrochemical detection of short DNA sequences related to the hepatitis-B virus. *Anal. Chim. Acta* 422:139–149.
290. Kim, E.J., T. Haruyama, Y. Yanagida, E. Kobatake, and M. Aizawa. 1999. Disposable creatinine sensor based on thick-film hydrogen peroxide electrode system. *Anal. Chim. Acta* 394:225–231.
291. Wang, J., J. Liu, and G. Cepra. 1997. Thermal stabilization of enzymes immobilized within carbon paste electrodes. *Anal. Chem.* 69:3124–3127.
292. Korbut, O., M. Buckova, P. Tarapcik, J. Labuda, and P. Grundler. 2001. Damage to DNA indicated by an electrically heated DNA-modified carbon paste electrode. *J. Electroanal. Chem.* 506:143–148.
293. Korbut, O., M. Buckova, J. Labuda, and P. Gruendler. 2003. Voltammetric detection of antioxidative properties of flavonoids using electrically heated DNA modified carbon paste electrode. *Sensors* 3:1–10.
294. Flechsig, G.-U., O. Korbout, S.B. Hocevar, S. Thongngamdee, B. Ogorevc, P. Gruendler, and J. Wang. 2002. Electrically heated bismuth-film electrode for voltammetric stripping measurements of trace metals. *Electroanalysis* 14:192–196.
295. Švancara, I., P. Kotzian, R. Metelka, M. Bartos, P. Foret, and K. Vytřas. 2002. Plastic bars with carbon paste: A new type of the working electrode in electroanalysis. In *Monitoring of Environmental Pollutants*, Vol. IV (in Czech), Eds. K. Vytřas, J. Kellner, and J. Fischer, pp. 145–158. Pardubice, Czech Republic: University of Pardubice.
296. Švancara, I., S. Hradilova, L. Nepejchalova, and M. Bartos. 2009. Temperature-controlled processes at carbon paste-based electrodes: Possibilities and limitations in electroanalytical measurements. In *Sensing in Electroanalysis*, Vol. 4, Eds. K. Vytřas, K. Kalcher, and I. Švancara, pp. 7–26. Pardubice, Czech Republic: University Press Center.
297. Lopes da Silva, W.T., C. Thobie Gautier, M.O.O. Rezende, and N. El Murr. 2002. Electrochemical behavior of Cu(II) on carbon paste electrode modified by humic acid. A cyclic voltammetry study. *Electroanalysis* 14:71–77.
298. Wang, J., A.N. Kawde, A. Erdem, and M. Salazar. 2001. Magnetic bead-based label-free electrochemical detection of DNA hybridization. *Analyst (U.K.)* 126:2020–2024.
299. Wang, J. and A.N. Kawde. 2002. Magnetic-field stimulated DNA oxidation. *Electrochem. Commun.* 4:349–352.
300. Dunwoody, D.C., M. Unlu, A.K.H. Wolf, W.L. Gellett, and J. Leddy. 2005. Magnet incorporated carbon electrodes: Methods for construction and demonstration of increased electrochemical flux. *Electroanalysis* 17:1487–1494.
301. Tang, D.-P., R. Yuan, and Chai Y.-Q. 2006. Direct electrochemical immunoassay based on protein-magnetic nanoparticle composites on to magnetic electrode surfaces by sterically enhanced magnetic field force. *Biotechnik. Lett.* 28:559–565.
302. Liu, Z.-M., H.-F. Yang, Y.-F. Li, Y.-L. Liu, G.-L. Shen, and R.-Q. Yu. 2006. Core-shell magnetic nanoparticles applied for immobilization of antibody on carbon paste electrode and amperometric immunosensing. *Sens. Actuators B* 113:956–962.

References

303. Tang, D.-P., R. Yuan, and Y.-Q. Chai. 2006. Magnetic core-shell Fe$_3$O$_4$@Ag nano-particles coated carbon paste interface for studies of carcinoembryonic antigen in clinical immunoassay. *J. Phys. Chem. B* 110:11640–11646.
304. Gaillochet, P., D. Bauer, and M.-C. Hennion. 1974. Rapid determination of uranium in ores by means of a carbon paste electrode (in French). *Analusis (Bruxelles)* 3:513–516.
305. Barikov, V.G., Z.B. Rozhdestvenskaya, and O.A. Songina. 1969. Phase electrochemical analysis with a carbon paste electrically-active electrode (in Russian). *Zavod. Lab. (U.S.S.R.)* 35:776–778.
306. Brainina, K.Z. and M.B. Vydrevich. 1981. Stripping analysis of solids. *J. Electroanal. Chem.* 121:1–28.
307. Grygar, T., F. Marken, U. Schroeder, and F. Scholz. 2002. Electrochemical analysis of solids. A review. *Collect. Czech. Chem. Commun.* 67:163–208.
308. Nagy, G., Zs. Feher, and E. Pungor. 1970. Application of silicone rubber-based graphite electrodes for continuous flow measurements: Part II. Voltammetric study of active substances injected into electrolyte streams. *Anal. Chim. Acta* 52:47–54.
309. Siska, E. and E. Pungor. 1971. Indirect voltammetric determination of potassium and caesium and direct amperometric determination of potassium with tetraphenylborate using silicone rubber based graphite electrodes. *Fresenius Z. Anal. Chem.* 257:7–11.
310. Stulik, K., V. Pacakova, and B. Starkova. 1981. Carbon pastes for voltammetric detectors in high-performance liquid chromatography. *J. Chromatogr. A* 213:41–46.
311. Lee, B.-G., K.-B. Rhyu, and K.-J. Yoon. 2010. Amperometric study of hydrogen peroxide biosensor with butadiene rubber as immobilization matrix. *J. Industr. Engineer. Chem.* 16:340–343.
312. Rajendran, V., E. Csoregi, Y. Okamoto, and L. Gorton. 1998. Amperometric peroxide sensor based on horseradish peroxidase and toluidine blue O-acrylamide polymer in carbon paste. *Anal. Chim. Acta* 373:241–251.
313. Sandulescu, R. 1999. Personal communication.
314. Sandulescu, R.V., S.M. Mirel, R.N. Oprean, and S. Lotrean. 2000. Comparative electrochemical study of some Phenothiazines with carbon paste, solid carbon paste, and glass-like carbon electrodes. *Collect. Czech. Chem. Commun.* 65:1014–1028.
315. Blankert, B., O. Dominguez, W. El Ayyas, J. Arcos, and J.-M. Kauffmann. 2004. Horseradish peroxidase electrode for the analysis of clozapine. *Anal. Lett.* 37:903–913.
316. Mascarenhas, R.J., A.K. Satpati, S. Yellappa, B.S. Sherigara, and A.K. Bopiah. 2006. Wax-impregnated carbon paste electrode modified with mercuric oxalate for the simultaneous determination of heavy metal ions in medicinal plants and ayurvedic tablets. *Anal. Sci. (Jpn.)* 22:871–875.
317. Niranjana, E., R.R. Naik, B.E.K. Swamy, B.S. Sherigara, and H. Jayadevappa. 2007. Studies on adsorption of Triton X-100 at carbon paste and ceresin wax carbon paste electrodes and enhancement effect in dopamine oxidation by cyclic voltammetry. *Int. J. Electrochem. Sci.* 2:923–934.
318. Wangsa, J. and N.D. Danielson. 1990. Electrochemical detection for high-performance liquid chromatography using a Kel-F wax-graphite electrode. *J. Chromatogr. A* 514:171–178.
319. Wangsa, J. and N.D. Danielson. 1991. Enzymatic determination of ethanol by flow injection analysis using a Kel-F wax carbon paste electrode. *Electroanalysis* 3:625–630.
320. Sikora, E. and K. Vytřas. 1997. Voltammetric determinations at polymeric colloidal gold paste electrodes. *Sci. Pap. Univ. Pardubice, Ser. A* 3:333–338.
321. Compagnone, D., P. Schweicher, J.-M. Kauffman, and G.G. Guilbault. 1998. Sub-micromolar detection of hydrogen peroxide at a peroxidase/tetramethylbenzidine solid carbon paste electrode. *Anal. Lett.* 31:1107–1120.
322. Cui, G., J.-H. Yoo, J.-S. Lee, J. Yoo, J.-H. Uhm, G.-S. Cha, and H. Nam. 2001. Effect of pretreatment on the surface and electrochemical properties of screen-printed carbon paste electrodes. *Analyst (U.K.)* 126:1399–1403.
323. Morrin, A., A.J. Killard, and M.R. Smyth. 2003. Electrochemical characterization of commercial and home-made screen-printed carbon electrodes. *Anal. Lett.* 36:2021–2039.
324. Huang, X. and W.T. Kok. 1993. Conductive carbon cement as electrode matrix for cobalt phthalocyanine-modified electrodes for detection in flowing solutions. *Anal. Chim. Acta* 273:245–253.
325. Niwa, O. 2005. Electroanalytical chemistry with carbon film electrodes and micro- and nano-structured carbon film-based electrodes (a review). *Bull. Chem. Soc. Jan.* 78:555–571.
326. Schachl, K., H. Alemu, K. Kalcher, J. Jezkova, I. Švancara, and K. Vytřas. 1997. Determination of hydrogen peroxide with sensors based on heterogeneous carbon materials modified with MnO$_2$. *Sci. Pap. Univ. Pardubice Ser. A* 3:41–55.

327. Fanjul Bolado, P., D. Hernandez Santos, P.J. Lamas Ardisana, A. Martin Pernia, and A. Costa Garcia. 2008. Electrochemical characterization of screen-printed and conventional carbon paste electrodes. *Electrochim. Acta* 53:3635–3642.
328. Xu, L.-J., N.-Y. He, J.-J. Du, Y. Deng, S. Li, and H.-N. Liu. 2008. Fabrication of porous pseudo-carbon paste electrode as a novel high-sensitive electrochemical biosensor. *Anal. Lett.* 41:2402–2411.
329. Xu, L.-J., N.-Y. He, J.-J. Du, Y. Deng, Z.-Y. Li, and T. Wang. 2009. Detailed investigation on the determination of tannic acid by using anodic stripping voltammetry and a porous electrochemical sensor. *Anal. Chim. Acta* 634:49–53.
330. Xu, L.-J., J.-J. Du, Y. Deng, and N.-Y. He. 2010. Fabrication of magnetic porous pseudo-carbon paste electrode electrochemical biosensor and its application in detection of Schistosoma egg antigen. *Electrochem. Commun.* 12:1329–1332.
331. Xu, L.-J., J.-J. Du, Y. Deng, Z.-Y. Li, C.-X. Xu, and N.-Y. He. 2011. Fabrication and characterization of nanoporous pseudo-carbon paste electrode. *Adv. Sci. Lett.* 4:104–107.
332. Wang, Z.-H. and H. Zhang. 2005. An in-laid super-thin carbon paste film modified electrode for simultaneous determination of xanthine and hypoxanthine. *Fenxi Huaxue* 33:671–674.
333. Yan, Q.-P., F.-Q. Zhao, G.-Z. Li, and B.-Z. Zeng. 2006. Voltammetric determination of uric acid with a glassy carbon electrode coated by paste of multiwalled carbon nanotubes and ionic liquid. *Electroanalysis* 18:1075–1080.
334. Fan, S.-H., F. Xiao, L.-Q. Liu, F.-Q. Zhao, and B.-Z. Zeng. 2008. Sensitive voltammetric response of methylparathion on single-walled carbon nanotube paste coated electrodes using ionic liquid as binder. *Sens. Actuators B* 132:34–39.
335. Song, J.-C., J. Yang, J.-F. Zeng, J. Tan, and L. Zhang. 2011. Graphite oxide film-modified electrode as an electrochemical sensor for acetaminophen. *Sens. Actuators B* 155:220–225.
336. Chen, L.-Y., Y.-H. Tang, K. Wang, C.-B. Liu, and S.-L. Luo. 2011. Direct electro-deposition of reduced graphene oxide on glassy carbon electrode and its electrochemical application. *Electrochem. Commun.* 13:133–137.
337. Compton, R.G., J.S. Foord, and F. Marken. 2003. Electroanalysis at diamond-like and doped-diamond electrodes (a review). *Electroanalysis* 15:1349–1363.
338. Kraft, A. 2007. Doped diamond: A compact review on a new, versatile electrode material. *Int. J. Electrochem. Sci.* 2:355–385.
339. Luong, J.H.T., K.B. Male, and J.D. Glennon. 2009. Boron-doped diamond electrode: Synthesis, characterization, functionalization, and analytical applications (a review). *Analyst (U.K.)* 134:1965–1979.
340. Mueller, T.R. and R.N. Adams. 1960. Voltammetry at inert electrodes. Part 1: Analytical applications of boron carbide electrodes. *Anal. Chim. Acta* 23:467–479.
341. Mueller T.R. and R.N. Adams. 1961. Voltammetry at inert electrodes. Part 2: Correlation of experimental results with theory for voltage and controlled potential scanning, controlled potential electrolysis, and chronopotentiometric techniques—Oxidation of ferrocyanide and o-dianisidine at boron carbide electrodes. *Anal. Chim. Acta* 25:482–497.
342. Stefan, R.I., S.G. Bairu, and J.F. van Staden. 2003. Diamond paste electrodes for the determination of iodide in vitamins and table salt. *Anal. Lett.* 36:1493–1500.
343. Stefan, R.I. and R.M. Nejem. 2003. Determination of L- and D-pipecolic acid using diamond paste based amperometric biosensors. *Anal. Lett.* 36:2635–2644.
344. Stefan, R.I., S.G. Bairu, J.F. van Staden. 2003. Diamond paste-based electrodes for the determination of Cr(III) in pharmaceuticals. *Anal. Bioanal. Chem.* 376:844–847.
345. Stefan, R.I. and R.G. Bokretsion. 2003. Determination of creatine and creatinine using a diamond paste based electrode. *Instrum. Sci. Technol.* 31:183–188.
346. Stefan, R.I. and S.G. Bairu. 2003. Diamond paste-based electrodes for the determination of Cr(VI) at trace levels. *Instrum. Sci. Technol.* 31:261–267.
347. Stefan, R.I., S.G. Bairu, and J.F. van Staden. 2003. Determination of Fe(III) in water samples using diamond paste based electrodes. *Instrument. Sci. Technol.* 31:411–416.
348. Stefan, R.I. and R.G. Bokretsion. 2003. Diamond paste-based immunosensor for rapid determination of azidothymidine. *J. Immunoassay Immunochem.* 24:319–324.
349. Stefan, R.I. and S.G. Bairu. 2004. Diamond paste based electrodes for determination of Pb(II) at trace concentration level. *Talanta* 63:605–608.
350. Stefan, R.I., R.M. Nejem, J.F. van Staden, and H.Y. Aboul Enein. 2004. New amperometric biosensors based on diamond paste for the assay of L- and D-pipecolic acids in serum samples. *Prep. Biochem. Biotechnol.* 34:135–143.

References

351. Stefan, R.I. and R.M. Nejem. 2004. Determination of L- and D-pipecolic acids using a diamond paste based electrode. *Instrum. Sci. Technol.* 32:311–320.
352. Stefan-van Staden, R.I. and R.G. Bokretsion. 2006. Simultaneous determination of creatine and creatinine using monocrystalline diamond paste based amperometric biosensors. *Anal. Lett.* 39:2227–2233.
353. Stefan-van Staden, R.I., S.G. Bairu, and J.F. van Staden. 2010. Diamond paste based electrodes for the determination of Ag(I) ion. *Anal. Methods* 2:650–652.
354. Stefan-van Staden, R.I., J.F. van Staden, and H.Y. Aboul Enein. 2010. Diamond paste-based electrodes for the determination of Sildenafil citrate (Viagra). *J. Solid State Electrochem.* 14:997–1000.
355. Van Staden, J.F. and R.I. Stefan-van Staden. 2010. Personal communication.
356. Gong, K.-P., Y.-M. Yang, M.-N. Zhang, L. Su, S.-X. Xiong, and L.-Q. Mao. 2005. Electrochemistry and electroanalytical applications of carbon nanotubes: A review. *Anal. Sci. (Jpn.)* 21:1383–1393.
357. Dumitrescu, I., R. Unwin Patrick, and J.V. MacPherson. 2009. Electrochemistry at carbon nanotubes: Perspective and issues. *Chem. Commun.* 45:6886–6901.
358. Pumera, M. 2009. The electrochemistry of carbon nanotubes: Fundamentals and applications. *Chem. Eur. J.* 15:4970–4978.
359. Britto, P.J., K.S.V. Santhanam, and P.M. Ajayan. 1996. Carbon nanotube electrode for oxidation of dopamine. *Bioelectrochem. Bioenerg.* 41:121–125.
360. Ricci, F., A. Amine, D. Moscone, and G. Palleschi. 2003. Prussian blue modified carbon nanotube paste electrodes: A comparative study and a biochemical application. *Anal. Lett.* 36:1921–1938.
361. Pedano, M.L. and Rivas, G.A. 2004. Adsorption and electrooxidation of nucleic acids at carbon nanotubes paste electrodes. *Electrochem. Commun.* 6:10–16.
362. Antiochia, R., I. Lavagnini, F. Magno, F. Valentini, and G. Palleschi. 2004. Single-wall carbon nanotube paste electrodes: A comparison with carbon paste, platinum and glassy carbon electrodes via cyclic voltammetric data. *Electroanalysis* 16:1451–1458.
363. Lawrence, N.S., R.P. Deo, and J. Wang. 2004. Detection of homocysteine at carbon nanotube paste electrodes. *Talanta* 63:443–449.
364. Rubianes, M.D. and G.A. Rivas. 2005. Enzymatic biosensors based on carbon nanotubes paste electrodes. *Electroanalysis* 17:73–78.
365. Chicharro, M., E. Bermejo, M. Moreno, A. Sanchez, A. Zapardiel, and G. Rivas. 2005. Adsorptive stripping voltammetric determination of amitrole at a multi-wall carbon nanotubes paste electrode. *Electroanalysis* 17:476–482.
366. He, J.-B., X.-Q. Lin, and J. Pan. 2005. Multi-wall carbon nanotube paste electrode for adsorptive stripping determination of quercetin: A comparison with a CPE via voltammetry and chronopotentiometry. *Electroanalysis* 17:1681–1686.
367. Chicharro, M., A. Sanchez, E. Bermejo, A. Zapardiel, M.D. Rubianes, and G.A. Rivas. 2005. Carbon nanotubes paste electrodes as new detectors for capillary electrophoresis. *Anal. Chim. Acta* 543:84–91.
368. Antiochia, R., I. Lavagnini, and F. Magno. 2005. Electrocatalytic oxidation of NADH at single-wall carbon-nanotube-paste electrodes: Kinetic considerations for use of redox mediator in solution and dissolved in paste. *Anal. Bioanal. Chem.* 381:1355–1361.
369. Ly, S.Y., S.K. Kim, T.H. Kim, Y.S. Jung, and S.M. Lee. 2005. Measuring mercury ion concentration with a carbon nanotube paste electrode using the cyclic voltammetry method. *J. Appl. Electrochem.* 35:567–571.
370. Arribas, A.S., E. Bermejo, M. Chicharro, A. Zapardiel, G.L. Luque, N.F. Ferreyra, and G.A. Rivas. 2006. Analytical applications of a carbon nanotubes composite modified with copper micro-particles as detector in flow systems. *Anal. Chim. Acta* 577:183–189.
371. Kurusu, F.S. Koide, I. Karube, and M. Gotoh. 2006. Electrocatalytic activity of bamboo-structured carbon nanotube paste electrode toward hydrogen peroxide. *Anal. Lett.* 39:903–911.
372. Antiochia, R. and I. Lavagnini. 2006. Alcohol biosensor based on immobilization of Meldola blue and alcohol dehydrogenase into a carbon nanotube paste electrode. *Anal. Lett.* 39:1643–1655.
373. Tian, X.-J., J.-F. Song, X.-J. Luan, Y.-Y. Wang, and Q.-Z. Shi. 2006. Determination of *Metformin* based on amplification of its voltammetric response by a combination of molecular wire and carbon nanotubes. *Anal. Bioanal. Chem.* 386:2081–2086.
374. Lin, X.-Q., J.-B. He, and Z.-G. Zha. 2006. Simultaneous determination of quercetin and rutin at a multi-wall carbon-nanotube paste electrodes by reversing differential pulse voltammetry. *Sens. Actuators B* 119:608–614.
375. Luque, G.L., N.F. Ferreyra, and G.A. Rivas. 2006. Glucose biosensor based on the use of carbon nanotube paste electrode modified with metallic particles. *Microchim. Acta* 152:277–283.

376. Ly, S.-Y. 2006. Detection of dopamine in the pharmacy with a carbon nanotube paste electrode using voltammetry. *Bioelectrochemistry* 68:227–231.
377. Ly, S.-Y. 2006. Real-time voltammetric assay of cadmium ions in plant tissue and fish brain core. *Bull. Korean Chem. Soc.* 27:1613–1617.
378. Nie, L.-B., J.-R. Chen, Y.-Q. Miao, and N.-Y. He. 2006. Gold nanoparticle-based layer-by-layer enhancement of DNA hybridization electrochemical signal at carbon nanotube modified carbon paste electrode. *Chin. Chem. Lett.* 17:795–798.
379. Zheng, L. and J. Song. 2007. Voltammetric behavior of Urapidil and its determination at multi-wall carbon nanotube paste electrode. *Talanta* 73:943–947.
380. Qu, J., X. Zou, B. Liu, and S. Dong. 2007. Assembly of polyoxometalates on carbon nanotubes paste electrode and its catalytic behaviors. *Anal. Chim. Acta* 599:51–57.
381. Timur, S., U. Anik, D. Odaci, and L. Gorton. 2007. Development of a microbial biosensor based on carbon nanotube modified electrodes. *Electrochem. Commun.* 9:1810–1815.
382. Ly, S.-Y., Y.-S. Jung, S.-K. Kim, and H.-K. Lee. 2007. Trace analysis of lead and copper ions in fish tissue using paste electrodes. *Anal. Lett.* 40:2683–2692.
383. Shahrokhian, S. and M. Amiri. 2007. Multi-walled carbon nanotube paste electrode for selective voltammetric detection of isoniazid. *Microchim. Acta* 157:149–158.
384. Shahrokhian, S. and L. Fotouhi. 2007. Carbon paste electrode incorporating multi-walled carbon nanotube/cobalt salophen for sensitive voltammetric determination of tryptophan. *Sens. Actuators B* 123:942–949.
385. Antiochia, R. and L. Gorton. 2007. Development of a carbon nanotube paste electrode osmium polymer-mediated biosensor for determination of glucose in alcoholic beverages. *Biosens. Bioelectron.* 22:2611–2617.
386. Chen, J.-H., Z.-Y. Lin, and G.-N. Chen. 2007. An electrochemiluminescent sensor for glucose employing a modified carbon nanotube paste electrode. *Anal. Bioanal. Chem.* 388:399–407.
387. Abbaspour, A. and R. Mirzajani. 2007. Electrochemical monitoring of piroxicam in different pharmaceutical forms with multi-walled carbon nanotubes paste electrode. *J. Pharm. Biomed. Anal.* 44:41–48.
388. Nie, L.-B., H.-S. Guo, Q.-G. He, J.-R. Chen, and Y.-Q. Miao. 2007. Enhanced electrochemical detection of DNA hybridization with carbon nanotube modified paste electrode. *J. Nanosci. Nanotechnol.* 7:560–564.
389. Chen, Y., Z. Lin, J. Chen, J. Sun, L. Zhang, and G. Chen. 2007. New capillary electrophoresis-electrochemiluminescence detection system equipped with an electrically heated Ru(bpy)$_3^{2+}$/multi-wall-carbon-nanotube paste electrode. *J. Chromatogr. A* 1172:84–91.
390. Ly, S.-Y., C.-H. Lee, and Y.-S. Jung. 2007. Measuring oxytetracycline using a simple prepared DNA immobilized on a carbon nanotube paste electrode in fish tissue. *J. Korean Chem. Soc.* 51:412–417.
391. Kachoosangi, R.T., L. Xiao, G.G. Wildgoose, F. Marken, P.C.B. Page, and R.G. Compton. 2007. A new method of studying ion transfer at liquid|liquid phase boundaries using a carbon nanotube paste electrode with a redox active binder. *J. Phys. Chem. C* 111:18353–18360.
392. Zheng, L. and J.F. Song. 2007. Adsorptive voltammetric determination of cisapride on carbon nanotubes paste electrode. *Fenxi Huaxue* 35:1018–1020.
393. Deng, P.-H., J.-J. Fei, J. Zhang, and J.-N. Li. 2008. Determination of trace copper by adsorptive voltammetry using a multiwalled carbon nanotube modified carbon paste electrode. *Electroanalysis* 20:1215–1219.
394. Zhang, X.-Z., K. Jiao, and X. Wang. 2008. Paste electrode based on short single-walled carbon nanotubes and room temperature ionic liquid: Preparation, characterization and application in DNA detection. *Electroanalysis* 20:1361–1366.
395. Li, X.-L. and J.-S. Ye. 2008. Comparison of the electrochemical reactivity of carbon nanotubes paste electrodes with different types of multiwalled carbon nanotubes. *Electroanalysis* 20:1917–1924.
396. Ly, S.-Y. 2008. Diagnosis of Cu(II) ions in vascular tracts by a fluorine-doped carbon nanotube sensor. *Talanta* 74:1635–1641.
397. Deng, P.H., J. Zhang, S.F. Liao, and J.N. Li. 2008. Adsorption voltammetry of the zirconium-alizarin red S complex at a multi-walled carbon nanotubes modified carbon paste electrode. *Microchim. Acta* 161:123–128.
398. Zhuang, Q., J.-H. Chen, J. Chen, and X.-H. Lin. 2008. Electrocatalytical properties of bergenin on a multi-wall carbon nanotubes modified carbon paste electrode and its determination in tablets. *Sens. Actuators B* 128:500–506.
399. Wang, Z.-H., X.-Y. Dong, and J. Li. 2008. An inlaying ultra-thin carbon paste electrode modified with functional single-wall carbon nanotubes for simultaneous determination of three purine derivatives. *Sens. Actuators B* 131:411–416.

References

400. Odaci, D., A. Telefoncu, and S. Timur. 2008. Pyranose oxidase biosensor based on carbon nanotube-modified carbon paste electrode. *Sens. Actuators B* 132:159–165.
401. Dong, S.-Q., S. Zhang, L.-Z. Chi, P.-A. He, Q.-J. Wang, and Y.-Z. Fang. 2008. Electrochemical behaviors of amino acids at multiwall carbon nanotubes and Cu_2O modified carbon paste electrode. *Anal. Biochem.* 381:199–204.
402. Ly, S.-Y. 2008. Voltammetric analysis of DL-alpha-tocopherol with a paste electrode. *J. Sci. Food Agricult.* 88:1272–1276.
403. Shahrokhian, S., Z. Kamalzadeh, A. Bezzaatpour, and D.M. Boghaei. 2008. Differential pulse voltammetric determination of N-acetylcysteine by the electro-catalytic oxidation at the surface of carbon nanotube-paste electrode modified with cobalt salophen complexes. *Sens. Actuators B* 133(2):599–606.
404. Kim, T.-H., S.-J. Kwon, H.-H. Yoon, C.-W. Chung, and J.-S. Kim. 2008. Fabrication of carbon nanotube paste using photosensitive polymer for field emission display. *Colloids Surf. A* 313:448–451.
405. Xie, J.-N., S.-Y. Wang, L. Aryasomayajula, and V.K. Varadan. 2008. Effect of nano-materials in platinum-decorated carbon nanotube paste-based electrodes for amperometric glucose detection. *J. Mater. Res.* 23:1457–1465.
406. Zhang, H., G.-P. Cao, Y.-S. Yang, and Z.-N. Gu. 2008. The capacitive performance of an ultralong aligned carbon nanotube electrode in an ionic liquid at 60°C. *Carbon* 46:30–34.
407. Deng, P.-H., J.-J. Fei, J. Zhang, and J.-N. Li. 2008. Trace determination of zirconium using anodic adsorptive voltammetry at a carbon paste electrode modified with multi-waited carbon nanotubes. *Chin. Sci. Bull.* 53:1665–1670.
408. Deng, P.-H., J. Zhang, and J.-N. Li. 2008. Anodic adsorption voltammetry of zirconium-alizarin red S complex at a multi-walled carbon nanotube modified carbon paste electrode. *Fenxi Huaxue* 36:691–694.
409. Ly, S.-Y., Y.-S. Jung, C.-H. Lee, and B.-W. Lee. 2008. Administering pesticide assays in *in-vivo* implanted biosensors. *Aust. J. Chem.* 61:826–832.
410. Ganjali, M.R., H. Khoshsafar, F. Faridbod, A. Shirzadmehr, M. Javanbakht, and P. Norouzi. 2009. Room temperature ionic liquids and multiwalled carbon nanotubes as modifiers for improvement of carbon paste ion selective electrode response. A comparison study with PVC membrane. *Electroanalysis* 21:2175–2178.
411. Zheng, L. and J.-F. Song. 2009. Ni(II)-*baicalein* complex modified multi-wall carbon nanotube paste electrode toward electrocatalytic oxidation of hydrazine. *Talanta* 79:319–326.
412. Sun, W., Y. Li, Y. Duan, and K. Jiao. 2009. Direct electrochemistry of guanosine on multi-walled carbon nanotubes modified carbon ionic liquid electrode. *Electrochim. Acta* 54:4105–4110.
413. Sun, W., X. Li, Y. Wang, R. Zhao, and K. Jiao. 2009. Electrochemistry and electrocatalysis of hemoglobin on multi-walled carbon nanotubes modified carbon ionic liquid electrode with hydrophilic EMIMBF4 as the modifier. *Electrochim. Acta* 54:4141–4148.
414. Zheng, L., J.-Q. Zhang, and J.-F. Song. 2009. Ni(II)-quercetin complex modified multiwall carbon nanotube ionic liquid paste electrode and its electrocatalytic activity toward the oxidation of glucose. *Electrochim. Acta* 54:4559–4565.
415. Faridbod, F., M.R. Ganjali, B. Larijani, and P. Nurouzi. 2009. Multi-walled carbon nanotubes (MWCNTs) and room temperature ionic liquids (RTILs) carbon paste Er(III) sensor based on a new derivative of dansyl chloride. *Electrochim. Acta* 55:234–239.
416. Tashkhourian, J., M.R. Hormozi Nezhad, J. Khodavesi, and S. Javadi. 2009. Silver nanoparticles modified carbon nanotube paste electrode for simultaneous determination of dopamine and ascorbic acid. *J. Electroanal. Chem.* 633:85–91.
417. Sun, W., Z. Zhai, X. Li, L. Qu, T. Zhan, and K. Jiao. 2009. Direct electrochemistry of hemoglobin in chitosan/multiwalled carbon nanotubes/ionic liquid-modified carbon paste electrode. *Anal. Lett.* 42:2460–2473.
418. Ganjali, M.R., H. Khoshsafar, A. Shirzadmehr, M. Javanbakht, and F. Faridbod. 2009. Improvement of carbon paste ion-selective electrode response by using room temperature ionic liquids and multi-walled carbon nanotubes. *Int. J. Electrochem. Sci.* 4:435–443.
419. Ganjali, M.R., N. Motakef-Kazemi, P. Norouzi, and S. Khoee. 2009. A Ho(III)-modified carbon paste electrode based on multiwalled carbon nanotubes (MWCNTs) and nanosilica. *Int. J. Electrochem. Sci.* 4:906–913.
420. Yaghoubian, H., H.M. Karimi, M.A. Khalilzadeh, and F. Karimi. 2009. Electrocatalytic oxidation of *Levodopa* at a ferrocene modified carbon nanotube paste electrode. *Int. J. Electrochem. Sci.* 4:993–1003.

421. Raoof, J.-B., M.S. Hejazi, R. Ojani, and E.H. Asl. 2009. A comparative study of carbon nanotube paste electrode for the development of indicator-free DNA sensors by differential pulse voltammetry and electrochemical impedance spectroscopy: Human iont-erleukin-2 oligonucleotide as a model. *Int. J. Electrochem. Sci.* 4:1436–1451.
422. Beitollahi, H., M. Mazloum Ardakani, H. Naeimi, and B. Ganjipour. 2009. Electrochemical characterization of 2,2′-[1,2-ethanediyl-bis-(nitriloethylidyne)]-bis-hydroquinone carbon nanotube paste electrode and its application to simultaneous voltammetric determination of ascorbic acid and uric acid. *J. Solid. State Electrochem.* 13:353–363.
423. Xiong, H., H. Xu, L. Wang, and S. Wang. 2009. Characterization and sensing properties of a CNTPE for acetaminophen. *Microchim. Acta* 167:129–133.
424. Zheng, L. and J.-F. Song. 2009. Electrocatalytic oxidation of methanol and other short chain aliphatic alcohols at Ni(II)-quercetin complex modified multi-wall carbon nanotube paste electrode. *J. Solid. State Electrochem.* 14:43–50.
425. Janegitz, B.C., L.H. Marcolino-Junior, S.P. Campana-Filho, R.C. Faria, and O. Fatibello Filho. 2009. Anodic stripping voltammetric determination of Cu(II) using a functionalized carbon nanotubes paste electrode modified with crosslinked chitosan. *Sens. Actuators B* 142:260–266.
426. Dai, H., Y.-M. Wang, X.-P. Wu, L. Zhang, and G. Chen. 2009. An electrochemiluminescent sensor for methamphetamine hydrochloride based on multi-wall carbon nanotube/ionic liquid composite electrode. *Biosens. Bioelectron.* 24:1230–1234.
427. Mazloum Ardakani, M., H. Beitollahi, B. Ganjipour, H. Naeimi, and M. Nejati. 2009. Electrochemical and catalytic investigations of dopamine and uric acid by modified carbon nanotube paste electrode. *Bioelectrochemistry* 75:1–8.
428. Santos, V.S., W.d.J. Rodrigues Santos, L.T. Kubota, T. Teixeira, and R. Cesar. 2009. Speciation of Sb(III) and Sb(V) in meglumine antimoniate pharmaceutical formulations by PSA using carbon nanotube electrode. *J. Pharm. Biomed. Anal.* 50:151–157.
429. Zheng, L. and J.-F. Song. 2009. Nickel(II)-*Baicalein* complex modified multiwall carbon nanotube paste electrode and its electrocatalytic oxidation toward glycine. *Anal. Biochem.* 391:56–63.
430. Ly, S.-Y., C.-H. Lee, and Y.-S. Jung. 2009. Voltammetric bioassay of caffeine using sensor implant. *Neuromol. Med.* 11:20–27.
431. Zhai, Z.-Q., J. Wu, W. Sun, and K. Jiao. 2009. Direct electrochemistry of hemoglobin and its electrocatalysis based on a carbon nanotube paste electrode. *J. Chin. Chem. Soc. (Taiwan)* 56:561–567.
432. Yaghoubian, H., M.H. Karimi, M.A. Khalilzadeh, and F. Karimi. 2009. Electrochemical detection of *Carbidopa* using a ferrocene-modified CNTPE. *J. Serb. Chem. Soc.* 74:1443–1453.
433. Lozano, M.L., M.C. Rodriguez, P. Herrasti, L. Galicia, and G.A. Rivas. 2010. Amperometric response of hydrogen peroxide at carbon nanotubes paste electrodes modified with an electrogenerated poly(Fe(III)-5-amino-phenantroline). Analytical applications for glucose biosensing. *Electroanalysis* 22:128–134.
434. Gligor, D., I. Craciunescu, I.C. Popescu, and L. Gorton. 2010. Influence of the electrode material on the electrochemical behavior of CPEs modified with meldola blue and methylene green adsorbed on a synthetic zeolite. *Electroanalysis* 22:509–512.
435. Suresh, S., A.K. Gupta, V.K. Rao, O. Kumar, and R. Vijayaraghavan. 2010. Amperometric immunosensor for ricin by using on graphite and carbon nanotube paste electrodes. *Talanta* 81:703–708.
436. Alamo, L.S.T., T. Tangkuaram, and S. Satienperakul. 2010. Determination of sulfite by pervaporation-flow injection with amperometric detection using copper hexacyano-ferrate-carbon nanotube modified carbon paste electrode. *Talanta* 81:793–1799.
437. Merisalu, M., J. Kruusma, and C.E. Banks. 2010. Metallic impurity free carbon nanotube paste electrodes. *Electrochem. Commun.* 12:144–147.
438. Shahrokhian, S. and E. Asadian. 2010. Simultaneous voltammetric determination of ascorbic acid, acetaminophen, and isoniazid using thionine immobilized multi-walled carbon nanotube modified carbon paste electrode. *Electrochim. Acta* 55:666–672.
439. Fotouhi, L., F. Raei, M.M. Heravi, and D. Nematollahi. 2010. Electrocatalytic activity of 6,7-dihydroxy-3-methyl-9-thia-4,4-diazafluoren-2-one/multi-wall carbon nano-tubes immobilized on carbon paste electrode for NADH oxidation: Application to the trace determination of NADH. *J. Electroanal. Chem.* 639:15–20.
440. Ensafi, A.A. and H. Karimi Maleh. 2010. Modified multiwall CNPE as a sensor for simultaneous determination of 6-thioguanine and folic acid using ferrocene-dicarboxylic acid as a mediator. *J. Electroanal. Chem.* 640:75–83.

441. Khalilzadeh, M.A. and H. Karimi Maleh. 2010. Sensitive and selective determination of phenylhydrazine in the presence of hydrazine at a ferrocene monocarboxylic acid modified carbon nanotube paste electrode. *Anal. Lett.* 43:186–196.
442. Mazloum Ardakani, M., H. Beitollahi, B Ganjipour, and H. Naeimi. 2010. Novel carbon nanotube paste electrode for simultaneous determination of norepinephrine, uric acid, and d-penicillamine. *Int. J. Electrochem. Sci.* 5:531–546.
443. Zare, H.R. and N. Nasirizadeh. 2010. Application of hematoxylin multi-wall carbon nanotube modified carbon paste electrode as a sensor for simultaneous determination of dopamine and acetaminophen. *Int. J. Electrochem. Sci.* 4:1691–1705.
444. Ensafi, A.A. and H. Karimi Maleh. 2010. Ferrocenedicarboxylic acid modified multiwall carbon nanotubes paste electrode for voltammetric determination of sulfite. *Int. J. Electrochem. Sci.* 5:392–406.
445. Ensafi, A.A., M. Taei, T. Khayamian, H. Karimi Maleh, and F. Hasanpour. 2010. Voltammetric measurement of trace amount of glutathione using multiwall carbon nano-tubes as a sensor and chlorpromazine as a mediator. *J. Solid State Electrochem.* 14:1415–1423.
446. Faridbod, F., M.R. Ganjali, B. Larijani, M. Hosseini, and P. Norouzi. 2010. Ho$^{(3+)}$-carbon paste sensor based on multi-walled carbon nanotubes for determination of the content of holmium in biological and environmental samples. *Mater. Sci. Engineer. C* 30:555–560.
447. Ganjali, M.R., N. Motakef Kazami, F. Faridbod, and S. Khoee. 2010. Determination of Pb(II) ions by a modified carbon paste electrode based on multi-walled carbon nanotubes and nanosilica. *J. Hazard. Mater.* 173:415–419.
448. Khani, H., M.K. Rofouei, P. Arab, V.K. Gupta, and Z. Vafaei. 2010. Multi-walled carbon nanotubes-ionic liquid-carbon paste electrode as a super selectivity sensor: Application to potentiometric monitoring of mercury ion(II). *J. Hazard. Mater.* 183:402–409.
449. Gligor, D., C. Varodi, A. Maicaneanu, and L.M. Muresan. 2010. Carbon nanotubes-graphite paste electrode modified with Cu(II)-exchanged zeolite for the detection of H_2O_2. *Stud. Univ. Babes-Bolyai Chem. (Romania)* 55:293–302.
450. Tiyapiboonchaiya, C., D.R. MacFarlane, J.-Z. Sun, and M. Forsyth. 2002. Polymer-in-ionic-liquid electrolytes. *Macromol. Chem. Phys.* 203:1906–1911.
451. Wei, D. and A. Ivaska. 2008. Applications of ionic liquids in electrochemical sensors. *Anal. Chim. Acta* 607:126–135.
452. Weidlich, T. and I. Švancara. 2010. Possibilities and limitations of (room temperature) ionic liquids in electrochemistry and electroanalytical measurements. In *Sensing in Electroanalysis*, Vol. 5, Eds. K. Vytřas, K. Kalcher, and I. Švancara, pp. 33–56. Pardubice, Czech Republic: University Press Center.
453. Maleki, N., A. Safavi, and F. Tajabadi. 2006. High-performance carbon composite electrode based on an ionic liquid as a binder. *Anal. Chem.* 78:3820–3826.
454. Shul, G., J. Sirieix Plenet, L. Gaillon, and M. Opallo. 2006. Ion transfer at carbon paste electrode based on ionic liquid. *Electrochem. Commun.* 8:1111–1114.
455. Niedziolka J., E. Rozniecka, J.-Y. Chen, and M. Opallo. 2006. Changing the direction of ion transfer across o-nitrophenyloctylether/water interface coupled to electrochemical redox reaction. *Electrochem. Commun.* 8:941–945.
456. Safavi, A., N. Maleki, O. Moradlou, and F. Tajabadi. 2006. Simultaneous determination of dopamine, ascorbic acid, and uric acid using carbon ionic liquid electrode. *Anal. Biochem.* 359:224–229.
457. Safavi, A., N. Maleki, F. Honarasa, F. Tajabadi, and F. Sedaghatpour. 2007. Ionic liquids for modifying the performance of carbon based potentiometric sensors. *Electroanalysis* 19:582–586.
458. Sun, W., M. Yang, R. Gao, and K. Jiao. 2007. Electrochemical determination of ascorbic acid in room temperature ionic liquid BPPF6 modified carbon paste electrode. *Electroanalysis* 19:1597–1602.
459. Maleki, N., A. Safavi, and F. Tajabadi. 2007. Investigation of the role of ionic liquids in imparting electrocatalytic behavior to carbon paste electrode. *Electroanalysis* 19:2247–2250.
460. Zheng, J.-B., Y. Zhang, and P.-P. Yang. 2007. An ionic liquid-type carbon paste electrode for electrochemical investigation and the determination of calcium dobesilate. *Talanta* 73:920–925.
461. Safavi, A., N. Maleki, and F. Tajabadi. 2007. Highly stable electrochemical oxidation of phenols at carbon ionic liquid electrode. *Analyst (U.K.)* 132:54–58.
462. Li, J., M. Huang, X. Liu, H. Wei, Y. Xu, G. Xu, and E. Wang. 2007. Enhanced electrochemiluminescence sensor from tris(2,2′-bipyridyl)Ru(II) incorporated into MCM-41 and ionic liquid-based carbon paste electrode. *Analyst (U.K.)* 132:687–691.

463. Wang, S.-F., H.-Y. Xiong, and Q.-X. Zeng. 2007. Design of carbon paste biosensors based on the mixture of ionic liquid and paraffin oil as a binder for high performance and stabilization. *Electrochem. Commun.* 9:807–812.
464. Sun, W., D.-D. Wang, R.-F. Gao, and K. Jiao. 2007. Direct electrochemistry and electrocatalysis of hemoglobin in sodium alginate film on a BMIMPF6 modified carbon paste electrode. *Electrochem. Commun.* 9:1159–1164.
465. Zhang, Y. and J.-B. Zheng. 2007. Comparative investigation on electrochemical behavior of hydroquinone at carbon ionic liquid electrode, ionic-liquid modified-, and the bare carbon paste electrode. *Electrochim. Acta* 52:7210–7216.
466. Sun, W., M. Yang, and K. Jiao. 2007. Electrocatalytic oxidation of dopamine at an ionic liquid modified CPE and its analytical application. *Anal. Bioanal. Chem.* 389:1283–1291.
467. Sun, W., M.-X. Yang, R.-F. Gao, and K. Jiao. 2007. Electrochemistry and electro-catalysis of hemoglobin in Nafion/nano-$CaCO_3$ film on a new ionic liquid BPPF6 modified carbon paste electrode. *J. Phys. Chem. B* 111:4560–4567.
468. Zhang, Y. and J.-B. Zheng. 2007. An ionic liquid bulk-modified carbon paste electrode and its electrocatalytic activity toward p-aminophenol. *Chin. J. Chem.* 25:1652–1657.
469. Sun, W., R.-F. Gao, R.-F. Bi, and K. Jiao. 2007. Preparation and characteristics of room temperature ionic liquid N-butylpyridinium hexafluorophosphate modified carbon paste electrode. *Fenxi Huaxue* 35:567–570.
470. Sun, W., R.-F. Gao, D.-D. Wang, and K. Jiao. 2007. Direct electrochemistry of hemoglobin at room temperature ionic liquid [BMIM]PF6 modified carbon paste electrode. *Yaoxue Xuebao* 23:1247–1251.
471. Safavi, A., N. Maleki, S. Momeni, and F. Tajabadi. 2008. Highly improved electro-catalytic behavior of sulfite at carbon ionic liquid electrode: Application to the analysis of some real samples. *Anal. Chim. Acta* 625:8–12.
472. Musameh, M.M. and J. Wang. 2008. Sensitive and stable amperometric measurements at ionic liquid-carbon paste micro-electrodes. *Anal. Chim. Acta* 606:45–49.
473. Musameh, M.M., R.T. Kachoosangi, and R.G. Compton. 2008. Enhanced stability and sensitivity of ionic liquid-carbon paste electrodes at elevated temperatures. *Analyst (U.K.)* 133:133–138.
474. Sun, W., Y. Li, M. Yang, S. Liu, and K. Jiao. 2008. Direct electrochemistry of single-stranded DNA on an ionic liquid modified carbon paste electrode. *Electrochem. Commun.* 10:298–301.
475. Shang, G.-X., H.-F. Zhang, and J.-B. Zheng. 2008. Direct electrochemistry of glucose oxidase based on its direct immobilization on carbon ionic liquid electrode and glucose sensing. *Electrochem. Commun.* 10:1140–1143.
476. Sun, W., Y.-Z. Li, M.-X. Yang, J. Li, and K. Jiao. 2008. Application of carbon ionic liquid electrode for the electrooxidative determination of catechol. *Sens. Actuators B* 133:387–392.
477. Shang Guan, X., H. Zhang, and J. Zheng. 2008. Electrochemical behavior and differential pulse voltammetric determination of Paracetamol at a carbon ionic liquid electrode. *Anal. Bioanal. Chem.* 391:1049–1055.
478. Musameh, M.M., R.T. Kachoosangi, L. Xiao, A. Ruseell, and R.G. Compton. 2008. Ionic liquid-carbon composite glucose biosensor. *Biosens. Bioelectron.* 24:87–92.
479. Sun, W., D.-D. Wang, J.-H. Zhong, and K. Jiao. 2008. Electrocatalytic activity of hemoglobin in sodium alginate/SiO_2 nanoparticle/ionic liquid BMIM-PF_6 composite film. *J. Solid State Electrochem.* 12:655–661.
480. Sun, W., Q. Jiang, M.-X. Yang, and K. Jiao. 2008. Electrochemical behaviors of hydroquinone on a carbon paste electrode with ionic liquid as the binder. *Bull. Korean Chem. Soc.* 29:915–920.
481. Sun, W., X.-Q. Li, and K. Jiao. 2009. Direct electrochemistry of myoglobin in a Nafion-ionic composite film modified carbon ionic liquid electrode. *Electroanalysis* 21:959–964.
482. Haghighi, B. and H. Hamidi. 2009. Electrochemical characterization and application of carbon ionic liquid electrodes containing 1:12 phosphomolybdic acid. *Electroanalysis* 21:1057–1065.
483. Sun, W., Y.-Y. Duan, Y.-H. Li, T.-R. Zhan, and K. Jiao. 2009. Electrochemistry and voltammetric determination of adenosine with N-hexyl-pyridinium hexafluorophosphate modified electrode. *Electroanalysis* 21:2667–2673.
484. Sun, W., Y. Duan, Y. Li, H. Gao, and K. Jiao. 2009. Electrochemical behaviors of guanosine on carbon ionic liquid electrode and its determination. *Talanta* 78:695–699.
485. Ding, C.-F., F. Zhao, R. Ren, and J.-M. Lin. 2009. An electrochemical biosensor for alpha-fetoprotein based on carbon paste electrode constructed of room temperature ionic liquid and gold nanoparticles. *Talanta* 78:1148–1154.

486. Dai, H., H. Xu, X. Wu, Y. Chi., and G. Chen. 2009. Fabrication of a new electrochemiluminescent sensor for Fentanyl citrate based on glassy carbon microspheres and ionic liquid composite paste electrode. *Anal. Chim. Acta* 647:60–65.
487. Ji, H., L. Zhu, D. Liang, Y. Liu, L. Cai, S. Zhang, and S. Liu. 2009. Use of a 12-molybdovanadate(V) modified ionic liquid carbon paste electrode as a bifunctional electrochemical sensor. *Electrochim. Acta* 54:7429–7434.
488. Gao, R. and J. Zheng. 2009. Direct electrochemistry of myoglobin based on DNA accumulation on a carbon ionic liquid electrode. *Electrochem. Commun.* 11:1527–1529.
489. Lin, Z., X. Chen, H. Chen, B. Qiu, and G. Chen. 2009. Electrochemiluminescent behavior of N6-isopentenyl-adenine/Ru(bpy)$_{(3)}^{(2+)}$ system on an electrically heated ionic liquid/carbon paste electrode. *Electrochem. Commun.* 11:2056–2059.
490. Sun, W., C.-X. Guo, Z. Zhu, and C.-M. Li. 2009. Ionic liquid/mesoporous carbon/protein composite microelectrode and its possible applications in biosensing. *Electrochem. Commun.* 11:2105–2108.
491. Sun, W., Y. Li, H. Gao, and K. Jiao. 2009. Direct electrochemistry of double stranded DNA on ionic liquid modified carbon paste electrode (CILE). *Microchim. Acta* 165:313–317.
492. Li, Y., X. Liu, X. Zeng, Y. Liu, X. Liu, W. Wei, and S. Luo. 2009. Non-enzymatic sensor for hydrogen peroxide based on a Prussian blue-modified carbon ionic liquid electrode. *Microchim. Acta* 165:393–398.
493. Sun, W., Q. Jiang, and K. Jiao. 2009. Electrochemical behaviors of metal on ionic-liquid-modified carbon paste electrode and its analytical application. *J. Solid State Electrochem.* 13:1193–1199.
494. Sun, W., Q. Jiang, Y. Wang, and K. Jiao. 2009. Electrochemical behaviors of metol on hydrophilic ionic liquid 1-ethyl-3-methyl-imidazolium tetrafluoroborate-modified electrode. *Sens. Actuators B* 136:419–424.
495. Li, Y.-H., X.-Y. Liu, X.-D. Zeng, Y. Liu, X.-T. Liu, W.-Z. Wei, and S.-L. Luo. 2009. Simultaneous determination of ultra-trace lead and cadmium at a hydroxyapatite-modified carbon ionic liquid electrode by square-wave stripping voltammetry. *Sens. Actuators B* 139:604–610.
496. Sun, W., Z.-Q. Zhai, D.-D. Wang, S.-F. Liu, and K. Jiao. 2009. Electrochemistry of hemoglobin entrapped in a Nafion/nano-ZnO film on carbon ionic liquid electrode. *Bioelectrochem.* 74:295–300.
497. Safavi, A., N. Maleki, and E. Farjami. 2009. Fabrication of a glucose sensor based on a novel nanocomposite electrode. *Biosens. Bioelectron.* 24:1655–1660.
498. Sun, W., M.-X. Yang, Y.-Z. Li, Q. Jiang, S.-F. Liu, and K. Jiao. 2009. Electrochemical behavior and determination of rutin on a pyridinium-based ionic liquid modified carbon paste electrode. *J. Pharm. Biomed. Anal.* 48:1326–1331.
499. Sun, W., X.-Q. Li, S.-F. Liu, and K. Jiao. 2009. Electrochemistry of hemoglobin in the chitosan and TiO$_2$ nanoparticles composite film modified carbon ionic liquid electrode and its electrocatalysis. *Bull. Korean Chem. Soc.* 30:582–588.
500. Sun, W., M.-X. Yang, Q. Jiang, and K. Jiao. 2009. Direct electrocatalytic reduction of *p*-nitrophenol at room temperature ionic liquid modified electrode. *Chin. Chem. Lett.* 19:1156–1158.
501. Zhao, R.-J., Q. Jiang, W. Sun, and K. Jiao. 2009. Electropolymerization of methylene blue on carbon ionic liquid electrode and its electrocatalysis to 3,4-dihydroxybenzoic acid. *J. Chin. Chem. Soc. (Taiwan)* 56:158–163.
502. Sun, B., H.-L. Qi, C. Ling, and C.-Q. Zhang. 2009. Sensitive electrogenerated chemiluminescence sensor for determination of *Heroin*. *Fenxi Huaxue* 37:1601–1605.
503. Sun, W., D. Wang, Z. Zhai, R. Gao, and K. Jiao. 2009. Direct electrochemistry of hemoglobin immobilized in the sodium alginate and SiO$_2$ nanoparticles bio-nanocomposite film on a carbon ionic liquid electrode. *J. Iran. Chem. Soc.* 6:412–419.
504. Yu, Q., Y. Liu, X.-Y. Liu, X.-D. Zeng, S.-L. Luo, and W.-Z. Wei. 2010. Simultaneous determination of dihydroxybenzene isomers at MWCNTs/beta-cyclodextrin modified CILE in the presence of cetylpyridinium bromide. *Electroanalysis* 22:1012–1018.
505. Wang, Y., H.-Y. Xiong, X.-H. Zhang, and S.-F. Wang. 2010. Electrochemical study of *Aloe-emodin* on an ionic liquid-type carbon paste electrode. *Microchim. Acta* 169:255–260.
506. Li, X.-Q., R.-J. Zhao, Y. Wang, X.-Y. Sun, W. Sun, C.-Z. Zhao, and K. Jiao. 2010. An electrochemical biosensor based on Nafion-ionic liquid and a myoglobin-modified carbon paste electrode. *Electrochim. Acta* 55:2173–2178.
507. Guo, C.-X., Z.-S. Lu, Y. Lei, and C.-M. Li. 2010. Ionic liquid-graphene composite for ultratrace explosive trinitrotoluene detection. *Electrochem. Commun.* 12:1237–1240.
508. Zhu, Z.-H., X. Li, Y. Zeng, and W. Sun. 2010. Ordered mesoporous carbon modified carbon ionic liquid electrode for the electrochemical detection of double-stranded DNA. *Biosens. Bioelectron.* 25:2313–2317.

509. Norouzi, P., S.Z. Rafiei, F. Faridbod, M. Adibi, and M.R. Ganjali. 2010. Er(3+)-carbon paste electrode based on new nano-composite. *Int. J. Electrochem. Sci.* 5(2010):367–376.
510. Ganjali, M.R., H. Ganjali, M. Hosseini, and P. Norouzi. 2010. A novel nano-composite Tb(3+) carbon paste electrode. *Int. J. Electrochem. Sci.* 5:967–977.
511. Faridbod, F., M.R. Ganjali, M. Pirali Hamedani, and P. Norouzi. 2010. MWCNTs-ionic liquids-ionophore-graphite nanocomposite based sensor for selective determination of ytterbium(III) ion. *Int. J. Electrochem. Sci.* 5:1103–1112.
512. Li, X.-Q., Y. Wang, X.-Y. Sun, T.-R. Zhan, and W. Sun. 2010. Direct electrochemistry of hemoglobin entrapped in dextran film on a carbon ionic liquid electrode. *J. Chem. Sci.* 122:271–278.
513. Xia, F.-Q, X. Zhang, C.-L. Zhou, D.-Z. Sun, Y.-M. Dong, and Z. Liu. 2010. Simultaneous determination of copper, lead, and cadmium at hexagonal mesoporous silica immobilized quercetin modified carbon paste electrode. *J. Automat. Methods Manag. Chem.* Article No. 824197.
514. Huang, K.-J., J.-Y. Sun, D.-J. Niu, W.-Z. Xie, and W. Wang. 2010. Direct electrochemistry and electrocatalysis of hemoglobin on carbon ionic liquid electrode. *Colloids Surf. B* 78:69–74.
515. Dong, T., J.-B. Du, M.-H. Cao, and C.-W. Hu, Chan. 2010. The electrochemical properties of 12-molybdophosphoric acid modified ionic liquid carbon paste electrode. *J. Cluster Sci.* 21:155–162.
516. Brezina, M. 1967. Polarographic reduction of oxygen on carbon electrode (in German). *Fresenius Z. Anal. Chem.* 224:74–84.
517. Bobbitt, J.M., J.F. Colaruotolo, and S.-J. Huang. 1973. Preparative graphite-paste electrode. *J. Electrochem. Soc.* 120:773–773.
518. Peng, T.-Z., H.-P. Li, and S.-W. Wang. 1993. Selective extraction and voltammetric determination of gold at a chemically modified carbon paste electrode. *Analyst (U.K.)* 118:1321–1324.
519. Sotomayor, M.D.P.T., A.A. Tanaka, and L.T. Kubota. 2002. Development of an enzymeless biosensor for the determination of phenolic compounds. *Anal. Chim. Acta* 455:215–223.
520. Efron, C. and M. Ariel. 1979. Lead dioxide-coated carbon paste electrodes. *Anal. Chim. Acta* 108:395–399.
521. Zielinska, R., E. Mulik, A. Michalska, S. Achmatowicz, and M. Maj Zurawska. 2002. All-solid-state planar miniature chloride ion-selective electrode. *Anal. Chim. Acta* 451:243–249.
522. Hernandez, L., P. Hernandez, M.H. Blanco, and M. Sanchez. 1988. Determination of copper(II) with a carbon paste electrode modified with an ion-exchange resin. *Analyst* 113:41–43.
523. Ulakhovich, N.A., D.Z. Zakieva, Y.G. Galyametdinov, and G.K. Budnikov. 1993. Voltammetric determination of cyanide by use of preconcentration at a carbon-paste electrode modified with liquid crystals. *J. Anal. Chem. (Russ.)*/transl. of *Zh. Anal. Khim.* 48:1048–1052.
524. Pandey, P.C. and H.H. Weetall. 1995. Peroxidase- and tetracyanoquinodimethane-modified graphite paste electrode for the measurement of glucose/lactate/glutamate using enzyme-packed bed reactor. *Anal. Biochem.* 224:428–433.
525. Prabhu, S.V., R.P. Baldwin, and L. Kryger. 1989. Preconcentration and determination of lead(II) at crown ether and cryptand containing chemically modified electrodes. *Electroanalysis* 1:13–21.
526. Yao, C.-L., K.-H. Park, and J.L. Bear. 1989. Chemically modified carbon paste electrode for chronoamperometric studies. Reduction of oxygen by tetrakis(m-2-anilinopyridinato) dirhodium(II,III) chloride. *Anal. Chem.* 61:279–282.
527. Švancara, I. 1987. The applicability of a carbon paste electrode as the integral part of a polarographic analyzer prototype proposed for stripping voltammetric techniques. *Report "SVOC"* (Student's scientific research activity in Czech). Pardubice, Czech Republic: University of Chemical Technology.
528. Galus, Z. and R.N. Adams. 1963. The investigation of the kinetics of moderately rapid electrode reactions using rotated disc electrodes (RDEs). *J. Phys. Chem.* 67:866–869.
529. Monien, H. and P. Jacob. 1971. Rotating carbon paste electrode for the inverse voltammetry in the picogram range. *Fresenius Z. Anal. Chem.* 255:33–34.
530. Neubert, G. and K.B. Prater. 1974. Electrooxidation of N,N-dimethylaniline: Electrochemical study at rotating disc electrodes. *J. Electrochem. Soc.* 121:745–752.
531. Itaya, K., I. Uchida, and S. Toshima. 1983. Mediated electron transfer reactions between redox centers in Prussian blue and reactants in solution. *J. Phys. Chem.* 87:105–112.
532. Alonso Vante, N., H. Tributsch, and O. Solorza Feria. 1995. Kinetics studies of oxygen reduction in acid medium on novel semiconducting transition metal chalcogenides. *Electrochim. Acta* 40:567–576.
533. Walcarius, A. and L. Lamberts L. 1997. Voltammetric response of the hexammino-ruthenium complex incorporated in zeolite-modified carbon paste electrode. *J. Electroanal. Chem.* 422:77–89.
534. Castellani, A.M. and Y. Gushikem. 2000. Electrochemical properties of a porphyrin-cobalt(II) adsorbed on silica-titania-phosphate composite surface prepared by the sol-gel method. *J. Coll. Interface Sci.* 230:195–199.

535. Pessoa, C.A., Y. Gushikem, and L.T. Kubota. 2001. Ferrocenecarboxylic acid adsorbed on Nb_2O_5 film grafted on a SiO_2 surface: NADH oxidation study. *Electrochim. Acta* 46:2499–2505.
536. Shamsipur, M., M. Najafi, M.R.M. Hosseini, and H. Sharghi. 2007. Electrocatalytic reduction of dioxygen, O_2, at carbon paste electrode modified with a novel cobalt(III) Schiffs base complex. *Electroanalysis* 19:1661–1667.
537. Mariame, C., M. El Rhazi, and I. Adraoui. 2009. Determination of traces of copper(II) by anodic stripping voltammetry at a rotated carbon paste disk electrode modified with poly(1,8-diaminonaphtalene). *J. Anal. Chem. (Russ.)* 64:632–636.
538. Ali Qureshi, G. and J. Lindquist. 1973. Liquid ion-exchange nitrate-selective electrode based on carbon paste. *Anal. Chim. Acta* 67:243–245.
539. Alonso Sedano, A.B., M.L. Tascon Garcia, M.D. Vazquez Barbado, and P. Sanchez Batanero. 2003. Electrochemical study of copper and bismuth compounds in the solid state by using voltammetry of immobilized microparticles: Application to $YBa_2Cu_3O_{7-x}$ and $Bi_2Sr_2CaCu_2O_{8-x}$ high transition temperature superconductors. *J. Solid State Electrochem.* 7:301–308.
540. Švancara, I., R. Metelka, and K. Vytřas. 2005. Piston-driven carbon paste electrode holders for electrochemical measurements. In *Sensing in Electroanalysis*, Eds. K. Vytřas and K. Kalcher, pp. 7–18. Pardubice, Czech Republic: University of Pardubice.
541. University of Pardubice, I. Švancara, R. Metelka, and K. Vytřas. 2010. Casing for carbon pastes for electrochemical measurements. *CZ Patent* registered at the Industrial Property Office of the Czech Rep., No. 301714/22.4.2010.
542. Ogorevc, B., X.-H. Cai, and I. Grabec. 1995. Determination of traces of copper by anodic stripping voltammetry after its preconcentration via an ion-exchange route at carbon paste electrodes modified with vermiculite. *Anal. Chim. Acta* 305:176–182.
543. Metrohm: Metrohm products: The whole world of ion analysis, http://products.metrohm.com/prod-62801020.aspx (accessed January 15, 2011).
544. BASi: Stationary voltametric electrodes & carbon paste, http://www.basinc.com/products/ec/sve.php and http://www.basinc.com/mans/carbon paste.pdf (both accessed January 15, 2011).
545. ALS–E-measuring chemical electrode and accessory/4. carbon paste, http://www.als-japan.com/1039.html (accessed January 15, 2011).
546. Rios, A., M.D. Luque de Castro, M. Valcarcel, and H.A. Mottola. 1987. Electrochemical determination of sulfur dioxide in air samples in closed-loop flow injection system. *Anal. Chem.* 59:666–670.
547. Bonakdar, M. and H.A. Mottola. 1989. Electrocatalysis at chemically modified electrodes. Detection/determination of redox gaseous species in continuous-flow systems. *Anal. Chim. Acta* 224:305–313.
548. Karolczak, M., R. Dreiling, R.N. Adams, L.J. Felice, and P.T. Kissinger. 1976. Electrochemical techniques for study of natural phenolic products and drugs in microliter volumes. *Anal. Lett.* 9:783–793.
549. Thomsen, E.J., L. Kryger, and R.P. Baldwin. 1988. Voltammetric determination of traces of nickel(II) with a medium exchange flow system and a chemically modified carbon paste electrode containing dimethylglyoxime. *Anal. Chem.* 60:151–155.
550. Stefan, R.I., R.G. Bokretsion, J.F. van Staden, and H.Y. Aboul-Enein. 2003. Immunosensor for the determination of azidothymidine. Its utilization as detector in a sequential injection analysis system. *Talanta* 59:883–887.
551. Yantasee, W., C. Timchalk, G.E. Fryxell, B.P. Dockendorff, and Y. Lin. 2005. Automated portable analyzer for lead(II) based on sequential flow injection and nano-structured electrochemical sensors. *Talanta* 68:256–261.
552. Wang, Y., Z.-Q. Liu, X.-Y, Hu, J.-L. Cao, F. Wang, Q. Xu, and C. Yang. 2009. On-line coupling of the sequential injection lab-on-valve to differential pulse anodic stripping voltammetry for determination of Pb(II) in water samples. *Talanta* 77:1203–1207.
553. Hoogvliet, J.C., J.M. Reijn, and W.P. Van Bennekom. 1991. Multichannel amperometric detection system for liquid chromatography and flow injection analysis. *Anal. Chem.* 63:2418–2423.
554. Rogers, K.R., J.Y. Becker, and J. Cembrano. 2000. Improved selective electrocatalytic oxidation of phenols by tyrosinase-based carbon paste electrode biosensor. *Electrochim. Acta* 45:4373–4379.
555. O'Shea, T.J. and S.M. Lunte. 1994. Chemically modified microelectrodes for capillary electrophoresis/electrochemistry. *Anal. Chem.* 66:307–311.
556. Labuda, J., A. Meister, P. Glaeser, and G. Werner. 1998. Metal oxide-modified carbon paste electrodes and microelectrodes for the detection of amino acids and their application to capillary electrophoresis. *Fresenius J. Anal. Chem.* 360:654–658.
557. Fu, C., L. Wang, and Y. Fang. 1999. Determination of oxalic acid in urine by co-electro-osmotic capillary electrophoresis with amperometric detection. *Talanta* 50:953–958.

558. Martin, R.S., A.J. Gawron, B.A. Fogarty, F.B. Regan, E. Dempsey, and S.M. Lunte. 2001. Carbon paste-based electrochemical detectors for microchip capillary electrophoresis/electrochemistry. *Analyst (U.K.)* 126:277–280.
559. Sun, X.-H., X.-R. Yang, and E.-K. Wang. 2003. Evaluation of a sol-gel derived carbon composite electrode as an amperometric detector for capillary electrophoresis. *J. Chromatogr. A* 991:109–116.
560. Yang, B.-Y., J.-Y. Mo, and R. Lai. 2005. Determination of nitrophenols of environment by dual-electrode and dual-channel electrochemical detection of capillary electrophoresis with polyvinylpyrrolidone modified carbon paste electrode. *Gaodeng Xuexiao Huaxue Xuebao* 26:227–230.
561. Cheng, X., Q.-J. Wang, S. Zhang, W.-D. Zhang, P.-G. He, and Y.-Z. Fang. 2007. Determination of four kinds of carbamate pesticides by capillary zone electrophoresis with amperometric detection at a polyamide-modified carbon paste electrode. *Talanta* 71:1083–1087.
562. Lee, K.-S., S.-H. Park, S.-Y. Won, and Y.-B. Shim. 2009. Electrophoretic total analysis of Tetracycline antibiotics with an amperometric microchip. *Electrophoresis* 30:3219–3227.
563. Wu, W.-M., X.-M. Yuan, X.-P. Wu, X.-C. Lin, and Z.-H. Xie. 2010. Analysis of phenolic xenoestrogens by pressurized capillary electrochromatography (CEC) with amperometric detection. *Electrophoresis* 31:1011–1018.
564. Dai, J. and G.R. Helz. 1988. Liquid chromatographic determination of nitrilotriacetic acid, ethylenediamine-tetraacetic acid, and related aminopolycarboxylic acids using an amperometric detector. *Anal. Chem.* 60:301–305.
565. Dominguez Sanchez, P., C.K. O'Sullivan, A.J. Miranda Ordierez, P. Tunon Blanco, and M.R. Smyth. 1994. Flow injection amperometric detection of aniline with a peroxidase modified carbon paste electrode. *Anal. Chim. Acta* 291:349–356.
566. Marcolino, J.L.H., M.F. Bergamini, M.F.S. Teixeira, E.T.G. Cavalheiro, and O.F. Filho. 2003. Flow injection amperometric determination of dipyrone in pharmaceutical formulations using a carbon paste electrode. *Farmaco (Brazil)* 58:999–1004.
567. Xu, Z.-M., Q. Yue, Z.-J. Zhuang, and D. Xiao. 2009. Flow injection amperometric determination of acetaminophen at a gold nanoparticle modified carbon paste electrode. *Microchim. Acta* 164:387–393.
568. Hynes, C.J., M. Bonakdar, and H.A. Mottola. 1989. Carbon paste electrode chemically modified by direct admixing of tris[4,7-diphenyl-1,10-phenanthroline]iron(II). *Electroanalysis* 1:155–160.
569. Ibrahim, H., Y.M. Issa, and H.M. Abu Shawish. 2004. Chemically modified CPE for the potentiometric determination of dicyclomine hydrochloride under batch and in FIA conditions. *Anal. Sci. (Jpn.)* 20:911–916.
570. Issa, Y.M., H. Ibrahim, and H.M. Abu Shawish. 2005. Carbon paste electrode for the potentiometric flow injection analysis of drotaverine hydrochloride in serum and urine. *Microchim. Acta* 150:47–54.
571. Chen, Y.-T., Z.-Y. Lin, J.-J. Sun, and G.-N. Chen. 2007. A new electrochemi-luminescent detection system equipped with an electrically heated carbon paste electrode for capillary electrophoresis. *Electrophoresis* 28:3250–3259.
572. Lin, Z.-Y., J.-J. Sun, J.-H. Chen, L. Guo, Y.-T. Chen, and G.-N. Chen. 2008. Electro-chemiluminescent biosensor for hypoxanthine based on the electrically heated carbon paste electrode modified with xanthine oxidase. *Anal. Chem.* 80:2826–2831.
573. Bonakdar, M., J.L. Vilchez, and H.A. Mottola. 1989. Bioamperometric sensors for phenol based on carbon paste electrodes. *J. Electroanal. Chem.* 266:47–55.
574. J. Kulys, P. Klitgaard, and H.E. Hansen. 1996. Bioelectrode response modulation caused by periodical carbon paste filling process. *Mater. Sci. Engineer. C* 4:39–44.
575. Miertus, S., J. Katrlik, A. Pizzariello, M. Stredansky, J. Svitel, and J. Svorc. 1998. Amperometric biosensors based on solid binding matrices applied in food monitoring. *Biosens. Bioelectron.* 13:911–923.
576. Ruzicka, J. and E.H. Hansen. 1980. Flow injection analysis. Principles, applications, and trends. *Anal. Chim. Acta* 114:19–44.
577. Stulik, K. and V. Pacakova. 1981. Electrochemical detection techniques in HPLC. *J. Electroanal. Chem.* 129:1–24.
578. Wang, J. 1990. Modified electrodes for electrochemical detection in flowing streams. *Anal. Chim. Acta* 234:41–48.
579. Baldwin, R.P. and K.N. Thomsen. 1991. Chemically modified electrodes in liquid chromatography detection. A review. *Talanta* 38:1–16.
580. Luque de Castro, M.D. and A. Izquierdo. 1991. Flow injection stripping analysis (a review). *Electroanalysis* 3:457–467.
581. Ruzicka, J. and E.H. Hansen. 2008. Retro-review of flow-injection analysis. *Trends Anal. Chem.* 27:390–393.
582. Desideri, P.G., L. Lepri, and D. Heimler. 1973. Electrochemical behavior of the aminobenzenesulfonic acids in aqueous solutions. *J. Electroanal. Chem.* 43:387–396.

References

583. Desideri, P.G., L. Lepri, and D. Heimler. 1974. Electrochemical behavior of the toluidines in aqueous solutions. Part II: p-Toluidine. *J. Electroanal. Chem.* 52:105–114.
584. Lvova, L., S.-S. Kim, A. Legin, Y. Vlasov, J.-S. Yang, G.S. Cha, and H. Nam. 2002. All-solid-state electronic tongue and its application for beverage analysis. *Anal. Chim. Acta* 468:303–314.
585. Švancara, I., P. Kotzian, M. Bartos, and K. Vytřas. 2005. Groove electrodes: A new alternative of using carbon paste in electroanalysis. *Electrochem. Commun.* 7:657–662.
586. Metelka, R., M. Zeravik, and K. Vytřas. 2008. The groove carbon paste electrode as a detector for FIA. In *Monitoring of Environmental Pollutants* (in Czech), Eds. K. Vytřas, J. Kellner, and J. Fischer, pp. 153–158. Pardubice, Czech Republic: University Press Center.
587. Metelka, R., M. Zeravik, and K. Vytřas. 2010. Carbon paste electrode containing dispersed bismuth powder for pH-measurements. In *Sensing in Electroanalysis*, Eds. K. Vytřas, K. Kalcher, and I. Švancara, pp. 257–267. Pardubice, Czech Republic: University Press Center.
588. Galus, Z., J.O. Schenk, and R.N. Adams. 1982. Electrochemical behavior of very small electrodes in solution. Double potential step, cyclic voltammetry, and chronopotentiometry with current reversal. *J. Electroanal. Chem.* 135:1–11.
589. Gasparini, R., M. Scarpa, F. Vianello, B. Mondovi, and A. Rigo. 1994. Renewable miniature enzyme-based sensing devices. *Anal. Chim. Acta* 294:299–304.
590. Dutra, R.F., K.A. Moreira, M.I.P. Oliveira, A.N. Araujo, M. Montenegro, J.L.L. Filho, and V.L. Silva. 2005. An inexpensive mini-biosensor for uric acid determination in human serum by flow-injection analysis. *Electroanalysis* 17:701–705.
591. Hason, S. and V. Vetterl. 2006. Amplified oligonucleotide sensing in microliter volumes containing copper ions by solution streaming. *Anal Chem.* 78:5179–5183.
592. Baldrianova, L., I. Švancara, and S. Sotiropoulos. 2007. Anodic stripping voltammetry at a new type of disposable bismuth-plated carbon paste mini-electrodes. *Anal. Chim. Acta* 599:249–255.
593. Conti, J.C., E. Strope, R.N. Adams, and C.A. Marsden. 1978. Voltammetry in brain tissue: Chronic recording of stimulated dopamine and 5-hydroxytryptamine. *Life Sci.* 23:2705–2715.
594. Cheng, H.Y., J. Schenk, R. Huff, and R.N. Adams. 1979. In-vivo electrochemistry: Behavior of microelectrodes in brain tissue. *J. Electroanal. Chem.* 100:23–31.
595. Lane, R.F., A.T. Hubbard, and C.D. Blaha. 1978. Brain dopaminergic neurons: In vivo electrochemical information concerning storage, metabolism, and release processes. *Bioelectrochem. Bioenerg.* 5:504–525.
596. Lane, R.F., A.T. Hubbard, and C.D. Blaha. 1979. Application of semidifferential electroanalysis to studies of neurotransmitters in the central nervous system. *J. Electroanal. Chem.* 95:117–122.
597. Brazell, M.P. and C.A. Marsden. 1982. Differential pulse voltammetry in the anesthetized rat: Identification of ascorbic acid, catechol, and indoleamine oxidation peaks in the striatum and frontal cortex. *Br. J. Pharmacol.* 75:539–547.
598. O'Neill, R.D., R.A. Gruenewald, M. Fillenz, and W.J. Albery. 1982. Linear sweep voltammetry with carbon paste electrodes in the rat striatum. *Neuroscience (U.K.)* 7:1945–1954.
599. O'Neill, R.D., M. Fillenz, and W.J. Albery. 1983. The development of linear sweep voltammetry with carbon paste electrodes in vivo. *J. Neurosci. Methods* 8:263–273.
600. Wang, J., L.D. Hutchins, S. Selim, and L.B. Cummins. 1984. Long voltammetric microelectrode for in-vivo monitoring of acetaminophen in primate species. *Biochem. Bioenerg.* 12:193–203.
601. Marsden, C.A., M.H. Joseph, Z.L. Kruk, N.T. Maidment, R.D. O'Neill, J.O. Schenk, and J.A. Stamford. 1988. In vivo voltammetry: Present electrodes and methods (a review). *Neuroscience (U.K.)* 25:389–400.
602. Blaha, C.D. 1996. A critical assessment of electrochemical procedures applied to the measurement of dopamine and its metabolites during drug-induced and species-typical behaviours (a review). *Behav. Pharmacol.* 7:585–714.
603. Di Ciano, P., C.D. Blaha, and A.G. Phillips. 2002. Inhibition of dopamine efflux in the rat nucleus accumbens during abstinence after free access to d-amphetamine. *Behav. Brain Res.* 128:1–12.
604. O'Neill, R.D., J.P. Lowry, and M. Mas. 1998. Monitoring brain chemistry in vivo: Voltammetric techniques, sensors, and behavioral applications (a review). *Crit. Rev. Neurobiol.* 12:69–127.
605. O'Neill, R.D. 2005. Long-term monitoring of brain dopamine metabolism in vivo with carbon paste electrodes (a review). *Sensors* 5:317–342.
606. Lowry, J.P. and R.D. O'Neill. 2006. Neuroanalytical chemistry in-vivo using electrochemical sensors. In *The Encyclopedia of Sensors*, Vol. 6, Eds. C.A. Grimes, E.C. Dickey, and M.V. Pishko, pp. 501–524. Stevenson Ranch: American Scientific Publishers (ASP).
607. Zou, Y. and J.-W. Mo. 1999. Ensembles of carbon paste microelectrodes. *Anal. Chim. Acta* 382:145–150.

608. Wang, J., X. Zhang, C. Parrado, and G. Rivas. 1999. Controlled release of DNA from carbon-paste microelectrodes. *Electrochem. Commun.* 1:197–202.
609. Wang J., L. Chen, M. Jiang, and F. Lu. 1999. Myoglobin-containing carbon-paste enzyme microelectrodes for the biosensing of glucose under oxygen-deficit conditions. *Anal. Chem.* 71:5009–5011.
610. Wang, J., X. Zhang, and M. Prakash. 1999. Glucose microsensors based on carbon paste enzyme electrodes modified with cupric hexacyanoferrate. *Anal. Chim. Acta* 395:11–16.
611. Wang, J., L. Chen, and M.P. Chatrathi. 2000. Evaluation of different fluorocarbon oils for their internal oxygen supply in glucose microsensors operated under oxygen-deficit conditions. *Anal. Chim. Acta* 411:187–192.
612. Wang, J. and X. Zhang. 2001. Needle-type dual microsensor for the simultaneous monitoring of glucose and insulin. *Anal. Chem.* 73:844–847.
613. Wang, J. and J.M. Zadeii. 1988. Ensembles of carbon paste ultramicroelectrodes. *J. Electroanal. Chem.* 249:339–345.
614. Cheng, I.F, L.D. Whiteley, and C.R. Martin. 1989. Ultramicroelectrode ensembles. Comparison of experimental and theoretical responses and evaluation of electroanalytical detection limits. *Anal. Chem.* 61:762–766.
615. Oni, J., P. Westbroek, and T. Nyokong. 2001. Construction and characterization of carbon paste ultramicroelectrodes. *Electrochem. Commun.* 3:524–528.
616. Chouaib, F. and M.A. Dosal. 1998. The utilisation of a carbon paste electrode in analytical chemistry. *Rev. Soc. Quim. Mex.* 24:119–124.
617. Metelka, R., K. Vytřas, and A. Bobrowski. 2000. Effect of the modification of mercuric oxide on the properties of mercury films at HgO-modified carbon paste electrodes. *J. Solid State Electrochem.* 4:348–352.
618. Harak, D.W. and H.A. Mottola. 1995. Backscattered electron imaging of ligand-modified carbon paste surfaces by complexation with iron(II) and/or chemical deposition of gold. *Langmuir* 11:1605–1611.
619. Wang, J., L.H. Wu, and R. Li. 1989. Scanning electrochemical microscopic monitoring of biological processes. *J. Electroanal. Chem.* 272:285–292.
620. Flechsig, G.U., M. Kienbaum, and P. Gruendler. 2005. Ex situ atomic force microscopy of bismuth film deposition at carbon paste electrodes. *Electrochem. Commun.* 7:1091–1097.
621. Lubert, K.H., M. Guttmann, L. Beyer, and K. Kalcher. 2001. Experimental indications for the existence of different states of palladium(0) at the surface of carbon paste electrodes. *Electrochem. Commun.* 3:102–106.
622. Švancara, I., L. Baldrianova, M. Vlcek, R. Metelka, and K. Vytřas. 2005. A role of the plating regime in the deposition of bismuth films onto a carbon paste electrode. Microscopic study. *Electroanalysis* 17:120–126.
623. Švancara, I., R. Metelka, M. Stiburkova, J. Seidlova, G. Jansova, K. Vytřas, and B. Pihlar. 2002. Carbon paste electrodes and screen-printed sensors plated with mercury- and bismuth films in stripping voltammetry of heavy metals. *Sci. Pap. Univ. Pardubice Ser. A* 8:19–33.
624. Khoo, S.B. and S.X. Guo. 2002. Rapidly renewable and reproducible mercury film coated carbon paste electrode for anodic stripping voltammetry. *Electroanalysis* 14:813–822.
625. Švancara, I., M. Fairouz, K. Ismail, R. Metelka, and K. Vytřas. 2003. A contribution to the characterisation of mercury- and bismuth-film carbon paste electrodes in stripping voltammetry. *Sci. Pap. Univ. Pardubice Ser. A* 9:31–48.
626. Vytřas, K., I. Švancara, and R. Metelka 2002. A novelty in potentiometric stripping analysis: Total replacement of mercury by bismuth. *Electroanalysis* 14:1359–1364.
627. Švancara, I., M. Fairouz, K. Ismail, J. Sramkova, R. Metelka, and K. Vytřas. 2004. Applicability of electrochemical stripping analysis at mercury- and bismuth-film carbon paste electrodes to crude oil digests. *Sci. Pap. Univ. Pardubice Ser. A* 10:5–20.
628. Kizek, R., M. Masarik, K.J. Kramer, D. Potesil, M. Bailey, J.A. Howard, B. Klejdus et al. 2005. An analysis of avidin, biotin and their interaction at attomole levels by voltammetric and chromatographic techniques. *Anal. Bioanal. Chem.* 381:1167–1178.
629. Wang, J., X. Cai, C. Jonsson, and M. Balakrishnan. 1996. Adsorptive stripping potentiometry of DNA at electrochemically pretreated carbon paste electrodes. *Electroanalysis* 8:20–24.
630. Cai, X., G. Rivas, P.A.M. Farias, H. Shiraishi, J. Wang, and E. Palecek. 1996. Potentiometric stripping analysis of bioactive peptides at carbon electrodes down to subnanomolar concentrations. *Anal. Chim. Acta* 332:49–57.
631. Kaplan, J.D., J.H. Marsh, and S.W. Orchard. 1993. Voltammetric characterization of activated carbons in modified carbon paste electrodes. *Electroanalysis* 5:509–516.

References

632. Biniak, S., M. Pakula, W. Darlewski, A. Swiatkowski, and P. Kula. 2009. Powdered activated carbon and carbon paste electrodes: Comparison of electrochemical behaviour. *J. Appl. Electrochem.* 39:593–600.
633. Huang, X., J.J. Pot, and W.T. Kok. 1995. Electrochemical characteristics of conductive carbon cement as matrix for chemically modified electrodes. *Anal. Chim. Acta* 300:5–14.
634. El Maslout, A., A. Benayada, H. Oudda, J. Bessiere, and Y. Pillet. 1996. In situ determination of hydrogenophosphate ion activity in concentrated H_3PO_4 solutions. *Analusis (Bruxelles)* 24:182–185.
635. Cai, X., K. Kalcher, G. Koelbl, C.G. Neuhold, W. Diewald, and B. Ogorevc. 1995. Electrocatalytic reduction of hydrogen peroxide on a palladium-modified carbon paste electrode. *Electroanalysis* 7:340–345.
636. Xu, G.B. and S.J. Dong. 1999. Chemiluminescent determination of luminol and hydrogen peroxide using hematin immobilized in the bulk of a carbon paste electrode. *Electroanalysis* 11:1180–1184.
637. Elsuccary, S.A.A., I. Švancara, R. Metelka, L. Baldrianova, M.E.M. Hassouna, and K. Vytřas. 2003. Applicability of bismuth film carbon paste electrodes in highly alkaline media. *Sci. Pap. Univ. Pardubice Ser. A* 9:5–17.
638. Konvalina, J., I. Švancara, K. Vytřas, and K. Kalcher. 1997. Study of conditions for voltammetric determinations of iodine at carbon paste electrodes. *Sci. Pap. Univ. Pardubice Ser. A* 3:153–162.
639. Švancara, I., J. Konvalina, K. Schachl, K. Kalcher, and K. Vytřas. 1998. Stripping voltammetric determination of iodide with synergistic accumulation at a carbon paste electrode. *Electroanalysis* 10:435–441.
640. Kitagawa, T., H. Takao, and Y. Fujikawa. 1966. Voltammetry at solid electrodes. Part I: Fundamental studies with a carbon paste electrode. *Bunseki Kagaku* 15:446–451.
641. Kitagawa, T. and S. Tsushima. 1966. Voltammetry at solid electrodes. Part II: Anodic oxidation of ethylenediaminetetraacetic acid and its related compounds at a carbon paste electrode. *Bunseki Kagaku* 15:452–458.
642. Taylor, R.J. and A.A. Humffray. 1973. Electrochemical studies on glassy carbon electrodes. I. Electron transfer kinetics. *J. Electroanal. Chem.* 42:347–354.
643. Taliene, V.R., T. Ruzgas, V. Razumas, and J. Kulys. 1994. Chronoamperometric and cyclic voltammetric study of carbon paste electrodes using ferricyanide and ferrocenemonocarboxylic acid. *J. Electroanal. Chem.* 372:85–89.
644. Kulys, J., L. Gorton, E. Dominguez, J. Emneus, and H. Jarskog. 1994. Electrochemical characterization of carbon pastes modified with proteins and polycations. *J. Electroanal. Chem.* 372:49–55.
645. Kulys, J. 1999. Personal communication.
646. Chicharro, M., A. Zapardiel, E. Bermejo, J.A. Perez, and L. Hernandez. 1994. Ephedrine determination in human urine using a carbon paste electrode modified with C18 bonded silica gel. *Anal. Lett.* 27:1809–1831.
647. Wang, J., Q. Chen, and M. Pedrero. 1995. Highly selective biosensing of lactate at lactate oxidase containing rhodium-dispersed carbon paste electrodes. *Anal. Chim. Acta* 304:41–46.
648. O'Shea, T.J., D. Leech, M.R. Smyth, and J.G. Vos. 1992. Determination of nitrite based on mediated oxidation at a carbon paste electrode modified with a ruthenium polymer. *Talanta* 39:443–447.
649. Pravda, M., O. Adeyoju, E.I. Iwuoha, J.G. Vos, M.R. Smyth, and K. Vytřas. 1995. Amperometric glucose biosensors based on an osmium (2+/3+) redox polymer-mediated electron transfer at carbon paste electrodes. *Electroanalysis* 7:619–625.
650. Paredes, P.A., J. Parellada, V.M. Fernandez, I. Katakis, and E. Dominguez. 1997. Amperometric mediated carbon paste biosensor based on D-fructose dehydrogenase for the determination of fructose in food analysis. *Biosens. Bioelectron.* 12:1233–1243.
651. Ruiz, M.A., P. Yanez Sedeno, and J.M. Pingarron. 1994. Voltammetric determination of the antioxidant tert-butylhydroxytoluene (BHT) at a carbon paste electrode modified with nickel phthalocyanine. *Electroanalysis* 6:475–479.
652. Cookeas, E.G. and C.E. Efstathiou. 1994. Flow injection amperometric determination of thiocyanate and selenocyanate at a cobalt phthalocyanine modified carbon paste electrode. *Analyst (U.K.)* 119:1607–1612.
653. Novakova, M., K. Kalcher, K. Schachl, A. Komersova, M. Bartos, and K. Vytřas. 1997. Voltammetric determination of ascorbic acid in food-stuff using modified carbon paste electrodes. *Sci. Pap. Univ. Pardubice Ser. A* 3:139–151.
654. Alvarez Gonzalez, M.I., S.B. Saidman, M.J. Lobo Castanon, A.J. Miranda Ordieres, and P. Tunon Blanco 2000. Electrocatalytic detection of NADH and glycerol by NAD+-modified carbon electrodes. *Anal. Chem.* 72:520–527.
655. Amini, M.K., S. Shahrokhian, S. Tangestaninejad, and V. Mirkhani. 2001. Iron(II) phthalocyanine-modified carbon-paste electrode for potentiometric detection of ascorbic acid. *Anal. Biochem.* 290:277–282.
656. Chen, H.Y., A.-M. Yu, J. Han, and M. Yz. 1995. Catalytic oxidation of NADH at a methylene green chemically modified electrode and FIA applications. *Anal. Lett.* 28:1579–1591.

657. Gosser, D.K., Jr. 1993. The cyclic Voltammetric Experiment. In *Cyclic Voltammetry: Simulation and Analysis of Reaction Mechanisms*, Chapter 2, pp. 27–69. New York: VCH Publishers.
658. Galus, Z. 1969. *Fundamentals of Electrochemical Analysis*, 2nd Rev. Edn., New York: Ellis Horwood.
659. Nishiyama, K., K. Inada, and I. Taniguchi. 2010. Selective determination of uric acid, tryptophan, and serotonin at glassy carbon paste electrode. *Electrochemistry (Tokyo, Jpn.)* 78:165–169.
660. Liu, Y., J. Huang, H. Hou, and T. You. 2008. Simultaneous determination of dopamine, ascorbic acid and uric acid with electrospun carbon nanofibers modified electrode. *Electrochem. Commun.* 10:1431–1434.
661. Rivas, G.A., M.D. Rubianes, M.L. Pedano, N.F. Ferreyra, G.L. Luque, M.C. Rodriguez, and S.A. Miscoria. 2007. Carbon nanotubes paste electrodes. A new alternative for the development of electrochemical sensors. *Electroanalysis* 19:823–831.
662. Wang, J. 2005. Carbon-nanotube based electrochemical biosensors: A review. *Electroanalysis* 17:7–14.
663. Liu, H., P. He, Z. Li, Y. Liu, J. Li, L. Zheng, and J. Li. 2005. The inherent capacitive behavior of imidazolium-based room-temperature ionic liquids at carbon paste electrode. *Electrochem. Solid State Lett.* 8:J17–J19.
664. Sun, W., R. Gao, X. Li, D. Wang, M. Yang, and K. Jiao. 2008. Fabrication and electrochemical behavior of hemoglobin modified carbon ionic liquid electrode. *Electroanalysis* 20:1048–1054.
665. Jiang, Q., W. Sun, and K. Jiao. 2010. Electrochemical behavior and determination of L-tryptophan on carbon ionic liquid electrode. *J. Anal. Chem.* 65:648–651.
666. Ikeda, T., H. Hamada, and M. Senda. 1986. Electrocatalytic oxidation of glucose at a glucose oxidase-immobilized benzoquinone-mixed carbon paste electrode. *Agric. Biol. Chem.* 50:883–890.
667. Ikeda, T., K. Miki, F. Fushimi, and M. Senda. 1988. Amperometric D-gluconate sensor using D-gluconate dehydrogenase from bacterial membranes. *Agric. Biol. Chem.* 52:1557–1563.
668. Andrieux, C.P., P. Audebert, P. Bacchi, and B. Divisia-Blohorn. 1996. Parameters regulating kinetics of amperometric biosensor made of enzymatic carbon paste and Nafion gel. *Sens. Mater.* 8:137–146.
669. Švancara, I., M. Fairouz, K. Ismail, R. Metelka, and K. Vytřas. 2004. A contribution to the characterization of mercury- and bismuth-film carbon paste electrodes in stripping voltammetry. *Sci. Pap. Univ. Pardubice Ser. A* 9:31–47.
670. Švancara, I., L. Baldrianova, E. Tesarova, S.B. Hocevar, S.A.A. Elsuccary, A. Economou, S. Sotiropoulos, B. Ogorevc, and K. Vytřas. 2006. Recent advances in anodic stripping voltammetry with bismuth-modified carbon paste electrodes. *Electroanalysis* 18:177–185.
671. Švancara, I., R. Metelka, and E. Tesarova. 2006. Stripping voltammetry at mercury film plated carbon paste- and screen-printed electrodes. Ten years of advanced laboratory practice for students at the University of Pardubice. *Sci. Pap. Univ. Pardubice Ser. A* 11:343–361.
672. Švancara, I., L. Baldrianova, E. Tesarova, T. Mikysek, and K. Vytřas. 2007. Anodic stripping voltammetry at bismuth-modified electrodes in ammonia-buffered media. *Sci. Pap. Univ. Pardubice Ser. A* 12:5–19.
673. Vytřas, K. and I. Švancara. 1994. Applications of carbon paste electrodes in electroanalysis. *Chem. Listy* 88:412–422.
674. Paneli, M.G. and A. Voulgaropoulos. 1993. Applications of adsorptive stripping voltammetry in the determination of trace and ultratrace metals. *Electroanalysis* 5:355–373.
675. Lovric, M. 2002. Stripping voltammetry. In *Electroanalytical Methods; Guide to Experiments and Applications*, Ed. F. Scholz, pp. 200–203. Berlin, Germany: Springer.
676. Wang, J. and B.A. Freiha. 1984. Preconcentration of uric acid at a carbon paste electrode. *Bioelectrochem. Bioenerg.* 12:225–234.
677. Ding, Y., J. Li, and J. Fei. 2005. Adsorptive catalytic voltammetry of physcion in the presence of dissolved oxygen at a carbon paste electrode. *Microchim. Acta* 150:125–130.
678. El Ries, M.A., A.A. Wassel, N.T. Abdel Ghani, and M.A. El Shall. 2005. Electrochemical adsorptive behavior of some fluoroquinolones at carbon paste electrode. *Anal. Sci.* 21:1249–1254.
679. Girousi, S.T., D.K. Alexiadou, and A.K. Ioannou. 2008. An electroanalytical study of the drug proflavine. *Microchim. Acta* 160:435–439.
680. Nemcova, L., J. Zima, and J. Barek. 2009. Determination of 5-amino-6-nitroquinoline at a carbon paste electrode. *Collect. Czech. Chem. Commun.* 74:1477–1488.
681. Jemelkova, Z., J. Zima, and J. Barek. 2009. Voltammetric and amperometric determination of doxorubicin using carbon paste electrodes. *Collect. Czech. Chem. Commun.* 74:1503–1515.
682. Ruiz Barrio, M.A. and J.M. Pingarron Corrazon. 1992. Voltammetric determination of pentachlorophenol with a silica gel-modified carbon paste electrode. *Fresenius J. Anal. Chem.* 344:34–38.
683. Baldwin, R.P., J.K. Christensen, and L. Kryger. 1986. Voltammetric determination of traces of nickel(II) at a chemically modified electrode based on dimethylglyoxime-containing carbon paste. *Anal. Chem.* 58:1790–1798.

684. Trojanowicz, M. and W. Matuszewski. 1989. Potentiometric stripping determination of nickel at a dimethylglyoxime-containing graphite paste electrode. *Talanta* 36:680–682.
685. Li, J., S. Liu, Z. Yan, X. Mao, and P. Gao. 2006. Adsorptive voltammetric studies on the cerium(III)-alizarin complexon complex at a carbon paste electrode. *Microchim. Acta* 154:241–246.
686. Li, Y.H., Q.L. Zhao, and M.H. Huang. 2007. Adsorptive anodic stripping voltammetry of zirconium(IV)-alizarin red S complex at a carbon paste electrode. *Microchim. Acta* 157:245–249.
687. Liu, N. and J.F. Song. 2005. Catalytic adsorptive stripping voltammetric determination of copper(II) on a carbon paste electrode. *Anal. Bioanal. Chem.* 383:358–364.
688. Li, Y.H., Y.X. Wang, and M.H. Huang. 2008. Determination of trace vanadium by adsorptive stripping voltammetry at a carbon paste electrode. *Electroanalysis* 20:1440–1444.
689. Li, Y.H., Q.L. Zhao, and M.H. Huang. 2005. Cathodic adsorptive voltammetry of the gallium-alizarin red S complex at a carbon paste electrode. *Electroanalysis* 17:343–347.
690. Li, Y.H., H.Q. Xie, and F.Q. Zhou. 2005. Alizarin violet modified carbon paste electrode for the determination of trace silver(I) by adsorptive voltammetry. *Talanta* 67:28–33.
691. Li, J., S. Liu, X. Mao, P. Gao, and Z. Yan. 2004. Trace determination of rare earths by adsorption voltammetry at a carbon paste electrode. *J. Electroanal. Chem.* 561:137–142.
692. Guo, H., Y. Li, P. Xiao, and N. He. 2005. Determination of trace amount of bismuth(III) by adsorptive anodic stripping voltammetry at carbon paste electrode. *Anal. Chim. Acta* 534:143–147.
693. Guo, H., Y. Li, X. Chen, L. Nie, and N. He. 2005. Determination of trace antimony(III) by adsorption voltammetry at carbon paste electrode. *Sensors* 5:284–292.
694. Li, Y.H., H.Q. Xie, F.Q. Zhou, and H.S. Guo. 2006. Determination of trace tin by anodic stripping voltammetry at a carbon paste electrode. *Electroanalysis* 18:976–980.
695. Švancara, I., P. Foret, and K. Vytřas. 2004. A study on the determination of chromium as chromate at a carbon paste electrode modified with surfactants. *Talanta* 64:844–852.
696. Lu, G., X. Wu, Y. Lan, and S. Yao. 1999. Studies on 1:12 phosphomolybdic heteropoly anion film modified carbon paste electrode. *Talanta* 49:511–515.
697. Švancara, I. and K. Vytřas. 2001. Determination of iodide in potassium iodide dosage tablets using cathodic stripping voltammetry with a carbon paste electrode. *Sci. Pap. Univ. Pardubice Ser. A* 7:5–15.
698. Sousa, M.d.F., O.E.S. Godinho, and L.M. Aleixo. 1995. An indirect voltammetric approach for the determination of cyanide at a chemically modified electrode. *Electroanalysis* 7:1095–1097.
699. Lefèvre, G., A. Walcarius, and J. Bessière. 1999. Voltammetric investigation of iodide sorption on cuprite dispersed into a carbon paste electrode. *Electrochim. Acta* 44:1817–1826.
700. Walcarius, A., T. Barbaise, and J. Bessière. 1997. Factors affecting the analytical applications of zeolite-modified electrodes. Preconcentration of electroactive species. *Anal. Chim. Acta* 340:61–76.
701. Walcarius, A., N. Luthi, J.L. Blin, B.L. Su, and L. Lamberts. 1999. Electrochemical evaluation of polysiloxane-immobilized amine ligands for the accumulation of copper(II) species. *Electrochim. Acta* 44:4601–4610.
702. Walcarius, A., M. Etienne, and C. Delacote. 2004. Uptake of inorganic HgII by organically modified silicates: Influence of pH and chloride concentration on the binding pathways and electrochemical monitoring of the processes. *Anal. Chim. Acta* 508:87–98.
703. Attia, A.K. 2010. Determination of antihypertensive drug moexipril hydrochloride based on the enhancement effect of sodium dodecyl sulfate at carbon paste electrode. *Talanta* 81:25–29.
704. Stadlober, M., K. Kalcher, G. Raber, and C. Neuhold. 1996. Anodic stripping voltammetric determination of titanium(IV) using a carbon paste electrode modified with cetyltrimethylammonium bromide. *Talanta* 43:1915–1924.
705. Stadlober, M., K. Kalcher, and G. Raber. 1997. A new method for the voltammetric determination of molybdenum(VI) using carbon paste electrodes modified in situ with cetyltrimethylammonium bromide. *Anal. Chim. Acta* 350:319–328.
706. Stadlober, M., K. Kalcher, and G. Raber. 1997. Anodic stripping voltammetric determination of vanadium(V) using a carbon paste electrode modified in situ with cetyltrimethylammonium bromide. *Electroanalysis* 9:225–230.
707. Abbas, M.N.E.-D. 2003. Chemically modified carbon paste electrode for iodide determination on the basis of cetyltrimethylammonium iodide ion-pair. *Anal. Sci.* 19:229–233.
708. Wang, J., B. Greene, and C. Morgan. 1984. Carbon paste electrodes modified with cation-exchange resin in differential pulse voltammetry. *Anal. Chim. Acta* 158:15–22.
709. Jezkova, J., J. Musilova, and K. Vytřas. 1997. Potentiometry with perchlorate and fluoroborate ion-selective carbon paste electrodes. *Electroanalysis* 9:1433–1436.

710. Konvalina, J. and K. Vytřas. 2001. Reductive determination of gold at a carbon paste electrode using constant-current stripping analysis. *Chem. Listy* 95:505–508.
711. Vytřas, K. and J. Konvalina. 1998. New possibilities of potentiometric stripping analysis based on ion-pair formation and accumulation of analyte at carbon paste electrodes. Preliminary note. *Electroanalysis* 10:787–790.
712. Metelka, R., S. Slavikova, and K. Vytřas. 2002. Determination of arsenate and organic arsenic via potentiometric titration of its heteropoly anions. *Talanta* 58:147–151.
713. Gellon, H., P.S. Gonzalez, and C.A. Fontan. 2003. Square wave anodic stripping determination of silver using a carbon paste electrode modified with a strong acid ion-exchanger. *Anal. Lett.* 36:2749–2765.
714. Bergamini, M.F., S.I. Vital, A.L. Santos, and N.R. Stradiotto. 2006. Lead ions determination in ethanol fuel by differential pulse anodic stripping voltammetry using a carbon paste electrode modified with ion-exchange resin Amberlite IR120. *Ecletica Quim.* 31:45–52.
715. Navratilova, Z. and P. Kula. 2003. Clay modified electrodes: Present applications and prospects. *Electroanalysis* 15:837–846.
716. Navratilova, Z. and P. Kula. 2000. Cation and anion exchange on clay modified electrodes. *J. Solid State Electrochem.* 4:342–347.
717. Walcarius, A. 1999. Zeolite-modified electrodes in electroanalytical chemistry. *Anal. Chim. Acta* 384:1–16.
718. Walcarius, A. 2008. Electroanalytical applications of microporous zeolites and mesoporous (organo) silicas: Recent trends. *Electroanalysis* 20:711–738.
719. Ganesan, V. and A. Walcarius. 2008. Ion exchange and ion exchange voltammetry with functionalized mesoporous silica materials. *Mater. Sci. Eng. B* 149:123–132.
720. Labuda, J. 1992. Chemically modified electrodes as sensors in analytical chemistry. *Selective Electrode Rev.* 14:33–86.
721. Rezaei, B. and S. Damiri. 2008. Multiwalled carbon nanotubes modified electrode as a sensor for adsorptive stripping voltammetric determination of hydrochlorothiazide. *IEEE Sens. J.* 8:1523–1529.
722. Anik, U. and M. Cubukcu. 2008. Examination of the electroanalytic performance of carbon nanotube (CNT) modified carbon paste electrodes as xanthine biosensor transducers. *Turk. J. Chem.* 32:711–719.
723. Sun, W., Y. Li, Y. Duan, and K. Jiao. 2008. Direct electrocatalytic oxidation of adenine and guanine on carbon ionic liquid electrode and the simultaneous determination. *Biosens. Bioelectron.* 24:988–993.
724. Chulkina, L.S., S.I. Sinyakova, and E.K. Vul'fson. 1970. Optimum composition of carbon paste electrodes for inverse voltammetric determinations of nanogram amounts of substances. *Zh. Anal. Khim.* 25:1268–1272.
725. Shul, G. and M. Opallo. 2005. Ion transfer across liquid–liquid interface coupled to electrochemical redox reaction at carbon paste electrode. *Electrochem. Commun.* 7:194–198.
726. Barek, J., J. Cvacka, A. Muck, V. Quaiserova, and J. Zima. 2001. Electrochemical methods for monitoring of environmental carcinogens. *Fresenius J. Anal. Chem.* 369:556–562.
727. Abu Zuhri, A.Z. and W. Voelter. 1998. Applications of adsorptive stripping voltammetry for the trace analysis of metals, pharmaceuticals, and biomolecules. *Fresenius J. Anal. Chem.* 360:1–9.
728. Ozkan, S.A., B. Uslu, and H.Y. Aboul-Enein. 2003. Analysis of pharmaceuticals and biological fluids using modern electroanalytical techniques. *Crit. Rev. Anal. Chem.* 33:155–181.
729. Radi, A. 1998. Preconcentration and voltammetric determination of indomethacin at carbon paste electrodes. *Electroanalysis* 10:103–106.
730. Shamsipur, M. and K. Farhadi. 2000. Adsorptive stripping voltammetric determination of ketoconazole in pharmaceutical preparations and urine using carbon paste electrodes. *Analyst* 125:1639–1643.
731. Peng, T., Z. Yang, and R. Lu. 1990. Adsorptive/extractive voltammetry for determining clozapine at carbon paste electrodes. *Gaodeng Xuexiao Huaxue Xuebao* 11:1067–1071.
732. Wang, J., B.A. Freiha, and B.K. Deshmukh. 1985. Adsorptive/extractive stripping voltammetry of phenothiazine compounds at carbon paste electrodes. *Bioelectrochem. Bioenerg.* 14:457–467.
733. Neuhold, C.G., K. Kalcher, X. Cai, and G. Raber. 1996. Catalytic determination of perchlorate using a modified carbon paste electrode. *Anal. Lett.* 29:1685–1704.
734. Kalcher, K. 1986. A new method for the voltammetric determination of nitrite. *Talanta* 33:489–494.
735. He, Q., J. Fei, and S. Hu. 2003. Voltammetric method based on an ion-pairing reaction for the determination of trace amount of iodide at carbon-paste electrodes. *Anal. Sci.* 19:681–686.
736. Galik, M., M. Cholota, I. Švancara, A. Bobrowski, and K. Vytřas. 2006. A study on stripping voltammetric determination of osmium(IV) at a carbon paste electrode modified in situ with cationic surfactants. *Electroanalysis* 18:2218–2224.

References

737. Švancara, I., I. Cermakova, K. Vytřas, W. Gossler, and K. Kalcher. 1999. Cationic surfactants as modifiers for carbon paste electrodes. Application to the determination of iodide. *Sci. Pap. Univ. Pardubice Ser. A* 5:95–108.
738. Mascarenhas, R.J., K.V. Reddy, B.E. Kumaraswamy, B.S. Sherigara, and V. Lakshminarayan. 2005. Electrochemical oxidation of L-dopa at a carbon paste electrode. *Bull. Electrochem.* 21:341–345.
739. Mascarenhas, R.J., I.N. Namboothiri, B.S. Sherigara, and B.E. Kumaraswamy. 2006. Electro-reductive study of beta-nitrostyrene in surfactant medium at wax impregnated carbon paste electrode. *Bull. Electrochem.* 22:253–256.
740. Yi, H. and C. Li. 2007. Voltammetric determination of ciprofloxacin based on the enhancement effect of cetyltrimethylammonium bromide (CTAB) at carbon paste electrode. *Russ. J. Electrochem.* 43:1377–1381.
741. Hegde, R.N., R.R. Hosamani, and S.T. Nandibewoor. 2009. Electrochemical oxidation and determination of theophylline at a carbon paste electrode using cetyltrimethyl ammonium bromide as enhancing agent. *Anal. Lett.* 42:2665–2682.
742. Teixeira, M.F.S., F.C. Moraes, O. Fatibello-Filho, and N. Bocchi. 2001. Voltammetric determination of lithium ions in pharmaceutical formulation using a l-MnO_2-modified carbon-paste electrode. *Anal. Chim. Acta* 443:249–255.
743. Lefèvre, G., J. Bessière, and A. Walcarius. 1999. Cuprite-modified electrode for the detection of iodide species. *Sens. Actuators B Chem.* B59:113–117.
744. Ojani, R., J.B. Raoof, and P.S. Afagh. 2004. Electropolymerization of 1-naphthylamine at the surface and bulk of CPE. *Bull. Electrochem.* 20:251–259.
745. Stefan-van Staden, R.I., J.F. van Staden, and H.Y. Aboul Enein. 2007. Enantioselective Biosensors. In *Chiral Separation Techniques: A Practical Approach*, Chapter 13, Ed. G. Subramanian. Weinheim, Germany: Wiley/VCH.
746. Stefan, R.I., R.G. Bokretsion, J.F. van Staden, and H.Y. Aboul Enein. 2003. Determination of L- and D-enantiomers of methotrexate using amperometric biosensors. *Talanta* 60:983–990.
747. Aboul Enein, H.Y., R.I. Stefan, and G.L. Radu. 1999. Biosensor for enantioselective analysis of S-Cilazapril, S-Trandolapril, and S-Pentopril. *Pharm. Develop. Technol.* 4:251–255.
748. Maistrenko, V.N., G.K. Budnikov, and V.N. Gusakov. 1996. New possibilities in voltammetry: Extraction into the electrode bulk. *J. Anal. Chem.* 51:942–948.
749. Abelairas, A., M.J. Puertollano, R.M. Alonso, M.P. Elizalde, R.M. Jimenez, and M. Huebra. 1994. Voltammetric study of the extractant 4-methyl-N-quinolin-8-yl-benzenesulfonamide at a carbon paste electrode. *Analyst* 119:303–306.
750. Zhang, Y., X. Lu, K. Zhu, Z. Wang, and J. Kang. 2002. Voltammetric detection of traces of copper using a carbon paste electrode modified with tetraphenylporphyrin. *Anal. Lett.* 35:369–381.
751. Wang, J. and B.A. Freiha. 1983. Preconcentration and differential pulse voltammetry of butylated hydroxyanisole at a carbon paste electrode. *Anal. Chim. Acta* 154:87–94.
752. Shangguan, X. and J. Zheng. 2009. Direct electron transfer and electrocatalysis of myoglobin based on its direct immobilization on carbon ionic liquid electrode. *Electroanalysis* 21:881–886.
753. Walcarius, A., L. Lamberts, and E.G. Derouane. 1993. The methyl viologen incorporated zeolite modified carbon paste electrode. Part 1. Electrochemical behavior in aqueous media. Effects of supporting electrolyte and immersion time. *Electrochim. Acta* 38:2257–2266.
754. Wang, J. and A. Walcarius. 1996. Zeolite containing oxidase-based carbon paste biosensors. *J. Electroanal. Chem.* 404:237–242.
755. Walcarius, A., P. Mariaulle, and L. Lamberts. 2003. Zeolite-modified solid carbon paste electrodes. *J. Solid State Electrochem.* 7:671–677.
756. Murray, R.W. 1980. Chemically modified electrodes. *Acc. Chem. Res.* 13:135–141.
757. Murray, R.W. 1984. Chemically modified electrodes. In *Electroanalytical Chemistry*, Vol. 13, Ed. A.J. Bard, pp. 191–368. New York: Marcel Dekker.
758. Murray, R.W., A.G. Ewing, and R.A. Durst. 1987. Chemically modified electrodes. Molecular design for electrochemistry. *Anal. Chem.* 59:379A–385A.
759. Wang, J. 1988. Voltammetry following non-electrolytic preconcentration. In *Electroanalytical Chemistry*, Vol. 16, Chapt. 1, Ed. A.J. Bard. pp. 1–89, New York: Marcel Dekker.
760. Watkins, B.F., J.R. Behling, E. Kariv, and L.L. Miller. 1975. Chiral electrode. *J. Am. Chem. Soc.* 97:3549–3550.
761. Horner, L. and W. Brich. 1977. Studies on the occurrence of hydrogen transfer, 42. Electroreduction of prochiral aryl ketones on surface-modified carbon electrodes. *Justus Liebigs Ann. Chem.* 8:1354–1364.

762. Cai, X., K. Kalcher, C. Neuhold, and B. Ogorevc. 1990. An improved voltammetric method for the determination of trace amounts of uric acid with electrochemically pretreated carbon paste electrodes. *Talanta* 41:407–413.
763. Barek, J., K. Peckova, and V. Vyskocil. 2008. Adsorptive stripping voltammetry of environmental carcinogens. *Curr. Anal. Chem.* 4:242–249.
764. Parra, V., T. Del Cano, M.L. Rodriguez-Mendez, J.A. de Saja, and R.F. Aroca. 2004. Electrochemical characterization of two perylenetetracarboxylic diimides: Langmuir-Blodgett films and carbon paste electrodes. *Chem. Mater.* 16:358–364.
765. Arrieta, A., M.L. Rodriguez-Mendez, and J.A. de Saja. 2003. Langmuir-Blodgett film and carbon paste electrodes based on phthalocyanines as sensing units for taste. *Sens. Actuators B* B95:357–365.
766. Xu, G.-Y., K. Jiao, J.-S. Fan, and B. Zhang. 2007. Immobilizing DNA on cetyltrimethyl ammonium bromide cationic membrane for the detection of specific gene related to NPT II. *Asian J. Chem.* 19:4161–4172.
767. Han, X., J. Ma, and Z. Gao. 2007. Electrocatalytic oxidation of ascorbic acid at cetyl pyridinium bromide-self-assembly monolayers in situ modified carbon paste electrodes. *Ningxia Daxue Xuebao, Ziran Kexueban* 28:147–150.
768. Grundler, P. 2007. *Chemical Sensors. An Introduction for Scientists and Engineers*, 1st Edn., Berlin, Germany: Springer. ISBN-10 3540457429; ISBN-13 978-3540457428.
769. Yantasee, W., G.E. Fryxell, M.M. Conner, and Y. Lin. 2005. Nanostructured electrochemical sensors based on functionalized nanoporous silica for voltammetric analysis of lead, mercury, and copper. *J. Nanosci. Nanotechnol.* 5:1537–1540.
770. Yantasee, W., Y.-H. Lin, G.E. Fryxell, and Z.-M. Wang. 2004. Carbon paste electrode modified with carbamoylphosphonic acid functionalized mesoporous silica: A new mercury-free sensor for uranium detection. *Electroanalysis* 16:870–873.
771. Yantasee, W., Y.-H. Lin, G.E. Fryxell, and B.J. Busche. 2004. Simultaneous detection of cadmium, copper, and lead using a carbon paste electrode modified with carbamoyl-phosphonic acid self-assembled monolayer on mesoporous silica (SAMMS). *Anal. Chim. Acta* 502:207–212.
772. Yantasee, W., Y.H. Lin, T.S. Zemanian, and G.E. Fryxell. 2003. Voltammetric detection of lead(II) and mercury(II) using a carbon paste electrode modified with thiol self-assembled monolayer on mesoporous silica (SAMMS). *Analyst (U.K.)* 128:467–462.
773. Ferancova, A. and J. Labuda. 2001. Cyclodextrins as electrode modifiers. *Fresenius J. Anal. Chem.* 370:1–10.
774. Mashhadizadeh, M.H., K. Eskandari, A. Foroumadi, and A. Shafiee. 2008. Self-assembled mercapto-compound-gold-nanoparticle-modified carbon paste electrode for potentiometric determination of cadmium(II). *Electroanalysis* 20:1891–1896.
775. Granero, A.M., H. Fernandez, E. Agostini, and M.A. Zon. 2008. An amperometric biosensor for trans-resveratrol determination in aqueous solutions by means of carbon paste electrodes modified with peroxidase basic isoenzymes from Brassica napus. *Electroanalysis* 20:858–864.
776. Takatsy, A., B. Csoka, L. Nagy, and G. Nagy. 2006. Periodically interrupted amperometry at membrane coated electrodes: A simplified pulsed amperometry. *Talanta* 69:281–285.
777. Kinoshita, H. 2005. Amperometric assay of glutathione in whole blood and serum. *Bunseki Kagaku* 54:71–74.
778. Ciucu, A.A., C. Negulescu, and R.P. Baldwin. 2003. Detection of pesticides using an amperometric biosensor based on ferrophthalocyanine chemically modified carbon paste electrode and immobilized bienzymatic system. *Biosens. Bioelectron.* 18:303–310.
779. Forzani, E.S., M.E. Bernardi, A.M. Sisti, J.A. Zarzur, and V.M. Solis. 2002. Indirect determination of pepsin activity in human seric protein samples by polyphenol oxidase electrodes. *Anal. Chim. Acta* 460:163–175.
780. Ikeda, T., K. Kato, M. Maeda, H. Tatsumi, K. Kano, and K. Matsushita. 1997. Electrocatalytic properties of *Acetobacter aceti* cells immobilized on electrodes for the quinone-mediated oxidation of ethanol. *J. Electroanal. Chem.* 430:197–204.
781. Bergmann, W., R. Rudolph, and U. Spohn. 199. A bienzyme modified carbon paste electrode for amperometric detection of pyruvate. *Anal. Chim. Acta* 394:233–241.
782. Bu, H.Z., S.R. Mikkelsen, and A.M. English. 1998. NAD(P)H sensors based on enzyme entrapment in ferrocene-containing polyacrylamide-based redox gels *Anal. Chem.* 70:4320–4325.
783. Mullor, S.G., M. Sanchez-Cabezudo, A.J.M. Ordieres, and B.L. Ruiz. 1996. Alcohol biosensor based on alcohol dehydrogenase and Meldola Blue immobilized into a carbon paste electrode. *Talanta* 43:779–784.
784. Lobo, M.J., A.J. Miranda, and P. Tunon. 1996. A comparative study of some phenoxazine and phenothiazine modified carbon paste electrodes for ethanol determination. *Electroanalysis* 8:591–596.

References

785. Bu, H.Z., S.R. Mikkelsen, and A.M. English. 1995. Characterization of a ferrocene-containing polyacrylamide-based redox gel for biosensor use. *Anal. Chem.* 67:4071–4076.
786. Takayama, K., T. Kurosaki, T. Ikeda, and T. Nagasawa. 1995. Bioelectrocatalytic hydroxylation of nicotinic acid at an electrode modified with immobilized bacterial cells of *Pseudomonas fluorescens* in the presence of electron transfer mediators. *J. Electroanal. Chem.* 381:47–53.
787. Amako, K., H. Yanai, T. Ikeda, T. Shiraishi, M. Takahashi, and K. Asada. 1993. Dimethyl-benzoquinone-mediated photoelectrochemical oxidation of water at a carbon paste electrode coated with photosystem II membranes. *J. Electroanal. Chem.* 362:71–77.
788. Ikeda, T., Y. Hashimoto, M. Senda, and Y. Isono. 1991. Nucleoside oxidase-modified carbon paste electrode containing quinone: Cathodic current response to nucleosides. *Electroanalysis* 3:891–897.
789. Bontempelli, G., N. Comisso, R. Toniolo, and G. Schiavon. 1997. Electroanalytical sensors for nonconducting media based on electrodes supported on perfluorinated ion-exchange membranes. *Electroanalysis* 9:433–443.
790. Harth, R., D. Ozer, J. Hayon, R. Ydgar, and A. Bettelheim. 1994. Application of ionically conductive polymers of perfluorosulfonic acid coatings and membranes in solid-state electrochemistry. *Curr. Top. Electrochem.* 3:531–543.
791. Ugo, P., L.M. Moretto, and F. Vezza. 2002. Ionomer-coated electrodes and nanoelectrode ensembles as electrochemical environmental sensors: Recent advances and prospects. *Chem. Phys. Chem.* 3:917–925.
792. Leech, D. 1996. Analytical applications of polymer-modified electrodes. *Electroact. Polym. Electrochem.* 2:269–296.
793. Mount, A.R. 1997. Transmission line models for modified electrodes. *Res. Chem. Kinet.* 4:1–29.
794. Bott, A.W. 2001. Electrochemical techniques for the characterization of redox polymers. *Curr. Sep.* 19:71–75.
795. Deronzier, A. and J.C. Moutet. 1994. Functionalized polypyrroles as versatile molecular materials for electrode modification. *Curr. Top. Electrochem.* 3:159–200.
796. Cosnier, S., C. Gondran, and A. Senillou. 1999. Functionalized polypyrroles: A sophisticated glue for the immobilization and electrical wiring of enzymes. *Synth. Met.* 102:1366–1369.
797. Cosnier, S. 2000. Biosensors based on immobilization of biomolecules by electrogenerated polymer films. New perspectives. *Appl. Biochem. Biotechnol.* 89:127–138.
798. Bobacka, J., A. Ivaska, and A. Lewenstam. 2003. Potentiometric ion sensors based on conducting polymers. *Electroanalysis* 15:366–374.
799. Kalcher, K. 1986. Voltammetric behavior of gold at a dithizone-modified carbon-paste electrode. *Fresenius J. Anal. Chem.* 325:181–185.
800. Cai, X., K. Kalcher, W. Diewald, C. Neuhold, and R.J. Magee. 1993. Voltammetric determination of trace amounts of mercury with a carbon paste electrode modified with an anion-exchanger. *Fresenius J. Anal. Chem.* 345:25–31.
801. Talaber, C., K. Kalcher, G. Koelbl, G. Raber, and X.-H. Cai. 1992, Voltammetric determination of gold with a carbon paste electrode modified with Rhodamine B hexadecyl ester. *Sci. Int. (Lahore)* 4:347–350.
802. Kalcher, K. 1985. Voltammetric determination of trace amounts of gold with a chemically modified carbon paste electrode. *Anal. Chim. Acta* 177:175–182.
803. Kalcher, K. 1986. Voltammetry of hexacyanoferrates using a chemically modified carbon-paste electrode. *Analyst (U.K.)* 111:625–630.
804. Neuhold, C.G., K. Kalcher, W. Diewald, X.-H. Cai, and G. Raber. 1994. Voltammetric determination of nitrate with a modified carbon paste electrode. *Electroanalysis* 6:227–236.
805. Wu, W.S., M.S. Uddin, H. Chi, and K. Hidajat. 1994. Electrochemically assisted metal uptake by cation exchange based chemically modified electrodes. *J. Appl. Electrochem.* 24:548–553.
806. Kalcher, K. 1985. Voltammetrische bestimmung von Iodid mit einer mit Ionenaustauschern modifizierten Kohlepasteelektrode. *Fresenius Z. Anal. Chem.* 321:666–670.
807. Hernandez, L., J.M. Melguizo, M.H. Blanco, and P. Hernandez. 1989. Determination of cadmium(II) with a carbon paste electrode modified with an ion-exchange resin. *Analyst* 114:397–399.
808. Escribano, M.T.S., J.R. Prokopio, J.M.P. Macias, and L. Hernandez. 1989. Determination of lead in rainwater and human urine with a carbon paste electrode modified with a chelating resin. *Int. J. Environ. Anal. Chem.* 37:107–115.
809. Kalcher, K., H. Greschonig, and R. Pietsch. 1987. Extractive preconcentration of gold with carbon paste electrodes modified with organophosphorus compounds. *Fresen. Z. Anal. Chem.* 327:513–517.
810. K. Kalcher. 1986. Voltammetric determination of bismuth with carbon paste electrodes modified with alkylmercaptans. *Fresenius Z. Anal. Chem.* 325:186–190.

811. Cai, X., B. Ogorevc, G. Tavcar, and K. Kalcher. 1995. Insoluble inorganic salts as carbon paste electrode modifiers for preconcentration and voltammetric determination of oxalic acid. *Electroanalysis* 7:639–643.
812. Renger, F., I. Švancara, and M. Suska. 1992/93. Voltammetric determination of nickel(II) and cobalt(II) at a carbon paste electrode and its application under expedition conditions. *Sb. Ved. Pr., Vys. Sk. Chemickotechnol., Pardubice* 56:5–19.
813. Li, J., Z. Yan, F. Yan, J. Fei, and L. Yi. 2007. Study on anodic adsorptive voltammetry of palladium-dimethyl glyoxime complex at a carbon paste electrode. *Yejin Fenxi* 27:23–26.
814. Holgado, T.M., J.M.P. Macias, and L.H. Hernandez. 1995. Voltammetric determination of lead with a chemically modified carbon paste electrode with diphenyl-thiocarbazone. *Anal. Chim. Acta* 309:117–122.
815. Vasquez, M.D., M.L. Tascon, and L. Deban. 2006. Determination of Pb(II) with a dithizone-modified carbon paste electrode. *J. Environ. Sci.* 41:2735–2742.
816. Raber, G., K. Kalcher, C.G. Neuhold, C. Talaber, and G. Koelbl. 1995. Adsorptive stripping voltammetry of palladium(II) with thioridazine in situ modified carbon paste electrodes. *Electroanalysis* 7:138–142.
817. Stadlober, M., K. Kalcher, and G. Raber. 1997. Voltammetric determination of titanium, vanadium, and molybdenum using a carbon paste electrode modified with cetyl-trimethylammonium bromide. *Sci. Pap. Univ. Pardubice, Ser. A* 3:103–137.
818. Koelbl, G., K. Kalcher, and A. Voulgaropoulos. 1992. Voltammetric determination of gold with a Rhodamine B-modified carbon paste electrode. *Fresenius J. Anal. Chem.* 342:83–86.
819. Brainina, Kh.Z., A.V. Chernysheva, N.Y. Stozhko, and L.N. Kalnyshevskaya. 1989. In situ modified electrodes in stripping voltammetry. *Analyst* 114:173–180.
820. Brainina, Kh.Z., N.A. Malakhova, and N.Y. Stojko. 2000. Stripping voltammetry in environmental and food analysis. *Fresenius J. Anal. Chem.* 368:307–325.
821. Abu-Shawish, H.M. and S.M. Saadeh. 2007. Chemically modified carbon paste electrode for potentiometric analysis of cyproheptadine hydrochloride in serum and urine. *Anal. Chem. Indian J.* 4:20–27.
822. Abu Shawish, H.M. 2008. Potentiometric response (to chlorpheniramine maleate) at modified carbon paste electrode based on mixed ion-exchangers. *Electroanalysis* 20:491–497.
823. Lorenzo, E., E. Alda, P. Hernandez, M.H. Blanco, and L. Hernandez. 1988. Voltammetric determination of nitrobenzene with a chemically modified carbon paste electrode. Application to wines, beers, and cider. *Fresenius Z. Anal. Chem.* 330:139–142.
824. Hernandez, L., P. Hernandez, and Z. Sosa. 1988. Determination of phenol by differential-pulse voltammetry with a sepiolite-modified carbon paste electrode. *Fresenius Z. Anal. Chem* 331:525–527.
825. Hernandez, P., J. Vicente, and L. Hernandez. 1989. Determination of tetramethrin (neo-pynamin) by differential pulse voltammetry with a carbon paste electrode modified with sepiolite. *Fresenius Z. Anal. Chem* 334:550–553.
826. Hernandez, L., P. Hernandez, E. Lorenzo, and Z. Sosa Ferrera. 1988. Comparative study of the electrochemical behavior of sepiolite- and hectorite-modified carbon paste electrodes in the determination of dinocap. *Analyst* 113:621–623.
827. Hernandez, L., E. Gonzales, and P. Hernandez. 1988. Determination of Clozapine by adsorptive anodic voltammetry using glassy carbon and modified carbon paste electrodes. *Analyst (U.K.)* 113:1715–1718.
828. Hernandez, P., E. Alda, and L. Hernandez. 1987. Determination of mercury(II) using a modified electrode with zeolite. *Fresenius Z. Anal. Chem.* 327:676–678.
829. Murr, N.E., M. Kerkeni, A. Sellami, and Y.B. Taarit. 1988. The zeolite-modified carbon paste electrode. *J. Electroanal. Chem.* 246:461–465.
830. Shaw, B.R., K.E. Creasy, C.J. Lanzcyncki, J.A. Sargeant, and M. Tirhado. 1988. Voltammetric response of zeolite-modified electrodes. *J. Electrochem. Soc.* 135:869–875.
831. Walcarius, A. 1998. Analytical applications of silica-modified electrodes. A comprehensive review. *Electroanalysis* 10:1217–1235.
832. Sugawara, K., A. Terauchi, N. Kamiya, G. Hirabayashi, and H. Kuramitz. 2008. Voltammetric behaviors of wheat-germ agglutinin on a chitin-modified carbon-paste electrode. *Anal. Sci.* 24:583–587.
833. Lan, Y., G. Lu, X. Wu, S. Yao, F. Song, and H. Xu. 1999. Determination of trace nitrite with a chitin-containing carbon paste electrode. *Fenxi Shiyanshi* 18:27–30.
834. Lan, Y., G. Lu, S. Yao, and F. Song. 1998. Determination of trace copper by chitin-modified carbon paste electrode. *Fenxi Huaxue* 26:1192–1195.
835. Rodrigues, C.A., V.T. de Favere, E. Stadler, and M.C.M. Laranjeira. 1994. Pre-concentration of anions on poly(N-acetyl-D-glucosamine) derivative-modified carbon paste electrode. *J. Brazil. Chem. Soc.* 4:14–16.

836. Bai, Z.P., T. Nakamura, and K. Izutsu. 1990. Enhanced voltammetric waves of iron(III)-EDTA at a chitin-containing carbon paste electrode and its analytical application. *Electroanalysis* 2:75.
837. Navratilova, Z. and P. Kula. 1998. Carbon paste electrodes modified with clays and humic acids. *Sci. Pap. Univ. Pardubice Ser. A Faculty Chem. Techn.* 3:195–206.
838. Wang, J., N. Naser, L. Angnes, H. Wu, and L. Chen. Metal-dispersed carbon paste electrodes. *Anal. Chem.* 64:1285–1288.
839. Sanchez, A., A. Zapardiel, F.L. de Prado, E. Bermejo, M. Moreno, J.A. Perez Lopez, and M. Chicharro. 2007. Flow injection analysis of Maleic hydrazide using an electrochemical sensor based on palladium-dispersed carbon paste electrode. *Electroanalysis* 19:1683–1688.
840. Matakova, R.N., R.F. Yulmetova, and A.I. Zebreva. 1984. Carbon paste electrodes for the study of metallic catalysts. *Zhurnal Analiticheskoi Khimii* 39:1200–1205.
841. Raoof, J.B., M.A. Karimi, S.R. Hosseini, and S. Mangelizadeh. 2010. Cetyltrimethylammonium bromide effect on highly electrocatalysis of methanol oxidation based on nickel particles electrodeposited into poly(m-toluidine) film on the carbon paste electrode. *J. Electroanal. Chem.* 638:33–38.
842. Beyene, N.W., P. Kotzian, K. Schachl, H. Alemu, E. Turkusic, A. Copra, H. Moderegger, I. Švancara, K. Vytřas, and K. Kalcher. 2004. (Bio)sensors based on manganese dioxide-modified carbon substrates: Retrospections, further improvements and applications. *Talanta* 64:1151–1159.
843. Garjonyte, R. and A. Malinauskas. 1998. Electrocatalytic reactions of hydrogen peroxide at carbon paste electrodes modified by some metal hexacyanoferrates. *Sens. Actuators B* B46:236–241.
844. Itaya, K., I. Uchida, and S. Toshima. 1984. Catalysis of the reduction of molecular oxygen at mixed valence complex modified electrodes of Prussian blue analogs. *Nippon Kagaku Kaishi* 11:1849–1853.
845. Boyer, A., K. Kalcher, and R. Pietsch. 1990. Voltammetric behavior of perborate on Prussian blue-modified carbon paste electrodes. *Electroanalysis* 2:155–161.
846. Weissenbacher, M., K. Kalcher, H. Greschonig, W. Ng, W.-H. Chan, and A.N. Voulgaropoulos. 1992. Electrochemical behavior of persulfate on carbon paste electrodes modified with Prussian blue and analogous compounds. *Fresenius J. Anal. Chem.* 344:87–92.
847. Gomez de Rio, M.I., C. De la Fuente, J.A. Acuna, M.D. Vazquez Barbado, M. Tascon Garcia, S. de Vicente Perez, and P. Sanchez Batanero. 1995. Determination of nitrites by using an electrocatalytic method on Prussian Blue chemically carbon paste electrode. *Quim Anal. (Barcelona)* 14:108–111.
848. Moscone, D., D. D'Ottavi, D. Compagnone, G. Palleschi, and A. Amine. 2001. Construction and analytical characterization of Prussian blue-based carbon paste electrodes and their assembly as oxidase enzyme sensors. *Anal. Chem.* 73:2529–2535.
849. Ivama, V.M. and S.H.P. Serrano. 2003. Rhodium-Prussian blue modified carbon paste electrode (Rh-PBMCPE) for amperometric detection of hydrogen peroxide. *J. Brazil. Chem. Soc.* 14:551–555.
850. Li, J., X. Wei, and Y. Yuan. 2009. Synthesis of magnetic nanoparticles composed by Prussian blue and glucose oxidase for preparing highly sensitive and selective glucose biosensor. *Sens. Actuators B* B139:400–406.
851. Lin, M.S., J.S. Lai, and J.S. Wang. 2002. Nickel(II) hexacyanoferrate based glucose biosensor. *Huaxue* 60:483–493.
852. Zaldivar, G.A.P. and Y. Gushikem. 1992. Tin(IV) oxide grafted on a silica gel surface as a conducting substrate base for nickel hexacyanoferrate. *J. Electroanal. Chem.* 337:165–174.
853. Nossol, E. and A.J.G. Zarbin. 2009. A simple and innovative route to prepare a novel carbon nanotube/Prussian blue electrode and its utilization as a highly sensitive H_2O_2 amperometric sensor. *Adv. Funct. Mater.* 19:3980–3986.
854. Dicks, J.M., W.J. Aston, G. Davis, and A.P.F. Turner. 1986. Mediated amperometric biosensors for D-galactose, glycolate and L-amino acids based on a ferrocene-modified carbon paste electrode. *Anal. Chim. Acta* 182:103–112.
855. Miki, K., T. Ikeda, S. Todoriki, and M. Senda. 1989. Bioelectrocatalysis at NAD-dependent dehydrogenase and diaphorase-modified carbon paste electrodes containing mediators. *Anal. Sci.* 5:269–274.
856. Hale, P.D., T. Inagaki, H.S. Lee, H.I. Karan, Y. Okamoto, and T.A. Skotheim. 1990. Amperometric glycolate sensors based on glycolate oxidase and polymeric electron transfer mediators. *Anal. Chim. Acta* 228:31–37.
857. Wang, J., L.H. Wu, R. Li, and J. Sanchez. 1990. Mixed ferrocene-glucose oxidase-carbon-paste electrode for amperometric determination of glucose. *Anal. Chim. Acta* 228:251–257.
858. Okuma, H., H. Takahashi, S. Sekimukai, K. Kawahara, and R. Akahoshi. 1991. Mediated amperometric biosensor for hypoxanthine based on a hydroxymethyl ferrocene-modified carbon paste electrode. *Anal. Chim. Acta* 244:161–164.

859. Beh, S.K., G.J. Moody, and J.D.R. Thomas. 1991. Studies on enzyme electrodes with ferrocene and carbon paste bound with cellulose triacetate. *Analyst* 116:459–462.
860. Dominguez Sanchez, P., A.J. Miranada Ordieres, A. Costa Garcia, and P. Tunon Blanco. 1991. Peroxidase-ferrocene modified carbon paste electrode as an amperometric sensor for the hydrogen peroxide assay. *Electroanalysis* 3:281–285.
861. Mizutani, F., A. Okuda, S. Yabuki, and T. Katsura. 1991. Glucose-sensing electrode based on carbon paste containing PEG-attached enzyme and ferrocene. *Chem. Sens.* 7(Suppl. A):17–20.
862. Hale, P.D., H.L. Lan, L.I. Boguslavsky, H.I. Karan, Y. Okamoto, and T.A. Skotheim. 1991. Amperometric glucose sensors based on ferrocene-modified poly(ethylene oxide) and glucose oxidase. *Anal. Chim. Acta* 251:121–128.
863. Kulys, J., W. Schuhmann, and H.L. Schmidt. 1992. Carbon-paste electrodes with incorporated lactate oxidase and mediators. *Anal. Lett.* 25:1011–1024.
864. Lin, M.S., M. Hare, and G.A. Rechnitz. 1992. Multienzyme containing tissue-based and ferrocene-mediated bioelectrode for the determination of polyamines. *Electroanalysis* 4:521–525.
865. Andrieux, C.P., P. Audebert, B. Divisia-Blohorn, and S. Linquette-Maillet. 1993. A new glucose sensor made of a Nafion-gel-layer covered carbon paste. *J. Electroanal. Chem.* 353:289–296.
866. Ciszewski, A. and Z. Gorski. 1995. Stripping measurements of hydrogen peroxide based on biocatalytic accumulation at a horseradish peroxidase/ferrocene/carbon paste electrode. *Electroanalysis* 7:495–497.
867. Oungpipat, W., P.W. Alexander, and P. Southwell-Keely. 1995. A reagentless amperometric biosensor for hydrogen peroxide determination based on asparagus tissue and ferrocene mediation. *Anal. Chim. Acta* 309:35–45.
868. Chi, Q.J., W. Gopel, T. Ruzgas, L. Gorton, and P. Heiduschka. 1997. Effects of pretreatments and modifiers on electrochemical properties of carbon paste electrodes. *Electroanalysis* 9:357–365.
869. Nakabayashi, Y., M. Wakuda, and H. Imai. 1998. Amperometric glucose sensors fabricated by electrochemical polymerization of phenols on carbon paste electrodes containing ferrocene as an electron transfer mediator. *Anal. Sci.* 14:1069–1076.
870. Guorong, Z., W. Xiaolei, S. Xingwang, and S. Tianling. 2000. β-Cyclodextrin-ferrocene inclusion complex modified carbon paste electrode for amperometric determination of ascorbic acid. *Talanta* 51:1019–1025.
871. Nakabayashi, Y. and H. Yoshikawa. 2000. Amperometric biosensors for sensing of hydrogen peroxide based on electron transfer between horseradish peroxidase and ferrocene as a mediator. *Anal. Sci.* 16:609–613.
872. Boujtita, M. and N. El Murr. 2000. Ferrocene-mediated carbon paste electrode modified with D-fructose dehydrogenase for batch mode measurement of D-fructose. *Appl. Biochem. Biotechnol.* 89:55–66.
873. Wang, J., J.-W. Mo, S. Li, and J. Porter. 2001. Comparison of oxygen-rich and mediator-based glucose-oxidase carbon-paste electrodes. *Anal. Chim. Acta* 441:183–189.
874. Raoof, J.B., R. Ojani, and A. Kiani. 2001. Carbon paste electrode spiked with ferrocene carboxylic acid and its application to the electrocatalytic determination of ascorbic acid. *J. Electroanal. Chem.* 515:45–51.
875. Ramirez-Molina, C., M. Boujtita, and N. El Murr. 2003. New strategy for dehydrogenase amperometric biosensors using surfactant to enhance the sensitivity of diaphorase/ferrocene modified carbon paste electrodes for electrocatalytic oxidation of NADH. *Electroanalysis* 15:1095–1100.
876. Serban, S. and N. El Murr. 2004. Synergetic effect for NADH oxidation of ferrocene and zeolite in modified carbon paste electrodes. New approach for dehydrogenase based biosensors. *Biosens. Bioelectron.* 20:161–166.
877. Raoof, J.B., R. Ojani, and M. Kolbadinezhad. 2005. Differential pulse voltammetry determination of L-cysteine with ferrocene-modified carbon paste electrode. *Bull. Chem. Soc. Jpn.* 78:818–826.
878. Delacote, C., J.-P. Bouillon, and A. Walcarius. 2006. Voltammetric response of ferrocene-grafted mesoporous silica. *Electrochim. Acta* 51:6373–6383.
879. Raoof, J.B., R. Ojani, and H. Karimi-Maleh. 2007. Electrocatalytic determination of sulfite at the surface of a new ferrocene derivative-modified carbon paste electrode. *Int. J. Electrochem. Sci.* 2:257–269.
880. Raoof, J.B., R. Ojani, and F. Chekin. 2007. Electrochemical analysis of D-penicillamine using a carbon paste electrode modified with ferrocene carboxylic acid. *Electroanalysis* 19:1883–1889.
881. Kamyabi, M.A. and F. Aghajanloo. 2009. Electrocatalytic response of dopamine at a carbon paste electrode modified with ferrocene. *Croat. Chem. Acta* 82:599–606.
882. Korfhage, K.M., K. Ravichandran, and R.P. Baldwin. 1984. Phthalocyanine-containing chemically modified electrodes for electrochemical detection in liquid chromatography/flow injection systems. *Anal. Chem.* 56:1514–1517.

883. Halbert, M.K. and R.P. Baldwin. 1985. Electrocatalytic and analytical response of cobalt phthalocyanine containing carbon paste electrodes toward sulfhydryl compounds. *Anal. Chem.* 57:591–595.
884. Halbert, M.K. and R.P. Baldwin. 1985. Determination of cysteine and glutathione in plasma and blood by liquid chromatography with electrochemical detection using a chemically modified electrode containing cobalt phthalocyanine. *J. Chromatogr. Biomed. Appl.* 345:43–49.
885. Santos, L.M. and R.P. Baldwin. 1986. Electrocatalytic response of cobalt phthalocyanine chemically modified electrodes toward oxalic acid and α-keto acids. *Anal. Chem.* 58:848–852.
886. Linders, C.R., G.J. Patriarche, J.M. Kauffmann, and G.G. Guilbault. 1986. Electrochemical study of thiols and disulfides using modified electrodes. *Anal. Lett.* 19:193–203.
887. Linders, C.R., B.J. Vincke, and G.J. Patriarche. 1986. Catalase like activity of iron phthalocyanine incorporated in a carbon paste electrode. *Anal. Lett.* 19:1831–1837.
888. Linders, C.R., J.M. Kauffmann, and G.J. Patriarche. 1986. Modified electrodes based on cobalt(II) phthalocyanine. Applications to the determination of organic sulfur compounds. *J. Pharm. Belgique* 41:373–379.
889. Santos, L.M. and R.P. Baldwin. 1987. Liquid chromatography/electrochemical detection of carbohydrates at a cobalt phthalocyanine containing chemically modified electrode. *Anal. Chem.* 59:1766–1770.
890. Santos, L.M. and R.P. Baldwin. 1988. Electrochemistry and chromatographic detection of monosaccharides, disaccharides, and related compounds at an electrocatalytic chemically modified electrode. *Anal. Chim. Acta* 206:85–96.
891. Tolbert, A.M., R.P. Baldwin, and L.M. Santos. 1989. Electrochemical detection of ribonucleotides at a carbon paste electrode containing cobalt phthalocyanine. *Anal. Lett.* 22:683–702.
892. Tolbert, A.M. and R.P. Baldwin. 1989. Liquid chromatography and electrochemical detection of alditols and acidic sugars at a cobalt phthalocyanine-containing chemically modified electrode. *Electroanalysis* 1:389–395.
893. Skladal, P. 1991. Determination of organophosphate and carbamate pesticides using a cobalt phthalocyanine-modified carbon paste electrode and a cholinesterase enzyme membrane. *Anal. Chim. Acta* 252:11–15.
894. Wang, J., L. Angnes, C. Liang, and O. Evans. 1991. Electrocatalysis and amperometric detection of organic peroxides at modified carbon-paste electrodes. *Talanta* 38:1077–1081.
895. Ruiz, M.A., M.P. Calvo, and J.M. Pingarron. 1994. Catalytic-voltammetric determination of the antioxidant tert-butylhydroxyanisole (BHA) at a nickel phthalocyanine modified carbon paste electrode. *Talanta* 41:289–294.
896. Qi, X.-H. and R.P. Baldwin. 1993. Liquid chromatography and electrochemical detection of organic peroxides by reduction at an iron phthalocyanine chemically modified electrode. *Electroanalysis* 5:547–554.
897. Fernandez Sanchez, C., A.J. Reviejo, and J.M. Pingarron. 1995. Voltammetric determination of the herbicides Thiram and Disulfiram with a cobalt phthalocyanine modified carbon paste electrode. *Analusis (Bruxelles)* 23:319–324.
898. Ozsoz, M., A. Erdem, E. Kilinc, and L. Gokgunnec. 1996. Mushroom-based cobalt phthalocyanine dispersed amperometric biosensor for the determination of phenolic compounds. *Electroanalysis* 8:147–150.
899. Mafatle, T.J. and T. Nyokong. 1996. Electrocatalytic oxidation of cysteine by molybdenum(V) phthalocyanine complexes. *J. Electroanal. Chem.* 408:213–218.
900. Santamaria, M.C., M.D.V. Barbado, M.L.T. Garcia, and P.S. Batanero. 1998. Determination of hydrogen peroxide by voltammetric techniques at carbon paste electrodes modified with transition metal phthalocyanines. *Quim. Anal. (Barcelona)* 17:147–152.
901. Nunes, G.S., D. Barcelo, B.S. Grabaric, J.M. Diaz-Cruz, and M.L. Ribeiro. 1999. Evaluation of a highly sensitive amperometric biosensor with low cholinesterase charge immobilized on a chemically modified carbon paste electrode for trace determination of carbamates in fruit, vegetable and water samples. *Anal. Chim. Acta* 399:37–49.
902. Cookeas, E.G. and C.E. Efstathiou. 2000. Flow injection-pulse amperometric detection of ephedrine at a cobalt phthalocyanine modified carbon paste electrode. *Analyst* 125:1147–1150.
903. Oni, J. and T. Nyokong. 2001. Simultaneous voltammetric determination of dopamine and serotonin on carbon paste electrodes modified with iron(II) phthalocyanine complexes. *Anal. Chim. Acta* 434:9–21.
904. Chicharro, M., A. Zapardiel, E. Bermejo, M. Moreno, and E. Madrid. 2002. Electro-catalytic amperometric determination of Amitrole using a cobalt phthalocyanine-modified carbon paste electrode. *Anal. Bioanal. Chem.* 373:277–283.
905. Chicharro, M., A. Zapardiel, E. Bermejo, E. Madrid, and C. Rodriguez. 2002. Flow injection analysis of Aziprotryne using an electrochemical sensor based on cobalt phthalocyanine modified carbon paste. *Electroanalysis* 14:892–898.

906. Oni, J., P. Westbroek, and T. Nyokong. 2002. Voltammetric detection of vitamin B1 at carbon paste electrodes and its determination in tablets. *Electroanalysis* 14:1165–1168.
907. Abbaspour, A., M.A. Mehrgardi, and R. Kia. 2004. Electrocatalytic oxidation of guanine and ss-DNA at a cobalt (II) phthalocyanine modified carbon paste electrode. *J. Electroanal. Chem.* 568:261–266.
908. Shahrokhian, S., A. Hamzehloei, A. Thaghani, and S.R. Mousavi. 2004. Electrocatalytic oxidation of 2-thiouracil and 2-thiobarbituric acid at a carbon-paste electrode modified with cobalt phthalocyanine. *Electroanalysis* 16:915–921.
909. Ozoemena, K.I., R.I. Stefan, and T. Nyokong. 2004. Determination of 2′,3′-dideoxyinosine using iron (II) phthalocyanine modified carbon paste electrode. *Anal. Lett.* 37:2641–2648.
910. Tahirovic, I., K. Kalcher, A. Sapcanin, Z. Pujic, and S. Muzaferovic. 2004. Utilization of butyrylthiocholine enzyme sensor for the detection of cholinesterase inhibitors in natural samples. *Pharmacia* 15:45–57.
911. Ozoemena, K.I., R.I. Stefan-van Staden, and T. Nyokong. 2009. Metallophthalocyanine based carbon paste electrodes for the determination of 2′,3′-dideoxyinosine. *Electroanalysis* 21:1651–1654.
912. Santos, W.J.R., A.L. Sousa, M.P.T. Sotomayor, F.S. Damos, S.M.C.N. Tanaka, L.T. Kubota, and A.A. Tanaka. 2009. Manganese phthalocyanine as a biomimetic electrocatalyst for phenols in the development of an amperometric sensor. *J. Braz. Chem. Soc.* 20:1180–1187.
913. Shahrokhian, S., M. Ghalkhani, and M.K. Amini. 2009. Application of carbon-paste electrode modified with iron phthalocyanine for voltammetric determination of epinephrine in the presence of ascorbic acid and uric acid. *Sens. Actuators B* B137:669–675.
914. Debnath, C., P. Saha, and A. Ortner. 2009. Electrocatalytic and analytical response of cobalt phthalocyanine modified carbon paste electrodes towards anti-malarial endoperoxide Artemisinin. *Electroanalysis* 21:657–661.
915. Yin, H.-S., Y.-L. Zhou, and S.-Y. Ai. 2009. Preparation and characteristic of cobalt phthalo-cyanine modified carbon paste electrode for bisphenol-A detection. *J. Electroanal. Chem.* 626:80–88.
916. Naik, R.R., E. Niranjana, B.E.K. Swamy, B.S. Sherigara, and H. Jayadevappa. 2008. Surfactant induced iron (II) phthalocyanine modified carbon paste electrode for simultaneous detection of ascorbic acid, dopamine and uric acid. *Int. J. Electrochem. Sci.* 3:1574–1583.
917. Conceicao, C.D.C., R.C. Faria, O. Fatibello, and A.A. Tanaka. 2008. Electrocatalytic oxidation and voltammetric determination of hydrazine in industrial boiler feed water using a cobalt phthalocyanine-modified electrode. *Anal. Lett.* 41:1010–1021.
918. Parra, V., A.A. Arrieta, J.A. Fernandez-Escudero, M.L. Rodriguez-Mendez, and J.A. De Saja. 2006. Electronic tongue based on chemically modified electrodes and voltammetry for the detection of adulterations in wines. *Sens. Actuators B* B118:448–453.
919. Parra, V., A.A. Arrieta, J.A. Fernandez-Escudero, H. Garcia, C. Apetrei, M.L. Rodriguez-Mendez, and J.A. de Saja. 2006. E-tongue based on a hybrid array of voltammetric sensors based on phthalocyanines, perylene derivatives and conducting polymers: Discrimination capability towards red wines elaborated with different varieties of grapes. *Sens. Actuators B* B115:54–61.
920. Raoof, J.B., R. Ojani, and M. Baghayeri. 2009. Simultaneous electrochemical determination of glutathione and tryptophan on a nano-TiO$_2$/ferrocene carboxylic acid modified carbon paste electrode. *Sens. Actuators B* B143:261–269.
921. Li, J., L.T. Xiao, X.M. Liu, G.M. Zeng, G.H. Huang, G.L. Shen, and R.Q. Yu. 2003. Amperometric biosensor with HRP immobilized on a sandwiched nano-Au/polymerized m-phenylenediamine film and ferrocene mediator. *Anal. Bioanal. Chem.* 376:902–907.
922. Gorton, L., H.I. Karan, P.D. Hale, T. Inagaki, Y. Okamoto, and T.A. Skotheim. 1990. A glucose electrode based on carbon paste chemically modified with a ferrocene-containing siloxane polymer and glucose oxidase, coated with a poly(ester-sulfonic acid) cation-exchanger. *Anal. Chim. Acta* 228:23–30.
923. Milagres, B.G., L.T. Kubota, and G. de Oliveira Neto. 1996. Immobilized ferrocene and glucose oxidase on titanium(IV) oxide grafted onto a silica gel surface and its application as an amperometric glucose biosensor. *Electroanalysis* 8:489–493.
924. Patel, H., X. Li, and H.I. Karan. 2003. Amperometric glucose sensors based on ferrocene containing polymeric electron transfer systems—A preliminary report. *Biosens. Bioelectron.* 18:1073–1076.
925. Qi, X., R.P. Baldwin, H. Li, and T.F. Guarr. 1991. Electrocatalytic amperometric detection at polymeric cobalt phthalocyanine electrodes. *Electroanalysis* 3:119–124.
926. Bashkin, J.K. and P.J. Kinlen. 1990. Oxygen-stable ferrocene reference electrodes. *Inorg. Chem.* 29:4507–4509.
927. Hammouti, B., A. Benayada, and A. Elmaslout. 1997. Realization of ferrocene reference electrode. Part I: In concentrated phosphoric acid media. *Bull. Electrochem.* 13:466–469.

References

928. Hammouti, B., A. Elmaslout, A. Benayada, and H. Oudda. 2004. Realisation of ferrocene reference electrode in isoacidic H_3PO_4, H_2SO_4 and HCl mixtures. *Trans. SAEST* 39:81–83.
929. Shahrokhian, S., M.K. Amini, I. Mohammadpoor-Baltork, and S. Tangestaninejad. 2000. Potentiometric detection of 2-mercaptobenzimidazole and 2-mercapto-benzothiazole at cobalt phthalocyanine modified carbon-paste electrode. *Electroanalysis* 12:863–867.
930. Amini, M.K., S. Shahrokhian, and Tangestaninejad. 1999. Cobalt phthalocyanine modified carbon paste electrode as a potentiometric sensor for determination of ascorbic acid. *J. Sci. Islam. Rep. Iran* 10:218–222.
931. Shahrokhian, S. and J. Yazdani. 2003. Electrocatalytic oxidation of thioglycolic acid at carbon paste electrode modified with cobalt phthalocyanine: Application as a potentiometric sensor. *Electrochim. Acta* 48:4143–4148.
932. Yang, G.-J., K. Wang, J.-J. Xu, and H.-Y. Chen. 2004. Determination of Theophylline in drugs and tea on nanosized cobalt phthalocyanine particles modified carbon paste electrode. *Anal. Lett.* 37:629–643.
933. Yang, G.J., J.J. Xu, K. Wang, and H.Y. Chen. 2006. Electrocatalytic oxidation of dopamine and ascorbic acid on carbon paste electrode modified with nanosized cobalt phthalocyanine particles: Simultaneous determination in the presence of CTAB. *Electroanalysis* 18:282–290.
934. Siswana, M., K.I. Ozoemena, and T. Nyokong. 2006. Electrocatalytic behaviour of carbon paste electrode modified with iron(II) phthalocyanine (FePc) nanoparticles towards the detection of Amitrole. *Talanta* 69:1136–1142.
935. Persson, B., H.L. Lan, L. Gorton, Y. Okamoto, P.D. Hale, L.I. Boguslavsky, and T. Skotheim. 1993. Amperometric biosensors based on electrocatalytic regeneration of NAD+ at redox polymer-modified electrodes. *Biosens. Bioelectron.* 8:81–88.
936. Habermuller, K., M. Mosbach, and W. Schuhmann. 2000. Electron-transfer mechanisms in amperometric biosensors. *Fresenius J. Anal. Chem.* 366:560–568.
937. Shu, H.C., B. Mattiasson, B. Persson, G. Nagy, L. Gorton, S. Sahni, L. Geng, L. Boguslavsky, and T. Skotheim. 1995. A reagentless amperometric electrode based on carbon paste, chemically modified with D-lactate dehydrogenase, NAD+, and mediator containing polymer for D-lactic acid analysis. Part I. Construction, composition, and characterization. *Biotechnol. Bioeng.* 46:270–279.
938. Huan, Z., B. Persson, L. Gorton, S. Sahni, T. Skotheim, and P. Barlett. 1996. Redox polymers for electrocatalytic oxidation of NADH—Cationic styrene and ethylenimine polymers. *Electroanalysis* 8:575–581.
939. Jeffries, C., N. Pasco, K. Baronian, and L. Gorton. 1997. Evaluation of a thermophile enzyme for a carbon paste amperometric biosensor: L-glutamate dehydrogenase. *Biosens. Bioelectron.* 12:225–232.
940. Jezkova, J., E.I. Iwuoha, M.R. Smyth, and K. Vytřas. 1997. Stabilization of an osmium bis-bipyridyl polymer-modified carbon paste amperometric glucose biosensor using polyethyleneimine. *Electroanalysis* 9:978–984.
941. Adachi, N., D. Ishikawa, N. Sato, and H. Okuma. 2009. Electrochemical behavior of GOD-modified AuNPs/carbon paste electrode. *Mater. Technol. (Jpn.)* 27:75–80.
942. Huan, D.L., S.H. Zuo, and M.B. Lan. 2007. Glucose biosensors based on carbon paste electrodes modified by ferrocene and Au nanoparticles. *Huadong Ligong Daxue Xuebao, Ziran Kexueban* 33:822–826.
943. Cubukcu, M., S. Timur, and U. Anik. 2007. Examination of performance of glassy carbon paste electrode modified with gold nanoparticle and xanthine oxidase for xanthine and hypoxanthine detection. *Talanta* 74:434–439.
944. Wang, F.-C., R. Yuan, and Y.-Q. Chai. 2007. A new amperometric biosensor for hydrogen peroxide determination based on HRP-nanogold-PTH-nanogold-modified carbon paste electrodes. *Eur. Food Res. Technol.* 225:95–104.
945. Aguei, L., J. Manso, P. Yanez-Sedeno, and J.M. Pingarron. 2006. Amperometric biosensor for hypoxanthine based on immobilized xanthine oxidase on nanocrystal gold-carbon paste electrodes. *Sens. Actuators B* 113:272–280.
946. Miscoria, S.A., G.D. Barrera, and G.A. Rivas. 2005. Enzymatic biosensor based on carbon paste electrodes modified with gold nanoparticles and polyphenol oxidase. *Electroanalysis* 17:1578–1582.
947. Aguei, L., C. Pena-Farfal, P. Yanez-Sedeno, and J.M. Pingarron. 2007. Electrochemical determination of homocysteine at a gold nanoparticle-modified electrode. *Talanta* 74:412–420.
948. Liu, S.-Q. and H.-X. Ju. 2003. Nitrite reduction and detection at a carbon paste electrode containing hemoglobin and colloidal gold. *Analyst* 128:1420–1424.
949. Liu, S. and H. Ju. 2003. Electrocatalysis via direct electrochemistry of myoglobin immobilized on colloidal gold nanoparticles. *Electroanalysis* 15:1488–1493.
950. Huang, K.J., J.Y. Sun, C.X. Xu, D.J. Niu, and W.Z. Xie. 2010. A disposable immunosensor based on gold colloid modified chitosan nanoparticles-entrapped carbon paste electrode. *Microchim. Acta* 168:51–58.

951. Du, D., X. Xu, S. Wang, and A. Zhang. 2007. Reagentless amperometric carbohydrate antigen 19-9 immunosensor based on direct electrochemistry of immobilized horseradish peroxidase. *Talanta* 71:1257–1262.
952. Tang, D., R. Yuan, and Y. Chai. 2006. Electrochemical immuno-bioanalysis for carcinoma antigen 125 based on thionine and gold nanoparticles-modified carbon paste interface. *Anal. Chim. Acta* 564:158–165.
953. Huang, D., H. Liu, B. Zhang, K. Jiao, and X. Fu. 2009. Highly sensitive electrochemical detection of sequence-specific DNA of 35S promoter of cauliflower mosaic virus gene using CdSe quantum dots and gold nanoparticles. *Microchim. Acta* 165:243–248.
954. Zhang, Y., T. Yang, N. Zhou, W. Zhang, and K. Jiao. 2008. Nano Au/TiO$_2$ hollow microsphere membranes for the improved sensitivity of detecting specific DNA sequences related to transgenes in transgenic plants. *Sci. China B: Chem.* 51:1066–1073.
955. Pan, J. 2007. Voltammetric detection of DNA hybridization using a non-competitive enzyme linked assay. *Biochem. Eng. J.* 35:183–190.
956. Du, D., S. Liu, J. Chen, H. Ju, H. Lian, and J. Li. 2005. Colloidal gold nanoparticle modified carbon paste interface for studies of tumor cell adhesion and viability. *Biomaterials* 26:6487–6495.
957. Mashhadizadeh, M.H., H. Khani, A. Foroumadi, and P. Sagharichi. 2010. Comparative studies of mercapto thiadiazoles self-assembled on gold nanoparticle as ionophores for Cu(II) carbon paste sensors. *Anal. Chim. Acta* 665:208–214.
958. Mashhadizadeh, M.H. and H. Khani. 2010. Sol-gel-Au nanoparticle modified carbon paste electrode for potentiometric determination of Al(III) at the sub-ppb level. *Anal. Methods* 2:24–31.
959. Mashhadizadeh, M.H., K. Eskandari, A. Foroumadi, and A. Shafiee. 2008. Copper(II) modified carbon paste electrodes based on self-assembled mercapto compounds-gold-nanoparticle. *Talanta* 76:497–502.
960. Heli, H., F. Faramarzi, A. Jabbari, A. Parsaei, and A.A. Moosavi Movahedi. 2010. Electrooxidation and determination of Etidronate using copper nanoparticles and microparticles modified carbon paste electrodes. *J. Braz. Chem. Soc.* 21:16–24.
961. Heli, H., M. Zarghan, A. Jabbari, A. Parsaei, and A.A. Moosavi Movahedi. 2010. Electrocatalytic oxidation of the antiviral drug Acyclovir on a copper nanoparticles-modified carbon paste electrode. *J. Solid State Electrochem.* 14:787–795.
962. Heli, H., A. Jabbari, M. Zarghan, and A.A. Moosavi-Movahedi. 2009. Copper nanoparticles-carbon microparticles nanocomposite for electrooxidation and sensitive detection of sotalol. *Sens. Actuators B* B140:245–251.
963. Heli, H., M. Hajjizadeh, A. Jabbari, and A.A. Moosavi-Movahedi. 2009. Fine steps of electrocatalytic oxidation and sensitive detection of some amino acids on copper nanoparticles. *Anal. Biochem.* 388:81–90.
964. Heli, H., M. Hajjizadeh, A. Jabbari, and A.A. Moosavi-Movahedi. 2009. Copper nanoparticles-modified carbon paste transducer as a biosensor for determination of acetylcholine. *Biosens. Bioelectron.* 24:2328–2333.
965. Valentini, F., V. Biagiotti, C. Lete, G. Palleschi, and J. Wang. 2007. The electrochemical detection of ammonia in drinking water based on multi-walled carbon nanotube/copper nanoparticle composite paste electrodes. *Sens. Actuators B* B128:326–333.
966. Zhang, X. and X. Hu. 2005. Electrochemical determination of amino acid with carbon paste electrode modified by copper nanoparticles. *Yangzhou Daxue Xuebao, Ziran Kexueban* 8:16–20.
967. Brondani, D., C. Weber Scheeren, J. Dupont, and I. Cruz Vieira. 2009. Biosensor based on platinum nanoparticles dispersed in ionic liquid and laccase for determination of adrenaline. *Sens. Actuators B* B140:252–259.
968. Yoon, J.-H., G. Muthuraman, S.-B. Yoon, and M.-S. Won. 2007. Pt-nanoparticle incorporated carbon paste electrode for the determination of Cu(II) ion by anodic stripping voltammetry. *Electroanalysis* 19:1160–1166.
969. Huang, J., Y. Liu, H. Hou, and T. You. 2008. Simultaneous electrochemical determination of dopamine, uric acid and ascorbic acid using palladium nanoparticle-loaded carbon nanofibers modified electrode. *Biosens. Bioelectron.* 24:632–637.
970. Remita, H., P.F. Siril, I.M. Mbomekalle, B. Keita, and L. Nadjo. 2006. Activity evaluation of carbon paste electrodes loaded with platinum nanoparticles prepared in different radiolytic conditions. *J. Solid State Electrochem.* 10:506–511.
971. Shaidarova, L.G., I.A. Chelnokova, A.V. Gedmina, G.K. Budnikov, S.A. Ziganshina, A.A. Mozhanova, and A.A. Bukharaev. 2006. Electrooxidation of oxalic acid at a carbon-paste electrode with deposited palladium nanoparticles. *J. Anal. Chem.* 61:375–381.

972. Li, W. and S. Luo. 2009. Preparation of Ni nanoparticle-ionic liquid modified carbon paste electrode and detection of dopamine. *Fenxi Ceshi Xuebao* 28:1287–1290.
973. You, T. and Y. Liu. 2009. Method for preparing nonenzymatic glucose sensor from nickel nanoparticle/carbon nanofiber composite by electrospinning. *Faming Zhuanli Shenqing Gongkai Shuomingshu*. 10 pp. CODEN: CN 101603941 A 20091216.
974. Sun, W., X. Li, P. Qin, and K. Jiao. 2009. Electrodeposition of Co nanoparticles on the carbon ionic liquid electrode as a platform for myoglobin electrochemical biosensor. *J. Phys. Chem. C* 113:11294–11300.
975. Liu, Y., H. Teng, H. Hou, and T. You. 2009. Nonenzymatic glucose sensor based on renewable electrospun Ni nanoparticle-loaded carbon nanofiber paste electrode. *Biosens. Bioelectron.* 24:3329–3334.
976. Lai, G.S., H.L. Zhang, and D.Y. Han. 2008. Electrocatalytic response of dopamine at an iron nanoparticles-Nafion-modified carbon paste electrode. *Anal. Lett.* 41:3088–3099.
977. Wang, C.-Y. and X.-Y. Hu. 2005. Determination of Benorilate in pharmaceutical formulations and its metabolite in urine at a carbon paste electrode modified by silver nanoparticles. *Talanta* 67:625–633.
978. Wen, W., Y.-M. Tan, H.-Y. Xiong, and S.-F. Wang. 2010. Voltammetric and spectroscopic investigations of the interaction between colchicine and bovine serum albumin. *Intern. J. Electrochem. Sci.* 5:232–241.
979. Gao, H.-L. and J.-P. Li. 2008. Amperometric immunosensor based on magnetic inorganic bio-nanoparticles sensing films. *Fenxi Huaxue* 36:1614–1618.
980. Li, J. and H. Gao. 2008. A renewable potentiometric immunosensor based on Fe_3O_4 nanoparticles immobilized anti-IgG. *Electroanalysis* 20:881–887.
981. Yuan, Y.-H. and J.-P. Li. 2007. Biosensor based on immobilizing horseradish peroxidase on Fe_3O_4 nanoparticles. *Fenxi Huaxue* 35:1078–1082.
982. Fu, X.H. 2008. Magnetic-controlled non-competitive enzyme-linked voltammetric immunoassay for carcinoembryonic antigen. *Biochem. Eng. J.* 39:267–275.
983. Heli, H., S. Majdi, and N. Sattarahmady. 2010. Ultrasensitive sensing of N-acetyl–cysteine using an electrocatalytic transducer of nanoparticles of iron(III) oxide core-cobalt hexacyanoferrate shell. *Sens. Actuators B* B145:185–193.
984. Pham, T.T.H. and S.J. Sim. 2010. Electrochemical analysis of gold-coated magnetic nanoparticles for detecting immunological interaction. *J. Nanopart. Res.* 12:227–235.
985. Tang, D., R. Yuan, and Y. Chai. 2008. Magneto-controlled bioelectronics for the antigen-antibody interaction based on magnetic-core/gold-shell nanoparticles functionalized biomimetic interface. *Bioprocess Biosyst. Eng.* 31:55–61.
986. Qiu, J.D., M. Xiong, R.P. Liang, H.P. Peng, and F. Liu. 2009. Synthesis and characterization of ferrocene modified Fe_3O_4@Au magnetic nanoparticles and its application. *Biosens. Bioelectron.* 24:2649–2653.
987. Zhang, Y., G.M. Zeng, L. Tang, H.Y. Yu, and J.B. Li. 2007. Catechol biosensor based on immobilizing laccase to modified core-shell magnetic nanoparticles supported on carbon paste electrode. *Huanjing Kexue* 28:2320–2325.
988. Qiu, J., H. Peng, and R. Liang. 2007. Ferrocene-modified Fe_3O_4@SiO_2 magnetic nanoparticles as building blocks for construction of reagentless enzyme-based biosensors. *Electrochem. Commun.* 9:2734–2738.
979. Zhang, Y., G.-M. Zeng, L. Tang, D.-L. Huang, X.-Y. Jiang, and Y.-N. Chen. 2007. A hydroquinone biosensor using modified core-shell magnetic nanoparticles supported on carbon paste electrode. *Biosens. Bioelectron.* 22:2121–2126.
990. Liu, Z., Y. Liu, H. Yang, Y. Yang, G. Shen, and R. Yu. 2005. A phenol biosensor based on immobilizing tyrosinase to modified core-shell magnetic nanoparticles supported at a carbon paste electrode. *Anal. Chim. Acta* 533:3–9.
991. Mazloum-Ardakani, M., H. Beitollahi, M.A. Sheikh-Mohseni, H. Naeimi, and N. Taghavinia. 2010. Novel nanostructure electrochemical sensor for electrocatalytic determination of norepinephrine in the presence of high concentrations of acetaminophene and folic acid. *Appl. Catal. A* 378:195–201.
992. Mazloum-Ardakani, M., H. Rajabi, H. Beitollahi, B.B.F. Mirjalili, A. Akbari, and N. Taghavinia. 2010. Voltammetric determination of dopamine at the surface of TiO_2 nanoparticles modified carbon paste electrode. *Int. J. Electrochem. Sci.* 5:147–157.
993. Mazloum-Ardakani, M., H. Beitollahi, M.A.S. Mohseni, A. Benvidi, H. Naeimi, M. Nejati-Barzoki, and N. Taghavinia. 2010. Simultaneous determination of epinephrine and acetaminophen concentrations using a novel carbon paste electrode prepared with 2,2′-[1,2 butanediylbis(nitriloethylidyne)]-bishydroquinone and TiO_2 nanoparticles. *Colloids Surf. B* 76:82–87.
994. Lin, H., X. Ji, Q. Chen, Y. Zhou, C.E. Banks, and K. Wu. 2009. Mesoporous-TiO_2 nanoparticles based carbon paste electrodes exhibit enhanced electrochemical sensitivity for phenols. *Electrochem. Commun.* 11:1990–1995.

995. Mazloum Ardakani, M., A. Talebi, H. Naeimi, M. Nejati Barzoky, and N. Taghavinia. 2009. Fabrication of modified TiO_2 nanoparticle carbon paste electrode for simultaneous determination of dopamine, uric acid, and L-cysteine. *J. Solid State Electrochem.* 13:1433–1440.
996. Sun, W., X.Q. Li, and K. Jiao. 2008. Direct electron transfer of hemoglobin on a carbon ionic liquid electrode with TiO_2 nanoparticle as enhancer. *J. Chin. Chem. Soc. (Taiwan)* 55:1074–1079.
997. Mazloum Ardakani, M., Z. Taleat, H. Beitollahi, M. Salavati-Niasari, B.B.F. Mirjalili, and N. Taghavinia. 2008. Electrocatalytic oxidation and nanomolar determination of guanine at the surface of a molybdenum (VI) complex-TiO_2 nanoparticle modified carbon paste electrode. *J. Electroanal. Chem.* 624:73–78.
998. Parham, H. and N. Rahbar. 2010. Square-wave voltammetric determination of Methyl-parathion at ZrO_2-nanoparticles modified carbon paste electrode. *J. Hazard. Mater.* 177:1077–1084.
999. Sun, D., X. Xie, and H. Zhang. 2010. Surface effects of mesoporous silica modified electrode and application in electrochemical detection of dopamine. *Colloids Surf. B* 75:88–92.
1000. Wang, J. and J. Liu. 1993. Fumed-silica containing carbon-paste dehydrogenase biosensors. *Anal. Chim. Acta* 284:385–391.
1001. Hrbac, J., V. Halouzka, R. Zboril, K. Papadopoulos, and T. Triantis. 2007. Carbon electrodes modified by nanoscopic iron(III) oxides to assemble chemical sensors for the amperometric detection of hydrogen peroxide. *Electroanalysis* 19:1850–1854.
1002. Kang, J.W., Z.F. Li, and X.Q. Lu. 2005. Electrocatalytic redox of 3-nitrobenzaldehyde thiosemicarbazone at nano-γ-Al_2O_3 modified electrodes. *Xibei Shifan Daxue Xuebao, Ziran Kexueban* 41:47–51.
1003. Reddy, S., B.E.K. Swamy, U. Chandra, B.S. Sherigara, and H. Jayadevappa. 2010. Synthesis of CdO nanoparticles and their modified carbon paste electrode for determination of dopamine and ascorbic acid by using cyclic voltammetry technique. *Int. J. Electrochem. Sci.* 5:10–17.
1004. Zhang, B., H. Liu, K. Jiao, X. Fu, and W. Sun. 2009. Detection for DNA specific sequences using the polarographic complex adsorptive wave of lead. *Qingdao Keji Daxue Xuebao, Ziran Kexueban* 30:95–100.
1005. Xie, J.K., K. Jiao, H. Liu, Q.X. Wang, S.F. Liu, and X. Fu. 2008. DNA electrochemical sensor based on PbSe nanoparticle for the sensitive detection of CaMV35S transgene gene sequence. *Fenxi Huaxue* 36:874–878.
1006. Xia, Q., X. Chen, K. Zhao, and J. Liu. 2008. Synthesis and characterizations of polycrystalline walnut-like CdS nanoparticle by solvothermal method with PVP as stabilizer. *Mater. Chem. Phys.* 111:98–105.
1007. Li, J.-P. and Y.-H. Yuan. 2006. Synthesis of magnetic prussian blue nanoparticles and the fabrication of chemically modified electrode. *Huaxue Xuebao* 64:261–265.
1008. Sheng, Q., Y. Shen, H. Zhang, and J. Zheng. 2006. Neodymium (III) hexacyanoferrate (II) nanoparticles induced by enzymatic reaction and their use in biosensing of glucose. *Electrochim. Acta* 53:4687–4692.
1009. Wang, J. and A.S. Arribas. 2006. Biocatalytically induced formation of cupric ferrocyanide nanoparticles and their application for electrochemical and optical biosensing of glucose. *Small* 2:129–134.
1010. Arribas, A.S., T. Vazquez, J. Wang, A. Mulchandani, and W. Chen. 2005. Electrochemical and optical bioassays of nerve agents based on the organophosphorus-hydrolase mediated growth of cupric ferrocyanide nanoparticles. *Electrochem. Commun.* 7:1371–1734.
1011. He, Q., D. Zheng, and S. Hu. 2009. Development of a novel serotonin sensor based on Na-montmorillonite nanoparticles modified carbon paste electrode. *Anal. Chem. Indian J.* 8:464–470.
1012. Khalilzadeh, B., M. Hasanzadeh, B. Bakhoday, A. Babaei, M. Hajjizadeh, and M. Zendehdel. 2009. Zeolite nanoparticle modified carbon paste electrode as a biosensor for simultaneous determination of dopamine and tryptophan. *J. Chin. Chem. Soc. (Taiwan)* 56:789–796.
1013. Li, X., X. Wang, G. Liu, and H. Lin. 2005. Study on preparation, electrochemistry actions of $[(C_2H_5)_4N]_4GeMo_{12}O_{40}$ nanoparticles bulk-modified carbon paste electrode. *Dongbei Shida Xuebao, Ziran Kexueban* 37:36–40.
1014. Wang, X., Z. Kang, Y. Lan, and E. Wang. 2003. Molybdovanadophosphate tetraethylammounium nanoparticle bulk-modified carbon paste electrode and its electrocatalysis toward the reduction of hydrogen peroxide. *Fenxi Huaxue* 31:941–944.
1015. Wang, X., Z. Kang, E. Wang, and C. Hu. 2002. Preparation, electrochemical property and application in chemically bulk-modified electrode of a hybrid inorganic–organic silicomolybdate nanoparticles. *Mater. Lett.* 56:393–396.
1016. Wang, X.-L., Z.-H. Kang, E.-B. Wang, and C.-W. Hu. 2002. Inorganic–organic hybrid 18-molybdodiphosphate nanoparticles bulk-modified carbon paste electrode and its electrocatalysis. *Chin. J. Chem.* 20:777–783.

1017. Bi, L.H., W.H. Zhou, H.Y. Wang, and S.Y. Dong. 2008. Preparation of nanoparticles of bipyridineruthenium and silicotungstate and application as electrochemiluminescence sensor. *Electroanalysis* 20:996–1001.
1018. Fritzen-Garcia, M.B., I.R.W.Z. Oliveira, B.G. Zanetti-Ramos, O. Fatibello-Filho, V. Soldi, A.A. Pasa, and T.B. Creczynski-Pasa. 2009. Carbon paste electrode modified with pine kernel peroxidase immobilized on pegylated polyurethane nanoparticles. *Sens. Actuators B* 139:570–575.
1019. de Oliveira, I.R.W.Z., M.B. Fritzen-Garcia, B.G. Zanetti-Ramos, O. Fatibello-Filho, A.A. Pasa, and T.B. Creczynski-Pasa. 2008. Pegylated polyurethane nanoparticles as support for a biosensor construction based on peroxidase extracted from pine kernel (*Araucaria angustifolia*). *ECS Trans.* 16:483–490.
1020. Zhu, Y.-C., J.-J. Guan, L. Cao, and J. Hao. 2010. Determination of trace iodide in iodised table salt on silver sulfate-modified carbon paste electrode by differential pulse voltammetry with electrochemical solid phase nano-extraction. *Talanta* 80:1234–1238.
1021. Liu, K.-E. and H.D. Abruna. 1989. Electroanalysis of aromatic aldehydes with modified carbon paste electrodes. *Anal. Chem.* 61:2599–2602.
1022. Blaha, C.D. and R.F. Lane. 1983. Chemically modified electrode for in vivo monitoring of brain catecholamines. *Brain Res. Bull.* 10:861–864.
1023. Broderick, P.A. 1985. In vivo electrochemical studies of rat striatal dopamine and serotonin release after morphine. *Life Sci.* 36:2269–2275.
1024. Lane, R.F., C.D. Blaha, and A.G. Phillips. 1986. In vivo electrochemical analysis of cholecystokinin-induced inhibition of dopamine release in the nucleus accumbens. *Brain Res.* 397:200–204.
1025. Lane, R.F., C.D. Blaha, and S.P. Hari. 1987. Electrochemistry in vivo: Monitoring dopamine release in the brain of the conscious, freely moving rat. *Brain Res. Bull.* 19:19–27.
1026. Glynn, G.E. and B.K. Tamamoto. 1989. In vivo neurochemical and anatomical heterogeneity of the dopamine uptake system in the rat caudate putamen. *Brain Res.* 481:235–241.
1027. Ormonde, D.E. and R.D. O'Neill. 1989. Altered response of carbon paste electrodes after contact with brain tissue. Implications for modified electrode use in vivo. *J. Electroanal. Chem.* 261:463–469.
1028. Lyne, P.D. and R.D. O'Neill. 1989. Selectivity of stearate-modified carbon paste electrodes for dopamine and ascorbic acid. *Anal. Chem.* 61:2323–2324.
1029. Gelbert, M.B. and D.J. Curran. 1986. Alternating current voltammetry of dopamine and ascorbic acid at carbon paste and stearic acid modified carbon paste electrodes. *Anal. Chem.* 58:1028–1032.
1030. Khodari, M., J.-M. Kauffmann, G.J. Patriarche, and M.A. Ghandour. 1989. Preconcentration and determination of promethazine at lipid-modified carbon paste electrodes. *Electroanalysis* 1:501–505.
1031. Itaya, K., I. Uchida, and V.D. Neff. 1986. Electrochemistry of polynuclear transition metal cyanides: Prussian blue and its analogues. *Acc. Chem. Res.* 19:162–168.
1032. Zen, J.M., A.S. Kumar, and H.W. Chen. 2001. Electrochemical behavior of stable cinder/prussian blue analogue and its mediated nitrite oxidation. *Electroanalysis* 13:1171–1178.
1033. Zen, J.M., A.S. Kumar, and H.W. Chen. 2000. Electrochemical formation of Prussian blue in natural iron-intercalated clay and cinder matrixes. *Electroanalysis* 12:542–545.
1034. Teixeira, M.F.S., A. Segnini, F.C. Moraes, L.H. Marcolino-Junior, O. Fatibello-Filho, and E.T.G. Cavalheiro. 2003. Determination of vitamin B6 (pyridoxine) in pharmaceutical preparations by cyclic voltammetry at a copper(II) hexacyan-ferrate(III) modified carbon paste electrode. *J. Braz. Chem. Soc.* 14:316–321.
1035. Wang, J., X. Zhang, and L. Chen. 2000. Comparison of glucose enzyme electrodes based on dispersed rhodium particles and cupric hexacyanoferrate within carbon paste transducers. *Electroanalysis* 12:1277–1281.
1036. Toito Suarez, W., L.H. Marcolino, Jr., and O. Fatibello-Filho. 2006. Voltammetric determination of N-acetylcysteine using a carbon paste electrode modified with copper(II) hexacyanoferrate(III). *Microchem. J.* 82:163–167.
1037. Ojani, R., J.B. Raoof, and B. Norouzi. 2008. Cu(II) hexacyanoferrate(III) modified carbon paste electrode: Application for electrocatalytic detection of nitrite. *Electroanalysis* 20:1996–2002.
1038. Abbaspour, A. and M.A. Mehrgardi. 2004. Electrocatalytic oxidation of guanine and DNA on a carbon paste electrode modified by cobalt hexacyanoferrate films. *Anal. Chem.* 76:5690–5696.
1039. Abbaspour, A. and M.A. Kamyabi. 2005. Electrocatalytic oxidation of hydrazine on a carbon paste electrode modified by hybrid hexacyanoferrates of copper and cobalt films. *J. Electroanal. Chem.* 576:73–83.
1040. Abbaspour, A. and A. Ghaffarinejad. 2008. Electrocatalytic oxidation of L-cysteine with a stable copper-cobalt hexacyanoferrate electrochemically modified carbon paste electrode. *Electrochim. Acta* 53:6643–6650.

1041. Siroueinejad, A., A. Abbaspour, and M. Shamsipur. 2009. Electrocatalytic oxidation and determination of sulfite with a novel copper-cobalt hexacyanoferrate modified carbon paste electrode. *Electroanalysis* 21:1387–1393.
1042. Shaidarova, L.G., S.A. Ziganshina, L.N. Tikhonova, and G.K. Budnikov. 2003. Electrocatalytic oxidation and flow-injection determination of sulfur-containing amino acids at graphite electrodes modified with a ruthenium hexacyanoferrate film. *J. Anal. Chem.* 58:1144–1150.
1043. Hosseinzadeh, R., R.E. Sabzi, and K. Ghasemlu. 2009. Effect of cetyltrimethyl ammonium bromide (CTAB) in determination of dopamine and ascorbic acid using carbon paste electrode modified with tin hexacyanoferrate. *Colloids Surf. B* 68:213–217.
1044. Farhadi, K., F. Kheiri, and M. Golzan. 2008. Th(IV)-hexacyanoferrate modified carbon paste electrode as a new electrocatalytic probe for simultaneous determination of ascorbic acid and dopamine from acidic media. *J. Braz. Chem. Soc.* 19:1405–1412.
1045. Sabzi, R.E., A. Hasanzadeh, K. Ghasemlu, and P. Heravi. 2007. Preparation and characterization of carbon paste electrode modified with tin and hexacyanoferrate ions. *J. Serb. Chem. Soc.* 72:993–1002.
1046. Sadakane, M. and E. Steckhan. 1998. Electrochemical properties of polyoxometalates as electrocatalysts. *Chem. Rev.* 98:219–237.
1047. Wang, X., E. Wang, and C. Hu. 2001. Hybrid inorganic–organic material containing 12-molybdophosphate bulk-modified carbon paste electrode. *Chem. Lett.* 1030–1031.
1048. Wang, X.L., E.B. Wang, Y. Lan, and C.W. Hu. 2002. Renewable PMo12-based inorganic–organic hybrid material bulk-modified carbon paste electrode: Preparation, electrochemistry and electrocatalysis. *Electroanalysis* 14:1116–1121.
1049. Han, Z., Y. Zhao, J. Peng, Y. Feng, J. Yin, and Q. Liu. 2005. The electrochemical behavior of Keggin polyoxometalate modified by tricyclic, aromatic entity. *Electroanalysis* 17:1097–1102.
1050. Han, Z., Y. Gao, J. Wang, and C. Hu. 2009. Preparation and electrochemistry properties of a new polyoxometalate-based supramolecular assembly. *Z. Anorg. Allg. Chem.* 635:2665–2670.
1051. Han, Z., Y. Zhao, J. Peng, Q. Liu, and E. Wang. 2005. Inorganic–organic hybrid polyoxometalate containing supramolecular helical chains: Preparation, characterization and application in chemically bulk-modified electrode. *Electrochim. Acta* 51:218–224.
1052. Ojani, R., M.S. Rahmanifar, and P. Naderi. 2008. Electrocatalytic reduction of nitrite by phosphotungstic heteropolyanion. Application for its simple and selective determination. *Electroanalysis* 20:1092–1098.
1053. Wang, X., J. Li, H. Hu, H. Lin, A. Tian, and Y. Chen. 2010. Two new organic–inorganic hybrids based on [SiW$_{12}$O$_{40}$]$_4$-anion: Hydrothermal syntheses, crystal structures, and electrochemical properties. *J. Inorg. Organomet. Polym. Mater.* 20:361–368.
1054. Zhang, L., H. Xue, N. Li, W. You, Z. Sun, and Z. Zhu. 2009. Hydrothermal synthesis and characterization of 3H$_2$[H$_2$PMo$_8$O$_{30}$]·4H$_2$O and its electrocatalysis toward the reduction of hydrogen peroxide. *Huaxue Yanjiu Yu Yingyong* 21:1408–1413.
1055. Mousty, C. 2004. Sensors and biosensors based on clay-modified electrodes—new trends. *Appl. Clay Sci.* 27:159–177.
1056. Mousty, C. 2010. Biosensing applications of clay-modified electrodes: A review. *Anal. Bioanal. Chem.* 396:315–325.
1057. Macha, S.M and A. Fitch. 1998. Clays as architectural units at modified electrodes. *Mikrochim. Acta* 128:1–18.
1058. Wang, J. and T. Martinez. 1989. Trace analysis at clay-modified carbon paste electrodes. *Electroanalysis* 1:167–172.
1059. Kula, P. and Z. Navratilova. 1996. Voltammetric copper(II) determination with a montmorillonite-modified carbon paste electrode. *Fresenius J. Anal. Chem.* 354:692–695.
1060. Raber, G., K. Kalcher, and M. Stadlober. 1998. New voltammetric methods for the determination of heavy metals using a montmorillonite modified carbon paste electrode. *Sci. Pap. Univ. Pardubice Ser. A* 3:163–193.
1061. Kula, P., Z. Navratilova, P. Kulova, and M. Kotoucek. 1999. Sorption and determination of Hg(II) on clay modified carbon paste electrodes. *Anal. Chim. Acta* 385:91–101.
1062. Navratilova, Z. and P. Kula. 2000. Determination of gold using clay modified carbon paste electrode. *Fresenius J. Anal. Chem.* 367:369–372.
1063. Kula, P. and Z. Navratilova. 2001. Anion exchange of gold chloro complexes on carbon paste electrode modified with montmorillonite for determination of gold in pharmaceuticals. *Electroanalysis* 13:795–798.
1064. Huang, W., C. Yang, and S. Zhang. 2002. Anodic stripping voltammetric determination of mercury by use of a sodium montmorillonite-modified carbon-paste electrode. *Anal. Bioanal. Chem.* 374:998–1001.

1065. Huang, W. 2004. Voltammetric determination of bismuth in water and nickel metal samples with a sodium montmorillonite (SWy-2) modified carbon paste electrode. *Microchim. Acta* 144:125–129.
1066. Huang, W., S. Zhang, and Y. Wu. 2006. Electrochemical behavior and detection of guanine using a sodium montmorillonite-modified carbon paste electrode. *Russ. J. Electrochem.* 42:153–156.
1067. Sun, D., C. Wan, G. Li, and K. Wu. 2007. Electrochemical determination of lead(II) using a montmorillonite calcium-modified carbon paste electrode. *Microchim. Acta* 158:255–260.
1068. Lin, H., G. Li, and K. Wu. 2007. Electrochemical determination of Sudan I using montmorillonite calcium modified carbon paste electrode. *Food Chem.* 107:531–536.
1069. Xu, G.Y., J.S. Fan, and K. Jiao. 2008. Immobilizing DNA on clay mineral modified carbon paste electrodes. *Appl. Clay Sci.* 40:119–123.
1070. Beltagi, A.M. 2009. Utilization of a montmorillonite-Ca-modified carbon paste electrode for the stripping voltammetric determination of diflunisal in its pharmaceutical formulations and human blood. *J. Appl. Electrochem.* 39:2375–2384.
1071. Hernandez, P., J. Vicente, M. Gonzalez, and L. Hernandez. 1990. Voltammetric determination of linuron at a carbon-paste electrode modified with sepiolite. *Talanta* 37:789–794.
1072. Hernandez, L., P. Hernandez, and E. Lorenzo. 1990. Direct determination of bentazepam in a biological sample with a sepiolite-modified carbon paste electrode. *Electroanalysis* 2:643–646.
1073. Chicharro, M., A. Zapardiel, E. Bermejo, J.A. Perez-Lopez, and L. Hernandez. 1995. Determination of ephedrine in human urine by square wave voltammetry with a sepiolite-modified carbon paste electrode. *Analusis (Bruxelles)* 23:131–134.
1074. Hernandez, L., P. Hernandez, M.H. Blanco, E. Lorenzo, and E. Alda. 1988. Determination of flunitrazepam by differential-pulse voltammetry using a bentonite-modified carbon paste electrode. *Analyst (U.K.)* 113:1719–1722.
1075. Naranjo Rodriguez, I., M. Barea Zamora, J.M. Barbera Salvador, J.A. Munoz Leyva, M.P. Hernandez Artiga, and J.L.H.H. de Cisneros. 1997. Voltammetric determination of 2-nitrophenol at a bentonite-modified carbon paste electrode. *Mikrochim. Acta* 126:87–92.
1076. Naranjo Rodriguez, I., J.A. Munoz Leyva, and J.L.H.H. de Cisneros. 1997. Use of a carbon paste modified electrode for the determination of 2-nitrophenol in a flow system by differential pulse voltammetry. *Anal. Chim. Acta* 344:167–173.
1077. Naranjo Rodriguez, I., J.A. Munoz Leyva, and J.L.H.H. de Cisneros. 1997. Use of a bentonite-modified carbon paste electrode for the determination of some phenols in a flow system by differential-pulse voltammetry. *Analyst (U.K.)* 122:601–604.
1078. Marchal, V., F. Barbier, F. Plassard, R. Faure, and O. Vittori. 1999. Determination of cadmium in bentonite clay mineral using a carbon paste electrode. *Fresenius J. Anal. Chem.* 363:710–712.
1079. Rezaei, B., M. Ghiaci, and M.E. Sedaghat. 2008. A selective modified bentonite-porphyrin carbon paste electrode for determination of Mn(II) by using anodic stripping voltammetry. *Sens. Actuators B* B131:439–447.
1080. Kalcher, K., I. Grabec, G. Raber, X. Cai, G. Tavcar, and B. Ogorevc. 1995. The vermiculite-modified carbon paste electrode as a model system for preconcentrating mono- and divalent cations. *J. Electroanal. Chem.* 386:149–156.
1081. Svegl, I.G., B. Ogorevc, and V. Hudnik. 1996. A methodological approach to the application of a vermiculite-modified carbon paste electrode in interaction studies: Influence of some pesticides on the uptake of Cu(II) from a solution to the solid phase. *Fresenius J. Anal. Chem.* 354:770–773.
1082. Svegl, I.G., M. Kolar, B. Ogorevc, and B. Pihlar. 1998. Vermiculite clay mineral as an effective carbon paste electrode modifier for the preconcentration and voltammetric determination of Hg(II) and Ag(I) ions. *Fresenius J. Anal. Chem.* 361:358–362.
1083. Oropeza, M.T., I. Gonzalez, M.M. Teutli-Leon, and L.F. Chazaro. 2005. Voltammetric study of cadmium-kaolinite system using modified carbon paste electrodes. In *Applications of Analytical Chemistry in Environmental Research*, Ed. M. Palomar, pp. 67–77. Trivandrum, India: Research Signpost.
1084. El Mhammedi, M.A., M. Bakasse, R. Najih, and A. Chtaini. 2009. A carbon paste electrode modified with kaolin for the detection of diquat. *Appl. Clay Sci.* 43:130–134.
1085. El Mhammedi, M.A., M. Achak, M. Bakasse, R. Bachirat, and A. Chtaini. 2010. Accumulation and trace measurement of paraquat at kaolin-modified carbon paste electrode. *Mater. Sci. Eng. C* 30:833–838.
1086. Ozkan, D., K. Kerman, B. Meric, P. Kara, H. Demirkan, M. Polverejan, T.J. Pinnavaia, and M. Ozsoz. 2002. Heterostructured fluorohectorite clay as an electrochemical sensor for the detection of 2,4-dichlorophenol and the herbicide 2,4-D. *Chem. Mater.* 14:1755–1761.
1087. Dilgin, Y., Z. Dursun, and G. Nisli. 2003. Flow injection amperometric determination of ascorbic acid using a photoelectrochemical reaction after immobilization of methylene blue on muscovite. *Turk. J. Chem.* 27:167–180.

1088. Tonle, I.K., E. Ngameni, D. Njopwouo, C. Carteret, and A. Walcarius. 2003. Functionalization of natural smectite-type clays by grafting with organosilanes: Physico-chemical characterization and application to mercury(II) uptake. *Phys. Chem. Chem. Phys.* 5:4951–4961.

1089. Tonle, I.K., E. Ngameni, and A. Walcarius. 2005. Preconcentration and voltammetric analysis of mercury(II) at a carbon paste electrode modified with natural smectite-type clays grafted with organic chelating groups. *Sens. Actuators B* B110:195–203.

1090. Tonle, I.K., E. Ngameni, H.L. Tcheumi, V. Tchieda, C. Carteret, and A. Walcarius. 2008. Sorption of methylene blue on an organoclay bearing thiol groups and application to electrochemical sensing of the dye. *Talanta* 74:489–497.

1091. Shan, D., S. Cosnier, and C. Mousty. 2003. Layered double hydroxides: An attractive material for electrochemical biosensor design. *Anal. Chem.* 75:3872–3879.

1092. Therias, S. and C. Mousty. 1995. Electrodes modified with synthetic anionic clays. *Appl. Clay Sci.* 10:147–162.

1093. Therias, S., C. Mousty, C. Forano, and J.P. Besse. 1996. Electrochemical transfer at anionic clay modified electrodes. Case of 2,2′-Azinobis(3-ethylbenzothiazoline-6-sulfonate). *Langmuir* 12:4914–4920.

1094. Labuda, J. and M. Hudakova. 1997. Hexacyanoferrate-anion exchanger-modified carbon paste electrodes. *Electroanalysis* 9:239–242.

1095. Domenech, A., A. Ribera, A. Cervilla, and E.L. Lopis. 1998. Electrochemistry of hydrotalcite-supported bis(2-mercapto-2,2-diphenyl-ethanoate) dioxymolybdate complexes. *J. Electroanal. Chem.* 458:31–41.

1096. Walcarius, A., G. Lefevre, J.P. Rapin, G. Renaudin, and M. Francois. 2001. Voltammetric detection of iodide after accumulation by Friedel's salt. *Electroanalysis* 13:313–320.

1097. Svegl, I.G. and B. Ogorevc. 2000. Soil-modified carbon paste electrode: A useful tool in environmental assessment of heavy metal ion binding interactions. *Fresenius J. Anal. Chem.* 367:701–706.

1098. Reddy, T.M. and S.J. Reddy. 2004. Differential Pulse Adsorptive stripping voltammetric determination of Nifedipine and Nimodipine in pharmaceutical formulations, urine, and serum samples by using a clay-modified carbon-paste electrode. *Anal. Lett.* 37:2079–2098.

1099. Hu, S. 1999. Electrocatalytic reduction of molecular oxygen on a sodium montmorillonite-methyl viologen carbon paste chemically modified electrode. *J. Electroanal. Chem.* 463:253–257.

1100. Yang, H., X. Zheng, W. Huang, and K. Wu. 2008. Modification of montmorillonite with cationic surfactant and application in electrochemical determination of 4-chlorophenol. *Colloids Surf. B* 65:281–284.

1101. Huang, W., D. Zhou, X. Liu, and X. Zheng. 2009. Electrochemical determination of phenol using CTAB-functionalized montmorillonite electrode. *Environ. Technol.* 30:701–706.

1102. Dias Filho, N.L., D.R. do Carmo, F. Gessner, and A.H. Rosa. 2005. Preparation of a clay-modified carbon paste electrode based on 2-thiazoline-2-thiol-hexadecylammonium sorption for the sensitive determination of mercury. *Anal. Sci.* 21:1309–1316.

1103. Dias Filho, N.L., D.R. do Carmo, and A.H. Rosa. 2006. Selective sorption of mercury(ii) from aqueous solution with an organically modified clay and its electroanalytical application. *Sep. Sci. Technol.* 41:733–746.

1104. Rolison, D.R. 1990. Zeolite-modified electrodes and electrode-modified zeolites. *Chem. Rev.* 90:867–878.

1105. Walcarius, A. 1995. Zeolite-modified electrodes: Analytical applications and prospects. *Electroanalysis* 8:971–986.

1106. Creasy, K.E. and B.R. Shaw. 1988. Simplex optimization of electroreduction of oxygen mediated by methyl viologen supported on zeolite-modified carbon paste electrode. *Electrochim. Acta* 33:551–556.

1107. Walcarius, A., L. Lamberts, and E.G. Derouane. 1993. The methyl viologen incorporated zeolite modified carbon paste electrode. Part 2. Ion exchange and electron transfer mechanism in aqueous medium. *Electrochim. Acta* 38:2267–2276.

1108. Wang, J. and A. Walcarius. 1996. Zeolite-modified carbon paste electrode for selective monitoring of dopamine. *J. Electroanal. Chem.* 407:183–187.

1109. Marko Varga, G., E. Burestedt, C.J. Svensson, J. Emneus, L. Gorton, T. Ruzgas, M. Lutz, and K.K. Unger. 1996. Effect of HY-zeolites on the performance of tyrosinase-modified carbon paste electrodes. *Electroanalysis* 8:1121–1126.

1110. Zou, M.-Z., H.-D. Xu, J. Lu, and Q.-H. Ru. 1997. Studies on zeolite modified electrode. Part I. Ascorbic acid sensor based on carbon paste electrode containing Fe(III)Y zeolite. *Chin. Chem. Lett.* 8:247–250.

1111. Walcarius, A. and L. Lamberts. 1998. The methylviologen-doped zeolite modified electrode as a new detector for suppressor free ion chromatography. *Anal. Lett.* 31:585–599.

1112. Guerra, S.V., C.R. Xavier, S. Nakagaki, and L.T. Kubota. 1998. Electrochemical behavior of copper porphyrin synthesized into zeolite cavity. A sensor for hydrazine. *Electroanalysis* 10:462–466.

1113. Guerra, S.V., L.T. Kubota, C.R. Xavier, and S. Nakagaki. 1999. Experimental optimization of selective hydrazine detection in flow injection analysis using a carbon paste electrode modified with copper porphyrin occluded into zeolite cavity. *Anal. Sci.* 15:1231–1234.
1114. Walcarius, A., P. Mariaulle, and L. Lamberts. 1999. Use of a zeolite-modified electrode for the study of the methylviologen-sodium ion-exchange in zeolite Y. *J. Electroanal. Chem.* 463:100–108.
1115. Walcarius, A. 1999. Factors affecting the analytical applications of zeolite modified electrodes: Indirect detection of nonelectroactive cations. *Anal. Chim. Acta* 388:79–91.
1116. Walcarius, A., P. Mariaulle, C. Louis, and L. Lamberts. 1999. Amperometric detection of nonelectroactive cations in electrolyte-free flow systems at zeolite modified electrodes. *Electroanalysis* 11:393–400.
1117. Walcarius, A., S. Rozanska, J. Bessiere, and J. Wang. 1999. Screen-printed zeolite-modified carbon electrodes. *Analyst* 124:1185–1190.
1118. Serban, S. and N. El Murr. 2003. Use of ferricinium exchanged zeolite for mediator stabilization and analytical performances enhancement of oxidase-based carbon paste biosensors. *Anal. Lett.* 36:1739–1753.
1119. Zhuang, Y., D. Zhang, and H. Ju. 2005. Sensitive determination of heroin based on electrogenerated chemiluminescence of tris(2,2′-bipyridyl)ruthenium(II) immobilized in zeolite Y modified carbon paste electrode. *Analyst* 130:534–540.
1120. Zendehdel, M., A. Babaei, and S. Alami. 2007. Intercalation of xylenol orange, morin and calmagite into NaY zeolite and their application in dye/zeolite modified electrode. *J. Inclusion Phenom. Macrocyclic Chem.* 59:345–349.
1121. Ardakani, M.M., Z. Akrami, H. Kazemian, and H.R. Zare. 2008. Accumulation and voltammetric determination of cobalt at zeolite-modified electrodes. *J. Anal. Chem.* 63:184–191.
1122. Babaei, A., M. Zendehdel, B. Khalilzadeh, and A. Taheri. 2008. Simultaneous determination of tryptophan, uric acid and ascorbic acid at iron(III) doped zeolite modified carbon paste electrode. *Colloids Surf. B* 66:226–232.
1123. Senthilkumar, S. and R. Saraswathi. 2009. Electrochemical sensing of cadmium and lead ions at zeolite-modified electrodes: Optimization and field measurements. *Sens. Actuators B* B141:65–75.
1124. Babaei, A., S. Mirzakhani, and B. Khalilzadeh. 2009. A sensitive simultaneous determination of epinephrine and tyrosine using an iron(III) doped zeolite-modified carbon paste electrode. *J. Braz. Chem. Soc.* 20:1862–1869.
1125. Babaei, A., B. Khalilzadeh, and M. Afrasiabi. 2010. A new sensor for the simultaneous determination of paracetamol and mefenamic acid in a pharmaceutical preparation and biological samples using copper(II) doped zeolite modified carbon paste electrode. *J. Appl. Electrochem.* 40:1537–1543.
1126. Teixeira, M.F.S., M.F. Bergamini, C.M.P. Marques, and N. Bocchi. 2004. Voltammetric determination of L-dopa using an electrode modified with trinuclear ruthenium ammine complex (Ru-red) supported on Y-type zeolite. *Talanta* 63:1083–1088.
1127. Chen, B., N.K. Goh, and L.S. Chia. 1996. Determination of copper by zeolite molecular sieve modified electrode. *Electrochim. Acta* 42:595–604.
1128. Gligor, D., L.M. Muresan, A. Dumitru, and I.C. Popescu. 2007. Electrochemical behavior of carbon paste electrodes modified with methylene green immobilized on two different X type zeolites. *J. Appl. Electrochem.* 37:261–267.
1129. Varodi, C., D. Gligor, and L.M. Muresan. 2007. Carbon paste electrodes modified with methylene blue immobilized on a synthetic zeolite. *Rev. Chim. (Roum.)* 52:81–88.
1130. Varodi, C., D. Gligor, A. Maicaneanu, and L.M. Muresan. 2007. Carbon paste electrode incorporating calcium-exchanged zeolite modified with methylene blue for amperometric detection of NADH. *Rev. Chim. (Roum.)* 58:890–894.
1131. Wang, J. and T. Martinez. 1988. Accumulation and voltammetric measurement of silver at zeolite-containing carbon-paste electrodes. *Anal. Chim. Acta* 207:95–102.
1132. Domenech, A., I. Casades, and H. Garcia. 1999. Electrochemical evidence for an impeded attack of water at anthracene and thianthrene radical ions located on the outermost layers of zeolites. *J. Org. Chem.* 64:3731–3735.
1133. Arvand, M., S. Sohrabnezhad, M.F. Mousavi, M. Shamsipur, and M.A. Zanjanchi. 2003. Electrochemical study of methylene blue incorporated into mordenite type zeolite and its application for amperometric determination of ascorbic acid in real samples. *Anal. Chim. Acta* 491:193–201.
1134. Arvand, M., M. Vaziri, and M. Vejdani. 2010. Electrochemical study of atenolol at a carbon paste electrode modified with mordenite type zeolite. *Mater. Sci. Eng. C* 30:709–714.
1135. Walcarius, A., V. Vromman, and J. Bessière. 1999. Flow injection indirect amperometric detection of ammonium ions using a clinoptilolite-modified electrode. *Sens. Actuators B* B56:136–143.

1136. Alpat, S., S.K. Alpat, and A. Telefoncu. 2005. A sensitive determination of dopamine in the presence of ascorbic acid using a nafion-coated clinoptilolite-modified carbon paste electrode. *Anal. Bioanal. Chem.* 383:695–700.
1137. Ardakani, M.M., M.A. Karimi, M.H. Mashhadizadeh, M. Pesteh, M.S. Azimi, and H. Kazemian. 2007. Potentiometric determination of monohydrogen arsenate by zeolite-modified carbon-paste electrode. *Int. J. Environ. Anal. Chem.* 87:285–294.
1138. Castro Martins, S., S. Khouzami, A. Tuel, Y. Ben Taarit, N. El Murr, and A. Sellami. 1993. Characterization of titanium silicalite using TS-1-modified carbon paste electrodes. *J. Electroanal. Chem.* 350:15–28.
1139. Castro Martins, S., A. Tuel, and Y. Ben Taarit. 1994. Characterization of titanium silicalites using cyclic voltammetry. *Stud. Surf. Sci. Catal.* 84:501–508.
1140. Arvand, M., M.A. Zanjanchi, and A. Islamnezhad. 2009. Zeolite-modified carbon-paste electrode as a selective voltammetric sensor for detection of tryptophan in pharmaceutical preparations. *Anal. Lett.* 42:727–738.
1141. Gligor, D., L. Muresan, and I.C. Popescu. 2004. Carbon paste electrodes incorporating methylene green-modified zeolite. *Acta Univ. Cibiniensis Ser. F (Roum.)* 7:29–35.
1142. Kilinc Alpat, S., U. Yuksel, and H. Akcay. 2005. Development of a novel carbon paste electrode containing a natural zeolite for the voltammetric determination of copper. *Electrochem. Commun.* 7:130–134.
1143. Zacahua Tlacuatl, G., J.J.C. Arellano, and A. Manzo-Robledo. 2009. Electrochemical characterization of carbon paste electrodes modified with natural zeolite. *Chem. Eng. Commun.* 196:1178–1188.
1144. Gligor, D., A. Maicaneanu, and A. Walcarius. 2010. Iron-enriched natural zeolite modified carbon paste electrode for H_2O_2 detection. *Electrochim. Acta* 55:4050–4056.
1145. Del Mar Cordero Rando, M., M. Barea Zamora, J.M. Barbera Salvador, I. Naranjo Rodriguez, J.A. Munoz Leyva, and J.L.H. de Cisneros. 1999. Electrochemical study of 4-nitrophenol at a modified carbon paste electrode. *Mikrochim. Acta* 132:7–11.
1146. Del Mar Cordero Rando, M., I. Naranjo Rodriguez, and J. H.H. de Cisneros. 1998. Voltammetric study of 2-methyl-4,6-dinitrophenol at a modified carbon paste electrode. *Anal. Chim. Acta* 370:231–238.
1147. Naranjo-Rodriguez, I., J.A. MunozLeyva, and J.L.H. de Cisneros. 2003. Flow injection study of 2,4,6-trichlorophenol by differential pulse voltammetry at a zeolite-modified carbon paste electrode. *Bull. Electrochem.* 19:289–294.
1148. Mazloum Ardakani, M., Z. Akrami, H. Kazemian, and H.R. Zare. 2009. Preconcentration and electroanalysis of copper at zeolite modified carbon paste electrode. *Int. J. Electrochem. Sci.* 4:308–319.
1149. Ejhieh, A.N. and N. Masoudipour. 2010. Application of a new potentiometric method for determination of phosphate based on a surfactant-modified zeolite carbon-paste electrode (SMZ-CPE). *Anal. Chim. Acta* 658:68–74.
1150. Stara, V. and M. Kopanica. 1989. Chemically modified carbon paste and carbon composite electrodes. *Electroanalysis* 1:251–256.
1151. Kopanica, M. and V. Stara. 1991. Silica gel-modified carbon composite electrodes. *Electroanalysis* 3:13–16.
1152. Barrio, R.J., Z. Gomez de Balugera, and M. Aranzazu Goicolea. 1993. Utilization of a silica-modified carbon paste electrode for the direct determination of todralazine in biological fluids. *Anal. Chim. Acta* 273:93–99.
1153. Wang, J. and N. Naser. 1994. Improved performance of carbon paste amperometric biosensors through the incorporation of fumed silica. *Electroanalysis* 6:571–575.
1154. Fernandez de Betono, S., A. Arranz, A. Cid, J.M. Moreda, and J.F. Arranz. 1995. Voltammetric determination of chloridazone with a silica-modified carbon paste electrode. *Quim. Anal.* 14:112–116.
1155. Arranz, A., M.F. Villalba, S. de Betono, J.M. Moreda, and J.F. Arranz. 1997. Anodic voltammetric assay of the herbicide Metamitron on a carbon paste electrode. *Fresenius J. Anal. Chem.* 357:768–772.
1156. Arranz, A., S.F. De Betono, J.M. Moreda, A. Cid, and J.F. Arranz. 1997. Preconcentration and voltammetric determination of the herbicide metamitron with a silica-modified carbon paste electrode. *Mikrochim. Acta* 127:273–279.
1157. Walcarius, A. and J. Bessiere. 1997. Silica-modified carbon paste electrode for copper determination in ammoniacal medium. *Electroanalysis* 9:707–713.
1158. Walcarius, A., C. Despas, and J. Bessière. 1999. Selective monitoring of Cu(II) species using a silica modified carbon paste electrode. *Anal. Chim. Acta* 385:79–89.
1159. Borgo, C.A., R.T. Ferrari, L.M.S. Colpini, C.M.M. Costa, M.L. Baesso, and A.C. Bento. 1999. Voltammetric response of a copper(II) complex incorporated silica modified carbon-paste electrode. *Anal. Chim. Acta* 385:103–109.

References

1160. Bond, A.M., W. Miao, T.D. Smith, and J. Jamis. 1999. Voltammetric reduction of mercury(II), silver(I), lead(II) and copper(II) ions adsorbed onto a new form of mesoporous silica. *Anal. Chim. Acta* 396:203–213.
1161. Walcarius, A., J. Devoy, and J. Bessière. 2000. Silica-modified electrode for the selective detection of mercury. *J. Solid State Electrochem.* 4:330–336.
1162. Meyer, A., S. Hoeffler, and K. Fischer. 2006. Influence of the physical properties and handling of silica gel modified carbon paste electrodes on the phase transfer of solved Cu(II) ions. *Anal. Chim. Acta* 571:105–112.
1163. Nasri, Z. and E. Shams. 2009. Application of silica gel as an effective modifier for the voltammetric determination of dopamine in the presence of ascorbic acid and uric acid. *Electrochim. Acta* 54:7416–7421.
1164. Brinker, C. and G. Scherer. 1990. *Sol-Gel Science: The Physics and Chemistry of Sol-Gel Processing.* New York: Academic Press.
1165. Li, J.-N., S.-N. Tan, and H. Ge. 1996. Silica sol-gel immobilized amperometric biosensor for hydrogen peroxide. *Anal. Chim. Acta* 335:137–145.
1166. Li, J.-N., L.-S. Chia, N-K. Goh, and S.-N. Tan. 1998. Silica sol-gel immobilized amperometric biosensor for the determination of phenolic compounds. *Anal. Chim. Acta* 362:203–211.
1167. De Jesus, D.S., C.M.C.M. Couto, A.N. Araujo, and M.C.B.S. Montenegro. 2003. Amperometric biosensor based on monoamine oxidase (MAO) immobilized in sol-gel film for benzydamine determination in pharmaceuticals. *J. Pharm. Biomed. Anal.* 33:983–990.
1168. Wang, Q., G. Lu, and B. Yang. 2004. Myoglobin/sol-gel film modified electrode: Direct electrochemistry and electrochemical catalysis. *Langmuir* 20:1342–1347.
1169. Wang, Q., G. Lu, and B. Yang. 2004. Direct electrochemistry and electrocatalysis of hemoglobin immobilized on carbon paste electrode by silica sol-gel film. *Biosens. Bioelectron.* 19:1269–1275.
1170. Wang, Q., G. Lu, and B. Yang. 2004. Hydrogen peroxide biosensor based on direct electrochemistry of hemoglobin immobilized on carbon paste electrode by a silica sol-gel film. *Sens. Actuators B* B99:50–57.
1171. Ribeiro, E.S., Y. Gushikem, J.C. Biazzotto, and O.A. Serra. 2002. Electrochemical properties and dissolved oxygen reduction study on an iron(III)-tetra(o-ureaphenyl) porphyrinosilica matrix surface. *J. Porphyrins Phthalocyanines* 6:527–532.
1172. Toledo, M., A.M.S. Lucho, and Y. Gushikem. 2004. In situ preparation of Co phthalocyanine on a porous silica gel surface and the study of the electrochemical oxidation of oxalic acid. *J. Mater. Sci.* 39:6851–6854.
1173. Garcia, C.A.B., G. Oliveira Neto, L.T. Kubota, and L.A. Grandin. 1996. A new amperometric biosensor for fructose using a carbon paste electrode modified with silica gel coated with Meldola's Blue and fructose 5-dehydrogenase. *J. Electroanal. Chem.* 418:147–151.
1174. Kubota, L.T., B.G. Milagres, F. Gouvea, and G. de Oliveira. 1996. A modified carbon paste electrode with silica gel coated with Meldola's Blue and salicylate hydroxylase as a biosensor for salicylate. *Anal. Lett.* 29:893–910.
1175. Liu, H.-Y., Z.-N. Zhang, Y.-B. Fan, M. Dai, X.L. Zhang, J.-J. Wei, Z.-N. Qiu, H.-B. Li, and X.-X. Wu. 1997. Reagentless amperometric biosensor highly sensitive to hydrogen peroxide based on the incorporation of Meldola Blue, fumed silica, and horseradish peroxidase into carbon paste. *Fresenius J. Anal. Chem.* 357:297–301.
1176. Ferrari, R.T., L.M.S. Colpini, and C.M.M. Costa. 2003. Electrochemical characterization of CoHP-Si incorporated into a carbon paste electrode. *Microchim. Acta* 142:213–217.
1177. Gushikem, Y. and S.S. Rosatto. 2001. Metal oxide thin films grafted on silica gel surfaces: Recent advances on the analytical application of these materials. *J. Braz. Chem. Soc.* 12:695–705.
1178. Rocha, R.F., S.S. Rosatto, R.E. Bruns, and L.T. Kubota. 1997. Factorial design optimization of redox properties of methylene blue adsorbed on a modified silica gel surface. *J. Electroanal. Chem.* 433:73–76.
1179. Perez, E.F., L.T. Kubota, A.A. Tanaka, and G. Oliveira Neto. 1998. Anodic oxidation of cysteine catalyzed by nickel tetrasulfonated phthalocyanine immobilized on silica gel modified with titanium(IV) oxide. *Electrochim. Acta* 43:1665–1673.
1180. Perez, E.F., G. Oliveira Neto, A.A. Tanaka, and L.T. Kubota. 1998. Electrochemical sensor for hydrazine based on silica modified with nickel tetrasulfonated phthalocyanine. *Electroanalysis* 10:111–115.
1181. Rosatto, S.S., L.T. Kubota, and G. Oliveira Neto. 1999. Biosensor for phenol based on the direct electron transfer blocking of peroxidase immobilizing on silica-titanium. *Anal. Chim. Acta* 390:65–72.
1182. Perez, E.F., G. Oliveira Neto, and L.T. Kubota. 2001. Bi-enzymatic amperometric biosensor for oxalate. *Sens. Actuators B* B72:80–85.

1183. Castellani, A.M., J.E. Goncalves, and Y. Gushikem. 2002. The use of carbon paste electrodes modified with cobalt tetrasulfonated phthalocyanine adsorbed in silica/titania for the reduction of oxygen. *J. New Mater. Electrochem. Syst.* 5:169–172.

1184. Dias, S.L.P., S.T. Fujiwara, Y. Gushikem, and R.E. Bruns. 2002. Methylene blue immobilized on cellulose surfaces modified with titanium dioxide and titanium phosphate: Factorial design optimization of redox properties. *J. Electroanal. Chem.* 531:141–146.

1185. Ferreira, C.U., J.E. Goncalves, R.A.C. Goncalves, and A.J.B. De Oliveira. 2003. Electrochemical properties of Meldola's blue immobilized on silica/titania/antimonate prepared by sol-gel method. *J. New Mater. Electrochem. Syst.* 6:251–257.

1186. Arenas, L.T., D.S.F. Gay, C.C. Moro, S.L.P. Dias, D.S. Azambuja, T.M.H. Costa, E.V. Benvenutti, and Y. Gushikem. 2008. Brilliant yellow dye immobilized on silica and silica/titania based hybrid xerogels containing bridged positively charged 1,4-diazoniabicyclo[2.2.2]octane: Preparation, characterization and electrochemical properties study. *Microporous Mesoporous Mater.* 112:273–283.

1187. Peixoto, C.R.M., L.T. Kubota, and Y. Gushikem. 1995. Use of ruthenium-(ethylenedinitrito)-tetraacetic acid monohydrate ion immobilized on zirconium(IV) oxide coated silica gel surface as an amperometric sensor for oxygen in water. *Anal. Proc.* 32:503–505.

1188. Yamashita, M., S.S. Rosatto, and L.T. Kubota. 2002. Electrochemical comparative study of riboflavin, FMN and FAD immobilized on the silica gel modified with zirconium oxide. *J. Braz. Chem. Soc.* 13:635–641.

1189. Yamashita, M., C.A. Pessoa, and L.T. Kubota. 2003. Electrochemical behavior of pyrroloquinoline quinone immobilized on silica gel modified with zirconium oxide. *J. Coll. Interface Sci.* 263:99–105.

1190. Zaitseva, G., Y. Gushikem, E.S. Ribeiro, and S.S. Rosatto. 2002. Electrochemical property of methylene blue redox dye immobilized on porous silica-zirconia-antimonia mixed oxide. *Electrochim. Acta* 47:1469–1474.

1191. Pessoa, C.A. and Y. Gushikem. 2001. Cobalt porphyrins immobilized on niobium(V) oxide grafted on a silica gel surface: Study of the catalytic reduction of dissolved dioxygen. *J. Porphyrins Phthalocyanines* 5:537–544.

1192. Rosatto, S.S., P.T. Sotomayor, L.T. Kubota, and Y. Gushikem. 2002. SiO_2/Nb_2O_5 sol-gel as a support for HRP immobilization in biosensor preparation for phenol detection. *Electrochim. Acta* 47:4451–4458.

1193. Santos, A.d.S., L. Gorton, and L.T. Kubota. 2002. Nile blue adsorbed onto silica gel modified with niobium oxide for electrocatalytic oxidation of NADH. *Electrochim. Acta* 47:3351–3360.

1194. Santos, A.d.S., L. Gorton, and L.T. Kubota. 2002. Electrocatalytic NADH oxidation using an electrode based on meldola blue immobilized on silica coated with niobium oxide. *Electroanalysis* 14:805–812.

1195. Santos, A.d.S., A.C. Pereira, and L.T. Kubota. 2002. Electrochemical and electrocatalytic studies of toluidine blue immobilized on a silica gel surface coated with niobium oxide. *J. Braz. Chem. Soc.* 13:495–501.

1196. De Lucca, A.R., A.d.S. Santos, A.C. Pereira, and L.T. Kubota. 2002. Electrochemical behavior and electrocatalytic study of the methylene green coated on modified silica gel. *J. Colloid Interface Sci.* 254:113–119.

1197. Santos, A.d.S., R.S. Freire, and L.T. Kubota. 2003. Highly stable amperometric biosensor for ethanol based on Meldola's blue adsorbed on silica gel modified with niobium oxide. *J. Electroanal. Chem.* 547:135–142.

1198. Pereira, A.C., A.d.S. Santos, and L.T. Kubota. 2003. Electrochemical behavior of riboflavin immobilized on different matrices. *J. Colloid Interface Sci.* 265:351–358.

1199. Pereira, A.C., A.d.S. Santos, and L.T. Kubota. 2003. o-Phenylenediamine adsorbed onto silica gel modified with niobium oxide for electrocatalytic oxidation. *Electrochim. Acta* 48:3541–3550.

1200. Santos, A.d.S., N. Duran, and L.T. Kubota. 2005. Biosensor for H_2O_2 response based on horseradish peroxidase: Effect of different mediators adsorbed on silica gel modified with niobium oxide. *Electroanalysis* 17:1103–1111.

1201. Francisco, M.S., W.S. Cardoso, L.T. Kubota, and Y. Gushikem. 2007. Electrocatalytic oxidation of phenolic compounds using an electrode modified with Ni(II) porphyrin adsorbed on SiO_2/Nb_2O_5-phosphate synthesized by the sol-gel method. *J. Electroanal. Chem.* 602:29–36.

1202. Skeika, T., C. Marcovicz, S. Nakagaki, S.T. Fujiwara, K. Wohnrath, N. Nagata, and C.A. Pessoa. 2007. Electrochemical studies of an iron porphyrin immobilized on Nb_2O_5/SiO_2 and its application for simultaneous determination of dopamine and ascorbic acid using multivariate calibration methodology. *Electroanalysis* 19:2543–2550.

1203. Francisco, M.S., W.S. Cardoso, and Y. Gushikem. 2005. Carbon paste electrodes of the mixed oxide SiO_2/Nb_2O_5 prepared by the sol-gel method: Dissolved dioxygen sensor. *J. Electroanal. Chem.* 574:291–297.

References

1204. Ribeiro, E.S., S.L.P. Dias, S.T. Fujiwara, Y. Gushikem, and R.E. Bruns. 2003. Electrochemical study and complete factorial design of Toluidine Blue immobilized on SiO_2/Sb_2O_3 binary oxide. *J. Appl. Electrochem.* 33:1069–1075.
1205. Ribeiro, E.S., S.S. Rosatto, Y. Gushikem, and L.T. Kubota. 2003. Electrochemical study of Meldola's blue, methylene blue and toluidine blue immobilized on a SiO_2/Sb_2O_3 binary oxide matrix obtained by the sol-gel processing method. *J. Solid State Electrochem.* 7:665–670.
1206. Ribeiro, E.S., S.L.P. Dias, Y. Gushikem, and L.T. Kubota. 2004. Cobalt(II) porphyrin complex immobilized on the binary oxide SiO_2/Sb_2O_3: Electrochemical properties and dissolved oxygen reduction study. *Electrochim. Acta* 49:829–834.
1207. Cardoso, W.S., M.S. Francisco, R. Landers, and Y. Gushikem. 2005. Co(II)-porphyrin adsorbed on SiO_2/SnO_2/phosphate prepared by the sol-gel method. *Electrochim. Acta* 50:4378–4384.
1208. Cardoso, W.S. and Y. Gushikem. 2005. Electrocatalytic oxidation of nitrite on a carbon paste electrode modified with Co(II) porphyrin adsorbed on SiO_2/SnO_2/Phosphate prepared by the sol-gel method. *J. Electroanal. Chem.* 583:300–306.
1209. Ferrari de Castilho, R., E.B. Ribeiro de Souza, R.V.d.S. Alfaya, and A.A.d.S. Alfaya. 2008. Meldola's blue immobilized on a SiO_2/SnO_2/phosphate Xerogel, a new sensor for determination of ascorbic acid in medicine and commercial fruit juice. *Electroanalysis* 20:157–162.
1210. Ghiaci, M., B. Rezaei, and R.J. Kalbasi. 2007. Highly selective SiO_2-Al_2O_3 mixed-oxide modified carbon paste electrode for anodic stripping voltammetric determination of Pb(II). *Talanta* 73:37–45.
1211. Ghiaci, M., B. Rezaei, and M. Arshadi. 2009. Characterization of modified carbon paste electrode by using Salen Schiff base ligand immobilized on SiO_2-Al_2O_3 as a highly sensitive sensor for anodic stripping voltammetric determination of copper(II). *Sens. Actuators B* B139:494–500.
1212. Kubota, L.T., F. Gouvea, A.N. Andrade, B.G. Milagres, and G. Oliveira Neto. 1996. Electrochemical sensor for NADH based on Meldola's blue immobilized on silica gel modified with titanium phosphate. *Electrochim. Acta* 41:1465–1469.
1213. Kooshki, M. and E. Shams. 2007. Selective response of dopamine in the presence of ascorbic acid on carbon paste electrode modified with titanium phosphated silica gel. *Anal. Chim. Acta* 587:110–115.
1214. Shams, E., F. Alibeygi, and R. Torabi. 2006. Determination of nanomolar concentrations of Pb(II) using carbon paste electrode modified with zirconium phosphated amorphous silica. *Electroanalysis* 18:773–778.
1215. Shams, E. and R. Torabi. 2006. Determination of nanomolar concentrations of cadmium by anodic-stripping voltammetry at a carbon paste electrode modified with zirconium phosphated amorphous silica. *Sens. Actuators B* B117:86–92.
1216. Shams, E., A. Babaei, A.R. Taheri, and M. Kooshki. 2009. Voltammetric determination of dopamine at a zirconium phosphated silica gel modified carbon paste electrode. *Bioelectrochemistry* 75:83–88.
1217. Pereira, A.C., D.V. Macedo, A.d.S. Santos, and L.T. Kubota. 2006. Amperometric biosensor for lactate based on meldola's blue adsorbed on silica gel modified with niobium oxide. *Electroanalysis* 18:1208–1214.
1218. Wang, J. and Z. Taha. 1990. Catalytic oxidation and flow detection of carbohydrates at ruthenium dioxide modified electrodes. *Anal. Chem.* 62:1413–1416.
1219. Leech, D., J. Wang, and M.R. Smyth. 1990. Electrocatalytic detection of streptomycin and related antibiotics at ruthenium dioxide modified graphite-epoxy composite electrodes. *Analyst* 115:1447–1450.
1220. Leech, D., J. Wang, and M.R. Smyth. 1991. Electrocatalysis and flow detection of alcohols at ruthenium dioxide-modified electrodes. *Electroanalysis* 3:37–42.
1221. Chen, L., Z. Taha, and J. Wang. 1992. Modified graphite composite electrodes for fixed-potential amperometric detection of carbohydrates. *Curr. Sep.* 11:13–15.
1222. Lyons, M.E.G., C.A. Fitzgerald, and M.R. Smyth. 1994. Glucose oxidation at ruthenium dioxide based electrodes. *Analyst* 119:855–861.
1223. Wang, J. and Y. Lin. 1994. Electrocatalytic flow detection of amino acids at ruthenium dioxide-modified carbon electrodes. *Electroanalysis* 6:125–129.
1224. Dousikou, M.F., M.A. Koupparis, and C.E. Efstathiou. 2006. Determination of catalase-like activity in plants based on the amperometric monitoring of hydrogen peroxide consumption using a carbon paste electrode modified with ruthenium(IV) oxide. *Phytochem. Anal.* 17:255–261.
1225. Xie, Y. and C.O. Huber. 1991. Electrocatalysis and amperometric detection using an electrode made of copper oxide and carbon paste. *Anal. Chem.* 63:1714–1719.
1226. Kano, K., M. Torimura, Y. Esaka, M. Goto, and T. Ueda. 1994. Electrocatalytic oxidation of carbohydrates at copper(II)-modified electrodes and its application to flow-through detection. *J. Electroanal. Chem.* 372:137–143.

1227. Carjonyte, R. and A. Malinauskas. 1998. Amperometric sensor for hydrogen peroxide, based on Cu_2O or CuO modified carbon paste electrodes. *Fresenius J. Anal. Chem.* 360:122–123.
1228. Xu, J.-Z., J-J. Zhu, H. Wang, and H.-Y. Chen. 2003. Nano-sized copper oxide modified carbon paste electrodes as an amperometric sensor for Amikacin. *Anal. Lett.* 36:2723–2733.
1229. Dong, S., S. Zhang, X. Cheng, P. He, Q. Wang, and Y. Fang. 2007. Simultaneous determination of sugars and ascorbic acid by capillary zone electrophoresis with amperometric detection at a carbon paste electrode modified with polyethylene glycol and Cu_2O. *J. Chromatogr. A* 1161:327–333.
1230. Mannino, S., M.S. Cosio, and S. Ratti. 1993. Cobalt(II, III) oxide chemically modified electrode as amperometric detector in flow-injection systems. *Electroanalysis* 5:145–148.
1231. Cheng, X., S. Zhang, H. Zhang, Q. Wang, P. He, and Y. Fang. 2007. Determination of carbohydrates by capillary zone electrophoresis with amperometric detection at a nano-nickel oxide modified carbon paste electrode. *Food Chem.* 106:830–835.
1232. Schachl, K., H. Alemu, K. Kalcher, J. Jezkova, I. Švancara, and K. Vytřas. 1997. Flow injection determination of hydrogen peroxide using a carbon paste electrode modified with a manganese dioxide film. *Anal. Lett.* 30:2655–2673.
1233. Schachl, K., H. Alemu, K. Kalcher, J. Jezkova, I. Švancara, and K. Vytřas. 1997. Amperometric determination of hydrogen peroxide with a manganese dioxide-modified carbon paste electrode using flow injection analysis. *Analyst* 122:985–989.
1234. Schachl, K., E. Turkusic, A. Komersova, M. Bartos, H. Moderegger, I. Švancara, H. Alemu, K. Vytřas, M. Jimenez-Castro, and K. Kalcher. 2002. Amperometric determination of glucose with a carbon paste biosensor. *Coll. Czech. Chem. Commun.* 67:302–313.
1235. Zheng, X. and Z. Guo. 2000. Potentiometric determination of hydrogen peroxide at MnO_2-doped carbon paste electrode. *Talanta* 50:1157–1162.
1236. Teixeira, M.F.S., F.C. Moraes, E.T.G. Cavalheiro, and N. Bocchi. 2003. Differential pulse anodic voltammetric determination of lithium ions in pharmaceutical formulations using a carbon paste electrode modified with spinel-type manganese oxide. *J. Pharm. Biomed. Anal.* 31:537–543.
1237. Teixeira, M.F.S., M.F. Bergamini, and N. Bocchi. 2004. Lithium ions determination by selective preconcentration and differential pulse anodic stripping voltammetry using a carbon paste electrode modified with a spinel-type manganese oxide. *Talanta* 62:603–609.
1238. Teixeira, M.F.S., E.T.G. Cavalheiro, M.F. Bergamini, F.C. Moraes, and N. Bocchi. 2004. Use of carbon paste electrode modified with spinel-type manganese oxide as potentiometric sensor for lithium ions in flow injection analysis. *Electroanalysis* 16:633–639.
1239. Martinez, M.T., A.S. Lima, N. Bocchi, and M.F.S. Teixeira. 2009. Voltammetric performance and application of a sensor for sodium ions constructed with layered birnessite-type manganese oxide. *Talanta* 80:519–525.
1240. Teixeira, M.F.S., M.F. Bergamini, and N. Bocchi. 2010. Application of the potentiometric stripping analysis with constant current for the determination of lithium ions using a spinel-type manganese(IV) oxide-modified carbon paste electrode. *Curr. Anal. Chem.* 6:161–165.
1241. Correa de Melo, H., A.P. Seleghim, W.L. Polito, O. Fatibello-Filho, and I.C. Vieira. 2007. Simultaneous differential pulse voltammetric determination of L-dopa and carbidopa in pharmaceuticals using a carbon paste electrode modified with lead dioxide immobilized in a polyester resin. *J. Braz. Chem. Soc.* 18:797–803.
1242. Walcarius, A., J. Devoy, and J. Bessiere. 1999. Electrochemical recognition of selective mercury adsorption on minerals. *Environ. Sci. Technol.* 33:4278–4284.
1243. Navarro, P., C. Jambon, and O. Vittori. 1996. Electrochemical response of a carbon paste electrode containing an adsorbed species on incorporated alumina. *Fresenius J. Anal. Chem.* 356:476–479.
1244. Rani, S., A. Jayaraman, L.D. Sharma, G. Murali Dhar, and T.S.R. Prasada Rao. 2000. Cyclic voltammetric studies of alumina supported monometallic Pt and bimetallic Pt-Sn catalysts using carbon paste electrodes. *J. Electroanal. Chem.* 495:62–70.
1245. Zuo, S.-H., L.-F. Zhang, H.-H. Yuan, M-.B. Lan, G.A. Lawrance, and G. Wei. 2009. Electrochemical detection of DNA hybridization by using a zirconia modified renewable carbon paste electrode. *Bioelectrochemistry* 74:223–226.
1246. Wu, J., H. Liu, and Z. Lin. 2008. Electrochemical performance of a carbon nanotube/La-doped TiO_2 nanocomposite and its use for preparation of an electrochemical nicotinic acid sensor. *Sensors* 8:7085–7096.
1247. Wang, L.-H. 2000. Determination of zinc pyrithione in hair care products on metal oxides modified carbon electrodes. *Electroanalysis* 12:227–232.
1248. Shimizu, Y., A. Ishikawa, K. Iseki, and S. Takase. 2000. Perovskite-type oxide-based electrode: A new sensor for hydrogen-phosphate ion. *J. Electrochem. Soc.* 147:3931–3934.

1249. Luque, G.L., N.F. Ferreyra, A.G. Leyva, and G.A. Rivas. 2009. Characterization of carbon paste electrodes modified with manganese based perovskites-type oxides from the amperometric determination of hydrogen peroxide. *Sens. Actuators B* 142:331–336.
1250. Wang, Y., Y. Xu, L. Luo, Y. Ding, and X. Liu. 2010. Preparation of perovskite-type composite oxide LaNi$_{0.5}$Ti$_{0.5}$O$_3$-NiFe$_2$O$_4$ and its application in glucose biosensor. *J. Electroanal. Chem.* 642:35–40.
1251. Wang, L.H. and Z.S. Chen. 1997. Determination of thioglycolic acid and cysteine in hair-treatment products at ceramic carbon composite electrodes. *Electroanalysis* 9:1294–1297.
1252. Domenech Carbo, A., M.T. Domenech Carbo, J.V. Gimeno Adelantado, M. Moya Moreno, and F. Bosch Reig. 2000. Voltammetric identification of lead(II) and (IV) in medieval glazes in abrasion-modified carbon paste and polymer film electrodes. Application to the study of alterations in archaeological ceramic. *Electroanalysis* 12:120–127.
1253. Domenech Carbo, A., S. Sanchez Ramosa, M.T. Domenech Carbo, J.V. Gimeno Adelantado, F. Bosch Reig, D.J. Yusa Marco, and M.C. Sauri Peris. 2002. Electrochemical determination of the Fe(III)/Fe(II) ratio in archeological ceramic materials using carbon paste and composite electrodes. *Electroanalysis* 14:685–696.
1254. Wang, J., F. Lu, L. Angnes, J. Liu, H. Sakslund, Q. Chen, M. Pedrero, L. Chen, and O. Hammerich. 1995. Remarkably selective metalized-carbon amperometric biosensors. *Anal. Chim. Acta* 305:3–7.
1255. Pedrero, M., P. Mateo, C. Parrado, and J.M. Pingarron. 2000. Metallized graphite-ethylene/propylene/diene terpolymer composite electrodes as electrocatalytic amperometric detectors in flowing systems. *Quim. Anal.* 19:171–177.
1256. Wang, J., L. Fang, D. Lopez, and H. Tobias. 1993. Highly selective and sensitive amperometric biosensing of glucose at ruthenium-dispersed carbon paste enzyme electrodes. *Anal. Lett.* 26:1819–1830.
1257. Wang, J., E. Gonzalez Romero, and A.J. Reviejo. 1993. Improved alcohol biosensor based on ruthenium-dispersed carbon paste enzyme electrodes. *J. Electroanal. Chem.* 353:113–120.
1258. Wang, J., F. Lu, and D. Lopez. 1994. Tyrosinase-based ruthenium dispersed carbon paste biosensor for phenols. *Biosens. Bioelectron.* 9:9–15.
1259. Katsu, T., X. Yang, and G.A. Rechnitz. 1994. Amperometric biosensor for adenosine-5'-triphosphate based on a platinum-dispersed carbon paste enzyme electrode. *Anal. Lett.* 27:1215–1224.
1260. Wang, J., L. Chen, and J. Liu. 1997. Critical comparison of metaled and mediator-based carbon paste glucose biosensors. *Electroanalysis* 9:298–301.
1261. Wang, J., L. Chen, J. Liu, and F. Lu. 1996. Enhanced selectivity and sensitivity of first-generation enzyme electrodes based on the coupling of rhodinized carbon paste transducers and permselective poly(o-phenylenediamine) coatings. *Electroanalysis* 8:1127–1130.
1262. Wang, J., J. Liu, L. Chen, and F. Lu. 1994. Highly selective membrane-free, mediator-free glucose biosensor. *Anal. Chem.* 66:3600–3603.
1263. Wang, J., G. Rivas, and M. Chicharro. 1996. Iridium-dispersed carbon paste enzyme electrodes. *Electroanalysis* 8:434–437.
1264. Rivas, G.A. and B. Maestroni. 1997. Iridium-dispersed carbon paste amino acid oxidase electrodes. *Anal. Lett.* 30:489–501.
1265. Rubianes, M.D. and G.A. Rivas. 2000. Amperometric biosensor for phenols and catechols based on iridium-polyphenol oxidase-modified carbon paste. *Electroanalysis* 12:1159–1162.
1266. Liu, J., F. Lu, and J. Wang. 1999. Metal-alloy-dispersed carbon-paste enzyme electrodes for amperometric biosensing of glucose. *Electrochem. Commun.* 1:341–344.
1267. Liu, J. and J. Wang. 2001. A novel improved design for the first-generation glucose biosensor. *Food Technol. Biotechnol.* 39:55–58.
1268. Cai, X. and K. Kalcher. 1994. Studies on the electrocatalytic reduction of aliphatic aldehydes on palladium-modified carbon paste electrodes. *Electroanalysis* 6:397–404.
1269. Cai, X., K. Kalcher, J. Lintschinger, C.G. Neuhold, J. Tykarski, and B. Ogorevc. 1995. Electrocatalytic amperometric detection of hydroxylamine with a palladium-modified carbon paste electrode. *Electroanalysis* 7:556–559.
1270. Celej, M.S. and G.A. Rivas. 1998. Amperometric glucose biosensor based on gold-dispersed carbon paste. *Electroanalysis* 10:771–775.
1271. Ju, H., S. Liu, B. Ge, F. Lisdat, and F.W. Scheller. 2002. Electrochemistry of cytochrome c immobilized on colloidal gold modified carbon paste electrodes and its electrocatalytic activity. *Electroanalysis* 14:141–147.
1272. Liu, S.-Q. and H.-X. Ju. 2002. Renewable reagentless hydrogen peroxide sensor based on direct electron transfer of horseradish peroxidase immobilized on colloidal gold-modified electrode. *Anal. Biochem.* 307:110–116.

1273. Lei, C.-X., S.-Q. Hu, G.-L. Shen, and R.-Q. Yu. 2003. Immobilization of horseradish peroxidase to a nano-Au monolayer modified chitosan-entrapped carbon paste electrode for the detection of hydrogen peroxide. *Talanta* 59:981–988.
1274. Liu, S. and H. Ju. 2003. Reagentless glucose biosensor based on direct electron transfer of glucose oxidase immobilized on colloidal gold modified carbon paste electrode. *Biosens. Bioelectron.* 19:177–183.
1275. Liu, S., J. Yu, and H. Ju. 2003. Renewable phenol biosensor based on a tyrosinase-colloidal gold modified carbon paste electrode. *J. Electroanal. Chem.* 540:61–67.
1276. Chen, Z.-P., Z.-F. Peng, P. Zhang, X.-F. Jin, J.-H. Jiang, X.-B. Zhang, G.-L. Shen, and R.-Q. Yu. 2007. A sensitive immunosensor using colloidal gold as electrochemical label. *Talanta* 72:1800–1804.
1277. Mao, X., M. Baloda, A.S. Gurung, Y. Lin, and G. Liu. 2008. Multiplex electrochemical immunoassay using gold nanoparticle probes and immuno-chromatographic strips. *Electrochem. Commun.* 10:1636–1640.
1278. Xu, L., J. Du, N. He, Y. Deng, S. Li, and T. Wang. 2009. Anodic stripping voltammetry for detection of DNA hybridization with porous pseudo-carbon paste electrode by gold nanoparticle-catalyzed silver enhancement. *J. Nanosci. Nanotechnol.* 9:2698–2703.
1279. Behpour, M., E. Honarmand, and S.M. Ghoreishi. 2010. Nanogold-modified carbon paste electrode for the determination of atenolol in pharmaceutical formulations and urine by voltammetric methods. *Bull. Korean Chem. Soc.* 31:845–849.
1280. Xu, Y., C. Hu, and S. Hu. 2010. A reagentless nitric oxide biosensor based on the direct electrochemistry of hemoglobin adsorbed on the gold colloids modified carbon paste electrode. *Sens. Actuators, B* B148:253–258.
1281. Rodriguez, M.C. and G.A. Rivas. 2001. Highly selective first generation glucose biosensor based on carbon paste containing copper and glucose oxidase. *Electroanalysis* 13:1179–1184.
1282. Rubianes, M.D. and G.A. Rivas. 2003. Use of a melanin-type polymer to improve the selectivity of glucose biosensors. *Anal. Lett.* 36:1311–1323.
1283. Miscoria, S.A., G.D. Barrera, and G.A. Rivas. 2002. Analytical performance of a glucose biosensor prepared by immobilization of glucose oxidase and different metals into a carbon paste electrode. *Electroanalysis* 14:981–987.
1284. Beristain Ortiz, G., E. Hernandez Guevara, E. Reynoso Soto, I. Rivero Espejel, and M. Oropeza-Guzman 2009. Electrochemical characterization of CPEs modified with gold nanoparticles deposited by immersion. *ECS Trans.* 20:251–258.
1285. Pessoa, C.A., Y. Gushikem, and L.T. Kubota. 1997. Electrochemical study of methylene blue immobilized in zirconium phosphate. *Electroanalysis* 9:800–803.
1286. Pessoa, C.A., Y. Gushikem, L.T. Kubota, and L. Gorton. 1997. Preliminary electrochemical study of phenothiazines and phenoxazines immobilized on zirconium phosphate. *J. Electroanal. Chem.* 431:23–27.
1287. Munteanu, F.D., L.T. Kubota, and L. Gorton. 2001. Effect of pH on the catalytic electrooxidation of NADH using different two-electron mediators immobilized on zirconium phosphate. *J. Electroanal. Chem.* 509:2–10.
1288. Dilgin, Y., Z. Dursun, G. Nisli, and L. Gorton. 2005. Photoelectrochemical investigation of methylene blue immobilized on zirconium phosphate modified carbon paste electrode in flow injection system. *Anal. Chim. Acta* 542:162–168.
1289. Malinauskas, A., T. Ruzgas, and L. Gorton. 2000. Electrochemical study of the redox dyes Nile Blue and Toluidine Blue adsorbed on graphite and zirconium phosphate modified graphite. *J. Electroanal. Chem.* 484:55–63.
1290. Malinauskas, A., T. Ruzgas, and L. Gorton. 2000. Electrocatalytic oxidation of coenzyme NADH at carbon paste electrodes, modified with zirconium phosphate and some redox mediators. *J. Colloid Interface Sci.* 224:325–332.
1291. Malinauskas, A., T. Ruzgas, L. Gorton, and L.T. Kubota. 2000. A reagentless amperometric carbon paste based sensor for NADH. *Electroanalysis* 12:194–198.
1292. Malinauskas, A., T. Ruzgas, and L. Gorton. 1999. Tuning the redox potential of riboflavin by zirconium phosphate in carbon paste electrodes. *Bioelectrochem. Bioenerg.* 49:21–27.
1293. Kubota, L.T. and L. Gorton. 1999. Electrochemical investigations of the reaction mechanism and kinetics between NADH and riboflavin immobilized on amorphous zirconium phosphate. *J. Solid State Electrochem.* 3:370–379.
1294. Munteanu, F.D., M. Mosbach, A. Schulte, W. Schuhmann, and L. Gorton. 2002. Fast-scan cyclic voltammetry and scanning electrochemical microscopy studies of the pH-dependent dissolution of 2-electron mediators immobilized on zirconium phosphate containing carbon pastes. *Electroanalysis* 14:1479–1487.

1295. Munteanu, F.D., N. Mano, A. Kuhn, and L. Gorton. 2002. Mediator-modified electrodes for catalytic NADH oxidation: High rate constants at interesting overpotentials. *Bioelectrochemistry* 56:67–72.
1296. Munteanu, F.D., N. Mano, A. Kuhn, and L. Gorton. 2004. NADH electrooxidation using carbon paste electrodes modified with nitro-fluorenone derivatives immobilized on zirconium phosphate. *J. Electroanal. Chem.* 564:167–178.
1297. Munteanu, F.D., D. Dicu, I.C. Popescu, and L. Gorton. 2003. NADH oxidation using carbonaceous electrodes modified with dibenzo-dithia-diazapentacene. *Electroanalysis* 15:383–391.
1298. Dicu, D., F.D. Munteanu, I. Catalin Popescu, and L. Gorton. 2003. Indophenol and o-quinone derivatives immobilized on zirconium phosphate for NADH electro-oxidation. *Anal. Lett.* 36:1755–1779.
1299. Santiago, M.E., M.M. Velez, S. Borrero, A. Diaz, C.A. Casillas, C. Hofmann, A.R. Guadalupe, and J.L. Colon. 2006. NADH electrooxidation using bis(1,10-phenanthroline-5,6-dione) (2,2′-bipyridine)ruthenium(II)-exchanged zirconium phosphate modified carbon paste electrodes. *Electroanalysis* 18:559–572.
1300. Santiago, M.B., G.A. Daniel, A. David, B. Casanas, G. Hernandez, A.R. Guadalupe, and J.L. Colon. 2010. Effect of enzyme and cofactor immobilization on the response of ethanol oxidation in zirconium phosphate modified biosensors. *Electroanalysis* 22:1097–1105.
1301. El Mhammedi, M.A., M. Bakasse, and A. Chtaini. 2007. Electrochemical studies and square wave voltammetry of paraquat at natural phosphate modified carbon paste electrode. *J. Hazard. Mater.* 145:1–7.
1302. El Mhammedi, M.A., M. Bakasse, and A. Chtaini. 2007. Square-wave voltammetric determination of paraquat at carbon paste electrode modified with hydroxyapatite. *Electroanalysis* 19:1727–1733.
1303. El Mhammedi, M.A., M. Bakasse, and A. Chtaini. 2008. Investigation of square wave voltammetric detection of diquat at carbon paste electrode impregnated with $Ca_{10}(PO_4)_6F_2$: Application in natural water samples. *Mater. Chem. Phys.* 109:519–525.
1304. El Mhammedi, M.A. and A. Chtaini. 2008. Electrochemical studies of adsorption of paraquat onto $Ca_{10}(PO_4)_6(OH)_2$ from aqueous solution. *Leonardo J. Sci.* 25–34.
1305. El Mhammedi, M.A., M. Achak, M. Bakasse, and A. Chtaini. 2009. Electrochemical determination of para-nitrophenol at apatite-modified carbon paste electrode: Application in river water samples. *J. Hazard. Mater.* 163:323–328.
1306. El Mhammedi, M.A., M. Achak, and A. Chtaini. 2009. $Ca_{10}(PO_4)_6(OH)_2$-modified carbon-paste electrode for the determination of trace lead(II) by square-wave voltammetry. *J. Hazard. Mater.* 161:55–61.
1307. El Mhammedi, M.A., M. Achak, R. Najih, M. Bakasse, and A. Chtaini. 2009. Microextraction and trace determination of cadmium by square wave voltammetry at the carbon paste electrode impregnated with $Ca_{10}(PO_4)_6(OH)_2$. *Mater. Chem. Phys.* 115:567–571.
1308. Abdollahi, S. 1995. Preconcentration and determination of Pb^{2+} at an $AlPO_4$ containing carbon paste electrode. *Anal. Chim. Acta* 304:381–388.
1309. Bai, Y., Y.D. Wang, W.J. Zheng, and Y.S. Chen. 2008. Study on coordination of selenoamino acids with Ag+ at silver nitrate-modified carbon paste electrode. *Colloids Surf. B* 63:110–115.
1310. Guo, S.X. and S.B. Khoo. 1999. Copper(II) diethyldithiocarbamate modified carbon paste electrode for highly selective accumulation of cysteine. *Anal. Lett.* 32:689–700.
1311. Yeom, J.-S., M.-S. Won, and Y.-B. Shim. 1999. Voltammetric determination of the iodide ion with a quinine copper(II) complex modified carbon paste electrode. *J. Electroanal. Chem.* 463:16–23.
1312. Abbas, M.N. and G.A.E. Mostafa. 2003. New triiodomercurate-modified carbon paste electrode for the potentiometric determination of mercury. *Anal. Chim. Acta* 478:329–335.
1313. Tarmure, C., R. Sandulescu, and C. Ionescu. 2000. Voltammetric determination of lactate dehydrogenase using a carbon paste electrode. *J. Pharm. Biomed. Anal.* 22:355–361.
1314. Zhu, Y., S. Zhang, Y. Tang, M. Guo, C. Jin, and T. Qi. 2010. Electrochemical solid-phase nanoextraction of copper(II) on a magnesium oxinate-modified carbon paste electrode by cyclic voltammetry. *J. Solid State Electrochem.* 14:1609–1614.
1315. Bonifacio, V.G., L.H. Marcolino, M.F.S. Teixeira, and O. Fatibello-Filho. 2004. Voltammetric determination of isoprenaline in pharmaceutical preparations using a copper(II) hexacyanoferrate(III) modified carbon paste electrode. *Microchem. J.* 78:55–59.
1316. Dursun, Z. and G. Nisli. 2004. Voltammetric behavior of copper(I) oxide modified carbon paste electrode in the presence of cysteine and ascorbic acid. *Talanta* 63:873–878.
1317. Lazarin, A.M. and C. Airoldi. 2005. Synthesis and electrochemical properties of meldola blue intercalated into barium and calcium phosphates. *Sens. Actuators B* B107:446–453.
1318. Lazarin, A.M. and C. Airoldi. 2008. Methylene blue intercalated into calcium phosphate—Electrochemical properties and an ascorbic acid oxidation study. *Solid State Sci.* 10:1139–1144.
1319. Navratilova, Z. 1991. Mercury(II) voltammetry on a 1,5-diphenylcarbazide containing carbon paste electrode. *Electroanalysis* 3:799–802.

1320. Cai, X., K. Kalcher, J. Lintschinger, and C. Neuhold. 1993. Stripping voltammetric determination of trace amounts of mercury using a carbon paste electrode modified with 2-mercapto-4(3H)-quinazolinone. *Mikrochim. Acta* 112:135–146.

1321. Kannan, R., S.R. Rajagopalan, and V. Lakshminarayanan. 1995. Imidazole-modified C-paste electrode in the determination of trace amounts of mercury. *J. Electrochem. Soc. India* 44:90–96.

1322. Won, M.S., D.W. Moon, and Y.B. Shim. 1995. Determination of mercury and silver at a modified carbon paste electrode containing glyoxal bis(2-hydroxyanil). *Electroanalysis* 7:1171–1176.

1323. Khoo, S.B. and Q. Cai. 1996. Metal displacement at the zinc-diethyldithiocarbamate-modified carbon paste electrode for the selective preconcentration and stripping analysis of mercury(II). *Electroanalysis* 8:549–556.

1324. Gismera, M.J., D. Hueso, J.R. Procopio, and M.T. Sevilla. 2004. Ion-selective carbon paste electrode based on tetraethyl thiuram disulfide for copper(II) and mercury(II). *Anal. Chim. Acta* 524:347–353.

1325. Colilla, M., M.A. Mendiola, J.R. Procopio, and M.T. Sevilla. 2005. Application of a carbon paste electrode modified with a Schiff base ligand to mercury speciation in water. *Electroanalysis* 17:933–940.

1326. Mashhadizadeh, M.H., M. Talakesh, M. Peste, A. Momeni, H. Hamidian, and M. Mazlum. 2006. A novel modified carbon paste electrode for potentiometric determination of mercury(II) ion. *Electroanalysis* 18:2174–2179.

1327. Rizea, M.C., A.F. Danet, and S. Kalinowski. 2007. Determination of mercury (II) after its preconcentration on a carbon paste electrode modified with Cadion A. *Rev. Chim.* 58:266–269.

1328. Yeom, J.S., M.S. Won, S.N. Choi, and Y.B. Shim. 1990. Determination of silver(I) at a chemically modified electrode based on 2-iminocyclopentanedithiocarboxylic acid. *Bull. Korean Chem. Soc.* 11:200–205.

1329. Sugawara, K., S. Tanaka, and M. Taga. 1991. Voltammetry of silver(I) using a carbon-paste electrode modified with 2,2′-dithiodipyridine. *J. Electroanal. Chem.* 304:249–255.

1330. Cai, X., K. Kalcher, C. Neuhold, W. Goessler, I. Grabec, and B. Ogorevc. 1994. Studies on the voltammetric behavior of a 2-mercaptoimidazole containing carbon paste electrode: Determination of traces of silver. *Fresenius J. Anal. Chem.* 348:736–741.

1331. Khodari, M., M.M. Abou, and R. Fandy. 1994. Determination of Ag(I) with chemically modified carbon paste electrode based on 2,3-dicyano-1,4-naphthoquinone. *Talanta* 41:2179–2182.

1332. Guttmann, M., K.H. Lubert, and L. Beyer. 1996. Preconcentration and voltammetric behavior of Ag$^+$ at carbon paste electrodes modified by N-benzoyl-N′,N′-di-iso-butylthiourea. *Fresenius J. Anal. Chem.* 356:263–266.

1333. Won, M.S., J.S. Yeom, J.H. Yoon, E.D. Jeong, and Y.B. Shim. 2003. Determination of Ag(I) ion at a modified carbon paste electrode containing N,N′-diphenyl oxamide. *Bull. Korean Chem. Soc.* 24:948–952.

1334. Gismera, M.J., J.R. Procopio, M.T. Sevilla, and L. Hernandez. 2003. Copper(II) ion-selective electrodes based on dithiosalicylic and thiosalicylic acids. *Electroanalysis* 15:126–132.

1335. Mohadesi, A. and M.A. Taher. 2007. Stripping voltammetric determination of silver(I) at carbon paste electrode modified with 3-amino-2-mercaptoquinazolin-4(3H)-one. *Talanta* 71:615–619.

1336. Lee, I.C. and M.H. Cho. 1997. Stripping voltammetric determination of Ag(I) with carbon paste electrode modified with podand compounds. *J. Korean Chem. Soc.* 41:557–560.

1337. Prabhu, S.V., R.P. Baldwin, and L. Kryger. 1987. Chemical preconcentration and determination of copper at a chemically modified carbon-paste electrode containing 2,9-dimethyl-1,10-phenanthroline. *Anal. Chem.* 59:1074–1078.

1338. Sugawara, K., S. Tanaka, and M. Taga. 1991. Accumulation voltammetry of copper(II) using a carbon paste electrode modified with diquinolyl-8,8′-disulfide. *Analyst* 116:131–134.

1339. Sugawara, K., S. Tanaka, and M. Taga. 1992. Accumulation voltammetry of copper(II) at carbon-paste electrode containing salicylideneamino-2-thiophenol. *Fresenius J. Anal. Chem.* 342:65–69.

1340. Peng, T., L. Sheh, and G. Wang. 1996. Linear scan stripping voltammetry of copper(II) at the chemically modified carbon paste electrode. *Mikrochim. Acta* 122:125–132.

1341. Safavi, A., M. Pakniat, and N. Maleki. 1996. Design and construction of a flow system for determination of Cu(II) ions in water by means of a chemically modified carbon paste electrode. *Anal. Chim. Acta* 335:275–282.

1342. Qu, J., L. Meng, and K. Liu. 1999. Simultaneous determination of lead and copper by carbon paste electrode modified with pyruvaldehyde bis(N,N′-dibutyl thiosemicarbazone). *Anal. Lett.* 32:1991–2006.

1343. Abbaspour, A. and S.M.M. Moosavi. 2002. Chemically modified carbon paste electrode for determination of copper(II) by potentiometric method. *Talanta* 56:91–96.

1344. Danet, A.F., D. Neagu, M.P. Dondoi, and N. Iliescu. 2004. Anodic stripping voltammetric determination of copper(II) with salicylaldoxime carbon paste electrodes. *Rev. Chim.* 55:1–4.

1345. Jureviciute, I. and A. Malinauskas. 2004. Preparation of 2-mercaptobenzothiazole modified carbon paste electrode and its application to the stripping analysis of copper. *Chem. Anal. (Warsaw, Pol.)* 49:339–349.
1346. Chaisuksant, R., L. Pattanarat, and K. Grudpan. 2008. Naphthazarin modified carbon paste electrode for determination of copper(II). *Microchim. Acta* 162:181–188.
1347. Won, M.-S., J.-H. Park, and Y.-B. Shim. 1993. Determination of copper(I) ion with a chemically modified carbon paste electrode based on di(2-iminocyclopentylidinemercaptomethyl) disulfide. *Electroanalysis* 5:421–426.
1348. Won, M.-S., H.-J. Kim, and Y.-B. Shim. 1996. Differential pulse voltammetric determination of copper(I) ion with a rubeanic acid-modified carbon paste electrode. *Bull. Korean Chem. Soc.* 17:1142–1146.
1349. Peng, T., T. Zhe, G. Wang, and B. Shen. 1994. Differential pulse voltammetric determination of lead(II) with benzoin oxime-modified carbon paste electrodes. *Electroanalysis* 6:597–603.
1350. Liu, J.-G., F.-X. Xie, S.-Y. Zhang, Y.-P. Tian, and S.-S. Ni. 1998. Design, synthesis and characterization of a novel Schiff base bis(iminoantipyrine) and its modified electrode. *Chin. Chem. Lett.* 9:245–248.
1351. Rahmani, A., M.F. Mousavi, S.M. Golabi, M. Shamsipur, and H. Sharghi. 2004. Voltammetric determination of lead(II) using chemically modified carbon paste electrode with bis[1-hydroxy-9,10-anthraquinone-2-methyl]sulfide. *Chem. Anal.* 49:359–368.
1352. Adraoui, I., M. El Rhazi, A. Amine, L. Idrissi, A. Curulli, and G. Palleschi. 2005. Lead determination by anodic stripping voltammetry using a p-phenylenediamine modified carbon paste electrode. *Electroanalysis* 17:685–693.
1353. Vazquez, M.D., M.L. Tascon, and L. Deban. 2006. Determination of Pb(II) with a dithizone-modified carbon paste electrode. *J. Environ. Sci. Health A* 41:2735–2746.
1354. Gao, Z., P. Li, and Z. Zhao. 1991. Voltammetric determination of traces of cobalt(II) with a chemically modified carbon paste electrode. *Fresenius J. Anal. Chem.* 339:137–141.
1355. Gao, Z., G. Wang, P. Li, and Z. Zhao. 1991. Differential pulse voltammetric determination of cobalt with a perfluorinated sulfonated polymer-2,2-bipyridyl modified carbon paste electrode. *Anal. Chem.* 63:953–957.
1356. Suska, M. 1989. A study on suitability of solid electrodes for the determination of low concentrations of Ni and Co using stripping voltammetry method (in Czech). MSc dissertation, University of Chemical Technology, Pardubice, Czech Republic.
1357. Khan, M.R. 1998. 1-(2-Pyridylazo)-2-naphthol modified carbon paste electrode for trace cobalt(II) determination by differential pulse cathodic voltammetry. *Analyst* 123:1351–1357.
1358. Lu, X., Z. Wang, Z. Geng, J. Kang, and J. Gao. 2000. 2,4,6-tri(3,5-Dimethylpyrazoyl)-1,3,5-triazine modified carbon paste electrode for trace Cobalt(II) determination by differential pulse anodic stripping voltammetry. *Talanta* 52:411–416.
1359. Nateghi, M.R., A. Hadji-Shabani, and M. Fakheri. 2008. Determination of cobalt(II) at a 5-[(4-chlorophenyl)azo-N-(4'-methyl phenyl)]salicyl aldimine modified carbon paste electrode by differential pulse cathodic voltammetry. *Anal. Chem., Indian J.* 7:89–94.
1360. Bing, C., R. Deen, G.-N. Khang, C.-L. Sai, and L. Kryger. 1999. Chemical accumulation and voltammetric determination of traces of nickel(II) at glassy carbon electrodes modified with dimethyl glyoxime containing polymer coatings. *Talanta* 49:651–659.
1361. Tartarotti, F.O., M. Firmino de Oliveira, V.R. Balbo, and N.R. Stradiotto. 2006. Determination of nickel in fuel ethanol using a carbon paste modified electrode containing dimethylglyoxime. *Microchim. Acta* 155:397–401.
1362. Cai, X., K. Kalcher, C. Neuhold, W. Diewald, and R.J. Magee. 1993. Voltammetric determination of gold using a carbon paste electrode modified with thiobenzanilide. *Analyst* 118:53–57.
1363. Dos Santos Mattos, C., D. Ribeiro do Carmo, M. Firmino de Oliveira, and N.R. Stradiotto. 2008. Voltammetric determination of total iron in fuel ethanol using a 1,10 phenantroline/Nafion carbon paste-modified electrode. *Int. J. Electrochem. Sci.* 3:338–345.
1364. Gao, Z., P. Li, G. Wang, and Z. Zhao. 1990. Preconcentration and differential-pulse voltammetric determination of iron(II) with Nafion-1,10-phenantroline-modified carbon paste electrodes. *Anal. Chim. Acta* 241:137–146.
1365. Gao, Z., P. Li, and Z. Zhao. 1991. Determination of iron(II) with chemically-modified carbon-paste electrodes. *Talanta* 38:1177–1184.
1366. Paniagua, A.R., M.D. Vazquez, M.L. Tascon, and P. Sanchez Batanero. 1993. Determination of chromium(VI) and chromium(III) by using a diphenylcarbazide-modified carbon paste electrode. *Electroanalysis* 5:155–163.
1367. Xu, J., Y. Kong, W. Wang, Z. Chen, and S. Yao. 2009. Determination of chromium(VI) in electronics materials using trioctylamine modified carbon paste electrode. *Anal. Sci.* 25:1427–1430.

1368. Ghaedi, M., A. Shokrollahi, A.R. Salimibeni, S. Noshadi, and S. Joybar. 2010. Preparation of a new chromium(III) selective electrode based on 1-[(2-hydroxy ethyl) amino]-4-methyl-9H-thioxanthen-9-one as a neutral carrier. *J. Hazard. Mater.* 178:157–163.

1369. Khoo, S.B., M.K. Soh, Q. Cai, M.R. Khan, and S.X. Guo. 1997. Differential pulse cathodic stripping voltammetric determination of manganese(II) and manganese(VII) at the 1-(2-pyridylazo)-2-naphthol-modified carbon paste electrode. *Electroanalysis* 9:45–51.

1370. Khoo, S.B. and J. Zhu. 1996. Determination of trace amounts of antimony(III) by differential-pulse anodic stripping voltammetry at a phenylfluorone-modified carbon paste electrode. *Analyst* 121:1983–1988.

1371. Wang, J., P. Tuzhi, R. Li, and J. Zadeii. 1989. Tropolone modified carbon paste electrodes for trace measurements of tin. *Anal. Lett.* 22:719–727.

1372. Cai, Q. and S.B. Khoo. 1995. Differential pulse stripping voltammetric determination of thallium with an 8-hydroxyquinoline-modified carbon paste electrode. *Electroanalysis* 7:379–385.

1373. Cai, Q. and S.B. Khoo. 1995. Determination of trace thallium after accumulation of thallium(III) at a 8-hydroxyquinoline-modified carbon paste electrode. *Analyst* 120:1047–1053.

1374. Gao, Z., K.S. Siow, and A. Ng. 1996. Catalytic voltammetric determination of molybdenum at a chemically modified carbon paste electrode. *Electroanalysis* 8:1183–1187.

1375. Ferri, T., F. Guidi, and R. Morabito. 1994. Carbon paste electrode and medium-exchange procedure in adsorptive cathodic stripping voltammetry of selenium(IV). *Electroanalysis* 6:1087–1093.

1376. Estevez Hernandez, O., I. Naranjo Rodriguez, J.L.H.H. de Cisneros, and E. Reguera. 2007. Evaluation of carbon paste electrodes modified with 1-furoylthioureas for the analysis of cadmium by differential pulse anodic stripping voltammetry. *Sens. Actuators B* B123:488–494.

1377. Javanbakht, M., H. Khoshsafar, M.R. Ganjali, P. Norouzi, and M. Adib. 2009. Adsorptive stripping voltammetric determination of nanomolar concentration of cerium(III) at a carbon paste electrode modified by N′-[(2-hydroxyphenyl) methylidene]-2-furohydrazide. *Electroanalysis* 21:1605–1610.

1378. Kwak, M.-K., D.-S. Park, M.-S. Won, and Y.-B. Shim. 1996. Anodic differential pulse voltammetric determination of iodide with a cinchonine-modified carbon paste electrode. *Electroanalysis* 8:680–684.

1379. Wang, J., J. Lu, D.D. Larson, and K. Olsen. 1995. Voltammetric sensor for uranium based on the propyl gallate-modified carbon paste electrode. *Electroanalysis* 7:247–250.

1380. Dick, R., H. Ruf, and H.J. Ache. 1989. Analytically useful stripping voltammetric behavior of technetium at a thenoyltrifluoroacetone-modified carbon paste electrode. *Electroanalysis* 1:81–85.

1381. Li, J., F. Yi, D. Shen, and J. Fei. 2002. Adsorptive stripping voltammetric study of scandium-alizarin complexan complex at a carbon paste electrode. *Anal. Lett.* 35:1361–1372.

1382. Zhang, J., J. Li, and P. Deng. 2001. Adsorption voltammetry of the scandium-alizarin red S complex onto a carbon paste electrode. *Talanta* 54:561–566.

1383. Shamsipur, M., G. Khayatian, S.Y. Kazemi, K. Niknam, and H. Sharghi. 2001. The synthesis of 1,4-diaza-2,3;8,9-dibenzo-7,10-dioxacyclododecane-5,12-dione and its use in calcium-selective carbon paste electrodes. *J. Inclusion Phenom. Macrocycl. Chem.* 40:303–307.

1384. Alemu, H. and B.S. Chandravanshi. 1998. Differential pulse anodic stripping voltammetric determination of copper(II) with N-phenylcinnamohydroxamic acid modified carbon paste electrodes. *Anal. Chim. Acta* 368:165–173.

1385. Alemu, H. and B.S. Chandravanshi. 1998. Electrochemical behavior of N-phenylcinnamohydroxamic acid incorporated into carbon paste electrode and adsorbed metal ions. *Electroanalysis* 10:116–120.

1386. Refera, T., B.S. Chandravanshi, and H. Alemu. 1998. Differential pulse anodic stripping voltammetric determination of cobalt(II) with N-p-chlorophenylcinnamo-hydroxamic acid modified carbon paste electrode. *Electroanalysis* 10:1038–1042.

1387. Degefa, T.H., B.S. Chandravanshi, and H. Alemu. 1999. Differential pulse anodic stripping voltammetric determination of lead(II) with N-p-chlorophenyl-cinnamohydroxamic acid modified carbon paste electrode. *Electroanalysis* 11:1305–1311.

1388. Fanta, K. and B.S. Chandravanshi. 2001. Differential pulse anodic stripping voltammetric determination of cadmium(II) with N-p-chlorophenylcinnamo-hydroxamic acid modified carbon paste electrode. *Electroanalysis* 13:484–492.

1389. Ghoneim, E.M. 2010. Simultaneous determination of Mn(II), Cu(II) and Fe(III) as 2-(5′-bromo-2′-pyridylazo)-5-diethylaminophenol complexes by adsorptive cathodic stripping voltammetry at a carbon paste electrode. *Talanta* 82:646–652.

1390. Wang, J. and Q. Chen. 1993. In situ elimination of metal inhibitory effects using ligand-containing carbon paste enzyme electrodes. *Anal. Chem.* 65:2698–2700.

1391. Wang, J. and M. Bonakdar. 1988. Preconcentration and voltammetric measurement of mercury with a crown ether-modified carbon-paste electrode. *Talanta* 35:277–280.

References

1392. Shaidarova, L.G., N.A. Ulakhovich, M.A. Al Gakhri, and G.K. Budnikov. 1993. Use of dibenzo-18-crown-6 as a modifier of a carbon-paste electrode for lead determination. *Izv. Vyss. Ucheb. Zaved., Khim. Khim. Tekhnol.* 36:51–54.
1393. Shaidarova, L.G., N.A. Ulakhovich, M.A. Al'-Gakhri, G.K. Budnikov, and A.N. Glebov. 1995. Use of carbon-paste electrodes modified with macrocyclic compounds in voltammetric analysis. *J. Anal. Chem.* 50:692–697.
1394. Shaidarova, L.G., I.L. Fedorova, N.A. Ulakhovich, and G.K. Budnikov. 1997. Stripping-voltammetric determination of some amino acids on a carbon-paste electrode modified by crown ethers. *J. Anal. Chem.* 52:238–241.
1395. Shaidarova, L.G., I.L. Fedorova, N.A. Ulakhovich, and G.K. Budnikov. 1998. Stripping voltammetry of biologically active organic compounds as host-guest complexes at electrodes modified with crown ether. *J. Anal. Chem.* 53:52–58.
1396. Kotkar, R.M. and A.K. Srivastava. 2006. Voltammetric determination of para-aminobenzoic acid using carbon paste electrode modified with macrocyclic compounds. *Sens. Actuators B* B119:524–530.
1397. Kotkar, R.M. and A.K. Srivastava. 2008. Electrochemical behavior of nicotinamide using carbon paste electrode modified with macrocyclic compounds. *J. Inclusion Phenom. Macrocyclic Chem.* 60:271–279.
1398. Desai, P.B., R.M. Kotkar, and A.K. Srivastava. 2008. Electrochemical behaviour of pyridoxine hydrochloride (vitamin B6) at carbon paste electrode modified with crown ethers. *J. Solid State Electrochem.* 12:1067–1075.
1399. Ijeri, V.S. and A.K. Srivastava. 2000. Voltammetric determination of copper at chemically modified electrodes based on crown ethers. *Fresenius J. Anal. Chem.* 367:373–377.
1400. Ijeri, V.S. and A.K. Srivastava. 2001. Voltammetric determination of lead at chemically modified electrodes based on crown ethers. *Anal. Sci.* 17:605–608.
1401. Srivastava, A.K. and R.R. Gaichore. 2010. Macrocyclic compounds based chemically modified electrodes for voltammetric determination of L-tryptophan using electrocatalytic oxidation. *Anal. Lett.* 43:1933–1950.
1402. Agrahari, S.K., S.D. Kumar, and A.K. Srivastava. 2009. Development of a carbon paste electrode containing benzo-15-crown-5 for trace determination of the uranyl ion by using a voltammetric technique. *J. AOAC Int.* 92:241–247.
1403. Tanaka, S. and H. Yoshida. 1989. Stripping voltammetry of silver(I) with a carbon-paste electrode modified with thiacrown compounds. *Talanta* 36:1044–1046.
1404. Ruiperez, J., M.A. Mendiola, M.T. Sevilla, J.R. Procopio, and L. Hernandez. 2002. Application of a macrocyclic thiohydrazone modified carbon paste electrode to copper speciation in water samples. *Electroanalysis* 14:532–539.
1405. Taraszewska, J., G. Roslonek, and W. Darlewski. 1994. Electrochemical behavior of nickel tetraazamacrocyclic complexes incorporated into carbon paste electrodes: Application in H_2O_2 electrocatalysis. *J. Electroanal. Chem.* 371:223–230.
1406. Jesus Gismera, M., M. Antonia Mendiola, J. Rodriguez Procopio, and M. Teresa Sevilla. 1999. Copper potentiometric sensors based on copper complexes containing thiohydrazone and thiosemicarbazone ligands. *Anal. Chim. Acta* 385:143–149.
1407. Pouretedal, H.R. and M.H. Keshavarz. 2005. Cyclam modified carbon paste electrode as a potentiometric sensor for determination of cobalt(II) ions. *Chem. Res. Chin. Univ.* 21:28–31.
1408. Kotkar, R.M., P.B. Desai, and A.K. Srivastava. 2007. Behavior of riboflavin on plain carbon paste and aza macrocycles based chemically modified electrodes. *Sens. Actuators, B* B124:90–98.
1409. Chung, T.D. and H. Kim. 1998. Electrochemistry of calixarene and its analytical applications. *J. Inclusion Phenom. Mol. Recogn. Chem.* 32:179–193.
1410. O'Connor, K.M., D.W.M. Arrigan, and G. Svehla. 1995. Calixarenes in electroanalysis. *Electroanalysis* 7:205–215.
1411. Kim, S.-H., M.-S. Won, and Y.-B. Shim. 1996. Determination of derivatives of phenol with a modified electrode containing b-cyclodextrin. *Bull. Korean Chem. Soc.* 17:342–347.
1412. Ferancova, A., E. Korgova, R. Miko, and J. Labuda. 2000. Determination of tricyclic antidepressants using a carbon paste electrode modified with b-cyclodextrin. *J. Electroanal. Chem.* 492:74–77.
1413. Ferancova, A., E. Korgova, J. Labuda, J. Zima, and J. Barek. 2002. Cyclodextrin modified carbon paste based electrodes as sensors for the determination of carcinogenic polycyclic aromatic amines. *Electroanalysis* 14:1668–1673.
1414. Yanez, C., L.J. Nunez Vergara, and J.A. Squella. 2002. Determination of nitrendipine with b-cyclodextrin modified carbon paste electrode. *Electroanalysis* 14:559–562.

1415. Roa Morales, G., T. Ramirez Silva, and L. Galicia. 2003. Carbon paste electrodes electrochemically modified with cyclodextrins. *J. Solid State Electrochem.* 7:355–360.

1416. Roa Morales, G., M.T. Ramirez Silva, M.A. Romero Romo, and L. Galicia. 2003. Determination of lead and cadmium using a polycyclodextrin-modified carbon paste electrode with anodic stripping voltammetry. *Anal. Bioanal. Chem.* 377:763–769.

1417. Reddy, T.M., M. Sreedhar, and S.J. Reddy. 2003. Electrochemical determination of sparfloxacin in pharmaceutical formulations and urine samples using a b-cyclodextrin modified carbon paste electrode. *Anal. Lett.* 36:1365–1379.

1418. Reddy, T.M., K. Balaji, and S.J. Reddy. 2007. Voltammetric behavior of some fluorinated quinolone antibacterial agents and their differential pulse voltammetric determination in drug formulations and urine samples using a beta-cyclodextrin-modified carbon-paste electrode. *J. Anal. Chem.* 62:168–175.

1419. El Hady, D. and N. El Maali. 2008. Selective square wave voltammetric determination of (+)-catechin in commercial tea samples using beta-cyclodextrin modified carbon paste electrode. *Microchim. Acta* 161:225–231.

1420. Ramirez Berriozabal, M., L. Galicia, S. Gutierrez-Granados, J.S. Cortes, and P. Herrasti. 2008. Selective electrochemical determination of uric acid in the presence of ascorbic acid using a carbon paste electrode modified with beta-cyclodextrin. *Electroanalysis* 20:1678–1683.

1421. Abbaspour, A. and A. Noori. 2008. Electrochemical studies on the oxidation of guanine and adenine at cyclodextrin modified electrodes. *Analyst* 133:1664–1672.

1422. Ortega, D.R., M.T.R. Silva, M.P. Pardave, G. Alarcon-Angeles, A.R. Hernandez, and M.R. Romo. 2009. Development a boron potentiometric determination methodology using a carbon paste electrode modified with a beta-cyclodextrin-azomethine-H inclusion complex. *ECS Trans.* 20:13–19.

1423. Roa Morales, G., M.T. Ramirez-Silva, R.L. Gonzalez, L. Galicia, and M. Romero-Romo. 2005. Electrochemical characterization and determination of mercury using carbon paste electrodes modified with cyclodextrins. *Electroanalysis* 17:694–700.

1424. Ferancova, A., M. Buckova, E. Korgova, O. Korbut, P. Gruendler, I. Waernmark, R. Stepan, J. Barek, J. Zima, and J. Labuda. 2005. Association interaction and voltammetric determination of 1-aminopyrene and 1-hydroxypyrene at cyclodextrin and DNA based electrochemical sensors. *Bioelectrochem.* 67:191–197.

1425. Ozoemena, K.I., R.I. Stefan, J.F. van Staden, and H.Y. Aboul-Enein. 2004. Utilization of maltodextrin based enantioselective, potentiometric membrane electrodes for the enantioselective assay of S-perindopril. *Talanta* 62:681–685.

1426. Barsoum, B.N., W.M. Watson, I.M. Mahdi, and E. Khalid. 2004. Electrometric assay for the determination of acetylcholine using a sensitive sensor based on carbon paste. *J. Electroanal. Chem.* 567:277–281.

1427. Roa Morales, G., M. T. Ramirez Silva, M. M. Romero Romo, and L. Galicia. 2005. Heavy metal determination by anodic stripping voltammetry with a carbon paste electrode modified with alpha-cyclodextrin. In *Applications of Analytical Chemistry in Environmental Research*, Ed. M. Palomar, pp. 57–66, Trivandrum, India: Research Signpost.

1428. Arrigan, D.W.M., G. Svehla, S.J. Harris, and M.A. McKervey. 1992. Stripping voltammetry with a polymeric calixarene modified carbon paste electrode. *Anal. Proc.* 29:27–29.

1429. Arrigan, D.W.M., G. Svehla, S.J. Harris, and M.A. McKervey. 1994. Use of calixarenes as modifiers of carbon paste electrodes for voltammetric analysis. *Electroanalysis* 6:97–106.

1430. Ijeri, V.S., M. Algarra, and A. Martins. 2004. Electrocatalytic determination of vitamin C using calixarene modified carbon paste electrodes. *Electroanalysis* 16:2082–2086.

1431. Canpolat, E.C., E. Sar, N.Y. Coskun, and H. Cankurtaran. 2007. Determination of trace amounts of copper in tap water samples with a calix[4]arene modified carbon paste electrode by differential pulse anodic stripping voltammetry. *Electroanalysis* 19:1109–1115.

1432. Vaze, V.D. and A.K. Srivastava. 2007. Electrochemical behavior of folic acid at calixarene based chemically modified electrodes and its determination by adsorptive stripping voltammetry. *Electrochim. Acta* 53:1713–1721.

1433. Raoof, J.B., R. Ojani, A. Alinezhad, and S.Z. Rezaie. 2010. Differential pulse anodic stripping voltammetry of silver(I) using p-isopropylcalix[6]arene modified carbon paste electrode. *Monatsh. Chem.* 141:279–284.

1434. Ikeda, T., I. Katasho, and M. Senda. 1985. Glucose oxidase-immobilized benzoquinone-mixed carbon paste electrode with pre-minigrid. *Anal. Sci.* 1:455–457.

1435. Ikeda, T., F. Matsushita, and M. Senda. 1990. D-Fructose dehydrogenase-modified carbon paste electrode containing p-benzoquinone as a mediated amperometric fructose sensor. *Agric. Biol. Chem.* 54:2919–2924.

1436. Ikeda, T., T. Shibata, and M. Senda. 1989. Amperometric enzyme electrode for maltose based on an oligosaccharide dehydrogenase-modified carbon paste electrode containing p-benzoquinone. *J. Electroanal. Chem.* 261:351–362.
1437. Ikeda, T., T. Shibata, S. Todoriki, M. Senda, and H. Kinoshita. 1990. Amperometric response to reducing carbohydrates of an enzyme electrode based on oligosaccharide dehydrogenase. Detection of lactose and a-amylase. *Anal. Chim. Acta* 230:75–82.
1438. Karan, H.I., P.D. Hale, H.-L. Lan, H.-S. Lee, L.-F. Liu, T.A. Skotheim, and Y. Okamoto. 1991. Quinone-modified polymers as electron transfer relay systems in amperometric glucose sensors. *Polym. Adv. Technol.* 2:229–235.
1439. Tessema, M., T. Ruzgas, L. Gorton, and T. Ikeda. 1995. Flow injection amperometric determination of glucose and some other low molecular weight saccharides based on oligosaccharide dehydrogenase mediated by benzoquinone systems. *Anal. Chim. Acta* 310:161–171.
1440. Kulys, J. 1999. The carbon paste electrode encrusted with a microreactor as glucose biosensor. *Biosens. Bioelectron.* 14:473–479.
1441. Takayama, K., T. Kurosaki, and T. Ikeda. 1993. Mediated electrocatalysis at a biocatalyst electrode based on a bacterium, *Gluconobacter industrius*. *J. Electroanal. Chem.* 356:295–301.
1442. Takayama, K., T. Ikeda, and T. Nagasawa. 1996. Mediated amperometric biosensor for nicotinic acid based on whole cells of Pseudomonas fluorescens. *Electroanalysis* 8:765–768.
1443. Kondo, T. and T. Ikeda. 2000. Rapid detection of substrate-oxidizing activity of Hiochi bacteria using benzoquinone-mediated amperometric method. *J. Biosci. Bioeng.* 90:217–219.
1444. Aklilu, M., M. Tessema, and M. Redi-Abshiro. 2008. Indirect voltammetric determination of caffeine content in coffee using 1,4-benzoquinone modified carbon paste electrode. *Talanta* 76:742–746.
1445. Raoof, J.B., R. Ojani, and M. Ramine. 2009. Voltammetric sensor for nitrite determination based on its electrocatalytic reduction at the surface of p-duroquinone modified carbon paste electrode. *J. Solid State Electrochem.* 13:1311–1319.
1446. Zare, H.R., N. Nasirizadeh, and M. Mazloum Ardakani. 2005. Electrochemical properties of a tetrabromo-p-benzoquinone modified carbon paste electrode. Application to the simultaneous determination of ascorbic acid, dopamine and uric acid. *J. Electroanal. Chem.* 577:25–33.
1447. Raoof, J.B., R. Ojani, and M. Ramine. 2006. Electrocatalytic oxidation and voltammetric determination of L-cysteic acid at the surface of p-bromanil modified carbon paste electrode. *Electroanalysis* 18:1722–1726.
1448. Motta, N. and A.R. Guadalupe. 1994. Activated carbon paste electrodes for biosensors. *Anal. Chem.* 66:566–571.
1449. Raoof, D.B. and S.M. Golabi. 1995. Electrochemical properties of carbon-paste electrodes spiked with some 1,4-naphthoquinone derivatives. *Bull. Chem. Soc. Jpn.* 68:2253–2261.
1450. El Atrash, S.S. and R.D. O'Neill. 1995. Characterization in vitro of a naphthoquinone-mediated glucose oxidase-modified carbon paste electrode designed for neurochemical analysis in vivo. *Electrochim. Acta* 40:2791–2797.
1451. Golabi, S.M. and J.B. Raoof. 1996. Catalysis of dioxygen reduction to hydrogen peroxide at the surface of carbon paste electrodes modified by 1,4-naphthoquinone and some of its derivatives. *J. Electroanal. Chem.* 416:75–82.
1452. Raoof, J.B. and R. Ojani. 2002. The aqueous electrochemical behavior of carbon paste electrodes spiked with some 9,10-anthraquinone derivatives. *Int. J. Chem.* 12:227–239.
1453. Manisankar, P. and A. Gomathi. 2005. Electrocatalytic reduction of dioxygen at the surface of carbon paste electrodes modified with 9,10-anthraquinone derivatives and dyes. *Electroanalysis* 17:1051–1057.
1454. Mousavi, M.F., A. Rahmani, S.M. Golabi, M. Shamsipur, and H. Sharghi. 2001. Differential pulse anodic stripping voltammetric determination of lead(II) with a 1,4-bis(prop-2'-enyloxy)-9,10-anthraquinone modified carbon paste electrode. *Talanta* 55:305–312.
1455. Shamsipur, M., A. Salimi, S.M. Golabi, H. Sharghi, and M.F. Mousavi. 2001. Electrochemical properties of modified carbon paste electrodes containing some amino derivatives of 9,10-anthraquinone. *J. Solid State Electrochem.* 5:68–73.
1456. Kawakami, M., N. Uriuda, H. Koya, and S. Gondo. 1995. Glucose sensor based on ubiquinone-modified carbon paste electrode. *Anal. Lett.* 28:1555–1569.
1457. Kawakami, M., K. Tanaka, N. Uriuda, and S. Gondo. 2000. Effects of nonionic surfactants on electrochemical behavior of ubiquinone and menaquinone incorporated in a carbon paste electrode. *Bioelectrochemistry* 52:51–56.
1458. Laurinavicius, V., B. Kurtinaitiene, V. Liauksminas, A. Ramanavicius, R. Meskys, R. Rudomanskis, T. Skotheim, and L. Boguslavsky. 1999. Oxygen insensitive glucose biosensor based on PQQ-dependent glucose dehydrogenase. *Anal. Lett.* 32:299–316.

1459. Mazloum Ardakani, M., Z. Taleat, H. Beitollahi, and H. Naeimi. 2010. Selective determination of cysteine in the presence of tryptophan by carbon paste electrode modified with quinizarine. *J. Iran. Chem. Soc.* 7:251–259.
1460. Taleat, Z., M. Mazloum Ardakani, H. Naeimi, H. Beitollahi, M. Nejati, and H. Reza Zare. 2008. Electrochemical behavior of ascorbic acid at a 2,2′-[3,6-dioxa-1,8-octanediylbis(nitriloethylidyne)]-bis-hydroquinone carbon paste electrode. *Anal. Sci.* 24:1039–1044.
1461. Mazloum Ardakani, M., F. Habibollahi, H.R. Zare, H. Naeimi, and M. Nejati. 2009. Electrocatalytic oxidation of ascorbic acid at a 2,2′-(1,8-octanediylbisnitriloethylidine)-bis-hydroquinone modified carbon paste electrode. *J. Appl. Electrochem.* 39:1117–1124.
1462. Mazloum Ardakani, M., Z. Taleat, H. Beitollahi, and H. Naeimi. 2010. Electrocatalytic oxidation of dopamine on 2,2′-[3,6-dioxa-1,8-octanediylbis (nitriloethylidyne)]-bis-hydroquinone modified carbon paste electrode. *Anal. Methods* 2:149–153.
1463. Erdem, A., K. Kerman, B. Meric, and M. Ozsoz. 2001. Methylene blue as a novel electrochemical hybridization indicator. *Electroanalysis* 13:219–223.
1464. Yang, W., M. Ozsoz, D.B. Hibbert, and J.J. Gooding. 2002. Evidence for the direct interaction between methylene blue and guanine bases using DNA-modified carbon paste electrodes. *Electroanalysis* 14:1299–1302.
1465. Kara, P., K. Kerman, D. Ozkan, B. Meric, A. Erdem, P.E. Nielsen, and M. Ozsoz. 2002. Label-free and label based electrochemical detection of hybridization by using methylene blue and peptide nucleic acid probes at chitosan modified carbon paste electrodes. *Electroanalysis* 14:1685–1690.
1466. Ozkan, D., P. Kara, K. Kerman, B. Meric, A. Erdem, F. Jelen, P.E. Nielsen, and M. Ozsoz. 2002. DNA and PNA sensing on mercury and carbon electrodes by using methylene blue as an electrochemical label. *Bioelectrochemistry* 58:119–126.
1467. Kara, P., K. Kerman, D. Ozkan, B. Meric, A. Erdem, Z. Ozkan, and M. Ozsoz. 2002. Electrochemical genosensor for the detection of interaction between methylene blue and DNA. *Electrochem. Commun.* 4:705–709.
1468. Kulys, J., L. Wang, and N. Daugvilaite. 1992. Amperometric methylene green-mediated pyruvate electrode based on pyruvate oxidase entrapped in carbon paste. *Anal. Chim. Acta* 265:15–20.
1469. Kulys, J., G. Gleixner, W. Schuhmann, and H.L. Schmidt. 1993. Biocatalysis and electrocatalysis at carbon paste electrodes doped by diaphorase-methylene green and diaphorase-meldola blue. *Electroanalysis* 5:201–207.
1470. Kulys, J., L. Wang, and A. Maksimoviene. 1993. L-Lactate oxidase electrode based on methylene green and carbon paste. *Anal. Chim. Acta* 274:53–58.
1471. Chi, Q. and S. Dong. 1994. Electrocatalytic oxidation of reduced nicotinamide coenzymes at methylene Green-modified electrodes and fabrication of amperometric alcohol biosensors. *Anal. Chim. Acta* 285:125–133.
1472. Yu, A.-M., C.-X. He, J. Zhou, and H.-Y. Chen. 1997. Flow injection analysis of ascorbic acid at a methylene green chemically modified electrode. *Fresenius J. Anal. Chem.* 357:84–85.
1473. Dominguez, E., H.-L. Lan, Y. Okamoto, P.D. Hale, T.A. Skotheim, and L. Gorton. 1993. A carbon paste electrode chemically modified with a phenothiazine polymer derivative for electrocatalytic oxidation of NADH. Preliminary study. *Biosens. Bioelectron.* 8:167–175.
1474. Lobo Castanon, M.J., A.J. Miranda Ordieres, and P. Tunon Blanco. 1996. Flow-injection analysis of ethanol with an alcohol dehydrogenase-modified carbon paste electrode. *Electroanalysis* 8:932–937.
1475. Ramirez Molina, C., M. Boujtita, and N. El Murr. 1999. A carbon paste electrode modified by entrapped toluidine blue-O for amperometric determination of L-lactate. *Anal. Chim. Acta* 401:155–162.
1476. Munteanu, F.D., Y. Okamoto, and L. Gorton. 2003. Electrochemical and catalytic investigation of carbon paste modified with toluidine blue O covalently immobilised on silica gel. *Anal. Chim. Acta* 476:43–54.
1477. Thenmozhi, K. and S.S. Narayanan. 2007. Amperometric determination of hydrogen peroxide using a carbon paste electrode with immobilized toluidine blue. *Bull. Electrochem.* 23:13–18.
1478. Razola, S.S., E. Aktas, J.C. Vire, and J.-M. Kauffmann. 2000. Reagentless enzyme electrode based on phenothiazine mediation of horseradish peroxidase for subnanomolar hydrogen peroxide determination. *Analyst* 125:79–85.
1479. Serradilla Razola, S., B. Blankert, G. Quarin, and J.-M. Kauffmann. 2003. Phenothiazine drugs as redox mediators in horseradish peroxidase bioelectrocatalysis. *Anal. Lett.* 36:1819–1833.
1480. Wang, J., G. Rivas, X. Cai, H. Shiraishi, P.A.M. Farias, N. Dontha, and D. Luo. 1996. Accumulation and trace measurements of phenothiazine drugs at DNA-modified electrodes. *Anal. Chim. Acta* 332:139–144.

References

1481. Bremle, G., B. Persson, and L. Gorton. 1991. An amperometric glucose electrode based on carbon paste, chemically modified with glucose dehydrogenase, nicotinamide adenine dinucleotide, and a phenoxazine mediator, coated with a poly(ester sulfonic acid) cation exchanger. *Electroanalysis* 3:77–86.
1482. Kulys, J., H.E. Hansen, T. Buch-Rasmussen, J. Wang, and M. Ozsoz. 1994. Glucose biosensor based on the incorporation of Meldola Blue and glucose oxidase within carbon paste. *Anal. Chim. Acta* 288:193–196.
1483. Garjonyte, R. and A. Malinauskas. 2003. Investigation of baker's yeast *Saccharomyces cerevisiae*- and mediator-based carbon paste electrodes as amperometric biosensors for lactic acid. *Sens. Actuators, B* 96:509–515.
1484. Luo, W., H. Liu, H. Deng, K. Sun, C. Zhao, D. Qi, and J. Deng. 1997. Biosensing of hydrogen peroxide at carbon paste electrode incorporating N-methyl phenazine methosulfate, fumed-silica and horseradish peroxidase. *Anal. Lett.* 30:205–220.
1485. Garjonyte, R., V. Melvydas, and A. Malinauskas. 2006. Mediated amperometric biosensors for lactic acid based on carbon paste electrodes modified with baker's yeast *Saccharomyces cerevisiae*. *Bioelectrochemistry* 68:191–196.
1486. Gunasingham, H. and C.-H. Tan. 1990. Carbon paste-tetrathiafulvalene amperometric enzyme electrode for the determination of glucose in flowing systems. *Analyst* 115:35–39.
1487. Hale, P.D., L.F. Liu, and T.A. Skotheim. 1991. Enzyme-modified carbon paste: Tetrathiafulvalene electrodes for the determination of acetylcholine. *Electroanalysis* 3:751–756.
1488. Almeida, N.F. and A.K. Mulchandani. 1993. A mediated amperometric enzyme electrode using tetrathiafulvalene and L-glutamate oxidase for the determination of L-glutamic acid. *Anal. Chim. Acta* 282:353–361.
1489. Mulchandani, A., A.S. Bassi, and A. Nguyen. 1995. Tetrathiafulvalene-mediated biosensor for L-lactate in dairy products. *J. Food Sci.* 60:74–78.
1490. Mulchandani, A. and A.S. Bassi. 1996. Determination of glutamine and glutamic acid in mammalian cell cultures using tetrathiafulvalene modified enzyme electrodes. *Biosens. Bioelectron.* 11:271–280.
1491. Murthy, A.S.N. and Anita. 1996. Tetrathiafulvalene as a mediator for the electrocatalytic oxidation of L-ascorbic acid. *Biosens. Bioelectron.* 11:191–193.
1492. Saby, C., F. Mizutani, and S. Yabuki. 1995. Glucose sensor based on carbon paste electrode incorporating poly(ethylene glycol)-modified glucose oxidase and various mediators. *Anal. Chim. Acta* 304:33–39.
1493. Sun, C., W. Song, D. Zhao, Q. Gao, and H. Xu. 1996. Tetrabutylammonium-tetracyanoquinodimethane as electron-transfer mediator in amperometric glucose sensor. *Microchem. J.* 53:296–302.
1494. Bassi, A.S., D. Tang, and M.A. Bergougnou. 1999. Mediated, amperometric biosensor for glucose-6-phosphate monitoring based on entrapped glucose-6-phosphate dehydrogenase, Mg^{2+} ions, tetracyanoquinodimethane, and nicotinamide adenine dinucleotide phosphate in carbon paste. *Anal. Biochem.* 268:223–228.
1495. Forzani, E.S., G.A. Rivas, and V.M. Solis. 1999. Kinetic behavior of dopamine-polyphenol oxidase on electrodes of tetrathiafulvalenium tetracyanoquinodimethanide and tetracyanoquinodimethane species. *J. Electroanal. Chem.* 461:174–183.
1496. Calvo Marzal, P., K.Y. Chumbimuni Torres, N.F. Hoeehr, G. Oliveira Neto, and L.T. Kubota. 2004. Determination of reduced glutathione using an amperometric carbon paste electrode chemically modified with TTF-TCNQ. *Sens. Actuators B* 100:333–340.
1497. Hu, S., C. Xu, J. Luo, J. Luo, and D. Cui. 2000. Biosensor for detection of hypoxanthine based on xanthine oxidase immobilized on chemically modified carbon paste electrode. *Anal. Chim. Acta* 412:55–61.
1498. Hale, P.D., L.I. Boguslavsky, H.I. Karan, H.L. Lan, H.S. Lee, Y. Okamoto, and T.A. Skotheim. 1991. Investigation of viologen derivatives as electron-transfer mediators in amperometric glucose sensors. *Anal. Chim. Acta* 248:155–161.
1499. Dhesi, R., T.M. Cotton, and R. Timkovich. 1983. Reaction of redox proteins at a 4,4'-bipyridyl-modified carbon paste electrode. *J. Electroanal. Chem.* 154:129–139.
1500. Pandurangachar, M., B.E.K. Swamy, B.N. Chandrashekar, and B.S. Sherigara. 2009. Cyclic voltammetric investigation of dopamine at p-aminobenzoic acid modified carbon paste electrode. *Int. J. Electrochem. Sci.* 4:1319–1328.
1501. Xu, G., J. Zhang, and S. Dong. 1999. Chemiluminescent determination of glucose with a modified carbon paste electrode. *Microchem. J.* 62:259–265.
1502. Zheng, N., Y. Zeng, P.G. Osborne, Y. Li, W. Chang, and Z. Wang. 2002. Electrocatalytic reduction of dioxygen on hemin based carbon paste electrode. *J. Appl. Electrochem.* 32:129–133.
1503. Zare, H.R., N. Nasirizadeh, M. Mazloum-Ardakani, and M. Namazian. 2006. Electrochemical properties and electrocatalytic activity of hematoxylin modified carbon paste electrode toward the oxidation of reduced nicotinamide adenine dinucleotide (NADH). *Sens. Actuators B* B120:288–294.

1504. Niranjana, E., R.R. Naik, B.E.K. Swamy, B.S. Sherigara, and H. Jayadevappa. 2008. Electrocatalytic oxidation of amlodipinebesylate with phenyl hydrazine as a mediator at carbon paste electrode. *Res. Rev. Electrochem.* 1:42–48.

1505. Chitravathi, S., B.E.K. Swamy, U. Chandra, G.P. Mamatha, and B.S. Sherigara. 2010. Electrocatalytic oxidation of sodium levothyroxine with phenyl hydrazine as a mediator at carbon paste electrode: A cyclic voltammetric study. *J. Electroanal. Chem.* 645:10–15.

1506. Mirmomtaz, E., A.A. Ensafi, and H. Karimi-Maleh. 2008. Electrocatalytic determination of 6-thioguanine at a p-aminophenol modified carbon paste electrode. *Electroanalysis* 20:1973–1979.

1507. Kachoosangi, R.T., G.G. Wildgoose, and R.G. Compton. 2008. Using capsaicin modified multiwalled carbon nanotube based electrodes and p-chloranil modified carbon paste electrodes for the determination of amines: Application to benzocaine and lidocaine. *Electroanalysis* 20:2495–2500.

1508. Kulys, J., A. Drungiliene, U. Wollenberger, and F. Scheller. 1998. Membrane covered carbon paste electrode for the electrochemical determination of peroxidase and microperoxidase in a flow system. *Bioelectrochem. Bioenerg.* 45:227–232.

1509. Shahrokhian, S. and M. Ghalkhani. 2006. Simultaneous voltammetric detection of ascorbic acid and uric acid at a carbon-paste modified electrode incorporating thionine-Nafion ion-pair as an electron mediator. *Electrochim. Acta* 51:2599–2606.

1510. Portaccio, M., D. Di Tuoro, F. Arduini, M. Lepore, D.G. Mita, N. Diano, L. Mita, and D. Moscone. 2010. A thionine-modified carbon paste amperometric biosensor for catechol and bisphenol A determination. *Biosens. Bioelectron.* 25:2003–2008.

1511. Lima, P.R., W.J.R. Santos, A.B. Oliveira, M.O.F. Goulart, and L.T. Kubota. 2008. Electrocatalytic activity of 4-nitrophthalonitrile-modified electrode for the L-glutathione detection. *J. Pharm. Biomed. Anal.* 47:758–764.

1512. Lima, P.R., W.d.J.R. Santos, A. Bof de Oliveira, M.O.F. Goulart, and L.T. Kubota. 2008. Electrochemical investigations of the reaction mechanism and kinetics between NADH and redox-active $(NC)_2C_6H_3$-NHOH/$(NC)_2C_6H_3$-NO from 4-nitrophthalonitrile-$(NC)_2C_6H_3$-NO_2-modified electrode. *Biosens. Bioelectron.* 24:448–454.

1513. Shahrokhian, S., H.R. Zare-Mehrjardi, and H. Khajehsharifi. 2009. Modification of carbon paste with congo red supported on multi-walled carbon nanotube for voltammetric determination of uric acid in the presence of ascorbic acid. *J. Solid State Electrochem.* 13:1567–1575.

1514. Sabzi, R.E., E. Minaei, and K. Farhadi. 2010. Congo red as a modifier for amperometric determination of ascorbic acid. *Asian J. Chem.* 22:2165–2171.

1515. Ensafi, A.A., A. Arabzadeh, and H. Karimi-Maleh. 2010. Simultaneous determination of dopamine and uric acid by electrocatalytic oxidation on a carbon paste electrode using Pyrogallol Red as a mediator. *Anal. Lett.* 43:1976–1988.

1516. Faguy, P.W., R.P. Baldwin, and R.M. Buchanan. 1997. Polymerized phthalocyanines for electrochemical detection, EOY report 1997808052; Office of Naval research. Freely available from: http://www.dtic.mil/cgi-bin/GetTRDoc? Location = U2&doc = GetTRDoc. pdf&AD = ADA327966 (accessed April 15, 2011).

1517. Amine, A., J.-M. Kauffmann, and G.J. Patriarche. 1991. Amperometric biosensors for glucose based on carbon paste modified electrodes. *Talanta* 38:107–110.

1518. Chen, L., M.S. Lin, M. Hara, and G.A. Rechnitz. 1991. Kohlrabi-based amperometric biosensor for hydrogen peroxide measurement. *Anal. Lett.* 24:1–14.

1519. Mizutani, F., S. Yabuki, A. Okuda, and T. Katsura. 1991. Glucose-sensing electrode based on carbon paste containing ferrocene and polyethylene glycol-modified enzyme. *Bull. Chem. Soc. Jpn.* 64:2849–2851.

1520. Kalab, T. and P. Skladal. 1994. Evaluation of mediators for development of amperometric microbial bioelectrodes. *Electroanalysis* 6:1004–1008.

1521. Katrlik, J., R. Brandsteter, J. Svorc, M. Rosenberg, and S. Miertus. 1997. Mediator type of glucose microbial biosensor based on *Aspergillus niger*. *Anal. Chim. Acta* 356:217–224.

1522. Rondeau, A., N. Larsson, M. Boujtita, L. Gorton, and N. El Murr. 1999. The synergetic effect of redox mediators and peroxidase in a bienzymatic biosensor for glucose assays in FIA. *Analusis* 27:649–656.

1523. Guemas, Y., M. Boujtita, and N. El Murr. 2000. Biosensor for determination of glucose and sucrose in fruit juices by flow injection analysis. *Appl. Biochem. Biotechnol* 89:171–181.

1524. Szuwarski, N., S. Gueguen, M. Boujtita, and N. El Murr. 2001. Use of thermally simulated aging to evaluate the contribution of constituents to the activity of mediated carbon paste glucose biosensors to improve their shelf-life. *Electroanalysis* 13:1237–1241.

1525. Kong, Y.-T., S.I. Imabayashi, K. Kano, T. Ikeda, and T. Kakiuchi. 2001. Peroxidase-based amperometric sensor for the determination of total phenols using two-stage peroxidase reactions. *Am. J. Enol. Vitic.* 52:381–385.

1526. Matsumoto, R., M. Mochizuki, K. Kano, and T. Ikeda. 2002. Unusual response in mediated biosensors with an oxidase/peroxidase bienzyme system. *Anal. Chem.* 74:3297–3303.

1527. Zaydan, R., M. Dion, and M. Boujtita. 2004. Development of a new method, based on a bioreactor coupled with an L-lactate biosensor, toward the determination of a nonspecific inhibition of L-lactic acid production during milk fermentation. *J. Agric. Food Chem.* 52:8–14.
1528. Ghasemi, V., J.B. Raoof, R. Ojani, and R. Hosseinzadeh. 2005. Voltammetric determination of ascorbic acid on a ferrocene derivative-modified carbon paste electrode. *Bull. Electrochem.* 21:115–122.
1529. Raoof, J.B., R. Ojani, and M. Kolbadinezhad. 2009. Voltammetric sensor for glutathione determination based on ferrocene-modified carbon paste electrode. *J. Solid State Electrochem.* 13:1411–1416.
1530. Amine, A., J.M. Kauffmann, G.C. Guilbault, and S. Bacha. 1993. Characterization of mixed enzyme-mediator-carbon paste electrodes. *Anal. Lett.* 26:1281–1299.
1531. Yabuki, S., F. Mizutani, and T. Katsura. 1994. Choline-sensing electrode based on polyethylene glycol-modified enzyme and mediator. *Sens. Actuators B* 20:159–162.
1532. Kulys, J., L. Tetianec, and P. Schneider. 2000. The development of an improved glucose biosensor using recombinant carbohydrate oxidase from *Microdochium nivale*. *Analyst* 125:1587–1590.
1533. Kulys, J., L. Tetianec, and P. Schneider. 2001. Recombinant *Microdochium nivale* carbohydrate oxidase and its application in an amperometric glucose sensor. *Biosens. Bioelectron.* 16:319–324.
1534. Charpentier, L. and N. El Murr. 1995. Amperometric determination of cholesterol in serum with use of a renewable surface peroxidase electrode. *Anal. Chim. Acta* 318:89–93.
1535. Raoof, J.B., R. Ojani, and M. Kolbadinezhad. 2005. Electrocatalytic characteristics of ferrocenecarboxylic acid modified carbon paste electrode in the oxidation and determination of L-cysteine. *Electroanalysis* 17:2043–2051.
1536. Raoof, J.B., R. Ojani, and H. Beitollahi. 2007. L-cysteine voltammetry at a carbon paste electrode bulk-modified with ferrocenedicarboxylic acid. *Electroanalysis* 19:1822–1830.
1537. Karimi Maleh, H., A.A. Ensafi, and H.R. Ensafi. 2009. Ferrocenedicarboxylic acid modified carbon paste electrode: A sensor for electrocatalytic determination of hydrochlorothiazide. *J. Braz. Chem. Soc.* 20:880–887.
1538. Khalilzadeh, M.A., F. Gholami, and H. Karimi-Maleh. 2009. Electrocatalytic determination of ampicillin using carbon-paste electrode modified with ferrocendicarboxylic acid. *Anal. Lett.* 42:584–599.
1539. Karimi Maleh, H., A.A. Ensafi, and A.R. Allafchian. 2010. Fast and sensitive determination of captopril by voltammetric method using ferrocenedicarboxylic acid modified carbon paste electrode. *J. Solid State Electrochem.* 14:9–15.
1540. Akhgar, M.R., M. Salari, H. Zamani, A. Changizi, and H. Hosseini-Mahdiabad. 2010. Electrocatalytic and simultaneous determination of phenylhydrazine and hydrazine using carbon paste electrode modified with carbon nanotubes and ferrocenedicarboxylic acid. *Int. J. Electrochem. Sci.* 5:782–796.
1541. Hattori, T. and S. Tanaka. 2003. Cyclic voltammetry examination for effect of counter anion on adsorptive voltammetry of 11-ferrocenytrimethylundecyl-ammonium at carbon paste electrode. *Electroanalysis* 15:1522–1528.
1542. Raoof, J.B., R. Ojani, R. Hosseinzadeh, and V. Ghasemi. 2003. Electrocatalytic characteristics of a 1-[4-(ferrocenylethynyl)phenyl]-1-ethanone modified carbon-paste electrode in the oxidation of ascorbic acid. *Anal. Sci.* 19:1251–1258.
1543. Raoof, J.B., R. Ojani, H. Beitollahi, and R. Hosseinzadeh. 2006. Electrocatalytic oxidation and highly selective voltammetric determination of L-cysteine at the surface of a 1-[4-(ferrocenyl ethynyl)phenyl]-1-ethanone modified carbon paste electrode. *Anal. Sci.* 22:1213–1220.
1544. Raoof, J.B., R. Ojani, F. Chekin, and R. Hossienzadeh. 2007. Carbon paste electrode incorporating 1-[4-(ferrocenyl ethynyl) phenyl]-1-ethanone for voltammetric determination of D-penicillamine. *Int. J. Electrochem. Sci.* 2:848–860.
1545. Raoof, J.B., R. Ojani, and H. Karimi-Maleh. 2008. Carbon paste electrode incorporating 1-[4-(ferrocenyl ethynyl) phenyl]-1-ethanone for electrocatalytic and voltammetric determination of tryptophan. *Electroanalysis* 20:1259–1262.
1546. Raoof, J.B., R. Ojani, and H. Karimi-Maleh. 2008. Electrocatalytic determination of sulfite using 1-[4-(ferrocenylethynyl)phenyl]-1-ethanone modified carbon paste electrode. *Asian J. Chem.* 20:483–494.
1547. Raoof, J.B., R. Ojani, and H. Karimi-Maleh. 2008. Voltammetric determination of L-cysteic acid on a 1-[4-(ferrocenyl-ethynyl)phenyl]-1-ethanone modified carbon paste electrode. *Bull. Chem. Soc. Ethiop.* 22:173–182.
1548. Raoof, J.B., R. Ojani, H. Beitollahi, and R. Hossienzadeh. 2006. Electrocatalytic determination of ascorbic acid at the surface of 2,7-bis(ferrocenyl ethyl)fluoren-9-one modified carbon paste electrode. *Electroanalysis* 18:1193–1201.

1549. Raoof, J.B., R. Ojani, and H. Beitollahi. 2007. Electrocatalytic determination of ascorbic acid at chemically modified carbon paste electrode with 2,7-bis (ferrocenylethynyl) fluoren-9-one. *Int. J. Electrochem. Sci.* 2:534–548.

1550. Raoof, J.B., R. Ojani, and H. Karimi Maleh. 2009. Electrocatalytic oxidation of glutathione at carbon paste electrode modified with 2,7-bis (ferrocenyl ethyl) fluoren-9-one: Application as a voltammetric sensor. *J. Appl. Electrochem.* 39:1169–1175.

1551. Losada, J., I. Cuadrado, M. Moran, C.M. Casado, B. Alonso, and M. Barranco. 1997. Ferrocenyl silicon-based dendrimers as mediators in amperometric biosensors. *Anal. Chim. Acta* 338:191–198.

1552. Casero, E., F. Pariente, E. Lorenzo, L. Beyer, and J. Losada. 2001. Electrocatalytic oxidation of nitric oxide at 6,17-diferrocenyldibenzo[b,i]5,9,14,18-tetraaza[14]annulen-nickel(II) modified electrodes. *Electroanalysis* 13:1411–1416.

1553. Ojani, R., J.B. Raoof, and B. Norouzi. 2008. Acetylferrocene modified carbon paste electrode; a sensor for electrocatalytic determination of hydrazine. *Electroanalysis* 20:1378–1382.

1554. Hale, P.D., L.I. Boguslavsky, T. Inagaki, H.I. Karan, H.S. Lee, T.A. Skotheim, and Y. Okamoto. 1991. Amperometric glucose biosensors based on redox polymer-mediated electron transfer. *Anal. Chem.* 63:677–682.

1555. Calvo, E.J., C. Donilowicz, and L. Diaz. 1993. Enzyme catalysis at hydrogel-modified electrodes with redox polymer mediator. *J. Chem. Soc. Faraday Trans.* 89:377–384.

1556. Rapicault, S., B. Limoges, and C. Degrand. 1996. Renewable perfluorosulfonated ionomer carbon paste electrode for competitive homogeneous electrochemical immunoassays using a redox cationic labeled hapten. *Anal. Chem.* 68:930–935.

1557. Rapicault, S., F. Paday, and C. Degrand. 1996. Redox-catalytic reduction of dioxygen by cobaltocene at a carbon paste electrode: Application to trace analysis. *J. Organomet. Chem.* 525:139–144.

1558. Bordes, A.L., B. Limoges, P. Brossier, and C. Degrand. 1997. Simultaneous homogeneous immunoassay of phenytoin and phenobarbital using a Nafion-loaded carbon paste electrode and two redox cationic labels. *Anal. Chim. Acta* 356:195–203.

1559. Bordes, A.L., B. Schollhorn, B. Limoges, and C. Degrand. 1998. Redox labeling of two antiepileptic drugs with metallocenes and their simultaneous detection by a nafion-modified electrode. *Appl. Organomet. Chem.* 12:59–65.

1560. Wring, S.A., J.P. Hart, and B.J. Birch. 1989. Development of an improved carbon electrode chemically modified with cobalt phthalocyanine as a reusable sensor for glutathione. *Analyst* 114:1563–1570.

1561. Fujiwara, S.T. and Y. Gushikem. 1999. Cobalt(II) phthalocyanine bonded to 3-n-propylimidazole immobilized on silica gel surface: Preparation and electrochemical properties. *J. Braz. Chem. Soc.* 10:389–393.

1562. Shaidarova, L.G., G.K. Budnikov, and S.A. Zaripova. 2001. Electrocatalytic determination of dithio-carbamate-based pesticides using electrodes modified with metal phthalocyanines. *J. Anal. Chem.* 56:748–753.

1563. Andronic, V., L. Oniciu, and C. Popescu. 2003. Electrochemical behavior of cobalt phthalocyanine (CoPc) modified carbon paste electrodes in aqueous solutions. *Rev. Roum. Chim.* 47:505–511.

1564. Shaidarova, L.G., S.A. Ziganshina, E.P. Medyantseva, and G.K. Budnikov. 2004. Amperometric cholinesterase biosensors with carbon paste electrodes modified with cobalt phthalocyanine. *Russ. J. Appl. Chem.* 77:241–248.

1565. Siangproh, W., O. Chailapakul, R. Laocharoensuk, and J. Wang. 2005. Microchip capillary electrophoresis/electrochemical detection of hydrazine compounds at a cobalt phthalocyanine modified electrochemical detector. *Talanta* 67:903–907.

1566. Mizutani, F., S. Yabuki, and S. Iijima. 1995. Amperometric glucose-sensing electrode based on carbon paste containing poly(ethylene glycol)-modified glucose oxidase and cobalt octaethoxyphthalocyanine. *Anal. Chim. Acta* 300:59–64.

1567. Mizutani, F., S. Yabuki, and S. Iijima. 1995. Carbon paste electrode incorporated with cobalt(II) octaethoxyphthalocyanine for the amperometric detection of hydrogen peroxide. *Electroanalysis* 7:706–709.

1568. Ribeiro, E.S. and Y. Gushikem. 1999. Iron(II) tetrasulphophthalocyanine complex adsorbed on a silica gel surface chemically modified by 3-n-propylpyridinium chloride. *Electrochim. Acta* 44:3589–3592.

1569. Shaidarova, L.G., A.Y. Fomin, S.A. Ziganshina, E.P. Medyantseva, and G.K. Budnikov. 2002. Electrocatalytic response of electrodes modified with metal phthalocyanines in the cholinesterase-thiocholine ester system. *J. Anal. Chem.* 57:150–156.

1570. Oni, J., P. Westbroek, and T. Nyokong. 2003. Electrochemical behavior and detection of dopamine and ascorbic acid at an iron(II)tetrasulfophthalocyanine modified carbon paste microelectrode. *Electroanalysis* 15:847–854.

1571. Parra, V., T. Hernando, M.L. Rodriguez-Mendez, and J.A. de Saja. 2004. Electrochemical sensor array made from bisphthalocyanine modified carbon paste electrodes for discrimination of red wines. *Electrochim. Acta* 49:5177–5185.
1572. Takeuchi, E.S. and R.W. Murray. 1985. Metalloporphyrin containing carbon paste electrodes. *J. Electroanal. Chem.* 188:49–57.
1573. Campanella, L., G. Favero, and M. Tomassetti. 1997. A modified amperometric electrode for the determination of free radicals. *Sens. Actuators B* B44:559–565.
1574. Kasanuki, T., M. Kamiyama, H. Noguchi, S. Ishii, and Y. Yosuhida. 2001. Hydrogen peroxide sensor based on carbon paste electrode containing a metal porphyrin complex. *Chem. Sens.* 17:427–429.
1575. Shamsipur, M., M. Najafi, M.R. Milani Hosseini, H. Sharghi, and S.H. Kazemi. 2009. Electrocatalytic behavior of a carbon paste electrode modified with iron(III)tetracyanophenylporphyrin chloride towards dioxygen reduction. *Pol. J. Chem.* 83:1173–1183.
1576. Sugawara, K., F. Yamamoto, S. Tanaka, and H. Nakamura. 1995. Electrochemical behavior of sugar investigated using a carbon paste electrode modified with copper(II)-porphyrin. *J. Electroanal. Chem.* 394:263–265.
1577. Abbaspour, A., A. Ghaffarinejad, and E. Safaei. 2004. Determination of L-histidine by modified carbon paste electrode using tetra-3,4-pyridinoporphirazinatocopper(II). *Talanta* 64:1036–1040.
1578. Joseph, R. and K. Girish Kumar. 2010. Differential pulse voltammetric determination and catalytic oxidation of sulfamethoxazole using [5,10,15,20-tetrakis (3-methoxy-4-hydroxy phenyl) porphyrinato] Cu (II) modified carbon paste sensor. *Drug Test. Anal.* 2:278–283.
1579. Abbaspour, A., M. Asadi, A. Ghaffarinejad, and E. Safaei. 2005. A selective modified carbon paste electrode for determination of cyanide using tetra-3,4-pyridinoporphyrazinatocobalt(II). *Talanta* 66:931–936.
1580. Gong, F.-C., X.-B. Zhang, C.-C. Guo, G.-L. Shen, and R.-Q. Yu. 2003. Amperometric metronidazole sensor based on the supermolecular recognition by metalloporphyrin incorporated in carbon paste electrode. *Sensors* 3:91–100.
1581. Amini, M.K., J.H. Khorasani, S.S. Khaloo, and S. Tangestaninejad. 2003. Cobalt(II) salophen-modified carbon-paste electrode for potentiometric and voltammetric determination of cysteine. *Anal. Biochem.* 320:32–38.
1582. Shahrokhian, S., A. Souri, and H. Khajehsharifi. 2004. Electrocatalytic oxidation of penicillamine at a carbon paste electrode modified with cobalt salophen. *J. Electroanal. Chem.* 565:95–101.
1583. Shahrokhian, S. and H.R. Zare-Mehrjardi. 2007. Cobalt salophen-modified carbon-paste electrode incorporating a cationic surfactant for simultaneous voltammetric detection of ascorbic acid and dopamine. *Sens. Actuators B* B121:530–537.
1584. Jamasbi, E.S., A. Rouhollahi, S. Shahrokhian, S. Haghgoo, and S. Aghajani. 2007. The electrocatalytic examination of cephalosporins at carbon paste electrode modified with CoSalophen. *Talanta* 71:1669–1674.
1585. Teixeira, M.F.S., G. Marino, E.R. Dockal, and E.T.G. Cavalheiro. 2004. Voltammetric determination of pyridoxine (Vitamin B6) at a carbon paste electrode modified with vanadyl(IV)-Salen complex. *Anal. Chim. Acta* 508:79–85.
1586. Shahrokhian, S. and M. Karimi. 2004. Voltammetric studies of a cobalt(II)-4-methylsalophen modified carbon-paste electrode and its application for the simultaneous determination of cysteine and ascorbic acid. *Electrochim. Acta* 50:77–84.
1587. Shahrokhian, S., M. Karimi, and H. Khajehsharifi. 2005. Carbon-paste electrode modified with cobalt-5-nitrolsalophen as a sensitive voltammetric sensor for detection of captopril. *Sens. Actuators B* B109:278–284.
1588. Shahrokhian, S. and M.J. Jannat Rezvani. 2005. Voltammetric studies of propylthiouracil at a carbon-paste electrode modified with cobalt(II)-4-chlorosalophen: Application to voltammetric determination in pharmaceutical and clinical preparations. *Microchim. Acta* 151:73–79.
1589. Shahrokhian, S. and M. Amiri. 2006. Voltammetric determination of thiocytosine based on its electrocatalytic oxidation on the surface of carbon-paste electrode modified with cobalt Schiff base complexes. *J. Solid State Electrochem.* 11:1133–1138.
1590. Shahrokhian, S. and H.R. Zare-Mehrjardi. 2007. Simultaneous voltammetric determination of uric acid and ascorbic acid using a carbon-paste electrode modified with multi-walled carbon nanotubes/Nafion and cobalt(II)-nitrosalophen. *Electroanalysis* 19:2234–2242.
1591. Shahrokhian, S. and M. Ghalkhani. 2008. Voltammetric determination of methimazole using a carbon paste electrode modified with a Schiff base complex of cobalt. *Electroanalysis* 20:1061–1066.
1592. Kamyabi, M.A. and F. Aghajanloo. 2008. Electrocatalytic oxidation and determination of nitrite on carbon paste electrode modified with oxovanadium(IV)-4-methylsalophen. *J. Electroanal. Chem.* 614:157–165.

1593. Shamsipur, M., A. Soleymanpour, M. Akhond, H. Sharghi, and M.A. Naseri. 2001. Iodide-selective carbon paste electrodes based on recently synthesized Schiff base complexes of Fe(III). *Anal. Chim. Acta* 450:37–44.

1594. Teixeira, M.F.S., E.R. Dockal, and E.T.G. Cavalheiro. 2005. Sensor for cysteine based on oxovanadium(IV) complex of Salen modified carbon paste electrode. *Sens. Actuators B* 106:619–625.

1595. Kamyabi, M.A., S. Shahabi, and H. Hosseini-Monfared. 2007. Electrocatalytic oxidation of hydrazine at a cobalt(II) Schiff-base-modified carbon paste electrode. *J. Electrochem. Soc.* 155:F8-F12.

1596. Nigam, P., S. Mohan, S. Kundu, and R. Prakash. 2009. Trace analysis of cefotaxime at carbon paste electrode modified with novel Schiff base Zn(II) complex. *Talanta* 77:1426–1431.

1597. Bonakdar, M., J. Yu, and H.A. Mottola. 1989. Continuous-flow performance of carbon electrodes modified with immobilized iron(II)/iron(III) centers. *Talanta* 36:219–225.

1598. Sun, G. and H.A. Mottola. 1991. Chemical species produced in the reaction between ethanedial (glyoxal) and 5-amino-1,10-phenanthroline and their iron(II) complexes: Electrochemical studies and analytical applications. *Anal. Chim. Acta* 242:241–247.

1599. Tobalina, F., F. Pariente, L. Hernandez, H.D. Abruna, and E. Lorenzo. 1999. Integrated ethanol biosensors based on carbon paste electrodes modified with [Re(phen-dione)(CO)$_3$Cl] and [Fe(phen-dione)$_3$] (PF$_6$)$_2$. *Anal. Chim. Acta* 395:17–26.

1600. Wang, H., G. Xu, and S. Dong. 2002. Electrochemiluminescence of dichlorotris (1,10-phenanthroline) ruthenium(II) with peroxydisulfate in purely aqueous solution at carbon paste electrode. *Microchem. J.* 72:43–48.

1601. Ivanova, E.V., V.S. Sergeeva, J. Oni, C. Kurzawa, A.D. Ryabov, and W. Schuhmann. 2003. Evaluation of redox mediators for amperometric biosensors: Ru-complex modified carbon-paste/enzyme electrodes. *Bioelectrochemistry* 60:65–71.

1602. Motonaka, J., Y. Mishima, K. Maruyama, K. Minagawa, and S. Ikeda. 2000. Interaction between immobilized DNA and tris(bpy) osmium(II/III) complex. *Sens. Actuators B* 66:234–236.

1603. Fojta, M., L. Havran, S. Billova, P. Kostecka, M. Masarik, and R. Kizek. 2003. Two-surface strategy in electrochemical DNA hybridization assays: Detection of osmium-labeled target DNA at carbon electrodes. *Electroanalysis* 15:431–440.

1604. El Maali, N.A. and J. Wang. 2001. Tris(2,2′-bipyridyl)dichloro-ruthenium(II) modified carbon paste electrodes for electrocatalytic detection of DNA. *Sens. Actuators B* B76:211–214.

1605. Sugawara, K., S. Tanaka, and M. Taga. 1991. Voltammetric behavior of cysteine by a carbon-paste electrode containing a cobalt(II) cyclohexylbutyrate. *Bioelectrochem. Bioenerg.* 26:469–474.

1606. Sugawara, K., S. Tanaka, K. Hasebe, and M. Taga. 1993. Electroanalysis for lipoic acid using a carbon paste electrode modified with nickel[II]-cyclohexylbutyrate. *J. Electroanal. Chem.* 347:393–398.

1607. Sugawara, K., S. Tanaka, and M. Taga. 1993. Determination of amino acids by a chemically modified carbon paste electrode with copper(II) cyclohexylbutyrate. *Bioelectrochem. Bioenerg.* 31:229–234.

1608. Nalini, B. and S.S. Narayanan. 1998. Electrocatalytic oxidation of sulfhydryl compounds at ruthenium(III) diphenyldithiocarbamate modified carbon paste electrode. *Electroanalysis* 10:779–783.

1609. Nalini, B. and S.S. Narayanan. 2000. Amperometric determination of ascorbic acid based on electrocatalytic oxidation using a ruthenium(III) diphenyldithiocarbamate-modified carbon paste electrode. *Anal. Chim. Acta* 405:93–97.

1610. Ramos, L.A., E.T.G. Cavalheiro, and G.O. Chierice. 2005. Evaluation of the electrochemical behavior and analytical potentialities of a carbon paste electrode modified with a ruthenium (III) piperidinedithiocarbamate complex. *Farmaco* 60:149–155.

1611. Cavalheiro, E.T.G., A. Segnini, A.B. Couto, F.C. Moraes, and M.F.S. Teixeira. 2007. Electrochemical behavior and electrocatalytic study of the manganese (II) pyrrolidine dithiocarbamate modified carbon paste electrode: Use in a flow injection system. *Anal. Chem. Indian J.* 3:124–132.

1612. Santos, P.M., B. Sandrino, T.F. Moreira, K. Wohnrath, N. Nagata, and C.A. Pessoa. 2007. Simultaneous voltammetric determination of dopamine and ascorbic acid using multivariate calibration methodology performed on a carbon paste electrode modified by a mer-[RuCl$_3$(dppb)(4-pic)] complex. *J. Braz. Chem. Soc.* 18:93–99.

1613. Zen, J.-M., A.S. Kumar, and C.-R. Chung. 2003. A glucose biosensor employing a stable artificial peroxidase based on ruthenium purple anchored cinder. *Anal. Chem.* 75:2703–2709.

1614. Gil, E.D. and L.T. Kubota. 2000. Electrochemical behavior of rhodium acetamidate immobilized on a carbon paste electrode: A hydrazine sensor. *J. Braz. Chem. Soc.* 11:304–310.

1615. Do Carmo, D.R., R.M. da Silva, and N.R. Stradiotto. 2003. Electrocatalytic and voltammetric determination of sulfhydryl compounds through iron nitroprusside modified graphite paste electrode. *J. Braz. Chem. Soc.* 14:616–620.

1616. Rezaei, B., S. Meghdadi, and M. Rezazadeh. 2009. Electrocatalytic determination of dopamine in pharmaceutical and human serum samples by using [N,N'-bis(2-pyridine carboxamido)-1,2-benzene] nickel(II) modified carbon paste electrode. *J. Anal. Chem.* 64:513–517.
1617. Kamyabi, M.A., Z. Asgari, H. Hosseini Monfared, and A. Morsali. 2009. Electrocatalytic oxidation of ascorbic acid and simultaneous determination of ascorbic acid and dopamine at a bis(4'-(4-pyridyl)-2,2':6',2''-terpyridine)iron(II) thiocyanate carbon past modified electrode. *J. Electroanal. Chem.* 632:170–176.
1618. Zhuang, R.-R. and F.-F. Jian. 2010. Electrocatalysis for the hydrogen peroxide and nitrite at carbon paste electrode modified with a new zinc complex of 1-pentyl-1H-benzo[d][1,2,3]triazole. *J. Solid State Electrochem.* 14:747–750.
1619. Zhuang, R.-R., F.-F. Jian, and K.-F. Wang. 2010. An electrochemical sensing platform based on a new zinc complex for the determination of hydrogen peroxide and nitrite. *Sens. Lett.* 8:228–232.
1620. Zhuang, R.-R., F.-F. Jian, and K. Wang. 2010. Synthesis, characterization and performance of a new copper complex as electrocatalyst of hydrogen peroxide and nitrite. *Sci. Adv. Mater.* 2:151–156.
1621. Zhuang, R.-R., F.-F. Jian, and K-.F. Wang. 2010. New cobalt(II)-based electrocatalyst for reduction of trichloroacetic acid and bromate. *Collect. Czech. Chem. Commun.* 75:637–647.
1622. Wang, L.-H. and S.-W. Tseng. 2001. Direct determination of d-panthenol and salt of pantothenic acid in cosmetic and pharmaceutical preparations by differential pulse voltammetry. *Anal. Chim. Acta* 432:39–48.
1623. Kauffmann, J.-M. 1997. Lipid based enzyme electrodes for environmental pollution control. *NATO ASI Ser. Ser. 2* 38:107–114.
1624. Lu, G.-H., M. Yang, Q.-F. Zheng, A.-L. Wang, and Z.-X. Jin. 1995. Anodic stripping voltammetry for the determination of trace iodide. *Electroanalysis* 7:591–593.
1625. Hu, C. and S. Hu. 2004. Electrochemical characterization of cetyltrimethyl ammonium bromide modified carbon paste electrode and the application in the immobilization of DNA. *Electrochim. Acta* 49:405–412.
1626. He, Q., C. Hu, X. Dang, Y. Wei, and S. Hu. 2004. Electrocatalytic reduction of dioxygen at cetyltrimethylammonium bromide modified carbon paste electrode. *Electrochemistry* 72:5–8.
1627. Amor Garcia, I., M.C. Blanco Lopez, M.J. Lobo Castanon, A.J. Miranda Ordieres, and P. Tunon Blanco. 2005. Flufenamic acid determination in human serum by adsorptive voltammetry with in situ surfactant-modified carbon paste electrodes. *Electroanalysis* 17:1555–1562.
1628. Huang, W. 2005. Voltammetric determination of bisphenol A using a carbon paste electrode based on the enhancement effect of cetyltrimethylammonium bromide (CTAB). *Bull. Korean Chem. Soc.* 26:1560–1564.
1629. Liu, S., J. Li, S. Zhang, and J. Zhao. 2005. Study on the adsorptive stripping voltammetric determination of trace cerium at a carbon paste electrode modified in situ with cetyltrimethylammonium bromide. *Appl. Surf. Sci.* 252:2078–2084.
1630. Li, C. 2007. Electrochemical determination of dipyridamole at a carbon paste electrode using cetyltrimethyl ammonium bromide as enhancing element. *Colloids Surf. B* 55:77–83.
1631. Char, M.P., E. Niranjana, B.E.K. Swamy, B.S. Sherigara, and K.V. Pai. 2008. Electrochemical studies of amaranth at surfactant modified carbon paste electrode: A cyclic voltammetry. *Int. J. Electrochem. Sci.* 3:588–596.
1632. Sharath Shankar, S., B.E.K. Swamy, U. Chandra, J.G. Manjunatha, and B.S. Sherigara. 2009. Simultaneous determination of dopamine, uric acid and ascorbic acid with CTAB modified carbon paste electrode. *Int. J. Electrochem. Sci.* 4:592–601.
1633. Manjunatha, J.G., B.E.K. Swamy, R. Deepa, V. Krishna, G.P. Mamatha, U. Chandra, S. Sharath Shankar, and B.S. Sherigara. 2009. Electrochemical studies of dopamine at Eperisone and cetyl trimethyl ammonium bromide surfactant modified carbon paste electrode: A cyclic voltammetric study. *Int. J. Electrochem. Sci.* 4:662–671.
1634. Manjunatha, J.G., B.E.K. Swamy, O. Gilbert, G.P. Mamatha, and B.S. Sherigara. 2010. Sensitive voltammetric determination of dopamine at salicylic acid and TX-100, SDS, CTAB modified carbon paste electrode. *Int. J. Electrochem. Sci.* 5:682–695.
1635. Li, C. 2007. Voltammetric determination of ethinylestradiol at a carbon paste electrode in the presence of cetyl pyridine bromide. *Bioelectrochemistry* 70:263–268.
1636. Švancara, I., M. Galik, and K. Vytřas. 2007. Stripping voltammetric determination of platinum metals at a carbon paste electrode modified with cationic surfactants. *Talanta* 72:512–518.
1637. Egashira, N., S. Aragaki, H. Iwanaga, and K. Ohga. 1991. Selective determination of pyridoxal with an octadecylamine-modified carbon-paste electrode. *Anal. Sci.* 7:691–694.

1638. Millan, K.M., A. Saraullo, and S.R. Mikkelsen. 1994. Voltammetric DNA biosensor for cystic fibrosis based on a modified carbon paste electrode. *Anal. Chem.* 66:2943–2948.
1639. Amine, A., J. Deni, and J.M. Kauffmann. 1994. Preparation and characterization of octadecylamine-containing carbon paste electrodes. *Anal. Chem.* 66:1595–1599.
1640. Girousi, S.T., A.A. Pantazaki, and A.N. Voulgaropoulos. 2001. Mitochondria-based amperometric biosensor for the determination of L-glutamic acid. *Electroanalysis* 13:243–245.
1641. Egashira, N., H. Iwanaga, K. Okabe, and K. Ohga. 1992. Selective determination of pyridoxal in nonfat dry milk using a carbon paste electrode modified with an amine having a long alkyl chain. *Anal. Sci.* 8:705–707.
1642. Ormonde, D.E. and R.D. O'Neill. 1990. The oxidation of ascorbic acid at carbon paste electrodes. Modified response following contact with surfactant, lipid and brain tissue. *J. Electroanal. Chem.* 279:109–121.
1643. Acuna, J.A., C. De La Fuente, M.D. Vazquez, M.L. Tascon, and P. Sanchez-Batanero. 1993. Voltammetric determination of piroxicam in micellar media by using conventional and surfactant chemically-modified carbon paste electrodes. *Talanta* 40:1637–1642.
1644. Posac, J.R., M.D. Vazquez, M.L. Tascon, J.A. Acuna, C. De La Fuente, E. Velasco, and P. Sanchez-Batanero. 1995. Determination of aceclofenac using adsorptive stripping voltammetric techniques on conventional and surfactant chemically modified carbon paste electrodes. *Talanta* 42:293–304.
1645. Manjunatha, J.G., B.E.K. Swamy, G.P. Mamatha, U. Chandra, E. Niranjana, and B.S. Sherigara. 2009. Cyclic voltammetric studies of dopamine at lamotrigine and TX-100 modified carbon paste electrode. *Int. J. Electrochem. Sci.* 4:187–196.
1646. Yuan, C.J., Y.C. Wang, and O. Reiko. 2009. Improving the detection of hydrogen peroxide of screen-printed carbon paste electrodes by modifying with nonionic surfactants. *Anal. Chim. Acta* 653:71–76.
1647. Mahanthesha, K.R., B.E.K. Swamy, U. Chandra, Y.D. Bodke, K.V.K. Pai, and B.S. Sherigara. 2009. Cyclic voltammetric investigations of alizarin at carbon paste electrode using surfactants. *Int. J. Electrochem. Sci.* 4:1237–1247.
1648. Manjunatha, J.G., B.E.K. Swamy, G.P. Mamatha, S. Sharath Shankar, O. Gilbert, B.N. Chandrashekar, and B.S. Sherigara. 2009. Electrochemical response of dopamine at phthalic acid and Triton X-100 modified carbon paste electrode: A cyclic voltammetry study. *Int. J. Electrochem. Sci.* 4:1469–1478.
1649. Manjunatha, J.G., B.E.K. Swamy, G.P. Mamatha, O. Gilbert, M.T. Shreenivas, and B.S. Sherigara. 2009. Electrocatalytic response of dopamine at Mannitol and Triton X-100 modified carbon paste electrode: A cyclic voltammetric study. *Int. J. Electrochem. Sci.* 4:1706–1718.
1650. Huang, S.-S., Z.-G. Chen, B.-F. Li, H.-G. Lin, and R.-Q. Yu. 1994. Preconcentration and voltammetric measurement of silver(I) with a carbon paste electrode modified with 2,9-dichloro-1,10-phenanthroline-surfactant. *Analyst* 119:1859–1862.
1651. Jaiswal, P.V., V.S. Ijeri, and A.K. Srivastava. 2001. Voltammetric behavior of menadione in surfactant media and its determination in sodium dodecyl sulfate surfactant system. *Bull. Chem. Soc. Jpn.* 74:2053–2057.
1652. Gutierrez Fernandez, S., M.C. Blanco Lopez, M.J. Lobo Castanon, A.J. Miranda Ordieres, and P. Tunon Blanco. 2004. Adsorptive stripping voltammetry of rifamycins at unmodified and surfactant-modified carbon paste electrodes. *Electroanalysis* 16:1660–1666.
1653. Blanco Lopez, M.C., M.J. Lobo Castanon, A.J. Miranda Ordieres, and P. Tunon Blanco. 2007. Electrochemical behavior of catecholamines and related compounds at in situ surfactant modified carbon paste electrodes. *Electroanalysis* 19:207–213.
1654. Alarcon Angeles, G., S. Corona Avendano, M. Palomar Pardave, A. Rojas Hernandez, M. Romero Romo, and M.T. Ramirez Silva. 2008. Selective electrochemical determination of dopamine in the presence of ascorbic acid using sodium dodecyl sulfate micelles as masking agent. *Electrochim. Acta* 53:3013–3020.
1655. Chandra, U., O. Gilbert, B.E.K. Swamy, Y.D. Bodke, and B.S. Sherigara. 2008. Electrochemical studies of Eriochrome Black T at carbon paste electrode and immobilized by SDS surfactant: A cyclic voltammetric study. *Int. J. Electrochem. Sci.* 3:1044–1054.
1656. Chitravathi, S., B.E. Kumaraswamy, E. Niranjana, U. Chandra, G.P. Mamatha, and B.S. Sherigara. 2009. Electrochemical studies of sodium levothyroxine at surfactant modified carbon paste electrode. *Int. J. Electrochem. Sci.* 4:223–237.
1657. Chowdappa, N., B.E.K. Swamy, E. Niranjana, and B.S. Sherigara. 2009. Cyclic voltammetric studies of serotonin at sodium dodecyl sulfate modified carbon paste electrode. *Int. J. Electrochem. Sci.* 4:425–434.
1658. Chandrashekar, B.N., B.E.K. Swamy, K.R. Vishnu Mahesh, U. Chandra, and B.S. Sherigara. 2009. Electrochemical studies of bromothymol blue at surfactant modified carbon paste electrode by cyclic voltammetry. *Int. J. Electrochem. Sci.* 4:471–480.

1659. Niranjana, E., B.E.K. Swamy, R. Raghavendra Naik, B.S. Sherigara, and H. Jayadevappa. 2009. Electrochemical investigations of potassium ferricyanide and dopamine by sodium dodecyl sulphate modified carbon paste electrode: A cyclic voltammetric study. *J. Electroanal. Chem.* 631:1–9.
1660. Ojani, R., J.B. Raoof, and B. Norouzi. 2010. Carbon paste electrode modified by cobalt ions dispersed into poly (N-methylaniline) preparing in the presence of SDS: Application in electrocatalytic oxidation of hydrogen peroxide. *J. Solid State Electrochem.* 14:621–631.
1661. Kauffmann, J.M. 2004. Personal communication.
1662. Lyne, P.D. and R.D. O'Neill. 1990. Stearate-modified carbon paste electrodes for detecting dopamine in vivo: Decrease in selectivity caused by lipids and other surface-active agents. *Anal. Chem.* 62:2347–2351.
1663. El Maali, N.A. 1992. Carbon paste electrodes modified with palmitic acid and stearic acid for the determination of folic acid (vitamin B9) in both aqueous and biological media. *Bioelectrochem. Bioenerg.* 27:465–473.
1664. Petit, C., A. Gonzalez-Cortes, and J.M. Kauffmann. 1996. On the origin of the differences between stearic-acid-modified carbon paste electrode performances after exposure to surfactant and brain tissues. *Bioelectrochem. Bioenerg.* 41:101–106.
1665. Xu, G. and S. Dong. 2000. Electrochemiluminescent detection of chlorpromazine by selective preconcentration at a lauric acid-modified carbon paste electrode using tris(2,2′-bipyridine)ruthenium(II). *Anal. Chem.* 72:5308–5312.
1666. Khodari, M., J.M. Kauffmann, G.J. Patriarche, and M.A. Ghandour. 1989. Applications in drug analysis of carbon paste electrodes modified by fatty acids. *J. Pharm. Biomed. Anal.* 7:1491–1497.
1667. Chastel, O., J.M. Kauffmann, G.J. Patriarche, and G.D. Christian. 1990. Electrochemical behavior of marcellomycin at lipid-modified carbon-paste electrodes. *Talanta* 37:213–217.
1668. Kauffmann, J.M., O. Chastel, G. Quarin, G.J. Patriarche, and M. Khodari. 1990. Electrochemical behavior of drugs at lipid modified carbon paste electrodes. *Bioelectrochem. Bioenerg.* 23:167–175.
1669. Arcos, J., J.M. Kauffmann, G.J. Patriarche, and P. Sanchez Batanero. 1990. Voltammetric determination of celiptium with carbon paste and lipid-modified carbon paste electrodes. *Anal. Chim. Acta* 236:299–305.
1670. Arcos, J., B. Garcia, A. Munguia, J. Lopez Palacios, J.-M. Kauffmann, and G.J. Patriarche. 1991. Spectrophotometric and electroanalytical study of minoxidil. *Anal. Lett.* 24:357–376.
1671. Khodari, M. 1993. Voltammetric determination of the antidepressant trimipramine at a lipid-modified carbon paste electrode. *Electroanalysis* 5:521–523.
1672. Amine, A., J. Deni, and J.-M. Kauffmann. 1994. Amperometric biosensor based on carbon paste mixed with enzyme, lipid and cytochrome c. *Bioelectrochem. Bioenerg.* 34:123–128.
1673. Mizutani, F., S. Yabuki, and T. Katsura. 1994. Amperometric glucose-sensing electrodes with the use of modified enzymes. *ACS Symp. Ser.* 556:41–46.
1674. Hu, S., M. Guo, G. Hu, and M. Jiang. 1995. Electrochemical oxidation and analytical applications of estrogens at modified carbon paste electrode. *Anal. Lett.* 28:1993–2003.
1675. Khodari, M., H. Mansour, and H.S. El Din. 1997. Preconcentration and determination of the tricyclic antidepressant drug imipramine at modified carbon paste electrode. *Anal. Lett.* 30:1909–1921.
1676. Murray, R.W. 1992. Molecular design of electrode surfaces. In *Techniques of Chemistry*, Vol. 22, Ed. R.W. Murray, pp. 1–48. New York: Wiley.
1677. Li, P., Z. Gao, Y. Xu, G. Wang, and Z. Zhao. 1990. Determination of trace amounts of silver with a chemically modified carbon paste electrode. *Anal. Chim. Acta* 229:213–219.
1678. Gao, Z., P. Li, S. Dong, and Z. Zhao. 1990. Voltammetric determination of trace amounts of gold(III) with a carbon paste electrode modified with chelating resin. *Anal. Chim. Acta* 232:367–376.
1679. Compagnone, D., J.V. Bannister, and G. Federici. 1992. Electrochemical sensors for the determination of metal ions. *Sens. Actuators B* B7:549–552.
1680. Agraz, R., M.T. Sevilla, and L. Hernandez. 1995. Voltammetric quantification and speciation of mercury compounds. *J. Electroanal. Chem.* 390:47–57.
1681. Agraz, R., J. de Miguel, M.T. Sevilla, and L. Hernandez. 1996. Application of a Chelite P modified carbon paste electrode to copper determination and speciation. *Electroanalysis* 8:565–570.
1682. Mikysek, T., I. Švancara, K. Vytřas, and F.G. Banica. 2008. Functionalised resin-modified carbon paste sensor for the voltammetric determination of Pb(II) within a wide concentration range. *Electrochem. Commun.* 10:242–245.
1683. Sar, E., H. Berber, B. Asci, and H. Cankurtaran. 2008. Determination of some heavy metal ions with a carbon paste electrode modified by poly(glycidyl methacrylate-methylmethacrylate-divinylbenzene) microspheres functionalized by 2-aminothiazole. *Electroanalysis* 20:1533–1541.
1684. Mikysek, T., I. Švancara, A. Banica, F.G. Banica, and K. Vytřas. 2010. Carbon paste electrode modified with thiourea-functionalized resin for determination of lead. *Sci. Pap. Univ. Pardubice Ser. A* 15:5–27.

1685. Ianniello, R.M. 1992. Crosslinked poly(N-vinylpyrrolidone) coated carbon paste electrodes prepared via plasma polymerization. *Anal. Lett.* 25:125–135.
1686. Yang, X., F. Wang, and S. Hu. 2006. The electrochemical oxidation of troxerutin and its sensitive determination in pharmaceutical dosage forms at PVP modified carbon paste electrode. *Colloids Surf. B* 52:8–13.
1687. Franzoi, A.C., A. Spinelli, and I.C. Vieira. 2008. Rutin determination in pharmaceutical formulations using a carbon paste electrode modified with poly(vinylpyrrolidone). *J. Pharm. Biomed. Anal.* 47:973–977.
1688. Alizadeh, T. 2009. High selective parathion voltammetric sensor development by using an acrylic based molecularly imprinted polymer-carbon paste electrode. *Electroanalysis* 21:1490–1498.
1689. Alizadeh, T., M.R. Ganjali, P. Norouzi, M. Zare, and A. Zeraatkar. 2009. A novel high selective and sensitive para-nitrophenol voltammetric sensor, based on a molecularly imprinted polymer-carbon paste electrode. *Talanta* 79:1197–1203.
1690. Alizadeh, T., M. Zare, M.R. Ganjali, P. Norouzi, and B. Tavana. 2010. A new molecularly imprinted polymer (MIP)-based electrochemical sensor for monitoring 2,4,6-trinitrotoluene (TNT) in natural waters and soil samples. *Biosens. Bioelectron.* 25:1166–1172.
1691. Alizadeh, T., M.R. Ganjali, M. Zare, and P. Norouzi. 2010. Development of a voltammetric sensor based on a molecularly imprinted polymer (MIP) for caffeine measurement. *Electrochim. Acta* 55:1568–1574.
1692. Alizadeh, T. and M. Akhoundian. 2010. Promethazine determination in plasma samples by using carbon paste electrode modified with molecularly imprinted polymer (MIP): Coupling of extraction, preconcentration and electrochemical determination. *Electrochim. Acta* 55:5867–5873.
1693. Navratilova, Z. and P. Kula. 1992. Determination of mercury on a carbon paste electrode modified with humic acid. *Electroanalysis* 4:683–687.
1694. Navratilova, Z. and P. Kula. 1993. Modified carbon paste electrodes for the study of metal-humic substances complexation. *Anal. Chim. Acta* 273:305–311.
1695. Jeong, E.D., M.S. Won, and Y.B. Shim. 1994. Simultaneous determination of lead, copper, and mercury at a modified carbon paste electrode containing humic acid. *Electroanalysis* 6:887–893.
1696. Kula, P. and Z. Navratilova. 1994. Humic acids interactions with metal salts incorporated into the carbon paste electrode. *Electroanalysis* 6:1009–1013.
1697. Wang, C.M., Q.Y. Sun, and H.L. Li. 1997. Voltammetric behavior and determination of bismuth on sodium humate-modified carbon paste electrode. *Electroanalysis* 9:645–649.
1698. Wang, C.M. and H.L. Li. 1998. Voltammetric behavior of mercury(I,II) ions at an amide-functionalized humic acids modified carbon paste electrode. *Electroanalysis* 10:44–49.
1699. Wang, C., H. Zhang, Y. Sun, and H. Li. 1998. Electrochemical behavior and determination of gold at chemically modified carbon paste electrode by the ethylenediamine fixed humic acid preparation. *Anal. Chim. Acta* 361:133–139.
1700. Sun, Q.Y., C.M. Wang, L.X. Li, and H.L. Li. 1999. Preconcentration and voltammetric determination of palladium(II) at sodium humate modified carbon paste electrodes. *Fresenius J. Anal. Chem.* 363:114–117.
1701. Wang, C.M., B. Zhu, and H.L. Li. 1999. Theoretical analysis and determination of the heterogeneous stability constant of copper(II)-humic acids complex at chemically modified carbon paste electrode. *Electroanalysis* 11:183–187.
1702. Thobie Gautier, C., W.T. Lopes da Silva, M.O.O. Rezende, and N. El Murr. 2003. Sensitive and reproducible quantification of Cu^{2+} by Stripping with a carbon paste electrode modified with humic acid. *J. Environ. Sci. Health A* A38:1811–1823.
1703. Airoldi, F.P.S., W.T.L. Da Silva, F.N. Crespilho, and M.O.O. Rezende. 2007. Evaluation of the electrochemical behavior of pentachlorophenol by cyclic voltammetry on carbon paste electrode modified by humic acids. *Water Environ. Res.* 79:63–67.
1704. Silva, L.S., T.N. Oliveira, M.A. Ballin, and C.R.M. Peixoto. 2006. Ascorbic acid determination using a carbon paste electrode modified with iron(III) ions adsorbed on humic acid. *Ecletica Quim.* 31:39–42.
1705. Bai, Z.P., T. Nakamura, and K. Izutsu. 1990. Enhanced voltammetric waves of iron(III)-EDTA at a chitin-containing carbon paste electrode and its analytical application. *Electroanalysis* 2:75–79.
1706. Bai, Z.P., T. Nakamura, and K. Izutsu. 1990. Voltammetric behavior of various anions at a chitin-containing carbon paste electrode. *Anal. Sci.* 6:443–447.
1707. Sugawara, K., T. Miyasita, S. Hoshi, and K. Akatsuka. 1997. Voltammetric detection of MoO_4^{2-} at a modified carbon paste electrode containing chitin. *Anal. Chim. Acta* 353:301–306.
1708. Miao, Y. and S.-N. Tan. 2000. Amperometric hydrogen peroxide biosensor based on immobilization of peroxidase in chitosan matrix crosslinked with glutaraldehyde. *Analyst* 125:1591–1594.

References

1709. Marcolino-Junior, L.H., B.C. Janegitz, B.C. Lourencao, and O. Fatibello-Filho. 2007. Anodic stripping voltammetric determination of mercury in water using a chitosan-modified carbon paste electrode. *Anal. Lett.* 40:3119–3128.
1710. Hassan, R., I.H.I. Habib, and H.N.A. Hassan. 2008. Voltammetric determination of lead(II) in medical lotion and biological samples using chitosan-carbon paste electrode. *Int. J. Electrochem. Sci.* 3:935–945.
1711. Sugawara, K., A. Yugami, N. Terui, and H. Kuramitz. 2009. Electrochemical study of functionalization on the surface of a chitin/platinum-modified glassy carbon paste electrode. *Anal. Sci.* 25:1365–1368.
1712. Dempsey, E., M.R. Smyth, and D.H.S. Richardson. 1992. Application of lichen-modified carbon paste electrodes to the voltammetric determination of metal ions in multielement and speciation studies. *Analyst* 117:1467–1470.
1713. Wang, J., Z. Taha, and N. Naser. 1991. Electroanalysis at modified carbon-paste electrodes containing natural ionic polysaccharides. *Talanta* 38:81–88.
1714. Perlado, J.C., A. Zapardiel, E. Bermejo, J.A. Perez, and L. Hernandez. 1995. Determination of phenylephrine with a modified carbon paste electrode. *Anal. Chim. Acta* 305:83–90.
1715. Gao, Z., A. Ivaska, and P. Li. 1992. Determination of trace amounts of copper(I) with a chemically modified carbon paste electrode. *Anal. Sci.* 8:337–343.
1716. Lee, G.-J., C.-K. Kim, M.-K. Lee, and C.-K. Rhee. 2010. Advanced use of nanobismuth/Nafion electrode for trace analyses of zinc, cadmium, and lead. *J. Electrochem. Soc.* 157:J241-J244.
1717. Mueller, K. 1986. In vivo voltammetric recording with nafion-coated carbon paste electrodes: Additional evidence that ascorbic acid release is monitored. *Pharm. Biochem. Behavior* 25:325–328.
1718. Boyd, D., J.R. Barreira Rodriguez, P. Tunon Blanco, and M.R. Smyth. 1994. Application of a Nafion-modified carbon paste electrode for the adsorptive stripping voltammetric determination of fenoterol in pharmaceutical preparations and biological fluids. *J. Pharm. Biomed. Anal.* 12:1069–1074.
1719. Boyd, D., J.R. Barreira Rodriguez, A.J. Miranda Ordieres, P. Tunon Blanco, and M.R. Smyth. 1994. Voltammetric study of salbutamol, fenoterol and metaproterenol at unmodified and nafion-modified carbon paste electrodes. *Analyst* 119:1979–1984.
1720. Moane, S., J.R. Barreira Rodriguez, O. Miranda, P. Tunon Blanco, and M.R. Smyth. 1995. Electrochemical behavior of clenbuterol at Nafion-modified carbon-paste electrodes. *J. Pharm. Biomed. Anal.* 14:57–63.
1721. Moane, S., M.R. Smyth, and M. O'Keeffe. 1996. Differential-pulse voltammetric determination of clenbuterol in bovine urine using a Nafion-modified carbon paste electrode. *Analyst* 121:779–784.
1722. Zen, J.M. and C.T. Hsu. 1998. A selective voltammetric method for uric acid detection at Nafion-coated carbon paste electrodes. *Talanta* 46:1363–1369.
1723. De Betono, S.F., A.A. Garcia, and J.F.A. Valentin. 1999. UV-spectrophotometry and square wave voltammetry at Nafion-modified carbon-paste electrode for the determination of doxazosin in urine and formulations. *J. Pharm. Biomed. Anal.* 20:621–630.
1724. Arranz, A., S.F. de Betono, C. Echevarria, J.M. Moreda, A. Cid, and J.F. Arranz Valentin. 1999. Voltammetric and spectrophotometric techniques for the determination of the antihypertensive drug Prazosin in urine and formulations. *J. Pharm. Biomed. Anal.* 21:797–807.
1725. Andrieux, C.P., P. Audebert, P. Bacchi, and B. Divisia Blohorn. 1995. Kinetic behavior of an amperometric biosensor made of an enzymic carbon paste and a Nafion gel investigated by rotating electrode studies. *J. Electroanal. Chem.* 394:141–148.
1726. Rapicault, S., B. Limoges, and C. Degrand. 1996. Alkaline phosphatase assay using a redox procationic labeled substrate and a renewable Nafion-loaded carbon paste electrode. *Electroanalysis* 8:880–884.
1727. Yamamoto, K., T. Ohgaru, M. Torimura, H. Kinoshita, K. Kano, and T. Ikeda. 2000. Highly sensitive flow injection determination of hydrogen peroxide with a peroxidase-immobilized electrode and its application to clinical chemistry. *Anal. Chim. Acta* 406:201–207.
1728. Bordes, A.L., B. Schollhorn, B. Limoges, and C. Degrand. 1999. Simultaneous detection of three drugs labeled by cationic metal complexes at a nafion-loaded carbon paste electrode. *Talanta* 48:201–208.
1729. Shahrokhian, S. and H.R. Zare Mehrjardi. 2007. Application of thionine-Nafion supported on multi-walled carbon nanotube for preparation of a modified electrode in simultaneous voltammetric detection of dopamine and ascorbic acid. *Electrochim. Acta* 52:6310–6317.
1730. Sun, W., R.F. Gao, and K. Jiao. 2007. Electrochemistry and electrocatalysis of a Nafion/nano-$CaCO_3$/Hb film modified carbon ionic liquid electrode using $BMIMPF_6$ as binder. *Electroanalysis* 19:1368–1374.
1731. Labuda, J., H. Korgova, and M. Vanickova. 1995. Theory and application of chemically modified carbon paste electrode to copper speciation determination. *Anal. Chim. Acta* 305:42–48.
1732. Labuda, J., E. Korgova, M. Vanickova, and P. Tarapcik. 1996. Comparison of copper speciation data obtained by voltammetric and ion-exchange methods. *Chem. Anal.* 41:865–871.

1733. Labuda, J., M. Buckova, and L. Halamova. 1997. Sensor-analyte interaction kinetics as a metal speciation criterion. *Electroanalysis* 9:1129–1131.
1734. Gonzalez, P., V.A. Cortinez, and C.A. Fontan. 2002. Determination of nickel by anodic adsorptive stripping voltammetry with a cation exchanger-modified carbon paste electrode. *Talanta* 58:679–690.
1735. Smolander, M., G. Marko Varga, and L. Gorton. 1995. Aldose dehydrogenase-modified carbon paste electrodes as amperometric aldose sensors. *Anal. Chim. Acta* 302:233–240.
1736. Mariaulle, P., F. Sinapi, L. Lamberts, and A. Walcarius. 2001. Application of electrodes modified with ion-exchange polymers for the amperometric detection of non-redox cations and anions in combination to ion chromatography. *Electrochim. Acta* 46:3543–3553.
1737. Agraz, R., M.T. Sevilla, J.M. Pinilla, and L. Hernandez. 1991. Voltammetric determination of cadmium on a carbon paste electrode modified with a chelating resin. *Electroanalysis* 3:393–397.
1738. Agraz, R., M.T. Sevilla, and L. Hernandez. 1993. Chemically modified electrode for the simultaneous determination of trace metals and speciation analysis. *Anal. Chim. Acta* 273:205–212.
1739. Agraz, R., M.T. Sevilla, and L. Hernandez. 1993. Copper speciation analysis using a chemically modified electrode. *Anal. Chim. Acta* 283:650–656.
1740. Ortiz Viana, M.M., M.P. da Silva, R. Agraz, J.R. Procopio, M.T. Sevilla, and L. Hernandez. 1999. Comparison of two kinetic approaches for copper speciation using ion-exchange columns and ion-exchange modified carbon paste electrodes. *Anal. Chim. Acta* 382:179–188.
1741. Alvarez, E., M.T. Sevilla, J.M. Pinilla, and L. Hernandez. 1992. Cathodic stripping voltammetry of paraquat on a carbon paste electrode modified with Amberlite XAD-2 resin. *Anal. Chim. Acta* 260:19–23.
1742. Geno, P.W., K. Ravichandran, and R.P. Baldwin. 1985. Chemically modified carbon paste electrodes. Part IV. Electrostatic binding and electrocatalysis at poly(4-vinylpyridine)-containing electrodes. *J. Electroanal. Chem.* 183:155–166.
1743. Guadalupe, A.R., S.S. Jhaveri, K.E. Liu, and H.D. Abruna. 1987. Electroanalysis of primary amines with chemically modified carbon paste electrodes. *Anal. Chem.* 59:2436–2438.
1744. Biryol, I., B. Uslu, and Z. Kucukyavuz. 1998. Voltammetric determination of amoxicillin using a carbon paste electrode modified with poly(4-vinyl pyridine). *S. T. P. Pharma Sci.* 8:383–386.
1745. Diewald, W., K. Kalcher, C. Neuhold, X. Cai, and R.J. Magee. 1993. Voltammetric behavior of thallium(III) on carbon paste electrodes chemically modified with an anion exchanger. *Anal. Chim. Acta* 273:237–244.
1746. Ahamad, R., J. Barek, A.R. Yusoff, S.M. Sinaga, and J. Zima. 2000. Determination of roxarsone using carbon paste and amberlite LA2 modified carbon paste electrodes. *Electroanalysis* 12:1220–1226.
1747. Chandra, U., B.E.K. Swamy, O. Gilbert, M. Pandurangachar, and B.S. Sherigara. 2009. Voltammetric resolution of dopamine in presence of ascorbic acid at polyvinyl alcohol modified carbon paste electrode. *Int. J. Electrochem. Sci.* 4:1479–1488.
1748. Saito, T. and M. Watanabe. 1998. Characterization of poly(vinylferrocene-co-2-hydroxyethyl methacrylate) for use as electron mediator in enzymic glucose sensor. *React. Funct. Polym.* 37:263–269.
1749. Chuang, C.-L., Y.-J. Wang, and H.-L. Lan. 1997. Amperometric glucose sensors based on ferrocene-containing B-polyethylenimine and immobilized glucose oxidase. *Anal. Chim. Acta* 353:37–44.
1750. Dominguez, E., H.-L. Lan, Y. Okamoto, P.D. Hale, T.A. Skotheim, L. Gorton, and B. Hahn-Haegerdal. 1993. Reagentless chemically modified carbon paste electrode based on a phenothiazine polymer derivative and yeast alcohol dehydrogenase for the analysis of ethanol. *Biosens. Bioelectron.* 8:229–237.
1751. Pasco, N., C. Jeffries, Q. Davies, A.J. Downard, A.D. Roddick-Lanzilotta, and L. Gorton. 1999. Characterization of a thermophilic L-glutamate dehydrogenase biosensor for amperometric determination of L-glutamate by flow injection analysis. *Biosens. Bioelectron.* 14:171–178.
1752. Yao, Q., S. Yabuki, and F. Mizutani. 2000. Preparation of a carbon paste/alcohol dehydrogenase electrode using polyethylene glycol-modified enzyme and oil-soluble mediator. *Sens. Actuators B* B65:147–149.
1753. Yang, M., Y. Yang, Y. Yang, G. Shen, and R. Yu. 2004. Bienzymatic amperometric biosensor for choline based on mediator thionine in situ electropolymerized within a carbon paste electrode. *Anal. Biochem.* 334:127–134.
1754. Hedenmo, M., A. Narvaez, E. Dominguez, and I. Katakis. 1996. Reagentless amperometric glucose dehydrogenase biosensor based on electrocatalytic oxidation of NADH by osmium phenanthrolinedione mediator. *Analyst* 121:1891–1895.
1755. Vijayakumar, A.R., E. Csoeregi, A. Heller, and L. Gorton. 1996. Alcohol biosensors based on coupled oxidase–peroxidase systems. *Anal. Chim. Acta* 327:223–234.
1756. Hale, P.D., L.I. Boguslavsky, T.A. Skotheim, H.I. Karan, H.L. Lan, and Y. Okamoto. 1990. Poly(xylylviologen) electron transfer mediators in amperometric glucose sensors. *Mol. Cryst. Liq. Cryst.* 190:259–264.

References

1757. Kaku, T., H.I. Karan, and Y. Okamoto. 1994. Amperometric glucose sensors based on immobilized glucose oxidase-polyquinone system. *Anal. Chem.* 66:1231–1235.
1758. Lan, H.L., T. Kaku, H.I. Karan, and Y. Okamoto. 1994. Poly(ether amine quinone)s as electron-transfer relay systems in amperometric glucose sensors. *ACS Symp., Ser. A* 556:124–136.
1759. Kaku, T., Y. Okamoto, L. Charles, W. Holness, and H.I. Karan. 1995. The effect of structure on poly(quinone) systems for amperometric glucose sensors. *Polymer* 36:2813–2818.
1760. Uslu, B. and I. Biryol. 1999. Voltammetric determination of amoxicillin using a poly(N-vinylimidazole) modified carbon paste electrode. *J. Pharm. Biomed. Anal.* 20:591–598.
1761. Sode, K., Y. Takahashi, S. Ohta, W. Tsugawa, and T. Yamazaki. 2001. A new concept for the construction of an artificial dehydrogenase for fructosylamine compounds and its application for an amperometric fructosylamine sensor. *Anal. Chim. Acta* 435:151–156.
1762. Zhou, Y.Z., L.J. Zhang, S.L. Chen, S.Y. Dong, and X.H. Zheng. 2009. Electroanalysis and simultaneous determination of dopamine and epinephrine at poly(isonicotinic acid)-modified carbon paste electrode in the presence of ascorbic acid. *Chin. Chem. Lett.* 20:217–220.
1763. Gilbert, O., B.E.K. Swamy, U. Chandra, and B.S. Sherigara. 2009. Electrocatalytic oxidation of dopamine and ascorbic acid at poly(Eriochrome black-T) modified carbon paste electrode. *Int. J. Electrochem. Sci.* 4:582–591.
1764. Gilbert, O., B.E.K. Swamy, U. Chandra, and B.S. Sherigara. 2009. Simultaneous detection of dopamine and ascorbic acid using polyglycine modified carbon paste electrode: A cyclic voltammetric study. *J. Electroanal. Chem.* 636:80–85.
1765. Ojani, R., J.B. Raoof, and P. Salmany Afagh. 2004. Electrocatalytic oxidation of some carbohydrates by poly(1-naphthylamine)/nickel modified carbon paste electrode. *J. Electroanal. Chem.* 571:1–8.
1766. Ojani, R., J.B. Raoof, and S. Fathi. 2008. Electrocatalytic oxidation of some carbohydrates by nickel/poly(o-aminophenol) modified carbon paste electrode. *Electroanalysis* 20:1825–1830.
1767. Ojani, R., J.B. Raoof, and S.R.H. Zavvarmahalleh. 2009. Preparation of Ni/poly(1,5-diaminonaphthalene)-modified carbon paste electrode; application in electrocatalytic oxidation of formaldehyde for fuel cells. *J. Solid State Electrochem.* 13:1605–1611.
1768. Ojani, R., J.B. Raoof, and S. Zamani. 2009. Electrocatalytic oxidation of folic acid on carbon paste electrode modified by nickel ions dispersed into poly(o-anisidine) film. *Electroanalysis* 21:2634–2639.
1769. Ojani, R., J.B. Raoof, and S. Zamani. 2010. A novel sensor for cephalosporins based on electrocatalytic oxidation by poly(o-anisidine)/SDS/Ni modified carbon paste electrode. *Talanta* 81:1522–1528.
1770. Ojani, R., J.B. Raoof, and R. Babazadeh. 2010. Electrocatalytic oxidation of hydrogen peroxide on poly(m-toluidine)-nickel modified carbon paste electrode in alkaline medium. *Electroanalysis* 22:1607–1616.
1771. Lobo Castanon, M.J., A.J. Miranda Ordieres, and P. Tunon Blanco. 1997. A bienzyme-poly-(o-phenylenediamine)-modified carbon paste electrode for the amperometric detection of L-lactate. *Anal. Chim. Acta* 346:165–174.
1772. Lobo Castanon, M.J., A.J. Miranda Ordieres, and P. Tunon Blanco. 1997. Amperometric detection of ethanol with poly-(o-phenylenediamine)-modified enzyme electrodes. *Biosens. Bioelectron.* 12:511–520.
1773. Alvarez Crespo, S.L., M.J. Lobo Castanon, A.J. Miranda Ordieres, and P. Tunon Blanco. 1997. Amperometric glutamate biosensor based on poly(o-phenylenediamine) film electrogenerated onto modified carbon paste electrodes. *Biosens. Bioelectron.* 12:739–747.
1774. Lowry, J.P., M. Miele, R.D. O'Neill, M.G. Boutelle, and M. Fillenz. 1998. An amperometric glucose-oxidase/poly(o-phenylenediamine) biosensor for monitoring brain extracellular glucose: In vivo characterization in the striatum of freely-moving rats. *J. Neurosci. Methods.* 79:65–74.
1775. Bassi, A.S., E. Lee, and J.X. Zhu. 1999. Carbon paste-mediated, amperometric, thin-film biosensors for fructose monitoring in honey. *Food Res. Int.* 31:119–127.
1776. Vidal, J.C., E. Garcia, S. Mendez, P. Yarnoz, and J.R. Castillo. 1999. Three approaches to the development of selective bilayer amperometric biosensors for glucose by in situ electropolymerization. *Analyst* 124:319–324.
1777. Wang, J., L. Chen, S.B. Hocevar, and B. Ogorevc. 2000. One-step electropolymeric co-immobilization of glucose oxidase and heparin for amperometric biosensing of glucose. *Analyst* 125:1431–1434.
1778. Saidman, S.B., M.J. Lobo-Castanon, A.J. Miranda-Ordieres, and P. Tunon-Blanco. 2000. Amperometric detection of D-sorbitol with NAD+-D-sorbitol dehydrogenase modified carbon paste electrode. *Anal. Chim. Acta* 424:45–50.
1779. Rodriguez Granda, P., M.J. Lobo Castanon, A.J. Miranda Ordieres, and P. Tunon Blanco. 2002. Modified carbon paste electrodes for flow injection amperometric determination of isocitrate dehydrogenase activity in serum. *Anal. Biochem.* 308:195–203.

1780. Wang, J., L. Chen, and D.-B. Luo. 1997. Electrocatalytic detection of hydrogen peroxide at a poly(m-phenylenediamine)-modified carbon paste electrode and its use for biosensing of glucose. *Anal. Commun.* 34:217–219.
1781. Yabuki, S., F. Mizutani, and T. Katsura. 1992. Glucose-sensing carbon paste electrode containing polyethylene glycol-modified glucose oxidase. *Biosens. Bioelectron.* 7:695–700.
1782. Yabuki, S., F. Mizutani, and T. Katsura. 1993. Electrical communication of polyethylene glycol-modified glucose oxidase in carbon paste and its application to the assay of glucose. *Sens. Actuators B* 13:166–168.
1783. Yabuki, S. and F. Mizutani. 1995. Modifications to a carbon paste glucose-sensing enzyme electrode and a reduction in the electrochemical interference from L-ascorbate. *Biosens. Bioelectron.* 10:353–358.
1784. Csoeregi, E., T. Laurell, I. Katakis, A. Heller, and L. Gorton. 1995. Online glucose monitoring by using microdialysis sampling and amperometric detection based on 'wired' glucose oxidase in carbon paste. *Mikrochim. Acta* 121:31–40.
1785. Vijayakumar, A., E. Csoeregi, T. Ruzgas, and L. Gorton. 1996. Comparison of carbon paste electrodes modified with native and polyethylene glycol derivatized horseradish peroxidases for the amperometric monitoring of H_2O_2. *Sens. Actuators B* B37:97–102.
1786. Lobo Castanon, M.J., A.J. Miranda Ordierez, J.M. Lopez Fonseca, and P. Tunon Blanco. 1996. Electrocatalytic detection of nicotinamide coenzymes by poly(o-aminophenol)- and poly(o-phenylenediamine)-modified carbon paste electrodes. *Anal. Chim. Acta* 325:33–42.
1787. Parellada, J., E. Dominguez, and V.M. Fernandez. 1996. Amperometric flow injection determination of fructose in honey with a carbon paste sensor based on fructose dehydrogenase. *Anal. Chim. Acta* 330:71–77.
1788. Kulys, J., K. Krikstopaitis, T. Ruzgas, and V. Razumas. 1995. Electrochemical and bioelectrocatalytical properties of polydimethylsiloxane carbon paste doped with glucose oxidase. *Mater. Sci. Eng. C* 3:51–56.
1789. Wang, J., S. Li, J.W. Mo, J. Porter, M.M. Musameh, and P.K. Dasgupta. 2002. Oxygen-independent poly(dimethylsiloxane)-based carbon-paste glucose biosensors. *Biosens. Bioelectron.* 17:999–1003.
1790. Hernandez, P., F. Galan, O. Nieto, and L. Hernandez. 1994. Direct determination of indole-3-acetic acid in plant tissues by electroanalytical techniques using a carbon paste modified with OV-17 electrode. *Electroanalysis* 6:577–583.
1791. Schuhmann, W. 1995. Conducting polymer based amperometric enzyme electrodes. *Mikrochim. Acta* 121:1–29.
1792. Gerard, M., A. Chaubey, and B.D. Malhotra. 2002. Application of conducting polymers to biosensors. *Biosens. Bioelectron.* 17:345–359.
1793. Vidal, J.C., E. Garcia-Ruiz, and J.R. Castillo. 2003. Recent Advances in Electropolymerized conducting polymers in amperometric biosensors. *Microchim. Acta* 143:93–111.
1794. Borole, D.D., U.R. Kapadi, P.P. Malhulikar, and D.G. Hundiwale. 2006. Conducting polymers: An emerging field of biosensors. *Design. Monomers Polym.* 9:1–11.
1795. Wang, X., H. Zhang, E. Wang, Z. Han, and C. Hu. 2004. Phosphomolybdate-polypyrrole composite bulk-modified carbon paste electrode for a hydrogen peroxide amperometric sensor. *Mater. Lett.* 58:1661–1664.
1796. Audebert, P. and G. Bidan. 1986. Comparison of carbon paste electrochemistry of polypyrroles prepared by chemical and electrochemical oxidation paths. Some characteristics of the chemically prepared polyhalopyrroles. *Synth. Met.* 14:71–80.
1797. Chen, Z. K. Okamura, M. Hanaki, and T. Nagaoka. 2002. Selective determination of tryptophan by using a carbon paste electrode modified with an overoxidized polypyrrole film. *Anal. Sci.* 18:417–421.
1798. Mailley, P., E.A. Cummings, S.C. Mailley, B.R. Eggins, E. McAdams, and S. Cosnier. 2003. Composite carbon paste biosensor for phenolic derivatives based on in situ electrogenerated polypyrrole binder. *Anal. Chem.* 75:5422–5428.
1799. Raoof, J.B., R. Ojani, and S. Rashid-Nadimi. 2004. Preparation of polypyrrole/ferrocyanide films modified carbon paste electrode and its application on the electrocatalytic determination of ascorbic acid. *Electrochim. Acta* 49:271–280.
1800. Mailley, P., E.A. Cummings, S. Mailley, S. Cosnier, B.R. Eggins, and E. McAdams. 2004. Amperometric detection of phenolic compounds by polypyrrole-based composite carbon paste electrodes. *Bioelectrochemistry* 63:291–296.
1801. Wu, X., J. Wu, L. Fu, X. Li, G. Shen, and R. Yu. 2005. Determination of DNA based on polypyrrole-intercalated graphite oxide nanocomposite modified carbon paste electrode. *Asian J. Phys.* 14:173–182.
1802. Raoof, J.B., R. Ojani, and S. Rashid-Nadimi. 2005. Voltammetric determination of ascorbic acid and dopamine in the same sample at the surface of a carbon paste electrode modified with polypyrrole/ferrocyanide films. *Electrochim. Acta* 50:4694–4698.

References

1803. Szymanska, I., H. Radecka, J. Radecki, P.A. Gale, and C.N. Warriner. 2006. Ferrocene-substituted calix[4] pyrrole modified carbon paste electrodes for anion detection in water. *J. Electroanal. Chem.* 591:223–228.
1804. Lori, J.A., A. Morrin, A.J. Killard, and M.R. Smyth. 2006. Development and characterization of Nickel-NTA-polyaniline modified electrodes. *Electroanalysis* 18:77–81.
1805. Consolin Filho, N., E.C. Venancio, E.S. de Medeiros, S.T. Tanimoto, S.A.S. Machado, and L.H.C. Mattoso. 2006. Voltammetric determination of imazaquin using polyaniline modified carbon paste electrode (CPE). *Sens. Lett.* 4:11–16.
1806. Gilbert, O., U. Chandra, B.E.K. Swamy, M.P. Char, C. Nagaraj, and B.S. Sherigara. 2008. Poly(alanine) modified carbon paste electrode for simultaneous detection of dopamine and ascorbic acid. *Int. J. Electrochem. Sci.* 3:1186–1195.
1807. Majid, S., M. El Rhazi, A. Amine, A. Curulli, and G. Palleschi. 2003. Carbon paste electrode bulk-modified with the conducting polymer poly(1,8-diaminonaphthalene): Application to lead determination. *Microchim. Acta* 143:195–204.
1808. Walcarius, A. 2001. Electrochemical applications of silica-based organic–inorganic hybrid materials. *Chem. Mater.* 13:3351–3372.
1809. Gonzalez, E., P. Hernandez, and L. Hernandez. 1990. Cyclic voltammetry of tifluadom at a C18-modified carbon-paste electrode. *Anal. Chim. Acta* 228:265–272.
1810. Hernandez, P., E. Lorenzo, J. Cerrada, and L. Hernandez. 1992. Preconcentration and determination of bentazepam at C18-modified carbon paste electrodes. *Fresenius J. Anal. Chem.* 342:429–432.
1811. Hernandez, L., P. Hernandez, and J. Vicente. 1993. Voltammetric determination of methyl parathion, ortho, meta and para nitrophenol with a carbon paste electrode modified with C18. *Fresenius J. Anal. Chem.* 345:712–715.
1812. Hernandez, P., O. Nieto, F. Galan, and L. Hernandez. 1993. Use of an electrode of carbon paste modified with C18 in the determination of bendiocarb. *Quim. Anal. (Barcelona)* 12:18–23.
1813. Faller, C., A. Meyer, and G. Henze. 1996. Voltammetric determination of ioxynil and 2-methyl-3-nitroaniline using C18 modified carbon paste electrodes. *Fresenius J. Anal. Chem.* 356:279–283.
1814. Arranz, A., J.M. Moreda, and J.F. Arranz. 2000. Anodic stripping voltammetric assay for the determination of prazosin at carbon paste electrodes. *Quim. Anal. (Barcelona)* 19:31–37.
1815. Arranz, A., J.M. Moreda, and J.F. Arranz. 2000. Preconcentration and voltammetric determination of the antihypertensive doxazosin on a C8 modified carbon paste electrode. *Mikrochim. Acta* 134:69–75.
1816. Etienne, M., J. Bessiere, and A. Walcarius. 2001. Voltammetric detection of copper(II) at a carbon paste electrode containing an organically modified silica. *Sens. Actuators B* 76:531–538.
1817. Li, L., W. Li, C. Sun, and L. Li. 2002. Fabrication of carbon paste electrode containing 1:12 phosphomolybdic anions encapsulated in modified mesoporous molecular sieve MCM-41 and its electrochemistry. *Electroanalysis* 14:368–375.
1818. Hamidi, H., E. Shams, B. Yadollahi, and F.K. Esfahani. 2008. Fabrication of bulk-modified carbon paste electrode containing alpha-PW12O403-polyanion supported on modified silica gel: Preparation, electrochemistry and electrocatalysis. *Talanta* 74:909–914.
1819. Hamidi, H., E. Shams, B. Yadollahi, and F.K. Esfahani. 2009. Fabrication of carbon paste electrode containing [PFeW11O39]4-polyoxoanion supported on modified amorphous silica gel and its electrocatalytic activity for H_2O_2 reduction. *Electrochim. Acta* 54:3495–3500.
1820. Lorencetti, L.L. and Y. Gushikem. 1993. Cyclic voltammetry study of hexacyanoferrate complex immobilized on silica gel surface chemically modified with pyridinium ion. *J. Braz. Chem. Soc.* 4:88–92.
1821. Fujiwara, S.T., C.A. Pessoa, and Y. Gushikem. 2003. Hexacyanoferrate ion adsorbed on propylpyridiniumsilsesquioxane polymer film-coated SiO_2/Al_2O_3: Use in an electrochemical oxidation study of cysteine. *Electrochim. Acta* 48:3625–3631.
1822. Lucho, A.M.S., E.C. Oliveira, H.O. Pastore, and Y. Gushikem. 2004. 3-n-Propylpyridinium chloride silsesquioxane polymer film-coated aluminum phosphate and adsorption of cobalt(II) tetrasulfophthalocyanine: An electrocatalytic oxidation study of oxalic acid. *J. Electroanal. Chem.* 573:55–60.
1823. Fujiwara, S.T., Y. Gushikem, C.A. Pessoa, and S. Nakagaki. 2005. Electrochemical studies of a new iron porphyrin entrapped in a propylpyridiniumsilsesquioxane polymer immobilized on a SiO_2/Al_2O_3 surface. *Electroanalysis* 17:783–788.
1824. Magosso, H.A., R.C.S. Luz, and Y. Gushikem. 2010. Preparation and properties of the hybrid material n-propyl(3-methylpyridinium)silsesquioxane chloride. Application in electrochemical determination of nitrite. *Electroanalysis* 22:216–222.
1825. Sayen, S., M. Etienne, J. Bessière, and A. Walcarius. 2002. Tuning the sensitivity of electrodes modified with an organic–inorganic hybrid by tailoring the structure of the nanocomposite material. *Electroanalysis* 14:1521–1525.

1826. Cavalheiro, E.T.G., G. Marino, and I. Cesarino. 2006. Differential pulse anodic stripping voltammetric determination of mercury(II) in natural water at a carbon paste electrode modified with organofunctionalized amorphous silica. *Anal. Chem., Indian J.* 2:37–44.
1827. Aleixo, L.M., B.S. de Fatima, O.E.S. Godinho, G. Oliveira Neto, Y. Gushikem, and J.C. Moreira. 1992. Development of a chemically modified electrode based on carbon paste and functionalized silica gel for preconcentration and voltammetric determination of mercury(II). *Anal. Chim. Acta* 271:143–148.
1828. Dias Filho, N.L. and D.R. do Carmo. 2005. Stripping voltammetry of mercury(II) with a chemically modified carbon paste electrode containing silica gel functionalized with 2,5-dimercapto-1,3,4-thiadiazole. *Electroanalysis* 17:1540–1546.
1829. Dias Filho, N.L., D.R. do Carmo, and A.H. Rosa. 2006. An electroanalytical application of 2-aminothiazole-modified silica gel after adsorption and separation of Hg(II) from heavy metals in aqueous solution. *Electrochim. Acta* 52:965–972.
1830. Takeuchi, R.M., A.L. Santos, P.M. Padilha, and N.R. Stradiotto. 2007. Copper determination in ethanol fuel by differential pulse anodic stripping voltammetry at a solid paraffin-based carbon paste electrode modified with 2-aminothiazole organofunctionalized silica. *Talanta* 71:771–777.
1831. Sayen, S., C. Gerardin, L. Rodehuser, and A. Walcarius. 2003. Electrochemical detection of copper(II) at an electrode modified by a carnosine-silica hybrid material. *Electroanalysis* 15:422–430.
1832. Walcarius, A., S. Sayen, C. Gerardin, F. Hamdoune, and L. Rodehuser. 2004. Dipeptide-functionalized mesoporous silica spheres. *Colloids Surf. A* 234:145–151.
1833. Ganesan, V. and A. Walcarius. 2004. Surfactant templated sulfonic acid functionalized silica microspheres as new efficient ion exchangers and electrode modifiers. *Langmuir* 20:3632–3640.
1834. Abu-Shawish, H.M., S.M. Saadeh, and A.R. Hussien. 2008. Enhanced sensitivity for Cu(II) by a salicylidine-functionalized polysiloxane carbon paste electrode. *Talanta* 76:941–948.
1835. Javanbakht, M., A. Badiei, M.R. Ganjali, P. Norouzi, A. Hasheminasab, and M. Abdouss. 2007. Use of organofunctionalized nanoporous silica gel to improve the lifetime of carbon paste electrode for determination of copper(II) ions. *Anal. Chim. Acta* 601:172–182.
1836. Torabi, R., E. Shams, M.A. Zolfigol, and S. Afshar. 2006. Anodic stripping voltammetric determination of lead(II) with a 2-aminopyridinated-silica modified carbon paste electrode. *Anal. Lett.* 39:2643–2655.
1837. Dias Filho, N.L., L. Caetano, D.R. do Carmo, and A.H. Rosa. 2006. Preparation of a silica gel modified with 2-amino-1,3,4-thiadiazole for adsorption of metal ions and electroanalytical application. *J. Braz. Chem. Soc.* 17:473–481.
1838. Takeuchi, R.M., A.L. Santos, P.M. Padilha, and N.R. Stradiotto. 2007. A solid paraffin-based carbon paste electrode modified with 2-aminothiazole organo-functionalized silica for differential pulse adsorptive stripping analysis of nickel in ethanol fuel. *Anal. Chim. Acta* 584:295–301.
1839. Tian, A., Z. Han, J. Peng, J. Zhai, and Y. Zhao. 2007. Inorganic–organic microporous solid of Wells-Dawson type polyoxometalate: Synthesis, characterization, and electrochemical properties. *Z. Anorg. Allg. Chem.* 633:495–503.
1840. Wang, X., Y. Bi, B. Chen, H. Lin, and G. Liu. 2008. Self-assembly of organic–inorganic hybrid materials constructed from eight-connected coordination polymer hosts with nanotube channels and polyoxometalate guests as templates. *Inorg. Chem.* 47:2442–2448.
1841. Li, C., R. Cao, K.P. O'Halloran, H. Ma, and L. Wu. 2008. Preparation, characterization and bifunctional electrocatalysis of an inorganic–organic complex with a vanadium-substituted polyoxometalate. *Electrochim. Acta* 54:484–489.
1842. Lin, S., X. Zhang, and M. Luo. 2009. A novel inorganic–organic hybrid compound constructed from copper(II)-monosubstituted polyoxometalates and poly(amidoamine). *J. Solid State Electrochem.* 13:1585–1589.
1843. Walcarius, A. 2005. Impact of mesoporous silica-based materials on electrochemistry and feedback from electrochemical science to the characterization of these ordered materials. *C. R. Chim.* 8:693–712.
1844. Guo, H., N. He, S. Ge, D. Yang, and J. Zhang. 2005. MCM-41 mesoporous material modified carbon paste electrode for the determination of cardiac troponin I by anodic stripping voltammetry. *Talanta* 68:61–66.
1845. Guo, H.S., N.Y. He, S.X. Ge, D. Yang, and J.N. Zhang. 2005. Determination of cardiac troponin I by anodic stripping voltammetry at SBA-15 modified carbon paste electrode. *Stud. Surf. Sci. Catal.* 156:695–702.
1846. He, N.-Y., H.-S. Guo, D. Yang, C.-R. Gu, and J.-N. Zhang. 2006. SBA-15 modified carbon paste electrode for rapid cTnI detection with enhanced sensitivity. *Chin. Chem. Lett.* 17:235–238.
1847. Zeng, Y., J. Xu, and K. Wu. 2008. Electrochemical determination of uric acid using a mesoporous SiO_2-modified electrode. *Microchim. Acta* 161:249–253.

1848. Xie, X., D. Zhou, X. Zheng, W. Huang, and K. Wu. 2009. Electrochemical sensing of rutin using an MCM-41 modified electrode. *Anal. Lett.* 42:678–688.
1849. Zhou, C., Z. Liu, Y. Dong, and D. Li. 2009. Electrochemical behavior of o-nitrophenol at hexagonal mesoporous silica modified carbon paste electrodes. *Electroanalysis* 21:853–858.
1850. Sun, D., F. Wang, K. Wu, J. Chen, and Y. Zhou. 2009. Electrochemical determination of hesperidin using mesoporous SiO_2 modified electrode. *Microchim. Acta* 167:35–39.
1851. Zhao, J., W. Huang, and X. Zheng. 2009. Mesoporous silica-based electrochemical sensor for simultaneous determination of honokiol and magnolol. *J. Appl. Electrochem.* 39:2415–2419.
1852. Song, J., J. Yang, and X. Yang. 2009. Electrochemical determination of 5-hydroxytryptamine using mesoporous SiO_2 modified carbon paste electrode. *Russ. J. Electrochem.* 45:1346–1350.
1853. Walcarius, A., C. Despas, P. Trens, M.J. Hudson, and J. Bessiere. 1998. Voltammetric in situ investigation of a MCM-41-modified carbon paste electrode—A new sensor. *J. Electroanal. Chem.* 453:249–252.
1854. Walcarius, A. and J. Bessiere. 1999. Electrochemistry with Mesoporous Silica: Selective Mercury(II) Binding. *Chem. Mater.* 11:3009–3011.
1855. Zeng, Y., J. Yang, and K. Wu. 2008. Electrochemistry and determination of epinephrine using a mesoporous Al-incorporated SiO_2 modified electrode. *Electrochim. Acta* 53:4615–4620.
1856. Lin, H., T. Gan, and K. Wu. 2009. Sensitive and rapid determination of catechol in tea samples using mesoporous Al-doped silica modified electrode. *Food Chem.* 113:701–704.
1857. Walcarius, A., M. Etienne, S. Sayen, and B. Lebeau. 2003. Grafted silicas in electroanalysis: Amorphous versus ordered mesoporous materials. *Electroanalysis* 15:414–421.
1858. Walcarius, A., C. Delacote, and S. Sayen. 2004. Electrochemical probing of mass transfer rates in mesoporous silica-based organic–inorganic hybrids. *Electrochim. Acta* 49:3775–3783.
1859. Ojani, R., E. Ahmadi, J.B. Raoof, and F. Mohamadnia. 2009. Characterization of a carbon paste electrode containing organically modified nanostructure silica: Application to voltammetric detection of ferricyanide. *J. Electroanal. Chem.* 626:23–29.
1860. Cesarino, I., G. Marino, J. do Rosario Matos, and E.T.G. Cavalheiro. 2007. Using the organofunctionalized SBA-15 nanostructured silica as a carbon paste electrode modifier: Determination of cadmium ions by differential anodic pulse stripping voltammetry. *J. Braz. Chem. Soc.* 18:810–817.
1861. Cesarino, I., G. Marino, J.R. Matos, and E.T.G. Cavalheiro. 2007. Evaluation of a carbon paste electrode modified with organofunctionalized SBA-15 silica in the determination of copper. *Ecletica Quim.* 32:29–34.
1862. Cesarino, I., G. Marino, J.d.R. Matos, and E.T.G. Cavalheiro. 2008. Evaluation of a carbon paste electrode modified with organofunctionalised SBA-15 nanostructured silica in the simultaneous determination of divalent lead, copper and mercury ions. *Talanta* 75:15–21.
1863. Morante Zarcero, S., A. Sanchez, M. Fajardo, I. Hierro, and I. Sierra. 2010. Voltammetric analysis of Pb(II) in natural waters using a carbon paste electrode modified with 5-mercapto-1-methyltetrazol grafted on hexagonal mesoporous silica. *Microchim. Acta* 169:57–64.
1864. Ganjali, M.R., M. Asgari, F. Faridbod, P. Norouzi, A. Badiei, and J. Gholami. 2010. Thiomorpholine-functionalized nanoporous mesopore as a sensing material for Cd^{2+} carbon paste electrode. *J. Solid State Electrochem.* 14:1359–1366.
1865. Popa, D.E., M. Buleandra, M. Mureseanu, M. Ionica, and I.G. Tanase. 2010. Carbon paste electrode modified with organofunctionalized mesoporous silica for electrochemical detection and quantitative determination of cadmium(II) using square wave anodic stripping voltammetry. *Rev. Chim. (Roum.)* 61:162–167.
1866. Zhou, W., Y. Chai, R. Yuan, J. Guo, and X. Wu. 2009. Organically nanoporous silica gel based on carbon paste electrode for potentiometric detection of trace Cr(III). *Anal. Chim. Acta* 647:210–214.
1867. Javanbakht, M., F. Divsar, A. Badiei, M.R. Ganjali, P. Norouzi, G.M. Ziarani, M. Chaloosi, and A.A. Jahangir. 2009. Potentiometric detection of mercury(II) ions using a carbon paste electrode modified with substituted thiourea-functionalized highly ordered nanoporous silica. *Anal. Sci. (Jpn.)* 25:789–794.
1868. Javanbakht, M., F. Divsar, A. Badiei, F. Fatollahi, Y. Khaniani, M.R. Ganjali, P. Norouzi, M. Chaloosi, and G.M. Ziarani. 2009. Determination of picomolar silver concentrations by differential pulse anodic stripping voltammetry at a carbon paste electrode modified with phenylthiourea-functionalized high ordered nanoporous silica gel. *Electrochim. Acta* 54:5381–5386.
1869. Javanbakht, M., M.R. Ganjali, P. Norouzi, A. Badiei, A. Hasheminasab, and M. Abdouss. 2007. Carbon paste electrode modified with functionalized nanoporous silica gel as a new sensor for determination of silver ion. *Electroanalysis* 19:1307–1314.

1870. Javanbakht, M., H. Khoshsafar, M.R. Ganjali, P. Norouzi, A. Badei, and A. Hasheminasab. 2008. Stripping voltammetry of cerium(III) with a chemically modified carbon paste electrode containing functionalized nanoporous silica gel. *Electroanalysis* 20:203–206.
1871. Javanbakht, M., H. Khoshsafar, M.R. Ganjali, A. Badiei, P. Norouzi, and A. Hasheminasab. 2009. Determination of nanomolar mercury(II) concentration by anodic-stripping voltammetry at a carbon paste electrode modified with functionalized nanoporous silica gel. *Curr. Anal. Chem.* 5:35–41.
1872. Lin, Y., X. Cui, and L. Li. 2005. Low-potential amperometric determination of hydrogen peroxide with a carbon paste electrode modified with nanostructured cryptomelane-type manganese oxides. *Electrochem. Commun.* 7:166–172.
1873. Cui, X., G. Liu, L. Li, W. Yantasee, and Y. Lin. 2005. Electrochemical sensor based on carbon paste electrode modified with nanostructured cryptomelane-type manganese oxides for detection of heavy metals. *Sens. Lett.* 3:16–21.
1874. Cui, X., G. Liu, and Y. Lin. 2005. Amperometric biosensors based on carbon paste electrodes modified with nanostructured mixed-valence manganese oxides and glucose oxidase. *Nanomedicine* 1:130–135.
1875. Xie, X., K. Yang, and D. Sun. 2008. Voltammetric determination of hypoxanthine based on the enhancement effect of mesoporous TiO_2-modified electrode. *Colloids Surf. B* 67:261–264.
1876. Tan, X., B. Li, G. Zhan, and C. Li. 2010. Sensitive voltammetric determination of methyl parathion using a carbon paste electrode modified with mesoporous zirconia. *Electroanalysis* 22:151–154.
1877. Manso, J., L. Aguei, P. Yanez-Sedeno, and J.M. Pingarron. 2004. Development and Characterization of colloidal gold-cysteamine-carbon paste electrodes. *Anal. Lett.* 37:887–902.
1878. Agui, L., J. Manso, P. Yanez-Sedeno, and J.M. Pingarron. 2004. Colloidal-gold cysteamine-modified carbon paste electrodes as suitable electrode materials for the electrochemical determination of sulphur-containing compounds Application to the determination of methionine. *Talanta* 64:1041–1047.
1879. Kalanur, S.S., J. Seetharamappa, and S.N. Prashanth. 2010. Voltammetric sensor for buzepide methiodide determination based on TiO_2 nanoparticle-modified carbon paste electrode. *Colloids Surf. B* 78:217–221.
1880. Mazloum Ardakani, M., H. Beitollahi, Z. Taleat, H. Naeimi, and N. Taghavinia. 2010. Selective voltammetric determination of D-penicillamine in the presence of tryptophan at a modified carbon paste electrode incorporating TiO_2 nanoparticles and quinizarine. *J. Electroanal. Chem.* 644:1–6.
1881. Sabzi, R.E., F. Ebrahimzadeh, and H. Sedghi. 2008. Electrochemical and electrocatalytic properties of cobalt oxide nanoparticles. *Asian J. Chem.* 20:3364–3372.
1882. Daneshgar, P., P. Norouzi, M.R. Ganjali, A. Ordikhani-Seyedlar, and H. Eshraghi. 2009. A dysprosium nanowire modified carbon paste electrode for determination of levodopa using fast Fourier transformation square-wave voltammetry method. *Colloids Surf. B* 68:27–32.
1883. Daneshgar, P., P. Norouzi, M.R. Ganjali, and F. Dousty. 2009. A dysprosium nanowire modified carbon paste electrode for determination of nanomolar level of Diphenhydramin by continuous square wave voltammetry in flow injection system. *Int. J. Electrochem. Sci.* 4:444–457.
1884. Daneshgar, P., P. Norouzi, F. Dousty, M.R. Ganjali, and A.A. Moosavi-Movahedi. 2009. Dysprosium hydroxide nanowires modified electrode for determination of rifampicin drug in human urine and capsules by adsorptive square wave voltammetry. *Curr. Pharm. Anal.* 5:246–255.
1885. Daneshgar, P., P. Norouzi, M.R. Ganjali, R. Dinarvand, and A.A. Moosavi-Movahedi. 2009. Determination of diclofenac on a dysprosium nanowire-modified carbon paste electrode accomplished in a flow injection system by advanced filtering. *Sensors* 9:7903–7918.
1886. Daneshgar, P., P. Norouzi, A.A. Moosavi-Movahedi, M.R. Ganjali, E. Haghshenas, F. Dousty, and M. Farhadi. 2009. Fabrication of carbon nanotube and dysprosium nanowire modified electrodes as a sensor for determination of curcumin. *J. Appl. Electrochem.* 39:1983–1992.
1887. Norouzi, P., F. Dousty, M.R. Ganjali, and P. Daneshgar. 2009. Dysprosium nanowire modified carbon paste electrode for the simultaneous determination of naproxen and paracetamol: Application in pharmaceutical formulation and biological fluid. *Int. J. Electrochem. Sci.* 4:1373–1386.
1888. Mashhadizadeh, M.H. and M. Akbarian. 2009. Voltammetric determination of some anti-malarial drugs using a carbon paste electrode modified with $Cu(OH)_2$ nano-wire. *Talanta* 78:1440–1445.
1889. Sun, W., D. Wang, G. Li, Z. Zhai, R. Zhao, and K. Jiao. 2008. Direct electron transfer of hemoglobin in a CdS nanorods and Nafion composite film on carbon ionic liquid electrode. *Electrochim. Acta* 53:8217–8221.
1890. Economou, E., and P.R. Fielden. 2003. Mercury film electrodes: Developments, trends and potentialities for electroanalysis. *Analyst* 128:205–212.
1891. Farghaly, O.A. 2004. A novel method for determination of magnesium in urine and water samples with mercury film-plated carbon paste electrode. *Talanta* 63:497–501.

1892. Sherigara, B.S., Y. Shivaraj, R.J. Mascarenhas, and A.K. Satpati. 2007. Simultaneous determination of lead, copper and cadmium onto mercury film supported on wax impregnated carbon paste electrode. *Electrochim. Acta* 52:3137–3142.
1893. Chadim, P., I. Švancara, B. Pihlar, and K. Vytřas. 2000. Gold-plated carbon paste electrodes for anodic stripping determination of arsenic. *Collect. Czech. Chem. Commun.* 65:1035–1046.
1894. Tesarova, E., L. Baldrianova, S.B. Hocevar, I. Švancara, K. Vytřas, and B. Ogorevc. 2009. Anodic stripping voltammetric measurement of trace heavy metals at antimony film carbon paste electrode. *Electrochim. Acta* 54:1506–1510.
1895. Bobrowski, A., A. Krolicka, and E. Lyczkowska. 2007. Carbon paste electrode plated with lead film. Voltammetric characteristics and application in adsorptive stripping voltammetry. *Electroanalysis* 20:61–67.
1896. Tesarova-Svobodova, E., L. Baldrianova, M. Stoces, I. Švancara, K. Vytřas, S.B. Hocevar, and B. Ogorevc. 2010. Antimony powder-modified carbon paste electrodes for electrochemical stripping determination of trace heavy metals. *Electrochim. Acta* 56:6673–6677.
1897. Pauliukaite, R. and K. Kalcher. 2001. Using of CPE and SPCE modified by Bi_2O_3 and Sb_2O_3 for trace analysis of some heavy metals. In *YISAC-01: 8th Young Investigators' Seminar on Analytical Chemistry, Book of Abstracts*, Pardubice, Czech Republic, July 2–5, 2001, pp. 10–11. Pardubice, Czech Republic: University of Pardubice.
1898. Švancara, I., M. Florescu, M. Stoces, L. Baldrianova, E. Svobodova, M. Badea. 2010. Carbon paste electrodes modified with a reaction product obtained by hydrolysis of an antimony(III) salt. In *Sensing in Electroanalysis*, Vol. 5, Eds. K. Vytřas, K. Kalcher, and I. Švancara, pp. 109–125. Pardubice, Czech Republic: University Press Center.
1899. Stoces, M. and I. Švancara. 2010. Carbon paste electrodes modified with bismuth- and antimony fluoride as a novel non-mercury metallic film-based sensors for electrochemical stripping analysis. Article in preparation.
1900. Cai, X., K. Kalcher, and C. Neuhold. 1994. Simultaneous determination of uric acid, xanthine and hypoxanthine with an electrochemically pretreated carbon paste electrode. *Fresenius J. Anal. Chem.* 348:660–665.
1901. Cai, X., B. Ogorevc, and K. Kalcher. 1995. Synergistic electrochemical and chemical modification of carbon paste electrodes for open-circuit preconcentration and voltammetric determination of trace adenine. *Electroanalysis* 7:1126–1131.
1902. Wang, J., X. Cai, J. Wang, C. Jonsson, and E. Palecek. 1995. Trace measurements 91 of RNA by potentiometric stripping analysis at carbon paste electrodes. *Anal. Chem.* 67:4065–4070.
1903. Wang, H., A. Zhang, H. Cui, D. Liu, and R. Liu. 1998. Adsorptive stripping voltammetric determination of phenol at an electrochemically pretreated carbon-paste electrode with solid paraffin as a binder. *Microchem. J.* 59:448–456.
1904. Saidman, S.B. 2002. Effects of electrochemical pretreatments on NADH oxidation at NAD^+-modified carbon paste electrodes. *Electroanalysis* 14:449–454.
1905. Dunwoody, D.C., A.K.H. Wolf, W.L. Gellett, and J. Leddy. 2004–2005. Magnetically modified carbon paste electrodes: Method for construction and demonstration of increased electrochemical flux over unmodified carbon paste electrodes. *Proc. Electrochem. Soc.* 18:181–191.
1906. Hulanicki, A., S. Glab, and F. Ingman A. 1991. Chemical sensors—Definitions and classification. *Pure Appl. Chem.* 63:1247–1250.
1907. Thevenot, D.R., K. Toth, R.A. Durst, and G.S. Wilson. 1999. Electrochemical biosensors: Recommended definitions and classification. *Pure Appl. Chem.* 71:2333–2348.
1908. Clark, L.C. and C. Lyons. 1962. Electrode systems for continuous monitoring in cardiovascular surgery. *Ann. N.Y. Acad. Sci.* 102:29–45.
1909. Clark L.C., R. Wolf, D. Granger, and Z. Taylor. 1953. Continuous recording of blood oxygen tensions by polarography. *J. Appl. Physiol.* 6:189–193.
1910. Severinghaus J.W. and P.B. Astrup. 1986. History of blood gas analysis. IV. Leland Clark's oxygen electrode. *J. Clin. Monit.* 2:125–139.
1911. Lubert, K.H. and K. Kalcher. 2010. History of electroanalytical methods. *Electroanalysis* 22:1937–1946.
1912. Patolsky, F., Y. Weizmann, and I. Willner. 2004. Long-range electrical contacting of redox enzymes by SWCNT connectors. *Angew. Chem. Int. Ed.* 43:2113–2117.
1913. Xiao, Y., F. Patolsky, E. Katz, J.F. Hainfeld, and I. Willner. 2003. "Plugging into Enzymes": Nanowiring of redox enzymes by a gold nanoparticle. *Science* 299:1877–1881.
1914. Razumas, V., J. Kazlauskaite, and R. Vidziunaite. 1996. Electrocatalytic reduction of hydrogen peroxide on the microperoxidase-11 modified carbon paste and graphite electrodes. *Bioelectrochem. Bioenerg.* 39:139–143.

1915. Wang, J. 1994. *Analytical Electrochemistry*, 1st Edn., New York: VCH Publishers.
1916. Wang, J. 1985. *Stripping Analysis. Principles, Instrumentation, and Application*, 1st Edn., New York: VCH Publishers.
1917. Kissinger, P.T and W.R. Heineman. 1996. *Laboratory Techniques in Electroanalytical Chemistry*, 2nd Rev. and Exp. Edn., New York: Marcel Dekker.
1918. Bockris, J.O'M. and A.K.N. Reddy. 1977. *Modern Electrochemistry 1*. 3rd Print, New York: Plenum Press.
1919. Bockris, J.O'M. and A.K.N. Reddy. 1977. *Modern Electrochemistry 2*. 3rd Print, New York: Plenum Press.
1920. Bard, A.J. and L.R. Faulkner. 1980. *Electrochemical Methods. Fundamentals and Applications*. New York: John Wiley.
1921. Monk, P.M.S. 2001. *Fundamentals of Electroanalytical Chemistry*. Chichester, U.K.: John Wiley.
1922. Scholz F. Ed. 2002. *Electroanalytical Methods*. Berlin, Germany: Springer-Verlag.
1923. Pletcher, D. 1991. *A First Course in Electrode Processes*. Hants, U.K.: The Electrochemical Consultancy.
1924. Bagotsky, V.S. 2006. *Fundamentals of Electrochemistry*. Hoboken, NJ: John Wiley.
1925. Mirceski, V., S. Komorski-Lovric, and M. Lovric. 2007. *Square Wave Voltammetry. Theory and Application*. Berlin, Germany: Springer-Verlag.
1926. Zhao, J., L. Zhang, X. Yao, and Q. Yang. 2002. Adsorptive voltammetric determination of cobalt in Chinese herbal medicine. *Lihua Jianyan, Huaxue Fence* 38:387–388.
1927. Lu, J., W. Jin, and S. Wang. 1990. Adsorption voltammetry of the vanadium-2-(5′-bromo-2′-pyridylazo)-5-diethylaminophenol (5-Br-PADAP) system. *Anal. Chim. Acta* 238:375–381.
1928. Jin, W. and X. Li. 1990. 1.5th- and 2.5th-order derivative adsorption voltammetry of the cobalt(II)-2-(5′-bromo-2′-pyridylazo)-diethylaminophenol system. *Anal. Chim. Acta* 236:453–458.
1929. Goto, M. and D. Ishii. 1975. Semidifferential electroanalysis. *J. Electroanal. Chem. Interf. Electrochem.* 61:361–365.
1930. Oldham, K.B. 1972. Signal-independent electroanalytical method. *Anal. Chem.* 44:196–198.
1931. Grennes, M. and K.B. Oldham. 1972. Semiintegral electroanalysis. Theory and verification. *Anal. Chem.* 44:1121–1129.
1932. Goto, M., T. Hirano, and D. Ishii. 1978. 2.5th Order differential electroanalysis. *Bull. Chem. Soc. Jan.* 51:470–475.
1933. Jin, W., H. Cui, and S. Wang. 1992. Linear potential sweep adsorption voltammetry for a reversible interfacial reaction: Comparison of conventional and derivative measuring techniques. *Anal. Chim. Acta* 268:301–306.
1934. Jin W., H. Cui, L. Zhu, and S. Wang. 1992. Linear potential sweep adsorption voltammetry for an irreversible interfacial reaction: Theory and sensitivity. *J. Electroanal. Chem.* 340:315–324.
1935. Jin W., H. Cui, and S. Wang. 1991. On the theory of the integer and half-integer integral and derivative linear potential sweep voltammetry for a reversible interfacial reaction. *J. Electroanal. Chem.* 297:37–47.
1936. Trnkova, L. and O. Dracka. 1996. Elimination voltammetry. Experimental verification and extension of theoretical results. *J. Electroanal. Chem.* 413:123–129.
1937. Trnkova, L. 2001. Electrochemical elimination methods. *Chem. Listy* 95:518–527.
1938. Trnkova, L., R. Kizek, and O. Dracka. 2000. Application of elimination voltammetry to adsorptive stripping of DNA. *Electroanal.* 12:905–911.
1939. Trnkova, L., J. Friml, and O. Dracka. 2001. Elimination voltammetry of adenine and cytosine mixtures. *Bioelectrochem.* 54:131–136.
1940. Trnkova, L. 2002. Electrochemical behavior of DNA at a silver electrode studied by cyclic and elimination voltammetry. *Talanta* 56:887–894.
1941. Trnkova, L., R. Kizek, and O. Dracka. 2002. Elimination voltammetry of nucleic acids on silver electrodes. *Bioelectrochemistry* 55:131–133.
1942. Trnkova, L., F. Jelen, and I. Postbieglova. 2003. Application of elimination voltammetry to the resolution of adenine and cytosine signals in oligonucleotides. I. Homooligodeoxynucleotides dA(9) and dC(9). *Electroanalysis* 15:1529–1535.
1943. Trnkova, L., R. Kizek, and J. Vacek. 2004. Square wave and elimination voltammetric analysis of azidothymidine in the presence of oligonucleotides and chromosomal DNA. *Bioelectrochemistry* 63:31–36.
1944. Trnkova, L., I. Postbieglova, and M. Holik. 2004. Electroanalytical determination of d(GCGAAGC) hairpin. *Bioelectrochemistry* 63, 25–30.
1945. Trnkova, L., F. Jelen, J. Petrlova, V. Adam, D. Potesil, and R. Kizek. 2005. Elimination voltammetry with linear scan as a new detection method for DNA sensors. *Sensors* 5:448–464.

1946. Palecek E., F. Jelen, C. Teijeiro, V. Fucik, and T.M. Jovin. 1993. Biopolymer-modified electrodes in the voltammetric determination of nucleic acids and proteins at the submicrogram level. *Anal. Chim. Acta* 273:175–186.
1947. Ruzicka, J. and E.H. Hansen. 1988. *Flow Injection Analysis*, 2nd Edn., New York: John Wiley & Sons, Inc.
1948. Wang, J., L. Chen, L. Angnes, and B. Tian. 1992. Computerized pipettes with programmable dispension for batch injection analysis. *Anal. Chim. Acta* 267:171–177.
1949. Wang, J. and Z. Taha. 1991. Batch injection analysis. *Anal. Chem.* 63:1053–1056.
1950. Quintino, M.S.M. and L. Angnes. 2004. Batch injection analysis: An almost unexplored powerful tool. *Electroanalysis* 16:513–523.
1951. Brett, C.M.A., A.M. Oliveira Brett, and L. Tugulea. 1996. Anodic stripping voltammetry of trace metals by batch injection analysis. *Anal. Chim. Acta* 322:151–157.
1952. Brett, C.M.A., A.M.O. Brett, and L.C. Mitoseriu. Amperometric batch injection analysis: Theoretical aspects of current transients and comparison with wall-jet electrodes in continuous flow. 1995. *Electroanalysis* 7:225–229.
1953. Brett, C.M.A., A.M. Oliveira Brett, and L.C. Mitoseriu. Amperometric and voltammetric detection in batch injection analysis. 1994. *Anal. Chem.* 66:3145–3150.
1954. Backofen, U., W. Hoffmann, and F.M. Matysik. 1998. Capillary batch injection analysis—A novel approach for analyzing nanoliter samples. *Anal. Chim. Acta* 362:213–220.
1955. Yamada, J. and H. Matsuda. 1973. Limiting diffusion currents in hydrodynamic voltammetry. III. Wall jet electrodes. *J. Electroanal. Chem.* 44:189–198.
1956. Bates, R.G. 1964. *Determination of pH, Theory and Practice*. New York: Wiley.
1957. Eiseman, G. 1967. *Glass Electrodes for Hydrogen and Other Cations*. New York: Marcel Dekker.
1958. Vytřas, K. 1995. Potentiometry, In *Encyclopedia of Pharmaceutical Technology*, Vol. 12, Eds. J. Swarbrick and J.C. Boylan, pp. 347–388. New York: Marcel Dekker.
1959. Lakshminarayanaiah, N. 1976. *Membrane Electrodes*. New York: Academic Press.
1960. Koryta, J. and K. Stulik. 1983. *Ion-Selective Electrodes*, 2nd Edn., Cambridge, U.K.: University Press.
1961. Ruzicka, J. 2009. *Flow-Injection Analysis*. Bellevue, WA: FIALab Instruments.
1962. Ruzicka, J., J.C. Tjell, and C.G. Lamm. 1972. Selectrode—Universal ion-selective electrode. Concept, construction and materials. *Anal. Chim. Acta* 62:15–28.
1963. Ruzicka, J. 1997. The seventies: Golden age for ion selective electrodes. *J. Chem. Educ.* 74:167–170.
1964. Filipenko, A.T., E.M. Skobets, O.P. Ryabushko, and Yu.S. Savin. 1984. Silver iodide carbon-paste ion-selective electrode. *Ukr. Khim. Zh. (Russ. Ed.)* 50:490–493.
1965. Hoffmann, C.R., M.R. Haskard, and D.E. Mulcahy. 1984. Carbon-filled polymer paste ion-selective probes. *Anal. Lett.* 17:1499–1509.
1966. Vytřas, K., J. Jezkova, and J. Kalous. 1997. Automated potentiometry as an ecologic alternative to two-phase titrations of surfactants. *Egypt. J. Anal. Chem.* 6:107–123.
1967. Vytřas, K., J. Jezkova, V. Dlabka, and J. Kalous. 1997. Studies on potentiometric titrations using simple liquid membrane-based electrodes: Coated wires vs. carbon pastes. *Sci. Pap. Univ. Pardubice Ser. A* 3:307–321.
1968. Vytřas, K., J. Jezkova, and J. Skorepa. 1998. Some aspects of the use of heteropoly anions in elemental analysis by simple potentiometric ion-pair formation-based titration. *Talanta* 46:1619–1622.
1969. Lee, Y.-K., C.-K. Kim, J.-T. Park, K.-S. Kim, and K.-J. Whang. 1985. Potentiometry with carbon paste-based ion-selective electrode for the determination of sulphate. *J. Korean Air Pollut. Res. Assoc.* 1:99–103.
1970. Lee, Y.-K., J.-T. Park, C.-K. Kim, and K.-J. Whang. 1986. Carbon paste-coated wire ion-selective electrode for nitrate ion. *Anal. Chem.* 58:2101–2103.
1971. Lee, Y.-K., S.-K. Rhim, and K.-J. Whang. 1989. Construction of carbon paste-coated wire ion-selective electrode for chloride and its application to environmental analysis of water. *Bull. Korean Chem. Soc.* 10:485–488.
1972. Vytřas, K. 1989. The use of ion-selective electrodes in the determination of drug substances. *J. Pharm. Biomed. Anal.* 7:789–812.
1973. Vytřas, K., T. Capoun, E. Halamek, J. Soucek, and B. Stajerova. 1990. Potentiometric ion-pair formation titrations of N-alkyl-N-ethylpyrrolidium cations using plastic membrane electrodes. *Collect. Czech. Chem. Commun.* 55:941–950.
1974. Fogg, A.G., M. Duzinkewycz, and A.S. Pathan. 1973. Ion-selective electrodes based on brilliant green tetrathiocyanatezincate(II). *Anal. Lett.* 6:1101–1106.
1975. Vytřas, K., J. Jezkova, J. Kalous, I. Švancara. 1994. A low ohmic resistance sensor for potentiometric titrations of tensides. In *XXVII. Seminar on Tensides and Detergents. Proceedings (in Czech)*, Novaky, Slovakia, pp. 35–48.
1976. Rejhonová, H. 2011. Development and testing of the dual sensors based on carbon pastes modified with Bi + Bi_2O_3 and Sb + Sb_2O_3 mixtures. MSc dissertation (in Czech). University of Pardubice, Pardubice, Czech Republic.

1977. Fogg, A.G. and J. Wang. Terminology and convention for electrochemical stripping analysis (technical report). 1999. *Pure Appl. Chem.* 71:891–897.
1978. Delahay, P. and T. Berzins. 1953. Theory of electrolysis at constant current with partial control by diffusion: Application to the study of complex ions. *J. Am. Chem. Soc.* 75:2486–2486.
1979. Berzins, T. and P. Delahay. 1953. Theory of electrolysis at constant current in unstirred solution. 2. Consecutive electrochemical reactions. *J. Am. Chem. Soc.* 75:4205–4213.
1980. Delahay, P. and C.C. Mattax. 1954. Theory of electrolysis at constant current in unstirred solution. 3. Experimental study of potential-time curves. *J. Am. Chem. Soc.* 76:874–878.
1981. Heineman, W.R. and P.T. Kissinger. 1996. Large-Amplitude Controlled-Current Techniques. In *Laboratory Techniques in Electroanalytical Chemistry*, 2nd Rev. Edn. Eds. P.T. Kissinger and W.R. Heineman, pp. 127–139. New York: Marcel Dekker.
1982. Jagner, D. and A. Graneli. 1976. Potentiometric stripping analysis. *Anal. Chim. Acta* 83:19–26.
1983. Ostapczuk, P. and M. Froning. 1992. Advanced electrochemical techniques for the determination of heavy metals in specimen bank materials. In *Specimen Banking—Environmental Monitoring & Modern Analytical Approaches*, Eds. M. Rossbach, J.D. Schladot, and P. Ostapczuk, pp. 153–165. Berlin, Germany: Springer Verlag.
1984. Ostapczuk, P. 1993. Present potentials and limitations in the determination of trace elements by potentiometric stripping analysis. *Anal. Chim. Acta* 273:35–40.
1985. Estela, J.M., C. Tomas, A. Cladera, and V. Cerda. 1995. Potentiometric stripping analysis—A review. *Crit. Rev. Anal. Chem.* 25:91–141.
1986. Christensen, J.K., L. Kryger, J. Mortensen, and J. Rasmussen. 1980. Reductive potentiometric stripping analysis for elements forming sparingly soluble mercury compounds with amalgamed metal as a reducting agent. *Anal. Chim. Acta* 121:71–83.
1987. Konvalina, J. and K. Vytřas. 2001. The present use of (chrono)potentiometric stripping analysis. *Chem. Listy* 95:344–351.
1988. Wang, J., G. Rivas, X.-H. Cai, M. Chicharro, N. Dontha, D.-B. Luo, E. Palecek, and P.E. Nielsen. 1997. Adsorption and detection of peptide nucleic acids at carbon paste electrodes. *Electroanalysis* 9:120–124.
1989. Vytřas, K., K. Kalcher, I. Švancara, K. Schachl, E. Khaled, J. Jezkova, J. Konvalina, and R. Metelka. 2001. Recent applications of carbon paste electrodes in potentiometry and stripping analysis. In *Chemical and Biological Sensors and Analytical Methods II*, Eds. M. Butler, P. Vanysek, and N. Yamazoe, pp. 277–283. Pennington, NJ: Electrochem. Soc.
1990. Konvalina, J. and K. Vytřas. 1999. Determination of thallium(III) at a carbon paste electrode using potentiometric stripping analysis. In *Monitoring of Environmental Pollutants*, Eds. K. Vytřas, J. Kellner, and J. Fischer, pp. 99–104. Pardubice, Czech Republic: University of Pardubice.
1991. Mortensen, J., E. Ouziel, H.J. Skov, and L. Kryger. 1979. Multiple-scanning potentiometric stripping analysis. *Anal. Chim. Acta* 112:297–312.
1992. Konvalina, J., E. Khaled, and K. Vytřas. 2000. Carbon paste electrode as a support for mercury film in potentiometric stripping determination of heavy metals. *Collect. Czech. Chem. Commun.* 65:1047–1054.
1993. Khaled, E., J. Konvalina, K. Vytřas, and H.N.A. Hassan. 2003. Investigation of carbon paste electrodes as supports for gold films in potentiometric stripping determination of copper(II) and mercury(II) traces. *Sci. Pap. Univ. Pardubice Ser. A* 9:19–29.
1994. Tesarova, E. and K. Vytřas. 2009. Potentiometric stripping analysis at antimony film electrodes. *Electroanalysis* 21:1075–1080.
1995. Hocevar, S.B., I. Švancara, B. Ogorevc, and K. Vytřas. 2007. Antimony film electrode for electrochemical stripping analysis. *Anal. Chem.* 79:8639–8643.
1996. Švancara, I., S.B. Hocevar, L. Baldrianova, E. Tesarova, B. Ogorevc, and K. Vytřas. 2007. Antimony-modified carbon paste electrodes: Initial studies and prospects. *Sci. Pap. Univ. Pardubice Ser. A* 13:5–19.
1997. Sopha, H., L. Baldrianova, E. Tesarova, S.B. Hocevar, I. Švancara, B. Ogorevc, and K. Vytřas. 2010. Insights into the simultaneous chronopotentiometric stripping measurements of indium(III), thallium(I) and zinc(II) in acidic medium at the in-situ prepared antimony film carbon paste electrode. *Electrochim. Acta* 55:7929–7933.
1998. Brett, M.A. and A.M. Oliveira Brett. 1998. *Electroanalysis*. Oxford, U.K.: University Press.
1999. Ruby, W.R. and C.G. Tremmel. 1968. Electrochemical properties of a carbon paste-silver halide electrode. *J. Electroanal. Chem.* 18:231–238.
2000. Hu, C., X. Dang, and S. Hu. 2004. Studies on adsorption of cetyltrimethylammonium bromide at carbon paste electrode and the enhancement effect in thyroxine reduction by voltammetry and electrochemical impedance spectroscopy. *J. Electroanal. Chem.* 572:161–171.

References

2001. He, J.B., Y. Zhou, and F.S. Meng. 2009. Time-derivative cyclic voltabsorptometry for voltammetric characterization of catechin film on a carbon-paste electrode: One voltammogram becomes four. *J. Solid State Electrochem.* 13:679–685.
2002. Sandulescu, R., S. Mirel, and R. Oprean. 2000. The development of spectrophotometric and electroanalytical methods for ascorbic acid and acetaminophen and their applications in the analysis of effervescent dosage forms. *J. Pharm. Biomed. Anal.* 23:77–87.
2003. Egashira, N., H. Kumasako, Y. Kurauchi, and K. Ohga. 1993. Determination of oxalate with fiber-optic electrochemiluminescence sensors. *Proc. Electrochem. Soc.* 93–7:674–679.
2004. Buckley, A.N. and R. Woods. 1991. Electrochemical and XPS studies of the surface oxidation of synthetic heazlewoodite (Ni_3S_2). *J. Appl. Electrochem.* 21:575–582.
2005. Nava, D., I. Gonzalez, D. Leinen, and J.R. Ramos-Barrado. 2008. Surface characterization by X-ray photoelectron spectroscopy and cyclic voltammetry of products formed during the potentiostatic reduction of chalcopyrite. *Electrochim. Acta* 53:4889–4899.
2006. Simic, N. and E. Ahlberg. 1999. Electrochemical, spectroscopic and structural investigations of the Cd/Cd(II) system in alkaline media 2. Concentration effects. *J. Electroanal. Chem.* 462:34–42.
2007. Ahlberg, E. and J. Asbjoernsson. 1993. Carbon paste electrodes in mineral processing: An electrochemical study of galena. *Hydrometallurgy* 34:171–185.
2008. Lara Castro, R.H., R. Briones, M. Monroy, M. Dossot, M. Mullet, and R. Cruz. 2010. Electrochemical and spectroscopic analysis of arsenopyrite (FeAsS) oxidation under calcareous soil conditions. *ECS Trans.* 28:105–116.
2009. Galus, Z. and R.N. Adams. 1962. Anodic oxidation studies of N,N-dimethylaniline. Part II: Stationary and rotated disk electrode studies. *J. Am. Chem. Soc.* 84:2061–2064.
2010. Galus, Z. and R.N. Adams. 1962. Anodic oxidation of triphenylmethane dyes. Part I. *J. Am. Chem. Soc.* 84:3207–3208.
2011. Galus, Z. and R.N. Adams. 1964. Anodic oxidation of triphenylmethane dyes. Part II. *J. Am. Chem. Soc.* 86:1666–1671.
2012. Hawley, D. and R.N. Adams. 1964. Anodic oxidation of p-methoxyphenol. *J. Electroanal. Chem.* 8:163–166.
2013. Leedy, D.W. and R.N. Adams. 1967. Reduction of N,N-dimethyl-*p*-nitrosoaniline. *J. Electroanal. Chem.* 14:119–122.
2014. Papouchado, L., G. Petrie, J.H. Sharp, and R.N. Adams. 1968. Anodic hydroxylation of aromatic compounds. *J. Am. Chem. Soc.* 90:5620–5622.
2015. Bacon, A.J. and R.N. Adams. 1968. Anodic oxidation of aromatic amines. Part III: Substituted anilines in aqueous media. *J. Am. Chem. Soc.* 90:6596–6597.
2016. Melicharek, M. and R.F. Nelson. 1970. Electrochemical oxidation of N,N-dimethyl-p-toluidine. *J. Electroanal. Chem.* 26:201–209.
2017. Grandi, G., R. Andreoli, and G.B. Gavioli. 1970. Effect of specific adsorption of some anions on the polarographic reduction. *J. Electroanal. Chem.* 27:177–183.
2018. Farsang, Gy., V. Vass, L. Ladanyi, and T.H.M. Saber. 1973. Electrochemical oxidation of 2,4-diaminodiphenylamine in aqueous solutions. *J. Electroanal. Chem.* 43:397–403.
2019. Sternson, L.A. 1974. Electrochemical determination of arylhydroxylamines in aqueous solutions and liver microsomal suspensions. *Anal. Chem.* 46:2228–2230.
2020. Sternson, L.A. 1974. Detection of arylhydroxylamines as intermediates in the metabolic reduction of nitro compounds. *Cell. Mol. Life Sci.* 31:268–270.
2021. Sternson, L.A. and J. Hes. 1975. Electrochemical method for the determination of aniline hydroxylation in liver. *Anal. Biochem.* 67:74–80.
2022. Brainina, Kh.Z. 1990. Personal communication.
2023. Gaillochet, P. 1975. Redox properties of oleums and concentrated or pure sulfuric acid. Electrochemical behavior of solids dispersed in carbon paste. Application to the determination of uranium in ores. PhD dissertation (in French), University of Paris 6.
2024. Bachiller, P.E., M.L.T. Garcia, M.D.V. Barbado, and P. Sanchez Batanero. 1997. Voltammetric method for distinguishing between electrochemical reactions involving solid or dissolved analytes within carbon paste electrodes. *J. Electroanal. Chem.* 424:217–219.
2025. Lecuire, J-M. 1975. Electrochemical reduction of iron oxides. Application to the measurement of nonstoichiometry. *J. Electroanal. Chem.* 66:195–205.
2026. Lecuire, J-M. and O. Evrard. 1977. Electrochemical reduction of iron oxides. Comparison of the electrochemical and crystallographic characteristics. *J. Electroanal. Chem.* 78:331–339.
2027. Mouhandess, M.T., F. Chassagneux, O. Vittori, A. Accary, and R.M. Reeves. 1984. Some theoretical aspects of electrodissolution of iron oxide α-Fe_2O_3 in carbon paste electrodes with acidic binder. *J. Electroanal. Chem.* 181:93–105.

2028. Ramirez, M.T., M.E. Palomar, I. González, and A. Rojas Hernandez. 1995. Carbon paste electrodes with electrolytic binder: Influence of the preparation method. *Electroanalysis* 7:184–188.
2029. Lamache, M. 1979. Carbon paste electrode with electrolytic binder and the incorporated poorly soluble electroactive compounds: Quantitative aspects (in French). *Electrochim. Acta* 24:79–84.
2030. Gruner, W., J. Kunath, L.N. Kalnishevskaja, J.V. Posokin, and K.Z. Brainina. 1993. Fundamentals and limitations for the application of the carbon paste electroactive electrode in the electroanalysis of solids. *Electroanalysis* 5:243–250.
2031. Roizenblat, E.M., A.A. Kosogov, E.Y. Sapozhnikova, and V.Y. Kolmanovich. 1988. Voltammetry of poorly soluble and non-conducting oxides in carbon paste electrode. *Electrokhimiya* 24:1352–1358.
2032. Moskzin, L.N., E.A. Koc, and M. Grigorieva. 1988. Electrode for determination of active Fe_2O_3 in magnesite samples. *Zh. Anal. Chim.* 43:1025–1028.
2033. Encinas, P.B., M.L.G. Tascon, M.D.B. Vazquez, and P. Sanchez Batanero. 1994. Electroanalytical study of copper and iron compounds in the solid state: Application to copper ferrite characterization. *J. Electroanal. Chem.* 367:99–108.
2034. Encinas Lorenzo, L.P., M.L. Tascon, M.D. Vazquez, C. de Francisco, and P. Sanchez Batanero. 1996. Electrochemical study of manganese and iron compounds at carbon paste electrodes with electrolytic binder. Application to the characterization of manganese ferrite. *J. Solid State Electrochem.* 1:232–240.
2035. Freour, R. 1985. The oxidation and reduction of γ-Fe_2O_3 and Fe_3O_4 in powdered form in carbon paste electrode with electroactive binder. *Electrochim. Acta* 30:795–798.
2036. Cepria, G., J.J. Cepria, and J. Ramajo. 2004. Fast and simple electroanalytical identification of iron oxides in geological samples without sample pretreatment. *Microchim. Acta* 144:139–145.
2037. Agasyan, P.K., A.I. Kamenev, and A.M. Troshenkov. 1987. Design of electroactive carbon paste electrodes for studying CuO, Ag, and Ag_2O. *Zavod. Lab.* 53:14–26.
2038. San Jose, M.T., A.M. Espinosa, M.L. Tascon, M.D. Vasquez, and P. Sanchez Batanero. 1991. Electrochemical behaviour of copper oxides at a carbon paste electrode. Application to the study of the superconductor YBaCuO. *Electrochim. Acta* 36:1209–1218.
2039. Zakharchuk, N.F. and N.F. Borisova. 1992. Peculiarities of the electrochemical behavior of a copper oxide graphite paste electrode in solutions of HCl. *Elektrokhimiya* 28:1787–1799.
2040. Zakharchuk, N., S. Meyer, B. Lange, and F. Scholz. 2000. A comparative study of lead oxide modified graphite paste electrodes and solid graphite electrodes with mechanically immobilized lead oxides. *Croat. Chem. Acta* 73:667–704.
2041. Centeno, B., M.L. Tascon, M.D. Vazquez, and P. Sanchez Batanero. 1991. Electrochemical study of PbO_2 at a carbon paste electrode with electrolytic binder. *Electrochim. Acta* 36:277–282.
2042. Liang, B., P. Li, and L. Gu. 1995. Carbon paste electrode with Mn-compounds. *Dianchi (Shuangyuekan)* 25:275–276.
2043. Fetisov, V.B., G.A. Kozhina, A.N. Ermakov, A.V. Fetisov, and E.G. Miroshnikova. 2007. Electrochemical dissolution of Mn_3O_4 in acid solutions. *J. Solid State Electrochem.* 11:1205–1210.
2044. Gonzalez, C., J.I. Gutierrez, J.R. Gonzalez Velasco, A. Cid, A. Arranz, and J.F. Arranz, 1996. Transformations of manganese oxides under different thermal conditions. *J. Therm. Anal.* 47: 93–102.
2045. Bezdicka, P., T. Grygar, B. Klapste, and J. Vondrak. MnOx/C composites as electrode materials. Part I. Synthesis, XRD and cyclic voltammetric investigation. 1999. *Electrochim. Acta* 45:913–920.
2046. El Sherief, A.E. 2000. A study of the electroleaching of manganese ore. *Hydrometallurgy* 55:311–326.
2047. Strappazzon, R. and L. Lamberts. 1994. Electrochemical study of the vanadium-oxygen system with a carbon paste electrode. *Analusis* 22:263–267.
2048. Elouadseri, M.M., O. Vittori, and B. Durand. 1986. The electrochemical reactivity of some vanadium compounds in an acidic medium of 1 M HCl. *Electrochim. Acta* 31:1335–1339.
2049. Grigoreva, M.F. and M.E. Dushina. 1992. Method of masking during analysis with the use of a carbon paste electroact. electrode. *Vestn. S. Peterb. Univ., Ser. 4: Fiz., Khim. (Ukraine)* 2:95–97 + 119.
2050. Grigoreva, M.F. and M.G. Ivanko. 1997. Simultaneous determination of vanadium(III,IV,V) in chloride melts. *Vestn. S. Peterb. Univ. Ser. 4: Fiz., Khim. (Ukraine)* 1:98–102.
2051. Barrado, E., R. Pardo, Y. Castrillejo, and M. Vega. 1997. Electrochemical behaviour of vanadium compounds at a carbon paste electrode. *J. Electroanal. Chem.* 427:35–42.
2052. Zakharchuk, N.F., N.A. Valisheva, N.R. Vitsina, T.P. Smirnova, K.P. Lel'kin, I.G. Yudelevich, and I.S. Illarionova. 1984. Phase analysis of oxide films on indium antimonide by voltammetry on a carbon-paste electrode. Part III: Electrochemical transformations of indium oxide (In_2O_3). *Izv. Sib. Otd. Akad. Nauk S.S.S.R Ser. Khim. Nauk* 2:83–88.

2053. Espinosa, A.M., M.T.S. Jose, M.L.T. Tascon, M.D. Vazquez, and P. Sanchez Batanero. 1991. Electrochemical behaviour of bismuth(V) and bismuth(III) compounds at a carbon paste electrode. Application to the study of the superconductor BiSrCaCuO. *Electrochim. Acta* 36:1561–1571.
2054. Vivier, V., A. Regis, G. Sagon, J.-Y. Nedelec, L.T. Yu, and C. Cachet-Vivier. 2001. Cyclic voltammetry study of bismuth oxide powder (Bi_2O_3) by means of a cavity micro-electrode coupled with Raman microspectrometry. *Electrochim. Acta* 46:907–914.
2055. Eguren, M., M.L. Tascon, M.D. Vazquez, and P. Sanchez Batanero. 1988. Study of the electrochemical behavior of solid tin dioxide using a carbon paste electrode. *Electrochim. Acta* 33:1009–1011.
2056. Kozhina, G.A., A.N. Ermakov, V.B. Fetisov, A.V. Fetisov, and K.Y. Shunyaev. 2009. Electrochemical dissolution of Co_3O_4 in acidic solutions. *Russ. J. Electrochem.* 45:1170–1175.
2057. Adekola, F.A., C. Colin, and D. Bauer. 1992. A study of the electrochemical behaviour of some platinum compounds at a carbon paste electrode with electrolytic binder. *Electrochim. Acta* 37:507–512.
2058. Hodos, M.Y., E.V. Bazarova, A.P. Palkin, K.Z. Brainina. 1984. The electrochemical behaviour of the $V_2O_5 + MoO_3$, system and the defect structure of $(Mo_xV1-x)_2O_5$ solid solution. *J. Electroanal. Chem.* 164:121–128.
2059. Grygar, T. 1996. The electrochemical dissolution of iron(III) and chromium(III) oxides and ferrites under conditions of abrasive stripping voltammetry. *J. Electroanal. Chem.* 405:117–125.
2060. Barrado, E., F. Prieto, M. Vega, R. Pardo, and J. Medina. 2000. Characterization and electrochemical behavior of a copper ferrite obtained by in situ precipitation from aqueous solutions. *Electroanalysis* 12:383–389.
2061. Barrado, E., F. Prieto, Y. Castrillejo, and J. Medina. 1999. Chemical and electrochemical characterization of lead ferrites produced in the purification of lead-bearing wastewater. *Electrochim. Acta* 45:1105–1111.
2062. Barrado, E., F. Prieto, J. Medina, and R. Pardo. 2001. Purification of cadmium waste water: Characterization and electrochemical behaviour of ferrites bearing cadmium(II). *Quim. Anal. (Barcelona)* 20:47–53.
2063. Barrado, E., F. Prieto, F.J. Garay, J. Medina, and M. Vega. 2002. Characterization of nickel-bearing ferrites obtained as by-products of hydrochemical wastewater purification processes. *Electrochim. Acta* 47:1959–1965.
2064. Mao, Q., S. Wu, and H. Zhang. 1995. A new kind of carbon paste electroactive electrode with paraffin as a binder. *Fenxi Huaxue* 23:648–651.
2065. Espinosa, A.M., M.L. Tascon, P. Encinas, M.D. Vázquez, and P. Sanchez Batanero. 1995. Electrochemical behaviour of copper, lead, and bismuth solid compounds and their ternary mixtures at solid state at a carbon paste electrode with electrolytic binder. Application to the study of the high transition temperature superconductor $Bi_{0.7}Pb_{0.3}SrCaCu_{1.8}Ox$. *Electrochim. Acta* 40:1623–1632.
2066. Domenech, A. and J. Alarcon. 2002. Electrochemistry of vanadium-doped tetragonal and monoclinic ZrO_2 attached to graphite/polyester composite electrodes. *J. Solid. State Electrochem.* 6:443–450.
2067. Ahlberg, E. and J. Ásbjörnsson. 1994. Carbon paste electrodes in mineral processing: An electrochemical study of sphalerite. *Hydrometallurgy* 36:19–37.
2068. Srinivasan, G.N. and S.V. Iyer. 2000. Cyclic voltammetric studies on sphalerite. *Bull. Electrochem.* 16:5–9.
2069. Nava, J.L., M.T. Oropeza, and I. Gonzalez. 2004. Oxidation of mineral species as a function of the anodic potential of zinc concentrate (sphalerite with various metal sulphides) in sulfuric acid. *J. Electrochem. Soc.* 151:B387–B393.
2070. Shi, S.-Y., Z.-H. Fang, and J.-R. Ni. 2006. Comparative study on the bioleaching of zinc sulphides. *Process Biochem.* 41:438–446.
2071. Shi, S.-H., Z.-H. Fang, and J.-R. Ni. 2006. Electrochemistry of marmatite-containing carbon paste electrode in the presence of bacterial strains. *Bioelectrochemistry* 68:113–118.
2072. Gonzalez, I.C., T.M.O. Guzmán, and I. González. 1999. Cyclic voltammetry applied to the characterisation of galena. *Hydrometallurgy* 53:133–144.
2073. Gonzalez, I.G., M.T.O. Guzmán, and I. González. 2000. An electrochemical study of galena concentrate in perchlorate medium at pH 2.0: The influence of chloride ions. *Electrochim. Acta* 45:2729–2741.
2074. Nava, J.L., M.T. Oropeza, and I. Gonzalez. 2002. Electrochemical characterisation of sulfur species formed during anodic dis-solution of galena concentrate in perchlorate medium. *Electrochim. Acta* 47:1513–1525.
2075. Nava, J.L. and I. González. 2005. The role of the carbon paste electrodes in the electrochemical study of metallic minerals. *Quim. Nova (Barcelona)* 28:901–909.
2076. Medvedeva, E.P., Z.B. Rozhdestvenskaya, and A.K. Kulikovskii. 1976. Kinetics of electro-oxidation of copper and chalcocite on a mineral-carbon paste electrode. *Izv. Akad. Nauk Kazakh. S.S.R. Ser. Khim.* 26:80–84.

2077. Lu, Z.Y., M.I. Jeffrey, and F. Lawson. 2000. An electrochemical study of the effect of chloride ions on the dissolution of chalcopyrite in acidic solutions. *Hydrometallurgy* 56:145–155.
2078. El Sherief, A.E. 2002. The influence of cathodic reduction, Fe^{2+} and Cu^{2+} ions on the electrochemical dissolution of chalcopyrite in acidic solution. *Miner. Eng.* 15:215–223.
2079. Nava, D. and I. Gonzalez. 2005. Electrochemical characterization of chemical species formed during the electrochemical treatment of chalcopyrite in sulfuric acid. *Electrochim. Acta* 51:5295–5303.
2080. Horta, D.G., H.A. Acciari, D. Bevilaqua, A.V. Benedetti, and O. Garcia Jr. 2009. The effect of chloride ions and A. ferrooxidans on the oxidative dissolution of the chalco-pyrite evaluated by electrochemical noise analysis (ENA). *Adv. Mater. Res. (Book Ser.)* 71–73:397–400.
2081. Almeida, C.M.V.B. and B.F. Giannetti. 2003. Electrochemical study of arsenopyrite weathering. *Phys. Chem. Chem. Phys.* 5:604–610.
2082. Liu, A. and J.-L. Wong. 2000. Chemical speciation of nickel in fly ash by phase separation and carbon paste electrode voltammetry. *J. Hazard. Mater.* 74:25–35.
2083. Wong, J.-L., J. Wenrui, and H. Yanan. 2001. Subspeciation of nickel disulfide by carbon paste electrode voltammetry catalyzed by Ni^{2+} ions. *Fresenius J. Anal. Chem.* 369:587–588.
2084. Holley, E.A., A.J. McQuillan, D. Craw, J.P. Kim, and S.G. Sander. 2007. Mercury mobilization by oxidative dissolution of cinnabar (α-HgS) and meta-cinnabar (β-HgS). *Chem. Geol.* 240:313–325.
2085. Luna Sanchez, R.M., I. Gonzalez, and G.T. Lapidus. 2002. An integrated approach to evaluate the leaching behaviour of silver from sulfide concentrates. *J. Appl. Electrochem.* 32:1157–1165.
2086. Laganovsky, A.V., Z.O. Kormosh, A.O. Fedorchuk, V.P. Sachanyuk, and O.V. Parasyuk. 2008. $AgCrTiS_4$: Synthesis, properties, and analytical application. *Metall. Mater. Trans. Ser. B* 39:155–159.
2087. Gobert, E. and V. Olivier. 1988. Electrochemical study of PbSe using the carbon paste electrode technique. *Electrochim. Acta* 33:245–250.
2088. Carbonnelle, P. and L. Lamberts. 1992. Electrochemical study of the copper-selenium system using carbon paste electrode. *Electrochim. Acta* 37:1321–1325.
2089. Espinosa, A.M., M.L. Tascon, M.D. Vazquez, and P. Sanchez Batanero. 1992. Electroanalytical study of selenium(+IV) at a carbon paste electrode with electrolytic binder and electroactive compound incorporated. *Electrochim. Acta* 37:1165.
2090. Perdicakis, M., N. Grosselin, and J. Bessiere. 1997. Voltammetric behaviour of single particles of silver chalcogenides in the micrometric range. Comparison with measurements at modified carbon paste electrodes. *Electrochim. Acta* 42:3351–3358.
2091. Lamache, M. and A. Adriamanana. 1986. The electrochemistry of chalcogenides containing molybdenum at carbon paste electrode with electrolytic binder. *Electrochim. Acta* 31:79–82.
2092. Matakova, R.N. and R.F. Yulmetova. 1984. Use of carbon paste electroactive electrode in the investigations of some metallic catalyst. *Zh. Anal. Khim.* 39:1200–1205.
2093. Tkach, A.V. and Y.B. Paderno. 1984. Electrochemical behavior of powdered rare earth borides incorporated in carbon paste electrodes. *Dop. Ak. Nauk Ukrain. R.S.R. Ser. B* 9:50–52.
2094. Gruner, W. 1986. Electrochemical characterization of solid materials in the volume range. Part III: Voltammetric studies on isolated iron carbide (Fe_3C). *Microchim. Acta* 1:301–309.
2095. Bennouna, A., B. Durand, and O. Vittori. 1987. The influence of the nature of the electrolyte on the electrochemical reactivity of barium chromate incorporated in a carbon paste electrode. *Electrochim. Acta* 32:713–721.
2096. Zakharchuk, N.F., G.F. Mustafina, T.P. Smirnova, and I.S. Illarionova. 1988. Voltammetry of a graphite paste electrode containing amorphous and crystalline arsenic. *Elektrokhimiya* 24:29–36.
2097. Adekola, F.A., C. Colin, and D. Bauer. 1993. Electrochemical study of some rhodium compounds at a carbon paste electrode with electrolytic binder. *Electrochim. Acta* 38:1331–1335.
2098. Marinovich, Y., S. Bailey, J. Avraamides, and S. Jayasekera. 1995. An electrochemical study of reduced ilmenite carbon paste electrodes. *J. Appl. Electrochem.* 25:823–832.
2099. Jaya, K.E., S. Berckman, V. Yegnaraman, and P.N. Mohandes. 2002. Electrochemical investigation of the rusting reaction of ilmenite using CV-studies *Hydrometallurgy* 65:217–225.
2100. Ibris, N., J.C.M. Rosca, A. Santana, and T. Visan. 2002. Comparative EIS study of a paste electrode containing zinc powder in neutral and near neutral solutions. *J. Solid State Electrochem.* 6:119–125.
2101. Carmo, D.R., R.M. da Silva, and N.R. Stradiotto. 2002. Electrochemical study of $Fe[Fe(CN)_5NO]$ in a graphite paste electrode. *Ecletica Quim.* 27:197–210.
2102. Covington, J.R. and R.J. Lacoste. 1965. Voltammetric determination of monoethylether of hydroquinone with carbon-ceresine wax paste electrode. *Anal. Chem.* 37:420–421.
2103. Dione, G., M. Guene, M. Toure, and M.M. Dieng. 1995. Influence of substitution on the electrochemical properties of anthraquinone in the solid state. *Bull. Chem. Soc. Ethiop.* 9:119–124.

References

2104. Gholabi, S.M., R. Davarkhah, and D. Nematollahi. 1997. Modified carbon paste electrode: An electroanalytical tool for estimation of thermodynamic parameters of water insoluble quinones. *Sci. Iran* 4:112–120.
2105. Martinez, R., M.T. Ramirez, and I. Gonzales. 1998. Voltammetric characterization of carbon paste electrodes with a nonconducting binder. Part I: Evidence of the influence of electroactive species dissolution into the paste on the voltammetric response. *Electroanalysis* 10:336–342.
2106. Perdicakis, M., N. Grosselin, and J. Bessière. 1999. Interaction of pyrite pulps with Ag^+ and Hg^{2+} ions. Electrochemical characterization of micrometric grains. *Anal. Chim. Acta* 385:467–485.
2107. Vlasa, A., S. Varvara, and L.M. Muresan. 2007. Electrochemical investigation of the influence of two thiadiazole derivatives on the patina of an archaeological bronze artefact using a CPE. *Stud. Univ. Babes-Bolyai Chem. (Roum.)* 52:63–71.
2108. Tascon, M.L., M.D. Vazquez, R. Pardo, P. Sanchez Batanero, and J. Iza. 1989. Determination of tin(IV) oxide with a mixed carbon paste electrode and using computerized instrumentation. *Electroanalysis* 1:363–366.
2109. Gruner, W., R. Stahlberg, K.Z. Brainina, N.L. Akselrod, and V.M. Kamyshov. 1990. Electrochemical determination of free CuO in YBaCuO superconductor materials. *Electroanalysis* 2:397–400.
2110. Adekola, F.A., M. Diaw, C. Colin, and D. Bauer. 1992. Electrochemical study of some palladium compounds at a carbon paste electrode—Application to the determination of palladium in oxidation automotive catalysts. *Electrochim. Acta* 37:2491–2495.
2111. Osipova, E.A., G.V. Prokhorova, O.I. Gurentsova, and N.E. Kopytova. 1994. Determination of germanium in semiconductor-materials by the adsorption inversion voltamperometric method. *Ind. Labor. (Russ.)* 60:82–83.
2112. Cepria, G., N. Alexa, E. Cordos, and R.J. Castillo. 2005. Electrochemical screening procedure for arsenic contaminated soils. *Talanta* 66:875–881.
2113. Bennouna, A., A. Kheribech, J.P. Scharff, F. Chassagneux, B. Durand, and O. Vittori. 1997. Evaluation of preconcentration response of a carbon paste electrode with conducting binder. Application to lead. *Electrochim. Acta* 42:2659–2665.
2114. Clark, L.C., Jr., G. Misrahy, and R.P. Fox. 1958. Chronically implanted polarographic electrodes. *J. Appl. Physiol.* 13:85–91.
2115. Clark, L.C., Jr. and C. Lyons. 1958. Studies of a glassy carbon electrode for brain polarography with observations on the effect of carbonic anhydrase inhibition. *Ala. J. Med. Sci.* 2:353–359.
2116. Adams, R.N. 1990. In vivo electrochemical measurements in the CNS. *Prog. Neurobiol.* 35:297–311.
2117. Mitchell, K. and R.N. Adams. 1993. Comparison of the effects of voltage-sensitive calcium channel antagonism on electrically stimulated release of dopamine and norepinephrine monitored in-vivo. *Brain. Res.* 604:349–353.
2118. Adams, R.N., A.A Boulton, and G.B. Baker (Eds). 1996. *Voltammetric Methods in Brain Systems.* New York: Humana Press.
2119. Al Mulla, I., J.P. Lowry, P.A. Serra, and R.D. O'Neill. 2009. Development of a voltammetric technique for monitoring brain dopamine metabolism: Compensation for interference caused by DOPAC electrogenerated during homovanillic acid detection. *Analyst (U.K.)* 134:893–898.
2120. Bhaskaran, D. and C.R. Freed. 1988. Changes in arterial blood pressure lead to baroreceptor-mediated changes in norepinephrine and 5-hydroxyindoleacetic acid in rat nucleus tractus solitarius. *J. Pharmacol. Exp. Ther.* 245:356–363.
2121. Broderick, P.A. 1988. Distinguishing in vitro electrochemical signatures for norepinephrine and dopamine. *Neurosci. Lett.* 95:275–280.
2122. Joseph, M.H. and H. Hodges. 1990. Lever pressing for food reward and changes in dopamine turnover and uric acid in rat caudate and nucleus accumbens studied chronically by in vivo voltammetry. *J. Neurosci. Methods* 34:143–149.
2123. Lin Z., W. Qiao, and M. Wu. 1992. The preparation of a spinach tissue-based carbon paste microelectrode and its performance in pharmacokinetic experiments in vivo. *Anal. Lett.* 25:1171–1181.
2124. Pantano, P. and W.G. Kuhr. 1995. Enzyme-modified microelectrodes for in vivo neurochemical measurements. *Electroanalysis* 7:405–416.
2125. Sokolovskaya, L.G., L.V. Sigolaeva, A.V. Eremenko, I.V. Gachok, G.F. Makhaeva, N.N. Strakhova, R.J. Richardson, and I.N. Kurochkin. 2005. Improved electrochemical analysis of neuropathy target esterase activity by a tyrosinase carbon paste electrode modified by 1-methoxyphenazine methosulfate. *Biotechnol. Lett.* 27:1211–1218.
2126. O'Neill, R.D., M. Fillenz, W.J. Albery, and N.J. Goddard. 1983. The monitoring of ascorbate and monoamine transmitter metabolites in the striatum of unanaesthetised rats using microprocessor-based voltammetry. *Neuroscience (U.K.)* 9:87–93.

2127. Blaha, C.D. and R.F. Lane. 1984. Direct in vivo electrochemical monitoring of dopamine release in response to neuroleptic drugs. *Eur. J. Pharmacol.* 98:113–117.
2128. Marrocco, R.T., R.F. Lane, J.W. McClurkin, C.D. Blaha, and M.F. Alkire. 1987. Release of cortical catecholamines by visual stimulation requires activity in thalamocortical afferents of monkey and cat. *J. Neurosci.* 7:2756–2767.
2129. Blaha, C.D. and A.G. Phillips. 1990. Application of in vivo electrochemistry to the measurement of changes in dopamine release during intracranial self-stimulation. *J. Neurosci. Methods* 34:125–133.
2130. Lane, R.F. and C.D. Blaha. 1990. Detection of catecholamines in brain tissue: Surface-modified electrodes enabling in vivo investigations of dopamine function. *Langmuir* 6:56–65.
2131. Blaha, C.D. and M.E. Jung. 1991. Electrochemical evaluation of stearate-modified graphite paste electrodes: Selective detection of dopamine is maintained after exposure to brain tissue. *J. Electroanal. Chem.* 310:317–334.
2132. Coury, A., C.D. Blaha, L.J. Atkinson, and A.G. Phillips. 1992. Cocaine-induced changes in extracellular levels of striatal dopamine measured concurrently by microdialysis with HPLC-EC and chronoamperometry. *Ann. N.Y. Acad. Sci.* 654:424–427.
2133. Blaha, C.D. 1996. Evaluation of stearate-graphite paste electrodes for chronic measurement of extracellular dopamine concentrations in the mammalian brain. *Pharmacol. Biochem. Behav.* 55:351–364.
2134. Blaha, C.D., D. Liu, and A.G. Phillips. 1996. Improved electrochemical properties of stearate-graphite paste electrodes after albumin and phospholipid treatments. *Biosens. Bioelectron.* 11:63–79.
2135. O'Neill, R.D., M. Fillenz, L. Sundstrom, and J.N. Rawlins. 1984. Voltammetrically monitored brain ascorbate as an index of excitatory amino acid release in the unrestrained rat. *Neurosci. Lett.* 52:227–233.
2136. O'Neill, R.D., M. Fillenz, R.A. Gruenewald, M.R. Bloomfield, W.J. Albery, C.M. Jamieson, J.H. Williams, and J.A. Gray. 1984. Voltammetric carbon paste electrodes monitor uric acid and not 5-HIAA at the 5-hydroxyindole potential in the rat brain. *Neurosci. Lett.* 45:39–46.
2137. O'Neill, R.D. and M. Fillenz. 1985. Simultaneous monitoring of dopamine release in rat frontal cortex, nucleus accumbens and striatum. Effect of drugs, circadian changes and correlations with motor activity. *Neuroscience (U.K.)* 16:49–55.
2138. O'Neill, R.D. and M. Fillenz. 1986. Microcomputer-controlled voltammetry in the analysis of transmitter release in rat brain. *Ann. N.Y. Acad. Sci.* 473:337–348.
2139. Brose, N., R.D. O'Neill, M.G. Boutelle, and M. Fillenz. 1989. The effects of anxiolytic and anxiogenic benzodiazepine receptor ligands on motor activity and levels of ascorbic acid in the nucleus accumbens and striatum of the rat. *Neuropharmacology* 28:509–514.
2140. O'Neill, R.D. 1990. Uric acid levels and dopamine transmission in rat striatum: Diurnal changes and effects of drugs. *Brain Res.* 507:267–272.
2141. Duff, A. and R.D. O'Neill. 1994. Effect of probe size on the concentration of brain extracellular uric acid monitored with carbon paste electrodes. *J. Neurochem.* 62:1496–1502.
2142. Lowry, J.P., M.G. Boutelle, R.D. O'Neill, and M. Fillenz. 1996. Characterization of carbon paste electrodes in vitro for simultaneous amperometric measurement of changes in oxygen and ascorbic acid concentrations in vivo. *Analyst (U.K.)* 121:761–766.
2143. Kane, D.A. and R.D. O'Neill. 1998. Major differences in the behavior of carbon paste and carbon fiber electrodes in a protein-lipid matrix: Implications for voltammetry in vivo. *Analyst (U.K.)* 123:2899–2903.
2144. Lowry, J.P. and M.Fillenz. 2001. Real-time monitoring of brain energy metabolism in vivo using microelectrochemical sensors: The effects of anesthesia. *Bioelectrochemistry* 54:39–47.
2145. Dixon, B.M., J.P. Lowry, and R.D. O'Neill. 2002. Characterization in vitro and in vivo of the oxygen dependence of an enzyme/polymer biosensor for monitoring brain glucose. *J. Neurosci. Methods* 119:135–142.
2146. Kennett, G.A. and M.H. Joseph. 1982. Does in vivo voltammetry in the hippocampus measure 5-HT release? *Brain Res.* 236:305–316.
2147. Echizen, H. and C.R. Freed. 1984. Measurement of serotonin turnover rate in rat dorsal raphe nucleus by in vivo electrochemistry. *J. Neurochem.* 42:1483–1486.
2148. Echizen, H. and C.R. Freed. 1986. Factors affecting in vivo electrochemistry: Electrode-tissue interaction and the ascorbate amplification effect. *Life Sci.* 39:77–89.
2149. Mueller, K. and C. Haskett. 1987. Effects of haloperidol on amphetamine-induced increases in ascorbic acid and uric acid as determined by voltammetry in vivo. *Pharmacol., Biochem. Behav.* 27:231–234.
2150. Mueller, K. 1989. Repeated administration of high doses of amphetamine increases release of ascorbic acid in caudate but not nucleus accumbens. *Brain Res.* 494:30–35.
2151. Joseph, M.H. and A.M.J. Young. 1991. Pharmacological evidence, using in vivo dialysis, that substances additional to ascorbic acid, uric acid and homovanillic acid contribute to the voltammetric signals obtained in unrestrained rats from chronically implanted carbon paste electrodes. *J. Neurosci. Methods* 36:209–218.

References

2152. Shi, G.-Y., K. Yamamoto, T.-.S. Zhou, F. Xu, T. Kato, J.-Y. Jin, and L.-T. Jin. 2003. On-line biosensors for simultaneous determination of glucose, choline, and glutamate integrated with a microseparation system. *Electrophoresis* 24:3266–3272.
2153. Congestri, F., F. Formenti, V. Sonntag, G. Hdou, and F. Crespi. 2008. Selective D3 receptor antagonist SB-277011-A potentiates the effect of cocaine on extracellular dopamine in the nucleus accumbens: A dual core-shell voltammetry in anesthetized rats. *Sensors* 8:6936–6951.
2154. Bolger F.B., S.B. McHugh, R. Bennett, J. Li, K. Ishiwari, J. Francois, M.W. Conway et al. 2011. Characterisation of carbon paste electrodes for real-time amperometric monitoring of brain tissue oxygen. *J. Neurosci. Methods* 195:135–142.
2155. Baker, G.A., Jr. and J.E. Brolley. 1983. Deconvolution of noisy experimental data. *J. Comput. Phys.* 51:227–240.
2156. Pizeta, I. 1994. Deconvolution of non-resolved voltammetric signals. *Anal. Chim. Acta* 285:95–102.
2157. Goto, M., K. Ikenoya, M. Kajihara, and D. Ishii. 1978. Application of semidifferential electroanalysis to anodic stripping voltammetry. *Anal. Chim. Acta* 101:131–138.
2158. Bustin D., J. Mocak, and J. Garaj. 1987. The utilization of fractional differentiation and integration in electrochemistry (in Slovak). *Chem. Listy* 81:1009–1033.
2159. Mocak, J., I. Janiga, M. Rievaj, and D. Bustin. 2007. The use of fractional differentiation or integration for signal improvement. *Meas. Sci. Rev.* 7:39–42.
2160. Cobos Murcia, J.A., L. Galicia, A. Rojas Hernandez, M.T. Ramirez Silva, R. Alvarez Bustamante, M. Romero Romo, G. Rosquete Pina, and M. Palomar Pardave. 2005. Electrochemical polymerisation of 5-amino-1,10-phenanthroline at different substrates: Experimental and theoretical study. *Polymer* 46:9053–9063.
2161. Parra, V., A.A. Arrieta, J.A. Fernández Escudero, M. Iniguez, J.A. de Saja, and M.L. Rodriguez Mendez. 2006. Monitoring of the ageing of red wines in oak barrels by an hybrid electronic tongue. *Anal. Chim. Acta* 563:229–237.
2162. Rodriguez Mendez, M.L., V. Parra, C. Apetrei, S. Villanuevam, M. Gay, N. Prieto, J. Martinez, and J.A. de Saja. 2008. Electronic tongue based on voltammetric electrodes modified with materials showing complementary electroactive properties. Characterization and applications. *Microchim. Acta* 163:23–31.
2163. Apetrei, C., F. Gutieréz, M.L. Rodriguez Mendez, and J.A. de Saja. 2007. Novel method based on carbon paste electrodes for the evaluation of bitterness in extra virgin olive oils. *Sens. Actuators B* 121:567–575.
2164. Kalvoda, R. 1999. Electronic noses and tongues. In *Modern Electroanalytical Methods—XVIII*. Book of Abstracts (in Czech), p. 8. Usti nad Labem (Czech Republic): SES Logis.
2165. Wang, X.-L., H.-Y. Lin, Y.-F. Bi, B.-K. Chen, and G.-C. Liu. 2008. An unprecedented extended architecture constructed from a 2-D interpenetrating cationic coordination framework templated by SiW12O40(4-) anion. *J. Solid State Chem.* 181:556–561.
2166. Cao, X.-G., L.-W. He, B.-Z. Lin, Z.-J. Chen, and P.-D. Liu. 2009. Synthesis and characterization of a polyoxotungstate-supported metal compound [{Cu(enMe)(2) (H$_2$O)}{Cu(enMe)(2)}(3)P2W18O62] × n H(2)O. *Inorg. Chim. Acta* 362:2505–2509.
2167. Wang, K.-F., F.-F. Jian, and R.-R. Zhuang. 2009. A new ionic liquid comprising lanthanum(III) bulk-modified carbon paste electrode: Preparation, electrochemistry and electrocatalysis. *Dalton Trans.* 23:4532–4537.
2168. Wolter, K.D. and J.T. Stock. 1978. Preparative electrolyses at graphite paste anodes. *J. Electrochem. Soc.* 125:531–533.
2169. Fitt, A.D. and P.D. Howell. 1998. The manufacture of continuous smelting electrodes from carbon-paste briquettes. *J. Eng. Math.* 33:353–376.
2170. Pettichini, S., H. Boggetti, B.A. Lopez de Mishima, H.T. Mishima, J. Rodriguez, and E. Pastor. 2003. Electrochemical reduction of carbon dioxide on copper alloy. *J. Argen. Chem. Soc.* 91:107–112.
2171. Roa Morales, G., L. Galicia, A. Rojas Hernandez, and M.T. Ramirez Silva. 2005. Electrochemical study on the selective formation of [Pb(cyclodextrin)$^{(2+)}$] (surface) inclusion complexes at the carbon paste electrode/ClO$_4$/1 M interphase. *Electrochim. Acta* 50:1925–1930.
2172. Oh, T.-S., J.-H. Lee, S.-E. Lee, K.-W. Min, S.-K. Kang, J.-B. Yoo, C.-Y. Park, and J.-M. Kim. 2005. A field-emission display with an asymmetric electrostatic-quadrupole lens structure. *Jpn. J. Appl. Phys.* 44:8692–8697.
2173. Cepria, G., L. Irigoyen, and J.R. Castillo. 2006. A microscale procedure to test the metal sorption properties of biomass sorbents: A time and reagents saving alternative to conventional methods. *Microchim. Acta* 154:287–295.
2174. Krizkova, S., P. Ryant, O. Krystofova, V. Adam, V. Galiova, M. Beklova, P. Babula et al. 2008. Multi-instrumental analysis of tissues of sunflower plants treated with silver(I) ions: Plants as bioindicators of environmental pollution. *Sensors* 8:445–463.

2175. Khotseng, L.E., S. Feng, G. Vaivars, and V. Linkov. 2008. Nickel/carbon nano-structured electrodes synthesized using template method. *Integr. Ferroelectr.* 103:72–79.
2176. Liang, Y., P. He, Y.-J. Ma, Y. Zhou, C.-H. Pei, and X.-B. Li. 2009. A novel bacterial cellulose-based carbon paste electrode and its polyoxometalate-modified properties. *Electrochem. Commun.* 11:1018–1021.
2177. Lee, C.-P., P.-Y. Chen, R. Vittal, and K.-C. Ho. 2010. Iodine-free high efficient quasi solid-state dye-sensitized solar cell containing ionic liquid and polyaniline-loaded carbon black. *J. Mater. Chem.* 20:2356–2361.
2178. Zuman, P. 1999. The use of polarography in the initial stages of investigations of mechanisms of organic electrode processes in aqueous solutions. *Sci. Pap. Univ. Pardubice Ser. A* 5:5–40.
2179. Barek, J., J. Costa Moreira, and J. Zima. 2005. Modern electrochemical methods for monitoring of chemical carcinogens (a review). *Sensors* 5:148–158.
2180. Barek, J., J. Fischer, T. Navratil, K. Peckova, B. Yosypchuk, and J. Zima. 2007. Nontraditional electrode materials in environmental analysis of biologically active organic compounds (a review). *Electroanalysis* 19:2003–2014.
2181. Badihi Mossberg, M., V. Buchner, and J. Rishpon. 2007. Electrochemical biosensors for pollutants in the environment (a review). *Electroanalysis* 19:2015–2028.
2182. Švancara, I. and J. Zima. 2011. Possibilities and limitations of carbon paste electrodes in organic electrochemistry (a review). *Curr. Org. Chem.* 15:3043–3058.
2183. Uslu, B. and S.A. Ozkan. 2007. Electroanalytical application of carbon based electrodes to the pharmaceuticals (mini-review). *Anal. Lett.* 40:817–853.
2184. Fojta, M. 2002. Electrochemical sensors for DNA interactions and damage (a review). *Electroanalysis* 14:1449–1463.
2185. Girousi, S. and Z. Stanic. 2011. The last decade of carbon paste electrodes in DNA electrochemistry. *Curr. Anal. Chem.* 7:80–100.
2186. Skladal, P. 1997. Advances in electrochemical immunosensors (a review). *Electroanalysis* 9:737–745.
2187. Liu, G.-D. and Y.-H. Lin. 2007. Nanomaterial labels in electrochemical immunosensors and immunoassays (a review; dedicated to Prof. Joseph Wang on the occasion of his 60th birthday). *Talanta* 74:308–317.
2188. Monien, H. 1968. Inverse voltammetric determination of small amounts of gold by peak potential measurement. *Fresenius Z. Anal. Chem.* 237:409–419.
2189. Vasilyeva, L.N. and T.A. Koroleva. 1971. Joint determination of elements by method 2205 of inversion voltamperometry on a graphite (paste) electrode. *J. Anal. Chem. (Russ.)*/transl. of *Zh. Anal. Chem.* 237:1682–1685.
2190. Vasilyeva, L.N., E.N. Vinogradova, T.A. Koroleva, and Z.L. Justus. 1975. Determination of gold on carbon paste electrode containing nujol oil in samples of metallic copper. *Zavod. Lab.* 41:1199–2003.
2191. Zarinskii, V.A., L.S. Chulkina, and N.N. Baranova. 1977. Inverse voltammetric determination of gold(III) in the presence of iron(III). *J. Anal. Chem. (Russ.)*/transl. of *Zh. Anal. Khim.* 32:530–534.
2192. Alexander, R., B. Kinsella, and A. Middleton. 1978. Cathodic voltammetric determination of gold. *J. Electroanal. Chem.* 93:19–27.
2193. Peng, T.-Z., Q. Shi, and R. Lu. 1990. Determination of gold at an extractive CPE modified with N235. *Yingyong Kexue Xuebao* 8:366–370.
2194. Wang, J., B.-M. Tian, and G.D. Rayson. 1992. Bioaccumulation and voltammetry of gold at flower-biomass modified electrodes. *Talanta* 39:1637–1642.
2195. Peng, T.-Z., G.-S. Wang, and P.-L. Zhu. 1992. Electrostatic adsorption of gold(III) on carbon paste electrode and its voltammetric determination. *Yingyong Huaxue* 9:76–78.
2196. Wang, G.-S., T.-Z. Peng, P. Wang, P.-L. Zhu, and L.-M. Qu, 1994. Determination of gold with 18-crown-6 modified carbon paste electrode. *Fenxi Shiyanshi* 13:54–56.
2197. Vytřas, K., I. Švancara, F. Renger, M. Srey, R. Vankova, and M. Hvizdalova. 1993. Voltammetric and potentiometric determination of gold in gold-plated electro-technical components. *Collect. Czech. Chem. Commun.* 58:2039–2046.
2198. Hu, R.-Z., H. Xu, J.-K. Fu, W.-Y. Hu, and Y.-Z. Chen. 1998. Electrochemical determination of trace amounts of gold(III) by cathodic stripping voltammetry using a carbon paste modified with bacteria. *Dian Huaxue* 4:323–327.
2199. Hu, R.-Z., W. Zhang, J.-K. Fu, W. Zhang, and Y.-Y. Liu. 1999. Determination of trace amounts of gold(III) by cathodic stripping voltammetry using a bacteria-modified carbon paste electrode. *Anal. Commun.* 36:147–148.
2200. Chulkina, L.S. and S.I. Sinyakova. 1969. Inverse voltammetric determination of nanogram amounts of silver using a carbon paste electrode. *J. Anal. Chem (Russ.)*/transl. of *Zh. Anal. Khim.* 24:247–250.

References

2201. Rubel, S., E. Stryjewska, and J. Golimowski. 1979. Voltammetric determination of trace amounts of silver in metallic copper. *Chem. Anal.* 24:247–254.
2202. Pei, J.-H., Q. Jin, and J.-Y. Zhong. 1991. Potentiometric determination of trace silver based on the use of a carbon paste electrode. *Talanta* 38:1185–1189.
2203. Nikitin, A.A., L.P. Sherbakova, and A.G. Zhulanova. 1991. Determination of trace silver in mercury electrolytes by stripping voltammetry. *Izv. Akad. Nauk Kaz. S.S.R. Sek. Khim.* 1:38–41.
2204. Huang, S.-S., B.-F. Li, H.-G. Lin, S. Liao, Z.-Q. Zhang, and R.-Q. Yu. 1993. Determination of trace Ag(I) on carbon paste electrode modified with 2,9-dichlor-1,10-phenanthroline. *Fenxi Huaxue* 21:1423–1427.
2205. Kim, S.-H., M.-S. Won, and Y.-B. Shim. 1994. Determination of Ag(I) ion with chemically modified carbon paste electrode containing cinchonidine. *J. Korean Chem. Soc.* 38:734–740.
2206. Rivas, G.A. and P.I. Ortiz. 1994. Electrochemical determination of Ag(I) and Cu(II) using activated carbon paste electrodes. *Anal. Lett.* 27:751–778.
2207. Labar, C. and L. Lamberts. 1997. Anodic stripping voltammetry with carbon paste electrodes for rapid Ag(I) and Cu(II) determinations. *Talanta* 44:733–742.
2208. Schildkraut, D.E., P.-T. Dao, J.P. Twist, A.T. Davis, and K.A. Robillard. 1998. Determination of silver ions at submicrogram-per-liter levels using anodic square-wave stripping voltammetry. *Environ. Toxic. Chem.* 17:642–649.
2209. Ha, K.-S., J.-H. Kim, Y.-S. Ha, S.-S. Lee, and M.-L. Seo. 2001. Anodic stripping voltammetric determination of silver(I) at a carbon paste electrode modified with S_2O_2-donor podand. *Anal. Lett.* 34:675–686.
2210. Zhang, S.-B., X.-J. Zhang, and X.-Q. Lin. 2002. An ethylenediaminetetraacetic acid modified carbon paste electrode for the determination of silver ion. *Fenxi Huaxue* 30:745–747.
2211. Yang, C.-H., W.-S. Huang, and S.-H. Zhang. 2003. Highly sensitive electrochemical determination of trace Pb^{2+} and Ag^+ in the presence of cetyltrimethylammonium bromide. *Fenxi Huaxue* 31:794–798.
2212. Hocevar, S.B. and I. Švancara. 2006. Unpublished results.
2213. Mashhadizadeh, M.H.A. Mostafavi, H.A. Abadi, and I. Sheikh Shoai. 2006. New Schiff base modified carbon paste and coated wire PVC membrane for silver ion. *Sens. Actuators, B* 113:930–936.
2214. Emmot, P. 1965. Application of anodic stripping voltammetry to the determination of mercury in lithium sulfate. *Talanta* 12:651–656.
2215. Ulrich, L. and P. Ruegsegger. 1975. Use of the carbon paste electrode for the inverse voltammetric determination of small quantities of mercury. *Fresenius Z. Anal. Chem.* 277:349–353.
2216. Jeong, E.-D., M.-S. Won, and Y.-B. Shim. 1991. Determination of mercury(II) ion at chemically modified carbon paste electrode containing l-sparteine. *J. Korean Chem. Soc.* 35:545–552.
2217. Švancara, I., K. Vytřas, C. Hua, and M.R. Smyth. 1992. Voltammetric determination of mercury(II) at a carbon paste electrode in aqueous solutions containing tetraphenyl-borate ion. *Talanta* 39:391–396.
2218. Huang, W., C. Yang, and S. Zhang. 2002. Anodic stripping voltammetric determination of mercury by use of a sodium montmorillonite-modified carbon-paste electrode. *Anal. Bioanal. Chem.* 274:998–1001.
2219. Kong, Y.-T., G.-H. Choi, M.-S. Won, and Y.-B. Shim. 2002. Determination of $Hg_2^{(2+)}$ ions using the specific reaction with a picolinic acid N-oxide modified electrode. *Chin. Chem. Lett.* 31:54–55.
2220. Filho, N.L.D., D.R. do Carmo, and A.H. Rosa. 2006. Selective sorption of mercury(II) from aqueous solution with an organically modified clay and its electroanalytical application. *Sep. Sci. Technol.* 41:733–746.
2221. Gismera, M.J., J.R. Procopio, and M.T. Sevilla. 2007. Characterization of mercury-humic acids interaction by potentiometric titration with a modified carbon paste mercury sensor. *Electroanalysis* 19:1055–1061.
2222. Zejli, H., J. de Cisneros, I.N. Rodriguez, H. Elbouhouti, M. Choukairi, D. Bouchta, and K.R. Temsamani. 2007. Electrochemical analysis of mercury using a *Cryptofix* carbon-paste electrode. *Anal. Lett.* 40:2788–2798.
2223. Cesarino, I., G. Marino, J.D. Matos, and E.T.G. Cavalheiro. 2008. Evaluation of a carbon paste electrode modified with organofunctionalised SBA-15 nanostructured silica in the simultaneous determination of lead, copper and mercury ions. *Talanta* 15:15–21.
2224. Shawish, A. and M. Hazem. 2009. A mercury(II) selective sensor based on N,N′-bis(salicyl-aldehyde)-phenylenediamine as neutral carrier for potentiometric analysis in water samples. *J. Hazard. Mat.* 167:602–608.
2225. Monien, H., U. Gerlach, and P. Jacob. 1981. Inverse voltammetry of some copper chelates in the carbon paste electrode. Determination of copper in drinking water by oxidation of copper dithiooxamide. *Fresenius Z. Anal. Chem.* 306:136–143.
2226. Torresday, J.G., D. Darnall, and J. Wang. 1988. Bioaccumulation and measurement of copper at an alga-modified carbon paste electrode. *Anal. Chem.* 60:72–76.

2227. Yarnitzky, M.N. and M. Ariel. 1990. Possibilities of determination of copper(II) in biological fluids using FIA-wall jet detector on a carbon paste. *Anal. Chim. Acta* 228:117–122.

2228. Zhang, G.-R. and C.-G. Fu. 1991 Adsorptive voltammetric determination of copper with a benzoin oxime graphite paste electrode. *Talanta* 38:1481–1485.

2229. Qiao, W.-J, Z.-H. Lin, and M. Wu. 1992. The chelating-resin modified electrodes and its application in the detection of copper in human samples. *J Tongji Med. Univ. (China)* 12:80–84.

2230. Peng, T.-Z., Z. Tang, and G.-S. Wang. 1993. Linear scan stripping voltammetry of copper(II) at Cupron-modified carbon paste electrode. *Fenxi Huaxue* 21:221–223.

2231. Bae, Z.-U., H.-S. Jun, and H.-Y. Chang. 1993. Voltammetric determination of Cu(II) at chemically modified carbon paste electrode containing 1-(2-pyridylazo-2-naftol)-PAN. *J. Korean Chem. Soc.* 37:723–727.

2232. Wang, G.-S., T.-Z. Peng, B.-E. Shen, L.-M. Qu, and P.-L. Zhu. 1993. Determination of copper with hectorite carbon paste electrode. *Fenxi Ceshi Xuebao* 12:67–70.

2233. Qiao, W.-J. and H. Ding. 1993. The catalytic stripping voltammetry of trace copper(II) at a chemically modified electrode based on oxine containing carbon paste. *J. Tongji Med. Univ. (China)* 13:90–193.

2234. Wang, G.-S., T.-Z. Peng, X.-H. Yang, and T.-H. Ding. 1994. Catalytická stripping voltammetry of Cu(II) traces based on oxin-modified carbon paste. *Metall. Anal. (China)/Yejin Fenxi* 12:8–12.

2235. Hu, X.-Y. and Z.-Z. Leng. 1995. Highly selective and super-Nernstian potentiometry for determination of Cu^{2+} using carbon paste electrode. *Anal. Lett.* 28:979–989.

2236. Bae, Z.-U., Y.-L. Kim, and H.-Y Chang. 1995. Voltammetric determination of copper(II) at chemically modified carbon paste electrodes containing algae. *Anal. Sci. Technol.* 8:611–615.

2237. Chang, C.-H. and C.-Y. Li. 1997. Preconcentration and determination of Cu(II) at a chemically modified electrode containing salicylaldehyde thiosemicarbazone. *Kexue Xuebao* 44:231–236.

2238. Tymecki, L., M. Jakubovska, S. Achamatowicz, R. Koncki, and S. Glab, Potentiometric thick-film graphite electrodes with improved response to copper ions. *Anal. Lett.* 34:71–78.

2239. Yang, S., X.-Q. Lu, Y.-H. Xue, X.-Q. Feng, and X.-F. Wang. 2003. 4-Methoxy-2,5-bis(3,5-dimethylpyrazoyl)-1,3,5-triazine modified carbon paste electrode for trace Cu(II) determination by differential pulse voltammetry. *Rare Met.* 22:250–253.

2240. Vytřas, K., L. Baldrianova, E. Tesarova, A. Bobrowski, and I. Švancara. 2005. Comments to stripping voltammetric determination of copper(II) at bismuth-modified carbon substrate electrodes. In *Sensing in Electroanalysis*, Eds. K. Vytřas and K. Kalcher, pp. 49–58. Pardubice, Czech Republic: University of Pardubice.

2241. Abu-Shawish, H.M. and S.M. Saadeh. 2007. A new chemically modified carbon paste electrode for determination of copper based on N,N'-disalicylidenehexamethylene-diaminate copper(II) complex. *Sens. Lett.* 5:565–571.

2242. Janegitz, B.C., L.H. Marcolino, and O.F. Filho. 2007. Anodic stripping voltammetric determination of copper (II) in wastewaters using a carbon paste electrode modified with chitosan. *Quim. Nova (Barcelona)* 30:1673–1676.

2243. Taher, M.A., M. Esfandyarpour, S. Abbasi, and A. Mohadesi. 2008. Indirect determination of trace copper(II) by adsorptive stripping voltammetry with Zincon at a carbon paste electrode. *Electroanalysis* 20:374–278.

2244. Gismera, M.J, M.T. Sevilla, and J.R. Procopio. 2008. Flow and batch systems for copper(II) potentiometric sensing. *Talanta* 14:190–197.

2245. Alpat, S.K., S. Alpat, B. Kutlu, O. Ozbayrak, and H.B. Buyukisik. 2008. Development of biosorption-based algal biosensor for Cu(II) using *Tetraselmis chuii*. *Sens. Actuators B* 128:273–278.

2246. Rohani, T. and M.A. Taher. 2008. A new method for application of the water-soluble dye SPADNS inside a carbon paste electrode for determination of trace amounts of copper. *J. AOAC Int.* 91:1478–1482.

2247. Lubert, K.H. and L. Beyer. 2008. Carbon paste electrode modified with the copper(II) complex of N-benzoyl-N',N'-Di-N-butyl-thiourea. Voltammetric behavior and response to copper(II). *Solvent Extr. Ion Exch.* 26:321–331.

2248. Stephanie, G.R., M. Etienne, Y. Rousselin, F. Denat, B. Lebeau, and A. Walcarius. 2009. Cyclam-functionalized silica-modified electrodes for selective determination of Cu(II). *Electroanalysis* 21:280–289.

2249. Csoka, B. and Z. Mekhalif. 2009. Carbon paste-based and dual function ion-selective microelectrodes for scanning electrochemical microscopic measurements. *Electrochim. Acta* 54:3225–3232.

2250. Yuce, M., H. Nazir, and G. Donmez. 2010. A voltammetric microbial biosensor modified with *Rhodotorula mucilaginosa* sp. for the determination of Cu(II). *Bioelectrochemistry* 79:66–70.

2251. Kamenev, A.I., M.I. Lunev, P.K. Agasyan, and Z.I. Lisichkina. 1977. Determination of lead in steel and cast iron by inverse voltammetry. *Zh. Anal. Khim (Russ.)* 32:1955–1960.

References

2252. Kopanica, M. and V. Stara. 1991. Determination of amalgam-forming metals by anodic stripping voltammetry in solutions with dissolved oxygen. *Electroanalysis* 3:925–928.
2253. Shaidarova, L.G., N.A. Ulakhovich, M.A. El Gakhri, and G.K. Budnikov. 1993. Use of dibenzo-18-crown-6 as a modifier of a carbon-paste electrode for lead determination. *Izv. Vysch. Uchebn. Zaved., Khim. Technol. (Russ.)* 35:51–54.
2254. Peng, T.-Z., Z. Tang, G.-S. Wang, and B.-E. Shen. 1995. A cuproin-modified carbon paste electrode for the determination of lead(II) using differential pulse voltammetry. *Curr. Sep.* 13:119–124.
2255. Glebov, A.N., L.G. Shaidarova, M. Al-Gahri, N.A. Ulakhovich, and A.B. Vepritskaya. 1993. Lead(II) macrocyclic complexes with alkali metal cations in voltammetry. *Koordinatsion. Khim. (Russ.)* 19:122–124.
2256. Shaidarova, L.G., N.A. Ulakhovich, M.A. El Gakhri, G.K. Budnikov, and A.N. Glebov. 1995. Inversion voltammetry of lead(II) on a carbon paste electrode modified by macrocyclic complexants. *J. Anal. Chem. (Russ.)*/transl. of *Zh. Anal. Khim.* 50:692–696.
2257. Zhang, Z.-Q., H. Liu, H. Zhang, and Y.-F. Li. 1997. Determination of trace cadmium with 8-hydroxyquinoline mixed adhersion agent-carbon paste electrode. *Lihua Jianyan Huaxue Fence* 33:506–507.
2258. Iliadou, E., S.T. Girousi, A.N. Voulgaropoulos, and K. Vytřas. 1997. Voltammetric determination of heavy metals in natural waters, biological samples, and food-stuff by using chemically modified carbon paste electrodes. *Sci. Pap. Univ. Pardubice Ser. A* 3:87–101.
2259. Huang, S.-S., Y.-D. Cheng, B.-F. Li, and G.-D. Liu. Simultaneous anodic stripping voltammetric determination of lead and cadmium with a carbon paste electrode modified by tributyl phosphate. *Microchim. Acta* 130:97–101.
2260. Bae, Z.-U., J.-E. Choi, and H.-Y. Chang. Anodic stripping voltammetric determination of cadmium(II) using algae-modified carbon paste electrodes. *J. Korean Chem. Soc.* 42:28–35.
2261. Huang, S.-S., Y.-D. Cheng, B.-F. Li, and X.-S. Shi. 1998. A carbon paste electrode containing tributyl phosphate and 1-phenyl-3-methyl-4-benzoyl-pyrazolone and its use for determination of lead traces. *Fenxi Huaxue* 41:126–130.
2262. Pazdera, R. 2001. Stripping potentiometry with carbon paste electrodes in practical analysis. MSc dissertation (in Czech), University of Pardubice, Pardubice, Czech Republic.
2263. Hu, C.-G., K.-B. Wu, X. Dai, and S.-S. Hu. 2003. Simultaneous determination of lead(II) and cadmium(II) at a diacetyldioxime modified carbon paste electrode by differential pulse stripping voltammetry. *Talanta* 60:17–24.
2264. Ismail, K. 2004. Carbon paste electrodes plated with a bismuth film. Some contribution to their voltammetric characterisation and applicability to the determination of heavy metals in crude oil. MSc dissertation, University of Pardubice, Pardubice, Czech Republic.
2265. Švancara, I., L. Baldrianova, E. Tesarova, S.A.A. Elsuccary, A. Economou, S. Sotiropoulos, A. Bobrowski, and K. Vytřas. 2004. Stripping voltammetry of mixtures of heavy ions at electrodes with bismuth film. In *Monitoring of Environmental Pollutants*, Vol. VI, Eds. K. Vytřas, J. Kellner, and J. Fischer, pp. 229–246. Pardubice, Czech Republic: University of Pardubice.
2266. Baldrianova, L., I. Švancara, M. Vlcek, A. Economou, and S. Sotiropoulos. 2006. Effect of BiIII concentration on the stripping voltammetric response of *in-situ* bismuth-coated carbon paste and gold electrodes. *Electrochim. Acta* 52:481–490.
2267. Baldrianova, L., I. Švancara, K. Vytřas, and S. Sotiropoulos. 2006. Variation of the metal analyte-to-bismuth ratio with deposition time in anodic stripping voltammetry at *in-situ* bismuth-coated carbon paste electrodes. In *Sensing in Electroanalysis*, Vol. 2, Eds. K. Vytřas and K. Kalcher, pp. 59–74. Pardubice, Czech Republic: University of Pardubice.
2268. Gismera, M.J., M.T. Sevilla, and J.R. Procopio. 2006. Potentiometric carbon paste sensors for lead(II) based on dithiodibenzoic and mercaptobenzoic acids. *Anal. Sci. (Jpn.)* 22:405–410.
2269. Adraoui, I., M.E. Rhazi, and A. Amine. 2007. Fibrinogen-coated bismuth film electrodes for voltammetric analysis of lead and cadmium using the batch injection analysis. *Anal. Lett.* 40:349–368.
2270. Mojica, E.-R.E., J.M. Vidal, A.B. Pelegrina, and J.R.L. Micor. 2007. Voltammetric determination of lead (II) ions at carbon paste electrode modified with banana tissue. *J. Appl. Sci.* 7:1286–1292.
2271. Gholivand, M.B. and M. Malekian. 2008. Determination of trace amount of lead(II) in sweet fruit-flavored powder drinks by differential pulse adsorptive stripping voltammetry at carbon paste electrode. *Electroanalysis* 20:367–373.
2272. Rievaj, M., P. Tomcik, M. Cernanska, Z. Janosikova, and D. Bustin. 2008. Trace determination of lead in environmental and biological samples by anodic stripping voltammetry on carbon paste electrode. *Chem. Anal.* 53:717–723.
2273. Kuang, Y.F., J.L. Zou, L.Z. Ma, Y.J. Feng, and P.H. Deng. 2008. Determination of trace Cd(II) in water sample using 1,10-phenanthroline-5,6-dione modified carbon paste electrode. *Fenxi Huaxue* 36:103–106.

2274. Švancara, I., T. Mikysek, K. Vytřas, and F.G. Banica. 2008. Carbon paste electrode modified with macroporous thiourea functionalised resin and its use for the determination of lead. In *Monitoring of Environmental Pollutants*, Vol. X, Eds. K. Vytřas, J. Kellner, and J. Fischer, pp. 231–256. Pardubice, Czech Republic: University Centre Press.
2275. Rodriguez, J.A., I.S. Ibarra, C.A.G. Vidal, M. Vega, and E. Barrado. 2009. Multicommutated anodic stripping voltammetry at tubular bismuth film electrode for lead determination in gunshot residues. *Electroanalysis* 21:452–458.
2276. Goubert Renaudin, S., M. Moreau, C. Despas, M. Meyer, F. Denat, B. Lebeau, and A. Walcarius. 2009. Voltammetric detection of Pb(II) using amide-cyclam-functionalized silica-modified carbon paste electrodes. *Electroanalysis* 21:1731–1742.
2277. Ghanjaoui, M.E.A., M. Srij, and M. El Rhazi. 2009. Assessment of lead and cadmium in canned foods by square-wave stripping voltammetry. *Anal. Lett.* 42:1294–1309.
2278. Abbastabar Ahangar, H., A. Shirzadmehr, K. Marjani, H. Khoshsafar, M Chaloosi, and L. Mohammadi. 2009. Ion-selective carbon paste electrode based on a new tripodal ligand for determination of cadmium(II). *J. Inclusion Phenom. Macrocyclic. Chem.* 63:287–293.
2279. Chen, J.-Y, G.-Z. Yang, W.-Y. Li, and J.-S. Sun. 2009. Determination of lead (II) at nanomolar level using H_2O_2-oxidized activated carbon modified electrode. *J. Electrochem. (Russ.)* 45:908–912.
2280. Hradilova, S. 2009. Non-electrolytic processes at carbon paste electrodes and utilisation in electroanalysis. MSc dissertation (in Czech), University of Pardubice, Pardubice, Czech Republic.
2281. Papp, Z., V. Guzsvany, I. Švancara, K. Vytřas, F. Gaal, L. Bjelica, and B. Abramovic. 2009. New applications of tricresyl phosphate-based carbon paste electrodes in voltammetric analysis. In *Sensing in Electroanalysis*, Vol. 4, Eds. K. Vytřas, K. Kalcher, and I. Švancara, pp. 47–58. Pardubice, Czech Republic: University Press Center.
2282. Bing, C. and L. Kryger. 1996. Accumulation and voltammetric determination of complexed metal ions at zeolite-modified sensor electrodes. *Talanta* 43:153–160.
2283. Metelka, R., I. Švancara, and K. Vytřas. 1999. Carbon paste electrodes modified with mercuricoxide. In *Monitoring of Environmental Pollutants*, Eds. K. Vytřas, J. Kellner, and J. Fischer, pp. 113–117. Pardubice, Czech Republic: University of Pardubice.
2284. Konvalina, J. and K. Vytřas. 1999. Determination of thallium(III) at a carbon paste electrode with the aid of potentiometric stripping analysis. In *Monitoring of Environmental Pollutants*, Eds. K. Vytřas, J. Kellner, and J. Fischer, pp. 99–104. Pardubice, Czech Republic: University of Pardubice.
2285. Konvalina, J. 2001. Carbon paste electrodes in stripping potentiometry (in Czech). PhD dissertation, pp. 75–85. University of Pardubice, Pardubice, Czech Republic.
2286. Monien, H. and K. Zinke. 1970. Inverse-voltammetric determination of tin and lead in presence of both at a carbon-paste electrode. *Fresenius Z. Anal. Chem.* 250:178–185.
2287. Yang, Z.-P., M. Alafandy, K. Boutakhrit, J.-M. Kauffmann, and J. Arcos. 1996. Electrochemical oxidation of 8-hydroxyquinoline and selective determination of tin(II) at solid electrodes. *Electroanalysis* 8:25–29.
2288. Xie, H.-Q., Y.-H. Li, F. Q. Zhou, H.-S. Guo, and B. Yi. 2001. Determination of trace tin by adsorptive voltammetry at an alizarin violet modified carbon paste electrode. *Fenxi Huaxue* 29:822–824.
2289. Švancara, I., L. Baldrianova, E. Tesarova, T. Mikysek, and K. Vytřas. 2005. Determination of tin(II) at bismuth-modified carbon paste electrodes: An initial study. In *Monitoring of Environmental Pollutants*, Eds. K. Vytřas, J. Kellner, and J. Fischer, pp. 139–148. Pardubice, Czech Republic: University of Pardubice.
2290. Tesarova, E., L. Baldrianova, A. Krolicka, I. Švancara, A. Bobrowski, and K. Vytřas. 2005. Role of supporting electrolyte in anodic stripping voltammetry of In(III) in the presence of Cd(II) and Pb(II) using bismuth film electrodes. In *Sensing in Electroanalysis*, Eds. K. Vytřas and K. Kalcher, pp. 75–87. Pardubice, Czech Republic: University of Pardubice.
2291. Nghi, T.-V. and F. Vydra. 1975. Voltammetry with disk electrodes and its analytical application. Part X. *J. Electroanal. Chem.* 64:163–173.
2292. Cai, X.-H., K. Kalcher, and R.J. Magee. 1993. Studies on the voltammetric behavior of a bismuthiol I-containing carbon paste electrode: Determination of traces of bismuth. *Electroanalysis* 5:413–419.
2293. Ferri, T., S. Paci, and R. Morabito. 1996. Chemically-modified carbon paste electrode for the voltammetric determination of bismuth. *Ann. Chim.* 86:245–256.
2294. Guo, H.-S. and Y.-H. Li. 2000. Determination of trace bismuth by adsorptive stripping voltammetry with pyrogallol red-modified carbon paste electrode. *Fenxi Huaxue* 28:1527–1530.
2295. Huang, W.-S. 2004. Voltammetric determination of bismuth in water and nickel metal samples with a sodium montmorillonite (SWy-2) modified carbon paste electrode. *Microchim. Acta* 14:125–129.

References

2296. Mostafa, G.A.E. and A.M. Homoda. 2008. Potentiometric carbon paste electrodes for the determination of bismuth in some pharmaceutical preparations. *Bull. Chem. Soc. Jpn.* 81:257–261.
2297. Watanabe, D., T. Furuike, M. Midorikawa, and T. Tanaka. 2005. Simultaneous determination of copper and antimony by differential pulse anodic stripping voltammetry with a carbon-paste electrode. *Bunseki Kagaku* 54:907–912.
2298. Santos, V.S., W.J.R. Santos, L.T. Kubota, and C.R.T. Tarley. 2009. Speciation of Sb(III) and Sb(V) in meglumine antimoniate pharmaceutical formulations by PSA using carbon nanotube electrode. *J. Pharm. Biomed. Anal.* 50:151–157.
2299. Cepria, G., S. Hamida, F. Laborda, and J.R. Castillo. 2009. Electroanalytical determination of arsenic(III) and total arsenic in 1 M HCl using a carbonaceous electrode without a reducing agent. *Anal. Lett.* 42:1971–1985.
2300. Gevorgyan, A.M., S.V. Vakhnenko, and A.T. Artykov. 2004. Thick film graphite-containing electrodes for determining selenium by stripping voltammetry. *J. Anal. Chem. (Russ.)* 59:374–380.
2301. Ibrahim, H., Y.M. Issa, and O.R. Shehab. 2010. New selenite ion-selective electrodes based on 5,10,15,20-tetrakis-(4-methoxyphenyl)-21H,23H-porphyrin-Co(II). *J. Hazard. Mater.* 181:857–867.
2302. Monien, H. and P. Jacob. 1972. Voltammetric determination of small amounts of iron without separation from the matrix. *Fresenius Z. Anal. Chem.* 260:195–202.
2303. Arrigan, D.W.M., J.D. Glennon, and G. Svehla. 1993. Approaches to electrode modification with hydroxamic acids. *Anal. Proc.* 30:141–142.
2304. Hu, X.-Y., Z.-Z. Leng, N.-N. Xu, and S.-H. Yu. 1996. Determination of iron(III) by potentiometric and oscillographic-potentiometric titration with highly sensitive carbon paste electrode. *Fenxi Huaxue* 24:595–598.
2305. Komersova, A., M. Bartos, K. Kalcher, and K. Vytřas. 1998. Trace iron determination in aminoisophthalic acid using differential-pulse cathodic stripping voltammetry at carbon paste electrodes. *J. Pharm. Biomed. Anal.* 16:1373–1379.
2306. Mikysek, T. M. Stoces, I. Švancara, and J. Ludvik. 2010. Relation between the composition and properties of carbon nanotubes paste electrodes (CNTPEs). In *Sensing in Electroanalysis*, Vol. 5, Eds. K. Vytřas, K. Kalcher, and I. Švancara, pp. 69–75. Pardubice, Czech Republic: University Press Center.
2307. Kasem, K.K. and H. D. Abruna. 1988. Electroanalysis with modified carbon paste electrodes. Coordination trends, selectivity, and sensitivity. *J. Electroanal. Chem.* 242:87–95.
2308. Švancara, I. 1988. Applications of a carbon paste electrode for voltammetric determination of nickel and cobalt at low concentrations (in Czech). MSc dissertation, University of Pardubice, Pardubice, Czech Republic.
2309. Jeong, E.-D., M.-S. Won, D.-S. Park, S.-N. Choi, and Y.-B. Shim. 1993. Differential pulse voltammetric determination of cobalt(II) ion with a chemically modified carbon paste electrode containing l-sparteine. *J. Korean Chem. Soc.* 37:881–888.
2310. Bae, Z.-U., Y.-C. Park, S.-H. Lee, W.-S. Jeon, and H.-Y. 1996. Chang. Voltammetric determination of cobalt(II) using carbon paste electrodes modified with 1-(2-pyridylazo)-2-naphthol. *Bull. Korean Chem. Soc.* 17:995–999.
2311. Huang, S.-S., Y. Chen, X.-F. Li, and F.-M. Li. 1997. Determination of trace cobalt in natural water by ternary complex system with phenanthroline chemically modified carbon paste electrode. *Guangdong Gongye Daxue Xuebao* 14:128–130.
2312. Shaidarova, L.G., N.A. Ulakhovich, I.L. Fedorova, and Yu. G. Galyametdinov. 1996. Stripping-voltammetric determination of transition metals using electrodes modified by azacrown compounds. *J. Anal. Chem. (Russ.)* 51:687–692.
2313. Wu, J.-R., X. Zhang, and Y.-T. Sun. 1996. Determination of microamount of nickel by potentiometry with dimethylglyoxime-modified carbon paste electrode. *Shandong Jiancai Xueyuan Xuebao* 10:14–18.
2314. Gonzalez, P.S., C.A. Fontan, and V.A. Cortinez. 1997. Determination of nickel by direct automatic potentiometric titration with EDTA and a chemically modified electrode based on a strong acid ion exchanger containing 4-(3,5-dichloro-2-pyridylazo)-1,3-diaminobenzene. *Talanta* 44:23–30.
2315. Deng, P.-H., J.-J. Fei, and Y.-L. Feng. 2010. Determination of trace vanadium(V) by adsorptive anodic stripping voltammetry on an acetylene black paste electrode in the presence of alizarin violet. *J. Electroanal. Chem.* 648:85–91.
2316. Hu, X.-Y. and Z.-Z. Leng. 1995. Highly sensitive potentiometry for determination of chromium(VI) with carbon paste electrode. *Anal. Proc.* 32:521–522.
2317. Gevorgyan, A.M., S.V. Vakhnenko, and A.T. Artykov. 2004. Determination of chromium in natural waters by stripping voltammetry. *J. Anal. Chem. (Russ.)* 59:371–373.
2318. Fernandez, M.E.S., L.M.C. Aguilera, and J.M.P. Satander, I.N. Rodriguez, and J.L.H. de Cisneros. 2005. An oxidative procedure for the electrochemical determination of Cr(VI) using modified carbon paste electrodes. *Bull. Electrochem.* 21:529–535.

2319. Moreno, R.A.S., M.J. Gismera, M.T. Sevilla, and J.R. Procopio. 2009. Chromium(III) determination without sample treatment by batch and flow injection potentiometry. *Anal. Chim. Acta* 634:68–74.
2320. Abu Shawish, H.M., S.M. Saadeh, K. Hartani, and H.M. Dalloul. 2009. Comparative study of chromium(III) ion-selective electrodes based on N,N-bis(salicylidene)-o-phenyl-enediaminate-chromium(III). *J. Iran. Chem. Soc.* 6:729–737.
2321. Ghaedi, M., A. Shokrollahi, A.R. Salimibeni, S. Noshadi, and S. Joybar. 2010. Preparation of a new chromium(III) selective electrode based on 1-[(2-hydroxy ethyl) amino]-4-methyl-9H-thioxanthen-9-one as a neutral carrier. *J. Hazard. Mater.* 178:157–163.
2322. Monien, H., P. Jacob, and B. Jaenisch. 1973. Voltammetric determination of molybdenum in the presence of tungsten. *Fresenius Z. Anal. Chem.* 267:108–114.
2323. Zheng, X.-W., Z.-J. Zhang, Q. Wang, and H.-C. Ding. 2003. Electrogenerated chemiluminescence determination of Mo(VI) based on its sensitizing effect in electrochemical reduction of luminol. *Fenxi Huaxue* 31:1076–1078.
2324. Shumilova, M.A., A.V. Trubachev, and D.I. Kurbatov. 1997. Voltammetry of tungsten(VI) on a carbon-paste electrode modified with 8-mercaptoquinoline and dimethylsulfoxide. *J. Anal. Chem. (Russ.)*/transl. of *Zh. Anal. Khim.* 52:753–755.
2325. Farsang, Gy. and L. Tomcsanyi. 1967. Voltammetric behavior of manganese(II) and its determination or a carbon paste electrode. *Acta Chim. Acad. Sci. Hung.* 52:123–132.
2326. Narayanan, A. and R. Neeb. 1974. Adsorptive enrichment of azo dye complexes on a carbon paste electrode and its application to the determination of small concentrations of cobalt and manganese. *Fresenius Z. Anal. Chem.* 269:344–348.
2327. Smit, M.H. and G.A. Rechnitz. 1992. Reagentless enzyme electrode for the determination of manganese through biocatalytic enhancement. *Anal. Chem.* 64:245–249.
2328. Rievaj, M., P. Tomcik, Z. Janosikova, D. Bustin, and R.G. Compton. 2008. Determination of trace Mn(II) in pharmaceutical diet supplements by cathodic strip-ping voltammetry on bare carbon paste electrode. *Chem. Anal. (Warsaw)* 53:153–161.
2329. Ruf, H. and K. Schorb. 1989. Stripping voltammetry of technetium using a TOA-modified carbon paste electrode. *Kernforschungszent. Karlsruhe (Ber.)* p. 18. KfK: KFK 4634.
2330. Galik, M., I. Švancara, and K. Vytřas. 2005. Stripping voltammetric determination of platinum metals at carbon paste electrodes modified with cationic surfactants. In *Sensing in Electroanalysis*, Eds. K. Vytřas and K. Kalcher, pp. 89–107. Pardubice, Czech Republic: University of Pardubice.
2331. Zakieva, D.Z., N.A. Ulakhovich, Y.G. Galyametdinov, and G.K. Budnikov. 1991. Selective determination of Pd(II) using a carbon-paste electrode modified with liquid-crystal azomethines. *Zh. Anal. Khim. (Russ.)* 46:2093–2095.
2332. Raber, G., K. Kalcher, C. Neuhold, C. Talaber, and G. Koelbl. 1995. Adsorptive stripping voltammetry of palladium(II) with thioridazine in situ modified carbon paste electrodes. *Electroanalysis* 7:137–142.
2333. Qiu, D.-F., Q. Zhao, and K.-Z. Liu. 1997. Determination of palladium(II) using flow injection analysis with a detector based on carbon paste chemically modified with dimethylglyoxime. *Huaxue Yanjiu* 8:41–44.
2334. Galik, M., M. Cholota, I. Švancara, A. Bobrowski, and K. Vytřas. 2006. Stripping voltammetry of osmium at carbon paste electrodes. In *Monitoring of Environmental Pollutants* (in Czech), Eds. K. Vytřas, J. Kellner, and J. Fischer, pp. 75–85. Pardubice, Czech Republic: University of Pardubice.
2335. Kalcher, K. 1986. Voltammetric behavior of hexachloroiridate(IV) solutions on chemically modified carbon paste electrodes. *Fresenius Z. Anal. Chem.* 324:47–51.
2336. Zarinskii, V.A. and L.S. Chulkina. 1977. Inverse voltammetric method for determining platinum in industrial solutions. *Zavod. Lab.* 43:148–150.
2337. Hoshi, T., J.-I. Anzai, and T. Osa. 1992. Potentiometric sensors based on carbon paste electrode. *Denki Kagaku* 60:1146–1147.
2338. Wangsa, J., M.A. Targove, and N.D. Danielson. 1990. Indirect electrochemical detection of cations with cerium(III) in the mobile phase. *Talanta* 37:1151–1154.
2339. Walcarius, A., L. Lamberts, and E.G. Derouane. 1995. Cation determination in aqueous solution using the methyl viologen-doped Zeolite-modified carbon paste electrode. *Electroanalysis* 7:120–128.
2340. Lima, A.S., N. Bocchi, H.M. Gomes, M.F.S. and Teixeira. 2009. An electrochemical sensor based on nanostructured hollandite-type manganese oxide for the detection of potassium(I) ions. *Sensors* 9:6613–6625.
2341. Teixeira, M.F.S., B.H. Freitas, P.M. Seraphim, L.O. Salmazo, M.A. Nobre, and S. Lanfredi. 2009. Development of an electrochemical sensor for potassium ions based on the use of KSr2Nb5O15 modified electrode. In *Proceedings of The Eurosensors XXIII Conference*, Lausanne, Switzerland, Eds. J. Brugger and D. Briand, *Proc. Chem.* 1:293–296

References

2342. Blasius, E. and K.-P. Janzen. 1972. o-(2-Hydroxy-5-methyl-phenylazo)benzoic acid as reagent for polarographic, inverse voltammetric, and photometric determination of Be(II) ions. *Fresenius Z. Anal. Chem.* 258:257–263.
2343. Liu, N. and J.F. Song. 2005. Determination of free calcium at a carbon paste electrode adsorptive stripping voltammetric method. *Fenxi Huaxue* 33:1261–1264.
2344. Specker, H., H. Monien, and B. Lendermann. 1971. Inverse voltammetric determination of aluminum and vanadium on a carbon paste electrode. *Chem. Anal. (Warsaw)* 17:1003–1014.
2345. Monien, H. and H. Specker. 1972. Inverse voltammetric determination of traces of aluminum in beryllium. *Anal. Lett.* 5:837–842.
2346. Cai, Q.T. and S.B. Khoo. 1993. Differential pulse voltammetric determination of aluminum at a 8-hydroxyquinoline modified carbon paste electrode. *Bull. Singapore Natl. Inst. Chem.* 21:157–170.
2347. Liu, H.-L. 2003. Chemically modified carbon paste sensor for aluminium(III) and its application. *Fenxi Huaxue* 31:1511–1513.
2348. Kurbatov, D.I. and L.Y. Buldakova. 1996. Use of a modified carbon-paste electrode for voltammetric determination of gallium and zinc. *Zavod. Lab. (Russ.)* 62:17–18.
2349. Kurbatov, D.I. and L.Y. Buldakova. 1996. Voltammetric determination of gallium(III) and zinc(II) in aqueous-organic supporting electrolytes. *J. Anal. Chem. (Russ.)* 51:386–390.
2350. Liu, S.-M., L.-H. Yi, and J.-N. Li. 2003. Studies on anodic adsorptive stripping voltammetry of gallium(III)-alizarin complexone at carbon paste electrodes and its application. *Fenxi Huaxue* 31:1489–1492.
2351. Wu, J.-R., W.-Z. Xu, and Y.-T. Sun. 1995. Study and application of chemically modified α-nitroso-β-naphthol-Zr(IV)-carbon paste electrodes. *Lihua Jianyan Huaxue Fence* 31:157–158.
2352. Li, J.-N., J. Zhang, P.-H. Deng, and J.-J. Fei. 2001. Carbon paste electrode for trace zirconium(IV) determination by adsorption voltammetry. *Analyst* 126:2032–2035.
2353. Li, J.-N., J. Zhang, P.-H. Deng, and Y.-Q. Peng. 2001. Adsorption voltammetry of the mix-polynuclear complex of zirconium-calcium-Alizarin Red S at a carbon paste electrode. *Anal. Chim. Acta* 431:81–87.
2354. Liu, S.-M., J.-N. Li, and X. Mao. 2003. Stripping voltammetric determination of zirconium with complexing preconcentration of zirconium(IV) at a morin-modified carbon paste electrode. *Electroanalysis* 15:1751–1755.
2355. Liu, S.-M., J.-N. Li, and X. Mao. 2004. Determination of zirconium by 2nd derivative adsorptive voltammetry of zirconium (IV)-*Morin* complex at a carbon paste electrode. *Fenxi Huaxue* 32:195–197.
2356. Li, J.-N., S.-M. Liu, Z.-H. Yan, X. Mao, and P. Gao. 2006. Adsorptive voltammetric studies on the cerium(III)-alizarin complexon complex at a carbon paste electrode: Development of a method for the determination of Ce(III). *Microchim. Acta* 154:241–243.
2357. Liu, S.-M., J.-N. Li, and P. Gao. 2003. Anodic adsorptive stripping voltammetry at a carbon paste electrode for determination of trace thorium. *Anal. Lett.* 36:1381–1392.
2358. Li, J.-N., F.-Y. Yi, Z.-M. Jiang, and J.-J. Fei. 2003. Adsorptive voltammetric study of Th(IV) alizarin complex at a carbon paste electrode. *Microchim. Acta* 143:287–292.
2359. Norouzi, P., Z.R. Sarmazdeh, F. Faridbod, M. Adibi, and M.R. Ganjali. 2010. Er(3+) carbon paste electrode based on new nano-composite. *Int. J. Electrochem. Sci.* 5:367–376.
2360. Ji, K.-B. and S.-S. Hu. 2004. Square wave voltammetric determination of trace amounts of europium(III) at montmorillonite-modified carbon paste electrodes. *Collect. Czech. Chem. Commun.* 69:1590–1599.
2361. Li, J.-N., S.-M. Liu, X. Mao, P. Gao, and Z.-H. Yan. 2004. Trace determination of rare earths by adsorption voltammetry at a carbon paste electrode. *J. Electroanal. Chem.* 56:137–142.
2362. Tomita, S., K. Sato, J.-I. Anzai. 2008. pH-Sensitive thin films composed of poly(methacrylic acid) and carboxyl-terminated dendrimer. *Sens. Lett.* 6:250–252.
2363. Brezina, M. and A. Hofmanova Matejkova. 1973. Electrochemical generation of superoxide ion on carbon paste electrodes. *J. Electroanal. Chem.* 44:460–462.
2364. Guzman, S.Q., O.M. Baudino, and V.A. Cortinez. 1987. Design and evaluation of an electrochemical sensor for determination of dissolved oxygen in water. *Talanta* 34:551–554.
2365. Jalali, F., A.M. Ashrafi, and D. Nematollahi. 2009. Measurement of dissolved oxygen in biological fluids by a modified carbon paste electrode. *Electroanalysis* 21:201–205.
2366. Xu, G.-B and S.-J. Dong. 1999. Chemiluminescent determination of luminol and hydrogen peroxide using hematin immobilized in the bulk of a carbon paste electrode. *Electroanalysis* 11:1180–1184.
2367. Li, C.-Y., Y. Chen, C.-F. Wang, H.-B. Li, and Y.-Y. Chen. 2003. Electrocatalytic oxidation of H_2O_2 at a carbon paste electrode modified with a nickel (II)-5,11,17,23-tetra-tert-butyl-25,27-bis(diethylcarbamoylmethoxy) calix[4]arene complex and its application. *Wuhan Univ. J. Nat. Sci.* 8:857–860.

2368. Kozan, J.V.B., R.P. Silva, S.H.P. Serrano, A.W.O. Lima, and L. Angnes. 2007. Biosensing hydrogen peroxide utilizing carbon paste electrodes containing peroxidases naturally immobilized on coconut fibers. *Anal. Chim. Acta* 591:200–207.
2369. Jian, F.-F., Y.-B. Qiao, H.-Q. Yu, and R.-R. Zhuang. 2007. Hydrogen peroxide biosensor based on the electrochemistry of the myoglobin-TATP composite film. *Anal. Lett.* 40:2664–2672.
2370. Yu, H., Q.-L. Sheng, and J.-B. Zheng. 2007. Preparation, electrochemical behavior and performance of gallium hexacyanoferrate as electrocatalyst of H_2O_2. *Electrochim. Acta* 52:4403–4410.
2371. Alpat, S., S.K. Alpat, Z. Dursun, and A. Telefoncu. 2009. Development of a new biosensor for mediatorless voltammetric determination of H_2O_2 and its application in milk samples. *J. Appl. Electrochem.* 39:971–977.
2372. Nyasulu, F.W. and H.A. Mottola. 1987. Amperometric determination of nitrogen dioxide in air samples by flow injection and reaction at a gas-liquid interface. *J. Automat. Chem.* 9:46–49.
2373. Suye, S.I., Y. Nakamura, S. Inuta, T. Ikeda, and M. Senda. 1996. Mediated amperometric determination of ammonia with a methanol dehydrogenase immobilized carbon paste electrode. *Biosens. Bioelectron.* 11:529–534.
2374. Coutinho, C.F.B., A.A. Muxel, C.G. Rocha, D.A. de Jesus, R.V.S. Alfaya, F.A.S. Almeida, Y. Gushikem, and A.A.S. Alfaya. 2007. Ammonium ion sensor based on SiO_2/ZrO_2/phosphate-NH_4^+ composite for quantification of ammonium ions in natural waters. *J. Brazil. Chem. Soc.* 18:189–194.
2375. Zare, H.R. and A. Nasirizadeh. 2006. Electrocatalytic characteristics of hydrazine and hydroxylamine oxidation at coumestan modified carbon paste electrode. *Electroanalysis* 18:507–512.
2376. Cepria, G. and J.R. Castillo. 1998. Electrocatalytic behavior of several cobalt complexes: Determination of hydrazine at neutral pH. *J. Appl. Electrochem.* 28:65–70.
2377. Pessoa, C.A., Y. Gushikem, and S. Nagasaki. 2002. Cobalt porphyrin immobilized on a niobium(V) oxide grafted-silica gel surface: Study of the catalytic oxidation of hydrazine. *Electroanalysis* 14:1072–1076.
2378. Raoof, J.B., R. Ojani, and M. Ramine. 2007. Electrocatalytic oxidation and voltammetric determination of hydrazine on the tetrabromo-p-benzoquinone modified carbon paste electrode. *Electroanalysis* 19:597–603.
2379. Kamyabi, M.A., S. Shahabi, and H.H. Monfared. 2007. Electrocatalytic oxidation of hydrazine at a cobalt(II) Schiff-base-modified carbon paste electrode. *J. Electrochem. Soc.* 155:F08–F12.
2380. Liawruangrath, S., W. Oungpipat, S. Watanesk, B. Liawruangrath, C. Dongduen, and P. Purachat. 2001. Asparagus-based amperometric sensor for fluoride determination. *Anal. Chim. Acta.* 448:37–46.
2381. Shamsipur, M., S. Ershad, N. Samadi, A. Moghimi, and H. Aghabozorg. 2005. A novel chemically modified carbon paste electrode based on a new mercury(II) complex for selective potentiometric determination of bromide ion. *J. Solid. State Electrochem.* 9:788–793.
2382. Malongo, T.K., S. Patris, P. Macours, F. Cotton, J. Nsangu, and J.-M. Kauffmann. 2008. Highly sensitive determination of iodide by ion chromatography with amperometric detection at a silver-based carbon paste electrode. *Talanta* 76:540–547.
2383. Amine, A., M. Alafandy, J.-M. Kauffmann, and M.N. Pekli. 1995. Cyanide determination using an amperometric biosensor based on cytochrome oxidase inhibition. *Anal. Chem.* 67:2822–2827.
2384. Tan, J., J.H. Bergantini, A. Merkoci, S. Alegret, and F. Sevilla. 2004. Oil dispersion of AgI/Ag_2S salts as a new electroactive material for potentiometric sensing of iodide and cyanide. *Sens. Actuators B* 101:57–62.
2385. Samo, A.R., M.Y. Khahawer, S.A. Arbani, and G.A. Qureshi. 1993. Quantitation of azide and lead in lead azide by voltammetric method. *J. Chem. Soc. Pak.* 15:187–190.
2386. Ghaedi, M., A. Shokrollahi, M. Montazerozohori, and S. Derki. 2010. Design and construction of azide carbon paste selective electrode based on a new Schiff's base complex of Iron. *IEEE Sens. J.* 10:814–819.
2387. Sarwar, M. and G.G. Willems. 1980. Voltammetric determination of nitrite: Comparison of carbon paste and glassy carbon electrodes. *J. Chem. Soc. Pak.* 2:123–125.
2388. Asplund, J. 1986. Voltammetric determination of nitrous acid in mixtures of nitric and sulfuric acids by oxidation of nitrite at a carbon paste electrode. *Anal. Chem. Symp. Ser.* 25:91–96.
2389. Lu, G.-H., D.-W. Long, T. Zhan, and H.-Y. Zhao. 2002. Electrochemical behavior of a ruthenium(II) polypyridine complex and its electrocatalysis of nitrite. *Fenxi Huaxue* 30:1115–1118.
2390. Badea, M., A. Amine, M. Benzine, A. Curulli, D. Moscone, A. Lupu, G. Volpe, and G. Palleschi. 2004. Rapid and selective electrochemical determination of nitrite in cured meat in the presence of ascorbic acid. *Mikrochim. Acta* 147:51–58.
2391. Idrissi, L., A. Amine, M. El Rhazi, and F.E. Cherkaoui. 2005. Electrochemical detection of nitrite based on the reaction with 2,3-diaminonaphthalene. *Anal. Lett.* 38:1943–1955.
2392. Ojani, R., J.B. Raoof, and E. Zarei. 2006. Electrocatalytic reduction of nitrite using ferricyanide: Application for its simple and selective determination. *Electrochim. Acta* 52:753–759.

References

2393. Ojani, R., J.B. Raoof, and E. Zarei. 2008. Poly(o-toluidine) modified carbon paste electrode: A sensor for electrocatalytic reduction of nitrite. *Electroanalysis* 20:379–385.
2394. Shiddiky, M.A., K.-S. Lee, J.-G. Son, D.-S. Park, and Y.-B. Shim. 2009. Development of extraction and analytical methods for nitrite ion from food samples: Microchip electrophoresis with a modified electrode. *J. Agricult. Food Chem.* 57:4051–4057.
2395. Lee, Y.-K., D.-H. Kim, H.-J. Mang, and K.-J. Whang. 1988. Construction of high performance carbon paste coated wire sulfate ion-selective electrode. *Punsok Kwahak* 1:10–18.
2396. Gonzales Alvarez, M.J., A.J. Miranda Ordieres, A. Costa Garcia, and P. Tunon Blanco. 1990. A carbon paste electrode modified with p-dimethyl-N-aniline and lauryl sulphate and its use in electroanalysis. In *Electrospain Analysis '90 in Asturias, An International Conference on Electroanalytical Chemistry*, Book of Abstracts, D-PP-20. Gijon (Spain).
2397. Amini, M.K., S. Shahrokhian, S. Tangestaninejad, and I.M. Baltork. 2001. Voltammetric and potentiometric behavior of 2-pyridinethiol, 2-mercaptoethanol and sulfide at iron(II) phthalocyanine modified carbon-paste electrode. *Iran. J. Chem. Chem. Eng.* 20:29–36.
2398. Nader, P.A., S.S. Vives, and H.A. Mottola. 1990. Studies with a sulfite oxidase-modified carbon paste electrode for detection/determination of sulfite ion and sulfur dioxide (g) in continuous-flow systems. *J. Electroanal. Chem.* 284:323–333.
2399. Kumar, S.S. and S.S. Narayanan. 2008. Electrocatalytic oxidation of sulfite on a nickel aquapentacyanoferrate modified electrode: Application for simple and selective determination. *Electroanalysis* 20:1427–1433.
2400. Soleymanpour, A., E.H. Asl, and M.A. Nasseri. 2006. Chemically modified carbon paste electrode for determination of sulfate ion, SO_4^{2-} by potentiometric method. *Electroanalysis* 18:1598–1604.
2401. Fernandez, J.J., J.R Lopez, X. Correig, and I. Katakis. 1998. Reagentless carbon paste phosphate biosensors: Preliminary studies. *Sens. Actuators B* 47:13–20.
2402. Quintana, J.C, L. Idrissi, G. Palleschi, P. Albertano, A. Amine, M.El Rhazi, and D. Moscone. 2004. Investigation of amperometric detection of phosphate: Application in seawater and cyanobacterial biofilm samples. *Talanta* 63:567–574.
2403. Xue, Y., X.-W. Zheng, and G.-X. Li. 2007. Determination of phosphate in water by means of a new electrochemiluminescence technique based on the combination of liquid–liquid extraction with benzene-modified carbon paste electrode. *Talanta* 72:450–456.
2404. Gurentsova, O.I., G.V. Prokhorova, and E.A. Osipova. 1992. Modified carbon paste electrode for voltammetric determination of trace silicon. *J. Anal. Chem. (Russ.)*/transl. of *Zh. Anal. Khim.* 47:1671–1675.
2405. Kotrly, S. and L. Sucha. 1988. *Handbook of Chemical Equilibria in Analytical Chemistry. Tables and Diagrams*, Orig. Edn. (in Czech), Prague, Czech Republic: SNTL.
2406. Hoegfeldt, E. 1982. *Stability Constants of Metal-Ion Complexes*. Oxford, U.K.: Pergamon.
2407. Heyrovsky, J. and J. Kuta. 1962. *Principles of Polarography*, Orig. Edn. (in Czech), pp. 111–123. Prague, Czech Republic: NCSAV.
2408. Heyrovsky, J. 1922. Electrolysis with mercury drop cathode (in Czech). *Chem. Listy* XVI:258–264.
2409. Burestedt, E., J. Emneus, L. Gorton, G. Marko Varga, E. Dominguez, F. Ortega, A. Narvaez et al. 1995. Optimization and validation of an automated solid phase extraction technique coupled on-line to enzyme-based biosensor detection for the determination of phenolic compounds in surface water samples. *Chromatographia* 41:207–215.
2410. Rogers, K., J.Y. Becker, J. Wang, and F. Lu. 1999. Determination of phenols in environmentally relevant matrices with the use of liquid chromatography with an enzyme electrode detector. *Field Anal. Chem. Technol.* 3:161–169.
2411. Rosatto S.S., G. de Oliveira Neto, and L.T. Kubota. 2001. Effect of DNA on the peroxidase based biosensor for phenol determination in waste waters. *Electroanalysis* 13:445–450.
2412. Chen, Z.-D. and M. Hojo. 2007. Determination of phenol using a carbon paste electrode modified with overoxidized polypyrrole/polyvinylpyrrolidone films. *Jpn. Analyst*/transl. of *Bunseki Kagaku* 56:669–673.
2413. Mulazimoglu, I.E. and E. Yilmaz. 2010. Quantitative determination of phenol in natural decayed leaves using procaine modified carbon paste electrode surface by cyclic voltammetry. *Desalination* 256:64–69.
2414. Eskinja, I., Z. Grabaric, and B.S. Grabaric. 1995. Monitoring of pyrocatechol indoor air pollution. *Atmosph. Environ.* 29:1165–1170.
2415. Vieira da Cruz, I., O. Fatibello Filho, and L. Angnes. 1999. Zucchini crude extract-palladium-modified carbon paste electrode for the determination of hydroquinone in photographic developers. *Anal. Chim. Acta* 398:145–151.

2416. Wang, L.-H. 1995. Simultaneous determination of hydroquinone ethers in cosmetics after preconcentration at a carbon paste electrode. *Analyst* 120:2241–2244.
2417. Yan, Z.-H., J.-N. Li, J.-J. Fei, X. Mao, P. Gao, and Y.-L. Ding. 2005. Study on the adsorptive catalytic voltammetry of emodin at a carbon paste electrode. *Anal. Lett.* 38:1641–1650.
2418. Li, J.-N., P. Gao, X.-L. Li, Z.-H. Yan, and X. Mao. 2005. Study on the adsorptive catalytic voltammetry of aloe-emodin at a carbon paste electrode. *Sci. Chin. Ser. B* 48:442–448.
2419. Babula, P., D. Huska, P. Hanustiak, J. Baloun, S. Krizkova, V. Adam, J. Hubalek et al. 2006. Flow injection analysis coupled with carbon electrodes as the tool for analysis of naphthoquinones. *Sensors* 6:1466–1482.
2420. Stobiecka, A., H. Radecka, and J. Radecki. 2007. Novel voltammetric biosensor for determining acrylamide in food samples. *Biosens. Bioelectron.* 22:2165–2170.
2421. Wang, L.-H., S.-Y. Jiang, and Y.-Z. Lan. 2004. Voltammetric behavior of Coumarins and Psoralens at a carbon fiber ultramicroelectrode and their determination in citrus essential oils. *Bull. Electrochem.* 20:445–451.
2422. De la Fuente, E., G. Villagra, and S. Bollo. 2007. Electrochemical nucleic acid biosensors for the detection of interaction between peroxynitrite and DNA. *Electroanalysis* 19:1518–1523.
2423. Kim, O.-S., K. Maekawa, and K. Kusuda. 1995. Cyclic voltammetry of lipophilic compounds in oil: Direct determination of lipid peroxide with a carbon paste electrode. *J. Am. Oil Chem. Soc.* 72:299–303.
2424. Hernandez, L., P. Hernandez, and Z. Sosa Ferrera. 1988. Differential pulse voltammetric determination of aniline with carbon paste electrode modified by sepiolite. *Fresenius Z. Anal. Chem.* 329:756–759.
2425. Zoulis, N.E. and C.E. Efstathiou. 1988. Voltammetric determination of N-alkylated anilines by adsorption/extraction at a carbon-paste electrode. *Anal. Chim. Acta* 204:201–211.
2426. Vitova, V. 2001. The electrochemistry of amines at carbon paste electrodes. MSc dissertation (in Czech), University of Pardubice, Pardubice, Czech Republic.
2427. Wang, L.-H. and Z.-S. Chen. 1998. Determination of phenylendiamines in oxidative hair dyes by differential pulse voltammetry. *J. Chin. Chem. Soc. (Taipei)* 45:53–58.
2428. Dvorakova, J. 2004. Determination of 1-aminonaphtalene at carbon paste electrodes using voltammetry and HPLC-EC. MSc dissertation (in Czech), Charles University, Prague, Czech Republic.
2429. Zima, J., H. Dejmkova, and J. Barek. 2007. HPLC determination of naphthalene amino derivatives using electrochemical detection at carbon paste electrodes. *Electroanalysis* 19:185–190.
2430. Dejmkova, H. 2005. Determination of naphthalene amino derivatives using HPLC with electrochemical detection. MSc dissertation (in Czech), Charles University, Prague, Czech Republic.
2431. Armalis, S., N. Novikova, E. Kubiliene, J. Zima, and J. Barek. 2002. Voltammetric determination of 2-aminofluorene and 2,7-diaminofluorene using carbon paste electrode. *Anal. Lett.* 35:1551–1559.
2432. German, N., S. Armalis, J. Zima, and J. Barek. V. Voltammetric determination of fluoren-9-ol and 2-acetamidofluorene using carbon paste electrodes. *Collect. Czech. Chem. Commun.* 70:292–304.
2433. Zitova, A. 2004. Determination of aminoquinolines at carbon paste electrodes using voltammetry and HPLC-EC. MSc dissertation (in Czech), Charles University, Prague, Czech Republic.
2434. Stoica, A.I., J. Zima, and J. Barek. 2005. Differential pulse voltammetric determination of 8-aminoquinoline at carbon paste electrode. *Anal. Lett.* 38:149–156.
2435. Zima, J., A.I. Stoica, A. Zitova, and J. Barek. 2006. Voltammetric determination of selected aminoquinolines using a carbon paste electrode. *Electroanalysis* 18:158–162.
2436. Cizek, K., J. Barek, S. Kuecuekkolbasi, M. Ersoez, and J. Zima. 2007. Voltammetric determination of 3-aminofluoranthene at different types of carbon electrodes. *Chem. Anal. (Warsaw)* 52:1003–1013.
2437. Luo, L.-Q., X. Wang, Y.-P. Ding, Q.-X. Li, J.-B. Jia, and D.-M. Deng. 2010. Electrochemical determination of nitrobenzene using bismuth-film modified carbon paste electrode in the presence of cetyltrimethylammonium bromide. *Anal. Methods* 2:1095–1100.
2438. Hranicka, Z. 2010. Determination of selected nitrophenols using modified carbon paste electrodes. MSc dissertation (in Czech), Charles University, Prague, Czech Republic.
2439. Lei, Y., P. Mulchandani, W. Chen., J. Wang, and A. Mulchandani. 2004. *Arthrobacter* sp. JS443-based whole cell amperometric biosensor for p-nitrophenol. *Electroanalysis* 16:2030–2034.
2440. Mulchandani, P., C.M. Hangarter, Y. Lei, W. Chen, and A. Mulchandani. 2005. Amperometric microbial biosensor for p-nitrophenol using *Moraxella* sp.-modified carbon paste electrode. *Biosens. Bioelectron.* 21:523–527.
2441. Jeon, Y.-G., B.-W. Kim, H.-J. Kim, Y.-D. Cho, and C. Chung. 1996. Characteristics of a-cyclodextrin modified carbon paste electrode. *Anal. Sci. Technol. (Korea)* 9:235–243.
2442. Maistrenko, V.N., S.V. Sapelnikova, F.A. Amirkhanova, V.N. Gusakov, and F.Kh. Kudasheva. 1997. Voltammetry of organic nitro-compounds at modified carbon paste electrodes. *Baskhir. Khim. Zh. (Russ.)* 4:54–56.

2443. Dejmkova, H., J. Zima, and J. Barek. 2008. Application of carbon paste electrodes with admixed bismuth powder for the determination of 4-amino-3-nitrophenol. In *Sensing in Electroanalysis*, Vol. 3, Eds. K. Vytřas, K. Kalcher, and I. Švancara, pp. 83–89. Pardubice, Czech Republic: University Press Center.
2444. Maistrenko, V.N., S.V. Sapelnikova, F.Kh. Kudasheva, and F.A. Amirkhanova. 2000. Isomer-selective carbon paste electrodes for the determination of nitrophenol, nitroaniline, and nitrobenzoic acid by adsorptive stripping voltammetry. *J. Anal. Chem. (Russ.)*/transl. of *Zh. Anal. Khim.* 55:586–589.
2445. Sapelnikova, S.V., N.V. Kuzmina, V.N. Maystrenko, and F.Kh. Kudasheva. 2002. Preconcentration and voltammetric determination of some nitro-compounds on carbon paste electrodes. *J. Anal. Chem. (Russ.)*/transl. of *Zh. Anal. Khim.* 57:443–447.
2446. Kacur, M. 2003. Voltammetric determination of fluorene nitro derivatives using carbon paste electrodes. MSc dissertation (in Czech), Charles University, Prague, Czech Republic.
2447. Nemcova, L. 2008. Determination of 5-amino-6-nitroquinoline at carbon paste electrode. MSc dissertation (in Czech), Charles University, Prague, Czech Republic.
2448. Wang, J. and L. Chen. 1995. Hydrazine detection using a tyrosinase-based inhibition biosensor. *Anal. Chem.* 67:3824–3827.
2449. Wang, J., M. Chicharro, G. Rivas, X.-H. Cai, N. Dontha, A.M. Percio Farias, and H. Shiraishi. 1996. A DNA-biosensor for the detection of hydrazines. *Anal. Chem.* 68:2251–2254.
2450. Chybova, O. and G. Beranova. 1991. Electrochemical reductions of some azo-compounds at a carbon paste electrode. Parts I + II. Two joint MSc dissertation (in Czech), University of Pardubice, Pardubice, Czech Republic.
2451. Vanickova, M., J. Labuda, M. Buckova, I. Surugiu, M. Mecklenburg, and B. Danielsson. 2000. Investigation of catechin and acridine derivatives using voltammetric and fluorimetric DNA-based sensors. *Collect. Czech. Chem. Commun.* 65:1055–1066.
2452. Yang, Y.-H., M.-H. Yang, H. Wang, L. Tang, G.-L. Shen, and R.-Q. Yu. 2004. Inhibition biosensor for determination of nicotine. *Anal. Chim. Acta* 509:151–157.
2453. Masini, J., S. Aragon, and F.W. Nyasulu. 1997. Electrochemistry of brucine. Part II: Brucine-based determination of nitrate. *Anal. Chem.* 69:1077–1081.
2454. Wan, C.-D., Y. Zhang, H.-G. Lin, K.-B. Wu, J.-W. Chen, and Y.-K. Zhou. 2009. Electrochemical determination of p-chlorophenol based on the surface enhancement effect of mesoporous TiO_2-modified electrode. *J. Electrochem. Soc.* 156:F151–F154.
2455. Diaz Goretti, D., C.M. Blanco Lopez, J.M. Lobo Castanon, A.J. Miranda Ordieres, and P. Tunon Blanco. 2009. Chloroperoxidase modified electrode for amperometric determination of 2,4,6-tri-chlorophenol. *Electroanalysis* 21:1348–1353.
2456. Fogg, A.G. and D. Bhanot. 1981. Further voltammetric studies of synthetic food coloring matters at glassy carbon and carbon paste electrodes using static and flowing systems. *Analyst (U.K.)* 106:883–889.
2457. Fanjul Bolado, P., M.B. Gonzales Garcia, and A. Costa Garcia. 2005. Voltammetric determination of Leucoindigo adsorbed on pretreated carbon paste electrodes: Its application in a flow system. *Electroanalysis* 17:148–154.
2458. Ma, M.-M. 2009. Voltammetric determination of dye-uptake for CI-Acid-Blue-120 dyeing silk. *Anal. Lett.* 42:3073–3084.
2459. Cheng, X.-H. and W. Guo. 2007. The oxidation kinetics of reduction intermediate product of Methyl red with hydrogen peroxide. *Dyes Pigm.* 72:372–377.
2460. Huang, W.-S., C.-H. Yang, W.-Y. Qu, and S.-H. Zhang. 2008. Voltammetric determination of Malachite green in fish samples based on enhancement effect of anionic surfactant. *Russ. J. Electrochem.* 44:946–951.
2461. Jezkova, J. 1994. The use of carbon paste electrodes in potentiometry. MSc dissertation (in Czech), University of Pardubice, Pardubice, Czech Republic.
2462. Ticha, I. 1996. Automated potentiometric titrations of surfactants. MSc dissertation (in Czech), University of Pardubice, Pardubice, Czech Republic.
2463. Xue, Y., X.-W. Zheng, and K.-G. Han. 2007. Electrogenerated chemiluminescence detecting sodium dodecyl benzene sulfonate based on its ion-associated complex effect. *Fenxi Huaxue* 35:370–374.
2464. Hattori, T. and K. Ooshima. 2002. Determination of anionic polyelectrolytes by adsorptive stripping voltammetry using ferrocenyltrimethylundecyl ammonium ion at a carbon paste electrode. *Bunseki Kagaku* 51:1171–1174.
2465. Medyantseva, E.P., N.A. Ulakhovich, G.K. Budnikov, and A.T. Groisberg. 1991. Voltammetric determination of the pesticides Carbathion and Amoben by using a carbon-paste electrode. *J. Anal. Chem (Russ)*/transl. of *Zh. Anal. Khim.* 46:962–966.
2466. Hernandez, L., P. Hernandez, and S. Garcia. 1990 (June 4–8). Determination of Benomyl fungicide using a carbon paste electrode containing silicone oil OV-17. In *Electrospainanalysis '90 in Asturias, An International Conference on Electroanalytical Chemistry,* Book of Abstracts, C-PP-07, Gijon.

2467. Simoes, F.R., L.H.C. Mattoso, and C.M.P. Vaz. 2006. Conducting polymers as sensor materials for the electrochemical detection of pesticides. *Sens. Lett.* 4:319–324.
2468. Medyantseva, E.P., N.A. Ulakhovich, and G.K. Budnikov. 1994. Application of modified carbon-paste electrodes for determination of Kaptax and BM-Ts thiazole. *Zavod. Lab. (Russ.)* 60:5–7.
2469. Beklova, M., S. Krizkova, V. Supalkova, R. Mikelova, V. Adam, J. Pikula, and R. Kizek. 2007. Determination of bromadiolone in pheasants and foxes by using differential pulse voltammetry. *Int. J. Environment. Anal. Chem.* 87:459–469.
2470. Ulakhovich, N.A., E.V. Priimak, E.P. Medyantseva, L.A. Anisimova, and M.A. Al Gahri 1998. Voltammetric determination of phosalone and carbophos pesticides using a modified carbon-paste electrode. *J. Anal. Chem. (Russ)*/transl. of *Zh. Anal. Khim.* 53:147–150.
2471. Sirisha, K., S. Mallipattu, and S.R.J. Reddy. 2007. Differential pulse adsorptive stripping voltammetric determination of chlorpyrifos at a sepiolite modified carbon paste electrode. *Anal. Lett.* 40:1939–1950.
2472. Papp, Zs., V. Guzsvany, I. Švancara, K. Vytřas, F. Gaal, L. Bjelica, and B. Abramovic. 2009. New applications of tricresyl phosphate-based carbon paste electrodes in voltammetric analysis. In *Sensing in Electroanalysis*, Vol. 4, Eds. K. Vytřas, K. Kalcher, and I. Švancara, pp. 47–58. Pardubice, Czech Republic: University Press Center.
2473. Simoes, F.R., L.H.C. Mattoso, and C.M.P. Vaz. 2004. Modified carbon paste poly-aniline electrodes for the electrochemical determination of the herbicide dichloro-phenoxyacetic acid (2,4-D). *Sens. Lett.* 2:221–225.
2474. Walcarius, A. and L. Lamberts. 1996. Square wave voltammetric determination of Paraquat and Diquat in aqueous solution. *J. Electroanal. Chem.* 406:59–68.
2475. El Bakouri, H., J.M. Palacios Santander, L. Cubillana Aguilera, A. Ouassini, I. Naranjo Rodriguez, and J. de Cisneros. 2005. Electrochemical analysis of Endosulfan using a C18-modified carbon-paste electrode. *Chemosphere* 60:1565–1571.
2476. Lei, Y., P. Mulchandani, W. Chen, and A. Mulchandani. 2007. Biosensor for direct determination of Fenitrothion and EPN using recombinant *Pseudomonas putida* JS444 with surface-expressed organophosphorous hydrolase. 2. Modified carbon paste electrode. *Appl. Biochem. Biotechnol.* 136:243–250.
2477. Papp, Zs., I. Švancara, V. Guzsvany, K. Vytřas, and F. Gaal. 2009. Voltammetric determination of Imidacloprid insecticide in selected samples using carbon paste electrodes. *Microchim. Acta* 166:169–175.
2478. Priyantha, N., A. Navaratne, S. Weliwegamage, and C.B. Ekanayake. 2007. Determination of MCPA through electrocatalysis by manganese species. *Int. J. Electrochem. Sci.* 2:433–443.
2479. Arribas, A.S., E. Bermejo, M. Chicharro, and A. Zapardiel. 2006. Voltammetric detection of the herbicide Metamitron at a bismuth film electrode in nondeaerated solution. *Electroanalysis* 18:2331–2336.
2480. Noguer, T., A.M. Balasoiu, A. Avramescu, and J.L. Marty. 2001. Development of a disposable biosensor for the detection of metamsodium and its metabolite MITC. *Anal. Lett.* 34:513–528.
2481. Mulchandani, P., W. Chen, and A. Mulchandani. 2001. Flow injection amperometric enzyme biosensor for direct determination of organophosphate nerve agents. *Environ. Sci. Technol.* 35:2562–2565.
2482. Mulchandani, P., W. Chen, A. Mulchandani, J. Wang, and L. Chen. 2001. Amperometric microbial biosensor for direct determination of organophosphate pesticides using recombinant microorganism with surface expressed organophosphorus hydrolase. *Biosens. Bioelectron.* 16:433–437.
2483. Lei, Y., P. Mulchandani, W. Chen, J. Wang, and A. Mulchandani. 2004. Whole cell-enzyme hybrid amperometric biosensor for determination of organo-phosphorous nerve agents with p-nitrophenyl substituent. *Biotechnol. Bioengineer.* 85:706–713.
2484. Lei, Y., P. Mulchandani, J. Wang, W. Chen, and A. Mulchandani. 2005. Highly sensitive and selective amperometric microbial biosensor for direct determination of p-nitrophenyl-substituted organophosphate nerve agents. *Environ. Sci. Technol.* 39:8853–8857.
2485. Liu, G.-D. and Y.-H. Lin. 2005. Electrochemical stripping analysis of organo-phosphate pesticides and nerve agents. *Electrochem. Commun.* 7:339–343.
2486. Alvarez, E., M. Teresa Sevilla, J. M. Pinilla, and L. Hernandez: Cathodic stripping voltammetry of Paraquat on a carbon paste electrode modified with amberlite XAD-2 resin. *Anal. Chim. Acta* 260 (1992):19–23.
2487. El Mhammedi, M.A., M. Bakasse, R. Bachirat, and A. Chtaini. 2008. Square wave voltammetry for analytical determination of Paraquat at carbon paste electrode modified with fluorapatite. *Food Chem.* 110:1001–1006.
2488. Alizadeh, T. 2009. High selective Parathion voltammetric sensor development by using an acrylic based molecularly imprinted polymer-modified carbon paste electrode. *Electroanalysis* 12:1490–1498.
2489. Papp, Zs.J., V.J. Guzsvany, S. Kubiak, A. Bobrowski, and L.J. Bjelica. 2010. Voltammetric determination of the neonicotinoid insecticide thiamethoxam using a tricresyl phosphate-based carbon paste electrode. *J. Serb. Chem. Soc.* 75:681–687.

References

2490. Noguer, T., A. Gradinaru, A. Ciucu, and J.L. Marty. 1999. A new disposable biosensor for the accurate and sensitive detection of ethylenebis(dithiocarbamate) fungicides. *Anal. Lett.* 32:1723–1738.
2491. Vytřas, K. Potentiometric titrations based on ion-pair formation. 1985. *Ion-Sel. Electrode Rev.* 7:77–164.
2492. Cullum, D.C. 1994. *Introduction to Surfactant Analysis*. London, U.K.: Chapman & Hall.
2493. Papp, Zs., V. Guzsvany, I. Švancara, and K. Vytřas. 2011. Carbon paste electrodes for the analysis of some agricultural pollutants and trace metals. *J. Agric. Sci. Technol.* 5:85–92.
2494. Acuna, J.A., C. de la Fuente, M.D. Vazquez, M.L. Tascon, M.I. Gomez del Rio, and P. Sanchez Batanero. 1998. Sensitive determination method of Aceclofenac by using voltammetric techniques on carbon paste electrode in aqueous-organic media. *Quim. Anal. (Barcelona)* 17:105–111.
2495. Miner, D.J., J.C. Price, R.M. Riggin, and P.T. Kissinger. 1981. Voltammetry of acetaminophen and its metabolites. *Anal. Chem.* 53:2258–2263.
2496. Wang, C.-Y., X.-Y. Hu, Z.-Z. Leng, G.-J. Yang, and G.-D. Jin. 2001. Differential pulse voltammetry for the determination of Paracetamol at a pumice mixed carbon paste electrode. *Anal. Lett.* 34:2747–2759.
2497. Yu, D.-H., O. Renedo, B. Blankert, V. Sima, R. Sandulescu, J. Arcos, and J.-M. Kauffmann. 2006. A peroxidase-based biosensor supported by nanoporous magnetic silica microparticles for Acetaminophen biotransformation and inhibition studies. *Electroanalysis* 18:1637–1642.
2498. Teixeira, M.F.S., L.H. Marcolino Junior, O. Fatibello Filho, F.C. Moraes, and R.S. Nunes. 2009. Determination of two analgesics (dipyrone and acetaminophen) in pharmaceutical preparations by cyclic voltammetry at a copper(II)-hexacyano-ferrate(III) modified carbon paste electrode. *Curr. Anal. Chem.* 5:303–310.
2499. Baldwin, R.P., D. Packett, and T.M. Woodcock. 1981. Electrochemical behavior of Adriamycin at carbon paste electrodes. *Anal. Chem.* 53:540–542.
2500. Chaney, E.N., Jr., and R.P. Baldwin. 1982. Electrochemical determination of Adriamycin compounds in urine by preconcentration at carbon paste electrodes. *Anal. Chem.* 54:2556–2560.
2501. Zhang, S.-H., K.-B. Wu, and S.-S. Hu. 2002. Carbon paste electrode based on surface activation for trace Adriamycin determination by a preconcentration and voltammetric method. *Anal. Sci. (Jpn.)* 18:1089–1092.
2502. Jalali, F. and R. Maghooli. 2010. Potentiometric determination of trace amounts of amantadine using a modified CPE. *Anal. Sci. (Jpn.)* 25:1227–1230.
2503. Jin, G.-D., X.-Y.Hu, Z.-Z. Leng, and C. Yao. 2002. Anodic voltammetric behavior of ambroxol at carbon paste electrode. *Fenxi Huaxue* 30:214–217.
2504. Kauffmann, J.-M., G.J. Patriarche, M. Chateau Gosselin, and B.L. Gallo Hermosa. 1986. Electrochemical oxidation of Khelline and Amikhelline. A cyclic voltammetry study. *Talanta* 33:733–738.
2505. Hermosa, B.L. Gallo, J.-M. Kauffmann, G.J. Patriarche, and G.G. Guilbault. 1986. Electrochemical behaviour of Benzofuran derivatives of pharmaceutical interest at solid electrodes. *Anal. Lett.* 19:2022–2021.
2506. Liu, N., W. Gao, and J.-F. Song. 2006. Catalytic adsorptive stripping voltammetry at a carbon paste electrode for the determination of Amiodarone. *Chin. J. Chem.* 24:1657–1661.
2507. Wang, L.-H. and C.-C. Wang. 2006. Electrochemical oxidation of Aminophylline at platinum film electrodes and its determination in cosmetic and pharmaceutical products. *Microchim. Acta* 153:95–100.
2508. Biryol, I., B. Uslu, and Z. Kucukyavuz. 1996. Voltammetric determination of Imipramine hydrochloride and amitriptyline hydrochloride using a polymer-modified carbon paste electrode. *J. Pharm. Biomed. Anal.* 15:371–381.
2509. Kazemipour, M., M. Ansari, A. Mohammadi, H. Beitollahi, and R. Ahmadi. 2009. Use of adsorptive square-wave anodic stripping voltammetry at carbon paste electrode for the determination of amlodipine besylate in pharmaceutical preparations. *J. Anal. Chem. (Russ.)* 64:65–70.
2510. Bergamini, M.F., M.F.S. Teixeira, E.R. Dockal, N. Bocchi, and E.T.G. Cavalheiro. 2006. Evaluation of different voltammetric techniques in the determination of Amoxicillin using a carbon paste electrode modified with [N,N′-ethylene-bis-(salicylidene-aminato)] oxovanadium(iv). *J. Electrochem. Soc.* 153:E94–E98.
2511. Cheng, H.-Y., E. Strope, and R.N. Adams. 1979. Electrochemical studies of the oxidation pathways of Apomorphine. *Anal. Chem.* 51:2243–2246.
2512. Supalkova, V., J. Petrek, L. Havel, S. Krizkova, J. Petrlova, V. Adam, D. Potesil et al. 2006. Electrochemical sensors for detection of acetylsalicylic acid (Aspirin). *Sensors* 6:1483–1497.
2513. Patil, R.H., R.N. Hegde, and S.T. Nandibewoor. 2009. Voltammetric oxidation and determination of Atenolol at a carbon paste electrode. *Ind. Eng. Chem. Res.* 48:10206–10210.
2514. Farghaly, O.A. and N.A.L. Mohamed. 2004. Voltammetric determination of Azithromycin at the carbon paste electrode. *Talanta* 62:531–538.

2515. El Maali, N.A. 1998. Electrochemical behavior of the monobactam antibiotic Aztreonam at different electrodes in biological fluids. *Bioelectrochem. Bioenerg.* 45:281–286.
2516. Wang, C.-Y., X.-Y. Hu, Q. Chen, and G. Rong. 2005. Differential pulse voltammetry for determination of Benorilate in pharmaceutical formulations at carbon paste electrode. *Anal. Lett.* 38:893–905.
2517. Kocourkova, M. 2008. Voltammetric determination of Benserazide drug at carbon paste electrodes (in Czech). MSc dissertation, Charles University, Prague, Czech Republic.
2518. Legorburu, M.J., R.M. Alonso, and R.M. Jimenez. 1993. Oxidative behavior of the sulfonamidic diuretic Bumetanide at carbon paste electrode. *Electroanalysis* 5:333–338.
2519. Garcia-Fernandez, M.A., M.T. Fernandez Abedul, and A. Costa Garcia. 1999. Voltammetric study and determination of buprenorphine in pharmaceuticals. *J. Pharm. Biomed. Anal.* 21:809–815.
2520. Garcia-Fernandez, M.A., M.T. Fernandez Abedul, and A. Costa Garcia. 2000. Determination of Buprenorphine in pharmaceuticals and human urine by adsorptive stripping voltammetry in batch and flow systems. *Electroanalysis* 12:483–489.
2521. Stefan-van Staden, R.I., R. Girmai Bokretsion, J.F. van Staden, and H.Y. Aboul Enein. 2009. Enantioanalysis of Butaclamol using enantioselective potentiometric electrodes. *Anal. Lett.* 42:1111–1118.
2522. Mebsout, F., J.-M. Kauffmann, and G.J. Patriarche. 1987. Redox behaviour of antitumor platinum(II) compound (Carboplatin) at solid electrodes. *J. Pharm. Biomed. Anal.* 5:223–231.
2523. Mebsout, F., J.-C. Vire, G.J. Patriarche, and G.D. Christian. 1988. Electrochemical study of carminomycin at solid electrodes. *Talanta* 35:993–996.
2524. El Maali, N.A., A.M.M. Ali, and M.A. Ghandour. 1993. Electrochemical reduction and oxidation of two cephalosporin antibiotics: Ceftriaxone (Rocephin) and Cefoperazone (Cefobid). *Electroanalysis* 5:599–604.
2525. El Maali, N.A., M.A. Ghandour, and J.-M. Kauffmann. 1995. Cephalosporin antibiotics at bare and modified carbon paste electrodes in both aqueous and biological media. *Bioelectrochem. Bioenerg.* 38:91–97.
2526. Yilmaz, N. and I. Biryol. 1998. Anodic voltammetry of Cefotaxime. *J. Pharm. Biomed. Anal.* 17:1335–1344.
2527. Rizk Nashwa, M.H., S.S. Abbas, F.A. El Sayed, and A. Abo Bakr. 2009. Novel ionophore for the potentiometric determination of cetirizine hydrochloride in pharmaceutical formulations and human urine. *Int. J. Electrochem. Sci.* 4:396–406.
2528. Torres, R.F., M.C. Mochon, C.J. Sanchez Jimenez, M.A.B. Lopez, and A.G. Perez. 2001. Electrochemical oxidation of Cisatracurium on carbon paste electrode and its analytical applications. *Talanta* 53:1179–1185.
2529. Mebsout, F., J-M. Kauffmann, and G.J. Patriarche. 1988. Redox behaviour of cisplatin at solid electrodes: Carbon paste, platinum. *J. Pharm. Biomed. Anal.* 6:441–448.
2530. Kotzian, P., I.Ch. Gherghi, S.T. Girousi, and K. Vytřas. 2005. Optimization of DNA accumulation onto carbon paste electrodes when applied in a study of its interaction with Cisplatin, In *Sensing in Electroanalysis*, Eds. K. Vytřas and K. Kalcher, pp. 109–118. Pardubice, Czech Republic: University of Pardubice.
2531. O'Dea, P., A. Costa Garcia, A.J. Miranda Ordierez, P. Tunon Blanco, and M.R. Smyth. 1991. Comparison of adsorptive stripping voltammetry at Hg and C paste electrodes for the determination of Ciprofloxacin in urine. *Electroanalysis* 3:337–342.
2532. Zhang, S.-H. and S. Wei. 2007. Electrochemical determination of Ciprofloxacin based on the enhancement effect of sodium dodecyl benzene sulfonate. *Bull. Korean Chem. Soc.* 28:543–546.
2533. Stefan-van Staden, RI, L. Holo, B. Moeketsi, J.F. van Staden, and H.Y. Aboul-Enein. 2009. Enantioselective determination of R-Clenbuterol using an enantioselective potentiometric membrane electrode based on b-cyclodextrin derivative. *Instrum. Sci. Technol.* 37:189–196.
2534. Pudilova, H. 2006. Determination of chloramphenicol at carbon paste electrodes (in Czech). MSc dissertation, Charles University, Prague, Czech Republic.
2535. Radi, A. 2005. Accumulation and trace measurement of chloroquine drug at DNA-modified carbon paste electrode. *Talanta* 65:271–275.
2536. Khodari, M. 1999. Electrometric studies on the ternary complexes of Me(II) with Chlorpromazine and some amino acids. *Microchim. Acta* 131:231–235.
2537. Peng, T.-Z., Z.-P. Yang, and R.-S. Lu. 1990 Stripping voltammetric determination of Perphenazine and some antipsychotic drugs at carbon paste electrodes. *Yaoxue Xuebao* 25:277–283.
2538. Angelikaki, A.G. and S.T. Girousi. 2008. Sensitive detection of tetracycline, oxytetracycline, and chlortetracycline in the presence of copper(II) ions using a DNA-modified carbon paste electrode. *Chem. Anal. (Warsaw)* 53:445–454.
2539. Ambrosi, A., R. Antiochia, L. Campanella, R. Dragone, and I. Lavagnini. 2005. Electrochemical determination of pharmaceuticals in spiked water samples. *J. Hazard. Mater.* 122:219–225.

2540. Yu, D.-H., B. Blankert, and J.-M. Kauffmann. 2007. Development of amperometric horse-radish peroxidase based biosensors for clozapine and for the screening of thiol compounds. *Biosens. Bioelectron.* 22:2707–2711.
2541. Farhadi, K., R.H. Yamchi, and R. Sabzi. 2007. Electrochemical study of interaction between Clozapine and DNA and its analytical application. *Anal. Lett.* 40:1750–1762.
2542. Fernandez Abedul, M.T., J.R. Barreira Rodriguez, A. Costa Garcia, and P. Tunon Blanco. 1991. Voltammetric determination of cocaine in confiscated samples. *Electroanalysis* 3:409–412.
2543. Abedul, M.T.F. and A.C. Garcia. 1996. Flow injection analysis with amperometric detection of cocaine in confiscated samples. *Anal. Chim. Acta* 328:67–71.
2544. Yu, S.-H., X.-Y. Hu, and Z.-Z. Leng. 1997. Measurement of colchicine by anodic voltammetry with carbon paste electrode. *Yaoxue Xuebao* 32:210–312.
2545. Stanic, Z., A. Voulgaropoulos, and S. Girousi. 2008. Electroanalytical study of the anti-oxidant and anti-tumor agent curcumin. *Electroanalysis* 20:1263–1266.
2546. Palaska, P., E. Aritzoglou, and S. Girousi. 2007. Sensitive detection of Cyclophosphamide using DNA-modified carbon paste, pencil graphite, and hanging mercury drop electrodes. *Talanta* 72:1199–1206.
2547. Rodriguez, R.J.B., A.C. Garcia, A.J.M. Ordierez, and P.T. Blanco. 1989. Electrochemical oxidation of dacarbazine and its major metabolite (AIC) on carbon electrodes. *Electroanalysis* 1:529–534.
2548. Wang, J., M-S. Lin, and V. Villa. 1987. Adsorptive stripping voltammetric determination of low levels of daunorubicin. *Analyst (U.K.)* 112:1303–1307.
2549. Wang, J., M. Ozsos, X.-H. Cai, G. Rivas, H. Shiraishi, D.H. Grant, M. Chicharro, J. Fernandes, and E. Palecek. 1998. Interactions of antitumor drug daunomycin with DNA in solution and at the surface. *Bioelectrochem. Bioenerg.* 45:33–40.
2550. Balaji, K., G.V.R. Reddy, T.M. Reddy, and S.J. Reddy. 2008. Determination of Prednisolone, Dexamethasone, and Hydrocortisone in pharmaceutical formulations and biological fluid samples by voltammetric techniques using b-cyclodextrin modified carbon paste electrode. *Afr. J. Pharm. Pharmacol.* 2:157–166.
2551. Khaled, E., H.N.A. Hassan, M. Mohamed, and A.A. Gehad Gseleim. 2010. Carbon paste and PVC electrodes for the flow injection potentiometric determination of dextromethorphan. *Talanta* 81:510–515.
2552. Li, Y.-F., W.-D. Li, and Z.-Q. Zhang. 1998. Electroanalytical characteristics of diazepam on carbon paste electrode. *Hunan Daxue Xuebao, Ziran Kexueban* 25:25–27,42.
2553. Vanickova, M., M. Buckova, and J. Labuda. 2000. Voltammetric determination of azepine and phenothiazine drugs with DNA biosensors. *Chem. Anal. (Warsaw)* 45:125–133.
2554. Oelschaeger, H., D. Rothley, and U. Dunzendorfer. 1986. Direct determination of diethylstilbestrol and its monoconjugates in plasma. *Arzneimittel Forschung* 36:759–763.
2555. Zhang, S.-H., K.-B. Wu, and S.-S. Hu. 2002. Voltammetric determination of diethylstilbestrol at a carbon paste electrode using cetylpyridine bromide as medium. *Talanta* 58:747–754.
2556. Teixeira, M.F.S., L.H. Marcolino, Jr., O. Fatibello Filho, E.R. Dockal, and E.T.G. Cavalheiro. 2004. Voltammetric determination of dipyrone using a N,N'-ethylene-bis-(salicylidene-aminato) oxovanadium(iv) modified carbon-paste electrode. *J. Braz. Chem. Soc.* 15:803–808.
2557. Liu, L., J.-F. Song, Y. Peng-Fei, and B. Cui. 2008. Enhancement action of lanthanum hydroxide nanowire towards voltammetric response of calcium dobesilate and its application. *Chin. J. Chem.* 26:220–224.
2558. Yang, G.-J., L.-T. Jin, and Z.-Z. Leng. 1998. Adsorptive voltammetry behavior of trace Dobutamine on carbon paste electrode and its determination. *Gaodeng Xuexiao Huaxue Xuebao* 19:1574–1577.
2559. Arranz, A., S.F. de Betono, J.M. Moreda, A. Cid, and J.F. Arranz. 1997. Cathodic stripping voltammetric determination of doxazosin in urine and pharmaceutical tablets using carbon paste electrodes. *Analyst (U.K.)* 122:849–854.
2560. Chaney, E.N. Jr. and R.P. Baldwin. 1985. Voltammetric determination of doxorubicin in urine by adsorptive preconcentration and flow injection analysis. *Anal. Chim. Acta* 176:105–112.
2561. Stefan, R.I., H.Y. Aboul Enein, and G.L. Radu. 1998. Biosensors for the enantioselective analysis of S-Enalapril and S-Ramipril. *Prep. Biochem. Biotechnol.* 28:305–312.
2562. Enein, H.Y. Aboul, R.I. Stefan, and J.F. van Staden. 1999. Potentiometric enantio-selective membrane electrode for S-Enalapril assay. *Analusis (Bruxelles)* 27:53–56.
2563. Stefan, R.I., J.F. van Staden, C. Bala, and H.Y. Aboul Enein. 2004. On-line assay of the S-enantiomers of enalapril, ramipril, and pentopril using a sequential injection analysis/amperometric biosensor system. *J. Pharm. Biomed. Anal.* 36:889–892.
2564. Connor, M.P., J. Wang, W. Kubiak, and M.R. Smyth. 1990. Tissue- and microbe-based electrochemical detectors for liquid chromatography. *Anal. Chim. Acta* 229:139–143.

2565. Erdem, A. and M. Ozsoz. 2001. Interaction of the anticancer drug epirubicin with DNA. *Anal. Chim. Acta* 437:107–114.
2566. Radi, A.E., N. Abd Elghany, and T. Wahdan. 2007. Electrochemical study of the antineoplastic agent etoposide at carbon paste electrode and its determination in spiked human serum by differential pulse voltammetry. *Chem. Pharm. Bull.* 55:1379–1382.
2567. Abbas, M.N. and G.A.E. Mostafa. 2003. Gallamine-tetraphenylborate modified carbon paste electrode for the potentiometric determination of gallamine triethiodide (Flaxedil). *J. Pharm. Biomed. Anal.* 31:819–826.
2568. Stefan-van Staden, R.I., R.G. Bokretsion, K.I. Ozoemena, J.F. van Staden, and H.Y. Aboul-Enein. 2006. Enantioselective, potentiometric membrane electrodes based on different cyclodextrins as chiral selectors for the assay of S-Flurbiprofen. *Electroanalysis* 18:1718–1721.
2569. Stefan-van Staden, R.I., R.G. Bokretsion, and K.I. Ozoemena. 2006. Utilization of maltodextrin-based enantioselective, potentiometric membrane electrodes for the enantioselective assay of S-Flurbiprofen. *Anal. Lett.* 39:1065–1073.
2570. Stefan-van Staden, R.I., J.F. van Staden, and H.Y. Aboul Enein. 2009. Macrocyclic antibiotics as chiral selectors to design enantioselective, potentiometric membrane electrodes for the determination of S-Flurbiprofen. *Anal. Bioanal. Chem.* 394:821–826.
2571. Radi, A. 2004. Voltammetric study of Glibenclamide at carbon paste and sephadex-modified carbon paste electrodes. *Anal. Bioanal. Chem.* 378:822–826.
2572. Radi, A, M.A. El Ries, and G.E. Bekhiet. 1999. Electrochemical oxidation of the hypoglycemic drug gliclazide. *Anal. Lett.* 32:1603–1612.
2573. Rodriguez, R.J.B., V. Cabal Diaz, A.C. Garcia, and P.T. Blanco. 1900. Voltammetric assay of heroin in illicit dosage forms. *Analyst (U.K.)* 115:209–212.
2574. Santamaria, M.d.C., M.D. Vazquez Barbado, M.L. Tascon Garcia, J.A. Acuna, and P.S. Batanero. 1995. Determination of indomethacin by using voltammetric techniques on carbon paste electrode. *Quim. Anal. (Barcelona)* 14:117–120.
2575. Radi, A. 2001. Stripping voltammetric determination of Indapamide in serum at castor oil-based carbon paste electrodes. *J. Pharm. Biomed. Anal.* 24:413–419.
2576. Liu, L., J.-F. Song, P.-F. Yu, and B. Cui. 2006. A novel electrochemical sensing system for inosine and its application for Inosine determination in pharmaceuticals and human serum. *Electrochem. Commun.* 8:1521–1526.
2577. Mebsout, F., J.-M. Kauffmann, G.J. Patriarche, J. Vereecken, and G.D. Christian. 1989. Electrochemical behavior of iproplatin at the carbon paste electrode. *Electroanalysis* 1:161–165.
2578. Hammam, E., A.M. Beltagi, and M.M. Ghoneim. 2004. Voltammetric assay of Rifampicin and Isoniazid drugs in pharmaceutical formulations and human serum at a carbon paste electrode. *Microchem. J.* 77:53–62.
2579. Wahdan, T. 2005. Voltammetric method for the simultaneous determination of Rifampicin and Isoniazid in pharmaceutical formulations. *Chem. Anal. (Warsaw)* 50:457–464.
2580. Radi, A. 1999. Voltammetric study of khellin at a DNA-coated carbon paste electrode. *Anal. Chim. Acta* 386:63–68.
2581. Radi, A. 2003. Anodic voltammetric assay of lansoprazole and omeprazole on a carbon paste electrode. *J. Pharm. Biomed. Anal.* 31:1007–1012.
2582. Radi, A., M.A. El Ries, and S. Kandil. 2003. Electrochemical study of the interaction of levofloxacin with DNA. *Anal. Chim. Acta* 495:61–67.
2583. Arenaza, M.J., B. Gallo, L.A. Berrueta, and F. Vicente. 1995. Electrooxidation and determination of the 1,4-benzodiazepine loprazolam at the carbon-paste electrode. *Anal. Chim. Acta* 305:91–95.
2584. Mebsout, F., J.-M. Kauffmann, and G.J. Patriarche. 1987. Electrochemical behavior of Marcellomycine on solid electrodes. *Analusis (Bruxelles)* 15:243–247.
2585. Liu, L. and J.-F. Song. 2006. voltammetric determination of mefenamic acid at lanthanum hydroxide nanowires modified carbon paste electrodes. *Anal. Biochem.* 354:22–27.
2586. Radi, A., M.A. El Ries, F. El Anwar, and Z. El Sherif. 2001. Electrochemical oxidation of meloxicam and its determination in tablet dosage form. *Anal. Lett.* 34:739–748.
2587. Tian, X.-J. and J.-F. Song. 2007. Catalytic action of copper(II) ion on electrochemical oxidation of metformine and its voltammetric determination in pharmaceuticals. *J. Pharm. Biomed. Anal.* 44:1192–1196.
2588. Khaled, E., H.N.A. Hassan, M.S. Kamel, and B.N. Barsoum. 2007. Novel etformin carbon paste and PVC electrodes. *Curr. Pharm. Anal.* 3:262–267.
2589. Ramirez Barreira, R.J., A. Costa Garcia, and P. Tunon Blanco. 1989. Electrooxidation of Methadone on carbon paste electrodes. *Electrochim. Acta* 34:957–961.

References

2590. Pekarova, Z. 2004. Voltammetric determination of methotrexate using a carbon paste electrode (in Czech). MSc dissertation, Charles University, Prague, Czech Republic.
2591. Zima, J., J. Pekarova, J. Barek, and I. Švancara. 2007. Possibilities and limitations of carbon paste electrodes in electroanalysis of pharmaceuticals: Voltammetric determination of Methotrexate. In *Sensing in Electroanalysis*, Vol. 2, Eds. K. Vytřas and K. Kalcher, pp. 141–158. Pardubice, Czech Republic: University of Pardubice.
2592. Abd El Hady, D., M.M. Selim, R. Gotti, and N.A. El Maali. 2006. Novel voltammetric method for enantioseparation of racemic methotrexate: Determination of its enantiomeric purity in some pharmaceuticals. *Sens. Actuators B* 113:978–988.
2593. Farghaly, O.A., M.A. Taher, A.H. Naggar, and A.Y. El Sayed. 2005. Square-wave anodic stripping voltammetric determination of metoclopramide in tablet and urine at carbon paste electrode. *J. Pharm. Biomed. Anal.* 38:14–20.
2594. Joseph, R. and K.G. Kumar. 2009. Electrochemical reduction and voltammetric determination of Metronidazole Benzoate at modified carbon paste electrode. *Anal. Lett.* 42:2309–2321.
2595. Gu, K., J.-J. Zhu, Y.-L. Zhu., J.-Z. Xu, and H.-Y. Chen. 2000. Voltammetric determination of mifepristone at a DNA-modified carbon paste electrode. *Fresenius J. Anal. Chem.* 368:832–835.
2596. Ozkan, D., H. Karadeniz, A. Erdem, M. Mascini, and M. Ozsoz. 2004. Electrochemical genosensor for Mitomycin-C/DNA interaction based on guanine signal. *J. Pharm. Biomed. Anal.* 35:905–912.
2597. Cerna, P. 2005. Voltammetric determination of mitoxantron on a carbon paste electrode (in Czech). MSc dissertation, Charles University, Prague, Czech Republic.
2598. Sandulescu, R., I. Marian, S. Mirel, R. Oprean, and L. Roman. 1998. Voltammetric study of 2-mercapto-5-phenylamino-1,3,4-thiadiazole using carbon paste electrodes. *J. Pharm. Biomed. Anal.* 18:75–81.
2599. Fernandez-Abedul, M.T., M.S. Velazquez Rodriguez, J.R.B. Rodriguez, and A.C. Garcia. 1997. Voltammetric determination of naltrexone in pharmaceuticals. *Anal. Lett.* 30:1491–1502.
2600. Fernandez-Abedul, M.T. and A.C. Garcia. 1997. Flow injection analysis with amperometric detection of Naltrexone in pharmaceuticals. *J. Pharm. Biomed. Anal.* 16:15–19.
2601. Aki, C., S. Yilmaz, Y. Dilgin, S. Yagmur, and E. Suren. 2005. Electrochemical study of Natamycin. Analytical application to pharmaceutical dosage forms by differential pulse voltammetry. *Pharmazie* 60:747–750.
2602. Wang, J., B.K. Desmukh, and M. Bonakdar. 1985. Electrochemical behavior and determination of nicardipine. *Anal. Lett.* 18:1087–1090.
2603. Radi, A. 1999. Preconcentration and voltammetric study of nicergoline at a carbon paste electrode. *Microchim. Acta* 132:49–53.
2604. Buchberger, W., G. Niessner, and R. Bakry. 1998. Determination of nifuroxazide with polarography and adsorptive stripping voltammetry at mercury and carbon paste electrodes. *Fresenius J. Anal. Chem.* 362:205–208.
2605. Radi, A. 1999. Voltammetric study of nifuroxazide at unmodified and sephadex-modified carbon paste electrodes. *Fresenius J. Anal. Chem.* 364:590–594.
2606. Radi, A. 2003. Determination of pantoprazole by adsorptive stripping voltammetry at carbon paste electrode. *Farmaco (Brazil)* 58:535–539.
2607. Stefan, R.I., J.F. van Staden, and H.Y. Aboul Enein. 1999. S-Perindopril assay using a potentiometric enantioselective membrane electrode. *Chirality* 11:631–634.
2608. Ozoemena, K.I., R.I. Stefan, J.F. van Staden, and H.Y. Aboul Enein. 2005. Enantioanalysis of S-Perindopril using different cyclodextrin-based potentiometric sensors. *Sens. Actuators B* 105:425–429.
2609. Raoof, J.B., R. Ojani, and F. Chekin. 2009. Voltammetric sensor for D-Penicillamine determination based on its electrocatalytic oxidation at the surface of ferrocenes modified carbon paste electrodes. *J. Chem. Sci.* 121:1083–1091.
2610. Raoof, J.B., R. Ojani, M. Majidian, and F. Chekin. 2009. Homogeneous electrocatalytic oxidation of D-Penicillamine with ferrocyanide at a carbon paste electrode: Application to voltammetric determination. *J. Appl. Electrochem.* 39:799–805.
2611. Radi, A., M.A. El Ries, and S. Kandil. 2005. Spectroscopic and voltammetric studies of Pefloxacin bound to calf thymus double stranded DNA. *Anal. Bioanal. Chem.* 381:451–455.
2612. Kubota, L.T. and L. Gorton. 1999. Electrochemical study of flavins, phenazines, phenoxazines, and phenothiazines immobilized on zirconium phosphate. *Electroanalysis* 1:719–728.
2613. Citak, M., S. Yilmaz, Y. Dilgin, G. Turker, S. Yagmur, H. Erdugan, and N. Erdugan. 2007. Osteryoung square wave voltammetric determination of phenazopyridine hydrochloride in human urine and tablet dosage forms based on electrochemical reduction at carbon paste electrode. *Curr. Pharm. Anal.* 3:141–145.

2614. Ibrahim, H. 2005. Chemically modified carbon paste electrode for the potentiometric flow injection analysis of piribedil in pharmaceutical preparation and urine. *J. Pharm. Biomed. Anal.* 38:624–632.
2615. Vire, J.-C., J.-M. Kauffmann, J. Braun, and G.J. Patriarche. 1985. Electrochemical characteristics of a novel non-steroidal anti-inflammatory drug: The piroxicam. *Analusis (Bruxelles)* 13:134–140.
2616. El Maali, N.A. and R.M. Hassan. 1990. Electrooxidation and determination of anti-inflammatory drugs piroxicam and tenoxicam at the CPE. *Bioelectrochem. Bioenerg.* 24:155–163.
2617. Panigua, A.R., M.D. Vazquez, M.L. Tascon, and P. Sanchez Batanero. 1994. Voltammetric determination of piroxicam after incorporation within carbon pastes. *Electroanalysis* 6:265–268.
2618. Rucki, R.J., A. Ross, and S.A. Moros. 1980. Application of an electrochemical detector to the determination of Procarbazine hydrochloride by high performance liquid chromatography. *J. Chromatogr. A* 190:359–365.
2619. Wang, C.-Y., X.-Y. Hu, G.-D. Jin, and Z.-Z. Leng. 2002. Differential pulse adsorption voltammetry for determination of procaine hydrochloride at a pumice modified carbon paste electrode in pharmaceutical preparations and urine. *J. Pharm. Biomed. Anal.* 30:131–139.
2620. Radi, A., A.A. Wassel, and M.A. El Ries. 2004. Adsorptive behaviour and voltammetric analysis of propranolol at a carbon paste electrode. *Chem. Anal. (Warsaw)* 49:51–58.
2621. El-Ries, M.A.N., G.G. Mohamed, and A.K. Attia. 2008. Electrochemical determination of the antidiabetic drug Repaglinide. *Yakugaku Zashi* 128:171–177.
2622. Girousi, S.T., I.C. Gherghi, and M.K. Karava. 2004. DNA-modified carbon paste electrode applied to the study of interaction between rifampicin and DNA in solution and at the electrode surface. *J. Pharm. Biomed. Anal.* 36:851–858.
2623. Mebsout, F., J-M. Kauffmann, G.J. Patriarche, J. Vereecken, and G.D. Christian. 1989. Electrochemical behavior of spiroplatin at carbon paste and platinum electrodes. *Electroanalysis* 1:257–261.
2624. Guo, X.-X., Z.-J. Song, X.-J. Tian, and J.-F. Song. 2008. Single-sweep voltammetric determination of tamoxifen at a carbon paste electrode. *Anal. Lett.* 41:1225–1235.
2625. Tian, X.-J. and J.-F. Song. 2006. Voltammetric behavior and determination of Tanshinone II-A at carbon paste electrode. *Fenxi Huaxue* 34:1283–1286.
2626. Wang, L.-H., H.-J. Tien, and C.-Y. Tai. 2006. Use of a disposable modified carbon paste electrode for liquid chromatography-amperometric detection of theophylline and three metabolites in human serum. *J. Chin. Chem. Soc. (Taiwan)* 53:1523–1530.
2627. Zima, J., M. Cienciala, J. Barek, and J.C. Moreira. 2007. Determination of thymol using HPLC-ED with a glassy CP-electrode. *Chem. Anal. (Warsaw)* 52:1049–1057.
2628. Kauffmann, J.-M., B. Lopez Ruiz, M.F. Gotor, and G.J. Patriarche. 1992. Electrochemical behaviour of Tizanidine at solid electrodes. *J. Pharm. Biomed. Anal.* 10:763–767.
2629. Abu Shawish, H.M., N. Abu Ghalwa, F.R. Zaggout, S.M. Saadeh, A.R. Al Dalou, and A.A. Abou Assi. 2010. Improved determination of tramadol hydrochloride in biological fluids and pharmaceutical preparations utilizing a modified carbon paste electrode. *Biochem. Eng. J.* 48:237–245.
2630. Kauffmann, J.-M., J.-C. Vire, G.J. Patriarche, L.J. Nunez Vergara, and J.A. Squella. 1987. Voltammetric oxidation of trazodone. *Electrochim. Acta* 32:1159–1162.
2631. Ortiz, M.C., J. Arcos, J.V. Juarros, J. Lopez Palacios, and L.A. Sarabia. 1993. Robust procedure for calibration and calculation of detection limit of trimipramine by adsorptive stripping voltammetry at a carbon paste electrode. *Anal. Chem.* 65:678–682.
2632. Zayed, S.I.M. 2004. New plastic membrane and carbon paste ion-selective electrodes for potentiometric determination of triprolidine. *Anal. Sci. (Jpn.)* 20:1043–1048.
2633. Guo, H.-S., N.Y. He, S.-X. Ge, D. Yang, and J.-N. Zhang. 2005. Molecular sieves materials modified carbon paste electrodes for the determination of cardiac Troponin I by anodic stripping voltammetry. *Micropor. Mesopor. Mater.* 85:89–95.
2634. He, N.-Y., H.-S. Guo, S.-X. Ge, D. Yang, and J.-N. Zhang. 2005. Determination of cardiac Troponin I by anodic stripping voltammetry over mesoporous materials modified carbon paste electrode. *Stud. Surf. Sci. Catal.* 158:2041–2048.
2635. Legorburu, M.J., R.M. Alonso, and R.M. Jimenez. 1993. Voltammetric study of the diuretic Xipamide. *Bioelectrochem. Bioenerg.* 32:57–66.
2636. Cosofret, V.V. 1982. *Membrane Electrodes in Drug-Substances Analysis*. Oxford, U.K.: Pergamon Press.
2637. Popkov, V.A. and V. Yu. Reshetnyak. 1983. Application of ion-selective electrodes in medicine and pharmacy. *Farmatsiya (Russ.)* 32:79–81.
2638. Kauffmann, J.-M., G.J. Patriarche, J.-C. Vire, and W.R. Heineman. 1985. Unusual electrochemical behaviour of a new phenothiazine. *Analyst (U.K.)* 110:349–351.

References

2639. Adams, R.N. 1969. Applications of modern electroanalytical techniques in pharmaceutical chemistry. *J. Pharm. Sci.* 58:1171–1184.
2640. Patriarche, G.J. and H. Zhang. 1990. Electroanalytical techniques for drug analysis (a review). *Electroanalysis* 2:573–579.
2641. Kauffmann, J.-M. and J.-C. Vire. 1993. Pharmaceutical and biomedical application of electroanalysis: A critical review. *Anal. Chim. Acta* 273:329–334.
2642. Lima, E.C., P.G. Fenga, J.R. Romero, and W.F. De Giovani. 1997. Electrochemical behavior of [Ru(4,4′-Me$_2$bpy)$_2$(PPh$_3$)(H$_2$O)](ClO$_4$)$_2$ in homogeneous solution and incorporated into a carbon paste electrode. Application to oxidations of benzylic compounds. *Polyhedron* 17:313–318.
2643. Kutner, W. 1989. A carbon molecular-sieve paste electrode modified with the ruthenium oxo-bridged dimer, diaqua-M-oxotetrakis(2,2′-bipyridine)diruthenium(4+), for electrocatalysis of benzyl alcohol oxidation. *J. Electroanal. Chem.* 259:99–111.
2644. Kutner, W., T.J. Meyer, and R.W. Murray. 1985. Electrochemical and electrocatalytic reactions of a ruthenium oxo complex in solution and in cation exchange beads in carbon paste electrodes. *J. Electroanal. Chem. Interfacial Electrochem.* 195:375–394.
2645. Razola, S.S., S. Pochet, K. Grosfils, and J.-M. Kauffmann. 2003. Amperometric determination of choline released from rat submandibular gland acinar cells using a choline oxidase biosensor. *Biosens. Bioelectron.* 18:185–191.
2646. Cardoso, W.S., V.L.N. Dias, W.M. Costa, I. Araujo Rodrigues, E.P. Marques, A.G. Sousa, J. Boaventura et al. 2009. Nickel-dimethylglyoxime complex modified graphite and carbon paste electrodes: Preparation and catalytic activity towards methanol/ethanol oxidation. *J. Appl. Electrochem.* 39:55–64.
2647. Bilitewski, U. and R.D. Schmid. 1989. Alcohol determination by modified carbon paste electrodes. *GBF Monographs (Biosens. Appl. Med. Environ. Prot. Process. Control)* 13:99–102.
2648. Boujtita, M. and N. El Murr. 1995. Biosensors for analysis of ethanol in foods. *J. Food Sci.* 60:201–204.
2649. Koyuncu, D., P.E. Erden, S. Pekyardimci, and E. Kilic. 2007. A new amperometric carbon paste enzyme electrode for ethanol determination. *Anal. Lett.* 40:1904–1922.
2650. Yao, Q., S. Yabuki, and F. Mizutani. 1998. Preparation of a carbon paste/alcohol dehydrogenase electrode based on oil-soluble mediator. *Chem. Sens.* 14 (Suppl. A, *Proceedings of the 26th Chemical Sensor Symposium*):153–156.
2651. Yabuki, S. and F. Mizutani. 1997. Preparation of carbon paste-enzyme electrode using polyethylene glycol-alcohol dehydrogenase hybrid. *Denki Kagaku Oyobi Kogyo Butsuri Kagaku* 65:471–473.
2652. Larsson, N. and L. Gorton. 1998. The effect of various polyelectrolytes on alcohol oxidase/horseradish peroxidase/ferrocene modified carbon paste electrodes. *Sci. Pap. Univ. Pardubice Ser. A* 3:31–39.
2653. Liden, H., A.R. Vijayakumar, L. Gorton, and G. Marko-Varga. 1998. Rapid alcohol determination in plasma and urine by column liquid chromatography with biosensor detection. *J. Pharm. Biomed. Anal.* 17:1111–1128.
2654. Marko Varga, G., K. Johansson, and L. Gorton. 1993. A reagentless coimmobilized alcohol biosensor as a detection unit in column liquid chromatography for the determination of methanol and ethanol. *Quim. Anal. (Barcelona)* 12:24–29.
2655. Marko Varga, G., K. Johansson, and L. Gorton. 1994. Enzyme-based biosensor as a selective detection unit in column liquid chromatography. *J. Chromatogr. A* 660:153–167.
2656. Marko Varga, G., K. Johansson, and L. Gorton. 1993. An enzyme electrode as a detection unit in column liquid chromatography for the determination of methanol and ethanol. In *Bioelectroanalysis, 2nd Symposium*, Ed. E. Pungor, pp. 377–391. Budapest: Akademiai Kiado.
2657. Gorton, L., G. Joensson-Pettersson, E. Csoregi, K. Johansson, E. Dominguez, and G. Marko-Varga. 1992. Amperometric biosensors based on an apparent direct electron transfer between electrodes and immobilized peroxidases. *Analyst* 117:1235–1241.
2658. Johansson, K., G. Joensson-Pettersson, L. Gorton, G. Marko-Varga, and E. Csoregi. 1993. A reagentless amperometric biosensor for alcohol detection in column liquid chromatography based on co-immobilized peroxidase and alcohol oxidase in carbon paste. *J. Biotechnol.* 31:301–316.
2659. Azevedo, A.M., D.M.F. Prazeres, J.M.S. Cabral, and L.P. Fonseca. 2005. Ethanol biosensors based on alcohol oxidase. *Biosens. Bioelectron.* 21:235–247.
2660. Ying, Q. and G. Huan. 1992. The primary study of yeast-carbon paste electrode for alcohol determination. *Shipin Yu Fajiao Gongye* 1:44–48.
2661. Kubiak, W.W. and J. Wang. 1989. Yeast-based carbon paste bioelectrode for ethanol. *Anal. Chim. Acta* 221:43–51.
2662. Lu, Y., J. Wang, P. Li, and W. Zhang. 1997. Study on yeast electrode for determining alcohol. *Weishengwuxue Tongbao* 24:218–220.

2663. Ozsoz, M. and J. Wang. 1991. Tomato seed-based amperometric sensor for the determination of alcohols. *Electroanalysis* 3:655–658.
2664. Kondo, T. and T. Ikeda. 1999. An electrochemical method for the measurements of substrate-oxidizing activity of acetic acid bacteria using a carbon-paste electrode modified with immobilized bacteria. *Appl. Microbiol. Biotechnol.* 51:664–668.
2665. Yazgan, I., T. Aydin, D. Odaci, and S. Timur. 2008. Use of pyranose oxidase enzyme in inhibitor biosensing. *Anal. Lett.* 41:2088–2096.
2666. Barranco, J. and A.R. Pierna. 2009. Kinetics of methanol electro-oxidation on Pt, Pt-Sn and Pt-Ru amorphous metallic alloys. *J. New Mater. Electrochem. Syst.* 12:69–76.
2667. Barranco, J. and A.R. Pierna. 2007. Bifunctional amorphous alloys more tolerant to carbon monoxide. *J. Power Sources* 169:71–76.
2668. Ojani, R., J.-B. Raoof, and B. Norouzi. 2009. Platinum-doped cobalt hexacyanoferrate film-carbon paste electrode for electrocatalytic oxidation of methanol. *Asian J. Chem.* 21:6752–6762.
2669. Ojani, R., J.-B. Raoof, and S. Fathi. 2009. Nickel-poly(o-aminophenol)-modified carbon paste electrode; an electrocatalyst for methanol oxidation. *J. Solid State Electrochem.* 13:927–934.
2670. Ojani, R., J.-B. Raoof, and S.R.H. Zavvarmahalleh. 2008. Electrocatalytic oxidation of methanol on carbon paste electrode modified by nickel ions dispersed into poly(1,5-diaminonaphthalene) film. *Electrochim. Acta* 53:2402–2407.
2671. Barranco, J. and A.R. Pierna. 2007. Amorphous $Ni_{59}Nb_{40}Pt_{(1-x)}Y_x$ (Y = Sn, Ru; x = 0%, 0.4%) modified carbon paste electrodes and their role in the electrochemical methanol deprotonation and CO oxidation process. *J. Non-Cryst. Solids* 353:851–854.
2672. Timur, S., D. Odaci, A. Dincer, F. Zihnioglu, A. Telefoncu, and L. Gorton. 2007. Sulfhydryl oxidase modified composite electrode for the detection of reduced thiolic compounds. *Sens. Actuators B* B125:234–239.
2673. Miki, K., H. Kinoshita, Y. Yamamoto, N. Taniguchi, and T. Ikeda. 1995. An amperometric pyruvate sensor based on a pyruvate dehydrogenase-immobilized carbon paste electrode containing vitamin K3 as a mediator. *Denki Kagaku Oyobi Kogyo Butsuri Kagaku* 63:1121–1127.
2674. Oungpipat, W. and P.W. Alexander. 1994. An amperometric bi-enzyme sensor for glycolic acid determination based on spinach tissue and ferrocene-mediation. *Anal. Chim. Acta* 295:37–46.
2675. Zhu, L., Y. Li, and G. Zhu. 2004. A novel renewable plant tissue-based electrochemiluminescent biosensor for glycolic acid. *Sens. Actuators B* B98:115–121.
2676. Liawrungrath, S., P. Purachat, W. Oungpipat, and C. Dongduen. 2008. Sunflower leaves tissue-based bioelectrode with amperometric flow-injection system for glycolic acid determination in urine. *Talanta* 77:500–506.
2677. Spohn, V.U. 1998. Enzyme-modified carbon paste electrodes—A new way to reagent-free enzymic analysis. *Nova Acta Leopold. Suppl.* 15:155–175.
2678. Shu, H.-C., L. Gorton, B. Persson, and B. Mattiasson. 1995. A reagentless amperometric electrode based on carbon paste, chemically modified with D-lactate dehydrogenase, NAD+, and mediator containing polymer for D-lactic acid analysis. II. Online monitoring of fermentation process. *Biotechnol. Bioeng.* 46:280–284.
2679. Shu, H.-C. and N.-P. Wu. 2001. A chemically modified carbon paste electrode with d-lactate dehydrogenase and alanine aminotransferase enzyme sequences for d-lactic acid analysis. *Talanta* 54:361–368.
2680. Pohanka, M. and P. Zboril. 2008. Amperometric biosensor for D-lactate assay. *Food Technol. Biotechnol.* 46:107–110.
2681. Ledru, S. and M. Boujtita. 2004. Electrocatalytic oxidation of ascorbate by heme-FeIII/heme-FeII redox couple of the HRP and its effect on the electrochemical behavior of an L-lactate biosensor. *Bioelectrochemistry* 64:71–78.
2682. Boujtita, M., M. Chapleau, and N. El Murr. 1996. Enzymic electrode for the determination of L-lactate. *Electroanalysis* 8:485–488.
2683. Spohn, U., D. Narasaiah, and L. Gorton. 1997. Reagentless hydrogen peroxide and L-lactate sensors based on carbon paste electrodes modified with different peroxidases and lactate oxidases. *J. Prakt. Chem.* 339:607–614.
2684. Spohn, U., D. Narasaiah, and L. Gorton. 1996. The influence of the carbon paste composition on the performance of an amperometric bienzyme sensor for L-lactate. *Electroanalysis* 8:507–514.
2685. Garjonyte R,V. Melvydas, and Malinauskas. 2008. Effect of yeast pretreatment on the characteristics of yeast-modified electrodes as mediated amperometric biosensors for lactic acid. *Bioelectrochemistry* 74:188–194.

2686. Kulys, J., L. Wang, and V. Razumas. 1992. Sensitive yeast bioelectrode to L-lactate. *Electroanalysis* 4:527–532.
2687. Gan, N. and C. Ge. 2006. Separation of organic acids by capillary zone electrophoresis and electrochemical detection in fruit juices. *Yingyang Xuebao* 28:255–258, 262.
2688. Liu, Y., J. Huang, D. Wang, H. Hou, and T. You. 2010. Electrochemical determination of oxalic acid using palladium nanoparticle-loaded carbon nanofiber modified electrode. *Anal. Methods* 2:855–859.
2689. Shaidarova, L.G., S.A. Zaripova, L.N. Tikhonova, G.K. Budnikov, and I.M. Fitsev. 2001. Electrocatalytic determination of oxalate ions on chemically modified electrodes. *Russ. J. Appl. Chem.* 74:750–754.
2690. Santos, L.M. and R.P. Baldwin. 1987. Profiling of oxalic acid and a-keto acids in blood and urine by liquid chromatography with electrochemical detection at a chemically modified electrode. *J. Chromatogr. Biomed. Appl.* 414:161–166.
2691. Wang, J. and N. Naser. 1995. Modified carbon-wax composite electrodes. *Anal. Chim. Acta* 316:253–259.
2692. Ramos, A.R., J. Arguello, H.A. Magosso, and Y. Gushikem. 2008. Al_2O_3 coated with 3-n-propyl-1-azonia-4-azabicyclo[2.2.2]octane silsesquioxane chloride and its use for immobilization of cobalt(II) tetrasulfonated phthalocyanine in oxalic acid electrooxidation. *J. Braz. Chem. Soc.* 19:755–761.
2693. Mishra, R., H. Yadav, and C.S. Pundir. 2010. An amperometric oxalate biosensor based on sorghum leaf oxalate oxidase immobilized on carbon paste electrode. *Anal. Lett.* 43:151–160.
2694. Zhou, D.-M., P. Nigam, J. Jones, and R. Marchant. 1995. Production of salicylate hydroxylase from *Pseudomonas putida* UUC-1 and its application in the construction of a biosensor. *J. Chem Technol. Biotechnol.* 64:331–338.
2695. Wimmerova, M. and L. Macholan. 1999. Sensitive amperometric biosensor for the determination of biogenic and synthetic amines using pea seedlings amine oxidase: A novel approach for enzyme immobilisation. *Biosens. Bioelectron.* 14:695–702.
2696. Rodriguez Mendez, M.L., M. Gay, C. Apetrei, and J.A. De Saja. 2009. Biogenic amines and fish freshness assessment using a multisensor system based on voltammetric electrodes. Comparison between CPE and screen-printed electrodes. *Electrochim. Acta* 54:7033–7041.
2697. Castilho, T.J., M.d.P.T. Sotomayor, and L.T. Kubota. 2005. Amperometric biosensor based on horseradish peroxidase for biogenic amine determinations in biological samples. *J. Pharm. Biomed. Anal.* 37:785–791.
2698. Zagorovskiy, G.M., I.G. Sydorenko, N.M. Vlasova, and V.V. Lobanov. 2008. Voltammetric determination of biogenic amines on the electrodes of carbon nanotubes modified with polyaniline. *Khim. Fiz. Tekhnol. Poverkh. (Ukr.)* 14:317–324.
2699. Kondo, T., J. Matsui, I. Takahashi, and S. Hatano. 2003. Construction of histamine sensor. *Aichi-ken Sangyo Gijutsu Kenkyusho Kenkyu Hokoku* 2:130–131.
2700. Sidorenko, I.G., O.V. Markitan, N.N. Vlasova, G.M. Zagorovskii, and V.V. Lobanov. 2007. Voltammetric determination of the content of biogenic amines in water using carbon nanotube electrode. *Khim. Fiz. Tekhnol. Poverkh. (Ukr.)* 13:195–200.
2701. Yang, J.-K., K.S. Ha, H.S. Baek, S.S. Lee, and M.L. Seo. 2004. Amperometric determination of urea using enzyme-modified carbon paste electrode. *Bull. Korean Chem. Soc.* 25:1499–1502.
2702. Wang, C. and Z. Gao. 2008. Electrocatalytic oxidation of N-acetyl-L-cysteine at caffeic acid modified carbon paste electrode. *Fenxi Ceshi Xuebao* 27:1326–1329.
2703. Bai, L. and Z.-N. Gao. 2008. Electrocatalytic oxidation of N-acetyl-L-cysteine at 10-methylphenothiazine modified carbon paste electrode and its practical analytical application. *Yingyong Huaxue* 25:702–705.
2704. Luque, G.L., N.F. Ferreyra, and G.A. Rivas. 2007. Electrochemical sensor for amino acids and albumin based on composites containing carbon nanotubes and copper microparticles. *Talanta* 71:1282–1287.
2705. Cui, S. 1993. Flow injection amperometric determination of cucurbitin and some other amino acids with copper-modified carbon paste electrode. *Yanbian Daxue Xuebao, Ziran Kexueban* 19:37–41.
2706. Chen, Q., J. Wang, G. Rayson, B. Tian, and Y. Lin. 1993. Sensor array for carbohydrates and amino acids based on electrocatalytic modified electrodes. *Anal. Chem.* 65:251–254.
2707. Liu, J. and J. Wang. 1999. Remarkable thermostability of bioelectrodes based on enzymes immobilized within hydrophobic semi-solid matrices. *Biotechnol. Appl. Biochem.* 30:177–183.
2708. Kacaniklic, V., K. Johansson, G. Marko-Varga, L. Gorton, G. Joensson-Pettersson, and E. Csoeregi. 1994. Amperometric biosensors for detection of L- and D-amino acids based on co-immobilized peroxidase and L- and D-amino acid oxidases in carbon paste electrodes. *Electroanalysis* 6:381–390.
2709. Wu, X., B.J. Van Wie, and D.A. Kidwell. 2004. An enzyme electrode for amperometric measurement of D-amino acid. *Biosens. Bioelectron.* 20:879–886.

2710. Gorton, L., G. Marko-Varga, B. Persson, Z. Huan, H. Linden, E. Burestedt, S. Ghobadi, M. Smolander, S. Sahni, and T. Skotheim. 1996. Amperometric biosensors based on carbon paste electrodes chemically modified with redox-enzymes. *Adv. Mol. Cell Biol.* 15B:421–450.
2711. Maleki, N., A. Safavi, F. Sedaghati, and F. Tajabadi. 2007. Efficient electrocatalysis of L-cysteine oxidation at carbon ionic liquid electrode. *Anal. Biochem.* 369:149–153.
2712. Wu, J., Y. Zou, N. Gao, J. Jiang, G. Shen, and R. Yu. 2005. Electrochemical performances of C/Fe nanocomposite and its use for mediator-free glucose biosensor preparation. *Talanta* 68:12–18.
2713. Baldrianova, L., P. Agrafiotou, I. Švancara, K. Vytřas, and S. Sotiropoulos. 2008. The determination of cysteine at Bi-powder carbon paste electrodes by cathodic stripping voltammetry. *Electrochem. Comm.* 10:918–921.
2714. Shaidarova, L.G., S.A. Ziganshina, and G.K. Budnikov. 2003. Electrocatalytic oxidation of cysteine and cystine at a carbon-paste electrode modified with ruthenium(IV) oxide. *J. Anal. Chem./*transl. of *Zh. Anal. Khim.* 58:577–582.
2715. Ojani, R., J.-B. Raoof, and E. Zarei. 2010. Preparation of poly N,N-dimethylaniline/ferrocyanide film modified carbon paste electrode: Application to electrocatalytic oxidation of L-cysteine. *J. Electroanal. Chem.* 638:241–245.
2716. Sugawara, K., S. Tanaka, and M. Taga. 1991. Voltammetric behavior of cysteine by a chemically modified carbon-paste electrode with copper(II) cyclohexylbutyrate. *J. Electroanal. Chem.* 316:305–314.
2717. Bai, L. and Z. Gao. 2008. Electrocatalytic oxidation of L-cysteine at acetylferrocene modified carbon paste electrode and its electrochemical kinetics. *Fenxi Ceshi Xuebao* 27:273–276.
2718. Khaloo, S.S., M.K. Amini, S. Tangestaninejad, S. Shahrokhian, and R. Kia. 2004. Voltammetric and potentiometric study of cysteine at cobalt(II) phthalocyanine modified carbon-paste electrode. *J. Iran. Chem. Soc.* 1:128–135.
2719. Zhang, Y.-J., X. Fang, J. Wang, and J.-D. Wang. 2008. Carbon paste electrode incorporating multi-walled carbon nanotube/a-(2,4-di-tert-butylphenoxy) phthalo-cyaninatocobalt for determination of L-cysteine. *Yingyong Huaxue* 25:909–912.
2720. Lima, P.R., W.J.R. Santos, R.d.C.S. Luz, F.S. Damos, A.B. Oliveira, M.O.F. Goulart, and L.T. Kubota. 2008. An amperometric sensor based on electrochemically triggered reaction: Redox-active Ar-NO/Ar-NHOH from 4-nitrophthalonitrile-modified electrode for the low voltage cysteine detection. *J. Electroanal. Chem.* 612:87–96.
2721. Huang, T.H., T. Kuwana, and A. Warsinke. 2002. Analysis of thiols with tyrosinase-modified carbon paste electrodes based on blocking of substrate recycling. *Biosens. Bioelectron.* 17:1107–1113.
2722. Amine, A. and J.M. Kauffmann. 1992. Preparation and characterization of a fragile enzyme immobilized carbon paste electrode. *Bioelectrochem. Bioenerg.* 28:117–125.
2723. Amine, A., J.M. Kauffmann, and G. Palleschi. 1993. Investigation of the batch injection analysis technique with amperometric biocatalytic electrodes using a modified small-volume cell. *Anal. Chim. Acta* 273:213–218.
2724. Ghobadi, S., E. Csoeregi, G. Marko-Varga, and L. Gorton. 1996. Bienzyme carbon paste electrodes for L-glutamate determination. *Curr. Sep.* 14:94–102.
2725. Wang, G., T. Peng, B. Shen, P. Zhu, and L. Qu. 1995. Determination of amino acid with carbon paste electrode. III. Determination of histidine. *Fenxi Huaxue* 23:703–706.
2726. Zou, Y., J. Wang, J. Mo, and R. Zhang. 1999. Voltammetric determination of amino acids by modified carbon paste electrodes: II. Study on the electrochemical behavior of histidine on alumina-modified carbon paste electrode. *Fenxi Ceshi Xuebao* 18:47–49.
2727. Stefan-van Staden, R.-I. and L. Holo. 2007. Enantioselective, potentiometric membrane electrodes based on cyclodextrins for the determination of L-histidine. *Sens. Actuators B* B120:399–402.
2728. Stefan-van Staden, R.-I., B. Lal, and L. Holo. 2007. Enantioselective potentiometric membrane electrodes based on C60 fullerene and its derivatives for the assay of L-Histidine. *Talanta* 71:1434–1437.
2729. Akalu G., A. Belay, E. Csoregi, L. Gorton, G. Johansson, K. Kalcher, N. Larsson et al. 2004. Glutamate oxidase advances the selective bioanalytical detection of the neurotoxic amino acid β-ODAP in grass pea: A decade of progress. *Pure Appl. Chem.* 76:765–775.
2730. Beyene, N.W., H. Moderegger, and K. Kalcher. 2004. Development of an amperometric biosensor for β-ODAP. *Electroanalysis* 16:268–274.
2731. Beyene, N.W., H. Moderegger, and K. Kalcher. 2004. Simple and effective procedure for immobilization of oxidases onto MnO_2-bulk-modified, screen-printed carbon electrodes. *S. Afr. J. Chem.* 57:1–7.
2732. Weiss, D.J., M. Dorris, A. Loh, and L. Peterson. 2007. Dehydrogenase based reagentless biosensor for monitoring phenylketonuria. *Biosens. Bioelectron.* 22:2436–2441.

2733. Huang, T., A. Warsinke, T. Kuwana, and F.W. Scheller. 1998. Determination of L-phenylalanine based on an NADH-detecting biosensor. *Anal. Chem.* 70:991–997.

2734. Ozoemena, K.I. and R.-I. Stefan. 2005. Enantioselective potentiometric membrane electrodes based on α-, β- and γ-cyclodextrins as chiral selectors for the assay of L-proline. *Talanta* 66:501–504.

2735. Linders, C.R., B.J. Vincke, J.C. Vire, J.M. Kauffmann, and G.J. Patriarche. 1985. Voltammetry of tryptophan using solid electrodes. *J. Pharm. Belg.* 40:27–33.

2736. Fiorucci, A.R. and E.T.G. Cavalheiro. 2002. The use of carbon paste electrode in the direct voltammetric determination of tryptophan in pharmaceutical formulations. *J. Pharm. Biomed. Anal.* 28:909–915.

2737. Zoulis, N.E., D.P. Nikolelis, and C.E. Efstathiou. 1990. Preconcentration of indolic compounds at a carbon paste electrode and indirect determination of L-tryptophan in serum by adsorptive stripping voltammetry. *Analyst (U.K.)* 115:291–295.

2738. Inada, K., K. Nishiyama, and I. Taniguchi. 2009. Voltammetric determination of tryptophan and serotonin at glassy carbon paste electrodes in a potential region free from interfering substances. *Chem. Lett.* 38:686–687.

2739. Wang, H., A. Zhang, and H. Cui. 1997. Electrochemical activation of a carbon-paste electrode with paraffin as a binder and its application to the determination of tryptophan. *Fenxi Huaxue* 25:85–88.

2740. Zou, Y.-D., J. Wang, J.-Y. Mo, and R.-J. Zhang. 1999. Voltammetric determination of amino acids by modified carbon paste electrodes. Part I. Electrochemical behavior of tryptophan and tyrosine. *Fenxi Ceshi Xuebao* 18:25–28.

2741. Wu, J., Q. Shi, G. Wang, and T. Peng. 1994. Potato-juice-modified carbon paste electrode for the determination of tryptophan. *Fenxi Huaxue* 22:599–601.

2742. Wang, G., T. Peng, B. Shen, P. Zhu, and L. Qu. 1993. Determination of an amino acid with montmorillonite-modified carbon paste electrode. I. Determination of tryptophan. *Fenxi Huaxue* 21:779–782.

2743. Kitagawa, T. and S. Tsushima. 1971. Simultaneous determination of tyrosine and dopa using rotating carbon paste disk electrode. *Bunseki Kagaku* 20:1561–1565.

2744. Toyokichi, K. and Y. Kanei. 1970. Voltammetry at solid electrode. VI. Anodic oxidation of tyrosine at carbon paste electrode. *Bunseki Kagaku* 19:642–649.

2745. Wang, G., T. Peng, B. Shen, P. Zhu, and L. Qu. 1994. Determination of amino acid with montmorillonite-modified carbon paste electrode. II. Determination of tyrosine. *Fenxi Huaxue* 22:590–592.

2746. Kong, Y., G.-T. Huang, Z.-D. Chen, and W.-C. Wang. 2007. Molecular-imprinted electrochemical sensor for recognition of enantiomorphs of D-tyrosine and L-tyrosine. *Lihua Jianyan, Huaxue Fence* 43:997–999, 1003.

2747. He, Q. 2003. Studies on the electrochemical behavior of tyrosine at a polyvinyl-pyrrolidone modified carbon paste electrode and its determination. *Fenxi Kexue Xuebao* 19:346–348.

2748. Reis, N.S., S.H.P. Serrano, R. Meneghatti, and E.d.S. Gil. 2009. Electrochemical methods used for the evaluation of the antioxidant activities of natural products. *Lat. Am. J. Pharm.* 28:949–953.

2749. Diego, E., L. Agui, A. Gonzalez-Cortes, P. Yanez-Sedeno, J.M. Pingarron, and J.M. Kauffmann. 1998. Critical comparison of paraffin carbon paste and graphite-poly(tetrafluorethylene) composite electrodes concerning the electroanalytical behavior of various antioxidants of different hydrophobicity. *Electroanalysis* 10:33–38.

2750. Jayasri, D. and S. Sriman Narayanan. 2006 (volume date 2007). Manganese(II) hexacyanoferrate based renewable amperometric sensor for the determination of butylated hydroxyanisole in food products. *Food Chem.* 101:607–614.

2751. Fernandes, S.C., I.R.W. Zwirtes de Oliveira, and I.C. Vieira. 2007. A green bean homogenate immobilized on chemically crosslinked chitin for determination of caffeic acid in white wine. *Enzyme Microb. Technol.* 40:661–668.

2752. Arduini, F., F. Di Giorgio, A. Amine, F. Cataldo, D. Moscone, and G. Palleschi. 2010. Electroanalytical characterization of carbon black nanomaterial paste electrode: Development of highly sensitive tyrosinase biosensor for catechol detection. *Anal. Lett.* 43:1688–1702.

2753. Gil, E.S., L. Muller, M.F. Santiago, and T.A. Garcia. 2009. Biosensor based on brut extract from laccase (*Pycnoporus sanguineus*) for environmental analysis of phenolic compounds. *Portugaliae Electrochim. Acta* 27:215–225.

2754. Szewczynska, M. and M. Trojanowicz. 2000. Amperometric enzymatic detection of phenols for HPLC. *Chem. Anal. (Warsaw)* 45:667–679.

2755. Dantoni, P., S.H.P. Serrano, A.M.O. Brett, and I.G.R. Gutz. 1998. Flow-injection determination of catechol with a new tyrosinase/DNA biosensor. *Anal. Chim. Acta* 366:137–145.

2756. Lima, A.W.O., E.K. Vidsiunas, V.B. Nascimento, and L. Angnes. 1998. Vegetable tissue from *Latania* sp.: An extraordinary source of naturally immobilized enzymes for the detection of phenolic compounds. *Analyst (U.K.)* 123:2377–2382.

2757. Tan, X.-C., Z.-W. Huang, J.-L. Zhang, Z.-R. Huang, and H. Liang. 2008. Determination of catechol in beer based on amperometric biosensor potato tissue. *Hubei Minzu Xueyuan Xuebao, Ziran Kexueban* 26:294–296.

2758. Cevik, S. and U. Anik. 2010. Banana tissue-nanoparticle/nanotube based glassy carbon paste electrode biosensors for catechol detection. *Sens. Lett.* 8:667–671.

2759. Grabaric, B.S., I. Kruhak, J. Velikonja, and M. Tkalcec. 1993. Electrochemical investigation of dopamine and pyrocatechol with unmodified and banana tissue modified carbon paste electrodes. *Prehrambeno-Tehnol. Biotehnol. Rev. (Serbo-Croat.)* 31:131–135.

2760. Eggins, B.R., C. Hickey, S.A. Toft, and D. Min Zhou. 1997. Determination of flavanols in beers with tissue biosensors. *Anal. Chim. Acta* 347:281–288.

2761. Adam, V., R. Mikelova, J. Hubalek, P. Hanustiak, M. Beklova, P. Hodek, A. Horna et al. 2007. Utilizing of square wave voltammetry to detect flavonoids in the presence of human urine. *Sensors* 7:2402–2418.

2762. Zoulis, N.E. and C.E. Efstathiou. 1996. Preconcentration at a carbon-paste electrode and determination by adsorptive-stripping voltammetry of rutin and other flavonoids. *Anal. Chim. Acta* 320:255–261.

2763. Volikakis, G.J. and C.E. Efstathiou. 2000. Determination of rutin and other flavonoids by flow-injection/adsorptive stripping voltammetry using nujol-graphite and diphenylether-graphite paste electrodes. *Talanta* 51:775–785.

2764. Hodek, P., P. Hanustiak, J. Krizkova, R. Mikelova, S. Krizkova, M. Stiborova, L. Trnkova, A. Horna, M. Beklova, and R. Kizek. 2006. Toxicological aspects of flavonoid interaction with biomacromolecules. *Neuroendocrinol. Lett.* 27(Suppl. 2):14–17.

2765. Ying, T., S. Hu, and T. Qi. 1994. Development of apple tissue-carbon paste electrode for hydroquinone determination. *Shanghai Gongye Daxue Xuebao* 15:280–282.

2766. Jenkins, S.H., Heineman W.R., and Halsall, H. 1988. Extending the detection limit of solid-phase electrochemical enzyme immunoassay to the attomole level. *Anal. Biochem.* 168:292–299.

2767. Lu, Y., J. Ma, S. Cao, C. Xu, and S. Hu. 1996. Electrochemical oxidation and determination of phenol. *Fenxi Kexue Xuebao* 12:323–325.

2768. Zou, Y-D. and J.-Y. Mo. 1997. The 2.5th order differential voltammetric determination of phenol with a composite carbon paste/polyamide electrode. *Anal. Chim. Acta* 353:71–78.

2769. Chen, W.-H., S. Yuan, and C.-G. Hu. 2005. Voltammetric detection of phenol at CTAB-modified carbon paste electrode. *Fenxi Kexue Xuebao* 21:54–56.

2770. Wang, G., T. Peng, and P. Zhu. 1993. Determination of phenol with montmorillonite-modified carbon paste electrode. *Fenxi Huaxue* 21:672–675.

2771. Tzang, C.H., C.-W. Li, J. Zhao, and M. Yang. 2005. Electrocatalytic phenol oxidation on mixed Pt-RuO2 nanoparticle modified electrode. *Anal. Lett.* 38:1735–1746.

2772. Pei, H.-L., H. Li, and W. Liu. 2007. Determination of phenol on a carbon paste electrode modified with cobalt phthalocyanine. *Fenxi Shiyanshi* 26:102–105.

2773. Tingry, S., C. Innocent, S. Touil, A. Deratani, and P. Seta. 2006. Carbon paste biosensor for phenol detection of impregnated tissue: Modification of selectivity by using β-cyclodextrin-containing PVA membrane. *Mater. Sci. Eng., C* 26:222–226.

2774. Kim, G.-Y., N.M. Cuong, S.-H. Cho, J. Shim, J.-J. Woo, and S.-H. Moon. 2007. Improvement of an enzyme electrode by poly(vinyl alcohol) coating for amperometric measurement of phenol. *Talanta* 71:129–135.

2775. Makhaeva, G., L. Sigolaeva, L. Zhuravleva, A. Eremenko, I. Kurochkin, V. Malygin, and R. Richardson. 2003. Biosensor detection of neuropathy target esterase in whole blood as a biomarker of exposure to neuropathic organophosphorus compounds. *J. Toxicol. Environ. Health Part A* 66:599–610.

2776. Sigolaeva, L.V., A. Makower, A.V. Eremenko, G.F. Makhaeva, V.V. Malygin, I.N. Kurochkin, and F.W. Scheller. 2001. Bioelectrochemical analysis of neuropathy target esterase activity in blood. *Anal. Biochem.* 290:1–9.

2777. Bassi, A.S. and C. McGrath. 1999. Carbon paste biosensor based on crude soybean seed hull extracts for phenol detection. *J. Agric. Food Chem.* 47:322–326.

2778. Ortega, F., E. Dominguez, E. Burestedt, J. Emneus, L. Gorton, and G. Marko-Varga. 1994. Phenol oxidase-based biosensors as selective detection units in column liquid chromatography for determination of phenolic compounds. *J. Chromatogr. A* 675:65–78.

2779. Petit, Chr., A. Nagy, G. Quarin, and J.-M. Kauffmann. 1996. Enzymic electrode for the analysis of phenol and catechol derivatives of pharmaceutical interest. *J. Pharm. Belg.* 51:1–8.

2780. Cui, L. and F. Meng. 2003. Study on determination of phenols with tyrosinase based biosensor activated by polysaccharide. *Lihua Jianyan, Huaxue Fence* 39:629–632.

2781. Cui, L. and F. Meng. 2003. Study of polysaccharide modified amperometric biosensor for the determination of phenols based on tyrosinase. *Huaxue Chuanganqi* 23:49–57.
2782. Lindgren, A., T. Ruzgas, J. Emneus, E. Csoeregi, L. Gorton, and G. Marko-Varga. 1996. Flow injection analysis of phenolic compounds with carbon paste electrodes modified with tyrosinase purchased from different companies. *Anal. Lett.* 29:1055–1068.
2783. Erdem, A., N. Altinigne, E. Kilinc, L. Gokgunnec, T. Dalbasti, and M. Ozsoz. 1998. Amperometric biosensor based on mushroom tissue tyrosinase for the determination of phenolic compounds. *FABAD Farmasotik Bilimler Dergisi* 23:1–6.
2784. Rogers K.R., F. Lu, and J. Wang. 1997. Liquid chromatography system with an enzyme biosensor for measurement of phenolics. In *Field Analytical Methods for Hazardous Wastes and Toxic Chemicals, Proceedings of a Specialty Conference*, Las Vegas, NV, pp. 861–869.
2785. Ruzgas, T., J. Emneus, L. Gorton, and G. Marko-Varga. 1995. The development of a peroxidase biosensor for monitoring phenol and related aromatic compounds. *Anal. Chim. Acta* 311:245–253.
2786. Wollenberger, U. and B. Neumann. 1997. Quinoprotein glucose dehydrogenase-modified carbon paste electrode for the detection of phenolic compounds. *Electroanalysis* 9:366–371.
2787. Apetrei, C., M.L. Rodriguez-Mendez, and J.A. de Saja. 2005. Modified carbon paste electrodes for discrimination of vegetable oils. *Sens. Actuators B* 111:403–409.
2788. Apetrei, C., I.M. Apetrei, I. Nevares, M. Del Alamo, V. Parra, M.L. Rodriguez-Mendez, and J.A. De Saja. 2007. Using an e-tongue based on voltammetric electrodes to discriminate among red wines aged in oak barrels or aged using alternative methods. *Electrochim. Acta* 52:2588–2594.
2789. Santhiago, M., R.A. Peralta, A. Neves, G.A. Micke, and I.C. Vieira. 2008. Rosmarinic acid determination using biomimetic sensor based on purple acid phosphatase mimetic. *Anal. Chim. Acta* 613:91–97.
2790. Franzoi, A.C., R.A. Peralta, A. Neves, and I.C. Vieira. 2009. Biomimetic sensor based on MnIII/MnII complex as manganese peroxidase mimetic for determination of rutin. *Talanta* 78:221–226.
2791. Zwirtes de Oliveira, I.R.W., S.C. Fernandes, and I.C. Vieira. 2006. Development of a biosensor based on gilo peroxidase immobilized on chitosan chemically crosslinked with epichlorohydrin for determination of rutin. *J. Pharm. Biomed. Anal.* 41:366–372.
2792. Arribas, A.S., E. Bermejo, A. Zapardiel, H. Tellez, J. Rodriguez-Flores, M. Zougagh, A. Rios, and M. Chicharro. 2009. Screening and confirmatory methods for the analysis of macrocyclic lactone mycotoxins by CE with amperometric detection. *Electrophoresis* 30:499–506.
2793. Smolander, M., L. Gorton, H.S. Lee, T. Skotheim, and H.-L. Lan. 1995. Ferrocene-containing polymers as electron transfer mediators in carbon paste electrodes modified with PQQ-dependent aldose dehydrogenase. *Electroanalysis* 7:941–946.
2794. Smolander, M. 1995. Electrochemical aldose detection with PQQ-dependent aldose dehydrogenase. *VTT Publ.* 229:105.
2795. Maestre, E., I. Katakis, A. Narvaez, and E. Dominguez. 2005. A multianalyte flow electrochemical cell: Application to the simultaneous determination of carbohydrates based on bioelectrocatalytic detection. *Biosens. Bioelectron.* 21:774–781.
2796. Chien, H.-C. and T.-C. Chou. 2010. Glassy carbon paste electrodes for the determination of fructosyl valine. *Electroanalysis* 22:688–693.
2797. Wang, Q.-L., B.-J. Yang, and G.-X. Lu. 2003. Direct electrochemistry of hemoglobin immobilized on carbon paste electrode by inorganic film. *Gaodeng Xuexiao Huaxue Xuebao* 24:1561–1566.
2798. Tsugawa, W., F. Ishimura, K. Ogawa, and K. Sode. 2000. Development of an enzyme sensor utilizing a novel fructosyl amine oxidase from a marine yeast. *Electrochemistry* 68:869–871.
2799. Liu, L., J.-F. Song, P.-F. Yu, and B. Cui. 2007. Voltammetric determination of glucose based on reduction of copper(II)-glucose complex at lanthanum hydroxide nanowire modified carbon paste electrodes. *Talanta* 71:1842–1848.
2800. Yang, T., K. Jiao, J. Yang, C. Zhao, and W. Qu. 2006. Electrocatalytic oxidation of glucose by poly(o-aminophenol)/nickel modified carbon paste electrode. *Fenxi Huaxue* 34:1415–1418.
2801. Gorton, L., G. Bremle, E. Csoeregi, G. Joensson-Pettersson, and B. Persson. 1991. Amperometric glucose sensors based on immobilized glucose-oxidizing enzymes and chemically modified electrodes. *Anal. Chim. Acta* 249:43–54.
2802. Cespedes, F., E. Martinez-Fabregas, J. Bartoli, and S. Alegret. 1993. Amperometric enzymatic glucose electrode based on an epoxy-graphite composite. *Anal. Chim. Acta* 273:409–417.
2803. Kurusu, F., H. Tsunoda, A. Saito, A. Tomita, A. Kadota, N. Kayahara, I. Karube, and M. Gotoh. 2006. The advantage of using carbon nanotubes compared with edge plane pyrolytic graphite as an electrode material for oxidase-based biosensors. *Analyst* 131:1292–1298.

2804. Savitri, D. and C.K. Mitra. 1998. Electrochemistry of reconstituted glucose oxidase on carbon paste electrodes. *Bioelectrochem. Bioenerg.* 47:67–73.
2805. Kotzian, P., P. Brazdilova, K. Kalcher, K. Handlir, and K. Vytřas. 2007. Oxides of platinum metal group as potential catalysts in carbonaceous amperometric biosensors based on oxidases. *Sens. Actuators B* B124:297–302.
2806. Ikeda, T. 1992. Electrochemical biosensors based on biocatalyst electrodes. *Bull. Electrochem.* 8:145–159.
2807. Ikeda, T., H. Hiasa, and M. Senda. 1988. In *Redox Chemistry and Interfacial Behavior of Biological Molecules*. Eds. G. Dryhurst and K. Niki. pp. 193–201. New York: Plenum Press.
2808. Inagaki, T., H.S. Lee, P.D. Hale, T.A. Skotheim, and Y. Okamoto. 1989. Synthesis and electrochemical characterization of siloxane polymers containing hydroquinone and 1,4-naphthohydroquinone. *Macromolecules* 22:4641–4643.
2809. Narvaez, A., J. Parellada, E. Dominguez, and I. Katakis. 1996. Effects of mediator and paste hydrophobicity on the response of amperometric carbon paste electrodes: Binder paste electrodes. *Quim. Anal. (Barcelona)* 15:75–82.
2810. Sakura, S. and R.P. Buck. 1992. Amperometric processes with glucose oxidase embedded in the electrode. *Bioelectrochem. Bioenerg.* 28:387–400.
2811. Slilam M., L. Charpentier, and N. El Murr. 1991. Glucose determination in must. *Analusis* 19:M49–M52.
2812. Turner A.P.F., I. Karube, and G.S. Wilson, Eds. 1987. *Biosensors. Fundamentals and Applications.* pp. 276–290. Oxford, U.K.: University Press.
2813. Amine, A., J.-M. Kauffmann, G.J. Patriarche, and A.E. Kaifer. 1991. Long-term operational stability of a mixed glucose oxidase-redox mediator-carbon paste electrode. *Anal. Lett.* 24:1293–1315.
2814. Hale, P.D., T. Inagaki, H.I. Karan, Y. Okamoto, and T.A. Skotheim. 1989. A new class of amperometric biosensor incorporating a polymeric electron-transfer mediator. *J. Am. Chem. Soc.* 111:3482–3484.
2815. Hale, P.D., L.I. Boguslavsky, T. Inagaki, H.S. Lee, T.A. Skotheim, H.I. Karan, and Y. Okamoto. 1990. Ferrocene-modified siloxane polymers as electron relay systems in amperometric glucose sensors. *Mol. Cryst. Liq. Cryst.* 190:251–258.
2816. Hale, P.D., T. Inagaki, H.S. Lee, T.A. Skotheim, H.I. Karan, and Y. Okamoto. 1990. In *Biosensor Technology—Fundamentals and Applications*. Eds. R.P. Buck, W.E. Hatfield, M. Umaiia, and E.F. Bowden, pp. 195–200. New York: Marcel Dekker.
2817. Hale, P.D., L.I. Boguslavsky, T.A. Skotheim, L.F. Liu, H.S. Lee, H.I. Karan, H.L. Lan, and Y. Okamoto. 1992. Polymer developments for biosensors. In *Biosensors and Chemical Sensors*. Eds. P.G. Edelman and J. Wang. *ACS Symposium Series No. 487*. pp. 111–124. Washington, D.C: American Chemical Society.
2818. Saito, T. and M. Watanabe. 1999. Electron transfer reaction from glucose oxidase to an electrode via redox copolymers. *Polym. J.* 31:1149–1154.
2819. Casado, C.M., M. Moran, J. Losada, and I. Cuadrado. 1995. Siloxane and organosilicon dimers, monomers, and polymers with amide-linked ferrocenyl moieties. synthesis, characterization, and redox properties. *Inorg. Chem.* 34:1668–1680.
2820. Hale, P.D., H.S. Lee, and Y. Okamoto. 1993. Electrical communication between glucose oxidase and novel ferrocene-containing siloxane-ethylene oxide copolymers: Biosensor applications. *Anal. Lett.* 26:1–16.
2821. Saito, T., N. Kurosawa, M. Watanae, Y. Iwasaki, and K. Ishihara. 1998. Design of enzymic glucose sensor using biocompatible polymeric mediator. *Kobunshi Ronbunshu* 55:200–206.
2822. Nagasaka, H., T. Saito, H. Hatakeyama, and M. Watanabe. 1995. Preparation of novel redox copolymers: Poly[ferrocenylmethyl methacrylate-co-methoxy-oligo(ethylene oxide) methacrylate], and their use as polymeric mediators in amperometric glucose sensors. *Denki Kagaku oyobi Kogyo Butsuri Kagaku* 63:1088–1094.
2823. Rosen-Margalit, I. and J. Rishpon. 1993. Novel approaches for the use of mediators in enzyme electrodes. *Biosens. Bioelectron.* 8:315–323.
2824. Sun, C., Y. Zhang, Q. Gao, and H. Xu. 1993. Amperometric biosensors for glucose based on the carbon paste electrode modified by 1,1′-dimethylferrocene and $(C_6H_5)_4B \cdot N(CH_3)_4$ ion associate complex. *Fenxi Huaxue* 21:882–886.
2825. Serban, S., A.F. Danet, and N. El Murr. 2004. Rapid and sensitive automated method for glucose monitoring in wine processing. *J. Agric. Food Chem.* 52:5588–5592.
2826. Gueguen, S., M. Boujtita, and N. El Murr. 1999. Effect of the mutarotase on the analytical behaviour of modified carbon paste type glucose biosensor. *Analusis (Brux.)* 27:587–591.
2827. Lawrence, N.S., R.P. Deo, and J. Wang. 2004. Biocatalytic carbon paste sensors based on a mediator pasting liquid. *Anal. Chem.* 76:3735–3739.

2828. Evans, R.G., C.E. Banks, and R.G. Compton. 2004. Amperometric detection of glucose using self-catalytic carbon paste electrodes. *Analyst (U.K.)* 129:428–431.
2829. Toyama, S., Y. Chisuwa, S. Yamauchi, and Y. Ikariyama. 1996. Less positive potential workable blood glucose sensor. *Chem. Sensors* 12 (Suppl. A):141–144.
2830. Rosen-Margalit, I., A. Bettelheim, and J. Rishpon. 1993. Cobalt phthalocyanine as a mediator for the electrooxidation of glucose oxidase at glucose electrodes. *Anal. Chim. Acta* 281:327–333.
2831. Turdean, G., I.C. Popescu, and L. Oniciu. 1998. Glucose oxidase-cobalt(II) phthalocyanine-carbon paste biosensor for amperometric detection of glucose. *Rev. Roum. Chim.* 43:203–208.
2832. Mizutani, F., S. Yabuki, and T. Katsura. 1992. Amperometric enzyme sensor for glucose with glucose oxidase and carbon paste electrode modified with catalyst for hydrogen peroxide oxidation. *Denki Kagaku oyobi Kogyo Butsuri Kagaku* 60:1141–1142.
2833. Luque, G.L., M.C. Rodriguez, and G.A. Rivas. 2005. Glucose biosensors based on the immobilization of copper oxide and glucose oxidase within a carbon paste matrix. *Talanta* 66:467–471.
2834. Wang, C.-J. and X.-Y. Hu. 2008. Potentiometric response of glucose on MnO_2-GOx carbon paste electrode. *Dianhuaxue* 14:76–82.
2835. Waryo, T.T., S. Begic, E. Turkusic, K. Vytřas, and K. Kalcher. 2005. Metal oxide-modified carbon amperometric H_2O_2-transducers and oxidase-biosensors. In *Sensing in Electroanalysis*, Eds. K. Vytřas and K. Kalcher, pp. 145–191. Pardubice, Czech Republic: University of Pardubice.
2836. Wang, J., M. Musameh, and J.-W. Mo. 2006. Acid stability of carbon paste enzyme electrodes. *Anal. Chem.* 78:7044–7047.
2837. Wang, J. and F. Lu. 1998. Oxygen-rich oxidase enzyme electrodes for operation in oxygen-free solutions. *J. Am. Chem. Soc.* 120:1048–1050.
2838. Sugawara, K., T. Kase, C. Kunugi, A. Shimodaira, N. Kamiya, and G. Hirabayashi. 2008. Development of glucose sensor using carbon paste electrode modified with keratin/glucose oxidase. *Gunma Daigaku Kyoikugakubu Kiyo, Shizenkagaku-hen* 56:97–103.
2839. Ming, L., X. Xi, and J. Liu. 2006. Electrochemically platinized carbon paste enzyme electrodes: A new design of amperometric glucose biosensors. *Biotechnol. Lett.* 28:1341–1345.
2840. Wang, J., N. Naser, and U. Wollenberger. 1993. Use of tyrosinase for enzymic elimination of acetaminophen interference in amperometric sensing. *Anal. Chim. Acta* 281:19–24.
2841. Gunasingham, H., C.H. Tan, and T.C. Aw. 1990. Clinical evaluation of amperometric enzyme electrodes in continuous-flow analysis for glucose in undiluted whole blood. *Clin. Chem.* 36:1657–1661.
2842. Gunasingham, H., C.-H. Tan, and T.C. Aw. 1990. Comparative study of first-, second- and third-generation amperometric glucose enzyme electrodes in continuous-flow analysis of undiluted whole blood. *Anal. Chim. Acta* 234:321–330.
2843. Gunasingham, H. and C.-H. Tan. 1990. Conducting organic salt amperometric glucose sensor in continuous-flow monitoring using a wall-jet cell. *Anal. Chim. Acta* 229:83–91.
2844. Pandey, P.C., A.M. Kayastha, and V. Pandey. 1992. Amperometric enzyme sensor for glucose based on graphite paste-modified electrodes. *Appl. Biochem. Biotechnol.* 33:139–144.
2845. Sun, C.-Q., W.-B. Song, D. Zhao, Q. Gao, and H.-D. Xu. 1995. Tetrabutylammonium-tetracyanoquinodimethane as electron-transfer mediator in amperometric glucose sensor. *Chin. Chem. Lett.* 6:141–142.
2846. Pandey, P.C., S. Upadhyay, H.C. Pathak, and C.M.D. Pandey. 1998. Sensitivity, selectivity and reproducibility of some mediated electrochemical biosensors/sensors. *Anal. Lett.* 31:2327–2348.
2847. Gruendig, B. and C. Krabisch. 1989. Electron mediator-modified electrode for the determination of glucose in fermentation media. *Anal. Chim. Acta* 222:75–81.
2848. Savitri, D. and C.K. Mitra. 1999. Modeling the surface phenomena in carbon paste electrodes by low frequency impedance and double-layer capacitance measurements. *Bioelectrochem. Bioenerg.* 48:163–169.
2849. Savitri, D. and C.K. Mitra. 1999. Role of mediators in electron transport from glucose oxidase redox center to electrode surface in a covalently coupled enzyme paste electrode. *J. Biosci.* 24:43–48.
2850. Baronas, R., F. Ivanauskas, and J. Kulys. 1999. Modeling a biosensor based on the heterogeneous microreactor. *J. Math. Chem.* 25:245–252.
2851. Kawakami, M., N. Uriuda, H. Koya, Y. Tajima, and S. Gondo. 1994. Amperometric glucose sensor based on ubiquinone-mixed carbon paste electrode. *Fukuoka Kogyo Daigaku Erekutoronikusu Kenkyusho Shoho* 11:35–40.
2852. Qiao, Y.-G., T.-L. Ying, and D.-Y. Qi. 1994. Glucose sensor based on carbon-paste electrode modified with ubiquinone-glucose oxidase. *Fenxi Huaxue* 22:709–711.
2853. Yabuli, S., F. Mizutani, and T. Katsura. 1992. In *Biosensor 92 Proceedings: The Second World Congress on Biosensors*. Eds. A.P.F. Turner, W.R. Heinemann, I. Karube, and R.D. Schmid, pp. 149–152. Amsterdam, the Netherlands: Elsevier Science.

2854. Okuda, A., F. Mizutani, and S. Yabuki. 1991. Glucose-sensing electrode based on carbon paste-containing polyethylene glycol-attached glucose oxidase and ferrocene. *Hokkaidoritsu Kogyo Shikenjo Hokoku* 290:173–177.

2855. Longchamp, S., M. Herrmann, J.-M. Nigretto, O. Gallet, L. Poulouin, and J.-M. Imhoff. 1998. Enhanced sensitivity of a first generation glucose amperometric biosensor based on fibrinogen-precoating to carbon paste electrode. *Electroanalysis* 10:1064–1066.

2856. El Rhazi, M. and H.N. Randriamahazaka. 1997. The pH-dependent permselectivity of the fibrinogen-modified carbon paste electrode. *Electroanalysis* 9:403–406.

2857. Kulys, J. and H.E. Hansen. 1995. Long-term response of an integrated carbon paste based glucose biosensor. *Anal. Chim. Acta* 303:285–294.

2858. Kulys, J. and H.E. Hansen. 1994. Carbon-paste biosensors array for long-term glucose measurement. *Biosens. Bioelectron.* 9:491–500.

2859. Gorton, L., M. Bardheim, G. Bremle, E. Csoregi, B. Persson, and G. Pettersson. 1991. In *Flow Injection Analysis (FIAJ Based on Enzymes or Antibodies)*. Eds. R.D. Schmid, pp. 305–314. Weinheim, Germany: VCH.

2860. Persson, B., H.L. Lan, L. Gorton, Y. Okamoto, P.D. Hale, L.I. Boguslavsky, and T. Skotheim. 1992. In *Biosensor 92 Proceedings: The Second World Congress on Biosensors*. Eds. A.P.F. Turner, W.R. Heinemann, I. Karube, and R.D. Schmid, pp. 149–152. Amsterdam, the Netherlands: Elsevier Science Ltd.

2861. Kurtinaitiene, B., V. Liauksminas, R. Meskys, R. Rudomanskis, and V. Laurinavicius. 1995. Mediated glucose biosensor based on PQQ-dependent glucose dehydrogenase. *Biologija (Vilnius)* 1995:50–52.

2862. Ivanauskas, F., I. Kaunietis, V. Laurinavicius, J. Razumiene, and R. Simkus. 2005. Computer simulation of the steady state currents at enzyme doped carbon paste electrodes. *J. Math. Chem.* 38:355–366.

2863. Babkina, E., E. Chigrinova, O. Ponamoreva, V. Alferov, and A. Reshetilov. 2006. Bioelectrocatalytic oxidation of glucose by immobilized bacteria Gluconobacter oxydans. Evaluation of water-insoluble mediator efficiency. *Electroanalysis* 18:2023–2029.

2864. Ikeda, T., T. Kurosaki, K. Takayama, and H. Kinoshita. 1993. Mediated bioelectrocatalysis based on a whole-cell at a carbon paste electrode coated with immobilized Gluconobacter industrius. *Denki Kagaku oyobi Kogyo Butsuri Kagaku* 61:889–890.

2865. Zhao, M., D.B. Hibbert, and J.J. Gooding. 2003. An oxygen-rich fill-and-flow channel biosensor. *Biosens. Bioelectron.* 18:827–833.

2866. Li, Y.-X., L.-D. Zhu, and G.-Y. Zhu. 2001. Optical fiber biosensor based on electrochemiluminescence for glucose analysis. *Fenxi Shiyanshi* 20:89–92.

2867. Katsu, T., X. Zhang, and G.A. Rechnitz. 1994. Simultaneous determination of lactose and glucose in milk using two working enzyme electrodes. *Talanta* 41:843–848.

2868. Liden, H., J. Volc, G. Marko-Varga, and L. Gorton. 1998. Pyranose oxidase-modified carbon paste electrodes for monosaccharide determination. *Electroanalysis* 10:223–230.

2869. Lin, L., J. Chen, L. Huang, and X. Lin. 2008. Electrochemical and UV-Vis spectrophotometric studies on the interaction between isorhamnetin and deoxyribonucleic acid. *Fenxi Ceshi Xuebao* 27:635–637, 640.

2870. Safavi, A., N. Maleki, and F. Tajabadi. 2010. SE-30 graphite composite electrode: An alternative for the development of electrochemical biosensors. *Electroanalysis* 22:2460–2466.

2871. Valentini, F., S. Orlanducci, M.L. Terranova, A. Amine, and G. Palleschi. 2004. Carbon nanotubes as electrode materials for the assembling of new electrochemical biosensors. *Sens. Actuators B* B100:117–125.

2872. Dai, H., H. Xu, Y. Lin, X. Wu, and G. Chen. 2009. A highly performing electrochemical sensor for NADH based on graphite/poly(methylmethacrylate) composite electrode. *Electrochem. Comm.* 11:343–346.

2873. Saidman, S.B. 2003. Electrochemical behavior of NAD+-modified carbon paste electrodes. *Indian J. Chem., Sect. A* 42:769–773.

2874. Saidman, S.B. and J.B. Bessone. 2000. The influence of electrode material on NAD+ oxidation. *Electrochim. Acta* 45:3151–3156.

2875. Huang, J., D. Wang, H. Hou, and T. You. 2008. Electrospun palladium nanoparticle-loaded carbon nanofibers and their electrocatalytic activities towards hydrogen peroxide and NADH. *Adv. Funct. Mater.* 18:441–448.

2876. Pinho da Silva, R. and S.H.P. Serrano. 2003. Electrochemical oxidation of biological molecules at carbon paste electrodes pre-treated in guanine solutions. *J. Pharm. Biomed. Anal.* 33:735–744.

2877. Zare, H.R., N. Nasirizadeh, S.-M. Golabi, M. Namazian, M. Mazloum-Ardakani, and D. Nematollahi. 2006. Electrochemical evaluation of coumestan modified carbon paste electrode: Study on its application as a NADH biosensor in presence of uric acid. *Sens. Actuators B* B114:610–617.

2878. Antiochia, R., I. Lavagnini, P. Pastore, and F. Magno. 2004. A comparison between the use of a redox mediator in solution and of surface modified electrodes in the electrocatalytic oxidation of nicotinamide adenine dinucleotide. *Bioelectrochemistry* 64:157–163.

2879. Delbem, M.F., W.J. Baader, and S.H.P. Serrano. 2002. Mechanism of 3,4-dihydroxybenzaldehyde electropolymerization at carbon paste electrodes-catalytic detection of NADH. *Quim. Nova (Barcelona)* 25:741–747.
2880. Lazarin, A.M. and C. Airoldi. 2004. Intercalation of methylene blue into barium phosphate-synthesis and electrochemical investigation. *Anal. Chim. Acta* 523:89–95.
2881. Wang, J., P.V.A. Pamidi, and M. Jiang. 1998. Low-potential stable detection of _-NADH at sol-gel derived carbon composite electrodes. *Anal. Chim. Acta* 360:171–178.
2882. Ladiu, C.I., R. Garcia, I.C. Popescu, and L. Gorton. 2007. NADH electro-catalytic oxidation at glassy carbon paste electrodes modified with Meldola Blue adsorbed onatitanium phosphate. *Rev. Chim. (Roum.)* 58:465–469.
2883. Ladiu, C.I., J.R. Garcia, I.C. Popescu, and L. Gorton. 2007. NADH electrocatalytic oxidation at glassy carbon paste electrodes modified with Meldola Blue adsorbed on acidic a-zirconium phosphate. *Rev. Roum. Chim.* 52:67–74.
2884. Ladiu, C.I., I.C. Popescu, and L. Gorton. 2005. Electrocatalytic oxidation of NADH at carbon paste electrodes modified with meldola blue adsorbed on zirconium phosphate: Effect of Ca^{2+} and polyethyleneimine. *J. Solid State Electrochem.* 9:296–303.
2885. Hoffmann, A.A., S.L.P. Dias, E.V. Benvenutti, E.C. Lima, F.A. Pavan, J.R. Rodrigues, R. Scotti, E.S. Ribeiro, and Y. Gushikem. 2007. Cationic dyes immobilized on cellulose acetate surface modified with titanium dioxide: Factorial design and an application as sensor for NADH. *J. Braz. Chem. Soc.* 18:1462–1472.
2886. Wu, J. 2001. Determination of reduced nicotinamide adenine dinucleotide with p-nitrophenylcarboxyltoluidine blue O carbon paste electrode and its electrochemical mechanism. *Fenxi Huaxue* 29:117.
2887. Gligor, D., F. Balaj, A. Maicaneanu, R. Gropeanu, I. Grosu, L. Muresan, and I.C. Popescu. 2009. Carbon paste electrodes modified with a new phenothiazine derivative adsorbed on zeolite and on mineral clay for NADH oxidation. *Mater. Chem. Phys.* 113:283–289.
2888. Safavi, A., O. Moradlou, and M. Saadatifar. 2010. Methylated azopyridine as a new electron transfer mediator for the electrocatalytic oxidation of NADH. *Electroanalysis* 22:1072–1077.
2889. Kinoshita, H., M. Torimura, K. Yamamoto, K. Kano, and T. Ikeda. 1999. Amperometric determination of NAD(P)H with peroxidase-based H_2O_2-sensing electrodes and its application to isocitrate dehydrogenase activity assay in serum. *J. Electroanal. Chem.* 478:33–39.
2890. Todoriki, S., K. Miki, T. Ikeda, and M. Senda. 1990. The chemically amplified response of a membrane-covered diaphorase and glucose-6-phosphate dehydrogenase co-immobilized carbon paste electrode to NADH and NAD. *Denki Kagaku Oyobi Kogyo Butsuri Kagaku* 58:1089–1096.
2891. Safavi, A., N. Maleki, E. Farjami, and F.A. Mahyari. 2009. Simultaneous electrochemical determination of glutathione and glutathione disulfide at a nanoscale copper hydroxide composite carbon ionic liquid electrode. *Anal. Chem.* 81:7538–7543.
2892. Kinoshita, H., T. Mira, and S. Kamihira. 1999. Amperometric determination of glutathione in yeast extract using a membrane-covered phthalocyanine embedded carbon-paste electrode. *Bunseki Kagaku* 48:117–120.
2893. Elyacoubi, A., S.I.M. Zayed, B. Blankert, and J.-M. Kauffmann. 2006. Development of an amperometric enzymatic biosensor based on gold modified magnetic nanoporous microparticles. *Electroanalysis* 18:345–350.
2894. Martinez-Montequin, S., C. Fernandez-Sanchez, and A. Costa-Garcia. 2000. Voltammetric monitoring of the interaction between streptavidin and biotinylated alkaline phosphatase through the enzymatic hydrolysis of 3-indoxyl phosphate. *Anal. Chim. Acta* 417:57–65.
2895. Kinoshita, H. 2002. Amperometric assay of γ-glutamyltransferase using glutathione as a substrate. *Bunseki Kagaku* 51:1183–1185.
2896. Rodriguez, M.C. and G.A. Rivas. 2009. Label-free electrochemical aptasensor for the detection of lysozyme. *Talanta* 78:212–216.
2897. Longchamp, S., H.N. Randriamahazaka, and J.-M. Nigretto. 1994. The catalytic activity and thermal denaturation of plasmin adsorbed to graphite carbon surfaces. *J. Colloid Interface Sci.* 166:444–450.
2898. Daraio de Peuriot, M. Nigretto, and J.M. Jozefowicz. 1981. Electrochemical activity determination of trypsin-like enzymes. IV. Coupled electrochemical and spectrophotometric assay of thrombin using the same H-D-Phe-Pip-Arg-4 MeO _ naphthylamide, 2HCl (S-2421) substrate. *Thrombosis Res.* 22:303–308.
2899. Reynaud, J.A., B. Malfoy, and A. Bere. 1980. The electrochemical oxidation of three proteins: RNAse A, bovine serum albumin and concanavalin A at solid electrodes. *Bioelectrochem. Bioenerg.* 7:595–606.
2900. Hernandez-Santos, D., M.B. Gonzalez-Garcia, and A. Costa-Garcia. 2000. Electrochemical determination of gold nanoparticles in colloidal solutions. *Electrochim. Acta* 46:607–615.

2901. Ito, S., S.-I. Yamazaki, K. Kano, and T. Ikeda. 2000. Highly sensitive electrochemical detection of alkaline phosphatase. *Anal. Chim. Acta* 424:57–63.

2902. Krizkova, S., V. Hrdinova, V. Adam, E.P.J. Burgess, K.J. Kramer, M. Masarik, and R. Kizek. 2008. Chip-Based CE for avidin determination in transgenic tobacco and its comparison with square-wave voltammetry and standard gel electrophoresis. *Chromatographia* 67(Suppl.):S75–S81.

2903. Tang, D. and B. Xia. 2008. Electrochemical immunosensor and biochemical analysis for carcinoembryonic antigen in clinical diagnosis. *Microchim. Acta* 163:41–48.

2904. Erabi T., Higuti, T., Sakata, K., Kakuno, T., Yamashita, J., Tanaka, M., and Horio, T. 1976. Polarographic studies in presence of Triton X-100 on oxidation-reduction components bound with chromatophores from *Rhodospirillum rubrum*. *J. Biochem.* 79:497–503.

2905. Lee, T.-Y., H.-J. Kim, J.-O. Moon, and Y.-B. Shim. 2004. Determination of cytochrome C with cellulose—DNA modified carbon paste electrodes. *Electroanalysis* 16:821–826.

2906. Merino, M., L.J. Nunez-Vergara, and J.A. Squella. 2000. Cytochrome c reductase immobilized on carbon paste electrode and its electrocatalytic effect on the reduction of cytochrome c. *Bol. Soc. Chil. Quim.* 45:433–439.

2907. Arkoub, I.A., H. Randriamahazaka, and J.-M. Nigretto. 1997. Effect of surface activation on charge and mass transfer rates of the hexacyanoferrate(III)/(II) redox probe at fibrinogen-modified carbon paste electrodes. *Anal. Chim. Acta* 340:99–108.

2908. Lu, Q., C. Hu, R. Cui, and S. Hu. 2007. Direct electron transfer of hemoglobin founded on electron tunneling of CTAB monolayer. *J. Phys. Chem. B* 111:9808–9813.

2909. Xu, Y., C. Hu, and S. Hu. 2009. Single-chain surfactant monolayer on carbon paste electrode and its application for the studies on the direct electron transfer of hemoglobin. *Bioelectrochemistry* 74:254–259.

2910. Jian, F., Y. Qiao, and R. Zhuang. 2007. Direct electrochemistry of hemoglobin in TATP film: Application in biological sensor. *Sens. Actuators B* B124:413–420.

2911. Sun, W., Z. Zhai, and K. Jiao. 2008. Hemoglobin modified carbon paste electrode: Direct electrochemistry and electrocatalysis. *Anal. Lett.* 41:2819–2831.

2912. Guan, L. and Q. Wang. 2006. Study on direct electrochemistry and peroxidase-like electrocatalysis of hemoglobin in organic solvents. *Huaxue Yanjiu, Yu Yingyong* 18:1035–1040.

2913. Sun, W., D.D. Wang, R.F. Gao, J. Sun, and K. Jiao. 2006. Direct electrochemistry of hemoglobin immobilized in sodium alginate film on the ionic liquid [BMIM]PF6 modified carbon paste electrode. *Chin. Chem. Lett.* 17:1589–1591.

2914. Li, Y., H. Cao, and Y. Zhang. 2005. Reductive dechlorination of trichloroacetic acid by bioelectrochemically catalytic method. *Huanjing Kexue* 26:55–58.

2915. Li, Y.-P., H.-B. Cao, and Y. Zhang. 2005. Direct electrochemistry of hemoglobin immobilized on carbon paste electrode modified by carbon nanotubes. *Wuli Huaxue Xuebao* 21:187–191.

2916. Sun, W., X. Li, Z. Zhai, and K. Jiao. 2008. Direct electrochemistry of hemoglobin on single-walled carbon nanotubes modified carbon ionic liquid electrode and its electrocatalysis. *Electroanalysis* 20:2649–2654.

2917. Qiao, L., R. Gao, and J. Zheng. 2010. Direct electrochemistry of hemoglobin immobilized on hydrophilic ionic liquid-chitosan-ZrO$_2$ nanoparticles composite film with carbon ionic liquid electrode as the platform. *Anal. Sci.* 26:1181–1186.

2918. Han, J., J. Wei, and H. Chen. 1993. Electrocatalysis of biomolecules at modified electrodes. (III). Catalytic reduction of hemoglobin at chemically modified carbon paste electrode containing cresyl fast violet. *Gaodeng Xuexiao Huaxue Xuebao* 14:332–333.

2919. Fernandez Sanchez, C., M.B. Gonzalez Garcia, and A. Costa Garcia. 2000. AC voltammetric carbon paste-based enzyme immunosensors. *Biosens. Bioelectron.* 14:917–924.

2920. Fernandez Sanchez, C. and A. Costa Garcia. 1999. Competitive enzyme immunosensor developed on a renewable carbon paste electrode support. *Anal. Chim. Acta* 402:119–127.

2921. Fernandez Sanchez, C. and A. Costa Garcia. 1997. Adsorption of immunoglobulin G on carbon paste electrodes as a basis for the development of immunoelectrochemical devices. *Biosens. Bioelectron.* 12:403–413.

2922. Fernandez Sanchez, C. and A. Costa Garcia. 1999. Inhibition of adsorbed alkaline phosphatase activity by an anti-enzyme antibody. An approach to carbon paste immunoelectrodes. *Electroanalysis* 11:1350–1354.

2923. Liu, G. and Y. Lin. 2005. A renewable electrochemical magnetic immunosensor based on gold nanoparticle labels. *J. Nanosci. Nanotechnol.* 5:1060–1065.

2924. Adam, V., O. Zitka, P. Dolezal, L. Zeman, A. Horna, J. Hubalek, J. Sileny, S. Krizkova, L. Trnkova, and R. Kizek. 2008. Lactoferrin isolation using monolithic column coupled with spectrometric or microamperometric detector. *Sensors* 8:464–487.

2925. Qiao, Y., F. Jian, H. Yu, and L. Hu. 2007. Composite films of lecithin and heme proteins with electrochemical and electrocatalytic activities. *J. Colloid Interface Sci.* 315:537–543.
2926. Bobes, C.F., M.T.F. Abedul, and A. Costa-Garcia. 2001. Pneumolysin ELISA with adsorptive voltammetric detection of indigo in a flow system. *Electroanalysis* 13:559–566.
2927. Bengoechea Alvarez, M.J., C. Fernandez Bobes, M.T. Fernandez Abedul, and A. Costa-Garcia. 2001. Sensitive detection for enzyme-linked immunosorbent assays based on the adsorptive stripping voltammetry of indigo in a flow system. *Anal. Chim. Acta* 442:55–62.
2928. Bartosik, M., V. Ostatna, and E. Palecek. 2009. Electrochemistry of riboflavin-binding protein and its interaction with riboflavin. *Bioelectrochemistry* 76:70–75.
2929. Lappin, A.G., C.A. Lewis, and W.J. Ingledew. 1985. Kinetics and mechanisms of reduction of rusticyanin, a blue copper protein from Thiobacillus ferrooxidans, by inorganic cations. *Inorg. Chem.* 24:1446–1450.
2930. Masarik, M., A. Stobiecka, R. Kizek, F. Jelen, Z. Pechan, W. Hoyer, T.M. Jovin, V. Subramaniam, and E. Palecek. 2004. Sensitive electrochemical detection of native and aggregated a-synuclein protein involved in Parkinson's disease. *Electroanalysis* 16:1172–1181.
2931. Batista, I.V., M.R.V. Lanza, I.L.T. Dias, S.M.C.N. Tanaka, A.A. Tanaka, and M.D.P.T. Sotomayor. 2008. Electrochemical sensor highly selective for estradiol valerate determination based on a modified carbon paste with iron tetrapyridinoporphyrazine. *Analyst (U.K.)* 133:1692–1699.
2932. Wu, K.-B., Q. He, and S.-S. Hu. 2002. Voltammetric determination of estrogens based on the enhancement effect of surfactant at carbon paste electrode. *Wuhan Univ. J. Natl. Sci.* 7:463–469.
2933. Raba, J., C.A. Fontan, and V.A. Cortinez. 1994. 4-(3,5-Dichloro-2-pyridylazo)-1,3-diaminobenzene-graphite paste electrode for end-point detection in the automatic potentiometric titration of zinc(II) with EDTA. *Talanta* 41:273–278.
2934. Wang, J., G. Rivas, X. Cai, M. Chicharro, P.A.M. Farias, and E. Palecek. 1996. Trace measurements of insulin by potentiometric stripping analysis at carbon paste electrodes. *Electroanalysis* 8:902–906.
2935. Bekheit, G.E. 2000. Determination of melatonin by differential pulse voltammetry at modified carbon paste electrode. *Asian J. Chem.* 12:541–547.
2936. Wang, S. and T. Peng. 2000. Adsorptive/extractive behavior of melatonin at carbon paste electrodes and its electrochemical determination. *Fenxi Huaxue* 28:1350–1354.
2937. Radi, A. and G.E. Bekhiet. 1998. Voltammetry of melatonin at carbon electrodes and determination in capsules. *Bioelectrochem. Bioenerg.* 45:275–279.
2938. Corujo-Antuna, J.L., E.M. Abad-Villar, M.T. Fernandez-Abedul, and A. Costa-Garcia. 2003. Voltammetric and flow amperometric methods for the determination of melatonin in pharmaceuticals. *J. Pharm. Biomed. Anal.* 31:421–429.
2939. Corujo-Antuna, J.L., S. Martinez-Montequin, M.T. Fernandez-Abedul, and A. Costa-Garcia. 2003. Sensitive adsorptive stripping voltammetric methodologies for the determination of melatonin in biological fluids. *Electroanalysis* 15:773–778.
2940. He, Q., X. Dang, C. Hu, and S. Hu. 2004. The effect of cetyltrimethyl ammonium bromide on the electrochemical determination of thyroxine. *Colloids Surf. B* 35:93–98.
2941. Dang, X.-P., W.-H. Chen, and S.-S. Hu. 2004. Enhanced effects of surfactants on electrooxidation of two species with opposite charges. *Wuhan Daxue Xuebao, Lixueban* 50:141–145.
2942. Hu, C., Q. He, Q. Li, and S. Hu. 2004. Enhanced reduction and determination of trace thyroxine at carbon paste electrode in the presence of trace cetyltrimethylammonium bromide. *Anal. Sci. (Jpn.)* 20:1049–1054.
2943. Sun, Y. 2005. Study on voltammetric determination of thyroxine by using carbon paste electrode based on the activity effect of surfactant. *Lihua Jianyan, Huaxue Fence* 41:229–231.
2944. Stefan-van Staden, R.-I., J.F. van Staden, H.Y. Aboul-Enein, M.C. Mirica, I. Balcu, and N. Mirica. 2008. Determination of (+)-3,3′,5,5′-tetraiodo-L-thyronine (L-T4) in serum and pharmaceutical formulations using a sequential injection analysis/immunosensor system. *J. Immunoassay Immunochem.* 29:348–355.
2945. Hooda, V., A. Gahlaut, H. Kumar, and C.S. Pundir. 2009. Biosensor based on enzyme coupled PVC reaction cell for electrochemical measurement of serum total cholesterol. *Sens. Actuators B* B136:235–241.
2946. Zhu, Q., M. Han, H. Wang, L. Liu, J. Bao, Z. Dai, and J. Shen. 2010. Electrogenerated chemiluminescence from CdS hollow spheres composited with carbon nanofiber and its sensing application. *Analyst* 135:2579–2584.
2947. Kinoshita, H., M. Torimura, K. Kano, and T. Ikeda. 1998. Amperometric determination of high-density lipoprotein cholesterol using polyethylene glycol-modified enzymes and a peroxidase-entrapped electrode. *Ann. Clin. Biochem.* 35:739–744.
2948. Kinoshita, H., T. Chijiwa, M. Torimura, K. Kano, and T. Ikeda. 1998. Direct determination of high-density lipoprotein- and total cholesterol in serum using a peroxidase-entrapped electrode and polyethylene glycol-modified enzymes. *Bunseki Kagaku* 47:233–238.

2949. Sweetser, P.B. and D.G. Swartzfager. 1978. Indole-3-acetic acid levels of plant tissue as determined by a new high performance liquid chromatographic method. *Plant Physiol.* 61:254–258.
2950. Zhang, S. and K. Wu. 2004. Square wave voltammetric determination of indole-3-acetic acid based on the enhancement effect of anionic surfactant at the carbon paste electrode. *Bull. Korean Chem. Soc.* 25:1321–1325.
2951. Li, C.-X., J. Li, L.-T. Xiao, G.-L. Shen, and R.-Q. Yu. 2003. Study on the electrochemical biosensor for photohormone indole 3-acetic acid. *Fenxi Kexue Xuebao* 19:205–208.
2952. Ballesteros, Y., M.J. Gonzalez De La Huebra, M.C. Quintana, P. Hernandez, and L. Hernandez. 2003. Voltamperometric determination of kinetin with a carbon paste modified electrode. *Microchem. J.* 74:193–202.
2953. Leite, O.D., O. Fatibello-Filho, and A.d.M. Barbosa. 2003. Determination of catecholamines in pharmaceutical formulations using a biosensor modified with a crude extract of fungi laccase (*Pleurotus ostreatus*). *J. Braz. Chem. Soc.* 14:297–303.
2954. Leite, O.D., K.O. Lupetti, O. Fatibello-Filho, I.C. Vieira, and A.d.M. Barbosa. 2003. Synergic effect studies of the bi-enzymatic system laccase-peroxidase in a voltammetric biosensor for catecholamines. *Talanta* 59:889–896.
2955. Ensafi, A.A., A. Arabzadeh, and H. Karimi-Maleh. 2010. Sequential determination of benserazide and levodopa by voltammetric method using chloranil as a mediator. *J. Braz. Chem. Soc.* 21:1572–1580.
2956. Serpentini, C.-L., C. Gauchet, D. De Montauzon, M. Comtat, J. Ginestar, and N. Paillous. 2000. First electrochemical investigation of the redox properties of DOPA-melanins by means of a carbon paste electrode. *Electrochim. Acta* 45:1663–1668.
2957. Wang, J. and B.A. Freiha. 1983. Evaluation of differential pulse voltammetry at carbon electrodes. *Talanta* 30:317–322.
2958. Liu, D., Y. Liu, H. Hou, and T. You. 2010. A new strategy to pretreat carbon nanofiber and its application in determination of dopamine. *J. Nanomater.*, ID 659207, 6 pages; DOI: 10.1155/2010/659207.
2959. Li, Z., J.-M. Xu, H.-D. Xu, N. Lu, and J.-R. Qi. 2007. Determination of dopamine using rice hull activated carbon modified carbon paste electrode by voltammetry. *Jilin Daxue Xuebao, Lixueban* 45:98–102.
2960. Pandurangachar, M., B.E.K. Swamy, B.N. Chandrashekar, O. Gilbert, S. Reddy, and B.S. Sherigara. 2010. Electrochemical investigations of potassium ferricyanide and dopamine by 1-butyl-4-methylpyridinium tetrafluoro borate modified carbon paste electrode: A cyclic voltammetric study. *Int. J. Electrochem. Sci.* 5:1187–1202.
2961. Zhang, Y., H.-F. Zhang, and J.-B. Zheng. 2008. A nafion/ionic liquid/modified carbon paste electrode for selective detection of dopamine in the presence of ascorbic acid and uric acid. *Fenxi Shiyanshi* 27:34–37.
2962. Shankar, S.S., B.E.K. Swamy, M. Pandurangachar, U. Chandra, B.N. Chandrashekar, J.G. Manjunatha, and B.S. Sherigara. 2010. Electrocatalytic oxidation of dopamine on acrylamide modified carbon paste electrode: A voltammetric study. *Int. J. Electrochem. Sci.* 5:944–954.
2963. Shreenivas, M.T., B.E.K. Swamy, U. Chandra, S.S. Shankar, J.G. Manjunatha, and B.S. Sherigara. 2010. Electrochemical investigations of dopamine at chemically modified Losartan carbon paste electrode: A cyclic voltammetric study. *Int. J. Electrochem. Sci.* 5:774–781.
2964. Wei, J., J.-B. He, S.-Q. Cao, Y.-W. Zhu, Y. Wang, and G.-P. Hang. 2010. Enhanced sensing of ascorbic acid, dopamine and serotonin at solid carbon paste electrode with a nonionic polymer film. *Talanta* 83:190–196.
2965. Chandrashekar, B.N., B.E.K. Swamy, M. Pandurangachar, S.S. Shankar, O. Gilbert, J.G. Manjunatha, and B.S. Sherigara. 2010. Electrochemical oxidation of dopamine at polyethylene glycol modified carbon paste electrode: A cyclic voltammetric study. *Int. J. Electrochem. Sci.* 5:578–592.
2966. Zheng, J. and X. Zhou. 2007. Sodium dodecyl sulfate-modified carbon paste electrodes for selective determination of dopamine in the presence of ascorbic acid. *Bioelectrochemistry* 70:408–415.
2967. Corona-Avendano, S., G. Alarcon-Angeles, M.T. Ramirez-Silva, G. Rosquete-Pina, M. Romero-Romo, and M. Palomar-Pardave. 2007. On the electrochemistry of dopamine in aqueous solution. Part I: The role of [SDS] on the voltammetric behavior of dopamine on a carbon paste electrode. *J. Electroanal. Chem.* 609:17–26.
2968. Patrascu, D.G., V. David, I. Balan, A. Ciobanu, I.G. David, P. Lazar, I. Ciurea, I. Stamatin, and A.A. Ciucu. 2010. Selective DPV method of dopamine determination in biological samples containing ascorbic acid. *Anal. Lett.* 43:1100–1110.
2969. Han, X.-X. and Z.-N. Gao. 2007. Electrochemical behaviors of dopamine and ascorbic acid at CPB modified carbon paste electrodes. *Fenxi Ceshi Xuebao* 26:612–616.
2970. Han, X.-X. and Z.-N. Gao. 2007. Electrochemical behavior of dopamine and ascorbic acid at CPB/CPE and its application. *Yingyong Huaxue* 24:770–773.

2971. Han, X.-X. and Z.-N. Gao. 2007. Electrochemical behaviors of uric acid and ascorbic acid at cetylpyridinium bromide in situ modified carbon paste electrodes. *Huaxue Yanjiu, Yu Yingyong* 19:254–257.
2972. Han, X.-X., B. Liang, Z.-N. Gao, and Z.-X. Zheng. 2007. Electrochemical behaviors of ascorbic acid and uric acid at cetylpyridinium bromide in situ modified carbon paste electrodes and their selective determination. *Fenxi Shiyanshi* 26:30–33.
2973. Shangguan, X.-D., J.-B. Zheng, Q.-L. Sheng, and Y.-P. He. 2010. Electrocatalytic oxidation and determination of dopamine at a carbon ionic liquid electrode modified with nafion-L-aspartic acid composite film. *Acad. J. Xi'an Jiatong Univ.* 22:1–6.
2974. Ramirez-Silva, M.T., S. Corona-Avendano, G. Alarcon-Angeles, A. Rojas-Hernandez, M. Romero-Romo, and M. Palomar-Pardave. 2009. Dopamine electrochemical behavior onto an electrode modified with a β-cyclodextrin polymer. *ECS Trans.* 20:151–157.
2975. Alarcon-Angeles, G., S. Corona-Avendano, M.E. Palomar-Pardave, M.A. Romero-Romo, A. Rojas-Hernandez, and M.T. Ramirez-Silva. 2008. Electrochemical and spectrophotometric evaluation of the formation constants of the AA-βCD and DA-βCD inclusion complexes. *ECS Trans.* 15:507–516.
2976. Alarcon-Angeles, G., S. Corona-Avendano, M.E. Palomar-Pardave, A. Merkoci, M.A. Romero-Romo, A. Rojas-Hernandez, and M.T. Ramirez-Silva. 2008. Electrochemical study of dopamine and ascorbic acid by means of supramolecular systems. *ECS Trans.* 15:325–334.
2977. Rubianes, M.D., A.S. Arribas, E. Bermejo, M. Chicharro, A. Zapardiel, and G. Rivas. 2010. Carbon nanotubes paste electrodes modified with a melanic polymer: Analytical applications for the sensitive and selective quantification of dopamine. *Sens. Actuators B* 144:274–279.
2978. Atta, N.F., A. Galal, F.M. Abu-Attia, and S.M. Azab 2010. Carbon paste gold nanoparticles sensor for the selective determination of dopamine in buffered solutions. *J. Electrochem. Soc.* 157:F116–F123.
2979. Sivanesan, A. and A. John. 2009. Gold nanoparticles modified electrodes for biosensors. In *Nanostructured Materials for Electrochemical Biosensors*. Eds. Y. Umasankar, S.A. Kumar, and S.M. Chen, pp. 97–128. New York: Nova Science Publishers.
2980. Chi, Y., Y. Wang, and Z. Zhu. 2009. Voltammetric behavior of dopamine at poly (o-aminophenol)/nickel modified carbon paste electrode. *Jisuanji Yu Yingyong Huaxue* 26:566–570.
2981. Navaratne, A. and G.A. Rechnitz. 1992. Use of tin oxide electrodes in flow injection analysis with application to plant tissue-based biosensors. *Anal. Lett.* 25:191–203.
2982. Wei, P. and G. Li. 2005. Electrochemical oxidation of dopamine at a carbon paste electrode modified with 3-mercaptopropyltrimethoxysilane copper encapsulated in mesoporous crystalline molecular sieve. *Fenxi Huaxue* 33:703–706.
2983. Rekha, B.E., K. Swamy, R. Deepa, V. Krishna, O. Gilbert, U. Chandra, and B.S. Sherigara. 2009. Electrochemical investigations of dopamine at chemically modified alcian blue carbon paste electrode: A cyclic voltammetric study. *Int. J. Electrochem. Sci.* 4:832–845.
2984. Naik, R.R., E. Niranjana, B.E.K. Swamy, B.S. Sherigara, K.R.V. Reddy, and H. Jayadevappa. 2010. Simultaneous detection of dopamine and uric acid at cobalt (II) octanitro phthalocyanine modified carbon paste electrode. *Res. Rev. Electrochem.* 2:33–36.
2985. Wohnrath, K., C.A. Pessoa, P.M. dos Santos, J.R. Garcia, A.A. Batista, and O.N. Oliveira. 2006. Electrochemical properties of a ruthenium complex immobilized as thin films and in carbon paste electrodes. *Progr. Solid State Chem.* 33:243–252.
2986. Raghavendra Naik, R., B.E.K. Swamy, U. Chandra, E. Niranjana, B.S. Sherigara, and H. Jayadevappa. 2009. Separation of ascorbic acid, dopamine and uric acid by acetone/water modified carbon paste electrode: A cyclic voltammetric study. *Int. J. Electrochem. Sci.* 4:855–862.
2987. Yoo, J.H., B.W. Woo, S.S. Kim, J.H. Uhm, H. Nam, and G.S. Cha. 2001. Voltammetric determination of dopamine with the N-hydroxysuccinimide modified carbon paste electrode. *J. Korean Electrochem. Soc.* 4:109–112.
2988. Deng, Z., Z. Zhu, T. Yao, and T. Li. 1999. Electrocatalytic oxidation determination of dopamine by tetracyanoquinodimethane modified carbon paste electrode. *Lihua Jianyan, Huaxue Fence* 35:8–9, 11.
2989. Chandra, U., B.E.K. Swamy, O. Gilbert, and B.S. Sherigara. 2010. Determination of dopamine in presence of ascorbic acid at Eriochrome Black T modified carbon paste electrode: A voltammetric study. *Int. J. Electrochem. Sci.* 5:1475–1483.
2990. Mazloum-Ardakani, M., R. Arazi, H. Beitollahi, and H. Naeimi. 2010. 2,2′-(1,3-propanediylbisnitriloethylidine)bis-hydroquinone/TiO_2 nanoparticles modified carbon paste electrode for selective determination of dopamine in the presence of uric acid and tryptophan. *Anal. Meth.* 2:1078–1084.
2991. Manjunatha, J.G., B.E.K. Swamy, G.P. Mamatha, O. Gilbert, B.N. Chandrashekar, and B.S. Sherigara. 2010. Electrochemical studies of dopamine and epinephrine at a poly (tannic acid) modified carbon paste electrode: A cyclic voltammetric study. *Int. J. Electrochem. Sci.* 5:1236–1245.

2992. Chandra, U., B.E.K. Swamy, O. Gilbert, and B.S. Sherigara. 2010. Voltammetric resolution of dopamine in the presence of ascorbic acid and uric acid at poly (calmagite) film coated carbon paste electrode. *Electrochim. Acta* 55:7166–7174.

2993. Petit, C., A. Gonzalez-Cortes, and J.-M. Kauffmann. 1995. Preparation and characterization of a new enzyme electrode based on solid paraffin and activated graphite particles. *Talanta* 42:1783–1789.

2994. Petit, C. and J.-M. Kauffmann. 1995. New carbon paste electrode for the development of biosensors. *Anal. Proc.* 32:11–12.

2995. Ding, H. and W. Qiano. 1992. Mixed plant tissue-carbon paste biomicroelectrode and its application to determination of dopamine in vitro and in vivo. *Chem. Res. Chin. Univ.* 8:147–151.

2996. Ding, H. and W. Qiao. 1992. Mixed banana tissue-carbon paste biomicroelectrode and its application in in vivo dopamine detection. *Fenxi Huaxue* 20:556–559.

2997. Ding H and Qiao W. 1992. Measurement of dopamine in rat striatum in vivo with a biomicroelectrode made from mixed plant tissue-carbon paste. *Yao xue xue bao/Acta Pharm. Sin.* 27:86–89.

2998. Raoof, J.-B., R. Ojani, and A. Kiani. 2005. Apple-modified carbon paste electrode: A biosensor for selective determination of dopamine in pharmaceutical formulations. *Bull. Electrochem.* 21:223–228.

2999. Qi, D., T. Ying, and S. Hu. 1993. A new bio-electrode based on mixed potato tissue-carbon paste. *Fenxi Huaxue* 21:212–214.

3000. Forzani, E.S., G.A. Rivas, and V.M. Solis. 1995. Amperometric determination of dopamine on an enzymically modified carbon paste electrode. *J. Electroanal. Chem.* 382:33–40.

3001. Wang, J., N. Naser, and M. Ozsoz. 1990. Plant tissue-based amperometric electrode for eliminating ascorbic acid interferences. *Anal. Chim. Acta* 234:315–320.

3002. Wu, M. and Z. Lin. 1996. A new two plant tissue-based carbon paste microelectrode for neurotransmitter detection. *J. Tongji Med. Univ.* 16:152–154.

3003. Caruso, C.S., I. Da Cruz Vieira, and O. Fatibello-Filho. 1999. Determination of epinephrine and dopamine in pharmaceutical formulations using a biosensor based on carbon paste modified with crude extract of cara root (*Dioscorea bulbifera*). *Anal. Lett.* 32:39–50.

3004. Cai, P., H. Huang, X. Yu, and B. Nie. 1996. Plant tissue-modified carbon paste electrode for determination of dopamine. *Chuangan Jishu Xuebao* 9:75–78.

3005. Frattini, P., G. Santagostino, S. Schinelli, M.L. Cucchi, and G. Corona. 1983. Assay of urinary vanilmandelic, homovanillic, and 5-hydroxyindole acetic acids by liquid chromatography with electrochemical detection. *J. Pharmacol. Meth.* 10:193–198.

3006. Joseph, M.H., B.V. Kadam, and D. Risby. 1981. Simple high-performance liquid chromatographic method for the concurrent determination of the amine metabolites vanillylmandelic acid, 3-methoxy-4-hydroxyphenylglycol, 5-hydroxyindoleacetic acid, dihydroxyphenylacetic acid and homovanillic acid in urine using electrochemical detection. *J. Chromatogr.* 226:361–368.

3007. Yang, G., L. Jin, and Z. Leng. 1998. Study on adsorptive voltammetry for adrenaline on carbon paste electrode. *Yaoxue Xuebao* 33:534–537.

3008. Ye, X., J. Hou, S. Gao, X. Gao, and D. Wang. 1991. Voltammetric behaviors of adrenaline and its Be(II) complex. *Chin. Chem. Lett.* 2:731–734.

3009. Salimi, A., H. MamKhezri, and R. Hallaj. 2006. Simultaneous determination of ascorbic acid, uric acid and neurotransmitters with a carbon ceramic electrode prepared by sol-gel technique. *Talanta* 70:823–832.

3010. Corona-Avendano, S., G. Alarcon-Angeles, M.T. Ramirez-Silva, M. Romero-Romo, A. Cuan, and M. Palomar-Pardave. 2009. Simultaneous electrochemical determination of adrenaline and ascorbic acid: Influence of [CTAB]. *J. Electrochem. Soc.* 156:J375–J381.

3011. Avendano, S.C., G.A. Angeles, M.R. Romo, M.T.R. Silva, and M.P. Pardave. 2008. Effect of CTAB interfacial supramolecular systems on the voltammetry signals of adrenalin and ascorbic acid. *ECS Trans.* 15:489–498.

3012. Xie, P., X. Chen, F. Wang, C. Hu, and S. Hu. 2006. Electrochemical behaviors of adrenaline at acetylene black electrode in the presence of sodium dodecyl sulfate. *Colloids Surf. B* 48:17–23.

3013. Lupetti, K.O., I.C. Vieira, H.J. Vieira, and O. Fatibello-Filho. 2002. Electroregenerable anion-exchange resin with triiodide carbon paste electrode for the voltammetric determination of adrenaline. *Analyst (U.K.)* 127:525–529.

3014. Chernyshov, D.V., N.V. Shvedene, E.R. Antipova, and I.V. Pletnev. 2008. Ionic liquid-based miniature electrochemical sensors for the voltammetric determination of catecholamines. *Anal. Chim. Acta* 621:178–184.

3015. Yu, A., J. Zhou, and H. Chen. 1997. Electrocatalytic oxidation of epinephrine at a methylene green chemically modified carbon paste electrode and its flow injection analysis. *Fenxi Huaxue* 25:959–961.

3016. Beitollahi, H., M.M. Ardakani, B. Ganjipour, and H. Naeimi. 2008. Novel 2,2'-[1,2-ethanediylbis(nitrilo ethylidyne)]-bis-hydroquinone double-wall carbon nanotube paste electrode for simultaneous determination of epinephrine, uric acid and folic acid. *Biosens. Bioelectron.* 24:362–368.

3017. Beitollahi, H., H. Karimi-Maleh, and H. Khabazzadeh. 2008. Nanomolar and selective determination of epinephrine in the presence of norepinephrine using carbon paste electrode modified with carbon nanotubes and novel 2-(4-Oxo-3-phenyl-3,4-dihydro-quinazolinyl)-N?-phenyl-hydrazinecarbothioamide. *Anal. Chem.* 80:9848–9851.

3018. Mataveli, L.R.V., N.d.J. Antunes, M.R.P.L. Brigagao, C. Schmidt de Magalhaes, C. Wisniewski, and P.O. Luccas. 2010. Evaluation of a simple and low cost potentiometric biosensor for pharmaceutical and in vivo adrenaline determination. *Biosens. Bioelectron.* 26:798–802.

3019. Felix, F.S., M. Yamashita, and L. Angnes. 2006. Epinephrine quantification in pharmaceutical formulations utilizing plant tissue biosensors. *Biosens. Bioelectron.* 21:2283–2289.

3020. Albery, W.J., T.W. Beck, W.N. Brooks, and M. Fillenz. 1981. The determination of noradrenaline using differential double pulse voltammetry. *J. Electroanal. Chem.* 125:205–217.

3021. Brent, P.J., S. Hall, A.J. Smith, and J. Aylward. 1985. A simplified liquid chromatography and electrochemical detection method for measurement of plasma noradrenaline. *J. Pharmacol. Methods* 14:243–248.

3022. Yaghoubian, H., V. Soltani-Nejad, and S. Roodsaz. 2010. Simultaneous voltammetric determination of norepinephrine, uric acid, and folic acid at the surface of modified chloranil carbon nanotube paste electrode. *Int. J. Electrochem. Sci.* 5:1411–1421.

3023. Verbiese-Genard, N., J.-M. Kauffmann, M. Hanocq, and L. Molle. 1984. Study of the electrooxidative behavior of 5-hydroxyindole-3-acetic acid, 5-hydroxytryptophan and serotonine in the presence of sodium ethylenediaminetetraacetic acid. *J. Electroanal. Chem.* 170:243–254.

3024. Cheung, W., P.-L. Chiu, R.-R. Parajuli, Y. Ma, S.-R. Ali, and H. He. 2009. Fabrication of high performance conducting polymer nanocomposites for biosensors and flexible electronics: Summary of the multiple roles of DNA dispersed and functionalized single walled carbon nanotubes. *J. Mater. Sci.* 19:6465–6480.

3025. Wang, J., G. Rivas, X. Cai, E. Palecek, P. Nielsen, H. Shiraishi, N. Dontha et al. 1997. DNA electrochemical biosensors for environmental monitoring. A review. *Anal. Chim. Acta* 347:1–8.

3026. Diculescu, V.C., A.-M.C. Paquim, and A.M.O. Brett. 2005. Electrochemical DNA sensors for detection of DNA damage. *Sensors* 5:377–393.

3027. Bowater, R.P., R.J.H. Davies, E. Palecek, and M. Fojta. 2009. Sensitive electrochemical assays of DNA structure: Electrochemical analysis of DNA. *Chim. Oggi* 27:50–54.

3028. Palecek, E. 2009. Fifty years of nucleic acid electrochemistry. *Electroanalysis* 21:239–251.

3029. Chang, X., D. Zhou, and Y. Xie. 1989. A comparative study on the anodic behavior of adenine, adenosine, and 5'-AMP at carbon paste electrode. *Huaxue Xuebao* 47:158–162.

3030. Zhang, L., S. Zuo, Y. Zhao, Y. Teng, H. Yuan, and M. Lan. 2008. Differential pulse voltammetry method for the determination of adenine and ssDNA concentration using an electrochemically pretreated carbon paste electrode. *Xiandai Yiqi* 1:19–22, 26.

3031. Gilmartin, M.A.T. and J.P. Hart. 1992. Comparative study of the voltammetric behavior of guanine at carbon paste and glassy carbon electrodes and its determination in purine mixtures by differential-pulse voltammetry. *Analyst* 117:1613–1618.

3032. Liu, L., J.-F. Song, P.-F. Yu, and B. Cui. 2007. Sensing system integrating lanthanum hydroxide nanowires with Copper(II) ion for uracil and its application. *Anal. Lett.* 40:2562–2573.

3033. Stempkowska, I., M. Ligaj, J. Jasnowska, J. Langer, and M. Filipiak. 2007. Electrochemical response of oligonucleotides on carbon paste electrode. *Bioelectrochemistry* 70:488–494.

3034. Wang, J., L. Chen, and M. Chicharro. 1996. Trace measurements of nucleic acids using flow injection amperometry. *Anal. Chim. Acta* 319:347–352.

3035. Cai, X., G. Rivas, P.A.M. Farias, H. Shiraishi, J. Wang, M. Fojta, and E. Palecek. 1996. Trace measurements of plasmid DNAs by adsorptive stripping potentiometry at carbon paste electrodes. *Bioelectrochem. Bioenerg.* 40:41–47.

3036. Pedano, M.L. and G.A. Rivas. 2010. Adsorption and electrooxidation of DNA at glassy carbon paste electrodes. *Anal. Lett.* 43:1703–1712.

3037. Swaile, B.A.H. and J.Q. Chambers. 1991. Electrochemical release of ethidium into and fluorescence detection of DNA-ethidium complexes in the diffusion layer at a carbon paste electrode. *Anal. Biochem.* 196:415–420.

3038. Chambers, J.Q., M.A. Lange, D.S. Trimble, and R.D. Mounts. 1989. Electrochemically controlled binding of ethidium to calf thymus DNA at a carbon-paste/ethidium tetracyanoquinodimethane electrode. *J. Electroanal. Chem.* 266:277–285.

3039. Li, J. and D. Fei. 2007. Determination of DNA using methylene blue modified magnetic electrode. *Chuangan Jishu Xuebao* 20:1940–1944.
3040. Wang, J., A.-N. Kawde, E. Sahlin, C. Parrado, and G. Rivas. 2000. Factors influencing the adsorptive stripping potentiometric response of synthetic oligonucleotides. *Electroanalysis* 12:917–920.
3041. Jiang, M. and J. Wang. 2001. Recognition and detection of oligonucleotides in the presence of chromosomal DNA based on entrapment within conducting-polymer networks. *J. Electroanal. Chem.* 500:584–589.
3042. Wang, J., G. Rivas, J.R. Fernandes, M. Jiang, J.L. Lopez Paz, R. Waymire, T.W. Nielsen, and R.C. Getts. 1998. Adsorption and detection of DNA dendrimers at carbon electrodes. *Electroanalysis* 10:553–556.
3043. Arguelho, M.L.P.M., J.D.P.H. Alves, N.R. Stradiotto, V. Lacerda, Jr., J.M. Pires, and A. Beatriz. 2010. Electrochemical and theoretical evaluation of the interaction between DNA and amodiaquine. Evidence of the guanine adduct formation. *Quim. Nova (Barcelona)* 33:1291–1296.
3044. Stanic, Z. and S. Girousi. 2008. Electrochemical study of the interaction between dsDNA and copper(I) using carbon paste and hanging mercury drop electrode. *Talanta* 76:116–121.
3045. Stanic, Z. and S. Girousi. 2009. Electrochemical study of the interaction between dsDNA and copper(II) using carbon paste and hanging mercury drop electrodes. *Microchim. Acta* 164:479–485.
3046. Panagoulis, D., E. Pontiki, E. Skeva, C. Raptopoulou, S. Girousi, D. Hadjipavlou-Litina, and C. Dendrinou-Samara. 2007. Synthesis and pharmacochemical study of new Cu(II) complexes with thiophen-2-yl saturated and alpha,beta-unsaturated substituted carboxylic acids. *J. Inorg. Biochem.* 101:623–634.
3047. Jasnowska, J., M. Ligaj, B. Stupnicka, and M. Filipiak. 2004. DNA sensor for o-dianisidine. *Bioelectrochemistry* 64:85–90.
3048. Yang, T., K. Jiao, J. Yang, C.-Z. Zhao, and W.-Y. Qu. 2006. Voltammetric behavior of dopamine at dsDNA-modified Ni2+/poly(o-aminophenol) carbon paste electrode. *Gaodeng Xuexiao Huaxue Xuebao* 27:2294–2296.
3049. Radi, A.-E. 2010. Application of electrochemical methods for analysis of fluoroquinolones antibacterial agents and fluoroquinolones-DNA interactions. *Open Chem. Biomed. Meth. J.* 3:27–36.
3050. Sun, W., Y. Wang, X. Li, J. Wu, T. Zhan, and K. Jiao. 2009. Electrochemical biosensor based on carbon ionic liquid electrode modified with Nafion/hemoglobin/DNA composite film. *Electroanalysis* 21:2454–2460.
3051. Du, M., T. Yang, and K. Jiao. 2010. Carbon nanotubes/(pLys/dsDNA)n layer-by-layer multilayer films for electrochemical studies of DNA damage. *J. Solid State Electrochem.* 14:2261–2266.
3052. Gherghi, I.Ch., S.Th. Girousi, A. Voulgaropoulos, and R. Tzimou-Tsitouridou. 2004. Adsorptive transfer stripping voltammetry applied to the study of the interaction between DNA and actinomycin D. *Int. J. Environ. Anal. Chem.* 84:865–874.
3053. Gherghi, I.Ch., S.Th. Girousi, A.N. Voulgaropoulos, and R. Tzimou-Tsitouridou. 2004. DNA modified carbon paste electrode applied to the voltammetric study of the interaction between DNA and acridine orange. *Chem. Anal. (Warsaw)* 49:467–480.
3054. Ozsoz, M., A. Erdem, P. Kara, K. Kerman, and D. Ozkan. 2003. Electrochemical biosensor for the detection of interaction between arsenic trioxide and DNA based on guanine signal. *Electroanalysis* 15:613–619.
3055. Kerman, K., B. Meric, D. Ozkan, P. Kara, A. Erdem, and M. Ozsoz. 2001. Electrochemical DNA biosensor for the determination of benzo[a]pyrene-DNA adducts. *Anal. Chim. Acta* 450:45–52.
3056. Serpi, C., Z. Stanic, and S. Girousi. 2010. Electroanalytical study of the interaction between double stranded DNA and antitumor agent curcumin. *Anal. Lett.* 43:1491–1506.
3057. Serpi, C., Z. Stanic, and S. Girousi. 2010. Electroanalytical study of the interaction between dsDNA and curcumin in the presence of copper(II). *Talanta* 81:1731–1734.
3058. Erdem, A., K. Kerman, D. Ozkan, O. Kucukoglu, E. Erciyas, and M. Ozsoz. 2001. Electrochemical determination of interaction between an alkylating anticancer agent and DNA in solution and at the surface. *Proc. Electrochem. Soc.* 18:563–575.
3059. Meric, B., K. Kerman, D. Ozkan, P. Kara, A. Erdem, O. Kucukoglu, E. Erciyas, and M. Ozsoz. 2002. Electrochemical biosensor for the interaction of DNA with the alkylating agent 4,4′-dihydroxy chalcone based on guanine and adenine signals. *J. Pharm. Biomed. Anal.* 30:1339–1346.
3060. Ioannou, A.K., A.A. Pantazaki, S.Th. Girousi, M.-C. Millot, C. Vidal-Madjar, and A.N. Voulgaropoulos. 2006. DNA biosensor based on carbon paste electrodes modified by polymer multilayer. *Electroanalysis* 18:456–464.
3061. Gherghi, I.Ch., S.Th. Girousi, A.N. Voulgaropoulos, and R. Tzimou-Tsitouridou. 2004. Differentiations in the electrochemical behavior of the interactions between DNA and compounds with affinity for DNA. *Anal. Lett.* 37:957–966.

References

3062. Gherghi, I.Ch., S.Th. Girousi, A. Voulgaropoulos, and R. Tzimou-Tsitouridou. 2004. Interaction of the mutagen ethidium bromide with DNA, using a carbon paste electrode and a hanging mercury drop electrode. *Anal. Chim. Acta* 505:135–144.
3063. Gherghi, I.Ch., S.Th. Girousi, A.A. Pantazaki, A.N. Voulgaropoulos, and R. Tzimou-Tsitouridou. 2003. Electrochemical DNA biosensors applicable to the study of interactions between DNA and DNA intercalators. *Int. J. Environm. Anal. Chem.* 83:693–700.
3064. Gherghi, I.Ch., S.T. Girousi, A.N. Voulgaropoulos, and R. Tzimou-Tsitouridou. 2003. Study of interactions between actinomycin D and DNA on carbon paste electrode (CPE) and on the hanging mercury drop (HMDE) surface. *J. Pharm. Biomed. Anal.* 31:1065–1078.
3065. Naik, T.R.R. and H.S.B. Naik. 2008. Electrochemical investigation of DNA binding on carbaldehyde oxime by cyclic voltammetry. *Int. J. Electrochem. Sci.* 3:409–415.
3066. Kara, P., K. Dagdeviren, and M. Ozsoz. 2007. An electrochemical DNA biosensor for the detection of DNA damage caused by radioactive iodine and technetium. *Turkey J. Chem.* 31:243–249.
3067. Diopan, V., P. Babula, V. Shestivska, V. Adam, M. Zemlicka, M. Dvorska, J. Hubalek, L. Trnkova, L. Havel, and R. Kizek. 2008. Electrochemical and spectrometric study of antioxidant activity of pomiferin, isopomiferin, osajin and catalposide. *J. Pharm. Biomed. Anal.* 48:127–133.
3068. Karadeniz, H., B. Gulmez, F. Sahinci, A. Erdem, G.I. Kaya, N. Unver, B. Kivcak, and M. Ozsoz. 2003. Disposable electrochemical biosensor for the detection of the interaction between DNA and lycorine based on guanine and adenine signals. *J. Pharm. Biomed. Anal.* 33:295–302.
3069. Kerman, K., D. Oezkan, P. Kara, H. Karadeniz, Z. Oezkan, A. Erdem, F. Jelen, and M. Oezsoez. 2004. Electrochemical detection of specific DNA sequences from PCR amplicons on carbon and mercury electrodes using meldola's blue as an intercalator. *Turk. J. Chem.* 28:523–533.
3070. Erdem, A. and M. Ozsoz. 2001. Voltammetry of the anticancer drug mitoxantrone and DNA. *Turk. J. Chem.* 25:469–475.
3071. Zhu, J.-J., K. Gu, J.-Z. Xu, and H.-Y. Chen. 2001. DNA modified carbon paste electrode for the detection of 6-mercaptopurine. *Anal. Lett.* 34:329–337.
3072. Ly, S-Y. 2008. Voltammetric assay of mercury ion in fish kidneys. *Toxicol. Res.* 24:23–28.
3073. Qi, H., P. Chen, and C. Zhang. 2007. Determination of ofloxacin and its interaction with DNA. *Yaowu Fenxi Zazhi* 27:96–99.
3074. Wang, Y., H.-Y. Xiong, X.-H. Zhang, and S.-F. Wang. 2010. Detection of rutin at DNA modified carbon paste electrode based on a mixture of ionic liquid and paraffin oil as a binder. *Microchim. Acta* 170:27–32.
3075. Snevajsova, P., L. Tison, I. Brozkova, J. Vytřas ova, R. Metelka, and K. Vytřas. 2010. Carbon paste electrode for voltammetric detection of a specific DNA sequence from potentially aflatoxigenic Aspergillus species. *Electrochem. Comm.* 12:106–109.
3076. Cai, X., G. Rivas, H. Shirashi, P. Farias, J. Wang, M. Tomschik, F. Jelen, and E. Palecek. 1997. Electrochemical analysis of formation of polynucleotide complexes in solution and at electrode surfaces. *Anal. Chim. Acta* 344:65–76.
3077. Li, J., W. Zhu, and H. Wang. 2009. Novel magnetic single-walled carbon nanotubes/methylene blue composite amperometric biosensor for DNA determination. *Anal. Lett.* 42:366–380.
3078. Hejazi, M.S., J.-B. Raoof, R. Ojani, S.M. Golabi, and E.H. Asl. 2010. Brilliant cresyl blue as electroactive indicator in electrochemical DNA oligonucleotide sensors. *Bioelectrochemistry* 78:141–146.
3079. Xu, G., J. Fan, and K. Jiao. 2008. Studies on the electrochemical property of dinuclear copper(II) complex containing dimethylglyoxime and its interaction with DNA. *Electroanalysis* 20:1209–1214.
3080. Ioannou, A.K., D.K. Alexiadou, S.A. Kouidou, A.N. Voulgaropoulos, and S.T. Girousi. 2010. Electroanalytical study of SYBR Green I and ethidium bromide intercalation in methylated and unmethylated amplicons. *Anal. Chim. Acta* 657:163–168.
3081. Girousi, S. and V. Kinigopoulou. 2010. Detection of short oligonucleotide sequences using an electrochemical DNA hybridization biosensor. *Cent. Eur. J. Chem.* 8:732–736.
3082. Alexiadou, D.K., A.K. Ioannou, S. Kouidou-Andreou, A.N. Voulgaropoulos, and S.T. Girousi. 2008. Electrochemical study of the interaction mechanism of proflavine (PF) with DNA using carbon paste (CPE) and hanging mercury drop (HMDE) electrode. *Anal. Lett.* 41:1742–1750.
3083. Wang, J., G. Rivas, C. Parrado, X. Cai, and M.N. Flair. 1997. Electrochemical biosensor for detecting DNA sequences from the pathogenic protozoan Cryptosporidium parvum. *Talanta* 44:2003–2010.
3084. Wang, J., G. Rivas, J.R. Fernandes, J.L. Lopez Paz, M. Jiang, and R. Waymire. 1998. Indicator-free electrochemical DNA hybridization biosensor. *Anal. Chim. Acta* 375:197–203.
3085. Wang, J., X. Cai, G. Rivas, and H. Shirashi. 1996. Stripping potentiometric transduction of DNA hybridization processes. *Anal. Chim. Acta* 326:141–147.

3086. Wang, J., J.R. Fernandes, and L.T. Kubota. 1998. Polishable and renewable DNA hybridization biosensors. *Anal. Chem.* 70:3699–3702.
3087. Ozkan, D., A. Erdem, P. Kara, K. Kerman, B. Meric, J. Hassmann, and M. Ozsoz. 2002. Allele-specific genotype detection of factor V Leiden mutation from polymerase chain reaction amplicons based on label-free electrochemical genosensor. *Anal. Chem.* 74:5931–5936.
3088. Kara, P., D. Ozkan, A. Erdem, K. Kerman, S. Pehlivan, F. Ozkinay, D. Unuvar, G. Itirli, and M. Ozsoz. 2003. Detection of Achondroplasia G380R mutation from PCR amplicons by using inosine modified carbon electrodes based on electrochemical DNA chip technology. *Clin. Chim. Acta* 336:57–64.
3089. Wang, J., G. Rivas, X. Cai, M. Chicharro, C. Parrado, N. Dontha, A. Begleiter, M. Mowat, E. Palecek, and P.E. Nielsen. 1997. Detection of point mutation in the p53 gene using a peptide nucleic acid biosensor. *Anal. Chim. Acta* 344:111–118.
3090. Wang, J., E. Palecek, P.E. Nielsen, G. Rivas, X. Cai, H. Shiraishi, N. Dontha, D. Luo, and P.A.M. Farias. 1996. Peptide nucleic acid probes for sequence-specific DNA biosensors. *J. Am. Chem. Soc.* 118:7667–7670.
3091. Kerman, K., D. Ozkan, P. Kara, A. Erdem, B. Meric, P.E. Nielsen, and M. Ozsoz. 2003. Label-free bioelectronic detection of point mutation by using peptide nucleic acid probes. *Electroanalysis* 15:667–670.
3092. Kerman, K., Y. Matsubara, Y. Morita, Y. Takamura, and E. Tamiya. 2004. Peptide nucleic acid modified magnetic beads for intercalator based electrochemical detection of DNA hybridization. *Sci. Technol. Adv. Mater.* 5:351–357.
3093. Wang, J., X. Cai, G. Rivas, H. Shiraishi, P.A.M. Farias, and N. Dontha. 1996. DNA electrochemical biosensor for the detection of short DNA sequences related to the human immunodeficiency virus. *Anal. Chem.* 68:2629–2634.
3094. Wang, J., G. Rivas, X. Cai, N. Dontha, H. Shiraishi, D. Luo, and F.S. Valera. 1997. Sequence-specific electrochemical biosensing of M. tuberculosis DNA. *Anal. Chim. Acta* 337:41–48.
3095. Erdem, A., K. Kerman, B. Meric, D. Ozkan, P. Kara, and M. Ozsoz. 2002. DNA biosensor for Microcystis spp. sequence detection by using methylene blue and ruthenium complex as electrochemical hybridization labels. *Turk. J. Chem.* 26:851–862.
3096. Wang, J., G.-U. Flechsig, A. Erdem, O. Korbut, and P. Gruendler. 2004. Label-free DNA hybridization based on coupling of a heated carbon paste electrode with magnetic separations. *Electroanalysis* 16:928–931.
3097. Masarik, M., R. Kizek, K.J. Kramer, S. Billova, M. Brazdova, J. Vacek, M. Bailey, F. Jelen, and J.A. Howard. 2003. Application of avidin-biotin technology and adsorptive transfer stripping square-wave voltammetry for detection of DNA hybridization and avidin in transgenic avidin maize. *Anal. Chem.* 75:2663–2669.
3098. Ligaj, M., T. Oczkowski, J. Jasnowska, W.G. Musial, and M. Filipiak. 2003. Electrochemical genosensors for detection of L. monocytogenes and genetically-modified components in food. *Pol. J. Food Nutr. Sci.* 12:61–63.
3099. Yang, T., C. Jiang, W. Zhang, and K. Jiao. 2010. Improved electrochemical performances of polyaniline nanotubes-poly-L-lysine composite for label-free impedance detection of DNA hybridization. *Sci. China Ser. B Chem.* 53:1371–1377.
3100. Qi, H., X. Li, P. Chen, and C. Zhang. 2007. Electrochemical detection of DNA hybridization based on polypyrrole/ss-DNA/multi-wall carbon nanotubes paste electrode. *Talanta* 72:1030–1035.
3101. Zhou, N., T. Yang, K. Jiao, and C.-X. Song. 2010. Electrochemical deoxyribonucleic acid biosensor based on multi-walled carbon nanotubes/Ag-TiO$_2$ composite film for label-free phosphinothricin acetyltransferase gene detection by electrochemical impedance spectroscopy. *Fenxi Huaxue* 38:301–306.
3102. Yin, C., H. Liu, K. Jiao, and X. Fu. 2009. Sensitive detection for DNA-specific sequences using Bi$_2$S$_3$-nanoparticle-labeled DNA probes. *Qingdao Keji Daxue Xuebao, Ziran Kexueban* 30:471–476.
3103. Erdem, A., K. Kerman, B. Meric, U.S. Akarca, and M. Ozsoz. 1999. DNA electrochemical biosensor for the detection of short DNA sequences related to the hepatitis B virus. *Electroanalysis* 11:586–588.
3104. Meric, B., K. Kerman, D. Ozkan, P. Kara, S. Erensoy, U.S. Akarca, M. Mascini, and M. Ozsoz. 2002. Electrochemical DNA biosensor for the detection of TT and Hepatitis B virus from PCR amplified real samples by using methylene blue. *Talanta* 56:837–846.
3105. Guo, M., Y. Li, H. Guo, X. Wu, and L. Fan. 2007. Electrochemical detection of short sequences related to the hepatitis B virus using MB on chitosan-modified CPE. *Bioelectrochemistry* 70:245–249.
3106. Cheng, L., Y. Li, Z. Liu, and F. Liu. 2003. Fast detection of human immunodeficiency virus gene with carbon paste electrode. *Fenxi Ceshi Xuebao* 22:75–77.
3107. Ligaj, M., M. Tichoniuk, and M. Filipiak. 2008. Detection of bar gene encoding phosphinothricin herbicide resistance in plants by electrochemical biosensor. *Bioelectrochemistry* 74:32–37.

3108. Jiang, C., T. Yang, K. Jiao, and H. Gao. 2008. A DNA electrochemical sensor with poly-L-lysine/single-walled carbon nanotubes films and its application for the highly sensitive EIS detection of PAT gene fragment and PCR amplification of NOS gene. *Electrochim. Acta* 53:2917–2924.

3109. Jiao, K., X.-Z. Zhang, G.-Y. Xu, and W. Sun. 2005. Fabrication and voltammetric characteristics of carbon paste electrode modified by ssDNA/stearic acid. *Huaxue Xuebao* 63:1100–1104.

3110. Ligaj, M., J. Jasnowska, W.G. Musial, and M. Filipiak. 2006. Covalent attachment of single-stranded DNA to carbon paste electrode modified by activated carboxyl groups. *Electrochim. Acta* 51:5193–5198.

3111. Zhang, Z.-Q., J.-P. Xie, J.-F. Xiong, M.-F. Zhou, and Z.-S. Zhang. 2005. Voltammetric biosensor for cucumber DNA. *Fenxi Kexue Xuebao* 21:5–8.

3112. Jiao, K., Y. Ren, G. Xu, and X. Zhang. 2005. Voltammetric study on deoxyribonucleic acid immobilization and hybridization on stearic acid/aluminum ion films and the detection of specific gene related to phosphinothricin acetyltransferase gene from Bacillus amyloliquefaciens gene. *Fenxi Huaxue* 33:1381–1384.

3113. Ren, Y., K. Jiao, G. Xu, W. Sun, and H. Gao. 2005. An electrochemical DNA sensor based on electrodepositing aluminum ion films on stearic acid-modified carbon paste electrode and its application for the detection of specific sequences related to bar gene and CP4 Epsps gene. *Electroanalysis* 17:2182–2189.

3114. Zhou, N., T. Yang, Y. Zhang, C. Jiang, and K. Jiao. 2009. Electrochemical characterization of deoxyribonucleic acid on aluminum(III)/poly(-glutamic acid) film and its application for the detection of phosphinothricin acetyltransferase gene-specific sequence. *Thin Solid Films* 518:338–342.

3115. Ikebukuro, K., Y. Kohiki, and K. Sode. 2002. Amperometric DNA sensor using the pyrroquinoline quinone glucose dehydrogenase-avidin conjugate. *Biosens. Bioelectron.* 17:1075–1080.

3116. Zhou, N., T. Yang, C. Jiang, M. Du, and K. Jiao. 2009. Highly sensitive electrochemical impedance spectroscopic detection of DNA hybridization based on Aunano-CNT/PAN-nano films. *Talanta* 77:1021–1026.

3117. Yang, T., N. Zhou, Y. Zhang, W. Zhang, K. Jiao, and G. Li. 2009. Synergistically improved sensitivity for the detection of specific DNA sequences using polyaniline nanofibers and multi-walled carbon nanotubes composites. *Biosens. Bioelectron.* 24:2165–2170.

3118. Wu, J., Y. Zou, X. Li, H. Liu, G. Shen, and R. Yu. 2005. A biosensor monitoring DNA hybridization based on polyaniline intercalated graphite oxide nanocomposite. *Sens. Actuators B* B104:43–49.

3119. Kerman, K., Y. Morita, Y. Takamura, M. Ozsoz, and E. Tamiya. 2004. DNA-directed attachment of carbon nanotubes for enhanced label-free electrochemical detection of DNA hybridization. *Electroanalysis* 16:1667–1672.

3120. Masarik, M., K. Cahova, R. Kizek, E. Palecek, and M. Fojta. 2007. Label-free voltammetric detection of single-nucleotide mismatches recognized by the protein MutS. *Anal. Bioanal. Chem.* 388:259–270.

3121. Ly, S.Y., Y.S. Jung, M.H. Kim, I.k. Han, W.W. Jung, and H.S. Kim. 2004. Determination of caffeine using a simple graphite pencil electrode with square-wave anodic stripping voltammetry. *Microchim. Acta* 146:207–213.

3122. Ly, S.Y., C.H. Lee, Y.S. Jung, O.M. Kwon, J.E. Lee, S.M. Baek, and K.J. Kwak. 2008. Simultaneous diagnostic assay of catechol and caffeine using an in vivo implanted neuro sensor. *Bull. Korean Chem. Soc.* 29:1742–1746.

3123. Sanghavi, B.J. and A.K. Srivastava. 2010. Simultaneous voltammetric determination of acetaminophen, aspirin and caffeine using an in situ surfactant-modified multiwalled carbon nanotube paste electrode. *Electrochim. Acta* 55:8638–8648.

3124. Stulik, K. and V. Pacakova. 1983. High-performance liquid chromatography of biologically important pyrimidine derivatives with ultraviolet—voltammetric—polarographic detection. *J. Chromatogr.* 273:77–86.

3125. Munson, J.W. and H. Abdine. 1978. Determination of theophylline in plasma by oxidation at the stationary carbon-paste electrode. *Talanta* 25:221–222.

3126. Inada, K., K. Nishiyama, and I. Taniguchi. 2009. Selective detection of uric acid in the presence of ascorbic acid and acetaminophen at a glassy carbon paste electrode in an alkaline solution. *Chem. Lett.* 38:814–815.

3127. Liao, H.-C., T.-P. Sun, C.-Y. Chen, J.-C. Chou, and S.-K. Hsiung. 2004. A multi-parameter biochemical sensor with a linear readout circuit. *Chem. Sensors* 20(Suppl. B):638–639.

3128. Xiao, F., C. Ruan, J. Li, L. Liu, F. Zhao, and B. Zeng. 2008. Voltammetric determination of xanthine with a single-walled carbon nanotube-ionic liquid paste modified glassy carbon electrode. *Electroanalysis* 20:361–366.

3129. Zhang, X.-L., Y.-F. Peng, and C.-G. Hu. 2008. Direct detection of uric acid in human urine at CTAB modified carbon paste electrodes by linear sweeping voltammetry. *Fenxi Kexue Xuebao* 24:641–644.

3130. Akbari, R., M. Noroozifar, M. Khorasani-Motlagh, and A. Taheri. 2010. Simultaneous determination of ascorbic acid and uric acid by a new modified carbon nanotube-paste electrode using chloromercuriferrocene. *Anal. Sci.* 26:425–430.
3131. Lima, P.R., P.R. Barbosa de Miranda, A. Bof de Oliveira, M.O.F. Goulart, and L. Tatsuo Kubota. 2009. Modified carbon paste electrode for kinetic investigation and simultaneous determination of ascorbic and uric acids. *Electroanalysis* 21:2311–2320.
3132. Fu, Z.-Q., M.-Y. Wan, and Y.-F. Tu. 2009. A uric acid biosensor based on screen printed carbon electrode modified by MnO_2 nano-particles and ferrocene. *Fenxi Kexue Xuebao* 25:79–82.
3133. Wu, X.-P., W.-C. Zhou, H. Dai, and S.-C. Sun. 2007. A new type of graphite/poly(methyl methacrylate) composite electrode for electrochemiluminescence. *Guangpu Shiyanshi* 24:1176–1180.
3134. Lorenzo, E., E. Gonzalez, F. Pariente, and L. Hernandez. 1991. Immobilized enzyme carbon paste electrodes as amperometric sensors. *Electroanalysis* 3:319–323.
3135. Sternson, L.A. 1976. Electrochemical determination of xanthine oxidase-catalyzed oxidation of xanthine. *Anal. Lett.* 9:641–652.
3136. Gonzalez, E., F. Pariente, E. Lorenzo, and L. Hernandez. 1991. Amperometric sensor for hypoxanthine and xanthine based on the detection of uric acid. *Anal. Chim. Acta* 242:267–273.
3137. Doblhoff-Dier, O. and G.A. Rechnitz. 1989. Amperometric enzyme based biosensor for the detection of xanthine via superoxide. *Anal. Lett.* 22:1047–1055.
3138. Anik, U. and S. Cevik. 2009. Double-walled carbon nanotube based carbon paste electrode as xanthine biosensor. *Microchim. Acta* 166:209–213.
3139. Kirgoez, U.A., S. Timur, J. Wang, and A. Telefoncu. 2004. Xanthine oxidase modified glassy carbon paste electrode. *Electrochem. Comm.* 6:913–916.
3140. Lindquist, J. 1975. Voltammetric determination of ascorbic acid using a carbon paste electrode. *Analyst* 100:339–348.
3141. Deakin, M.R., P.M. Kovach, K.J. Stutts, and R.M. Wightman. 1986. Heterogeneous mechanisms of the oxidation of catechols and ascorbic acid at carbon electrodes. *Anal. Chem.* 58:1474–1480.
3142. Wang, Q., A. Geiger, R. Frias, and T.D. Golden. 2000. An introduction to electrochemistry for undergraduates: Detection of vitamin C (ascorbic acid) by inexpensive electrode sensors. *Chem. Educ.* 5:58–60.
3143. Wang, C., J. Yang, B. Hua, X. Zhai, and H. Zhang. 1998. Development and application of polypyrrole carbon paste electrode. *Fenxi Huaxue* 26:847–849.
3144. Hu, X. and Z. Leng. 1995. Determination of L-ascorbic acid by adsorption potentiometry with carbon paste electrode. *Anal. Lett.* 28:2263–2274.
3145. Sandulescu, R., R. Oprean, and L. Roman. 1997. Carbon-paste electrodes in the quantitative determination of ascorbic acid in pharmaceutical forms. *Farmacia (Bucharest)* 45:23–35.
3146. Marian, I.O., R. Sandulescu, and N. Bonciocat. 2000. Diffusion coefficient (or concentration) determination of ascorbic acid using carbon paste electrodes in Fredholm alternative. *J. Pharm. Biomed. Anal.* 23:227–230.
3147. Yang, M., Q. Jiang, Y. Li, W. Sun, and K. Jiao. 2009. Direct electrochemistry of ascorbic acid on the [BMIM]PF6 modified carbon paste electrode. *Qingdao Keji Daxue Xuebao, Ziran Kexueban* 30:6–8.
3148. Sandulescu, R., S. Lotrean, and V. Mirel. 2003. Electrochemical and spectrophotometric study of interactions of β-cyclodextrin with some pharmaceuticals. *Farmacia (Bucharest)* 51:31–41.
3149. Xiang, B. 2009. Optimal condition for determination of ascorbic acid adsorptive stripping voltammetry. *Huagong Shikan* 23:41–43.
3150. Xiang, B. 2009. The determination of ascorbic acid by adsorptive stripping voltammetry using mesoporous SiO_2 modified carbon paste electrode. *Huagong Shikan* 23:20–22.
3151. Teixeira, M.F.S., L.A. Ramos, O. Fatibello-Filho, and E.T.G. Cavalheiro. 2003. Carbon paste electrode modified with copper(II) phosphate immobilized in a polyester resin for voltammetric determination of L-ascorbic acid in pharmaceutical formulations. *Anal. Bioanal. Chem.* 376:214–219.
3152. Fujiwara, S.T., C.A. Pessoa, and Y. Gushikem. 2002. Copper (II) tetrasulfophthalocyanine entrapped in a propylpyridiniumsilsesquioxane polymer immobilized on a SiO_2/Al_2O_3 surface: Use for electrochemical oxidation of ascorbic acid. *Anal. Lett.* 35:1117–1134.
3153. Nezamzadeh, A., M.K. Amini, and H. Faghihian. 2007. Square-wave voltammetric determination of ascorbic acid based on its electrocatalytic oxidation at zeolite-modified carbon-paste electrodes. *Int. J. Electrochem. Sci.* 2:583–594.
3154. Liu, S., G. Dai, Y. Zhao, and X. Hu. 2009. Voltammetric behavior and electrochemical determination of ascorbic acid at ferrous pentacyanonitrosylferrate film modified carbon paste electrode. *Anal. Lett.* 42:2914–2927.
3155. do Carmo, D.R., R.M. da Silva, and N.R. Stradiotto. 2004. Electrocatalysis and determination of ascorbic acid through graphite paste electrode modified with iron nitroprusside. *Portugaliae Electrochim. Acta* 22:71–79.

3156. Raoof, J.B., R. Ojani, and A. Kiani. 2003. Ferrocene spiked carbon paste electrode and its application to electrocatalytic determination of ascorbic acid. *Bull. Electrochem.* 19:17–22.
3157. Bai, L., Z. Zheng, and Z. Gao. 2007. Electrocatalytic oxidation of ascorbic acid at ferrocene modified carbon paste electrode and its practical analytical application. *Huaxue Chuanganqi* 27:53–57.
3158. Wang, X., G. Zhang, Y. Li, and T. Sun. 1999. Electrocatalytic oxidation of ascorbic acid on ferrocene modified carbon paste electrode. *Huaxue Chuanganqi* 19:39–43.
3159. Sun, W., D.-Q. Lu, Y.-J. Li, and X.-W. He. 2003. Preparation and application of modified PVC carbon paste electrodes. *Fenxi Shiyanshi* 22:40–43.
3160. Wang, X.-L., G.-R. Zhang, X.-W. Shi, and T.-L. Sun. 2000. Studies on electrocatalytic oxidation of ascorbic acid at β-cyclodextrin-ferrocene inclusion complex-modified carbon paste electrode. *Gaodeng Xuexiao Huaxue Xuebao* 21:1383–1385.
3161. Liu, C.-Y., M.-Q. Huang, and G.-H. Lu. 2005. Voltammetric determination of ascorbic acid at a carbon paste electrode modified with iron(I) phthalocyanine and dioctyl phthalate. *Wuhan Huagong Xueyuan Xuebao* 27:3–6.
3162. Liu, C.-Y., J.-F. Hu, X.-M. Li, and G.-H. Lu. 2005. Voltammetric determination of ascorbic acid at a carbon paste electrode modified with Iron(II) phthalocyanine and dioctyl phthalate. *Xinan Shifan Daxue Xuebao, Ziran Kexueban* 30:289–292.
3163. Ijeri, V.S., P.V. Jaiswal, and A.K. Srivastava. 2001. Chemically modified electrodes based on macrocyclic compounds for determination of Vitamin C by electrocatalytic oxidation. *Anal. Chim. Acta* 439:291–297.
3164. Goreti, M., F. Sales, M.S.A. Castanheira, R.M.S. Ferreira, M. Carmo, V.G. Vaz, and C. Delerue-Matos. 2008. Chemically modified carbon paste electrodes for ascorbic acid determination in soft drinks by flow injection amperometric analysis. *Portugaliae Electrochim. Acta* 26:147–157.
3165. Qi, Y.-J., X.-D. Yuan, and L. Tian. 2009. Synthesis, structure and electrochemical properties of three-dimensional complex [C4H11N2]3[PMo12O40]. *Wuji Huaxue Xuebao* 25:1110–1114.
3166. Ojani, R., J.-B. Raoof, and S. Zamani. 2005. Electrochemical behavior of chloranil chemically modified carbon paste electrode. Application to the electrocatalytic determination of ascorbic acid. *Electroanalysis* 17:1740–1745.
3167. Liu, B., L. Chen, H. Liu, and J. Deng. 1995. Tetracyanoquinodimethane modified carbon paste electrodes for oxidation of ascorbic acid. *Fenxi Huaxue* 23:206–210.
3168. Faguy, P.W., W. Ma, J.A. Lowe, W.P. Pan, and T. Brown. 1994. Conducting polymer-clay composites for electrochemical applications. *J. Mater. Sci.* 4:771–772.
3169. Wang, Z.-H., A.-P. Huang, H.-Y. Qiao, T.-W. Chen, and D. Wang. 2009. Fabrication of inlaid super-thin microcrystalline cellulose modified carbon paste film electrode and its application. *Yingyong Huaxue* 26:840–844.
3170. Dilgin, Y. and G. Nisli. 2006. Flow injection photoamperometric investigation of ascorbic acid using methylene blue immobilized on titanium phosphate. *Anal. Lett.* 39:451–465.
3171. Hoffmann, A.A., S.L.P. Dias, J.R. Rodrigues, F.A. Pavan, E.V. Benvenutti, and E.C. Lima. 2008. Methylene blue immobilized on cellulose acetate with titanium dioxide: An application as sensor for ascorbic acid. *J. Braz. Chem. Soc.* 19:943–949.
3172. Varodi, C., D. Gligor, and L.M. Muresan. 2007. Carbon paste electrodes incorporating synthetic zeolites and methylene blue for amperometric detection of ascorbic acid. *Stud. Univ. Babes-Bolyai Chem.* 52:109–117.
3173. Canevari, T.d.C., R. Ferrari de Castilho, R.V.d.S. Alfaya, and A.A.d.S. Alfaya. 2006. Development of a new ascorbic acid sensor. *Semina Cienc. Exat. Tecnol.* 27:129–138.
3174. Pavan, F.A., E.S. Ribeiro, and Y. Gushikem. 2005. Congo red immobilized on a silica/aniline xerogel: Preparation and application as an amperometric sensor for ascorbic acid. *Electroanalysis* 17:625–629.
3175. Ying, T., Y. Qiao, S. Hu, and D. Qi. 1994. Development and application of green pepper seed tissue-carbon paste bioelectrode. *Fenxi Huaxue* 22:492–494.
3176. Wang, Q. 2007. Gooseberry modified carbon paste electrode for the determination of ascorbic acid. *Jinan Daxue Xuebao, Ziran Kexueban* 21:151–153.
3177. Dai, H., X. Wu, Y. Wang, W. Zhou, and G. Chen. 2008. An electrochemiluminescent biosensor for vitamin C based on inhibition of luminol electrochemiluminescence on graphite/poly(methyl methacrylate) composite electrode. *Electrochim. Acta* 53:5113–5117.
3178. Zeng, G.-M., Z.-G. Cjhen, F.-Y. Zhou, and Z.-Q. Zhangm. 1994. Preparation of voltammetric sensors for drug analysis and its application. IV. Carbon paste electrode for the determination of nicotinic acid. *Huaxue Chuanganqi* 14:66–68, 62.
3179. Hart, J.P., S.A. Wring, and I.C. Morgan. 1989. Preconcentration of vitamin K1 (phylloquinone) at carbon paste electrodes and its determination in plasma by adsorptive stripping voltammetry. *Analyst (U.K.)* 114:933–937.

3180. Soderhjelm, P. and J. Lindquist. 1975. Voltammetric determination of pyridoxine using a carbon paste electrode. *Analyst (U.K.)* 100:349–354.
3181. Liu, Y., Y. Bai, T. Cheng, W. Zheng, and Y. Zhou. 2004. Electrochemical behavior of riboflavin (VB2) on the liquid paraffin carbon paste electrode and determination of VB2 by differential pulse voltammetry. *Guangzhou Huagong* 32:46–49, 22.
3182. Percira, A.C. and L.T. Kubota. 2004. Optimization of the carbon paste electrode preparation containing riboflavin immobilized on an inorganic support. *Quim. Nova (Barcelona)* 27:725–729.
3183. Jaiswal, P.V., V.S. Ijeri, and A.K. Srivastava. 2001. Voltammetric behavior of a-tocopherol and its determination using surfactant+ethanol+water and surfactant+acetonitrile+water mixed solvent systems. *Anal. Chim. Acta* 441:201–206.
3184. Kim, O.S., S. Shiragami, and K. Kusuda. 1994. Increase of the redox current in cyclic voltammetry of a compound dissolved in Nujol by contacting with a surface active compound in an aqueous solution. *J. Electroanal. Chem.* 367:271–273.
3185. Kim, O.-S. and K. Kusuda. 1994. Electrochemical behavior of a-tocopherol in a thin film of Nujol, a model of adipose tissue. *Bioelectrochem. Bioenerg.* 33:61–65.
3186. Atuma, S.S., K. Lundstrom, and J. Lindquist. 1975. Electrochemical determination of vitamin A. II. Further voltammetric determination of vitamin A and initial work on determination of vitamin D in the presence of vitamin A. *Analyst* 100:827–834.
3187. Atuma, S.S., J. Lindquist, and K. Lundstroem. 1974. Electrochemical determination of vitamin A. I. Voltammetric determination of vitamin A in pharmaceutical preparations. *Analyst* 99:683–689.
3188. Lesunova, R.P., O.A. Fursik, and N.A. Khodyreva. 1974. Use of carbon paste electrodes in vitamin analysis. *Dostizh. Razvit. Nov. Metodov Khim. Anal. (Ukr.)* 1974:27–28.
3189. Shaikh, A.A., M. Begum, A.H. Khan, and M.Q. Ehsan. 2006. Cyclic voltammetric studies of the redox behavior of iron(III)-vitamin B6 complex at carbon paste electrode. *Russ. J. Electrochem.* 42:620–625.
3190. Xu, H., C. Wang, J. Zhai, G. Ke, and H. Li. 2004. Extractive stripping voltammetric behavior of vitamin E at carbon paste electrode and its electrochemical determination. *Lihua Jianyan, Huaxue Fence* 40:284–287.
3191. Zhai, J., H. Xu, and C. Liu. 2000. Voltammetric method for determination of vitamin E with paste electrode extraction. *J. Gansu Univ. Technol.* (Engl. Edn.) E-4:123–126.
3192. Gu, L., J. Wang, and H.-L. Zhuang. 2009. Electrochemical determination of vitamin E in vitamin E capsules at carbon paste electrode with cationic surfactant as sensitizing agent. *Lihua Jianyan, Huaxue Fence* 45:885–888.
3193. Hu, R.-Z., D.-P. Wei, G. Wei, and Y.-Y. Liu. 2001. Simultaneous determination of gold(III) and copper(II) with two-way stripping voltammetry. *Dian Huaxue* 7:339–344.
3194. Hu, R., Y. Ruan, H. Xu, J. Fu, W. Zhang, and W. Hu. 1999. Preparation and application of gold film on the surface of bacterial-modified carbon paste electrode. *Dian Huaxue* 5:231–235.
3195. Wang, J., T. Martinez, and D. Darnall. 1989. Electrocatalysis at algae modified electrodes. *J. Electroanal. Chem.* 259:295–300.
3196. Yuece, M., H. Nazir, and G. Doenmez. 2010. An advanced investigation on a new algal sensor determining Pb(II) ions from aqueous media. *Biosens. Bioelectron.* 26:321–326.
3197. Kato, K., N. Taniguchi, Y. Yamamoto, K. Kano, and T. Ikeda. 1996. A whole cell electrode for ethanol based on the respiratory reaction in cytoplasmic membrane of *Acetobacter aceti*. *Denki Kagaku Oyobi Kogyo Butsuri Kagaku* 64:1259–1260.
3198. Yeni, F., D. Odaci, and S. Timur. 2008. Use of eggshell membrane as an immobilization platform in microbial sensing. *Anal. Lett.* 41:2743–2758.
3199. Yao, H. and G.J. Ramelow. 1998. Biomass-modified carbon paste electrodes for monitoring dissolved metal ions. *Talanta* 45:1139–1146.
3200. Fatibello-Filho, O., K.O. Lupetti, O.D. Leite, and I.C. Vieira. 2007. Electrochemical biosensors based on vegetable tissues and crude extracts for environmental, food and pharmaceutical analysis. *Compr. Anal. Chem.* 49:357–377.
3201. Wang, J., L.H. Wu, S. Martinez, and J. Sanchez. 1991. Tissue bioelectrode for eliminating protein interferences. *Anal. Chem.* 63:398–400.
3202. Wang, J. and M.S. Lin. 1989. Horseradish-root-modified carbon paste bioelectrode. *Electroanalysis* 1:43–48.
3203. Wang, J., A. Ciszewski, and N. Naser. 1992. Stripping measurements of hydrogen peroxide based on biocatalytic accumulation at mediatorless peroxidase/carbon paste electrodes. *Electroanalysis* 4:777–782.
3204. Lee, B.-G., K.-J. Yoon, and H.-S. Kwon. 2000. Spinach root-tissue based amperometric biosensor for the determination of hydrogen peroxide. *Anal. Sci. Technol.* 13:315–322.

References

3205. Kwon, H.-S., K.-K. Kim, and C.-G. Lee. 1996. Cabbage root tissue-based amperometric biosensor for determination of hydrogen peroxide. *J. Korean Chem. Soc.* 40:278–282.
3206. Lee, B.-G., K.-B. Rhyu, and K.-J. Yoon. 2009. Electrochemical investigation of carbon paste biosensor bound with natural rubber solution. *Bull. Korean Chem. Soc.* 30:2457–2460.
3207. Yoon, K.-J., K.-J. Kim, and H.-S. Know. 1999. Electrochemical properties of the chicken small intestinal tissue based enzyme electrode for the determination of hydrogen peroxide. *J. Korean Chem. Soc.* 43:271–279.
3208. Wang, J., B. Freiha, N. Naser, E.G. Romero, U. Wollenberger, M. Ozsoz, and O. Evans. 1991. Amperometric biosensing of organic peroxides with peroxidase-modified electrodes. *Anal. Chim. Acta* 254:81–88.
3209. Erdem, A., D. Ozkan, B. Meric, K. Kerman, and M. Ozsoz. 2001. Incorporation of EDTA for the elimination of metal inhibitory effects in an amperometric biosensor based on mushroom tissue polyphenol oxidase. *Turkey J. Chem.* 25:231–239.
3210. Liu, J. and J. Wang. 2001. Plant tissue bioelectrode for environmental monitoring of tyrosinase inhibitor. *Huanjing Huaxue* 20:398–404.
3211. Wang, J., S.A. Kane, J. Liu, M.R. Smyth, and K. Rogers. 1996. Mushroom tissue-based biosensor for inhibitor monitoring. *Food Technol. Biotechnol.* 34:51–55.
3212. Morisaki, H., M. Sugimoto, and H. Shiraishi. 2000. Attachment of bacterial cells to carbon electrodes. *Bioelectrochem.* 51:21–25.
3213. Obuchowska, A. 2008. Quantitation of bacteria through adsorption of intracellular biomolecules on carbon paste and screen-printed carbon electrodes and voltammetry of redox-active probes. *Anal. Bioanal. Chem.* 390:1361–1371.
3214. Che, Y.H., Y. Li, M. Slavik, and D. Paul. 2000. Rapid detection of Salmonella typhimurium in chicken carcass wash water using an immunoelectrochemical method. *J. Food. Prot.* 63:1043–1048.
3215. Yang, Z., Y. Li, C. Balagtas, M. Slavik, and D. Paul. 1998. Immunoelectrochemical assay in combination with homogeneous enzyme-labeled antibody conjugation for rapid detection of Salmonella. *Electroanalysis* 10:913–916.
3216. O'Neill R.D. 1986. The effects of intracerebral and systemic administration of drugs on transmitter release in the unrestrained rat measured using microcomputer-controlled voltammetry. *Anal. Chem. Symp. Ser.* 25 (*Electrochem. Sens. Anal.*) pp. 215–220. Amsterdam: Elsevier.
3217. Bolger, F.B. and J.P. Lowry. 2005. Brain tissue oxygen: In vivo monitoring with carbon paste electrodes. *Sensors* 5:473–487.
3218. Boutelle, M.G., L. Svensson, and M. Fillenz. 1989. Rapid changes in striatal ascorbate in response to tail-pinch monitored by constant potential voltammetry. *Neuroscience* 30:11–17.
3219. Justice, J.B. Jr., S.A. Wages, A.C. Michael, R.D. Blakely, and D.B. Neill. 1983. Interpretations of voltammetry in the striatum based on chromatography of striatal dialyzate. *J. Liq. Chromatogr.* 6:1873–1896.
3220. Fillenz, M. and R.D. O'Neill. 1986. Effects of light reversal on the circadian pattern of motor activity and voltammetric signals recorded in rat forebrain. *J. Physiol.* 374:91–101.
3221. Joseph, M.H., H. Hodges, and J. Gray. 1989. Lever pressing for food reward and in vivo voltammetry: Evidence for increases in extracellular homovanillic acid, the dopamine metabolite, and uric acid in the rat caudate nucleus. *Neuroscience* 32:195–201.
3222. Mason, P.A., J.A. Durr, D. Bhaskaran, and C. Freed. 1988. Plasma osmolality predicts extracellular fluid catechol concentrations in the lateral hypothalamus. *J. Neurochem.* 51:552–560.
3223. Haskett, C. and K. Mueller. 1987. The effects of serotonin depletion on the voltammetric response to amphetamine. *Pharmacol. Biochem. Behav.* 28:381–384.
3224. O'Neill, R.D. 1986. Adenosine modulation of striatal neurotransmitter release monitored in vivo using voltammetry. *Neurosci. Lett.* 63:11–16.
3225. O'Neill, R.D. and M. Fillenz. 1985. Circadian changes in extracellular ascorbate in rat cortex, accumbens, striatum and hippocampus: Correlations with motor activity. *Neurosci. Lett.* 60:331–336.
3226. Clemens, J.A. and L. Phebus. 1984. Brain dialysis in conscious rats confirms in vivo electrochemical evidence that dopaminergic stimulation releases ascorbate. *Life Sci.* 35:671–677.
3227. Kovach, P.M., A.G. Ewing, R.L. Wilson, and R. Wightman. 1984. In vitro comparison of the selectivity of electrodes for in vivo electrochemistry. *J. Neurosci. Methods* 10:215–227.
3228. Echizen, H. and C. Freed. 1983. In vivo electrochemical detection of extraneuronal 5-hydroxyindole acetic acid and norepinephrine in the dorsal raphe nucleus of urethane-anesthetized rats. *Brain Res.* 277:55–62.
3229. O'Neill, R.D., M. Fillenz, and W. Albery. 1982. Circadian changes in homovanillic acid and ascorbate levels in the rat striatum using microprocessor-controlled voltammetry. *Neurosci. Lett.* 34:189–193.

3230. Saponjic, R.M., K. Mueller, D. Krug, and P.M. Kunko. 1994. The effects of haloperidol, scopolamine, and MK-801 on amphetamine-induced increased in ascorbic and uric acid as determined by voltammetry in vivo. *Pharmacol. Biochem. Behav.* 48:161–168.
3231. Miele, M. and M. Fillenz. 1996. In vivo determination of extracellular brain ascorbate. *J. Neurosci. Methods* 70:15–19.
3232. Fillenz, M., M.G. Boutelle, L. Fellows, A. Fray, and M. Miele. 1994. The use of quantitative microdialysis to study the dynamics of glucose, lactate and ascorbate in rat brain. In *Monitoring Molecules in Neuroscience, Proceedings of the 6th International Conference on In-Vivo Methods*, Ed. A. Louilot, pp. 79–80. London, U.K.
3233. Boutelle, M.G., L. Svensson, and M. Fillenz. 1990. Effect of diazepam on behavior and associated changes in ascorbate concentration in rat brain areas: Striatum, n. accumbens and hippocampus. *Psychopharmacology* 100:230–236.
3234. Bhaskaran, D. and C.R. Freed. 1989. Catechol and indole metabolism in rostral ventrolateral medulla change synchronously with changing blood pressure. *J. Pharmacol. Exper. Ther.* 249:660–666.
3235. Brose, N., R.D. O'Neill, M.G. Boutelle, and M. Fillenz. 1988. Dopamine in the basal ganglia and benzodiazepine-induced sedation. *Neuropharmacology* 27:589–595.
3236. Brose, N., R.D. O'Neill, M.G. Boutelle, S.M. Anderson, and M. Fillenz. 1987. Effects of an anxiogenic benzodiazepine receptor ligand on motor activity and dopamine release in nucleus accumbens and striatum in the rat. *J. Neurosci.* 7:2917–2926.
3237. Bhaskaran, D. and C. Freed. 1987. Nucleus tractus solitarius: An evaluation by in vivo voltammetry. *Life Sci.* 41:323–331.
3238. Mulchahey, J.J. and J. Neill. 1986. Dopamine levels in the anterior pituitary gland monitored by in vivo electrochemistry. *Brain Res.* 386:332–340.
3239. O'Neill, R.D. 1986. Effects of intranigral injection of taurine and GABA on striatal dopamine release monitored voltammetrically in the unanaesthetized rat. *Brain Res.* 382:28–32.
3240. Michael, A.C., J.B. Justice Jr., and O'Neill. 1985. In vivo voltammetric determination of the kinetics of dopamine metabolism in the rat. *Neurosci. Lett.* 56:365–369.
3241. O'Neill, R.D. and M. Fillenz. 1985. Detection of homovanillic acid in vivo using microcomputer-controlled voltammetry: Simultaneous monitoring of rat motor activity and striatal dopamine release. *Neuroscience* 14:753–763.
3242. Trulson, M. 1985. Simultaneous recording of dorsal raphe unit activity and serotonin release in the striatum using voltammetry in awake, behaving cats. *Life Sci.* 37:2199–2204.
3243. Echizen, H. and C. Freed. 1984. Altered serotonin and norepinephrine metabolism in rat dorsal raphe nucleus after drug-induced hypertension. *Life Sci.* 34:1581–1589.
3244. O'Neill, R.D. and M. Fillenz. 1986. Microcomputer-controlled voltammetry in the analysis of transmitter release in rat brain. *Ann. N. Y. Acad. Sci.* 473:337–348.
3245. Miller, A.D. and C.D. Blaha. 2004. Nigrostriatal dopamine release modulated by mesopontine muscarinic receptors. *Neuroreport* 15:1805–1808.
3246. Di Ciano, P., C.D. Blaha, and A. Phillips. 1996. Changes in dopamine oxidation currents in the nucleus accumbens during unlimited-access self-administration of d-amphetamine by rats. *Behav. Pharmacol.* 7:714–729.
3247. Coury, A., C.D. Blaha, L.J. Atkinson, and A.G. Phillips. 1992. Cocaine-induced changes in extracellular levels of striatal dopamine measured concurrently by microdialysis with HPLC-EC and chronoamperometry. *Ann. N. Y. Acad. Sci.* 654:424–427.
3248. Lowry, J.P., M.G. Boutelle, and M. Fillenz. 1997. Measurement of brain tissue oxygen at a carbon paste electrode can serve as an index of increases in regional cerebral blood flow. *J. Neurosci. Methods* 71:177–182.
3249. Lowry, J.P., Demestre M, and Fillenz. 1998. Relation between cerebral blood flow and extracellular glucose in rat striatum during mild hypoxia and hyperoxia. *Dev. Neurosci.* 20:52–58.
3250. O'Neill, R.D., J.L. Gonzalez Mora, M.G. Boutelle, D.E. Ormonde, J.P. Lowry, A. Duff, B. Fumero, M. Fillenz, and M. Mas. 1991. Anomalously high concentrations of brain extracellular uric acid detected with chronically implanted probes: Implications for in vivo sampling techniques. *J. Neurochem.* 57:22–29.
3251. Mueller, K., R. Palmour R, C.D. Andrews, and P. Knott. 1985. In vivo voltammetric evidence of production of uric acid by rat caudate. *Brain Res.* 335:231–235.
3252. Yabuki, S. 2008. Development of high-performance biosensors based on functional matrices. *Chem. Sens.* 24:84–89.
3253. Wang, J. 1991. Modified electrodes for electrochemical sensors. *Electroanalysis* 3:255–259.

References

3254. Ates, M. and A.S. Sarac. 2009. Conducting polymer coated carbon surfaces and biosensor applications. *Progr. Org. Coatings* 66:337–358.
3255. Zhao, W., H. Dong, W. Zhang, and G. Cui. 2008. Application of carbon nanotubes in modified oxidase biosensors. *Zhongguo Zuzhi Gongcheng Yanjiu Yu Linchuang Kangfu* 12:3793–3795.
3256. Cremer, M. 1906. On the origin of the electromotive properties of the tissues contributing to the knowledge on the formation of the polyphase electrolyte chains (in German). *Z. Biol.* 47:562–608.
3257. Haber, F. and Z. Klemensiewicz. 1909. Über elektrische Phasengrenzkräfte/About electrical interfacial forces. *Z. Phys. Chem.* 67:385–431.
3258. Palecek, E. 1958. Oscillographic polarography of nucleic acids and their building blocks. *Naturwiss* 45:186–187.
3259. Palecek, E. 1960. Oscillographic polarography of highly polymerized deoxyribonucleic acid. *Nature* 188:656–657.
3260. Janata, J. 1975. Immunoelectrode. *J. Am. Chem. Soc.* 97:2914–2916.
3261. Guilbault, G.G. and J.G. Montalvo. 1969. Urea-specific enzyme electrode. *J. Am. Chem. Soc.* 91:2164–2165.
3262. Egholm, M., O. Buchardt, P.E. Nielsen, and R.H. Berg. 1992. Peptide nucleic acids (PNA). Oligonucleotide analogs with an achiral peptide backbone. *J. Am. Chem. Soc.* 114:1895–1897.
3263. Wang, J. 1988. DNA biosensors based on peptide nucleic acid (PNA) recognition layers. A review. *Biosens. Bioelectron.* 13:757–762.
3264. Rechnitz, G.A. 1981. Bioselective membrane electrode probes. *Science* 214:287–291.
3265. Chen, Z.-P., J.-H. Jiang, X.-B. Zhang, G.-L. Shen, and R.-Q. Yu. 2007. An electrochemical amplification immunoassay using bi-electrode signal transduction system. *Talanta* 71:2029–2033.
3266. Navaratne, A., M.-S. Lin, S.-L. Meng, and G.A. Rechnitz. 1990. Eggplant-based bioamperometric sensor for the detection of catechol. *Anal. Chim. Acta* 237:107–113.
3267. Yan, Q.-P., F.-Q. Zhao, and B.-Z. Zeng. 2006. Voltammetric behavior of epinephrine on carbon nanotubes-ionic liquid paste modified glassy carbon electrodes. *Fenxi Kexue Xuebao* 22:523–526.
3268. Zari, N., H. Mohammedi, A. Amine, and M.M. Ennaji. 2007. DNA hydrolysis and voltammetric determination of guanine and adenine using different electrodes. *Anal. Lett.* 40:1698–1713.
3269. Erdem, A., B. Meric, K. Kerman, T. Dalbasti, and M. Ozsoz. 1999. Detection of interaction between metal complex indicator and DNA by using electrochemical biosensor. *Electroanalysis* 11:1372–1376.
3270. Comba, F. N., M.D. Rubianes, P. Herrasti, and G.A. Rivas. 2010. Glucose biosensing at carbon paste electrodes containing iron nanoparticles. *Sens. Actuators B* 149:306–309.
3271. Liu, X.-H., Z. Ding, Y.-H. He, Z.-H. Xue, X.-P. Zhao, and X.-Q. Lu. 2010. Electrochemical behavior of hydroquinone at multi-walled carbon nanotubes and ionic liquid composite film modified electrode. *Colloids Surf. B* 79:27–32.
3272. Amani, V., H.R.L.Z. Zhad, A. Ebadi, O. Sadeghi, E. Najafi, and N. Tavassoli. 2011. Modification of carbon paste electrodes using pyridine functionalized SBA-15 and multi-walled carbon nanotubes (MWCNTs) for determination of lead ions in real aqueous solutions. *J. New Mater. Electrochem. Syst.* 14:39–42.
3273. Apetrei, C., I. Mirela Apetrei, J. Antonio de Saja, and M.L. Rodriguez Mendez. 2011. Carbon paste electrodes made from different carbonaceous materials: Application in the study of antioxidants. *Sensors* 11:1328–1344.
3274. Du, M., T. Yang, S.-Y. Ma, C.-Z. Zhao, and K. Jiao. 2011. Ionic liquid-functionalized graphene as modifier for electrochemical and electrocatalytic improvement: Comparison of different carbon electrodes. *Anal. Chim. Acta* 690:169–174.
3275. Li, X., Q.-J. Niu, G.-J. Li, T.-R. Zhan, and W. Sun. 2011. Electropolymerization of Brilliant cresyl blue on carbon ionic liquid electrode and electrocatalytic application for the voltammetric determination of ascorbic acid. *J. Brazil. Chem. Soc.* 22:422–427.

Authors

Ivan Švancara received his MSc and PhD from the Department of Analytical Chemistry at the University of Pardubice in 1988 and 1995, respectively, the latter under the supervision of Professor Karel Vytřas.

Since then, Dr. Švancara has been working at the same institute, first as an assistant professor (1995–2002), then as an associate professor (2002–2007), and finally as a full professor of analytical chemistry (2008). During his early research and pedagogical career, he went for a series of visits abroad, namely, at the D.C.U. Dublin (with the group of Dr. M.R. Smyth in 1991), at the KFU Graz (with Professor K. Kalcher in 1993), at the Research Centre Jülich (with Dr. H. Emons and Dr. P. Ostapczuk in 1995), at the National Institute of Chemistry (with Dr. B. Ogorevc in 1999), and at AGH Krakow (with Professor Bobrowski in 2000). His major research interests include the development and application of new types of electrodes and sensors as well as electroanalytical techniques and inorganic analyses, focusing specifically on carbon paste and nonmercury metallic electrodes. Dr. Švancara has published approximately 130 scientific papers (nearly 70 articles being registered by the Web of Science [WoS] database), including numerous reviews and 2 book chapters, and has presented his work in nearly 200 contributions at conferences and seminars locally and across Europe. These activities have resulted in more than 1200 citations (1000 references at the WoS) and a Hirsch index of $I_H = 24$. He is also a coauthor of three textbooks for students and has reviewed around 250 manuscripts submitted to impacted international journals. From the late 1990s, he has been a member of the Czech Chemical Society, working in the Committee for Analytical Chemistry. Finally, Dr. Švancara has been a member of the organizing committee for several international conferences.

Kurt Kalcher completed his studies at the Karl-Franzens University (KFU) with a dissertation in inorganic chemistry entitled "Contributions to the Chemistry of Cyantrichloride, $CINCCI_2$"; he also received his PhD in 1980 from the same institution. He then went for a postdoctoral stay at the Nuclear Research Center in Jülich (Germany) under the supervision of Professor Nürnberg and Dr. Valenta, and conducted intensive electroanalytical research while he was there.

Dr. Kalcher continued his academic career at KFU with habilitation on chemically modified carbon paste electrodes in analytical chemistry in 1988. Since then, he has been employed there as an associate professor. His research interests include the development of electrochemical sensors for the determination of inorganic and biological analytes on the basis of carbon paste and screen-printed electrodes, as well as design, automation, and data handling with small analytical devices using microprocessors. He has published over 160 full papers and has presented almost 200 contributions at international conferences (mainly oral presentations). These activities have resulted in more than 3000 citations and a Hirsch factor of I_H 30. Dr. Kalcher has received numerous guest professor position offers in Bosnia and Herzegovina, Poland, Slovenia, and Thailand.

Alain Walcarius received his MSc and PhD from the Facultés Universitaires Notre-Dame de la Paix, Namur (Belgium), in 1989 and 1994, respectively, under the supervision of Professor Luc Lamberts. He then went to New Mexico State University for a postdoctoral stay with the group of Joe-Wang in Las Cruces (NM). After a second postdoctoral stay at Nancy University in France (with the group of Jacques Bessière), he got a permanent position at the National Center of Scientific Research (CNRS, France) in 1996 as a research associate. He is currently a research director in the Laboratory of Physical Chemistry and Microbiology for the Environment (LCPME, Nancy, France), where his analytical and electroanalytical chemistry group works in the area of reactions at solid/liquid interfaces, at the confluence between analytical chemistry, electrochemistry, and chemistry of materials. He has also been the director of the LCPME since January 2011. His research interests include the intersection between the chemistry of silica-based organic–inorganic hybrid materials and electroanalysis. Dr. Walcarius has published 144 peer-reviewed manuscripts and 8 book chapters and has presented his work in more than 200 contributions at conferences and seminars (53 invited lectures). These activities have resulted in about 3500 citations (according to the WoS) and a Hirsch index of $I_H = 32$, reflected also in his regular reviews of renowned periodicals. Dr. Walcarius is a recipient of the Tajima Prize 2006 from the International Society of Electrochemistry.

Karel Vytřas graduated from the University of Chemical Technology in 1966 under the supervision of Professor Stanislav Kotrlý. His early research was associated with studies of metallochromic indicators and objective color measurements (which was also the major subject of his PhD in 1970).

Dr. Vytřas worked in the Department of Analytical Chemistry of the aforementioned institution as an assistant professor (1969–1982), an associate professor (1982–1991), and finally as a professor of analytical chemistry (1991). From the 1970s, he had focused his research on ion-selective electrodes, especially coated-wire types, and applied them in determining organic substances (for which he received his DSc in 1990). He then started to build his electroanalytical team oriented to modern techniques of electrochemical stripping analysis, biosensing, or development of new types of electrodes, among which the carbon paste–based ones have always played a dominant role. Dr. Vytřas has published approximately 350 scientific papers (nearly 160 being registered by WoS) and has presented his research results at numerous conferences. These activities have resulted in 1460 citations and a Hirsch index of $I_H = 27$. In 1976–1977, young Dr. Vytřas worked as a visiting scientist at the Institute of Analytical Chemistry, University of Oslo, for six months. He later received a number of guest professorships, namely, at universities in Bulgaria (1988), Spain (1991), Egypt (1993, 2000), and Portugal (1995). He has also served in high academic positions at his home university: as a vice-rector (1990–1994), vice-dean (1997–2002), and head of the Department of Analytical Chemistry (1994–2010). In 2004, Dr. Vytřas was awarded the Hanuš medal by the Czech Chemical Society.

Author Index

A

Abadi, H.A., 216
Abad-Villar, E.M., 191, 202, 381
Abbas, M.N.E.-D., 65, 88, 219, 275, 312
Abbas, S.S., 307
Abbasi, S., 227
Abbaspour, A., 32, 79, 82, 89–90, 96, 195, 204, 225, 280, 320, 333, 338, 341, 398–399
Abbastabar Ahangar, H., 239
Abd El Hady, D., 316
Abd Elghany, N., 312
Abdel Ghani, N.T., 313, 315, 318, 322
Abdine, H., 191, 206, 413
Abdollahi, S., 88
Abdouss, M., 102, 216
Abedul, M.T.F., 191, 202, 309, 379
Abelairas, A., 70
Abo Bakr, A., 307
Abou Assi, A.A., 323
Abou, M.M., 89, 214
Aboul Enein, H.Y., 28, 40, 67–68, 90, 191, 202, 305–306, 308, 311, 313, 318, 321, 323, 382
Aboul, R., 311
Abramovic, B., 188, 240, 297, 299–300
Abruna, H.D., 81, 93, 96, 99, 185, 193–194, 198, 250, 285, 326, 329, 334
Abu Ghalwa, N., 323
Abu Shawish, H.M., 78, 101, 227–228, 255, 267, 308, 310–311, 323
Abu Zuhri, A.Z., 67
Abu-Attia, F.M., 191, 203, 388
Accary, A., 160–162
Acciari, H.A., 161–162
Achak, M., 84, 86, 88, 209, 239, 257
Achamatowicz, S., 225
Ache, H.J., 90, 257
Achmatowicz, S., 36, 42
Acuna, J.A., 79, 82, 97, 277, 303, 314, 320
Adachi, N., 80, 199, 361
Adams, R.N., 1–4, 11, 16–21, 23, 27–28, 36, 38, 40–45, 47, 49, 52–53, 55–58, 61, 65–66, 96, 104, 142, 151–153, 155, 157–158, 161–163, 165, 169, 183, 191, 197, 207, 305, 346, 436, xv
Adam,V., 131, 168, 185, 191, 196, 201–202, 205, 285, 296, 305, 345, 375, 379, 404
Adekola, F.A., 161–162
Adeyoju, O., 59, 80, 99–100, 198, 354
Adibi, M., 34–35, 90, 263–264
Adraoui, I., 38–39, 89, 229, 234, 236
Adriamanana, A., 161
Afagh, P.S., 67
Afrasiabi, M., 84
Afshar, S., 101, 174, 235
Agasyan, P.K., 161, 230
Aghabozorg, H., 274

Aghajani, S., 96, 307
Aghajanloo, F., 79, 95–96, 206, 278, 389
Agostini, E., 75, 197, 350
Agrafiotou, P., 191, 195, 337
Agrahari, S.K., 90, 256
Agraz, R., 99, 218, 223, 230
Aguei, L., 80, 82, 103, 195, 201, 206, 341–342, 371, 412
Agui, L., 103, 191, 195–196, 342, 344
Aguilera, L.M.C., 254
Ahamad, R., 99, 202, 383
Ahlberg, E., 150, 161–162
Ahmadi, E., 102, 250
Ahmadi, R., 304
Ahmady, A., 166'
Ahmed, M.F., 166, 192, 328
Ai, S.-Y., 79, 96, 185, 284, 333
Airoldi, C., 89, 191, 200–201, 206–207, 369, 425
Airoldi, F.P.S., 99, 292
Aizawa, M., 25
Ajayan, P.M., 29, 31–32
Akahoshi, R., 79, 206, 412
Akalu, G., 191, 196, 342
Akarca, U.S., 25, 93, 191, 205, 404, 409
Akatsuka, K., 25, 99, 199, 215, 255, 360
Akbari, R., 191, 206–207, 415, 421
Akbarian, M., 103, 308, 320
Akcay, H., 85, 226
Akhgar, M.R., 96, 201
Akhond, M., 96
Akhoundian, M., 99, 447
Akhtar, M.S., 9, 161–162
Aki, C., 317
Aklilu, M., 92, 206, 306, 412
Akrami, Z., 84–85, 229, 251, 277
Akselrod, N.L., 162
Aktas, E., 93
Al Dalou, A.R., 323
Al Gahri, M.A., 296, 301
Al Gahri, M.M., 230
Al Gakhri, M.A., 90
Al Mulla, I., 163, 165, 209, 437
Alafandy, M., 244, 277
Alami, S., 84–85
Alamo, L.S.T., 30, 32, 82, 280
Alarcon, J., 161
Alarcon-Angeles, G., 90, 191, 203, 387–388, 395, 418
Albahadily, F.N., 4, 40, 59, 97, 163, 200, 366
Albertano, P., 184, 281
Albery, W.J., 163, 165, 191, 204, 209, 397, 436–437, 598–599
Alda, E., 78, 84–85, 185, 217, 288, 312
Alegret, S., 191, 198, 243, 277, 353
Aleixo, L.M., 64, 101, 217, 276
Alemu, H., 5, 9, 19, 27, 51, 79, 87, 90, 196, 199, 208–209, 251, 267, 342, 359, 427, 438
Alexa, N., 162

603

Alexander, P.W., 79, 191, 193, 209, 329, 433
Alexander, R., 210
Alexiadou, D.K., 191, 205, 321, 406
Alfaya, A.A.D.S., 87, 191, 206–207, 271, 429
Alfaya, R.V.D.S., 87, 191, 206–207, 271, 429
Alferov, V., 191, 200, 208, 364, 429
Algarra, M., 90, 206–207, 424
Ali, A.M.M., 307
Ali Qureshi, G., 13, 135, 261, 279
Ali, S.-R., 191, 204, 398
Ali, Z., 357, 398
Alibeygi, F., 87, 235
Alinezhad, A., 90, 92
Alizadeh, T., 99, 185, 188, 206, 289, 301, 412, 447
Alkire, M.F., 163, 165, 209
Allafchian, A.R., 96, 307
Almeida, C.M.V.B., 161
Almeida, F.A.S., 271
Almeida, N.F., 93, 195, 341
Alonso, B., 96, 198, 357
Alonso, R.M., 70, 306, 324, 435
Alonso Sedano, A.B., 38, 161–162
AlonsoVante, N., 38
Alpat, S.K., 85, 203, 208, 228, 269, 387, 428
Alvarez Bustamante, R., 166
Alvarez Crespo, S.L., 9, 75, 99–100, 116, 195, 341
Alvarez, E., 99, 301
Alvarez Gonzalez, M.I., 9, 59, 75, 116, 192, 201, 328, 367
Alves, J.D.P.H., 191, 205, 398, 402
Amako, K., 75
Amani,V., 447
Ambrosi, A., 309–310, 318, 321
Amine, A., 19, 29–31, 49, 63, 79–80, 82, 89, 95, 97, 100, 148, 184, 191, 193, 195–196, 198, 200, 202, 204, 233–234, 236, 249, 277–278, 281, 330–331, 340–341, 344, 354–356, 358–359, 366, 385, 398
Amini, M.K., 59, 79–80, 96, 191, 195, 203, 206–207, 279, 333, 339–340, 396, 420, 422
Amiri, M., 32, 96, 314
Amirkhanova, F.A., 289
Amor Garcia, I., 97
Andrade, A.N., 87, 200–201, 370
Andreoli, R., 152, 158, 286
Andrews, C.D., 191, 209, 437
Andrieux, C.P., 99, 198, 356
Andronic, V., 96
Angeles, G.A., 191, 203, 395
Angelikaki, A.G., 205, 309, 318, 322, 405
Angnes, L.L., 25, 79, 87, 96, 135, 185, 191, 196, 199, 203, 209, 233, 268, 283, 285, 333, 345, 359, 395, 397, 430–431, 433
Anik, U., 31, 66, 80, 185, 191, 196, 206, 209, 282, 345, 417, 429, 431
Anisimova, L.A., 296, 301
Anita, 93, 206–207, 424
Ansari, M., 304
Antiochia, R., 30, 32, 63, 191–192, 200–201, 309–310, 318, 321, 326, 364, 368
Antipova, E.R., 191, 203, 396
Antonia Mendiola, M., 90, 224
Antonio de Saja, J., 447
Antunes, N.d.J., 191, 203, 397, 433, 436
Anzai, J.-I., 260, 266

Apetrei, C., 79, 166, 191, 194, 197, 333, 335, 348, 350, 447
Apetrei, I.M., 191, 350
Arab, P., 25
Arabzadeh, A., 95, 191, 202–203, 384, 391
Aragaki, S., 97, 208, 426
Aragon, S., 291
Aranzazu Goicolea, M., 86, 323
Araujo, A.N., 25, 43–44, 86, 93, 101, 206, 334, 393, 416, 431
Araujo Rodrigues, I., 191, 325
Arazi, R., 191, 392
Arbani, S.A., 183, 277
Arcos, J., 97, 201, 244, 303, 307, 309, 316, 323, 373
Ardakani, M.M., 84–86, 191, 203, 206, 247, 251, 396, 416
Arduini, F., 95, 191, 196, 202, 284, 344, 385
Arellano, J.J.C., 85
Arenas, L.T., 87, 149, 206–207, 425
Arenaza, M.J., 315
Arguelho, M.L.P.M., 191, 205, 398, 402
Arguello, J., 191, 193, 333
Ariel, M., 36, 222, 279
Aritzoglou, E., 205, 309, 402
Arkoub, I.A., 191, 202, 376
Armalis, S., 185, 287
Aroca, R.F., 74
Arranz, A., 86, 99, 101, 150, 161, 296, 299, 311, 320
Arranz Valentin, J.F., 86, 99, 101, 150, 161, 296, 299, 311, 320
Arribas, A.S., 30–32, 81, 191, 197, 199, 203, 299, 350, 359, 388
Arrieta, A.A., 74, 79, 166, 197, 204, 333, 350
Arrigan, D.W.M., 90, 249
Arshad, M., 87, 101, 229
Artykov, A.T., 248, 254
Arvand, M., 85, 196, 206, 305, 343, 425
Aryasomayajula, L., 31–32, 185, 283
Asada, K., 75
Asadi, M., 96
Asadian, E., 95, 207, 303, 314, 424
Asbjoernsson, J., 150, 162
Ásbjörnsson, J., 161–162
Asci, B., 99, 172, 221, 227, 237
Asgari, M., 102, 240
Asgari, Z., 203, 206, 390, 421
Ashpur, V., 9, 26, 159
Ashrafi, A.M., 267
Asl, E.H., 191, 205, 280
Asplund, J., 277
Aston, W.J., 5, 79, 95, 193, 195, 273, 329, 337, 352
Astrup, P.B., 112
Ates, M., 191, 209
Atkinson, L.J., 163, 165, 191, 209, 437
Atta, N.F., 191, 203, 388
Attia, A.K., 65, 317, 321
Atuma, S.S., 3, 26, 165, 191, 208, 427
Audebert, P., 99–100, 198, 356
Avendano, S.C., 191, 203, 395
Avraamides, J., 162
Avramescu, A., 188, 299
Aw, T.C., 191, 199, 361, 434
Aydin, T., 191, 201, 328, 373
Aylward, J., 191, 204
Azab, S.M., 191, 203, 388

Author Index

Azambuja, D.S., 87, 149, 206–207, 425
Azevedo, A.M., 191–192, 327
Azimi, M.S., 85–86, 247

B

Baader, W.J., 191, 200–201, 368
Babaei, A., 81, 84–87, 196, 203, 206–207, 343, 387, 389, 396, 415, 420
Babazadeh, R., 99–100
Babkina, E., 191, 200, 208, 364, 429
Babula, P., 168, 185, 191, 205, 285, 404
Bacchi, P., 99, 198, 356
Bacha, S., 95, 354
Bachiller, P.E., 159–160
Bachirat, R., 84, 86, 301
Backofen, U., 135
Bacon, A.J., 151–152, 155
Bacon, J., 3
Badea, M., 104, 240, 242, 244–245, 278
Badiei, A., 101–102, 216, 221, 227, 240, 263
Badihi Mossberg, M., 169, 186
Bae, Z.-U., 208, 223, 231, 237, 251, 428
Baek, H.S., 191, 194, 335
Baek, S.M., 191, 206, 209, 412, 437
Baesso, M.L., 86
Baghayeri, M., 80, 103, 149, 196, 201, 333, 343, 372
Bagotsky, V.S., 123
Bai, L., 191, 194–195, 206–207, 335, 339, 421
Bai, Y., 88, 191, 196, 342, 426
Bai, Z.P., 78, 99, 249, 253
Bailey, M., 54, 58, 191, 205, 408
Bailey, S., 162
Bairu, S.G., 8, 14, 28, 216, 250, 254, 275
Bakasse, M., 84, 86, 88, 101, 209, 257, 297, 301
Baker, G.A., 163
Baker, G.B., 162–163
Bakhoday, B., 81, 84, 86, 196, 203, 343, 389
Bakry, R., 317
Bala, C., 311, 318, 321
Balagtas, C., 191, 209
Balaj, F., 191, 200–201, 371
Balaji, K., 90, 310, 313, 315, 318, 320
Balakrishnan, M., 104, 143, 204, 400
Balan, I., 191, 387
Balasoiu, A.M., 188, 299
Balbo, V.R., 90, 252
Balcu, I., 191, 202, 382
Baldrianova, L., 25, 43, 49, 54–55, 59, 63, 103–104, 135, 143, 146–147, 149, 175, 191, 195, 226, 233–237, 239–240, 242–245, 337
Baldwin, R.P., 4, 36, 64, 76, 79, 89–90, 92, 95–96, 99–100, 185, 188, 191, 193, 195, 197, 201, 203, 206–207, 222, 230, 251, 285, 296, 301, 304, 311, 333, 335, 339, 341, 351, 372, 391, 420, 423
Ballesteros, Y., 191, 202, 383
Ballin, M.A., 99, 206–207, 420
Baloda, M., 88, 103, 202, 378
Baloun, J., 185, 285
Baltork, I.M., 279
Banica, A., 99, 238
Banica, F.G., 19, 99, 238
Banks, C.E., 19, 29, 80, 102, 191, 197–198, 348, 358, 427
Bannister, J.V., 99, 249

Bao, J., 191, 202, 383
Baranova, N.N., 210
Barbado, M.D.V., 79, 96, 159–160, 268, 333
Barbaise, T., 64, 84, 203, 387
Barbera Salvador, J.M., 84–85, 185, 289
Barbier, F., 84, 168
Barbosade Miranda, P.R., 191, 206–207, 415, 424
Barcelo, D., 79, 96, 333
Bard, A.J., 123
Bardheim, M., 191, 200, 363
Barea Zamora, M., 84–85, 185, 289
Barek, J., 8–9, 14, 17, 40–41, 51–52, 64–65, 67, 74, 90, 99, 105, 142, 169–170, 185–188, 194, 197, 202, 205, 285, 287–290, 311, 316, 323, 335, 350, 383, 401, 447
Barikov, V.G., 26, 158–159, 161
Barlett, P., 80, 99–100, 193, 200, 332, 364, 369
Baronas, R., 191, 200, 362, 364
Baronian, K., 80, 99–100, 195, 340
Barrado, E., 161–162, 174, 238
Barranco, J., 191–192, 328
Barranco, M., 96, 198, 357
Barreira Rodriguez, J.R., 99, 308–309, 312, 315, 321
Barrera, G.D., 80, 88, 103, 199, 361, 393
Barrio, R.J., 86, 323
Barsoum, B.N., 90, 97, 140, 243, 315
Bartoli, J., 191, 198, 353
Bartos, M., 19, 25, 42–43, 51, 59, 62, 87, 96, 135, 142, 183, 199, 206–207, 239, 249–250, 359, 364, 422, 585
Bartosik, M., 191, 202, 380
Bashkin, J.K., 80, 333
Bassi, A.S., 93, 99–100, 191, 197, 331, 348
Batanero, P.S., 79, 96, 268, 314, 333
Bates, R.G., 135–136, 140
Batista, A.A., 191, 206–207, 422
Batista, I.V., 191, 202, 381
Baudino, O.M., 266
Bauer, D., 3, 16, 26, 158, 160–162
Bazarova, E.V., 161
Bear, J.L., 46, 59, 97
Beatriz, A., 191, 205, 398, 402
Beck, T.W., 191, 204, 397
Becker, J.Y., 40, 185, 196–197, 215, 282, 345, 349
Begic, S., 191, 199, 359
Begleiter, A., 191, 205, 407
Begum, M., 191, 208, 427
Beh, S.K., 79, 198, 354
Behling, J.R., 73
Behpour, M., 88, 103, 305
Beilby, A.L., 3, 51
Beitollahi, H., 32, 80, 93, 96, 103, 191, 194–196, 203–204, 206–207, 304, 319, 335, 339–340, 343, 391–392, 396–397, 415–416, 421, 423
Bekhiet, G.E., 191, 202, 313, 381
Beklova, M., 168, 191, 196–197, 205, 296, 345–346, 404
Belay, A., 191, 196, 342
Bell, J.M., 5, 273
Beltagi, A.M., 84, 314, 321
Ben Taarit, Y., 85
Benayada, A., 7, 54, 80, 281, 333
Benedetti, A.V., 161–162
Bengoechea Alvarez, M.J., 191, 202, 379
Bennett, R., 163, 165, 209, 436, 447
Bennouna, A., 162

Bento, A.C., 86
Benvenutti, E.V., 87, 149, 191, 200–201, 206–207, 370, 425
Benvidi, A., 80, 93, 103, 203, 396
Benzine, M., 278
Beranova, G., 186, 290, 292
Berber, H., 99, 172, 221, 227, 237
Berckman, S., 162
Bere, A., 191, 374
Berg, R.H., 205
Bergamini, M.F., 65, 84, 86–87, 101, 202, 232, 260, 305, 310, 384
Bergantini, J.H., 243, 277
Bergmann, W., 193, 329
Bergougnou, M.A., 93, 99
Beristain Ortiz, G., 88
Bermejo, E., 25, 30–32, 58–59, 79, 84, 88, 96, 99, 101, 104, 191, 197, 203, 230, 295, 299, 311, 319, 333, 350, 385, 388
Bernardi, M.E., 201, 374
Berrueta, L.A., 315
Berzins, T., 143
Besse, J.P., 84, 149
Bessiere, J., 161
Bessière, J., 54, 64, 67, 80, 84–87, 101, 161–162, 199, 203, 224–225, 275, 281, 360, 387
Bessière, V., 86
Bessone, J.B., 191, 200–201, 367
Bettelheim, A., 75, 191, 198, 358
Bevilaqua, D., 161–162
Beyene, N.W., 79, 191, 196, 209, 342, 359, 438
Beyer, L., 49, 63, 89, 96, 103, 149, 228
Bezdicka, P., 161–162
Bezerra, V.S., 25, 93, 393, 431
Bezzaatpour, A., 32, 194, 303, 335
Bhanot, D., 292
Bhaskaran, D., 163, 165, 191, 209, 436–437
Bi, L.H., 81, 193, 333
Bi, R.-F., 34–35
Bi, Y.-F., 101, 167, 437
Biagiotti, V., 80
Biazzotto, J.C., 86, 96
Bidan, G., 100
Bilitewski, U., 191–192, 326
Billova, S., 97, 191, 205, 408
Bing, C., 90
Biniak, S., 59, 63
Birch, B.J., 96
Biris, A.R., 357, 398
Biryol, I., 99–100, 304–305, 307, 313
Bjelica, L., 188, 240, 297, 299–300
Blaha, C.D., 81, 163, 165, 188, 191, 205, 209, 437
Blakely, R.D., 191, 209, 436
Blanco Lopez, C.M., 292
Blanco Lopez, M.C., 97
Blanco, M.H., 36, 67, 77–78, 84, 99, 174, 185, 222, 230, 288, 304, 312
Blanco, P.T., 309, 313
Blank, L., 3, 158
Blankert, B., 93, 191, 201, 303, 309, 373
Blasius, E., 180, 261
Blin, J.L., 101–102
Bloomfield, M.R., 163, 165, 209, 437
Boaventura, J., 191, 325
Bobacka, J., 75

Bobbitt, J.M., 24, 36, 54, 67, 135, 140, 158, 261, 274, 279, 289
Bobes, C.F., 191, 202, 379
Bobrowski, A., 7, 24, 49, 55, 59, 63, 67, 85, 87, 97, 103–104, 146–149, 174–175, 178, 198, 226, 232–233, 245, 247, 251, 258
Bobrowski, R., 7, 24, 55, 63, 87, 103, 174, 232, 241
Bocchi, N., 67, 84, 86–87, 202, 260–261, 305, 384
Bockris, J.O'M., 123
Bodke, Y.D., 97, 185, 284
Bof de Oliveira, A., 95, 191, 200–201, 206–207, 415, 424
Boggetti, H., 168
Boghaei, D.M., 32, 194, 303, 335
Boguslavsky, L.I., 80, 93, 96, 99–100, 191, 193, 198, 200, 209, 330, 355, 363–364, 438
Bokretsion, R.G., 28, 40, 68, 305, 313
Bolger, F.B., 163, 165, 191, 209, 436, 447
Bolger, G.W., 14, 16, 23, 210
Bollo, S., 185, 205, 286, 405
Bonakdar, M., 40–41, 56, 70–71, 90, 96, 183, 189, 197, 201, 234, 270, 317, 349
Bonciocat, N., 191, 206, 419
Bond, A.M., 86
Bonifacio,V.G., 89, 314
Bontempelli, G., 75
Bopiah, A.K., 236, 242
Bordes, A.L., 96, 99, 319–320
Borgo, C.A., 86
Borisova, N.F., 161
Borole, D.D., 100
Borrero, S., 88, 97, 192, 200–201, 325, 368
Bosch Reig, F., 87, 162
Bott, A.W., 75
Bouchta, D., 220
Bouillon, J.-P., 102
Boujtita, M., 93, 95, 191–193, 198, 200–201, 206–207, 326, 330–331, 352–353, 357, 367, 417, 425
Boulton, A.A., 162–163
Boutakhrit, K., 244
Boutelle, M.G., 99–100, 163, 165, 191, 209, 436–437
Bowater, R.P., 191, 204, 398
Boyd, D., 99, 312, 315, 321
Boyer, A., 79, 82, 281
Brainina, K.Z., 9–10, 26, 78, 158–162
Brandsteter, R., 95, 200, 208, 365, 429
Braun, J., 320
Brazdilova, P., 191, 198, 354
Brazdova, M., 191, 205, 408
Brazell, M.P., 98, 163, 165
Bremle, G., 93, 191, 198, 200, 353, 363
Brent, P.J., 191, 204
Brett, A.M.O., 135, 191, 196, 204, 345, 398
Brett, C.M.A., 135
Brett, M.A., 143–144
Brezina, M., 3, 36, 53, 55, 266
Brich, W., 73
Brigagao, M.R.P.L., 191, 203, 397, 433, 436
Brinker, C., 86
Briones, R., 150
Britto, P.J., 29, 31–32
Broderick, P.A., 81, 163
Brolley, J.E., 163
Brondani, D., 80, 203, 396
Brooks, W.N., 191, 204, 397

Author Index

Brose, N., 163, 165, 191, 209, 436–437
Brossier, P., 96, 99, 319–320
Brown, T., 191, 206–207, 417, 424
Brozkova, I., 191, 205, 406
Bruckenstein, S., 3
Bruns, R.E., 87, 93
Bu, H.Z., 79–80, 96, 100, 198, 200–201, 356, 368
Buchanan, R.M., 95, 100, 203, 206–207, 391, 423
Buchardt, O., 205
Buchberger, W., 317
Buchner, V., 169, 186
Buch-Rasmussen, T., 93, 200, 362–363
Buck, R.P., 191, 198, 354
Buckley, A.N., 150
Buckova, M., 43, 90, 99, 185, 197, 205, 286–287, 310, 319, 346, 401–405
Budnikov, G.K., 4, 9, 67, 69–70, 80, 82, 88, 90, 96, 103, 149, 187–188, 191, 193, 195, 201, 230, 258, 276, 292, 296–297, 299, 301, 332–333, 337–338, 372
Bukharaev, A.A., 80, 88, 103, 149, 193, 332
Buldakova, L.Y., 262
Buleandra, M., 102, 240
Burestedt, E., 24, 84, 185, 191, 195, 197, 201, 209, 282, 337, 341, 348
Burgess, E.P.J., 191, 201, 375
Busche, B.J., 101–102, 233
Bustin, D., 163, 238, 257, 262
Buyukisik, H.B., 208, 228, 428
Buzuk, M., 9

C

Cabal Diaz, V., 313
Cabral, J.M.S., 191–192, 327
Cachet-Vivier, C., 161
Caetano, L., 101, 220
Cahova, K., 191, 205, 411
Cai, L., 35, 82, 195, 338
Cai, P., 191, 203, 395, 433
Cai, Q.T., 89–90, 218, 243, 262
Cai, X.-H., 6–7, 9–10, 26, 40, 54, 67, 74, 76–78, 84, 88–90, 93, 99, 104, 106, 108, 143, 146, 185–186, 188, 191, 193, 202, 204–206, 211, 214, 217, 223, 242, 245, 267, 271, 276, 279, 285, 290, 309, 319, 329, 333, 381, 398, 400–402, 404, 406–408, 414
Calvo, E.J., 96, 100, 198, 356
Calvo Marzal, P., 93, 201, 372
Calvo, M.P., 79, 96, 196, 333, 344
Campana-Filho, S.P., 31, 229
Campanella, L., 96, 309–310, 318, 321
Canevari, T.D.C., 191, 206–207
Cankurtaran, H., 90, 99, 172, 221, 226–227, 237
Canpolat, E.C., 90, 226
Cao, G.-P., 29, 32, 34
Cao, H.-B., 191, 202, 377
Cao, J.-L., 239
Cao, L.-Y., 24, 81, 88, 237, 276
Cao, M.-H., 33, 35, 66
Cao, R., 101
Cao, S.-Q., 191, 197, 202–203, 206–207, 346, 386, 424
Cao, X.-G., 167
Capoun, T., 135, 139
Carbonnelle, P., 161
Cardoso, W.S., 87, 191, 196, 267, 278, 325, 344

Carjonyte, R., 87
Carmo, D.R., 162
Carmo, M., 191, 206–207, 423
Carteret, C., 84, 101, 219, 293
Caruso, C.S., 191, 203, 395, 433
Casades, I., 85
Casado, C.M., 96, 191, 198, 355, 357
Casanas, B., 88
Casero, E., 96
Casillas, C.A., 88, 97, 192, 200–201, 325, 368
Castanheira, M.S.A., 191, 206–207, 423
Castellani, A.M., 87, 96
Castilho, T.J., 191, 194, 203–204, 335, 393, 397
Castillo, J.R., 99–100, 168, 198, 248, 271, 355
Castillo, R.J., 162
Castrillejo, Y., 161–162
Castro Martins, S., 85
Cataldi, T.R.I., 6, 26, 87, 162
Cataldo, F., 191, 196, 202, 344, 385
Catalin Popescu, I., 88, 92, 95, 200–201, 368
Cavalheiro, E.T.G., 82, 87, 96–98, 101–102, 172, 185, 191, 195–196, 201, 206–208, 221, 227, 232, 236, 260, 289, 300, 305, 310, 338, 342, 371, 419, 422, 426
Celej, M.S., 88, 199
Cembrano, J., 40, 185, 196, 215, 282, 345
Centeno, B., 161
Centonze, D., 6, 26, 87, 162
Cepra, G., 25, 199, 201, 336, 360, 374
Cepria, G., 161–162, 168, 248, 271
Cepria, J.J., 161–162
Cerda, V., 143, 145
Cermakova, I., 67, 97, 183, 275
Cerna, P., 316
Cernanska, M., 238
Cerrada, J., 101, 306
Cervilla, A., 84
Cesar, R., 32, 247
Cesarino, I., 101–102, 172, 221, 227, 236
Cespedes, F., 191, 198, 353
Cevik, S., 191, 196, 206, 209, 345, 417, 431
Cha, G.S., 26, 166, 191, 203, 391
Chadim, P., 49, 103, 149, 175, 247
Chai, Y.-Q., 26, 68, 80, 102, 202, 254, 269, 375–376, 438
Chailapakul, O., 96, 272
Chaisuksant, R., 89, 228
Chaloosi, M., 102, 216, 221, 239
Chambers, C.A.H., 3, 17–18, 37, 61, 69–70, 99
Chambers, J.A., 46–47
Chambers, J.Q., 3, 66, 191, 204, 398, 401
Chan, C.-W., 33, 35, 66
Chan, W.-H., 79, 281
Chandra, U., 80, 95, 97, 99–100, 103, 166, 185, 191, 202–203, 206–207, 284, 293, 386–393, 414, 416, 418, 423–425, 447
Chandrashekar, B.N., 93, 97, 191, 202–203, 293, 386, 388, 392
Chandravanshi, B.S., 90, 208, 231, 251, 427
Chaney, E.N., 311
Chang, C.-H., 224
Chang, H.-Y., 208, 223, 231, 237, 428
Chang, W., 95
Chang, X., 191, 204, 398
Changizi, A., 96, 201

Chapleau, M., 191, 193, 326, 331
Char, M.P., 97, 100, 203, 206–207, 392, 424
Charles, L., 99–100, 199, 362
Charpentier, L., 95, 191, 198, 202, 354, 382
Chassagneux, F., 160–162
Chastel, O., 97, 315
Chateau Gosselin, M., 304, 314
Chatrathi, M.P., 163, 165, 199, 360
Chaubey, A., 100
Chazaro, L.F., 84
Che, Y.H., 191, 209, 435
Cheek, G.T., 3, 17, 73–74, 213
Chekin, F., 96, 319
Chelnokova, I.A., 80, 88, 103, 149, 193, 332
Chen, B.-K., 84, 101, 167, 205, 437
Chen, C.-Y., 191, 206, 413
Chen, G.-N., 31–35, 41, 43, 150, 182, 191, 200–201, 206–207, 286, 295, 297, 312, 316, 363, 367, 399, 412, 425
Chen, H., 35, 43, 182, 191, 202–203, 378, 396, 399
Chen, H.-W., 82, 84, 166, 204, 399
Chen, H.-Y., 59, 80–81, 87, 93, 96, 103, 191, 203, 205–207, 304, 316, 323, 333, 369, 390, 405, 413, 422, 425
Chen, J., 31–32, 41, 43, 66, 80, 101, 191, 200, 205, 295, 297, 306, 366, 404
Chen, J.-C., 149
Chen, J.-H., 31–32, 66, 150, 306, 363, 412
Chen, J.-R., 32, 81, 411
Chen, J.-W., 291
Chen, J.-Y., 67, 239
Chen, L., 44, 79, 82, 87–88, 95, 99–100, 121, 135, 163, 165, 191, 197–200, 204, 206–207, 209, 290, 300–301, 352, 359–360, 363, 398, 400, 424
Chen, L.-R., 166
Chen, L.-Y., 27
Chen, P.-Y., 168, 191, 205, 405, 408
Chen, Q., 59, 80, 87, 90, 102, 191, 193–194, 197, 209, 305, 331, 336, 348, 352
Chen, S.L., 99–100, 203, 392
Chen, T.-W., 191, 206–207, 424
Chen, W.-H., 81, 191, 197, 202, 208, 289, 298, 300–301, 347, 382, 429
Chen, X., 35, 43, 64, 81, 182, 191, 203, 246, 396, 399
Chen, Y., 82, 251, 268
Chen, Y.-N., 80, 202
Chen, Y.S., 88, 196, 342
Chen, Y.-T., 31–32, 41, 43, 150, 286, 295, 297, 412
Chen, Y.-Y., 268
Chen, Y.-Z., 208, 212, 428
Chen, Z.-D., 185, 191, 196–197, 282, 343, 347
Chen, Z.-G., 97, 437
Chen, Z.-J., 167
Chen, Z.K., 100, 196, 343
Chen, Z.-P., 88, 202, 378–379
Chen, Z.-S., 87, 287
Cheng, H.-Y., 44, 158, 305
Cheng, I.F., 12, 45
Cheng, L., 191, 205, 409
Cheng, T., 191, 426
Cheng, X.-H., 87, 100, 188, 197, 206–207, 293, 296, 298–300, 351, 419
Cheng, Y.-D., 174, 231
Cherkaoui, F.E., 278
Chernysheva, A.V., 78
Chernyshov, D.V., 191, 203, 396
Cheung, W., 191, 204, 398
Chi, H., 77–78
Chi, L.-Z., 32, 87, 194, 336
Chi, Q.J., 93, 104, 192, 202, 327, 380
Chi, Y., 33–35, 191, 203, 312, 389
Chia, L.-S., 7, 84, 86, 198, 356
Chicharro, M., 30–32, 58–59, 79, 84, 87–88, 96, 99, 101, 104, 143, 146, 186, 191, 197, 199, 202–205, 290, 295, 299, 309, 311, 333, 350, 381, 385, 388, 398, 400–401, 404, 407
Chida, I.U., 82
Chien, H.-C., 191, 198, 353
Chierice, G.O., 97, 206–207, 422
Chigrinova, E., 191, 200, 208, 364, 429
Chijiwa, T., 191, 202, 383
Chisuwa, Y., 191, 358
Chitravathi, S., 95, 97
Chiu, P.-L., 191, 204, 398
Cho, M.H., 89, 99, 215
Cho, S.-H., 191, 197, 347
Cho, Y.-D., 289
Choi, G.-H., 7, 208, 219, 231, 428
Choi, S.-N., 251
Choi, Y.-K., 18, 185, 282
Cholota, M., 67, 85, 97, 178, 198, 258
Chou, J.-C., 191, 206, 413
Chou, T.-C., 191, 198, 353
Chouaib, F., 45, 161–162
Chough, S.H., 185, 196, 215, 282, 345
Choukairi, M., 220
Chowdappa, N., 97, 204, 397
Christensen, J.K., 64, 90, 143, 145
Christian, G.D., 4, 24, 55, 97, 245, 307, 314–315, 322
Chtaini, A., 84, 86, 88, 101, 209, 239, 257, 297, 301
Chuang, C.L., 99–100, 198, 356
Chulkina, L.S., 66, 210, 213, 259
Chumbimuni Torres, K.Y., 93, 201, 372
Chung, C., 289
Chung, C.-R., 97, 198, 359
Chung, C.-W., 168
Chung, T.D., 90
Chybova, O., 186, 290, 292
Cid, A., 86, 99, 150, 161, 296, 299, 311, 320
Cienciala, M., 197, 323, 350
Ciobanu, A., 191, 387
Ciszewski, A., 95, 191, 209, 430
Citak, M., 319
Citterio, D., 18
Ciucu, A.A., 96, 188, 191, 296, 301–302, 387
Ciurea, I., 191, 387
Cizek, K., 288
Cjhen, Z.-G., 191, 207, 426
Cladera, A., 143, 145
Clark, L.C., 112, 162, 191
Clemens, J.A., 191, 209, 436
Cline, K.K., 10
Cobos Murcia, J.A., 166
Colaruotolo, J.F., 24, 36, 54, 67, 135, 140, 158, 261, 274, 279, 289
Colby, J., 5, 273
Colilla, M., 89, 219
Colin, C., 161–162
Colon, J.L., 88, 97, 192, 200–201, 325, 368

Colpini, L.M.S., 86, 96
Comba, F.N., 447
Comisso, N., 75
Compagnone, D., 26, 79, 82, 93, 99, 198, 200, 249, 267, 363
Compton, R.G., 8, 19, 28–29, 31–35, 38, 43, 95, 191, 194, 198, 257, 262, 306, 315, 334, 358, 427
Conceicao, C.D.C., 79, 96, 273
Congestri, F., 163, 165, 209
Conner, M.M., 75
Connor, M.P., 25, 99, 197, 203, 209, 311, 349, 394, 429–430, 432–434
Consolin Filho, N., 100, 299
Conti, J.C., 44, 158, 163, 165
Conway, M.W., 163, 165, 209, 436, 447
Cookeas, E.G., 19, 59, 79, 96, 175, 185, 194–195, 248, 277, 285, 334, 337
Copra, A., 79, 196, 209, 342, 359, 438
Cordos, E., 162
Corona Avendano, S., 97, 191, 203, 387–388, 395, 418
Corona, G., 191, 203, 395
Correade Melo, H., 87, 100, 202, 260, 307, 314, 384
Correig, X., 281
Cortes, J.S., 90, 206, 415
Cortinez, V.A., 99, 191, 202, 252, 266, 381
Corujo-Antuna, J.L., 191, 202, 381
Cosio, M.S., 87, 195, 197, 203, 338, 351, 389
Coskun, N.Y., 90, 226
Cosnier, S., 75, 84, 100, 150, 197, 348
Cosofret, V.V., 189
Costa, C.M.M., 86, 96
Costa Garcia, A., 7, 49, 79, 87, 95, 191, 198, 200–202, 279, 292, 306, 308–309, 315, 356, 363, 373, 375, 378–381
Costa Moreira, J., 169–170, 186
Costa, T.M.H., 87, 149, 206–207, 425
Costa, W.M., 191, 325
Cotton, F., 241, 275
Cotton, T.M., 93, 202, 375
Coury, A., 163, 165, 191, 209, 437
Coutinho, C.F.B., 271
Couto, A.B., 97, 185, 289, 300
Couto, C.M.C.M., 86, 101, 334
Covington, J.R., 162
Craw, D., 161–162
Creasy, K.E., 78, 84, 86, 93, 266
Creczynski-Pasa, T.B., 81, 203, 395, 432
Cremer, M., 191
Crespi, F., 163, 165, 209
Crespilho, F.N., 99, 292
Crow, D.R., 6
Cruz, R., 150
Cruz Vieira, I., 80, 203, 396
Csoeregi, E., 99–100, 191–192, 195, 197–198, 200, 327, 337, 341, 348, 353, 363
Csoka, B., 75, 229
Csoregi, E., 80, 191–192, 196, 200, 327, 342, 363
Cuadrado, I., 96, 191, 198, 355, 357
Cuan, A., 191, 203, 395, 418
Cubillana Aguilera, L., 298
Cubukcu, M., 66, 80, 206, 417
Cucchi, M.L., 191, 203, 395
Cui, B., 191, 198, 311, 314, 353, 398–400
Cui, D., 93, 100, 206, 412

Cui, G., 26, 191, 209
Cui, H., 104, 131, 191, 196–197, 342, 346
Cui, L., 191, 197, 348
Cui, R., 191, 202, 376
Cui, S., 191, 194, 201, 334, 375
Cui, X.-J., 102, 199–200, 206, 268, 359, 367, 385, 413
Cullum, D.C., 186
Cummings, E.A., 25, 100, 196–197, 209, 345, 348, 433
Cummins, L.B., 600
Cuong, N.M., 191, 197, 347
Curran, D.J., 4, 16, 53, 81, 97, 203, 206–207, 386, 418
Curulli, A., 89, 100, 233–234, 278
Cvacka, J., 194, 335

D

da Silva, M.P., 99
da Silva, R.M., 97, 162, 191, 206–207, 421
Da Silva, W.T.L., 99, 292
DaCruz Vieira, I., 191, 203, 395, 433
Dagdeviren, K., 191, 205, 404
Dahmen, E.M.F., 26, 36, 135, 140, 213, 274
Dai, G., 191, 206–207, 420
Dai, H., 32–35, 191, 200–201, 206–207, 312, 316, 367, 425
Dai, J., 185, 188, 287
Dai, X., 232
Dai, Z., 191, 202, 383
Dalbasti, T., 205, 406
Dalloul, H.M., 255
Damiri, S., 66, 81
Damos, F.S., 191, 195, 340
Daneshgar, P., 103, 202, 303, 310, 314, 317, 321, 384
Danet, A.F., 89, 191, 198, 226, 356
Dang, X.-P., 97, 149, 191, 202–203, 267, 382, 391
Daniel, G.A., 88
Danielson, N.D., 4, 17, 26, 52, 97, 202, 260, 381
Danielsson, B., 185, 205, 286, 405
Dantoni, P., 191, 196, 345
Dao, P.-T., 215
Daraio de Peuriot, M., 191, 201, 374
Darlewski, W., 59, 63, 90
Darnall, D., 25, 191, 208, 211, 222, 428
Dasgupta, P.K., 99, 199, 360
Daugvilaite, N., 93, 193, 329
Davarkhah, R., 162
David, A., 88
David, I.G., 191, 387
David, V., 191, 387
Davies, G., 5, 273
Davies, Q., 99–100, 195, 340
Davies, R.J.H., 191, 204, 398
Davies, T.J., 19, 427
Davis, A.T., 215
Davis, D.G., 2, 56, 143–144, 146, 274
Davis, G., 79, 95, 193, 195, 329, 337, 352
De Betono, S.F., 86, 99, 150, 299, 311, 320
de Cisneros, J.L.H.H., 84–85, 90, 185, 220, 254, 288–289, 292, 298
de Fatima, B.S., 101, 217
de Favere, V.T., 99
de Francisco, C., 161–162
De Giovani, W.F., 191–192, 325
de Jesus, D.A., 271
de Jesus, D.S., 86, 101, 334

De la Fuente, C., 79, 82, 97, 277, 303, 320
De la Fuente, E., 185, 205, 286, 405
de Lima Filho, J.L., 25, 93, 393, 431
De Lucca, A.R., 87, 93, 200–201, 369
de Medeiros, E.S., 100, 299
de Miguel, J., 99, 223
De Oliveira, A.J.B., 87, 93
de Oliveira, G., 86, 93, 194, 333
de Oliveira, I.R., 81
de Oliveira Neto, G., 80, 87, 96, 185, 198, 282, 333, 355
de Prado, F.L.E., 79, 88, 299
De Saja, J.A., 74, 79, 96, 166, 191, 194, 197, 204, 333, 335, 348, 350
de Vicente Perez, S., 79, 82, 277
Deakin, M.R., 191, 206, 418
Deban, L., 89, 235
Debnath, C., 79, 96, 305, 333
Deen, R., 90
Deepa, R., 97, 191, 203, 390, 392
Degrand, C., 96, 99, 319–320
Dejmkova, H., 188, 285, 287–289
Del Alamo, M., 191, 350
Del Cano, T., 74
Del Mar Cordero Rando, M., 85, 289
Delacote, C., 64, 101–102, 149, 219
Delahay, P., 143
Delbem, M.F., 191, 200–201, 368
Delerue-Matos, C., 191, 206–207, 423
Delplancke, J.L., 49, 59, 97, 148, 203, 387
Demestre, M., 191, 209, 437
Demirkan, H., 84, 297
Dempsey, E., 25, 41, 99, 195, 201, 209, 337, 373, 429–430
Denat, F., 172, 229, 238
Dendrinou-Samara, C., 191, 205, 398, 402
Deng, D.-M., 288, 447
Deng, H., 93
Deng, J., 93, 191, 206–207, 424
Deng, N., 166
Deng, P.H., 30–32, 64, 90, 228, 238, 253, 263–264
Deng, Y., 27, 88, 103, 204, 398, 406
Deng, Z., 191, 203, 391
Deni, J., 97, 148, 193, 249, 330
Deo, R.P., 32, 191, 198, 357
Deocaris, C.C., 25, 235
Deratani, A., 191, 197, 347
Derki, S., 277
Deronzier, A., 75
Derouane, E.G., 71, 84, 260–261
Desai, P.B., 90, 208, 426–427
Deshmukh, B.K., 19, 56, 67, 69–71, 189, 308, 317, 319
Desideri, P.G., 42, 152, 158
Despas, C., 86, 101, 238
Devoy, J., 86–87
Dhesi, R., 93, 202, 375
Di Ciano, P., 163, 165, 191, 209, 437
Di Giorgio, F., 191, 196, 202, 344, 385
Di Tuoro, D., 95, 284
Diano, N., 95, 284
Dias, I.L.T., 191, 202, 381
Dias, S.L.P., 87, 93, 149, 191, 200–201, 206–207, 266, 370, 425
Dias, V.L.N., 191, 325
DiasFilho, N.L., 84, 96, 101, 219–220
Diaw, M., 162

Diaz, A., 88, 97, 192, 200–201, 325, 368
Diaz Goretti, D., 292
Diaz, L., 96, 100, 198, 356
Diaz-Cruz, J.M., 79, 96, 333
Dick, R., 90, 257
Dicks, J.M., 79, 95, 193, 195, 329, 337, 352
Dicu, D., 88, 92, 95, 200–201, 368, 370
Diculescu, V.C., 191, 204, 398
Diego, E., 191, 196, 344
Diewald, W., 6, 26, 53–55, 58, 67, 76–78, 88, 90, 99, 211, 214, 217, 242, 267, 279
Digua, K., 49, 59, 97, 148, 203, 387
Dilgin, Y., 84, 88, 93, 191, 206–207, 293, 317, 319, 425
Dinarvand, R., 103
Dincer, A., 191–192, 201, 328, 372
Ding, C.-F., 35, 80, 88, 103
Ding, H.-C., 191, 203, 209, 223, 256, 394, 432, 437
Ding, T.-H., 223
Ding, Y., 87, 198, 320, 361
Ding, Y.-L., 185, 284
Ding, Y.-P., 288, 447
Ding, Z., 447
Dion, M., 95, 193, 331
Diopan, V., 191, 205, 404
Divisia-Blohorn, B., 99, 198, 356
Divsar, F., 102, 216, 221
Dixon, B.M., 163, 165, 209
Dlabka, V., 135, 139–140
do Carmo, D.R., 84, 96–97, 101, 191, 206–207, 219–220, 421
do Rosario Matos, J., 102, 236
Doblhoff-Dier, O., 191, 206, 417
Dockal, E.R., 96, 98, 195, 201, 305, 338, 371, 426
Dockal, O.E.R., 310
Dockendorff, B.P., 75
Doenmez, G., 191, 208, 428
Dolezal, P., 191, 202, 379
Domenech, A., 84–85, 161
Domenech Carbo, A., 87, 162
Domenech Carbo, M.T., 87, 162
Dominguez, E., 20, 27, 54, 58–59, 93, 99–100, 185, 191–192, 197–198, 200, 202, 209, 282, 327, 338, 348, 352, 354, 363–364, 369, 380
Dominguez, O., 309
Dominguez Sanchez, P., 79, 95, 185, 188, 191, 198, 200, 287, 356, 363
Dondoi, M.P., 89, 226
Dong, B.-X., 207, 423
Dong, H., 191, 209
Dong, S., 32, 87, 93, 95, 100, 192, 197, 206–207, 327, 351, 365, 419
Dong, S.-J., 54, 95, 97, 99, 183, 211, 268
Dong, S.-Q., 32, 87, 194, 336
Dong, S.Y., 81, 97, 99–101, 185, 193, 203, 281, 288, 333, 392
Dong, T., 33, 35, 66
Dong, X.-Y., 31–32
Dong, Y.-M., 35, 166, 229, 240
Dongduen, C., 25, 191, 193, 209, 274, 329, 430, 433
Donilowicz, C., 96, 100, 198, 356
Donmez, G., 229
Dontha, N., 93, 143, 146, 186, 191, 204–205, 290, 319, 398, 401–402, 407–408
Dorris, M., 191, 196, 342

Author Index

Dos Santos Mattos, C., 90, 250
dos Santos, P.M., 191, 206–207, 422
Dosal, M.A., 45, 161–162
Dossot, M., 150
D'Ottavi, D., 79, 82, 198
Dousikou, M.F., 87, 201, 373
Dousty, F., 103, 303, 310, 317, 321
Downard, A.J., 99–100, 195, 340
Dracka, O., 131
Dragone, R., 309–310, 318, 321
Dreiling, R., 40–41, 43–44, 158, 162–163, 197, 346
Dryhust, G., 10
Du, D., 80, 352
Du, J., 88, 103, 406
Du, J.-B., 33, 35, 66
Du, J.-J., 27, 204, 398
Du, M., 191, 205, 398, 403, 411, 447
Duan, Y.-Y., 35, 66, 204, 398–399
Duff, A., 163, 165, 191, 209, 437
Dumitrescu, I., 29
Dumitru, A., 84, 86, 269
Dunwoody, D.C., 26, 104
Dunzendorfer, U., 310
Dupont, J., 80, 203, 396
Duran, N., 7, 87, 268
Durand, B., 161–162
Durr, J.A., 191, 209, 436
Durst, R.A., 73, 109
Dursun, Z., 84, 88–89, 93, 195, 206–207, 269, 293, 338, 419, 425
Dushina, M.E., 161
Dutra, R.F., 43–44, 206, 416
Duzinkewycz, M., 135, 140, 174, 241
Dvorakova, J., 287

E

Ebadi, A., 447
Ebrahimzadeh, F., 103
Echevarria, C., 99, 150, 320
Echizen, H., 163, 165, 191, 209, 436–437
Economou, A., 24, 103–104, 233–235, 243
Economou, E., 103–104
Efron, C., 36, 279
Efstathiou, C.E., 19, 59, 79, 87, 96, 175, 185, 191, 194–196, 201, 204, 248, 277, 285, 334, 337, 342, 345–346, 373, 397
Egashira, N., 97, 150, 208, 426
Eggins, B.R., 25, 100, 191, 196–197, 209, 345, 348, 433
Egholm, M., 205
Eguren, M., 161
Ehsan, M.Q., 191, 208, 427
Eiseman, G., 135–136, 140
Ejhieh, A.N., 86, 281
Ekanayake, C.B., 299
El Anwar, F., 315
El Atrash, S.S., 92, 163, 199
El Ayyas, W., 309
El Bakouri, H., 298
El Din, H.S., 97, 313
El Gakhri, M.A., 230
El Hady, D., 90, 196, 344
El Maali, N.A., 90, 97, 196, 205, 207, 305, 307, 316, 320, 322, 344, 401, 425

El Mhammedi, M.A., 84, 86, 88, 101, 209, 239, 257, 297, 301
El Murr, N., 25, 84–86, 93, 95–96, 99, 149, 191–193, 198, 200–202, 206, 225, 326, 330–331, 352–357, 367–368, 382, 417
El Rhazi, M., 38–39, 89, 100, 184, 191, 200, 202, 229, 233–234, 239, 278, 281, 363, 376
El Sayed, A.Y., 316
El Sayed, F.A., 307
El Shall, M.A., 313, 315, 318, 322
El Sherief, A.E., 161–162
El Sherif, Z., 315
Elbouhouti, H., 220
Elizalde, M.P., 70
Elmaslout, A., 7, 54, 80, 281, 333
Elouadseri, M.M., 161
El-Ries, M.A.N., 205, 313, 315, 318–319, 321–322, 405
Elsuccary, S.A.A., 54, 63, 103, 233–234, 243
Elyacoubi, A., 191, 201, 373
Emmot, P., 217
Emneus, J., 24, 58, 84, 185, 191, 197, 201–202, 282, 348–349, 380
Encinas, P.B., 161–162
Enein, H.Y., 311
English, A.M., 79, 96, 100, 198, 356
Ennaji, M.M., 204, 398
Ensafi, A.A., 32, 95–96, 191, 201–204, 280, 307, 313, 372, 384, 391
Ensafi, H.R., 96, 313
Erabi, T., 191, 202, 375
Erciyas, E., 191, 205, 404
Erdem, A., 25–26, 79, 93, 96, 99, 191, 197, 205, 209, 312, 316, 333, 349, 398, 403–409, 411, 434
Erden, P.E., 191–192, 326
Erdugan, H., 319
Erdugan, N., 319
Eremenko, A.V., 163, 191, 197, 201, 209, 347, 374
Erensoy, S., 191, 205, 409
Ermakov, A.N., 161–162
Ershad, S., 274
Ersoez, M., 288
Esaka, Y., 87, 197, 351
Escribano, M.T.S., 78, 99, 230
Esfahani, F.K., 101, 276
Esfandyarpour, M., 227
Eshraghi, H., 103, 202, 314, 384
Eskandari, K., 75, 80, 88, 103, 227, 237
Eskinja, I., 185, 283
Espinosa, A.M., 161–162
Estela, J.M., 143, 145
Estevez Hernandez, O., 90
Etienne, M., 64, 101–102, 172, 219, 225, 229
Evans, O., 79, 96, 185, 191, 209, 285, 333, 434
Evans, R.G., 191, 198, 358
Everhart, M.E., 2, 56, 143–144, 146, 274
Evrard, O., 160
Ewing, A.G., 73, 191, 209, 436–437

F

Fagain, C.O., 99, 206, 416
Faghihian, H., 191, 206–207, 420
Faguy, P.W., 95, 100, 191, 203, 206–207, 391, 417, 423–424
Fairouz, M., 53–54, 233, 252

Fajardo, M., 102, 240
Fakheri, M., 90
Faller, C., 101, 289, 299
Fan, J.S., 74, 84, 191, 205, 401, 406, 409
Fan, L.-S., 15, 32, 191, 205, 409
Fan, S.-H., 35, 300
Fandy, R., 89, 214
Fang, L., 87, 199, 359
Fang, X., 191, 195, 339
Fang, Y.-Z., 32, 41, 79, 87, 100, 188, 193–194, 197, 206–207, 296, 298–300, 333, 336, 351, 419
Fang, Z.-H., 161
Fanjul Bolado, P., 49, 292
Fanta, K., 231
Faramarzi, F., 80, 88, 103, 312
Farghaly, O.A., 103, 261, 305, 316
Farhadi, K., 67, 82, 95, 203, 205–207, 309, 314, 389, 402, 420
Farhadi, M., 103
Faria, R.C., 31, 79, 96, 229, 273
Farias, P.A.M., 7, 24, 93, 103, 191, 202, 204–205, 319, 381, 398, 400, 402, 406–408
Faridbod, F., 30–35, 66, 102, 180, 240, 263–265, 302
Farjami, E., 34–35, 191, 197, 201, 359, 371
Farsang, Gy., 3, 37, 152, 158, 213, 256
Fathi, S., 99–100, 166, 191–192, 328
Fatibello Filho, L.H., 310
Fatibello Filho, O., 20, 25, 31, 67, 81–82, 87, 89, 99–100, 185, 191, 202–203, 206–209, 220, 229, 260, 283, 303, 307, 310, 314, 384, 395–396, 419, 426, 430, 432–433
Fatibello, O., 79, 96, 273
Fatollahi, F., 102, 216
Faulkner, L.R., 123
Faure, R., 84, 168
Favero, G., 96
Federici, G., 99, 249
Fedorchuk, A.O., 161
Fedorova, I.L., 90, 188, 195, 201, 252, 296–297, 299, 301, 337, 372
Feher, Zs., 52
Fei, D., 191, 205, 398, 401
Fei, J.-J., 30–31, 64, 67, 90, 185, 228, 253, 263–264, 275, 284, 320
Feinberg, B., 3, 158
Felice, L.J., 40–41, 43–44, 158, 197, 346
Felix, F.S., 191, 203, 209, 397, 433
Fellows, L., 191, 209, 437
Feng, A.-X., 82, 206
Feng, S., 168
Feng, X.-Q., 225
Feng, Y., 82–83, 149
Feng, Y.J., 238
Feng, Y.-L., 253
Fenga, P.G., 191–192, 325
Ferancova, A., 75, 90, 105, 185, 194, 205, 287, 323, 335, 401
Fernandes, J.R., 191, 205, 309, 398, 401, 404, 407
Fernandes, S.C., 191, 196, 209, 344, 350, 431
Fernandez Alvarez, J.H., 16, 96, 322
Fernandez Bobes, C., 191, 202, 379
Fernandez de Betono, S., 86, 296
Fernández Escudero, J.A., 79, 166, 197, 204, 333, 350
Fernandez, H., 75, 197, 350, 430
Fernandez, J.J., 281
Fernandez, M.E.S., 254
Fernandez Sanchez, C., 79, 96, 191, 201–202, 297, 302, 333, 373, 378
Fernandez, V.M., 59, 99, 338
Fernandez-Abedul, M.T., 191, 202, 306, 309, 317, 379, 381
Ferrari de Castilho, R., 87, 191, 206–207, 429
Ferrari, R.T., 86, 96
Ferreira, C.U., 7, 87, 93
Ferreira, R.M.S., 191, 206–207, 423
Ferreyra, N.F., 30–32, 63, 66, 87, 93, 191, 193–195, 198–199, 201, 203–204, 269, 331, 336, 341, 353, 359, 361, 374, 393, 400
Ferri, T., 90, 245, 248
Fetisov, A.V., 161–162
Fetisov, V.B., 161–162
Fielden, P.R., 103–104
Filho, J.L.L., 43–44, 206, 416
Filho, N.L.D., 220
Filho, O.F., 227, 310
Filipenko, A.T., 135, 140, 213
Filipiak, M., 191, 204–205, 398–399, 402, 408–410
Fillenz, M., 99–100, 163–165, 191, 204, 209, 397, 435–437, 598–599
Fiorucci, A.R., 191, 196, 342
Firmino de Oliveira, M., 90, 252
Fischer, J., 169–170, 186–187
Fischer, K., 86
Fitch, A., 83
Fitsev, I.M., 191, 193, 333
Fitt, A.D., 168
Fitzgerald, C.A., 87, 198, 353
Flair, M.N., 191, 205, 407
Flechsig, G.U., 7, 25, 42, 49, 147, 149, 191, 205, 232, 241, 401
Florescu, M., 104, 240, 242, 244–245
Fogarty, B.A., 41, 195, 201, 337, 373
Fogg, A.G., 135, 140, 143, 145, 174, 241, 292
Fojta, M., 7, 97, 170, 191, 201, 204–205, 375, 398, 400, 408, 411
Fomin, A.Y., 96
Fonseca, L.P., 191–192, 327
Fontan, C.A., 65, 99, 191, 202, 252, 381
Foord, J.S., 28
Forano, C., 84, 149
Foret, P., 25, 42–43, 51, 64, 67, 97, 254
Formenti, F., 163, 165, 209
Foroumadi, A., 75, 80, 88, 103, 237
Forzani, E.S., 25, 93, 191, 196, 201, 203, 209, 345, 374, 393–395, 431–432
Fotouhi, L., 31–32, 93, 96, 196, 201, 343, 349, 369
Fox, R.P., 162
Francisco, M.S., 87, 196, 267, 344
Francois, J., 163, 165, 209, 436, 447
Francois, M., 84, 275
Franzoi, A.C., 99, 191, 197, 350
Frattini, P., 191, 203, 395
Fray, A., 191, 209, 437
Freed, C.R., 163, 165, 191, 209, 436–437
Freiha, B.A., 64, 67, 69–70, 188, 191, 203, 206–207, 209, 308, 319, 385, 413, 418, 434
Freire, R.S., 7, 87, 192, 326
Freitas, B.H., 261
French, W.G., 2, 4, 26, 73, 151, 158, 162

Author Index

Freour, R., 161
Frias, R., 191, 206, 418
Friml, J., 131
Fritzen-Garcia, M.B., 81, 203, 395, 432
Froning, M., 143, 145–146
Fryxell, G.E., 7, 75, 101–102, 179, 219, 233, 256
Fu, C.-G., 41, 79, 193, 222, 333
Fu, J.-K., 191, 208, 212, 428
Fu, L., 100, 149, 205, 411
Fu, X.H., 80–81, 92, 103, 191, 205, 409
Fu, Z.-Q., 191, 206, 416
Fujikawa, Y., 56, 152, 158, 291
Fujiwara, S.T., 20, 87, 93, 96, 101, 191, 193, 195, 203, 206–207, 271, 333, 338, 391, 420–421
Fukushi, H., 25, 99, 199, 360
Fumero, B., 191, 209, 437
Fursik, O.A., 191, 208, 427
Furuike, T., 247
Fushimi, F., 63

G

Gaal, F., 188, 240, 297, 299–300
Gachok, I.V., 163, 197, 201, 209, 347, 374
Gahlaut, A., 191, 202, 382
Gaichore, R.R., 90
Gaillochet, P., 3, 16, 26, 158, 160, 162
Gaillon, L., 35, 67
Galal, A., 191, 203, 388
Galan, F., 99, 101, 202, 295, 383
Gale, P.A., 100, 274
Galicia, L., 29, 32, 90, 166, 168, 206, 233, 354, 415
Galik, M., 67, 85, 97, 178, 198, 257–259
Galiova, V., 168
Gallet, O., 191, 200, 363
Gallo, B., 304, 306, 315
Gallo Hermosa, B.L., 304, 314
Galus, Z., 2, 23, 27, 36, 38, 43, 49, 53, 56–58, 61, 63, 65–66, 96, 104, 142, 151–153, 207
Galyametdinov, Y.G., 252, 258, 276
Gan, N., 191, 193, 198, 332, 352
Gan, T., 101
Ganesan, V., 65, 77–78, 101–102, 106, 149
Ganjali, H., 35
Ganjali, M.R., 30–35, 66, 86, 90, 99, 101–103, 180, 185, 202, 206, 216, 221, 227, 240, 263–265, 289, 302–303, 310, 314, 317, 321, 384, 412
Ganjipour, B., 32, 93, 191, 203–204, 206, 392, 396–397, 415–416, 423
Gao, F., 99, 149, 202, 377
Gao, H.-L., 80, 103, 185, 191, 197, 205, 283, 346, 348, 378, 399–400, 410
Gao, J., 90
Gao, M.-X., 99
Gao, N., 191, 195, 199, 337, 361
Gao, P., 64, 180, 185, 263–265, 284
Gao, Q., 93, 191, 198–199, 356, 362
Gao, R.F., 34–35, 71, 80, 99, 149, 191, 202, 376–377, 419
Gao, S., 191, 203, 395
Gao, W., 304
Gao, X., 191, 203, 395
Gao, Y., 82–83
Gao, Z.-N., 89–90, 99, 190–191, 194–195, 203, 206–207, 211, 213, 222, 249–250, 255, 335, 339, 387, 414, 418, 421

Garaj, J., 163
Garay, F.J., 162
Garcia, A.A., 99, 150, 311
Garcia, A.C., 309, 313, 317
Garcia, B., 97, 316
Garcia, C.A.B., 86, 93
Garcia, E., 99–100, 198, 355
Garcia, H., 79, 85, 333
Garcia, J.R., 191, 200–201, 206–207, 370, 422
Garcia, M.L.T., 79, 96, 159–160, 268, 333
Garcia, O., 161–162
Garcia, S., 188, 295
Garcia, T.A., 191, 196, 345
Garcia-Fernandez, M.A., 306
Gardea Torresdey, J., 25, 208, 211, 428
Garjonyte, R., 79, 82, 93, 191, 193, 209, 332, 435
Gasparini, R., 43, 194, 335
Gavioli, G.B., 152, 158, 286
Gawron, A.J., 41, 195, 201, 337, 373
Gay, D.S.F., 87, 149, 206–207, 425
Gay, M., 166, 191, 194, 197, 335, 348
Ge, B., 88, 202, 375
Ge, C., 191, 193, 198, 332, 352
Ge, H., 86, 93
Ge, S.X., 101, 324
Gedmina, A.V., 80, 88, 103, 149, 193, 332
Gehad Gseleim, A.A, 310
Geiger, A., 191, 206, 418
Gelbert, M.B., 16, 81, 97, 203, 206–207, 386, 418
Gellett, W.L., 26, 104
Gellon, H., 65, 99
Geng, L., 80, 193, 330
Geng, Z., 90
Geno, P.W., 99, 206–207, 420
Gerard, M., 100
Gerardin, C., 101–102
Gerardin, L., 101, 225
Gerlach, U., 222
German, N., 185, 287
Gessner, F., 84, 101, 219
Getts, R.C., 191, 205, 398, 401
Gevorgyan, A.M., 248, 254
Ghaedi, M., 90, 255, 277
Ghaffarinejad, A., 82, 96, 195, 338, 341
Ghalkhani, M., 79, 95–96, 203, 206–207, 316, 333, 396, 416, 424
Ghandour, M.A., 81, 97, 307, 321
Ghanjaoui, M.E.A., 239
Ghasemi, V., 95–96, 206–207, 421
Ghasemlu, K., 82, 203–204, 206–207, 389, 399, 420
Gherghi, I.Ch., 191, 205, 307, 321, 398, 403–405
Ghiaci, M., 84, 87, 101, 229, 236, 257
Ghobadi, S., 191, 195, 209, 337, 341
Gholabi, S.M., 162
Gholami, F., 96, 305
Gholami, J., 102, 240
Gholivand, M.B., 186, 237
Ghoneim, E.M., 90
Ghoneim, M.M., 314, 321
Ghoreishi, S.M., 88, 103, 305
Giannetti, B.F., 161
Gil, E.S., 7, 97, 191, 196, 205–206, 344–345, 402, 418

Gilbert, O., 95, 97, 99–100, 166, 191, 203, 206–207, 386, 388, 390–393, 416, 424–425
Gilmartin, M.A.T., 191, 204, 398
Gimeno Adelantado, J.V., 87, 162
Girish Kumar, K., 96
Girmai Bokretsion, R., 306
Girousi, S.T., 97, 121, 170, 191, 205, 231, 307, 309, 318, 321–322, 398, 402–406, 447
Gismera, M.J., 89, 220, 228, 235, 254
Glab, S., 109, 225
Glaeser, P., 79, 87, 96, 194, 336
Glebov, A.N., 230
Gleixner, G., 93, 200–201, 369
Glennon, J.D., 28, 249
Gligor, D., 84–86, 191, 200–201, 206–207, 269–270, 369, 371, 425
Glynn, G.E., 81, 163
Gobert, E., 161
Gobhadi, S., 24
Goddard, N.J., 163, 165, 209, 436
Godinho, O.E.S., 64, 101, 217, 276
Goessler, W., 89, 214
Goh, N.-K., 7, 84, 86, 198, 356
Gokgunnec, L., 79, 96, 197, 209, 333, 349, 434
Golabi, S.M., 89, 92, 162, 191, 200–201, 205, 231, 253, 368
Golden, T.D., 191, 206, 418
Golimowski, J., 213
Golzan, M., 82, 203, 206–207, 389, 420
Gomathi, A., 92, 267
Gomes, H.M., 261
Gomez de Rio, M.I., 79, 82, 277, 303
Gomez, S.P., 25, 235
Gomezde Balugera, Z., 86, 323
Goncalves, C., 19, 82, 193, 195, 198, 331, 341, 358
Goncalves, J.E., 87, 93
Goncalves, R.A.C., 87, 93
Gondo, S., 93, 191, 199, 362
Gondran, C., 75
Gong, F.-C., 25, 68, 96, 316
Gong, K.-P., 29
Gonzales Alvarez, M.J., 279
Gonzales, E., 78, 84, 309
Gonzales Garcia, M.B., 292
Gonzales, I., 162
Gonzalez, C., 161
Gonzalez De La Huebra, M.J., 191, 202, 383
Gonzalez, E., 101, 162, 191, 206, 323, 417
Gonzalez Garcia, M.B., 7, 87, 191, 202, 375, 378, 380
Gonzalez, I., 8, 15, 63, 84, 149
Gonzalez, I.C., 150, 160–162
Gonzalez, I.G., 161–162
Gonzalez, M., 84, 188, 299
Gonzalez Mora, J.L., 191, 209, 437
Gonzalez, P.S., 65, 99, 252
Gonzalez, R.L., 90
Gonzalez Romero, E., 87
Gonzalez Velasco, J.R., 161
Gonzalez-Cortes, A., 97, 191, 196, 203, 209, 344, 393, 436
Gooding, J.J., 93, 191, 200, 205, 365, 405
Goreti, M., 191, 206–207, 423
Gorski, Z., 95

Gorton, L., 5, 9, 13, 19, 24, 31, 40, 58, 67, 80, 82, 84, 87–89, 92–93, 95–96, 99–101, 110, 113, 185, 191–193, 195–198, 200–202, 206–209, 282, 293, 319, 327–328, 330–333, 337, 340–342, 348–349, 351, 353, 355, 357–358, 363–365, 368–372, 380, 425, 427, 429–430, 438
Gosser, D.K. Jr., 41, 61
Gossler, W., 67, 97, 183, 275
Goto, M., 87, 131, 163, 197, 351
Gotoh, M., 191, 198, 353
Gotor, M.F., 323
Gotti, R., 316
Goubert Renaudin, S., 238
Goulart, M.O.F., 95, 104, 191, 195, 200–201, 206–207, 340, 372, 415, 424
Gouvea, F., 86–87, 93, 194, 200–201, 333, 370
Grabaric, B.S., 79, 96, 185, 191, 196, 203, 209, 283, 333, 345, 394
Grabaric, Z., 185, 283
Grabec, I., 40, 84, 89, 106, 108, 214, 223
Gradinaru, A., 188, 302
Grandi, G., 152, 158, 286
Grandin, L.A., 86, 93
Graneli, A., 143–145
Granero, A.M., 75, 197, 350
Granger, D., 112
Grant, D.H., 205, 309, 404
Gray, J.A., 163, 165, 191, 209, 436–437
Greene, B., 99
Grennan, K., 7
Grennes, M., 131
Greschonig, H., 77–79, 211, 281
Grigoreva, M.F., 161–162
Grinciene, G., 25, 104, 240
Groisberg, A.T., 292
Gropeanu, R., 191, 200–201, 371
Grosfils, K., 191, 325
Grosselin, N., 161–162
Grosu, I., 191, 200–201, 371
Grudpan, K., 89, 228
Gruendig, B., 191, 198–199, 362
Gruendler, P., 7, 25, 42–43, 49, 90, 147, 149, 185, 191, 197, 205, 232, 241, 287, 346, 401, 404
Gruenewald, R.A., 163, 165, 209, 437, 598
Grundler, P., 43, 74, 80, 205, 403
Gruner, W., 160, 162
Grygar, T., 159, 161–162, 202, 379
Gu, C.-R., 101, 324
Gu, K., 191, 205, 316, 405
Gu, L., 161, 191, 208, 427
Gu, Z.-N., 29, 32, 34
Guadalupe, A.R., 88, 97, 99, 192, 194, 200–201, 325, 334, 368
Guadalupe, R., 92, 104, 362
Guan, J.-J., 81, 88, 276
Guan, L., 191, 202, 376
Gueguen, S., 95, 191, 198, 206, 357, 417
Guemas, Y., 95, 198, 352, 357
Guerra, S.V., 84, 86, 96, 271
Guidi, F., 90, 248
Guilbault, G.C., 95, 354
Guilbault, G.G., 6, 26, 79, 93, 191, 195, 200, 267, 304, 306, 333, 339, 363, 366
Gulmez, B., 191, 205, 404

Author Index

Gunasingham, H., 93, 191, 199, 208, 361, 427, 434
Guo, C.-C., 96, 316
Guo, C.-X., 34–35
Guo, H.-S., 64, 81, 101, 191, 205, 244, 246, 324, 409
Guo, J., 102, 254
Guo, L., 412
Guo, M., 89, 97, 191, 202, 205, 209, 381, 409, 447
Guo, Q.-H., 8, 80, 192, 325
Guo, S.X., 24, 88, 90, 97, 103, 195, 338
Guo, W., 293
Guo, X.-X., 322
Guo, Z., 87
Guorong, Z., 80, 90, 93, 207
Gupta, A.K., 32, 186, 294
Gupta, V.K., 25
Gurentsova, O.I., 162, 281
Gurung, A.S., 88, 103, 202, 378
Gusakov, V.N., 69–70, 289
Gushikem, Y., 7, 38, 79, 86–88, 93, 96, 101, 149, 191, 193, 195–197, 200–201, 206–207, 217, 266–267, 271, 278, 319, 333, 338, 344, 347, 367, 369–370, 420, 425
Gutiéréz, F., 166, 197, 350
Gutierrez Fernandez, S., 97
Gutierrez, J.I., 161
Gutierrez-Granados, S., 90, 206, 415
Guttmann, M., 49, 63, 89, 103, 149
Gutz, I.G.R., 191, 196, 345
Guzman, S.Q., 266
Guzmán, T.M.O., 161–162
Guzsvany, V., 187–188, 240, 297, 299–300, 447

H

Ha, K.S., 191, 194, 215, 335
Ha, Y.-S., 215
Haber, F., 191
Habermuller, K., 80, 209, 438
Habib, I.H.I., 99, 238
Habibollahi, F., 93, 206–207, 423
Hadjipavlou-Litina, D., 191, 205, 398, 402
Hadji-Shabani, A., 90
Haghgoo, S., 96, 307
Haghighi, B., 35, 276
Haghshenas, E., 103
Hahn-Haegerdal, B., 99–100, 192, 327
Hainfeld, J.F., 115
Hajjizadeh, M., 80–81, 84, 86, 88, 103, 194, 196, 203, 336, 343, 389
Halamek, E., 135, 139
Halamova, L., 99
Halbert, M.K., 79, 96, 195, 201, 339, 341, 372
Hale, P.D., 79–80, 92–93, 96, 99–100, 191–193, 198–200, 202, 209, 327, 329, 333, 354–355, 362–364, 369, 384, 438
Hall, S., 191, 204
Halouzka, V., 80, 198, 357
Halsall, H., 191, 197, 346
Hamada, H., 63, 92, 198, 354
Hamdoune, F., 101–102
Hamida, S., 248
Hamidi, H., 35, 101, 276
Hamidian, H., 89, 220
Hammam, E., 314, 321

Hammerich, O., 87, 209
Hammouti, B., 7, 80, 333
Hamzehloei, A., 79, 96, 206, 333, 413
Han, I.k., 191, 206, 412
Han, J., 59, 191, 202, 369, 378
Han, K.-G., 187, 294
Han, M., 191, 202, 383
Han, S., 15, 32
Han, X.-X., 74, 191, 203, 206–207, 387, 414, 418
Han, Z., 82–83, 100–101, 149
Han, Z.-B., 83, 97, 149, 167
Han, Z.G., 82, 206
Hanaki, M., 100, 196, 343
Hand, H.-Y., 251
Handlir, K., 191, 198, 354
Hang, G.-P., 191, 202–203, 206–207, 386, 424
Hangarter, C.M., 185, 197, 208, 289, 346, 429
Hanocq, M., 191, 204, 397
Hansen, E.H., 3, 53, 134, 140, 576, 581
Hansen, H.E., 41, 93, 191, 200, 362–363
Hanustiak, P., 185, 191, 196–197, 205, 285, 345–346, 404
Hao, J., 81, 88, 276
Hara, M., 16, 19, 96–97, 187, 293
Harak, D.W., 49, 59, 147, 149
Hare, M., 95, 194, 209, 335, 434
Hari, S.P., 81, 163, 165, 205
Harris, G., 90
Harris, S.J., 90
Hart, J.B., 44, 162–163, 165
Hart, J.P., 4, 96, 110, 120–121, 170, 191, 204, 208, 398, 426
Hartani, K., 255
Harth, R., 75
Haruyama, T., 25
Hasanpour, F., 32, 201, 372
Hasanzadeh, A., 82
Hasanzadeh, M., 81, 84, 86, 196, 203, 343, 389
Hasebe, K., 97, 200, 366
Hasheminasab, A., 102, 216, 221, 263
Hashimoto, Y., 75, 92, 204, 399
Haskard, M.R., 135, 140
Haskett, C., 163, 165, 191, 209, 436
Hason, S., 43, 205, 401
Hassan, H.N.A., 97, 99, 140, 143, 146, 219, 226, 238, 243, 310, 315
Hassan, R.M., 99, 238, 320, 322
Hassmann, J., 191, 205, 407
Hassouna, M.E.M., 54, 63, 103, 233, 243
Hatakeyama, H., 191, 198, 355
Hatano, S., 191, 194, 335
Hattori, T., 16, 19, 96–97, 186–187, 293–294
Havel, L., 305
Havran, L., 7, 97, 201, 205, 375, 408
Hawley, D., 3, 42–43, 151–152, 157
Hawley, M.D., 3
Hayon, J., 75
Hazem, M., 221
Hdou, G., 163, 165, 209
He, C.-X., 93, 206–207, 425
He, H., 191, 204, 398
He, J.-B., 31–32, 149, 166, 191, 197, 202–203, 206–207, 349, 386, 424
He, L.-W., 167
He, N.Y., 27, 32, 64, 88, 101, 103, 204, 246, 324, 398, 406, 411

He, P., 8, 20, 32–35, 54, 63, 66, 82, 87, 100, 168, 197, 206–207, 351, 419
He, P.-A., 32, 87, 194, 336
He, P.-G., 188, 296, 298–300
He, Q.-G., 67, 81, 84, 97, 191, 196, 202, 204, 267, 275, 343, 381–382, 397
He, X.-W., 191, 206–207, 421
He, Y.-H., 447
He, Y.-P., 191, 388
Hedenmo, M., 99–100, 200, 209, 364
Hegde, R.N., 67, 206, 305, 413
Heimler, D., 42, 158
Heineman, W.R., 123, 143–144, 189, 191, 197, 346
Hejazi, M.S., 191, 205, 408
Heli, H., 80, 88, 103, 194, 304, 312, 335–336
Heller, A., 99–100, 192, 327, 363
Helz, G.R., 185, 188, 287
Hennion, M.-C., 162
Henze, G., 101, 289, 299
Heravi, M.M., 31, 93, 201, 369
Heravi, P., 82
Hermosa, B.L., 304, 306
Hernandez, A.R., 90
Hernandez Artiga, M.P., 84, 185
Hernandez, G., 88
Hernandez Guevara, E., 88
Hernandez, L.H., 25, 36, 58–59, 67, 77–78, 84–85, 89–90, 96, 99, 101, 104, 162, 174, 185, 188, 191, 194, 197–198, 202, 206, 217–218, 222–223, 225, 230, 287–288, 295, 299, 301–302, 304, 306, 309, 311–312, 319, 323, 326, 334, 347, 383, 417
Hernandez, P., 36, 67, 77–78, 84–85, 99, 101, 162, 174, 185, 188, 191, 194, 197, 202, 217, 222, 230, 287–288, 295, 299, 302, 304, 306, 309, 312, 323, 334, 347, 383
Hernandez Santos, D., 7, 49, 87
Hernandez-Santos, D., 191, 202, 375
Hernando, T., 96, 197, 350
Herrasti, P., 29, 32, 90, 206, 354, 415, 447
Herrmann, M., 191, 200, 363
Hes, J., 158, 185, 286
Heyrovsky, J., 172, 174
Hiasa, H., 191, 198, 354
Hibbert, D.B., 93, 191, 200, 205, 365, 405
Hickey, C., 191, 196, 209, 345, 433
Hidajat, K., 77–78
Hierro, I., 102, 240
Higgins, I.J., 5, 273
Higuti, T., 191, 202, 375
Hill, H.A.O., 5, 273
Hirabayashi, G., 78, 99, 191, 201, 360, 365, 375
Hirano, T., 131
Ho, K.-C., 168
Hocevar, S.B., 99–100, 103–104, 121, 135, 143, 146, 215, 237, 239, 242–243, 245
Hodek, P., 191, 196–197, 205, 345–346, 404
Hodges, H., 163, 191, 209, 436–437
Hodos, M.Y., 161
Hoeehr, N.F., 93, 201, 372
Hoeffler, S., 86
Hoegfeldt, E., 172
Hočevar, S.B., 7–8, 12, 16, 24–25, 41, 45, 103–104, 146, 174, 183, 232, 234, 240–241, 243, 275
Hoffmann, A.A., 191, 200–201, 206–207, 370, 425

Hoffmann, C.R., 135, 140
Hoffmann, W., 135
Hofmann, C., 88, 97, 192, 200–201, 325, 368
Hofmanova Matejkova, A., 266
Hojo, M., 185, 197, 282, 347
Holgado, T.M., 78, 89, 230
Holik, M., 131
Holley, E.A., 161–162
Holness, W., 99–100, 199, 362
Holo, L., 191, 195, 308, 341
Homoda, A.M., 246
Honarasa, F., 33, 35
Honarmand, E., 88, 103, 305
Hooda, V., 191, 202, 382
Hoogvliet, J.C., 40, 203, 284, 395
Horie, H., 25, 200, 209, 365, 434
Horio, T., 191, 202, 375
Hormozi Nezhad, M.R., 80, 88, 203, 207, 389, 423
Horna, A., 191, 196–197, 202, 205, 345–346, 379, 404
Horner, L., 73
Horta, D.G., 161–162
Hosamani, R.R., 67, 206, 413
Hoshi, S., 25, 99, 199, 215, 255, 360
Hoshi, T., 260, 266
Hosseini, M.R.M., 31–32, 35, 38, 96, 267
Hosseini, S.R., 79, 99–100, 166, 192, 328
Hosseini-Mahdiabad, H., 96, 201
Hosseini-Monfared, H., 96, 203, 206, 390, 421
Hossienzadeh, R., 82, 95–96, 195, 203–204, 206–207, 319, 339, 389, 399, 420–421
Hou, H.-Q., 8, 16, 34, 63, 80, 103, 191–193, 195, 198, 200, 203, 206–207, 325, 332, 337, 367, 373, 385, 389, 414–415, 419, 423
Hou, J., 191, 203, 395
Howard, J.A., 54, 58, 191, 205, 408
Howell, P.D., 168
Hoyer, W., 191, 201–202, 380
Hradilova, S., 43, 183, 239
Hranicka, Z., 188, 288–289
Hrbac, J., 80, 87, 103, 268
Hrdinova, V., 191, 201, 375
Hsiung, S.-K., 191, 206, 413
Hsu, C.T., 99, 206, 414
Hu, C., 81–83, 88, 97, 100–101, 103, 149, 191, 202–203, 205, 267, 376, 382, 391, 396, 423
Hu, C.-G., 99, 191, 197, 206, 232, 293, 347, 414
Hu, C.-W., 81–83, 97, 149, 167
Hu, G., 97, 202, 209, 381
Hu, H.-L., 82, 205
Hu, J.-F., 191, 206–207, 421
Hu, L., 191, 202, 379
Hu, R.-Z., 191, 208, 212, 428
Hu, S., 67, 81, 84, 93, 97, 100, 191, 197, 202–207, 209, 266–267, 275, 346, 376, 382, 394, 396–397, 412, 423, 425, 431–433
Hu, S.-Q., 88, 103, 357, 398
Hu, S.-S., 97, 99, 149, 191, 202–203, 209, 232, 264, 293, 304, 310, 324, 381–382, 391
Hu, W., 191, 208, 428
Hu, X.-Y., 80, 191, 194, 199, 206–208, 212, 223, 239, 249, 254, 303–305, 309, 336, 359, 418, 420, 428
Hua, B., 191, 206, 418
Hua, C., 16, 24, 77–78, 96, 99, 217, 223, 322
Huan, D.L., 80, 198, 357

Author Index

Huan, G., 191–192, 209, 327, 435
Huan, Z., 80, 99–100, 191, 193, 195, 200, 209, 332, 337, 341, 364, 369
Huang, A.-P., 191, 206–207, 424
Huang, D.-L., 80, 92, 103, 202, 205, 409
Huang, G.H., 80, 95, 99, 333
Huang, G.-T., 191, 196, 343
Huang, H., 191, 203, 395, 433
Huang, J., 63, 80, 103, 191, 193, 200, 203, 206–207, 332, 367, 385, 389, 414–415, 419, 423
Huang, J.-S., 34
Huang, K.-J., 34–35, 80, 88, 99, 103, 202, 377
Huang, L., 191, 200, 205, 366, 404
Huang, M., 35, 102, 286
Huang, M.H., 64, 253, 263
Huang, M.-Q., 191, 206–207, 421
Huang, S.-J., 36, 54, 135, 140, 158, 289
Huang, S.-S., 97, 174, 214, 231, 251, 437
Huang, T.H., 24, 191, 195–196, 199, 201, 340, 342, 373
Huang, W.-S., 84, 97, 101, 185, 197, 204, 209, 215, 218–219, 246, 282, 284, 293, 347, 398, 438
Huang, X., 27, 54, 96, 195, 197–198, 201–202, 339, 346, 353, 372, 382
Huang, Z.-R., 191, 196, 209, 345, 431
Huang, Z.-W., 191, 196, 209, 345, 431
Hubalek, J., 185, 191, 196, 202, 285, 345, 379
Hubbard, A.T., 163
Huber, C.O., 87, 192, 195, 197, 325, 334, 336, 351
Hudakova, M., 84, 206–207, 420
Hudnik, V., 84
Hudson, M.J., 101
Huebra, M., 70
Hueso, D., 89
Huff, R., 44, 158
Hulanicki, A., 109
Humffray, A.A., 56, 58
Hundiwale, D.G., 100
Huska, D., 185, 285
Hussien, A.R., 101, 228
Hutchins, L.D., 600
Hvizdalova, M., 9, 12, 14, 17, 19–20, 24, 49, 51–52, 54–55, 103, 147, 212, 241
Hynes, C.J., 201

I

Ianniello, R.M., 99
Ibarra, I.S., 174, 238
Ibrahim, H., 19, 135, 208, 248, 267, 310–311, 320
Ibris, N., 162
Idrissi, L., 89, 184, 234, 278, 281
Iijima, S., 8, 16, 29, 32, 96, 99, 198, 358
Ijeri, V.S., 90, 97, 191, 206–208, 224, 231, 423–424, 427
Ikariyama, Y., 191, 358
Ikebukuro, K., 191, 205, 411
Ikeda, S., 97, 205, 405
Ikeda, T., 7, 25, 63, 75, 79, 92, 95, 99, 149, 168, 191–193, 197–198, 200–202, 204, 208, 270, 328–329, 331, 349, 352, 354, 357, 364–365, 371, 375, 383, 399, 426, 428–429
75, 208, 426, 429
Ikenoya, K., 163
Iliadou, E., 231
Iliescu, N., 89, 226

Illarionova, I.S., 161–162
Imabayashi, S.I., 95, 197, 349
Imai, H., 95, 198, 355
Imhoff, J.-M., 191, 200, 363
Inada, K., 63, 191, 196, 204, 206, 342, 397, 413
Inagaki, H.S., 79–80, 96, 100, 193, 329
Inagaki, T., 80, 96, 99–100, 191, 198, 333, 354–355
Ingledew, W.J., 191, 202, 380
IngmanA, F., 109
Iniguez, M., 166
Innocent, C., 191, 197, 347
Inuta, S., 270
Ioannou, A.K., 191, 205, 321, 404, 406
Ion, A., 19
Ionescu, C., 88, 201, 374
Irigoyen, L., 168
Iseki, K., 87, 281
Ishihara, K., 191, 198
Ishii, D., 131, 163
Ishii, S., 96
Ishikawa, A., 87, 281
Ishikawa, D., 80, 199, 361
Ishimura, F., 191, 198, 353
Ishiwari, K., 163, 165, 209, 436, 447
Islamnezhad, A., 85, 196, 343
Ismail, K., 53–54, 233, 252
Isono, Y., 75, 92, 204, 399
Issa, Y.M., 248, 267, 310–311
Itaya, K., 38, 79, 82, 259
Itirli, G., 191, 205, 407
Ito, S., 191, 202, 375
Ito, Y., 25, 200, 208, 429
Ivama, V.M., 79, 82, 268
Ivanauskas, F., 191, 200, 362, 364
Ivanko, M.G., 161–162
Ivanova, E.V., 97, 192, 200, 326, 364
Ivaska, A., 32–33, 75, 99, 222
Iwanaga, H., 97, 208, 426
Iwasaki, Y., 191, 198
Iwuoha, E.I., 59, 80, 88, 99–100, 198, 354
Iyer, S.V., 161
Iza, J., 162
Izquierdo, A., 42
Izutsu, K., 78, 99, 249, 253

J

Jabbari, A., 80, 88, 103, 194, 304, 312, 336
Jacob, P., 222, 249, 253, 255
Jacobs, E.S., 2, 38, 210, 213
Jaenisch, B., 255
Jagner, D., 143–145
Jahangir, A.A., 102, 221
Jaiswal, P.V., 97, 191, 206–208, 423, 427
Jakubovska, M., 225
Jalali, F., 267, 304
Jamasbi, E.S., 96, 307
Jambon, C., 87
Jamieson, C.M., 163, 165, 209, 437
Jamis, J., 86
Janata, J., 191
Janegitz, B.C., 31, 99, 220, 227, 229
Janiga, I., 163
Jannat Rezvani, M.J., 96, 321

Janosikova, Z., 238, 257, 262
Jansova, G., 51, 87, 103, 174, 232, 241
Janzen, K.-P., 180, 261
Jarskog, H., 58, 202, 380
Jasinski, M., 7, 42, 205, 401
Jasnowska, J., 191, 204–205, 398–399, 402, 408, 410
Javadi, S., 80, 88, 203, 207, 389, 423
Javanbakht, M., 30, 33, 35, 66, 90, 101–102, 216, 221, 227, 263, 265
Jaya, K.E., 162
Jayadev, S., 386
Jayadevappa, H., 80, 95, 97, 103, 188, 191, 203, 206–207, 250, 294, 387–391, 423
Jayadevappa, J., 79, 96, 206–207, 333, 390, 415, 421
Jayaraman, A., 87
Jayasekera, S., 162
Jayasri, D., 191, 196, 344
Jeffrey, M.I., 161–162
Jeffries, C., 80, 99–100, 195, 340
Jelen, F., 93, 131, 191, 201–202, 205, 380, 404, 406–408
Jemelkova, Z., 64, 311
Jenkins, S.H., 191, 197, 346
Jeon, W.-S., 251
Jeon, Y.-G., 289
Jeong, E.D., 89, 99, 215, 217–218, 251
Jesus Gismera, M., 90, 224
Jezkova, J., 14, 16, 19, 21, 25–27, 40, 49, 54, 59, 65, 80, 87–88, 97, 99–100, 135, 139–140, 142–143, 146, 186–187, 243, 245, 267, 276, 294
Jhaveri, S.S., 99, 194, 334
Ji, H., 35, 82, 195, 338
Ji, K.-B., 264
Ji, X., 80, 102, 197, 348
Ji, Z.-M., 8, 13, 54, 235, 239
Jia, J.-B., 24, 237, 288, 447
Jian, F.-F., 97, 167, 191, 202, 268, 279, 376–377, 379
Jiang, C., 191, 205, 408, 410–411
Jiang, J.-H., 88, 191, 195, 199, 202, 337, 361, 378–379
Jiang, M., 44, 97, 191, 199–202, 205, 209, 360, 370, 381, 398, 401
Jiang, Q., 33, 35, 63, 66, 191, 206–207, 283–285, 289, 419
Jiang, S.-Y., 185, 285
Jiang, X.-Y., 80, 202
Jiang, Z.-M., 264
Jiao, K., 31–35, 63, 66, 71, 74, 80–81, 84, 92, 99, 103, 149, 191, 198, 202, 204–207, 283–285, 289, 337, 353, 376–379, 385, 398–401, 403, 406, 408–411, 419, 447
Jimenez, R.M., 70, 306, 324, 435
Jimenez-Castro, M., 87, 199, 359
Jin, C., 89, 447
Jin, G.-D., 303–304
Jin, H.-J., 8, 149, 167
Jin, J.-Y., 163, 165, 209
Jin, L.-T., 15, 163, 165, 185, 191, 203, 209, 288, 311, 395
Jin, Q., 214
Jin, W., 131
Jin, X.-F., 88, 202, 378
Jin, Z.-X., 97, 274
Joensson-Pettersson, G., 191–192, 195, 198, 200, 327, 337, 353, 363
Johansson, E., 13, 195, 337
Johansson, G., 191, 196, 342
Johansson, K., 191–192, 195, 327, 337, 363

John, A., 191, 203, 388
Jones, J., 191, 194, 333
Jonsson, C., 7, 10, 104, 143, 204–205, 400–401
Jose, M.T.S., 161–162
Joseph, M.H., 163, 165, 191, 203, 209, 395, 436–437
Joseph, R., 96, 316
Jovin, T.M., 191, 201–202, 380
Joybar, S., 90, 255
Jozefowicz, J.M., 191, 201, 374
Ju, H.-X., 80, 84, 86, 88, 150, 197, 202, 278, 348, 375
Ju, S., 80, 88, 202
Juarros, J.V., 323
Julakova, E., 10
Jun, H.-S., 223, 237
Jung, M.E., 163, 209
Jung, W.W., 191, 206, 412
Jung, Y.S., 31–32, 165, 191, 206, 209, 219, 227, 237, 298, 318, 402–403, 405, 412, 437
Jureviciute, I., 89, 225
Justice, J.B., 191, 209, 436–437
Justus, Z.L., 210

K

Kacaniklic, V., 191, 195, 337
Kachoosangi, R.T., 8, 29, 31–35, 38, 43, 95, 194, 306, 315, 334
Kacur, M., 290
Kadam, B.V., 191, 203, 395
Kadota, A., 191, 198, 353
Kaifer, A.E., 191, 198, 355, 359
Kajihara, M., 163
Kakiuchi, T., 95, 197, 349
Kaku, T., 99–100, 199, 362
Kakuno, T., 191, 202, 375
Kalab, T., 95, 193, 331
Kalanur, S.S., 103, 306
Kalbasi, R.J., 87, 236
Kalcher, K., 1–2, 4–12, 14, 16–22, 24–27, 29, 32, 40, 45, 49–56, 58–59, 61–67, 70, 74, 76–82, 84, 87–90, 96–97, 99, 101, 103–104, 106, 108, 112, 143, 146–147, 149, 151, 166, 169–171, 174–175, 181, 183–186, 188, 191–193, 196, 198–199, 204, 206–207, 209–211, 214, 217–218, 232, 241–242, 245, 247, 249–250, 253, 255, 258–259, 263, 267, 271, 274–277, 279, 281, 285, 329, 333, 342, 354, 359, 364, 398, 414, 422, 438, 447
Kalnishevskaja, L.N., 160
Kalnyshevskaya, L.N., 78
Kalous, J., 135, 139–140, 142, 187
Kalvoda, R., 166
Kamalzadeh, Z., 32, 194, 303, 335
Kamel, M.S., 315
Kamenev, A.I., 161, 230
Kamihira, S., 191, 201, 372
Kamiya, N., 78, 99, 191, 201, 360, 365, 375
Kamiyama, M., 96
Kamyabi, M.A., 79, 82, 95–96, 203, 206, 272, 278, 389–390, 421
Kamyshov, V.M., 162
Kandil, S., 205, 315, 319, 405
Kane, D.A., 163, 209, 436
Kane, S.A., 18, 185, 191, 209, 282, 435
Kanei, Y., 191, 196, 202, 343, 384

Author Index

Kang, J., 70, 89–90, 225, 400
Kang, J.W., 80
Kang, S.-K., 168
Kang, Z.-H., 81
Kannan, R., 89
Kano, K., 7, 25, 87, 92, 95, 99, 149, 168, 191–192, 197–198, 200–202, 208, 328, 349, 351, 357, 371, 375, 383, 428–429
Kapadi, U.R., 100
Kaplan, J.D., 54, 59
Kara, P., 84, 93, 99, 191, 205, 297, 398, 403–405, 407–409
Karadeniz, H., 191, 205, 316, 404–405
Karan, H.I., 79–80, 92, 96, 99–100, 191, 193, 198–200, 329, 333, 355–356, 362–363
Karava, M.K., 205, 321, 405
Karimi, F., 31–32, 95, 161, 202, 307, 314, 384
Karimi, H.M., 32, 95, 202, 314, 384
Karimi, M.A., 79, 85–86, 96, 99–100, 192, 195, 206–207, 247, 307, 328, 340, 422
Karimi, M.H., 31–32, 161, 307
Karimi-Maleh, H., 32, 95–96, 191, 196, 201–204, 280, 290, 305, 307, 313, 345, 372, 384, 391, 396
Karimi-Maleh, T.H., 32, 201, 372
Kariv, E., 73
Karolczak, M., 3, 40–41, 43–44, 158, 197, 346
Karube, I., 191, 198, 353, 355
Kasanuki, T., 96
Kase, T., 191, 360
Kasem, K.K., 250
Katakis, I., 20, 27, 54, 59, 99–100, 191, 198, 200, 209, 281, 338, 352, 354, 363–364
Katasho, I., 92, 198, 354
Kato, K., 92, 191–192, 208, 328, 428
Kato, M., 16, 19, 96–97, 187, 293
Kato, T., 163, 165, 209
Katrlik, J., 95, 200, 208, 280, 365, 429
Katsu, T., 87, 191, 200, 365
Katsura, T., 79, 95, 97, 99, 191–192, 198, 200, 236, 325, 356, 358, 363
Katz, E., 115
Kauffmann, J.-M., 4–6, 12–13, 22, 24, 26, 40–41, 49, 55, 59, 79, 81, 93, 95, 97, 148, 188–189, 191, 193, 195–198, 200–201, 203–204, 206–207, 209, 241, 244–245, 249, 267, 275, 277, 303–307, 309, 311, 314–317, 319–323, 325, 330, 333, 339–340, 342, 344, 348, 354–355, 359, 363, 366, 373, 387, 393–394, 397, 418, 436
Kaunietis, I., 191, 200, 364
Kawahara, K., 79, 206, 412
Kawakami, M., 93, 191, 199, 362
Kawde, A.N., 26, 40, 104, 191, 205, 398, 401, 411
Kaya, G.I., 191, 205, 404
Kayahara, N., 191, 198, 353
Kayastha, A.M., 191, 199, 361
Kazemi, S.H., 96, 267
Kazemi, S.Y., 90, 261
Kazemian, H., 84–86, 229, 247, 251, 277
Kazemipour, M., 304
Ke, G., 191, 208, 427
Keita, B., 80, 103
Kennett, G.A., 163, 165, 209, 437
Kerkeni, M., 85, 222
Kerman, K., 25, 84, 93, 99, 191, 205, 297, 398, 403–409, 411, 434

Keshavarz, M.H., 90, 251
Khabazzadeh, H., 191, 203, 396
Khahawer, M.Y., 183, 277
Khajehsharifi, H., 95–96, 206, 307, 319, 416
Khaled, E., 97, 140, 143, 146, 219, 226, 231, 243, 310, 315
Khalid, E., 90
Khalilzadeh, B., 81, 84–86, 196, 203, 206–207, 343, 389, 396, 415, 420
Khalilzadeh, M.A., 31–32, 95–96, 161, 202, 290, 305, 307, 314, 384
Khaloo, S.S., 96, 191, 195, 339–340
Khan, A.H., 191, 208, 427
Khan, M.R., 89–90, 251
Khang, G.-N., 90
Khani, H., 25, 80, 88, 103, 229, 262
Khaniani, Y., 102, 216
Khayamian, T., 32, 201, 372
Khayatian, G., 90, 261
Kheiri, F., 82, 203, 206–207, 389, 420
Kheribech, A., 162
Khodari, M., 81, 89, 97, 214, 308, 313, 315, 321
Khodavesi, J., 80, 88, 203, 207, 389, 423
Khodyreva, N.A., 191, 208, 427
Khoee, S., 31–32, 86, 103, 240, 265
Khoo, S.B., 24, 88–90, 97, 103, 195, 218, 243, 246, 262, 338
Khorasani, J.H., 96, 195, 340
Khorasani-Motlagh, M., 191, 206–207, 415, 421
Khoshsafar, H., 30, 33, 35, 66, 90, 102, 221, 239, 263, 265
Khotseng, L.E., 168
Khouzami, S., 85
Kia, R., 79, 191, 195, 204, 333, 339, 399
Kiani, A., 96, 191, 203, 206–207, 209, 394, 421, 432
Kidwell, D.A., 191, 195, 337
Kiehnast, I., 3, 36, 53
Kienbaum, M., 49, 147, 149
Kilic, E., 191–192, 326
Kilinc Alpat, S., 85, 226
Kilinc, E., 79, 96, 197, 209, 333, 349, 434
Killard, A.J., 7, 51, 93, 100, 103
Kim, B.-W., 289
Kim, C.-K., 135, 140, 279–280
Kim, D.-H., 279
Kim, E.J., 25
Kim, G.-Y., 191, 197, 347
Kim, H., 90
Kim, H.-J., 191, 202, 224, 289, 375
Kim, H.S., 191, 206, 412
Kim, J.-H., 215
Kim, J.-M., 168
Kim, J.P., 161–162
Kim, J.-S., 168
Kim, K.-J., 191, 209, 430
Kim, K.-K., 191, 209, 430
Kim, K.-S., 135, 140, 280
Kim, M.H., 191, 206, 412
Kim, O.S., 185, 191, 208, 286, 427
Kim, S.-H., 90, 197, 214, 348
Kim, S.-K., 31–32, 219, 227, 237, 402
Kim, S.-S., 166
Kim, T.-H., 31–32, 168, 219
Kim, Y.-L., 208, 223, 428
Kingsley, E.D., 4, 16, 53
Kinigopoulou, V., 191, 205

Kinlen, P.J., 80, 333
Kinoshita, H., 92, 99, 191, 193, 198, 200–202, 208, 329, 352, 357, 364, 371–373, 383, 429
Kinsella, B., 210
Kirgoez, U.A., 191, 206
Kirgoz, U.A., 7–8, 14, 27, 51
Kissinger, P.T., 2, 40–41, 43–44, 123, 143–144, 158, 162–163, 165, 197, 303, 319, 346
Kitagawa, T., 56, 152–153, 158, 191, 196, 202, 291, 343, 384
Kivcak, B., 191, 205, 404
Kizek, R., 54, 58, 97, 131, 191, 197, 201–202, 205, 296, 346, 379–380, 404, 408, 411
Klapste, B., 161–162
Klejdus, B., 54, 58
Klemensiewicz, Z., 191
Klitgaard, P., 41, 363
Knott, P., 191, 209, 437
Know, H.-S., 191, 209, 430
Kobatake, E., 25
Koc, E.A., 161
Kocourkova, M., 305
Koelbl, G., 54, 76–78, 88, 90, 101, 211, 258, 267
Kohiki, Y., 191, 205, 411
Koide, S., 353
Kok, W.T., 27, 54, 96, 195, 197–198, 201–202, 339, 346, 353, 372, 382
Kokkinos, C., 24, 103–104
Kolar, M., 84, 215, 218
Kolbadinezhad, M., 95–96, 195, 201, 270, 338–339, 372, 421
Kolbl, G., 9
Kolmanovich, V.Y., 161
Komersova, A., 59, 87, 96, 199, 206–207, 249, 359, 422
Komorski-Lovric, S., 123, 132
Koncki, R., 225
Kondo, T., 92, 191–192, 194, 200, 208, 328, 335, 428–429
Kong, Y.-T., 7, 90, 95, 191, 196–197, 219, 255, 343, 349
Konvalina, J., 55, 64–65, 67, 70, 74, 143, 145–146, 183, 212, 219, 226, 231, 243, 275
Kooshki, M., 87, 203, 387
Kopanica, M., 86, 197, 230, 347
Kopytova, N.E., 162
Korbout, O., 25, 232, 241
Korbut, O., 43, 90, 185, 191, 197, 205, 287, 346, 401, 403–404
Korfhage, K.M., 79, 96, 197, 351
Korgova, E., 90, 105, 185, 194, 205, 287, 323, 335, 401
Korgova, H., 99
Kormosh, Z.O., 161
Koroleva, T.A., 210
Koroljova Skorobogatko, O.V., 24
Koryta, J., 135, 140, 186
Kosogov, A.A., 161
Kostecka, P., 97, 205, 408
Kotkar, R.M., 90, 207–208, 426–427
Kotoucek, M., 84, 172
Kotrly, S., 172
Kotzian, P., 14, 19, 21, 25–26, 40, 42–43, 49, 51, 54, 59, 79, 191, 196, 198, 209, 245, 307, 342, 354, 359, 438, 585
Kouidou, S.A., 191, 205, 406
Kouidou-Andreou, S., 191, 205
Koupparis, M.A., 87, 201, 373

Kovach, P.M., 191, 206, 209, 418, 436–437
Koya, H., 93, 191, 199, 362
Koyuncu, D., 191–192, 326
Kozan, J.V.B., 25, 185, 209, 268, 285, 430
Kozhina, G.A., 161–162
Krabisch, C., 191, 198–199, 362
Kraft, A., 28
Kramer, K.J., 54, 58, 191, 201, 205, 375, 408
Krikstopaitis, K., 99, 198, 358
Krishna, V., 97, 191, 203, 390, 392
Krizkova, J., 191, 197, 205, 346, 404
Krizkova, S., 168, 185, 191, 197, 201–202, 205, 285, 296, 305, 346, 375, 379, 404
Krolicka, A., 7, 24, 55, 63, 87, 103–104, 174, 232, 241, 245, 251
Krug, D., 191, 209, 436
Kruhak, I., 191, 196, 203, 209, 345, 394
Kruk, Z.L., 165
Kruusma, J., 29
Kryger, L., 36, 64, 89–90, 143, 145, 222, 230, 251
Krystofova, O., 168
Kuang, Y.F., 238
Kubiak, S., 188
Kubiak, W.W., 191–192, 197, 203, 209, 311, 327, 349, 394, 430, 432–435
Kubiliene, E., 185, 287
Kubota, L.T., 32, 36, 38, 80, 84, 86–88, 93, 95–97, 104, 185, 191–198, 200–201, 203–208, 247, 266, 268, 271, 282, 319, 326, 330, 333, 335, 340, 344, 347–348, 355, 366–372, 390, 393, 397, 402, 407, 425–427
Kucukoglu, O., 191, 205, 404
Kucukyavuz, Z., 99, 304–305, 313
Kudasheva, F.Kh., 185, 196, 289
Kuecuekkolbasi, S., 288
Kuhn, A., 88, 93, 200–201, 369
Kuhr, W.G., 163, 209, 436
Kula, P., 59, 63, 65, 78, 83–84, 99, 106, 172, 212, 217, 224
Kulikovskii, A.K., 161
Kulova, P., 84, 172
Kulys, J., 41, 58, 71, 92–93, 95, 99, 191, 193, 198–202, 209, 329, 331–332, 355, 358, 362–364, 369, 380, 435
Kumar, A.S., 82, 84, 97, 198, 204, 359, 399
Kumar, H., 191, 202, 382
Kumar, K.G., 316
Kumar, O., 32, 186, 294
Kumar, S.D., 90, 256
Kumar, S.S., 280
Kumaraswamy, B.E., 67, 97, 202, 384
Kumasako, H., 150
Kunath, J., 160
Kundu, S., 96, 307
Kunko, P.M., 191, 209, 436
Kunugi, C., 191, 360
Kuramitz, H., 25, 78, 99, 199, 201, 360, 365, 375
Kurauchi, Y., 150
Kurbatov, D.I., 177, 256, 262
Kureishi, Y., 7, 15, 25, 168
Kurochkin, I.N., 163, 191, 197, 201, 209, 347, 374
Kurosaki, T., 75, 92, 191, 200, 208, 364, 426, 429
Kurtinaitiene, B., 93, 191, 200, 364
Kurusu, F., 191, 198, 353
Kurzawa, C., 97, 192, 200, 326, 364
Kusuda, K., 185, 191, 208, 286, 427

Author Index

Kuta, J., 172
Kutlu, B., 208, 228, 428
Kutner, W., 191–192, 325
Kuwana, T., 1–2, 4, 24, 26, 73, 151, 153, 158, 162, 191, 195–196, 199, 201, 340, 342, 373, xv
Kuzmina, N.V., 185, 196, 289
Kwak, K.J., 191, 206, 209, 412, 437
Kwak, M.-K., 90, 274
Kwon, H.-S., 191, 209, 430
Kwon, O.M., 191, 206, 209, 412, 437
Kwon, S.-J., 168

L

Labar, C., 214
Laborda, F., 248
Labuda, J., 43, 66, 75, 79, 84, 87, 90, 96, 99, 105, 185–186, 188, 194, 197, 205–207, 287, 291, 310, 319, 323, 335–336, 346, 401–404, 420
Lacerda, V., 191, 205, 398, 402
Lacoste, R.J., 162
Ladanyi, L., 152, 158
Ladiu, C.I., 191, 200–201, 370
Laganovsky, A.V., 161
Lai, J.S., 79, 359
Lai, R., 560
Lakshmeinarayan, V., 67, 202, 384
Lakshminarayanaiah, N., 135, 140, 186
Lakshminarayanan, V., 89
Lal, B., 191, 195, 341
Lamache, M., 160–161
Lamas Ardisana, P.J., 49
Lamberts, L., 71, 84, 86, 99, 101–102, 106, 161, 181, 214, 225, 260–261, 297, 301
Lamm, C.G., 135, 140
Lan, H.-L., 80, 92–93, 96, 99–100, 191–192, 197–200, 209, 327, 351, 355–356, 362–364, 369, 438
Lan, M.-B., 80, 87, 191, 198, 204, 357, 398, 406
Lan, Y.-Z., 64, 81–82, 185, 281, 285
Landers, R., 87
Landsberg, R., 3
Lane, R.F., 81, 163, 165, 188, 205, 209, 437, 596
Lanfredi, S., 261
Lange, B., 161–162
Lange, M.A., 191, 204, 398, 401
Langer, J., 191, 204, 398–399
Lan,Y., 224, 277
Lanza, M.R.V., 191, 202, 381
Lanzcyncki, C.J., 78, 84, 93
Laocharoensuk, R., 96, 272
Lapidus, G.T., 161
Lappin, A.G., 191, 202, 380
Lara Castro, R.H., 150
Laranjeira, M.C.M., 99
Larcon Angeles, G., 97, 203, 387
Larijani, B., 31–32, 35, 180, 264
Larson, D.D., 90, 179, 256
Larsson, N., 95, 191–192, 196, 198, 327, 342, 357
Laudet, A., 4, 24, 55, 245
Laurell, T., 99, 363
Laurinavicius, V., 93, 191, 200, 364
Lavagnini, I., 30, 32, 63, 191–192, 200–201, 309–310, 318, 321, 326, 368
Lawrance, G.A., 87, 406

Lawrence, N.S., 32, 191, 198, 357
Lawrence, R.J., 46–47
Lawson, F., 161–162
Lazar, P., 191, 387
Lazarin, A.M., 89, 191, 200–201, 206–207, 369, 425
Lebeau, B., 102, 172, 229, 238
Lecuire, J-M., 160
Leddy, J., 26, 104
Ledru, S., 191, 193, 206–207, 331, 425
Lee, B.-G., 26, 191, 209, 270, 430
Lee, B.-W., 31–32, 165, 298, 403, 437
Lee, C.-G., 191, 209, 430
Lee, C.-H., 31–32, 165, 191, 206, 209, 298, 318, 402–403, 405, 412, 437
Lee, C.-P., 168
Lee, E., 99–100
Lee, G.-J., 99
Lee, H.-K., 32, 227, 237, 402
Lee, H.-S., 79–80, 92, 96, 100, 191, 193, 197–200, 329, 351, 354–355, 362–363
Lee, H.Y., 2, 36, 38, 207
Lee, I.C., 89, 99, 215
Lee, J.E., 191, 206, 209, 412, 437
Lee, J.-H., 168
Lee, J.K., 17–18, 37, 61, 69–70, 99
Lee, J.-S., 26
Lee, K.-S., 41, 279, 309, 311, 318, 322
Lee, S.-E., 168
Lee, S.-H., 251
Lee, S.M., 31–32, 219
Lee, S.-S., 191, 194, 215, 335
Lee, T.-Y., 191, 202, 375
Lee, Y.-K., 135, 140, 274, 279–280
Leech, D., 59, 75, 87, 99–100, 192, 277, 317–318, 322, 325
Leedy, D.W., 151–152
Lefèvre, G., 64, 67, 84, 87, 275
Legin, A., 166
Legorburu, M.J., 306, 324, 435
Lei, C.-X., 25, 68, 88, 99
Lei, Y., 34–35, 185, 197, 208, 289, 298, 300–301, 346, 429
Leinen, D., 150
Leite, O.D., 191, 202, 209, 384, 430–431
Lelasattarathkul, T., 25
Lel'kin, K.P., 161
Lendermann, B., 262
Leng, Z.-Z., 191, 203, 206, 223, 249, 254, 303–304, 311, 395, 418
Lepore, M., 95, 284
Lepri, L., 42, 158
Lesunova, R.P., 191, 208, 427
Lete, G., 80
Lewenstam, A., 75
Lewis, C.A., 191, 202, 380
Leyva, A.G., 87, 93, 199, 269, 359
Li, B., 99
Li, B.-F., 97, 174, 214, 231, 437
Li, B.H.L., 99, 246
Li, C., 67, 97, 99, 308, 311–312
Li, C.-M., 34–35
Li, C.-W., 191, 197, 347
Li, C.-X., 42, 191, 202, 209, 383, 433
Li, C.-Y., 101, 224, 268
Li, F.-M., 251

Li, G., 8, 13, 54, 84, 103, 191, 202–203, 235–236, 239, 293, 378, 390
Li, G.-J., 447
Li, G.-X., 281
Li, G.-Z., 27, 29, 35
Li, H., 99, 191, 197, 208, 212, 347, 427
Li, H.-B., 268
Li, H.-J., 15, 32
Li, H.L., 99, 106, 218, 258
Li, H.-P., 97, 211
Li, J., 8, 20, 31–32, 34–35, 54, 63–64, 66, 78, 102, 163, 165, 191, 202, 204–206, 209, 286, 320, 383, 398, 401, 406, 414, 417, 433, 436, 447
Li, J.-H., 8, 20, 32–35, 54, 63, 66, 82
Li, J.-N., 30–32, 64, 79–80, 82, 84, 86, 90, 95, 97, 99, 103, 163, 180, 185, 228, 262–265, 284, 333, 358, 378
Li, J.-P., 80–81, 103, 185, 197, 269, 283, 346, 348
Li, L.X., 99, 101–102, 207, 258, 268, 276, 423
Li, N., 82–83
Li, P., 89–90, 99, 161, 190–192, 209, 211, 213, 222, 249–250, 327
Li, Q.-X., 191, 202, 288, 382, 447
Li, R., 49, 79, 90, 95, 149, 198, 209, 244, 355, 435
Li, S., 27, 88, 93, 95, 99, 103, 199, 360, 406
Li, T., 191, 203, 391
Li, W., 80, 101, 203, 276, 389
Li, W.-D., 310
Li, W.-Y., 239
Li, X., 32–35, 63, 66, 71, 80–81, 96, 100–101, 131, 149, 185, 191, 198, 202, 205, 288, 333, 356, 376–377, 379, 398, 403, 408, 411, 447
Li, X.-B., 168
Li, X.-F., 251
Li, X.-L., 185, 284, 366
Li, X.-M., 191, 206–207, 421
Li, X.-Q., 34–35, 80, 99, 103, 202, 377
Li, Y., 34–35, 63–64, 66, 95, 191, 193, 202, 204–207, 209, 246, 329, 377, 398–400, 409, 419, 421, 433, 435
Li, Y.-F., 19, 24, 77–78, 80, 90, 103, 202, 231, 258, 308, 310
Li, Y.H., 64, 244, 253, 262–263
Li, Y.-H., 35, 63–64, 89, 215, 239, 244, 246
Li, Y.-J., 191, 206–207, 421
Li, Y.-P., 191, 202, 377
Li, Y.-X., 191, 200, 365
Li, Y.-Z., 35, 66, 204
Li, Z., 8, 63, 191, 203, 386
Li, Z.-F., 6, 80
Li, Z.-Y., 8, 20, 27, 32–35, 54, 63, 66, 82, 204, 398
Lian, H., 80
Liang, B., 161, 191, 203, 206–207, 387, 414, 418
Liang, C., 79, 96, 185, 285, 333
Liang, D., 35, 82, 195, 338
Liang, H., 191, 196, 209, 345, 431
Liang, R.P., 80, 103, 149, 198, 203, 356, 389
Liang, Y., 168
Liao, H.-C., 191, 206, 413
Liao, S.F., 30, 32, 64, 214
Liauksminas, V., 93, 191, 200, 364
Liawruangrath, B., 209, 274, 430
Liawruangrath, S., 25, 209, 274, 430
Liawrungrath, S., 191, 193, 209, 329, 433
Liden, H., 24, 191–192, 200, 327, 365
Ligaj, M., 191, 204–205, 398–399, 402, 408–410

Lima, A.S., 87, 261
Lima, A.W.O., 25, 185, 191, 196, 203, 209, 233, 268, 285, 345, 395, 430–431
Lima, E.C., 191, 200–201, 206–207, 370, 425
Lima, P.R., 95, 104, 191, 195, 200–201, 206–207, 340, 372, 415, 424
Limoges, B., 96, 99, 319–320
Lin, B.-Z., 167
Lin, H., 80–82, 84, 101–102, 197, 293, 348
Lin, H.-G., 97, 214, 291, 437
Lin, H.-Y., 167, 205, 437
Lin, J.-M., 35, 80, 88, 103
Lin, L., 191, 200, 205, 366, 404
Lin, M.-S., 24–25, 79, 95, 191, 194, 203, 209, 309, 335, 359, 394, 430, 432, 434
Lin, S., 101
Lin, X., 191, 200, 205, 366, 404
Lin, X.-C., 563
Lin, X.-H., 32, 66, 306
Lin, X.-Q., 31–32, 197, 215, 349
Lin, Y.-H., 7, 75, 87–88, 101–103, 170, 179, 191, 194, 197, 199–202, 219, 233, 256, 268, 300, 336, 352, 359, 367, 378
Lin, Z., 31–32, 35, 41, 43, 87, 163, 182, 191, 203, 207, 209, 295, 297, 394–395, 399, 426, 432, 437
Lin, Z.-H., 223
Lin, Z.-Y., 31, 150, 286, 363, 412
Linden, H., 191, 195, 209, 337, 341
Linders, C.R., 6, 79, 96, 191, 195–196, 200, 333, 339, 342, 366
Lindgren, A., 191, 197, 348
Lindquist, J., 3, 6, 13, 19, 26, 38, 55, 61, 158, 165, 191, 206–208, 279, 412, 418, 426–427
Ling, C., 35, 313
Linkov, V., 168
Linquette-Maillet, S., 99, 198, 356
Linquette-Mailley, S., 25, 196, 209, 345, 433
Lins da Silva, V., 25, 93, 393, 431
Lintschinger, J., 88–89, 217, 271
Lisdat, F., 88, 202, 375
Lisichkina, Z.I., 230
Liu, A., 161–162
Liu, B., 32, 191, 206–207, 424
Liu, C., 191, 208, 427
Liu, C.-B., 27
Liu, C.-Y., 191, 206–207, 421
Liu, D., 104, 163, 191, 197, 203, 209, 346, 385
Liu, F., 80, 103, 149, 191, 203, 205, 389, 409
Liu, G., 81, 88, 101–103, 191, 199, 202, 359, 378
Liu, G.-C., 167, 205, 437
Liu, G.-D., 170, 174, 231, 300
Liu, H., 6, 19, 24, 63, 77–78, 80–81, 87, 90, 92–93, 103, 191, 205–207, 258, 409, 411, 424, 426
Liu, H.-L., 231, 262
Liu, H.-N., 27
Liu, H.S., 207, 423
Liu, H.-T., 8, 20, 32–35, 54, 63, 66, 82
Liu, H.-Y., 86, 93
Liu, J., 25, 81, 86–87, 99, 191, 199–201, 209, 326, 330, 336, 360–361, 363, 374, 435
Liu, K., 89, 224, 231
Liu, K.-E., 81, 93, 99, 185, 193–194, 285, 329, 334
Liu, K.-Z., 258

Author Index

Liu, L., 191, 195, 198, 202, 206, 311, 314–315, 342, 353, 383, 398–400, 414, 417
Liu, L.-F., 92–93, 191, 198–199, 202, 355, 362, 384
Liu, L.-Q., 35, 300
Liu, N., 64, 226, 262, 304
Liu, P.-D., 167
Liu, Q., 82–83, 149, 206
Liu, R., 104, 196–197, 342, 346
Liu, S., 35, 63–64, 82, 191, 195, 204, 206–207, 338, 420
Liu, S.-F., 35, 80–81, 103, 205, 377–378, 409
Liu, S.-M., 180, 262–265
Liu, S.-Q., 80, 88, 97, 163, 197, 202, 263, 278, 348, 375
Liu, W., 191, 197, 347
Liu, X., 34–35, 84, 87, 102, 185, 197–198, 282, 286, 347, 361
Liu, X.-H., 447
Liu, X.M., 80, 95, 99, 333
Liu, X.-Q., 15, 32
Liu, X.-T., 35, 63, 239
Liu, X.-Y., 19, 24, 35, 63, 239, 283, 308
Liu, Y., 8, 16, 20, 32–35, 54, 63, 66, 80, 82, 103, 191–193, 195, 198, 203, 206–207, 239, 283, 325, 332, 337–338, 373, 385, 389, 414–415, 419, 423, 426
Liu, Y.-L., 80, 103, 202
Liu, Y.-Y., 191, 208, 212, 428
Liu, Z., 35, 80, 101, 103, 166, 185, 191, 205, 229, 240, 288, 409
Liu, Z.-M., 80, 103, 202
Liu, Z.-Q., 239
Lobanov, V.V., 191, 194, 335
Lobo Castanon, J.M., 292
Lobo Castanon, M.J., 9, 59, 75, 93, 97, 99–100, 116, 192–193, 195, 200–201, 326, 328, 330, 341, 367, 369, 371, 373
Lobo, M.J., 93, 192, 326
Loh, A., 191, 196, 342
Lokesh, S.V., 386
Long, D.-W., 278
Longchamp, S., 191, 200–201, 363, 374
Lopes da Silva, W.T., 25, 99, 149, 225
Lopez, D., 87, 197, 199, 203, 349, 359, 393
Lopez de Mishima, B.A., 168
Lopez Fonseca, J.M., 99, 192, 195, 200–201, 326, 330, 369
Lopez, J.R, 281
Lopez, M.A.B., 307
Lopez Palacios, J., 97, 316, 323
Lopez Paz, J.L., 7, 42, 191, 205, 398, 401
Lopez Ruiz, B., 323
Lopis, E.L., 84
Lorencetti, L.L., 101, 206–207, 420
Lorenzo, E., 78, 84–85, 96, 101, 185, 191, 198, 206, 288, 306, 312, 326, 417
Lori, J.A., 100
Losada, J., 96, 191, 198, 355, 357
Lotrean, S., 93, 191, 206–207, 319, 321, 419
Louis, C., 84, 86, 181, 260–261
Lourencao, B.C., 99, 220
Lovric, M., 64, 104, 123, 132
Lowe, J.A., 191, 206–207, 417, 424
Lowry, J.P., 44, 99–100, 162–163, 165, 191, 209, 436–437, 604
Lozano, M.L., 29–30, 32, 166, 198, 354, 358
Lu, D.-Q., 191, 206–207, 421
Lu, F., 18, 44, 87, 99, 185, 191, 197, 199–200, 203, 209, 282, 349, 360, 363, 393
Lu, G., 64, 82, 224, 277, 281
Lu, G.-H., 86, 97, 149, 191, 202, 206–207, 274, 278, 376, 379, 421
Lu, G.-X., 191, 198, 202, 353, 376
Lu, J.-M., 7, 24, 84, 90, 103, 131, 179, 206–207, 256, 420
Lu, N., 191, 203, 386
Lu, Q., 191, 202, 376
Lu, R.-S., 67, 211, 308–309, 312, 319, 323
Lu, X.-Q., 70, 80, 89–90, 225, 400, 447
Lu, Y., 191–192, 197, 209, 327, 346
Lu, Z.-S., 34–35
Lu, Z.Y., 161–162
Luan, X.-J., 30, 32, 315
Lubert, K.H., 49, 63, 89, 103, 112, 149, 192, 228
Luccas, P.O., 191, 203, 397, 433, 436
Lucho, A.M.S., 86, 101, 193, 333
Ludvik, J., 19–20, 29, 32, 51, 62, 250, 364
Luna Sanchez, R.M., 161
Lundstroem, K., 4, 18, 56–57, 191, 208, 427
Lundstrom, K., 191, 208, 427
Lunev, M.I., 230
Lunte, S.M., 41, 96, 195, 201, 337, 339, 372–373
Luo, D.-B., 77, 93, 99–100, 143, 146, 191, 196, 200, 205, 319, 344, 363, 401–402, 407–408, 413
Luo, J., 93, 100, 206, 412
Luo, L.-Q., 87, 198, 288, 361, 447
Luo, M., 101
Luo, S.-L., 27, 34–35, 63, 80, 203, 239, 283, 389
Luo, W., 93
Luong, J.H.T., 28
Lupetti, K.O., 20, 25, 191, 203, 209, 303, 396, 430
Lupu, A., 278
Lupu, D., 357, 398
Luque de Castro, M.D., 40, 42, 96, 273
Luque, G.L., 30–32, 63, 66, 87, 93, 191, 193–195, 198–199, 201, 203–204, 269, 331, 336, 341, 353, 359, 361, 374, 393, 400
Luthi, N., 101–102
Lutz, M., 24, 84, 201
Luz, R.d.C.S., 101, 149, 191, 195, 340
Lvova, L., 166
Ly, S.-Y., 30–32, 161, 163, 165, 191, 203, 205–206, 209, 219, 227, 236–237, 298, 318, 393, 402–403, 405, 412, 437
Lyczkowska, E., 104, 251
Lyne, P.D., 81, 97, 163, 209, 436–437
Lyons, C., 112, 162, 191
Lyons, M.E.G., 87, 198, 353

M

Ma, G.-H., 149
Ma, H.-Y., 42, 101
Ma, J., 191, 197, 346
Ma, L.Z., 238
Ma, M.-M., 293
Ma, S.-Y., 447
Ma, W., 191, 206–207, 417, 424
Ma, Y.-J., 168, 191, 204, 398
Macedo, D.V., 87, 193, 330
MacFarlane, D.R., 32
Macha, S., 83

Machado, S.A.S., 100, 299
Macholan, L., 191, 194, 334
Macias, J.M.P., 78, 89, 99, 230
Macours, P., 241, 275
MacPherson, J.V., 29
Madrid, E., 79, 96, 99, 295, 333
Maeda, M., 92, 192, 208, 328, 428
Maekawa, K., 185, 286
Maestre, E., 191, 198, 352
Maestroni, B., 87, 195, 336
Mafatle, T.J., 79, 96, 195, 333, 339
Magee, R.J., 76–78, 90, 99, 211, 217, 242, 245
Maghooli, R., 304
Magno, F., 32, 63, 191, 200–201, 368
Magosso, H.A., 101, 149, 191, 193, 333
Mahanthesha, K.R., 95, 97, 185, 203, 284, 391
Mahdi, I.M., 90
Mahesh, H.M., 386
Mahyari, F.A., 191, 201, 371
Maicaneanu, A., 84–86, 191, 200–201, 270, 369, 371
Maidment, N.T., 165
Mailley, P., 25, 100, 196–197, 209, 345, 348, 433
Mailley, S.C., 100, 197, 348
Maistrenko, V.N., 69–70, 196, 289
Maj Zurawska, M., 36, 42
Majdi, S., 80, 103, 194, 335
Majid, S., 100, 233
Majidian, M., 319
Makhaeva, G.F., 163, 191, 197, 201, 209, 347, 374
Makower, A., 24, 191, 197, 201, 347, 374
Maksimoviene, A., 93, 193, 331
Malachesky, P.A., 3
Malakhova, N.A., 78
Male, K.B., 28
Maleki, N., 33–35, 49, 66, 89, 191, 195, 197, 200–201, 206–207, 224, 280, 337, 346, 359, 366–367, 371, 385, 414, 419
Malekian, M., 186, 237
Malfoy, B., 191, 374
Malhotra, B.D., 100
Malhulikar, P.P., 100
Malinauskas, A., 79, 82, 87–89, 93, 191, 193, 200–201, 208–209, 225, 332, 369, 371, 427, 435
Mallipattu, S., 296
Malongo, T.K., 241, 275
Malygin, V.V., 191, 197, 201, 347, 374
Mamatha, G.P., 95, 97, 191, 203, 388, 392
MamKhezri, H., 191, 203, 206–207, 395, 414, 419
Mang, H.-J., 279
Mangelizadeh, S., 79, 99–100, 192, 328
Manisankar, P., 92, 267
Manjunatha, J.G., 97, 191, 202–203, 206, 386–388, 392, 414, 418
Mannino, S., 87, 195, 197, 203, 338, 351, 389
Mano, N., 88, 93, 200–201, 369
Manso, J., 80, 82, 103, 195, 206, 342, 412
Mansour, H., 97, 313
Manzo-Robledo, A., 85
Mao, L.-Q., 29
Mao, Q., 161–162
Mao, X.-Y., 64, 88, 103, 180, 185, 200, 202, 206, 263, 265, 284, 367, 378, 385, 413
Marchal,V., 84, 168
Marchant, R., 191, 194, 333

Marcolino-Junior, L.H., 31, 82, 89, 99, 208, 220, 227, 229, 303, 310, 314, 426
Marcoux, L.S., 3–4, 17, 52
Marcovicz, C., 20, 87, 203, 206, 391, 421
Marcus, M.F., 3
Mariame, C., 38–39, 229
Marian, I.O., 191, 206, 317, 419
Mariaulle, P., 84, 86, 99, 106, 181, 225, 260–261
Marino, G., 96, 98, 101–102, 172, 221, 227, 232, 236, 426
Marinovich, Y., 162
Marjani, K., 239
Marken, F., 28–29, 32, 159, 162, 202, 379
Markitan, O.V., 191, 194, 335
Marko-Varga, G., 13, 24, 84, 99, 185, 191–192, 195, 197, 200–201, 209, 282, 325, 327, 337, 341, 348–349, 351, 363, 365
Marques, C.M.P., 84, 86, 202, 384
Marques, E.P., 191, 325
Marrocco, R.T., 163, 165, 209
Marsden, C.A., 44, 98, 158, 163, 165
Marsh, J.H., 54, 59
Martin, C.R., 12, 45
Martin Pernia, A., 49
Martin, R.S., 41, 195, 201, 337, 373
Martinez, J., 166
Martinez, M.T., 87
Martinez, R., 162
Martinez, T., 49, 51, 84–85, 147, 149, 191, 208, 213, 249, 428
Martinez-Fabregas, E., 191, 198, 353
Martinez-Montequin, S., 191, 201–202, 209, 373, 381, 430, 433
Martins, A., 90, 206–207, 424
Marty, J.L., 188, 299, 302
Maruyama, K., 15, 97, 205, 405
Mas, M., 191, 209, 437, 604
Masarik, M., 54, 58, 97, 191, 201–202, 205, 380, 408, 411
Mascarenhas, R.J., 67, 103, 202, 236, 242, 384, 386
Mascini, M., 191, 205, 316, 405, 409
Mashhadizadeh, M.H.A., 75, 80, 85–86, 88–89, 103, 216, 220, 227, 229, 237, 247, 262, 308, 320
Mashkina, S.V., 19, 24, 259
Masini, J., 291
Mason, P.A., 191, 209, 436
Masoudipour, N., 86, 281
Matakova, R.N., 79, 162
Mataveli, L.R.V., 191, 203, 397, 433, 436
Mateo, P., 87
Mather, B.R., 3, 51
Matos, J.D., 172, 221, 227
Matousek, M., 63, 218
Matsubara, Y., 191, 205, 407
Matsuda, H., 135
Matsui, H., 25, 99, 215
Matsui, J., 191, 194, 335
Matsumoto, R., 95, 198, 357
Matsushita, F., 92
Matsushita, K., 92, 192, 208, 328, 428
Mattax, C.C., 143
Mattiasson, B., 80, 191, 193, 330
Mattoso, L.H.C., 100, 295, 297–299
Matuszewski, W., 5, 64, 116, 143, 146, 191, 198, 252, 353
Matysik, F.M., 135
Maystrenko, V.N., 185, 289

Author Index

Mazloum Ardakani, M., 32, 80, 85, 93, 95, 103, 191, 194–196, 200–201, 203–204, 206–207, 229, 277, 319, 335, 340, 343, 368, 390–392, 396–397, 399, 415, 423
Mazlum, M., 89, 220
Mbomekalle, I.M., 80, 103
McAdams, E.T., 25, 100, 196–197, 209, 345, 348, 433
McAllister, D.L., 10
McClurkin, J.W., 163, 165, 209
McCormick, L., 49, 51, 147, 149
McCreery, R.L., 3, 10, 44, 158, 162–163
McFadden, S., 25, 196, 209, 345, 433
McGrath, C., 191, 197, 348
McHugh, S.B., 163, 165, 209, 436, 447
McKervey, M.A., 90
McKervey, S.J., 90
McQuillan, A.J., 161–162
Mebsout, F., 307, 314–315, 322
Mecklenburg, M., 185, 205, 286, 405
Medina, J., 161–162
Medvedeva, E.P., 161
Medyantseva, E.P., 4, 9, 19, 24, 67, 96, 187–188, 259, 295–296, 301
Meghdadi, S., 97, 203, 389
Mehrgardi, M.A., 79, 82, 204, 333, 399
Meier, E.P., 3, 66
Meister, A., 79, 87, 96, 194, 336
Meittes, L., 10
Mekhalif, Z., 229
Melguizo, J.M., 99, 174, 230
Melicharek, M., 151–152
Melvydas, V., 93, 191, 193, 209, 332, 435
Memon, K.Z., 9, 161–162
Memon, M., 9, 161–162
Mendez, S., 99–100, 198, 355
Mendiola, M.A., 89–90, 219, 225
Meneghatti, R., 191, 196, 206, 344, 418
Meng, F.S., 149, 191, 197, 348
Meng, L., 89, 224, 231
Meng, S.-L., 203, 209, 394, 432
Meric, B., 25, 84, 93, 99, 191, 205, 297, 398, 403–409, 434
Merino, M., 191, 375
Merisalu, M., 29
Merkoci, A., 243, 277
Merkoci, M.E., 191, 203, 388
Mesaric, S., 26, 36, 135, 140, 213, 274
Meskys, R., 93, 191, 200, 364
Metelka, A., 7, 24, 55, 63, 87, 103, 174, 232, 241
Metelka, R., 6, 9–10, 14, 16, 19, 21, 24–27, 39–40, 42–43, 45, 49–51, 53–55, 59, 61, 63, 65, 87, 103–104, 135, 140–143, 146–149, 151, 174, 191, 205, 231–234, 241, 243, 245, 247, 252, 266, 406
Meyer, A., 86, 101, 289, 299
Meyer, M., 238
Meyer, S., 161–162
Meyer, T.J., 191–192, 325
Miao, W., 86
Miao, Y.-Q., 7, 32, 99, 198, 356, 411
Michael, A.C., 191, 209, 436–437
Michalska, A., 36, 42
Michotte, Y., 6, 26, 41, 203, 394
Micke, G.A., 191, 350
Micor, J.R.L., 25, 209, 235, 237, 430
Middleton, A., 210

Midorikawa, M., 247
Miele, M., 99–100, 165, 191, 209, 437
Miertus, S., 95, 200, 208, 280, 365, 429
Mihailescu, G., 357, 398
Mikelova, R., 191, 196–197, 205, 296, 345–346, 404
Miki, A., 7, 25, 149, 168
Miki, K., 63, 79, 92, 191, 193, 198, 200–201, 329, 331, 354, 371
Mikkelsen, S.R., 79, 96–97, 100, 198, 205, 356, 407
Miko, R., 90, 323
Mikysek, T., 19–20, 29, 32, 51, 62–63, 99, 175, 236, 238, 243, 250, 364
Milagres, B.G., 80, 86–87, 93, 96, 194, 198, 200–201, 333, 355, 370
Miland, E., 99, 206, 416
Milani Hosseini, M.R., 96, 267
Millan, K.M., 97, 205, 407
Miller, A.D., 191, 209, 437
Miller, L.L., 73
Millot, M.-C., 191, 205, 404
Min, K.-W., 168
Min Zhou, D., 191, 196, 209, 345, 433
Minaei, E., 95
Minagawa, K., 97, 205, 405
Miner, D.J., 303, 319
Ming,L., 361
Mira, T., 191, 201, 372
Miranda Hernandez, A., 8, 15, 63, 149
Miranda, O., 99, 308
Miranda-Ordieres, A.J., 9, 59, 75, 79, 93, 95, 97, 99–100, 116, 185, 188, 191–193, 195, 198, 200–201, 206, 279, 287, 308, 312, 315, 321, 326, 328, 330, 341, 356, 363, 367, 369, 371, 373, 416
Mirceski, V., 123, 132
Mirel, S.M., 6, 93, 150, 206–207, 303, 317, 319, 321, 419
Mirel, V., 191, 206–207, 419
Mirela Apetrei, I., 447
Mirica, M.C., 191, 202, 382
Mirica, N., 191, 202, 382
Mirjalili, B.B.F., 80, 103, 203, 391
Mirkhani, V., 59, 422
Mirmomtaz, E., 95
Miroshnikova, E.G., 162
Mirzajani, R., 32, 320
Mirzakhani, S., 84, 86, 196, 203, 343, 396
Miscoria, S.A., 63, 66, 80, 88, 103, 193, 198–199, 203–204, 331, 353, 361, 393, 400
Mishima, H.T., 168
Mishima, Y., 97, 205, 405
Mishra, R., 191, 193, 333
Misrahy, G., 162
Mita, D.G., 95, 284
Mita,L., 95, 284
Mitchell, K., 161–162, 165
Mitoseriu, L.C., 135
Mitra, C.K., 14, 191, 198–199, 353, 362
Miyasita, T., 99, 255
Mizutani, F., 79, 93, 95–97, 99–100, 191–193, 198, 200, 236, 325, 327, 331, 356–358, 363
Mo, J., 191, 195, 341
Mo, J.-W., 7–8, 14, 27, 51, 93, 95, 99, 191, 199, 201, 360, 373, 607
Mo, J.-Y., 191, 196–197, 343, 347, 560
Moane, S., 99, 308

Mocak, J., 163
Mochizuki, M., 95, 198, 357
Mochon, M.C., 307
Moderegger, H., 79, 87, 191, 196, 199, 209, 342, 359, 438
Moeketsi, B., 308
Moghimi, A., 274
Mohadesi, A., 89, 216, 227
Mohamadnia, F., 102, 250
Mohamed, G.G., 321
Mohamed, M., 310
Mohamed, N.A.L., 305
Mohammadi, A., 304
Mohammadi, L., 239
Mohammadpoor-Baltork, I., 80, 96, 333
Mohammedi, H., 204, 398
Mohan, S., 96, 307
Mohandes, P.N., 162
Mohseni, M.A.S., 80, 93, 103, 203, 396
Mojica, E.-R.E., 25, 209, 235, 237, 430
Molle, L., 191, 204, 397
Momeni, A., 89, 220
Momeni, S., 35, 280
Mondovi, B., 43, 194, 335
Monfared, F., 166, 185, 285
Monfared, H.H., 272
Monien, H., 37–38, 70, 210, 213, 222, 244, 249, 253, 255, 262
Monk, P.M.S., 123, 133, 205, 402
Monroy, M., 150
Montalvo, J.G., 191
Montazerozohori, M., 277
Montenegro, M.C.B.S., 25, 43–44, 86, 93, 101, 206, 334, 393, 416, 431
Moody, G.J., 79, 198, 354
Moon, D.W., 89, 172, 214, 218
Moon, J.-O., 191, 202, 375
Moon, S.-H., 191, 197, 347
Moosavi Movahedi, A.A., 80, 88, 103, 194, 304, 312, 321, 336
Moosavi, S.M.M., 89, 225
Morabito, R., 90, 245, 248
Moradlou, O., 66, 191, 200–201, 371
Moraes, F.C., 67, 82, 87, 97, 185, 208, 260, 289, 300, 303, 310, 426
Moran, M., 96, 191, 198, 355, 357
Morante Zarcero, S., 102, 240
Moreau, M., 238
Moreda, J.M., 86, 99, 101, 150, 296, 299, 311, 320
Moreira, J.C., 101, 197, 217, 323, 350
Moreira, K.A., 43–44, 206, 416
Moreira, T.F., 97, 203, 206–207, 391, 422
Moreno, M., 31–32, 79, 88, 96, 99, 295, 299, 333
Moreno, R.A.S., 254
Moressi, M.B., 430
Moretto, L.M., 75
Morgan, C., 99
Morgan, I.C., 191, 208, 426
Morisaki, H., 191, 209, 435
Morita, Y., 191, 205, 407, 411
Moro, C.C., 87, 149, 206–207, 425
Moros, S.A., 320
Morrin, A., 51, 93, 100, 103
Morsali, A., 203, 206, 390, 421
Mortensen, J., 143, 145

Mosbach, M., 88, 93, 208, 427
Mosbach, S., 80, 209, 438
Moscone, D., 19, 29–31, 63, 79, 82, 95, 184, 191, 193, 195–196, 198, 202, 278, 281, 284, 331, 341, 344, 358, 385
Moskzin, L.N., 161
Mostafa, G.A.E., 88, 216, 219, 246, 312
Motakef-Kazemi, N., 31–32, 86, 103, 240, 265
Motonaka, J., 97, 205, 405
Motta, N., 92, 104, 362
Mottola, H.A., 4, 40–41, 49, 59, 96–97, 147, 149, 163, 183, 197, 200–201, 234, 270, 273, 279, 349, 366
Mouhandess, M.T., 160–162
Mount, A.R., 75
Mounts, R.D., 191, 204, 398, 401
Mousavi, M.F., 85, 89, 92, 162, 206, 231, 253, 425
Mousavi, S.R., 79, 96, 206, 333, 413
Mousty, C., 83–84, 149–150
Moutet, J.C., 75
Mowat, M., 191, 205, 407
Moya Moreno, M., 87, 162
Mozhanova, A.A., 80, 88, 103, 149, 193, 332
Muck, A., 8, 14, 17, 40–41, 194, 335
Mueller, K., 99, 163, 165, 191, 209, 436–437
Mueller, T.R., 28
Mulazimoglu, I.E., 185, 197, 282, 347
Mulcahy, D.E., 135, 140
Mulchahey, J.J., 191, 209, 437
Mulchandani, A.K., 81, 93, 195, 208, 289, 298, 300–301, 331, 341, 429
Mulchandani, P., 185, 197, 208, 289, 298, 300–301, 346, 429
Mulik, E., 36, 42
Muller, L., 191, 196, 345
Mullet, M., 150
Mullor, S.G., 75, 93, 192, 206–207, 326, 418
Munden, R., 16, 96, 322
Munguia, A., 97, 316
Munoz Leyva, J.A., 84–85, 185, 288–289, 292
Munson, J.W., 191, 206, 413
Munteanu, F.D., 88, 92–93, 95, 101, 200–201, 208, 368–370, 427
Murali Dhar, G., 87
Muresan, L.M., 84–86, 162, 191, 200–201, 206–207, 269–270, 369, 371, 425
Murr, N.E., 85, 222
Murray, R.W., 73, 96–99, 191–192, 325
Murrill, E., 3, 158
Murthy, A.S.N., 93, 206–207, 424
Musameh, M.M., 34–35, 99, 191, 199, 201, 273, 360, 367, 373
Musha, S., 3, 5, 192–193, 326, 330
Musial, W.G., 191, 205, 408, 410
Musilova, J., 65, 97, 135, 140, 142, 276
Mustafina, G.F., 162
Muthuraman, G., 80, 88, 226
Muxel, A.A., 271
Muzaferovic, S., 79, 333

N

Nacimi, H., 93, 203, 391
Nader, P.A., 279
Naderi, P., 82–83, 272

Author Index

Nadjo, L., 80, 103
Naeimi, H., 32, 80, 93, 103, 191, 194–196, 203–204, 206–207, 319, 335, 340, 343, 390, 392, 396–397, 415–416, 423
Nagaoka, T., 100, 196, 343
Nagaraj, C., 100, 203, 206–207, 392, 424
Nagasaka, H., 191, 198, 355
Nagasaki, S., 271
Nagasawa, T., 75, 92, 208, 426, 429
Nagashree, K.L., 166, 192, 328
Nagata, N., 20, 87, 97, 203, 206–207, 391, 421–422
Naggar, A.H., 316
Nagy, A., 191, 197, 203, 348, 393
Nagy, G., 52, 75, 80, 193, 330
Nagy, L., 75
Naik, H.S.B., 191, 205, 404
Naik, R.R., 79, 95–97, 188, 191, 203, 206–207, 294, 333, 388, 390, 415, 421
Naik, T.R.R., 191, 205, 404
Najafi, E., 447
Najafi, M., 38, 96, 267
Najih, R., 84, 88, 257, 297
Nakabayashi, Y., 95, 198, 202, 355, 357, 382
Nakagaki, S., 20, 84, 86–87, 96, 101, 203, 206, 271, 391, 421
Nakamura, H., 96, 198, 351
Nakamura, T., 78, 99, 249, 253
Nakamura, Y., 270
Nalini, B., 97, 195, 201, 206–207, 338, 371, 422
Nam, H., 26, 166, 191, 203, 391
Namazian, M., 95, 191, 200–201, 368
Namboothiri, I.N., 67
Nandibewoor, S.T., 67, 206, 305, 413
Naranjo Rodriguez, I., 84–85, 90, 185, 288–289, 292, 298
Narasaiah, D., 24, 99, 191, 193, 330–331
Narayanan, A., 3, 36, 53, 257, 261
Narayanan, S.S., 93, 97, 195, 201, 206–207, 280, 338, 371, 422
Narvaez, A., 20, 27, 54, 99–100, 185, 191, 198, 200, 209, 282, 352, 354, 364
Nascimento, V.B., 191, 196, 203, 209, 233, 345, 395, 431
Naser, N., 79, 86–87, 99, 191, 193, 198–199, 203, 209, 333, 357, 359, 361, 394, 430, 432–434
Naseri, M.A., 96
Nasirizadeh, A., 271
Nasirizadeh, N., 95, 191, 200–201, 203, 206–207, 368, 391–392, 423
Nasri, Z., 86, 203, 387
Nasseri, M.A., 280
Nateghi, M.R., 90
Nava, D., 150
Nava, J.L., 161–162
Navaratne, A., 25, 191, 203, 209, 299, 389, 394, 406, 432
Navarro, P., 87
Navratil, T., 169–170, 186–187
Navratilova, Z., 14, 24, 65, 78, 83–84, 89, 99, 103, 106, 172, 188, 212, 217, 224, 247, 288–289
Nazir, H., 191, 208, 229, 428
Neagu, D., 89, 226
Nedelec, J.-Y., 161
Neeb, R., 3, 10, 36, 53, 257, 261
Neff, V.D., 82
Negulescu, C., 96, 296, 301
Neill, D.B., 191, 209, 436

Neill, J., 191, 209, 437
Neiman, E., 10
Neiman, V., 10
Nejati Barzoky, M., 80, 93, 103, 195, 203, 206, 340, 390, 396, 415
Nejati, M., 93, 206–207, 392, 415, 423
Nejem, R.M., 28
Nelson, R.F., 3, 17, 73–74, 151–152, 213
Nematollahi, D., 31, 93, 162, 191, 200–201, 267, 368–369
Nemcova, L., 185, 290
Nepejchalova, L., 43, 183, 239
Neubert, G., 38, 151–152
Neuhold, C.G., 5–6, 12, 22, 24, 26, 54, 65, 67, 74, 76–78, 88–90, 97, 99, 101, 104, 206, 211, 214, 217, 242, 258, 263, 267, 271, 276, 279, 414
Neumann, B., 191, 203, 349, 393
Nevares, I., 191, 350
Neves, A., 191, 197, 350
Nezamzadeh, A., 191, 206–207, 420
Ng, A., 90, 255
Ng, W., 79, 281
Ngameni, E., 84, 101, 219, 293
Nghi, T.-V., 245
Nguyen, A., 93, 331
Ni, J.-R., 161
Nie, B., 191, 203, 395, 433
Nie, L.-B., 32, 64, 81, 246, 411
Niedziolka, J., 67
Nielsen, P.E., 93, 99, 143, 146, 191, 204–205, 398, 401, 407, 409
Nielsen, T.W., 191, 205, 398, 401
Niessner, G., 317
Nieto, O., 99, 101, 202, 295, 383
Nigam, P., 96, 191, 194, 307, 333
Nigretto, J.-M., 191, 200–202, 363, 374, 376
Nigretto, M., 191, 201, 374
Nikitin, A.A., 214
Niknam, K., 90, 261
Nikolelis, D.P., 191, 196, 204, 342, 397
Niranjana, E., 79, 95–97, 188, 191, 203–204, 206–207, 250, 294, 333, 387–388, 390–392, 397, 415, 421
Nishiyama, K., 63, 191, 196, 204, 206, 342, 397, 413
Nisli, G., 84, 88–89, 93, 191, 195, 206–207, 293, 338, 419, 425
Niu, D.-J., 34–35, 80, 88, 99, 103, 202, 377
Niu, Q.-J., 447
Niwa, O., 27, 89
Njopwouo, D., 84, 101, 219
Nobre, M.A., 261
Noguchi, H., 96
Noguer, T., 188, 299, 302
Noori, A., 90, 204, 398
Norkus, E., 7, 16, 24, 53, 55, 63, 87, 103, 174, 231–232, 241
Noroozifar, M., 191, 206–207, 415, 421
Norouzi, B., 82, 96–97, 166, 191–192, 214, 269, 272, 279, 328
Norouzi, P., 30–35, 86, 90, 99, 102–103, 180, 185, 202, 206, 216, 221, 240, 263–265, 289, 302–303, 310, 314, 317, 321, 384, 412
Noshadi, S., 90, 255
Nossol, E., 8, 15, 79, 102
Novakova, M., 59, 96, 206–207, 422
Novic, M., 58, 65, 146
Novikova, N., 185, 287

Novotny, R., 9, 12, 14, 20, 49, 51, 55, 147
Nsangu, J., 241, 275
Nunes, G.S., 79, 96, 333
Nunes, R.S., 303, 310
Nunez Vergara, L.J., 90, 191, 317, 323, 375
Nurouzi, P., 32, 35, 180, 264
Nyasulu, F.W., 183, 186, 270, 291
Nyokong, T., 45, 79–81, 96, 103, 195, 203, 206–208, 295, 333, 339, 390, 422, 427

O

Obuchowska, A., 191, 209, 435
O'Connor, K.M., 90
Oczkowski, T., 191, 205, 408
Odaci, D., 31, 80, 185, 191–192, 200–201, 208, 282, 328, 365, 372–373, 429
O'Dea, P., 308
Oelschaeger, H., 310
Oezkan, D., 191, 205, 404
Oezkan, Z., 191, 205, 404
Oezsoez, M., 191, 205, 404
Ogawa, K., 191, 198, 353
Ogorevc, B., 7–8, 12, 16, 24–25, 40–41, 45, 54, 58, 65, 74, 77–78, 84, 88–89, 99–100, 103–104, 106, 108, 121, 135, 143, 146, 174, 183, 193, 204, 214–215, 218, 223, 232, 234, 237, 239, 241–243, 245, 267, 271, 275, 333, 398
Oh, T.-S., 168
O'Halloran, K.P., 101
Ohga, K., 97, 150, 208, 426
Ohgaru, T., 99, 198, 357
Ohta, S., 99–100
Ojani, R.P., 67, 80, 82–83, 90, 92, 95–97, 99–100, 102–103, 149, 166, 185, 191–192, 195–196, 198, 201, 203, 205–207, 209, 214, 250, 269–270, 272, 278–280, 285, 319, 328, 333, 338–339, 343, 345, 351, 372, 389, 394, 408, 420–421, 423, 432
Okabe, K., 97, 208, 426
Okamoto, Y., 79–80, 92–93, 96, 99–101, 191–193, 198–201, 209, 327, 329, 333, 354–355, 362–364, 369–370, 438
Okamura, M., 100, 196, 343
O'Keeffe, M., 99, 308
Okuda, A., 79, 95, 99, 191, 198, 356
Okuma, H., 79–80, 199, 206, 361, 412
Oldham, K.B., 131
Olenic, L., 357, 398
Oliveira, A.B., 95, 104, 191, 195, 201, 340, 372
Oliveira Brett, A.M., 135, 143–144
Oliveira, E.C., 101, 193, 333
Oliveira, M.I.P., 43–44, 206, 416
Oliveira Neto, G., 86–87, 93, 96, 101, 193, 197, 200–201, 217, 271, 333, 347, 370, 372
Oliveira, O.N., 191, 206–207, 422
Oliveira, T.N., 99, 206–207, 420
Oliveira, W.Z., 81, 203, 395, 432
Olivier, V., 161
Olsen, K., 90, 179, 256
Olson, C.L., 2, 16, 18, 21, 36, 38, 45, 52–53, 55, 152–153, 155, 207
Omrani, A., 166, 185, 285

O'Neill, R.D., 9, 13, 44, 81, 92, 97, 99–100, 162–165, 170, 191, 199, 209, 435–437, 598–599, 604
Oni, J.P., 45, 79, 96–97, 192, 200, 203, 206–208, 326, 333, 364, 390, 422, 427
Oniciu, L., 96, 191, 198, 358
Ooshima, K., 186, 294
Opallo, M., 35, 67
Oprean, R., 150, 191, 206–207, 303, 317, 418–419
Oprean, R.N., 93, 319, 321
Orchard, S.W., 54, 59
Ordieres, A.J.M., 75, 93, 192, 206–207, 309, 326, 418
Ordikhani-Seyedlar, A., 103, 202, 314, 384
Orlanducci, S., 29, 49, 80, 191, 200, 366, 385
Ormonde, D.E., 81, 97, 163, 165, 191, 209, 436–437
Oropeza, M.T., 84, 88, 161–162
Ortega, D.R., 90
Ortega, F., 185, 191, 197, 282, 348
Ortiz, M.C., 323
Ortiz, P.I., 214
Ortiz Viana, M.M., 99
Ortner, A., 79, 96, 305, 333
Osa, T., 260, 266
Osborne, P.G., 95
O'Shea, T.J., 41, 59, 96, 99–100, 195, 201, 277, 339, 372
Osipova, E.A., 162, 281
Ostapczuk, P., 143, 145–146
Ostatna, V., 191, 202, 380
O'Sullivan, C.K., 185, 188, 287
Ouangpipat, W., 25
Ouassini, A., 298
Oudda, H., 7, 54, 80, 281, 333
Oungpipat, W., 79, 191, 193, 209, 274, 329, 430, 433
Ouziel, E., 143, 145
Ozbayrak, O., 208, 228, 428
Ozer, D., 75
Ozkan, D., 84, 93, 99, 191, 205, 297, 316, 398, 403–405, 407–409, 434
Ozkan, S.A., 67, 170, 191
Ozkan, Z., 93, 205, 405
Ozkinay, F., 191, 205, 407
Ozoemena, K.I., 79–81, 90, 96, 103, 191, 196, 295, 313, 318, 333, 342
Ozsoz, M., 25, 79, 84, 93, 96, 99, 191–192, 197, 200, 203, 205, 209, 297, 309, 316, 327, 333, 349, 362–363, 394, 398, 403–409, 411, 430, 432, 434

P

Pacakova, V., 26, 191, 577
Paci, S., 245
Packett, D., 188, 304
Paday, F., 96
Paderno, Y.B., 162
Padilha, P.M., 101, 226, 252
Page, P.C.B., 29, 32
Pai, K.V.K., 97, 185, 284
Pakniat, M., 89, 224
Pakula, M., 59, 63
Palacios Santander, J.M., 298
Palaska, P., 205, 309, 402
Palecek, E., 7, 10, 143, 146, 191, 201–202, 204–205, 309, 375, 380–381, 398, 400–401, 404, 406–407, 411
Palkin, A.P., 161

Author Index

Palleschi, G., 19, 29–32, 49, 63, 79–80, 82, 89, 100, 184, 191, 193, 195–196, 198, 200, 202, 233–234, 278, 281, 331, 340–341, 344, 358, 366, 385
Palmour, R. R., 191, 209, 437
Palomar, M.E., 160, 162
Palomar-Pardave, A., 191, 203, 388
Palomar-Pardave, M.E., 97, 166, 191, 203, 387–388, 395, 418
Pamidi, P.V.A., 191, 200–201, 370
Pan, J., 32, 80, 197, 205, 349, 409
Pan, W.P., 191, 206–207, 417, 424
Panagoulis, D., 191, 205, 398, 402
Pandey, C.M.D., 191, 198–199, 362
Pandey, P.C., 36, 191, 198–199, 361–362
Pandey, V., 191, 199, 361
Pandurangachar, M., 93, 99–100, 166, 191, 202–203, 206–207, 386, 388, 393, 416, 424
Paneli, M.G., 64
Paniagua, A.R., 90, 253
Panigua, A.R., 320
Pantano, P., 163, 209, 436
Pantazaki, A.A., 97, 121, 191, 205, 404
Pap, S., 22, 40, 188, 305, 311, 322
Papadopoulos, K., 80, 198, 357
Papouchado, L., 3, 151–152, 155
Papp, Zs., 187–188, 240, 297, 299–300, 447
Paquim, A.-M.C., 191, 204, 398
Parajuli, R.-R., 191, 204, 398
Parasyuk, O.V., 161
Pardave, M.P., 90, 191, 203, 395
Pardo, R., 161–162
Paredes, P.A., 59, 338
Parellada, J., 20, 27, 54, 59, 99, 191, 198, 338, 354
Parham, H., 80, 103, 300
Pariente, F., 96, 191, 198, 206, 326, 417
Park, C.-Y., 168
Park, D.-S., 90, 251, 274, 279
Park, J.-H., 89, 223
Park, J.-T., 135, 140, 279–280
Park, K.-H., 46, 59, 97
Park, S.-H., 41, 309, 311, 318, 322
Park, Y.-C., 251
Parra, V., 79, 96, 166, 191, 197, 204, 333, 350
Parrado, C., 87, 191, 205, 398, 401, 405, 407
Parra, V., 74
Parsaei, A., 80, 88, 103, 304, 312
Pasa, A.A., 81, 203, 395, 432
Pasco, N., 80, 99–100, 195, 340
Pastor, E., 168
Pastore, H.O., 101, 193, 333
Pastore, P., 191, 200–201, 368
Patel, H., 80, 96, 100, 198, 333, 356
Pathak, H.C., 191, 198–199, 362
Pathan, A.S., 135, 140, 174, 241
Patil, R.H., 305
Patolsky, F.F., 115
Patrascu, D.G., 191, 387
Patriarche, G.J., 4, 6, 13, 22, 24, 40, 55, 79, 81, 95–97, 188–189, 191, 195–196, 198, 200, 245, 304–307, 309, 311, 314–316, 319–323, 333, 339, 342, 354–355, 359, 366
Patris, S., 241, 275
Pattanarat, L., 89, 228
Paul, D., 191, 209, 435
Pauliukaite, R., 7, 18, 24, 55, 63, 87, 103–104, 174, 232, 241
Pavan, F.A., 191, 200–201, 206–207, 370, 425
Pazdera, R., 16, 53, 231
Pechan, Z., 191, 201–202, 380
Peckova, K., 9, 74, 169–170, 186–187, 447
Pedano, M.L., 32, 63, 66, 191, 193, 198, 203–204, 331, 353, 393, 398, 400
Pedrero, M., 59, 87, 193, 209, 331
Pehlivan, S., 191, 205, 407
Pei, C.-H., 168
Pei, H.-L., 191, 197, 347
Pei, J.-H., 214
Peixoto, C.R.M., 87, 99, 206–207, 420
Pekarova, J., 316
Pekarova, Z., 316
Pekli, M.N., 277
Pekyardimci, S., 191–192, 326
Pelegrina, A.B., 209, 237, 430
Pena-Farfal, C., 80, 195, 201, 341, 371
Peng, H.P., 80, 103, 149, 198, 203, 356, 389
Peng, J., 8, 82–83, 101, 149, 206–207, 423
Peng, T.-Z., 67, 89–90, 97, 191, 195, 197, 202, 211–212, 223, 230, 308–309, 312, 319, 323, 341, 347, 381
Peng, Y.-F., 191, 206, 414
Peng, Y.-Q., 263
Peng, Z.-F., 88, 202, 378
Peng-Fei, Y., 311
Peralta, R.A., 191, 197, 350
Percio Farias, A.M., 186, 290
Percira, A.C., 191, 208, 426
Perdicakis, M., 161–162
Pereira, A.C., 87, 93, 96, 193, 200–201, 208, 330, 369–370, 426
Perez, A.G., 307
Perez, E.F., 87, 96, 193, 271, 333
Perez-Lopez, J.A., 25, 58–59, 79, 84, 88, 99, 101, 104, 230, 299, 311, 319
Perlado, J.C., 99, 319
Persson, B., 80, 93, 99–100, 191, 193, 195, 198, 200, 209, 330, 332, 337, 341, 353, 363–364, 369, 438
Pessoa, C.A., 20, 38, 87–88, 96–97, 101, 191, 195, 200–201, 203, 206–207, 271, 319, 338, 367–368, 391, 420–422, 425
Pesteh, M., 85–86, 89, 220, 247
Petek, M., 3
Peterson, L., 191, 196, 342
Petit, C., 6, 26, 41, 97, 191, 197, 203, 209, 348, 393–394, 436
Petrek, J., 305
Petrie, G., 3, 151–152, 155
Petrlova, J., 131, 305
Pettersson, G., 191, 200, 363
Pettichini, S., 168
Pfeiffer, D., 24, 99, 193, 331
Pham, T.T.H., 80, 380
Phebus, L., 191, 209, 436
Phillips, A.G., 81, 163, 165, 191, 209, 437
Piekarski, S., 42–43
Pierna, A.R., 191–192, 328
Pietsch, R., 77–79, 82, 211, 281
Pihlar, B., 7, 49, 51, 84, 87, 103, 149, 174–175, 215, 218, 232, 241, 247
Pikula, J., 296

Pillet, Y., 54, 80, 281
Pingarron, J.M., 59, 64, 79–80, 82, 86–87, 96, 103, 191, 195–196, 201, 206, 292, 297, 302, 333, 341–342, 344, 371, 412
Pinho da Silva, R., 191, 200, 206, 368, 416
Pinilla, J.M., 99, 230, 301
Pinnavaia, T.J., 84, 297
Pirali Hamedani, M., 35, 180, 265, 302
Pires, J.M., 191, 205, 398, 402
Pizeta, I., 163
Pizzariello, A., 280
Plassard, F., 84, 168
Pletcher, D., 123
Pletnev, I.V., 191, 203, 396
Pochet, S., 191, 325
Pohanka, M., 191, 193, 331
Polito, W.L., 87, 100, 202, 260, 307, 314, 384
Polverejan, M., 84, 297
Ponamoreva, O., 191, 200, 208, 364, 429
Pontiki, E., 191, 205, 398, 402
Popa, D.E., 102, 240
Popescu, C., 96
Popescu, I.C., 84–86, 88, 191, 198, 200–201, 269, 358, 370–371
Popkov, V.A., 189
Portaccio, M., 95, 284
Porter, J., 93, 95, 99, 199, 360
Posac, J.R., 97, 303
Posokin, J.V., 160
Postbieglova, I., 131
Pot, J.J., 54, 197–198, 202, 346, 353, 382
Potesil, D., 54, 58, 131, 305
Poulouin, L., 191, 200, 363
Pouretedal, H.R., 90, 251
Prabhu, S.V., 36, 89–90, 222, 230
Prakash, M., 82
Prakash, R., 96, 307
Prasada Rao, T.S.R., 87
Prashanth, S.N., 103, 306
Prater, B.G., 3–4, 17, 38, 52
Prater, K.B., 3, 38, 151–152
Prater, K.G., 3–4, 17, 52
Pravda, M., 6, 24, 26, 41, 59, 80, 99–100, 103, 198, 203, 241, 354, 394
Prazeres, D.M.F., 191–192, 327
Prete, M.P., 13, 22, 40, 188–189, 191, 305, 309, 311, 319–320, 322
Price, J.C., 303, 319
Prieto, F., 161–162
Prieto, N., 166
Priimak, E.V., 296, 301
Prior, C., 25, 104, 174
Priyantha, N., 299
Procopio, J.R., 89–90, 99, 219–220, 225, 228, 235, 254
Prokhorova, G.V., 162, 281
Prokopio, J.R., 78, 99, 230
Pruneanu, S., 357, 398
Pudilova, H., 308
Puertollano, M.J., 70
Pujic, Z., 79, 333
Pumera, M., 29
Pundir, C.S., 191, 193, 202, 333, 382
Pungor, E., 3, 36, 52, 260
Purachat, P., 191, 193, 209, 274, 329, 430, 433

Q

Qi, D.-Y., 93, 191, 199, 203, 206–207, 209, 362, 394, 425, 431–432
Qi, F., 166
Qi, H.-L., 35, 191, 205, 313, 405, 408
Qi, J.-R., 191, 203, 386
Qi, T., 89, 191, 197, 209, 346, 433, 447
Qi, X.-H., 79–80, 96, 185, 202, 285, 333
Qi, Y.-J., 191, 206–207, 423
Qian, H., 15, 185, 288
Qiano, W., 191, 203, 209, 394, 432, 437
Qiao, H.-Y., 191, 206–207, 424
Qiao, L., 191, 202, 377
Qiao, W.-J., 163, 191, 203, 209, 223, 394, 432, 437
Qiao, Y., 191, 202, 206–207, 376–377, 379, 425, 431
Qiao, Y.-B., 202, 268, 379
Qiao, Y.-G., 191, 199, 362
Qin, P., 80, 191, 379
Qiu, B., 35, 43, 182, 399
Qiu, D.-F., 258
Qiu, J.D., 80, 103, 149, 198, 203, 356, 389
Qu, J., 32, 89, 224, 231
Qu, L.-M., 32, 35, 63, 66, 191, 195, 212, 223, 341, 377
Qu, W.-Y., 191, 205, 209, 293, 398, 403, 438
Quaiserova, V., 194, 335
Quarin, G., 93, 97, 191, 197, 203, 348, 393
Quintana, J.C., 184, 281
Quintana, M.C., 191, 202, 383
Quintino, M.S.M., 135
Qureshi, G.A., 183, 277

R

Raba, J., 191, 202, 381
Raber, G., 65, 67, 76–78, 84, 90, 97, 99, 101, 106, 108, 211, 253, 255, 258, 263, 276, 279
Radecka, H., 100, 194, 274, 285, 334
Radecki, J., 100, 194, 274, 285, 334
Radi, A.-E., 9, 20, 67, 171, 191, 202, 205, 308, 313–315, 317–319, 321, 381, 398, 402–403, 405
Radi, E., 312
Radu, G.L., 68, 308, 311, 321, 323
Raei, F., 31, 93, 201, 369
Rafiei, S.Z., 34–35
Raghavendra Naik, R., 97, 191, 203, 250, 387, 391
Rahbar, N., 80, 103, 300
Rahmani, A., 89, 92, 231, 253
Rahmanifar, M.S., 82–83, 272
Rajabi, H., 80, 103, 203, 391
Rajagopalan, S.R., 89
Rajendran, V., 80
Ramajo, J., 161–162
Ramanavicius, A., 93
Ramelow, G.J., 191, 209, 429
Ramine, M., 92, 95, 272, 278
Ramirez Barreira, R.J., 315
Ramirez Berriozabal, M., 90, 206, 415
Ramirez-Molina, C., 93, 95, 193, 200–201, 330, 367
Ramirez-Silva, M.T., 90, 97, 160, 162, 166, 168, 191, 203, 233, 387–388, 395, 418
Ramos, A.R., 191, 193, 333
Ramos, J.A., 25, 230
Ramos, L.A., 97, 191, 206–207, 419, 422

Author Index

Ramos-Barrado, J.R., 150
Randriamahazaka, H.N., 191, 200–202, 363, 374, 376
Rani, S., 87
Rao, V.K., 32, 186, 294
Raoof, D.B., 92, 162
Raoof, J.B., 67, 79–80, 82, 90, 92, 95–97, 99–100, 102–103, 149, 166, 185, 191–192, 195–196, 201, 203, 205–207, 209, 214, 250, 269–270, 272, 278–280, 285, 319, 328, 333, 338–339, 343, 345, 372, 389, 394, 408, 420–421, 423, 432
Rapicault, S., 96, 99, 320
Rapin, J.P., 84, 275
Raptopoulou, C., 191, 205, 398, 402
Rashid-Nadimi, S., 100, 203, 206–207, 389, 420
Rasmussen, J., 143, 145
Ratti, S., 87, 195, 197, 203, 338, 351, 389
Ravichandran, K., 4, 38, 58–59, 76, 79, 92, 95–96, 99, 104, 163, 197, 206–207, 351, 418, 420, 423
Raviraj, N.H., 166, 192, 328
Rawlins, J.N., 163, 165, 209, 436
Rayson, G.D., 191, 194, 197, 211, 336, 352
Razola, S.S., 93, 191, 325
Razumas, V., 58, 71, 99, 191, 193, 198, 209, 332, 358, 435
Razumiene, J., 191, 200, 364
Rechnitz, G.A., 24–25, 87, 95, 191, 194, 200, 203, 206, 209, 257, 334–335, 365, 389, 394, 406, 417, 431–432, 434
Reddy, A.K.N., 123
Reddy, G.V.R., 310, 313, 320
Reddy, K.R.V., 191, 203, 390
Reddy, K.V., 67, 202, 384
Reddy, L.M., 7, 84, 297
Reddy, S., 80, 103, 191, 203, 206–207, 386, 389, 423
Reddy, S.J., 84, 90, 310, 313, 315, 318, 320, 322
Reddy, S.R.J., 7, 84, 296–297
Reddy, T.M., 84, 90, 310, 313, 315, 318, 320, 322
Redi-Abshiro, M., 92, 206, 306, 412
Reeves, R.M., 160–162
Refera, T., 208, 251, 427
Refshauge, C., 40–41, 44
Regan, F.B., 41, 195, 201, 337, 373
Regis, A., 161
Reguera, E., 90
Reijn, J.M., 40, 203, 284, 395
Reiko, O., 97
Reis, N.S., 191, 196, 206, 344, 418
Rejhonova, H., 135, 142
Rekha, B.E., 191, 203, 390
Remita, H., 80, 103
Ren, R., 35, 80, 88, 103
Ren, Y., 191, 205, 410
Renaudin, G., 84, 275
Renedo, O., 201, 303, 373
Renger, F., 17–18, 39, 52, 55, 69, 77–78, 104, 189, 212, 250, 252, 308, 319
Ren,Y., 191, 205, 410
Reshetilov, A., 191, 200, 208, 364, 429
Reshetnyak, V.Yu., 189
Reviejo, A.J., 79, 87, 96, 297, 302, 333
Reynaud, J.A., 191, 374
ReynosoSoto, E., 88
Reza Zare, H., 93, 206–207, 423
Rezaei, B., 66, 81, 84, 87, 97, 101, 203, 229, 236, 257, 389
Rezaie, S.Z., 90, 92

Rezazadeh, M., 97, 203, 389
Rezende, M.O.O., 25, 99, 149, 225, 292
Rhazi, M.E., 236
Rhim, S.-K., 135, 140, 274
Rhyu, K.-B., 26, 191, 209, 270, 430
Ribeiro de Souza, E.B., 87, 206–207, 429
Ribeiro do Carmo, D., 90, 250
Ribeiro, E.S., 86–87, 93, 96, 191, 200–201, 206–207, 266, 369–370, 425
Ribeiro, M.L., 79, 96, 333
Ribera, A., 84
Ricci, F., 19, 29–31, 63, 79, 82, 193, 195, 198, 331, 341, 358
Rice, M.E., 23, 27, 49, 53, 56–58, 61, 65–66, 96, 104, 142
Richardson, D.H.S., 25, 99, 209, 429–430
Richardson, R.J., 163, 191, 197, 201, 209, 347, 374
Rievaj, M., 163, 238, 257, 262
Riggin, R.M., 303, 319
Rigo, A., 43, 194, 335
Rincon, M.E., 8, 15, 63, 149
Rios, A., 40, 96, 191, 197, 273, 350
Risby, D., 191, 203, 395
Rishpon, J., 169, 186, 191, 198, 358
Rivas, G.A., 7–8, 16, 20, 25, 29–32, 42, 63, 66, 80, 87–88, 93, 103, 143, 146, 186, 191, 193–199, 201–205, 207, 209, 214, 269, 290, 295, 309, 319, 326, 331, 336, 341, 344–345, 349, 353–354, 359, 361, 374, 381, 385, 387–388, 393–395, 398, 400–402, 404–408, 419, 430–432, 434, 447
Rivero Espejel, I., 88
Rizea, M.C., 89, 221
Rizk Nashwa, M.H., 307
Roa Morales, G., 90, 168, 233
Robillard, K.A., 215
Rocha, C.G., 271
Rocha, R.F., 87, 93
Roddick-Lanzilotta, A.D., 99–100, 195, 340
Rodehuser, L., 101–102, 225
Rodrigues, C.A., 97, 99, 207, 422
Rodrigues, J.R., 191, 200–201, 206–207, 370, 425
Rodrigues Santos, W.d.J., 32, 247
Rodriguez, C., 79, 96, 295, 333
Rodriguez Granda, P., 99, 200–201, 371, 373
Rodriguez, I.N., 220, 254
Rodriguez, J.A., 174, 238
Rodriguez, J.R.B., 317
Rodriguez, M.C., 29, 32, 63, 66, 88, 103, 191, 193, 197–199, 201, 203–204, 209, 331, 349, 353–354, 359, 374, 393, 400, 434
Rodriguez Procopio, J., 90, 224
Rodriguez, R.J.B., 309, 313
Rodriguez-Flores, J., 191, 197, 350
Rodriguez-Mendez, M.L., 74, 79, 96, 166, 168, 191, 194, 197, 204, 333, 335, 348, 350, 447
Rofouei, M.K., 25
Rogers, K.R., 18, 40, 185, 191, 196–197, 209, 215, 282, 345, 349, 435
Rohani, T., 228
Roizenblat, E.M., 161
Rojas-Hernandez, A., 97, 160, 162, 166, 168, 191, 203, 387–388
Rolison, D.R., 84
Roman, L., 6, 191, 206–207, 317, 418
Romero, E.G., 191, 209, 434
Romero, J.R., 191–192, 325

Romero-Romo, M.A., 90, 97, 166, 191, 203, 233, 387–388, 395, 418
Romo, M.R., 90, 191, 203, 395
Rondeau, A., 95, 198, 357
Rong, G., 305
Roodsaz, S., 191, 204, 206, 397, 416
Rosa, A.H., 84, 96, 101, 219–220
Rosatto, S.S., 86–87, 93, 95, 185, 197, 200–201, 282, 347, 366, 369–370, 427
Rosca, J.C.M., 162
Rosenberg, M., 95, 200, 208, 365, 429
Rosen-Margalit, I., 191, 198, 358
Roslonek, G., 90
RosquetePina, G., 166
Ross, A., 320
Rothley, D., 310
Rott, A., 10
Rouhollahi, A., 96, 307
Rousselin, Y., 172, 229
Rozanska, S., 84–85
Rozhdestvenskaya, Z.B., 26, 158–159, 161
Rozniecka, E., 67
Ru, Q.-H., 84, 206–207, 420
Ruan, C., 191, 206, 414, 417
Ruan, Y., 191, 208, 428
Rubel, S., 213
Rubianes, M.D., 8, 16, 20, 29–32, 63, 66, 87, 191, 193, 197–199, 203–204, 207, 209, 326, 331, 344, 349, 353, 385, 387–388, 393, 400, 419, 434, 447
Ruby, W.R., 143–144, 274
Rucki, R.J., 320
Rudolph, R., 193, 329
Rudomanskis, R., 93, 191, 200, 364
Ruegsegger, P., 217
Ruf, H., 90, 257
Ruiperez, J., 90, 225
Ruiz Barrio, M.A., 59, 64, 79, 86, 96, 196, 292, 333, 344
Ruiz, B.L., 75, 93, 192, 206–207, 326, 418
Ruseell, A., 34
Ruzgas, T., 58, 71, 84, 88–89, 92–93, 99, 191, 197–198, 200–201, 208, 348–349, 358, 365, 369, 371, 427
Ruzicka, J., 3, 53, 134–135, 140, 576, 581
Ryabov, A.D., 97, 192, 200, 326, 364
Ryabushko, O.P., 135, 140, 213
Ryant, P., 168

S

Saadatifar, M., 191, 200–201, 371
Saadeh, S.M., 78, 101, 227–228, 255, 323
Saber, T.H.M., 152, 158
Saby, C., 93, 99, 198, 357
Sabzi, R.E., 82, 95, 103, 203–207, 309, 389, 399, 402, 420
Sachanyuk, V.P., 161
Sadakane, M., 82
Sadeghi, O., 447
Safaei, E., 96, 195, 341
Safavi, A., 33–35, 66, 89, 191, 195, 197, 200–201, 224, 280, 337, 346, 359, 366, 371
Sagar, K., 16, 96, 322
Sagon, G., 161
Saha, P., 79, 96, 305, 333
Sahinci, F., 191, 205, 404
Sahlin, E., 7, 42, 191, 205, 398, 401

Sahni, S., 80, 99–100, 191, 193, 195, 200, 209, 330, 332, 337, 341, 364, 369
Sai, C.-L., 90
Saidman, S.B., 9, 59, 75, 99–100, 104, 116, 191–192, 200–201, 328, 367
Saito, A., 191, 198, 353
Saito, T., 99–100, 191, 198, 355
Sakata, K., 191, 202, 375
Sakslund, H., 87, 209
Sakura, S., 191, 198, 354
Salari, M., 96, 201
Salazar, M., 26, 205, 411
Sales, F., 191, 206–207, 423
Salimi, A., 92, 162, 191, 203, 206–207, 395, 414, 419
Salimibeni, A.R., 90, 255
Salmany Afagh, P., 99–100
Salmazo, L.O., 261
Samadi, N., 274
Samo, A.R., 183, 277
San Jose, M.T., 161
Sanchez, A., 31–32, 79, 88, 102, 240, 295, 299, 385
Sanchez, J., 79, 95, 191, 198, 209, 355, 430, 433
Sanchez Jimenez, C.J., 307
Sanchez, M., 36, 67, 77–78, 99, 222, 304
Sanchez Ramosa, S., 87, 162
Sanchez-Batanero, P., 79, 82, 97, 159–162, 277, 303, 307, 320
Sanchez-Cabezudo, M., 75, 93, 192, 206–207, 326, 418
Sander, S.G., 161–162
Sandford, R.W., 3
Sandrino, B., 97, 203, 206–207, 391, 422
Sandulescu, R., 6, 26, 88, 150, 191, 201, 206–207, 303, 317, 373–374, 418–419
Sandulescu, R.V., 93, 319, 321
Sanghavi, B.J., 191, 206, 413
Santagostino, G., 191, 203, 395
Santamaria, M.d.C., 79, 96, 268, 314, 333
Santana, A., 162
Santhanam, K.S.V., 29, 31–32
Santhiago, M., 191, 350
Santiago, M.B., 88
Santiago, M.E., 88, 97, 192, 200–201, 325, 368
Santiago, M.F., 191, 196, 345
Santos, A.D.S., 87, 93, 96, 192–193, 200–201, 208, 268, 326, 330, 369–370, 426
Santos, A.L., 65, 101, 226, 252
Santos, L.M., 79, 96, 191, 193, 195, 197, 333, 335, 351
Santos, P.M., 97, 203, 206–207, 391, 422
Santos, V.S., 32, 247
Santos, W.J.R., 79, 95, 104, 191, 195, 197, 200–201, 203, 247, 333, 340, 349, 372, 391
Sapcanin, A., 79, 333
Sapelnikova, S.V., 185, 196, 289
Sapio, J.P., 24, 67, 158, 261, 274, 279
Saponjic, R.M., 191, 209, 436
Sapozhnikova, E.Y., 161
Sar, E., 90, 99, 172, 221, 226–227, 237
Sarabia, L.A., 323
Sarac, A.S., 191, 209
Saraswathi, R., 84–85, 239, 271
Saraullo, A., 97, 205, 407
Sargeant, J.A., 78, 84, 93
Sarmazdeh, Z.R., 264
Sarwar, M., 277

Author Index

Satander, J.M.P., 254
Satienperakul, S., 30, 32, 82, 280
Sato, K., 266
Sato, N., 80, 199, 361
Satpati, A.K., 103, 236, 242
Sattarahmady, N., 80, 103, 194, 335
Sauri Peris, M.C., 87, 162
Savin, Yu.S., 135, 140, 213
Savitri, D., 191, 198–199, 353, 362
Sayen, S., 101–102, 149, 225
Scarpa, M., 43, 194, 335
Schachl, K., 5, 9, 12–13, 17–22, 27, 38, 40, 51–53, 55, 59, 61, 63–65, 67, 70, 74, 79, 87, 96, 104, 143, 146, 148, 183, 196, 199, 206–207, 209, 218, 267, 275, 342, 359, 422, 438
Scharff, J.P., 162
Scheller, F.W., 24, 88, 191, 196, 199, 202, 342, 375
Schenk, J.O., 43–44, 158, 165
Scherer, G., 86
Schiavon, G., 75
Schildkraut, D.E., 215
Schinelli, S., 191, 203, 395
Schmid, G.M., 14, 16, 23, 210
Schmid, R.D., 191–192, 326
Schmidt de Magalhaes, C., 191, 203, 397, 433, 436
Schmidt, H.L., 93, 193, 200–201, 331, 369
Schneider, P., 95, 198, 355
Schollhorn, B., 96, 99, 319–320
Scholz, F., 123, 159, 161–162, 202, 379
Schorb, K., 257
Schroeder, U., 159, 162, 202, 379
Schuhmann, W., 80, 88, 93, 97, 100, 192–193, 200–201, 208–209, 326, 331, 364, 369, 427, 438
Schulte, A., 88, 93, 208, 427
Schultz, F.A., 2, 4, 153, 158
Schweicher, P., 26, 93, 200, 267, 363
Scotti, R., 191, 200–201, 370
Sedaghat, M.E., 84, 257
Sedaghati, F., 191, 195, 337
Sedaghatpour, F., 33, 35
Sedghi, H., 103
Seetharamappa, J., 103, 306
Segnini, A., 82, 97, 185, 208, 289, 300, 426
Seidlova, J., 51, 87, 103, 174, 232, 241
Sekimukai, S., 79, 206, 412
Seleghim, A.P., 87, 100, 202, 260, 307, 314, 384
Selim, M.M., 316
Selim, S., 600
Sellami, A., 85, 222
Senda, M., 63, 75, 79, 92, 191, 193, 198, 200–201, 204, 270, 331, 352, 354, 371, 399
Sendam, M., 92, 198, 354
Senillou, A., 75
Senthilkumar, S., 84–85, 239, 271
Seo, M.-L., 191, 194, 215, 335
Seraphim, P.M., 261
Serban, S., 84, 86, 96, 191, 198, 200–201, 355–356, 368
Sergeeva, V.S., 97, 192, 200, 326, 364
Serpi, C., 191, 205, 398, 403
Serra, O.A., 86, 96
Serra, P.A., 163, 165, 209, 437
Serradilla Razola, S., 93
Serrano, S.H.P., 25, 79, 82, 185, 191, 196, 200–201, 206, 209, 268, 285, 344–345, 368, 416, 418, 430

Seta, P., 191, 197, 347
Severinghaus, J.W., 112
Sevilla, F., 279
Sevilla, M.T., 89–90, 99, 217–220, 223, 225, 228, 230, 235, 254
Sha, J.-Q., 207, 423
Shafiee, A., 75, 80, 88, 103, 237
Shahabi, S., 96, 272
Shahrokhian, S., 32, 59, 79–80, 91, 95–96, 99, 104, 191, 194–196, 203, 206–207, 279, 303, 307, 314, 316, 319, 321, 333, 335, 339–340, 343, 349, 390, 392, 396, 413, 415–416, 422, 424
Shaidarova, L.G., 80, 82, 88, 90, 96, 103, 149, 188, 191, 193, 195, 201, 230, 252, 296–297, 299, 301, 332–333, 337–338, 372
Shaikh, A.A., 191, 208, 427
Shams, E., 86–87, 101, 174, 203, 235, 276, 387
Shamsipur, M., 38, 67, 82, 85, 89–90, 92, 96, 162, 206, 231, 253, 261, 267, 274, 280, 314, 425
Shan, D., 84, 150
Shang Guan, X.D., 35, 67, 191, 303, 388
Shang, G.-X., 34–35
Shankar, S.S., 95, 191, 202–203, 386, 391
Sharath Shankar, S., 97, 203, 206, 387–388, 392, 414, 418
Sharghi, H., 38, 89–90, 92, 96, 162, 231, 253, 261, 267
Sharma, L.D., 87
Sharp, J.H., 151–152
Shaw, B.R., 78, 84, 86, 93, 266
Shawish, A., 221
Sheh, L., 89–90, 223
Shehab, O.R., 248
Sheikh Shoai, I., 216
Sheikh-Mohseni, M.A., 80, 103, 204, 397
Shen, B.-E., 89–90, 191, 195, 223, 230, 341
Shen, D., 90, 264
Shen, G.-L., 25, 68, 80, 88, 95–96, 99–100, 103, 149, 191, 195, 199, 202, 205, 209, 293, 316, 333, 337, 361, 378–379, 383, 411, 433, 438
Shen, J., 191, 202, 383
Shen, Y., 81, 103, 199, 203, 359
Sheng, Q.-L., 81, 103, 191, 199, 203, 269, 359, 388
Sherbakova, L.P., 214
Sherigara, B.S., 67, 79–80, 93, 95–97, 99–100, 103, 166, 185, 188, 191, 202–204, 206–207, 236, 242, 250, 284, 293–294, 333, 384, 386–393, 397, 414–416, 418, 421, 423–425
Shestivska, V., 191, 205, 404
Shi, G.-Y., 163, 165, 209
Shi, L.-H., 8, 15, 32–35, 63, 66, 82
Shi, Q.-Z., 30, 32, 191, 196, 211, 315, 343, 434
Shi, S.-H., 161
Shi, S.-Y., 161
Shi, X.-S., 231
Shi, X.-W., 191, 206–207, 421
Shibata, T., 92, 198, 352
Shiddiky, M.A., 279
Shim, J., 191, 197, 347
Shim, Y.-B., 7, 41, 88–90, 99, 172, 191, 197, 202, 214–215, 217–219, 223–224, 251, 274–275, 279, 309, 311, 318, 322, 348, 375
Shimizu, Y., 87, 281
Shimodaira, A., 191, 360

Shiragami, S., 191, 208, 427
Shiraishi, H., 7, 15, 25, 93, 168, 186, 191, 202, 204–205, 209, 290, 309, 319, 381, 398, 400, 402, 404, 406–408, 435
Shiraishi, T., 75
Shirzadmehr, A., 30, 33, 35, 66, 239, 263, 265
Shivaraj, Y., 103
Shokrollahi, A., 90, 255, 277
Shou, C.-Q., 166
Shreenivas, M.T., 97, 191, 202–203, 386, 388
Shu, H.-C., 80, 191, 193, 330
Shul, G., 35, 67
Shumilova, M.A., 177, 256
Shunyaev, K.Y., 161–162
Shvedene, N.V., 191, 203, 396
Siangproh, W., 96, 272
Sidorenko, I.G., 191, 194, 335
Sierra, I., 102, 240
Sigolaeva, L.V., 163, 191, 197, 201, 209, 347, 374
Sikora, E., 63, 218
Sileny, J., 191, 202, 379
Silva, L.S., 99, 206–207, 420
Silva, M.T.R., 90, 191, 203, 395
Silva, R.P., 25, 185, 209, 268, 285, 430
Silva, V.L., 43–44, 206, 416
Sim, S.J., 80, 380
Sima, V., 201, 303, 373
Simic, N., 150
Simkus, R., 191, 200, 364
Simoes, F.R., 295, 297–298
Sinaga, S.M., 99, 202, 383
Sinapi, F., 99, 181
Sinyakova, S.I., 66, 213
Siow, K.S., 90, 255
Sirieix Plenet, J., 35, 67
Siril, P.F., 80, 103
Sirisha, K.R., 7, 84, 296–297
Siroueinejad, A., 82, 280
Siska, E., 260
Sisti, A.M., 201, 374
Siswana, M., 80–81, 103, 295
Sivanesan, A., 191, 203, 388
Skeika, T., 20, 87, 203, 206, 391, 421
Skeva, E., 191, 205, 398, 402
Skladal, P., 79, 95–96, 170, 193, 298, 300, 302, 331, 333, 349, 434
Skobets, E.M., 135, 140, 213
Skorepa, J., 135, 140
Skotheim, T.A., 79–80, 92–93, 96, 99–100, 191–193, 195, 197–200, 202, 209, 327, 329–330, 332–333, 337, 341, 351, 354–355, 362–364, 369, 384, 438
Skov, H.J., 143, 145
Slavik, M., 191, 209, 435
Slavikova, S., 65, 135, 140, 247
Slepuschkin, V., 10
Slilam, M., 191, 198, 354
Smirnova, T.P., 161–162
Smit, M.H., 257
Smith, A.J., 191, 204
Smith, T.D., 86
Smolander, M., 99, 191, 195, 197, 209, 337, 341, 351

Smyth, M.R., 7, 16–18, 25, 39, 51–52, 55, 59, 69, 80, 87–88, 93, 96, 99–100, 103, 185, 188–189, 191–192, 197–198, 203, 206, 209, 217, 277, 282, 287, 308, 311–312, 315, 317–319, 321–322, 325, 349, 353–354, 394, 416, 429–430, 432–435
Snevajsova, P., 191, 205, 406
Sode, K., 99–100, 191, 198, 205, 353, 411
Soderhjelm, P., 18, 191, 196, 208, 345, 426
Soh, M.K., 90
Sohrabnezhad, S., 85, 206, 425
Sokolovskaya, L.G., 163, 197, 201, 209, 347, 374
Soldi, V., 81, 203, 395, 432
Soleymanpour, A., 96, 280
Solis, V.M., 25, 93, 191, 196, 201, 203, 209, 345, 374, 393–395, 430–432
Solorza Feria, O., 38
Soltani-Nejad, V., 191, 204, 206, 397, 416
Son, J.-G., 279
Song, C.-X., 191, 205, 409
Song, F., 224, 277
Song, J., 32, 101, 204, 324, 397
Song, J.-C., 27, 447
Song, J.-F., 30, 32, 34, 63–64, 97, 191–192, 195, 198, 226, 262, 273, 304, 308, 311, 314–315, 322, 328, 342, 353, 359, 398–400
Song, W.-B., 93, 191, 199, 362
Song, Z.-J., 322
Songina, O.A., 26, 158–159, 161
Sonntag, V., 163, 165, 209
Sopha, H., 25, 104, 143, 146, 240
Sosa Ferrera, Z., 84–85, 194, 287, 334
Sosa, Z., 84, 197, 347
Sotiropoulos, S., 43, 49, 59, 149, 191, 195, 233–236, 243, 337
Sotomayor, M.D.P.T., 36, 38, 96, 191, 194, 197, 202–204, 335, 348, 381, 390, 393, 397
Sotomayor, P.T., 87, 197, 347
Soucek, J., 135, 139
Souri, A., 96, 319
Sousa, A.G., 191, 325
Sousa, M.d.F., 64, 276
Southwell-Keely, P., 79
Specker, H., 37–38, 70, 213, 262
Spichiger Keller, U.E., 18
Spinelli, A., 99
Spohn, V.U., 24, 99, 191, 193, 329–331
Squella, J.A., 90, 191, 317, 323, 375
Sramkova, J., 53, 252
Sreedhar, M., 7, 84, 90, 297, 322
Srey, M., 212
Srij, M.M., 239
Sriman Narayanan, S., 191, 196, 344
Srinivasan, G.N., 161
Srivastava, A.K., 90, 97, 191, 206–208, 224, 231, 256, 413, 423, 426–427
Stadler, E., 97, 99, 207, 422
Stadlober, M., 65, 77–78, 84, 97, 263
Stahlberg, R., 162
Stajerova, B., 135, 139
Stamatin, I., 191, 387
Stamford, J.A., 165
Stanic, Z., 170, 191, 205, 309, 398, 402–403, 447
Stara, V., 86, 197, 230, 347

Author Index

Starkova, B., 26
Steckhan, E., 82
Stefan, I., 311
Stefan, R.-I., 8, 14, 28, 40, 68, 79, 90, 96, 191, 196, 233, 250, 254, 275, 305, 308, 311, 318, 321, 323, 333, 342, 746
Stefan-van Staden, R.-I., 28–29, 67, 79, 96, 191, 195, 202, 216, 306, 308, 313, 321, 333, 341, 382
Stempkowska, I., 191, 204, 398–399
Stepan, R., 90, 185, 205, 287, 401
Stephanie, G.R., 172, 229
Sternson, A.W., 158
Sternson, L.A., 158, 185–186, 191, 206, 286, 290, 417
Stiburkova, M., 51, 87, 103, 174, 232, 241
Stobiecka, A., 191, 194, 201–202, 285, 334, 380
Stoces, M., 19–20, 29, 32, 51, 62, 104, 135, 142, 166, 240, 242, 244–245, 250, 447
Stoces, T.M., 250
Stock, J.T., 167
Stoica, A.I., 185, 188, 288
Stojko, N.Y., 78
Stozhko, N.Y., 78
Stradiotto, N.R., 65, 90, 97, 101, 162, 191, 205–207, 226, 250, 252, 398, 402, 421
Strakhova, N.N., 163, 197, 201, 209, 347, 374
Strappazzon, R., 161
Stredansky, M., 280
Strope, E., 44, 158, 163, 165, 305
Stryjewska, E., 213
Stuben, D., 9, 161–162
Stulik, K., 4, 10, 26, 41, 135, 140, 186, 191, 577
Stupnicka, B., 191, 205, 398, 402
Stutts, K.J., 191, 206, 418
Su, B.L., 101–102
Su, L., 29
Su, Z.-H., 8, 42, 149, 167
Subramaniam, V., 191, 201–202, 380
Sucha, L., 172
Sugawara, K., 25, 78, 89, 96–97, 99, 191, 194–195, 198–201, 214–215, 222, 255, 336, 338, 351, 360, 365–366, 375
Sugimoto, M., 191, 209, 435
Sun, B., 35, 313
Sun, C., 93, 101, 191, 198–199, 276, 356, 362
Sun, C.-N., 8, 33–35, 63, 66, 82
Sun, C.-Q., 191, 199, 362
Sun, D.-Z., 35, 80, 84, 101–102, 206, 229, 236, 240, 388, 412
Sun, G., 96
Sun, J., 31–32, 41, 43, 191, 202, 295, 297, 376
Sun, J.-J., 150, 286, 412
Sun, J.-S., 239
Sun, J.-Y., 34–35, 80, 88, 99, 103, 202, 377
Sun, K., 93
Sun, Q.Y., 99, 246, 258
Sun, S.-C., 191, 206
Sun, T., 191, 206–207, 421
Sun, T.-L., 191, 206–207, 421
Sun, T.-P., 191, 206, 413
Sun, W.R., 31–35, 63, 66, 71, 80–81, 99, 103, 149, 191, 202, 204–207, 283–285, 289, 337, 376–379, 385, 398–400, 403, 409–410, 419, 421, 447
Sun, X.-H., 559

Sun, X.-Y., 34–35, 99, 377
Sun, Y.-T., 99, 191, 202, 212, 252, 263, 382
Sun, Z., 82–83
Sundstrom, L., 163, 165, 209, 436
Supalkova, V., 296, 305
Suren, E., 317
Suresh, S., 32, 186, 294
Surugiu, I., 185, 205, 286, 405
Suska, M., 77–78, 89, 104, 250, 252
Suye, S.I., 270
Švancara, I., 1–2, 4–10, 12–14, 16–22, 24–27, 29, 32–36, 38–43, 45, 49–59, 61–67, 69–70, 74, 77–79, 81, 85, 87, 97, 99, 102–104, 135, 142–143, 146–149, 151, 159, 166, 169, 171–172, 174–175, 178, 183–184, 186–189, 191, 195–196, 198–199, 209, 211–212, 214–215, 217–218, 226, 231–245, 247, 250–252, 254, 257–259, 267, 275, 297, 299–300, 308, 316, 319, 337, 342, 359, 364, 438, 447, 585
Svegl, I.G., 84, 215, 218
Svehla, G., 90, 249
Svehla, M., 90
Svensson, C.J., 84, 201
Svensson, L., 191, 209, 436–437
Svitel, J., 280
Svobodova, E., 104, 240, 242, 244–245
Svorc, J., 95, 200, 208, 365, 429
Swaile, B.A.H., 191, 204, 398, 401
Swamy, B.E.K., 79–80, 93, 95–97, 99–100, 103, 166, 185, 188, 191, 202–204, 206–207, 250, 284, 293–294, 333, 386–393, 397, 414–416, 418, 421, 423–425
Swamy, K., 191, 203, 390
Swartzfager, D.G., 191, 202, 383
Sweetser, P.B., 191, 202, 383
Swiatkowski, A., 59, 63
Sydorenko, I.G., 191, 194, 335
Szepesvary, E., 3, 36
Szewczynska, M., 191, 196, 209, 345
Szuwarski, N., 95, 198, 357
Szymanska, I., 100, 274

T

Taarit, Y.B., 85, 222
Taei, M., 32, 201, 372
Taga, M., 89, 97, 191, 194–195, 200, 214, 222, 336, 338, 366
Taghavinia, N., 80, 93, 103, 194–196, 203–204, 206, 319, 335, 340, 343, 390–391, 396–397, 415
Taha, Z., 87, 99, 135, 197–199, 352, 354
Taher, M.A., 89, 216, 227–228, 316
Taheri, A.R., 84–85, 87, 191, 196, 206–207, 343, 387, 415, 420–421
Tahirovic, I., 79, 333
Tai, C.-Y., 206, 323, 413
Tajabadi, F., 33, 35, 66, 191, 195, 197, 200, 280, 337, 346, 366
Tajima, Y., 191, 362
Takahashi, H., 79, 206, 412
Takahashi, I., 191, 194, 335
Takahashi, M., 75
Takahashi, Y., 99–100

Takamura, Y., 191, 205, 407, 411
Takano, T., 25, 99, 199, 360
Takao, H., 56, 152, 158, 291
Takase, S., 87, 281
Takatsy, A., 75
Takayama, K., 75, 92, 191, 200, 208, 364, 426, 429
Takeuchi, E.S., 96
Takeuchi, R.M., 101, 226, 252
Talaber, C., 76–78, 90, 101, 211, 258
Talakesh, M., 89, 220
Taleat, Z., 93, 103, 194–196, 203, 206–207, 319, 335, 340, 343, 391, 423
Talebi, A., 80, 103, 195, 203, 206, 340, 390, 415
Taliene, V.R., 58, 71, 198, 358
Tamamoto, B.K., 81, 163
Tamiaki, H., 7, 15, 25, 168
Tamiya, E., 191, 205, 407, 411
Tan, C.-H., 93, 191, 199, 208, 361, 427, 434
Tan, J., 27, 243, 277, 447
Tan, S.-N., 7, 86, 198, 356
Tan, X.-C., 99, 102, 191, 196, 209, 300, 345, 431
Tan, Y.-M., 80, 88, 226
Tanaka, A.A., 36, 38, 79, 87, 96, 191, 197, 202–203, 271, 273, 348, 381, 390
Tanaka, K., 93
Tanaka, M., 191, 202, 375
Tanaka, S.M.C.N., 16, 19, 25, 89–90, 96–97, 99, 187, 191, 194–195, 198–200, 202, 213–214, 222, 293, 336, 338, 351, 360, 366, 381
Tanaka, T., 247
Tang, D.-P., 26, 68, 80, 93, 99, 191, 202, 375–376, 438
Tang, L., 80, 202, 209, 293, 438
Tang, X.-F., 8, 16, 63, 195, 337
Tang, Y.-H., 27, 89, 447
Tang, Z., 223, 230
Tangestaninejad, S., 59, 80, 96, 191, 195, 206–207, 279, 333, 339–340, 422
Tangkuaram, T., 30, 32, 82, 280
Taniguchi, I., 63, 191, 196, 204, 206, 342, 397, 413
Taniguchi, N., 191, 193, 208, 329, 428
Tanimoto, S.T., 100, 299
Tarapcik, P., 43, 99, 205, 403
Taraszewska, J., 90
Targove, M.A., 4, 17, 26, 52, 202, 260, 381
Tarley, C.R.T., 247
Tarmure, C., 88, 201, 374
Tartarotti, F.O., 90, 252
Tascon Garcia, M., 79, 82, 277
Tascon Garcia, M.D., 38, 161–162
Tascon Garcia, M.L., 314
Tascon, M.L., 89–90, 97, 161–162, 235, 253, 303, 320
Tascon, M.L.G., 161–162
Tascon, M.L.T., 161–162
Tashkhourian, J., 80, 88, 203, 207, 389, 423
Tatawawadi, S.V., 42–43
Tatsumi, H., 92, 192, 208, 328, 428
Tatsuo Kubota, L., 191, 206–207, 415, 424
Tavana, B., 99, 185, 289
Tavassoli, N., 447
Tavcar, G., 77–78, 84, 88, 106, 108, 193, 333
Taylor, R.J., 56, 58
Taylor, Z., 112
Tcheumi, H.L., 84, 293
Tchieda, V., 84, 293

Teixeira, M.F.S., 67, 82, 84, 86–87, 89, 96–98, 101, 185, 191, 195, 201–202, 206–208, 232, 260–261, 289, 300, 303, 305, 310, 314, 338, 371, 384, 419, 426
Teixeira, T., 32, 247
Telefoncu, A., 85, 191–192, 200–201, 203, 206, 269, 328, 365, 372, 387
Tellez, H., 191, 197, 350
Temsamani, K.R., 220
Teng, H., 80, 103, 198, 203, 373, 389
Teng, Y., 191, 204, 398
Terauchi, A., 78, 99, 201, 365, 375
Teresa Sevilla, M., 90, 224, 301
Terranova, M.L., 29, 49, 80, 191, 200, 366, 385
Terui, N., 99, 199, 360
Tesarova-Svobodova, E., 25, 49, 59, 63, 103–104, 135, 143, 146, 149, 175, 226, 233–234, 236–240, 242–245
Tessema, M., 92, 200, 206, 306, 365, 412
Tetianec, L., 95, 198, 355
Teutli-Leon, M.M., 84
Thaghani, A., 79, 96, 206, 333, 413
Thenmozhi, K., 93
Therias, S., 84, 149
Thevenot, D.R., 109
Thiele, R., 3
Thobie Gautier, C., 25, 99, 149, 225
Thomas, J.D.R., 79, 198, 354
Thomsen, E.J., 251
Thomsen, K.N., 90
Thongngamdee, S., 25, 232, 241
Tian, A.-X., 82, 101, 207, 423
Tian, B.-M., 135, 191, 194, 197, 211, 336, 352
Tian, C., 8, 15, 29, 63, 66, 198, 200, 203, 206, 354, 367, 386, 414
Tian, L., 191, 206–207, 423
Tian, X.-J., 30, 32, 315, 322
Tian, Y.-H., 82, 206
Tianling, S., 80, 90, 93, 207
Ticha, I., 186, 294
Tichoniuk, M., 191, 205, 409
Tien, H.-J., 206, 323, 413
Tikhonova, L.N., 82, 191, 193, 195, 333, 338
Timchalk, C., 75
Timkovich, R., 93, 202, 375
Timur, S., 31, 80, 185, 191–192, 200–201, 206, 208, 282, 328, 365, 372–373, 417, 429
Tingry, S., 191, 197, 347
Tirhado, M., 78, 84, 93
Tison, L., 191, 205, 406
Tiyapiboonchaiya, C., 32
Tjell, J.C., 3, 53, 135, 140
Tkach, A.V., 162
Tkalcec, M., 191, 196, 203, 209, 345, 394
Tobalina, F., 96, 198, 326
Tobias, H., 87, 199, 359
Todoriki, S., 79, 92, 191, 193, 198, 200–201, 331, 352, 371
Toft, S.A., 191, 196, 209, 345, 433
Toito Suarez, W., 82
Tolbert, A.M., 79, 96, 333
Toledo, M., 86, 193, 333
Tomas, C., 143, 145
Tomassetti, M., 96
Tomcik, P., 19, 238, 257, 262, 427
Tomcsanyi, L., 256

Author Index

Tomita, A., 191, 198, 353
Tomita, S., 266
Tomschik, M., 7, 191, 201, 205, 375, 406
Toniolo, R., 75
Tonle, I.K., 84, 101, 219, 293
Torabi, R., 87, 101, 174, 235
Torimura, M., 7, 25, 87, 99, 149, 168, 191, 197–198, 200–202, 351, 357, 371, 383
Torres, R.F., 307
Torresday, J.G., 208, 222, 428
Toshima, S., 38, 79, 259
Toth, K., 109
Touil, S., 191, 197, 347
Toyama, S., 191, 358
Toyokichi, K., 191, 196, 202, 343, 384
Tremmel, C.G., 143–144, 274
Trens, P., 101
Triantis, T., 80, 198, 357
Tributsch, H., 38
Trimble, D.S., 191, 204, 398, 401
Trnkova, L., 131, 163, 191, 197, 202, 205, 272, 346, 379, 404
Trojanowicz, M., 5, 64, 116, 143, 146, 191, 196, 198, 209, 252, 345, 353
Troshenkov, A.M., 161
Trubachev, A.V., 177, 256
Trulson, M., 191, 209, 437
Tseng, S.-W., 97, 208, 426
Tsugawa, W., 99–100, 191, 198, 353
Tsunoda, H., 191, 198, 353
Tsushima, S., 153, 158, 191, 196, 202, 291, 343, 384
Tu, Y.-F., 191, 206, 416
Tuel, A., 85
Tugulea, L., 135
Tunon Blanco, P., 9, 59, 75, 79, 93, 95, 97, 99–100, 116, 185, 188, 191–193, 195, 198, 200–201, 206, 279, 287, 292, 308–309, 312, 315, 321, 326, 328, 330, 341, 356, 363, 367, 369, 371, 373, 416
Tunon, P., 93, 192, 326
Turdean, G., 191, 198, 358
Turker, G., 319
Turkusic, E., 79, 87, 191, 196, 199, 209, 342, 359, 438
Turner, A.P.F., 5, 79, 95, 191, 193, 195, 198, 273, 329, 337, 352, 355
Tuzhi, P., 90, 244
Twist, J.P., 215
Tykarski, J., 88, 271
Tymecki, L., 225
Tzang, C.H., 191, 197, 347
Tzimou-Tsitouridou, R., 191, 205, 398, 403–404

U

Uchida, I., 38, 79, 259
Uddin, M.S., 24, 77–78, 99, 223
Ueda, T., 87, 197, 351
Ugo, P., 75
Uhm, J.-H., 26, 191, 203, 391
Ulakhovich, N.A., 4, 9, 19, 24, 67, 90, 187–188, 195, 201, 230, 252, 258–259, 276, 292, 296–297, 299, 301, 337, 372
Ulrich, L., 217
Unger, K.K., 84, 201
Unlu, M., 26

Unuvar, D., 191, 205, 407
Unver, N., 191, 205, 404
Unwin Patrick, R., 29
Upadhyay, S., 191, 198–199, 362
Urbaniczky, C., 4, 18, 56–57
Uriuda, N., 93, 191, 199, 362
Uslu, B., 67, 99–100, 170, 191, 304–305, 313

V

Vacek, J., 131, 191, 205, 408
Vaculikova., 24, 103
Vafaei, Z., 25
Vaivars, G., 168
Vakhnenko, S.V., 248, 254
Valcarcel, M., 40, 96, 273
Valentin, J.F.A., 99, 150, 311
Valentini, F., 29, 32, 49, 63, 80, 191, 200, 366, 385
Valera, F.S., 191, 205, 408
Valisheva, N.A., 161
Van Bennekom, W.P., 40, 203, 284, 395
van Staden, J.F., 28–29, 40, 67–68, 90, 191, 202, 216, 254, 275, 305–306, 308, 311, 313, 318, 321, 382
Vanickova, M., 99, 186, 188, 205, 291, 310, 319, 401–402
Vankova, R., 212
VanWie, B.J., 191, 195, 337
Varadan, V.K., 31–32, 185, 283
Varma, S., 14, 54
Varodi, C., 84, 86, 191, 200–201, 206–207, 269–270, 369, 425
Varvara, S., 162
Vasilyeva, L.N., 210
Vasquez, M.D., 161, 235
Vass, V., 152, 158
Vaz, C.M.P., 295, 297–298
Vaz, V.G., 191, 206–207, 423
Vaze, V.D., 90
Vaziri, M., 85, 305
Vazquez Barbado, M.D., 79, 82, 89–90, 97, 161–162, 253, 277, 303, 314, 320
Vazquez, T., 81
Vega, M., 161–162, 174, 238
Vejdani, M., 85, 305
Velasco, E., 97, 303
Velazquez Rodriguez, M.S., 317
Velez, M.M., 88, 97, 192, 200–201, 325, 368
Velikonja, J., 191, 196, 203, 209, 345, 394
Venancio, E.C., 100, 299
Vepritskaya, A.B., 230
Verbiese-Genard, N., 191, 204, 397
Vereecken, J., 314, 322
Vetterl, V., 43, 205, 401
Vezza, F., 75
Vianello, F., 43, 194, 335
Vicente, F., 315
Vicente, J., 84, 101, 188, 299, 302
Vidal, C.A.G., 174, 238
Vidal, J.C., 99–100, 198, 355
Vidal, J.M., 209, 237, 430
Vidal-Madjar, C., 191, 205, 404
Vidsiunas, E.K., 191, 196, 203, 209, 233, 345, 395, 431
Vieira, H.J., 191, 203, 396
Vieira, I.C., 87, 99–100, 191, 196–197, 202–203, 209, 260, 307, 314, 344, 350, 384, 396, 430–431

Vieira, L.D., 20, 25, 303
Vieirada Cruz, I., 185, 283
Vijayakumar, A.R., 99, 191–192, 327
Vijayaraghavan, R., 32, 186, 294
Vilchez, J.L., 41, 197, 349
Villa, V., 309
Villalba, M.F., 86, 299
Villanuevam, S., 166
Vincke, B.J., 79, 96, 191, 196, 333, 342
Vinogradova, E.N., 210
Vire, J.-C., 13, 22, 40, 93, 188–189, 191, 196, 305, 307, 309, 311, 319–320, 322–323, 342
Visan, T., 162
Vishnu Mahesh, K.R., 97, 293
Vital, S.I., 65
Vitova, V., 185, 287
Vitsina, N.R., 161
Vittal, R., 168
Vittori, O., 84, 87, 160–162, 168
Vives, S.S., 279
Vivier, V., 161
Vlasa, V., 162
Vlasov, Y., 166
Vlasova, N.M., 191, 194, 335
Vlasova, N.N., 191, 194, 335
Vlcek, M., 49, 55, 59, 147, 149, 233, 235
VMitra, C.K., 54
Voelter, W., 67
Volc, J., 191, 200, 365
Volikakis, G.J., 191, 196, 346
Volpe, G., 278
Vondrak, J., 161–162
Vos, J.G., 59, 80, 99–100, 198, 277, 354
Voulgaropoulos, A.N., 64, 77–79, 90, 97, 121, 191, 205, 211, 231, 281, 309, 398, 403–404, 406
Vromman, V., 85, 271
Vul'fson, E.K., 66
Vydra, F., 10, 245
Vydrevich, M.B., 159
Vyskocil, V., 74
Vytřas, K., 1–2, 4–10, 12, 14, 16–22, 24–27, 29, 32, 39–43, 45, 49–59, 61–67, 69–70, 74, 77–81, 85, 87–88, 96–97, 99–100, 103–104, 135–136, 139–140, 142–143, 145–149, 151, 166, 169, 171–172, 174–175, 178, 183–189, 191, 195–196, 198–199, 203, 205–207, 209, 212, 214, 217–219, 226, 231–245, 247, 249–250, 252, 254, 257–259, 267, 275–276, 297, 299–300, 307–308, 319, 337, 342, 354, 359, 364, 394, 406, 422, 438, 447, 585
Vytrasova, J., 191, 205, 406

W

Wadano, A., 7, 25, 149, 168
Waernmark, I., 90, 185, 205, 287, 401
Wages, S.A., 191, 209, 436
Wahdan, T., 312, 314, 321
Wakuda, M., 95, 198, 355
Walcarius, A., 1–2, 4–10, 14, 16, 18, 20–22, 25–27, 29, 40, 45, 50–52, 56, 59, 61, 64–67, 71, 74, 77–78, 81, 84–87, 99, 101–102, 104, 106, 117, 146, 149, 151, 163, 165, 171–172, 181, 184, 186, 199, 203, 217, 219, 224–225, 229, 238, 260–261, 271, 275, 293, 297, 301, 360, 387, 447

Wan, C.-D., 13, 291
Wan, G., 84, 236
Wan, M.-Y., 191, 206, 416
Wang, A.-L., 97, 274
Wang, C., 99, 191, 194, 206, 208, 212, 335, 418, 427
Wang, C.-C., 191, 199, 201, 304, 336, 360, 374
Wang, C.-F., 268
Wang, C.-J., 191, 199, 359
Wang, C.M., 99, 106, 218, 246, 258
Wang, C.-Y., 80, 303, 305
Wang, D., 34–35, 71, 80, 103, 149, 191, 193, 200, 202–203, 206–207, 332, 367, 376, 378, 395, 424
Wang, D.-D., 34–35, 80, 99, 191, 202, 376–378
Wang, D.-W., 34
Wang, E., 35, 100, 102, 286
Wang, E.-B., 8, 81–83, 97, 149, 167
Wang, E.-K., 559
Wang, F.-C., 80, 99, 101, 191, 203, 239, 269, 324, 396
Wang, G.-S., 89–90, 99, 190–191, 195–197, 211–213, 223, 230, 249–250, 341, 343, 347
Wang, H.Y., 81, 87, 97, 104, 191, 193, 196–197, 202, 205, 209, 281, 293, 304, 333, 342, 346, 383, 406, 438
Wang, J., 1–2, 4–10, 12, 14, 18, 21–22, 24–27, 32, 35, 40, 42, 44–45, 49, 51–54, 56, 58–59, 63–64, 66–67, 69–71, 73–74, 77, 81–82, 87, 93, 99–100, 103–104, 113, 121, 123, 133, 135, 143, 145–147, 149, 163, 165, 171, 184–186, 188–189, 191–209, 211, 222, 232, 241, 273, 282, 289–290, 300–301, 308–309, 311, 317, 319, 327, 331, 333, 336, 339, 341, 343–344, 349, 352, 354, 357, 360–363, 367, 370, 373–374, 381, 385, 394, 398, 400–401, 404–408, 411, 413, 418, 427–428, 430–435, 447, 578, 600
Wang, J.-D., 191, 195, 339
Wang, J.J., 191–192, 209, 327
Wang, J.S., 8, 79, 81–88, 90, 93, 95–97, 99, 163, 165, 179, 185, 192, 194, 197–200, 203, 205, 209, 213, 244, 249, 256, 272, 285, 317–319, 322, 325–326, 330, 333, 336, 349, 352, 355, 357, 359–361, 363, 387, 393, 401–402
Wang, K.-F., 27, 80–81, 96–97, 103, 167, 203, 206–207, 323, 333, 390, 413, 422
Wang, L.-H., 8, 16, 27, 32, 34–35, 41, 79, 87, 93, 97, 185, 191, 193, 200, 206, 208–209, 283, 285–287, 303–304, 323, 329, 331–333, 362, 413, 426, 435
Wang, P., 212
Wang, Q., 86–87, 100, 149, 191, 197, 202, 206–207, 256, 351, 376, 379, 418–419, 425, 431
Wang, Q.-J., 32, 87, 188, 194, 296, 298–300, 336
Wang, Q.-L., 191, 198, 202, 353, 376
Wang, Q.X., 81, 205, 409
Wang, S., 32, 80, 131, 191, 202, 303, 352, 381
Wang, S.-F., 16, 27, 32, 34–35, 80, 88, 185, 191, 205, 226, 269, 284, 286, 349, 405
Wang, S.-W., 97, 211
Wang, S.-Y., 31–32, 185, 283
Wang, T., 27, 88, 103, 204, 398, 406
Wang, W.-C., 34–35, 90, 191, 196, 202, 255, 343, 377
Wang, X., 33–35, 81–82, 100–101, 191, 206–207, 288, 400, 421, 447
Wang, X.E., 82
Wang, X.-F., 225
Wang, X.-L., 81–83, 97, 149, 167, 191, 205–207, 421, 437

Author Index

Wang, Y., 33–35, 66, 87, 99, 149, 166, 185, 191, 198, 202–203, 205–207, 239, 284, 361, 377, 386, 389, 398, 403, 405, 424–425
Wang, Y.C., 97
Wang, Y.D., 88, 196, 342
Wang, Y.-F., 81
Wang, Y.J., 99–100, 198, 356
Wang, Y.-M., 32, 35, 316
Wang, Y.X., 64, 253
Wang, Y.-Y., 30, 32, 315
Wang, Z.-H., 24, 27, 31–32, 70, 89–90, 95, 191, 206–207, 225, 237, 400, 424
Wang, Z.-M., 102, 179, 256
Wangsa, J., 4, 17, 26, 52, 97, 99, 202, 260, 326, 381
Warriner, C.N., 100, 274
Warsinke, A., 24, 191, 195–196, 199, 201, 340, 342, 373
Waryo, T.T., 191, 199, 359
Wassel, A.A., 313, 315, 318, 321–322
Watanabe, D., 247
Watanabe, M., 99–100, 191, 198, 355
Watanae, M., 191, 198
Watanesk, S., 209, 274, 430
Watkins, B.F., 73
Watson, G., 357, 398
Watson, W.M., 90
Waymire, R., 191, 205, 398, 401
Weber Scheeren, C., 80, 203, 396
Weetall, H.H., 36
Wei, D.-P., 32–33, 191, 208, 428
Wei, G., 87, 191, 208, 406, 428
Wei, H., 35, 102, 286
Wei, J., 191, 202–203, 206–207, 378, 386, 424
Wei, P., 191, 203, 390
Wei, S., 308
Wei, W.-Z., 34–35, 63, 239, 283
Wei, X., 79–80, 82, 84, 358
Wei, Y., 97, 267
Weidlich, T., 25, 32–35, 104, 240
Weiss, D.J., 191, 196, 342
Weissenbacher, M., 79, 281
Weizmann, Y., 115
Wen, W., 80, 149, 309
Wenrui, J., 161
Werner, G., 79, 87, 96, 194, 336
Westbroek, P., 45, 79, 96, 203, 206–208, 333, 390, 422, 427
Whang, K.-J., 135, 140, 274, 279–280
Whiteley, L.D., 45
Wightman, R.M., 191, 206, 209, 418, 436–437
Wijesuriya, D., 25, 194, 209, 334, 431
Wildgoose, G.G., 8, 29, 31–35, 38, 43, 95, 194, 306, 315, 334
Willems, G.G., 277
Williams, J.H., 163, 165, 209, 437
Willner, I., 115
Wilson, G.S., 109, 191, 198, 355
Wilson, R.L., 191, 209, 436–437
Wimmerova, M., 191, 194, 334
Wisniewski, C., 191, 203, 397, 433, 436
Wohnrath, K., 20, 87, 97, 191, 203, 206–207, 391, 421–422
Wolf, A.K.H., 26, 104
Wolf, R., 112
Wollenberger, U., 191, 203, 209, 349, 361, 393, 433–434
Wolter, K.D., 167

Won, M.-S., 7, 80, 88–90, 99, 172, 197, 214–215, 217–219, 223–224, 226, 251, 274–275, 348
Won, S.-Y., 41, 309, 311, 318, 322
Wong, J.-L., 161–162
Woo, B.W., 191, 203, 391
Woo, J.-J., 191, 197, 347
Woodcock, T.M., 188, 304
Woods, R., 150
Wring, S.A., 96, 191, 208, 426
Wu, H., 79, 87, 199, 359
Wu, J., 191, 200–201, 370
Wu, J.-R., 31, 87, 100, 149, 191, 195–196, 199, 205, 207, 252, 263, 337, 343, 361, 398, 403, 411, 426, 434
Wu, K.-B., 8, 13, 54, 80, 84, 101–102, 191, 197, 202–203, 206, 232, 235–236, 239, 291, 293, 304, 310, 348, 381, 383, 396, 415
Wu, L.H., 49, 79, 95, 101, 149, 191, 198, 209, 355, 430, 433, 435
Wu, M., 163, 191, 203, 209, 223, 394–395, 432, 437
Wu, N.-P., 191, 193, 330
Wu, S., 161–162
Wu, W.-S., 24, 77–78, 99, 223
Wu, X.-P., 32–35, 64, 82, 100, 102, 149, 191, 195, 200–201, 205–207, 254, 277, 281, 312, 316, 337, 367, 409, 411, 417, 425, 563
Wu, Y.-H., 200, 206, 367, 385, 413

X

Xavier, C.R., 84, 86, 96, 271
Xi, X., 361
Xia, B., 191, 202, 375
Xia, F.-Q., 35, 229, 240
Xia, Q., 81
Xiang, B., 191, 206–207, 419
Xiao, D., 567
Xiao, F., 35, 191, 206, 300, 414, 417
Xiao, L.-T., 29, 32, 34, 80, 95, 99, 191, 202, 209, 333, 383, 433
Xiao, P., 64, 246
Xiao, Y., 115
Xiao, Z.-L., 19, 24, 308
Xiaolei, W., 80, 90, 93, 207
Xie, H.-Q., 64, 89, 215, 244
Xie, J.K., 81, 205, 409
Xie, J.-N., 31–32, 185, 283
Xie, J.-P., 191, 205, 410
Xie, P., 191, 203, 396
Xie, W.-Z., 34–35, 80, 88, 99, 103, 202, 377
Xie, X., 80, 101–102, 206, 388, 412
Xie, Y., 87, 191–192, 195, 197, 204, 325, 334, 336, 351, 398
Xie, Z.-H., 563
Xingwang, S., 80, 90, 93, 207
Xiong, H.-Y., 16, 27, 32, 34–35, 80, 88, 185, 191, 205, 226, 269, 284, 286, 303, 349, 405
Xiong, J.-F., 191, 205, 410
Xiong, M., 80, 103, 149, 203, 389
Xiong, S.-X., 29
Xu, C.-X., 27, 80, 88, 93, 99–100, 103, 191, 197, 206, 346, 412
Xu, F., 163, 165, 209
Xu, G., 35, 97, 102, 191, 205, 286, 406, 410
Xu, G.-B., 15, 32, 54, 95, 183, 268
Xu, G.-Y., 74, 84, 95, 97, 191, 205, 281, 365, 401, 409–410

Xu, H.-D., 32–35, 84, 93, 191, 198–201, 203, 206–208, 212, 277, 303, 312, 356, 362, 367, 386, 420, 427–428
Xu, J., 101, 206, 415
Xu, J.-H., 99, 293
Xu, J.J., 80–81, 90, 96, 103, 203, 206–207, 255, 323, 333, 390, 413, 422
Xu, J.-M., 191, 203, 386
Xu, J.-Z., 87, 191, 205, 304, 316, 405
Xu, L.-J., 8, 27, 88, 103, 204, 398, 406
Xu, N.-N., 249
Xu, Q., 239
Xu, W.-Z., 263
Xu, X., 80, 352
Xu, Y., 35, 87–88, 99, 102–103, 191, 198, 202, 213, 286, 361, 376
Xu, Z.-M., 567
Xue, H., 82–83
Xue, Y.-H., 187, 225, 281, 294
Xue, Z.-H., 447

Y

Ya, E., 10
Yabuki, S., 79, 93, 95–97, 99–100, 191–192, 198, 200, 209, 236, 325, 327, 356–358, 363, 438
Yabuli, S., 191, 200, 363
Yadav, H., 191, 193, 333
Yadollahi, B., 101, 276
Yaghoubian, H., 31–32, 95, 161, 191, 202, 204, 206, 307, 314, 384, 397, 416
Yagmur, S., 317, 319
Yamada, J., 135
Yamamoto, F., 96, 198, 351
Yamamoto, K., 99, 163, 165, 191, 198, 200–201, 209, 357, 371
Yamamoto, Y., 191, 193, 208, 329, 428
Yamashita, J., 191, 202, 375
Yamashita, M., 87, 95, 191, 200–201, 203, 209, 366, 368, 397, 427, 433
Yamauchi, S., 191, 358
Yamazaki, S.-I., 25, 191, 200, 202, 208, 375, 429
Yamazaki, T., 99–100
Yamchi, R.H., 205, 309, 402
Yan, F., 78
Yan, Q.-P., 27, 29, 35, 395
Yan, Z.-H., 64, 78, 180, 185, 263, 265, 284
Yanagida, Y., 25
Yanai, H., 75
Yanan, H., 161
Yanez, C., 90, 317
Yanez-Sedeno, P., 59, 79–80, 82, 96, 103, 191, 195–196, 201, 206, 341–342, 344, 371, 412
Yang, B., 86, 149, 202, 376, 379
Yang, B.-J., 191, 198, 202, 353, 376
Yang, B.-Y., 560
Yang, C.-H., 84, 99, 209, 215, 218, 239, 293, 438
Yang, D., 101, 324
Yang, G.-J., 80–81, 96, 103, 191, 203, 206–207, 303, 311, 323, 333, 390, 395, 413, 422
Yang, G.-Z., 239
Yang, H.-F., 80, 84, 103, 202
Yang, J., 27, 101, 191, 198, 203–206, 353, 396–398, 403, 418, 447
Yang, J.-K., 191, 194, 335
Yang, J.-S., 166
Yang, K., 102, 206, 412
Yang, M., 35, 63, 71, 97, 99–100, 149, 191, 197, 202, 204, 206–207, 274, 347, 376, 385, 419
Yang, M.-H., 186, 206, 291, 413
Yang, M.-X., 35, 66, 99, 204, 283, 289
Yang, P.-P., 35, 311
Yang, Q., 131
Yang, R., 8, 15, 29, 63, 66, 198, 200, 203, 206, 354, 367, 386, 414
Yang, S., 225
Yang, T., 80, 191, 198, 205, 353, 398, 403, 408–411, 447
Yang, W., 93, 205, 405
Yang, X., 87
Yang, X.-H., 99, 101, 204, 223, 324, 397
Yang, X.-R., 559
Yang, Y., 80, 103
Yang, Y.-H., 99–100, 186, 206, 291, 413
Yang, Y.-M., 29
Yang, Y.-S., 29, 32, 34
Yang, Z.-P., 5, 12, 22, 24, 67, 191, 209, 244, 308–309, 312, 319, 323
Yaniv, D.R., 49, 51, 147, 149
Yantasee, W., 7, 75, 101–102, 179, 219, 233, 256
Yao, C.-L., 46, 59, 97, 304
Yao, H., 191, 209, 429
Yao, Q., 99–100, 191–192, 327
Yao, S., 64, 82, 224, 277, 281
Yao, T., 3, 5, 191–193, 203, 326, 330, 391
Yao, X., 131
Yao, Y., 149
Yarnitzky, M.N., 222
Yarnoz, P., 99–100, 198, 355
Yazdani, J., 80, 96, 333
Yazgan, I., 191, 201, 328, 373
Ydgar, R., 75
Ye, J.-N., 15, 185, 288
Ye, J.-S., 366
Ye, X., 191, 203, 395
Yegnaraman, V., 162
Yellappa, S., 236, 242
Yeni, F., 191, 208, 429
Yeom, J.-S., 88–89, 214–215, 275
Yi, B., 244
Yi, F.-Y., 90, 264
Yi, H., 67, 308
Yi, L.-H., 78, 262
Yilmaz, E., 185, 197, 282, 347
Yilmaz, N., 307
Yilmaz, S., 317, 319
Yin, C., 191, 205, 409
Yin, H.-S., 79, 96, 185, 284, 333
Yin, J., 82–83, 149
Ying, Q., 191–192, 209, 327, 435
Ying, T.-L., 191, 197, 199, 203, 206–207, 209, 346, 362, 394, 425, 431–433
Yoo, J., 26
Yoo, J.-B., 168
Yoo, J.-H., 26, 191, 203, 391
Yoon, H.-H., 168
Yoon, J.H., 80, 88–89, 215, 226
Yoon, K.-J., 26, 191, 209, 270, 430
Yoon, S.-B., 80, 88, 226

Author Index

Yoshida, H., 90, 213
Yoshikawa, H., 95, 198, 202, 357, 382
Yosuhida, Y., 96
Yosypchuk, B., 169–170, 186–187
You, T.-Y., 8, 16, 34, 63, 80, 103, 191, 193, 195, 198, 200, 203, 206–207, 332, 337, 367, 373, 385, 389, 414–415, 419, 423
You, W., 82–83
You, Y., 8, 80, 192, 325
Young, A.M.J., 163, 165, 209, 436
Yu, A.-M., 59, 93, 191, 203, 206–207, 369, 396, 425
Yu, D.-H., 201, 303, 309, 373
Yu, H.-Q., 191, 202, 268–269, 379
Yu, J., 88, 96, 197, 348
Yu, P.-F., 191, 198, 314, 353, 398–400
Yu, Q., 35, 283
Yu, R.-Q., 25, 68, 80, 88, 95–97, 99–100, 103, 149, 191, 195, 199, 202, 205, 209, 214, 293, 316, 333, 337, 361, 378–379, 411, 437–438
Yu, S.-H., 249, 309
Yu, .T., 161
Yu, X., 191, 203, 395, 433
Yuan, C.J., 97
Yuan, H.-H., 87, 191, 204, 398, 406
Yuan, R., 26, 68, 80, 102, 202, 254, 269, 375–376, 438
Yuan, S., 191, 197, 347
Yuan, X.-D., 191, 206–207, 423
Yuan, X.-M., 563
Yuan, Y.-H., 79–82, 84, 269, 358
Yuce, M., 229
Yudelevich, I.G., 161
Yue, Q., 567
Yuece, M., 191, 208, 428
Yugami, A., 99, 199, 360
Yuksel, U., 85, 226
Yulmetova, R.F., 79, 162
Yusa Marco, D.J., 87, 162
Yusoff, A.R., 99, 202, 383
Yz, M., 59, 369

Z

Zacahua Tlacuatl, G., 85
Zadeii, J.M., 45, 90, 244
Zaggout, F.R., 323
Zagorovskii, G.M., 191, 194, 335
Zaitseva, G., 87, 93, 200, 369
Zakharchuk, N.F., 13, 16, 20–21, 161–162
Zakieva, D.Z., 258, 276
Zaldivar, G.A.P., 79, 87
Zamani, H., 96, 201
Zamani, S., 99–100, 191, 206–207, 423
Zanetti-Ramos, B.G., 81, 203, 395, 432
Zanjanchi, M.A., 85, 196, 206, 343, 425
Zapardiel, A., 25, 30–32, 58–59, 79, 84, 88, 96, 99, 101, 104, 191, 197, 203, 230, 295, 299, 311, 319, 333, 350, 385, 388
Zarbin, A.J.G., 8, 15, 79, 102
Zare, M., 99, 185, 206, 289, 412
Zare Mehrjardi, H.R., 84–85, 91, 93, 95–96, 99, 104, 191, 200–201, 203, 206–207, 229, 251, 271, 277, 368, 390–392, 415–416, 422–424
Zarei, E., 100, 166, 191, 195, 278, 338
Zarghan, M., 80, 88, 103, 304

Zari, N., 204, 398
Zarinskii, V.A., 210, 259
Zaripova, S.A., 96, 191, 193, 333
Zarzur, J.A., 201, 374
Zavvarmahalleh, S.R.H., 99–100, 191–192, 198, 328, 351
Zaydan, R., 95, 193, 331
Zayed, S.I.M., 191, 201, 324, 373
Zboril, P., 191, 193, 331
Zboril, R., 80, 198, 357
Zebreva, A.I., 79
Zejli, H., 220
Zeman, L., 191, 202, 379
Zemanian, T.S., 7, 75, 101, 219, 233
Zemlicka, M., 191, 205, 404
Zen, J.-M., 82, 84, 97, 99, 198, 204, 206, 359, 399, 414
Zendehdel, M., 81, 84–86, 196, 203, 206–207, 343, 389, 415, 420
Zeng, B.-Z., 27, 29, 35, 191, 206, 300, 395, 414, 417
Zeng, G.-M., 80, 95, 99, 191, 202, 207, 333, 426
Zeng, J.-F., 27, 447
Zeng, Q.-X., 35, 269, 349
Zeng, X.-D., 34–35, 63, 239, 283
Zeng, Y., 33–34, 95, 101, 203, 206, 396, 415
Zeraatkar, A., 99, 289
Zeravik, M., 42
Zha, Z.-G., 31
Zhad, H.R.L.Z., 447
Zhai, J., 101, 191, 208, 427
Zhai, X., 191, 206, 418
Zhai, Z.-Q., 31–32, 34–35, 63, 66, 80, 103, 191, 202, 337, 376–378
Zhan, G., 99
Zhan, T.-R., 32, 34–35, 63, 66, 99, 191, 205, 278, 377, 398, 403, 447
Zhang, A., 80, 104, 191, 196–197, 342, 346, 352
Zhang, B., 74, 80–81, 92, 103, 205, 409
Zhang, C.-Q., 35, 191, 205, 313, 405, 408
Zhang, D., 84, 86, 150
Zhang, G.-R., 191, 206–207, 222, 421
Zhang, H.-F., 19, 24, 27, 29, 32, 34–35, 77–78, 80–81, 83, 87, 90, 97, 99–101, 103, 149, 161–162, 167, 191, 197, 199, 203, 206, 212, 258, 303, 351, 359, 386, 388, 418
Zhang, J., 30–32, 64, 90, 95, 228, 264, 365
Zhang, J.-L., 191, 196, 209, 345, 431
Zhang, J.-N., 101, 263, 324
Zhang, J.-Q., 34, 198, 359
Zhang, L., 8, 27, 31–32, 35, 41, 43, 80, 82–83, 191–192, 204, 295, 297, 316, 325, 398, 447
Zhang, L.-F., 87, 406
Zhang, L.J., 99–100, 131, 203, 392
Zhang, M.-N., 29
Zhang, P., 88, 202, 378
Zhang, R.-J., 191, 195–196, 341, 343
Zhang, S., 32, 35, 82, 84, 87, 89, 97, 100, 163, 188, 191, 194–195, 197, 202, 204, 206–207, 263, 296, 298–300, 336, 338, 351, 383, 398, 419, 447
Zhang, S.-B., 215, 218
Zhang, S.-H., 215, 231, 304, 308, 310
Zhang, W.-D., 80, 188, 191–192, 205, 208–209, 212, 296, 298–300, 327, 408, 410–411, 428
Zhang, X., 24, 35, 44, 80, 82, 88, 191, 194, 198–200, 202, 205, 229, 240, 336, 360, 365, 381, 405, 410
Zhang, X.-B., 88, 96, 202, 316, 378–379

Zhang, X.-H., 16, 27, 32, 34–35, 185, 191, 205, 284, 286, 405
Zhang, X.-J., 101, 215, 252
Zhang, X.-L., 191, 206, 414
Zhang, X.-Z., 33–35, 191, 205, 400, 410
Zhang, Y.-J., 33, 35, 63, 70, 80, 89, 191, 195–196, 198, 202–203, 205, 225, 283, 291, 311, 339, 345, 356, 377, 386, 400, 410–411
Zhang, Z.-J., 256
Zhang, Z.-Q., 6, 19, 24, 77–78, 90, 191, 205, 214, 231, 258, 308, 310, 410
Zhang, Z.-S., 191, 205, 410
Zhangm, Z.-Q., 191, 207, 426
Zhao, C.-Z., 35, 93, 191, 198, 205, 353, 398, 403, 447
Zhao, D., 93, 191, 199, 362
Zhao, F., 35, 80, 88, 103, 191, 206, 414, 417
Zhao, F.-Q., 27, 29, 35, 300, 395
Zhao, H.-Y., 81, 205, 278
Zhao, J., 97, 99, 101, 131, 163, 191, 197, 263, 293, 347
Zhao, K., 81
Zhao, M., 191, 200, 365
Zhao, Q.L., 64, 258, 263
Zhao, R.-J., 35, 103, 202, 285, 378
Zhao, W., 191, 209
Zhao, X.-P., 447
Zhao, Y.-L., 82–83, 101, 149, 191, 204, 206–207, 398, 420
Zhao, Z.-F., 8, 42, 89–90, 99, 149, 167, 211, 213, 249–250
Zhe, T., 89–90, 230
Zheng, D., 81, 84, 204, 397
Zheng, J.-B., 33–35, 63, 67, 81, 103, 191, 199, 202–203, 269, 283, 303, 311, 359, 377, 386–388
Zheng, L.-Z., 8, 20, 30, 32, 34–35, 54, 63, 97, 192, 198, 273, 308, 324, 328, 359
Zheng, N., 95
Zheng, Q.-F., 97, 274
Zheng, W.J., 88, 191, 196, 342, 426
Zheng, X., 84, 87, 185, 197, 282, 347
Zheng, X.H., 99–100, 203, 392
Zheng, X.-W., 101, 187, 256, 281, 294
Zheng, Y.-X., 166
Zheng, Z.-X., 191, 203, 206–207, 387, 414, 418, 421
Zhong, J.-H., 34–35, 80
Zhong, J.-Y., 214
Zhou, B.-B., 8, 42, 149, 167
Zhou, C.-L., 35, 101, 166, 185, 229, 240, 288
Zhou, D.-M., 84, 101, 185, 191, 194, 197, 204, 282, 333, 347, 398
Zhou, F.Q., 64, 89, 215, 244
Zhou, F.-Y., 191, 207, 426
Zhou, J., 93, 191, 203, 206–207, 396, 425
Zhou, M.-F., 191, 205, 410
Zhou, N., 80, 191, 205, 409–411
Zhou, T.-S., 163, 165, 209
Zhou, W., 102, 191, 206–207, 254, 425
Zhou, W.-C., 191, 206
Zhou, W.H., 81, 193, 333
Zhou, X., 191, 203, 387
Zhou, Y., 191, 426
Zhou, Y.-K., 101, 149, 168, 291
Zhou, Y.L., 79–80, 96, 102, 185, 197, 284, 333, 348

Zhou, Y.Z., 99–100, 203, 392
Zhu, B., 99, 106
Zhu, D., 8, 15, 29, 63, 66, 198, 200, 203, 206, 354, 367, 386, 414
Zhu, G.-Y., 8, 33–35, 63, 66, 82, 191, 193, 200, 209, 329, 365, 433
Zhu, J., 90, 246
Zhu, J.-J., 87, 191, 205, 304, 316, 405
Zhu, J.X., 99–100
Zhu, K., 70, 89, 225, 400
Zhu, L., 8, 15, 29, 35, 63, 66, 82, 191, 193, 195, 198, 200, 203, 206, 209, 329, 338, 354, 367, 386, 414, 433
Zhu, L.-D., 8, 15, 29, 63, 66, 191, 198, 200, 203, 206, 354, 365, 367, 385–386, 413–414
Zhu, P.-L., 191, 195, 197, 211–212, 223, 341, 347
Zhu, Q., 191, 202, 383
Zhu, S.-Y., 15, 32
Zhu, W., 191, 205, 406
Zhu, Y., 89, 447
Zhu, Y.-C., 81, 88, 276
Zhu, Y.-L., 205, 316, 405
Zhu, Y.-W., 191, 202–203, 206–207, 386, 424
Zhu, Z.-H., 33–35, 82–83, 191, 203, 389, 391
Zhuan, R.-R., 167
Zhuang, H.-L., 191, 208, 427
Zhuang, Q., 32, 66, 306
Zhuang, R.-R., 97, 191, 202, 268, 279, 376–377, 379
Zhuang, Y., 84, 86, 150
Zhuang, Z.-J., 567
Zhulanova, A.G., 214
Zhuravleva, L., 191, 197, 201, 347, 374
Zhylyak, G., 18
Ziarani, G.M., 102, 216, 221
Zielinska, R., 36, 42
Ziganshina, S.A., 80, 82, 88, 96, 103, 149, 191, 193, 195, 332, 338
Zihnioglu, F., 191–192, 201, 328, 372
Zima, J., 8–9, 14, 17, 19–21, 25–26, 40–41, 49, 51–54, 59, 64–65, 67, 90, 99, 104–105, 142, 169–170, 185–188, 194, 197, 202, 205, 245, 285, 287–290, 311, 316, 323, 335, 350, 383, 401, 447
Zinke, K., 37–38, 70, 213, 244
Zitka, O., 191, 202, 379
Zitova, A., 185, 188, 288
Zolfigol, M.A., 101, 174, 235
Zon, A., 430
Zon, M.A., 75, 197, 350
Zou, J.L., 238
Zou, M.-Z., 84, 206–207, 420
Zou, X., 32
Zou, Y.-D., 191, 195–196, 199, 205, 337, 341, 343, 361, 411, 607
Zougagh, M., 191, 197, 350
Zoulis, N.E., 191, 196, 204, 287, 342, 345, 397
Zou,Y-D., 191, 197, 347
Zstathiou, C.E., 287
Zuman, P., 10, 169, 185–186
Zuo, S.H., 80, 87, 191, 198, 204, 357, 398, 406
Zwirtes de Oliveira, I.R.W., 191, 196, 209, 344, 350, 431

Subject Index

A

Adams, Ralph Norman, 449–451
Adams's investigation, aromatic compounds
    aminophenols, 157
    aromatic amines, 154–155
    catecholamines, 157–158
    CPEs measurement, 153–154
    phenolic compounds, 155–156
Amino acids
    cathodic stripping voltammetry, 195
    cysteine, 195
    glutamate and histidine, 195
    mucolytic agent, 194
    L-proline, 196
    tryptophan and tyrosine, 196
    voltammetric and amperometric investigation, 194

B

Binary mixture
    carbon paste definition, 11–12
    carbon powder/graphite
        carbonaceous materials, 13–14
        forms of carbon, 14–16
        spectroscopic (spectral) graphite, 12–13
    handling and storage, 22
    mixing and preparation
        carbon-to-pasting-liquid ratio variation, 20
        chemically and biologically modified CPEs, 22
        equipment and accessories, 21
        homogenization, 21
        "ready-to-use" carbon paste with accessories, 21–22
        solid electrodes, 20
    pasting liquid/binder
        aliphatic and aromatic hydrocarbons, 18
        carbon paste binders types, 20
        chemical inertness and electroinactivity, 16
        controlled miscibility, organic solvents, 17
        halogenated hydrocarbons and similar derivatives, 19
        low volatility, 16
        minimal solubility, water, 17
        paraffin (mineral) oils, 17–18
        silicone oils and greases, 18–19
        transient element, 19
Biologically important compounds (BIC) determination
    alcohols, 192, 325–328
    aldehydes, ketones, and carboxylic acids, 193–194, 329–333
    amino compounds
        amides and amines, 194, 334–343
        amino acids (*see* Amino acids)
    antioxidants and phenolic compounds, 196–197, 344–350

brain electrochemistry/*in vivo* voltammetry, CPm/μEs application, 209, 436–437
carbohydrates and related compounds, 351–365
    alditols and saccharides, 197
    CP-biosensor testing, GOD, 198–199
    FAD, 198
    hemoglobin HbA$_{1c}$, 198
    pectins, 200
    pyranose oxidase, 200
    third-generation glucose biosensor, 199
coenzymes, enzymes, proteins, and related compounds, 366–380
    enzymes activity estimation, 201
    FAD, 200
    glutathione, 201
    nicotine adenine dinucleotide, 200
    silver deposition and re-oxidation, 202
compounds and bonus, 209, 438
hormones, phytohormones, and related compounds, 202, 381–383
neurotransmitters, 384–397
    amperometric detectors, 203
    brain electrochemistry, 203
    differential pulse voltammetry, 202
    dual-tissue modification, 202
    epinephrine/adrenaline, 203
    serotonin, accumulation and oxidation, 204
nucleic acid, nucleic bases, and related compounds, 204–205, 398–411
purines, pyridines, and pyrimidines, 205–206, 412–417
vitamins, 418–427
    ascorbic acid, 206–207
    L-ascorbic acid oxidation, 207
    B group vitamins, 207
    cyclodextrine and mesoporous silicagel, 207
    panthenol and pyridoxal, 208
    tocopherols, 208
whole cells, microorganisms, tissues, and tissue extracts as modifiers, 208–209, 428–435
Butler–Volmer equation, 126–127

C

Carbon ionic liquid electrodes (CILEs), 8–9
Carbon nanotube paste electrodes (CNTPEs), 8–9
    CMCNTPEs, 30
    CNTP-biosensors, 31
    definition, 29
Carbon nanotubes
    categories, 29
    CNT-CPEs, 31
    CNTFEs, 31
    CNTPEs, 29–31
    current-flow measurements, 29–30

643

electrochemical techniques, 31–32
inorganic and organic species spectrum, 32
three-component systems, 30
Carbon paste (CP), xv–xvii
  binary mixture (*see* Binary mixture)
  bulk interactions
    bulk and transport models, chemical equilibrium, 70–71
    extraction and reextraction, 69–70
  individual polarizability
    acidic and mild acidic media, 54
    alkaline and highly alkaline media, 54
    coating/packaging effect, 55
    graphite powder quality, 55
    highly acidic media, 54
    measurements, 53–54
    neutral and basic media, 54
    type of binder, 55
  microstructure
    electrochemical behavior, 51
    liquid binder, 50
    microstructures and surface morphology, 50
    modern microscopic techniques, 49
    theoretical models, 49
  special study themes
    electrocatalysts electrochemical characterization, CPEEs, 167
    electronic tongues, 166–167
    electropolymerized films, substrates, 166
    industrial use, 167–168
  specific reaction kinetics
    anodization, 56, 58
    cathodization, 58
    degree of hydrophobicity/lipophilicity, 56
    electrocatalyst, 59
    electrolytic activation, 57
    pH-metric neutralization titration, 59
    pinhole mechanism, 58
    potential cycling, 59
    pseudo-reversible pathway, 56
    redox processes, 56
    repelling effect, 56–57
    surface treatment effect, 58
Carbon paste electrodes (CPEs), xv
  biological analysis, biosensors and electrodes, 109–110
  carbon nanotubes
    categories, 29
    CNT-CPEs, 31
    CNTFEs, 31
    CNTPEs, 29–31
    current-flow measurements, 29–30
    electrochemical techniques, 31–32
    inorganic and organic species spectrum, 32
    three-component systems, 30
  education and practical training, 447–448
  ionic liquids
    electroanalytical applications, 35
    faradic experiments, 33
    room temperature, 32–34
    voltammetric measurements, 34–35
  miniaturized variants
    microelectrodes, 44–45
    minielectrodes, 43–44
    ultramicroelectrodes, 44–45

modification strategies, biosensors, 121
modifiers and working principles
  electrocatalysis and mediation, 110–111
  enzymes (*see* Enzymes)
  organic compounds, 169–170
  pharmaceutical and clinical analysis, 170
planar constructions, 42–43
special constructions
  CMCPE design, 46
  CP-RWE, 46
  electromotor, 47
  Heyrovsky's dropping mercury electrode, 47
  "U"-shaped electrode, 45
types
  biological modifications, 25
  carbon nanotubes (*see* Carbon nanotubes)
  carbon paste electroactive electrodes, 26
  chemical modification, 24–25
  common/classical type, 23
  CPFEs, 27–28
  diamond as electrode material and diamond paste electrodes, 28–29
  dry/wet CPEs, 23
  ionic liquids (*see* Ionic liquids)
  modified and unmodified CPEs, 24
  physical modifications, 25–26
  soft/hard (desiccated) CPEs, 24
  solid, solid-like, and pseudo carbon paste electrodes, 26–27
Carbon paste film electrodes (CPFEs)
  carbon paste-based electrodes, types, 27–28
  planar configurations, 43
Carbon paste holders
  carbon paste-based detectors
    advantages, 41
    capillary electrophoresis, 40–41
    constructions, 41
    electrochemical detection, flowing streams, 41–42
  commercially available carbon paste electrode bodies, 40
  CPEs miniaturized variants
    microelectrodes, 44–45
    minielectrodes, 43–44
    ultramicroelectrodes, 44–45
  CPEs planar constructions, 42–43
  CPEs special constructions
    CMCPE design, 46
    CP-RWE, 46
    electromotor, 47
    Heyrovsky's dropping mercury electrode, 47
    "U"-shaped electrode, 45
  piston-driven electrode holders
    CP-holders, 39
    Monien's design, 38–39
    piston-driven carbon paste holders, designs, 39
    voltammetric experiments, 39
    wall-jet CP-detector, 40
  tubings and rods (plugs) with hollow ends
    Adams's construction, 36–37
    Adams's laboratories, electrodes variants, 36, 38
    electrochemical stripping analysis, 37
    rotated disc electrodes, hydrodynamic measurements, 38
    Teflon "plug," 36–37

Subject Index

Carbon paste inventor, 449–451
Carbon paste surface interactions
  electrocatalysis-assisted detection, 65–66
  electrode reactions with charge transfer/electrolytic processes, 63
  ion exchange and ion-pair formation, 64–65
  nonelectrolytic character, adsorption and related electrode reactions, 63–64
  synergistic processes and interactions
    adsorption/extraction combination, 67
    electropolymerization and enantioselectivity, 67
    extraction, 66
    incorporation/ejection, 67
    intercalation/precipitation, 67
    voltammetric method, 67–68
Carbon powder/graphite
  carbonaceous materials, 13–14
  forms of carbon, 14–16
  spectroscopic (spectral) graphite, 12–13
Chemically modified carbon nanotube paste electrodes (CMCNTPEs), 30
Chemically modified carbon paste electrodes (CMCPEs)
  activation and restoration, 77
  vs. bare CPEs
    sorption processes, 105–106
    vermiculite-modified, 106
    voltammetric current, 107–108
  biosensors, 5
  bulk modification, 76
  electrocatalysis (*see* Electrocatalysis)
  electrochemically reversible system, 104–105
  era, 4–5
  green chemistry, 7
  immobilization and preconcentration, 77–78
  inorganic materials (*see* Inorganic materials)
  international journals publishing spectrum, 9
  intrinsic modification, 73
  in vivo measurement, 81
  modifier criteria, 76–77
  modifier-substrate interaction, 76–77
  nanomaterial
    nanoparticles, 103
    ordered mesoporous materials, 101–102
  organic catalysts (*see* Organic catalysts)
  organic–inorganic hybrid material, 100–101
  organic ligands
    macrocyclic compounds, 90–91
    molecular ligands, 89–90
  organic polymers and macromolecules (*see* Organic polymers and macromolecules)
  organometallic complexes (*see* Organometallic complexes)
  surface modification
    monolayers, 74–75
    polylayers, 75–76
    submonolayers, 74
  surface treatments and alteration (*see* Surface treatments and alteration)
  surfactants, amphiphilic and lipophilic, 97
Chlorpromazine (Clp) anodic oxidation, 189–190
Chronopotentiometry
  anodic stripping voltammograms, 146–147
  antimony film electrode, 146
  E-t curve, 143
  PSA definition, 144
  Sand's equation, 144
  stripping potentiogram, 145
  three-electrode configuration, 143
CMCPEs, *see* Chemically modified carbon paste electrodes
Constant current stripping analysis (CCSA), 145–146
CP, *see* Carbon paste
CP-biosensor testing, glucose oxidase (GOD), 198–199
CPEs, *see* Carbon paste electrodes

D

Diffusion layer
  diffusion coefficients, 127
  macrosized stationary electrodes, 127–128
  microelectrodes, 129
  rotating disk electrodes, 128–129
Direct potentiometry and potentiometric titrations
  advantages, 139
  chemical analysis completion, 135
  Donnan potential, 140
  extraction constant, $K_{ex}(QX)$, 139
  hydrogen ion-selective electrode, 141
  ion-selective electrode, 136–137
  liquid anion exchangers, anion-selective electrodes, 142
  Nernst equation, 136, 141
  Nikolskii–Eisenman equation, 138
  pH-measurement, calibration curves, 142, 144
  pH scale definition, 136
  potentiometric ion-pair formation-based titrations, 142–143
  Selectrodes, 140
  slope factor determination, 138
Direct voltammetry
  at faster scan rates, 131–133
  at slow scan rates, 130–131
Dropping carbon electrode (DCE), 1
Dropping mercury electrode (DME), xv

E

Electrocatalysis
  mediation oxidation, 78–79
  metal and metal oxides, 79
  metal complexes and salts, 79–80
  nanosized materials, 80–81
  nanosized polymeric resins, 81
  redox polymers, 80
Electrochemical investigation
  CPs special studies
    minor research, 165
    selected themes (*see* Carbon paste, special study themes)
    survey examples, 168
  electrochemistry *in vivo*
    application and prospects survey, 165
    measurement characterization (*see* Measurement characterization)
    position, 162–163
  electrochemistry of solids (*see* Electrochemistry, solids)

organic compounds electrode reaction
and mechanisms
Adams's investigation, aromatic compounds
(*see* Adams's investigation, aromatic
compounds)
Adams's legacy, 158
historical circumstances, 151–153
Electrochemical stripping analysis (ESA)
nucleic acid, nucleic bases, and related
compounds, 205
organic substances and environmental pollutants
determination, 187
synergictic accumulation mechanism, iodine
determination, 183–184
Electrochemistry
anodic polarography, 1
carbon paste-based electrodes
carbon pastes and new technology, 6–7
carbon pastes with screen-printed sensors
and other carbon composites, 5–6
carbon powder and bromoform mixture, 2
chemical modifications, 3–4
CILEs, 8–9
CMCPE, 4–5
CNTPEs, 8–9
with enzymes, biosensors, 5
expansion, electrochemical laboratories, 3
first modification, 2–3
green chemistry, 7
proposals, characterization and applications, 2
DCE, 1
in publication and literature, 9–10
solids
bromonaphthalene-based CPE, 159
chalcocite oxidation, 160–161
CPEE configurations, 159–160
cyclic voltammetry, 160
Faraday law, 160
practical applications, 161–162
voltammetry of microparticles, 159
Electrode material, carbon paste
electrochemical characteristics
individual polarizability (*see* Carbon paste,
individual polarizability)
specific reaction kinetics (*see* Carbon paste,
specific reaction kinetics)
very low background, 52–53
interactions
carbon paste bulk (*see* Carbon paste, bulk
interactions)
at carbon paste surface (*see* Carbon paste surface
interactions)
physicochemical properties
aging, 52
hydrophobicity, 52
microstructure (*see* Carbon paste, microstructure)
ohmic resistance, 51
organic solvents, instability, 52
unmodified CPEs testing
carbon-to-pasting-liquid ratio, 61
CMCPEs/CP biosensors, 61
CPE-index, χ_{CPE}, 61–62
degree of irreversibility, 62
testing guide, flow chart, 59–60

Enzymes
auxiliary enzymes, 117
biomolecules, 121
dehydrogenases, 116–117
enzyme kinetics, 118
hydrolases, 117
immunosensors, 120
nucleic acids
accumulation and interaction studies, 118
bovine serum albumin, 120
hybridization studies, 119
oxidation current production, 118–119
preconcentration, 118
oxidases
biosensors, 113
CMCPE, 113
electrochemical detection and sensor
generation, 112
electron transfer cofactors, 113–115
glucose oxidase, 116
hydrogen peroxide detection, 113
redox polymers, wiring, 115–116
oxidoreductases and hydrolases, 111–112
tissues and cells, 120–121

F

Faradic processes
Butler–Volmer equation, 126–127
kinetics, 125–126
Randles-Sevcik equation, 126
reaction, 125
Tafel equation, 127
Flavin adenine dinucleotide (FAD)
carbohydrates and related compounds, 198
coenzymes, enzymes, proteins, and related
compounds, 200
vitamins, 208

G

Gouy–Chapman–Stern model, 123–124

I

Inorganic ions, complex species, and molecules
determination
alkaline earth metals and alkaline metals
host-guest interactions, 180
indirect amperometric detection, 181
ion chromatography, 181–182
potentiometry techniques, determination, 180,
260–265
redox-active cations, 181
electrolytic deposition/anodic reoxidation scheme, 171
heavy metals
ammonia buffer, masking, 175–176
bismuth paste electrode, 174
cadmium and lead determination, 174, 230–240
direct voltammetry/anodic stripping
voltammetry, 173
electrolytic deposition-anodic reoxidation
scheme, 175
metallic film electrodes, 174
Tl, Sn, In, Bi, and Sb determination, 175, 241–248

Subject Index

inorganic analysis, 170–171
metalloids, 175, 177
metals of fourth and third groups, rare earth metals
    Alizarin derivatives, 179–180
    differential pulse voltammetry of titanium(IV), 180
    lanthanides/actinides analysis, 179, 260–265
    potentiometric detection, 179–180
metals of iron, manganese, chromium, and vanadium groups
    differential pulse voltammetry, vanadium, 177–178
    dimethylglyoxime, 177
    iron, cobalt, and nickel determination, 175, 249–252
    phenanthroline ligands, 176
    V, VI, VII, and VIII group metals, determination, 175, 177, 253–259
noble metals
    copper, 172, 222–229
    cyclam-functionalized silica samples, copper analysis, 172–173
    gold, 172, 210–212
    mercury, 172, 217–221
    silver, 172, 213–216
    tricresyl phosphate-based CPE testing, 172–173
non-metallic ions, complexes, and neutral molecules
    amperometric/voltammetric detection, 183
    electrochemiluminescence measurements, 181
    iodine determination, synergictic accumulation mechanism, 183–184
    ion-exchanger-modified carbon paste, 181–182
    non-metal anions and complex structures determination, 183, 274–281
    non-metal cations and inorganic molecules determination, 183, 266–273
platinum metals and uranium, 178–179, 253–259
voltammetric analysis, 171
"white sites" mapping, 184–185
Inorganic materials
    metal element, 87–88
    metal oxides, 87
    natural clays, 83–84
    POMs, 82–83
    Prussian-blue derivatives, 82
    silica-based materials, 86–87
    sparingly soluble or insoluble complexes and salts, 88–89
    zeolite-based molecular sieves, 84–86
Instrumental measurements
    amperometry, 134–136
    electrochemical techniques
        diffusion layer (*see* Diffusion layer)
        double layer concept and capacitive current, 123–124
        Faradic processes (*see* Faradic processes)
        mass transport, 124–125
    nonelectrochemical techniques
        microscopic observations, 147–149
        physicochemical techniques, special characterizations, 149–150
    potentiometry
        chronopotentiometry and stripping chronopotentiometry (*see* Chronopotentiometry)
        direct potentiometry and potentiometric titrations (*see* Direct potentiometry and potentiometric titrations)

    voltammetry
        direct voltammetry (*see* Direct voltammetry)
        stripping voltammetry, 133
Ionic liquids
    electroanalytical applications, 35
    faradic experiments, 33
    room temperature, 32–34
    voltammetric measurements, 34–35

M

Measurement characterization
    brain electrochemistry, 165
    CPm/μEs recording characteristics, 163
    deconvolution, 163
    linear-sweep voltammetric signal, homovanillic acid, 164
    voltammetric measurements, 163
Medium exchange (MEX) approach
    definition, 171
    pharmaceutical and clinical analysis, 189
Metallic-film-plated carbon paste electrodes (MeF-CPEs), 103–104

N

Nanomaterial
    nanoparticles, 103
    ordered mesoporous material, 101–102
Nernst equation, 136, 141
Nikolskii–Eisenman equation, 138

O

Organic catalysts
    charge transfer mediators, 91
    mediated electrocatalysis, 92
    organic mediators, 93–95
    phenothiazine, phenoxazine, and phenazine derivatives, 93
    quinone compounds, 92–93
Organic ligands
    macrocyclic compounds, 90–91
    molecular ligands, 89–90
Organic polymers and macromolecules
    chelating resins and similarly functioning, 98–99
    conducting polymers, 100
    ion exchangers, 99
    permselective or protective coatings, 99
    redox polymers, 100
Organic substances and environmental pollutants
    aliphatic and aromatic amines, 185
    amino- and nitro-functional group derivatives, 186–187
    C/TCP electrode, environmental investigation, 187–188
    environmental pollutants, synthetic substances, and industrially important products, 185–186, 282–294
    ESA, 187
    green-chemistry approaches, 187
    ISO 2271, ISO 2871-1, and ISO 2871-2, 186
    mechanical regeneration/electrolytic activation, 188

microbial biosensor, 185
potentiometry advantages, 186
preparations and products marketed under commercial names, 187, 295–302
time-controlled biodegradability, 187

Organometallic complexes
ferrocene derivatives, 95–96
phthalocyanine and porphyrin derivatives, 96
Schiff base complexes, 96–97

P

Pharmaceutical and clinical analysis
Clp anodic oxidation, 189–190
Clp determination, 191
and drugs determination, 189, 303–324
ion-pairing principles, 189
MEX approach, 189
substances determination, 188
surface removal, 191

Piston-driven electrode holders
CP-holders, 39
Monien's design, 38–39
piston-driven carbon paste holders, designs, 39
voltammetric experiments, 39
wall-jet CP-detector, 40

Polyoxometallates (POMs), 82–83

Potentiometric stripping analysis (PSA)
chemical oxidation and CCSA, 145
definition, 144
extraction and ion-pairing processes, 146
PSA curves, 146–147

R

Randles-Sevcik equation, 126, 131

S

Specific reaction kinetics
anodization, 56, 58
cathodization, 58
degree of hydrophobicity/lipophilicity, 56
electrocatalyst, 59
electrolytic activation, 57
pH-metric neutralization titration, 59
pinhole mechanism, 58
potential cycling, 59
pseudo-reversible pathway, 56
redox processes, 56
repelling effect, 56–57
surface treatment effect, 58

Stripping chronopotentiometry, *see* Chronopotentiometry

Surface treatments and alteration
electrolytic modification, 104
MeF-CPEs, 103–104

T

Tafel equation, 127